Nanodispersions

Eli Ruckenstein · Marian Manciu

Nanodispersions

Interactions, Stability, and Dynamics

Springer

Eli Ruckenstein
Department of Chemical
and Biological Engineering
University at Buffalo
The State University of New York
303 Furnas Hall
Buffalo NY 14260-4200
USA
feaeliru@buffalo.edu

Marian Manciu
Department of Physics
University of Texas at El Paso
500 W. University Avenue
El Paso TX 79968
Physical Science Bldg. 210
USA
mmanciu@utep.edu

ISBN 978-1-4419-1414-9 e-ISBN 978-1-4419-1415-6
DOI 10.1007/978-1-4419-1415-6
Springer New York Dordrecht Heidelberg London

Library of Congress Control Number: 2009939271

Printed on acid-free paper

Springer is part of Springer Science+Business Media (www.springer.com)

Preface

This book contains a number of papers published by Ruckenstein and coworkers on the topic of nanodispersions. Aerosols are the focus of the first chapter which features a model for the sticking probability as the main contribution. One concludes that, when the particles are small enough, the dissociation rate can become sufficiently large for doublets to reach equilibrium with single particles. However, above a critical radius for the particles, the doublets become stable and their concentration increases with time, providing nuclei for aerosol growth.

The second chapter examines the deposition of Brownian particles on surfaces when the interaction forces between particles and collector play a role. When the range of interactions between the two (which can be called the interaction force boundary layer) is small compared to the thickness of the diffusion boundary layer of the particles, the interactions can be replaced by a boundary condition. This has the form of a first order chemical reaction, and an expression is derived for the reaction rate constant. Although cells are larger than the usual Brownian particles, the deposition of cancer cells or platelets on surfaces is treated similarly but on the basis of a Fokker-Plank equation.

Micellar aggregates are considered in chapter 3 and a critical concentration is defined on the basis of a change in the shape of the size distribution of aggregates. This is followed by the examination, via a second order perturbation theory, of the phase behavior of a sterically stabilized non-aqueous colloidal dispersion containing free polymer molecules. This chapter is also concerned with the thermodynamic stability of microemulsions, which is treated via a new thermodynamic formalism. In addition, a molecular thermodynamics approach is suggested, which can predict the structural and compositional characteristics of microemulsions. Thermodynamic approaches similar to that used for microemulsions are applied to the phase transition in monolayers of insoluble surfactants and to lamellar liquid crystals.

Stern has noted that the traditional Poisson-Boltzmann approach leads to an ionic density in the vicinity of the interface which exceeds the available volume. This anomaly can be corrected by taking into account the hydration of ions. This issue is examined in Chapter 4. This chapter also examines the Helfrich force [Helfrich, W. Z. *Naturforsch.* **1978**, 33a, 305], a repulsion generated by thermally undulating interfaces. Together with the hydration force, this force is responsible for the stability of neutral lipid bilayers. In contrast to the Helfrich theory, which is valid only at large separations between bilayers, the present theory also provides the exponential behaviour that is observed experimentally at small separations. An equation is derived for the force generated between two charged plates immersed in an electrolyte solution, which contains small charged particles, by coupling double layer and depletion forces. It is also shown via Monte Carlo simulations that in concentrated colloidal systems, pairwise repulsive interactions between particles can generate a collective attractive interaction.

Chapter 5 is concerned with specific ion effects due to ion hydration forces and is based on the Structure Making / Structure Breaking (SM/SB) model. The structure making ions prefer the bulk water and are repelled by the surface, while the structure breaking ions disturb the order of bulk water and are expelled toward the interface. The model was used to calculate the force between two parallel plates. There are numerous experiments that point out that the cations (except H^+) are repelled by the air-water interface, whereas the anions (except F^-) are accumulated there. It is shown that the simple SM/SB model can account for these observations quantitatively.

Chapter 6 emphasizes that, as the double layer force, the hydration force is a consequence of electrostatic interactions. In the latter case, the interactions are between neighboring dipoles, and a theory in which the hydration and double layer forces are treated in a unified manner is proposed. The dipoles of the surface align the dipoles of the neighboring water molecules; the dipoles of the partially aligned water molecules, in turn, partially align the next layer of molecules, and so on. Thus a decaying polarization field is generated. The overlap of the polarization fields of two plates brought close enough generates repulsion. Whereas in the traditional continuum theory a uniform dipolar field on a surface does not generate an electric field outside, the discrete nature of the water molecules is considered to be responsible for the generation of a local field, which is calculated by assuming that water is organized in ice-like layers. In addition, in our "polarization model" the surface charge is increasingly neutralized with increasing electrolyte concentration and replaced with dipoles. As a consequence, the repulsive force due to charges is diminished whereas that due to dipoles is increased as the ionic strength increases. At low ionic strengths the repulsion is decreased as the ionic strength increases (particularly because of the screening of the electric field). At high ionic strengths the increasing number of surface dipoles generated by the increasing ionic strength causes an increase in repulsion.

Chapter 7 examines a number of consequences of the polarization model presented above. One of the most significant is the existence of a minimum in the repulsive interactions between colloidal particles as a function of electrolyte concentration, an old experimental result, which remained unexplained for 75 years. Voet reported that stable solutions of various metals (Pt, Pd) and salts (sulfides, halides) in highly concentrated solutions of acids (sulfuric, phosphoric) coagulated upon the dilution with water, hence by decreasing the electrolyte concentration [Voet, A. Thesis, Amsterdam, 1935 See also: Kruyt, H. R. *Colloid Science*; Elsevier: Amsterdam, 1952]. This effect can explain (i) the light scattering observations regarding the apoferritin molecules at various NaCH3COO concentrations which reveal high repulsion at high ionic strengths, as well as (ii) the restabilization of protein-covered latexes at high ionic strengths. The polarization model is combined with thermal undulations to explain the behaviour of common and Newton black films as well as the swelling of neutral lipid bilayers induced by simple salts. This theory, combined with ion-hydration forces, is extended to rough silica surfaces, which disorganize the structure of the interfacial water.

The last chapter, Chapter 8, is devoted to the interactions between grafted polyelectrolyte brushes, which exhibit peculiar repulsive and attractive interactions. Several approaches are suggested. First, a general theory is developed, based on a simple cubic lattice model and matrix formalism, which is used to calculate the steric interaction between two grafted polyelectrolyte brushes. Another approach based on an approximate partition function is suggested to treat the steric repulsion between grafted polymer brushes, leading to simple equations for both neutral and charged grafted brushes. A general formalism for double layer interactions between polyelectrolyte brushes is proposed and it is shown that the net interactions can become attractive, due to the bridging generated by the polyelectrolyte chains between plates. Consequently, the repulsive steric interactions compete with attractive bridging forces. A simple model to calculate the available area for competitive adsorption of molecules is also proposed.

Contents

Introduction to CHAPTER 1

Coagulation, dissociation and growth of aerosols

1.1 G. Narsimhan, E. Ruckenstein: "*The Brownian Coagulation of Aerosols over the Entire Range of Knudsen Numbers: Connection between the Sticking Probability and the Interaction Forces*" JOURNAL OF COLLOID AND INTERFACE SCIENCE 104: 2 (1985) 344–369.

1.2 G. Narsimhan, E. Ruckenstein: "*Monte Carlo Simulation of Brownian Coagulation over the Entire Range of Particle Sizes from Near Molecular to Colloidal: Connection between Collision Efficiency and Interparticle Forces*," JOURNAL OF COLLOID AND INTERFACE SCIENCE 107: 1 (1985) 174–193.

1.3 G. Narsimhan, E. Ruckenstein: "*Dissociation Kinetics of Doublets of Aerosol Particles*," JOURNAL OF COLLOID AND INTERFACE SCIENCE 116: 1 (1987) 278–287.

1.4 G. Narsimhan, E. Ruckenstein: "*A Possible Nucleation Type of Mechanism for the Growth of Small Aerosol Particles*," JOURNAL OF COLLOID AND INTERFACE SCIENCE 116: 1 (1987) 288–295.

Traditional approaches to Brownian coagulation account for the interaction between particles through a phenomenological sticking probability, which is usually assumed to be unity. A model is proposed for the Brownian coagulation coefficient of electrically neutral aerosol particles, which takes into account the interparticle van der Waals attraction and Born repulsion [1.1]. A "sphere of influence", which has a thickness equal to the correlation length of the relative Brownian motion of two particles is suggested; within this region, the relative motion of the particles is considered free molecular, whereas outside this region it is described by a Fokker-Plank equation. For very small particles, the sticking probability becomes vanishingly small, while for sufficiently large particles it is virtually unity. In contrast to earlier models, the expression derived for the coagulation coefficient of large particles displays the proper continuum and free molecular limits and agrees well with the Fuchs empirical formula [1.1]. Upper and lower bounds of the ratio between the coagulation coefficient and the Smoluchowski coagulation coefficient are expressed as functions of the Knudsen number for various Hamaker constants [1.1].

The model was tested using Monte-Carlo simulations (for selected values of the Hamaker constant) over the entire range of the Knudsen number [1.2]. For large particles, the Brownian coagulation coefficient agrees well with the Fuchs interpolation formula and with the analytical expression obtained in [1.1]. For sufficiently small particles [1.2], the coagulation coefficient agrees very well with the lower bound obtained analytically [1.1]. For intermediate particle sizes, the range of interparticle attractive forces becomes comparable to the particle size. This enhances the coagulation coefficient, which acquires values even greater than for the free molecular limit.

The coagulation of aerosols is typically regarded as a result of the short-range van der Waals attraction between particles; when two particles come sufficiently close to each other, they can coagulate. This is true if the particles are large enough and hence the potential well due to van der Waals and Born interactions is sufficiently deep; it is not accurate if the particles are too small and the potential well so shallow that the particles are able to escape from it. In the latter case, the coagulated aerosol doublets can acquire enough energy from the collision with other molecules to dissociate. A novel theoretical treatment, based on the assumption that the time scale of oscillations within the well is much shorter than the time scale of Brownian motion, allows one to average the Fokker-Plank equation over the positions of the doublet and leads to a one-dimensional Fokker-Plank equation in terms of the energy of the relative motion of the constituents of the doublet [1.3]. The average lifetime of a doublet can thus be calculated. It is shown that in air at 298 K, the average dissociation time increases dramatically, from 10^{-7} to 10^{-1} seconds, as the radius changes from 15 to 50 Å. The doublets formed by particles with a radius larger than 50 Å are found to be extremely stable [1.3].

Very small aerosol particles can be generated via physical or chemical nucleation, and their subsequent growth is due either to condensation on the surface of the particles or to the Brownian coagulation of the particles themselves. If

E. Ruckenstein, M. Manciu, *Nanodispersions*, DOI 10.1007/978-1-4419-1415-6_1, © Springer Science+Business Media, LLC 2010

the particles are sufficiently large, the latter process is irreversible (i.e., the sticking probability is unity). However, as noted above, for very small particles there is a non-negligible chance for the doublet to dissociate, because the interaction between the two particles leads to a shallow potential well. The rate of formation of doublets is calculated as suggested in [1.1] and the rate of dissociation is calculated as indicated in [1.3]. The difference between the rates of coagulation and dissociation increases as the particle size increases, becoming zero at a critical particle radius. For particles that are sufficiently small compared to the critical radius, the doublets become unstable and can reach a dynamic equilibrium with the single particles. For particles that are large compared to the critical size, the doublets are stable and their concentration increases with time. They provide nuclei for aerosol growth [1.4].

The Brownian Coagulation of Aerosols over the Entire Range of Knudsen Numbers: Connection between the Sticking Probability and the Interaction Forces

GANESAN NARSIMHAN AND ELI RUCKENSTEIN[1]

Department of Chemical Engineering, State University of New York at Buffalo, Buffalo, New York 14260

Received June 6, 1984; accepted August 9, 1984

The existing models for Brownian coagulation of aerosols account for the effect of interparticle forces through a phenomenological sticking probability, i.e., the probability of coagulation upon collision. This probability is usually assumed to be equal to unity. A model for Brownian coagulation of equal-sized electrically neutral aerosol particles is proposed, which takes into account explicitly the van der Waals attraction and Born repulsion, instead of the phenomenological sticking probability. In this model, the relative motion between two particles in the vicinity of the sphere of influence is considered to be free molecular; the thickness of this region is taken to be equal to the average correlation length for the relative Brownian motion. The relative motion of the particles outside this region is described by the Fokker–Plank equation. The sticking probability predicted by the model, in terms of the interaction potential between two particles, becomes vanishingly small for very small particles. The collisions with the medium that a particle experiences during its escape from the potential well and the corresponding energy dissipation, neglected in this model, will increase the coagulation coefficient, in particular for the large particles because they stay longer in the region in which the interaction potential is acting. For this reason the model is likely to be valid for sufficiently small particles and to provide a lower bound for the coagulation coefficient of the larger particles. However, for large particles the interaction potential is deep and the decay length of the potential is small compared to the particle sizes. As a result, the region of the interaction potential can be replaced by a sink. This is equivalent to considering the sticking probability as equal to unity. In contrast to the earlier models, the expression derived for the coagulation coefficient in the latter case displays the proper continuum and free molecular limits and agrees well with the Fuchs empirical formula. It yields an upper bound for sufficiently small particles but can be exceeded for particles of intermediate sizes. Comparison with the experimental data seem to indicate the validity of this upper bound for particles as small as 0.01 μm.

1. INTRODUCTION

The evolution of the size distribution of aerosols due to coagulation can be described by the population balance equation

$$\frac{\partial n(m, t)}{\partial t} = \frac{1}{2}\int_0^m \beta(m - m', m') \times n(m', t)n(m - m', t)dm' - \int_0^\infty \beta(m, m')n(m, t)n(m', t)dm', \quad [1]$$

where $n(m, t)dm$ is the number of particles per unit volume with a mass between m and $m + dm$ and the rate of coagulation between particles of mass m and m' is characterized by the coagulation coefficient $\beta(m, m')$. In the case of a stationary medium, the collision and subsequent coagulation of particles is a result of their Brownian motion and of the mutual forces of interaction between them. The interactions between particles due to the dispersion forces are mainly responsible for the coagulation of two colliding particles. The effect of these short-range dispersion forces on coagulation is usually accounted for through the sticking probability, i.e., the probability of coagulation upon collision. It

[1] To whom all correspondence should be addressed.

0021-9797/85 $3.00

is customary to assume that every collision between particles leads to coagulation, i.e., that the sticking probability is equal to unity.

When the particle sizes are much larger than the mean free path of the suspending medium, the Knudsen number ($Kn = \lambda_g/a$, where a and λ_g are the radius of the particle and the mean free path of the medium, respectively) is extremely small. Under such conditions, the Brownian coagulation coefficient is well described by the Smoluchowski equation (1)

$$\beta \equiv \beta_s(a_1, a_2) = 4\pi(a_1 + a_2)(D_1 + D_2), \quad [2]$$

where the diffusion coefficient D for a particle of radius a is given by the expression

$$D = \frac{kT}{6\pi a\eta}, \quad [3]$$

where η is the viscosity of the medium, k is as usual the Boltzmann constant and T the absolute temperature of the medium.

Smoluchowski's equation was extended up to $Kn < 0.25$, by introducing a slip correction factor $C(Kn)$ in the expression for the diffusion coefficient. The coagulation coefficient is given for $Kn < 0.25$ by the expression

$$\beta \equiv \beta_c(a_1, a_2) = \beta_s(a_1, a_2)\left(\frac{C(Kn_1)}{a_1} + \frac{C(Kn_2)}{a_2}\right), \quad [4]$$

where the empirical slip correction factor C due to Millikan (2) has the form

$$C(Kn) = 1 + Kn\{1.257 + 0.4\exp(-1.1/Kn)\}, \quad [5]$$

while the slip correction factor derived by Philips (3), for an accommodation coefficient equal to unity, is given by

$$C(Kn) = \frac{5 + 4Kn + 6Kn^2 + 18Kn^3}{5 - Kn + (8 + \pi)Kn^2}. \quad [6]$$

In the free molecular limit ($Kn > 10$), the coagulation coefficient is obtained from the kinetic theory of rarefied gases as (4)

$$\beta \equiv \beta_M(a_1, a_2) = \alpha\pi(a_1 + a_2)^2(\bar{c}_1^2 + \bar{c}_2^2)^{1/2}, \quad [7]$$

where α is the sticking probability and $\bar{c}_i = (8kT/\pi m_i)^{1/2}$ is the average velocity of a particle of mass m_i.

For the transition region ($0.1 < Kn < 10$), Fuchs (5), assuming the sticking probability equal to unity, proposed the following semiempirical expression for the coagulation coefficient

$$\beta \equiv \beta_F(a_1, a_2) = \beta_c\left[\frac{a_1 + a_2}{a_1 + a_2 + g_{12}} + \frac{4(D_1 + D_2)}{(a_1 + a_2)(\bar{c}_1^2 + \bar{c}_2^2)^{1/2}}\right]^{-1}, \quad [8]$$

where

$$g_{12} = (g_1^2 + g_2^2)^{1/2},$$

$$g_i = \frac{1}{6a_i\lambda_i}[(2a_i + \lambda_i)^3 - (4a_i^2 + \lambda_i^2)^{3/2}] - 2a_i,$$

$$\lambda_i = 8D_i/\pi\bar{c}_i,$$

$$D_i = \frac{kT}{6\pi a\eta}C(Kn_i),$$

and

$$\bar{c}_i = \left(\frac{8kT}{\pi m_i}\right)^{1/2}.$$

Sitarski and Seinfeld (6) were the first to provide a theoretical basis for Fuch's semiempirical formula, by solving the Fokker–Plank equation by means of Grad's (7) 13-moment method. Their solution was further improved by Mork *et al.* (8). The Brownian coagulation coefficient predicted by these models agrees fairly well with the Fuchs interpolation formula. However, the model does not predict the proper free molecular limit. The validity of the Fuchs semiempirical formula was further reinforced, by the Monte Carlo simulations of Brownian coagulation, by Nowakowski and Sitarski (9).

Even though it has been recognized that

the sticking probability is less than unity for very small particles, there have been no attempts to explain the physical origin of the sticking probability. The previous treatments consider the sticking probability as an empirical parameter. The objective of this paper is to propose a model for Brownian coagulation of electrically neutral aerosol particles capable of calculating the coagulation coefficient without the use of the sticking probability. Instead, the model takes into account the van der Waals attraction and Born repulsion between two particles. The surface of the particles is assumed to be completely rough, with momentum and thermal accommodation coefficients of unity. The relative Brownian motion between two particles, in the vicinity of the sphere of influence (whose radius is equal to the sum of the radii of the colliding particles), is correlated over a distance of the order of the correlation length. In other words, the two colliding particles retain the memory of their relative motion when the shortest distance between them is smaller than this length. The simplifying assumption is made that the motion of the particles in this inner region is free molecular. The motion outside this region is described by the Fokker–Plank equation. It is to be noted that, in reality the coagulation coefficient is greater than that predicted by assuming that the motion is free molecular in the inner region. This is a result of the collisions that a particle experiences with the medium during its escape from the potential well. This enhancement due to the collisions with the medium is likely to be significant particularly for large particles since they stay longer in the region in which the interaction potential is acting. Therefore, the coagulation coefficient predicted by the model is likely to be valid for small particles and to provide a lower bound for the larger particles. However, for large particles, the potential well is deep due to strong interparticle forces, and the decay length of the potential is small compared to particle sizes. Therefore, the potential can be replaced by a sink at the sphere of influence. This is equivalent to a sticking probability equal to unity. The development of the model is presented in the first section. It provides, as already noted, a lower bound for the coagulation coefficient, which should constitute, however, a good approximation for small particles. An expression for the coagulation coefficient of large particles is derived in the subsequent section. This provides an upper bound for sufficiently small particles. Comparison of the model with other models and experimental data is presented in the last section.

2. MODEL FOR BROWNIAN COAGULATION—LOWER BOUND

Brownian motion of a particle can be described by the Langevin equation (10)

$$\frac{d\mathbf{c}}{dt} = -\zeta\mathbf{c} + \mathbf{B}(t), \qquad [9]$$

where $\mathbf{c}$ is the velocity of the particle, ζ is the friction coefficient, and $\mathbf{B}(t)$ is the random force due to collisions with the medium. The relaxation time, i.e., the time over which the particle momentum is correlated, is of the order of ζ^{-1}. For times sufficiently large compared to ζ^{-1}, the single-particle distribution function $f(\mathbf{r}, \mathbf{c}, t)$ is governed by the Fokker–Plank equation (10)

$$\frac{\partial f}{\partial t} + \mathbf{c}\cdot\nabla_{\mathbf{r}} f + \left(\frac{\mathbf{F}}{m_{\mathrm{p}}}\right)\cdot\nabla_{\mathbf{c}} f = \zeta\nabla_{\mathbf{c}}\cdot\left[\mathbf{c}f + \left(\frac{kT}{m_{\mathrm{p}}}\right)\nabla_{\mathbf{c}} f\right], \qquad [10]$$

where m_{p} is the mass of the particle, $\mathbf{r}$ is the position vector of the particle, T is the temperature of the medium, and $\mathbf{F}$ is the external force acting on the particle. For times very large compared to ζ^{-1}, the Fokker–Plank equation reduces to the classical diffusion equation

$$\frac{\partial n(\mathbf{r}, t)}{\partial t} = \nabla_{\mathbf{r}}\cdot\left(D\nabla_{\mathbf{r}} n(\mathbf{r}, t) - \frac{\mathbf{F}}{\zeta}\, n(\mathbf{r}, t)\right), \qquad [11]$$

$n(\mathbf{r}, t)$ and D being the number density and the diffusion coefficient of the particle, respectively. Since the viscosity of gases is small, ζ^{-1} (given by Eqs. [82] and [84]) is large and therefore the Brownian motion of aerosols must be described by Eq. [10].

The rate of Brownian coagulation between two particles of specified sizes can be predicted from (i) the rate of collisions between the particles, obtained through the analysis of their relative motion, and (ii) the probability that a collision will lead to coagulation. Existing treatments for Brownian coagulation of aerosols do not account for interparticle forces in their description of the relative motion of the particles, but use, instead, a phenomenological sticking probability. In the absence of interparticle forces, the Brownian motion of the particles can be considered to be independent of each other, i.e., the Brownian motion of a particle proceeds unaffected so long as no other particle comes within its sphere of influence whose radius is equal to the sum of radii of the particles. Consequently, it is customary to consider the Brownian motion of a particle with respect to a second stationary particle. The rate of coagulation can therefore be calculated by solving the Fokker–Plank equation for the single-particle distribution function with the boundary conditions, for mass and energy fluxes at the sphere of influence, which account for the sticking probability.

As already mentioned the present treatment attempts to clarify the connection between the sticking probability and the mutual forces of interaction between particles. The van der Waals attraction and Born repulsion forces are included in the calculation of the rate of collisions between two electrically neutral aerosol particles. The overall interaction potential between two particles is calculated through the integration of the intermolecular potential, modeled as the Lennard–Jones 6–12 potential, under the assumption of pairwise additivity. The expression for the overall interaction potential in terms of the Hamaker constant and the molecular diameter can be found in Appendix I. The motion of a particle can no longer be assumed to be independent of the motion of the other particle, because of the interaction force between the two. Therefore, it is necessary to describe the relative motion of two particles in order to predict the rates of collision and of subsequent coagulation.

Let us now consider the relative motion of two particles of the same radius R_p and mass m_p, and denote by $W(\mathbf{r}_1, \mathbf{c}_1; \mathbf{r}_2, \mathbf{c}_2)d\mathbf{r}_1 d\mathbf{c}_1 d\mathbf{r}_2 d\mathbf{c}_2$ the probability of finding the first particle between $\mathbf{r}_1$ and $\mathbf{r}_1 + d\mathbf{r}_1$, with the velocity between $\mathbf{c}_1$ and $\mathbf{c}_1 + d\mathbf{c}_1$, and the second particle between $\mathbf{r}_2$ and $\mathbf{r}_2 + d\mathbf{r}_2$, with the velocity between $\mathbf{c}_2$ and $\mathbf{c}_2 + d\mathbf{c}_2$. The distribution function W satisfies the steady-state Fokker–Plank equation

$$\mathbf{c}_1 \cdot \nabla_{\mathbf{r}_1} W + \mathbf{c}_2 \cdot \nabla_{\mathbf{r}_2} W + \frac{\mathbf{F}_1}{m_p} \cdot \nabla_{\mathbf{c}_1} W + \frac{\mathbf{F}_2}{m_p} \cdot \nabla_{\mathbf{c}_2} W = \zeta \nabla_{\mathbf{c}_1} \cdot (\mathbf{c}_1 W) + \zeta \nabla_{\mathbf{c}_2} \cdot (\mathbf{c}_2 W) + \frac{\zeta k T}{m_p} \nabla^2_{\mathbf{c}_1} W + \frac{\zeta k T}{m_p} \nabla^2_{\mathbf{c}_2} W, \quad [12]$$

where $\mathbf{F}_1$ and $\mathbf{F}_2$ are the forces acting on the particles 1 and 2 along their line of centers, i.e.,

$$\mathbf{F}_1 = -\mathbf{F}_2 = \frac{F(|\mathbf{r}_1 - \mathbf{r}_2|)}{|\mathbf{r}_1 - \mathbf{r}_2|} (\mathbf{r}_1 - \mathbf{r}_2), \quad [13]$$

where $|\mathbf{r}|$ denotes the magnitude of the vector $\mathbf{r}$.

Introducing the center of mass and the relative coordinates:

$$\mathbf{R} = (\mathbf{r}_1 + \mathbf{r}_2)/2, \quad \mathbf{r} = \mathbf{r}_1 - \mathbf{r}_2$$

$$\mathbf{C} = (\mathbf{c}_1 + \mathbf{c}_2)/2, \quad \mathbf{c} = \mathbf{c}_1 - \mathbf{c}_2, \quad [14]$$

Eq. [12] admits a separable solution of the form (11)

$$W(\mathbf{r}_1, \mathbf{c}_1; \mathbf{r}_2, \mathbf{c}_2) = W_C(\mathbf{R}, \mathbf{C}) f(\mathbf{r}, \mathbf{c}), \quad [15]$$

where $W_C(\mathbf{R}, \mathbf{C})$ satisfies the equation

$$\mathbf{C} \cdot \nabla_{\mathbf{R}} W_C = \zeta \nabla_C \cdot (\mathbf{C} W_C) + \frac{\zeta k T}{2 m_p} \nabla^2_C W_C \quad [16]$$

and $f(\mathbf{r}, \mathbf{c})$ is given by the equation

$$\mathbf{c}\cdot\nabla_{\mathrm{r}}f+\frac{2F}{m_{\mathrm{p}}}\frac{\mathbf{r}}{r}\cdot\nabla_{\mathrm{c}}f = \zeta\nabla_{\mathrm{c}}\cdot(\mathbf{c}f)+\frac{2\zeta kT}{m_{\mathrm{p}}}\nabla_{\mathrm{c}}^{2}f. \quad [17]$$

Since the rate of collisions between two particles depends only on the relative motion, only Eq. [17] needs to be considered in the subsequent treatment.

The relative motion can now be represented by the Brownian motion of a fictitious particle of reduced mass $m_{\mathrm{r}}(=m_{\mathrm{p}}/2)$, with the friction coefficient ζ, experiencing the same force **F**. Consequently,

$$\mathbf{c}\cdot\nabla_{\mathrm{r}}f+\frac{F}{m_{\mathrm{r}}}\frac{\mathbf{r}}{r}\cdot\nabla_{\mathrm{c}}f = \zeta\nabla_{\mathrm{c}}\cdot\left[\mathbf{c}f+\left(\frac{kT}{m_{\mathrm{r}}}\right)\nabla_{\mathrm{c}}f\right], \quad [18]$$

where **r** and **c** now denote the position vector and the velocity of the fictitious particle, respectively. However, the Brownian motion of the particle in the vicinity of the sphere of influence, in a region whose thickness is of the order of the average correlation length, cannot be described by the Fokker–Plank equation. This is because the time scale of motion of the particle in that region is not large compared to ζ^{-1}. The average correlation length (or, equivalently, the apparent mean free path) λ_{r} for the Brownian motion of the fictitious particle of mass m_{r} can be defined as

$$\lambda_{\mathrm{r}} = \bar{c}\zeta^{-1}, \quad [19]$$

where

$$\bar{c} = \left(\frac{16kT}{\pi m_{\mathrm{p}}}\right)^{1/2} \quad [20]$$

is the average relative velocity between two particles of mass m_{p}.

As pointed out earlier, the Fokker–Plank equation [18] describes the motion of the fictitious particle only outside a sphere of radius $R_{\mathrm{S}}+\lambda_{\mathrm{r}}$, where $R_{\mathrm{S}}(=2R_{\mathrm{p}})$ is the radius of the sphere of influence. The motion of the fictitious particle in the region of thickness λ_{r}, next to the sphere of influence, can be described by the Langevin equation. The decay length of the overall interaction potential due to the van der Waals attraction and Born repulsion, based on the shortest distance between two particles, is much smaller than the correlation length λ_{r}. For example, the decay length of the overall interaction potential between two particles of radius 100 Å is of the order of 50 Å, whereas the correlation length for particles of unit density and a radius of 100 Å, in air, is of the order of 300 Å.[2] Therefore, the motion of the fictitious particle of mass m_{r} in the inner region of thickness λ_{r} can be further subdivided into three regions, as shown in Fig. 1. The motion of the fictitious particle in the Region I will be represented by the Fokker–Plank equation (Eq. [18]). Even though the motion of the fictitious particle in the inner regions (Regions II, III, and IV) can only be described by the Langevin equation (since the fictitious particle does undergo collisions with the medium), in the present model, it is assumed that the change in the momentum of the fictitious particle due to the collisions with the medium is negligible in the inner regions, i.e., the motion of the fictitious particle in Regions II, III, and IV can be approximated by a

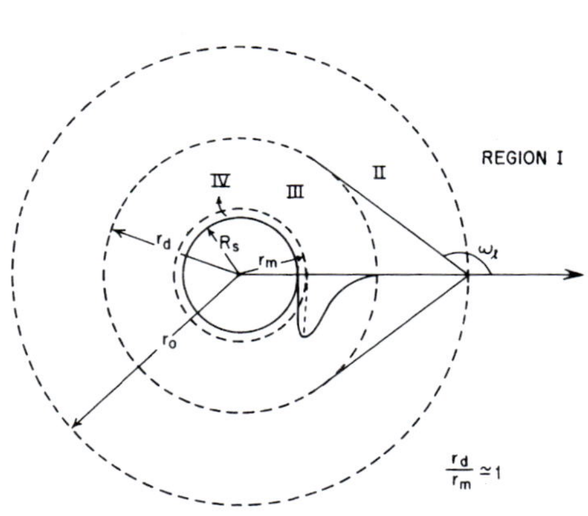

FIG. 1. Different regions for the motion of the fictitious particle of reduced mass.

[2] One may note that the correlation length for the relative Brownian motion in liquids is two orders of magnitude smaller because of a much higher viscosity.

collisionless motion. In Region II, the interaction potential between the two constituent particles of mass m_p is negligible, while in Regions III and IV the particles experience attractive and repulsive forces, respectively.

We now consider a dilute system of monodispersed aerosol particles; the volume of the system is considered to be very large compared to the characteristic lengths, such as the mean free path of the molecules of the medium and the correlation length λ_r. The motion of the fictitious particle in such a system can be considered quasi-steady, since the time scale of change in the particle concentration due to coagulation is large compared to the time scale of particle motion. In the following sections, we solve for the motion of the fictitious particle in the four regions and calculate the rate of collision and the rate of subsequent coagulation by imposing the continuity of concentration and of the fluxes of mass and energy at the interfaces between the regions.

Region I. Since the motion of the fictitious particle in this region takes place in the absence of the overall interaction potential between particles (see Fig. 1), the particle distribution function is governed by the steady-state Fokker–Plank equation

$$\mathbf{c}\cdot\nabla_{\mathbf{r}} f = \zeta\nabla_{\mathbf{c}}\cdot\left[\mathbf{c}f + \left(\frac{kT}{m_r}\right)\nabla_{\mathbf{c}} f\right]. \quad [21]$$

When the two particles of mass m_p are far apart, i.e., $\mathbf{r} \to \infty$, the distribution of the relative velocity $\mathbf{c}$ is Maxwellian and is therefore given by (11)

$$\lim_{\mathbf{r}\to\infty} f(\mathbf{c}, \mathbf{r}) = \left(\frac{m_p}{4\pi kT}\right)^{3/2} \exp\left(-\frac{m_p c^2}{4kT}\right). \quad [22]$$

Denoting

$$h^2 = m_r/kT, \quad \mathbf{v} = h\mathbf{c}, \quad g = f/h^3$$

and

$$\omega(\mathbf{v}) = \frac{1}{(2\pi)^{3/2}} \exp(-v^2/2), \quad [23]$$

$v = |\mathbf{v}|$ being the magnitude of $\mathbf{v}$, Eq. [21] can be recast in the form

$$\frac{1}{h}\mathbf{v}\cdot\nabla_{\mathbf{r}} g = \zeta\nabla_{\mathbf{v}}\cdot\{\omega(\mathbf{v})\nabla_{\mathbf{v}}[g/\omega(\mathbf{v})]\}. \quad [24]$$

The above equation can be solved by the Grad 13-moment method (7), by expanding the unknown distribution function g in terms of the multidimensional Hermite tensor polynomials, as

$$g(\mathbf{r}, \mathbf{v}) = \omega(\mathbf{v})[a^{(0)}H^{(0)} + a_i^{(1)}H_i^{(1)} + \tfrac{1}{2}a_{ij}^{(2)}H_{ij}^{(2)} + \tfrac{1}{10}a_i^{(3)}H_i^{(3)}], \quad [25]$$

where the summation convention is used for repeated indices and the Hermite tensor polynomials are orthogonal, with the weighting function $\omega(\mathbf{v})$. They are given by

$$H^{(0)} = 1, \quad H_i^{(1)} = v_i,$$
$$H_{ij}^{(2)} = v_i v_j - \delta_{ij} \quad [26]$$

and

$$H_i^{(3)} = v_i(v^2 - 5),$$

where δ_{ij} is the Kroneker delta. Denoting the moments of the distribution function $f(\mathbf{r}, \mathbf{c})$ by

$$n(\mathbf{r}) = \int f d\mathbf{c},$$
$$j_i(\mathbf{r}) = \int c_i f d\mathbf{c},$$
$$p_{ij}(\mathbf{r}) = \int c_i c_j f d\mathbf{c} - \frac{1}{h^2} n(\mathbf{r})\delta_{ij},$$
$$S_{ijk}(\mathbf{r}) = \int c_i c_j c_k f d\mathbf{c},$$
$$p(\mathbf{r}) = \frac{1}{3}\int c^2 f d\mathbf{c} \quad [27]$$

and

$$S_i(\mathbf{r}) = \tfrac{1}{2}S_{ijj}(\mathbf{r}),$$

where $n(\mathbf{r})$ is the number density, $j_i(\mathbf{r})$ is the ith component of the mass flux, $p_{ij}(\mathbf{r})$ is the pressure tensor (symmetric), $p(\mathbf{r})$ is the hydrostatic pressure, and $S_i(\mathbf{r})$ is the ith component of the flux of the kinetic energy per unit mass, all referring to the fictitious particle of reduced mass m_r.

The coefficients $a_i^{(n)}$ in the expansion [25] can be related to the moments of the distribution function by the expressions

$$a^{(0)} = n(\mathbf{r}),$$

$$a_i^{(1)} = hj_i(\mathbf{r}),$$

$$a_{ij}^{(2)} = h^2 p_{ij}(\mathbf{r}) \quad [28]$$

and

$$a_i^{(3)} = 2h^3 S_i(\mathbf{r}) - 5hj_i(\mathbf{r}).$$

Introducing the expansion [25] into the Fokker–Plank equation [24] and using the orthogonal properties of the Hermite polynomials, the following moment equations are obtained:

$$\frac{dj_i}{dr_i} = 0, \quad [29]$$

$$\frac{dn}{dr_i} + h^2 \frac{dp_{ij}}{dr_j} = -h^2 \zeta j_i, \quad [30]$$

$$\delta_{ij} \frac{dS_k}{dr_k} + 2\frac{dS_i}{dr_j} = -5\zeta p_{ij} \quad [31]$$

and

$$\frac{dp_{jj}}{dr_i} + 2\frac{dp_{ij}}{dr_j} = -3\zeta(2h^2 S_i - 5j_i). \quad [32]$$

The above equations are to be solved with the boundary conditions

As $r \longrightarrow \infty$,

$$n(\mathbf{r}) = n_\infty,$$

$$\mathbf{j}(\mathbf{r}) = \mathbf{o} \quad [33]$$

and

$$\mathbf{S}(\mathbf{r}) = \mathbf{o},$$

where n_∞ is the concentration at infinity of the fictitious particles of reduced mass m_r. Since the fictitious particle describes the relative motion of two particles of mass m_p,

$$n_\infty = \tfrac{1}{2}n_0^2, \quad [34]$$

where n_0 is the concentration of particles of mass m_p.

Assuming spherical symmetry, the mass and energy flux vectors can be expressed in terms of scalar functions as

$$\mathbf{j}(\mathbf{r}) = \frac{\mathbf{r}}{r} J(r) \quad [35]$$

and

$$\mathbf{S}(\mathbf{r}) = \frac{\mathbf{r}}{r} S(r). \quad [36]$$

Expressing the mass and energy flux vectors in terms of the scalar functions as given by Eqs. [35] and [36], the moment equations, Eqs. [29] to [33], can be solved (8) to give

$$J(r) = \frac{A}{r^2}, \quad [37]$$

$$S(r) = \frac{5}{2h^2}\frac{A}{r^2} + \frac{B}{qr}\left(1 + \frac{1}{qr}\right)e^{-qr}, \quad [38]$$

$$n(r) = n_\infty + \frac{h^2\zeta A}{r} - \frac{3h^2 B}{5\zeta r} e^{-qr} \quad [39]$$

and

$$p_{ij}(r) = -\frac{1}{5\zeta}\left[2\left(2\delta_{ij} - \frac{r_i r_j}{r^2}\right)\frac{S}{r} + \left(\delta_{ij} + 2\frac{r_i r_j}{r^2}\right)\frac{dS}{dr}\right], \quad [40]$$

where

$$\frac{1}{q^2} = \frac{11}{30}\frac{1}{h^2\zeta^2} \quad [41]$$

and A and B are unknown constants to be determined later. Substituting for the coefficients in the expansion [25], the expression for the distribution function in Region I is given by

$$g(\mathbf{r}, \mathbf{v}) = \omega(\mathbf{v})\left[n(r) + \frac{1}{2}hJ(r)v\mu(7 - v^2) - \frac{h^2}{10\zeta}\{v^2(1 - 3\mu^2)\mathcal{A}(r) - (v^2 + 2v^2\mu^2 - 5) \times \mathcal{B}(r)\} + h^3 S(r)v\mu\left(\frac{1}{5}v^2 - 1\right)\right], \quad [42]$$

where

$$\mu = \frac{\mathbf{r}\cdot\mathbf{v}}{rv}, \quad [43]$$

$$\mathcal{A}(r) = \frac{5A}{h^2 r^3} + \frac{2Be^{-qr}}{qr^2} + \frac{2Be^{-qr}}{q^2 r^3} \quad [44]$$

and

$$\mathcal{B}(r) = \frac{Be^{-qr}}{r}. \quad [45]$$

Region II. Since the motion is assumed to be collisionless and the interaction potential is negligible in this region, the particle path is rectilinear and the distribution function is described by

$$\mathbf{v} \cdot \nabla_r g = 0. \qquad [46]$$

Therefore, only those fictitious particles whose velocity vector at the interface of Regions I and II lies in the cone of angle $(\pi - \omega_1)$ (see Fig. 1) will enter into the region III. In other words, for a "particle" to enter into the Region III,

$$-1 \leqslant \mu \leqslant -\left(1 - \frac{r_d^2}{r_0^2}\right)^{1/2} \qquad [47]$$

at the interface between Regions I and II.

Region III. The motion of the fictitious "particle" in this region is influenced by the attractive force between two particles. Since the effect of collisions with the medium on the motion of the fictitious particle is assumed to be negligible, the distribution function is governed by the collisionless Boltzmann equation

$$\mathbf{c} \;\; \nabla_r f + \frac{\mathbf{F}}{m_r} \;\; \nabla_c f = 0 \qquad [48]$$

The distribution function f is therefore constant along the trajectory of the fictitious particle.

Since the fictitious particle moves in a central force field described by a spherically symmetric potential function $U(r)$, its angular momentum is conserved. Therefore, the motion of the fictitious particle will be in a plane defined by the velocity and the radius vectors. The Lagrangian may then be conveniently expressed in polar coordinates as

$$L = \tfrac{1}{2} m_r(\dot{r}^2 + r^2\dot{\theta}^2) - U(r). \qquad [49]$$

Since the Langrangian is independent of θ, the angular momentum conjugate to the coordinate θ is conserved and is given by

$$l = m_r r^2 \dot{\theta} = \text{constant} = m_r r_0 c_0 \sin \omega_0, \qquad [50]$$

where, as shown in Fig. 2, c_0 and ω_0 are the velocity and the angle between the velocity vector and the radius vector at the interface between Regions I and II, respectively. Since the motion is collisionless, the total energy

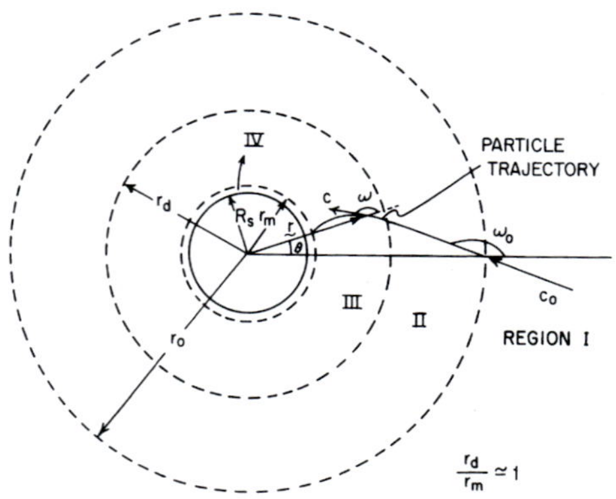

FIG. 2. Trajectory of the fictitious particle of reduced mass.

E of the fictitious particle is conserved and is given by the expression

$$E = \frac{1}{2} m_r \dot{r}^2 + \frac{1}{2} \frac{l^2}{m_r r^2} + U(r) \qquad [51]$$

From Eqs. [50] and [51], the equation of motion of the fictitious particle can be written as

$$\dot{r} = \frac{dr}{dt} = \left\{\frac{2}{m_r}(E - U) - \frac{l^2}{m_r^2 r^2}\right\}^{1/2} \qquad [52]$$

Solving the above equation, one can obtain the magnitude and direction of the velocity vector at the bottom of the potential well (i.e., at the interface between Regions III and IV) in terms of the velocity vector at the interface between regions I and II as

$$c_m^2 = c_0^2 + \frac{2\Phi_0}{m_r} \qquad [53]$$

and

$$\cos \omega_m = \mu_m = \left\{1 - \frac{r_0^2 c_0^2 \sin^2 \omega_0}{r_m^2\left(c_0^2 + \dfrac{2\Phi_0}{m_r}\right)}\right\}^{1/2}, \qquad [54]$$

where Φ_0 is the magnitude of the depth of the potential well, c_m and ω_m are the velocity and the angle between the velocity vector and the radius vector, respectively, at the bottom of the potential well, and r_0 and r_m are the radii of the interfaces between Regions I and II, and Regions III and IV, respectively (see Fig. 1).

In writing the above expressions for the magnitude and direction of the velocity vector at the bottom of the potential well, it has been tacitly assumed that every "particle" that enters Region III eventually reaches the bottom of the potential well. This, in general, need not be the case. In fact, the particle may be deflected before reaching the bottom of the potential well, due to the effect of the centrifugal force experienced by the particle because of its angular momentum. Therefore, there exists a limit on the magnitude of the velocity of the particle, $c_{01}(\omega_0)$, for a given direction at the interface between Regions I and II above which the particle will be deflected before reaching the bottom of the well. This upper limit has its minimum value when

$$\omega_0 = \cos^{-1}\left(1 - \frac{r_d^2}{r_0^2}\right)^{1/2}$$

The minimum value of the limit $c_{01}(\text{min})$ can be obtained from the equation of motion [52] as

$$c_{01}(\text{min}) = \left\{2\Phi_0 \Big/ m_r\left(\frac{r_d^2}{r_m^2} - 1\right)\right\}^{1/2} \quad [55]$$

However, when the decay length of the potential is sufficiently small compared to the radius of the sphere of influence, i.e.,

$$r_d/r_m \simeq r_d/R_S \simeq 1,$$

the minimum value of the upper limit on the magnitude of the velocity, $c_{01}(\text{min})$, is very large. Therefore, it can be assumed that all the particles that enter Region III eventually reach the bottom of the potential well. Of course, this happens because the effect of collisions with the medium is neglected in the inner regions.

Integrating Eq. [48] with respect to **c**, one obtains

$$\nabla_{\mathbf{r}} \cdot \mathbf{j} = 0. \quad [56]$$

Multiplying Eq. [48] by c^2 and integrating, one is led to

$$\nabla_{\mathbf{r}} \cdot \mathbf{S} = \frac{1}{m_r} \mathbf{F} \cdot \mathbf{J}. \quad [57]$$

Equations [56] and [57] give the differential equations for the mass and energy flux vectors in this region. Using the continuity of mass and energy fluxes at the interface between Regions I and II, one obtains the expressions for the scalar mass and energy flux functions at the bottom of the potential well in the form

$$J(r_m) = \frac{A}{r_m^2} \quad [58]$$

and

$$S(r_m) = \frac{r_0^2}{r_m^2} S(r_0) + \frac{\Phi_0}{m_r} J(r_m). \quad [59]$$

One may note that the energy flux in this region is increased by the attractive potential.

Region IV. In this region, the fictitious particle experiences collisionless motion in the presence of a repulsive force. Since the repulsive part of the overall interaction potential decays rapidly, this region is very thin. The fictitious particle will move from the bottom of the potential well until its radial velocity component becomes zero, at which point it will be reflected. The point at which the fictitious particle is reflected depends upon its kinetic energy at the bottom of the potential well. The motion of the fictitious particle in this region corresponds to the relative motion of the two particles of mass m_p when they are extremely close to each other, at the point of collision. Their relative motion will be strongly influenced by the interactions among the surface molecules and hence will depend on the surface roughness. It is, therefore, extremely difficult to describe the motion of the fictitious particle in this region. Because of this inherent difficulty, this region is replaced by a reflecting boundary and the effect of this region on the motion of the fictitious particle is accounted for through the boundary condition at a reflecting spherical surface. If there are no surface imperfections, the fictitious particle will be reflected specularly, i.e., the collision will be perfectly elastic with the normal component of the velocity reversed, while the components parallel to the surface remain unchanged. On the other hand, if the fictitious particle is momentarily trapped by the reflecting surface due to surface imperfections, i.e., the fictitious particle spends enough time on the surface, then the fictitious particle will be reflected diffusively, i.e, the velocity of the reflected particle will be independent of the velocity of the incident particle and the velocity of the reflected particle will be distributed according to the half-range Maxwellian distribution. A situation intermediate to the above extremes can be described using an empirical momentum accommodation coefficient α_c which is the fraction of the incident "particles" that is reflected diffusively.

The reflecting boundary condition, due to Maxwell, can be written in terms of the momentum accommodation coefficient α_c as (4)

$$f^{+*}(\mathbf{c}) = (1 - \alpha_c)f^{-*}\{\mathbf{c} - 2(\mathbf{c}\cdot\mathbf{n})\mathbf{n}\} + \alpha_c n_r\left(\frac{h_r}{\pi}\right)^{3/2} e^{-h_r c^2}, \quad [60]$$

where $f^{+*}(\mathbf{c})$ and $f^{-*}(\mathbf{c})$ are the velocity distribution of the reflected and incident particles at the reflecting surface, respectively, $\mathbf{n}$ is the unit normal vector to the reflecting surface,

$$h_r = \frac{m_r}{2kT_r} \quad [61]$$

and n_r and T_r are parameters to be calculated as shown below.

For an accommodation coefficient α_c equal to unity, the velocity distribution of the reflected particles will be given by the half-range Maxwellian distribution and the boundary condition at the reflecting surface becomes

$$f^{+*}(\mathbf{c}) = n_r\left(\frac{h_r}{\pi}\right)^{3/2} e^{-h_r c^2}. \quad [62]$$

The parameter n_r is determined from the condition that the reflecting spherical surface does not collect particles, i.e.,

$$-\int_{\mathbf{c}_i\cdot\mathbf{n}<0} (\mathbf{c}_i\cdot\mathbf{n})f^{-*}(\mathbf{c}_i)d\mathbf{c}_i = \int_{\mathbf{c}_r\cdot\mathbf{n}>0} (\mathbf{c}_r\cdot\mathbf{n})f^{+*}(\mathbf{c}_r)d\mathbf{c}_r, \quad [63]$$

where $\mathbf{c}_i$ and $\mathbf{c}_r$ refer to the velocity of the incident and reflected "particles," respectively.

Defining N_i as the flux of the particles incident upon the reflecting surface, one is led to

$$N_i = -\int (\mathbf{c}_i\cdot\mathbf{n})f^{-*}(\mathbf{c}_i)d\mathbf{c}_i. \quad [64]$$

From Eqs. [60], [61], [63] and [64], one obtains

$$n_r = 2N_i\left(\frac{m_r\pi}{2kT_r}\right)^{1/2}. \quad [65]$$

The parameter T_r is the temperature corresponding to the half-range Maxwellian distribution of the particles reflected diffusively. The temperature T_r is related to the temperature of the reflecting surface through the thermal accommodation coefficient α_E, defined as

$$\alpha_E = \frac{E_i - E_r}{E_i - E_s},$$

where E_i is the energy flux of the incident "particles," E_r is the energy flux of the reflected "particles," and E_s is the energy flux of the reflected "particles" if they were in equilibrium with the reflecting surface. The thermal accommodation coefficient describes the extent to which the fictitious particle accommodates itself to the surface temperature before reflection. The values of both the momentum and thermal accommodation coefficients are needed in order to describe the velocity distribution of the reflected particles. In the absence of enough information regarding the nature of the surface of colliding particles, we assume that the particles of mass m_p upon collision stay in contact long enough (due to surface roughness) to reach thermal equilibrium with the surface and be reflected diffusively. In other words, in the present model, the momentum and thermal accommodation coefficients are assumed to be unity. Therefore, the parameter T_r can be identified as the surface temperature and the reflecting boundary condition is given by Eq. [62]. The surface temperature T_r can then be defined as

$$kT_r = \tfrac{2}{3}e, \quad [66]$$

where

$$e = \frac{1}{n}\int \frac{1}{2} m_r C^2 f d\mathbf{c}, \quad [67]$$

$$n = \int f d\mathbf{c}, \quad [68]$$

$$\mathbf{C} = \mathbf{c} - \mathbf{u} \quad [69]$$

and

$$\mathbf{u} = \int \mathbf{c} f d\mathbf{c} \Big/ \int f d\mathbf{c}. \quad [70]$$

In the above equations, e, n, and $\mathbf{u}$ are the internal energy, the number density, and the average velocity of the fictitious particle, respectively, at the reflecting spherical surface. Since the flux of particles at the reflecting surface is zero (i.e., $\mathbf{u} = 0$), Eq. [66] can be rewritten as

$$kT_r\left\{\int_{\mu<0} g^-(r_m, v_m, \mu_m) d\mathbf{v}_m + \int_{\mu>0} g^{+*}(v, \mu) d\mathbf{v}\right\} = \frac{1}{3h^2} m_r \times \left\{\int_{\mu<0} v_m^2 g^-(r_m, v_m, \mu_m) d\mathbf{v}_m + \int_{\mu>0} v^2 g^{+*}(v, \mu) d\mathbf{v}\right\}, \quad [71]$$

where, the dimensionless velocity distribution as defined by Eq. [23],

$$g(r, v, \mu) = \begin{cases} g^+(r, v, \mu) & \text{for} \quad \mu \geq 0 \text{ (particles moving away from the reflecting surface)} \\ g^-(r, v, \mu) & \text{for} \quad \mu \leq 0 \text{ (particles moving toward the reflecting surface)} \end{cases}$$

and $g^{+*}(v, \mu)$ is the dimensionless velocity distribution of the reflected particles at the reflecting surface.

Because the motion of the fictitious particle is considered to be collisionless, in the Regions II and III, the distribution function of the incoming particles is conserved along their trajectories, i.e.,

$$g^-(r_m, v_m, \mu_m) = g^-(r_0, v, \mu) \quad [72]$$

along the particle trajectory, where v_m is the magnitude of the dimensionless velocity $h\mathbf{c}$ at the bottom of the potential well. Equation [71] can therefore be expressed in terms of the distribution function of the incoming particles at the interface between Regions I and II $g^-(r_0, v, \mu)$.

Let us introduce the notations

$$\kappa = \frac{\zeta^{-1}}{(R_s/2\bar{c})} = \frac{4}{(2\pi)^{1/2} h \zeta r_m} \simeq \frac{4}{(2\pi)^{1/2} h \zeta R_S},$$

$$R = \frac{r_m}{r_0} \simeq \frac{R_S}{r_0} = \frac{1}{1+\kappa}$$

$$\mu_0 \simeq (1 - R^2)^{1/2} = \cos \omega_0,$$

$$\Phi = \frac{\Phi_0}{kT},$$

$$\theta = \frac{T_r}{T} \quad [73]$$

and

$$\delta = \left(\frac{30}{11}\right)^{1/2} \frac{4}{(2\pi)^{1/2}}$$

Here, κ, the effective particle Knudsen number, is the ratio of the relaxation time ζ^{-1} for Brownian motion of the particle to the time needed for the fictitious particle to traverse the distance R_S with its average velocity, θ is the dimensionless surface temperature, and Φ is the dimensionless depth of the potential well.

Using the expression for the distribution function at the interface of Regions I and II given by Eq. [42] and the reflecting boundary condition (Eq. [62]), the surface temperature defined by Eq. [71] can be written in terms of dimensionless variables as

$$\theta\left[\frac{n_\infty R_S^2}{(2\pi)^{1/2}h}\frac{(1-\mu_0)}{2R^2} + A\left\{\frac{(1-\mu_0)}{\pi R\kappa} - \frac{R^2}{2\pi} + \frac{3}{16}(\mu_0 - \mu_0^3)R\kappa\right\} + Bh^2R_S^2 e^{-\delta/\kappa R} \times \left\{\left(-\frac{3}{40}\frac{(1-\mu_0)}{R^2} + \frac{3}{80}\frac{(\mu_0-\mu_0^3)}{R^2} + \frac{1}{10\pi\delta}\right)\kappa R + \left(\frac{3}{40}\frac{(\mu_0-\mu_0^3)}{R^2\delta} + \frac{1}{10\pi\delta^2}\right)\right.$$

$$
\times \kappa^2R^2 + \frac{3}{40}\frac{(\mu_0 - \mu_0^3)}{R^2\delta^2}\kappa^3R^3\Big\}\Big]
$$

$$
= \left(1 + \frac{2\Phi}{3}\right)\Big[\frac{n_\infty R_S^2}{(2\pi)^{1/2}h}\frac{(1-\mu_0)}{2R^2} + A\Big\{\frac{(1-\mu_0)}{\pi R\kappa} - \frac{2R^2}{3\pi} + \frac{5}{16}(\mu_0 - \mu_0^3)\kappa R\Big\} + Bh^2R_S^2e^{-\delta/\kappa R}
$$

$$
\times \Big\{\left(-\frac{3}{40}\frac{(1-\mu_0)}{R^2} + \frac{1}{24}\frac{(\mu_0-\mu_0^3)}{R^2} - \frac{2}{15\pi\delta}\right)\kappa R + \left(\frac{1}{8}\frac{(\mu_0-\mu_0^3)}{R^2\delta} - \frac{2}{15\pi\delta^2}\right)
$$

$$
\times \kappa^2R^2 + \frac{1}{8}\frac{(\mu_0-\mu_0^3)}{R^2\delta^2}\kappa^3R^3\Big\}\Big] \qquad [74]
$$

In obtaining the above equation, the following approximation was made

$$
v_m^2 \simeq v^2\left(1 + \frac{2\Phi}{\overline{v^2}}\right) \simeq v^2\left(1 + \frac{2\Phi}{3}\right),
$$

since the average of the square of the dimensionless relative velocity $\overline{v^2} = 3$ for a Maxwellian distribution.

Boundary Conditions

We have thus far obtained the distribution functions for the incoming and reflected particles of reduced mass in different regions in terms of the unknown constants A and B and the unknown dimensionless surface temperature θ. The surface temperature was then related to the internal energy and the number density of the fictitious particle at the reflecting surface. Now, in order to determine the constants A and B, we must specify the boundary conditions for the mass and the energy flux at the sphere of influence.

Since the motion of the fictitious particle in Regions II and III is assumed to be collisionless, a reflected particle of reduced mass m_r with kinetic energy less than the depth of the potential well Φ_0 will be captured by the potential well. Therefore, the boundary condition for the mass flux is given by

$$
\frac{A}{R_S^2} = -2\pi\int_0^{(2\Phi_0/m_r)^{1/2}}\int_0^1 c^3\mu n_r\left(\frac{h_r}{\pi}\right)^{3/2} \times e^{-h_rc^2}d\mu dc. \qquad [75]
$$

The energy of the fictitious particle is not conserved during collision, since the collision is not elastic. Therefore, the net energy flux will consist of (i) the energy lost during collision (the difference between the energy fluxes of incident and reflected particles), (ii) the energy of the captured particles (energy of particles with kinetic energy less than the depth of the potential well). Using Eqs. [59], [62], and [73], the boundary condition for the energy flux can therefore be written as

$$
\frac{r_0^2}{R_S^2}S(r_0) + \frac{A}{R_S^2}\frac{\Phi_0}{m_r} = \frac{1}{2}\int c_m^3\mu_m f^-(R_S, \mathbf{c}_m)d\mathbf{c}_m + \pi\int_{(2\Phi_0/m_r)^{1/2}}^{\infty}\int_0^1 c^5\mu n_r\left(\frac{h_r}{\pi}\right)^{3/2}e^{-h_rc^2}d\mu dc. \qquad [76]
$$

Using the expression for the distribution function of the incoming particles at the interface of Regions I and II (Eq. [42]) and the reflecting boundary condition (Eq. [62]) the above boundary conditions (Eqs. [75] and [76]) can be written in the form

$$
A = \Big[-\frac{n_\infty R_S^2}{(2\pi)^{1/2}h} + A\Big\{\frac{1}{2}(1-\mu_0^3) - \frac{2R}{\pi\kappa} - \frac{(1+2\mu_0^2-3\mu_0^4)}{4}\kappa R\Big\} + Bh^2R_S^2e^{-\delta/\kappa R}
$$

$$
\times\Big\{\frac{3}{20}\kappa R - \frac{1}{40}\left(\frac{4(1-\mu_0^4)}{R^2} - 1\right)\kappa R - \frac{1}{10}\frac{(1+2\mu_0^2-3\mu_0^4)}{R^2\delta}\kappa^2R^2 - \frac{1}{10}\frac{(1+2\mu_0^2-3\mu_0^4)}{R^2\delta^2}\kappa^3R^3\Big\}\Big]
$$

$$
\times\left[1 - e^{-\phi/\theta}\left(1 + \frac{\Phi}{\theta}\right)\right] \qquad [77]
$$

and

$$
\frac{5}{2}A + \frac{Bh^2R_S^2e^{-\delta/\kappa R}}{\delta R}\left(1 + \frac{\kappa R}{\delta}\right) + A\Phi
$$

$$
= \left(1 + \frac{2\Phi}{3}\right)\Big[-\frac{2n_\infty R_S^2}{(2\pi)^{1/2}h} + A\Big\{\frac{5}{4}(1-\mu_0^3) - \frac{4R}{\pi\kappa} - \frac{3}{4}(1+2\mu_0^2-3\mu_0^4)\kappa R\Big\}
$$

$$+ Bh^2R_S^2e^{-\delta/\kappa R}\left\{\frac{3}{10}\kappa R - \frac{3}{10} \times \frac{(1+2\mu_0^2-3\mu_0^4)}{R^2\delta}\kappa^2R^2 - \frac{3}{10} \times \frac{(1+2\mu_0^2-3\mu_0^4)}{R^2\delta^2}\kappa^3R^3 - \frac{1}{20} \times \left(1+\frac{6(1-\mu_0^4)}{R^2}\right)\kappa R + \frac{1}{2}\left(1-\mu_0^3\right)\frac{\kappa}{\delta R} \times \left(1+\frac{\kappa R}{\delta}\right)\right\}\Bigg] - 2\theta e^{-\Phi/\theta}\left(1+\frac{\Phi}{\theta}+\frac{1}{2}\frac{\Phi^2}{\theta^2}\right) \times \Bigg[-\frac{n_\infty R_S^2}{(2\pi)^{1/2}h} + A\left\{\frac{1}{2}(1-\mu_0^3) - \frac{2R}{\pi\kappa} - \frac{(1+2\mu_0^2-3\mu_0^4)}{4}\kappa R\right\} + Bh^2R_S^2e^{-\delta/\kappa R} \times \left\{\frac{3}{20}\kappa R - \frac{1}{40}\left(\frac{4(1-\mu_0^4)}{R^2}-1\right)\kappa R - \frac{1}{10}\frac{(1+2\mu_0^2-3\mu_0^4)}{R^2\delta}\kappa^2R^2 - \frac{1}{10}\frac{(1+2\mu_0^2-3\mu_0^4)}{R^2\delta^2}\kappa^3R^3\right\}\Bigg] \quad [78]$$

where the dimensionless quantities: R, κ, μ_0, Φ, and θ are defined by Eq. [73] and δ is a numerical constant also given in Eq. [73].

In order to evaluate the rate of coagulation, Eqs. [74], [77], and [78] must be solved for the constants A, B and the unknown dimensionless surface temperature θ. After considerable algebra, one can obtain the functional form of the expression for the constant A as

$$A = -\frac{n_\infty R_S^2\kappa}{(2\pi)^{1/2}h}\left[\left(1-e^{-\Phi/\theta}\left(1+\frac{\Phi}{\theta}\right)\right) \times (a_1+a_2\kappa+a_3\kappa^2)\Big/\left\{\left(1-e^{-\Phi/\theta} \times \left(1+\frac{\Phi}{\theta}\right)\right)(A_1+A_2\kappa+A_3\kappa^2+A_4\kappa^3) + (B_1\kappa R + B_2\kappa^2R^2 + B_3\kappa^3R^3)\right\}\right], \quad [79]$$

where the coefficients a_1, a_2, a_3, A_1, A_2, A_3, A_4, B_1, B_2, B_3 are functions of the dimensionless depth of the potential Φ, the dimensionless surface temperature θ, and μ_0 and R. The expressions for these coefficients in terms of the above variables can be found in Appendix II.

The rate of coagulation, ψ, of particles of mass m_p is equal to the net rate of flow of particles of reduced mass m_r across the sphere of influence of radius R_S, i.e.,

$$\psi = 4\pi R_S^2 J(R_S) = 4\pi A \quad [80]$$

Therefore, the coagulation coefficient β_1 can be calculated from

$$\beta_1 = \frac{\psi}{n_\infty} = \frac{4\pi A}{n_\infty} \quad [81]$$

The coagulation coefficient is a function of the radius of the particle R_p, its mass m_p, the effective particle Knudsen number κ, the temperature of the medium T, and the depth of the interaction potential well between two particles. Using the expression for the overall interaction potential given in Appendix I, the depth of the interaction potential well can be calculated from the knowledge of the Hamaker constant for the particle. The friction coefficient ζ is related to the diffusion coefficient of the particle, D, through the Einstein equation

$$D = \frac{kT}{m_p\zeta}. \quad [82]$$

The effective particle Knudsen number κ, defined by the Eq. [73], can be related to the diffusion coefficient of the particle as

$$\kappa = \frac{8hD}{(2\pi)^{1/2}R_S}. \quad [83]$$

Using the slip correction factor due to Philips (3) for an accommodation coefficient of unity, the expression for the diffusion coefficient of the particle in terms of the Knudsen number $Kn(=\lambda_g/R_p)$ and the viscosity of the medium η is given by

$$D = \frac{kT}{6\pi R\eta}\left(\frac{5+4Kn+6Kn^2+18Kn^3}{5-Kn+(8+\pi)Kn^2}\right). \quad [84]$$

From Eqs. [79], [81], and [83], it results that the coagulation coefficient β_1 is given by the expression

$$\beta_1 = 16R_SD\left[\left(1 - e^{-\Phi/\theta}\left(1 + \frac{\Phi}{\theta}\right)\right)\right.$$
$$\times (a_1 + a_2\kappa + a_3\kappa^2)\Big/\left\{\left(1 - e^{-\Phi/\theta}\left(1 + \frac{\Phi}{\theta}\right)\right)\right.$$
$$\times (A_1 + A_2\kappa + A_3\kappa^2 + A_4\kappa^3)$$
$$\left.\left. + (B_1\kappa R + B_2\kappa^2R^2 + B_3\kappa^3R^3)\right\}\right]. \quad [85]$$

The expressions for the coefficients a_1, a_2, a_3, A_1, A_2, A_3, A_4, B_1, B_2, and B_3 in the above equation can be found in Appendix II.

The limiting behavior of the coagulation coefficient β_1 in the continuum and the free molecular regimes can now be examined. In the continuum limit, where $Kn \to 0$, one has

$$\kappa \to 0, \quad \mu_0 \to 1, \quad \text{and} \quad R \to 1$$

and Eq. [85] reduces to the Smoluchowski's equation

$$\lim_{Kn\to 0} \beta_1 = 8\pi R_SD \quad [86]$$

In the free molecular limit, where $Kn \to \infty$, one has

$$\kappa \to \infty, \quad R \to 0, \quad \kappa R \to 1, \quad \Phi \to 0,$$
$$\theta \to 1, \quad \mu_0 \to \tfrac{1}{2}R^2$$

and Eq. [85] reduces to

$$\lim_{Kn\to\infty} \beta_1 = \pi R_S^2(2\bar{c}^2)^{1/2}\{1 - e^{-\Phi}(1 + \Phi)\}, \quad [87]$$

where $\bar{c} = (8kT/\pi m_p)^{1/2}$ is the average velocity of a particle of mass m_p.

The first factor in Eq. [87] is the free molecular limit for the rate of collisions between particles of mass m_p, whereas the second factor can be interpreted as the collision efficiency (sticking probability). The collision efficiency becomes negligible for very small particles, since in such cases the overall interaction potential becomes vanishingly small.

Range of Validity of the Model

As pointed out earlier, this model neglects the effect of collisions with the medium on the motion of the particle of reduced mass in Regions II and III. This is an idealization since the average distance travelled by the particle between the successive collisions with the molecules of the medium is much smaller (of the order of 1 Å) than the thickness of these regions. The collisions experienced by a reflected particle with the medium during its escape from the potential well increases the probability of capture. This is a consequence of the fact that the minimum threshold kinetic energy necessary for the escape becomes greater than the depth of the potential well, since the particle has also to overcome the friction due to the collisions with the medium. The enhancement of the coagulation coefficient as a result of collisions depends upon the time spent by a particle in the region in which the potential is acting. The deeper the potential well and the larger the decay length of the potential, the longer will a particle stay in the region of the interaction potential.[3] For small particles, the potential well is not deep, i.e., the dimensionless depth of the potential Φ is smaller than unity. Moreover, as the particle size decreases, the decay length of the potential also decreases. In other words, for small particles, the time spent in the region of the interaction potential is small and consequently the enhancement of the coagulation coefficient due to the collisions with the medium is also small. Therefore, the lower bound for the coagulation coefficient is likely to constitute a satisfactory approximation for sufficiently small particles. On the other hand, for large particles, the potential well is deep, i.e., the dimensionless depth of the potential Φ is much greater than unity. In addition,

[3] The effect of collisions with the medium on the flux of the incoming particles is not as important because their residence time in the inner regions, before collision with the other particle, is smaller than the relaxation time ζ^{-1}, because of the attractive interactions between the two particles.

the decay length of the potential while increasing with the size of the particles becomes small compared to the latter quantity. As a result, the interaction potential can be replaced by a sink at the sphere of influence. This is equivalent to a sticking probability equal to unity. The expression for the coagulation coefficient of large particles is derived in the next section.

3. COAGULATION COEFFICIENT OF LARGE PARTICLES

As already pointed out earlier, the potential well for the large particles is very deep. In addition, the decay length of the interaction potential is small compared to the radius of the particles. Consequently, the interaction potential can be replaced by a sink at the sphere of influence. Therefore, Regions III and IV in Fig. 1 will collapse into a sink at the sphere of influence and only the motion of the fictitious particle in Regions I and II need to be considered. The equations governing the motion of the fictitious particle in these regions are the same as those discussed earlier in Section 2.

Integrating Eq. [46] with respect to **c**, one obtains

$$\nabla_{\mathbf{r}}\ \mathbf{j} = 0 \qquad [88]$$

Multiplying Eq. [46] by c^2 and integrating, one is led to

$$\nabla_{\mathbf{r}}\ \mathbf{S} = 0 \qquad [89]$$

Equations [88] and [89] provide the differential equations for the mass and energy flux vectors in Region II. Invoking the continuity of mass and energy fluxes at the interface of Regions I and II, the scalar mass and energy flux functions at the sphere of influence are given by

$$J(R_S) = \frac{A}{R_S^2} \qquad [90]$$

and

$$S(R_S) = \frac{r_0^2}{R_S^2} S(r_0) \qquad [91]$$

The mass and energy fluxes at the sphere of influence are equal to the mass and energy fluxes of the incident particles because all the incident particles are captured by the sink. The boundary conditions for mass and energy fluxes can therefore be written as

$$\frac{A}{R_S^2} = \frac{1}{R_S} \int_{(\mathbf{R}_S\ \mathbf{c})<0} (\mathbf{R}_S\ \mathbf{c}) f^-(\mathbf{R}_S, \mathbf{c}) d\mathbf{c} \qquad [92]$$

and

$$\frac{r_0^2}{R_S^2} S(r_0) = \frac{1}{2R_S} \int_{(\mathbf{R}_S\ \mathbf{c})<0} (\mathbf{R}_S\ \mathbf{c}) c^2 f^-(\mathbf{R}_S, \mathbf{c}) d\mathbf{c} \qquad [93]$$

Using the expression for the distribution function of the incoming particles at the interface between Regions I and II (Eq. [42]), the above boundary conditions become

$$A = -\frac{n_\infty R_S^2}{(2\pi)^{1/2} h} + A\left\{\frac{1}{2}(1-\mu_0^3) - \frac{2R}{\pi\kappa} - \frac{(1+2\mu_0^2-3\mu_0^4)}{4}\kappa R\right\} + Bh^2 R_S^2 e^{-\delta/\kappa R} \times \left\{\frac{3}{20}\kappa R - \frac{1}{40}\left(\frac{4(1-\mu_0^4)}{R^2} - 1\right)\kappa R - \frac{1}{10}\frac{(1+2\mu_0^2-3\mu_0^4)}{R^2\delta}\kappa^2 R^2 - \frac{1}{10}\frac{(1+2\mu_0^2-3\mu_0^4)}{R^2\delta^2}\kappa^3 R^3\right\} \qquad [94]$$

and

$$\frac{5}{2}A + Bh^2 R_S^2 e^{-\delta/\kappa R}\frac{\kappa}{\delta R}\left(1+\frac{\kappa R}{\delta}\right) = -\frac{2n_\infty R_S^2}{(2\pi)^{1/2} h} + A\left\{\frac{5}{4}(1-\mu_0^3) - \frac{4R}{\pi\kappa} - \frac{3}{4}(1+2\mu_0^2-3\mu_0^4)\kappa R\right\} + Bh^2 R_S^2 e^{-\delta/\kappa R} \times \left\{\frac{3}{10}\kappa R - \frac{3}{10}\frac{(1+2\mu_0^2-3\mu_0^4)}{R^2\delta}\kappa^2 R^2 - \frac{3}{10}\frac{(1+2\mu_0^2-3\mu_0^4)}{R^2\delta^2}\kappa^3 R^3 - \frac{1}{20}\left(1+\frac{6(1-\mu_0^4)}{R^2}\right)\kappa R + \frac{1}{2}(1-\mu_0^3)\frac{\kappa}{\delta R}\left(1+\frac{\kappa R}{\delta}\right)\right\}. \qquad [95]$$

In order to calculate the coagulation coefficient β_u, Eqs. [94] and [95] must be solved for the unknown constants A and B. It results that the constant A is given by

$$A = -\frac{n_\infty R_S^2 \kappa}{(2\pi)^{1/2} h} \times \left[\frac{a_s + b_s\kappa + c_s\kappa^2}{A_s + B_s\kappa R + C_s\kappa^2 R^2 + D_s\kappa^3 R^3}\right], \quad [96]$$

where the coefficients a_s, b_s, c_s, A_s, B_s, C_s, D_s are functions of μ_0 and R. The expressions for these coefficients can be found in Appendix II.

From Eqs. [80] to [83], the coagulation coefficient β_u can be expressed in terms of the effective particle Knudsen number κ and the diffusion coefficient of the particle D as

$$\beta_u = 16 R_S D \times \left[\frac{a_s + b_s\kappa + c_s\kappa^2}{A_s + B_s\kappa R + C_s\kappa^2 R^2 + D_s\kappa^3 R^3}\right], \quad [97]$$

where the diffusion coefficient of the particle D is given by Eq. [84].

In the continuum limit, where $Kn \rightarrow 0$, one has

$$\kappa \rightarrow 0, \quad \mu_0 \rightarrow 1, \quad \text{and} \quad R \rightarrow 1,$$

Eq. [97] reduces to the Smoluchowski equation

$$\lim_{Kn \rightarrow 0} \beta_u = 8\pi R_S D. \quad [98]$$

In the free molecular limit, where $Kn \rightarrow \infty$, one has

$$\kappa \rightarrow \infty, \quad R \rightarrow 0, \quad \kappa R \rightarrow 1,$$

$$\text{and} \quad \mu_0 \rightarrow \tfrac{1}{2}R^2,$$

Eq. [97] reduces to

$$\lim_{Kn \rightarrow \infty} \beta_u = \pi R_S^2 (2\bar{c}^2)^{1/2}. \quad [99]$$

Therefore, the expression for the coagulation coefficient, which is valid for large particles, reduces to the free molecular limit for the rate of collisions when $Kn \rightarrow \infty$.

The coagulation coefficient given by Eq. [97] is referred to in what follows as the upper bound, even though, as explained later, it constitutes an upper bound only for sufficiently small particles.

4. COMPARISON OF THE MODEL WITH EXISTING MODELS AND EXPERIMENTAL DATA

The lower bound for the coagulation coefficient, β_l, is calculated from Eq. [81] using the values of the constant A determined from the numerical solution of Eqs. [74], [77], and [78]. The upper bound for the coagulation coefficient, β_u, is obtained from Eq. [97]. In the above calculations, the Philips slip correction factor (Eq. [6]), is used for the calculation of the diffusion coefficient of particles. The results are expressed in terms of the dimensionless coagulation coefficient γ defined as the ratio between the coagulation coefficient and the Smoluchowski coagulation coefficient

$$\gamma = \beta/\beta_S = \beta \frac{3\eta}{8kT}. \quad [100]$$

For particles of unit density, in air, at a pressure of 1 atm and temperature of 298°K, the upper and lower bounds of the dimensionless coagulation coefficients are plotted as a function of Knudsen number, for different Hamaker constants, in Fig. 3. Obviously, the upper bound for the coagulation coefficient is independent of the Hamaker constant. The upper and the lower bounds tend to the Smoluchowski expression for small Knudsen numbers. For large Knudsen numbers, the upper bound coincides with the free molecular limit, as can be seen from Fig. 3. The lower bound is found to decrease dramatically with a decrease in the Hamaker constant, for large Knudsen numbers. Both the lower and the upper bounds exhibit a maximum at intermediate values of the Knudsen number.

Table I compares the dimensionless coagulation coefficient predicted by the present model with other models. Since the Hamaker constant for most of the aerosol systems is of the order of 10^{-12} erg, this value is used in the calculation of the lower bound. Particle diffusion coefficients based on Philips slip correction factor for an accommodation coefficient of unity are used for the calculation of the coagulation coefficients β_F (the Fuchs interpolation formula) and β_{SS} (the Sitarski

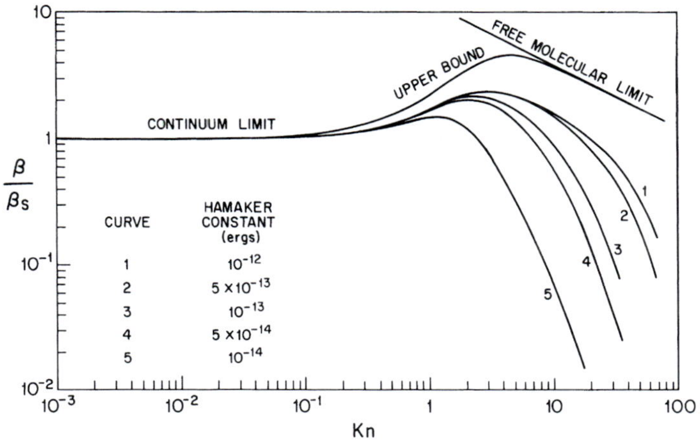

FIG. 3. Dimensionless coagulation coefficients predicted by the upper and the lower bounds for equal-sized particles of unit density, in air, at 1 atm and 298°K.

and Seinfeld (6) formula). The upper bound for the coagulation coefficient predicts higher coagulation coefficient than the Fuchs formula with a maximum deviation of about 13% in the transition regime. Compared to the treatment of Sitarski and Seinfeld (6), the upper bound predicts coagulation coefficients that are 20% higher in the transition regime, the deviation becoming as high as 50% for large Knudsen numbers. The upper and lower bounds for the coagulation coefficient predicted by the present model at different temperatures and for Hamaker constants of 10^{-12} and 10^{-13} erg are shown in Figs. 4 and 5, respectively. The upper bound for the coagulation coefficient β_u increases with an increase in the temperature, due to the increase in the diffusion coefficient of the particles. For a Hamaker constant of 10^{-12} erg, the lower bound β_l increases with temperature for particle sizes as small as 0.003 μm. However, for a Hamaker constant of 10^{-13} erg, the lower bound β_l increases with temperature for large particles, but decreases with an increase in the temperature for small particles. This behavior can be attributed to two contrary effects. On the one hand, an increase in temperature increases the rate of collisions

TABLE I

Comparison of the Coagulation Coefficients Predicted by Different Models for Equal-Sized Particles, of Unit Density, in Air, at 1 atm[a]

Particle radius (μm)	Kn	β_F/β_s (Fuchs)	β_u/β_s (Upper bound)	β_l/β_s (Lower bound)	β_{ss}/β_s (Sitarski and Seinfeld (6))
0.001	68.60	1.501	1.501	0.178	1.001
0.002	34.30	2.119	2.119	0.618	1.444
0.003	22.87	2.582	2.585	0.920	1.836
0.005	13.72	3.257	3.283	1.329	2.567
0.010	6.86	3.999	4.271	1.905	3.720
0.040	1.715	2.687	3.167	2.165	2.752
0.10	0.686	1.600	1.716	1.483	1.610
0.50	0.137	1.100	1.128	1.079	1.105
1.00	0.0686	1.049	1.067	1.036	1.052
5.00	0.0137	1.007	1.014	1.002	1.008

[a] The lower bound values reported are for a Hamaker constant of 10^{-12} erg.

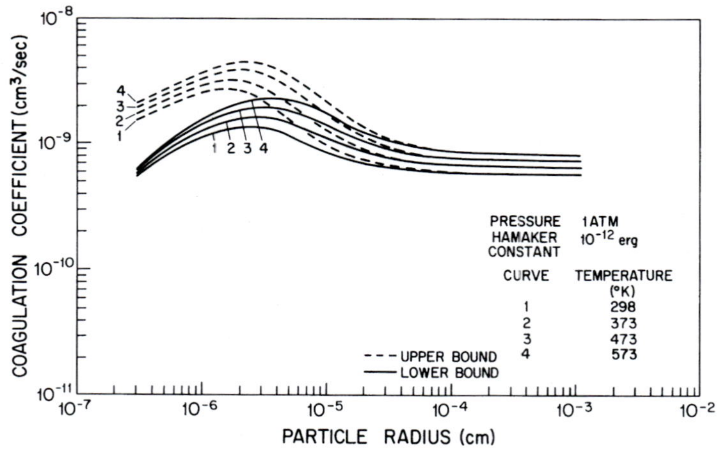

FIG. 4. Predicted coagulation coefficients for equal-sized particles of unit density, in air, at 1 atm, for different temperatures, and for a Hamaker constant of 10^{-12} erg.

between particles, due to larger diffusion coefficients. On the other hand, the fraction of colliding particles that escape from the potential well increases, resulting in a smaller sticking probability. The former effect increases, whereas the latter effect decreases the rate of coagulation. For a Hamaker constant of 10^{-13} erg, the depth of the potential well for very small particles is comparable in the temperature range of 298 to 573°K, to their mean kinetic energy. As a result, at a higher temperature, a larger number of particles can escape from the potential well leading to a lower coagulation rate. However, for a Hamaker constant of 10^{-12} erg, the depth of the potential well is much greater than the mean kinetic energy of the particles, even for particles as small as 0.003 μm. Therefore, an increase in the temperature does not appreciably decrease in this case the sticking probability and the coagulation coefficient increases with the temperature.

The effect of pressure on both the upper and lower bounds is shown in Figs. 6 and 7 for Hamaker constants of 10^{-12} and 10^{-13} erg, respectively. Both the upper and the lower bounds decrease with the increase in pressure because the diffusion coefficient becomes smaller. However, the effect of pressure on the upper and the lower bounds for

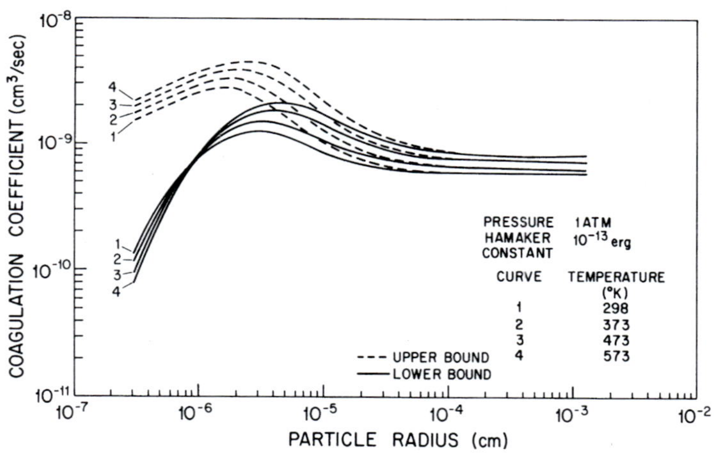

FIG. 5. Predicted coagulation coefficients for equal-sized particles of unit density, in air, at 1 atm, for different temperatures, and for a Hamaker constant of 10^{-13} erg.

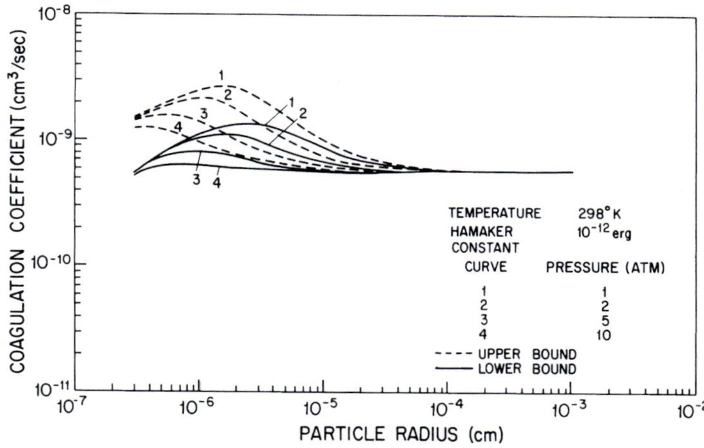

FIG. 6. Predicted coagulation coefficients for equal-sized particles of unit density, in air, at 298°K, for different pressures, and for a Hamaker constant of 10^{-12} erg.

particles of very small and very large sizes is negligible. This happens because the dependence of the diffusion coefficient on pressure occurs via the mean free path (which appears in the slip correction factor through the Knudsen number) and viscosity of the medium. For sufficiently large particles $Kn \rightarrow 0$ and for very small particles $Kn \rightarrow \infty$. In addition, the viscosity of the medium does not vary appreciably in the pressure range of 1 to 10 atm.

Wagner and Kerker (12) and Chatterjee *et al.* (13) have measured the coagulation coefficients for monodispersed aerosols of diethyl hexyl sebacate. In their experiments, the Knudsen number was varied by varying the pressure of the medium (He) for particles of almost constant size. Their experimental data are compared with different models in Tables II and III. Comparison of the experimental data with the upper bound, the lower bound (for a Hamaker constant of 10^{-12} erg), and the Fuchs interpolation formula is also shown in Fig. 8. Since the Knudsen number was varied by varying the pressure of the medium for particles of constant size, the upper and the lower bounds approach the limiting values for large Knudsen numbers. The experimental data of Wagner and Kerker (12) lie between the upper bound and the Fuchs for-

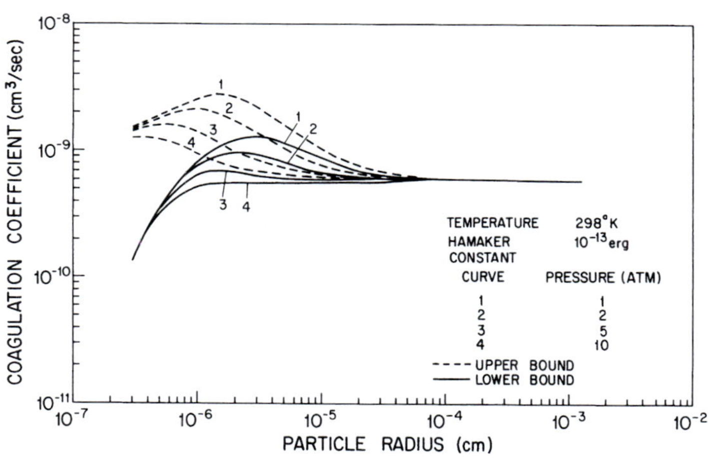

FIG. 7. Predicted coagulation coefficients for equal-sized particles of unit density, in air, at 298°K, for different pressures, and for a Hamaker constant of 10^{-13} erg.

TABLE II

Comparison of Different Models with the Experimental Data of Wagner and Kerker (12)[a]

Particle radius (μm)	Kn	β_{ex}/β_S (Experimental)	β_u/β_S (Upper bound)	β_l/β_S (Lower bound)	β_{SS}/β_S (Sitarski and Seinfeld)	β_F/β_S (Fuchs)
0.198	1.02	2.20	2.192	1.964	2.10	2.078
0.209	1.91	3.42	3.622	3.026	3.368	3.317
0.190	2.95	4.95	5.246	4.019	4.725	4.631
0.188	8.52	11.03	12.021	7.048	10.275	10.039
0.184	12.2	14.14	14.792	7.867	12.655	12.474

[a] The lower bound values reported are for a Hamaker constant of 10^{-12} erg.

mula. The maximum deviation between the experimental data and the upper bound is about 8% whereas the maximum deviation with the Fuchs formula is about 13%. The experimental data for Chatterjee *et al.* (13) seem to agree better with the lower bound with a maximum deviation of about 9%.

Experimental data of the coagulation coefficient for NaCl aerosols of different sizes obtained by Shon *et al.* (14) are compared with the upper bound, with the lower bound (for a Hamaker constant of 10^{-12} erg) and with the Fuchs interpolation formula in Fig. 9. The experimental data exhibit considerable scatter; the upper bound for the coagulation coefficient agrees somewhat better with the experimental data than the Fuchs interpolation formula.

Experimental data regarding the coagulation coefficients of Pt aerosols, measured by Nolan and Keenan (15) are compared with different models in Table IV. The corrections for polydispersity in the above data, as reported by Mercer (16), are accounted for in the experimental values of the coagulation coefficient. The apparent density of Pt aerosol of 0.62 gm/cm^3, reported by Nolan and Keenan (15), from the simultaneous measurements of the effective diffusivity and settling velocity, is used for the calculation of the coagulation coefficients for different models instead of the density of Pt. Comparison of the experimental data with the upper bound, with the lower bound (for a Hamaker constant of 10^{-12} erg) and with the Fuchs interpolation formula is shown in Fig. 10. The experimental data lie between the two bounds and agree with the Fuchs interpolation except for the particles with sizes of 0.006 and 0.0076 μm. The latter coagulation coefficients are by 1.4 to 3 times greater than those predicted by the upper bound.

In order to explain this behavior, one may note that, as the particle size decreases, the potential well becomes less deep, whereas the ratio of the decay length of the potential and the particle size increases. Consequently, there are two opposing effects: (1) the collisions become less effective because of the increase in the probability of escape of the particles

TABLE III

Comparison of Different Models with the Experimental Data of Chatterjee *et al.* (13)[a]

Particle radius (μm)	Kn	β_{ex}/β_S (Experimental)	β_u/β_S (Upper bound)	β_l/β_S (Lower bound)	β_{SS}/β_S (Sitarski and Seinfeld)	β_F/β_S (Fuchs)
0.23	0.83	1.59	1.938	1.746	1.843	1.827
0.23	1.10	2.06	2.373	2.081	2.222	2.20
0.23	1.66	2.78	3.237	2.764	3.028	2.987

[a] The lower bound values reported are for a Hamaker constant of 10^{-12} erg.

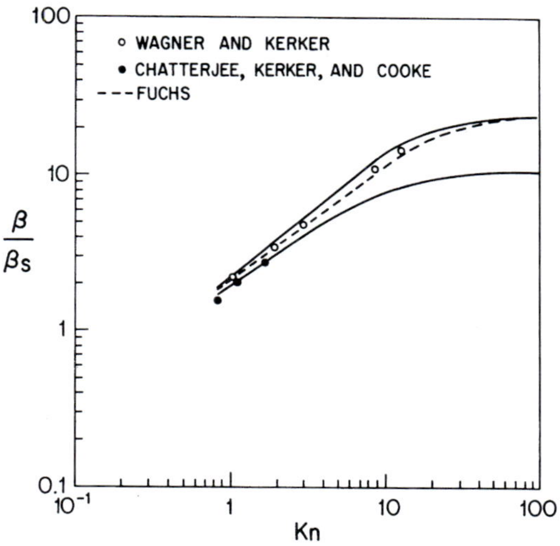

FIG. 8. Comparison of the upper bound, the lower bound (for a Hamaker constant of 10^{-12} erg), and the Fuchs interpolation formula with the experimental coagulation coefficients obtained by Wagner and Kerker (12) and Chatterjee *et al.* (13).

from the potential well, and (2) the rate of collisions between the particles increases because of the increase in the collision cross section due to the greater range of the van der Waals attraction relative to the particle size. The former effect decreases, whereas the latter increases the rate of coagulation. In a certain range of intermediate particle sizes, the latter effect may predominate so that the coagulation coefficient may even exceed the upper bound. In addition, the very low value of the apparent density of Pt aerosols compared to the density of Pt indicates that the aerosols were made up of loose aggregates (16). The small particles are, therefore, expected to be highly nonspherical. The coagulation coefficient of such nonspherical particles is higher than that predicted for the spherical particles of equivalent size, because of the increase in the effective area (capture area) for coagulation (16). Moreover, the smaller apparent densities result also in higher coagulation rates because of the increase in the diffusion coefficient of the particles. Fuchs and Sutugin (17) measured coagulation coefficients for NaCl particles of 0.0025 and 0.004 μm in size that were 2.3 to 2.8 times greater than those predicted by the free molecular limit. Assuming a sticking probability of unity, they explain this discrepancy through a correction factor which accounts for the increase in the collision cross section because of the interparticle van der Waals attraction. Marlow (18) extended their analysis to the entire range of Knudsen numbers and concluded that the effect of the van der Waals forces on the coagulation coefficient is negligible for large particles only. In order to explain the high experimental coagulation coefficients of very small Ag particles, Burtscher and Schmidt-Ott (19), assuming a sticking probability of unity, suggest a Hamaker constant of 10^{-8} erg. For a certain range of intermediate particle sizes, the potential well could be indeed sufficiently deep for the sticking probability to be near unity and yet the collision cross section could be enhanced by the long-range van der Waals attraction. However, such a large value of the Hamaker constant, as suggested by Burtscher and Schmidt-Ott (19), is probably an artifact due to the other factors discussed above. Even though it is reasonable to expect that the lower bound is a better approximation for particles of very small sizes, for reasons mentioned above, it is difficult to infer the range of particle sizes in which the lower bound is indeed valid. The comparison of the models with the experimental data seems to indicate that the upper bound,

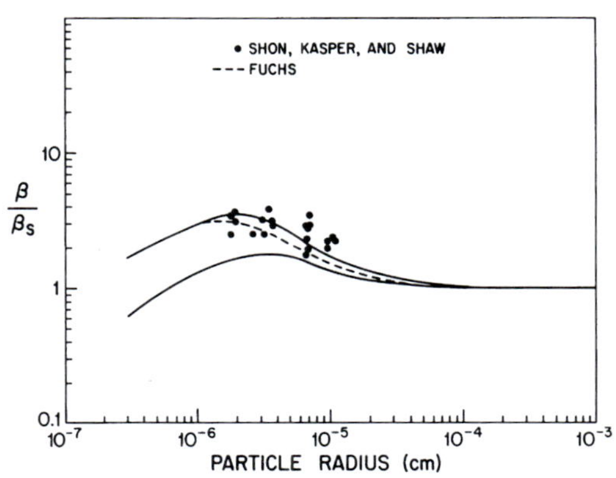

FIG. 9. Comparison of the upper bound, the lower bound (for a Hamaker constant of 10^{-12} erg), and the Fuchs interpolation formula with the experimental coagulation coefficients obtained by Shon *et al.* (14).

TABLE IV

Comparison of Different Models with the Experimental Data of Nolan *et al.* (15)[a]

Particle radius (μm)	*Kn*	β_{ex}/β_S (Experimental)	β_u/β_S (Upper bound)	β_l/β_S (Lower bound)	β_{SS}/β_S (Sitarski and Seinfeld)	β_F/β_S (Fuchs)
0.006	10.80	13.197	4.443	1.871	3.731	4.298
0.0076	8.50	7.210	4.834	2.105	4.163	4.539
0.010	6.70	4.670	5.193	2.362	4.511	4.696
0.013	5.00	4.623	5.333	2.549	4.512	4.530
0.020	3.30	3.743	4.920	2.698	4.075	3.989
0.030	2.30	2.54	3.971	2.605	3.432	3.347

[a] The lower bound values reported are for a Hamaker constant of 10^{-12} erg.

which is equivalent to a sticking probability of unity, is valid, for particles as small as 0.01 μm.

CONCLUSIONS

A model for Brownian coagulation of equal-sized electrically neutral aerosol particles is proposed. The model accounts for the van der Waals attraction and Born repulsion in the calculation of the rate of collisions and subsequent coagulation. In this model, the relative motion between two particles is considered to be free molecular in the neighborhood of the sphere of influence. The thickness of this region is taken to be equal to the correlation length of the relative Brownian motion. The relative motion of the particles outside this region is described by the Fokker–Plank equation. For very small particles, the collision efficiency predicted by the model becomes vanishingly small because of the negligible interaction forces between particles. This model neglects the effect of collisions which a particle experiences with the medium during its escape from the potential well on the coagulation coefficient. The enhancement of the coagulation coefficient due to these collisions is likely to be significant for large particles since they stay longer in the region in which the interaction potential is acting because the potential well is very deep. Therefore, the coagulation coefficient predicted by the model is valid only for small particles and will provide a lower bound for the larger particles. For large particles, the interaction potential can be replaced by a sink at the sphere of influence because the potential well is very deep and the decay length of the interaction potential is small compared to the particle size. This is equivalent to considering a sticking probability of unity. The expression derived for the coagulation coefficient of large particles exhibits the proper continuum and free molecular limits and provides an upper bound for sufficiently small particles. The upper bound predicts slightly higher coagulation coefficients than the Fuchs interpolation formula and than that derived by Sitarski and Seinfeld (6) and Mork *et al.* (8). Experimental data for the coagulation coefficients obtained by Wagner and Kerker (12), Shon *et al.* (14), Chatterjee *et al.* (13), and Nolan and Keenan (15) agree well with the upper bound except

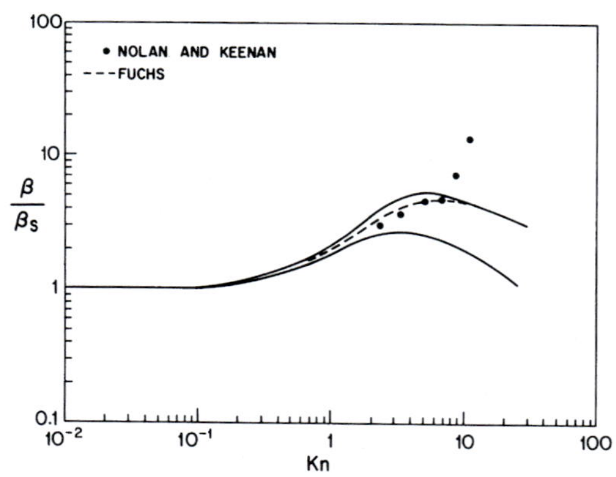

FIG. 10. Comparison of the upper bound, the lower bound (for a Hamaker constant of 10^{-12} erg), and the Fuchs interpolation formula with the experimental coagulation coefficients obtained by Nolan and Keenan (15).

for very small particles. The comparison with the experimental data seems to indicate that the upper bound constitutes a good approximation for particles as small as 0.01 μm. The size range of small particles for which the lower bound is valid could not be determined from the experimental data because the small particles are likely to be nonspherical and because of uncertainty in their apparent density.

The main result of the paper is shown in Fig. 3 in which the upper and lower bounds of the ratio between the coagulation coefficient and the Smoluchowski's coagulation coefficient are plotted as a function of the Knudsen number, for different Hamaker constants (for the lower bound), for particles of unit density.

APPENDIX I

Expression for the Overall Interaction Potential between Two Spherical Particles

Under the assumption of pairwise additivity, the energy of interaction U between two particles containing q molecules per unit volume is given by

$$U = \int_{V_1} dv_1 \int_{V_2} dv_2 q^2 \phi(r), \quad [I.1]$$

where V_1 and V_2 are the volumes of the particles and $\phi(r)$ is the intermolecular potential. For the Lennard–Jones 6–12 potential, the expression for the intermolecular potential $\phi(r)$ is

$$\phi(r) = -\phi_m\left[\left(\frac{r_m}{r}\right)^{12} - 2\left(\frac{r_m}{r}\right)^{6}\right], \quad [I.2]$$

where ϕ_m and r_m are the minimum value of the potential and the distance corresponding to this minimum, respectively.

Using the expression for the Lennard–Jones 6–12 potential in Eq. [I.1] and performing the integrations for two equal sized spherical particles of radius R_p, the expression for the overall interaction potential U can be written as

$$U = -A\left[\frac{1}{720 \times 21}\left\{\frac{(2R'/d')}{(d' - 2R')^6} - \frac{(2R'/d')}{(d' + 2R')^6}\right\} + \frac{1}{720 \times 105}\frac{1}{d'} \times \left\{\frac{2}{d'^5} - \frac{1}{(d' - 2R')^5} - \frac{1}{(d' + 2R')^5}\right\} - \frac{1}{90 \times 28}\left\{\frac{(R'^2/d')}{(d' - 2R')^7} + \frac{(R'^2/d')}{(d' + 2R')^7} + \frac{2(R'^2/d')}{d'^7}\right\} + \frac{1}{6}\left\{\ln\frac{(d' + 2R')(d' - 2R')}{d'^2} + \frac{2R'^2}{(d' + 2R')(d' - 2R')} + \frac{2R'^2}{d'^2}\right\}\right], \quad [I.3]$$

where $d' = d/r_m$, d being the distance between the centers of the two particles, $R' = R_p/r_m$, and A is the Hamaker constant defined as

$$A = -\left(\frac{\rho N \pi}{M}\right)^2 2 r_m^6 \phi_m,$$

where N is the Avogadro number and ρ and M are the density and the molecular weight of the particles, respectively.

From Eq. [I.3] it is obvious that the depth of the overall interaction potential well between two equal-sized particles is a function of the particle radius R_p and the Hamaker constant A only.

APPENDIX II

Expressions for the Coefficients in Eqs. (79) and (96)

The expressions for the coefficients in Eq. [79] in terms of the dimensionless depth of the potential Φ, the dimensionless surface temperature θ, μ_0, and R are given by

$$a_1 = -\frac{1}{\delta R} - \frac{1}{20}\left(1 + \frac{2\Phi}{3}\right)\left(1 + \frac{6(1 - \mu_0^4)}{R^2}\right)R + \frac{1}{2}\left(1 + \frac{2\Phi}{3}\right)\frac{(1 - \mu_0^3)}{\delta R} + \frac{1}{20}\left(\frac{4(1 - \mu_0^4)}{R^2} - 1\right)\left(1 + \frac{2\Phi}{3}\right)R, \quad [II.1]$$

$$a_2 = \frac{1}{2\delta}\left(1 + \frac{2\Phi}{3}\right)(1 + 2\mu_0^2 - 3\mu_0^4) - \frac{1}{2\delta^2}\left(1 + \frac{2\Phi}{3}\right)(1 - \mu_0^3) - \frac{1}{\delta^2}, \quad \text{[II.2]}$$

$$a_3 = -\frac{1}{10\delta^2}\left(1 + \frac{2\Phi}{3}\right)(1 + 2\mu_0^2 - 3\mu_0^4)R, \quad \text{[II.3]}$$

$$A_1 = -\frac{2}{\pi\delta} + \frac{1}{10\pi}\left(\frac{4(1-\mu_0^4)}{R^2} - 1\right)\left(1 + \frac{2\Phi}{3}\right)R^2 - \frac{1}{10\pi}\left(1 + \frac{6(1-\mu_0^4)}{R^2}\right)R^2\left(1 + \frac{2\Phi}{3}\right) + \frac{1}{\pi\delta}\left(1 + \frac{2\Phi}{3}\right)(1 - \mu_0^3) \quad \text{[II.4]}$$

$$A_2 = -\frac{3}{8}R + \frac{1}{16}\left(\frac{4(1-\mu_0^4)}{R^2} - 1\right)R - \frac{3}{40}\Phi R + \frac{\Phi}{80}\left(\frac{4(1-\mu_0^4)}{R^2} - 1\right)R$$
$$+ \frac{3}{80}(1-\mu_0^3)\left(1 + \frac{2\Phi}{3}\right)R - \frac{1}{32}(1-\mu_0^3)\left(1 + \frac{2\Phi}{3}\right)\left(\frac{4(1-\mu_0^4)}{R^2} - 1\right)R$$
$$+ \frac{1}{40}\left(1 + \frac{2\Phi}{3}\right)(1-\mu_0^3)\left(1 + \frac{6(1-\mu_0^4)}{R^2}\right)R + \frac{1}{\pi\delta}\left(1 + \frac{2\Phi}{3}\right)(1 + 2\mu_0^2 - 3\mu_0^4)R$$
$$- \frac{9}{5\pi\delta}\left(1 + \frac{2\Phi}{3}\right)(1 + 2\mu_0^2 - 3\mu_0^4)R - \frac{2R}{\pi\delta^2} + \frac{1}{2\delta R}(1-\mu_0^3) - \frac{1}{4\delta R}(1-\mu_0^3)^2, \quad \text{[II.5]}$$

$$A_3 = -\frac{9}{80}\left(1 + \frac{2\Phi}{3}\right)(1 + 2\mu_0^2 - 3\mu_0^4)R^2 + \frac{3}{160}\left(\frac{4(1-\mu_0^4)}{R^2} - 1\right)(1 + 2\mu_0^2 - 3\mu_0^4)$$
$$\times\left(1 + \frac{2\Phi}{3}\right)R^2 + \frac{3}{40}\left(1 + \frac{2\Phi}{3}\right)(1 + 2\mu_0^2 - 3\mu_0^4)R^2 - \frac{1}{80}\left(1 + \frac{2\Phi}{3}\right)(1 + 2\mu_0^2 - 3\mu_0^4)$$
$$\times\left(1 + \frac{6(1-\mu_0^4)}{R^2}\right)R^2 - \frac{3}{5\pi\delta}\left(1 + \frac{2\Phi}{3}\right)(1 + 2\mu_0^2 - 3\mu_0^4)R^2$$
$$+ \frac{2}{5\pi\delta^2}\left(1 + \frac{2\Phi}{3}\right)(1 + 2\mu_0^2 - 3\mu_0^4)R^2 + \frac{\Phi}{20\delta}(1 + 2\mu_0^2 - 3\mu_0^4) - \frac{1}{8\delta}(1-\mu_0^3)\left(1 + \frac{2\Phi}{3}\right)$$
$$+ \frac{1}{10\delta}(1 + 2\mu_0^2 - 3\mu_0^4)(1-\mu_0^3)\theta_1 - \frac{1}{40\delta}\left(1 + \frac{2\Phi}{3}\right)(1-\mu_0^3)(1 + 2\mu_0^2 - 3\mu_0^4)$$
$$+ \frac{1}{4\delta^2}\left(1 + \frac{2\Phi}{3}\right)(1-\mu_0^3)^2, \quad \text{[II.6]}$$

$$A_4 = \frac{3}{20\delta}\left(1 + \frac{2\Phi}{3}\right)(1 + 2\mu_0^2 - 3\mu_0^4)R - \frac{3}{20\delta}(1 + 2\mu_0^2 - 3\mu_0^4)\theta_1 R$$
$$+ \frac{\Phi}{20\delta}(1 + 2\mu_0^2 - 3\mu_0^4)R - \frac{1}{4\delta^2}(1-\mu_0^3)\left(1 + \frac{2\Phi}{3}\right)(1 + 2\mu_0^2 - 3\mu_0^4)R, \quad \text{[II.7]}$$

$$B_1 = -\frac{1}{\delta R^2} + \frac{3}{10}\left(1 + \frac{2\Phi}{3}\right) - \frac{1}{20}\left(1 + \frac{2\Phi}{3}\right)\left(1 + \frac{6(1-\mu_0^4)}{R^2}\right)$$
$$+ \frac{1}{2}\left(1 + \frac{2\Phi}{3}\right)\frac{(1-\mu_0^3)}{R^2\delta} - \frac{3}{10}\theta_1 + \frac{1}{20}\left(\frac{4(1-\mu_0^4)}{R^2} - 1\right)\theta_1, \quad \text{[II.8]}$$

$$B_2 = \frac{3}{10}\left(1 + \frac{2\Phi}{3}\right)\frac{(1 + 2\mu_0^2 - 3\mu_0^4)}{R^2\delta} - \frac{3}{20}\left(1 + \frac{2\Phi}{3}\right)\frac{(1 - \mu_0^3)}{R^2\delta^2} + \frac{1}{5}\frac{(1 + 2\mu_0^2 - 3\mu_0^2)}{R^2\delta}\theta_1 - \frac{1}{R^2\delta^2}, \quad \text{[II.9]}$$

and

$$B_3 = \frac{1}{5}\theta_1\frac{(1 + 2\mu_0^2 - 3\mu_0^4)}{R^2\delta^2} - \frac{3}{10}\left(1 + \frac{2\Phi}{3}\right)\frac{(1 + 2\mu_0^2 - 3\mu_0^4)}{R^2\delta^2}, \quad \text{[II.10]}$$

where

$$\theta_1 = \theta e^{-\Phi/\theta}\left(1 + \frac{\Phi}{\theta} + \frac{1}{2}\frac{\Phi^2}{\theta^2}\right).$$

The expressions for the coefficients in Eq. [96] in terms of μ_0 and R are given by

$$a_s = -\frac{1}{\delta R} + \frac{1}{2\delta R}(1 - \mu_0^3) + \frac{1}{20}\left(\frac{4(1 - \mu_0^4)}{R^2} - 1\right)R - \frac{1}{20}\left(1 + \frac{6(1 - \mu_0^4)}{R^2}\right)R, \quad \text{[II.11]}$$

$$b_s = -\frac{1}{\delta^2} - \frac{1}{10}\frac{(1 + 2\mu_0^2 - 3\mu_0^4)}{\delta} + \frac{1}{2\delta^2}(1 - \mu_0^3), \quad \text{[II.12]}$$

$$c_s = -\frac{1}{10}(1 + 2\mu_0^2 - 3\mu_0^4)R, \quad \text{[II.13]}$$

$$A_s = \frac{2}{\pi\delta} - \frac{3}{5\pi}R^2 + \frac{1}{10\pi}\left(1 + \frac{6(1 - \mu_0^4)}{R^2}\right)R^2 - \frac{(1 - \mu_0^3)}{\pi\delta} + \frac{3}{5\pi}R^2 - \frac{1}{10\pi}\left(\frac{4(1 - \mu_0^4)}{R^2} - 1\right)R^2, \quad \text{[II.14]}$$

$$B_s = \frac{1}{\delta R^2} - \frac{3}{10} + \frac{1}{20}\left(1 + \frac{6(1 - \mu_0^4)}{R^2}\right) - \frac{1}{2}\frac{(1 - \mu_0^3)}{\delta R^2} + \frac{2}{\pi\delta^2} + \frac{3}{5\pi\delta}(1 + 2\mu_0^2 - 3\mu_0^4) - \frac{(1 - \mu_0^3)}{\pi\delta^2} - \frac{(1 - \mu_0^3)}{2\delta R^2} + \frac{3}{20}(1 - \mu_0^3) + \frac{1}{32}(1 - \mu_0^3)\left(\frac{4(1 - \mu_0^4)}{R^2} - 1\right) - \frac{2}{5\pi\delta}(1 + 2\mu_0^2 - 3\mu_0^4) + \frac{3}{8} - \frac{1}{16}\left(\frac{4(1 - \mu_0^4)}{R^2} - 1\right), \quad \text{[II.15]}$$

$$C_s = \frac{1}{\delta^2 R^2} + \frac{3}{10}\frac{(1 + 2\mu_0^2 - 3\mu_0^4)}{R^2\delta} - \frac{1}{2}\frac{(1 - \mu_0^3)}{R^2\delta^2} + \frac{3}{5\pi\delta^2}(1 + 2\mu_0^2 - 3\mu_0^4) - \frac{1}{2\delta^2}\frac{(1 - \mu_0^3)}{R^2} - \frac{3}{20}(1 - \mu_0^3)\frac{(1 + 2\mu_0^2 - 3\mu_0^4)}{R^2\delta} + \frac{1}{4}\frac{(1 - \mu_0^3)^2}{R^2\delta^2} + \frac{(1 + 2\mu_0^2 - 3\mu_0^4)}{4\delta R^2} - \frac{3}{40}(1 + 2\mu_0^2 - 3\mu_0^4) + \frac{1}{80}(1 + 2\mu_0^2 - 3\mu_0^4)\left(1 + \frac{6(1 - \mu_0^4)}{R^2}\right) - \frac{(1 - \mu_0^3)}{8\delta R^2}(1 + 2\mu_0^2 - 3\mu_0^4) + \frac{1}{8}(1 - \mu_0^3)\frac{(1 + 2\mu_0^2 - 3\mu_0^4)}{R^2\delta} - \frac{2}{5\pi}\frac{(1 + 2\mu_0^2 - 3\mu_0^4)}{\delta^2} + \frac{9}{80}(1 + 2\mu_0^2 - 3\mu_0^4) - \frac{3}{160}(1 + 2\mu_0^2 - 3\mu_0^4)\left(\frac{4(1 - \mu_0^4)}{R^2} - 1\right) - \frac{1}{4}\frac{(1 + 2\mu_0^2 - 3\mu_0^4)}{R^2\delta} \quad \text{[II.16]}$$

and

$$D_s = \frac{3}{10}\frac{(1 + 2\mu_0^2 - 3\mu_0^4)}{R^2\delta^2} - \frac{3}{20}(1 - \mu_0^3)\frac{(1 + 2\mu_0^2 - 3\mu_0^4)}{R^2\delta^2} + \frac{(1 + 2\mu_0^2 - 3\mu_0^4)}{4R^2\delta^2}$$
$$+ \frac{3}{40\delta}\frac{(1 + 2\mu_0^2 - 3\mu_0^4)}{R^2} - \frac{1}{8}(1 - \mu_0^3)\frac{(1 + 2\mu_0^2 - 3\mu_0^4)}{R^2\delta} + \frac{1}{8}\frac{(1 + 2\mu_0^2 - 3\mu_0^4)}{R^2\delta^2}(1 - \mu_0^3)$$
$$- \frac{3}{40}\frac{(1 + 2\mu_0^2 - 3\mu_0^4)^2}{R^2\delta} - \frac{1}{4}\frac{(1 + 2\mu_0^2 - 3\mu_0^4)}{R^2\delta^2}. \quad [II.17]$$

ACKNOWLEDGMENT

We are indebted to Dr. J. Seinfeld for illuminating discussions.

REFERENCES

1. Smoluchowski, M. V., *Z. Phys. Chem.* **92,** 144 (1918).
2. Millikan, R. A., *Phys. Rev.* **22,** 1 (1923).
3. Philips, W. F., *Phys. Fluids* **18,** 1089 (1975).
4. Hidy, G. M., and Brock, J. R., "The Dynamics of Aerocolloidal Systems." Pergamon, Oxford, 1970.
5. Fuchs, N. A., "Mechanics of Aerosols." Pergamon, New York, 1964.
6. Sitarski, M., and Seinfeld, J. H., *J. Colloid Interface Sci.* **61,** 261 (1977).
7. Grad, H., *Commun. Pure Appl. Math.* **2,** 331 (1949).
8. Mork, K. J., Razi Naqvi, K., and Waldenstrom, S., *J. Colloid Interface Sci.* **98,** 103 (1984).
9. Nowakowski, B., and Sitarski, M., *J. Colloid Interface Sci.* **83,** 614 (1981).
10. Chandrasekhar, S., *Rev. Mod. Phys.* **15,** 1 (1943).
11. Sahni, D. C., *J. Colloid Interface Sci.* **91,** 418 (1983).
12. Wagner, P. E., and Kerker, M., *J. Chem. Phys.* **66,** 638 (1977).
13. Chatterjee, A., Kerker, M., and Cooke, D. D., *J. Colloid Interface Sci.* **53,** 71 (1975).
14. Shon, S., Kasper, G., and Shaw, D. T., *J. Colloid Interface Sci.* **73,** 233 (1980).
15. Nolan, P. J., and Keenan, E. L., *Proc. R. Ir. Acad. Sect. A.* **52,** 171 (1949).
16. Mercer, T. T., *in* "Fundamentals of Aerosol Science" (D. T. Shaw, Ed.). Wiley, New York, 1978.
17. Fuchs, N. A., and Sutugin, A. G., *J. Colloid Interface Sci.* **20,** 492 (1965).
18. Marlow, W. H., *J. Chem. Phys.* **73,** 6284, 6288 (1980).
19. Burtscher, H., and Schmidt-Ott, A., *Phys. Rev. Lett.* **48,** 1734 (1982).

Monte Carlo Simulation of Brownian Coagulation over the Entire Range of Particle Sizes from Near Molecular to Colloidal: Connection between Collision Efficiency and Interparticle Forces

GANESAN NARSIMHAN AND ELI RUCKENSTEIN[1]

Department of Chemical Engineering, State University of New York at Buffalo, Buffalo, New York 14260

Received November 7, 1984; accepted January 7, 1985

The effect of interparticle forces on Brownian coagulation of aerosols is usually accounted for through a phenomenological sticking probability, i.e., the probability of coagulation upon collision. This probability is customarily assumed to be equal to unity even though it is recognized that it is smaller than unity for very small particles. A Monte Carlo method for the simulation of Brownian coagulation of equal-sized electrically neutral spherical aerosol particles is presented, which takes into account the interparticle forces due to van der Waals attraction and Born repulsion instead of the phenomenological sticking probability. The particle motion is described by the generalized Brownian dynamics. The Brownian coagulation coefficient for particles of unit density, for a Hamaker constant of 10^{-12} erg, is calculated over the entire range of Knudsen numbers. For large particles, the results of the simulation agree very well with those provided by the model proposed previously by the authors (1) and also with the Fuchs interpolation formula. In this case the results indicate that the sticking probability is close to unity. For sufficiently small particles, the coagulation coefficient agrees with the lower bound provided by the model proposed by the authors (1). The extrapolation to sufficiently small particles of the model (1) proposed for large particles provides an upper bound with much larger values. For particles of intermediate sizes, the coagulation coefficient is found to be even higher than the values predicted by the model valid for large particles (1). The coagulation coefficients of the particles of intermediary size are greater than those corresponding to a sticking probability of unity whereas those of the sufficiently small particles are much smaller because of the following two opposite effects: (i) As the particle size decreases, the ratio of the decay length of the interaction potential (based on the shortest distance between the particles) and the particle size increases; this relative increase in the range of the van der Waals attraction increases the rate of collisions between particles. (ii) As the particle size decreases, the interaction potential becomes less deep and, as a result, the probability of escape of the particles from the potential well increases. In the intermediary size range, the former effect, which was not accounted for in the model for large particles (being negligible in that case), is more important, whereas for very small particle sizes, the latter effect predominates. The calculated values of the collision efficiency are small for sufficiently small particles, increase with particle size and eventually reach the value of unity for large particles. Extrapolating to particles of molecular dimensions, one can understand why in the cases in which only physical interactions are involved the collision efficiency is negligibly small.

INTRODUCTION

Dispersed systems consisting of particles suspended in gases are frequently encountered in nature and in industrial surroundings. Brownian coagulation is an important mechanism by which coarsening of particle sizes in such aerocolloidal systems occurs. It plays an important role in, for example, the formation of carbon particles (soot) in the combustion of hydrocarbons, growth of particles formed by physical or chemical nucleation, etc. In order to describe the evolution of particle size distribution in such systems, information regarding the coagulation rates of particles of different sizes is needed. Collision and subsequent coagulation of aerocolloidal particles in a stationary medium is

[1] To whom correspondence should be addressed.

0021-9797/85 $3.00

a result of their Brownian motion and of the mutual forces of interaction between them. The interaction between particles due to the dispersion forces are mainly responsible for the coagulation of two colliding particles. The effect of these short-range forces on coagulation is usually accounted for through a phenomenological sticking probability, i.e., the probability of coagulation upon collision. Since the dispersion forces of large particles are usually strong, it is customary to assume that this sticking probability is equal to unity, i.e., every collision leads to coagulation.

The nature of the Brownian motion of particles depends strongly on the relative magnitudes of the mean free path of the suspending medium λ_g and the particle size R, ratio known as the Knudsen number, Kn $(=\lambda_g/R)$. When the particle sizes are much larger than the mean free path, i.e., Kn is sufficiently small, the suspending medium can be considered as a continuum and the Brownian coagulation coefficient is well described, for a sticking probability of unity, by the Smoluchowski equation (2):

$$\beta_s(R_1, R_2) = 4\pi(R_1 + R_2)(D_1 + D_2), \quad [1]$$

where the diffusion coefficient D for a particle of radius R is given by the expression

$$D = \frac{kT}{6\pi R\eta}. \quad [2]$$

Here η is the viscosity of the medium, k is the Boltzmann constant, and T is the absolute temperature of the medium.

The Smoluchowski equation can also be extended to Knudsen numbers in the range $0.1 < \mathrm{Kn} < 0.25$ by introducing a slip correction factor $C(\mathrm{Kn})$ in the expression for the particle diffusion coefficient. The empirical slip correction factor due to Millikan (3) has the form

$$C(\mathrm{Kn}) = 1 + \mathrm{Kn}\{1.257 + 0.4\exp(-1.1/\mathrm{Kn})\} \quad [3]$$

and the slip correction factor derived by Philips (4), by assuming an accommodation coefficient of unity, is given by

$$C(\mathrm{Kn}) = \frac{5 + 4\,\mathrm{Kn} + 6\,\mathrm{Kn}^2 + 18\,\mathrm{Kn}^3}{5 - \mathrm{Kn} + (8 + \pi)\mathrm{Kn}^2}. \quad [4]$$

When the particle size is very small compared to the mean free path, i.e., Kn > 10, the particle can be regarded as a large spherical molecule undergoing independent binary collisions with the gas molecules and the coagulation coefficient is obtained from the kinetic theory of rarefied gases as (5)

$$\beta_M(R_1, R_2) = \alpha\pi(R_1 + R_2)^2(c_1^2 + c_2^2)^{1/2}, \quad [5]$$

where α is the sticking probability and $c_i = (8kT/\pi m_i)^{1/2}$ is the root mean square velocity of a particle of mass m_i.

For the transition region (0.1 < Kn < 10), Fuchs (6), assuming the sticking probability equal to unity, proposed an empirical interpolation expression for the coagulation coefficient (between the continuum and free molecular limits). His expression for the coagulation coefficient is given by:

$$\beta_F(R_1, R_2) = \beta_s(R_1, R_2)\left(\frac{C(\mathrm{Kn}_1)}{R_1} + \frac{C(\mathrm{Kn}_2)}{R_2}\right) \times \left[\frac{R_1 + R_2}{R_1 + R_2 + g_{12}} + \frac{4(D_1 + D_2)}{(R_1 + R_2)(c_1^2 + c_2^2)^{1/2}}\right]^{-1}, \quad [6]$$

where

$$g_{12} = (g_1^2 + g_2^2)^{1/2},$$

$$g_i = \frac{1}{6R_i\lambda_i}[(2R_i + \lambda_i)^3 - (4R_i^2 + \lambda_i^2)^{3/2}] - 2R_i,$$

$$\lambda_i = 8D_i/\pi c_i,$$

$$D_i = \frac{kT}{6\pi R_i\eta}C(\mathrm{Kn}_i),$$

and

$$c_i = \left(\frac{8kT}{\pi m_i}\right)^{1/2}, \quad (i = 1, 2).$$

A theoretical basis for the Fuchs semiempirical formula has been provided by Sitarski and Seinfeld (7), and by Mork, Razi Naqvi, and Waldenström (8), who obtained the coagulation coefficient through the solution of the Fokker–Plank equation. However, the Brownian coagulation coefficient predicted by their calculations does not exhibit the proper free molecular limit. In contrast, the Brownian coagulation coefficient, obtained by Nowakowski and Sitarski (9) through Monte Carlo simulation, by assuming a sticking probability of unity, agrees well with the Fuchs interpolation formula over the entire range of Knudsen numbers.

Even though it has been recognized that the sticking probability is much less than unity for very small particles, earlier investigations have not attempted to explain its physical origin, but, instead, consider it as an empirical parameter. The authors have proposed a model (1) for the Brownian coagulation of electrically neutral aerosol particles on the basis of which the coagulation coefficient was calculated without the use of an empirical sticking probability. Instead, the model accounts for the interparticle forces due to the van der Waals attraction and Born repulsion. The model employs the simplifying assumptions that the relative motion of two particles when they are close to each other (in the vicinity of the sphere of influence of radius $2R$) is free molecular and that the thickness of this region is equal to the average correlation length for the relative Brownian motion of the particles, (i.e., the shortest distance between the colliding particles below which they retain the memory of their relative motion). The relative motion outside this region is described by the Fokker–Plank equation. The repulsive part of the interaction potential is replaced by a reflecting boundary with momentum and thermal accommodation coefficients of unity. As a result of the simplifying assumption that the particle motion in the vicinity of the sphere of influence is free molecular, the reflected particles which have a kinetic energy less than the depth of the potential well are assumed to be captured in the potential well. One may note, however, that, in reality, the coagulation coefficient is greater than that predicted because the above assumption does not account for the collisions with the medium that a particle experiences during its escape from the potential well. This enhancement in the coagulation coefficient due to collision with the medium is likely to be significant particularly for the large particles since (i) they undergo a larger number of collisions with the medium because the mean free path of the medium is much smaller than their size; and (ii) they remain longer in the potential well because the potential well is deep. However, for sufficiently small particles, this effect is negligible since: (i) they undergo a very small number of collisions with the medium because the mean free path of the medium is larger than their size; and (ii) they do not stay long in the potential well because the potential well is shallow. Consequently, the coagulation coefficient predicted by the model is likely to be valid for sufficiently small particles and to provide a lower bound for the larger particles. The coagulation coefficient predicted by the model for very small particles, i.e., particles in the free molecular range, is much smaller than that provided by the free molecular limit for a sticking probability of unity. In other words, the model predicts vanishingly small sticking probabilities for particles of very small sizes. For large particles, however, the potential well due to the dispersion and Born interactions is deep and the decay length of the interaction potential (based on the shortest distance between particles) is small compared to the particle sizes. Therefore, the coagulation coefficient for large particles is calculated by replacing the interaction potential by a sink (i.e., the sticking probability is unity). However, as the particle size decreases the potential energy well becomes less deep whereas the ratio of the decay length of the potential and the particle size increases. This leads to two contrary effects: (i) The collisions become less effective because the probability of escape of particles from the potential well becomes higher, i.e., the sticking probability becomes less than unity and (ii) The rate of collisions between the particles increases because the effective collision cross section becomes larger as a result of the increased range of the van der Waals attraction relative to the particle size. The former effect decreases, while the latter increases the coagulation rate. It is important to note that in a certain range of intermediate particle sizes, the latter effect predominates so that the coagulation coefficient may even be greater than that calculated by replacing the interaction potential by a sink. For sufficiently large particles, the increase in the rate of collisions due to the increased range of the interaction potential is negligible because the ratio of the decay length of the interaction potential and the particle size is very small. In contrast, for sufficiently small particles, the effect of the collisions with the medium experienced by the particles escaping the potential well is negligible. The previous model (1) provides expressions for the coagulation coefficients of sufficiently large and sufficiently small particles by neglecting the former and the latter effects, respectively. For intermediate-size particles, however, both the above effects are important. In order to predict the coagulation coefficient of particles of intermediate size, it is, therefore, necessary to account for both these effects.

The objective of the present paper is to calculate through Monte Carlo simulations the coagulation coefficient for equal-sized electrically neutral particles over the entire range of Knudsen numbers, by accounting for the interparticle forces (instead of using the phenomenological sticking probability) in order to (i) verify the validity of the already derived expressions (1) for very large and very small particles; and, particularly (ii) obtain the coagulation coefficient of particles of intermediate sizes for which analytical expressions could not be derived. The simulation algorithm is based on the previous model (1). However, the present approach does not employ the above simplifying assumptions regarding either the particle motion or the interaction potential since it accounts for the effect of the increased range of the interaction potential relative to the particle size as well as the effect of collisions with the medium experienced by the particles in the vicinity of the sphere of influence. The particle motion is described by the generalized Brownian dynamics. The relative Brownian motion of two equal-sized particles is represented by that of a fictitious particle of reduced mass. The repulsive part of the interaction potential between two particles is replaced by a spherically reflecting boundary; the particles upon collision are assumed to be in contact long enough to reach thermal equilibrium between themselves (not with the medium), i.e., the momentum and thermal accommodation coefficients are assumed equal to unity. The model is formulated in the next section and the subsequent sections present the simulation of particle trajectory through the generalized Brownian dynamics and the algorithm for Monte Carlo simulation. Comparison of the simulation results with the previous model (1) is made in the last section.

MODEL FOR BROWNIAN COAGULATION

The Brownian motion of a particle of mass m_p is described by the Langevin equation (10):

$$m_p \frac{d\mathbf{v}}{dt} = -m_p \zeta \mathbf{v} + \mathbf{F} + \mathbf{B}(t), \qquad [7]$$

where $\mathbf{v}$ is the velocity of the particle, ζ is the friction coefficient, $\mathbf{F}$ is the interaction force acting on the particle and $\mathbf{B}(t)$ is the random force due to the collisions with the medium.

The rate of the Brownian coagulation between two particles of specified sizes can be predicted from: (i) the rate of collisions between the particles, obtained through the analysis of their relative motion; and (ii) the probability that a collision will lead to coagulation. It is customary to ignore the interparticle forces in the description of the relative motion between the two particles and to account for its effect on coagulation through a phenomenological sticking probability. The Brownian motions of the two particles are, therefore, usually considered to be independent. Moreover, since, excepting the very small particles, the mutual forces of interaction are fairly strong, every collision between particles is usually assumed to result in coagulation, i.e., the sticking probability is assumed to be unity. The rate of the Brownian coagulation can therefore be calculated from the rate of collisions between particles.

As pointed out earlier, the present treatment attempts to clarify the connection between the sticking probability and the mutual forces of interaction between particles. The van der Waals attraction and Born repulsion forces are included in the analysis of the relative motion between two electrically neutral aerosol particles. The overall interaction potential between two particles is calculated through the integration of the intermolecular potential, modelled as the Lennard–Jones 6–12 potential, under the assumption of pairwise additivity. The expression for the overall interaction potential in terms of the Hamaker constant and the molecular diameter can be found in Appendix I of (1). The Brownian motions of the two particles are no longer independent because of the interaction force between the two. It is, therefore, necessary to describe the relative motion between the two particles in order to predict the rate of collision and of subsequent coagulation.

Consider the relative motion between two particles of mass m_p and radius R. The Langevin equations for their Brownian motion can be written as

$$m_p \frac{d\mathbf{v}_1}{dt} = -m_p \zeta \mathbf{v}_1 + \mathbf{F}_1 + \mathbf{B}_1(t) \qquad [8]$$

$$m_p \frac{d\mathbf{v}_2}{dt} = -m_p \zeta \mathbf{v}_2 + \mathbf{F}_2 + \mathbf{B}_2(t), \qquad [9]$$

where the subscripts 1 and 2 refer to the particles 1 and 2, respectively. $\mathbf{F}_1$ and $\mathbf{F}_2$ are the interaction forces acting on the particles 1 and 2 along their line of centers, i.e.,

$$\mathbf{F}_1 = -\mathbf{F}_2 = \frac{F(\mathbf{r}_1 - \mathbf{r}_2)}{|\mathbf{r}_1 - \mathbf{r}_2|}(\mathbf{r}_1 - \mathbf{r}_2),$$

F being the magnitude of the interaction force and $|\mathbf{r}|$ denoting the magnitude of the vector $\mathbf{r}$. Taking the origin of the coordinate system as the center of mass of the particles, and defining the relative coordinates

$$\mathbf{r} = \mathbf{r}_1 - \mathbf{r}_2 \quad \text{and} \quad \mathbf{v} = \mathbf{v}_1 - \mathbf{v}_2,$$

the Langevin equation for the relative motion of the two particles can be written as

$$m_r \frac{d\mathbf{v}}{dt} = -m_r \zeta \mathbf{v} + \frac{\mathbf{r}}{r} F + \mathbf{B}(t), \qquad [10]$$

where m_r ($=m_p/2$) is the reduced mass and

$$\mathbf{B}(t) = \tfrac{1}{2}(\mathbf{B}_1(t) - \mathbf{B}_2(t)). \qquad [11]$$

Rewriting the above equation in terms of the momentum $\mathbf{p} = m_r\mathbf{v}$ of the fictitious particle of reduced mass, one has

$$\frac{d\mathbf{p}}{dt} = -\zeta \mathbf{p} - \nabla_r \phi + \mathbf{B}(t), \qquad [12]$$

where ϕ is the interaction potential between the two particles.

The motion of the fictitious particle (denoted hereafter as "particle") of reduced mass can be divided into three regions, as shown in Fig. 1. In region I, the "particle" experiences negligible interaction force since the interaction potential in this region is negli-

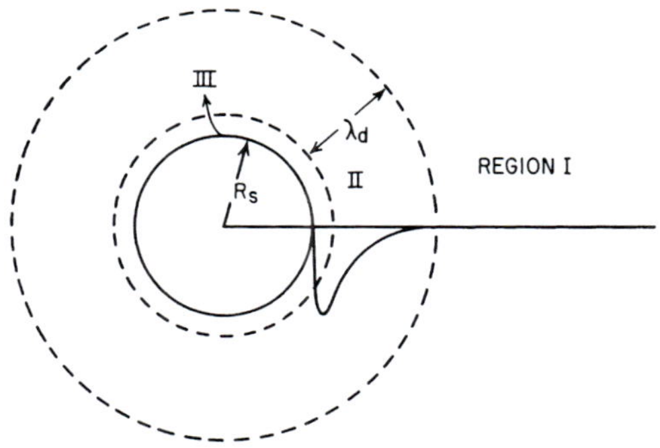

FIG. 1. Different regions for the motion of the fictitious particle of reduced mass.

gible. In region II, the "particle" experiences an attractive force, the thickness of this region being equal to the decay length λ_d of the interaction potential. In region III, the "particle" experiences a repulsive force. The motion of the fictitious particle in regions I and II can be described by the Langevin equation (Eq. [12]). The details of the simulation of the "particle" motion in these regions are described in the next section.

Since the repulsive part of the overall potential decays rapidly, region III is very thin. The fictitious particle will move from the bottom of the potential well until its radial velocity component becomes zero, at which point it will be reflected (i.e., the colliding particles are reflected). The point at which the fictitious particle is reflected depends upon its kinetic energy at the bottom of the potential well. The motion of the fictitious particle in this region corresponds to the relative motion of two particles when they are extremely close to each other, near the point of collision. Their relative motion will be strongly influenced by the interactions among the surface molecules and hence will depend on the surface roughness. It is, therefore, extremely difficult to describe the motion of the fictitious particle in this region. Because of this inherent difficulty, region III (refer to Fig. 1) is replaced by a reflecting boundary (sphere of radius R_s) and the effect of this region on the motion of the fictitious particle is accounted for through the boundary condition at a reflecting spherical surface. If there are no surface imperfections, the colliding particles will be reflected specularly, i.e., the collision will be perfectly elastic with the normal component of the velocity reversed, while the components parallel to the surface remain unchanged. On the other hand, if the colliding particles are momentarily trapped due to surface imperfections, i.e., they spend enough time in contact, then they will be reflected diffusively. The velocity of the reflected particles will be in this case independent of their incident velocities and their relative velocity will be distributed according to the half range Maxwellian distribution corresponding to the temperature of the colliding particles (referred to as the temperature of the reflecting spherical surface and calculated as shown below). A situation intermediate to the above extremes can be described using an empirical momentum accommodation coefficient α_c which is the fraction of the incident particles that is reflected diffusively. Since the fictitious particle exhibits random Brownian motion its velocity when incident upon the reflecting surface of radius R_s will be random.

The reflecting boundary condition, due to Maxwell, can be written in terms of the momentum accommodation coefficient α_c as (5)

$$f^{+*}(\mathbf{v}) = (1 - \alpha_c)f^{-*}(\mathbf{v})\{\mathbf{v} - 2(\mathbf{v}\cdot\mathbf{n})\mathbf{n}\} + \alpha_c n_r\left(\frac{m_r}{2\pi kT_r}\right)^{3/2}\exp\left[-\frac{m_r v^2}{2kT_r}\right], \quad [13]$$

where $f^{+*}(\mathbf{v})$ and $f^{-*}(\mathbf{v})$ are the velocity distributions of the reflected and incident fictitious particles at the reflecting surface, respectively, and $\mathbf{n}$ is the unit normal to the reflecting spherical surface.

The momentum accommodation coefficient α_c, as pointed out earlier, depends on the nature of the surface of the colliding particles. In the absence of information regarding the nature of the surface of the colliding particles, it will be assumed that they stay in contact long enough (due to surface roughness) to reach equilibrium and be reflected diffusively. In other words, the momentum accommodation coefficient is assumed to be unity. Therefore, the velocity distribution of the reflected "particle" will be given by the half-range Maxwellian distribution

$$f^{+*}(\mathbf{v}) = n_r\left(\frac{m_r}{2\pi kT_r}\right)^{3/2}\exp\left[-\frac{m_r v^2}{2kT_r}\right]. \quad [13a]$$

The parameter n_r in Eq. [13] is determined from the condition that every pair of colliding particles are reflected, i.e., the net flux of the fictitious particles at the reflecting spherical surface is zero. Therefore,

$$-\int_{\mathbf{v}_i\cdot\mathbf{n}<0}(\mathbf{v}_i\cdot\mathbf{n})f^{-*}(\mathbf{v}_i)d\mathbf{v}_i$$

$$=\int_{\mathbf{v}_r\cdot\mathbf{n}>0}(\mathbf{v}_r\cdot\mathbf{n})f^{+*}(\mathbf{v}_r)d\mathbf{v}_r, \quad [13b]$$

where $\mathbf{v}_i$ and $\mathbf{v}_r$ are the relative velocity of the incident and reflected particles respectively at the time of collision.

Using the zero flux condition [13b] and the boundary condition [13a], the parameter n_r can be expressed in terms of the flux of the fictitious particles incident upon the reflecting surface N_i and the "surface temperature" T_r as,

$$n_r = 2N_i\left(\frac{m_r\pi}{2kT_r}\right)^{1/2}, \quad [13c]$$

where $N_i = -\int_{\mathbf{v}_i\cdot\mathbf{n}<0}(\mathbf{v}_i\cdot\mathbf{n})f^{-*}(\mathbf{v}_i)d\mathbf{v}_i$.

The temperature of the colliding particle pairs (the "surface temperature") T_r is given by,

$$kT_r = \frac{2}{3}e, \quad [14]$$

where

$$e = \frac{1}{n}\int\frac{1}{2}m_rV^2 f d\mathbf{v},$$

$$n = \int f d\mathbf{v},$$

$$\mathbf{V} = \mathbf{v} - \mathbf{u},$$

$$\mathbf{u} = \int \mathbf{v} f d\mathbf{v}\Big/\int f d\mathbf{v}$$

and

$$f(\mathbf{v}) = f^{+*}(\mathbf{v}) + f^{-*}(\mathbf{v}),$$

$f(\mathbf{v})$ being the total velocity distribution of the incident and reflected "particles" at the reflecting surface.

In the above equation, e, n, and u are the internal energy, the number density and the average velocity of the fictitious particle, respectively, at the reflecting spherical surface. Since the net flux of the fictitious particles at the reflecting surface is zero (i.e., $u = 0$), Eq. [14] can be written in terms of the relative velocity distributions of the incident and reflected colliding particles as,

$$kT_r\left[\int_{\mathbf{v}_i\cdot\mathbf{n}<0} f^{-*}(\mathbf{v}_i)d\mathbf{v}_i + \int_{\mathbf{v}_r\cdot\mathbf{n}>0} f^{+*}(\mathbf{v}_r)d\mathbf{v}_r\right]$$

$$= \frac{m_r}{3}\left[\int_{\mathbf{v}_i\cdot\mathbf{n}<0} v_i^2 f^{-*}(\mathbf{v}_i)d\mathbf{v}_i + \int_{\mathbf{v}_r\cdot\mathbf{n}>0} v_r^2 f^{+*}(\mathbf{v}_r)d\mathbf{v}_r\right]. \quad [14a]$$

Using the relative velocity distribution of the reflected particles at the moment of collision for an accommodation coefficient of unity, as given by Eq. [13a], Eq. [14a] can be recast to express the "surface temperature" T_r in terms of the relative velocity distribution of the colliding particles $f^{-*}(\mathbf{v}_i)$ as,

$$\frac{kT_r}{m_r}\int_{\mathbf{v}_i\cdot\mathbf{n}<0} f^{-*}(\mathbf{v}_i)d\mathbf{v}_i = \frac{1}{3}\int_{\mathbf{v}_i\cdot\mathbf{n}<0} v_i^2 f^{-*}(\mathbf{v}_i)d\mathbf{v}_i. \quad [14b]$$

It may be noted that the "surface temperature" T_r, defined by Eq. [14b], is given by the ensemble average of the kinetic energy of the relative motion of the particle pairs at the time of collision. The ensemble average of the kinetic energy of the relative motion of the particle pairs when they are far apart, i.e., the ensemble average of the energy of their relative thermal motion, is governed by the temperature of the suspending medium. Since the colliding particle pairs are accelerated by the attractive interparticle force, the ensemble average of the kinetic energy of their relative motion will be greater than the ensemble average of the energy of their relative thermal motion when they are far apart. The "surface temperature" T_r is therefore greater than the temperature of the suspending medium, the deviation being larger as the interparticle attractive force increases. An ensemble of the colliding particle pairs can be obtained through the generation of the trajectories of a large number of particle pairs (or, equivalently, a large number of fictitious particles) initially separated by a large specified distance, until the particles collide. The relative velocity distribution of the colliding particle pairs $f^{-*}(\mathbf{v})$ and therefore the "surface temperature" T_r can be calculated from the above ensemble using Eq. [14b]. The distribution of the relative velocity of the reflected particle pairs will then be given by the half-range Maxwellian distribution Eq. [13a].

The trajectories of the reflected particle pairs can be calculated from the solution of the Langevin equation, using as the initial condition the relative velocity governed by the distribution function [13a]. This calculation will determine as to whether the reflected particle pair will escape from the potential well or not. If the kinetic energy of the relative motion of the reflected particles is sufficiently large compared to the depth of the potential well, the reflected particles would probably overcome the friction due to the collisions with the medium (during their stay in the potential well) and escape, i.e., the collision will not lead to coagulation. The meaning of a coagulation event is clarified later in the paper. On that basis, the ensemble average of the rate of coagulation over a large number of particle pairs should yield the rate of Brownian coagulation.

BROWNIAN DYNAMICS OF AEROSOL PARTICLES

The random Brownian motion of the fictitious particle of reduced mass m_r in terms of its momentum $\mathbf{p} = m_r\mathbf{v}$ is described by the Langevin equation:

$$\frac{d\mathbf{p}}{dt} = -\zeta\mathbf{p} - \frac{\mathbf{r}}{r}\frac{d\phi}{dr} + \mathbf{B}(t), \qquad [15]$$

where $\mathbf{B}(t) = \frac{1}{2}(\mathbf{B}_1(t) - \mathbf{B}_2(t))$.

The x component, for example, of the above equation is given by

$$\frac{dp_x}{dt} = -\zeta p_x - \frac{x}{r}\frac{d\phi}{dr} + B_x(t), \qquad [15a]$$

where p_x and $B_x(t)$ refer to the x component of the momentum of the fictitious particle and the x component of the random force, respectively. The above equation is supplemented with the conditions

$$\langle B_x(t)B_x(0)\rangle = \delta(t), \qquad [16]$$

$$\langle B_x(t)\rangle = 0, \qquad [17]$$

$$\langle B_x(t)p_x(0)\rangle = 0 \quad \text{for} \quad t > 0, \qquad [18]$$

$$W(B) = \exp\{-B^2/2\langle B^2\rangle\}/(2\pi\langle B^2\rangle)^{1/2} \qquad [19]$$

$$\langle p_x^2\rangle = m_r kT, \qquad [20]$$

where k is the Boltzmann constant, T is the temperature of the medium, $W(B)$ is the distribution function for the random force, and the brackets $\langle\cdot\rangle$ denote the ensemble average.

In view of the fact that the correlation function for the random force, as given by Eq. [16], is a Dirac δ function, the "strict" Langevin equation (Eq. [15]) is not amenable to computer simulation. In order to circumvent the above difficulty, it is convenient to describe the motion of the fictitious particle by the generalized Langevin equation. The generalized Langevin equation, which can be derived from the Liouville equation (11), along with the supplementary conditions are

$$\frac{dp_x}{dt} = -\int_0^t f(t-s)p_x(s)ds - \frac{x}{r}\frac{d\phi}{dr} + B_x(t), \qquad [21]$$

$$\langle B_x(t)p_x(0)\rangle = 0 \quad \text{for} \quad t > 0, \qquad [22]$$

$$\langle B_x(t)\rangle = 0 \qquad [23]$$

and

$$\langle p_x^2\rangle = m_r kT. \qquad [24]$$

The simple friction coefficient ζ in Eq. [15a] is replaced in the generalized Langevin equation (Eq. [21]) by a friction kernel containing the memory function $f(t)$. In addition, in place of the condition given by Eq. [16], the second fluctuation–dissipation theorem provides the more general condition (11)

$$\langle B_x(t)B_x(0)\rangle = \langle p_x^2\rangle f(t), \qquad [25]$$

which does not contain the Dirac δ function.

The Langevin equation [15a] is a special case of the generalized Langevin equation [21] when the memory function is given by

$$f(t) = \zeta\delta(t). \qquad [26]$$

Defining the autocorrelation times for the momentum, t_p, and for the random force, t_B, as

$$t_p = \langle p_x^2 \rangle^{-1} \int_0^\infty \langle p_x(t)p_x(0)\rangle dt \quad [27]$$

and

$$t_B = \langle B_x^2 \rangle^{-1} \int_0^\infty \langle B_x(t)B_x(0)\rangle dt, \quad [28]$$

the following relation (11)

$$t_p t_B = \langle p_x^2 \rangle / \langle B_x^2 \rangle \quad [29]$$

can be derived from Eqs. [21]–[24]. The generalized Langevin equation [21] reduces to the "strict" Langevin equation [15a] when $t_B \ll t_p$ (11, 12). Since for the "strict" Langevin equation one has $\langle p_x(t)p_x(0)\rangle = \langle p^2 \rangle \times \exp(-\zeta t)$, one obtains from Eq. [27], that $t_p = 1/\zeta$.

The "particle" trajectories can be simulated using a random force in the generalized Langevin equation that is constant during a small time step t_s with values given by a Gaussian distribution. The memory function for this form of random force is (12)

$$f(t) = 2(t_s - |t|)S(t_s - |t|)/t_p t_s^2 \quad [30]$$

and the time correlation of the random force is given by

$$\langle B_x(t)B_x(0)\rangle = \langle B_x^2\rangle[(t_s - |t|)/t_s]S(t_s - |t|), \quad [31]$$

where S is the unit step function. The correlation time for the random force t_B, as defined by Eq. [28], is given by (12)

$$t_B = t_s/2. \quad [32]$$

Therefore, from Eqs. [20] and [29], one has

$$\langle B_x^2 \rangle = 2m_r kT/t_p t_s. \quad [33]$$

In order to simulate the motion of the fictitious particle by an equation which constitutes a close approximation of the "strict" Langevin equation [15], we choose in the generalized Langevin equation $t_s \ll t_p\ (=1/\zeta)$ and the value of the random force is taken according to the distribution

$$W(B) = \exp\{-t_s B^2/4\zeta m_r kT\} \Big/ \left(2\pi \frac{2m_r kT\zeta}{t_s}\right)^{1/2}. \quad [19a]$$

Let us consider the fictitious particle of reduced mass m_r for which the location and the momentum at time $t_n = nt_s$ are x_n and p_n, respectively, and the random force $B_x^{(n)}$ has a constant value in the time interval $t_n < t < t_{n+1}$. The external force acting on the fictitious particle is given by,

$$F_x = -\frac{x}{r}\frac{d\phi}{dr}.$$

If the position $r^{(n)}$ of the particle (or, equivalently, the distance between two particles of mass m_p) is greater than the decay length λ_d of the interaction potential, the change in the interaction potential experienced by the fictitious particle during the $(n + 1)$th time interval is small. Therefore, it can be assumed that the "particle" experiences a constant external force $F_x^{(n)}$ during the time interval $t_n < t < t_{n+1}$. The generalized Langevin equation can then be integrated to give

$$p_x^{(n+1)} = p_x^{(n)} + \psi^{(n)} t_s \quad [34]$$

and

$$x^{(n+1)} = x^{(n)} + (p_x^{(n)} t_s + \tfrac{1}{2} t_s^2 \psi^{(n)})/m_r, \quad [35]$$

where

$$\psi^{(n)} = B_x^{(n)} - \frac{x^{(n)}}{r^{(n)}}\left(\frac{d\phi}{dr}\right) - \zeta p_x^{(n)}. \quad [36]$$

On the other hand, if the position $r^{(n)}$ of the particle is smaller than the decay length of the interaction potential, the external force experienced by the "particle" during the time interval $t_n < t < t_{n+1}$ cannot be assumed constant because the interaction potential is a strong function of position. In order to discretize the generalized Langevin equation, the time interval $t_n < t < t_{n+1}$ is further subdivided into K subintervals such that the external force experienced by the particle during the subinterval can be assumed constant.

Denoting by $p_{x,i}^{(n)}$ and $x_{,i}^{(n)}$ the momentum and position of the "particle" at time $t = t_n + it'_s$, t'_s $(=t_s/K)$ being the width of the subinterval, the generalized Langevin equation can be integrated to give

$$P_{x,i+1}^{(n)} = p_{x,i}^{(n)} + \chi_{,i}^{(n)} t'_s, \quad i = 0, 1, 2 \cdots K - 1 \qquad [37]$$

and

$$x_{,i+1}^{(n)} = x_{,i}^{(n)} + (p_{x,i}^{(n)} t'_s + \tfrac{1}{2} t_s'^2 \chi_{,i}^{(n)})/m_r, \quad i = 0, 1, 2 \cdots K - 1 \qquad [38]$$

where

$$\chi_{,i}^{(n)} = B_x^{(n)} - \frac{x_{,i}^{(n)}}{r_{,i}^{(n)}} \left(\frac{d\phi}{dr}\right)_i - \zeta p_{x,i}^{(n)}. \qquad [39]$$

It is to be noted that the random force $B_x^{(n)}$ is still assumed to be constant over the time interval $t_n < t < t_{n+1}$.

Introducing the dimensionless variables

$$P_x = \frac{h}{m_r} p_x, \quad h = (kT/m_r)^{1/2}$$

$$X = x/R_s, \quad R_s = 2R$$

$$R = r/R_s,$$

$$\kappa = \frac{4}{(2\pi)^{1/2} h \zeta R_s},$$

$$\theta = \frac{t_s}{t_p} = t_s \zeta,$$

$$\theta' = \frac{t'_s}{t_p} = t'_s \zeta = \frac{\theta}{K},$$

$$B'_x = \frac{B_x}{\langle B^2 \rangle^{1/2}} = \frac{B_x}{\left(\frac{2m_r \zeta kT}{t_s}\right)^{1/2}}$$

and

$$\Phi_0 = \frac{\phi}{kT}, \qquad [40]$$

the x component of the momentum and the position of the "particle" in regions I ($r > \lambda_d$) and II ($r < \lambda_d$) can be written as

Region I:

$$P_x^{(n+1)} = P_x^{(n)} - \theta P_x^{(n)} - \frac{X^{(n)}}{R^{(n)}} \frac{(2\pi)^{1/2}}{4} \kappa\theta \left(\frac{d\phi_0}{dR}\right)_n + 2^{1/2}\theta^{1/2} B'_x \qquad [41]$$

and

$$X^{(n+1)} = X^{(n)} + \kappa\theta P_x^{(n)} - \frac{1}{2}\kappa\theta^2 P_x^{(n)} - \frac{1}{2}\frac{X^{(n)}}{R^{(n)}} \kappa^2\theta^2 \left(\frac{d\phi_0}{dR}\right)_n + \frac{1}{2^{1/2}} \kappa\theta^{3/2} B'_x. \qquad [42]$$

Region II:

$$P_{x,i+1}^{(n)} = P_{x,i}^{(n)} - \theta' P_{x,i}^{(n)} - \frac{X_{,i}^{(n)}}{R_{,i}^{(n)}} \frac{(2\pi)^{1/2}}{4} \times \kappa\theta' \left(\frac{d\phi_0}{dR}\right)_i + 2^{1/2}\theta'^{1/2} \left(\frac{\theta'}{\theta}\right)^{1/2} B'_x \qquad [43]$$

and

$$X_{,i+1}^{(n)} = X_{,i}^{(n)} + \kappa\theta' P_{x,i}^{(n)} - \frac{1}{2}\kappa\theta'^2 P_{x,i}^{(n)} - \frac{1}{2}\frac{X_{,i}^{(n)}}{R_{,i}^{(n)}} \times \kappa^2\theta'^2 \left(\frac{d\phi_0}{dR}\right)_i + \frac{1}{2^{1/2}} \kappa\theta'^{3/2} \left(\frac{\theta'}{\theta}\right)^{1/2} B'_x. \qquad [44]$$

Similar equations can be written for the y and z components.

MONTE CARLO COMPUTER SIMULATION OF BROWNIAN COAGULATION

The Monte Carlo simulation of Brownian coagulation involves the evaluation of the ensemble average of the coagulation rate over a large number of particle pairs, through the generation of particle trajectories. The interparticle forces due to the van der Waals attraction and Born repulsion are accounted for in the description of the relative motion of two particles. The relative Brownian motion of two particles is described by the

Brownian motion of a fictitious particle of reduced mass m_r. The trajectory of the fictitious particle is generated according to the generalized Brownian dynamics described in the previous section. The relative velocity of the two colliding particles at the instant of reflection is determined according to the model for Brownian coagulation also described in the earlier section. The trajectories of the reflected particle pairs is calculated to determine whether the collision leads to coagulation or not. The flow chart for the simulation algorithm is shown in Fig. 2.

We start to follow the relative motion between two particles of mass m_p separated by a specified initial distance. This initial distance is chosen to be large compared to the correlation length λ_r (the average distance over which the relative velocity of the particle pair is correlated). Therefore, the initial relative motion of the particle pair, which is considered as the initial condition in the calculations, is uncorrelated and its initial relative velocity distribution is Maxwellian. The correlation length λ_r is given by (1)

$$\lambda_r = \bar{v}\zeta^{-1}, \quad [45]$$

where

$$\bar{v} = \left(\frac{16kT}{\pi m_p}\right)^{1/2} \quad [46]$$

is the average relative velocity between two particles of mass m_p.

The initial distance r_0 is chosen as

$$r_0 = R_s + s\lambda_r, \quad [47]$$

where R_s ($=2R$) is the radius of the sphere of influence and s is the distance parameter. The initial dimensionless distance R_0 ($=r_0/R_s$) is therefore given by

$$R_0 = 1 + s\kappa, \quad [47a]$$

κ being the modified particle Knudsen number defined by Eq. [40].

The initial dimensionless momentum $\mathbf{P}_0$ of the fictitious particle is governed by the Maxwellian distribution. The distribution of the x component P_{0x} of the initial dimensionless momentum of the fictitious particle is given by

$$F(P_x) = \frac{1}{(2\pi)^{1/2}} \exp\left(-\frac{P_x^2}{2}\right). \quad [48]$$

The initial dimensionless momentum $\mathbf{P}_0$ of the fictitious particle is chosen from the Maxwellian distribution such that

$$\mathbf{P}_0 \cdot \mathbf{R}_0 < 0, \quad [49]$$

i.e., the initial relative velocity of the two particles of mass m_p is chosen such that they approach each other. For convenience $\mathbf{R}_0$ is chosen as $(0, 0, R_0)$. The problem is spherically symmetric and the relative orientation of the two particles is not important.

Given the initial momentum of the fictitious particle, its trajectory can now be generated through the generalized Brownian dynamics. The calculation is carried out until either the fictitious particle collides with the reflecting spherical surface (or, equivalently, the particle pair collide) or the radial distance of the fictitious particle is equal to the initial distance (or, equivalently, the distance between the pair of particles is equal to the initial distance). If the fictitious particle collides with the reflecting surface, the generation of the velocity of the reflected fictitious particle from the half-range Maxwellian distribution (Eq. [13a]) requires the knowledge of the "surface temperature" T_r. As pointed out earlier, the "surface temperature" T_r is provided by the ensemble average of the kinetic energy of the relative motion of the colliding particle pairs. An ensemble of colliding particle pairs can be obtained through the generation of trajectories of a large number of fictitious particles from the initial position. The "surface temperature" T_r can then be calculated from the above ensemble using Eq. [14b]. One may note that the velocity of the reflected fictitious particle is independent of its incident velocity at the reflecting surface, because the momentum accommodation coefficient is assumed unity. After the calculation of the surface temperature, the reflected velocities of the ensemble of colliding fictitious particles can be generated from the half-range Maxwellian distribution given by Eq. [13a]. Further generation of the trajectory of a reflected fictitious particle would ascertain as to whether the "particle" escapes from the potential well or not. The incidence of the reflected "particle" upon the reflecting surface for the second time implies that the two

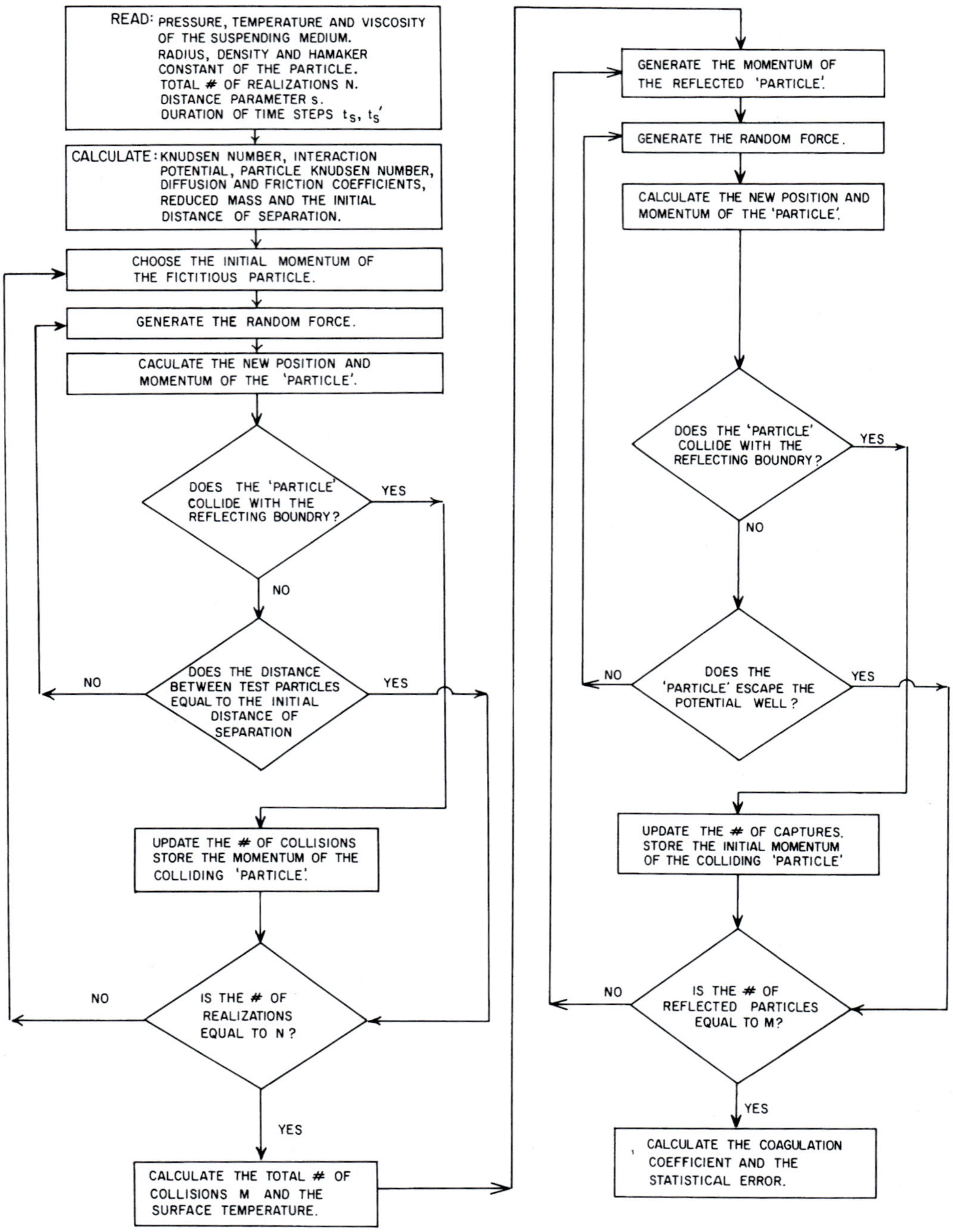

FIG. 2. Flow chart for the Monte Carlo simulation algorithm for Brownian coagulation.

colliding particles form a doublet. The "particle" incident upon the reflecting surface for the second time will be reflected with a velocity governed by the half range Maxwellian distribution. The "particle" can eventually escape the potential well after any number of reflections. The deeper the potential well, the larger is the number of reflections experienced by the "particle" and therefore the larger is the time spent by the "particle" in the potential well. Consequently, the average life of the doublet increases as the particle size increases. The fact that the two colliding particles do not collide again during their escape from the potential well implies that no coagulation act has occurred between them. However, if the particles reflected once collide at least once more before escaping the potential well, a coagulation act has occurred since a doublet with a lifetime greater than the time scale of collision is formed.[2] We outline below the details of the Monte-Carlo calculations.

The initial position of the fictitious particle is chosen such that its initial distance is given by Eq. [47a]. The initial dimensionless momentum of the fictitious particle is chosen according to the Maxwellian distribution given by Eq. [48] subject to the constraint [49]. The trajectory of the fictitious particle at subsequent times is calculated by means of Eqs. [41] and [42] or Eqs. [43] and [44] depending on whether the "particle" is in region I or II. The trajectory of the "particle" is generated until either: (i) the distance between the particle pair becomes again equal to the initial distance of separation, without any collision event; or (ii) the particle is incident upon the reflecting spherical surface (i.e., the two particles collide). The above calculation for the generation of particle trajectories is repeated for a specified N pairs of particles of mass m_p, i.e., for N fictitious particles. Out of these N pairs, $M(<N)$ will result in collision.

The dimensionless surface temperature τ defined as

$$\tau = T_r/T \qquad [50]$$

is then calculated from Eq. [14b] recast in terms of the dimensionless momentum of the colliding fictitious particles as,

$$\tau = \frac{1}{3M} \sum_{i=1}^{M} (P_{ix}^2 + P_{iy}^2 + P_{iz}^2), \qquad [51]$$

where P_{ix}, P_{iy}, and P_{iz} are the dimensionless components of the momentum of the ith fictitious particle (as defined by Eq. [40]), at the instant of collision. The components of the dimensionless momentum of the reflected "particles" at the instant of reflection are then chosen from the half-range Maxwellian distribution as given by Eq. [13a] recast in terms of dimensionless momentum as

$$F(P) = \frac{1}{(2\pi\tau)^{1/2}} \exp\left(-\frac{P^2}{2\tau}\right). \qquad [52]$$

The trajectories of the M reflected fictitious particles are further generated by means of Eqs. [43] and [44] until either: (i) the "particle" collides again with the reflecting spherical surface; or (ii) the "particle" escapes the potential well (Region II) and enters region I. If the fictitious particle is incident upon the reflecting surface for the second time (i.e., event (i) as described above), as explained earlier, a coagulation event has occurred.

From the ensemble of N particle pairs we can therefore calculate the integral

$$J = 4\pi r_0^2 \int \mathbf{v}^0 f^0(\mathbf{v}^0) g(\mathbf{v}^0) d\mathbf{v}^0 \qquad [53]$$

where J is the coagulation rate between a pair of particles of mass m_p, r_0 is the initial distance of separation between the particles, $g(\mathbf{v}^0)$ is the probability of coagulation, $\mathbf{v}^0$ is the initial velocity of the fictitious particle subject to the constraint [49], and $f^0(\mathbf{v}^0)$ is the Maxwellian distribution function for the relative velocity.

The coagulation rate between a pair of particles separated by an initial distance r_0 can be calculated as

$$J = 4\pi r_0^2 \bar{u}_z \qquad [54]$$

[2] The time spent by the "particle" after the second reflection in the potential well, before its eventual escape, determines the lifetime of the doublet. Calculations of the life time distribution of the doublets, to be published later, provide information regarding their rate of dissociation.

where

$$\bar{u_z} = \frac{1}{N} \sum_{i=1}^{L} u_{z,i}, \qquad [55]$$

N is the total number of particle pairs, L is the total number of captures, and $u_{z,i}$ is the initial z component of the relative velocity of the ith coagulating particle pair.

Thus far, we have calculated the coagulation rate J for a pair of particles separated by an initial distance r_0. However, the coagulation coefficient β is given by

$$\beta = \frac{J}{n_\infty} \qquad [56]$$

where n_∞ is the concentration of the fictitious particle of reduced mass at infinite distance, i.e., the concentration of particle pairs separated by infinite distance.

In order to infer the concentration of the fictitious particle at infinite distance n_∞ corresponding to the specified concentration $n(r_0)$ at the initial distance r_0, we need to obtain the concentration profile of the fictitious particle over a length scale very large compared to the average correlation length. The concentration profile under such condition is given by the Smoluchowski equation

$$\frac{dn}{dr} - n\frac{d\phi_0}{dr} = -\zeta h^2 J \qquad [57]$$

where ϕ_0 $(=\phi/kT)$ is the dimensionless interaction potential as defined by Eq. [40].

Integrating, we get

$$n(r)e^{-\phi_0(r)} = n_\infty - \frac{h^2\zeta J}{4\pi}\int_r^\infty \frac{e^{-\phi_0(r')}}{r'^2}\,dr'.$$

Therefore,

$$n_\infty = n(r)e^{-\phi_0(r)} + \frac{h^2\zeta J}{4\pi}\int_r^\infty \frac{e^{-\phi_0(r')}}{r'^2}\,dr'.$$

Since $n(r) = 1$ at $r = r_0 = R_s + s\lambda_r$,

$$n_\infty = e^{-\phi_0(r_0)} + \frac{h^2\zeta J}{4\pi R_s}\int_{(1+s\kappa)}^\infty \frac{e^{-\phi_0(r')}}{r'^2}\,dr'$$

i.e.,

$$n_\infty = e^{-\phi_0(r_0)} + \frac{J}{\frac{8kT}{3\eta}}\frac{1}{C}\int_{(1+s\kappa)}^\infty \frac{e^{-\phi_0(r')}}{r'^2}\,dr', \qquad [58]$$

where C is the slip correction factor for the particle diffusion coefficient given by Eq. [4].

Therefore, the coagulation coefficient β is given by

$$\beta = \frac{J}{e^{-\phi_0(r_0)} + \left(J\int_{(1+s\kappa)}^\infty \frac{e^{-\phi_0(r')}}{r'^2}\,dr' \Big/ \frac{8kT}{3\eta}C\right)}. \qquad [59]$$

The dimensionless coagulation coefficient γ, defined as the ratio of the coagulation coefficient and the Smoluchowski coagulation coefficient

$$\gamma = \beta/\beta_s = \beta\frac{3\eta}{8kT} \qquad [60]$$

is given by

$$\gamma = \frac{\gamma'}{e^{-\phi_0(r_0)} + \left(\gamma'\int_{(1+s\kappa)}^\infty \frac{e^{-\phi_0(r')}}{r'^2}\,dr'/C\right)}, \qquad [61]$$

where $\gamma' = J(3\eta/8kT)$.

Since the initial distance of the fictitious particle r_0 is much larger than the decay length λ_d of the interaction potential, the value of the dimensionless interaction potential at the initial distance $\phi_0(r_0)$ is much smaller than the dimensionless depth of the potential well. For large- and intermediary-size particles, the potential well, being deep, has a value several orders of magnitude greater than kT. For the values of the initial distance r_0 used in the calculations, it was found that the dimensionless interaction potential $\phi_0(r_0)$, though much smaller than the dimensionless depth of the potential, is not negligible. However, for small particles, the potential well being shallow (of the order of or smaller than kT), the dimensionless interaction potential $\phi_0(r_0)$ is found to be negligible. For small particles, the dimensionless coagulation coefficient γ is therefore given by

$$\gamma = \frac{\gamma'}{1 + \frac{\gamma'}{C(1+s\kappa)}}. \qquad [61a]$$

COMPARISON OF SIMULATION RESULTS WITH THE EQUATION VALID FOR LARGE PARTICLES AND WITH THE LOWER BOUND FOR THE COAGULATION COEFFICIENT

Computer simulations of the coagulation of equal-sized particles of unit density in air at 298 K and 1 atm were performed based on the algorithm discussed in the previous section. Since the Hamaker constant for most of the aerosol systems is of the order of 10^{-12} erg, this value was used for the calculation of the interaction potential between the particles. Computations were performed on a CDC 815 computer. In all the computations, the duration of the time step t_s for the random force was taken to be equal to one-tenth of the relaxation time for Brownian motion, i.e., $t_s = 0.1\zeta^{-1}$, so that the condition $t_s \ll \zeta^{-1}$ is satisfactorily fulfilled. The time step t'_s for the motion of the fictitious particle in the region of the potential well, i.e., region II, was taken to be $0.05t_s$. In other words, the values of the dimensionless times θ and θ' were taken as

$$\theta = 0.1$$

and

$$\theta' = 0.005.$$

It was assumed that the motion of the fictitious particle within the above time steps is rectilinear. This simplification, which accelerates the computer calculation of the trajectories of the fictitious particle, has been shown to be justified (9). The Philips slip correction factor for an accommodation coefficient of unity (Eq. [4]) was used in the calculation of the diffusion coefficients of particles. The values of the dimensionless coagulation coefficients γ, obtained by the computer simulation for different particle sizes, are given in Table I. The statistical errors of the Monte Carlo simulation were estimated by the standard 3σ method (corresponding to a probability of 0.997) (13). The number of particle pairs that must be generated in order to lower the error to a reasonable level depends both on the initial distance of separation between

TABLE I

Dimensionless Coagulation Coefficients Calculated through Monte Carlo Simulation

Particle radius (μm)	Knudsen number, Kn	Particle Knudsen number, κ	Distance parameter, s	No. of realizations, N	No. of collisions, M	No. of captures, L	Dimensionless coagulation coefficient, γ
0.0008	85.75	65.995	1.0	1×10^6	60	14	0.2235 ± 0.1854
0.001	68.60	47.28	1.5	1×10^6	77	25	0.490 ± 0.3146
0.0012	57.167	36.01	1.5	5×10^5	93	31	1.0475 ± 0.6080
0.0015	45.733	25.813	1.5	5×10^5	243	93	1.7588 ± 0.5954
0.002	34.30	16.816	1.5	5×10^5	808	362	3.0292 ± 0.5511
0.0025	27.44	12.068	1.5	5×10^5	1652	777	4.0047 ± 0.4976
0.003	22.867	9.208	1.5	3×10^5	1598	849	5.2164 ± 0.5746
0.005	13.72	4.329	1.5	2×10^5	4113	2450	6.8536 ± 0.4125
0.01	6.86	1.572	2.0	5×10^4	3193	2072	7.4214 ± 0.4390
0.02	3.43	0.583	4.0	2×10^4	1786	1136	6.0640 ± 0.4352
0.04	1.715	0.2237	4.0	1×10^4	2249	1521	4.0959 ± 0.2367
0.10	0.686	0.0727	4.0	5×10^3	2478	2205	2.2155 ± 0.1409
0.50	0.1372	0.02134	4.0	5×10^3	4317	4316	1.3855 ± 0.0503
1.0	0.0686	0.01428	4.0	5×10^3	4661	4661	1.2437 ± 0.0422
5.0	0.01372	0.00608	4.0	5×10^3	4988	4988	1.0934 ± 0.0352
10.0	0.00686	0.00427	4.0	5×10^3	5000	5000	1.0638 ± 0.0345

TABLE II

Dimensionless Coagulation Coefficients Calculated through Monte Carlo Simulation for Different Values of the Distance Parameter

	γ_{sim}					
Kn	$s = 2$	3	4	5	6	10
6.86	7.4214 ± 0.439	7.6281 ± 0.8787		7.2239 ± 1.306		
3.43		5.8306 ± 0.364	6.064 ± 0.4352		6.152 ± 0.3929	
0.686	2.216 ± 0.093		2.2155 ± 0.1409		2.300 ± 0.1709	2.525 ± 0.235
0.1372			1.3855 ± 0.0503			1.4145 ± 0.1017
0.0686			1.2437 ± 0.0422			1.2623 ± 0.0748

the particles and the particle size. The larger the initial distance of separation the larger should be the number of realizations in order to obtain a reasonable number of collisions (encounters) between particles. Moreover, for very small particles, the number of realizations has to be extremely high, in order to obtain a reasonable number of particle captures, since the collision efficiency is small for very small particles. The number of realizations ranges from 5000 for large particles to 10^6 for very small particles, as can be seen from Table I. Consequently, the computation time increases from a few minutes for the large particles to a few hours for the very small particles. The computed values of the dimensionless coagulation coefficient for different values of the distance parameter s are given in Table II. The effect of the distance parameter s on the computed dimensionless coagulation coefficient for different Knudsen numbers is shown in Fig. 3. The statistical errors of the simulation increase as the distance parameter s increases, as a result of the decrease in the probability of encounter between two particles. Therefore, a greater number of particle pairs must be generated at larger values of s, resulting in larger computation times.

Figure 4 compares the results of the Monte Carlo simulation with the equation valid for large particles[3] and the lower bound for the coagulation coefficient predicted by the previous models (1). Since the Fuchs semiempirical formula agrees very closely with our equation derived for large particles it is not possible to include the former in the Fig. 4. The statistical error in the simulation is found to increase with the decrease in the

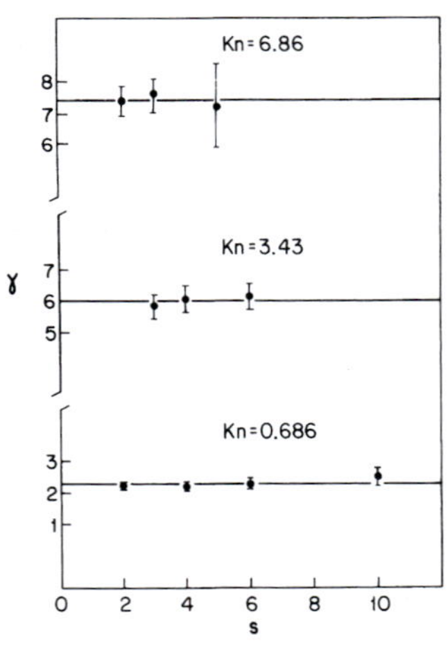

FIG. 3. Effect of the distance parameter s on the computed values of the dimensionless coagulation coefficient γ for equal-sized particles of unit density and Knudsen numbers of 6.86, 3.43, and 0.686, in air at 298 K and 1 atm, for a Hamaker constant of 10^{-12} erg.

[3] This equation was called in the previous paper "upper bound" because it indeed provides an upper bound for sufficiently small particles. It can be, however, exceeded for intermediate-size particles and for this reason we prefer to change its name.

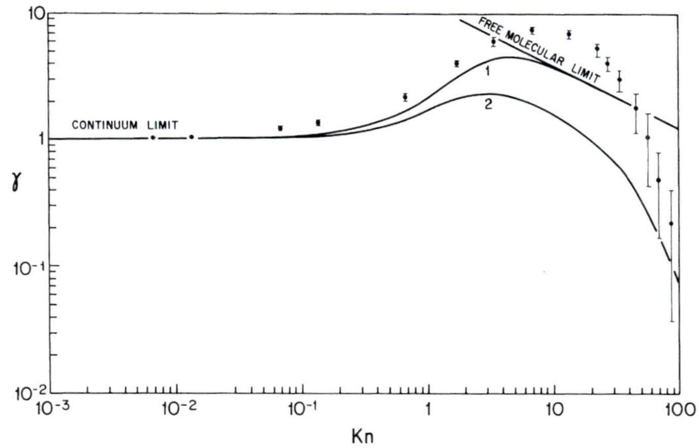

FIG. 4. Comparison of the dimensionless coagulation coefficient γ calculated through Monte Carlo simulation (Φ) with the previous models (1) for equal-sized particles of unit density, in air, at 298 K and 1 atm for a Hamaker constant of 10^{-12} erg. Curve 1 refers to the values of dimensionless coagulation coefficient γ calculated from the equation derived previously for large particles (1). Curve 2 refers to the values of the dimensionless coagulation coefficient γ calculated from the equation derived previously for sufficiently small particles (1) (the lower bound).

particle size. The simulation results agree very well for large particle sizes with the equation previously derived for large particles (1). Excepting for sufficiently small particles, the dimensionless coagulation coefficient, obtained through Monte Carlo simulations for particle sizes in the transition and free molecular ranges (i.e., $0.1 < \mathrm{Kn} < 50$), is found to be even higher than that calculated from the equation valid for large particles whereas, for sufficiently small particles, i.e., $\mathrm{Kn} > 50$, γ is found to be smaller than that calculated from the equation valid for large particles. As already noted in the Introduction, such a behavior can be attributed to two contrary effects. For small particles, i.e., particle sizes in the transition and free molecular range, the potential well is not deep whereas the decay length of the potential is greater than the particle size. Therefore, the collisions may be less effective because of the shallow potential well (i.e., the sticking probability may be less than unity), whereas the rate of collisions between the particles may increase because of the increase in the effective collision cross section caused by the increased range of the interaction potential relative to the particle size. The former effect decreases while the latter increases the coagulation rate. Excepting for sufficiently small particles, i.e., $0.1 < \mathrm{Kn} < 50$, the increase in the coagulation coefficient due to the increase in the effective collision cross section predominates over the decrease in the coagulation coefficient due to the decrease in the effectiveness of collisions, thus resulting in an enhancement in the coagulation coefficient. However, for sufficiently small particles, i.e., $\mathrm{Kn} > 50$, the increase in the rate of collisions is offset by the decrease in the effectiveness of collisions (i.e., collision efficiency) so that the coagulation coefficient becomes smaller than the free molecular limit.

The dimensionless coagulation coefficient γ for a particle size of 0.0012 μm lies between the equation valid for large particles and the lower bound, whereas, the values of γ for particle sizes of 0.001 and 0.0008 μm are found to agree with the lower bound within the statistical error of simulation. As pointed out earlier, the statistical error of the Monte Carlo simulation is high for very small particles because the number of coagulation events is very small, as a result of the small collision efficiency. More accurate estimates could, in principle, be obtained by ensuring a sufficiently large number of coagulation events. This, however, would require simulation of an exceedingly large number of particle pairs. In our calculations, the total number of realizations could not be increased beyond 10^6 because of prohibitive computation costs.

The enhancement factor e (notation not to be confused with the internal energy of the "particle" defined earlier), defined as the ratio of the coagulation coefficient to that predicted by the equation valid for large particles, is plotted as a function of the particle size in Fig. 5. The enhancement factor e is smaller than unity for sufficiently small particles, i.e., less than 0.0015 μm. However, for particle sizes in the range of 0.0015 to 0.05 μm, the enhancement factor e is greater than unity, with a maximum value around 2.0. The factor e reaches asymptotically the value of unity for large particle sizes. Nolan *et al.* (14) and Fuchs and Sutugin (15) measured coagulation rates of Pt and NaCl particles, respectively, in the transition and free molecular ranges, which are 2 to 3 times larger than those predicted by the Fuchs interpolation formula. Graham and Homer (16), Brockman *et al.* (17), and Burtscher and Schmidt-Ott (18) also report enhancement in the coagulation rates of particles in the transition range. Recently, Okuyama *et al.* (19) measured coagulation rates of NaCl, $ZnCl_2$, and Ag aerosol particles in the transition and free molecular ranges, by monitoring the inlet and outlet size distributions of aerosols in a laminar pipeflow. Their results indicate that the observed enhancement in the coagulation coefficient is a strong function of the interparticles forces. For NaCl particles (Hamaker constant of 8 $\times$ 10^{-13} ergs), the observed enhancement factor is around 2.0. This is in agreement with the present calculated values of the enhancement factor.

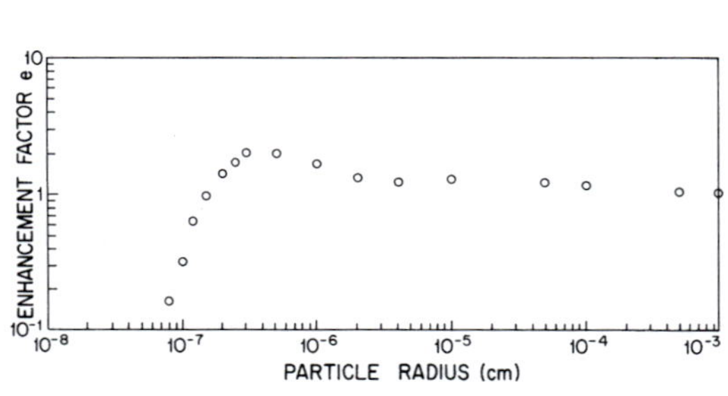

FIG. 5. Effect of the particle size on the enhancement factor e for equal-sized particles of unit density, in air, at 298 K and 1 atm, for a Hamaker constant of 10^{-12} erg.

Fuchs and Sutugin (15) explained the enhancement in the rate of coagulation of the small particles through a correction factor which accounts for the increase in the collision cross section because of interparticle van der Waals attraction. Their analysis assumed that every collision is effective, i.e., the sticking probability is unity. Marlow (20) extended their analysis to the entire range of Knudsen numbers and concluded that the effect of the van der Waals forces on the coagulation coefficient is negligible for the large particles. We have already noted that for large particles the potential energy well is deep and the decay length of the interaction potential (based on the minimum distance) is small compared to the particle size. For this reason, the sticking probability is unity and the collision cross section is equal to the cross section of the particle. For intermediary size particles, the potential well is no longer as deep and the decay length of the interaction potential becomes comparable to the particle size. Therefore, for a certain size range the potential well could be still sufficiently deep for the sticking probability to be near unity yet the collision cross section could be enhanced by the van der Waals attraction. However, for very small particles the potential well is so shallow that, even though the collision cross section is enhanced, the coagulation coefficient is very small because of the former effect.

Figure 6 shows the calculated values of the collision efficiency α (the ratio of the number of captures and the number of collisions) as a function of the particle size. The collision efficiency α increases with the particle size and eventually reaches the value of unity for large particles. Figure 7 shows the dimensionless surface temperature τ as a function of the particle size. The dimensionless surface temperature τ increases rapidly with the particle size. For very small particles, i.e., particles in the free molecular range, τ is of the order of unity. Since the interaction potential between such small particles is very small, the kinetic energy of the relative motion of the two particles at the time of collision is of the same order as the kinetic energy of the relative thermal motion of the particles. Consequently, the surface temperature, which is given by the ensemble average of the kinetic energy of the relative motion of the two particles at the time of collision, is of the same order as the temperature of the suspending medium. However, for large particles, the dimensionless surface temperature τ is extremely large, of the order of 10^2 or higher. This is a result of the increase in the depth of the potential energy well as the particle size increases. Large particles moving towards each other are subjected to great acceleration due to the large attractive potential. Therefore, the kinetic energy of the relative motion at the time of collision is very large.

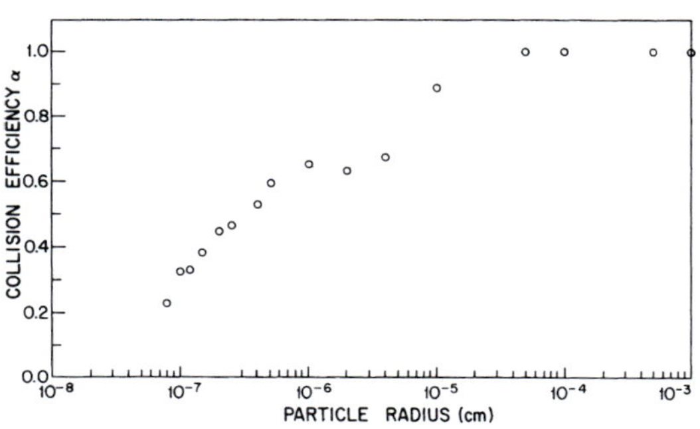

FIG. 6. Effect of the particle size on the collision efficiency α for equal sized particles of unit density, in air, at 298 K and 1 atm, for a Hamaker constant of 10^{-12} erg.

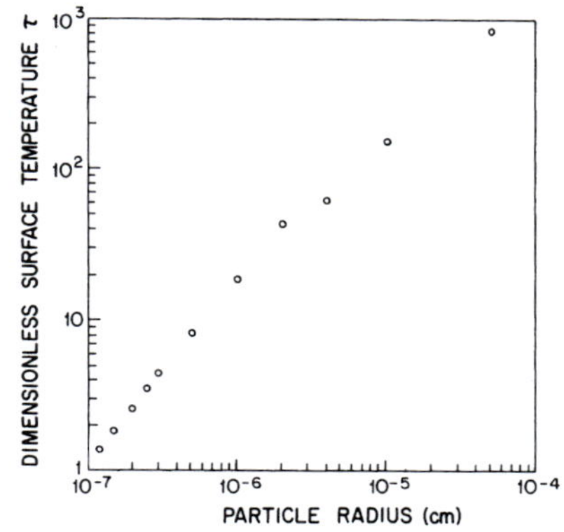

FIG. 7. Effect of the particle size on the dimensionless surface temperature τ for equal-sized particles of unit density, in air, at 298 K and 1 atm, for a Hamaker constant of 10^{-12} erg.

The Monte Carlo simulation can provide the values of the Brownian coagulation coefficient, without the use of empirical sticking probabilities, for particles of different densities and Hamaker constants, at different temperatures and pressures of the suspending medium. The method can easily be adapted to the evaluation of the Brownian coagulation coefficient between particles of different sizes. Since the relative motion of particles of different sizes cannot be reduced to the motion of a fictitious particle of reduced mass (21), the trajectories of both the particles must be generated, in order to calculate the Brownian coagulation coefficient. The method can also be easily extended to the calculation of the Brownian coagulation coefficient of charged particles by suitably modifying the interaction potential.

CONCLUSIONS

The Brownian coagulation coefficient for equal-sized electrically neutral particles, of unit density, for the entire range of Knudsen numbers, have been calculated through the Monte Carlo simulation. The interparticle forces due to van der Waals attraction and

Born repulsion have been taken into account instead of using a phenomenological sticking probability. The particle motion is described by the generalized Brownian dynamics. The relative Brownian motion of two equal-sized particles is represented by that of a fictitious particle of reduced mass. The repulsive part of the interaction potential between the particles is replaced by a spherically reflecting boundary with momentum and thermal accommodation coefficients of unity. For large particles, the results of the simulation agree very well with those predicted by the previous model (1) and with the Fuchs interpolation formula. For very small particles, the coagulation coefficients obtained through simulation are found to be much smaller than the free molecular limit but agree fairly well with the lower bound predicted by the previous model (1). The calculated values of the coagulation coefficient for particles of intermediary sizes are found to be even higher than those corresponding to a sticking probability of unity. Such an enhancement in the coagulation coefficient is consistent with the experimental observations and is attributed to the predominant effect of the increase in the rate of collisions (caused by the increased range of the van der Waals attraction relative to the particle size) over the decrease in the effectiveness of collisions (caused by the increase in the probability of escape of the particles from the potential well). The calculated values of the collision efficiency increase with the particle size and eventually reach the value of unity for large particles.

REFERENCES

1. Narsimhan, G., and Ruckenstein, E., *J. Colloid Interface Sci.* **104,** 344 (1985).
2. Smoluchowski, M. V., *Z. Phys. Chem.* **92,** 144 (1918).
3. Millikan, R. A., *Phys. Rev.* **22,** 1 (1923).
4. Philips, W. F., *Phys. Fluids* **18,** 1089 (1975).
5. Hidy, G. M., and Brock, J. R., "The Dynamics of Aerocolloidal Systems." Pergamon, Oxford, 1970.
6. Fuchs, N. A., "Mechanics of Aerosols." Pergamon, New York, 1964.
7. Sitarski, M., and Seinfeld, J. H., *J. Colloid Interface Sci.* **61,** 261 (1977).
8. Mork, K. J., Razi Naqvi, K., and Waldenström, S., *J. Colloid Interface Sci.* **98,** 103 (1984).
9. Nowakowski, B., and Sitarski, M., *J. Colloid Interface Sci.* **83,** 614 (1981).
10. Chandrasekhar, S., *Rev. Mod. Phys.* **15,** 1 (1943).
11. Kubo, R., *Rep. Progr. Phys.* **29,** 255 (1966).
12. Turq, P., Lantelme, F., and Friedman, H. L., *J. Chem. Phys.* **66,** 3039 (1977).
13. Sitarski, M., *J. Colloid Interface Sci.* **73,** 152 (1980).
14. Nolan, P. J., and Keenan, E. L., *Proc. Roy. Irish Acad. A* **52,** 171 (1949).
15. Fuchs, N. A., and Sutugin, A. G., *J. Colloid Interface Sci.* **20,** 492 (1965).
16. Graham S. C., and Homer, J. B., *Discuss. Faraday Soc.* **7,** 85 (1973).
17. Brockman, J. E., McMurry, P. H., and Liu, B. Y. H., *J. Colloid Interface Sci.* **88,** 522 (1982).
18. Burtscher, H., and Schmidt-Ott, A., *Phys. Rev. Lett.* **48,** 1734 (1982).
19. Okuyama, K., Kousaka, Y., and Hayashi, K., *J. Colloid Interface Sci.* **101,** 98 (1984).
20. Marlow, W. H., *J. Chem. Phys.* **73,** 6284, 6288 (1980).
21. Sahni, D. C., *J. Colloid Interface Sci.* **91,** 418 (1983).

Dissociation Kinetics of Doublets of Aerosol Particles

GANESAN NARSIMHAN[1] AND ELI RUCKENSTEIN[2]

Department of Chemical Engineering, State University of New York, Buffalo, New York 14260

Received April 14, 1986; accepted July 1, 1986

If the dispersion forces, responsible for coagulation, are not too strong (i.e., if the aerosol particles are sufficiently small), then the coagulated aerosol doublets can derive enough energy from the collisions with the molecules of the suspending medium to overcome the interaction potential well, thus leading to their dissociation. Because of the great disparity between the time scales of oscillation in the potential well and Brownian motion of the coagulated particle pairs, the Fokker–Planck equation can be averaged with respect to the relative position of the constituents of the doublet. One thus obtains a one-dimensional Fokker–Planck equation in terms of the energy of the relative motion of the two constituents. As a result, the average lifetime of a doublet can be calculated as a first passage time in the energy of the relative motion of its constituents. The average dissociation time of the doublets, in air at 1 atm and 298 K, for a Hamaker constant of 10^{-12} erg, has been thus calculated for different radii of the constituent particles. The average dissociation time is found to increase dramatically, from 10^{-7} to 10^{-1} s, as the radius changes from 15 to 50 Å. This is a result of the rapid increase in the depth of the interaction potential well with increasing radii. The doublets consisting of particles whose radii are greater than 50 Å are found to be extremely stable.

INTRODUCTION

The coagulation of aerosols is mainly a result of the short-range dispersion forces between particles. Since the dispersion forces between large particles are strong, it is customary to assume that every collision between two such particles results in irreversible coagulation. This assumption is reasonable because the interaction potential well (due to the van der Waals attraction and Born repulsion) is, in this case, very deep. For sufficiently small particles, however, not every collision leads to coagulation, since the dispersion forces between them are not strong enough (i.e., the interaction potential well is not sufficiently deep). The effectiveness of particle collisions, as characterized by the sticking probability, has been calculated in terms of the interaction potential between them (1, 2).

The coagulated particles are subjected, in addition, to collisions with the molecules of the suspending medium. If the dispersion forces, responsible for coagulation, are not sufficiently strong, i.e., if the particles are sufficiently small, the molecules of the suspending medium can impart sufficient energy to the coagulated particles to overcome the interaction potential well leading to their dissociation. In other words, the coagulation of sufficiently small particles is reversible. The aim of the present paper is to develop a theory capable of predicting the average dissociation time of a doublet. Information regarding the rate of dissociation of coagulated particle pairs is useful for the calculation of the evolution of particle size distribution. The effect of doublet dissociation rates on aerosol growth due to coagulation is discussed in a subsequent paper.

The average lifetime of a coagulated pair (doublet) of small particles can be calculated by examining the relative Brownian motion of the particle pair in the interaction potential well. If the energy imparted by the collisions of the molecules of the medium is smaller than the depth of the interaction potential well, the particle pair will exhibit an oscillatory motion within the potential well. It will be shown later that, for sufficiently small particles, the time scale of these oscillations is much smaller than

[1] G. Narsimhan is now associated with the Department of Agricultural Engineering, Purdue University, West Lafayette, Indiana.

[2] To whom correspondence should be addressed.

0021-9797/87 $3.00

the time scale of the Brownian motion. This disparity in the time scales enables one to describe the relative Brownian motion of the particle pair within the potential well only in terms of a stochastic differential equation for the energy of their relative motion. The particle pair will dissociate as soon as the energy of their relative motion equals the depth of the interaction potential well. The calculation of the average lifetime of a doublet of equal-size particles can, therefore, be formulated as a first passage time (the time at which the energy of the relative motion becomes equal to the depth of the potential well) in the particle energy space. The first passage time concept has been widely used in the dynamic description of a variety of physical problems (3).

The derivation of the dynamic equations describing the relative Brownian motion of the particle pair is presented in the next section. The subsequent section discusses the first passage time approach for the evaluation of the mean dissociation time of a doublet. The rates of dissociation of doublets of equal-size particles are calculated in the last section.

DYNAMIC EQUATIONS FOR THE RELATIVE BROWNIAN MOTION OF A PAIR OF EQUAL-SIZE PARTICLES

Let us consider a doublet consisting of two equal-size aerosol particles of mass m_p and radius R. The formation of the doublet is a result of the interparticle van der Waals attractive forces between its two constituents. These constituents are, however, subjected to random collisions with the molecules of the suspending medium. For large particles, the interparticle dispersion forces are strong. Therefore, the molecular collisions cannot impart sufficient energy to dissociate the doublet. On the other hand, for a doublet consisting of sufficiently small particles, the molecular collisions can impart sufficient energy to its constituents to overcome the interaction potential well, thus leading to dissociation. The lifetime of the doublet is determined by the time spent by its constituents in the potential well before their eventual escape. The overall interaction potential between the two constituents was calculated through the integration of the intermolecular potential, modeled as the Lennard Jones 6–12 potential, under the assumption of pairwise additivity (1). The two particles of the doublet experience Brownian motion in the interaction potential well. Their individual Brownian motions can, therefore, be described by the Langevin equations (4)

$$m_p \frac{d\mathbf{v}_1'}{dt'} = -m_p \zeta \mathbf{v}_1 + \mathbf{F}_1 + \sqrt{2m_p \zeta kT}\,\mathbf{I}\boldsymbol{\xi}_1'(t') \quad [1]$$

and

$$m_p \frac{d\mathbf{v}_2'}{dt'} = -m_p \zeta \mathbf{v}_2 + \mathbf{F}_2 + \sqrt{2m_p \zeta kT}\,\mathbf{I}\boldsymbol{\xi}_2'(t'), \quad [2]$$

where $\mathbf{v}'$ is the velocity, ζ is the friction coefficient, t' is the time, T is the absolute temperature of the suspending medium, k is the Boltzmann constant, $\mathbf{F}$ is the interparticle force, $\boldsymbol{\xi}(t')$ is a white noise vector, $\mathbf{I}$ is the identity matrix, and the subscripts 1 and 2 refer to the particles 1 and 2, respectively. The interaction forces $\mathbf{F}_1$ and $\mathbf{F}_2$ act on the particles 1 and 2, respectively, along their line of centers and are related via

$$\mathbf{F}_1 = -\mathbf{F}_2 = \frac{F(\mathbf{r}_1' - \mathbf{r}_2')}{|\mathbf{r}_1' - \mathbf{r}_2'|}(\mathbf{r}_1' - \mathbf{r}_2'), \quad [3]$$

where $\mathbf{r}_1'$ and $\mathbf{r}_2'$ are the position vectors of particles 1 and 2, respectively, $|\mathbf{r}|$ is the magnitude of the vector $\mathbf{r}$, and F is the magnitude of the interaction force. Taking the center of mass of the doublet as the origin of the coordinate system and introducing the relative coordinates

$$\mathbf{r}' = \mathbf{r}_1' - \mathbf{r}_2', \quad \mathbf{v}' = \mathbf{v}_1' - \mathbf{v}_2',$$

the Langevin equation for the relative motion of the two particles of the doublet can be written as

$$m_r \frac{d\mathbf{v}'}{dt'} = -m_r \zeta \mathbf{v}' - \frac{\mathbf{r}'}{r'}\frac{d\phi}{dr'} + \sqrt{2m_r \zeta kT}\,\mathbf{I}\boldsymbol{\xi}'(t'), \quad [4]$$

where $m_r(=m_p/2)$ is the reduced mass and ϕ is the interaction potential

$$\langle \boldsymbol{\xi}'(t) \rangle = \mathbf{0} \quad \text{and} \quad \langle \boldsymbol{\xi}(t')\boldsymbol{\xi}(t_0') \rangle = \mathbf{I}\delta(t' - t_0').$$

Here δ is the Dirac delta function and $\langle \cdot \rangle$ denotes the ensemble average.

The relative motion of the two particles of the doublet can, therefore, be described by the Brownian motion of a fictitious particle of reduced mass m_r in the interaction potential well with the same friction coefficient as that of the particles. The motion of the fictitious particle of reduced mass (referred to hereafter as the particle) around the sphere of influence of radius $R_s(=2R)$ is divided into two regions, which are shown in Fig. 1. The thickness of region I is equal to the decay length (expressed here as the shortest distance between the surfaces of the two particles) of the interaction potential. As long as the particle moves in this region, it remains a doublet. In region II, the interaction potential is negligible, i.e., the particle experiences negligible interaction force and, therefore, the particle will dissociate. This occurs if the particle gains sufficient energy from the colliding molecules of the suspending medium to overcome the potential well, i.e., to escape from region I and enter region II (Fig. 1). If the particle, on the other hand, does not gain sufficient energy, it will experience an oscillatory motion within the potential well. The relative magnitudes of the time scales of oscillation and of its Brownian motion can be evaluated by comparing the decay length λ_d of the interaction potential and the correlation length λ_r (the distance over which the Brownian motion of the particle is correlated). The correlation length λ_r is related to the time scale ζ^{-1} of Brownian motion via

$$\lambda_r \sim \zeta^{-1}\sqrt{\frac{kT}{m_r}}. \qquad [5]$$

For sufficiently small particles, the decay length λ_d is much smaller than the correlation length λ_r. Indeed, for particles of 100 Å radius and unit density, in air at 1 atm and 298 K, and for a Hamaker constant of 10^{-12} erg, λ_d is of the order of 50 Å, whereas λ_r is of the order of 300 Å. Therefore, the time scale of oscillation of a captured particle is much smaller than the time scale of its Brownian motion. Consequently, a captured particle will undergo very few collisions with the medium during one period of oscillation. As a result, the total energy of the captured particle (potential and kinetic) will nearly be conserved during one period of oscillation. Moreover, the depth of the overall interaction potential well, even for doublets of sufficiently small particles, can be of the order of or larger than kT. For instance, the depth of the potential well for two particles of 50 Å radius and of Hamaker constant 10^{-12} erg is of the order of 17 kT. Therefore, the particle in the potential well must undergo a very large number of collisions with the medium before acquiring sufficient energy to escape from the potential well. In other words, the time scale of dissociation of a doublet is expected to be much larger than the time scale of the Brownian motion and, therefore, even much larger than the time scale of oscillations. Because of this great disparity between the time scales of dissociation and oscillation, the lifetime of a doublet can be estimated without a detailed description of the particle motion during a period of oscillation. In other words, the lifetime of a doublet can be estimated by averaging the equation over the oscillatory motion.

Let us define the following dimensionless variables:

$$\mathbf{r} = \mathbf{r}'/\lambda_d, \quad t = t'/\tau_0, \quad \tau_0 = \frac{\lambda_d}{\sqrt{kT/m_r}}$$

$$\mathbf{v} = \mathbf{v}'/\sqrt{kT/m_r}, \quad \Phi = \phi/kT, \qquad [6]$$

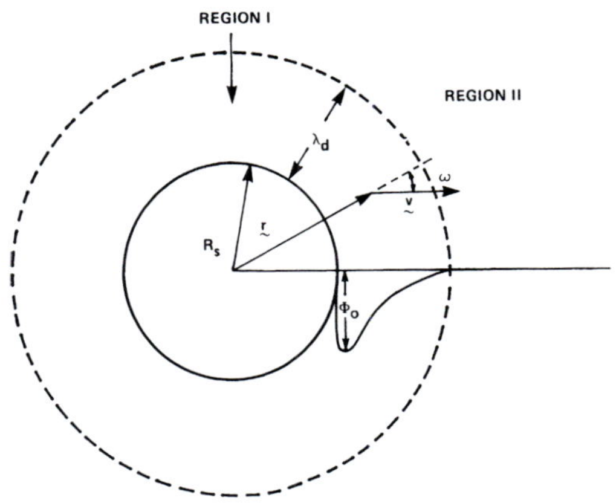

FIG. 1. Different regions for the motion of the fictitious particle of reduced mass.

and

$$\Lambda = \frac{\tau_0}{\zeta^{-1}} = \frac{\lambda_d}{\lambda_r},$$

where τ_0 is the time scale of oscillations. The Langevin equation (Eq. [4]) for the motion of the particle can, therefore, be written as

$$\frac{d^2\mathbf{r}}{dt^2} + \Lambda\frac{d\mathbf{r}}{dt} + \frac{\mathbf{r}}{r}\frac{d\Phi}{dr} = (2\Lambda)^{1/2}\mathbf{I}\boldsymbol{\xi}(t), \quad [7]$$

where the white noise vector $\boldsymbol{\xi}(t)$ satisfies the following conditions:

$$\langle\boldsymbol{\xi}(t)\rangle = \mathbf{0}, \quad \langle\boldsymbol{\xi}(t)\boldsymbol{\xi}(t_0)\rangle = \mathbf{I}\delta(t-t_0). \quad [8]$$

Taking the scalar product of Eq. [7] with the velocity vector $\dot{\mathbf{r}}$, one obtains

$$\frac{d}{dt}\left(\frac{1}{2}\dot{\mathbf{r}}\cdot\dot{\mathbf{r}}\right) + \Lambda(\dot{\mathbf{r}}\cdot\dot{\mathbf{r}}) + \frac{d\Phi}{dt} = (2\Lambda)^{1/2}\dot{\mathbf{r}}\cdot\mathbf{I}\boldsymbol{\xi}(t). \quad [9]$$

Since the right-hand side of the above equation is a linear combination of the three uncorrelated components of the white noise vector $\boldsymbol{\xi}(t)$, it represents a scalar uncorrelated white noise (5), i.e.,

$$\dot{\mathbf{r}}\cdot\mathbf{I}\boldsymbol{\xi}(t) = \dot{x}\xi_1(t) + \dot{y}\xi_2(t) + \dot{z}\xi_3(t),$$

where $\dot{x}$, $\dot{y}$, and $\dot{z}$ are the three velocity components and $\xi_1(t)$, $\xi_2(t)$, and $\xi_3(t)$ are the three components of the white noise vector $\boldsymbol{\xi}(t)$. Therefore,

$$\dot{\mathbf{r}}\cdot\mathbf{I}\boldsymbol{\xi}(t) = (\dot{\mathbf{r}}\cdot\dot{\mathbf{r}})^{1/2}\xi(t),$$

where $\langle\xi(t)\rangle = 0$, $\langle\xi(t)\xi(t-t_0)\rangle = \delta(t-t_0)$. Equation [9] can now be rewritten as

$$\frac{d}{dt}\left(\frac{1}{2}\dot{\mathbf{r}}\cdot\dot{\mathbf{r}} + \Phi\right) + \Lambda(\dot{\mathbf{r}}\cdot\dot{\mathbf{r}}) = (2\Lambda)^{1/2}(\dot{\mathbf{r}}\cdot\dot{\mathbf{r}})^{1/2}\xi(t). \quad [10]$$

Since the dimensionless velocity v and energy E of the particle are given by

$$\dot{\mathbf{r}}\cdot\dot{\mathbf{r}} = v^2 \quad [11]$$

and

$$E = \frac{1}{2}v^2 + \Phi,$$

Eq. [10] can be recast in the form

$$\frac{dE}{dt} = -2\Lambda(E-\Phi) + \sqrt{4\Lambda(E-\Phi)}\xi(t). \quad [12]$$

The associated equation for the evolution of the position r of the particle is

$$\frac{dr}{dt} = \mathbf{v}\cdot\frac{\mathbf{r}}{r} = v\cos\omega, \quad [13]$$

where ω is the angle between the velocity vector $\mathbf{v}$ and the position vector $\mathbf{r}$ (Fig. 1). Equations [12] and [13] are the stochastic differential equations which describe the motion of the particle in the potential well. Since the ratio of the time scales of oscillation and Brownian motion $\Lambda \ll 1$, Eqs. [12] and [13] show that the time scale of change of the position r is much shorter than the time scale of change of the energy E. Therefore, the particle experiences several oscillations during the time scale in which its energy changes. As already pointed out, the details of the particle motion over the time scale of oscillation do not appreciably influence the evolution of its energy. Consequently, in order to calculate the evolution of the energy E, one can approximate Eq. [13] by replacing $\cos\omega$ by the average value $\sqrt{\langle\cos^2\omega\rangle}$. This average value is calculated by assuming that the velocity vector can be oriented in any direction with equal probability. Therefore, Eq. [13] can be rewritten as

$$\frac{dr}{dt} \simeq v\sqrt{\langle\cos^2\omega\rangle} \quad [13a]$$

and, further, since $\langle\cos^2\omega\rangle = \frac{1}{3}$, as

$$\frac{dr}{dt} \simeq \{\tfrac{2}{3}(E-\Phi)\}^{1/2}. \quad [14]$$

FIRST PASSAGE TIME FORMULATION FOR THE CALCULATION OF THE MEAN DISSOCIATION TIME OF A DOUBLET

Consider a particle whose initial dimensionless position and energy are r_0 and E_0, respectively. Since this particle experiences Brownian motion in the potential well, both its position and its energy will be random at any time. Therefore, the evolution of its position and energy should be described by the conditional joint probability density $w(r, E; t|r_0, E_0; 0)$, defined as

$$w(r, E; t|r_0, E_0; 0)drdE,$$

which is the probability that at time t the position of the particle is between r and $r + dr$ and its energy between E and $E + dE$, when its initial position and energy are r_0 and E_0, respectively.

Employing the stochastic differential equations [12] and [14], the Fokker–Planck equation for the evolution of the conditional joint probability density $w(r, E; t|r_0, E_0; 0)$ has the form (6)

$$\frac{\partial}{\partial t}w(r,E;t|r_0,E_0,0) = -\frac{\partial}{\partial r}[\{\tfrac{2}{3}(E-\Phi)\}^{1/2} \times w(r,E;t|r_0,E_0,0)] + 2\Lambda\frac{\partial}{\partial E}[(E-\Phi) \times w(r,E;t|r_0,E_0;0)] + 2\Lambda\frac{\partial^2}{\partial E^2}[(E-\Phi) \times w(r,E;t|r_0,E_0;0)]. \quad [15]$$

As already noted, the time scale of oscillation of the particle is much smaller than its time scale of Brownian motion. Therefore, the particle undergoes very few collisions and its energy is nearly conserved during many periods of oscillations. Consequently, the conditional joint probability density can be decomposed as

$$w(r,E;t|r_0,E_0;0) = q(r|E)p(E;t|r_0,E_0;0), \quad [16]$$

where $q(r|E)dr$ is the conditional probability that the position of the particle (which is affected by oscillations) is between r and $r + dr$ when its energy is E, and $p(E; t|r_0, E_0; 0)dE$ is the conditional probability that the energy of the particle at time t is between E and $E + dE$ when its initial position and energy are r_0 and E_0, respectively. The evolution of the energy of the particle is weakly dependent on its initial position as the energy is almost conserved during many periods of oscillations, i.e., $p(E; t|r_0, E_0; 0) = p(E; t|E_0; 0)$.

The time spent by the particle at any position r is inversely proportional to its velocity (6) and is given by Eq. [14]. Therefore,

$$q(r|E) = \begin{cases} \dfrac{C(E)}{\sqrt{2(E-\Phi)/3}}, & \text{for } \Phi(r) < E \\ 0, & \text{for } \Phi(r) \geq E, \end{cases} \quad [17]$$

where $C(E)$ is a normalizing constant which is given by

$$C(E) = \frac{1}{\displaystyle\int_0^{\Phi^{-1}(E)} \frac{dr}{\sqrt{2(E-\Phi)/3}}}.$$

The conditional joint probability density can, therefore, be written as

$$w(r,E;t|r_0,E_0;0) = \frac{p(E;t|E_0;0)}{\psi'(E)\sqrt{2(E-\Phi)}}, \quad [18]$$

where

$$\psi'(E) = \int_0^{\Phi^{-1}(E)} \frac{1}{\sqrt{2(E-\Phi)}}dr. \quad [19]$$

Combining Eqs. [15] and [18], one obtains

$$\frac{1}{\psi'(E)\sqrt{2(E-\Phi)}}\frac{\partial}{\partial t}p(E;t|E_0;0) = \Lambda\frac{\partial}{\partial E}\left\{\frac{\sqrt{2(E-\Phi)}p(E;t|E_0;0)}{\psi'(E)}\right\} + \Lambda\frac{\partial^2}{\partial E^2}\left\{\frac{\sqrt{2(E-\Phi)}p(E;t|E_0;0)}{\psi'(E)}\right\}. \quad [20]$$

The Fokker–Planck equation for the conditional probability density $p(E; t|E_0; 0)$ can be obtained by averaging Eq. [20] with respect to the particle position to yield

$$\frac{1}{\psi'(E)}\left\langle\frac{1}{\sqrt{2(E-\Phi)}}\right\rangle\frac{\partial}{\partial t}p(E;t|E_0;0) = \Lambda\frac{\partial}{\partial E}\left\{\frac{\langle\sqrt{2(E-\Phi)}\rangle p(E;t|E_0;0)}{\psi'(E)}\right\} + \Lambda\frac{\partial^2}{\partial E^2}\left\{\frac{\langle\sqrt{2(E-\Phi)}\rangle p(E;t|E_0;0)}{\psi'(E)}\right\}, \quad [21]$$

where $\langle\cdot\rangle$ denotes the average with respect to the particle position r,

$$\left\langle\frac{1}{\sqrt{2(E-\Phi)}}\right\rangle = \psi'(E)$$

and

$$\langle\sqrt{2(E-\Phi)}\rangle = \int_0^{\Phi^{-1}(E)} \sqrt{2(E-\Phi)}dr = \psi(E). \quad [22]$$

Equation [21] can be recast in the form

$$\frac{\partial}{\partial t}p(E;t|E_0;0) = \Lambda\frac{\partial}{\partial E}\left\{\frac{\psi(E)}{\psi'(E)}p(E;t|E_0;0)\right\} + \Lambda\frac{\partial^2}{\partial E^2}\left\{\frac{\psi(E)}{\psi'(E)}p(E;t|E_0;0)\right\}. \quad [23]$$

The particle leaves the potential well as soon as its energy becomes equal to the depth of the potential well. In other words, the dimensionless energy E becomes equal to the dimensionless depth of the potential well Φ_0 at the time of dissociation. In order to calculate the average dissociation time, the probability that the dissociation time is τ must be calculated as a function of the initial conditions. Hence, the final energy and time of a dissociating doublet are fixed. It is, therefore, more convenient to deal with the backward evolution of the conditional probability density $p(E; t|E_0; t')$ (i.e. evolution with respect to t' for fixed E and t).

The corresponding backward Fokker–Planck equation (Kolmogorov equation) for the conditional density $p(E; t|E_0; t')$ is given by (5)

$$\frac{\partial}{\partial t'}p(E;t|E_0;t') = \Lambda\frac{\psi(E_0)}{\psi'(E_0)}\frac{\partial}{\partial E_0}p(E;t|E_0;t') - \Lambda\frac{\psi(E_0)}{\psi'(E_0)}\frac{\partial^2}{\partial E_0^2}p(E;t|E_0;t'). \quad [24]$$

Since the particle motion is time homogeneous (i.e., the autocorrelation function for the energy of the particle is independent of the initial time), the above equation can be rearranged in the form

$$-\frac{\partial}{\partial\tau}p(E;0|E_0;-\tau) = \Lambda\frac{\psi(E_0)}{\psi'(E_0)}\frac{\partial}{\partial E_0} \times p(E;0|E_0;-\tau) - \Lambda\frac{\psi(E_0)}{\psi'(E_0)}\frac{\partial^2}{\partial E_0^2} \times p(E;0|E_0;-\tau), \quad [24a]$$

where $\tau = t - t'$.

Further, $p(E; 0|E_0; -\tau) = p(E; \tau|E_0; 0)$ and Eq. [24a] becomes

$$\frac{\partial}{\partial\tau}p(E;\tau|E_0;0) = -\Lambda\frac{\psi(E_0)}{\psi'(E_0)}\frac{\partial}{\partial E_0}p(E;\tau|E_0;0) + \Lambda\frac{\psi(E_0)}{\psi'(E_0)}\frac{\partial^2}{\partial E_0^2}p(E;\tau|E_0;0). \quad [25]$$

Let τ be the dimensionless time at which the particle with initial dimensionless energy E_0 leaves the potential well (region I). Since the particle will remain in the potential well as long as its total energy is less than the depth of the potential well

$$\text{Prob}\,(\tau \geqslant t|E_0;0) = G(E_0,t) = \int_0^{\Phi_0} p(E;t|E_0;0)dE, \quad [26]$$

where Φ_0 is the dimensionless depth of the potential well.

Integrating Eq. [25] with respect to E, yields

$$\frac{\partial}{\partial\tau}G(E_0,\tau) = -\Lambda\frac{\psi(E_0)}{\psi'(E_0)}\frac{\partial}{\partial E_0}G(E_0,\tau) + \Lambda\frac{\psi(E_0)}{\psi'(E_0)}\frac{\partial^2}{\partial E_0^2}G(E_0,\tau). \quad [27]$$

Since $p(E; 0|E_0; 0) = \delta(E - E_0)$, it is clear that

$$G(E_0,0) = \begin{cases} 1 & \text{for } 0 \leqslant E_0 \leqslant \Phi_0 \\ 0 & \text{otherwise.} \end{cases} \quad [28]$$

The particle whose initial energy is equal to the depth of the potential well leaves the potential well immediately. Therefore,

$$G(\Phi_0,\tau) = 0 \quad \text{for all } \tau. \quad [29]$$

The flux of the particles that leave the potential well with zero initial energy is zero, i.e.,

$$\frac{\partial}{\partial E}G(E,\tau)|=0 \quad \text{for all } \tau$$

$$E=0. \qquad [30]$$

One may note that the boundary conditions [29] and [30] represent absorbing and reflecting boundary conditions in the particle energy space at $E_0 = \Phi_0$ and $E_0 = 0$, respectively.

Since $G(E_0, \tau)$ is the probability that the first passage time of the particle (with initial dimensionless energy E_0) is $\geq\tau$, the mean first passage time $T^*(E_0)$ is given by

$$T^*(E_0)=-\int_0^\infty \tau\frac{\partial}{\partial\tau}G(E_0,\tau)d\tau$$

$$=\int_0^\infty G(E_0,\tau)d\tau. \qquad [31]$$

Integrating Eq. [27] with respect to τ and taking into account that $G(E_0, \infty) = 0$, one obtains

$$-1=-\Lambda\frac{\psi(E_0)}{\psi'(E_0)}\frac{d}{dE_0}T^*(E_0)$$

$$+\Lambda\frac{\psi(E_0)}{\psi'(E_0)}\frac{d^2}{dE_0^2}T^*(E_0), \qquad [32]$$

with the associated boundary conditions

$$T^*(\Phi_0)=0 \qquad [33]$$

and

$$\frac{d}{dE_0}T^*(E_0)\bigg|_{E_0=0}=0. \qquad [34]$$

Solving Eq. [32] with the boundary conditions [33]–[34], yields

$$T^*(E_0)=\begin{cases}\dfrac{1}{\Lambda}\displaystyle\int_{E_0}^{\Phi_0}e^x\int_0^x e^{-y}\frac{\psi'(y)}{\psi(y)}dydx, \\ \qquad\qquad \text{for } 0\leq E_0\leq\Phi_0 \\ 0 \quad \text{for } E_0>\Phi_0.\end{cases} \qquad [35]$$

The above equation provides the average dissociation time of a doublet for a specified initial energy of relative motion of the two constituent particles. The initial energy of the relative motion of the particles of a doublet (or, the initial energy of the fictitious particle of reduced mass) is, however, random being a result of the collisions with the molecules of the medium. The initial energy, therefore, is described by the probability density $f(E_0)$ defined as

$f(E_0)dE_0$ = Probability that the initial dimensionless energy is between E_0 and $E_0 + dE_0$.

Let us consider an ensemble containing a large number of isolated doublets exposed to a heat bath of temperature T at time $t = 0$. If the time scale for the doublet to reach thermal equilibrium with the heat bath is much smaller than the time scale of dissociation of the doublet, the initial energy distribution can be assumed to be Maxwellian. Since the depth of the potential well is of the order of or greater than kT, even for doublets of sufficiently small particles, the time scale to reach thermal equilibrium (which is of the order of the relaxation time for Brownian motion) is much smaller than the time scale of dissociation. Consequently, it is reasonable to approximate the initial energy distribution as Maxwellian, i.e.,

$$f(E_0)=\frac{2}{\sqrt{\pi}}E_0^{1/2}e^{-E_0}. \qquad [36]$$

The dimensionless average dissociation time $\langle T^*\rangle$ of a doublet is therefore given by

$$\langle T^*\rangle=\frac{1}{\Lambda}\frac{2}{\sqrt{\pi}}\int_0^{\Phi_0}E_0^{1/2}e^{-E_0}\int_{E_0}^{\Phi_0}e^x\int_0^x e^{-y}$$

$$\times\frac{\psi'(y)}{\psi(y)}dydxdE_0, \qquad [37]$$

which, rewritten for the average dissociation time $\langle T\rangle$ of a doublet becomes

$$\langle T\rangle=\tau_0\langle T^*\rangle=\frac{2\zeta^{-1}}{\sqrt{\pi}}\int_0^{\Phi_0}E_0^{1/2}e^{-E_0}$$

$$\times\int_{E_0}^{\Phi_0}e^x\int_0^x e^{-y}\frac{\psi'(y)}{\psi(y)}dydxdE_0. \qquad [38]$$

THE MEAN LIFETIMES OF DOUBLETS CONSISTING OF EQUAL-SIZE AEROSOL PARTICLES

The average dissociation time $\langle T \rangle$ for doublets of equal-size particles, in air at 1 atm and 298 K, has been calculated using Eq. [38]. The overall interaction potential between the constituent particles has been obtained by the integration of the Lennard–Jones 6–12 intermolecular potential, under the assumption of pairwise additivity. The expression for this overall interaction potential in terms of the Hamaker constant can be found elsewhere (1). All the calculations have been performed for a Hamaker constant of 10^{-12} erg. The time scale of Brownian motion ζ^{-1} of the constituent particles was calculated from

$$\zeta^{-1} = \frac{kT}{m_p D}, \quad [39]$$

where D is the diffusion coefficient of the particle of mass m_p. Accounting for the Philips (7) slip correction factor, the diffusion coefficient D can be expressed in terms of the Knudsen number ($\mathrm{Kn} = \lambda_g/R$, λ_g being the mean free path of the suspending medium) and viscosity of the medium η as

$$D = \frac{kT}{6\pi R\eta}\left(\frac{5 + 4\mathrm{Kn} + 6\mathrm{Kn}^2 + 18\mathrm{Kn}^3}{5 - \mathrm{Kn} + (8+\pi)\mathrm{Kn}^2}\right). \quad [40]$$

The values of the time scale of Brownian motion ζ^{-1} for doublets are given in Table I, as a function of the radius R of the constituent particles.

The calculated values of the average dissociation time of the doublet for different values of the radius R are plotted in Fig. 2. One may note that the average dissociation time increases exponentially with increasing particle size. It changes by six orders of magnitude, i.e., from 10^{-7} to 10^{-1} s, as the particle radius increases from 15 to 50 Å. Such a dramatic increase in the average dissociation time with only a moderate increase in the particle size is a result of the rapid increase in the depth of the overall interaction potential well. As the potential well becomes deeper, the particles must acquire more energy (equal to the depth of the potential well) from the collisions with the medium in order to dissociate and, hence, they need more time to dissociate. The calculated values of the average dissociation times for particles of radius greater than 50 Å are found to be extremely large and, hence, are not reported. In other words, the calculations indicate that the doublets composed of particles greater than 50 Å in radius are extremely stable. The dimensionless average dissociation time (expressed in terms of the time scale of the Brownian motion) as well as the dimensionless depth of the overall interaction potential well are plotted as functions of the radius of the constituent particles in Fig. 3. The dimensionless depth of the potential well is found to increase from 2 to a value as large as 15 as the radius was varied from 15 to 50 Å. Except for the doublets containing particles of 15 Å radius, the average dissociation time is much greater than the time scale of Brownian motion. The values of the parameter Λ (the ratio of the time scales of oscillation and Brownian motion) are plotted against R in Fig. 4. Since the first passage time analysis involves the assumption that the time scale of oscillation is much smaller than the time scale of Brownian motion, the values of the parameter Λ must be much smaller than unity. The cal-

TABLE I

Radius (Å)	ζ^{-1} (s)	Φ_0	Λ	$\langle T \rangle$ (s)
15	2.01×10^{-9}	2.2368	5.67×10^{-2}	6.05×10^{-8}
20	2.69×10^{-9}	3.7674	7.51×10^{-2}	6.22×10^{-7}
25	3.37×10^{-9}	5.4679	9.30×10^{-2}	4.95×10^{-6}
30	4.06×10^{-9}	7.286	0.1096	3.79×10^{-5}
50	6.84×10^{-9}	15.2447	0.181	0.1729

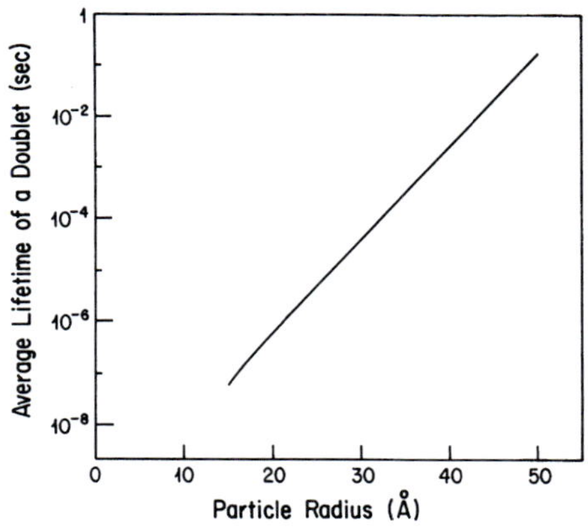

FIG. 2. The variation of the average dissociation time with the radius of the constituents of a doublet consisting of equal-size particles of unit density, in air at 1 atm and 298 K, for a Hamaker constant of 10^{-12} erg.

culated values of the parameter Λ satisfy this condition, since they vary from 0.057 to 0.181 as the particle radius increases from 15 to 50 Å (Table I).

The dimensionless dissociation time (expressed as the number of time scales of Brownian motion) is plotted as a function of the initial energy for different particle sizes in Fig. 5. The dissociation time is found to remain almost the same for initial energies smaller than $1kT$, and to decrease sharply for

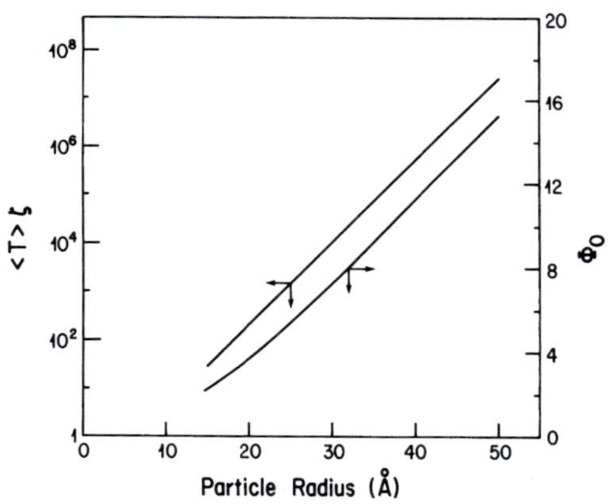

FIG. 3. The variation of the dimensionless average dissociation time as well as the dimensionless depth of the interaction potential well with the radius of the constituents of a doublet consisting of equal-size particles of unit density, in air at 1 atm and 298 K, for a Hamaker constant of 10^{-12} erg.

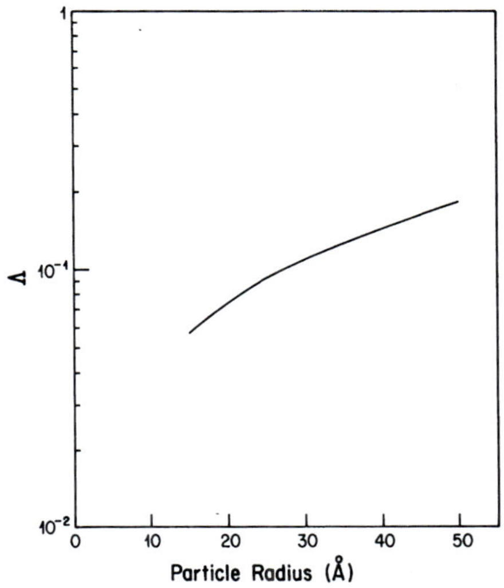

FIG. 4. The variation of the ratio of the time scales of oscillation and Brownian motion Λ with the radius of the constituents of a doublet consisting of equal-size particles of unit density, in air at 1 atm and 298 K, for a Hamaker constant of 10^{-12} erg.

larger values, becoming zero for an initial energy equal to the depth of the potential well. The range of the initial energies for which the dissociation time is constant increases with increasing particle sizes. Except for doublets consisting of particles of 15 Å radius, the dissociation time is independent of the initial energy when the latter is as large or even larger than $\frac{3}{2}kT$. In fact, for doublets composed of particles of 50 Å radius, the dissociation time is found to remain constant for initial energies as large as 10 kT. If the time scale that is necessary to attain thermal equilibrium is indeed much smaller than the time scale of dissociation, the particles with small initial energies will attain rapidly the average kinetic energy of the medium ($\frac{3}{2}kT$) through collisions with the molecules. As a result, the dissociation time is expected to be fairly independent of the initial energy, when the latter is less than $\frac{3}{2}kT$. Therefore, the insensitivity of the dissociation time to initial energies as large as the mean kinetic energy of the suspending medium is consistent with the assumption that the time scale necessary to attain thermal equilibrium is much shorter than the time

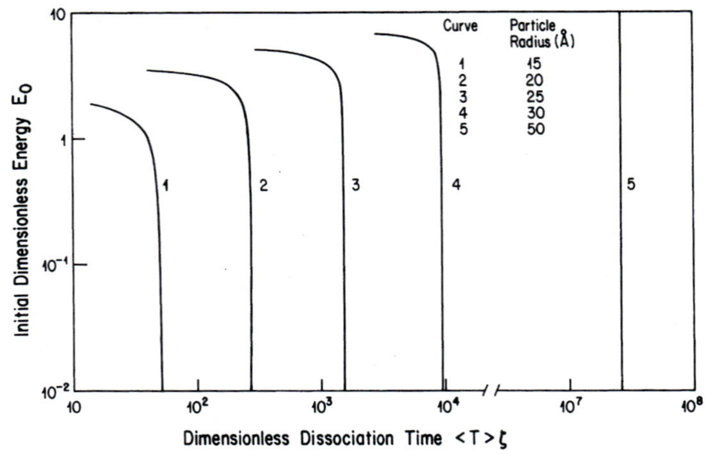

FIG. 5. The variation of the dimensionless dissociation time with the initial dimensionless energy for different values of the radius of the constituents of a doublet consisting of equal-size particles of unit density, in air at 1 atm and 298 K, for a Hamaker constant of 10^{-12} erg.

scale of dissociation. Consequently, the assumption that the initial energy distribution is Maxwellian is reasonable.

CONCLUSIONS

The relative Brownian motion between the constituents of doublets consisting of sufficiently small equal-size aerosol particles is described by a one-dimensional Fokker–Planck equation in the particle energy space. A first passage time approach is employed for the calculation of the average lifetime of the doublets. This calculation is based on the assumption that the initial distribution of the energy of the relative motion of the constituent particles is Maxwellian. The average dissociation time of doublets, in air at 1 atm and 298 K, for a Hamaker constant of 10^{-12} erg has been calculated for different sizes of the constituent particles. The calculations are found to be consistent with the assumption that the initial energy distribution of the particle pair is Maxwellian. The dissociation rate is found to decrease dramatically by several orders of magnitude with increasing particle size, as a result of the rapid increase in the depth of the interaction potential well. The doublets consisting of particles whose radii are greater than 50 Å are found to be extremely stable.

REFERENCES

1. Narsimhan, G., and Ruckenstein, E., *J. Colloid Interface Sci.* **104,** 344 (1985).
2. Narsimhan, G., and Ruckenstein, E., *J. Colloid Interface Sci.* **107,** 174 (1985).
3. Weiss, G. H., *Adv. Chem. Phys.* **13,** 1 (1967).
4. Chandrasekhar, S., *Rev. Mod. Phys.* **15,** 1 (1943).
5. Gardiner, C. W., "Handbook of Stochastic Methods," Springer–Verlag, New York/Berlin, 1983.
6. Stratonovich, R. L., "Topics in the Theory of Random Noise," Vol. I, Gordon and Breach, New York, 1963.
7. Philips, W. F., *Phys. Fluids* **18,** 1089 (1975).

A Possible Nucleation Type of Mechanism for the Growth of Small Aerosol Particles

GANESAN NARSIMHAN[1] AND ELI RUCKENSTEIN[2]

Department of Chemical Engineering, State University of New York, Buffalo, New York 14260

Received April 14, 1986; accepted July 1, 1986

The collisions between small aerosol particles do not necessarily result in coagulation, since the dispersion forces responsible for coagulation are relatively weak; i.e., the interaction potential well due to the van der Waals attraction and Born repulsion is not deep. In addition, because of weak dispersion forces, the constituents of a doublet of small particles can derive sufficient energy from the collisions with the molecules of the suspending medium to overcome the potential well and dissociate. The rate of formation of a doublet, i.e., the rate of coagulation per particle, is calculated by accounting for the sticking probability in terms of the depth of the interaction potential well. The average dissociation time of a doublet has been determined as a first passage time in the energy of the relative motion of the two constituent particles. For sufficiently small particles, the rate of dissociation of a doublet is much larger than its rate of formation. This unstable doublet reaches, therefore, an equilibrium with the single particles over times much shorter than the time scale of coagulation. Because of the dramatic decrease in the rate of dissociation of a doublet with an increase in the size of its constituents, there exists a critical particle radius above which the doublet is stable; i.e., its rate of formation becomes greater than its rate of dissociation. When the doublet is unstable, aerosol growth occurs via a nucleation mechanism, i.e., via successive equilibria of unstable agglomerates leading to a critical size aggregate, which constitutes a nucleus for growth. On the other hand, when the doublet is stable, the singlets themselves act as nuclei for aerosol growth.

INTRODUCTION

Aerosol particles of extremely small sizes (molecular clusters) can form via physical or chemical nucleation. The subsequent growth of these particles is brought about both by the condensation of monomers on the surface of the particles as well as by the Brownian coagulation of the particles themselves. Brownian coagulation of such small particles as a result of collisions has been shown to affect both aerosol formation and its subsequent growth (1). The short-range dispersion forces are mainly responsible for the coagulation of aerosol particles. It is usually assumed that every collision between two aerosol particles leads to irreversible coagulation. The above assumption is justified when the particles are large, since the interparticle dispersion forces are strong in this case; i.e., the interaction potential well (due to the van der Waals attraction and Born repulsion) is very deep. However, not every collision between two sufficiently small particles results in coagulation, since the interaction potential well is not sufficiently deep; i.e., the interparticle forces are relatively weak. The collision efficiency, i.e., the sticking probability, has been calculated in terms of the interaction potential between the particles (2, 3).

Moreover, the molecules of the suspending medium continually transfer energy through collisions to the constituents of the coagulated particle pairs. If the constituent particles are sufficiently small, the particle pair can derive sufficient energy from the molecules of the suspending medium to overcome the interaction potential well and, hence, dissociate. In

[1] G. Narsimhan is now associated with the Department of Agricultural Engineering, Purdue University, West Lafayette, IN 47907.

[2] To whom correspondence should be addressed.

0021-9797/87 $3.00

other words, the coagulation of sufficiently small particles is reversible. As shown previously (4), the time scale of oscillations in the potential well of the constituents of the doublet is much shorter than the time scale of Brownian motion. As a result, the relative motion between the two particles of the doublet in the interaction potential well can be described by a one-dimensional Fokker–Planck equation in the particle energy space. The average lifetime of a doublet consisting of equal-size particles has been calculated as a first passage time in the energy of the relative motion of the two constituent particles (4).

Therefore, the doublets of sufficiently small particles are continually formed (due to coagulation) and destroyed (because of their dissociation). If the rate of formation of a doublet (i.e., the rate of coagulation per particle) is greater than its rate of dissociation (i.e., the rate of dissociation per doublet), the doublet is stable. As a result, the concentration of doublets increases continuously with time. As will be shown later, the time scale of evolution of these stable doublets is of the order of the time scale of coagulation. Of course, these stable doublets interact with the particles to form triplets which, in turn, form quadruplets, etc., resulting in the growth of the particle size.

On the other hand, if the rate of formation of a doublet is much smaller than its rate of dissociation, the doublet is unstable. It will be shown later that the unstable doublets reach a dynamic equilibrium with the singlets in an extremely short time (of the order of the time scale of dissociation). The equilibrium concentration of these unstable doublets is small and depends upon the relative magnitudes of the rates of formation and dissociation. Since the dissociation rate of a doublet decreases rather dramatically with increasing particle size (because of the rapid increase in the depth of the interaction potential well with increasing particle size), there exists a critical particle size above which the coagulated particle pair is stable, i.e., its rate of formation is much greater than its rate of dissociation. This critical particle size is analogous to the critical cluster size for the growth of molecular clusters in nucleation. The unstable doublets undergo collisions with the particles to form triplets which, if unstable, attain equilibrium with singlets, again in a very short time, their concentration being much lower than that of the doublets. The triplets, in turn, form quadruplets and so on, until a stable agglomerate is formed which constitutes a nucleus for further growth. Hence, the aerosol growth occurs either only by coagulation or through the formation of a stable nucleus, depending on whether the doublets are stable or not.

A comparison between the rates of formation and dissociation of doublets as well as the calculation of the critical particle size are presented in the next section. The subsequent section discusses the implications of the above two mechanisms of aerosol growth on the calculation of the Brownian coagulation coefficient from the evolution of particle number concentration.

COMPARISON BETWEEN THE RATES OF FORMATION AND DISSOCIATION OF A DOUBLET CONSISTING OF EQUAL-SIZE AEROSOL PARTICLES

The Rate of Formation of a Doublet

Since the doublets are continually being formed by coagulation, their rate of formation depends on the rate of coagulation. The rate of formation of a doublet r_f (or, equivalently, the rate of coagulation per particle) can be expressed in terms of the coagulation coefficient β and the particle number concentration n as

$$r_f = \beta n \tag{1}$$

The initial particle number concentration n_0 is related to the volume fraction ϕ_p of the particles via the expression

$$n_0 = \frac{\phi_p}{(4/3)\pi R^3}, \tag{2}$$

where R is the radius of the particles.

When the particles are sufficiently small, i.e., the Knudsen number ($=\lambda_g/R$, where λ_g is the mean free path of the suspending medium) is large (free molecular limit), the particles can be considered as gigantic molecules and the rates of collision r_c between the particles can be obtained from the kinetic theory of gases as (5)

$$r_c = 4\pi R^2\left(\frac{16kT}{\pi m_p}\right)^{1/2}, \quad [3]$$

where k is the Boltzmann constant, T is the absolute temperature of the suspending medium, and m_p is the mass of the particle.

As pointed out earlier, not every collision between the small particles leads to coagulation because the interaction potential well (due to the van der Waals attraction and Born repulsion) is not deep. Consequently, the coagulation coefficient β can be written as

$$\beta = \gamma r_c = \gamma 4\pi R^2\left(\frac{16kT}{\pi m_p}\right)^{1/2}, \quad [4]$$

where γ is the sticking probability, i.e., the probability of coagulation upon collision.

Examining the relative Brownian motion between two singlets and accounting for the energy exchange between the two colliding particles through a reflecting boundary condition, the expression for the sticking probability γ, in the free molecular limit, is given by (2)

$$\gamma = 1 - e^{-\Phi_0}(1+\Phi_0), \quad [5]$$

where Φ_0 is the dimensionless depth of the interaction potential well ($=\phi_0/kT$, ϕ_0 being the depth of the potential well).

Combining Equations [4] and [5], the expression for the coagulation coefficient β becomes:

$$\beta = \{1 - e^{-\Phi_0}(1+\Phi_0)\}4\pi R^2\left(\frac{16kT}{\pi m_p}\right)^{1/2}. \quad [6]$$

The Rate of Dissociation of a Doublet

As pointed out earlier, when the constituent particles of a doublet are small, they can derive sufficient energy from the collisions with the molecules of the suspending medium to overcome the interaction potential well and thus dissociate. If the energy imparted by the collisions of the molecules of the suspending medium is smaller than the depth of the interaction potential well, the two particles of the doublet will exhibit an oscillatory motion within the potential well. For sufficiently small particles, the time scale of these oscillations is much smaller than the time scale of the Brownian motion of the constituent particles (4). Because of this disparity between the two time scales, the relative Brownian motion between the two particles of a doublet can be described by a one-dimensional Fokker–Planck equation in the energy space of the relative motion of the two. The average dissociation time $\langle T\rangle$ can, therefore, be calculated as a first passage time (i.e., the time at which the energy of the relative motion becomes equal to the depth of the interaction potential well) in the energy of the relative motion. The expression for the average dissociation time $\langle T\rangle$, in terms of the dimensionless energy E and the time scale of Brownian motion ζ^{-1}, is given by (4)

$$\langle T\rangle = \frac{2\zeta^{-1}}{\sqrt{\pi}}\int_0^{\Phi_0} E^{1/2}e^{-E}\int_E^{\Phi_0} e^{x}\int_0^{x} e^{-y} \times \frac{\psi'(y)}{\psi(y)}\,dy\,dx\,dE, \quad [7]$$

and the dimensionless energy E is given by

$$E = \tfrac{1}{2}m_r v^2/kT + \Phi, \quad [8]$$

Here $m_r(=m_p/2)$ is the reduced mass, v is the relative velocity between the two constituent particles of the doublet, Φ $(=\phi/kT)$ is the dimensionless interaction potential,

$$\psi(E) = \int_0^{\Phi^{-1}(E)} \sqrt{2(E-\Phi)}\,dr, \quad [9a]$$

and

$$\psi'(E) = \int_0^{\Phi^{-1}(E)} \frac{1}{\sqrt{2(E-\Phi)}}\,dr. \quad [9b]$$

The time scale of the Brownian motion ζ^{-1} is related to the diffusion coefficient D of the particle via the expression

$$\zeta^{-1} = \frac{kT}{m_p D}, \quad [10]$$

where the diffusion coefficient expressed in terms of the Knudsen number Kn and viscosity of the suspending medium η is

$$D=\frac{kT}{6\pi R\eta}\left(\frac{5+4\text{Kn}+6\text{Kn}^2+18\text{Kn}^3}{5-\text{Kn}+(8+\pi)\text{Kn}^2}\right). \quad [11]$$

The rate of dissociation r_d of a doublet is the reciprocal of the average dissociation time $\langle T\rangle$, i.e.,

$$r_d=\frac{1}{\langle T\rangle}. \quad [12]$$

The rates of formation and dissociation of a doublet consisting of equal-size aerosol particles, in air at 1 atm and 298 K, have been calculated for a Hamaker constant of 10^{-12} erg from Eqs. [6] and [7]–[12]. The rate of formation of a doublet r_f is plotted in Fig. 1 as a function of R for different values of ϕ_p. The rate of formation of a doublet is found to decrease with increasing particle size at constant ϕ_p because of the predominant effect of the decrease in the particle number concentration (Eq. [2]). The rate of formation of a doublet does not show as strong a dependence on particle size as does the rate of dissociation (Fig. 1). As expected, the rate of formation of a doublet is found to be higher at larger values of the particle volume fraction because of a greater number of collisions between the particles.

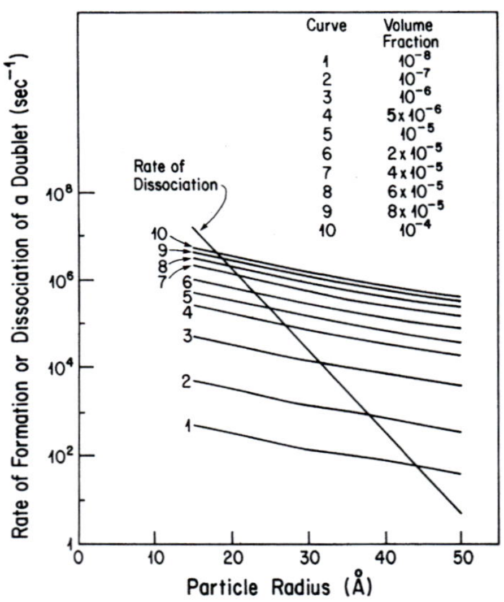

FIG. 1. Comparison between the rates of formation and dissociation of a doublet consisting of equal-size aerosol particles of unit density, in air at 1 atm and 298 K, for a Hamaker constant of 10^{-12} erg and for different particle volume fractions.

The rate of dissociation r_d is also plotted in Fig. 1 as a function of the radius R and it is found to be very sensitive to particle size. Indeed, it decreases dramatically with increasing R because of the rapid increase in the depth of the interaction potential well.

It is interesting to note that the rate of formation of a doublet is smaller than the rate of dissociation for sufficiently small particles. The difference between the rates of dissociation and formation decreases as the particle size increases, eventually becoming equal at a critical particle radius R^*. For particles smaller than R^*, the doublets are unstable and they reach a dynamic equilibrium with the single particles. For particles greater than R^*, however, the doublet is stable and, therefore, its concentration increases with time. The variation of the critical radius R^* with the particle volume fraction ϕ_p is plotted in Fig. 2. R^* is found to be a weak function of ϕ_p, decreasing from 44 to 18 Å as ϕ_p increases from 10^{-8} to 10^{-4}.

The time evolution of the number concentrations of particles and doublets is given by

$$\frac{dn}{dt}=-\beta n^2+\frac{2}{\langle T\rangle}n_d \quad [13]$$

and

$$\frac{dn_d}{dt}=\frac{1}{2}\beta n^2-\frac{n_d}{\langle T\rangle}, \quad [14]$$

where n and n_d are the number concentrations of particles and doublets, respectively, at time t.

Introducing the dimensionless variables, $n^* = n/n_0$, $n_d^* = n_d/n_0$, $\tau = \beta n_0 t$, $\alpha = 1/\beta n_0\langle T\rangle$, where n_0 is the initial number concentration of particles, $1/(\beta n_0)$ is the time scale of coagulation, and α is the ratio of the rates of dissociation and formation of a doublet.

Equations [13] and [14] can be rewritten as

$$\frac{dn^*}{d\tau}=-n^{*2}+2\alpha n_d^* \quad [13a]$$

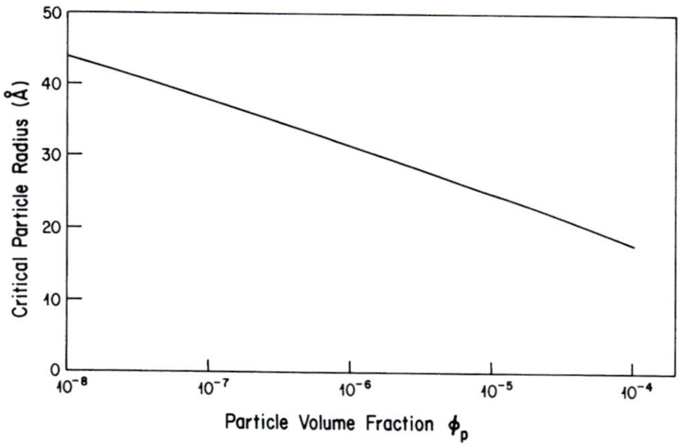

FIG. 2. The variation of the critical particle radius with the particle volume fraction for particles of unit density, in air at 1 atm and 298 K, and for a Hamaker constant of 10^{-12} erg.

and

$$\frac{dn_d^*}{d\tau} = \tfrac{1}{2}n^{*2} - \alpha n_d^*. \qquad [14a]$$

Equations [13a] and [14a] are solved for the initial conditions

$$n^* = 1 \quad n_d^* = 0 \quad \text{for } \tau = 0, \qquad [15]$$

to yield

$$n_d^* = \frac{[1 - \exp\{-\alpha^{1/2}(\alpha+2)^{1/2}\tau\}]}{\{(\alpha+2) + \alpha^{1/2}(\alpha+2)^{1/2}\} - [\{(\alpha+2) - \alpha^{1/2}(\alpha+2)^{1/2}\}\exp\{-\alpha^{1/2}(\alpha+2)^{1/2}\tau\}]}. \qquad [16]$$

The value of the parameter α determines whether the doublets are stable or unstable. For $\alpha > 1(<1)$, the doublets are unstable (stable). For $\alpha \gg 1$, the dimensionless time scale of evolution of the number concentration of the doublets is of the order of $1/\alpha$, whereas for $\alpha \ll 1$, the dimensionless time scale of evolution is of the order of $1/\alpha^{1/2}$. In other words, the time scale of evolution of the highly unstable doublets is much shorter than the time scale of coagulation, whereas the highly stable doublets evolve over much longer times. Therefore, the unstable doublets attain equilibrium in an extremely short time (of the order of the time scale of dissociation, i.e., in 10^{-7} to 10^{-4} s).

Since the number concentration of the stable doublets increases continuously with time, the number of collisions between doublets and single particles would become significant at large times, resulting in the formation of triplets, quadruplets, etc. Equation [16] accounts only for the presence of single particles and doublets. It can, therefore, describe the evolution of the number concentration of stable doublets only for short times.

The expression for the dimensionless equilibrium concentration of the doublets $n_{d_{eq}}^*$ is given by

$$n_{d_{eq}}^* = \frac{1}{(\alpha+2) + \alpha^{1/2}(\alpha+2)^{1/2}} \qquad \alpha > 1. \qquad [17]$$

For $\alpha \gg 1$, $n_{d_{eq}}^* \sim \dfrac{1}{\alpha}$.

The parameter α, being the ratio of the rates of dissociation and formation of the doublet, depends on the particle size as well as on the particle volume fraction. The equilibrium doublet concentration $n_{d_{eq}}^*$ is plotted in Fig. 3 against the particle size for different particle volume fractions. The dotted portions of the curves in Fig. 3 represent the hypothetical equilibrium concentrations of stable doublets.

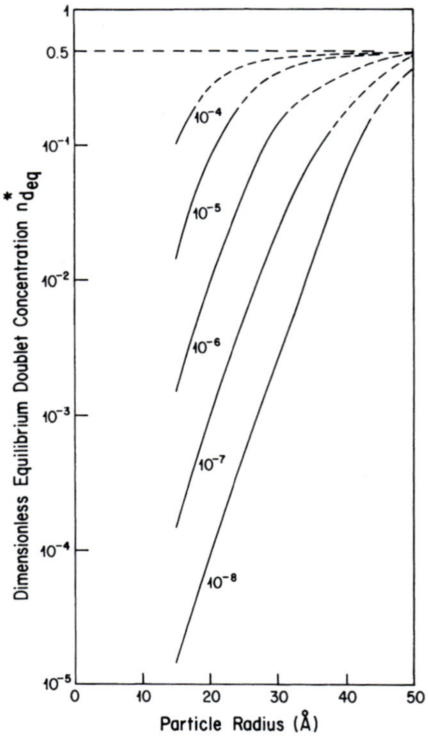

FIG. 3. The variation of the dimensionless equilibrium doublet concentration with the radius of particles of unit density, in air at 1 atm and 298 K, for a Hamaker constant of 10^{-12} erg and for different initial particle volume fractions.

The equilibrium doublet concentration is found to be extremely small for small particles and low volume fractions, to increase dramatically with increasing particle size, and to exhibit a weak dependence on the particle volume fraction.

TWO ALTERNATE MECHANISMS OF AEROSOL GROWTH

Let us consider a system of monodispersed aerosol particles. Because of the collisions between the particles, they will coagulate to form doublets. The rate of formation of a doublet depends, of course, on the particle radius, the particle density, the viscosity and temperature of the suspending medium, the Hamaker constant, and the particle number concentration. If the rate of formation of a doublet is much less than its rate of dissociation, the doublets are unstable. These unstable doublets attain equilibrium with the particles over a time scale which is much shorter than the time scale of coagulation. Their equilibrium concentrations are very small and depend on the relative magnitudes of their rates of formation and dissociation. The doublets continually undergo collisions with the particles to form triplets. The triplets, if unstable, quickly reach equilibrium with the particles, their concentrations being much smaller than those of the doublets. Over time scales of the order of the time scale of coagulation (i.e., much higher than the time scales of dissociation), the system will consist of extremely low concentrations of triplets, quadruplets, etc. in equilibrium with single particles. This continues until stable i_c-tuplets are formed, which grow in size. In other words, the formation and growth of the stable i_c-tuplet occur via the formation of extremely low concentrations of i-tuplets ($i < i_c$). Therefore, the smallest stable i_c-tuplet acts as a nucleus for aerosol growth. The rate of dissociation of agglomerates exhibit a very strong dependence on their size compared to their rate of formation. Therefore, α is expected to be much larger than unity for all unstable agglomerates (up to ($i_c - 1$)-tuplet) and becomes much smaller than unity for i_c-tuplet.

The system can be schematically represented as

$$[1]+[1] \rightleftharpoons [2]$$

$$[2]+[1] \rightleftharpoons [3]$$

$$\vdots$$

$$[i-1]+[1] \rightleftharpoons [i]$$

$$\vdots$$

$$[i_c-1]+[1] \rightarrow [i_c], \quad [18]$$

where $[i]$ denotes i-tuplet. The dimensionless equilibrium concentration of an i-tuplet ($i < i_c$) can be expressed in terms of the dimensionless single particle concentration as

$$n_i^* = K_2 K_3 \cdots K_i n_1^{*\,i}, \quad [19]$$

with the equilibrium constants $\{K_i\}$, $i = 1, 2, \ldots$, in Eq. [19] given by

$$K_i = \beta_{i-1,1} n_0 \langle T_i \rangle. \quad [20]$$

Here β_{i1} is the coagulation coefficient between an i-tuplet and a single particle, n_0 is the initial single particle concentration, and $\langle T_i \rangle$ is the average dissociation time for an i-tuplet to dissociate to an $(i - 1)$-tuplet.

The dynamic equations for the dimensionless concentrations of single particles and stable i_c-tuplets can be written as

$$\frac{dn^*_{i_c}}{d\tau} = -\left(\frac{\beta_{i_c 1}}{\beta_{11}}\right) n^*_{i_c} n^*_1 + \left(\frac{\beta_{i_c-1,1}}{\beta_{11}}\right) K n^{*i_c}_1 \quad [21]$$

and

$$\frac{dn^*_1}{d\tau} = -\sum_{j=2}^{i_c-1} L_j j^2 n^{*(j-1)}_1 \frac{dn^*_1}{d\tau} - \left(\frac{\beta_{i_c 1}}{\beta_{11}}\right) n^*_{i_c} n^*_1 - i_c \left(\frac{\beta_{i_c-1,1}}{\beta_{11}}\right) K n^{*i_c}_1. \quad [22]$$

Equation [22] is valid at short times since the growth of $(i_c + 1)$ and larger tuplets is neglected. In Eqs. [21] and [22] the constants L_j and K are given by

$$L_j = K_2 K_3 \cdots K_j, \quad K = L_{(i_c-1)}.$$

If the singlet concentration is sufficiently large, one can neglect the change in the singlet concentration at short times, to obtain

$$n_{i_c} = \left(\frac{\beta_{i_c-1,1}}{\beta_{i_c 1}}\right) K n_0 [1 - \exp(-\beta_{i_c 1} n_0 t)]. \quad [23]$$

For large i_c, $\beta_{i_c 1} \sim \beta_{i_c-1,1}$, and

$$n_{i_c} \simeq K n_0 [1 - \exp(-\beta_{i_c 1} n_0 t)]. \quad [23a]$$

The rate of nucleation at times is small compared to the time scale for coagulation. It is therefore given by

$$\frac{dn_{i_c}}{dt} \simeq \beta_{i_c 1} K n_0^2 \quad [24]$$

and the rate of change of singlet concentration is given by

$$\frac{dn_1}{dt} \simeq -i_c \frac{dn_{i_c}}{dt} = -i_c \beta_{i_c 1} K n_0^2. \quad [25]$$

On the other hand, if the rate of formation of a doublet is much greater than its rate of dissociation, the doublet will be stable. The doublet concentration increases continuously with time. The doublets will also interact with the particles resulting in the formation of triplets, quadruplets, etc. Therefore, the aerosol growth, in this case, does not require the formation of a nucleus. The single particles themselves act as nuclei for aerosol growth. The evolution of the concentration of the particles over a time of the order of the time scale of coagulation is given, in this case, by

$$\frac{dn}{dt} = -\beta_{11} n^2, \quad [26]$$

β_{11} being the coagulation coefficient of the single particles.

CONCLUSIONS

A comparison between the rates of formation and dissociation of a doublet consisting of sufficiently small equal-size aerosol particles of unit density, in air at 1 atm and 298 K, has been made for a Hamaker constant of 10^{-12} erg. The rate of formation of a doublet (the rate of coagulation per particle) has been calculated, accounting for the sticking probability in terms of the depth of the interaction potential well. The average dissociation time of a doublet has been obtained as a first passage time in the energy of the relative motion of the two constituent particles. The rate of dissociation of a doublet is found to decrease dramatically with increasing particle size because of the rapid increase in the depth of the interaction potential well. The rate of formation of a doublet, however, though a strong function of the particle volume fraction, does not exhibit as strong a dependence on particle size as does the rate of dissociation. For a doublet consisting of very small particles, the rate of dissociation is found to be greater than its rate of formation, thus rendering the doublet unstable. As the particle size increases, the difference between the rates of dissociation and formation of a doublet decreases. A critical particle size is found to exist above which the rate of formation of a doublet is greater than its rate of dissociation; i.e., the doublet is stable. The unstable doublets reach equilibrium with the singlets (particles) over times much shorter than the time scale of coagulation (i.e., in 10^{-7} to 10^{-4} s), their equilibrium concentration being dependent on the relative magnitudes of their rates of dissociation and formation. Because of successive collisions with particles, the doublets form agglomerates of different sizes. These agglomerates, if unstable, quickly reach equilibrium, their concentrations decreasing as the agglomerate size increases. The smallest stable agglomerate, formed via successive equilibria of smaller sized unstable agglomerates, constitutes a nucleus for the aerosol growth. When the doublets themselves are stable, however, the singlets (particles) themselves act as the nuclei for the aerosol growth. Therefore, the aerosol growth occurs by two different mechanisms depending on the stability of the doublet.

REFERENCES

1. Gelbard, F., and Seinfeld, J. H., *J. Colloid Interface Sci.* **68,** 363 (1979).
2. Narsimhan, G., and Ruckenstein, E., *J. Colloid Interface Sci.* **104,** 344 (1985).
3. Narsimhan, G., and Ruckenstein, E., *J. Colloid Interface Sci.* **107,** 174 (1985).
4. Narsimhan, G., and Ruckenstein, E., *J. Colloid Interface Sci.* **116,** 279 (1987).
5. Hidy, G. M., and Brock, J. R., "The Dynamics of Aerocolloidal Systems," Pergamon, Oxford, 1970.

Introduction to CHAPTER 2

Dynamics of deposition of Brownian particles or cells on surfaces

2.1 E. Ruckenstein, D. Prieve: "*Rate of deposition of Brownian particles under the action of London and double-layer forces,*" JOURNAL CHEM. SOC. FARADAY TRANSACTIONS II, 69 (1973) 1522–1536.

2.2 E. Ruckenstein, D.C Prieve: "*Adsorption and desorption of particles and their chromatographic separation,*" AIChE JOURNAL 22 (1976) 276–283

2.3 E. Ruckenstein, D.C. Prieve: "*On reversible adsorption of hydrosols and repeptization,*" AIChE JOURNAL 22 (1976) 1145–1147.

2.4 D.C. Prieve, E. Ruckenstein: "*Effect of London forces upon the rate of deposition of Brownian particles,*" AIChE JOURNAL 20(1974) 1178–1186.

2.5 D. C. Prieve, E. Ruckenstein: "*Role of surface chemistry in particle deposition,*" JOURNAL OF COLLOID AND INTERFACE SCIENCE 60 (1977) 337–348.

2.6 D.C. Prieve, E. Ruckenstein: "*The double-layer interaction between dissimilar ionizable surfaces and its effect on the rate of deposition,*" JOURNAL OF COLLOID AND INTERFACE SCIENCE 63 (1978) 317–329.

2.7 E. Ruckenstein: "*Reversible rate of adsorption or coagulation of Brownian particles- effects of the shape of the interaction potential,*" JOURNAL OF COLLOID AND INTERFACE SCIENCE 66 (1978) 531–543.

2.8 E. Ruckenstein, D.C Prieve: "*Dynamics of cell deposition on surfaces,*" JOURNAL OF THEORETICAL BIOLOGY 51 (1975) 429–438.

2.9 E. Ruckenstein, A. Marmur, W.N.Gill: "*Coverage dependent rate of cell deposition,*" JOURNAL OF THEORETICAL BIOLOGY 58(1976)439–454.

2.10 A. Marmur, W. N. Gill, E. Ruckenstein: "*Kinetics of cell deposition under the action of an external field,*" BULLETIN OF MATHEMATICAL BIOLOGY 38 (1976) 713–721.

2.11 E. Ruckenstein: "*Thermodynamic insights on macroemulsion stability,*" ADVANCES IN COLLOD AND INTERFACE SCIENCE 79 (1999) 59–76.

2.12 E. Ruckenstein: "*On the stability of concentrated, non-aqueous dispersions,*" COLLOIDS AND SURFACES 69 (1993) 271–275.

The rate of deposition of Brownian particles is predicted by taking into account the effects of diffusion and convection of single particles and interaction forces between particles and collector [2.1] -[2.6]. It is demonstrated that the interaction forces can be incorporated into a boundary condition that has the form of a first order chemical reaction which takes place on the collector [2.1], and an expression is derived for the rate constant. The rate of deposition is obtained by solving the convective diffusion equation subject to that boundary condition. The procedure developed for deposition is extended to the case when both deposition and desorption occur. In the latter case, the interaction potential contains the Born repulsion, in addition to the London and double-layer interactions [2.2]-[2.7]. Paper [2.7] differs from [2.2] because it considers the deposition at both primary and secondary minima. Papers [2.8], [2.9] and [2.10] treat the deposition of cancer cells or platelets on surfaces.

For a constant amount of nonionic surfactant, the interfacial tension at the planar oil-water interface, for the same amounts of oil and water, passes through a minimum when plotted against the hydrophilic-lipophilic balance (HLB). The emulsion stability passes through maxima in the W/O and O/W ranges and through a minimum between the two at the phase inversion point. The minima in the two cases coincide. These observations are explained on the basis of thermodynamics. The stability of macroemulsions can be correlated with the surface excess of surfactant, which also passes through two maxima and a minimum between them [2.11].

In paper [2.12] it is shown that the van der Waals interactions between two particles can be decreased if they are covered with layers that have a Hamaker constant which is near to that of the suspending liquid. It is also suggested that, at sufficiently high concentrations, the collective behavior of the colloidal particles can generate repulsion when the pairwise interactions are attractive. These two effects are suggested to be responsible for the kinetic stability of the system, and a methodology for achieving kinetic stability is provided.

E. Ruckenstein, M. Manciu, *Nanodispersions*, DOI 10.1007/978-1-4419-1415-6_2, © Springer Science+Business Media, LLC 2010

Rate of Deposition of Brownian Particles under the Action of London and Double-layer Forces

By Eli Ruckenstein*† and Dennis C. Prieve

Department of Chemical Engineering, University of Delaware, Newark, Delaware 19711

Received 9*th April*, 1973

The rate of collection of Brownian particles under the influence of interaction forces between the collector surface and the particles is calculated by (*a*) incorporating the interaction forces in the rate constant of a virtual, first order, chemical reaction taking place on the surface of the collector, and by (*b*) solving the convective diffusion equation subject to that chemical reaction as a boundary condition. Several geometries (sphere, cylinder, rotating disc) are considered for the collector.

An Arrhenius type equation is obtained for the apparent reaction rate constant. Equations for the apparent activation energy and for the frequency factor are established as functions of Hamaker's constant, ionic strength, surface potentials and particle radius.

Knowledge of effects which London and double-layer forces exert upon rates of deposition is useful for selecting surfaces that allow or prevent adhesion. For example, a coating on the wall of an artificial organ might be used to repel platelets in flowing blood, thus preventing their damage, whereas in filtration it would be advantageous to have a fibre surface which attracts particles in order to increase efficiency.

The deposition of non-Brownian particles under the influence of interaction forces was treated by Spielman and FitzPatrick [1] and Spielman and Cukor.[2] Theoretical predictions of the rate of deposition for Brownian particles have been made in two extreme situations. Levich [3] treated the case in which convection and diffusion control the rate of deposition. Hull and Kitchener [4] considered interaction forces and diffusion while neglecting convection.

The objective of the present research is to predict the rate of deposition of Brownian particles by considering the effects of diffusion, convection, and interaction forces between particle and collector. It will be shown that, when the repulsion due to the double-layer is sufficiently large, the interaction forces can be incorporated into a boundary condition for the convective-diffusion equation. This boundary condition takes the form of a virtual first-order chemical reaction which occurs on the surface of the collector.

INTERACTION POTENTIAL

Two classes of interactions are considered: (i) London–van der Waals attractions and (ii) electrical double-layer interactions. The unretarded London–van der Waals potential ϕ_{Ad} between a sphere and a plate is [5, 6]

$$\phi_{Ad} = A\left[\tfrac{1}{6}\ln\left(\frac{h+2a_p}{h}\right)-\tfrac{1}{3}\frac{a_p}{h}\frac{h+a_p}{h+2a_p}\right], \tag{1}$$

where h is the minimum distance of separation between the surface of the sphere and the plate, a_p is the radius of the sphere, and A is Hamaker's constant.

† present address: State University of New York at Buffalo, Faculty of Engineering and Applied Sciences, Buffalo, New York 14214, U.S.A.

When the sphere and plate are composed of similar material the double-layer interaction potential ϕ_{DL} may be approximated by [2, 7]

$$\phi_{DL} = \varepsilon a_p \Psi_1 \Psi_2 \ln(1+e^{-\kappa h}), \tag{2}$$

where ε is the dielectric constant of the fluid, Ψ_i is the value of the electrostatic potential of surface i, κ is Debye's reciprocal length parameter

$$\kappa^2 = \frac{4\pi}{\varepsilon kT} \sum_i n_i e_i^2, \tag{3}$$

n_i is the number of ions of species i per unit volume, e_i is the charge of species i, k is Boltzmann's constant, and T is the absolute temperature. Eqn (2) is valid only when $|\Psi_1| \approx |\Psi_2| < 50$ mV and $\kappa a_p > 5$. In addition Ψ_i is assumed to be independent of κh.

Adding eqn (1) and (2), one obtains the total interaction potential:

$$\phi = \tfrac{2}{3}A\left[\frac{3\varepsilon a_p \Psi_1 \Psi_2}{2A} \ln(1+e^{-\tau H}) + \tfrac{1}{4}\ln\left(\frac{H+2}{H}\right) - \tfrac{1}{2}\frac{H+1}{H(H+2)}\right] \tag{4}$$

where $H = h/a_p$ and $\tau = \kappa a_p$.

POSITION-DEPENDENT DIFFUSIVITY

The mobility, m, and diffusion coefficient, D, are related by the Nernst–Einstein equation:

$$D = mkT. \tag{5}$$

In the theory of diffusion in liquids, the mobility of a Brownian particle is determined using Stokes's law. Consequently,

$$m = 1/6\pi\mu a_p, \tag{6}$$

where μ is the viscosity of the fluid and a_p is the particle radius.

When the particle is very near to a solid wall, such as the surface of the collector, the mobility is decreased as a result of friction between the wall and fluid which increases the force required to push the fluid out of the path of the approaching particle. A correction factor was calculated by Brenner [2] which, incorporated into (6), gives:

$$m = \frac{f(h/a_p)}{6\pi\mu a_p}. \tag{7}$$

Brenner's correction factor, $f(H)$, tends to H when $H \ll 1$ and to $(H-\frac{1}{8})/(H+1)$ when $H \gg 1$. Values of $f(H)$ for intermediate values of H may be determined from an infinite series. Since $0 \leqslant f \leqslant 1$, the diffusion coefficient is always smaller in the region near the collector than in the bulk and it becomes vanishingly small when the particle touches the surface. An additional reduction in the diffusion coefficient can be caused by the deviation of the viscosity from its bulk value.

PARTICLE TRANSFER WITHIN THE INTERACTION-FORCE BOUNDARY LAYER

Because the surface–particle interaction forces are important only near to the surface of the collector, it is natural to divide the domain into two regions. The inner region, called the interaction-force boundary layer, has a thickness determined by the range over which interaction forces are important (less than about 10^{-5} cm). Because of the proximity of the surface, convection is negligible. Outside the interaction-force boundary layer, the usual convective-diffusion equation applies. It will be shown that, near the outer limit of the inner region, the radial flux is proportional to the concentration. Generally, the thickness of the inner region is small compared with the thickness of the region in which an appreciable variation of the concentration of particles takes place. This suggests that the convective-diffusion equation of the outer region should be solved subject to a virtual, first-order, chemical reaction occurring on the collector surface.

Consider a thin region, in contact with the collector surface, within which convection and tangential diffusion are negligible. Then the radial flux, $N_y(\theta)$, is independent of the distance between the particle centre and the surface of the collector, y (see fig. 1). The radial flux, considered negative if directed toward the collector, may be decomposed into contributions from diffusion and interaction forces:

$$N_y(\theta) = -D(y)\frac{\partial c}{\partial y} - m(y)\frac{\mathrm{d}\phi}{\mathrm{d}y}c, \qquad (8)$$

where $c(\theta, y)$ is the concentration of particles and $-\mathrm{d}\phi/\mathrm{d}y$ is the interaction force.

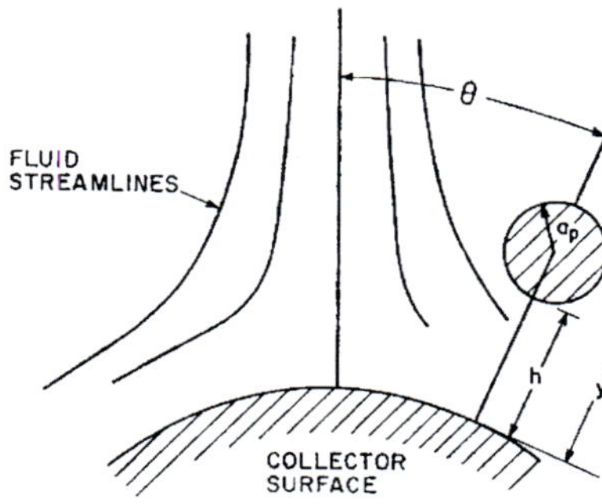

FIG. 1.—Colloidal particle near to the surface of the collector.

When the particles adhere to the collector surface they are no longer a part of the dispersed phase. Consequently, eqn (8) can be solved using the boundary condition

$$c(\theta, a_p) = 0. \qquad (9)$$

Combining (5), (7), and (8), and solving for $c(\theta, y)$ subject to (9), yields

$$c(\theta, y) = -\frac{N_y(\theta)}{D_\infty} \exp(-\phi(y)/kT) \int_{a_p}^{y} \exp(\phi(y')/kT) \frac{\mathrm{d}y'}{f\left(\dfrac{y'-a_p}{a_p}\right)}, \qquad (10)$$

where D_∞ is the value of D far from the surface.

When double-layer repulsions are sufficiently large, the total interaction potential exhibits a maximum at $y = y_m$ and tends to vanish at large distances. The thickness δ_F of the inner region (interaction-force boundary layer) will be taken as that value of y, greater than y_m, for which

$$\phi(\delta_F) \equiv 0.01\, \phi(y_m). \qquad (11)$$

For $y > \delta_F$ the interaction potential is essentially zero. Thus, near the outer limit of the inner region the concentration is given by:

$$c(\theta, \delta_F) = -\frac{N_y(\theta)}{D_\infty} \int_{a_p}^{\delta_F} \exp(\phi(y')/kT) \frac{dy'}{f\left(\dfrac{y'-a_p}{a_p}\right)}. \qquad (12)$$

Fig. 2 shows that, when

$$\exp\left(\phi(y_m)/kT\right)\Big/ f\left(\frac{y_m}{a_p}-1\right) \gg 1,$$

the value of the proportionality constant is relatively insensitive to how δ_F is defined. Additional approximations are possible if the above ratio is sufficiently large.

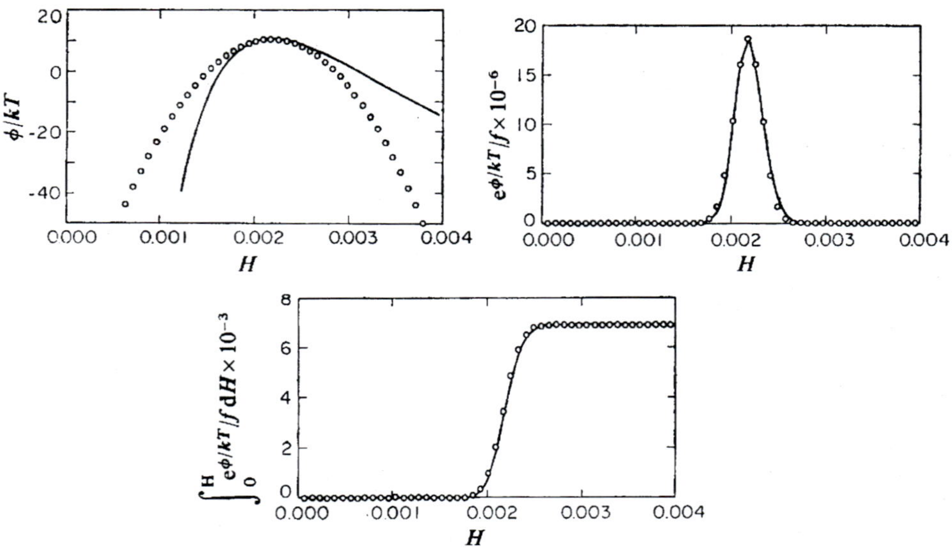

FIG. 2.—Validity of the parabolic approximation for the total interaction potential in the evaluation of the integral in eqn (12). Solid curves are exact values, while points represent quantities determined by using the parabolic approximation.

Expanding $\phi(y)$ in a Taylor series about $y = y_m$, truncating after the second-order term, and recognizing that $d\phi/dy|_{y_m} = 0$, we obtain

$$\phi(y) \approx E - \frac{\gamma}{2}(y-y_m)^2, \tag{13}$$

where $\gamma \equiv -d^2\phi/dy^2|_{y_m}$ and $E = \phi(y_m)$. Because $f(H)$ is a slowly varying function compared to an exponential, it is reasonable to replace it by its value $f(H_m)$ at $y = y_m$. Combining this approximation with (12) and (13), one obtains:

$c(\theta, y = \delta_F) =$

$$-\frac{N_y(\theta)}{D_\infty}\frac{\exp(E/kT)}{f(H_m)}\sqrt{\frac{\pi kT}{2\gamma}}\left\{\operatorname{erf}\left[(\delta_F - y_m)\sqrt{\frac{\gamma}{2kT}}\right] + \operatorname{erf}\left[(y_m - a_p)\sqrt{\frac{\gamma}{2kT}}\right]\right\}. \tag{14}$$

Whenever the arguments of both error-functions appearing in (14) are greater than about two, they can be replaced by infinity, which leads to

$$c(\theta, \delta_F) = -\frac{\sqrt{2\pi kT/\gamma}}{D_\infty f(H_m)}\exp(E/kT)N_y(\theta). \tag{15}$$

Thus, at the outer edge of the interaction-force boundary layer, the radial flux is proportional to the concentration. The transfer within this thin layer appears to the outer region as a virtual, first-order reaction occurring at the collector surface. The value of the apparent rate constant, K, can be deduced from (15):

$$K = D_\infty \sqrt{\frac{\gamma}{2\pi kT}} f(H_m) \exp(-E/kT). \tag{16}$$

This expression has the Arrhenius form and E is the maximum value of the potential energy, an " activation energy " for deposition. This is expected because the potential profile of fig. 2 resembles the plot of the energy against " reaction coordinate " used in the theory of rate processes. The factor $f(H_m)$ accounts for the dependence of the diffusion coefficient on the distance and evaluations show that it can decrease the frequency factor in eqn (16) by two orders of magnitude.

Because the diffusion coefficient depends upon y over a much larger region than the interaction forces, a transition region must also be considered. Neglecting convection in this region, one may write

$$-N_y(\theta, y) = \frac{c(\theta, \delta_D) - c(\theta, \delta_F)}{\int_{\delta_F}^{\delta_D} \frac{dy}{D(y)}}, \tag{17}$$

where δ_D is the thickness of the region in which the diffusion coefficient significantly depends upon y. For convection to be negligible, δ_D must be much smaller than the thickness of the diffusion boundary layer. When the above equation is valid, eqn (15) is relaced by

$$-N_y(\theta) = Kc(\theta, \delta_F) = \frac{c(\theta, \delta_D) - c(\theta, \delta_F)}{\int_{\delta_F}^{\delta_D} \frac{dy}{D(y)}}, \tag{18}$$

or, upon elimination of $c(\theta, \delta_F)$,

$$-N_y(\theta) = \frac{c(\theta, \delta_D)}{\frac{1}{K} + \frac{1}{\int_{\delta_F}^{\delta_D} \frac{dy}{D(y)}}}. \tag{19}$$

However, when $K \ll \int_{\delta_F}^{\delta_D} \frac{dy}{D(y)}$, this result reduces to

$$-N_y(\theta) = Kc(\theta, \delta_D). \tag{20}$$

The activation energy E and the frequency factor ν can be computed for a specific interaction potential profile as follows: maximizing eqn (4) with respect to H, one finds that E/A is a function of only two dimensionless groups:

$$E/A = F_1(N_\Psi, \tau), \tag{21}$$

where

$$N_\Psi = 3\varepsilon a_p \Psi_1 \Psi_2 / 2A, \tag{22a}$$

and

$$\tau = \kappa a_p. \tag{22b}$$

The values of E/A are plotted in fig. 4 against N_Ψ and τ. From eqn (16), the frequency factor can be identified as :

$$\nu = D_\infty f(H_m)\sqrt{\frac{\gamma}{2\pi kT}}. \tag{23}$$

Eqn (4) and (23) lead to :

$$\left(\frac{kT}{A}\right)^{\frac{1}{2}}\frac{a_p\nu}{D_\infty} = f(H_m)\left[-\frac{1}{2\pi}\frac{d^2(\phi/A)}{dH^2}\bigg|_{H=H_m}\right]^{\frac{1}{2}} = F_2(N_\Psi, \tau). \tag{24}$$

This equation is plotted in fig. 3.

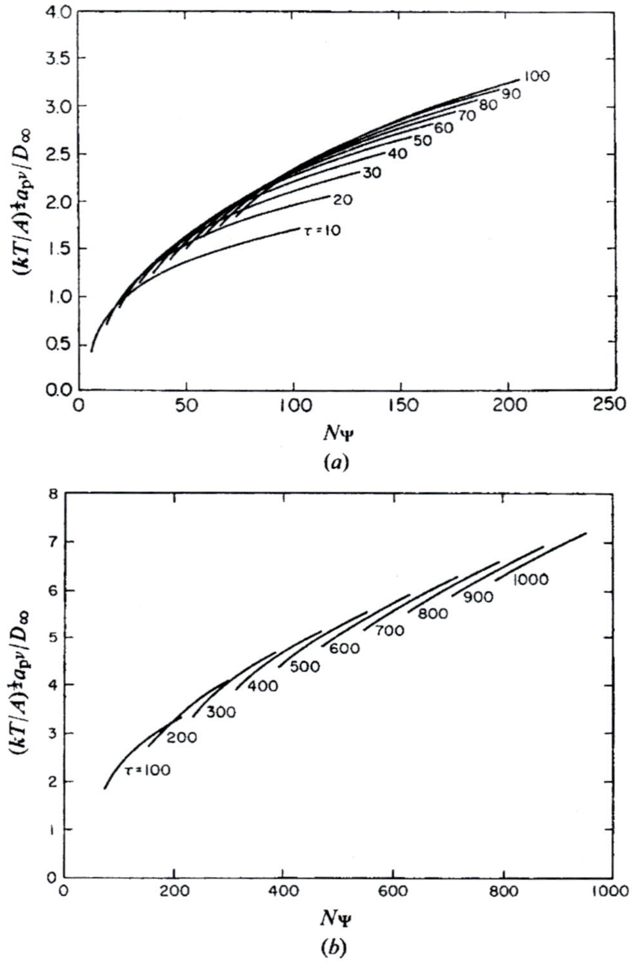

FIG. 3.—Plot of a dimensionless quantity containing the apparent frequency factor against N_Ψ and τ : (a) low ionic strength, (b) high ionic strength.

CONVECTIVE-DIFFUSION WITH SURFACE REACTION

In the previous section it was shown that, when double-layer repulsions are sufficiently strong, the resistance of the interaction-force boundary layer can be taken into account by imposing on the convective-diffusion equation a virtual, first-order, chemical reaction as a boundary condition on the surface of the collector. Because the diffusion coefficient for Brownian particles is very small, a diffusion boundary layer approximation can be used to solve the convective-diffusion equation. The method suggested by Ruckenstein [9] will be employed. Details of the analysis for the case of a spherical collector are presented. Results for a cylinder and rotating disc are summarized at the end of this section.

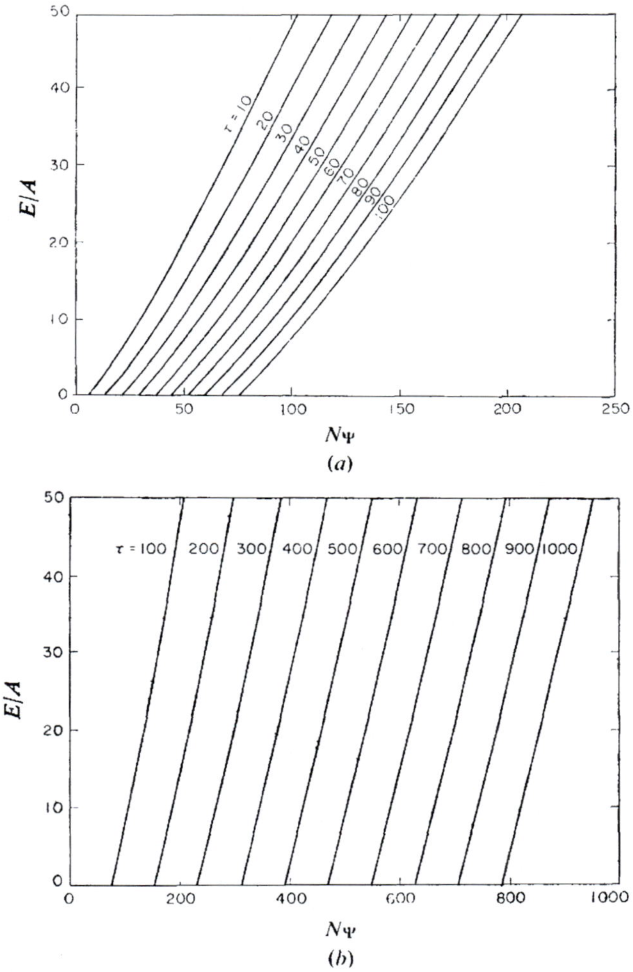

FIG. 4.—Ratio of the apparent activation energy to Hamaker's constant against N_Ψ and τ: (*a*) low ionic strength, (*b*) high ionic strength.

Consider a sphere of radius a held fixed in a creeping flow field with approach velocity U. The fluid contains Brownian particles having a diffusion coefficient D_∞. Should the Peclet number $2aU/D_\infty$ have a value much greater than unity, the diffusion boundary layer will be sufficiently thin so that curvature effects and tangential diffusion are negligible. Under these conditions the convective-diffusion equation assumes the following form:

$$u\frac{\partial c}{\partial x}+v\frac{\partial c}{\partial y} = D_\infty\frac{\partial^2 c}{\partial y^2}, \tag{25}$$

where x is the circumferential distance to the forward stagnation point. The tangential and radial components of velocity, u and v, are calculated from Stokes's solution for creeping flow. When the diffusion boundary layer is very thin, only the first terms in the Taylor expansion with respect to y are needed, hence

$$u = \tfrac{3}{2}U\frac{y}{a}\sin\frac{x}{a}, \text{ and} \tag{26a}$$

$$v = -\tfrac{3}{4}U\left(\frac{y}{a}\right)^2\cos\frac{x}{a}. \tag{26b}$$

If δ_D is very small compared with the thickness of the region in which there is an appreciable variation in concentration (diffusion boundary layer), the appropriate boundary conditions are

$$D_\infty\frac{\partial c}{\partial y}\bigg|_{y=0} = Kc(x, 0), \tag{27a}$$

$$c(x, \infty) = c_0, \text{ and} \tag{27b}$$

$$\frac{\partial c}{\partial x}\bigg|_{x=0} = 0. \tag{27c}$$

In order to solve (25) a change of variables is introduced which simplifies the equation and allows the construction of the solution for a non-uniform surface concentration from that valid for a uniform surface concentration by means of Duhamel's theorem. Since u and v are functions of x, Duhamel's theorem cannot be directly applied to (25).[10]

Consider an arbitrary change of variables, $X_1(x, y)$ and $Y(x, y)$, which transforms (25) into

$$\frac{\partial c}{\partial X_1}\left(u\frac{\partial X_1}{\partial x}+v\frac{\partial X_1}{\partial y}-D_\infty\frac{\partial^2 X_1}{\partial y^2}\right)+\frac{\partial c}{\partial Y}\left(u\frac{\partial Y}{\partial x}+v\frac{\partial Y}{\partial y}-D_\infty\frac{\partial^2 Y}{\partial y^2}\right) =$$

$$D_\infty\left[\left(\frac{\partial X_1}{\partial y}\right)^2\frac{\partial^2 c}{\partial X_1^2}+2\left(\frac{\partial X_1}{\partial y}\right)\left(\frac{\partial Y}{\partial y}\right)\frac{\partial^2 c}{\partial X_1\partial Y}+\left(\frac{\partial Y}{\partial y}\right)^2\frac{\partial^2 c}{\partial Y^2}\right]. \tag{28}$$

By choosing Y to be a particular solution of (25) and X_1 to be independent of y, eqn (28) is simplified. When u and v are given by (26), a solution of the form $Y = yg(x)$ is possible. Introducing $yg(x)$ in (25) wherever c appears and solving the resulting ordinary differential equation for g, one finds that

$$Y(x, y) = y\sin(x/a) \tag{29}$$

satisfies the transport equation. Substituting (29) into (28) and selecting $X_1 = x$ lead to

$$u(Y)\frac{\partial c}{\partial x} = D_\infty\sin^2\left(\frac{x}{a}\right)\frac{\partial^2 c}{\partial Y^2}. \tag{30}$$

Introducing the transformation:

$$\mathrm{d}X \equiv \sin^2\left(\frac{x}{a}\right)\mathrm{d}x, \quad \text{or} \quad X = \frac{x}{2}-\frac{a}{4}\sin\left(\frac{2x}{a}\right), \tag{31}$$

eqn (30) becomes

$$\tfrac{3}{2}U\frac{Y}{a}\frac{\partial c}{\partial X} = D_\infty\frac{\partial^2 c}{\partial Y^2}. \tag{32}$$

The boundary conditions (27) can be rewritten in the new coordinates as

$$c(X,\, Y=0) = \frac{D_\infty}{K}\sin\left(\frac{x}{a}\right)\frac{\partial c}{\partial Y}\bigg|_{Y=0}, \tag{33a}$$

$$c(X,\, Y\to\infty) = c_0, \text{ and} \tag{33b}$$

$$\frac{\partial c}{\partial X}\bigg|_{X=Y=0} = \text{finite}. \tag{33c}$$

In retrospect, the effect of the change of variables has been to deform the velocity field from axisymmetric stagnation flow over a sphere to linear shear flow along a flat plate. The main advantage of the new coordinates is that the coefficients of the derivatives in (32) are independent of X, and consequently Duhamel's theorem can be applied. Thus, the following procedure can be used: (1) solve equation (32) subject to a uniform surface concentration; (2) extend this solution to one valid for an arbitrary, nonuniform surface concentration by applying Duhamel's theorem; (3) select the surface concentration which satisfies (33*a*).

(1) For a uniform surface concentration the solution to (32) is readily obtained by performing a similarity transformation.

$$\psi(\eta) = c(X,\, Y), \quad \text{where } \eta = Y/X^{\frac{1}{3}}.$$

Using this combination of variables, the partial differential eqn (32) is transformed to an ordinary differential equation

$$\frac{d^2\psi}{d\eta^2} + \frac{U}{2aD_\infty}\eta^2\frac{d\psi}{d\eta} = 0. \tag{34}$$

For a uniform surface concentration the boundary conditions are:

$$\psi(0) = c_i = \text{constant}, \; \psi(\infty) = c_0.$$

The solution of (34) is

$$\frac{\psi(\eta)-c_i}{c_0-c_i} = \frac{1}{\Gamma(\frac{4}{3})}\int_0^{\alpha\eta}\exp(-s^3)\,ds, \tag{35}$$

where $\alpha^3 = U/6aD_\infty$ and $\Gamma(4/3) = 0.8930$. In terms of X and Y, eqn (35) can be written as

$$c(X,\, Y) = c_i + \frac{c_0-c_i}{\Gamma(\frac{4}{3})}\int_0^{\alpha Y/X^{\frac{1}{3}}}\exp(-s^3)\,ds. \tag{36}$$

(2) Now consider a nonuniform surface concentration such that the boundary condition becomes

$$c(X,\, Y=0) = c_i(X). \tag{37}$$

Applying Duhamel's theorem, one obtains

$$c(X,\, Y) = \frac{\partial}{\partial X}\int_0^X\left\{c_i(\lambda) + \frac{c_0-c_i(\lambda)}{\Gamma(\frac{4}{3})}\int_0^{\alpha Y/(X-\lambda)^{\frac{1}{3}}}\exp(-s^3)\,ds\right\}d\lambda. \tag{38}$$

Performing the differentiation of this expression using Leibnitz's formula leads to

$$c(X,\, Y) = c_0 - \tfrac{1}{3}\int_0^X\frac{c_0-c_i(\lambda)}{\Gamma(\frac{4}{3})}\frac{\alpha Y}{(X-\lambda)^{\frac{4}{3}}}\exp\left(-\frac{\alpha^3Y^3}{X-\lambda}\right)d\lambda. \tag{39}$$

(3) Finally the surface concentration profile is selected by introducing eqn (39) in the boundary condition (33*a*). The evaluation of $\lim_{Y\to 0}(\partial c/\partial Y)$ is accomplished by applying L'Hôpital's rule and assuming that dc_i/dX is bounded. One obtains:

$$\frac{\partial c}{\partial Y}(X, Y=0) = \frac{\alpha}{X^{\frac{1}{3}}}\frac{c_0 - c_i(0)}{\Gamma(\frac{2}{3})} - \alpha\int_0^X \frac{c_i'(\lambda)}{\Gamma(\frac{2}{3})}\frac{d\lambda}{(X-\lambda)^{\frac{1}{3}}}, \tag{40}$$

where $c_i'(X) = (dc_i/dX)(X)$. Substituting (40) into (33*a*), gives:

$$c_i(X) = \frac{D_\infty \alpha}{K\Gamma(\frac{2}{3})}\frac{\sin(x/a)}{X^{\frac{1}{3}}}\left[c_0 - c_i(0) - X^{\frac{1}{3}}\int_0^X c_i'(\lambda)\frac{d\lambda}{(X-\lambda)^{\frac{1}{3}}}\right]. \tag{41}$$

Since $c_i'(0)$ must be finite, according to (33*c*), the integral term vanishes as $X\to 0$. This allows one to explicitly determine $c_i(0)$ from:

$$c_i(0) = \frac{D_\infty \alpha}{K\Gamma(\frac{2}{3})}[c_0 - c_i(0)]\lim_{x\to 0}\left\{\frac{\sin(x/a)}{[X(x)]^{\frac{1}{3}}}\right\}. \tag{42}$$

Employing (31), the limit is found equal to $(3/a)^{\frac{1}{3}}$. Eqn (42) can subsequently be rearranged to give:

$$c_i(0) = \frac{B}{B+1}c_0, \tag{43}$$

where

$$B \equiv \left(\frac{3}{a}\right)^{\frac{1}{3}}\frac{\alpha D_\infty}{K\Gamma(\frac{2}{3})}. \tag{44}$$

Eliminating $c_i(0)$ between (41) and (43) yields:

$$c_i(X) = \left(\frac{a}{3}\right)^{\frac{1}{3}} B\frac{\sin(x/a)}{X^{\frac{1}{3}}}\left[\frac{1}{B+1} - X^{\frac{1}{3}}\int_0^X c_i'(\lambda)\frac{d\lambda}{(X-\lambda)^{\frac{1}{3}}}\right]. \tag{45}$$

A noniterative numerical solution for $c_i(X)$ is readily obtained by taking $c_i'(\lambda)$ to be constant over a small segment of the integration interval. The derivative is approximated by a finite difference.

Now that $c_i(X)$ is available, one can compute the overall rate of reaction by integrating the local rate per unit area over the entire surface of the sphere. This yields:

$$I = \int_0^\pi Kc_i[X(\theta)](2\pi a \sin\theta)\, a\, d\theta. \tag{46}$$

In the event that K becomes very small (B becomes very large) the process will be reaction controlled and the overall rate I will be equal to $4\pi a^2 K c_0$, denoted by I_∞. This rate may be used to nondimensionalize (46):

$$I/I_\infty = \tfrac{1}{2}\int_0^\pi \tilde{c}_i[\tilde{X}(\theta)]\sin\theta\, d\theta, \tag{47}$$

where

$$\tilde{c}_i(\tilde{X}) = c_i(X)/c_0 \quad \text{and} \quad \tilde{X} = X/a. \tag{48}$$

The procedure outlined above may also be used in the case of flow normal to a cylinder. The radial and tangential velocities near to the surface can be expressed as

$$u = 4\beta U(y/a)\sin(x/a) \tag{49a}$$

$$v = -2\beta U(y/a)^2\cos(x/a) \tag{49b}$$

where β is a dimensionless flow parameter. For slow flow around an isolated cylinder, Lamb [11] indicates $\beta = 1-\ln\sqrt{2aU\rho/\mu}$. Eqn (49) is very much like its spherical analogue, eqn (26). Thus it is not surprising that the same technique can be applied to both situations. Minor differences occur also in the expression of the differential surface area, and in the expression of I_∞. These differences will be summarized at the end of this section.

Ignoring edge effects, Levich [3] writes the convective diffusion equation for a rotating disc as:

$$v(y)\frac{\mathrm{d}c}{\mathrm{d}y} = D_\infty\frac{\mathrm{d}^2c}{\mathrm{d}y^2}, \tag{50}$$

where radial variations are neglected. Eqn (50) affords a particularly simple solution since it is already an ordinary differential equation. Near to the surface of the disc, Cochran's [12] expression for the normal velocity component becomes:

$$v(y) = -0.510\sqrt{\frac{\omega^3\rho}{\mu}}y^2, \tag{51}$$

where ω is the angular velocity of the disc and μ/ρ is the kinematic viscosity of the fluid. Requiring $c(0) = c_i$ and $c(\infty) = c_0$, the solution to (50) is [3]

$$c(y) = c_i+\frac{c_0-c_i}{\Gamma(\frac{4}{3})}\int_0^{\alpha y}\exp(-t^3)\,\mathrm{d}t, \tag{52}$$

where

$$\alpha^3 = \frac{0.170}{D_\infty}\sqrt{\frac{\omega^3\rho}{\mu}}.$$

To obtain the surface concentration when a chemical reaction occurs, eqn (52) is differentiated with respect to y and substituted into

$$Kc_i = D_\infty\left.\frac{\mathrm{d}c}{\mathrm{d}y}\right|_{y=0}.$$

Solving this equation for c_i leads to:

$$c_i = \frac{B}{B+1}c_0, \tag{53}$$

where $B = \alpha D_\infty/K\Gamma(4/3)$. If πa^2 is the area of the disc, the rate of collection is given by:

$$I = \pi a^2Kc_i. \tag{54}$$

Taking I_∞ to be πa^2Kc_0, the dimensionless rate is:

$$I/I_\infty = B/(B+1). \tag{55}$$

Because of the simplicity of analysis, the rotating disc is frequently employed to study the rate of deposition of particles.

For each case, the surface concentration profile can be found by solving the following integral equation:

$$\tilde{c}_i(\tilde{X}) = F(\tilde{X})\left[\frac{B}{B+1}-B\tilde{X}^{\frac{1}{3}}\int_0^{\tilde{X}}\frac{\tilde{c}_i'(\lambda)}{(\tilde{X}-\lambda)^{\frac{1}{3}}}\,\mathrm{d}\lambda\right], \tag{56}$$

where B, $F(\tilde{X})$, and $\tilde{X}(\theta)$ are summarized for each of the three cases in table 1. Once $\tilde{c}_i(\tilde{X})$ has been established, the rate of mass transfer is determined using the appropriate expression from table 2. The value of B uniquely determines the value of I/I_∞. Graphs of I/I_∞ against B for each of the three geometries are contained in fig. 5.

TABLE 1.—SUMMARY OF PARAMETERS USED IN EQN (56)

collector geometry	B	$F(\tilde{X})$	$\tilde{X}(\theta)$
sphere	$\frac{1}{K\Gamma(\frac{4}{3})}\left(\frac{UD_\infty^2}{2a^2}\right)^{\frac{1}{3}}$	$(\frac{1}{3})^{\frac{1}{3}}\frac{\sin\theta(\tilde{x})}{\tilde{X}^{\frac{1}{3}}}$	$\int_0^\theta \sin^2 t\,dt$
cylinder	$\frac{1}{K\Gamma(\frac{4}{3})}\left(\frac{2\beta UD_\infty^2}{3a^2}\right)^{\frac{1}{3}}$	$(\frac{2}{3})^{\frac{1}{3}}\frac{\sqrt{\sin(\theta(\tilde{X}))}}{\tilde{X}^{\frac{1}{3}}}$	$\int_0^\theta \sqrt{(\sin t)}\,dt$
disc	$\frac{1}{K\Gamma(\frac{4}{3})}\left(0.170D_\infty^2\sqrt{\frac{\omega^3\rho}{\mu}}\right)^{\frac{1}{3}}$	1	†

† Since concentration on the surface is uniform, no such transformation is required. Solution to (56) is given by $\tilde{c}_i = B/(B+1)$.

TABLE 2.—SUMMARY OF EXPRESSIONS FOR EVALUATING THE RATE OF COLLECTION

collector geometry	I	I_∞	I/I_∞
sphere	$2\pi a^2\int_0^\pi Kc_i \sin\theta\,d\theta$	$4\pi a^2 Kc_0$	$\frac{1}{2}\int_0^\pi \tilde{c}_i \sin\theta\,d\theta$
cylinder	$2aL\int_0^\pi Kc_i\,d\theta$	$2\pi aLKc_0$	$\frac{1}{\pi}\int_0^\pi \tilde{c}_i\,d\theta$
disc	$\pi a^2 Kc_i$	$\pi a^2 Kc_0$	$\tilde{c}_i$

Two limiting situations may be identified: (1) the rate constant K is very small compared to αD_∞, hence the process occurring in the interaction forces boundary layer controls the deposition rate, and (2) the rate constant K is very large hence the convective diffusion is the controlling factor. The first limiting case was treated by Hull and Kitchener (except for the variation of the diffusion coefficient) while the second was treated by Levich. In the present paper an equation is established which is valid for all values of the rate constant thus also incorporating both limiting situations.

DISCUSSION

In this paper it is shown that the rate of deposition of Brownian particles on the collector can be calculated by solving the convective diffusion equation subject to a virtual first order chemical reaction as a boundary condition at the surface. The boundary condition concentrates the surface–particle interaction forces. When the interaction potential between the particle and the collector experiences a sufficiently high maximum (see fig. 2) the apparent rate constant of the boundary condition has the Arrhenius form. Equations for the apparent activation energy and the apparent frequency factor are established for this case as functions of Hamaker's constant, dielectric constant, ionic strength, surface potentials and particle radius. The rate constant includes the effect of the variation in particle diffusion coefficient which occurs very near to the surface. This effect can decrease the value of the rate constant by a factor of 100.

The rate of deposition is calculated for a spherical, cylindrical, or rotating disc collector; the dimensionless rate of deposition was found to be a function of a single dimensionless group B, involving the rate constant, characteristic velocity, diffusion coefficient, and (except in the case of a disc) the collector size.

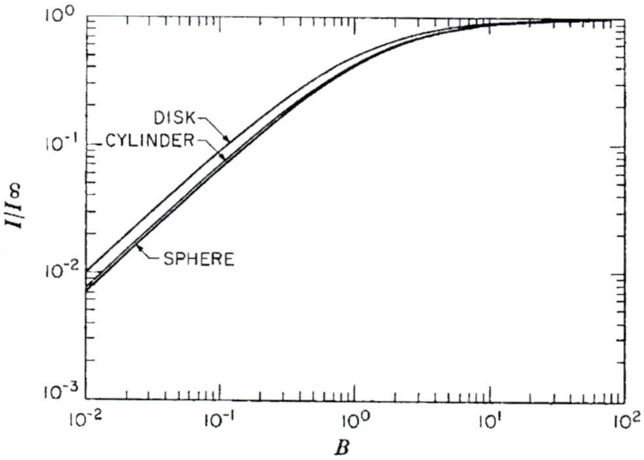

FIG. 5.—Ratio of actual rate of deposition to that for which the virtual reaction controls.

The present equations lead to the result of Levich when the apparent rate constant is large enough, and to a refined version (because it includes the effect of the distance dependent diffusion coefficient) of that of Hull and Kitchener when the apparent rate constant is small enough.

The emphasis in this paper is on the technique used to calculate the rate of deposition when London and double-layer forces are involved. For the purpose of illustration a simple expression for the interaction potential is considered. The main equations established here can however be readily applied to any expression for this potential. The expression for the double-layer potential which is available in the literature (eqn (2)) is not completely satisfactory because it implies that the surface potential is independent of separation distance. Calculations in progress,[13] which consider the degree of dissociation of surface groups, show that neither the surface potential nor the surface charge remains constant as the approach distance is changed. Because small changes in the surface potential lead to very large changes in the " activation energy ", and consequently catastrophic changes in the rate constant, one cannot expect agreement between predictions based on the approximate potential (2) and experiment. In addition to neglecting the variation in surface potential, the model used for the interaction potential implies a homogeneous collector surface. In reality all surfaces are geometrically and energetically heterogeneous. Surface roughness and a nonuniform distribution of " active sites " are always present. Also, the properties of the collector surface change during the deposition period, whereas the current analysis assumes a steady state.

A few experimental data are available in the literature, namely those of Hull and Kitchener,[4] and FitzPatrick and Spielman.[14] Hull and Kitchener observed a nonuniform distribution of the collected particles, which becomes more uniform when a surface active agent is present. Surfactants alter the nature of both collector and particle surfaces, thus modifying, in an essential manner, the surface-particle interaction potential. The profiles of potential energy which are available in the literature and used in the present paper do not consider the effect of surfactants.

Predictions of the deposition rate for the experimental conditions of Hull and Kitchener, based upon the analysis presented here are orders of magnitude smaller than those observed.† An explanation for this disagreement was suggested above. The rate of deposition is very sensitive to certain parameters. For instance, a 5 mV decrease in the surface potential can increase the deposition rate by six orders of magnitude (see fig. 6). A decrease of both surface potentials by a factor of four (from

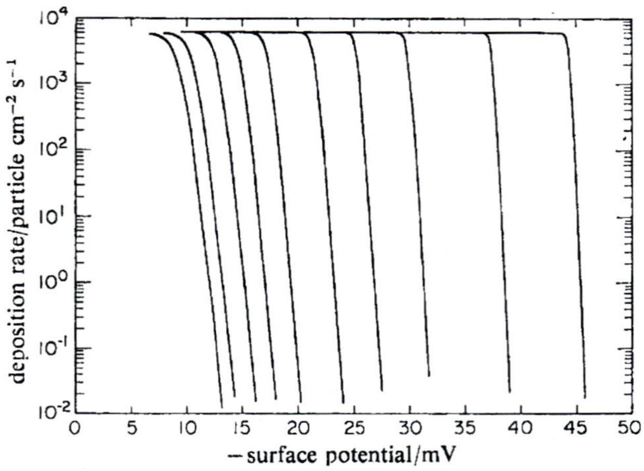

FIG. 6.—Effect of surface potential and ionic strength upon the rate of deposition to the surface of a rotating disc. Values of the fixed parameters were chosen from the conditions found in the experiments of Hull and Kitchener.[4] $a_p = 1.54 \times 10^{-5}$ cm; $\omega = 6$ rev./s; $c_0 = 1.28 \times 10^8$ particle/cm^3; $A = 1 \times 10^{-13}$ erg. Ionic strengths, from left to right are, 0.001, 0.002, 0.005, 0.01, 0.02, 0.05, 0.1, 0.2, 0.5 and 1.0 mol/dm^3.

−60 to −15 mV at low ionic strength) increases the theoretical rate of deposition by several hundred orders of magnitude and can result in the agreement between the predictions and the experimental data of Hull and Kitchener. Also small changes in the ionic strength and dielectric constant can have similar consequences.

NOMENCLATURE

A	Hamaker's constant (erg)
a	radius of collector (cm)
a_p	radius of particles (cm)
B	dimensionless group defined by table 1
c	local concentration of particles (particle/cm^3)
c_i	concentration of particles on collector surface (particle/cm^3)
c_i'	dc_i/dX (particle/cm^4)
$\bar{c}_i$	c_i/c_0 (dimensionless)
$\bar{c}_i'$	$d\bar{c}_i/dX$ (dimensionless)
c_0	concentration of particles far from collector (particle/cm^3)
D	local diffusion coefficient (cm^2/s)
D_∞	diffusion coefficient in an infinite medium (cm^2/s)
e_i	charge on ionic species i (e.s.u.)
E	maximum value of ϕ (erg)
f	Brenner's correction factor for mobility (dimensionless)
F	function defined by table 1 (dimensionless)
F_1	E/A (dimensionless)
F_2	$a_p\nu\sqrt{(kT/A)}/D_\infty$ (dimensionless)
h	minimum separation between collector and particle surfaces (cm)
H	h/a_p (dimensionless)
H_m	value of H for which ϕ is largest (dimensionless)

† For those conditions K is small and our equation gives nearly the same results as the limiting equation deduced by Hull and Kitchener.

I	rate of deposition (particle/s)
I_∞	rate of deposition when surface reaction dominates (see table 2) (particle/s)
k	Boltzmann constant (erg/K)
K	rate constant (cm/s)
L	length of cylinder (cm)
m	mobility (cm s^{-1} dyn^{-1})
n_i	concentration of ionic species i (ion/cm^3)
N_y	radial flux of particles (particle cm^{-2} s^{-1})
N_Ψ	$3\varepsilon a_p \Psi_1 \Psi_2 / 2A$ (dimensionless)
r	radial position of particle centre (cm)
T	absolute temperature (K)
u	tangential component of fluid velocity (cm/s)
U	approach velocity (cm/s)
v	radial component of fluid velocity (cm/s)
x	$a\theta$ (cm)
X	$a\check{X}$ (cm)
$\check{X}$	see table 1 (dimensionless)
X_1	x (cm)
y	$r-a$ (cm)
y_m	value of y for which ϕ is largest (cm)
Y	$y \sin(x/a)$ for a spherical collector (cm)
α	$(U/6aD_\infty)^{1/3}$ for spherical collector ($cm^{-2/3}$)
β	flow parameter for cylindrical collector (dimensionless)
γ	$-(d^2\phi/dy^2)_{y=y_m}$ (erg/cm^2)
δ_D	thickness of region over which D varies significantly with h (cm)
δ_F	defined by eqn (14) (cm)
ε	dielectric constant (dimensionless)
η	$Y/X^{1/3}$ ($cm^{2/3}$)
θ	angular position of particle centre measured from forward stagnation line (rad)
κ	reciprocal Debye length (cm^{-1})
μ	fluid viscosity (g cm^{-1} s^{-1})
ν	frequency factor (cm/s)
ρ	fluid density (g/cm^3)
τ	κa_p (dimensionless)
ϕ	total interaction potential (erg)
ϕ_{Ad}	London–van der Waals potential (erg)
ϕ_{DL}	double-layer potential (erg)
ψ	dependent variable of similarity transform (particle/cm^3)
Ψ_i	surface potential of surface i (e.s.u.)
ω	angular velocity of disc (rad/s)

[1] L. A. Spielman and J. A. FitzPatrick, *J. Colloid Sci.*, 1973, **43**, 51.
[2] L. A. Spielman and P. M. Cukor, *J. Colloid Sci.*, 1973, **42**, 607.
[3] V. G. Levich, *Physicochemical Hydrodynamics* (Prentice-Hall, N.J., 1962).
[4] M. Hull and J. A. Kitchener, *Trans. Faraday Soc.*, 1969, **65**, 3093.
[5] H. C. Hamaker, *Physica*, 1937, **4**, 1058.
[6] E. J. W. Verwey and J. Th. G. Overbeek, *Theory of the Stability of Lyophobic Colloids* (Elsevier, New York, 1948).
[7] R. Hogg, T. W. Healy and D. W. Fuerstenau, *Trans. Faraday Soc.*, 1966, **62**, 1638.
[8] H. Brenner, *Chem. Eng. Sci.*, 1961, **16**, 242.
[9] E. Ruckenstein, *Chem. Eng. Sci.*, 1971, **26**, 1795.
[10] H. S. Carslaw and J. C. Jaeger, *Conduction of Heat in Solids* (Clarendon, Oxford, 1959).
[11] H. Lamb, *Hydrodynamics* (Cambridge Univ. 1945).
[12] W. G. Cochran, *Proc. Cambridge Phil. Soc.*, 1934, **30**, 365.
[13] D. C. Prieve and E. Ruckenstein, paper to be published.
[14] J. A. FitzPatrick and L. A. Spielman, *J. Colloid Sci.*, 1973, **43**, 350.

Adsorption and Desorption of Particles and Their Chromatographic Separation

ELI RUCKENSTEIN
Faculty of Engineering and Applied Science
State University of New York at Buffalo
Buffalo, New York 14214
and
DENNIS C. PRIEVE
Department of Chemical Engineering
Carnegie-Mellon University
Pittsburgh, Pennsylvania 15213

Rates of adsorption/desorption are calculated by considering Born, double-layer, and van der Waals forces. Conditions exist where rates are highly sensitive to particle size, shape, and chemical constitution because diffusion over a potential barrier controls. On this basis, a new separation technique, called potential-barrier chromatography, is suggested.

SCOPE

Removal of suspended solids from a liquid may be required, either to recover valuable products or to remove obnoxious particulates. Clarification of dilute aqueous suspensions is frequently performed in deep-bed granular filters, where the grains provide a surface on which the particles may adsorb. If two types of particles differ in their adsorption rate, separation of them may be accomplished by selectively adsorbing one species on the grains. Furthermore, reversibility of adsorption would allow chromatographic separation of particle mixtures, analogous to gas chromatography.

Physical adsorption arises from physical interactions between the suspended particles and the collector, such as van der Waals attraction, double-layer repulsion, and Born repulsion. The total interaction energy, as a function of the particle-collector gap width, displays either one minimum and no maximum or two minima and one maximum. Several mechanisms for chromatographic sep-

aration of particle species may be classified by the shape of the interaction energy profile. At low ionic strengths, a large maximum in the profile is expected, which represents an insurmountable energy barrier, preventing particle adsorption. Particle residence times then are determined by the particle's velocity which, owing to hydrodynamics alone, is greater than the average fluid velocity by an amount dependent on particle size. This corresponds to hydrodynamic chromatography. Conversely, at high ionic strength no maximum is expected, and particles rapidly adsorb on the surface. The residence time is now determined by adsorption and desorption rates and is expected to be substantially greater than fluid residence time. Finally, at moderate ionic strengths, a surmountable energy barrier is developed which allows adsorption and desorption, but at reduced rates which depend on the height of the barrier.

The present paper focuses on the latter situation, where chromatographic separation of suspended particle species is expected to be most selective. Rates of adsorption and desorption over an energy barrier are calculated as a function of the particle radius and ionic strength in order to predict the sensitivity of the rate to these properties.

CONCLUSIONS AND SIGNIFICANCE

When the Born, double-layer, and van der Waals forces act over distances that are short compared to the diffusion boundary-layer thickness, and when these forces form an energy barrier, the adsorption and desorption rates may be calculated by lumping the effect of the interactions into a boundary condition on the usual convective-diffusion equation. This condition takes the form of a first-order, reversible reaction on the collector's surface. The apparent rate constants and equilibrium collector capacity are explicitly related to the interaction profile and are shown to have the Arrhenius form. They do not depend on the collector geometry or flow pattern. Under conditions where the apparent reaction controls the rate, an exponential sensitivity to particle size is predicted. Decreasing ionic strength is observed to enhance this sensitivity, but at the cost of lower rates. Similar sensitivity is also expected for properties related to chemical constitution. By giving careful attention to the choice of column packing and the ionic strength of the carrier solution, the particle's residence time may be controlled by the adsorption/desorption of particles over an energy barrier. This process is defined here as potential-barrier chromatography. It promises to be much more selective toward particle properties than gel or hydrodynamic chromatography.

Early experiments on flocculation revealed that it may be induced by adding electrolyte to the solution in which the sol is suspended and can sometimes be reversed by dialyzing the electrolyte from the floc. The DLVO theory (Verwey and Overbeek, 1948) adequately explains flocculation but does not allow for reversibility. If only van der Waals attractions and electrostatic double-layer repulsions are considered, the particles in the floc are held together in an infinitely deep energy well, regardless of the ionic strength. Hamaker (1937*a*) qualitatively explained reversibility by considering short-range Born repulsion among the interactions. Then the energy well is not infinitely deep, and its depth becomes dependent on ionic strength.

Measurements of the rate of deposition of particles, suspended in a moving phase, onto a surface also change dramatically with ionic strength (Marshall and Kitchener, 1966; Hull and Kitchener, 1969; Fitzpatrick and Spielman, 1973; Clint et al., 1973). This indicates that repulsive double-layer forces are also of importance to the transport rates of particulate solutes. When the interactions act over distances that are small compared to the diffusion boundary-layer thickness, the rate of transport can be computed (Ruckenstein and Prieve, 1973; Spielman and Friedlander, 1974) by lumping the interactions into a boundary condition on the usual convective-diffusion equation. This takes the form of an irreversible, first-order reaction on the surface. A similar analysis has also been performed for the case of unsteady deposition from stagnant suspensions (Ruckenstein and Prieve, 1975).

In the present paper, the previous analysis for moving suspensions is extended to allow reversible deposition by including the short-range Born repulsion. The apparent surface reaction is then shown to be reversible and first order in both directions and thus is analogous to gas adsorption at low surface coverage. Applications include solid-liquid separations, particle separation, and chromatography.

A SIMPLIFIED QUANTITATIVE MODEL OF PHYSICAL INTERACTIONS

The van der Waals interaction energy between a sphere and a semiinfinite plate has been estimated, as a function of the sphere-plate separation distance, by Hamaker

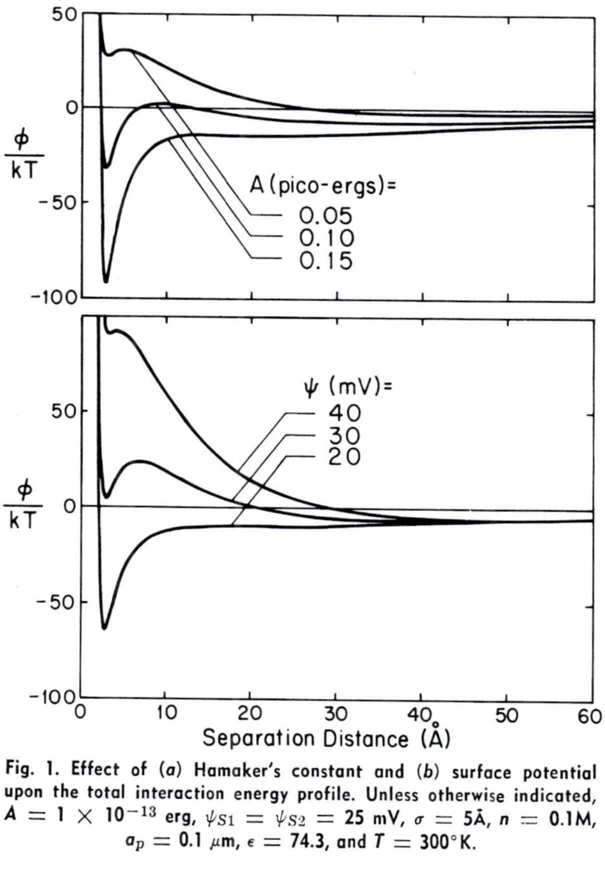

Fig. 1. Effect of (*a*) Hamaker's constant and (*b*) surface potential upon the total interaction energy profile. Unless otherwise indicated, $A = 1 \times 10^{-13}$ erg, $\psi_{S1} = \psi_{S2} = 25$ mV, $\sigma = 5$Å, $n = 0.1$M, $a_p = 0.1\ \mu$m, $\epsilon = 74.3$, and $T = 300$°K.

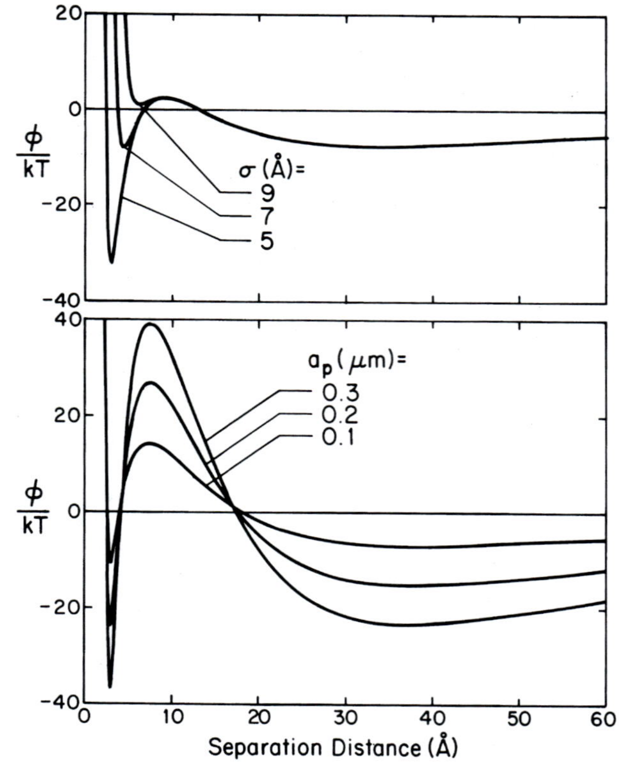

Fig. 2. Effect of (*a*) collision diameter and (*b*) particle radius upon the total interaction energy profile. Other properties have the same value as in Figure 1, except $\psi_{S1} = \psi_{S2} = 28$ mV in (*b*).

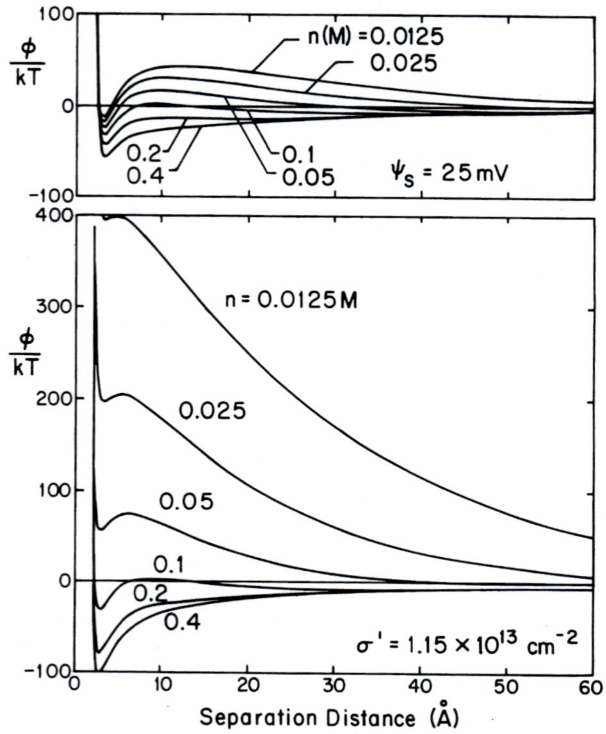

Fig. 3. Effect of ionic strength upon the total interaction energy profile when (*a*) surface potential or (*b*) surface charge density is held fixed. In (*b*), the charge density was chosen so that $\psi_{S1} = \psi_{S2} = 25$ mV when $n = 0.1$M. Other parameters have the same value as in Figure 1.

(1937*b*), who assumed that the molecular interactions could be linearly superimposed:

$$\phi_6(h) = \frac{A}{6}\left[\ln\left(\frac{h+2a_p}{h}\right) - 2\frac{a_p}{h}\frac{h+a_p}{h+2a_p}\right] \quad (1)$$

Lifshitz' more recent continuum analysis of the van der Waals interaction (Israelachvili and Tabor, 1973) leads to approximately the same distance dependence but to a more accurate physical interpretation of Hamaker's constant A. However, this distinction becomes unimportant to the present analysis if A is treated as a measurable property.

Born repulsion is a short-range molecular interaction, resulting from the overlap of electron orbitals. It is the twelfth-order term of the empirical Lennard-Jones 6-12 potential. To estimate the Born repulsion between a sphere and a plate, the authors of the present paper assumed these molecular interactions also could be linearly superimposed to obtain

$$\phi_{12}(h) = \frac{A\sigma^6}{7\,560}\left[\frac{8a_p+h}{(2a_p+h)^7} + \frac{6a_p-h}{h^7}\right] \quad (2)$$

where the collision diameter σ will be treated as a second characteristic property of the solids.

When macroscopic surfaces come in contact with polar solvents, such as water, an electrostatic charge can develop on the surface as a result of the loss or gain of a proton or another ion by the surface. This gives rise to a double-layer force between two opposing surfaces. Using the method suggested by Verwey and Overbeek (1948, p. 70) for weak interactions, the authors have calculated the double-layer interaction between two opposing, dissimilar planes (Prieve and Ruckenstein, 1975) and then applied Derjaguin's method (Hogg, Healy, and Fuerstenau, 1966) to obtain the sphere-plane interaction energy:

$$\phi_{DL}(h) =$$

$$16\epsilon\left(\frac{kT}{e}\right)^2 a_p \tanh\left(\frac{e\psi_{S1}}{4kT}\right)\tanh\left(\frac{e\psi_{S2}}{4kT}\right)e^{-\kappa h} \quad (3)$$

This expression is appropriate for monovalent electrolyte solutions when $\kappa h > 2$ and $\kappa a_p >> 1$.

Summing the contribution from each effect gives the total energy of interaction:

$$\phi(h) = \phi_{12}(h) + \phi_6(h) + \phi_{DL}(h) \quad (4)$$

Equations (1) to (4) form a simplified quantitative model for the interactions. A number of additional effects, such as surface roughness on the order of σ, solvation of the surface, and elastic deformations resulting from the interactions could also be incorporated in the model. For the present objectives the simplified model of Equations (1) through (4) will be used; however, any expression for $\phi(h)$ can be substituted in the final result of the next section.

Figures 1 to 3 illustrate the quantitative effect of various properties on the interaction profile $\phi(h)$. Four observations are noteworthy. First, there exists a narrow range of Hamaker's constant and surface potentials for which a maximum $[\phi_{max} = \phi(h_{max})]$ and two minima $[\phi_{mn1} = \phi(h_{mn1})$ and $\phi_{mn2} = \phi(h_{mn2})]$, exist where $h_{mn2} < h_{max} < h_{mn1}$. If A is too small (large) or ψ_S too large (small), ϕ_{max} and $\phi_{mn2}(\phi_{mn1})$ merge leaving only one extremum. Since a maximum is assumed in the rate analysis of the next section, the properties A, ψ_{S1}, and ψ_{S2} are restricted to those values (see Figure 4) where three extrema occur. Second, $\phi(h)$ at all h is roughly proportional to the particle radius. Third, holding the charge density fixed when the ionic strength is changed leads to more dramatic changes in $\phi(h)$ than when the surface potential is held fixed. Charge density, surface po-

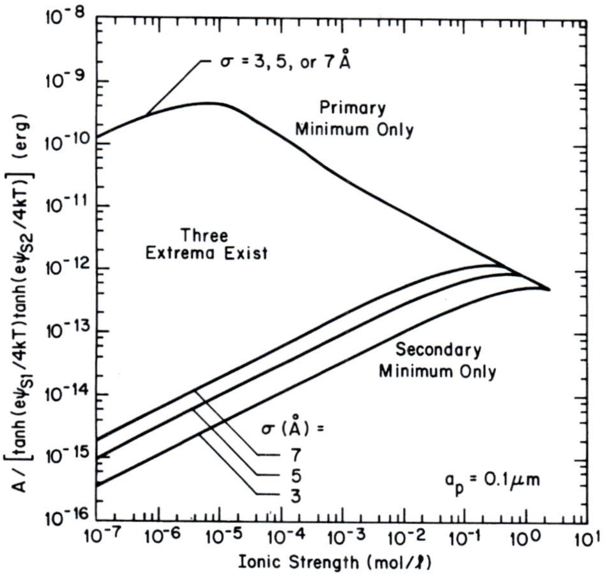

Fig. 4. Map of the region in which the potential energy profile possesses three extrema. Particle radius has practically no effect on the boundaries, except at very low ionic strength. This map was drawn for $a_p = 0.1$ μm.

tential, and ionic strength are related by the Gouy-Chapman equation (Verwey and Overbeek, 1948, p. 32)

$$\sigma' = (2\epsilon kTn/\pi)^{1/2} \sinh (e\psi_S/2kT) \quad (5)$$

provided the surfaces are widely separated and the electrolyte is univalent. Fourth, depending on ionic strength, the primary minimum (ϕ_{mn2}) may be above the secondary minimum (ϕ_{mn1}). Table 1 summarizes the effect of each parameter on the values of the extrema.

EVALUATION OF ADSORPTION/DESORPTION RATES

Transport through the Interaction Boundary Layer

Consider a thin region near to the collector, whose thickness δ_o is defined so that fluid motion can be neglected inside the region, and particle-collector interactions can be neglected outside it. This region is defined as the interaction boundary layer. By denoting by $c(x)dx$ the number of centers per unit surface area located between x and $x + dx$, where $x = h + a_p$, the net flux of particles through the region can be expressed as the sum of contributions from the total interaction force $-d\phi/dh$ and from Brownian motion

$$J = - D(h) \frac{dc}{dh} - m(h) \frac{d\phi}{dh} c \quad (6)$$

where the particle mobility can be evaluated from Stokes' law, as modified by Brenner (1961) to account for wall effects (which become important when the particle is near the collector)

$$m(h) = \frac{f(h/a_p)}{6\pi\mu a_p} \quad (7)$$

and where the diffusion coefficient is given by the Nernst-Einstein equation

$$D = mkT \quad (8)$$

The wall correction $f(H)$ tends to H when $H << 1$ and to unity when $H >> 1$.

Suppose that the interaction forces establish an energy barrier that retards the motion of particles both toward and away from the collector. If this barrier reduces the adsorption and desorption rates significantly, particles near the primary minimum will have time to achieve a balance between the interaction forces and Brownian motion, before their population changes. Integration of Equation (6) with $J \approx 0$ and $D = mkT$ leads to a Boltzmann distribution

$$c(h) = \frac{n_2}{I_2} \exp [- \phi(h)/kT] \text{ for } h < \delta_2 \quad (9a)$$

where $I_2 = \int_o^{\delta_2} \exp [- \phi(h)/kT]dh$, and δ_2 is the largest h in the interval (h_{mn2}, h_{max}) for which a Boltzmann distribution can be assumed. The integration constant n_2/I_2 was determined so that $\int_o^{\delta_2} c(h)dh = n_2$, where n_2 is the number of particles adsorbed per unit area. These arguments can also be applied near the secondary minimum. Here the equilibrium distribution is given by

$$c(h) = c_i \exp [- \phi(h)/kT] \text{ for } \delta_1 < h < \delta_o \quad (9b)$$

where δ_1 is the smallest h in the interval (h_{max}, h_{mn1}) for which a Boltzmann distribution can be assumed. In this region the integration constant is calculated as $c_i = c(\delta_o)$, noting that δ_o is defined so $\phi(\delta_o) \approx 0$.

A Boltzmann distribution cannot be obtained in the middle region ($\delta_2 < h < \delta_1$) containing the maximum because particles do not reside in this region long enough to establish an equilibrium. However, this zone is sufficiently thin so that J can be assumed to be constant throughout. Integration of Equation (6) then leads to

$$-J = \{c(h) \exp [\phi(h)/kT]\} \Big|_{\delta_2}^{\delta_1} \left\{ \int_{\delta_2}^{\delta_1} \frac{\exp [\phi(h)/kT]}{D(h)} dh \right\}^{-1} \quad (10)$$

Evaluating $c(\delta_2)$ from Equation (9a) and $c(\delta_1)$ from Equation (9b), we get

$$-J = K_f c_i - K_r n_2 \quad (11)$$

where $K_f^{-1} = \int_{\delta_2}^{\delta_1} D^{-1} \exp (\phi/kT)dh$ and $K_r = K_f/I_2$.

TABLE 1. EFFECT OF PHYSICAL PARAMETERS ON EXTREMA IN THE INTERACTION ENERGY PROFILE*

Physical parameter	Effect on extrema				
	ϕ_{mn2}	ϕ_{max}	ϕ_{mn1}	$\phi_{max} - \phi_{mn2}$	$\phi_{max} - \phi_{mn1}$
Hamaker's constant	−	−	−	+	−
Surface potential	+	+	+	−	+
Collision diameter	+	~0	0	−	~0
Particle radius	(+)	(+)	(+)	+	+
Ionic strength					
(constant potential)	−	−	−	−	−
(constant charge)	−	−	−	+	−

* + denotes increase in extremum results from an increase in value of parameter.
− denotes decrease in extremum results from an increase in value of parameter.
(+) denotes increase in absolute magnitude results from an increase in value of parameter.
0 denotes no significant effect.

To determine the net rate of adsorption of particles suspended in a fluid that is flowing over the collector, one may then solve the usual convective-diffusion equation subject to a reversible first-order reaction as the boundary condition, provided the diffusion boundary layer is much thicker than the interaction boundary layer.

The forward rate constant is evaluated as the integral of exp $[\phi(h)/kT]/D(h)$, which assumes its largest value at $h = h_{\max}$ and decays rapidly as $|h - h_{\max}|$ increases. Then it is reasonable, following Kramers (1940), to expand $\phi(h)$ in a Taylor series about $h = h_{\max}$

$$\phi(h) = \phi_{\max} - \frac{1}{2}\gamma_{\max}(h - h_{\max})^2 \quad (12)$$

truncating after the second-order term, with $\gamma_{\max} = -d^2\phi/dh^2|_{h=h_{\max}}$. Substituting this parabolic approximation into the definition of K_f and integrating, we get an expression containing error functions. If the maximum is a sharp spike, their arguments become sufficiently large so that the error functions tend to a maximum value of unity, whereupon the expression for the forward rate constant reduces to

$$K_f \approx D(h_{\max}) \left(\frac{\gamma_{\max}}{2\pi kT}\right)^{1/2} \exp(-\phi_{\max}/kT) \quad (13a)$$

Because $D(h)$ varies slowly compared to the exponential, it was replaced by its value $D(h_{\max})$ at the central point of the expansion. Similar arguments can be applied to the definition of I_2 to obtain

$$K_r \approx D(h_{\max}) \frac{(\gamma_{\max}\gamma_{mn2})^{1/2}}{2\pi kT} \exp\left(-\frac{\phi_{\max} - \phi_{mn2}}{kT}\right) \quad (13b)$$

where $\gamma_{mn2} = +d^2\phi/dh^2|_{h=h_{mn2}}$. Hence the apparent rate constants have the Arrhenius form, with $\phi_{\max}$ as the activation energy for adsorption, $\phi_{\max} - \phi_{mn2}$ as the activation energy for desorption, and ϕ_{mn2} as the heat of adsorption. When saturated, $J = 0$, and the grain capacity is $(K_f/K_r)c_o$. The ratio K_f/K_r is determined mainly by ϕ_{mn2}, which becomes more negative as ionic strength increases (Figure 3), thereby leading to larger K_f/K_r. Particle-particle interactions have implicitly been neglected, which requires the grain capacity to correspond to substantially less than monolayer coverage. This condition is met provided that the bulk particle concentration is sufficiently dilute.

In summary, the expression for the net rate of particle transport through the interaction boundary layer takes the same form as a first-order, reversible, heterogeneous chemical reaction, provided that $\phi_{\max}$ and $\phi_{\max} - \phi_{mn2}$ are large compared to kT, $\min\{\delta_1 - h_{\max}, h_{\max} - \delta_2\}$ $(\gamma_{\max}/2kT)^{1/2} > 2$, $\min\{h_{mn2}, \delta_2 - h_{mn2}\}$ $(\gamma_{mn2}/2kT)^{1/2} > 2$, and δ_o is small compared to the diffusion boundary-layer thickness. An additional assumption is that the particles are not carried over the barrier by their inertia. Therefore, the inertial force on the particle must be small compared to the applied force $-d\phi/dh$

$$|F_{\text{inertia}}| = M \left| V\frac{dV}{dh} \right| \approx Mm \left| \frac{d\phi}{dh}\frac{d}{dh}\left(m\frac{d\phi}{dh}\right) \right| << \left| \frac{d\phi}{dh} \right|$$

which is satisfied when $|d(md\phi/dh)/dh| << m^{-1}M^{-1}$. If inertia is not negligible, the problem must be solved by using the more general Fokker-Plank equation (Chandrasekhar, 1943, p. 65).

Transport to a Rotating Disk

As an example, suppose that a fluid containing particles is in contact with a rotating disk. What will be the net rate of adsorption onto the collector disk? Levich (1962) neglected radial variations and solved the usual convective-diffusion equation, taking the concentration as c_i at the disk surface and as c_o far away from the disk. If c_i is eliminated from his solution by imposing Equation (11) at this surface, with $J = -D\, dc/dy|_{y=0}$, the result for the net adsorption rate will be

$$-J = \frac{B}{B+1}(K_f c_o - K_r n_2) \quad (14)$$

where

$$B = \frac{\alpha D(\infty)}{K_f \Gamma(4/3)} \text{ and } \alpha^3 = \frac{0.170}{D(\infty)}\left(\frac{\omega^3 \rho_f}{\mu}\right)^{1/2}$$

As particles accumulate on the disk, n_2 and J can be expected to change slowly with time. Substituting dn_2/dt for - J and integrating, we get

$$n_{?}\ t) = \frac{K_f}{K_r}c_o + \left(n_{2i} - \frac{K_f}{K_r}c_o\right)\exp[-(B+1)K_r t/B] \quad (15a)$$

and

$$-J(t) = \frac{B}{B+1}(K_f c_o - K_r n_{2i})\exp[-(B+1)K_r t/B] \quad (15b)$$

At steady state, the number density of adsorbed particles is $K_f c_o/K_r$, regardless of the initial density n_{2i}. Thus, if $n_{2i} < K_f c_o/K_r$, a net adsorption will take place, whereas if $n_{2i} > K_f c_o/Kr$, a net desorption occurs. From Equation (13), K_f/K_r depends primarily on the depth ϕ_{mn2} of the primary minimum which in turn is influenced significantly by the ionic strength (Figure 3).

Transport in Creeping Flow over Cylinders and Spheres

Because the concentration must depend on both angular and radial positions, the exact analysis for cylinders and spheres is considerably more complex than for the rotating disk. However, rigorous calculations of the rate in these geometries with $K_r = 0$ (Ruckenstein and Prieve, 1973) revealed that a dimensionless parameter B always can be defined which uniquely determines the dimensionless area-average flux and, further, that the dependence of this flux on B is nearly identical ($\pm 20\%$) for all geometries examined. Extrapolating this experience to $K_r \neq 0$ one expects, as a first approximation, that Equation (15) may also be applied to cylinder and sphere geometries, provided that n_2 and J are area-average quantities and B is evaluated as $\alpha D(\infty)/K_f\Gamma(4/3)$, where α^3 is $U/2a^2D(\infty)$ or $2\beta U/3a^2D(\infty)$ for a sphere or a cylinder, respectively.

RESULTS AND DISCUSSION

Evaluation of Apparent Rate Constants

Figure 5 shows the dependence of the apparent adsorption rate constants upon particle radius and ionic strength for two extreme cases: surface potential or charge density held fixed as the ionic strength is changed, where the ψs, σ', and n are related by Equation (5). These results were obtained by using trial-and-error procedures on a digital computer to locate the three extrema in Equation (4) for each set of conditions. Then the necessary functions were evaluated and substituted into Equations (13).

One feature of Figure 5 is the enormous variation in the rate constants (tens of orders of magnitude) resulting from relatively small changes in particle size or ionic strength. Measurements of deposition rates, under condi-

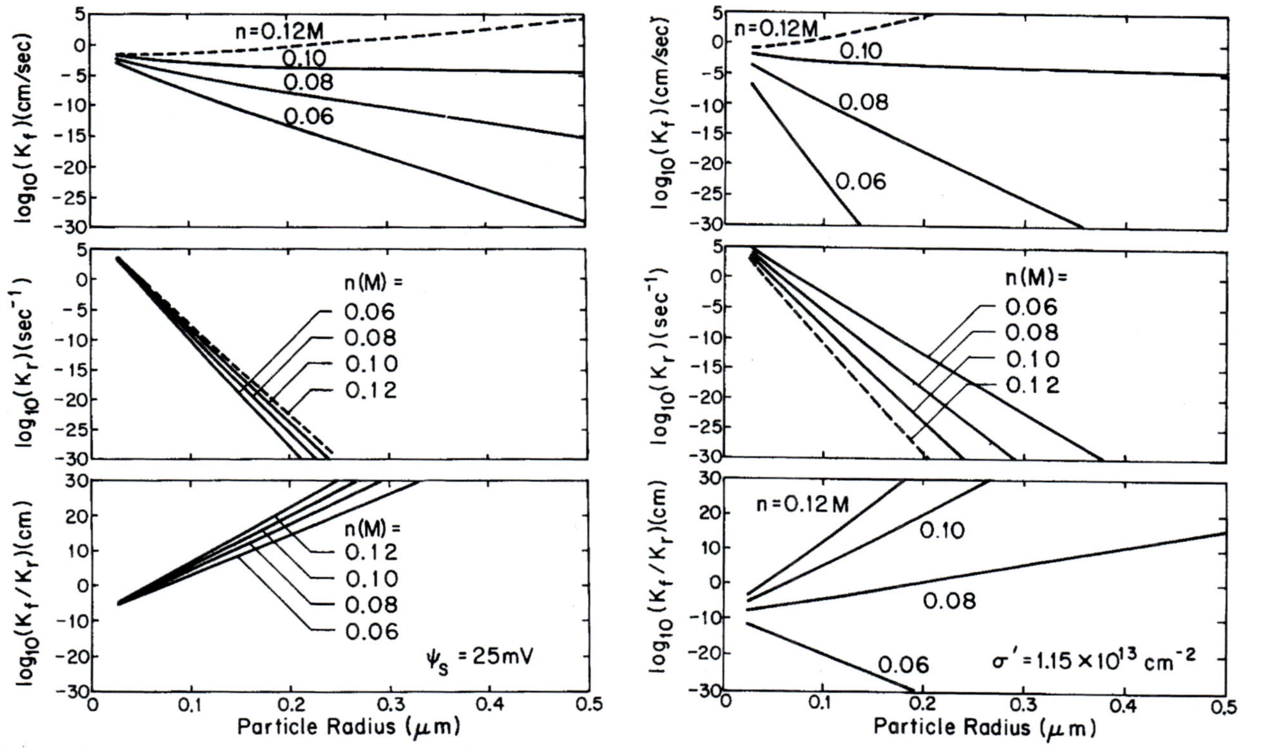

Fig. 5. Effect of particle radius and ionic strength upon the adsorption/desorption rate constants and the collector capacity when (a) surface potential is 25 mV at all ionic strengths or (b) surface charge density is the same at all ionic strengths. Charge density in (b) is chosen as in Figure 3b, while other parameters have the same values as in Figure 1.

tions where the interaction forces are rate limiting (rates then reflect K_f) by Clint et al. (1973) and Hull and Kitchener (1969) indeed do show great sensitivity to ionic strength and in the direction indicated but not as great as inferred from Figure 5. Oversensitivity in the predictions undoubtedly results from inaccuracies in the estimation of the interaction energy and its dependence upon ionic strength. Small errors in the interaction energy are greatly amplified by the exponentiation in Equations (13). Available expressions for the energy are not sufficiently accurate for quantitative rate predictions. However, they can be used, as in the present paper, to make qualitative predictions of the rate. Furthermore, this sensitivity may be exploited to study the nature of the total particle-surface interaction by measuring the rates of deposition and equilibrium surface coverage.

Because the rate constants have the Arrhenius form and the apparent activation energies are nearly proportional to particle radius (Figure 2*b*), the semilog plots of K_f and K_r against a_p are nearly linear, as shown in Figure 5. The slope of the K_f lines is negative at low ionic strengths, where $\phi_{\max}$ is positive, and positive at high ionic strength, where $\phi_{\max}$ is negative (Figure 3). Since $\phi_{\max} - \phi_{mn2}$ must be positive, the slope of the K_r lines is always negative. Similarly, the sign of the slope of the K_f/K_r lines is determined by the sign of ϕ_{mn2}. Ionic strength has a more profound effect on the rate when the surface charge is held fixed rather than the surface potential. This follows directly from the effect of ionic strength on the interaction energy.

Implications for Particle Removal

If a particular type of particle is to be removed from a suspension by adsorption onto the grains of a packed bed, then clearly the maximum rate will be achieved by choosing grains having a charge opposite to the particles so that no energy barrier exists. However, such grains may saturate quickly and be difficult to regenerate. Regeneration may be easier if the grains have the same charge as the particles. By adsorbing at high ionic strengths, Figure 5 shows that the rate of adsorption (K_f), as well as the grain's capacity ($n_{2f}/c_o = K_f/K_r$) to hold adsorbed particles, will be most favorable. When grains and particles have the same charge, regeneration of the saturated grains may be possible by purging the bed with a particle free solution having low ionic strength, since this corresponds to the maximum rate of desorption (K_r) and the minimum grain capacity (K_f/K_r). However, even at the lowest ionic strength, the desorption rate for a given system may be insufficient for practical regeneration. One solution might be to choose a different grain material to give the system a smaller Hamaker constant or larger collision diameter. A second anticipated problem is the sintering which may occur between the particles and grains. Such strong bonds would prevent desorption. Possibly, this may be averted by strongly adsorbing certain molecules on the grains which prevent sintering but still allow the particle to get close enough to physically adsorb.

Implications for Particle Separation

Suppose that two or more different types of particles are dispersed in water to form a mixed suspension. A means to separate one component of the mixture from the others is to selectively adsorb that component on the grains of a packed bed, while the remaining components pass through. Selectivity of physical adsorption results from differences in particle size, shape, or chemical constitution where the latter will affect the surface potential, Hamaker's constant, and collision diameter. Variation in the adsorption rate constant K_f with particle size can be seen in Figure 5. From these curves it can be concluded that selectivity, based on size, improves as the ionic strength becomes different from the value where $\phi_{\max}$ is changing sign (0.1M for the system of Figure 5). By

TABLE 2. CHROMATOGRAPHIC TECHNIQUES FOR SEPARATION OF SUSPENDED PARTICLES

Technique	Mechanism	Basis of separation
Permeation chromatography	Movement of particles into and out of pores in packing grains	Particle size
Hydrodynamic chromatography	Movement of particles into and out of regions having low fluid velocity external to the grains	Particle size
Potential-barrier chromatography	Brownian movement of particles over a potential-energy barrier to become physically attached to or detached from the surface of grains	Particle size, particle constitution (surface charge density, collision diameter of surface molecules, Hamaker constant)

decreasing the ionic strength below 0.1M, the separation factor ($d \ln K_f/d\,a_p$) is enhanced at the cost of decreasing adsorption rates and grain capacity. Although both the separation factor and K_f are enhanced above 0.1M, no energy barrier exists. Then, convective diffusion can be expected to control the rate, making the adsorption less selective. By operating at lower ionic strengths, the rate becomes controlled by the resistance to transport through the interaction boundary layer, which can be expected to be much more sensitive to particle size than the resistance of the diffusion boundary layer (decaying exponential vs. cube root).

Chromatographic Separation of Suspended Particles

If the desorption rate is sufficiently large, chromatographic separation and analyses of mixed suspensions become possible. Dependence of adsorption and desorption rates upon particle properties causes some components of a pulse of mixed particles carried through a packed bed to be eluted more quickly than others. This mechanism is analogous to conventional gas chromatography, and all the formalisms developed for the latter may be applied.

Gel and hydrodynamic chromatography (Determan, 1968; DeMarzio and Guttman, 1969) are two other methods for particle size analysis. They differ from the method proposed above in the mechanism by which particles are retained in the bed. When the packing contains pores which are permeable to some but not all of the particles, those particles which are small enough to enter the pores through Brownian motion will require a larger average residence time. This is the mechanism of gel (permeation) chromatography. Hydrodynamic chromatography results from the nonuniformity in the velocity profile between grains; pores are not required. Because the center of smaller particles is able to move closer to the surface of grains, these particles have a lower average velocity.

Physical interactions between particles and grains can greatly affect, and sometimes overwhelm, hydrodynamic chromatography. For example, if particles and grains bear like charges, so that a strong repulsive force develops, none of the particles will be able to get as near to the wall as in the absence of the force. Consequently, all particles move more rapidly through the bed. This effect has been observed at low ionic strengths by Small (1974). Conversely, strong attractive forces may cause rapid physical adsorption of the particles, whereupon the residence time becomes controlled by the time between adsorption and desorption. Then hydrodynamics are much less important. Finally, when both attractive and repulsive forces act, adsorption still occurs, but a surmountable energy barrier now can develop which will control the adsorption and desorption rates. Separation under these conditions is defined here as potential-barrier chromatography (see Table 2). At sufficiently high ionic strengths, where the energy barrier is surmountable, Figure 4 shows that the range of A, ψ_{S1}, and ψ_{S2} corresponding to the existence of a maximum is narrow. However, if a surmountable barrier can be achieved, the rates of adsorption and desorption will be very sensitive to particle size (see Figure 5). Sensitivity to the surface charge density and Hamaker constant likewise could be demonstrated. Consequently, the selectivity of the new proposed potential-barrier chromatography is anticipated to be substantially better than when rapid adsorption occurs in the absence of an energy barrier (ordinary diffusion controls rate) or when no adsorption takes place owing to an insurmountable barrier (hydrodynamic chromatography). Furthermore, separation of particles having the same size, but different chemical constitution, can also be performed by potential-barrier chromatography, due to the sensitivity to the charge on the particle's surface, its collision diameter and Hamaker constant. Finally, shape selectivity may also be possible, based on differences in the interaction between a plane collector and disks or spheres of the same volume and chemical constitution.

Applications to Other Problems

By including Born repulsion in the calculation of the interaction energy profile, the primary minimum ϕ_{mn2} is finite and depends on ionic strength. Allowing the primary to be above the secondary minimum ϕ_{mn1}, one is able to explain the large equilibrium separation (about 90 Å) which is known to occur between biological cells. Without short-range repulsion, the primary energy well becomes infinitely deep, and h_{mn2} (<10 Å) is the only stable equilibrium separation.

Hamaker (1937*a*) has suggested that short-range repulsion is responsible for repeptization (reverse of flocculation). By repeating the analysis of this paper for interactions between unequal spheres, one could calculate both flocculation and repeptization rates and, in principal, thereby predict the distribution of floc sizes, including the limiting equilibrium distribution.

NOTATION

a = radius of collector, cm
a_p = particle radius, cm
A = Hamaker's constant, erg
B = $\alpha D(\infty)/K_f\Gamma(4/3)$
c = local particle concentration, particles/cm^3
c_i = particle concentration at interface between the interaction and diffusion boundary layers (at $h = \delta_o$), particles/cm^3
c_o = particle concentration in bulk of solution, particles/cm^3
D = local diffusion coefficient, cm^2/s
e = protonic charge, statcoul
f = local wall correction factor for D and m
h = minimum separation distance between particle and collector surfaces, cm
H = h/a_p
I_2 = $\int_0^{\delta_2} \exp[-\phi(h)/kT]dh$, cm

J = particle flux, particles/(cm^2-s)
k = Boltzmann constant, erg/°K

$$K_f = \left\{ \int_{\delta_2}^{\delta_1} \frac{\exp[\phi(h)/kT]}{D(h)} dh \right\}^{-1}, \text{ cm/s}$$

K_r = K_f/I_2, s^{-1}
m = local particle mobility, (cm/s)/dyne
M = mass of a single particle, g
n = ionic strength of bulk solution, ions/cm^3*
n_2 = number of adsorbed particles, particles/cm^2
n_{2i} = initial value of n_2, particles/cm^2
n_{2f} = final value of n_2, particles/cm^2
T = absolute temperature, °K
U = approach velocity, cm/s
y = distance measured normal to collector, cm

Greek Letters

α = $[(0.170/D(\infty))(\omega^3\rho_f/\mu)^{1/2}]^{1/3}$, cm^{-1}
β = $1 - 0.5 \ln(2aU\rho_f/\mu)$
γ = $|d^2\phi/dh^2|$, erg/cm^2
Γ = Gamma function
δ_o = thickness of interaction boundary layer, cm
δ_1 = smallest h in the interval $h_{max} < h < h_{mn1}$ for which a Boltzmann distribution of particles may be assumed, cm
δ_2 = largest h in the interval $h_{mn2} < h < h_{max}$ for which a Boltzmann distribution of particles may be assumed, cm
ϵ = fluid dielectric constant
κ = $(8\pi ne^2/\epsilon kT)^{1/2}$, cm^{-1}
μ = fluid viscosity, g/(cm-s)
ρ_f = fluid density, g/cm^3
σ = collision diameter, cm
σ' = electrostatic surface charge density, statcoul*
ϕ = total potential energy of interaction, erg
ϕ_6 = van der Waals energy of interaction, erg
ϕ_{12} = Born energy of interaction, erg
ϕ_{DL} = double-layer energy of interaction, erg
ψ_{Si} = electrostatic potential of surface i, statvolt*
ω = disk rotation speed, rad/s

Subscripts

max = evaluated at the maximum of $\phi(h)$
mn1 = evaluated at the secondary minimum of $\phi(h)$
mn2 = evaluated at the primary minimum of $\phi(h)$

LITERATURE CITED

Brenner, H., "The Slow Motion of a Sphere through a Viscous Fluid Towards a Plane Surface," *Chem. Eng. Sci.*, **16**, 242 (1961).

Chandrasekhar, S., "Stochastic Problems in Physics and Astronomy," *Rev. Mod. Physics*, **15**, 1 (1943).

Clint, G. E., J. H. Clint, J. M. Corkill, and T. Walker, "Deposition of Latex Particles onto a Planar Surface," *J. Colloid Interface Sci.*, **44**, 121 (1973).

Determan, H., *Gel Chromatography*, Springer-Verlag, New York (1968).

DiMarzio, E. A., and C. M. Guttman, "Separation by Flow," *J. Polymer Sci.*, **B7**, 267 (1969).

FitzPatrick, J. A., and L. A. Spielman, "Filtration of Aqueous Latex Suspensions through Beds of Glass Spheres," *J. Colloid Interface Sci.*, **43**, 350 (1973).

Hamaker, H. C., "A General Theory of Lyophobic Colloids," *Rec. Trav. Chim.*, **56**, 3 (1937*a*).

———, "The London-van der Waals Attraction between Spherical Particles," *Physica*, **4**, 1058 (1937*b*).

Hogg, H., T. W. Healy, and D. W. Fuerstenau, "Mutual Coagulation of Colloidal Dispersions," *Trans. Faraday Soc.*, **62**, 1638 (1966).

Hull, M., and J. A. Kitchener, "Interaction of Spherical Colloidal Particles with Planar Surfaces," *ibid.*, **65**, 3093 (1969).

Israelachvili, J. N., and D. Tabor, "Van der Waals Forces: Theory and Experiment," in *Progress in Surface and Membrane Science*, J. F. Danielli, ed., Vol. 7, p. 1, Academic Press, New York (1973).

Kramers, H. A., "Brownian Motion in a Field of Force and the Diffusional Model of Chemical Reactions," *Physica*, **7**, 284 (1940).

Levich, V. G., *Physicochemical Hydrodynamics*, Prentice-Hall, Englewood Cliffs, N. J. (1962).

Marshall, J. K., and J. A. Kitchener, "The Deposition of Colloidal Particles on Smooth Solids," *J. Colloid Interface Sci.*, **22**, 342 (1966).

Prieve, D. C., and E. Ruckenstein, "The Surface Potential of and Double-Layer Interaction Force between Surfaces Characterized by Multiple Ionizable Groups," *J. Theoretical Biol.*, in press.

Ruckenstein, E., and D. C. Prieve, "The Rate of Deposition of Brownian Particles under the Action of London and Double-Layer Forces," *J.C.S. Faraday II*, **69**, 1522 (1973).

———, "Dynamics of Cell Deposition," *J. Theoretical Biol.*, **51**, 429 (1975).

Small, H., "Hydrodynamic Chromatography: A Technique for Size Analysis of Colloidal Particles," *J. Colloid Interface Sci.*, **48**, 147 (1974).

Spielman, L. A., and S. K. Friedlander, "Role of the Electrical Double Layer in Particle Deposition by Convective Diffusion," *ibid.*, **44**, 22 (1974).

Verwey, E. J. W., and J. Th. G. Overbeek, *Theory of the Stability of Lyophobic Colloids*, Elsevier, Amsterdam (1948).

* A consistent set of units has been specified above for each symbol, although more convenient units are sometimes employed in the text (for example, mV instead of statvolts).

Manuscript received July 8, 1975; revision received November 13 and accepted November 14, 1975.

On Reversible Adsorption of Hydrosols and Repeptization

ELI RUCKENSTEIN

Faculty of Engineering and Applied Sciences
State University of New York at Buffalo
Buffalo, NY 14214

and DENNIS C. PRIEVE

Department of Chemical Engineering
Carnegie-Mellon University
Pittsburgh, PA 15213

In a previous paper (Ruckenstein and Prieve, 1976) the reversible deposition of particles from a moving fluid was treated by lumping the effect of the interaction forces between particles and the deposition surface in a boundary condition to the usual convective-diffusion equation. The boundary condition was obtained by analyzing the flux of particles through a very thin layer (~ 100A) next to the collector's surface. Within this interaction force boundary layer, a quasi steady state was assumed up to a point between the maximum and primary minimum (in the profile of the potential energy of interaction), with quasi equilibrium assumed between that point and the wall (a region in which most of the accumulation is expected to occur). In what follows, this is called approximation A. The goals of this note are: (1) to compare approximation A with approximation B, which assumes quasi steady state over the entire thickness of the interaction force boundary layer, and (2) to extend the analysis to flocculation and repeptization of hydrosols.

In one dimension the equation of continuity for the particles within the thin layer is

$$\frac{\partial c}{\partial t} + \frac{\partial J}{\partial h} = 0$$

where the nomenclature is that of Ruckenstein and Prieve (1976) and J is given by Equation (6) of that same paper. To prevent particles from penetrating the wall at $h = 0$ and to match the concentration at $h = \delta_0$ with that outside the interaction force boundary layer, the following boundary conditions must be imposed:

$$J = 0 \quad \text{at} \quad h = 0$$
$$c = c_i \quad \text{at} \quad h = \delta_0$$

where δ_0 is the thickness of the interaction forces boundary layer. The solution of this unsteady problem would allow the accumulation of particles near the surface to be evaluated. More details will be found in Prieve and Ruckenstein (to be published)

When the interaction forces establish an energy barrier that reduces adsorption and desorption rates significantly, particles around the primary minimum ($0 \leq h \leq \delta_2$) have time to achieve quasi equilibrium before their population changes. Then, assuming a quasi steady state for the remainder of the region, we have (alternative A)

$$J = 0 \quad \text{for} \quad 0 \leq h \leq \delta_2$$
$$\partial J/\partial h = 0 \quad \text{for} \quad \delta_2 < h \leq \delta_0$$

where δ_2 must be chosen strictly less than h_{max}. This position (h_{max}), corresponding to a maximum in the interaction potential, is unstable; therefore, it is not reasonable to assume quasi equilibrium there.

Note that the boundary condition $J = 0$ at $h = 0$ is automatically satisfied. Using the other boundary condition and Equation (6), we previously obtained Equation (11)

$$-J = \frac{c_i \left(\int_0^{\delta_2} e^{-\phi/kT}\, ds \right) - n_2}{\left(\int_0^{\delta_2} e^{-\phi/kT}\, ds \right) \left(\int_{\delta_2}^{\delta_0} D^{-1} e^{\phi/kT}\, ds \right)} \quad \text{(A)}$$

for the net adsorption rate, where n_2 is the number of adsorbed particles per unit area, or

$$n_2 = \int_0^{\delta_2} c\, ds$$

One may be tempted to replace $h = \delta_0$ by $h = \infty$. However, this causes the integrals to diverge (Prieve and Ruckenstein, 1976) because of the way the diffusion coefficient (or particle mobility) tends to its bulk value. Prieve and Ruckenstein (1976) have shown that for a low potential barrier the integral depends on the value of δ_0. Because it is not possible to define a precise value of δ_0, the lumping of the effect of the interaction forces in a boundary condition is meaningless in this case. If the potential barrier is high enough, the integral is practically independent of δ_0;

provided δ_0 is assigned a physically reasonable, finite value. Then the value of the integral will be in agreement with that obtained using the parabolic approximation for the interaction potential around the maximum (Prieve and Ruckenstein, 1976).

If the quasi steady state approximation is imposed for all h, then (alternative B)

$$\frac{\partial J}{\partial h} = 0 \quad \text{for} \quad 0 \leq h \leq \delta_0$$

There is no reason to believe that this approximation is better than the previous one. It represents only another limiting case. The integration of Equation (6) under this assumption leads to a result very similar to Equation (10):

$$-J = (c(h) \exp[\phi(h)/kT])\Big|_h^{\delta_0} \left(\int_h^{\delta_0} \frac{\exp[\phi(s)/kT]}{D(s)} ds \right)^{-1}.$$

Substituting $c(\delta_0) = c_i$, $\phi(\delta_0) = 0$ and the above definition of n_2, we obtain

$$-J = \frac{c_i \left(\int_0^{\delta_2} e^{-\phi/kT} ds \right) - n_2}{\int_0^{\delta_2} e^{-\phi(h)/kT} \int_h^{\delta_0} \frac{e^{\phi(s)/kT}}{D(s)} ds dh} \qquad \text{(B)}$$

which differs from (A) in that the denominator is a double integral rather than the product of two integrals.

Obviously, Equations (A) and (B) are not identical. The difference stems from the assumption that $J = 0$ in (A) for $0 \leq h \leq \delta_2$ compared to the assumption $\partial J/\partial h = 0$ in (B) for the same range. But consider the case where $\phi(h)$ possesses a maximum ϕ_{max} and a primary minimum ϕ_{mn2} (a secondary minimum may also exist but is not relevant to the current discussion). Also suppose that $\phi_{max} - \phi_{mn2}$ and ϕ_{max} are large compared to kT. (The latter is a necessary condition for the lumping of the effect of the interaction forces into a boundary condition to the convective-diffusion equation and is implicitly assumed in our development.) Then the integrand in the denominator of (B), $e^{-\phi(h)/kT} \int_h^{\delta_0} D^{-1} e^{\phi(s)/kT} ds$, will take on its largest value near the primary minimum ($h = h_{mn2}$) and will decay rapidly as $|h - h_{mn2}|$ becomes non zero. This suggests that the variable lower limit h can be replaced by the constant h_{mn2} without significantly changing the value of the double integral. Now the integral can be factored into the product of two integrals. Since the integrand $D^{-1} e^{\phi/kT}$ assumes its maximum value at $h = h_{max}$ where $\phi = \phi_{max}$ and decays rapidly as $|h - h_{max}|$ differs from zero, we can compress the limits of integration from (h_{mn2}, δ_0) to (δ_2, δ_0) where $h_{mn2} < \delta_2 < h_{max} < \delta_0$. Under these conditions (A) and (B) are the same.

In the preceding discussion, a particle is said to be adsorbed when it is located near the primary minimum because it is here that the largest rate of accumulation occurs. Mathematically, this region was defined as the interval $0 \leq h \leq \delta_2$, where δ_2 can reasonably be assumed somewhere in the interval $h_{mn2} < \delta_2 < h_{max}$. Particles near the secondary minimum (if one exists) are not counted as adsorbed because generally the number of particles in its vicinity is not changing much with time compared to those near the primary minimum. For some applications, a definition of adsorbed particles which includes the secondary minimum may be more appropriate.

A similar analysis can also be performed in which flocculation is considered reversible. One must assume of course that elementary particles forming the floc maintain their individuality; thus they have not sintered together to form a single mass.

Suppose we have an electrolyte solution initially containing spherical flocs of radius a_f, each composed of many elementary particles of radius a, so that $a_f >> a$. Further, suppose that at $t = 0$ the electrolyte solution is suddenly diluted. The lowering of the ionic strength will increase the double layer repulsion between particles in the floc and thereby tend to disrupt it. Of course, some time is required for ions within the interior of the floc to diffuse out; this time is on the order of $t_e \sim a_f^2/D_e$, where D_e is the diffusion coefficient of the electrolyte. Once the ions have had time to diffuse out, the floc could disintegrate into two or more fragments of any size smaller than a_f. Considering the potential energy of interaction between two ideally spherical fragments of radii a_1 and a_2, where $a_1^3 + a_2^3 = a_f^3$, one finds that an energy barrier, roughly proportional to $a_1 a_2/(a_1 + a_2)$ opposes separation of the two fragments. This barrier to separation is smallest when $a_1 << a_2$; hence the detachment of elementary particles from the outer surface of the floc is the most probable event following a lowering of the ionic strength. This would also be true for $t < t_e$ since the exterior of the floc is exposed to lower ionic strength before the interior, and thus the particles of the exterior first experience the increased double layer repulsion.

As an illustration of how the method of Ruckenstein and Prieve can be extended to repeptization of flocculated sols, assume that the detachment of elementary particles is the only possible event following dilution of the electrolyte and further assume, for the sake of simplicity, that free elementary particles do not flocculate, except to reassociate with the original floc. Let $J(r)$ denote the flux of elementary particle centers through the imaginary spherical surface of radius r, concentric with the floc's center and let $4\pi r^2 c(r)dr$ denote the number of particle centers located between r and $r + dr$, relative to the floc's center. Using approximation B, but taking into account the spherical geometry, one obtains

$$N = 4\pi r^2 J(r) = \frac{n - c_\infty \int_0^{\delta_2} e^{-\phi/kT} dh}{\frac{1}{4\pi} \int_0^{\delta_2} e^{-\phi(h)/kT} \int_h^{\infty} D^{-1} (a + a_f + s)^{-2} e^{\phi(s)/kT} ds dh} = K_r' n - K_f' c_\infty \qquad \text{(C)}$$

where $n = \int_0^{\delta_2} c \, dh$ and c_∞ is the number of free elementary particles per unit volume. Because $a_f >> \delta_2$, the curvature effect was neglected in the definition of n.

When $a_f >> a$, the potential energy of interaction between an elementary particle and a floc is equivalent to the potential energy $\phi(h)$ between a particle and a flat plate; thus $\phi(h)$ used by Ruckenstein and Prieve (1976) still applies. Furthermore, if the conditions necessary for Equation (13) of that paper apply, then the rate constants for reversible flocculation may be calculated as $K_f' = 4\pi a_f^2 K_f$ and $K_r' = 4\pi a_f^2 K_r$, where K_f and K_r are given by Equation (13).

Multiplying Equation (C) by the number of flocs per unit volume, c_f, yields the net rate of increase in free elementary particles c_∞ per unit volume, or:

$$\frac{dc_\infty}{dt} = N c_f = K_r' n c_f - K_f' c_\infty c_f. \qquad \text{(D)}$$

As elementary particles detach from the floc, the radius a_f gradually decreases so that the sum of the number of free elementary particles and the number of elementary parti-

TABLE 1. REPEPTIZATION RATES AND EQUILIBRIA[a]

Ionic strength (mol/l)	$\frac{\phi_{mn2}}{kT}$	$\frac{\phi_{max}}{kT}$	K_r' (cm²/sec)	K_f' (cm³/sec)	τ (sec)	c_a^* (cm⁻³)
0.08	−21.4	2.93	6.1×10^{-10}	2.6×10^{-9}	3.8	6.8×10^{8}
0.07	−18.2	5.26	1.7×10^{-9}	2.8×10^{-10}	36	1.8×10^{10}
0.06	−13.8	8.55	5.1×10^{-9}	1.1×10^{-11}	900	1.3×10^{12}

[a] Values of parameters used in these calculations include $a = 10^{-5}$ cm, $a_f = 10^{-4}$ cm, $c_f = 10^{8}$ cm^{-3}, $\delta = 2.5$Å, $\epsilon = 74.3$, $T = 300$°K, $\psi = -10$mV at 0.1 mol/l (recomputed from Gouy-Chapman equation at lower ionic strengths, assuming surface charge is unchanged) and $A = 1.769 \times 10^{-14}$ erg (chosen so that $\phi_{max} = 0$ at 0.1 mol/l). For Notations see Ruckenstein and Prieve (1976).

cles in flocs equals a constant, c_o, per unit volume. Equation (D) can be integrated exactly, taking into account the changes of K_r' and K_f' with time. However, more insight into the problem can be obtained from the simple equation obtained in the limiting case when the amount of dilution of the electrolyte is small so that nearly all the elementary particles remain in the flocs. Then K_r' and K_f', which are proportional to a_f^2, remain nearly constant. Interpreting n as the number of elementary particles attached per unit area to the floc and assuming hexagonal close packing, then $n = (\sqrt{3})/6a^2$. As particles detach from the floc, new particles underneath are exposed so that n is practically independent of time. Then (D) integrates to yield

$$(c_a^* - c_a(t)) = (c_a^* - c_{ai})e^{-t/\tau}$$

where $c_a^* = K_r' n/K_f'$, $\tau = 1/K_f c_f$ and $c_{ai} = c_a(t = 0)$.

Table 1 shows some values of the time constant τ and the equilibrium concentration c_a^* obtained from these equations, employing the simplified model of the interaction forces developed by Ruckenstein and Prieve (1976). As might be expected, lowering the ionic strength makes the primary minimum, ϕ_{mn2}, less deep, thus shifting the equilibrium in favor of more detached elementary particles, c_a^*. However, decreasing the ionic strength increases the time τ to reach equilibrium, in spite of the fact that the repeptization rate, $K_r'n$, is increasing. This can be rationalized by observing that both the repeptization rate $K_r'n$ and the equilibrium concentration c_a^* are increased by dilution, but c_a^* increases faster so that a longer time τ is needed to reach equilibrium. Of course, the analysis here could be generalized to allow the recombination of singlets to form doublets, the combination of singlets and doublets to form triplets and quadruplets, etc. Finally, an equilibrium distribution of floc sizes should be obtainable.

In this note, an equation is established for the rate of reversible adsorption on the basis of the quasi steady state assumption over the entire thickness of the interaction force boundary layer. When an interaction force boundary layer occurs ($\phi_{max} >> kT$), we find that the approximation of quasi steady state over part of the region and quasi-equilibrium over the other part leads to practically the same rate as the approximation of quasi steady state over the entire region if $\phi_{max} - \phi_{mn2} >> kT$. Hence, either approximation is equally appropriate. A simple analysis of the repeptization of colloids is discussed on the basis of the equation established for reversible adsorption.

LITERATURE CITED

Prieve, D. C., and Eli Ruckenstein, "Rate of Deposition of Brownian Particles Calculated by Lumping Interaction Forces into a Boundary Condition," *J. Colloid Interface Sci.*, 1976 (in press).

———, "Kinetics of Repeptization" (to be published).

Ruckenstein, Eli, and D. C. Prieve, "Adsorption and Desorption of Particles and Their Chromatographic Separation," *AIChE J.*, 22, 276 (1976).

Manuscript received May 24, and accepted May 26, 1976.

Effect of London Forces upon the Rate of Deposition of Brownian Particles

Rates of deposition of particles onto a spherical collector, immersed in a creeping fluid, are calculated as a function of Peclet number, aspect ratio, and the ratio of Hamaker's constant to kT for all cases of practical importance involving convective-diffusion and London forces. Boundaries of regions where London forces or Brownian motion may be neglected are established in terms of the above dimensionless groups. Gravitational forces were included in the calculations for some representative conditions but may be neglected for small or neutrally buoyant particles or for high flow rates.

Close agreement was noted between deposition rates calculated by summing individual contributions from each transport mechanism and deposition rates calculated from a rigorous consideration of their combined effect. Thus the additivity rule was verified for the cases studied.

DENNIS C. PRIEVE

Department of Chemical Engineering
University of Delaware
Newark, Delaware 19711

and

ELI RUCKENSTEIN

Faculty of Engineering and Applied Science
State University of New York at Buffalo
Buffalo, New York 14214

SCOPE

Deep-bed granular filters are used to clarify fluids containing a small amount of fine particulate matter. Spaces between grains of the bed are generally quite large compared with the size of the suspended particles so that the bed does not act like a sieve. Instead the bed provides a surface on which the particles may stick by the action of London (van der Waals) forces. Besides holding the particles to the collector surface, London forces may act over a sufficiently long range to aid in the transport of particles to the surface.

The effectiveness of deep-bed filters in removing suspended particles is measured by the value of the filter coefficient which in turn is related to the capture efficiency of a single characteristic grain of the bed. Capture efficiencies are evaluated in the present paper for all cases of practical importance in which London forces and convective-diffusion serve to transport particles to the surface of a spherical collector immersed in a creeping flow field. Gravitational forces are considered in some cases, but the general results apply mainly to submicron or neutrally buoyant particles suspended in a viscous fluid such as water. Results obtained by linearly superimposing the individual contribution from each mechanism are compared with those obtained from a rigorous coupling of mechanisms in order to test the additivity rule.

Suspended particles are considered to have finite size; thus both the mobility coefficient and diffusion coefficient of the particles depend not only on the size of the particle but also upon the distance between the particle and the collector. A numerical finite-difference technique is used to solve the general transport equation.

Several limiting cases are discussed in order to obtain insight and perspective. To identify these limiting cases, some quantity is sought for each mechanism which will characterize its relative importance.

The objectives of the present paper are to: (1) compute the rate of deposition of particles onto a spherical collector in a creeping flow field for all situations in which London forces and convective-diffusion act as transport mechanisms, (2) identify limiting behaviors according to the relative values of the characteristic parameters for each mechanism, (3) establish the physical conditions in which each of the limiting cases is valid, and (4) test the accuracy of the additivity rule.

CONCLUSIONS AND SIGNIFICANCE

The capture efficiency or Sherwood number was shown to be a function of three dimensionless groups—the Peclet number, the aspect ratio (collector radius divided by particle radius), and the ratio of Hamaker's constant (indicating the intensity of London forces) to the thermal energy of the particles. Calculated values for the rate of deposition, expressed as the Sherwood number, are plotted in Figure 6 as a function of the three dimensionless groups.

The distance from the collector surface over which each mechanism or effect exerts an influence was found to be a useful characteristic parameter. Table 1 lists the five lengths used to identify each of the limiting cases summarized by Table 2. Three principal limiting cases were considered: (1) situations where London forces are negligible, (2) situations in which diffusion is negligible, and (3) situations where the fluid motion is negligible. Several subcases appear for various orderings of other lengths. Regions of applicability for each of the limiting cases, along with the corresponding rate, are outlined in Table 3.

For those cases in which gravitational forces are also included, values of the Sherwood number are summarized in Figure 8. The ratio G of the gravitational potential of a particle, located one particle radius above the collector, to the thermal energy was used to characterize the effect of gravity. For an aspect ratio of 100, gravitational forces were found to be negligible when $G < 10^{-2}$.

Superimposing the individual contribution of each mechanism was found, in those cases tested, to yield very nearly the same overall deposition rate as obtained by rigorous consideration of coupling.

Correspondence concerning this paper should be addressed to E. Ruckenstein. D. C. Prieve is with the Department of Chemical Engineering, Carnegie-Mellon University, Pittsburgh, Pennsylvania 15213.

Because colloidal particles have finite size, their mobility and diffusion coefficients depend not only upon their size but also upon the distance from the collector surface. This variation with distance stems from friction between the collector surface and the fluid which increases the force required to push the fluid out of the path of the approaching particle. In the usual transport equation containing only convective and diffusive terms, the size of the molecules is small enough for the thickness δ_m to be small compared to the length δ_D, where δ_m and δ_D are explained in Table 1. Other situations arise in which these conditions are not met, or in which London or gravitational forces are important. To identify the limiting cases, it is useful to seek some quantity for each mechanism which allows the ordering of its relative importance.

For the case of purely attractive forces (such as London-van der Waals forces) the length δ_F over which they act is a useful characteristic. An attractive force which acts over a distance which is much less than δ_D will not contribute substantially to the overall rate. When repulsive forces (such as the electrostatic double-layer forces) are also present, they may effectively prevent particles from arriving at the collector, even when they act only over a very short distance. For this reason the decay length alone cannot characterize the relative importance of the joint effect of attractive and repulsive forces. Useful characteristics of their combined effect may be obtained by considering the total potential energy of interaction between the particle and the collector.

When repulsive forces are acting together with attractive forces, the parameter which best characterizes the effect of the repulsive force is the height E of the energy barrier defined in Figure 1. If E is large enough, the particles at $h_{\min}$ will experience difficulty in overcoming the barrier. Should the thickness of the region in which London and double-layer forces contribute substantially to the flux also be small compared to δ_D, then the effect of the interaction forces can be lumped into a boundary condition of the transport equation which takes the form of an apparent, first-order, chemical reaction (Ruckenstein and Prieve, 1973; Spielman and Friedlander, 1974). On the other hand, should the barrier be small enough, the repulsive forces will have little effect.

The present paper deals with suspensions in nonpolar solvents or in aqueous solutions of moderate or high ionic strength, where double layer forces are negligible. Gravitational forces are considered in a number of representative cases; however, the emphasis is upon the mechanisms of convective-diffusion and London forces. Several limiting cases occur which can be classified by an ordering of the characteristic lengths summarized by Table 1. Three principal limiting behaviors may be identified: (1) situations where London forces are negligible, (2) situations where diffusion is negligible, and (3) situations where fluid motion is negligible. Table 2 identifies each case and lists the corresponding investigator and technique employed. Several subcases appear due to various orderings of the other lengths.

THEORETICAL DEVELOPMENT AND SOLUTION OF TRANSPORT EQUATION

When an isolated sphere is held stationary in a creeping flow field containing suspended particles (see Figure 2), the equation of continuity in spherical coordinates, allowing azimuthal symmetry, is given by

$$\frac{1}{r^2}\frac{\partial}{\partial r}(r^2 N_r) + \frac{1}{r\sin\theta}\frac{\partial}{\partial\theta}(N_\theta \sin\theta) = 0 \quad (1)$$

TABLE 1. CHARACTERISTIC LENGTHS

Length	Definition
δ_F	Length over which London forces contribute substantially to the radial particle flux.
δ_D	Length over which convective-diffusion contributes substantially to the radial particle flux.
δ_m	Length over which the mobility and diffusion coefficients vary appreciably with radial distance.
a	Collector radius.
a_p	Particle radius.

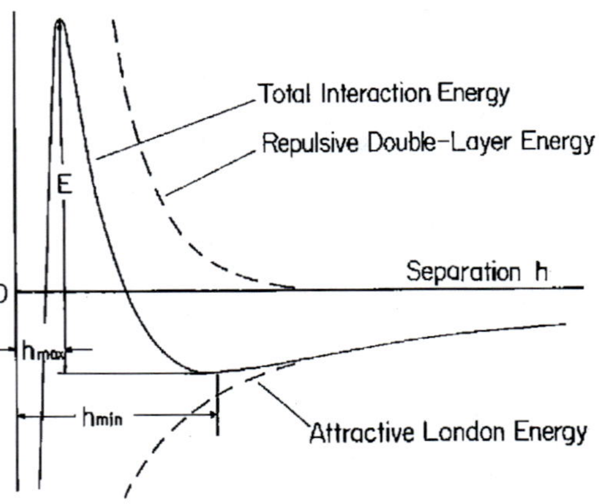

Fig. 1. Profile of the total energy of interaction between a colloidal particle and the collector surface arising from the double-layer repulsion and London attraction.

The radial and tangential components of the total particle flux vector, taken to be positive in the direction of increasing r or increasing θ, can be written as

$$N_r = v_r c - D\frac{\partial c}{\partial r}, \text{ and} \quad (2a)$$

$$N_\theta = v_\theta c, \quad (2b)$$

where the first term (proportional to c) represents the particle flux resulting from London forces and fluid motion; thus v_r and v_θ may be interpreted as the non-Brownian particle velocities. Except far from the collector, these are not equivalent to the corresponding components of the fluid velocity since particle trajectories do not coincide with fluid streamlines even in the absence of Brownian motion or interaction forces.

Non-Brownian Particle Velocities

By performing a radial force balance, Spielman and Fitzpatrick (1973) determined the radial velocity of a particle attracted to a spherical collector by London forces when particle inertia and Brownian motion are negligible:

$$\underbrace{\frac{v_r(h,\theta)}{m(h/a_p)}}_{\text{viscous drag}} = \underbrace{\frac{s_r(h,\theta)\, f_2(h/a_p)}{m_\infty}}_{\text{fluid motion force}} - \underbrace{\frac{2}{3}\frac{A a_p^3}{(h+2a_p)^2 h^2}}_{\text{London force}} \quad (3)$$

Hamaker's (1937) expression for the unretarded London force was employed, whereas the expression for the force exerted by fluid motion was taken from the results of

Table 2. Summary of Limiting Cases

Case	Subcase	Ordering of lengths	References	Comments
1		a_p and $\delta_F << \delta_D$		Negligible interaction forces. Continuum mechanics apply.
	a	$a_p, \delta_F, \delta_m << \delta_D << a$	Levich (1962) Lighthill (1950)	Diffusion boundary-layer approximation is valid. D is independent of position. Analytical solution possible.
	b	$a_p, \delta_F, \delta_m << \delta_D \sim a$	Yuge (1956)	D is independent of position. Transition between subcases *a* and *c*.
	c	$a_p, \delta_F, \delta_m, a << \delta_D$		Fluid considered stagnant. D is independent of position. Analytical solution possible.
	d	δ_F and $a_p << \delta_D$ and a	This paper	D depends on distance from surface. Numerical solution for rate.
2		$\delta_D << \delta_F$		Diffusion is negligible.
	a	$\delta_D << \delta_F << a$	Spielman and FitzPatrick (1973)	Trajectory analysis used. Approximate velocity profile, only valid near collector, is used.
	b	δ_D and $\delta_F << a_p$	Fuchs (1964)	Particle trajectories assumed to coincide with fluid streamlines. Capture by Interception. Analytic solution possible.
	c	$\delta_D << \delta_F$ and a	This paper	Exact fluid velocity profile used.
3		a_p and $a << \delta_D$	This paper	Fluid considered stagnant. Analytic solution possible.

Goren and O'Neill (1971), who calculated the normal force exerted by creeping flow over a sphere near a plane surface. Although the flow field in the present problem is somewhat more complex than that considered by Goren and O'Neill, the curvature of the plane surface as well as a tangential velocity component have only secondary effects upon the normal force, so their result is reasonably applicable here.

Since tangential diffusion has been neglected, only fluid motion contributes to the tangential particle velocity. Near to the collector the fluid appears to be experiencing uniform, linear, shear flow. Goldman et al. (1967b) studied the viscous motion of a sphere in Couette flow and found the particle velocity to be proportional to, but less than, the undisturbed fluid velocity at the particle's center:

$$v_\theta(h, \theta) = s_\theta(h, \theta)\, f_3(h/a_p) \tag{4}$$

In situations where separations comparable to the collector radius ($h \gtrsim a$) need to be considered, the particle no longer experiences Couette flow. However, Equation (4) correctly predicts that v_θ approaches s_θ at large separations, whereas at small separations, where the distinction between v_θ and s_θ is most important, the conditions of its derivation are practically satisfied.

Particle inertia has been assumed to be substantially smaller than the viscous drag force. This assumption will be valid provided the Reynolds number $\rho_p a_p \beta U/\mu$ is sufficiently small, which is more likely to be the case with liquids than gases.

Position-Dependent Mobility and Diffusion Coefficient

Suppose a spherical particle is moving under the action of a constant force toward a solid surface having a radius of curvature much larger than the particle. Friction between the fluid and the particle creates a drag force which opposes motion. As the particle approaches the plane solid surface, its velocity decreases as a result of an increase in the drag caused by additional friction between the fluid and the wall. This decrease in particle velocity under a constant applied force implies a decrease in particle mobility (velocity per unit applied force). When the particle Reynolds number is small, the mobility can be calculated using a modified form of Stokes's equation:

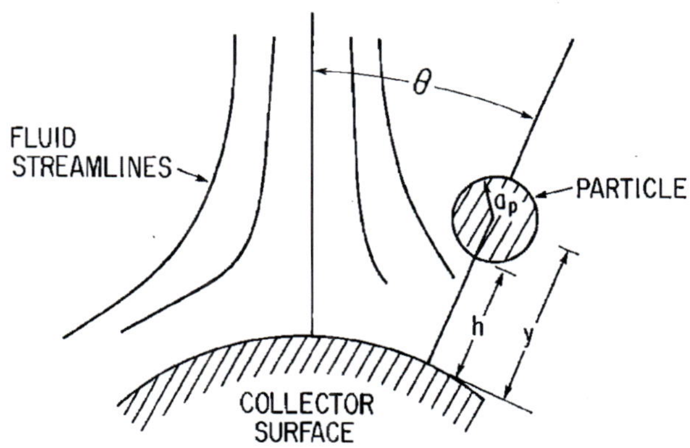

Fig. 2. Colloidal particle suspended in a fluid creeping over the surface of a spherical collector.

$$m = \frac{f_1\,(h/a_p)}{6\pi\mu a_p} \tag{5}$$

where the factor f_1 was calculated by Brenner (1961) to correct for the wall effect by considering the motion of a sphere through a quiescent fluid near a plane surface. At small separations ($h < a_p$) where f_1 becomes substantially less than unity, the fluid may be considered essentially quiescent.

According to the Nernst-Einstein equation ($D = mkT$) the mobility and diffusion coefficients are proportional. Then, using Equation (5), the diffusion coefficient of the particles can be expressed as

$$D = D_\infty f_1\,(h/a_p) \tag{6a}$$

where

$$D_\infty = kT/6\pi \mu a_p \tag{6b}$$

Brenner's correction factor $f_1(H)$ tends to H when $H << 1$ and to $(H - 1/8)/(H + 1)$ when $H >> 1$.

Undisturbed Fluid Velocities

When the flow over the collector is sufficiently slow ($\rho a \beta U/\mu < 1$), Stokes's expressions for the undisturbed fluid velocity can be used:

$$s_r = -\beta U\left[1 - \frac{3}{2}\left(\frac{a}{r}\right) + \frac{1}{2}\left(\frac{a}{r}\right)^3\right]\cos\theta \tag{7a}$$

and

$$s_\theta = \beta U\left[1 - \frac{3}{4}\left(\frac{a}{r}\right) - \frac{1}{4}\left(\frac{a}{r}\right)^3\right]\sin\theta \tag{7b}$$

For an isolated sphere β is unity and U is the approach velocity; for creeping flow around one grain of a packed bed, β is a semiempirical factor introduced to account for the effects of the bed porosity and U is the superficial velocity. Happel's (1958) model may be used (Spielman and Fitzpatrick, 1973) to estimate β. However, Happel's model assumes that flow occurs only in a narrow annulus near the surface of the grain; integration of the transport equation sometimes involves a larger region. A more adequate model does not presently exist.

Transport Equation and Its Boundary Conditions

Combining the continuity equation and flux expressions together with particle and fluid velocities, then writing the result in terms of dimensionless variables, yields the transport equation:

$$\frac{2R}{Pe} f_1(H)\frac{\partial^2 C}{\partial H^2} + \left[\frac{2R}{Pe}\left(\frac{df_1}{dH} + \frac{2f_1}{R+Y}\right) - v\right]\frac{\partial C}{\partial H}$$

$$-\left[\frac{2}{R+Y}\left(v + \frac{\partial u}{\partial \theta}\right) + \frac{\partial v}{\partial H}\right]C = \frac{u}{R+Y}\frac{\partial C}{\partial \theta} \tag{8}$$

with

$$u = \left[\frac{3}{2}R^2 + Y\left(\frac{9}{4}R + Y\right)\right]\frac{Yf_3(H)\sin\theta}{(R+Y)^3} \tag{9a}$$

and

$$v = -Y^2\left(\frac{3}{2}R + Y\right)\frac{f_1(H)\,f_2(H)\cos\theta}{(R+Y)^3}$$

$$-\frac{4}{3}\left(\frac{A}{kT}\right)\left(\frac{R}{Pe}\right)\frac{f_1(H)}{(Y^2-1)^2} \tag{9b}$$

Mobility and diffusion coefficients were evaluated from Equations (5) and (6). The associated boundary conditions are

$$C(H = 0, \theta) = 0, \tag{10a}$$

$$C(H = \infty, \theta) = 1, \tag{10b}$$

and

$$\left.\frac{\partial C}{\partial \theta}\right|_{\theta=0} = 0 \tag{10c}$$

The concentration of particles in contact with the collector is taken to be zero because these particles are no longer part of the disperse phase. At large distances from the collector the particle concentration must tend to that of the approaching fluid. The third boundary condition arises from the symmetry about the forward stagnation path ($\theta = 0$).

Numerical Solution of Transport Equation

The infinite radial domain ($0 \leq H < \infty$) was compressed into a finite domain ($0 \leq \eta \leq 1$) by use of the following transformation:

$$H = b \ln(1 - \eta) \tag{11a}$$

which is similar to a transformation employed by Vrentas et al. (1966). With condition (10c), the transport equation along $\theta = 0$ is an ordinary boundary value problem, for which finite differences with a uniform η-step could yield a solution. For $\theta > 0$ the technique of Crank and Nicolson (1947) was used.

When the Peclet number equals about R^3, or when $Ad < 1$ and Brownian motion is negligible, the radial concentration profile displays a maximum value greater than unity. Then a generalized form of transformation (11a)

$$H = b_1 \ln(1 - \eta) + b_2 \ln(1 - \eta^n) \tag{11b}$$

gave additional flexibility to distribute grid points of the H-domain both inside and outside the point corresponding to the maximum concentration.

Evaluation of the Deposition Rate

Once the concentration profile is determined that satisfies the transport equation, the radial flux can be evaluated from Equation (2a) and integrated over the collector surface to obtain the overall rate of deposition. When particles come in contact with the collector, the diffusion coefficient vanishes, leaving the radial particle flux equal to $v_r c$; however, the product $v_r c$ is indeterminant at $r = a$ ($H = 0$), since the particle concentration must vanish [Equation (10a)], while London forces are accelerating the particles to nearly infinite velocity at the moment of contact. One means of avoiding this indeterminancy is to integrate the flux over some surface other than $r = a$. From the Gauss Divergence Theorem and the continuity equation, the integral of the radial particle flux over any spherical surface concentric with the collector will have the same value regardless of the radius of the surface. Thus the following expression for the overall rate will be independent of r:

$$I = 2\pi r^2 \int_0^\pi - N_r(r, \theta)\sin\theta\, d\theta \tag{12}$$

In addition to avoiding the indeterminancy at $r = a$, the observation that this expression is independent of r provides a test of the accuracy of the numerical technique used to evaluate C.

Two dimensionless forms of the rate are useful. The ratio of the actual deposition rate to the rate of convection at large distances upstream from the collector is defined as the capture efficiency:

$$e = \frac{I}{\pi a^2 \beta U c_\infty}$$

$$= 2\left(1 + \frac{Y}{R}\right)^2 \int_0^\pi \left(\frac{2R}{Pe} f_1 \frac{\partial C}{\partial H} - vC\right)\sin\theta\, d\theta \tag{13a}$$

The Sherwood number may be interpreted as twice the ratio of the actual rate to the rate computed when only diffusion is considered:

$$Sh = \frac{2I}{4\pi a D_\infty c_\infty}$$

$$= \frac{1}{2}\left(1 + \frac{Y}{R}\right)^2 \int_0^\pi \left[2Rf_1\frac{\partial C}{\partial H} - (Pe)vC\right]\sin\theta\, d\theta \tag{13b}$$

Eliminating I between these two equations results in the following relation:

$$Sh = \frac{1}{4} e\,(Pe) \tag{14}$$

DIMENSIONAL ANALYSIS

Inspection of the dimensionless transport equation developed above reveals that three dimensionless groups—Pe, R, and A/kT—influence the dimensionless concentration profile. Since the expressions for the capture efficiency and Sherwood number involve no additional groups, it is possible to write

$$e = e\left(Pe, R, \frac{A}{kT}\right), \quad \text{or} \tag{15a}$$

$$Sh = Sh\left(Pe, R, \frac{A}{kT}\right) \tag{15b}$$

However, for large Peclet numbers and small Hamaker's constants, the quantity eR^2 can be expressed as a function of only two dimensionless groups.

For large Peclet numbers and small Hamaker's constants, appreciable concentration variation occurs only very near to the collector. Then Stokes's expressions for the fluid velocity may be expanded in a Taylor series about the collector surface and higher order terms together with curvature effects may be neglected, yielding

$$2\frac{R^3}{Pe}f_1\frac{\partial^2 C}{\partial H^2} + \left(2\frac{R^3}{Pe}\frac{df_1}{dH} - R^2 v\right)\frac{\partial C}{\partial H} - \left[\frac{\partial(Ru)}{\partial \theta} + \frac{\partial(R^2 v)}{\partial H}\right]C = Ru\frac{\partial C}{\partial \theta} \tag{16}$$

with

$$Ru = \frac{3}{2}Y f_3(H)\sin\theta \tag{17a}$$

and

$$R^2 v = -\frac{3}{2}Y^2 f_1(H)\, f_2(H)\cos\theta - \frac{4}{3}\left(\frac{A}{kT}\right)\left(\frac{R^3}{Pe}\right)\frac{f_1(H)}{(Y^2-1)^2} \tag{17b}$$

These equations involve only two independent dimensionless groups: Pe/R^3 and $\frac{4}{3}\left(\frac{A}{kT}\right)\left(\frac{R^3}{Pe}\right)$. In later discussions it will be convenient to give this second group a new name: Ad. Recalling Equation (12) for the capture efficiency, together with those immediately above, we can deduce

$$eR^2 = f(Ad, Pe/R^3) \tag{18}$$

whenever the concentration variations occur only very near to the collector surface.

RESULTS AND DISCUSSION

In the presentation of results, the principal limiting cases summarized in Table 2 will be discussed first. Then the general results will be presented, followed by the identification of the regions in which the rate predictions for each limiting case are valid. Finally, gravitational effects and the additivity rule are discussed.

Case 1: Brownian Particles with Weak London Forces

In this case London forces cause adhesion of the particles coming in contact with the collector but do not aid in the transport; hence A/kT is taken as zero. Because one of the subcases (Case 1c) is pure diffusion, it seems more reasonable to express the rate of deposition as the Sherwood number.

Figure 3 summarizes the results obtained when the contribution of London forces to the transport rate is negligible. When the boundary-layer analysis of Levich and Lighthill is valid (Case 1a) the Sherwood number is given by

$$Sh = 0.995(Pe)^{1/3} \tag{19}$$

whereas when pure diffusion occurs (Case 1c), the Sherwood number is, by definition, two. Deviations from the Levich-Lighthill equation occur when (1) δ_D is not small compared to the collector radius, or (2) when the particle size becomes large when compared with the diffusion boundary-layer thickness. To avoid condition (1) the Peclet number must be larger than 100, whereas to avoid condition (2) the Peclet number must be smaller than $10^{-4}R^3$. The form of this latter condition may be derived by taking $\delta_m = \alpha_1 a_p$, as suggested by Equation (6), and by taking $\delta_D = \alpha_2 a(Pe)^{-1/3}$. Then $\delta_m << \delta_D$ whenever $Pe^{1/3} << \frac{\alpha_2}{\alpha_1}R$. The ratio α_2/α_1 as well as the validity of this condition can be deduced by inspection of Figure 3. Thus Equation (19) is valid in the interval $100 \leqq Pe \leqq 10^{-4}R^3$ when A/kT is sufficiently small.

Use of Fick's law to describe the diffusion process requires the solute particle to be small compared with the diffusion boundary layer. The analysis presented above suggests that, for Peclet numbers greater than 100, the ratio δ_D/a_p is proportional to $(Pe)^{1/3}/R$. The solid curves in Figure 3 are truncated at the value of the Peclet number corresponding to $Pe/R^3 = 10^{-2}$, where an inspection of the radial concentration profile revealed that the ratio δ_D/a_p is about ten.

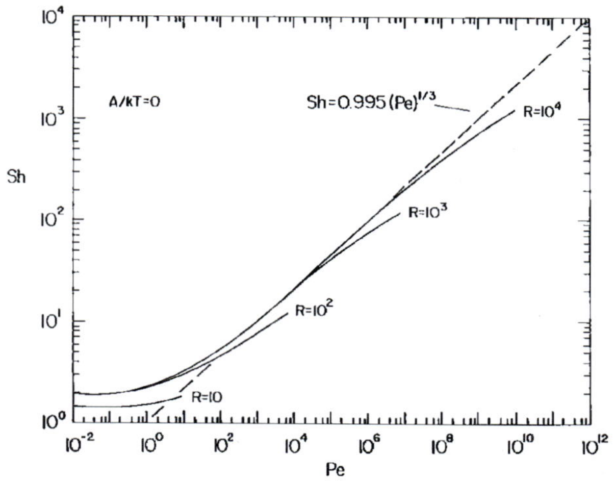

Fig. 3. Case 1. Sherwood numbers computed for the convective-diffusion of particles of finite size to the surface of a spherical collector by neglecting interaction forces. The dashed line is the Levich-Lighthill equation (19) which is valid when a diffusion boundary-layer exists and the particles are infinitesimal.

Case 2: Non-Brownian Particles

All particles, regardless of their size, undergo some degree of random or Brownian motion. However, the significance of this movement as a transport mechanism is less for large particles. In this paper *non-Brownian* refers to situations in which Brownian motion may be ignored when computing particle transport rates. Neglecting the diffusive terms in the transport Equation (8), a dimensional analysis shows that the product eR^2 depends only upon R and $\frac{4}{3}\left(\frac{A}{kT}\right)\frac{R^3}{Pe}$.

The group $\frac{4}{3}\left(\frac{A}{kT}\right)\frac{R^3}{Pe}$, denoted by Ad, can also be written as

$$Ad = \frac{4}{3}\left(\frac{A}{kT}\right)\frac{R^3}{Pe} = \frac{Aa^2}{9\pi\mu a_p^4 \beta U} \tag{20}$$

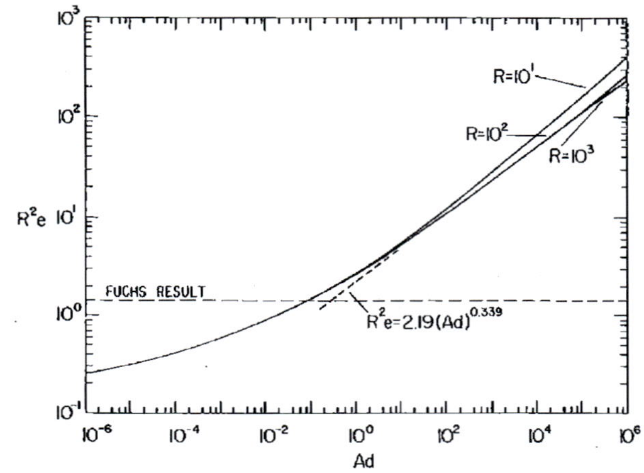

Fig. 4. Case 2. Single-grain capture efficiencies for the collection of non-Brownian particles by a spherical grain of a packed bed. A linear asymptote is noted by the dashed line which is valid for $10 < Ad < 10^{-6}R^7$.

This dimensionless group, introduced for conciseness in rate correlations, has no simple physical interpretation. It is the product of several ratios: A/kT represents the ratio of the characteristic London interaction energy to the thermal energy of the particle, R is the aspect ratio, while the Peclet number may be considered as the ratio of a characteristic energy for drag losses to the thermal energy possessed by the particle. This interpretation for the Peclet number becomes evident by using the relation $D = mkT$ to write

$$Pe = \frac{2a\beta U}{D_\infty} = \frac{2a\beta U/m_\infty}{kT} \tag{21}$$

Capture efficiencies are presented in Figure 4 for cases in which Brownian motion may be neglected. Small Ad implies weak London forces, hence $\delta_F << a$. From Equation (18) one then expects that eR^2 will depend only upon Ad for small Ad, as may be observed in Figure 4. Spielman and Fitzpatrick (1973) used a trajectory analysis to treat this limiting case.

Curves corresponding to large aspect ratios have a linear asymptote for large values of Ad:

$$eR^2 = 2.19\,(Ad)^{0.339}, \quad \text{for} \quad 10 < Ad < 10^{-6}R^7 \tag{22}$$

No linear asymptote is observable for small Ad. For $Ad \lessapprox 10^{-6}R^7$ the value of eR^2 was observed to be independent of R and $\delta_F << a$.

Fuchs (1964) computed the efficiency of capture by assuming the particles always follow fluid streamlines (Case 2b). Capture takes place when a streamline approaches one particle radius of the collector surface. Fuchs's analysis yields $eR^2 = 3/2$, which is plotted in Figure 4 as the dotted horizontal line. This result does not consider the effect of Brownian motion or London forces and therefore should yield the limiting value of eR^2 for non-Brownian particles when Ad approaches zero. However, Figure 4 shows that as Ad becomes small, eR^2 computed from a more rigorous analysis is considerably smaller than 3/2. Careful analysis of particle motion in flow fields near to a solid surface have shown that the particle velocity differs considerably from the undisturbed fluid velocity at the particle's center (Goren and O'Neill, 1971; Goldman et al., 1967). The radial component of the particle velocity, deduced from Equation (3), approaches zero as the particle approaches the surface in the absence of London forces, whereas the undisturbed radial fluid velocity at the particles center does not vanish even when the particle is in contact with the surface. Particle velocities coincide with fluid velocities only when the particle is several diameters away from a solid surface. For this reason Fuchs's analysis may yield only a qualitative indication of the efficiency. It does, for example, correctly and simply show that the efficiency for non-Brownian particles and weak London forces is proportional to R^{-2}.

Case 3: Stagnant Fluid

When the velocity at which the fluid approaches the collector becomes sufficiently slow the fluid may be considered to be essentially stagnant. The continuity equation for this situation takes on a particularly simple form which can be integrated to yield

$$N_r = -\frac{\alpha}{r^2} \tag{23}$$

If the London force is expressed as the gradient of the potential energy of interaction ϕ, the total flux may be written as the sum of the London and diffusive fluxes:

$$N_r = -m\frac{d\phi}{dr}c - D\frac{\partial c}{\partial r} \tag{24}$$

Equating these two equations for the radial flux yields a first-order ordinary differential equation in c, which may be integrated to yield

$$c(\infty) - c(a + a_p) = \alpha \exp[-\phi(a + a_p)/kT] \int_{a+a_p}^{\infty} \exp[\phi(r)/kT] \frac{dr}{r^2 D(r)} \tag{25}$$

Applying the boundary conditions of Equation (10), α may be evaluated. The rate of deposition per unit area may then be calculated as $-N_r|_{r=a+a_p}$ from Equation (23). In terms of dimensionless variables the result is

$$Sh = \frac{2}{R}\left\{\int_0^\infty \frac{\exp(\phi/kT)}{(R + H + 1)^2 f_1(H)}\,dH\right\}^{-1}, \tag{26}$$

which is plotted in Figure 5. Small A/kT and large R correspond to pure diffusion of a solute of infinitesimal dimensions through a stagnant fluid (Case 1c) for which a_p, δ_F, δ_m and a are much less than δ_D, and the Sherwood number takes on a value of two. Both δ_m and δ_F are proportional to a_p, whereas δ_D is proportional to a. If A/kT

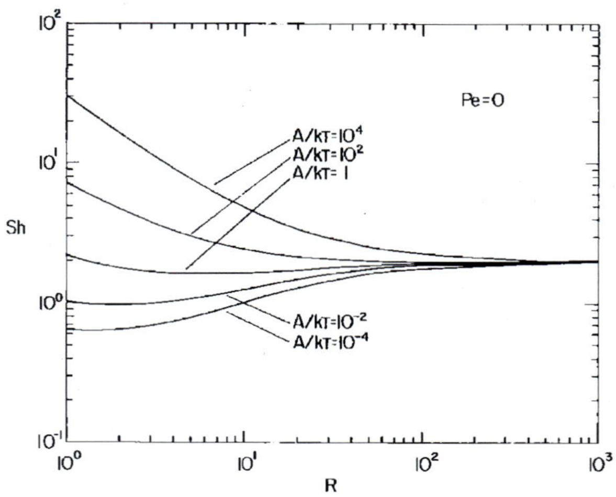

Fig. 5. Case 3. Sherwood numbers for the transport of finite size particles through a stagnant fluid to a spherical collector under the action of diffusion and London forces.

is small so that $\delta_F << \delta_m$, then as the aspect ratio R decreases at constant A/kT, δ_m becomes comparable to δ_D, and the variation in the diffusion coefficient begins to inhibit the transport. If A/kT is large so that $\delta_m << \delta_F$ then, decreasing R at constant A/kT, δ_F becomes comparable to δ_D and the transport rate is enhanced due to the action of the London forces.

Reported values of A/kT nearly always are in the range $10^{-2} < A/kT < 10^2$. Thus Figure 5 shows that, even in the best of circumstances ($R = 1$), this variation in Hamaker's constant by a factor of 10^4, results in a change in the Sherwood number by only a factor of 7. Particle transport rates in stagnant fluids are not highly sensitive to small changes in Hamaker's constant.

General Results

The results presented in Figure 6 do not contain any of the approximations which lead to the three special cases which have been discussed previously. They were obtained by solving the general transport Equation (8). Limiting behaviors can, however, be observed.

Dashed lines (- - -) represent the Levich-Lighthill equation (19). For Peclet numbers less than unity, the Sherwood numbers correspond to those computed for a stagnant fluid (Case 1c or Case 3) using Equation (26). At very large Peclet numbers the solid curves become tangent to the dotted lines (. . . . .) which represent the Sherwood numbers computed by neglecting Brownian motion (Case 2). The dotted curves were deduced from Figure 4, using Equation (14) to relate Sherwood number and capture efficiency, together with Equation (20) to relate Ad to A/kT and the Peclet number.

Criteria for Neglecting London forces or Brownian Motion

How small must the Peclet number be in order that $\delta_F << \delta_D$ for a given R and A/kT? How large must Peclet number be so that $\delta_D << \delta_F$ for the same R and A/kT? Answers to these questions are useful to design experiments that study London forces in the absence of Brownian motion or in analyzing a particular process to determine which effect is dominant.

Let the convective-diffusion-controlled region be defined as those sets of conditions (Pe, R, A/kT) for which the rate may be calculated to within 10% by ignoring London forces. Similarly, let the London-force-controlled region be those sets of conditions for which the rate may be calculated to within 10% by ignoring Brownian motion. These definitions suggest a method for determining limits for the regions.

In the convective-diffusion-controlled region, the rate should be independent of the value of A/kT. Selecting arbitrary values for the Peclet number and aspect ratio, the Sherwood number is first calculated for $A/kT = 0$. Then the value of the ratio A/kT is gradually increased,

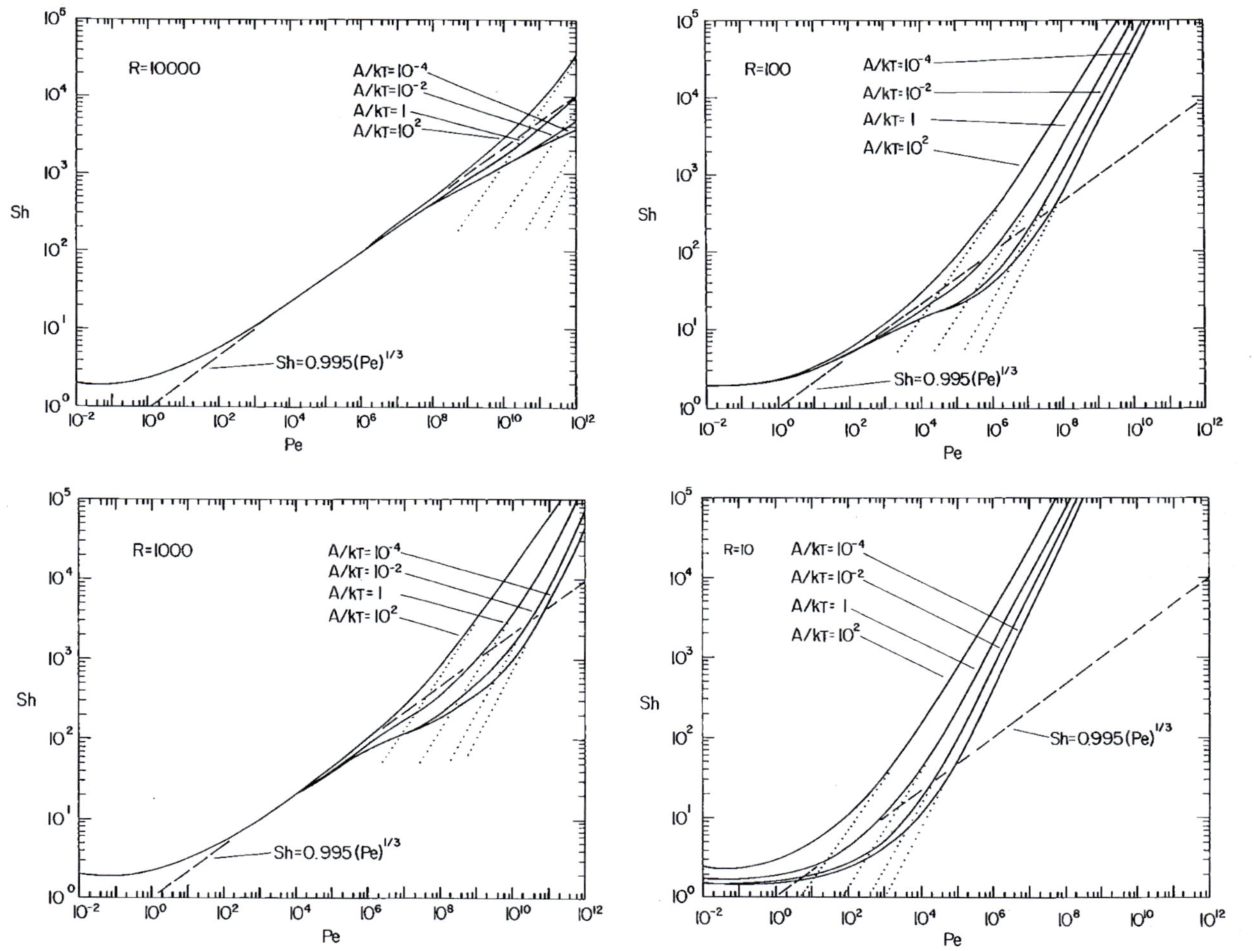

Fig. 6. Sherwood numbers computed for the transport of finite particles to a spherical collector under the combined action of convective-diffusion and London forces. Values of the aspect ratio are (a) $R = 10^4$, (b) $R = 10^3$, (c) $R = 10^2$, and (d) $R = 10$. For each aspect ratio, the value of A/kT was taken (upper curves to lower curves) as 10^2, 1, 10^{-2}, and 10^{-4}. Dashed lines represent the Levich-Lighthill equation (19), while the dotted curves represent Sherwood numbers deduced from Figure 4 which ignores the transport from diffusion.

holding Pe and R fixed, until the Sherwood number has increased to a value that is 1.1 times its initial value. In this manner one point (Pe, R, A/kT) on the boundary of the convective-diffusion region has been determined.

Within the London-force-controlled region the value of the collection efficiency depends only upon the values of R and the product $\frac{4}{3}\left(\frac{A}{kT}\right)\frac{R^3}{Pe}$ (denoted Ad). Arbitrarily selecting values for R and Ad, the efficiency is first computed for a value of the Peclet number that is sufficiently large so that no further change in the efficiency can be noted by changes in the Peclet number. Then the Peclet number and A/kT are gradually diminished, maintaining R and Ad fixed, until the efficiency increases by a factor of 1.1. In this manner one point on the boundary of the London-force controlled region is determined. Figure 7 summarizes the results. Selection of Pe/R^3 as the ordinate removes most of the R-dependence from the location of the boundaries.

Having established bounds on the regions where $\delta_F << \delta_D$ and $\delta_D << \delta_F$, it is now possible to summarize the conditions of validity for each of the limiting cases discussed. This summary is reported in Table 3 along with the corresponding rates of deposition.

Effect of Gravitational forces

Components of the gravitational force may be included in the radial and tangential force balances, given by Equations (3) and (4), to yield

$$\frac{v_r}{m} = \frac{s_r f_2}{m_\infty}$$

$$-\frac{2}{3}\,\frac{A a_p^3}{(h + 2a_p)^2 h^2} - \frac{4\pi}{3}\, a_p^3\,(\rho_p - \rho)\, g \cos\theta, \quad (27a)$$

and

$$v_\theta = s_\theta f_3 + \frac{4\pi}{3}\, a_p^3\,(\rho_p - \rho)\, g \sin\theta\, \frac{f_4(h/a_p)}{6\pi\mu a_p}, \quad (27b)$$

where $f_4(H)$ is the correction factor for the wall effect upon the tangential mobility. Goldman et al. (1967a) determined values for f_4 by considering the slow viscous motion of a sphere, through a quiescent fluid, parallel to a wall. A new dimensionless group—$G = \frac{4}{3}\pi a_p^4(\rho_p - \rho)g/kT$—appears upon nondimensionalization of Equation (27). This group may be interpreted as the ratio of the

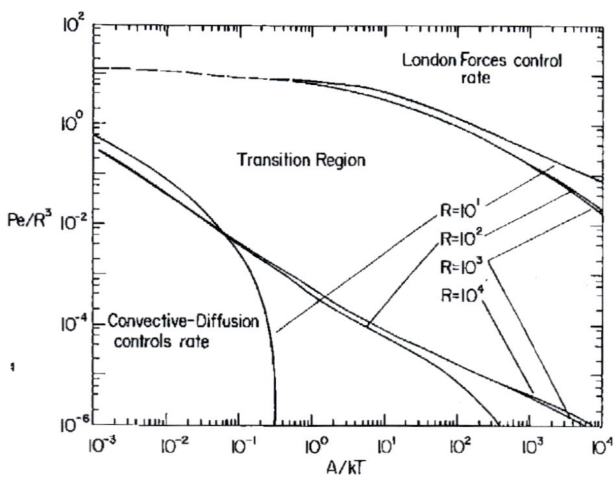

Fig. 7. Identification of regions in which either London forces or diffusion may be neglected when calculating the rate of deposition.

TABLE 3. TRANSPORT RATES AND REGIONS OF VALIDITY FOR LIMITING CASES

Case[a]	Region of validity[b]	Rate[b]
1a	$100 < Pe < \min\,[10^{-4}R^3,\ g_7(R, A/kT)]$	$Sh = 0.995\,(Pe)^{1/3}$
1b	$1 < Pe < \min\,[10^{-4}R^3,\ g_7(R, A/kT), 100]$	$Sh = g_3(Pe)$
1c	$Pe < \min\,[1, g_7(R, A/kT)]$ $R > 100$	$Sh = 2$
1d	$Pe < g_7(R, A/kT)$ $R > \max(Pe^{1/3}, 10)$	$Sh = g_3(Pe, R)$
2a	$g_7(R, Pe) < A/kT < 10^{-6}R^4Pe$	$eR^2 = g_4(Ad)$
2b	(see discussion)	$eR^2 = 1.5$
2c	$A/kT > g_7(R, Pe)$	$eR^2 = g_4(Ad, R)$
3	$Pe < 1$	$Sh = g_5(R, A/kT)$

[a] Refer to Table 2 for description of Cases.
[b] Value of the function $g_i(\ldots)$ may be evaluated from Figure i.

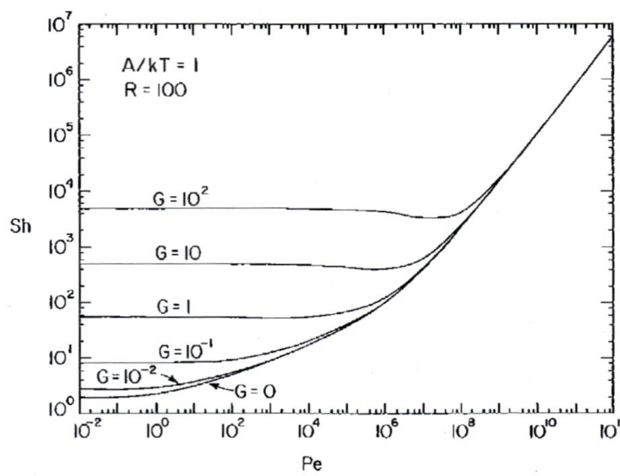

Fig. 8. Effect of gravitational forces upon the rate of deposition for a particular aspect ratio and Hamaker's constant.

gravitational potential of a particle, located one particle radius from the collector, to the characteristic thermal energy.

Figure 8 shows the effect which gravitational forces have upon the deposition rates for one typical situation, namely $A/kT = 1$ and $R = 100$. In the extreme case in which gravitational forces completely dominate the process, one could expect the particles to travel at their terminal velocity v_x along straight, vertical trajectories which leads to deposition at a rate $I = \pi a^2 v_x c_r$. In dimensionless terms this result is expressed by

$$Sh = \frac{1}{2}\,RG \quad (28)$$

which agrees well with the results of Figure 8 at small Peclet numbers and $G > 1$.

As an example, 0.25-micron (radius) particles with a density of 1.1 g/cm³ correspond to $G \sim 0.01$. Particles of this radius or smaller are not influenced significantly by gravity, according to Figure 8. However, deposition of larger particles is affected by gravity at small Peclet numbers.

Adequacy of Additivity Rule

In lieu of giving rigorous consideration to the effects of coupling between transport mechanisms, as was done in

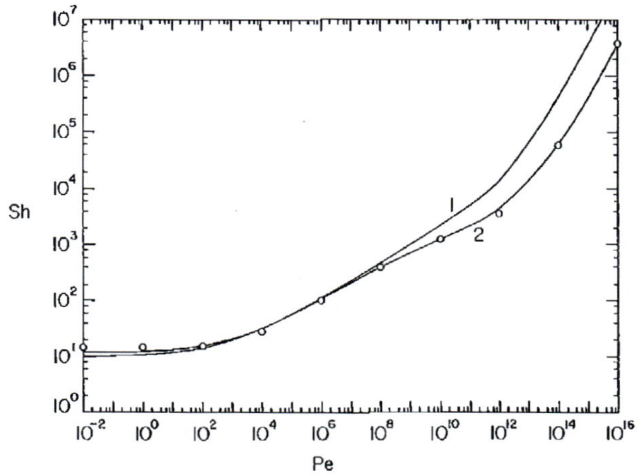

Fig. 9. Adequacy of the additivity rule for $R = 10^4$ and $A/kT = 10^4$. Lower curve represents results obtained by summing individual contributions from gravitational forces [Equation (28)], Brownian motion (Figure 3), and London forces (Figure 4). Points were obtained from simultaneously considering all effects, while upper curve is Eq. (29).

the present paper, the individual contributions to the deposition rate from each mechanism may be computed separately and subsequently summed to yield an approximation to the total deposition rate. This procedure, known as the additivity rule, was tested by Yao et al. (1971) and found to be adequate for practical calculations. Their transport equation, however, ignores the difference between non-Brownian particle velocity and fluid velocity as well as the variation of the diffusion coefficient with distance. It is of interest to retest the additivity rule after including the considerations ignored by them. Figure 9 compares the deposition rates computed by (1) a rigorous consideration of coupling effects, (2) summing individual contributions taken from Figures 3 and 4 together with Equation (28), and (3) the equation of Yao et al.:

$$Sh = \frac{1}{2} RG + 0.995 (Pe)^{1/3} + \frac{3}{8} \frac{Pe}{R^2} \qquad (29)$$

For the conditions selected, three regions may be distinguished. Gravitational forces, Brownian motion, or London forces dominate for small, moderate, and large Peclet numbers, respectively. Agreement between (1) and (2) is quite good, whereas (1) and (3) differ by an order of magnitude at high Peclet numbers. This difference stems from the distinction between particle velocities and fluid velocities ignored in the analysis of Fuchs (1964) and Yao et al. (1971). However, if this distinction is considered in each of the individual contributions, the additivity rule yields results, in an easy manner, which agree reasonably well with the results obtained from a rigorous consideration of coupling effects.

NOTATION

a = radius of collector, cm
a_p = radius of particle, cm
A = Hamaker's constant, erg
$Ad = \frac{4}{3}\left(\frac{A}{kT}\right)\frac{R^3}{Pe}$
b, b_1, b_2, n = parameters in transformation of the radial variable
c = local concentration of particles, particles/cm^3
c_∞ = concentration of particles far from collector, particles/cm^3
C = c/c_∞
D = local diffusion coefficient, cm^2/s
D_∞ = diffusion coefficient far from collector, cm^2/s
e = capture efficiency, $e = I/\pi a^2 \beta U c_\infty$
E = height of interaction energy barrier, erg
f_1, f_4 = correction factor for effect of wall proximity on radial and angular particle mobilities
f_2, f_3 = universal functions used to relate fluid and particle velocities
g_i = function whose value may be deduced from Figure i
G = $4/3\ \pi a_p^4(\rho_p - \rho)\ g/kT$
h = minimum separation between collector and particle surfaces, cm
h_{max} = separation at which ϕ is maximum, cm
h_{min} = separation at which ϕ is minimum, cm
H = h/a_p
I = rate of deposition, particles/s
k = Boltzmann's constant, erg/°K
m = mobility coefficient, cm/dyne-s
m_∞ = mobility coefficient far from collector, cm/dyne-s
N_r = radial particle flux, particles/cm^2-s
N_θ = angular particle flux, particles/cm^2-s
Pe = Peclet number, $Pe = 2a\ \beta\ U/D_\infty$
r = radial position of particle center, cm
R = aspect ratio, $R = a/a_p$
s_r = radial fluid velocity, cm/s
s_θ = angular fluid velocity, cm/s
Sh = Sherwood number, $Sh = 2I/4\pi a D_\infty c_\infty$
T = absolute temperature, °K
u = $v_\theta/\beta U$
U = approach velocity, cm/s
v = $v_r/\beta U$
v_r = radial particle velocity, cm/s
v_θ = angular particle velocity, cm/s
v_∞ = terminal velocity of particle far from collector, cm/s
Y = $H + 1$

Greek Letters

α = integration constant, particles/s
α_1, α_2 = proportionality constants
β = flow parameter
$\delta_D, \delta_F, \delta_m$ = characteristic lengths defined in Table 1, cm
η = transformed radial position
θ = angular position of particle center
μ = fluid viscosity, g/cm-s
ρ = fluid density, g/cm^3
ρ_p = particle density, g/cm^3
ϕ = total interaction energy, erg

LITERATURE CITED

Brenner, H., "The Slow Motion of a Sphere through a Viscous Fluid towards a Plane Surface," *Chem. Eng. Sci.*, **16**, 242 (1961).

Crank, J., and P. Nicolson, "A Practical Method for Numerical Evaluation of Partial Differential Equations of the Heat Conduction Type," *Proc. Camb. Phil. Soc.*, **43**, 50 (1947).

Fuchs, N. A., *The Mechanics of Aerosols*, MacMillan, New York (1964).

Goldman, A. J., R. G. Cox, and H. Brenner, "Slow Viscous Motion of a Sphere Parallel to a Plane Wall—I Motion through a Quiescent Fluid," *Chem. Eng. Sci.*, **22**, 637 (1967).

Goldman, A. J., R. G. Cox, and H. Brenner, "Slow Viscous Motion of a Sphere Parallel to a Plane Wall—II Couette Flow," *ibid.*, **22**, 653 (1967).

Goren, S. L., and M. E. O'Neill, "On the Hydrodynamic Resistance to a Particle of a Dilute Suspension when in the Neighborhood of a Large Obstacle," *ibid.*, **26**, 325 (1971).

Hamaker, H. C., "The London-van der Waals Attraction between Spherical Particles," *Physica*, **4**, 1058 (1937).

Happel, J., "Viscous Flow in Multiparticle Systems: Slow Motion of Fluids Relative to Beds of Spherical Particles," *AIChE J.*, **4**, 197 (1958).

Levich, V. G., *Physiocochemical Hydrodynamics*, Prentice-Hall, New Jersey (1962).

Lighthill, M. J., "Contributions to the Theory of Heat Transfer through a Laminar Boundary Layer," *Proc. Roy. Soc.*, **A202**, 359 (1950).

Ruckenstein, E., and D. C. Prieve, "Rate of Deposition of Brownian Particles under the Action of London and Double-Layer Forces," *J.C.S. Faraday II*, **69**, 1522 (1973).

Spielman, L. A., and J. A. Fitzpatrick, "Theory for Particle Collection under London and Gravity forces for Application to Water Filtration," *J. Colloid Interface Sci.*, **43**, 51 (1973).

Spielman, L. A., and S. K. Friedlander, "Role of the Electrical Double-Layer in Particle Deposition by Convective-Diffusion," *J. Colloid Interface Sci.*, **44**, 22 (1974).

Vrentas, J. S., J. L. Duda, and K. G. Bargeron, "Effect of Axial Diffusion of Vorticity on Flow Development in Circular Conduits: Part I. Numerical Solutions," *AIChE J.*, **12**, 837 (1966).

Yao, K. M., M. T. Habibian, and C. R. O'Melia, "Water and Waste-Water Filtration: Concepts and Applications," *Environ. Sci. Tech.*, **5**, 1105 (1971).

Yuge, T., "Theory of Heat Transfer of a Heated or Cool Body in High Speed Gas Flow," Rept. Inst. High Speed Mechanics of Tohoku Univ., **6**, 143 (1956).

Manuscript received May 29, 1974; revision received and accepted by September 16, 1974.

Role of Surface Chemistry in Particle Deposition

DENNIS C. PRIEVE

Department of Chemical Engineering, Carnegie-Mellon University, Pittsburgh, Pennsylvania 15213

AND

ELI RUCKENSTEIN

Faculty of Engineering and Applied Science, State University of New York at Buffalo, Buffalo, New York 14214

Received April 7, 1976; accepted August 20, 1976

The mechanism by which a solid surface acquires its charge in an electrolyte solution was found to affect calculated rates of particle deposition almost as dramatically as the charge itself. This is due to the mechanism's effect on changes in surface charge density, which occur as the surface approaches another charged surface. One class of mechanisms, based on the dissociation of acidic and basic surface groups, was considered. The changes in the surface charge density occurring in this case can affect the rate of deposition by several orders of magnitude relative to the rates for approaches at constant charge.

INTRODUCTION

Experimental studies of the rate of deposition of suspended particles onto a collector surface (1–5) suggest that particle–collector interaction forces have as profound an effect on rates of deposition as particle–particle interactions have upon flocculation rates. When particle and collector bear like charges of large magnitude, the resulting double-layer repulsion may effectively prevent deposition. Conversely, when particle and collector bear opposite charges, deposition occurs at the maximum rate permitted by convective diffusion (2). In some applications, such as platelet deposition prior to thrombus formation, prevention of deposition is desired; while in other applications, such as deep-bed granular filtration, enhancement of deposition is the goal.

Clearly, the magnitudes and relative sign of charges on the particles and collector have an important effect upon the rate of deposition. But surface charge densities may change as the particle approaches the collector. The amount of change depends on the mechanism which generates the charge. For example, one possible mechanism is the dissociation of a surface acid. Loss of a proton from acidic surface groups leaves a negative charge on the surface. As two such surfaces approach one another, overall electroneutrality requires the counterion concentration to increase in the solution between the plates to offset the decrease in volume. Since H^+ is a counterion, its concentration next to the surface tends to increase, which, in turn, tends to shift the dissociation equilibrium in favor of fewer dissociated acid groups and a lower surface charge density. If the surface acid is a strong one, this decrease in the absolute magnitude of the surface charge density will be much smaller than the decrease corresponding to a weak acid. In this and similar cases, the charge density at infinite separation, which may be deduced from ζ-potential measurements, is not adequate to fully characterize the electrostatic interactions. The number density and dissocia-

Journal of Colloid and Interface Science, Vol. 60, No. 2, June 15, 1977
ISSN 0021-9797

tion constant of ionogenic groups are also needed to determine how the charge will change upon interaction with other charged surfaces. Computations of the double-layer force between surfaces bearing single (6) or multiple (7–9) ionogenic groups have been performed previously.

In the present paper, the rate of deposition is calculated as a function of particle radius, ionic strength, surface potential, Hamaker's constant, and surface chemistry. By the last term, we mean the effect of the dissociation constant and the number density of acidic and basic groups on the surface, when the surface potential at large separations is held fixed. For the sake of simplicity, all calculations are done for deposition onto a rotating disk.

EVALUATION OF THE DEPOSITION RATE

When van der Waals and double-layer forces are effective over a distance which is short compared to the diffusion boundary-layer thickness, the rate of deposition may be calculated by lumping the effect of the particle–collector interactions into a boundary condition on the usual convective–diffusion equation. This condition takes the form of a first-order irreversible reaction (10, 11). Using this boundary condition to eliminate the solute concentration next to the disk from Levich's (12) boundary-layer solution of the convective–diffusion equation for a rotating disk, one obtains

$$N_y = [B/(B+1)]Kc_0 \qquad [1]$$

for the deposition flux, where

$$B = [0.170 D_\infty^2 (\omega^3 \rho/\mu)^{\frac{1}{2}}]^{\frac{1}{3}}/K\Gamma(\tfrac{4}{3})$$

and c_0 is the particle concentration far from the disk.

If the double-layer repulsion is sufficiently strong to cause the total potential energy of interaction $\phi(x)$ to assume a maximum (larger than kT) in its dependence on the particle–collector separation x, Ruckenstein and Prieve (10) have shown that the apparent rate constant K is related to the interaction profile by

$$K = D(x_{max})(\gamma/2\pi kT)^{\frac{1}{2}} \exp(-E/kT), \qquad [2]$$

where x_{max} is the location of the maximum, $\gamma = -d^2\phi/dx^2|_{x=x_{max}}$ and $E = \phi(x_{max})$. Position dependence in the particle diffusion coefficient $D(x)$ results from the increase in difficulty of pushing the fluid out of the path of the particle when it approaches a few radii of the collector (13). Equation [2] can be used to calculate the rate constant only if $E > 3kT$. However, inaccuracies in estimates of K for $E < 3kT$ are not significant in calculating the deposition flux, since K then will be sufficiently large that [1] reduces to Levich's result for the maximum rate:

$$N_{y,max} = 0.62 D_\infty^{\frac{2}{3}} (\rho/\mu)^{1/6} \omega^{\frac{1}{2}} c_0. \qquad [3]$$

This corresponds to mass-transfer limitation of the apparent surface reaction. Thus the combination of Eqs. [1] and [2] is expected to give reasonable estimates of the rate of deposition for all particle–collector interaction profiles, provided the interactions are confined to a region which is thin compared to the diffusion boundary layer.

EVALUATION OF THE INTERACTION ENERGY

Van der Waals Forces

The contribution from unretarded van der Waals forces to the total potential energy of interaction between a sphere of radius a, whose center is separated from the surface of a semi-infinite plate by a distance $x + a$, was calculated by Hamaker (14):

$$\phi_{VdW}(x) = A\left(\frac{1}{6}\ln\frac{x+2a}{x} - \frac{1}{3}\frac{a}{x}\frac{x+a}{x+2a}\right). \qquad [4]$$

While Lifshitz' continuum analysis leads to a more accurate physical interpretation of the Hamaker constant A, both Lifshitz' result and Eq. [4] predict that ϕ varies as x^{-1} for $x \ll a$, a result which has been confirmed experimentally (15). Since we treat A as an adjustable parameter, we use Hamaker's equation (Eq. [4]) below.

Double-Layer Force: Linearized Poisson–Boltzmann Equation

Two approaches for the calculation of the double-layer contribution are explored. Hogg *et al.* (16) linearized the Poisson–Boltzmann equation to compute the double-layer force between two dissimilar plane surfaces, then used Derjaguin's approximation to extend this result to the interaction of two spheres of different radii. When the radius of one sphere is infinite, their result becomes

$$\phi_{DL}(x) = \pm(\epsilon a/4)[(\Psi_1 + \Psi_2)^2 \ln(1 \pm e^{-\kappa x}) + (\Psi_1 - \Psi_2)^2 \ln(1 \mp e^{-\kappa x})], \quad [5]$$

where the upper sign applies for approach at constant surface potential and the lower sign applies for approach at constant surface charge density. Ψ_i denotes the surface potential of the ith surface at large separations. This result can be used provided $|\Psi_i| < 25$ mV and $\kappa a > 5$.

Double-Layer Force and Ionic Equilibria

A second and more rigorous approach to the calculation of the double-layer contribution is to solve the nonlinear Poisson–Boltzmann equation, taking into account the following dissociation equilibria to determine the surface charge and potential at each value of x:

acid sites:

$$G_{ai}\text{H (surf)} \rightleftarrows G_{ai}^{-} \text{ (surf)} + \text{H}^{+} \text{ (aq)} \quad \text{for } i = 1, 2, \ldots, m_a; \quad [6a]$$

basic sites:

$$G_{bj}\text{H}^{+} \text{ (surf)} \rightleftarrows G_{bj} \text{ (surf)} + \text{H}^{+} \text{ (aq)} \quad \text{for } j = 1, 2, \ldots, m_b. \quad [6b]$$

The method for calculating the force $p(l')$ between two plane surfaces, bearing multiple ionizable groups, separated by a distance l' has been presented previously (7–9). A summary is presented in Appendix I. To obtain the potential energy of the double-layer interaction between a sphere and a plate from the force between two plates, Derjaguin's method (16) is used:

$$\phi_{DL}(x) = 2\pi a \int_x^\infty \int_l^\infty p(l')dl'dl. \quad [7]$$

Bell *et al.* (17) have found that, when $\kappa a > 5$ and $\kappa x < 2$, the error in using Derjaguin's approximation is less than 10%. This was based on the interaction of two spherical particles. For sphere–plane interactions, the error should be even less. To calculate the rates according to Eqs. [1] through [3], one needs accurate estimates of ϕ_{DL} near x_{max}, which is typically of the order of κ^{-1}; hence $\kappa x_{max} \sim 1 < 2$.

When only numerical values of p are obtainable for a finite number of l', the integration in Eq. [7] must be performed numerically. Since considerable computational effort is needed for the calculation of p at each l', the number of these l' must be kept to a minimum, while maintaining the error below some tolerable value. Noting that $p(l')$ is an exponentially decaying function over most values of l', the authors approximated the integrand by $ae^{-bl'}$ over a small segment $[x_1, x_2]$ of the total integration interval $[x, \infty]$. Then the integral over this segment can be approximated by

$$\int_{x_1}^{x_2} f(x)dx \approx \frac{[f(x_2) - f(x_1)](x_2 - x_1)}{\ln [f(x_2)/f(x_1)]}. \quad [8]$$

This formula is applied successively to obtain the integral over a large interval. To obtain the results which follow, $p(l')$ was evaluated at 20 discrete, logarithmically spaced values of l' between 10^{-2} and 10^1 Debye lengths. This was found to keep the relative error in $\phi_{DL}(x)$, due to numerical integration, below 0.1%.

Total Interaction Energy

The total interaction energy is the sum of the van der Waals and double-layer contributions:

$$\phi(x) = \phi_{vdW}(x) + \phi_{DL}(x), \quad [9]$$

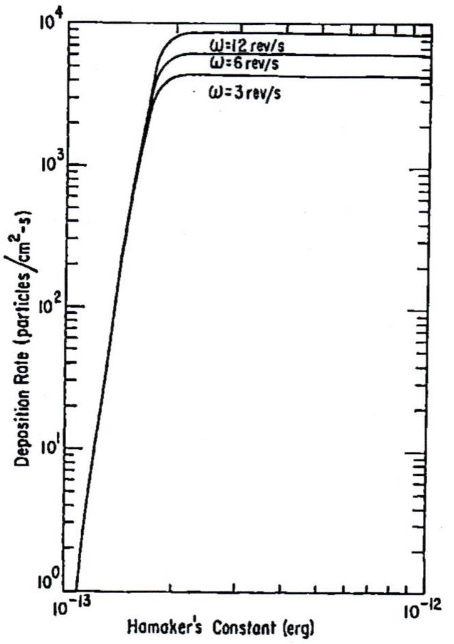

FIG. 1. Effect of disk rotation speed on particle deposition rate. Values of fixed parameters include $a = 10^{-5}$ cm, $\Psi_1 = \Psi_2 = -20$ mV, $c_0 = 10^9/\text{cm}^3$, $n = 0.01$ M, $\rho = 1$ g/cm³, $\mu = 0.01$ g/cm sec, $\epsilon = 74.3$, and $T = 300$°K.

where ϕ_{vdW} is given by [4] and ϕ_{DL} is given by [5] or ϕ_{DL} is evaluated numerically from Eq. [7], together with the nonlinear Poisson–Boltzmann equation. Other effects, such as discrete charge, surface roughness, and specific ion adsorption into the Stern layer, could also be included in Eq. [9], but they have been neglected here.

RESULTS AND DISCUSSION

Deposition of Constant-Potential Particles

Figures 1 through 4 illustrate the effect of Hamaker's constant, disk rotation speed, ionic strength, surface potentials, and particle radius on the rate of deposition of particles onto a rotating disk as computed from Eqs. [1] and [2]. For this set of figures, the interaction energy profiles $\phi(x)$ were evaluated from Eqs. [4], [5], and [9].

The curves of Fig. 1 display two distinct regions. For small values of the Hamaker constant, the van der Waals attraction is weak and a large energy barrier E develops, owing to the electrostatic repulsion. This makes K small and when it is small enough, Eq. [1] reduces to $N_y = Kc_0$, which corresponds to "reaction"-controlled deposition. Then the rate is insensitive to rotation rate. Conversely, when A is large, the strong van der Waals attraction overwhelms the double-layer repulsion, making the energy barrier small or nonexistent. This causes K to be large and Eq. [1] reduces to the mass-transfer limit given by Eq. [3]. Then the rate of deposition is proportional to $\omega^{1/2}$ and insensitive to A.

Like the rotation speed, the fluid density affects the deposition rate only for large Hamaker constants, where the flux will be proportional to $\rho^{1/6}$.

Ionic strength and surface potentials affect only the value of K; hence mass-transfer-

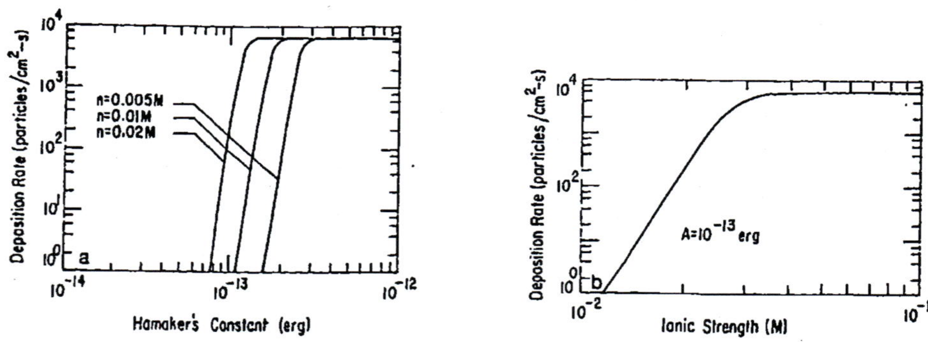

FIGS. 2a, b. Effect of ionic strength on particle deposition rate. Values of fixed parameters include $a = 10^{-5}$ cm, $\Psi_1 = \Psi_2 = -20$ mV, $c_0 = 10^9$ cm⁻³, $\omega = 6$ rev/sec, $\rho = 1$ g/cm³, $\mu = 0.01$ g/cm sec, $\epsilon = 74.3$, and $T = 300$°K.

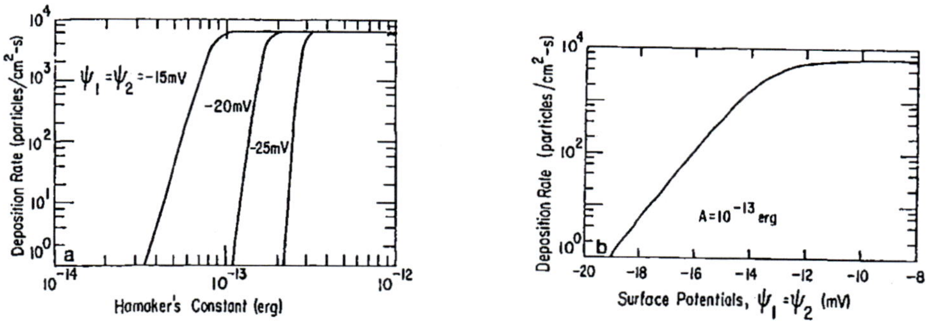

FIGS. 3a, b. Effect of surface potentials on particle deposition rate. Values of fixed parameters include $a = 10^{-5}$ cm, $n = 0.01$ *M*, $c_0 = 10^8/\text{cm}^3$, $\omega = 6$ rev/sec, $\rho = 1$ g/cm^3, $\mu = 0.01$ g/cm sec, $\epsilon = 74.3$, and $T = 300$°K.

limited rates (large Hamaker constant) are insensitive to these parameters, as shown in Figs. 2 and 3. However, they strongly influence the rate of reaction-controlled deposition. Note that halving the ionic strength lowers the rate by about three orders of magnitude, while increasing the absolute magnitude of the surface potential by 1 mV on each surface lowers the rate by a factor of about 300. Also, the rate becomes more sensitive to the Hamaker constant when the surface potential is increased.

Particle radius affects both the potential energy of interaction and the diffusion coefficient. Hence particle size influences the rate under all conditions. Under reaction-control conditions, Fig. 4 shows that a 1-μm increase in particle radius reduces the deposition rate by a factor of 5, while under mass-transfer control, the rate is proportional to $a^{-\frac{2}{3}}$. Note also that increasing particle size increases the sensitivity of the rate to the Hamaker constant. A similar increase in the sensitivity to surface potential and ionic strength could also be demonstrated when the particle radius is increased.

Constant Charge Compared with Constant Potential

All rate calculations presented above have used the double-layer interaction given by Eq. [5] as applied to approach at constant surface potential. Suppose the charge density had been held constant instead. How would this change affect the rate of deposition? The answer is illustrated in Fig. 5. When the sphere and plate are many Debye lengths apart, the

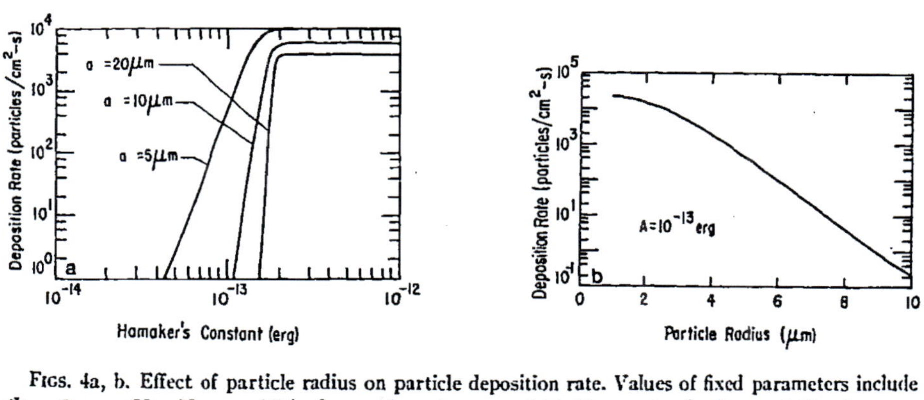

FIGS. 4a, b. Effect of particle radius on particle deposition rate. Values of fixed parameters include $\Psi_1 = \Psi_2 = -20$ mV, $c_0 = 10^8/\text{cm}^3$, $\omega = 6$ rev/sec, $n = 0.01$ *M*, $\rho = 1$ g/cm^3, $\mu = 0.01$ g/cm sec, $\epsilon = 74.3$, and $T = 300$°K.

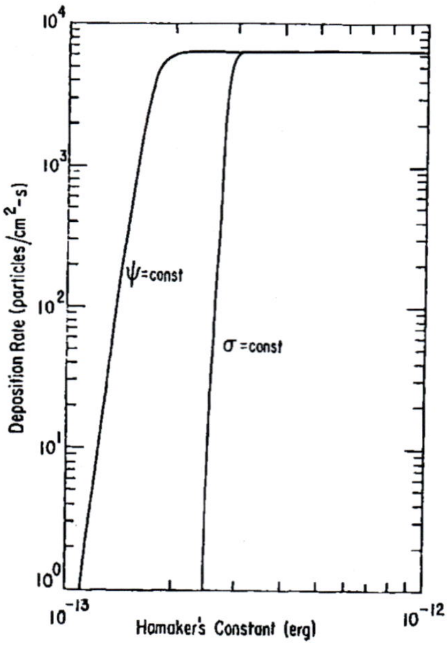

FIG. 5. Comparison of particle deposition rates calculated by assuming constant surface charge or constant surface potential. Values of fixed parameters include $a = 0.1\ \mu$m, $\Psi_1 = \Psi_2 = -20$ mV, $c_0 = 10^9$ cm^{-3}, $\omega = 6$rev/sec, $n = 0.01\ M$, $\rho = 1$ g/cm^3, $\mu = 0.01$ g/cm sec, $\epsilon = 74.3$, and $T = 300$°K.

surface potentials for the two curves are chosen to be identical. Note that for small Hamaker constants (reaction-controlled rate), the difference in the rates can be many orders of magnitude. As the sphere approaches the plate, the potentials on both surfaces will increase in order to maintain constant charge. Since this leads to a stronger double-layer repulsion, the constant-charge assumption leads to much smaller rates than the constant-potential assumption. Furthermore, the former yields rates which are more sensitive to the Hamaker constant than the latter. Greater sensitivity to the ionic strength, particle radius, and surface potential could also be demonstrated for the constant-charge case. Clearly the choice between constant charge and constant potential should not be made casually. Careful consideration must be given to the origin of surface charge.

Deposition of Particles Having Acidic Surfaces

If the surface acquires its charge through the dissociation of ionizable groups, neither the surface potential nor the surface charge density can be expected to remain constant. Changes in the separation distance will shift the ionic equilibria, which will affect both potential and charge. Figure 6 shows the effect of the dissociation constant (expressed as pK_a) when both surfaces bear a single type of acid. The number density (N_a) of acid groups on each surface was chosen for each pK_a so that the potential of widely separated surfaces is -20 mV, the same as in Fig. 5. For p$K_a = 5$, the surfaces are completely dissociated. Since the pH near the surface must decrease significantly to favor any reassociation of H^+ (which requires a large change in surface potential), surfaces with strong acids approach one another at constant surface charge density.

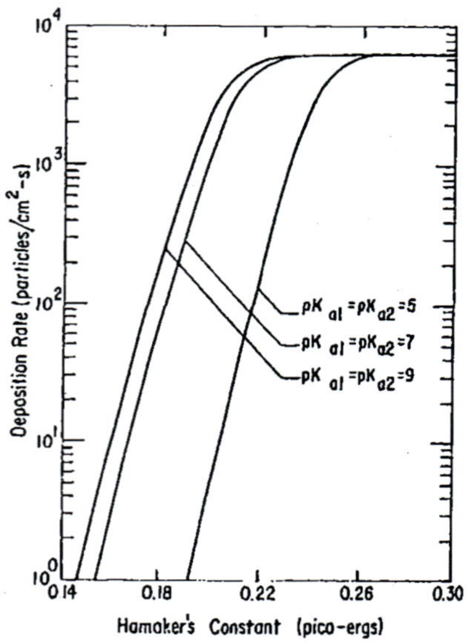

FIG. 6. Effect of dissociation constant on the particle deposition rate, when particle and collector surfaces bear identical numbers of the same acid. The number density of surface groups was adjusted to keep the potential of widely separated surfaces at -20 mV: $N_a = 2.922 \times 10^{12}, 9.135 \times 10^{12}$, and 6.304×10^{13} cm^{-2} for p$K_a = 5, 7$, and 9, respectively. Other fixed parameters are pH$_r = 7$, $n = 0.01\ M$, $a = 0.1\ \mu$m, $c_0 = 10^9$ cm^{-3}, $\omega = 6$ rev/sec, $\rho = 1$ g/cm^3, $\mu = 0.01$ g/cm sec, $\epsilon = 74.3$, and $T = 300$°K.

Other curves are for $pK_a = 7$ (partial dissociation) and $pK_a = 9$ (low degree of dissociation: weak acid). Significantly lower rates are obtained for surfaces bearing the stronger acid, although the potential of the noninteracting surface (infinite separation) is -20 mV in all cases.

Deposition of Particles Having Amphoteric Surfaces

Surfaces which bear both acidic and basic groups possess an isoelectric point: At some bulk pH the number of positively charged sites exactly balances the number of negative sites so that the net charge on the surface is zero. When equal numbers of one type of acidic and one type of basic group occur, it is easy to deduce the pH at which the net charge is zero (7):

$$[H^+]_{iso} = (K_a K_b)^{\frac{1}{2}}$$

or

$$pH_{iso} = \tfrac{1}{2}(pK_a + pK_b). \qquad [10]$$

Figure 7 shows the effect of proximity to the isoelectric pH upon the deposition rate. Particle and collector surfaces are identical in composition, both having equal numbers of acidic and basic groups ($N_a = N_b$). When widely separated, both surfaces were chosen to have a potential of -20 mV for all cases. This implies $pH_{s\infty} = 6.658$ if $pH_r = 7$. Next, pK_b was set at 10 for all cases and pK_a was selected with the help of Eq. [10] so that $pH_{s\infty} - pH_{iso}$ has the value indicated on each curve. Finally, the number of acidic and basic groups was chosen to give the desired surface potential of -20 mV when the surfaces do not interact.

Note that, as the isoelectric pH approaches the pH of the surface ($pH_{s\infty}$), the deposition rate increases for a given Hamaker constant. This is most easily explained in terms of changes in the surface potential and charge density which occur as the two surfaces are brought together. Consider the interaction between two identical *plane* surfaces which display an isoelectric point. Prieve and Ruckenstein (7) have noted that, as the separation distance is decreased, the ionic equilibria shift so that the net surface charge density tends to vanish. This corresponds to $pH_s \rightarrow pH_{iso}$ as the distance is decreased and, from Boltzmann's equation,

$$\Psi_s = 2.303(kT/e)(pH_s - pH_r) \rightarrow 2.303(kT/e)(pH_{iso} - pH_r). \qquad [11]$$

Subtracting $\Psi_{s\infty} = 2.303(kT/e)(pH_{s\infty} - pH_r)$ from the above, one obtains

$$\Delta\Psi_s = \Psi_s - \Psi_{s\infty} \rightarrow 2.303(kT/e)(pH_{iso} - pH_{s\infty}). \qquad [12]$$

Thus the maximum change in the surface potential, as the two surfaces are brought together, is proportional to the proximity to the isoelectric point: If the surfaces are very near to isoelectric conditions when widely separated, then the change in surface potential

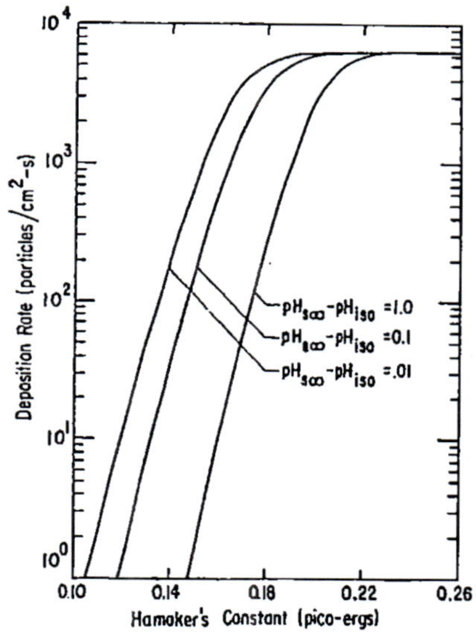

FIG. 7. Effect of isoelectric pH on particle deposition rate. Particle and collector surfaces are identical, each bearing equal numbers of acid and basic groups. The number density of surface groups was adjusted to keep the potential of widely separated surfaces at -20 mV: $N_a = N_b = 6.342 \times 10^{15}$, 1.702×10^{16}, and 1.404×10^{17} cm^{-2}, with $pK_a = 1.317$, 3.117, and 3.297, for $pH_{s\infty} - pH_{iso} = 1.0$, 0.1, and 0.01, respectively. Other fixed parameters include $pK_b = 10$, $pH_r = 7$, $n = 0.01$ *M*, $a = 0.1$ μm, $c_0 = 10^8$ cm^{-3}, $\omega = 6$ rev/sec, $\rho = 1$ g/cm^3, $\mu = 0.01$ g/cm sec, $\epsilon = 74.3$, and $T = 300$°K.

will be very small. Regarding Fig. 7, the curve having $pH_{s\infty} - pH_{iso} = 0.01$ corresponds to a change in Ψ_s from −20 to −20.6 mV, whereas $pH_{s\infty} - pH_{iso} = 1.0$ corresponds to a change from −20 to −78.5 mV. Clearly the latter situation corresponds to stronger double-layer repulsion at small separations and consequently, a lower rate of deposition.

The Importance of Surface Chemistry

Collectively, Figs. 5–7 illustrate that the surface potential, as determined from electrokinetic measurements, does not adequately characterize the double layer of that surface. If it did, all nine curves in Figs. 5–7 would coincide, since they all have a potential of −20 mV when the surface is not interacting with other surfaces. Instead, complete characterization of the double layer for rate calculations requires the number density of each type of chemical surface group (the N_{ai} and N_{bj}) together with the dissociation constant of each group (the K_{ai} and K_{bj}). These data for acid and basic groups may be evaluated from electrokinetic measurements over a wide range of pH. This is analogous to titrating the surface. Prieve and Ruckenstein (7) discuss and interpret the results of such experiments on human erythrocytes and sheep leucocytes. Of course, ions other than protons may also adsorb on the solid to change the charge and potential. Their dissociation characteristics must also be determined.

Comparison of Predicted and Observed Rates

Hull and Kitchener (2) measured the rate of deposition of 0.3-μm-diameter polystyrene latex particles onto a rotating disk coated with a film of polyvinyl formaldehyde. In electrolytes of high ionic strength (where the double-layer repulsion is negligible), they found close agreement between experiments and the prediction of Levich's boundary-layer analysis (Eq. [3]), indicating that a diffusion boundary layer exists and that its thickness is large compared to the domain of van der Waals and hydrodynamic interactions. These are necessary conditions for the analysis presented in the present paper. Measurements of the rate and surface potentials were also performed at five lower ionic strengths.

Measured rates, together with predictions, are summarized by Table I. Unfortunately, insufficient electrokinetic measurements were reported to characterize the surfaces chemically. However, the rate calculations of Figs. 5–7 are bounded by two extremes: approach at constant charge or approach at constant potential. Speculations 1 and 2 (in Table I) refer to predictions based on Eq. [5] for approach at constant potential or charge, respectively. Because the measured surface potentials lie outside the range needed for the linearized Poisson–Boltzmann equation used in Eq. [5], two additional predictions were based on solutions to the nonlinear Poisson–Boltzmann equation: speculations 3 and 4 are for approach at constant potential or charge, respectively. Measured surface potentials were used in all predictions, whereas the value of Hamaker's constant was obtained by forcing agreement between observed and predicted values at one ionic strength.

Predicted values of the rate for all four speculations in Table I are much more sensitive to ionic strength than observed values. At high ionic strengths, the energy barrier disappears and the rate becomes equal to the maximum value (6110 particles/cm^2 sec) predicted from Levich's equation (Eq. [3]). One possible explanation for the discrepancy in ionic strength dependence is that the Hamaker constant has a different value in each of the five electrolyte solutions tested. The Hamaker constant can be affected by adsorbed layers of surfactant (18). Since the concentration of surfactant used in solutions of different ionic strengths varied between 1×10^{-4} and 4×10^{-4} M/liter, the Hamaker constant could be affected differently. However, to obtain agreement between predicted and observed rates under speculation 1, the Hamaker constant would have to vary from 0.98×10^{-13} erg

TABLE I

Comparison of Experimental and Predicted Deposition Rates for the Conditions of the Hull and Kitchener (2) Experiment

n (M)	$\Psi_{particle}$ (mV)	Ψ_{disk} (mV)	$N_{y,\,exptl}$ (particles/cm² sec)	$N_{y,\,theory}$ (particles/cm² sec)[a]			
				Spec. 1	Spec. 2	Spec. 3	Spec. 4
0.03	−78	−51	26.3	$<10^{-40}$	$<10^{-40}$	$<10^{-40}$	$<10^{-40}$
0.04	−66	−47	69.4	68.65[b]	69.57[b]	68.77[b]	70.16[b]
0.05	−42	−42	202	6110	6110	6110	6110
0.10	−41	−37	481	6110	6110	6110	6110
0.30	−38	−32	625	6110	6110	6110	6110
Values of Hamaker's constant (pico-erg) =				0.5496[b]	0.6808[b]	1.1234[b]	0.7199[b]

[a] Spec. 1, etc., refer to different speculations on the nature of the surfaces and are identified in the text.
[b] The value of Hamaker's constant was adjusted to give close agreement between predicted and experimental rates for $n = 0.04$ M.

at the highest ionic strength to 10.71×10^{-13} erg at the lowest.

Discrepancies between predicted and observed ionic strength dependence may also be attributed either to uncertainties in electrokinetic measurements and their interpretation, or to inaccuracies in the current theory for predicting double-layer force. Evidence for the latter is given by Hough and Ottewill (19), who directly measured the double-layer repulsion between plane parallel surfaces as a function of separation distance and found that the current theory overpredicts the distance dependence. Since the separation distance is scaled in the current theory with respect to the Debye length, which is a function of ionic strength, this exaggerated distance dependence may also cause an exaggerated dependence on the ionic strength.

Clint *et al.* (4) measured the rate of deposition of 0.43-μm-diameter polystyrene latex particles onto a rotating disk coated with a polystyrene film in $Ba(NO_3)_2$ solutions of three different ionic strengths. Results are reported in Table II. Also reported in this table are the surface potentials of the disk which are needed to force agreement between predicted and observed rates. Speculations 1 and 2 again refer to approach at constant surface potential or charge, respectively, when the potential is small enough to linearize the Poisson–Boltzmann equation. However, when the charge is held constant and the separation is small, the surface potentials may become too large to use the linearized equation. So additional calculations were based on solutions to the nonlinear equation with constant surface charge (speculation 4). The existence of divalent cations was considered in these calculations. All of the calculations of the rate in Table II employed a Hamaker constant equal to 0.15×10^{-13} erg, which Clint *et al.* deduced.

Note that the disk potentials deduced from speculation 1 agree with observed values, to the extent allowed by the experimental uncertainty, while those deduced from speculation 2 or 4 do not agree. For some of the ionic strengths under speculation 2 the disk and particle have surface charges of opposite sign. Repulsion of oppositely charged surfaces has been discussed by Parsegian and Gingell (20), who solved the linear Poisson–Boltzmann equation assuming constant surface charge density. Further discussion on the repulsion between oppositely charged surfaces and its effect on deposition rates is given by Prieve and Ruckenstein (21), who numerically solved the nonlinear Poisson–Boltzmann equation, assuming the surfaces bear multiple ionizable sites. A necessary condition for repulsion is that, $-\Psi_1/\Psi_2$ must be different from unity.

Rates calculated in this paper show a very high sensitivity to every parameter affecting the potential energy of interaction, arising from

TABLE II

Comparison of Experimental and Predicted Deposition Rates for the Conditions of the Experiment by Clint *et al.* (4)

n (M)	$-\Psi_{\text{particles}}$ (mV)	$-\Psi_{\text{disk}}$ (mV)	$N_{\nu,\text{exptl}}$ (particles/cm² sec)	$-\Psi_{\text{disk, theory}}$ (mV)[a]		
				Spec. 1	Spec. 2	Spec. 4
0.015	26.0	7 ± 2	422	5.075	−4.062	<0.01
0.030	22.3	7 ± 2	900	5.639	−1.469	0.90
0.045	17.8	7 ± 2	1420	6.601	1.154	3.07

[a] Spec. 1, 2, and 4 refer to different speculations on the nature of the surfaces and are described in the text $\Psi_{\text{disk, theory}}$ values were adjusted to give agreement between observed and predicted rates.

the exponentiation in Eq. [2]. Consequently, rate measurements of deposition should provide a sensitive means of studying the interaction forces between nonidentical surfaces. For instance, existing models of interaction forces predict that $\phi(x)$ is nearly proportional to the particle radius, so that E should also be proportional. Semilog plots of rates versus particle radius should then be linear (as in Fig. 4) if the surface characteristics are the same for particles of all sizes. Experimental verification of this and similar predictions would provide insight into the accuracy of current models of interaction forces.

CONCLUSIONS

Calculations of the deposition rate show a dramatic dependence on the mechanism by which the surface charge is generated, in addition to the dependence on the charge itself. This reflects the importance of surface chemistry in particle deposition. Characterization of the electrostatic interactions, involving a given surface, require electrokinetic measurements on that surface under a broad spectrum of electrolyte conditions so that the number density and dissociation constant of ionizable surface groups can be deduced: a single electrokinetic measurement is not sufficient.

In one comparison of predicted and observed deposition rates, the predictions were too strongly dependent on ionic strength, while reasonable agreement was obtained in a second comparison. Inaccuracy in the predicted double-layer force, which would be greatly amplified in the deposition rates, probably accounts for the discrepancy between predicted and observed ionic-strength dependence. Additional careful measurements of deposition rates, involving fully characterized particles and collectors, are needed before a conclusive test of the theory can be performed. On the basis of the sensitivity of rate calculations to various parameters, measurements of deposition rates should provide a means of studying interaction forces between the particle and collector.

APPENDIX I: COMPUTING THE DOUBLE-LAYER FORCE BETWEEN PLANE SURFACES BEARING IONIZABLE GROUPS

The dissociation equilibria (see Eq. [6]) involving ionizable surface groups are characterized by their dissociation constants:

$$K_{ai} = [G_{ai}^-][H^+]_s/[G_{ai}H], \quad [A1a]$$

$$K_{bj} = [G_{bj}][H^+]_s/[G_{bj}H^+], \quad [A1b]$$

where $[H^+]_s$ denotes the volumetric concentration of H^+ next to the surface and the remaining bracketed quantities denote the number of each type of site per unit surface. The charge density on the surface is proportional to the net number of positive sites:

$$\sigma/e = \sum_{j=1}^{m_b} [G_{bj}H^+] - \sum_{i=1}^{m_a} [G_{ai}^-]. \quad [A2]$$

If

$$N_{aik} \equiv [G_{aik}^-] + [G_{aik}H]$$

and

$$N_{bjk} \equiv [G_{bjk}] + [G_{bjk}H^+],$$

then [A1] and [A2] yield

$$\frac{\sigma_k}{e} = \sum_{j=1}^{m_b} \frac{[H^+]_{sk}}{K_{bjk} + [H^+]_{sk}} N_{bjk} - \sum_{i=1}^{m_a} \frac{K_{aik}}{K_{aik} + [H^+]_{sk}} N_{aik}. \quad [A3]$$

Note that the coefficient of N_{aik} or N_{bjk} is the degree of dissociation of the ith acidic or jth basic group on surface k ($k = 1, 2$). The hydrogen ion concentration next to the kth surface is related to the electrostatic potential of that surface through the Boltzmann equation

$$[H^+]_{sk} = [H^+]_r \exp(-e\Psi_k/kT), \quad [A4]$$

where $[H^+]_r$ is the concentration within an infinite reservoir, having $\psi = 0$, with which the double layer is in equilibrium. As boundary conditions on the Poisson–Boltzmann equation, it is required that $d\psi/dx|_{x=0} = -4\pi\sigma_1/\epsilon$ and $d\psi/dx|_{x=l} = +4\pi\sigma_2/\epsilon$ or, after substituting [A4] into [A3]:

$$\left.\frac{d\psi}{dx}\right|_{x=0} = -\frac{4\pi e}{\epsilon}\left\{\sum_{j=1}^{m_b} \frac{[H^+]_r N_{bj1}}{[H^+]_r + K_{bj1}\exp[e\psi(0)/kT]} - \sum_{i=1}^{m_a} \frac{K_{ai1}N_{ai1}}{K_{ai1} + [H^+]_r \exp[-e\psi(0)/kT]}\right\}, \quad [A5a]$$

$$\left.\frac{d\psi}{dx}\right|_{x=l} = +\frac{4\pi e}{\epsilon}\left\{\sum_{j=1}^{m_b} \frac{[H^+]_r N_{bj2}}{[H^+]_r + K_{bj2}\exp[e\psi(l)/kT]} - \sum_{i=1}^{m_a} \frac{K_{ai2}N_{ai2}}{K_{ai2} + [H^+]_r \exp[-e\psi(l)/kT]}\right\}. \quad [A5b]$$

These boundary conditions explicitly give neither the potential nor the derivative of potential at either surface, but instead specify a complex relation between the potential and the derivative at each surface. An analytical solution (in terms of Jacobi θ functions) to the nonlinear Poisson–Boltzmann equation has been obtained, subject to boundary conditions [A5] for the case of two identical surfaces each bearing a single type of acid group (6). Numerical solutions are available (7–9) for the interaction of surfaces bearing both acidic and basic groups.

Once the electrostatic potential profile $\psi(x)$ between two parallel plates is known for $0 \le x \le l$, the repulsive force p per unit area of surface can be evaluated from

$$p = -\frac{\epsilon}{8\pi}\left(\frac{d\psi}{dx}\right)^2 + kT\sum_{i=1}^{I} n_{ri} \times [\exp(-z_i e\psi/kT) - 1], \quad [A6]$$

where the sum is over all ions in the solutions. Although both ψ and $d\psi/dx$ depend on x, their dependence is such that [A6] must yield the same p for all x. This can be proved by differentiating both sides by x: the right-hand side will identically vanish when $\psi(x)$ satisfies the general Poisson–Boltzmann equation.

APPENDIX II: NOMENCLATURE

a	particle radius (cm)
A	Hamaker's constant (ergs)
c_0	particle concentration far from collector (particles/cm^3)
D	local diffusion coefficient (cm^2/sec)
D_∞	diffusion coefficient far from collector (cm^2/sec)
e	magnitude of electron charge (statcoul)
E	$\phi(x_{max})$ (ergs)
k	Boltzmann's constant (ergs/°K)
$\dot{K}$	apparent first-order reaction rate constant (cm/sec)
K_{ai}	dissociation equilibrium constant for ith acid (moles/liter)
K_{bj}	dissociation equilibrium constant for jth base (moles/liter)
l	distance between two plane surfaces (cm)

m_a, m_b number of types of acids or bases, respectively (dimensionless)
n ionic strength (ions/cm³)
n_{ri} number density of ionic species i in reservoir (ions/cm³)
N_{aik} number density of acidic groups of type i on surface k (groups/cm²)
N_{bjk} number density of basic groups of type j on surface k (groups/cm²)
N_y flux of particles to collector (particles/cm² sec)
$N_{y,max}$ flux of particles to collector surface when convective–diffusion controls (particles/cm² sec)
p repulsive double-layer force between two plane surfaces (dyn/cm²)
T absolute temperature (°K)
x distance between particle and collector surfaces (cm)
x_{max} value of x corresponding to a maximum in $\phi(x)$ (cm)
z_i valence of ith ionic species (dimensionless)

Greek characters

γ $-d^2\phi/dx^2$ evaluated at $x = x_{max}$ (ergs/cm²)
Γ gamma function (dimensionless)
ϵ dielectric constant of fluid (dimensionless)
κ reciprocal Debye length, $\kappa = (8\pi ne^2/\epsilon kT)^{\frac{1}{2}}$ (cm⁻¹)
μ fluid viscosity (g/cm sec)
ρ density of fluid (g/cm³)
σ_k charge density on kth surface (statcoul/cm²)
ϕ total potential energy of interaction between a sphere and a plate (ergs)
ϕ_{vdW}, ϕ_{DL} contributions to ϕ from van der Waals or double-layer forces, respectively (ergs)
ψ local electrostatic potential (statvolt)
Ψ_k value of ψ at kth surface (statvolt)
ω rotational speed of disk (rad/sec)

REFERENCES

1. Marshall, J. R. and Kitchener, J. A., *J. Colloid Interface Sci.* **22**, 342 (1966).
2. Hull, M. and Kitchener, J. A., *Trans. Faraday Soc.* **65**, 3093 (1969).
3. FitzPatrick, J. A. and Spielman, L. A., *J. Colloid Interface Sci.* **43**, 350 (1973).
4. Clint, G. E., Clint, J. H., Corkill, J. M., and Walker, T., *J. Colloid Interface Sci.* **44**, 121 (1973).
5. Mádi, I., *Colloid Polym. Sci.* **252**, 22 (1974).
6. Ninham, B. W. and Parsegian, V. A., *J. Theor. Biol.* **31**, 405 (1971).
7. Prieve, D. C. and Ruckenstein, E., *J. Theor. Biol.* **56**, 205 (1976).
8. Prieve, D. C., Ph.D. Dissertation, University of Delaware, 1974.
9. Chan, D., Perram, J. W., White, L. R., and Healy, T. W., *J. Chem. Soc. Faraday I* **71**, 1046 (1975).
10. Ruckenstein, E. and Prieve, D. C., *J. Chem. Soc. Faraday II* **69**, 1522 (1973).
11. Spielman, L. A. and Friedlander, S. K., *J. Colloid Interface Sci.* **44**, 22 (1974).
12. Levich, V. G., "Physicochemical Hydrodynamics." Prentice-Hall, Englewood Cliffs, N.J., 1962.
13. Brenner, H., *Chem. Eng. Sci.* **16**, 242 (1961).
14. Hamaker, H. C., *Physica* **4**, 1058 (1937).
15. Tabor, D. and Winterton, R. H. S., *Nature* **219**, 1120 (1969).
16. Hogg, R., Healy, T. W., and Fuerstenau, D. W., *Trans. Faraday Soc.* **62**, 1638 (1966).
17. Bell, G. M., Levine, S., and McCartney, L. N., *J. Colloid Interface Sci.* **33**, 335 (1970).
18. Weiss, G. H., Kiefer, J. E., and Parsegian, V. A., *J. Colloid Interface Sci.* **45**, 615 (1973).
19. Hough, D. B. and Ottewill, R. H., "Colloid and Interface Science" (M. Kerker, ed.), Vol. 4, p. 45, Academic Press, New York, 1976.
20. Parsegian, V. A. and Gingell, D., *Biophys. J.* **12**, 1192 (1972).
21. Prieve, D. and Ruckenstein, E., Paper 122e presented at 69th Annual A.I.Ch.E. Meeting, Chicago, Nov. 28–Dec. 2, 1976.

The Double-Layer Interaction between Dissimilar Ionizable Surfaces and Its Effect on the Rate of Deposition[1]

DENNIS C. PRIEVE

Department of Chemical Engineering, Carnegie-Mellon University, Pittsburgh, Pennsylvania 15213

AND

ELI RUCKENSTEIN

Faculty of Engineering and Applied Science, State University of New York at Buffalo, Buffalo, New York 14214

Received February 11, 1977; accepted May 17, 1977

Rates of deposition were calculated for cases in which sol and collector surfaces are dissimilar and acquire their charge by the dissociation of multiple acidic and/or basic sites. The double-layer interaction between such dissimilar surfaces can induce a sign reversal in the surface charge or the force. These sign reversals are not observed between identical surfaces.

A qualitative analysis of the governing equations allows prediction of the double-layer interaction at very small separations without solving the complete equations. Two strongly acidic or basic surfaces practically always repel if the separation is small enough, regardless of the relative signs of their surface charges, whereas two amphoteric surfaces having high buffer capacity practically always attract in the limit of small separation.

The lowest rates of deposition are obtained when both particle and collector are strongly acidic or basic and of like charge, whereas the highest rates are obtained when both surfaces are amphoteric. These bounds on the rate are orders of magnitude apart. Of interest was the observation that a strongly acidic particle could be effectively prevented from depositing on a strongly basic collector by double-layer repulsion, even though the charges on the two surfaces have opposite signs. A necessary condition is that the ratio of the two surface potentials at infinite separation must be significantly different from -1.

Introduction

Biological surfaces, as well as metal oxides, acquire an electrostatic charge in water through the dissociation of acidic or basic groups which lie near the solid/aqueous interface. As two such surfaces, separated by an aqueous electrolyte, are brought together the volume between them decreases and consequently the counterion concentration of the solution must tend to increase in order to preserve overall electroneutrality. In particular, the concentration of H^+ near the interfaces will be altered, which in turn will shift the acid/base dissociation equilibria in favor of either more or less associated protons, thereby changing the surface charge density. At the same time, the increased counterion concentration reflects a change in the magnitude of the surface potential. Thus, in general neither the surface charge density nor the surface potential remains fixed during the approach of these surfaces. Instead, the Poisson–Boltzmann equation must be solved using boundary conditions which

[1] Paper 122e presented at 69th Annu. AIChE Meet., Chicago, Nov. 28–Dec. 2, 1976.

0021-9797/78/0632-0317$02.00/0

simultaneously consider the multiple dissociation equilibria of ionizable groups on each surface.

Such calculations were first performed by Ninham and Parsegian (1) for the case of two identical plane, acidic surfaces. Later, Prieve and Ruckenstein (2, 3) and Chan *et al.* (4, 5) independently published numerical results of similar calculations for the case of two identical, plane, amphoteric surfaces. Rates of hydrosol adsorption were calculated by Prieve and Ruckenstein (6), who used the double-layer interaction between the particle and collector corresponding to identical and acidic or amphoteric surfaces. However, in a real situation the particle and collector surfaces are different.

The double-layer force between nonidentical surfaces can depend upon separation distance in a manner that is much different from that between identical surfaces. For example, Derjaguin (7) predicted that charged surfaces bearing different surface potentials of the same sign may attract, rather than repel, across ionic solutions. Later calculations by Parsegian and Gingell (8) confirmed Derjaguin's predictions by assuming small surface potentials (in order to use the Debye–Huckel approximation) and either constant charge or constant potential. They concluded that surfaces of like *charge* must repel but, since charge and potential do not necessarily have the same sign, surfaces of like *potential* (as well as surfaces having opposite charges) may attract or repel, depending on the separation distance. Thus the interaction of dissimilar surfaces is qualitatively different from that of identical surfaces and cannot be predicted from the latter by replacing the dissociation characteristics of the different surfaces with some average values. Instead the Poisson–Boltzmann equation must be solved using the dissimilar surface characteristics. A few such calculations have been previously reported (9).

The purpose of the present paper is threefold. First, solutions of the nonlinear Poisson–Boltzmann equation are presented for a number of cases to illustrate some of the complex behavior displayed by the double-layer interaction between dissimilar surfaces. Second, a qualitative analysis of the fundamental equations is performed which allows conclusions to be deduced regarding the charge, surface potential, and double-layer force at small separations without solving the equations. Finally, the rate of adsorption of hydrosols is determined, based on the double-layer interactions computed above for dissimilar surfaces.

Calculation of the Double-Layer Force

To calculate the double-layer force, the nonlinear Poisson–Boltzmann equation was solved for the case of two plane parallel plates, subject to boundary conditions which arise from consideration of the simultaneous dissociation equilibria of multiple ionizable groups on each surface. Derjaguin's approximation is then used to extend these results to calculate the force between a sphere and a plane. Details of the method can be found in Ref. (6).

Two types of surfaces will be employed in the results which follow. A type "A" surface is defined as one having a single kind of strongly acidic or strongly basic site. This surface remains fully dissociated for any reasonable degree of double-layer overlap. A type "B" surface is defined as one having equal numbers of a single kind of acidic site and a single kind of basic site. The dissociation characteristics and number density of surface sites on an amphoteric type "B" surface were chosen to lend to the surface a significant buffer capacity. Thus the pH near a surface of type "B" tends to remain constant, except at very small separations. By Boltzmann's equation, the surface potential of a type "B" surface then is also nearly constant, whereas a type "A" surface maintains nearly constant charge. Interaction of weakly acidic surfaces was also computed (6, 9), but only interactions of type-A or type-B surfaces are reported in the present paper because these interactions seem to lead to bounds on the adsorption rates.

Two interacting surfaces may differ either in their type, in the sign of their charge, in the absolute magnitude of their charge, or any combination of the above. Considering two different types, two different charge magnitudes, and two different signs, there are eight different surfaces to be considered. Characteristics of these surfaces are summarized in Table I. In what follows, these surfaces will be referenced by type and surface potential. For example, a B + 10 surface is the amphoteric surface from Table I of type "B", having a surface potential of +10 mV when it is infinitely far from any other surface (of course the potential may not equal +10 mV when its double-layer overlaps with that of another surface).

Figures 1 and 2 summarize the results computed for the double-layer interaction between two parallel plates. Each figure presents the charge density and electrostatic potential of both interacting surfaces, together with the double-layer force per unit area ($p > 0$ denotes repulsion), as a function of the dimensionless separation distance κl, where κ is the reciprocal Debye length and l is the distance separating the parallel plane surfaces. Figure 1 summarizes the interaction between several pairs of surfaces which differ in type or in their surface potential at infinite separation. Figure 2 summarizes the interaction of two type-A surfaces having charges of opposite sign. Note that sign reversals in the force or charge occur in both figures. These sign reversals do not occur when the surfaces are identical.

The discussion which follows attempts to provide some physical explanations for the calculated behavior illustrated by Figs. 1 and 2. Secondly, a qualitative analysis of the equations is presented which allows estimation of the limiting behavior at very small separations without having to solve the equations.

Qualitative Behavior of Surface Potential and Charge Density at Small Separations

For every pair of surfaces in Figs. 1 and 2 the interaction is such that the two surface potentials tend to become equal as $\kappa l \to 0$. One may explain this tendency by arguing that the electric field is approximated by $d\psi/dx \approx [\psi(l) - \psi(0)]/l$ for small enough l. Since the charge density and

TABLE I

Characteristics of Surfaces[a]

Type	$\psi_{s\infty}$ (mV)	N_a (No./cm²)	N_b (No./cm²)	pK_a	pK_b	$pH_{iso} - pH_{s\infty}$	σ_∞/e (No./cm²)
A	−20	2.860×10^{12}	0	2	—	—	-2.860×10^{12}
	−10	1.403×10^{12}	0	2	—	—	-1.403×10^{12}
	+10	0	1.403×10^{12}	—	10	—	$+1.403 \times 10^{12}$
	+20	0	2.860×10^{12}	—	10	—	$+2.860 \times 10^{12}$
B	−20	1.404×10^{17}	1.404×10^{17}	3.297[b]	10	+0.01	-2.860×10^{12}
	−10	4.637×10^{16}	4.637×10^{16}	3.639[b]	10	+0.01	-1.403×10^{12}
	+10	4.637×10^{16}	4.637×10^{16}	3.639[b]	10	−0.01	$+1.403 \times 10^{12}$
	+20	1.404×10^{17}	1.404×10^{17}	3.297[b]	10	−0.01	$+2.860 \times 10^{12}$

[a] Characteristics of electrolyte are $n = 0.01$ moles/liter, $\epsilon = 74.3$, $T = 300°K$, and $pH_r = 7$.
[b] pK_a chosen to give value of $pH_{iso} - pH_{s\infty}$ indicated.

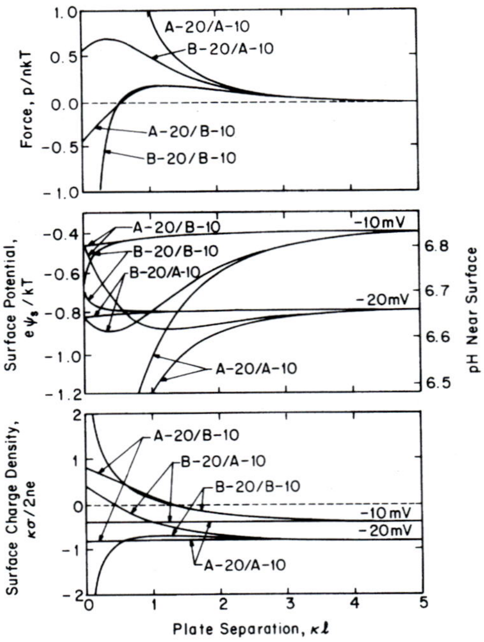

FIG. 1. Interaction of two surfaces having $\psi_{s\infty}$ of different magnitude, but the same sign ($\psi_{s\infty} = -20$ or -10 mV). Note sign changes in both charge and force. A − 20 is a strong acid-like surface, whereas B − 20 is a buffer-like amphoteric surface. For other details, see Table I.

hence $d\psi/dx$ remain bounded, $\psi(l) - \psi(0)$ must vanish as l vanishes. In other words, an infinitesimally thin film cannot maintain a finite potential drop.

During the interaction of two strongly acidic or two strongly basic surfaces, both surface potentials tend to become enormous in magnitude at small separations (see A − 20/A − 10 in Fig. 1, or A + 80/A − 5.84 in Fig. 2, for example). This can be simply explained by considering the requirement of overall electroneutrality:

$$\sigma_0 + \sigma_l + \int_0^l \rho(x)dx = 0, \qquad [1]$$

where ρ is the net number of positive charges per unit volume at some position x in the solution. Consider the limit as $\kappa l \rightarrow 0$. From the discussion above, $\psi(x)$ tends to a uniform constant, say ψ_{s0}. Writing $\rho = e\Sigma z_i c_i = (c_+ - c_-)e$, which is essentially independent of x at small separations, and using Boltzmann's equation to relate c_+ and c_- to ψ_{s0}, Eq. [1] becomes

$$\sinh\left(\frac{e\psi_{s0}}{kT}\right) = \frac{\kappa}{2ne}\lim_{\kappa l\rightarrow 0}\left(\frac{\sigma_0 + \sigma_l}{\kappa l}\right). \qquad [2]$$

Two limiting situations may occur. Suppose the interacting surfaces are both of type A and remain with fixed charge. Equation [2] would then correctly predict that $\psi_{s0} \rightarrow \pm\infty$, provided $\sigma_0 + \sigma_l \neq 0$. This behavior may also be rationalized by observing that as $\kappa l \rightarrow 0$ the volume of the intervening solution per unit area of plate is decreasing, so that in order to hold the same number of counterions per unit area, the average concentration of counterions in the solution between the

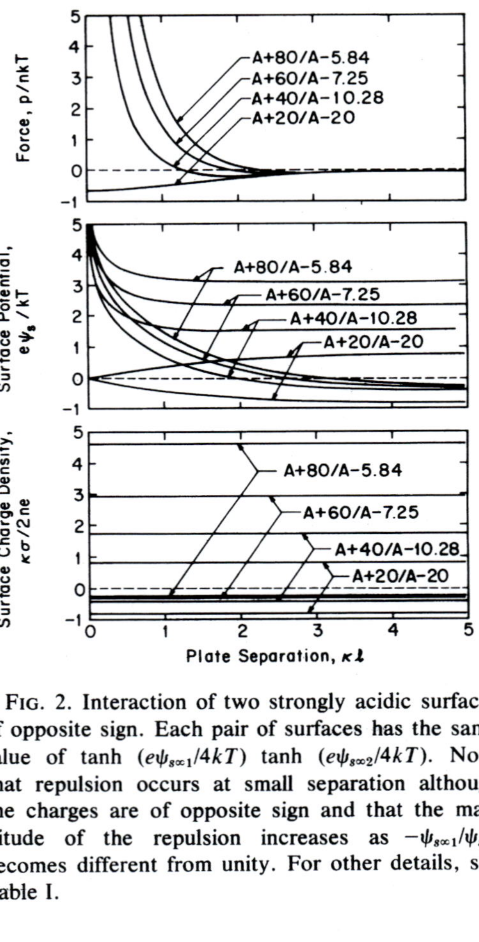

FIG. 2. Interaction of two strongly acidic surfaces of opposite sign. Each pair of surfaces has the same value of tanh $(e\psi_{s\infty 1}/4kT)$ tanh $(e\psi_{s\infty 2}/4kT)$. Note that repulsion occurs at small separation although the charges are of opposite sign and that the magnitude of the repulsion increases as $-\psi_{s\infty 1}/\psi_{s\infty 2}$ becomes different from unity. For other details, see Table I.

plates must increase. Boltzmann's equation, relating ion concentrations to electrostatic potential, then implies that the surface potentials must also increase in absolute magnitude without bound. On the other hand, if either surface is composed of ionizable groups whose degree of dissociation can be induced to change as $\kappa l \to 0$, then $\sigma_0 + \sigma_l \to 0$ and ψ_{s0} may remain bounded according to Eq. [2].

Much of the behavior of surface potentials and surface charge densities as the separation is changed can be rationalized by considering overall electroneutrality and the tendency for surface potentials to become equal as the separation vanishes, as discussed above. However, consider the B − 20/B − 10 interaction in Fig. 1. During the interaction of these two buffer-like surfaces, the surface potentials remain essentially constant as κl is decreased until the separation distance is much below a Debye length, whereupon the two surface potentials tend to become equal. At the same time, as $\kappa l \to 0$ the charge density on one surface undergoes a sign reversal and both surface charge densities tend to become enormous in absolute magnitude, but opposite in sign. Of course, then $\sigma_0 + \sigma_l \to 0$ so that ψ_{s0} can remain finite according to the electroneutrality condition of Eq. [2]. However, $\sigma_0 + \sigma_l \to 0$ could also be accomplished if the two surface charge densities individually tend to zero. Why do they instead tend to become large in absolute magnitude and opposite in sign?

The answer rests with the dissociation equilibria. In order for the two surface potentials to become equal as $\kappa l \to 0$, the potential on the more negative B − 20 surface must become less negative. Then the hydrogen ion concentration near this surface decreases, favoring the dissociation of additional protons from the surface and thus more negative surface charges. Likewise, the less negative B − 10 surface assumes a more negative surface potential as $\kappa l \to 0$ and a less negative surface charge density. In short, the two surface charge densities tend to become equal in magnitude but opposite in sign, rather than simultaneously tend to zero, because this behavior is consistent with the acid/base dissociation equilibria of the surface groups.

Qualitative Behavior of the Double-Layer Force at Small Separations

The double-layer force p acting between two parallel plane surfaces has two contributions—the Maxwell stress and the osmotic pressure:

$$p = -\frac{\epsilon}{8\pi}\left(\left.\frac{d\psi}{dx}\right|_{x=0}\right)^2 + 2nkT\{\cosh\,[e\psi(0)/kT] - 1\}. \quad [3]$$

The Maxwell stress arises because there exists a net charge on the surface. Since the net charge in solution next to the surface always bears a sign opposite to that of the charge on the surface, the Maxwell stress term is always attractive, tending to restore electroneutrality to the surface either by attracting the charges in solution to the surface or by bringing the surface closer to the charges in solution.

Osmotic pressure arises because the total concentration of all ions [irrespective of their charge, i.e., $c_+ + c_- = 2n \cosh (e\psi_s/kT)$] is always higher near the surface than in the reservoir (where $c_+ + c_- = 2n$). Thus the liquid in the reservoir tends to flow into the space between plates to dilute this concentration and thereby force the plates apart. So the osmotic pressure contribution is always repulsive. Of course, the sign of the net force will depend on whether the attractive Maxwell stress or the repulsive osmotic pressure is larger.

Note that the Maxwell stress next to one surface depends only on the charge density of that surface ($d\psi/dx|_{x=0} = -4\pi\sigma_0/\epsilon$ from Gauss' law), while the osmotic pressure on the same surface depends solely on the surface's potential $\psi(0)$.

Consider the limiting force at small separations. Will it be attractive or repulsive? Substituting $d\psi/dx|_{x=0} = -4\pi\sigma_0/\epsilon$ and $\psi(0) \rightarrow \psi_{s0}$ as given by Eq. [2] into Eq. [3], then applying $\cosh y = (1 + \sinh^2 y)^{1/2}$, one obtains:

$$\lim_{\kappa l\to 0} \frac{p}{nkT} = \lim_{\kappa l\to 0} \left\{ 2\left[1 + \left(\frac{\kappa}{2ne}\frac{\sigma_0 + \sigma_l}{\kappa l}\right)^2\right]^{1/2} - 2 - \left(\frac{\kappa\sigma_0}{2ne}\right)^2 \right\}. \quad [4]$$

Several limiting cases can now be discussed. Suppose that the surfaces are either strongly acidic or strongly basic (type A) so that σ_0 and σ_l are independent of κl. Then after introducing the definition of κ, Eq. [4] predicts

$$\lim_{\kappa l\to 0} p = -\frac{2\pi}{\epsilon}\sigma^2 < 0, \quad [5]$$

if σ_0 and σ_l are constant and $\sigma_0 = -\sigma_l = \sigma$, or

$$\lim_{\kappa l\to 0} p = \frac{kT}{e} \lim_{l\to 0} \frac{|\sigma_0 + \sigma_l|}{l} > 0, \quad [6]$$

if σ_0 and σ_l are constant and $\sigma_0 + \sigma_l \neq 0$. The absolute value arises from taking the principal square root in Eq. [4], which in turn is demanded since $\cosh y \geq 1$. Thus the force between constant charge surfaces becomes infinitely repulsive at small separation, regardless of the relative signs of the surface charges (see A − 20/A − 10 in Fig. 1, or A + 80/A − 5.84 in Fig. 2), except in the unlikely event that the two surface charges are precisely equal and opposite; in that case the limiting force is finite, attractive and given by Eq. [5] (see A + 20/A − 20 in Fig. 2).

Another limiting case is the interaction of two amphoteric surfaces with a high buffer capacity. In the case that the surfaces are identical, it was observed above that $\sigma_0 \rightarrow \sigma_l \rightarrow 0$ as $\kappa l \rightarrow 0$ to satisfy overall electroneutrality. Then the Maxwell stress term in Eq. [3] vanishes, leaving:

$$\lim_{\kappa l\to 0} p = 2nkT \times [\cosh(e\psi_{s0}/kT) - 1] \geq 0. \quad [7]$$

On the other hand, if the amphoteric surfaces are not identical $\sigma_0 + \sigma_l$ still tends to zero, but computations show that σ_0 and σ_l tend to become equal and large in magnitude and opposite in sign. Then the osmotic pressure term in Eq. [4] remains bounded, but the Maxwell stress can overwhelm the repulsive contribution (as in Eq. [5]) so that:

$$\lim_{\kappa l\to 0} p \approx -\frac{2\pi}{\epsilon}\sigma^2 = -\frac{2\pi e^2}{\epsilon} \times [\min(N_0, N_l)]^2 < 0. \quad [8]$$

To obtain the second relation above, it was assumed that each surface has an equal number of acidic and basic groups, which is N_0 on the surface at $x = 0$ and N_l on the surface at $x = l$. Since the two surface charge densities become equal in absolute value and since $|\sigma_i/e|$ cannot exceed the number of acidic or basic groups on either surface, it is the surface bearing the smaller number of groups which determines the limiting charge density σ in Eq. [8]. Equation [8] predicts a strongly attractive force between nonidentical highly buffered surfaces as $\kappa l \rightarrow 0$, as for B − 20/B − 10 in Fig. 1.

The final limiting case covered in this paper is the interaction between a strongly acidic or basic surface and a highly buffered amphoteric surface. For practical purposes this may be considered as the interaction of a constant charge surface with a constant potential surface. Then the limiting value of the surface potentials (ψ_{s0}) must be approximately equal to the potential ($\psi_{sB\infty}$) on the buffer-like surface at large separations and, to satisfy overall electroneutrality, the charge density on the buffer-like surface becomes equal and opposite to that on the constant-charge surface ($\sigma_{A\infty}$). Substituting these limiting values into Eq. [3]:

$$\lim_{\kappa l \to 0} \frac{p}{nkT} \approx 2[\cosh(e\psi_{sB\infty}/kT) - 1] - \left(\frac{\kappa\sigma_{A\infty}}{2ne}\right)^2. \quad [9]$$

Finally, it may be noted that, since $\lim_{\kappa l\to\infty} p = 0$, $\sigma_{A\infty}$ may be related to $\psi_{sA\infty}$ through Eq. [3] (which reduces to the Gouy—Chapman equation when $p = 0$). Substituting this relation into [9]:

$$\lim_{\kappa l \to 0} p \approx 2nkT[\cosh(e\psi_{sB\infty}/kT) - \cosh(e\psi_{sA\infty}/kT)]. \quad [10]$$

Thus the sign of the limiting force is determined solely by the absolute magnitude of surface potential on the type-B surface, relative to the absolute magnitude of the surface potential at large separation on the type-A surface. For $|\psi_{sB\infty}| < |\psi_{sA\infty}|$ the force becomes attractive at small separations (see A − 20/B − 10 in Fig. 1); for $|\psi_{sB\infty}| = |\psi_{sA\infty}|$ the force tends to vanish; and for $|\psi_{sB\infty}| > |\psi_{sA\infty}|$ the force becomes repulsive (see B − 20/A − 10 in Fig. 1).

Superficially, it may appear surprising that surfaces bearing opposite charges may either attract or repel each other (Fig. 2). However, these surfaces are not separated by a vacuum so that Coulomb's law is not directly applicable. Instead, the intervening space contains mobile charges (ions) which accumulate near an oppositely charged surface. This accumulation can mask the true charge on the surface, allowing either repulsion or attraction in the case of oppositely charged surfaces. On the other hand, in the case that the two surfaces have net charges of the same sign, they must share the same counterions according to the magnitude of their net charge density. This prevents either surface from becoming completely masked by counterions. Thus at a given separation, if the net charge densities of the two interacting surfaces are of the same sign, then the force is always repulsive, but when the net charges are of opposite sign and unequal magnitude the force may have either sign.

A summary of the behavior of the double-layer interaction in the limit of small separations is provided in Table II.

Calculation of the Deposition Rate

A true test of the significance of different forces computed between similar pairs of surfaces (e.g., those pairs in Fig. 1) is whether these different forces lead to different rates of deposition. Toward this end, the rate of deposition of spherical hydrosol particles onto a rotating disk collector

TABLE II

Limiting Behavior at Small Separations

Interactions[a]	Restriction	$\lim_{\kappa l\to 0} \psi_s \equiv \psi_{s0}$	$\lim_{\kappa l\to 0} \sigma$	$\lim_{\kappa l\to 0} p$
A-A	$\sigma_0 = -\sigma_l \equiv \sigma$	0	$+\sigma, -\sigma$	$-2\pi\sigma^2/\epsilon$
	$\sigma_0 \neq -\sigma_l$	$\pm\infty$	σ_0, σ_l	$kT\lvert\sigma_0 + \sigma_l\rvert/(el)$
A-B		$\approx\psi_{sB\infty}$	$\sigma_0 \to \sigma_l \approx \sigma_{A\infty}$	$\approx 2nkT[\cosh(e\psi_{sB\infty}/kT) - \cosh(e\psi_{sA\infty}/kT)]$
B-B	$\sigma_0 = \sigma_l$	Finite	0	$2nkT[\cosh(e\psi_{s0}/kT) - 1]$
	$\sigma_0 \neq \sigma_l$	Finite	$\sigma_0 \to -\sigma_l \equiv \sigma$ $\lvert\sigma\rvert = \min(N_0, N_l)e$	$-2\pi\sigma^2/\epsilon$

[a] Letters in this column denote the types of the two interacting surfaces, where the types have been defined by Table I.

was computed by the method of Ruckenstein and Prieve (10). The usual convective-diffusion equation is solved subject to a first-order, irreversible, reaction-like, boundary condition on the collector surface. The boundary condition contains all the effects of the interaction forces and of the decreasing diffusion coefficient and particle mobility near the collector. They compute the apparent reaction rate constant K according to:

$$K = D(h_{\max})(\gamma/2\pi kT)^{1/2} \times \exp(-E/kT), \quad [11]$$

where $h_{\max}$ is the normal distance from the collector at which the total potential energy of interaction $\phi(h)$ experiences a maximum $E = \phi(h_{\max})$, $\gamma \equiv -d^2\phi/dh^2|_{h=h_{\max}}$, and $D(h)$ is the local diffusion coefficient. Modifying Levich's (11) boundary-layer solution by imposing the apparent first-order reaction at the surface of the rotating disk, one obtains N_y, the rate of deposition per unit area of disk:

$$N_y = \left(\frac{1}{K} + \frac{1}{V}\right)^{-1} c_\infty, \qquad V \equiv 0.620 D_\infty{}^{2/3}\omega^{1/2}v^{-1/6}, \quad [12]$$

where c_∞ is the bulk volumetric concentration of particles, D_∞ is the diffusion coefficient of particles in the bulk, ω is the disk rotation speed, and v is the fluid's kinematic viscosity. The principal assumption in this approach is that the distance over which the forces act and over which the mobility varies is short compared to the thickness of the diffusion boundary layer. While Eq. [11] also requires $E \geq 3kT$, this assumption is not critical since smaller E leads to large K for which Eq. [12] simply reduces to $N_y = Vc_\infty$ (insensitive to K).

For the calculations discussed below, the total potential energy of interaction was decomposed into contributions from van der Waals forces and double-layer forces: $\phi(h) = \phi_{\mathrm{VdW}}(h) + \phi_{\mathrm{DL}}(h)$. The van der Waals contribution was computed from Hamaker's superposition principle (12)

$$\phi_{\mathrm{VdW}}(h) = A\left(\frac{1}{6}\ln\frac{h+2a}{h} - \frac{1}{3}\frac{a}{h}\frac{h+a}{h+2a}\right), \quad [13]$$

while Derjaguin's approximation (13) was used to obtain the double-layer energy:

$$\phi_{\mathrm{DL}}(h) = 2\pi a \int_h^\infty \int_{l'}^\infty p(l)\,dl\,dl', \quad [14]$$

where A is Hamaker's constant, a is the particle radius, $h + a$ is the distance between the particle's center and the collector surface, and $p(l)$ is the double-layer force per unit area between parallel plates separated by a distance l.

Rates of Deposition if Particle and Collector Initially Have Like Charge

Figure 3 summarizes the rates of deposition calculated for hydrosols depositing onto a rotating disk. The four curves correspond to the four pairs of surfaces whose interactions are characterized in Fig. 1. Surface characteristics of particle and collector are interchangeable in the calculation of the rates. Values of other parameters include $a = 0.1$ μm, $c_\infty = 10^8$ cm^{-3}, $\omega = 6$ rev/sec, and $v = 0.01$ cm^2/sec. Rates are presented as a function of Hamaker's constant, which characterizes the van der Waals attraction, because this parameter is most difficult to experimentally determine, and because the rate is sensitive to its value.

The curves of Fig. 3 display two distinct regions. For small values of Hamaker's constant, van der Waals attraction is weak and a large energy barrier E develops due to the electrostatic repulsion. This makes K small and when small enough, Eq. [12] reduces to $N_y = Kc_\infty$, which corresponds to "reaction"-controlled deposition. Conversely, when A is large,

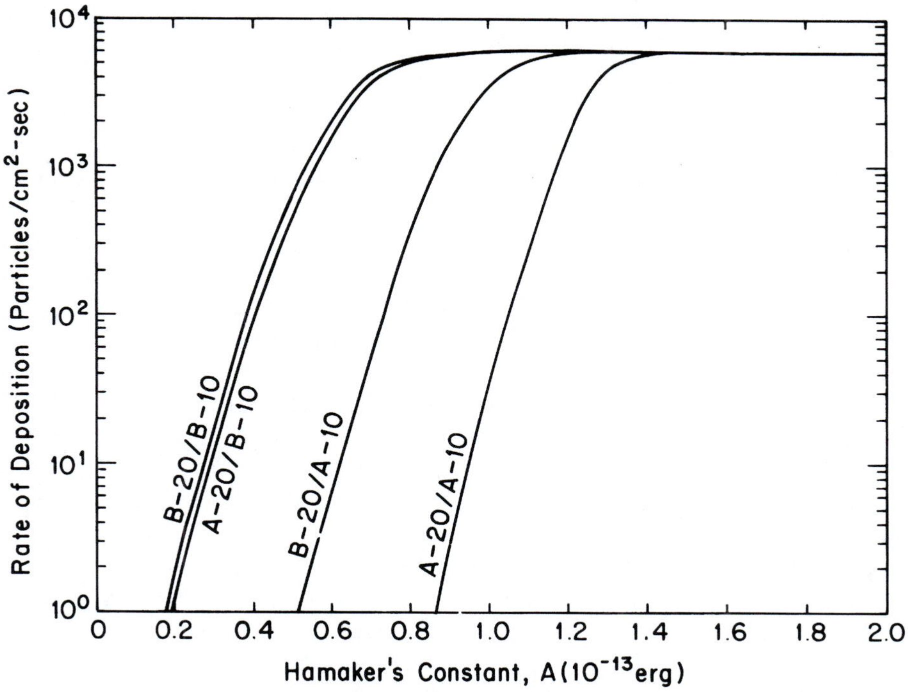

FIG. 3. A comparison of the rates of deposition for the four pairs of surfaces whose double-layer interactions are characterized by Fig. 1. Note that the mixed cases fall between the cases involving the same type of surface.

the strong van der Waals attraction overwhelms the double-layer repulsion, making the energy barrier small or nonexistent. This causes K to be large and Eq. [12] reduces to the mass-transfer limit $N_y = Vc_\infty$. Then the rate of deposition is insensitive to A.

For a given value of Hamaker's constant, the rate of deposition is lowest for the interaction of two type-A surfaces and highest for two type-B surfaces, with the interaction of mixed types corresponding to intermediate rates. Note that several orders of magnitude (at least 10^4) separate the lowest and highest rates for a given Hamaker constant if the deposition rate is "reaction"-controlled. Ordering the rates according to A − 20/A − 10 < B − 20/A − 10 < B − 20/B − 10 is consistent with Fig. 1 in that the A − 20/A − 10 interaction is more strongly repulsive at every separation than the B − 20/A − 10 interaction, which in turn is more strongly repulsive than the B − 20/B − 10 interaction. However, a comparison between A − 20/B − 10 and B − 20/B − 10 is more difficult since one pair experiences a greater repulsion than the other for some separations, while at other separations the opposite ordering of repulsive force is obtained. Although the limiting values of the double-layer force at very small separations are quite different for A − 20/B − 10 and B − 20/B − 10 (Fig. 1), the corresponding rates of deposition are nearly the same (Fig. 3).

Apparently, the asymptotic behavior of the double-layer interaction at very small separations does not strongly affect the rate. This may be rationalized as follows. The height E of the energy barrier opposing deposition is the single most important property of the interaction forces. The double-layer contribution to E equals the integral of the repulsive force over separations *larger* than h_{max}. Double-layer interactions at smaller separations may help locate $h_{\max}$, but have only a second order effect on E. For Fig. 3, the barrier was located at separations between 0.7 and 1.4 Debye lengths ($1/\kappa$), depending on the magnitude of Hamaker's constant.

Deposition of Hydrosols on Oppositely Charged Collectors

The double-layer force between the oppositely charged surfaces is attractive at all separations for many pairs of surfaces. In the absence of an energy barrier, the theory presented here then predicts that K will be very large and so the rate of deposition will always equal Vc_∞. This result has also been experimentally verified by Hull and Kitchener (14). If the purely attractive forces act over distances which are short compared to the diffusion boundary-layer thickness, then the rate will not exceed Vc_∞ regardless of the strength of the attraction: the resistance of the diffusion boundary-layer controls the rate. However, if the attractive forces act over distances comparable to the diffusion boundary-layer thickness (calculated by ignoring the interaction forces) then the rate may exceed Vc_∞, although the amount of the excess cannot be calculated from the theory presented here; instead the complete transport equations must be solved. An example of such a calculation is provided by Ref. (15), where the rate of deposition of hydrosols was calculated for the case of creeping flow over a large spherical collector in the presence of a purely attractive particle–collector interaction. Based on the results reported in Ref. (15), attractive van der Waals forces significantly affect the rate in the absence of double-layer repulsion whenever $(A/kT)^{6/7}$ (Pe/R^3) > 0.002, where Pe is the Peclet number $(2a_sU/D_\infty)$, R is the collector-to-particle radius ratio (a_s/a), and U is the approach velocity of the fluid. In order to extend this result to the rotating disk one might replace the characteristic hydrodynamic length a_s by $(\nu/\omega)^{1/2}$ and the characteristic velocity U by $(\nu\omega)^{1/2}$. This analogy would predict that London forces are important if $(A/kT)^{6/7}\omega^{3/2}\nu^{-1/2}D_\infty{}^{-1}a^3 > 0.002$.

Conversely, some pairs of oppositely charged, plane surfaces experience double-layer repulsion when the separation is sufficiently small (see Fig. 2). However, since the double-layer force is always attractive between oppositely charged surfaces at large separations, it is still possible that no barrier to deposition will arise. Particles at a given distance from the collector may have a lower potential energy of interaction than at infinite separation, although net repulsion occurs at that distance. A necessary condition for the existence of a significant barrier to deposition is that the double integral in Eq. [14] must become sufficiently positive at some separation so that, when added to the corresponding negative van der Waals energy ϕ_{VdW}, the sum is greater than a few kT.

As an example of how this might be accomplished, consider the four pairs of type-A surfaces whose double-layer interactions are illustrated in Fig. 2. Equation [6] states that the repulsion between constant-charge surfaces of opposite sign at small separations is proportional to $|\sigma_0 + \sigma_l|$, whereas Ref. (3) reports the attraction at large separations ($\kappa l \gg 1$) is proportional to $\tanh\,(e\psi_{s\infty 1}/4kT)\,\tanh\,(e\psi_{s\infty 2}/4kT) \equiv \beta$. Now $|\sigma_0 + \sigma_l|$ can be increased while holding β fixed by making $-\psi_{s\infty 1}/\psi_{s\infty 2}$ different from unity. The surface potentials at infinite separation were chosen in Fig. 2 so that each pair of surfaces has the same value of β; in other words, the double-layer attraction at large separations is the same for all pairs. The behavior of the force at small separations is in accordance with Eqs. [5] and [6]: at small separations the strength of the repulsion increases as the surface potentials at infinite separation (hence the surface charge densities) become more dissimilar in magnitude.

Figure 4 gives the calculated rates of deposition corresponding to the four pairs of surfaces whose double-layer interactions are reported in Fig. 2. Values of other parameters are the same as in Fig. 3. Note that if the two surface potentials at infinite separation are sufficiently dissimilar, the double-layer repulsion occurring at small separation can become sufficiently strong to lower the rate of deposition by many orders of magnitude, although the charge on particle and collector are of opposite sign. The principal condition is that the magnitude of the charge density or zeta-potential of the two surfaces be sufficiently dissimilar. No experimental rates of deposition of hydrosols on oppositely charged collector under these conditions have been reported in the literature.

SUMMARY AND CONCLUSIONS

Double-layer interactions were computed between dissimilar plane, parallel surfaces which are either strongly acidic, strongly basic, or amphoteric with a high buffer capacity. Sign reversals in both the double-layer force and surface charge density occurred as the separation was decreased, which would not have occurred if the interacting surfaces were identical.

A qualitative analysis of the governing equations was performed, which allowed prediction of the limiting behavior of the double-layer force, surface charge and surface potential at small separations without solving the complete equations. Between strongly acidic or strongly basic surfaces, this qualitative analysis predicts that the double-layer force always becomes strongly repulsive at small separations, regardless of the relative signs of the surface charges, unless the two surfaces have exactly equal but opposite charge densities. Conversely, the force always becomes attractive at small separations between amphoteric surfaces having a large buffer capacity, re-

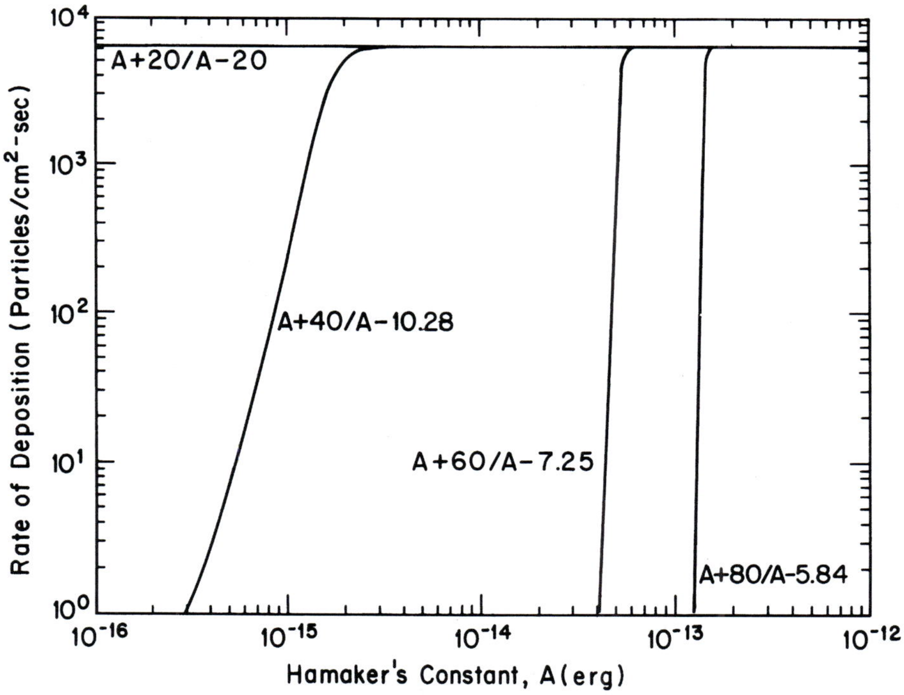

FIG. 4. A comparison of the rates of deposition corresponding to the four pairs of surfaces in Fig. 2. Note that as $-\psi_{s\infty 1}/\psi_{s\infty 2}$ becomes sufficiently different from unity the rate can be reduced by many orders of magnitude by the double-layer repulsion at small separations although particle and collector have opposite charges.

gardless of the relative signs of the surface charges at large separation, unless the two surfaces are identical. Finally, the sign of the limiting force at small separation between a strongly acidic (or basic) surface and an amphoteric surface depends solely on the absolute values of the two surface potentials at infinite separation: if the amphoteric surface has a larger potential than the acidic surface then the force becomes repulsive, whereas the force becomes attractive if the acidic surface has the larger potential. Again, the sign of the limiting force at small separation is independent of the relative signs of the two surface potentials at infinite separation. Table II quantitatively summarizes the limiting behavior described above. Solutions of the complete equations governing the double-layer interaction confirm this behavior predicted by the qualitative analysis.

Rates of adsorption of hydrosols onto a rotating-disk collector were also calculated for cases in which the particle and collector surfaces correspond to those pairs of surfaces for which the double-layer interaction had been computed. Van der Waals attraction and the position-dependence of the particle's diffusion coefficient were also considered in the analysis. Double-layer repulsion can lower the rate of adsorption by many orders of magnitude compared to rates calculated on the basis of convective-diffusion only. For given charge densities on the particle and collector surfaces at infinite separation, the lowest rates were obtained when both surfaces are strongly acidic, while the largest rates were calculated when both surfaces are amphoteric with a high buffer capacity. However, the rates in these two extremes can be orders of magnitude apart.

Finally, it was observed from the calculations that, if particle and collector surfaces bear fixed charge densities of opposite sign but sufficiently different magnitudes, then the double-layer repulsion which occurs at small separations is strong enough to lower the rate of deposition by orders of magnitude. Thus the calculations predict that if the absolute values of the two surface charge densities differ enough, then hydrosols will not deposit on a collector, even when particle and collector bear opposite charges.

NOMENCLATURE

a	particle radius (cm)
A	Hamaker's constant (erg)
c_i	local concentration of ith ion (ions/cm^3)
c_∞	particle concentration far from collector (particles/cm^3)
D	local diffusion coefficient (cm^2/sec)
D_∞	diffusion coefficient far from collector (cm^2/sec)
e	magnitude of electron charge (statcoul)
E	$\phi(h_{max})$(erg)
h	minimum distance between particle and collector surfaces (cm)
h_{max}	value of h corresponding to a maximum in $\phi(h)$(cm)
k	Boltzmann's constant (erg/°K)
K	apparent first-order reaction rate constant (cm/sec)
l	distance between two plane surfaces (cm)
n	ionic strength (ions/cm^3)
N_a	number density of acidic sites on surface (cm^{-2})
N_b	number density of basic sites on surface (cm^{-2})
N_y	flux of particles to collector (particles/cm^{-2}-sec)
p	repulsive double-layer force between two plane surfaces (dyn/cm^2)
pH_{iso}	pH at which σ is identically zero
pH_r	pH of the reservoir where $\psi \equiv 0$
pK_a	acid pK for acidic surface groups
pK_b	acid pK for basic surface groups

T absolute temperature (°K)

x distance measured normal from one of two plane surfaces (cm)

γ $-d^2\phi/dh^2$ evaluated at $h = h_{max}$ (erg/cm^2)

ϵ dielectric constant of fluid

κ reciprocal Debye length, $\kappa = (8\pi ne^2/\epsilon kT)^{1/2}$(cm^{-1})

ν kinematic viscosity of fluid (cm^2/sec)

ρ local net space charge density (statcoul/cm^3)

σ surface charge density (statcoul/cm^2)

ϕ total potential energy of interaction between a sphere and a plate (erg)

ϕ_{vdw}, ϕ_{DL} contributions to ϕ from van der Waals or double-layer forces, respectively (erg)

ψ local electrostatic potential between parallel plates (statvolts)

ω rotational speed of disk (radian/sec)

Subscripts

s surface

o limit as $\kappa l \rightarrow 0$

∞ limit as $\kappa l \rightarrow \infty$

0 surface at $x = 0$

l surface at $x = l$

A type-A surface

B type-B surface

REFERENCES

1. Ninham, B. W., and Parsegian, V. A., *J. Theor. Biol.* **31,** 405 (1971).
2. Prieve, D. C., Ph.D. dissertation, Univ. of Delaware, Newark, 1974.
3. Prieve, D. C., and Ruckenstein, E., *J. Theor. Biol.* **56,** 205 (1976).
4. Chan, D., and White, L. R., *J. Theor. Biol.* **48,** 253 (1974).
5. Chan, D., Perram, J. W., White, L. R., and Healy, T. W., *J.C.S. Faraday I* **71,** 1046 (1975).
6. Prieve, D. C., and Ruckenstein, E., *J. Colloid Interface Sci.* **60,** 337 (1977).
7. Derjaguin, B. V., *Discuss. Faraday Soc.* **18,** 85 (1954).
8. Parsegian, V. A., and Gingell, D., *Biophys. J.* **12,** 1192 (1972).
9. Prieve, D. C., and Ruckenstein, E., "Colloid and Interface Science," (M. Kerker, Ed.), p. 73. Academic Press, New York, 1976.
10. Ruckenstein, E., and Prieve, D. C., *J.C.S. Faraday II* **69,** 1522 (1973).
11. Levich, V. G., "Physiocochemical Hydrodynamics," Prentice-Hall, Englewood Cliffs, N.J., 1962.
12. Hamaker, H. C., *Physica* **4,** 1058 (1937).
13. Hogg, R., Healy, T. W., and Fuerstenau, D. W., *Trans. Faraday Soc.* **62,** 1638 (1966).
14. Hull, M., and Kitchener, J. A., *Trans. Faraday Soc.* **65,** 3093 (1969).
15. Prieve, D. C., and Ruckenstein, E., *AIChE J.* **20,** 1178 (1974).

Reversible Rate of Adsorption or Coagulation of Brownian Particles—Effect of the Shape of the Interaction Potential

E. RUCKENSTEIN[1]

Katholieke Universiteit Leuven, Instituut voor Chemie-ingenieurstechniek, de Croylaan 2, B-3030 Leuven (Heverlee), Belgium

Received January 9, 1978; accepted April 5, 1978

Kinetic equations for reversible adsorption and reversible coagulation are established when the interaction potential has primary and secondary minima of comparable depths. The process is assumed to occur in two successive steps. First the particles move from the bulk of the fluid to the secondary minimum. A fraction of the particles which have arrived at the secondary minimum move further to the primary minimum. Quasi-steady state is assumed for each of the steps separately. Conditions are identified under which rates of reversible adsorption or coagulation at the primary minimum can be computed by neglecting the rate of accumulation at the secondary minimum. The interaction force boundary layer approach has been improved by introducing the tangential velocity of the particles near the surface of the collector into the kinetic equations. To account for reversibility a short-range repulsion term is included in the interaction potential.

INTRODUCTION

Compared to small molecules the description of convective diffusion of particles of finite size in a fluid near a solid boundary has to account for both the interaction forces between particles and collector (such as van der Waals and double-layer forces) and for the hydrodynamic interactions between particles and fluid. The effect of the London–van der Waals forces and double-layer attractive forces is important if the range over which they act is comparable to the thickness over which the convective diffusion affects the transport of the particles. If, however, because of the competition between the double-layer repulsive forces and London attractive forces, a potential barrier is generated, then the effect of the interaction forces is important even when they act over distances much shorter than the thickness of the diffusion boundary layer. For the latter case Ruckenstein and Prieve (1) and Spielman and Friedlander (2) have suggested an approximate method to predict the rate of deposition of particles. In that method, fluid convection is neglected inside a thin layer near the collector, and the hydrodynamic interactions and the interaction forces are neglected outside the thin interaction force boundary layer. The interaction force boundary layer approximation reduces the computation of the rate of deposition to the solution of the convective-diffusion equation used for small molecules for a boundary condition at the surface of the collector which has the form of a first-order chemical reaction. By including short-range repulsive forces among the interaction forces, Ruckenstein and Prieve (3) allow for a finite depth of the primary minimum and therefore for reversible adsorption. In this case the boundary condition takes the form of a reversible first-order chemical reaction.

One of the main assumptions made in the treatment of the rate of deposition of par-

[1] On leave from State University of New York at Buffalo, Faculty of Engineering and Applied Sciences, Buffalo, N. Y., 14214.

0021-9797/78/0663-0531$02.00/0

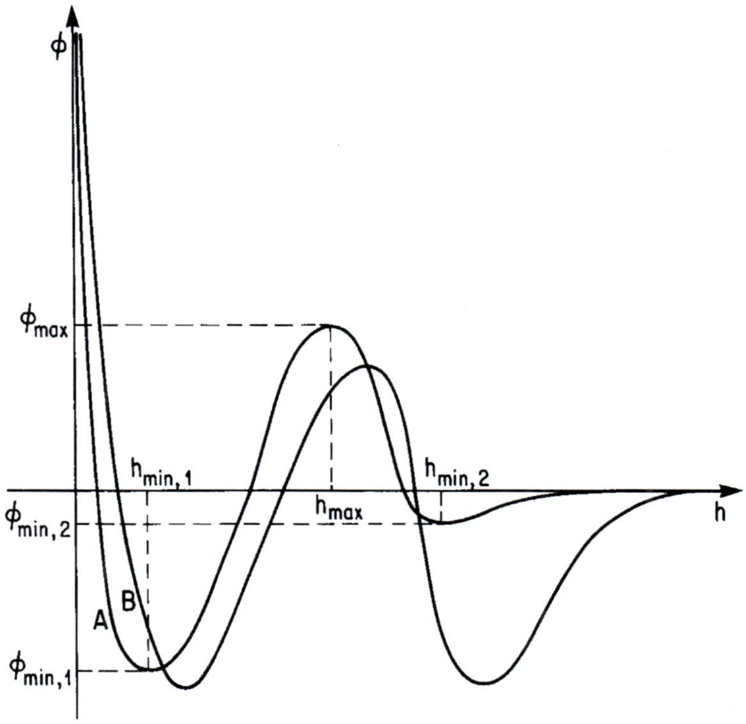

FIGURE 1

ticles on a collector is that accumulation of particles occurs mainly at the primary minimum. Intuition suggests that this happens if the secondary minimum is shallow compared to the primary minimum and if the potential barrier is not too high (Fig. 1, curve A). If, however, the secondary minimum is also sufficiently deep (Fig. 1, curve B), accumulation occurs in both places. The main goal of the present paper is to develop rate equations when accumulation occurs at both minima, and to identify the conditions under which, after a short transient, accumulation is restricted to the primary minimum. The theoretical considerations are carried out in the framework of the interaction force boundary layer approach assuming separately, quasi-steady-state diffusion in a force field between the bulk and the secondary minimum, and between the secondary and the primary minimum. Because similar problems arise in the coagulation kinetics, they are examined in the last part of the paper.

In the interaction force boundary layer approach the tangential velocity of the particles in the layer is neglected. Because the center of sufficiently large particles is far from the wall (even when the minimum distance h between wall and particle is small) this velocity may be appreciable. A second goal of the paper is to account for this effect.

Adsorption of particles on a collector, as well as their coagulation, is usually assumed irreversible. There are, however, numerous cases in which, by changing the ionic strength or the pH, the coagulated sols redisperse (4). This phenomenon, known under the name of repeptization was examined recently from a theoretical point of view (3). In addition Matijevic (5) has investigated the adsorption and desorption of monodisperse spherical colloidal particles of chromium hydroxide on glass as a function of pH and ionic strength, using a packed column of glass beads. He observed that the particles desorb for pH > 10 if the concentration of electrolyte is smaller than about 0.1 *M*. For these reasons the emphasis in the present paper is on reversible adsorption and reversible coagulation.

KINETICS OF REVERSIBLE ADSORPTION

It was shown in Ref. (3) that depending upon the values of the parameters several types of shapes are possible for the interaction potential. An interaction potential having two deep minima separated by a moderate potential barrier is assumed. Because the minima are deep, the relaxation times needed to reach equilibrium distribution at each is assumed short compared to the time during which the concentration of particles in these two "points" changes appreciably. Therefore near and on either side of the secondary minimum in a range $\delta_2 < h < \delta_2'$ a Boltzmann distribution is valid for the concentration $c(h)$ of particles

$$c(h) = C_2 \exp[-\phi(h)/kT] \quad \text{for} \quad \delta_2 < h < \delta_2'. \qquad [1]$$

Here $\phi(h)$ is the interaction potential between collector and particles, h is the minimum distance between them, k is Boltzmann's constant, and T is the temperature in degrees Kelvin. The number of particles n_2 accumulated per unit area at the secondary minimum is defined by

$$n_2 = \int_{\delta_2}^{\delta_2'} c(h)dh = C_2 \int_{\delta_2}^{\delta_2'} \times \exp[-\phi(h)/kT]dh \equiv C_2 I_2. \qquad [2]$$

When the integration constant C_2 is eliminated, Eq. [1] becomes

$$c(h) = \frac{n_2 \exp[-\phi(h)/kT]}{I_2}. \qquad [3]$$

For the region near the primary minimum, one obtains similarly that

$$c(h) = \frac{n_1 \exp[-\phi(h)/kT]}{I_1}, \qquad [4]$$

where n_1 is the number of particles accumulated per unit area at the primary minimum, defined by

$$n_1 = \int_{\delta_1}^{\delta_1'} c(h)dh = C_1 \int_{\delta_1}^{\delta_1'} \times \exp[-\phi(h)/kT]dh \equiv C_1 I_1 \quad \text{for} \quad \delta_1 < h < \delta_1'. \qquad [5]$$

Here δ_1, and δ_1' are values of h near and on either side of the primary minimum.

The flux of particles is the sum of a diffusional contribution and of a contribution due to the interaction forces between collector and particles [the interaction forces between particles are neglected here but are discussed elsewhere (6)]. Hence

$$J = -D(h)\frac{\partial c}{\partial h} - m(h)\frac{d\phi}{dh}c, \qquad [6]$$

where the mobility coefficient m and the diffusion coefficient D are distance dependent to account for the fact that as a particle approaches the collector the drainage of the liquid between particle and collector is increasingly difficult. Einstein's equation

$$D = mkT \qquad [7]$$

provides a link between the two coefficients, and the mobility coefficient is given by the equation

$$m(h) = \frac{f(h/a)}{6\pi\mu a}. \qquad [8]$$

Here a is the radius of the particle, μ is the dynamic viscosity of the liquid, and $f(h/a)$ is a friction factor correction calculated by Brenner (7), which has the following asymptotic behavior: $f(Z) \to Z$ when $Z \ll 1$, and $F(Z) \to 1$ when $Z \gg 1$. Assuming quasi-steady state between $h = \delta_1'$ and $h = \delta_2$, Eq. [6] leads to

$$-J = \left(\frac{n_2}{I_2} - \frac{n_1}{I_1}\right) \times \left\{\int_{\delta_1'}^{\delta_2} \frac{\exp[\phi(h)/kT]}{D(h)} dh\right\}^{-1}. \qquad [9]$$

If the minima are deep enough and the maximum is sufficiently high, one can expand $\phi(h)$ around the corresponding extremum as

$$\phi(h) = \phi(h_{\min}) + \tfrac{1}{2}\gamma_{\min}(h - h_{\min})^2, \quad \gamma_{\min} = \left(\frac{d^2\phi}{dh^2}\right)_{h=h_{\min}}, \quad [10a]$$

$$\phi(h) = \phi(h_{\max}) - \tfrac{1}{2}\gamma_{\max}(h - h_{\max})^2, \quad \gamma_{\max} = -\left(\frac{d^2\phi}{dh^2}\right)_{h=h_{\max}}. \quad [10b]$$

If the following conditions are satisfied

$$(\delta_i' - h_{\min,i})\left(\frac{\gamma_{\min,i}}{2kT}\right)^{1/2} \geqslant 2; \quad (h_{\min,i} - \delta_i)\left(\frac{\gamma_{\min,i}}{2kT}\right)^{1/2} \geqslant 2 \quad (i = 1, 2) \quad [11a]$$

$$(\delta_2 - h_{\max})\left(\frac{\gamma_{\max}}{2kT}\right)^{1/2} \geqslant 2; \quad (h_{\max} - \delta_1')\left(\frac{\gamma_{\max}}{2kT}\right)^{1/2} \geqslant 2, \quad [11b]$$

then using the extremum as the origin, the limits of integration can be extended from $-\infty$ to $+\infty$ to obtain

$$I_1 = \left(\frac{2\pi kT}{\gamma_{\min,1}}\right)^{1/2} \exp[-\phi(h_{\min,1})/kT], \quad [12a]$$

$$I_2 = \left(\frac{2\pi kT}{\gamma_{\min,2}}\right)^{1/2} \exp[-\phi(h_{\min,2})/kT], \quad [12b]$$

and

$$\left[\int_{\delta_1'}^{\delta_2} \frac{\exp[\phi(h)/kT]}{D(h)} dh\right]^{-1} = D(h_{\max})\left(\frac{\gamma_{\max}}{2\pi kT}\right)^{1/2} \times \exp[-\phi(h_{\max})/kT]. \quad [12c]$$

Since $D(h)$ varies slowly compared to the exponential and because the main contribution of the integral is near $h_{\max}$, $D(h)$ was replaced in [12c] by $D(h_{\max})$. With these approximations, Eq. [9] becomes

$$-J = K''n_2 - K'n_1, \quad [13]$$

where

$$K'' = D(h_{\max})\frac{(\gamma_{\max}\gamma_{\min,2})^{1/2}}{2\pi kT} \times \exp[-(\phi_{\max} - \phi_{\min,2})/kT] \quad [14]$$

and

$$K' = D(h_{\max})\frac{(\gamma_{\max}\gamma_{\min,1})^{1/2}}{2\pi kT} \times \exp[-(\phi_{\max} - \phi_{\min,1})/kT]. \quad [15]$$

Equations [13] to [15] are very similar to those already established by Ruckenstein and Prieve (3). Because the adsorption coefficient is defined here in terms of a concentration per unit area instead of a concentration per unit volume the expressions of the adsorption coefficient K'' and of the desorption coefficient K' are symmetrical.

Equations [13], [14], and [15] involve the assumption that the time scale of the process is large compared to the relaxation time τ of the velocity distribution of particles, hence that this distribution reaches equilibrium rapidly in each of the points of the system. A measure of this relaxation time is the reciprocal of the friction coefficient obtained from the Langevin equation for the Brownian motion of a free particle ($\tau = M/6\pi a\mu$), where M is the mass of the particle. If this condition is not satisfied, the Fokker–Planck equation (8) should be the starting point of the analysis.

RATES OF ACCUMULATION AT THE PRIMARY AND SECONDARY MINIMUM

Denoting by $-J'$ the rate of transfer, per unit area, of particles from the bulk of the fluid to the secondary minimum, the conservation of particles leads to

$$\partial n_2/\partial t = -J' - (K''n_2 - K'n_1) \quad [16]$$

and

$$\partial n_1/\partial t = K''n_2 - K'n_1, \quad [17]$$

where t represents the time. $-J'$ is computed from equations valid for the convective diffusion of small molecules

$$-J' = K(c_\infty - c_i), \quad [18]$$

where K is the mass transfer coefficient, c_∞

is the concentration of particles in the bulk of the liquid, and c_i is the concentration of particles at the border of the interaction force boundary layer. Because the resistance to the motion of the particles from the border of the interaction force boundary layer to the secondary minimum is small compared to the other resistances, thermodynamic equilibrium is assumed in that region. Consequently

$$c_i = \frac{n_2}{I_2} = n_2\left(\frac{\gamma_{\min,2}}{2\pi kT}\right)^{1/2} \times \exp[\phi(h_{\min,2})/kT] \equiv Hn_2. \quad [19]$$

The kinetic description just formulated is acceptable as long as particles accumulated at the primary minimum do not react (sinter) with the collector. If sintering occurs, then Eq. [17] has to be replaced by

$$\frac{\partial n_1}{\partial t} = K''n_2 - K'n_1 - \frac{\partial n_0}{\partial t},$$

where n_0 is the number of particles sintered per unit area. One may assume that $\partial n_0/\partial t = K_0 n_1$. If the time of sintering $(K_0)^{-1}$ is large compared to the desorption time, $(K')^{-1}$, the effect of sintering can be neglected.

CONDITIONS FOR ACCUMULATION TO OCCUR ONLY AT THE PRIMARY MINIMUM

Using Eq. [18] for $-J'$, Eqs. [16] and [17] become

$$\frac{dn_1}{dt} = K(c_\infty - Hn_2)$$

$$-\frac{dn_2}{dt} = K''n_2 - K'n_1. \quad [20]$$

Equations [20] are solved for the initial condition

$$n_1 = n_2 = 0 \quad \text{for} \quad t = 0. \quad [21]$$

One obtains

$$n_1 = \frac{c_\infty K''}{HK'}\Bigg[1 - \frac{(K' + K'' + KH) + \{(K' + K'' + KH)^2 - 4KHK'\}^{1/2}}{2\{(K' + K'' + KH)^2 - 4KHK'\}^{1/2}} \times \exp\left(-\frac{(K' + K'' + KH) - \{(K' + K'' + KH)^2 - 4KHK'\}^{1/2}}{2}t\right) - \frac{-(K' + K'' + KH) + \{(K' + K'' + KH)^2 - 4KHK'\}^{1/2}}{2\{(K' + K'' + KH)^2 - 4KHK'\}^{1/2}} \times \exp\left(-\frac{(K' + K'' + KH) + \{(K' + K'' + KH)^2 - 4KHK'\}^{1/2}}{2}t\right)\Bigg] \quad [22a]$$

and

$$n_2 = \frac{c_\infty}{\{(K' + K'' + KH)^2 - 4KHK'\}^{1/2}} \times \Bigg[\left(K + \frac{-(K' + K'' + KH) - \{(K' + K'' + KH)^2 - 4KHK'\}^{1/2}}{2H}\right) \times \exp\left(\frac{-(K' + K'' + KH) + \{(K' + K'' + KH)^2 - 4KHK'\}^{1/2}}{2}t\right) - \left(K + \frac{-(K' + K'' + KH) + \{(K' + K'' + KH)^2 - 4KHK'\}^{1/2}}{2H}\right) \times \exp\left(\frac{-(K' + K'' + KH) - \{(K' + K'' + KH)^2 - 4KHK'\}^{1/2}}{2}t\right)\Bigg] + \frac{c_\infty}{H}. \quad [22b]$$

As soon as the accumulation at the secondary minimum becomes negligibly small, Eqs. [20] simplify to

$$\frac{dn_1}{dt} = K''n_2 - K'n_1 = K(c_\infty - Hn_2), \qquad [23]$$

from which one obtains

$$n_1 = \frac{c_\infty K''}{HK'}\left[1 - \exp\left(-\frac{K'HK}{K'' + HK}t\right)\right] \qquad [24]$$

What conditions have to be satisfied for Eq. [24] to be a satisfactory approximation of Eq. [22]?

Observe first that if

$$\frac{4KHK'}{(K' + K'' + KH)^2} \ll 1, \qquad [25]$$

then

$$\{(K' + K'' + KH)^2 - 4KHK'\}^{1/2} \approx (K' + K'' + KH)\left(1 - 2\frac{KHK'}{(K' + K'' + KH)^2}\right)$$

and Eq. [22a] becomes

$$n_1 = \frac{c_\infty K''}{HK'}\left[1 - \left(1 + \frac{KHK'}{(K' + K'' + KH)^2}\right)\exp\left(-\frac{KHK'}{K' + K'' + KH}t\right) + \frac{KHK'}{(K' + K'' + KH)^2}\exp(-(K' + K'' + KH)t)\right] \qquad [26]$$

If in addition to [25]

$$K' \ll K'' + KH, \quad \text{and} \quad K'' + KH \qquad [27]$$

is large enough then Eq. [24] involving no further accumulation at the secondary minimum, is a good approximation of the exact equation [22a]. The condition that $K'' + KH$ should be large assures that after a short transient of duration $3/(K'' + KH)$, the last term in Eq. [26] becomes negligible. One may notice that if the depth of the secondary minimum increases, both H and K'' decrease, and hence conditions [25] and [27] may no longer be satisfied. Also if $\phi_{max} - \phi_{min,1}$ decreases, K' increases and hence the inequalities may also no longer be valid.

In essence, inequalities [25] and [27] give the conditions under which, after a short transient, the rate of accumulation at the primary minimum can be computed assuming steady state between the bulk and the primary minimum.

A SIMPLIFIED QUANTITATIVE MODEL OF THE INTERACTION POTENTIAL

The main difference between the interaction potential used here and those used previously consists in the inclusion of a short-range repulsion. This assures the existence of a primary minimum of finite depth and hence the occurrence of reversible adsorption. For illustration purposes the short-range repulsion between a sphere and a plate was computed by linearly superimposing the 12-order term of the empirical Lennard–Jones 6–12 potential to obtain (3)

$$\phi_{12}(h) = \frac{A\sigma^6}{7560} \times \left[\frac{8a + h}{(2a + h)^7} + \frac{6a - h}{h^7}\right], \qquad [28]$$

where σ is the collision diameter and A is

the Hamaker constant. The repulsive force is due not only to the Born repulsion but probably also to the disjoining pressure arising from the modification of the structure of water near the solid surface. Here we account for this by considering σ as an adjustable parameter. An exponential repulsion or a sum of exponentials might be more adequate than the 12th power law, because, inter alia, Marcèlja and Radic (9) suggest that the organization of water leads to such a repulsion.

The van der Waals interaction energy between a sphere and a semi-infinite plate has been estimated by Hamaker (4)

$$\phi_6(h) = \frac{A}{6}\left[\ln\left(\frac{h+2a}{h}\right) - 2\frac{a}{h}\frac{h+a}{h+2a}\right]. \quad [29]$$

Lifshitz' more recent continuum analysis of the dispersion interactions leads to approximately the same distance dependence. For the double-layer repulsive potential for the sphere–plane interaction energy we use the expression

$$\phi_{DL}(h) = 16\epsilon\left(\frac{kT}{e}\right)^2 a \tanh\left(\frac{e\psi_1}{4kT}\right) \times \tanh\left(\frac{e\psi_2}{4kT}\right)e^{-\kappa h}, \quad [30]$$

where ψ_1 and ψ_2 are the surface potentials, e is the protonic charge, ϵ the dielectric constant of the liquid, and κ the reciprocal Debye length. This equation is valid for 1–1 electrolytes, $\kappa h > 2$ and $\kappa a > 5$.

The total interaction potential is obtained by summing the three contributions

$$\phi(h) = \phi_{12}(h) + \phi_6(h) + \phi_{DL}(h). \quad [31]$$

Equations [28] to [31] form a simplified model of the interactions because neither solvation of the surface, nor surface roughness on the order of σ or larger have been included in a satisfactory manner. Depending upon the values of various parameters, Eqs. [28] to [31] lead to various shapes of the interaction potential. For instance if σ is too large the interaction potential has only a single minimum. A detailed parametric study of the interaction potential is given in Ref. (3). Computations concerning the adsorption and desorption coefficients, K'' and K', have been already carried out both for the cases when the surfaces are characterized by constant surface potentials (3) or when they acquire their charge through the dissociation of various acid and/or basic groups (10, 11).

NUMERICAL ILLUSTRATIONS

Some numerical values of the adsorption coefficients can give a better feeling of the conditions under which inequalities [25] and [27] are satisfied. The computations have been carried out by using Eq. [31] for the interaction potential. Only a few of the numerous computations are listed in Table I. The table shows that there are conditions under which accumulation at the secondary minimum cannot be neglected. For instance, conditions [25] and [27] are not satisfied for the first three values of the Hamaker constant in Table Ia. Consequently, in these cases, accumulation at the secondary minimum cannot be neglected. They are, however, satisfied for the fourth case of Table Ia and for the first three cases of Table Ib.

It may be noted that in all the cases examined in Table I, $K'' \ll KH$. This simplifies the two criteria immensely because in these conditions both inequalities [25] and [27] are satisfied if $4K'/H \ll K \approx 10^{-4}$. Substituting Eqs. [15] and [19], one obtains $D(h_{max})(\gamma_{max}\gamma_{min,1}/2\pi kT\gamma_{min,2})^{1/2} \exp[-(\phi_{max} + \phi_{min,2} - \phi_{min,1})/kT] \ll 10^{-5}$. This condition is satisfied if

$$(\phi_{max} + \phi_{min,2} - \phi_{min,1})/kT \gg 1.$$

From Table I, one can conclude that, at least for the range of values used, inequalities [25] and [27] are satisfied if

$$\phi_{max} + \phi_{min,2} - \phi_{min,1} \gtrsim 12kT. \quad [32]$$

TABLE I

Values of the Adsorption Coefficients[a]

	$\phi_{min,1}/kT$	ϕ_{max}/kT	$\phi_{min,2}/kT$	$[\phi_{max} - (\phi_{min,1} - \phi_{min,2})]/kT$	H (cm^{-1})	K' (sec^{-1})	K'' (sec^{-1})	$4KHK'$ (sec^{-2})	$K'' + KH$ (sec^{-1})	$(K'' + KH)^2$ (sec^{-2})
(a) A (ergs)										
5×10^{-14}	28.01	31.57	−2.90	0.66	1.67×10^5	3.85×10^4	2.68×10^{-11}	2.57×10^6	1.67×10^1	2.79×10^2
6×10^{-14}	16.15	24.37	−3.68	4.54	8.75×10^4	4.83×10^2	1.95×10^{-8}	1.69×10^4	8.75	7.66×10
7×10^{-14}	4.19	18.06	−4.52	9.35	4.25×10^4	2.02	5.22×10^{-6}	34.3	4.25	18.1
8×10^{-14}	−7.81	12.45	−5.41	14.85	1.93×10^4	3.79×10^{-3}	6.38×10^{-4}	0.0292	1.93	3.72
(b) ψ (mV)										
28	−18.47	15.12	−6.78	18.81	5.40×10^3	2.28×10^{-5}	1.39×10^{-5}	5.01×10^{-5}	0.549	0.301
29	−2.91	19.91	−6.62	16.20	6.33×10^3	3.70×10^{-4}	1.27×10^{-7}	9.40×10^{-4}	0.633	0.401
30	4.85	25.03	−6.47	13.71	7.23×10^3	5.21×10^{-3}	9.90×10^{-10}	0.0151	0.723	0.523
31	12.79	30.48	−6.34	11.35	8.12×10^3	6.35×10^{-2}	4.96×10^{-12}	0.206	0.812	0.659

[a] For (a) ionic strength = 0.1 mole/liter; $a = 10^{-5}$ cm, $\psi_1 = \psi_2 = 25$ mV; $K = 10^{-4}$ cm/sec; $\sigma = 5$ Å. For (b) ionic strength = 0.1 mole/liter; $a = 10^{-5}$ cm, $A = 10^{-13}$ erg; $K = 10^{-4}$ cm/sec; $\sigma = 5$ Å.

Observe, that for the values of the parameters used in Table I, $K'/K'' \gg 1$. This means that the amount accumulated, in conditions of thermodynamic equilibrium, is much larger at the secondary than at the primary minimum. However, when inequalities [25] and [27] are satisfied $dn_2/dt \approx 0$, after a short transient.

IMPROVED KINETIC EQUATIONS

Even if h is very small, the center of sufficiently large particles can have appreciable tangential velocities. A better approximation of the kinetic equations, accounting in part for this effect, is provided by the conservation equations,

$$\frac{\partial n_2}{\partial t} + \frac{\partial(n_2\bar{u}_2)}{\partial x} = -J' - (K''n_2 - K'n_1) \quad [33]$$

and

$$\frac{\partial n_1}{\partial t} + \frac{\partial(n_1\bar{u}_1)}{\partial x} = K''n_2 - K'n_1, \quad [34]$$

where $\bar{u}_1$ and $\bar{u}_2$ are the average velocities in the direction x of the flow of the particles located around the primary and the secondary minimum, respectively. Equations [33] and [34] represent an approximate formulation of the kinetic equations because the velocity distribution near the surface of the collector makes the flux J distance dependent and hence Eq. [13] approximate. However, only a numerical solution of the complete transport equations, such as that carried out in a particular case by Prieve and Ruckenstein (12) can take fully into account the mentioned effect.

For deposition of particles on a rotating disc, it is easy to use Eqs. [33] and [34]. Indeed, in this case, Eq. [33] becomes

$$\frac{\partial n_2}{\partial t} + \frac{1}{r}\frac{\partial}{\partial r} r\bar{v}_{r,2}n_2 = -J' - (K''n_2 - K'n_1), \quad [35]$$

where r is the radial distance and $\bar{v}_{r,2}$ the average radial velocity component of the particles around the secondary minimum. Because, for a rotating disc n_2 is practically independent of r, Eq. [35] can be written as

$$\frac{\partial n_2}{\partial t} = -J' - (K'''n_2 - K'n_1), \quad [36]$$

where

$$K''' \equiv K'' + \frac{1}{r}\frac{\partial}{\partial r} r\bar{v}_{r,2}. \quad [37]$$

The radial velocity component of the fluid, v_r, can be computed from the equation (13)

$$v_r = r\omega F(\xi),$$

where

$$F(\xi) = 0.510y(\omega/\nu)^{1/2} \quad \text{for} \quad y \ll (\nu/\omega)^{1/2}$$

Here ω is the angular velocity of the disc, ν is the kinematic viscosity, and y the distance from the disc.

The actual particle velocity is less than that of the fluid at the center of the particle

$$\bar{v}_{r,2} = f_3(h/a)v_r,$$

where f_3 was given by Goldman *et al.* (14). For $h/a \ll 1$ (15)

$$f_3 \to \frac{0.7431}{0.6376 - 0.2\ln(h/a)}.$$

Equations [18] and [19] show that the contribution of the radial motion can be neglected around the secondary minimum when

$$2\omega F\left(a\left(\frac{\omega}{\nu}\right)^{1/2}\right) f_3\left(\frac{h_{\min,2}}{a}\right) \ll KH + K''. \quad [38]$$

For $a = 10^{-5}$ cm, $h_{\min,2} \approx 10^{-6}$ cm, $\omega = 62.8$ sec^{-1}, and $\nu = 10^{-2}$ cm^2/sec, one obtains

$$\frac{1}{r}\frac{d}{dr}(rv_r) = 2\omega F = 5.58 \times 10^{-2}\ \text{sec}^{-1}$$

and

$$f_3 = 0.677.$$

Thus

$$\frac{1}{r}\frac{d}{dr} r\bar{v}_{r,2} \approx 0.0378\ \text{sec}^{-1}.$$

For these values of the parameters, D is 2×10^{-8} cm^2/sec and $K = 0.62\ D^{2/3}\ \nu^{-1/6}\ \omega^{1/2}$ is 1.3×10^{-4} cm/sec. Inequality [38] is satisfied for the cases given in Table I.

COAGULATION KINETICS

Smoluchowski (16) has treated the kinetics of coagulation assuming that the particles which collide during their Brownian motion coagulate. The rate of collision was computed assuming quasi-steady diffusion around each particle. Smoluchowski's result was improved by Fuchs (17) who assumed quasi-steady diffusion in the force field caused by the interaction forces between two particles. The main difference between the present treatment and those used previously consists in the splitting of the diffusion in a force field into two steps. The first step is a quasi-steady-state diffusion in a force field between the bulk of the fluid and the secondary minimum, while the second step is a quasi-steady-state diffusion in a force field between the secondary and the primary minimum. Accumulation of particles occurs at both minima. Only the rate of coagulation of isodispersed sols at short times is calculated here.

Denoting by N_2 the quasi-steady flow-rate of particles between the bulk of the fluid and the secondary minimum, one obtains

$$N_2 = \left(c_\infty - \frac{n_2}{I_2}\right)\left[\int_{\delta_2'}^{\infty} \frac{\exp[\phi(s)/kT)ds}{8\pi(2a+s)^2D(s)}\right]^{-1} \equiv Kc_\infty - KHn_2. \quad [39]$$

Here n_2 represents the number of particles at the secondary minimum given by

$$n_2 = \int_{\delta_2}^{\delta_2'} 4\pi(2a+s)^2c(s)ds \quad [40]$$

and

$$H^{-1} \equiv I_2 = \int_{\delta_2}^{\delta_2'} 4\pi(2a+s)^2 \times \exp[-\phi(s)/kT]ds. \quad [41]$$

Similarly, denoting by N_1 the quasi-steady-state flow-rate of particles between the secondary and the primary minimum, one can write

$$N_1 = \left(\frac{n_2}{I_2} - \frac{n_1}{I_1}\right) \times \left[\int_{\delta_1'}^{\delta_2} \frac{\exp[\phi(s)/kT]ds}{8\pi(2a+s)^2D(s)}\right]^{-1} \equiv K''n_2 - K'n_1, \quad [42]$$

where n_1 is the number of particles at the primary minimum given by

$$n_1 = \int_{\delta_1}^{\delta_1'} 4\pi(2a + s)^2 c(s) ds \quad [43]$$

and

$$I_1 = \int_{\delta_1}^{\delta_1'} 4\pi(2a + s)^2 \exp[-\phi(s)/kT] ds. \quad [44]$$

Keeping the notation similar to that used in the previous sections, although the quantities involved have somewhat different meanings, it is easy to extend some of the previous results to the kinetics of coagulation. The conservation equation can be written as

$$\frac{dn_1}{dt} = K''n_2 - K'n_1$$

$$= (Kc_\infty - KHn_2) - \frac{dn_2}{dt}. \quad [45]$$

RATE OF COAGULATION AT THE PRIMARY MINIMUM

Before the number of doublets, triplets, etc., becomes too large, one can write for the rate of coagulation at the primary minimum of isodispersed sols

$$-\left(\frac{dc_\infty}{dt}\right)_1 = c_\infty \frac{dn_1}{dt}. \quad [46]$$

Calculating dn_1/dt from Eq. [22a], one obtains

$$-\left(\frac{dc_\infty}{dt}\right)_1 = \frac{2K''Kc_\infty^2}{\{(K' + K'' + KH)^2 - 4KHK'\}^{1/2}} \exp\left(-\frac{K' + K'' + KH}{2} t\right) \times \sin h \frac{\{(K' + K'' + KH)^2 - 4KHK'\}^{1/2} t}{2}. \quad [47]$$

For small values of K', such that

$$4KHK' \ll (K' + K'' + KH)^2 \quad [48]$$

one obtains

$$-\left(\frac{dc_\infty}{dt}\right)_1 = \frac{K''Kc_\infty^2}{K' + K'' + KH}(1 - e^{-(K'+K''+KH)t}). \quad [49]$$

Comparing Eqs. [42] and [39], one can conclude that if the potential barrier is high enough and positive,

$$KH \gg K''.$$

As already mentioned, this considerably simplifies the two criteria [25] and [27] needed for negligible rate of accumulation at the secondary minimum. In these conditions both criteria are satisfied when

$$4K' \ll HK. \quad [50]$$

When both K' and K'' are much smaller than KH, Eq. [49] becomes

$$-\left(\frac{dc_\infty}{dt}\right)_1 = \frac{K''}{H} c_\infty^2 (1 - e^{-KHt}). \quad [51]$$

Since the ratio K''/H is given by

$$\frac{K''}{H} = \left[\int_{\delta_1'}^{\delta_2} \frac{\exp[\phi(s)/kT] ds}{8\pi(2a + s)^2 D(s)}\right]^{-1}, \quad [52]$$

one obtains

$$-\left(\frac{dc_\infty}{dt}\right)_1 = 8\pi c_\infty^2 \left[\int_0^\infty \frac{\exp[\phi(s)/kT] ds}{(2a + s)^2 D(s)}\right]^{-1} \times (1 - e^{-KHt}). \quad [53]$$

The limits of integration have been extended because the integrand takes its largest value at $h = h_{max}$ and decays rapidly as $|h - h_{max}|$ differs from zero.

Further, if KH is large enough, Eq. [53]

leads after a short transient to the well-known equation for the initial rate of coagulation at the primary minimum

$$-\left(\frac{dc_\infty}{dt}\right)_1 = 8\pi c_\infty^2\left[\int_0^\infty \frac{\exp[\phi(s)/kT]ds}{(2a+s)^2D(s)}\right]^{-1}. \quad [54]$$

Equation [54] is generally derived on the basis of the quasi-steady-state flow-rate of particles between the bulk and the primary minimum and involves no accumulation at the secondary minimum. One can now specify that Eq. [54] is valid if the potential barrier is high enough and positive and if $4K' \ll HK$.

RATE OF COAGULATION AT THE SECONDARY MINIMUM

Before the number of doublets, triplets, etc., becomes too large, one can write for the rate of coagulation at the secondary minimum

$$-\left(\frac{dc_\infty}{dt}\right)_2 = c_\infty\frac{dn_2}{dt} \quad [55]$$

and therefore

$$-\left(\frac{dc_\infty}{dt}\right)_2 = \frac{c_\infty^2K}{2\{(K''+K'+KH)^2-4KHK'\}^{1/2}}\{((K'-K''-KH)+[(K''+K'+KH)^2 - 4KHK']^{1/2})\exp\left(\frac{-(K'+K''+KH)+[(K''+K'+KH)^2-4KHK']^{1/2}}{2}t\right) - ((K'-K''-KH)-\{(K''+K'+KH)^2-4KHK'\}^{1/2}) \times \exp\left(\frac{-(K'+K''+KH)-[(K'+K''+KH)^2-4KHK']^{1/2}}{2}t\right). \quad [56]$$

When the potential barrier is very high, $K'' \approx K' \approx 0$, and accumulation occurs only at the secondary minimum. In this case, Eq. [56] becomes

$$-\left(\frac{dc_\infty}{dt}\right)_2 = Kc_\infty^2e^{-KHt} \quad [57]$$

One may notice that the value of H decreases and consequently the rate of accumulation at the secondary minimum increases as the secondary minimum becomes deeper.

EQUATIONS FOR VARIOUS COAGULATION COEFFICIENTS

Except for K, all the integrals are centered at extrema. This allows the use of the parabolic approximation for the interaction potential. One obtains

$$H = \frac{1}{4\pi(2a+h_{\min,2})^2}\times\left(\frac{\gamma_{\min,2}}{2\pi kT}\right)^{1/2}e^{\phi_{\min,2}/kT}, \quad [58]$$

$$K' = 2\frac{(2a+h_{\max})^2D(h_{\max})}{(2a+h_{\min,1})^2}\times\frac{(\gamma_{\max}\gamma_{\min,1})^{1/2}}{2\pi kT}e^{-(\phi_{\max}-\phi_{\min,1})/kT}, \quad [59]$$

and

$$K'' = 2\frac{(2a+h_{\max})^2D(h_{\max})}{(2a+h_{\min,2})^2}\times\frac{(\gamma_{\max}\gamma_{\min,2})^{1/2}}{2\pi kT}e^{-(\phi_{\max}-\phi_{\min,2})/kT}. \quad [60]$$

One may notice that when $h_{\min,1} < h_{\max}$

$< h_{min,2} \ll 2a$, this set of coefficients has nearly the same form as the set for reversible adsorption. The factor $16\pi a^2$ in H is the area for flocculation. The extra factor of 2 in the expressions of K' and K'' is due to the fact that both interacting particles undergo Brownian motion.

To estimate the remaining constant K, let us assume that for $h > \delta_2'$, $|\phi/kT| \ll 1$ and $D \approx D_\infty$. Then

$$K^{-1} = \frac{1}{8\pi D_\infty}\int_{\delta_2'}^{\infty}\frac{ds}{(2a+s)^2} = \frac{1}{8\pi D_\infty}\frac{1}{2a+\delta_2'} \quad [61]$$

Assuming, further, that $\delta_2' \ll 2a$, one obtains Smoluchowski's equation

$$K = 16\pi D_\infty a. \quad [62]$$

Since $D_\infty = kT/6\pi\mu a$,

$$K = \frac{8}{3}\frac{kT}{\mu}. \quad [63]$$

In water, at ambient temperature, $K = 10^{-11}$ cm^3/sec.

NUMERICAL ILLUSTRATIONS FOR COAGULATION

For illustration purposes H, K', and K'' are evaluated from the results obtained for reversible adsorption. Considering the first case in Table Ia one obtains for $a = 10^{-5}$ cm

H	K'	K''	$4KHK'$	$K'' + KH$	$(K'' + KH)^2$
0.33×10^{14}	7.7×10^4	5.36×10^{-11}	10^8	0.33×10^3	10^5

In this case inequalities [25] and [27] are not satisfied, and consequently accumulation at the secondary minimum cannot be neglected when computing the rate of accumulation at the primary minimum. Considering the last case in Table Ia, one obtains, also for $a = 10^{-5}$ cm,

H	K'	K''	$4KHK'$	$K'' + KH$	$(K'' + KH)^2$
3.86×10^{12}	7.58×10^{-3}	12.8×10^{-3}	2.34	3.86×10^1	10^3

The inequalities are this time satisfied, and therefore, accumulation at the secondary minimum can be neglected in the computation of the rate of deposition at the primary minimum.

If the potential barrier is high enough and positive, the two conditions reduce to inequality [50]. Using Eqs. [58] and [59] for H and K', inequality [50] becomes

$$D(h_{max})a^2\left(\frac{\gamma_{max}\gamma_{min,1}}{kT\gamma_{min,2}}\right)^{1/2} \times e^{-(\phi_{max}-\phi_{min,1}+\phi_{min,2})/kT} \ll 10^{-13}, \quad [64]$$

which is satisfied if

$$\phi_{max} - \phi_{min,1} + \phi_{min,2} \gg kT.$$

Computations showed that for the range of values used and for $a = 10^{-5}$ cm one has $4K' = 0.1HK$ when

$$\phi_{max} - \phi_{min,1} + \phi_{min,2} \approx 10kT.$$

SUMMARY AND DISCUSSION

By splitting the quasi-steady-state assumption of diffusion of particles under the action of the interaction force field into two parts, kinetic equations which account for accumulation at both the primary and secondary minimum are formulated. Conditions are established under which, after a short transient, reversible adsorption or coagulation can be treated by neglecting accumulation at the secondary minimum. The effect of tangential velocity of particles on the rate of reversible adsorption is analyzed and a criterion established when the effect can be neglected. Equations for the rate of coagulation at the secondary and primary minimum are derived. The main message of the paper is the observation that the rate of deposition on surfaces and the rate of coagulation are affected in some conditions by reversibility and by the rate of accumulation at the secondary minimum. Hence in computing rates of reversible adsorption in a packed column, one has to account in some circumstances for accumulation at both minima. It was already shown (18, 19) that in the deposition of platelets from a stagnant solution of plasma rich platelets on a horizontal surface, the rate of accumulation at the secondary minimum affects the relationship between the number of platelets deposited and time. This is caused by the low sedimentation velocity of the platelets. The disagreement between the classical theory of slow coagulation and experiment (20) may perhaps be explained in terms of reversibility and accumulation at the secondary minimum. This problem will be the subject of a forthcoming paper.

ACKNOWLEDGMENT

This work was supported by the National Science Foundation.

REFERENCES

1. Ruckenstein, E., and Prieve, D. C., *J. Chem. Soc. Faraday II* **69,** 1522 (1973).
2. Spielman, L. A., and Friedlander, S. K., *J. Colloid Interface Sci.* **48,** 147 (1974).
3. Ruckenstein, E., and Prieve, D. C., *AIChE J.* **22,** 276 (1976); **22,** 1145 (1976).
4. Verwey, E. J. W., and Overbeek, J. Th. G., "Theory of Stability of Lyophobic Colloids." Elsevier, Amsterdam, 1948.
5. Matijevic, E., *J. Colloid Interface Sci.* **58,** 314 (1977).
6. Ruckenstein, E., *J. Powder Technol.*, in press.
7. Brenner, H., *Chem. Eng. Sci.*, **16,** 242 (1961).
8. Chandrasekhar, S., *Rev. Mod. Phys.* **15,** 1 (1943).
9. Marcèlja, S., and Radic, N., *Chem. Phys. Lett.*, **42,** 129 (1976).
10. Prieve, D. C., and Ruckenstein, E., *in* "Colloid and Interface Science" (M. Kerker, Ed.), Vol. IV, p. 73. Academic Press, New York, 1976.
11. Prieve, D. C., and Ruckenstein, E., *J. Theor. Biol.* **56,** 205 (1976); Prieve, D., "Rate of Deposition of Brownian Particles under the action of London and Double-Layer Forces." Ph.D. Dissertation, University of Delaware, 1974.
12. Prieve, D. C., and Ruckenstein, E., *AIChE J.* **20,** 1178 (1974).
13. Levich, V. B., "Physicochemical Hydrodynamics." Prentice–Hall, Englewood Cliffs, N. J., 1962.
14. Goldman, A. J., Cox, R. G., and Brenner, H., *Chem. Eng. Sci.* **22,** 653 (1967).
15. Spielman, L. A., Fitzpatrick, J. A., *J. Colloid Interface Sci.* **43,** 51 (1973).
16. Smoluchowski, M., *Z. Phys. Chem.* **92,** 129 (1917).
17. Fuchs, N., *Z. Phys.* **89,** 736 (1934).
18. Marmur, A., Gill, W. N., and Ruckenstein, E., *Bull. Math. Biol.* **38,** 713 (1976).
19. Ruckenstein, E., Marmur, A., and Rakower, S. R., *Thrombosis and Haemostasis* **36,** 334 (1976).
20. Ottewill, R. H., and Shaw, J. N., *Discuss. Faraday Soc.* **42,** 154 (1966).

J. theor. Biol. (1975) **51,** 429–438

Dynamics of Cell Deposition on Surfaces

ELI RUCKENSTEIN

Faculty of Engineering and Applied Sciences, State University of New York at Buffalo, Buffalo, New York 14214, *U.S.A.*

AND

DENNIS C. PRIEVE†

Department of Chemical Engineering, University of Delaware, Newark, Delaware 19711, *U.S.A.*

(*Received* 29 *April* 1974, *and in revised form* 10 *September* 1974)

The rate of deposition of particles onto a surface, in the presence of London, double-layer, and gravitational forces, is calculated in terms of the energy of interaction between cell and surface by assuming that Brownian motion over a potential energy barrier is the rate-determining step of the process.

1. Introduction

During metastasis, malignant cells are carried in the bloodstream from the primary tumor to locations throughout the body. These cells adhere to the vascular endothelium as well as other cells and secondary growth subsequently occurs. Cells are attracted to other surfaces by the electrodynamic London–van der Waals forces. However, all cells and biological surfaces studied so far have a net negative surface charge; thus electrostatic repulsions also act. Previous studies considering these forces (Weiss, 1973) have shown that, to achieve intimate contact with another surface, a cell must overcome a substantial energy barrier. Cells carried by the bloodstream are unlikely to have sufficient radial momentum on the average to overcome this barrier; hence another mechanism is responsible. *In vitro* experiments, (Weiss & Harlos, 1972) in which cells settle through a stagnant fluid under the influence of gravity to a glass surface and then adhere to it, have shown that the rate at which cells adhere to the surface is a factor of ten less than the rate of gravitational settling. One means of explaining the low deposition rate is to

† Currently at Department of Chemical Engineering, Carnegie–Mellon University, Pittsburgh Pa. 15213, U.S.A.

assume that the cells are carried over the energy barrier by a stochastic process such as Brownian motion. In this manner only a small fraction of the total cells are, at any given time, in a state in which they can overcome the barrier. Waiting for that state to occur lowers the deposition rate.

Interaction and adhesion of biological surfaces are central considerations for other physiological conditions as well. Platelets, erythrocytes, the vascular endothelium and other tissues interact during thrombosis and hemostasis. Also, when erythrocytes come in contact with artificial surfaces, damage often occurs and blood trauma may result. Finally, the accumulation of cholesterol deposits on the interior walls of arteries is responsible for atherosclerosis.

Rates of deposition of cells in a gravitational field are calculated in the current paper by considering Brownian motion over a potential barrier formed by London and double-layer forces acting between the cells and the surface on which they deposit.

2. Approximate Model for Cell-surface Interactions

The potential energy of interaction ϕ between a very thick plate and a smooth sphere of radius a_p, whose center is located a distance a_p+x from the surface of the plate, is formed by summing the London (Hamaker, 1937), double-layer (Hogg, Healy & Fuerstenau, 1966), and gravitational contributions:

$$\phi = A\left[\tfrac{1}{6}\ln\left(\frac{x+2a_p}{x}\right) - \tfrac{1}{3}\frac{a_p}{x}\frac{x+a_p}{x+2a_p}\right] + \tfrac{1}{4}\varepsilon a_p\left[(\psi_1^2+\psi_2^2)\ln(1-e^{-2\kappa x}) + 2\psi_1\psi_2\ln\left(\frac{1+e^{-\kappa x}}{1-e^{-\kappa x}}\right)\right] + \tfrac{4}{3}\pi a_p^3(\rho_p-\rho_f)gx, \quad (1)$$

where A is Hamaker's constant, ε is the dielectric constant, ψ_1 and ψ_2 are the two surface potentials, $1/\kappa$ is the Debye length, g is the gravitational acceleration, and ρ_p and ρ_f are the densities of the cell and fluid, respectively. In the derivation of the double-layer interaction, the surface potentials have been assumed to be small (<25 mV) and independent of x. Figure 1 illustrates the effect of the surface potential and Hamaker's constant upon the potential energy profile. When the electrostatic repulsion terms are significant, $\phi(x)$ has a maximum at $x = x_{max}$, where x_{max} is 5–10 Å. At smaller values of x a very deep "primary" minimum (not shown in Fig. 1) is found, whereas a shallower "secondary" minimum is displayed at $x = x_{min}$, where x_{min} is 25–50 Å. In the current paper, cells "deposit" on the surface, when they have reached the primary minimum which is considered, for practical purposes,

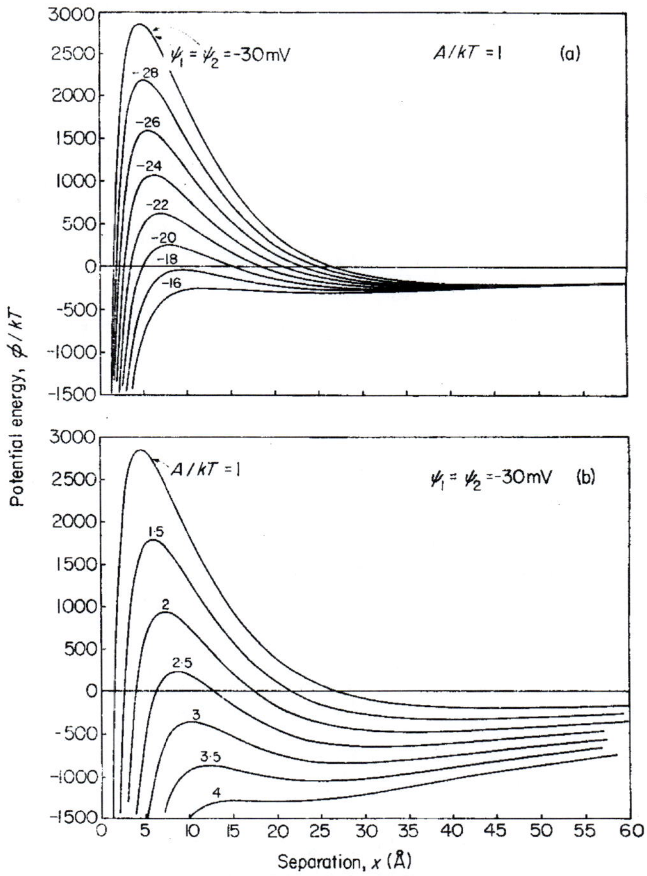

FIG. 1. Effect of surface potential (a) and Hamaker's constant (b) upon the total interaction energy profile for a sphere of radius 6·7 μm approaching a flat plate in a solution having $\varepsilon = 74{\cdot}3$ and $\kappa^{-1} = 8$Å.

to occur at $x = 0$ and to have a negatively infinite potential energy. It can be seen from Fig. 1 that reducing the surface potential diminishes the height of the maximum without greatly affecting the minimum, whereas increasing Hamaker's constant decreases the height of the maximum and increases the depth of the secondary minimum.

3. General Unsteady Analysis

Consider an initially uniform suspension of cells covering to a depth L the top surface of a plane, horizontal, glass coverslip. The distribution of cells $c(x, t)$ must satisfy the diffusion equation, modified by inclusion of the interaction force term:

$$\frac{\partial c}{\partial t} = \frac{\partial}{\partial x}\left(D\frac{\partial c}{\partial x} + m\frac{\mathrm{d}\phi}{\mathrm{d}x}c\right), \tag{2}$$

where $D(x)$ and $m(x)$ are the local diffusion coefficient and cellular mobility. Position variations in D and m arise from the finite size of the cells and the presence of the collecting surface. As a cell approaches a plane surface under the action of a constant force, its velocity decreases because friction between the fluid and the plane surface slow the rate at which fluid may be displaced from the path of the approaching cell. The cell velocity per unit applied force, or cell mobility, may be calculated from a modified Stokes equation when the Reynolds number is small:

$$m = \frac{f_1(x/a_p)}{6\pi\mu a_p} \tag{3}$$

where the factor f_1 accounts for the effect of the proximity of the plane surface and μ is the fluid viscosity. Values of f_1 calculated by Brenner (1961) tend to x/a_p when $x/a_p \ll 1$ and to unity when $x/a_p \gg 1$.

Since cells in contact with the surface are no longer part of the disperse phase, the concentration c must vanish when contact is made. Requiring the cellular flux to vanish at the upper surface of the suspension leads to the second boundary condition. Thus

$$c = 0 \qquad \text{at } x = 0, \tag{4a}$$

$$D\frac{\partial c}{\partial x} + m\frac{\mathrm{d}\phi}{\mathrm{d}x}c = 0 \qquad \text{at } x = L, \text{ and} \tag{4b}$$

$$c = c_0 \qquad \text{at } t = 0 \tag{4c}$$

constitute the boundary and initial conditions needed to complete the general formulation of the problem. However, if the rate of deposition is known to be substantially less than the rate of sedimentation and the London and double-layer forces act only very near to the surface (<100 Å), a simpler analysis is possible.

4. Rate-determining Step Approximation

The process of cell deposition in the presence of repulsive forces may be considered as a two-step sequence. First the cells move, primarily under the action of gravity, to a region very near to the surface. In order to move closer to the surface the particle must experience the energy barrier formed by the electrostatic double-layer repulsions and London attraction. Diffusion of cells over the energy barrier is the second step of the process. If the deposition rate is much smaller than the sedimentation rate the second step of the sequence is rate-determining. This situation is then analogous to a two-step chemical reaction sequence:

$$\begin{pmatrix}\text{Cells far}\\ \text{from surface}\end{pmatrix} \underset{\text{Diffusion}}{\overset{\text{Net force}}{\rightleftharpoons}} \begin{pmatrix}\text{Cells near}\\ \text{surface}\end{pmatrix} \xrightarrow{\text{Diffusion}} \begin{pmatrix}\text{Adhered}\\ \text{cells}\end{pmatrix}$$

in which the second step is irreversible and rate-determining. To evaluate the overall rate following Kramers (1940), one first determines the concentration of cells near the surface (the intermediate) by assuming the first step of the sequence is nearly in equilibrium, or that the rate of transport of cells to the secondary minimum by the action of the forces is nearly equal to the rate of diffusion of cells away from the secondary minimum:

$$-m(x)\frac{\mathrm{d}\phi}{\mathrm{d}x}c = D(x)\frac{\partial c}{\partial x}. \tag{5}$$

When integrated, taking $D = mkT$, this implies that the distribution of cells near the secondary minimum has the Maxwell–Boltzmann form:

$$c = a\exp(-\phi/kT) \quad \text{for} \quad x \geq \delta \tag{6}$$

where $a(t)$ is an integration constant and δ is the smallest value of x for which an equilibrium of the two opposing rates may be assumed. The integration constant may be evaluated in terms of the number of cells $n(t)$ remaining in the suspension per unit surface by noting that nearly all the cells remaining will accumulate outside the barrier so that:

$$n = \int_0^L c(x)\,\mathrm{d}x \approx a\int_\delta^L \exp(-\phi/kT)\,\mathrm{d}x. \tag{7}$$

Eliminating $a(t)$ between equations (6) and (7) leaves:

$$c = n\exp(-\phi/kT)\left\{\int_\delta^L \exp(-\phi/kT)\,\mathrm{d}x\right\}^{-1} \quad \text{for} \quad x \geq \delta. \tag{8}$$

For x smaller than δ, the cellular flux may be written as the sum of the contributions from Brownian motion and the total interaction force:

$$-J = D\frac{\partial c}{\partial x} + m\frac{\mathrm{d}\phi}{\mathrm{d}x}c. \tag{9}$$

Since the cells are quite large ($a_p \sim 10^{-3}$ cm) with respect to the distances ($\delta \sim 10^{-7}$ cm) involved, a diffusional representation is not entirely adequate; however equation (9) is expected to give the average behavior of a large number of cells deposited over a sufficiently large area of the coverslip. Since the inner region is very thin, the flux J may be taken to be nearly independent of x, and equation (9) when integrated, yields:

$$-J = [c\exp(\phi/kT)]\Big|^{\delta}\left\{\int^{\delta}\frac{\exp(\phi/kT)}{D}\,\mathrm{d}x\right\}^{-1}. \tag{10}$$

The concentration at the upper limit must coincide with the equilibrium value given by equation (8), while the value at the lower limit must vanish, leaving:

$$-J = n\left\{\int_{\delta}^{L} \exp(-\phi/kT)\,dx \int_{0}^{\delta} \frac{\exp(\phi/kT)}{D}\,dx\right\}^{-1}. \tag{11}$$

Since neither of these integrals is time-dependent and $J = dn/dt$, equation (11) implies that the number of cells $n(t)$ remaining in suspension decays exponentially with time:

$$n(t) = n(0)\exp(-t/\tau) \tag{12}$$

where the time constant τ is given by:

$$\tau = \int_{0}^{\delta} \frac{\exp(\phi/kT)}{D}\,dx \int_{\delta}^{L} \exp(-\phi/kT)\,dx. \tag{13}$$

When $\phi(x)$ resembles the profile in Fig. 1, the value of τ will be relatively insensitive to δ, provided it is chosen so that $x_{max} < \delta < x_{min}$. Thus if $\phi(x)$ is known, as in equation (1), and $D(x) = m(x)\,kT$ may be calculated from equation (3), then the time constant for the deposition process may be evaluated by performing the integrations indicated by equation (13). The time for half of the original cells to deposit (the half-time for the deposition process) is then $\tau \ln 2$.

5. Approximation to Equation (13) for the Case of Deep Secondary Minimum

In the special case in which the secondary minimum is deep, $\phi(\delta)-\phi(x_{min}) \gg kT$ and $-\phi(x_{min}) \gg kT$, so that one would expect that the principal contribution to the second integral of (13) will come from the region near $x = x_{min}$. This expectation arises from the observation that $\exp(-\phi/kT)$ takes on its maximum value within the integration interval at $x = x_{min}$ and decreases rapidly as $|x-x_{min}|$ becomes different from zero. Then it seems reasonable to expand $\phi(x)$ in a Taylor Series about $x = x_{min}$:

$$\phi(x) \approx \phi_{min} + (\tfrac{1}{2})\gamma_{min}(x-x_{min})^2 \tag{14}$$

where

$$\phi_{min} = \phi(x_{min}) \text{ and } \gamma_{min} = d^2\phi/dx^2|_{x=x_{min}}.$$

Substituting the parabolic approximation (14) into the integrand yields:

$$\int_{\delta}^{L} \exp(-\phi/kT)\,dx \approx \left(\frac{\pi kT}{2\gamma_{min}}\right)^{1/2}\left\{\operatorname{erf}\left[(L-x_{min})\left(\frac{\gamma_{min}}{2kT}\right)^{1/2}\right] + \operatorname{erf}\left[(x_{min}-\delta)\left(\frac{\gamma_{min}}{2kT}\right)^{1/2}\right]\right\}\exp(-\phi_{min}/kT). \tag{15}$$

If the secondary minimum is deep γ_{min}/kT may be expected to be large enough so that the arguments of both error functions will be greater than two and can thus be replaced by infinity to yield:

$$\int_{\delta}^{L} \exp(-\phi/kT) \approx \left(\frac{2\pi kT}{\gamma_{min}}\right)^{1/2} \exp(-\phi_{min}/kT). \quad (16)$$

In summary, approximation (16) requires $\phi(\delta)-\phi_{min} \gg kT$, $-\phi_{min} \gg kT$, and $(x_{min}-\delta)\sqrt{\gamma_{min}/2kT} > 2$. Using a similar analysis one can expand $\phi(x)$ about $x = x_{max}$ to show:

$$\int_{0}^{\delta} \exp(\phi/kT)\frac{dx}{D} \approx \frac{1}{D(x_{max})}\left(\frac{2\pi kT}{\gamma_{max}}\right)^{1/2} \exp(\phi_{max}/kT) \quad (17)$$

where $\phi_{max} = \phi(x_{max})$ and $\gamma_{max} = -d^2\phi/dx^2|_{x=x_{max}}$. Since $D(x)$ varies slowly with x compared to the exponential factor, equation (17) was obtained by approximating the function $D(x)$ by its value $D(x_{max})$ at the central point of expansion. Substituting approximations (16) and (17) into (13), the time constant for the deposition may be expressed as:

$$\tau \approx \frac{1}{D(x_{max})}\frac{2\pi kT}{(\gamma_{max}\gamma_{min})^{1/2}} \exp[(\phi_{max}-\phi_{min})/kT]. \quad (18)$$

This approximate analysis shows that the deposition time constant is exponentially sensitive to the height of the energy barrier $\phi_{max}-\phi_{min}$.

6. Approximation to Equation (13) for the Case of Shallow Secondary Minimum

Figure 1 suggests that as Hamaker's constant becomes very small compared to kT, ϕ_{min} vanishes. For this condition equation (16) is not valid. Instead, if the contribution to $\phi(x)$ from London and double-layer forces is much smaller than kT for $x \geq \delta$, then only the gravitational contribution needs to be considered:

$$\phi(x) \approx \tfrac{4}{3}\pi a_p^3(\rho_p-\rho_f)gx \quad (19)$$

for $x \geq \delta$. The new approximation of the integral becomes:

$$\int_{\delta}^{L} \exp(-\phi/kT)\,dx \approx \frac{3kT}{4\pi a_p^3(\rho_p-\rho_f)g} \times$$

$$\times\left[\exp\left(-\frac{4\pi a_p^3(\rho_p-\rho_f)g\delta}{3kT}\right) - \exp\left(-\frac{4\pi a_p^3(\rho_p-\rho_f)gL}{3kT}\right)\right]. \quad (20)$$

Recalling that $\delta \sim 10^{-7}$ cm whereas $L \sim 10^{-1}$ cm, the second exponential is certainly negligible compared to the first. Using reasonable values for a_p and $\rho_p-\rho_f$, one usually finds that δ is sufficiently small so that the first exponential may be taken as unity, whereupon:

$$\int_{\delta}^{L} \exp(-\phi/kT)\,\mathrm{d}x \approx \frac{3kT}{4\pi a_p^3(\rho_p-\rho_f)g}. \tag{21}$$

If $\phi_{max} \gg kT$ equation (17) is still valid. Substituting equations (17) and (21) into (13) yields a second approximation for the deposition time constant:

$$\tau \approx \frac{3kT\sqrt{2\pi kT/\gamma_{max}}}{4\pi a_p^3(\rho_p-\rho_f)gD(x_{max})} \exp(\phi_{max}/kT). \tag{22}$$

As in equation (18) this approximation also displays exponential sensitivity to the height of the energy barrier.

7. Results and Discussion

Weiss & Harlos (1972) have measured the rate of deposition of mouse tumor (EAT) cells onto a glass coverslip and found the rate to be much smaller than the calculated sedimentation rate. The relevant physical parameters reported were $\psi_1 = -27$ mV, $\psi_2 = -32$ mV, $\varepsilon = 74{\cdot}3$, $\kappa^{-1} = 8$ Å, $a_p = 6{\cdot}7\times10^{-4}$ cm, $T = 310$°K, $\mu = 7{\cdot}29\times10^{-3}$ g/cm-sec and $\rho_p-\rho_f = 0{\cdot}071$ g/cm^3. Equation (18) was found to yield the observed (Weiss & Harlos, 1972) deposition time constant $\tau = 2600$ sec provided the Hamaker constant could be chosen as $1{\cdot}60\times10^{-13}$ erg or $3{\cdot}81kT$†. This situation corresponds to $\phi_{max} = -1213{\cdot}2kT$ at $x_{max} = 15{\cdot}2$ Å, $\phi_{min} = -1227{\cdot}2kT$ at $x_{min} = 20{\cdot}6$ Å, $\gamma_{max} = 1800$ erg/cm^2, $\gamma_{min} = 801$ erg/cm^2, and $D(x_{max}) = 1{\cdot}04\times10^{-13}$ cm^2/sec, while $D(\infty) = 4{\cdot}57\times10^{-10}$ cm^2/sec. Each of the assumptions made to simplify equation (13) was tested and found to be valid.

This result, with a highly negative ϕ_{max}, raises the possibility that the cells might be carried over the potential energy barrier by their own inertia. Inertial forces, ignored in the analysis leading to equation (13), may be evaluated from:

$$F_{inertia} = M\frac{\mathrm{d}v}{dt} = M\frac{\mathrm{d}v}{\mathrm{d}x}\frac{\mathrm{d}x}{dt} = Mv\frac{\mathrm{d}v}{\mathrm{d}x} \tag{23}$$

where M is the mass of a cell and v is the local cell velocity or $-m\,\mathrm{d}\phi/\mathrm{d}x$. Calculations of the inertial force for the conditions indicated above, showed that it is less than an upper limit of 5% of the drag force, v/m, for all $x \geq x_{max}$. Inclusion of the fluid inertia in equation (23) could be expected to increase this percentage to about 7·5% (Fuchs, 1964). Thus kinetic energy which might be gained by moving down the potential hill into the secondary minimum will be nearly completely dissipated by viscous drag.

† Equation (18) is highly sensitive to the value of Hamaker's constant. A small fractional change in A/kT can change the calculated τ by many orders of magnitude.

While the value of Hamaker's constant required for agreement is typical for nonbiological materials (Gregory, 1969), it is more than an order of magnitude larger than values that are inferred from other experimental observations on cell-surface interactions (Weiss, 1968); however, both deductions contain assumptions and neither may be considered a direct measurement of Hamaker's constant. Measurements of colloid stability ratios for the flocculation of sheep leucocytes (Wilkins, Ottewill & Bangham, 1962) have been used to infer a Hamaker's constant of 5×10^{-15} erg. But these experiments are representative of a cell–water–cell system and not the cell–water–glass system currently being considered.

Calculations of deposition time constants from equation (18) were repeated, assuming lower values for the surface potential. Values of the Hamaker's constant necessary to obtain agreement with the experimental results of Weiss & Harlos, together with the corresponding values of other parameters are summarized by Table 1. Thus, to obtain agreement using a small Hamaker constant, the surface potential must be considerably lower.

TABLE 1

Surface potentials and Hamaker constants for which equation (18) *yields the* 2600 *sec deposition time* (τ) *observed under the conditions of the Weiss & Harlos Experiment*

$-\psi_1$ (mV)	$-\psi_2$ (mV)	A (10^{-14} erg)	ϕ_{max}/kT	x_{max} (Å)	γ_{max} (erg/cm²)	ϕ_{min}/kT	x_{min} (Å)	γ_{min} (erg/cm²)
27	32	16·0	−1213·2	15·2	1800	−1227·2	20·6	801
20	20	7·31	−539.3	14·5	1181	−552·7	21·4	420
10	10	1·71	−111·7	12·9	590	−124.0	23·6	118
5	5	0·361	−12·7	10·5	334	−23·8	27·4	26·9

Deviations between observed deposition time constants and those computed from equation (18), using surface potentials and Hamaker constants consistent with other experiments, do not imply that equation (18) is in error. The contribution from London and double-layer forces to the total interaction energy, expressed by equation (1) ignores a number of effects which may be of importance in biological interactions. First, the cell surface is not smooth but contains projections known as filopods (Weiss & Harlos, 1972), which may vary in size and number with the local environment. Second, the cell surface has been shown to be electrically heterogeneous (Weiss & Zeigel, 1972). Third, if the dissociation of ionizable groups on the surface is responsible for the charge, the degree of dissociation will vary as the cell approaches the glass such that neither the charge density nor the surface potential will remain constant. Finally, the cell is a complex organism: if the surface potential and hence the local pH change as the cell approaches the surface, some transformation may be triggered within the cell that either alters the electrical nature of the surface and/or generates some positively charged exudates.

In short, the rate of deposition of cells onto a surface has been evaluated in terms of the potential energy of interaction for cases in which Brownian motion of the cells over a potential barrier is the rate-determining step of the process. The rate is exponentially sensitive to the height of the barrier. However, accurate calculation of deposition rates in biological systems awaits an appropriate expression for the interaction forces that considers the complexities observed for such systems.

REFERENCES

BRENNER, H. (1961). *Chem. Engng Sci.* **16,** 242.
FUCHS, N. A. (1964). *The Mechanics of Aerosols.* New York: MacMillan.
GREGORY, J. (1969). *Adv. Colloid Interface Sci.* **2,** 369.
HAMAKER, H. C. (1937). *Physica* **4,** 1058.
HOGG, R., HEALY, T. W. & FUERSTENAU, D. W. (1966). *Trans. Faraday Soc.* **62,** 1938.
KRAMERS, H. A. (1940). *Physica* **7,** 284.
WEISS, L. (1968). *Expl Cell Res.* **53,** 603.
WEISS, L. (1973). *Chemotherapy of Cancer Dissemination and Metastasis* (S. Garattini & G. Franchi, eds). New York: Raven Press.
WEISS, L. & HARLOS, J. P. (1972). *J. theor. Biol.* **37,** 169.
WEISS, L. & ZEIGEL, R. (1972). *J. theor. Biol.* **34,** 21.
WILKINS, D. J., OTTEWILL, R. H. & BANGHAM, A. D. (1962). *J. theor. Biol.* **2,** 176.

J. theor. Biol. (1976) **58,** 439–454

Coverage Dependent Rate of Cell Deposition

ELI RUCKENSTEIN, ABRAHAM MARMUR AND WILLIAM N. GILL

Faculty of Engineering and Applied Sciences, State University of New York at Buffalo, Buffalo, NY 14214, U.S.A.

(*Received* 7 *May* 1975, *and in revised form* 18 *September* 1975)

A quantitative criterion is developed for the applicability of a previous diffusional model for cell deposition on surfaces, using as a starting point a modified Fokker-Planck equation.

The previous model for cell deposition through a stagnant liquid is extended to include the coverage-dependent effect of the deposited cells on subsequent deposition. This effect is shown to be responsible for an experimentally measured increase in the rate of cellular deposition with time, in contrast with the ever-decreasing rate of deposition predicted by the previous model. The present model also predicts that the decay in the fraction of cells in solution depends on the initial number of cells, again in contrast with the prediction of the previous model.

1. Introduction

Deposition and adhesion of cells on surfaces play an important part in many biological processes. For example, blood cell deposition and adhesion on surfaces are essential in the vital process of haemostasis and fundamental in the events leading to thrombus formation. Also, malignant cell adhesion on various tissues is responsible for the growth of secondary tumors, after being carried in the blood stream from the primary tumor location.

Some insight into the mechanism of cell deposition was provided by electrokinetic studies which have been conducted in relation with thrombus formation by Sawyer & Srinivasan (1967) and co-workers. They demonstrated that thrombosis occurred on a blood vessel wall which was connected to a positively charged electrode, and alternatively, that thrombosis was prevented by a negatively charged electrode. This was explained by the fact that all mammalian cells are negatively charged and, therefore, an electrical repulsion force exists between the blood cells and the vessel wall. While these experiments revealed the effect of the electrical repulsion force in preventing cell deposition, they did not throw light upon the attraction force that must exist in order to account for cell deposition which takes place even on negatively charged surfaces like, for example, glass.

The combined effect of attraction and repulsion forces has been treated by many investigators in terms borrowed from theories of colloidal stability (Weiss, 1972). These theories treat the adhesion of colloidal particles by taking into account three types of forces: (a) electrostatic repulsion force (Hogg, Healy & Fuerstenau, 1966); (b) London–Van der Waals molecular attraction force (Hamaker, 1937); (c) gravity force. The electrostatic repulsion force is due to the negative charges that exist on the cell membrane and on the deposition surface because of the development of electrostatic double layers when they are in contact with a solution. The London attraction force is due to the time distribution of the movement of electrons in each molecule and, therefore, it exists between each pair of molecules and consequently between each pair of particles. For example, this force is responsible, among other phenomena, for the condensation of vapors to liquids.

The potential energy of interaction ϕ between a sphere and a plate, which serve in this approach as a model for the cell and the contacted surface, can be obtained by summing up the individual contributions of the above mentioned forces:

$$\phi = A\left[\frac{1}{6}\ln\left(\frac{x+2R}{x}\right) - \frac{1}{3}\frac{R}{x}\frac{x+R}{x+2R}\right] + \\ +\frac{1}{4}\varepsilon R\left[(\psi_s^2+\psi_1^2)\ln(1-e^{-2\kappa x})+2\psi_s\psi_1\ln\frac{1+e^{-\kappa x}}{1-e^{-\kappa x}}\right] + \frac{4}{3}\pi R^3\Delta\rho\, gx, \quad (1)$$

where A is Hamaker's constant, R is the radius of the sphere, x is the separation between the sphere and the plate, ε is the dielectric constant, ψ_1 is the surface-potential of the cell, ψ_s is the surface-potential of the surface, $1/\kappa$ is the Debye length, $\Delta\rho$ is the difference between the density of the sphere and the density of the solution, and g is the gravitational acceleration.

This potential energy function is known to have two minima. The minimum closest to the surface is the "primary minimum" and the other is the "secondary minimum". The value of the potential energy of interaction is extremely different at these two minima, and therefore speculations have been made regarding the question of whether cell adhesion to the surface takes place at the primary or secondary minimum. To answer this question two experimental approaches have been suggested. Some investigators (Weiss, 1968; George, Weed & Reed, 1971) tried to measure the height of the potential barrier by measuring the centrifugal force needed to detach the adhering cells. This approach is not satisfactory because it is uncertain whether this information, concerning cell detachment, is directly related to cell adhesion, due to the changes in the cell caused by the contact with the surface.

A more adequate approach seems to be the measurement of the rate of cell deposition. In *in vitro* experiments, Weiss & Harlos (1972) measured the rate of tumor cell deposition through a stagnant solution onto glass and found it to be much slower than the rate of gravitational settling to the secondary minimum. They theorized that this delay in deposition is due to establishment of adhesion at the primary minimum. They regarded the potential barrier as impenetrable to the cells and, therefore, suggested two possible mechanisms for adhesion at the primary minimum, which do not involve escape of the whole cell over the potential barrier: protrusion of cellular probes from the secondary minimum to the primary minimum, and/or extrusion of an intercellular "glue".

However, Ruckenstein & Prieve (1975*a*) pointed out that cells are able to overcome the potential barrier between the secondary and the primary minima by a stochastic process similar to Brownian motion. The time delay in the experiments of Weiss & Harlos (1972) was explained by suggesting that this escape over the potential barrier is the time consuming step in the overall process of deposition, because only a small fraction of the cells are able, at a given time, to be carried over the barrier.

They developed a theoretical model, based on a diffusional approach, which leads to an exponential decay in the number of cells per unit area, n, that exist at time t in a solution which contained n_0 cells per unit area at time $t = 0$:

$$n = n_0 \, e^{-t/\tau}. \tag{2}$$

The time constant, τ, which characterizes this decay, was calculated as a function of the relevant properties of the cell, the surface, and the solution.

By using reasonable values for the physical parameters, equation (2) predicted well the average decay in the number of cells in solution, experimentally measured by Weiss and Harlos (1972). Figure 1 shows the experimental data as well as the theoretical curve. This figure demonstrates that, although the existing model predicts reasonably the average experimental decay, it does not predict its instantaneous shape. A possible explanation for this difference is that the time constant of deposition on the fraction of the surface already covered by cells is different from that which characterizes the bare surface, due to the changes in the surface potential and in Hamaker's constant, which are caused by the adhesion of the cells.

The purpose of the present paper is to reconsider the previous model from a more general point of view, starting with the Fokker-Planck equation for the motion of Brownian particles, to state the quantitive condition for the validity of that model and to extend it to include the effect of coverage-dependent rate of deposition. Quantitative and qualitative experimental

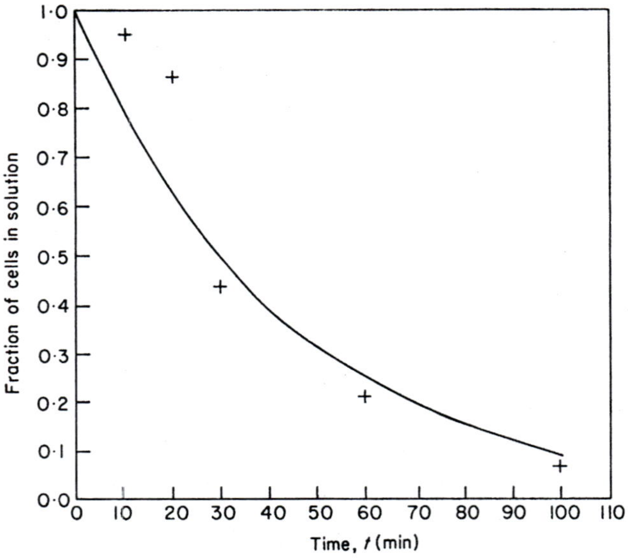

Fig. 1. Experimental data (Weiss & Harlos, 1972) and a theoretical curve (Ruckenstein & Prieve, 1975*a*) for the fraction of cells in solution as a function of time. $\tau = 2600$sec.

data concerning deposition of cancer cells as well as blood cells will be presented in order to demonstrate the usefulness of the extended theoretical model in explaining cell adhesion.

2. The Validity of the Diffusional Model for the Rate of Cell Deposition through a Stagnant Fluid

The Brownian motion of a particle under the influence of an external force field, and its consequent escape over a potential barrier has to be treated, in general, using the Fokker-Planck equation. This equation gives the distribution function W governing the probability that a particle will be after time t at a point x with velocity u (Chandrasekhar, 1943). In one dimension it has the form:

$$\frac{\partial W}{\partial t} + u\frac{\partial W}{\partial x} + K(x)\frac{\partial W}{\partial u} = \beta u\frac{\partial W}{\partial u} + \beta W + q\frac{\partial^2 W}{\partial u^2}, \tag{3}$$

where K is the acceleration produced by the external force field (equivalent to $-(1/M)\,\mathrm{d}\phi/\mathrm{d}x$, where M is the mass of the particle), β is the friction coefficient given by

$$\beta = \frac{6\pi R\mu}{M} \tag{4}$$

and q is related to β by:

$$q = \frac{\beta kT}{M}, \tag{5}$$

k is Boltzman's constant, and T is the absolute temperature. Equation (3) is correct only far away from any wall because the friction coefficient β is assumed constant. This assumption is not suitable for the case under consideration because as the cell approaches the surface, the viscous shear between the displaced solution and the surface increases the resistance to its motion. Therefore, equation (3) has to be modified to include the effect of a near-by wall by introducing the position-dependent cellular mobility instead of the constant friction coefficient. The cellular mobility is given by:

$$m = \frac{f(x/R)}{6\pi\mu R}, \tag{6}$$

where for $x \to \infty$, $f(x/R) \to 1$ and $m = 1/\beta M$. In equation (6) μ is the viscosity of the solution and the function $f(x/R)$ was calculated by Brenner (1961). For the case $x \ll R$, which is of interest for cell deposition, where $[x/R \sim 0(10^{-4})]$, $f(x/R) \cong x/R$. Therefore, the cellular mobility is finally expressed by:

$$m(x) = \frac{x}{6\pi\mu R^2}, \qquad x/R \ll 1. \tag{7}$$

Using equations (4), (5) and (7) the Fokker-Planck equation is modified to:

$$\frac{\partial W}{\partial t} + u\frac{\partial W}{\partial x} - \frac{1}{M}\frac{d\phi}{dx}\frac{\partial W}{\partial u} = \frac{1}{m(x)M}\left[u\frac{\partial W}{\partial u} + W + \frac{kT}{M}\frac{\partial^2 W}{\partial u^2}\right] \tag{8}$$

It is important to notice that both the original and the modified Fokker-Planck equations give the probability distribution of a particle as a function of time, position and velocity. However, if we are interested in time intervals large enough compared to β^{-1}, the Fokker-Planck equation, equation (3), can be reduced to a diffusional equation for the distribution function w, frequently called the Smoluchowski equation (Chandrasekhar, 1943):

$$\frac{\partial w}{\partial t} = \frac{\partial}{\partial x}\left[\frac{q}{\beta^2}\frac{\partial w}{\partial x} - \frac{K(x)}{\beta}w\right], \tag{9}$$

in which q/β^2 can be interpreted as a constant diffusion coefficient D. This equation gives the probability distribution of a particle as a function of time and position only. In order to take into account the proximity of the wall, the Smoluchowski equation can be modified similarly to the Fokker-Planck equation to give:

$$\frac{\partial w}{\partial t} = \frac{\partial}{\partial x}\left[D(x)\frac{\partial w}{\partial x} + m(x)\frac{d\phi}{dx}w\right], \tag{10}$$

where

$$D(x) = m(x)kT \tag{11}$$

is the diffusion coefficient. Equation (10) is identical in form with the diffusion equation, on which the previous model is based (Ruckenstein & Prieve, 1975*a*).

In order to establish the validity condition of a diffusion like equation for the probability of escape of a particle over a potential barrier, the solution of the modified Fokker-Planck equation is compared to the solution of the modified Smoluchowski equation. Since the main contribution to the determination of the escape probability comes from the neighborhood of the maximum in the potential energy ($x = x_{max}$), the potential energy function was approximated by a parabolic function and the original Fokker-Planck equation was approximated at the vicinity of x_{max} by (Chandrasekhar, 1943):

$$\frac{\partial W}{\partial t} + u\frac{\partial W}{\partial x} + \frac{\gamma_{max}}{M}(x - x_{max})\frac{\partial W}{\partial u} = \beta u\frac{\partial W}{\partial u} + \beta W + q\frac{\partial^2 W}{\partial u^2}, \tag{12}$$

where $\gamma_{max} = -d^2\phi/dx^2|_{x=x_{max}}$. Using the same reasoning, the value of the cellular mobility, $1/(\beta M)$, in the bulk is replaced by its value at the vicinity of the maximum in the potential energy, and equation (8) is approximated by:

$$\frac{\partial W}{\partial t} + u\frac{\partial W}{\partial x} + \frac{\gamma_{max}}{M}(x - x_{max})\frac{\partial W}{\partial u} = \frac{1}{m(x_{max})M}\left[u\frac{\partial W}{\partial u} + W + \frac{kT}{M}\frac{\partial^2 W}{\partial u^2}\right] \tag{13}$$

and equation (10) is approximated by:

$$\frac{\partial w}{\partial t} = \frac{\partial}{\partial x}\left[D(x_{max})\frac{\partial w}{\partial x} + m(x_{max})\left.\frac{d\phi}{dx}\right|_{x=x_{max}} w\right]. \tag{14}$$

Following the mathematical steps described by Chandrasekhar (1943), the probability per unit time, P, that a Brownian particle will escape over a potential barier $\Delta\phi$ is derived from equation (13) as:

$$P = [(\gamma_{min}/\gamma_{max})^{1/2}/2\pi M]\left[\left(\frac{1}{4m^2(x_{max})} + M\gamma_{max}\right)^{1/2} - \frac{1}{2m(x_{max})}\right]e^{-\Delta\phi/kT} \tag{15}$$

and from equation (14) as:

$$P = [(\gamma_{min}\gamma_{max})^{1/2} m(x_{max})/2\pi]e^{-\Delta\phi/kT}. \tag{16}$$

where $\gamma_{min} = d^2\phi/dx^2$ at the secondary minimum in ϕ, and $\Delta\phi$ is the difference between the maximum and the secondary minimum in ϕ.

The difference between these equations and those derived by Chandrasekhar (1943) is that here the value of the particle mobility is taken at the point of the maximum in the potential energy rather than the value in the bulk. Since the ratio of these values is about 10^4, this modification may be crucial for the case of cell deposition, as demonstrated below.

Equation (15) can be approximated by equation (16) if the following condition is fulfilled:

$$m(x_{max}) \ll (4M\gamma_{max})^{-1/2}. \tag{17}$$

Using the following typical numerical values (Ruckenstein & Prieve, 1975*a*): $R = 6{\cdot}7\times10^{-4}$ cm, $\mu = 7{\cdot}3\times10^{-3}$ g/cm-sec, $x_{max} = 15{\cdot}2$ Å, $\Delta\rho = 0{\cdot}07$ g/cm^3, $\gamma_{max} = 1800$ erg/cm^2 and taking $M = \frac{4}{3}\pi\Delta\rho R^3$, the value of $m(x_{max})$ is $2{\cdot}5$ sec/g and the value of $(4M\gamma_{max})^{-1/2}$ is $1{\cdot}25\times10^3$ sec/g. These values clearly satisfy equation (17). It is worth noting, however, that if the value of m had been used without the correction x/R for the proximity of the surface, its value would have been $1{\cdot}1\times10^4$ which would not have satisfied equation (17). Therefore, the diffusion-like equation (10) is expected to be suitable for the treatment of cell deposition.

The cellular flux over the potential barrier, J, is given by

$$J \equiv -\frac{dn}{dt} = Pn^*, \tag{18}$$

where n^* is the number of cells per unit area in the close vicinity of the secondary minimum of ϕ. For the case treated by Ruckenstein & Prieve (1975*a*), in which the gravitational sedimentation towards the secondary minimum is very rapid compared with the rate of escape over the potential barrier, most of the cells arrive rapidly to the secondary minimum and $n^* \approx n$. For this case then:

$$-\frac{dn}{dt} = \frac{n}{(1/P)} \equiv \frac{n}{\tau}, \tag{19}$$

in which $1/P$ is defined as a time-constant, τ. This equation and equation (16), which expresses P explicitly, have been derived on the basis of the modified Smoluchowski equation and are identical with the equations derived by Ruckenstein & Prieve (1975*a*), based on a diffusional approach.

Coverage-dependent rate of cell deposition

The previous model for the rate of cell deposition, which has been discussed in the preceding section, does not consider the continual change in the characteristics of the surface during the progression of the deposition process. The integration of equation (19), assuming uniform rate of deposition over the whole surface, leads to the exponential decay given by equation (2). As mentioned in the introduction, the available experimental results (Weiss & Harlos, 1972) do not agree with the theoretically predicted exponential shape of the decay curve, and a possible explanation is that the coverage-dependent change in the surface characteristics is affecting the rate of deposition.

As cells start to adhere to the surface, the uniformity in its characteristics is disturbed. The altered spots on the surface, which consist of adhering cells, present new characteristics concerning the surface potential and Hamaker's constant, and consequently, the rate of cellular deposition is not uniform over the surface. The present model is based on the simplified assumption that the overall rate of deposition is the arithmetic sum of two contributions: deposition on the bare surface and deposition on the altered surface. Each of them has its own time constant and depends also on the fraction of the area which is already covered, X. Therefore, by extending equation (19), the overall rate of deposition is given by:

$$J = -\frac{dn}{dt} = \frac{n(1-X)}{\tau_1} + \frac{nX}{\tau_2}, \tag{20}$$

where τ_1 is the time constant for deposition on the bare surface and τ_2 is associated with the covered surface.

Theoretically, the values of τ_1 and τ_2 should be computed using equation (16), keeping in mind that $\tau = 1/P$, with the appropriate values of the surface potentials and Hamaker's constants for each of the two regions. Practically, however, because of the difficulties in the direct measurement of Hamaker's constant and in determining the surface potential of the cells adhering to the surface, it is preferable to evaluate τ_1 and τ_2 empirically by comparison with experimental data for cellular deposition. This will be discussed in more detail along with the presentation of the results.

In postulating the additivity of the rates of deposition in equation (20), it is assumed that the distinct covered areas on the surface, of which the total covered area consists, are large enough compared to a cell so that the model of a plane still holds. This assumption is not valid at the beginning when only single cells are adhering, but then X is very small and the effect of the coverage is not yet pronounced. It becomes valid later when aggregates of cells are formed on the surface.

Deposition on the already covered surface does not increase, at first, the fraction that it occupies of the whole surface. However, the cells that form the second layer may migrate, depending on their wetting properties, and settle down on the surface near the cells of the first layer. Thus, if the rate of this migration is very slow, the increase in X is due solely to the deposition on the bare surface:

$$\frac{dX}{dt} = \frac{2\pi R^2}{\tau_1} n(1-X). \tag{21}$$

But, if the migration of the cells towards the nearby bare surface is rapid, each deposited cell will increase the fraction of the covered area and X will be given by:

$$X = 2\pi R^2(n_0 - n). \tag{21a}$$

In these equations it is assumed that the cell is "flattened" on the surface and occupies an area of $2\pi R^2$. First, the case presented by equation (21) will be analyzed and then the other case presented by equation (21a) will be treated.

Another theoretical possibility, which will, however, just be mentioned, is that cells which deposit on a surface move towards each other and form aggregates.

Slow migration on the surface

The system of equations (20) and (21) can be re-arranged in dimensionless form as follows:

$$\begin{cases} -\dfrac{dY}{d\theta} = \left[(1-X)(1+\alpha) + \left(1 + \alpha - \dfrac{\alpha}{\beta}\right) X\right] Y & (22) \\ \dfrac{dX}{d\theta} = (1+\alpha)\beta(1-X)Y & (23) \end{cases}$$

where

$$\beta \equiv 2\pi R^2 n_0, \tag{24}$$

$$\theta \equiv \frac{t}{\dfrac{\tau_1 \tau_2}{(1-\beta)\tau_2 + \beta\tau_1}}, \tag{25}$$

$$Y \equiv n/n_0 \tag{26}$$

and

$$\alpha \equiv \frac{\beta}{[\tau_2/(\tau_2 - \tau_1)] - \beta}. \tag{27}$$

The choice of these dimensionless parameters is justified by the simple form of the solutions of this system in the limiting cases. The initial conditions are:

$$\text{at } \theta = 0, \quad Y = 1, \tag{28}$$

$$\text{at } \theta = 0, \quad X = 0. \tag{28a}$$

The discussion and solution of these equations is first focused on some special and limiting cases, and then the general case is treated.

The Case $\tau_1 = \tau_2$: This is the trivial case in which the rate of deposition is uniform all over the surface. In this case $\theta = t/\tau_1 = t/\tau_2$ and $\alpha = 0$. Therefore, equations (22) and (23) are reduced to:

$$\begin{cases} -\dfrac{dY}{d\theta} = Y & (29) \\ \dfrac{dX}{d\theta} = \beta(1-X)Y. & (30) \end{cases}$$

Equation (29) is solved independently of equation (30) and its solution is

$$Y = e^{-\theta}, \tag{31}$$

which is, in fact, identical with the results given by the previous model, equation (2). For the fraction of the surface which is covered one obtains:

$$X = 1 - e^{-\beta(1-e^{-\theta})}. \tag{32}$$

The Case $\tau_1 \ll \tau_2$. In this case the cells tend to deposit on the bare surface only, because the rate of deposition on it is much higher than that on the covered surface. When τ_1 is neglected in comparison with τ_2 in equations (25) and (27) one gets $\theta \cong (1-\beta)t/\tau_1$ and $\alpha \cong \beta/(1-\beta)$. Then equations (22) and (23) are reduced to:

$$\begin{cases} -\dfrac{dY}{d\theta} \cong (1-X)(1+\alpha)Y & (33) \\ \dfrac{dX}{d\theta} \cong \alpha(1-X)Y. & (34) \end{cases}$$

Dividing equation (34) by equation (33) and integrating using the initial conditions equations (28) and (28a), results in

$$X \cong \frac{\alpha}{1+\alpha}(1-Y) \cong \beta(1-Y). \tag{35}$$

This equation states that the covered area is approximately proportional to the number of deposited cells which means, as expected, that deposition occurs practically only on the bare surface. Substituting equation (35) into equation (33) gives the following differential equation:

$$-\frac{dY}{d\theta} \cong Y + \alpha Y^2, \tag{36}$$

which by integration leads to

$$Y \cong \frac{e^{-\theta}}{1+\alpha(1-e^{-\theta})}. \tag{37}$$

Since in this case $\theta \cong t/[\tau_1(1+\alpha)]$, the time constant for the exponential factor, $\tau_1(1+\alpha)$, is larger than in the previous case, equation (31). Also, the factor $[1+\alpha(1-e^{-\theta})]^{-1}$ is always smaller than unity. Therefore, the decay in Y is slower than in the case in which $\tau_1 = \tau_2$.

The general case for slow migration: The case of $\tau_2 < \tau_1$, in which cell deposition is expected to be accelerated by its own progression, does not seem to be susceptible to a convenient analytical solution, and therefore, it is treated together with the general case.

Solving equation (22) for X, differentiating with respect to θ and substituting these expressions in equation (23) result in the following equation for Y:

$$\frac{d^2 \ln Y}{d\theta^2} + (1+\alpha)\beta \frac{dY}{d\theta} + (1+\alpha)[\beta(1+\alpha)-\alpha]Y = 0, \quad (38)$$

which has to be solved numerically.

Rapid migration on the surface

Referring to the case of cells migrating rapidly from the covered surface to the nearby bare surface, when X from equation (21a) is substituted into equation (22), the following equation is derived for Y:

$$\frac{dY}{d\theta} + Y + \alpha Y^2 = 0, \quad (39)$$

which is of the same formal form as equation (36) although the physical situation is different. In the present case α and θ depend on τ_2 as well as on τ_1. The solution of equation (39) is given by:

$$Y = \frac{e^{-\theta}}{1+\alpha(1-e^{-\theta})}. \quad (40)$$

This equation reduces to equation (31) when $\tau_1 = \tau_2$ because then $\alpha = 0$. It reduces also naturally to equation (37) when $\tau_1 \ll \tau_2$. For the case $\tau_2 < \tau_1$ it has to be compared with the numerical solution of equation (38).

Thus, in contrast with the previous model, equation (2), which is identical with the case $\tau_1 = \tau_2$, the present model predicts that the decay in the fraction of cells in solution, Y, is not, in general, exponential, and depends on the initial number of cells and on a time constant for the altered surface, in addition to the time constant for the bare surface.

Results and Discussion

As mentioned in the introduction, Fig. 1 demonstrates the usefulness of the previous model in predicting the average decay in the number of tumor cells in solution, although the shape of the experimental curve does not agree in detail with that of the theoretical one. Unfortunately, there are no similar quantitative data concerning blood cell deposition, but some of the qualitative features of this important phenomenon can be explained using this model.

For example, one of the most prominent features of blood cell deposition is that platelets are always the first to adhere to the surface and red blood cells (RBC) are deposited later (Dutton, Webber, Johnson & Baier, 1969). This is especially intriguing since the concentration of RBC in the blood is about ten times higher than that of the platelets. However, a simple explanation is offered by equations (16) and (19) which show that the rate of cellular deposition is most sensitive to the term $(\phi_{max}-\phi_{min})$. Inspection of equation (1) reveals that the potential energy is almost directly proportional to R, the cell radius, when $x \ll R$, and that it increases when the cellular surface potential increases. Therefore $(\phi_{max}-\phi_{min})$ is much larger for RBC (which are much greater than platelets and have a higher surface potential) than it is for platelets, and consequently the rate of escape over the potential energy barrier of platelets is higher than that of RBC. A more detailed discussion of blood cell deposition will be given elsewhere.

However, the previous model cannot predict the instantaneous shape of the tumor cell deposition curve, probably because it does not consider the development of non-uniformity on the surface.

The fraction of cells in solution, Y, as calculated using the present model, is presented in Fig. 2 as a function of the dimensionless time θ, α and β being parameters. The curve for $\alpha = 0$ is the exponential decay predicted by the previous model (and by the present model when $\tau_1 = \tau_2$), and is independent of the value of β.

The curves for $\alpha > 0$ correspond to $\tau_1 < \tau_2$. Equation (27) shows that for values of τ_1/τ_2 lower than, say 0·2, α is almost solely determined by β. If also the value of β is lower than, say 0·1, $\alpha \approx \beta$, and equation (37) shows that Y is decaying, practically, exponentially. For higher values of β $\alpha \approx \beta/(1-\beta)$, and the decay of Y deviates from the exponential shape, but as shown in Fig. 2, this deviation is never significantly large.

The results for $\alpha < 0$ correspond to $\tau_2 < \tau_1$. The full lines present the numerical solution of equation (38), and the dotted lines present the calculations using equation (40). These curves clearly show, for low values of θ, an increase in the rate of deposition. This increase is expected because of the higher rate of deposition on the covered surface and is more pronounced as the value of α approaches -1, which means that τ_2 becomes much smaller than τ_1. Three values of β were used in the numerical solutions of equation (38): $2{\cdot}25 \times 10^{-2}$, $4{\cdot}5 \times 10^{-2}$ (which corresponds to the experiments of Weiss & Harlos, 1972) and 9×10^{-2}. All the curves corresponding to these

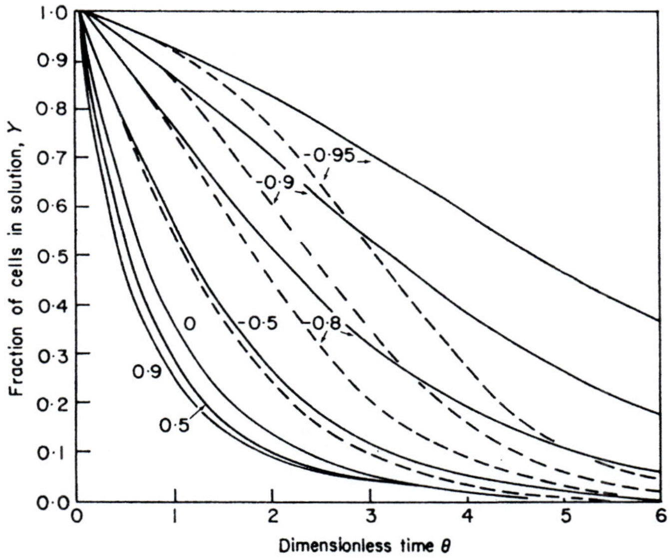

FIG. 2. The fraction of cells in solution as a function of the dimensionless time. The numbers give the value of α. ———, equation (38); – – – –, equation (40). For the values of β see text.

values of β, for a given value of α, were practically identical. However, β appears also in the definition of θ, and therefore, when Y is plotted against the real time t, the value of β affects the results as will be shown below. It is interesting to notice that the difference between equation (40) and the solution of equation (38) increases as α approaches -1 and, for a given value of α, it increases with time. In analyzing the curves shown in Fig. 2, it should be kept in mind that when the value of α is decreased, the value of $\tau_1 \tau_2/[(1-\beta)\tau_2+\beta\tau_1]$ is lowered too. Thus, for example, the curve which corresponds to $\alpha = -0{\cdot}95$, when drawn as a function of the real time t, will decay much faster than the curve for $\alpha = 0$, in contrast with the behavior when the co-ordinates Y, θ are used (Fig. 2).

By comparing the shapes of the curves in Fig. 2 with the experimental data of Weiss & Harlos (1972), which are given in Fig. 1, it is clear that this experiment can be represented by a theoretical curve for which $\tau_2 < \tau_1$.

Figure 3 shows again the experimental data of Weiss & Harlos (1972), but this time along with theoretical curves which are computed using equations (38) and (40). The full line is the solution of equation (38), and the dotted line is calculated using equation (40).

The numerical values of the properties of the cell and solution used in this calculation were as follows: (Weiss & Harlos, 1972; Ruckenstein &

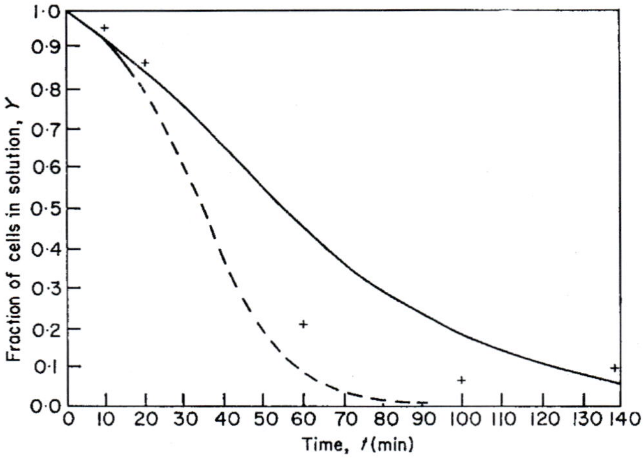

FIG. 3. Experimental data (Weiss & Harlos, 1972) and the present theoretical curves for the fraction of cells in solution as a function of time. ———, equation (38); – – – – equation (40). $\tau_1 = 1{\cdot}546 \times 10^4$ sec and $\tau_2 = 29{\cdot}86$ sec.

Prieve, 1975*a*) $\psi_1 = -12{\cdot}5$ mV, $R = 6{\cdot}7 \times 10^{-4}$ cm, $D(\infty) = 4{\cdot}57 \times 10^{-10}$ cm^2/sec, $n_0 = 1{\cdot}6 \times 10^4$ cells/cm^2, $T = 310°$ K, $\varepsilon = 74{\cdot}3$, $1/\kappa = 8 \times 10^{-8}$ cm. The cell surface potential was assumed to be identical with its zeta-potential as measured by Weiss (1968). In these experiments Weiss measured also the mobility of glass powder in the same medium in which cell deposition was later measured, with 5% calf serum. The zeta potential for glass in this medium after addition of 5% calf serum was $-9{\cdot}3$ mV. The value of Hamaker's constant for the deposition on the bare surface was fitted so that the initial rate of deposition as predicted by the theoretical calculations would be equal to its experimentally measured value. In this way, Hamaker's constant was evaluated using experimental data for the early stages of deposition in which the surface coverage is still minor, and its value was determined as $A = 0{\cdot}461kT$. Using this value and the values presented above for the properties of the cells, surface and solution, the value of the time constant of deposition on the bare surface was computed from equations (16) and (19) as $\tau_1 = 1{\cdot}546 \times 10^4$ sec. Because of the difficulties in determining the values of both the surface potential of the covered surface and the corresponding Hamaker's constant, it appears reasonable to fit the value of τ_2 directly as an empirical constant. The best agreement with the experimental results was found when the value of τ_2 was 29·86 sec. Then $\tau_1 \tau_2/[(1-\beta)\tau_2 + \beta\tau_1] = 6{\cdot}281 \times 10^2$ sec and $\alpha = -0{\cdot}96$.

Equations (38) and (40) present two limiting situations for the rate of cellular migration on the surface. Therefore, it is natural to fit the theoretical curves so the experimental data lie mostly between the two theoretical curves. The shape of the theoretical curves is seen to be in a very good agreement with the experimental curve, except for very large contact periods. The theoretical curves predict that deposition proceeds almost to completion, while the experimental data show a "saturation behavior" when the number of cells in solution is about 7–10% of the original number. This may be explained as follows: As the number of adhering cells on the surface increases, some of them are likely to escape over the potential barrier in the reverse direction and a dynamic equilibrium may be achieved between the adhering and detaching cells. Such an equilibrium is characterized by a constant concentration which is achieved with time in a manner similar to the trend of the experimental data. (For a quantitative treatment see Ruckenstein & Prieve, 1975*b*.)

Figure 4 demonstrates the dependence of the cell deposition process on the initial concentration of cells. The physical parameters are as in Fig. 3, except for n_0, which is for curve (a) 8×10^3 cells/cm² ($\beta = 2{\cdot}25\times10^{-2}$), for curve (b) $1{\cdot}6\times10^4$ cells/cm² ($\beta = 4{\cdot}5\times10^{-2}$), as in the experimental data of Fig. 3 and for curve (c) $3{\cdot}2\times10^4$ cells/cm² ($\beta = 9\times10^{-2}$). The full lines represent again equation (38) and the dotted lines are based on

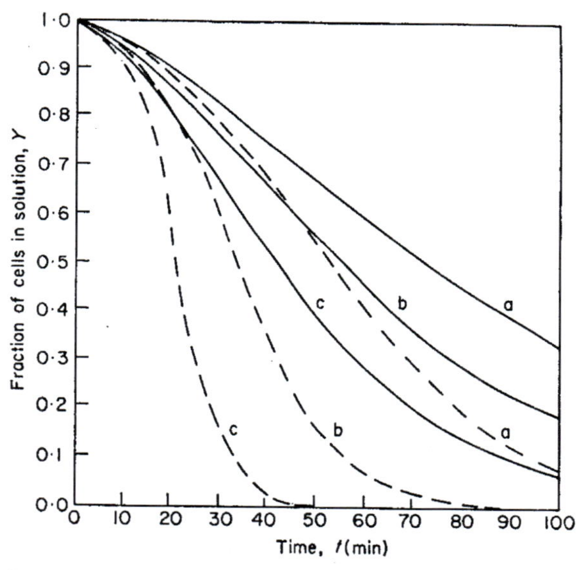

FIG. 4. The effect of initial cellular concentration on cell deposition: (a) $\beta = 2{\cdot}25\times10^{-2}$, (b) $\beta = 4{\cdot}5\times10^{-2}$, (c) $\beta = 9\times10^{-2}$. The other parameters as in Fig. 3. ———, equation (38); – – – –, equation (40).

equation (40). This figure shows that, as expected, an increase in the initial concentration increases the effect of the coverage of the surface, and therefore increases the rate of deposition.

REFERENCES

BRENNER, H. (1961). *Chem. Eng. Sci.* **16**, 242.
CHANDRASEKHAR, S. (1943). *Rev. Mod. Phys.* **15**, 1.
DUTTON, R. C., WEBBER, A. J., JOHNSON, S. A. & BAIER, R. E. (1969). *J. Biomed. Mater. Res.* **3**, 13.
GEORGE, J. N., WEED, R. I. & REED, C. F. (1971). *J. Cell. Physiol.* **77**, 51.
HAMAKER, H. C. (1937). *Physica* **4**, 1058.
HOGG, R., HEALY, J. W. & FUERSTENAU, D. W. (1966). *Trans. Faraday Soc.* **62**, 1638.
RUCKENSTEIN, E. & PRIEVE, D. C. (1975*a*). *J. theor. Biol.* **51**, 429.
RUCKENSTEIN, E. & PRIEVE, D. C. (1975*b*). *Los Angeles AIChE Mtg.* Manuscript. No 7795.
SAWYER, P. N. & SRINIVASAN, S., (1967). *Am. J. Surg.* **114**, 42.
WEISS, L. (1968). *Exp. Cell. Res.* **51**, 609.
WEISS, L. & HARLOS, J. P. (1972). *J. theor. Biol.* **37**, 169.

BULLETIN OF
MATHEMATICAL BIOLOGY
VOLUME 38, 1976

KINETICS OF CELL DEPOSITION UNDER THE ACTION OF AN EXTERNAL FIELD

■ A. Marmur, W. N. Gill and E. Ruckenstein†
Faculty of Engineering and Applied Sciences,
State University of New York at Buffalo,
Buffalo, N.Y. 14214, U.S.A.

The kinetics of cellular deposition from a stagnant solution to a surface are studied, taking into account the combined effect of an interaction field between the cells and the surface and of an external field. Since the forces involved in the adhesion of cells to a surface are short ranged, the cells are conveyed to the vicinity of the surface only by the external field. The equations developed are general, in the sense that they are independent of any particular form of the potential energy function, provided that it presents an appreciable potential barrier between the cells and the deposition surface. The characteristic shape of the curve representing the decay of the fraction of cells in solution with time is shown to be affected by the value of Pt^*, consisting of the probability per unit time, P, for the escape over the potential barrier, and the sedimentation time, t^*. A simple inspection procedure of this curve can disclose the relative significance of the external field and of the potential barrier in the overall kinetics of deposition. In addition, such an inspection can reveal the existence of alterations in the cellular adhesiveness with increasing coverage of the deposition surface. By matching the equation obtained to experimental results, the cellular adhesiveness, in terms of P, and the sedimentation rate (in the case of very slow sedimentation) can be evaluated.

Introduction. Cell deposition on various surfaces plays an important part in the initiation of many biological processes, such as hemostasis, thrombosis and growth of secondary tumors. It consists of two main stages: (a) the conveyance of cells to the vicinity of the interacting surface, and (b) the adhesion of cells to this surface. In most of the biological systems, the main field which conveys the cells to the surface is a flow field. However, for experimental studies of the process of adhesion *per se*, it is convenient to use a stagnant solution in which cells are conveyed to the surface by gravity (Weiss and Harlos, 1972) or by a

†Correspondence concerning this paper should be addressed to E. Ruckenstein.

centrifugal field (George, Weed and Reed, 1971). While the process of adhesion and its mechanisms have been studied extensively, as will be briefly reviewed below, the effect of the external field on the results of adhesion experiments has been overlooked. A failure to take this effect into account may lead, as will be shown in the present paper, to an inaccurate interpretation of experimental results.

The forces involved in cell adhesion are of two types: (a) attraction forces, of which the London–Van der Waals dispersion force is mostly considered, and (b) an electrostatic force, which is due to the development of double layers when a cell as well as an interacting surface are in contact with a solution. Mammalian cells are known to be negatively charged and so are most of the non-biologic surfaces. Therefore, the double layer force is in most of the cases a repulsion force. Theoretically, various forms of force–distance relationships are possible (Curtis, 1973). However, of special interest is the frequently encountered case, in which a potential energy barrier exists between the cells and the interacting surface. This occurs when the repulsive potential due to the double layer forces is large enough compared to the attractive potential. The potential energy function has in this case a very deep primary minimum close to the surface, followed by a maximum and a secondary minimum. The potential barrier is the difference between the maximum and the secondary minimum in the potential energy. Its existence was demonstrated in the experiment carried out by Weiss and Harlos (1972), who showed that the time needed for tumor cells to adhere from a stagnant solution to a horizontal surface of glass coverslip is much longer than the time needed for gravitational sedimentation to this surface. The existence of such a potential barrier raised speculations about the mechanism by which cells are able to overcome it and adhere to a surface at the primary minimum. Weiss and Harlos (1972) considered the potential barrier as impenetrable to the cells, and therefore speculated that adhesion involves protrusion of cellular probes from the secondary to the primary minimum, and/or extrusion of an intercellular "glue". However, Ruckenstein and Prieve (1975) suggested that cells are able to overcome the potential barrier by a stochastic motion, similar to Brownian motion. On this basis they established an equation for the rate of deposition for the case in which the conveyance of cells to the secondary minimum in a stagnant solution is very rapid compared to the rate of escape over the potential barrier. This condition was valid in the above mentioned experiment of Weiss and Harlos (1972). However, in other situations, such as platelet deposition in a gravity field, the rate of cellular transfer to the surface is slow and may considerably affect the overall rate of deposition.

The purpose of the present paper is to study theoretically the kinetics of cell deposition from stagnant solutions under conditions in which the cellular adhesion rate is determined both by the escape over the potential barrier and by the cellular conveyance rate, and to point out methods for analyzing experimental data measured under such circumstances. The theoretical analysis will be devoted to the case in which an appreciable potential barrier exists between the cells and the surface, but will be quite general in the sense that no particular potential energy function will be assumed.

Theory. The probability per unit time for a particle to overcome a potential barrier far away from any surface when the potential energy function, ϕ, has a deep minimum and a steep maximum is given by (Chandrasekhar, 1943):

$$P = [(\gamma_{\min}\gamma_{\max})^{1/2}m/2\pi]\exp(-\Delta\phi/kT), \tag{1}$$

where ϕ is the potential energy, $\gamma_{\max} = -\mathrm{d}^2\phi/\mathrm{d}x^2$ at the maximum in the potential energy, $\gamma_{\min} = \mathrm{d}^2\phi/\mathrm{d}x^2$ at the secondary minimum in the potential energy, x is the separation between the cell and the surface, $\Delta\phi = \phi_{\max} - \phi_{\min}$, $\phi_{\max}$ is the value of ϕ at the maximum, and $\phi_{\min}$ the value of ϕ at the secondary minimum, k is Boltzman's constant, T is the absolute temperature, and m is the mobility of the particle which is defined as the ratio between its velocity and the drag force acting on it.

However, the probability for a cell to overcome the potential barrier and adhere to a surface is affected by the presence of this surface through its influence on the cellular mobility and through the modifications that it induces in the cellular shape. The mobility of a particle near a solid surface decreases as it approaches the surface, because the viscous shear between the solution which is displaced by the particle and the surface increases the resistance to the motion of the particle. The changes in shape affect the interaction potential between the surface and the cells. Since the probability of escape over the potential barrier is mostly determined by the vicinity of the maximum in the potential energy, the mobility at this particular point is used to evaluate P in (1):

$$P = [(\gamma_{\min}\gamma_{\max})^{1/2}m(x_{\max})/2\pi]\exp(-\Delta\phi/kT), \tag{2}$$

where $x_{\max}$ is the separation which corresponds to $\phi_{\max}$. In addition to its control over the conveyance of the cells, the external field may influence the value of the probability of escape over the potential barrier through its contribution to the potential energy in the close vicinity of the surface. The effect of gravitational acceleration on the potential energy is usually negligible very near to the surface, and therefore, does not affect P. However, since the contribution of the external field to ϕ increases linearly with x, a substantial increase in the acceleration of the external field affects the value of P. In the extreme case of a very strong external field, the maximum and minimum in ϕ are eliminated and no potential barrier exists, i.e. $P \to \infty$. In the intermediary case, the potential barrier does exist, but P is affected by the external field.

Equation (1) holds for a particle or a cell of any shape, even when the shape changes with the distance x, provided that it is reaching the surface mainly by a translational motion. Equation (1) also assumes a quasi-steady state of the motion of particles over the potential barrier. This approximation can be made because the region over which the potential ϕ acts is very thin and consequently the flux of particles through it can be considered practically constant with respect to the distance x at a given time t.

Once a cell makes contact with a surface, cell spreading usually occurs, and this completes and strengthens the cellular adhesion. If the cell spreading is rapid, the probability for the cell to detach from the surface is low. If this spreading is slow, cells may detach from the surface and this will affect the measurement of the adhesion kinetics. In the present paper it is assumed that the spreading is rapid, i.e., that a cell which touches the surface is permanently adhering to it.

The flux of cells leaving the bulk of the solution, $-\mathrm{d}n/\mathrm{d}t$, and adhering to the surface by escaping over the potential barrier can be computed by multiplying the probability per unit time, P, by the number of cells per unit area, n^*, which are at a given time at the vicinity of the secondary minimum:

$$-\frac{\mathrm{d}n}{\mathrm{d}t} = Pn^*, \tag{3}$$

where n is the number of cells per unit area, which are in the solution at time t. Since P is a constant, only the time dependence of n^* needs to be known in order to integrate (3). In general, the value of n^* is determined by the external field and by the diffusional motion of the cells. The relative influence of the external field compared with that of the diffusional motion can be estimated from the ratio between the convective contribution caused by the external field and the diffusional contribution due to Brownian motion of the cells expressed by a Peclet number:

$$Pe = VL/D. \tag{4}$$

In (4), L is the distance that has to be traveled by the cells in the solution (the level of the solution above the horizontal surface), D is the diffusion coefficient, and V is the sedimentation velocity under the action of the external field. This velocity can be evaluated from the equation

$$V = \frac{2}{9}\frac{R^2\Delta\rho a}{\mu}, \tag{5}$$

where R is the "effective radius" of the cell, $\Delta\rho$ is the density difference between the cell and the solution, a is the acceleration produced by the external field, and μ is the viscosity of the solution. Equation (5) is valid for relatively dilute solutions of spherical particles and is used here for evaluation purposes only. The diffusion coefficient can be estimated by:

$$D = kT/6\pi R\mu. \tag{6}$$

Using (5) and (6) to evaluate V and D in (4), one gets for the Peclet number:

$$Pe = \frac{4}{3}\frac{\pi R^3 a \Delta\rho L}{kT}. \tag{7}$$

Using values as low as $R \approx 1\mu\text{m}$, $\Delta\rho = 10^{-2}\text{g/cm}^3$, $a = g$ (gravitational acceleration), and $L = 0.1$ cm at room temperature ($T = 300\text{K}$) one obtains a Peclet number of 10^2. This means that for most practical situations the diffusional motion in the bulk can be neglected compared with the motion induced by the external field.

When the diffusional motion in the bulk can be neglected and when the velocity of the cells towards the surface under the action of the external field is constant, three different regions can be observed: (a) the most distant region from the deposition surface, which is practically depleted of cells because of the action of the external field; (b) an intermediary region in which the concentration of cells is practically constant; (c) the layer of cells at the vicinity of the secondary minimum $\phi_{\min}$, near the surface, into which the cells are entering at a constant rate and leave by escaping over the potential barrier. If t^* is the time needed for the most distant cell to reach the vicinity of $\phi_{\min}$, then the rate at which cells are accumulating at the vicinity of the secondary minimum per unit area is n_0/t^*, where n_0 is the initial number of cells in solution per unit area. Therefore, the number of cells per unit area that have reached the vicinity of the surface at time t is $(n_0/t^*)t$. The number of cells per unit area that have left this layer and have adhered to the surface is (n_0-n). If $t > t^*$, all the cells in the solution have already reached the secondary minimum, and $n^* = n$. The number of cells per unit area in the vicinity of the secondary minimum is thus given by:

$$n^* = \begin{cases} n_0^* + n_0 t/t^* - (n_0 - n), & t/t^* \leqq 1 \\ n, & t/t^* > 1, \end{cases} \tag{8}$$

where n_0^* is the initial number of cells per unit area at the vicinity of the secondary minimum. n_0^* is usually negligible compared to n_0 and therefore, will not appear in subsequent equations. Substituting (8) into (3) and integrating gives:

$$\frac{n}{n_0} = \begin{cases} 1 - \dfrac{Pt + \exp(-Pt) - 1}{Pt^*}, & t/t^* \leqq 1 \\[2ex] \dfrac{\exp(Pt^*) - 1}{Pt^*} \exp(-Pt), & t/t^* > 1. \end{cases} \tag{9}$$

K

The value of Pt^* determines the relative significance of the transfer by the external field and the escape over the potential barrier in the process of cell deposition. When $Pt^* \ll 1$, that is when sedimentation by the external field is very rapid compared with the rate of overcoming the potential barrier, (9) transforms for $t/t^* > 1$ to an exponential expression, $n/n_0 = \exp(-Pt)$, previously derived specifically for this case (Ruckenstein and Prieve, 1975). This situation involves "low cellular adhesiveness" and/or large cells with appreciable density difference between them and the suspending solution, such as red blood cells. On the other hand, small cells with density close to that of the solution and/or not too low "adhesiveness", such as platelets, fall into the category of $Pt^* \gg 1$, which is especially emphasized in the present paper.

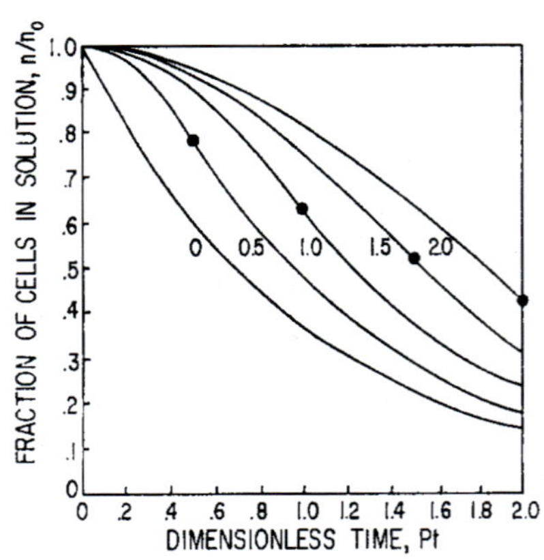

Figure 1. The effect of the sedimentation on the decay in the fraction of cells in solution. The numbers give the value of Pt^*. The dot shows the point of change in the curvature at $Pt = Pt^*$

Results and Discussion. Figure 1 shows the decay in the fraction of cells in solution with the dimensionless time, Pt for various values of the parameter Pt^*, as predicted by (9). For $Pt^* \to 0$, the shape of the curve for practically all finite values of Pt is exponential, as previously predicted by Ruckenstein and Prieve (1975). For $Pt^* > 0$, there exists a point on the decay curve, at $t = t^*$, before which ($t \leq t^*$), the rate of deposition increases with time and after which, ($t > t^*$), the rate of deposition decreases with time as in the exponential case. The characteristic shapes of these curves may serve as a useful tool to analyze experimental data. A simple inspection of an experimental curve presenting n/n_0 as a function of time and a comparison between it and the curves in Figure 1 can reveal the relative contribution of the external field and the potential barrier to the kinetics of deposition. This is important in order to avoid misinterpretation of low sedimentation rate as low adhesiveness in studies of cell adhesion. In the analysis of experimental data, the following possibilities should be checked:

(a) *No detectable point of change in the curvature of the curve.* This case is subdivided into two subcases, as follows:

(1) If the rate of deposition decreases with time, it means that $Pt^* \ll 1$, i.e. high rate of sedimentation compared with the rate of adhesion. For this case (9) is adequately described by:

$$n/n_0 \cong \exp(-Pt) \tag{10}$$

and the slope of the curve on a semi-log paper gives the probability P of escape over the potential barrier.

(2) If the rate of deposition increases with time, it means that $Pt^* \gg 1$, that is, sedimentation is very slow compared with the rate of adhesion. By finding the values of P and t^* that provide the best fit of (9) for $t/t^* < 1$ to

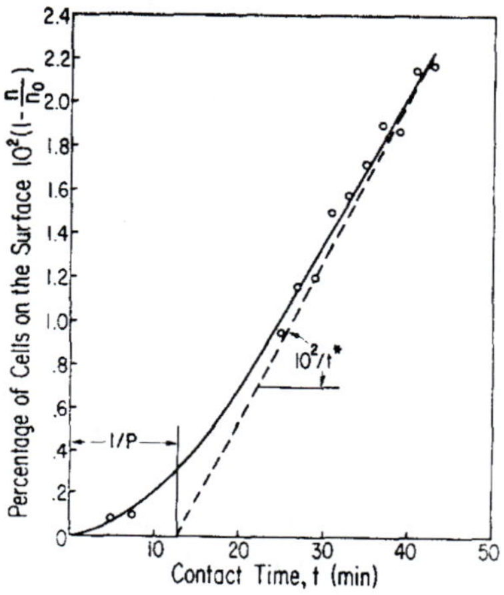

Figure 2. Experimental and theoretical results for platelet adhesion to siliconized glass

experimental results, both the cellular adhesiveness, in terms of P, and the sedimentation rate can be evaluated. This procedure can be made easier if experimental results are available for large values of Pt, say, $Pt \geqq 3$, because then (9) transforms to a linear equation:

$$n/n_0 \cong 1 - t/t^* + 1/Pt^*, \quad \text{for } t/t^* \leqq 1 \text{ and } Pt \geqq 3. \tag{11}$$

Since the platelets, for example, have a very low sedimentation rate and relatively large adhesiveness, one may expect that their behavior conforms to the category characterized by $Pt^* \gg 1$. Figure 2 presents an example of experimental data for the adhesion of platelets from a stagnant platelet-rich plasma onto a siliconized glass surface, under the action of gravity (Ruckenstein, Marmur, and Rakower, 1976) in terms of the percentage of cells adhering to the surface, $10^2 \times (1 - n/n_0)$, as a function of time. The solid line in Figure 2 represents the best fit of (9) to the experimental data, using the least-squares method. It can be seen that all the experimental points lie very close to the theoretical curve, a fact which confirms the qualitative features of (9). The values of P and t^* thus computed are 0.079 min^{-1} and 1.37×10^3 min, accordingly. The value of Pt^* is then 1.08×10^2 which, being much greater than unity, conforms to the presently discussed situation. It is worth noting that the straight line drawn according to (11) with the above values of P and t^* almost merges with the full curve at $Pt \cong 3$ to 4.

(b) *Detectable point of change in the curvature of decay curve.* The time t^*, at which the change in curvature is theoretically predicted, can be easily estimated on the basis of independent data, especially if the sedimentation rate is not very slow. If the experimental point of change in curvature occurs at $t \approx t^*$, it means that the conveyance of cells to the surface by the external field is indeed the mechanism that is responsible for the deviation from the exponential shape. It also means, that both the external field and the height of the potential barrier determine the kinetics of the cellular deposition. In this case the value of P may be found by matching the curve given by (9) to the experimental data.

However, a point of curvature change may occur at a time different from t^*, in addition to or in the absence of another point of curvature change at $t \approx t^*$. This means that P is not constant and there exists a mechanism, which is not taken into account in the present equations, which enhances adhesion. This possibility is demonstrated by the experiment of Weiss and Harlos (1972), in which they measured the rate of deposition of tumor cells in a gravity field. In their results, a point of curvature change occurs at $t \cong 30$ min, while the calculated time of sedimentation is $t^* \cong 6$ min. A possible mechanism which enhances deposition is the effect of the change in the characteristics of the surface due to its coverage by the cells. The surface is then divided into two regions: the bare surface and the covered surface, each of them characterized by its own value of P. The overall probability per unit time to escape over the potential barrier is given by $P = P_1X + P_2(1-X)$, where X is the fraction of the surface area covered by cells, P_1 is the probability related to the covered surface and P_2 to the bare surface. This dependence of P on X implies also a time-dependence of P. Computations done on the basis of this equation (Ruckenstein, Marmur, and Gill, 1976) show indeed, that for the case $P_1 > P_2$ an inflection point is theoretically predicted, in accordance with the experimental data (Weiss and Harlos, 1972).

Conclusions. An external field, such as gravity or a centrifugal field, affects the kinetics of cellular deposition by controlling the rate at which cells are conveyed to the close vicinity of the surface. The subsequent adhesion to the surface takes place via an escape over the potential barrier generated essentially by the London and double-layer forces. The extent of the effect of the external field can be easily estimated by a simple inspection of the shape of the curve representing the fraction of cells in solution as a function of time. Simple graphical methods can be used to evaluate the cellular adhesiveness, in terms of the probability per unit time for the escape over the potential barrier, if the inspection of the results does not reveal evidence for a time dependence of this probability. As a by-product, the rate of sedimentation can be easily determined in the case of very slow sedimentation, such as that of platelets under the action of gravity.

LITERATURE

Chandrasekhar, S. 1943. "Stochastic Problems in Physics and Astronomy," *Rev. Mod. Phys.*, **15,** 1–91.

Curtis, A. S. G. 1973. "Cell Adhesion," In *Biophysics and Molecular Biology* (Butler and Noble, eds.) **27,** pp. 315–386. Pergamon Press.

George, J. N., R. I. Weed and C. F. Reed. 1971. "Adhesion of Human Erythrocytes to Glass: The Nature of the Interaction and the Effect of Serum and Plasma." *J. Cell. Physiol.*, **77,** 51–60.

Hamaker, H. C. 1937. "The London–Van der Waals Attraction Between Spherical Particles," *Physica,* **4,** 1058–1072.

Hogg, R., T. W. Healy and D. W. Fuerstenau. 1966. "Mutual Coagulation of Colloidal Dispersion." *Trans. Faraday Soc.*, **62,** 1638–1651.

Ruckenstein, E., A. Marmur and S. R. Rakower. 1976. "Sedimentation and Adhesion of Platelets from Platelet Rich Plasma onto Glass Surface." submitted for publication.

Ruckenstein, E. and D. C. Prieve. 1975. "Dynamics of Cell Deposition on Surfaces." *J. Theor. Biol.*, **51,** 429–438.

Ruckenstein, E., A. Marmur and W. N. Gill. 1976. "Coverage Dependent Rate of Cell Deposition." *J. Theor. Biology*, **58.**

Weiss, L. and J. P. Harlos. 1972. "Some Speculations on the Rate of Adhesion of Cells to Coverslips," *J. Theor. Biol.* **37,** 169–179.

RECEIVED 8-18-75

REVISED 3-12-76

Advances in Colloid and Interface Science
79 (1999) 59–76

ADVANCES IN COLLOID AND INTERFACE SCIENCE

Thermodynamic insights on macroemulsion stability

Eli Ruckenstein*

Chemical Engineering Department, State University of New York at Buffalo, Clifford C. Furnas Hall, SUNY-Buffalo, Box 604200, Buffalo, NY 14260-4200, USA

Abstract

Experiments reported in the literature for nonionic surfactants have revealed that: (1) the interfacial tension at the planar surface between oil and water, at a constant amount of surfactant and equal volumes of water and oil, passes through a minimum when plotted against the HLB (the hydrophilic–lipophilic balance); (2) under the same conditions, the emulsion stability against HLB passes through two maxima, one in the W/O range and the other in the O/W range, and a minimum between the two at the phase inversion point; (3) the minima under (1) and (2) coincide; (4) the phase inversion temperatures of emulsions, formed of oil and water as distinct phases, or microemulsions and their excess phase, and the corresponding microemulsions are the same. These observations are explained on the basis of thermodynamics. It is also shown that the stability of emulsions can be correlated with the surface excess of surfactant, which also passes through two maxima and a minimum between the two when plotted against HLB or temperature. In addition, a two ratio approach is used to determine when the Bancroft rule is obeyed or violated in microemulsions. © 1999 Elsevier Science B.V. All rights reserved.

Keywords: Macroemulsion stability; Hydrophilic–hydrophobic balance (HLB); Interfacial tension vs. HLB; Surface excess vs. HLB; Surface excess vs. temperature; Phase inversion temperature; Bancroft rule

Contents

*Tel.: +1 716 645 2911; fax: +1 716 645 3822; e-mail: feaeliru@acsu.buffalo.edu

PII: S0001-8686(98)00079-7

1. Introduction

While kinetics are mostly responsible for the stability of macroemulsions, thermodynamics plays a role because adsorption of the surfactant or polymer on the interface of the globules affects the stability in a major way. In what follows it will be shown that thermodynamics can explain some of the overall features observed experimentally regarding the stability of emulsions.

Two attempts have been made to relate macroemulsions to thermodynamics. One group of authors tried to explain the surfactant performance through the interfacial tension [1–4]. They observed that a lowering of the interfacial tension sometimes increases the stability, but that a too-low interfacial tension has the opposite effect. Interesting experiments have been carried out by Boyd et al. [5], who prepared emulsions containing equal volumes of oil and water and a fixed weight amount of mixtures of nonionic dispersants. They observed that increasing the hydrophilic–lipophilic balance (HLB) of the dispersant, the initial rate of coalescence displayed two minima and a maximum at the phase inversion point. The first minimum was for a W/O emulsion and the second for an O/W emulsion. Berger et al. [4], using a series of polypropylene glycol ethoxylates as dispersants (2 wt.% with respect to water), mineral oil as the oil phase and water containing 2 wt.% NaCl as the water phase, repeated the kind of experiments reported by Boyd et al. [5]. Equal volumes of the two phases have been mixed and the time in which complete separation took place was used as a measure of stability. They noted that for the same weight amount of dispersant, the emulsion stability passed, with increasing HLB, through a maximum in the W/O range, followed by a minimum at the phase inversion point and again a maximum in the O/W range. The interfacial tension between the two phases before emulsification exhibited a minimum for the HLB corresponding to the phase inversion point. It is clear that the stability depends upon the HLB and is not directly related to the interfacial tension.

A second group of researchers has noted a parallelism between emulsions and microemulsions formed with surfactants possessing a poly(ethylene oxide) head group [6,7]. The emulsions were of the oil in water (O/W) kind at low temperatures and of the water in oil (W/O) kind at relatively high temperatures. Similarly, an O/W microemulsion was in equilibrium with excess oil at low temperatures; a W/O microemulsion coexisted with excess water at high temperatures and a middle phase microemulsion coexisted with both excess phases in an intermediate range of temperatures. Shinoda and Saito [6,7] noted that the phase inversion temperature provided by the water–oil–poly(ethylene oxide) surfactant phase diagram coincides with the phase inversion temperature of the emulsion formed by mixing a microemulsion with its excess phase, and that the emulsion was of the same kind as the microemulsion.

While microemulsions are thermodynamically stable, and the stability of emulsions has a kinetic origin, in both cases the adsorption of the dispersant upon the interface of the globules is responsible for stability. For this reason it appears natural to attempt to explain the above equality between the two inversion temperatures on the basis of surfactant adsorption. In addition, both the micro and macro-emulsions obey in many cases the Bancroft rule [8,9], which indicates that the phase in which a larger amount of dispersant is present becomes the continuous phase; there are, however, some violations of this rule which will be discussed later in the paper.

The scope of the present review is to emphasize that thermodynamics can explain the above experimental observations. The next section (Section 2), which is based on ref. [10], will be concerned with the effects of HLB (denoted in what follows h) on the interfacial tension and on the stability of macroemulsions, the goal being to explain the observations of Boyd et al. [5] and of Berger et al. [4]. Section 3, which is based on ref. [11], will examine the effect of temperature on the interfacial tension at the oil–water interface by assuming that no microemulsion or emulsion is formed, as well as its effect on the stability of emulsions. Shinoda and Saito's observations regarding the equality of the two inversion temperatures will be thus explained. Finally, the Bancroft rule [8,9], and some of the violations of this rule, will be examined in the spirit of ref. [12].

2. Effect of HLB (h) on the interfacial tension and the stability of macroemulsions [10]

2.1. The interfacial free energy

At constant temperature, the interfacial tension γ of a water–oil system containing a single surfactant solute can be calculated at thermodynamic equilibrium starting with the Gibbs adsorption equation

$$d\gamma = -\Gamma d\mu \tag{1}$$

where Γ is the surface excess of the surfactant and μ is its chemical potential. In order to characterize the tendency of dispersants to generate O/W or W/O emulsions, Griffin [13,14] suggested an empirical scale which he called hydrophilic–lipophilic balance (HLB) which will be denoted as h. Surfactants with low values of h are more compatible with the oil phase (are more hydrophobic) and tend to stabilize the water in oil (W/O) emulsions, while those with high values of h are more compatible with water (are more hydrophilic) and tend to stabilize the (O/W) emulsions. The surfactants with intermediary values of h are ineffective as stabilizers. Davies [15–17] proposed a group contribution method to evaluate h from the molecular structure of the surfactants. While the HLB scale does not account for the effect of temperature and the nature of the oil, it is useful at room temperature and as a qualitative concept and will be used in this section.

Consider a planar interface between water and oil, and a surfactant with a poly(ethylene oxide) head group distributed at equilibrium between the two phases. It is assumed that no macro or micro-emulsification nor solubilization takes place. For sufficiently small values of h, the surfactant is partitioned mostly in the oil phase, while for sufficiently large values of h, it is partitioned mostly in the water phase. The surfactant is mostly nonaggregated at small values of h, because it is present in oil where only small aggregates are formed [18,19], but becomes aggregated above the CMC for large values of h, when the surfactant is mostly partitioned in water. The case in which the concentration in the water phase is below the CMC will be considered first. For dilute systems, one can write

$$\mu = \mu_W^0 + kT\ln C_W = \mu_O^0 + kT\ln C_O, \tag{2}$$

where C is the concentration of the surfactant, the superscript 0 refers to the standard state, the subscripts W and O to water and oil, respectively, k is the Boltzmann constant and T is the absolute temperature. Consequently, the concentrations in the oil and water phases are related via the equation

$$C_O = C_W e^{\Delta\mu_0/kT}, \tag{3}$$

where

$$\Delta\mu_0 = \mu_W^0 - \mu_O^0. \tag{4}$$

A mass balance for equal volumes of water and oil leads to

$$C_O + C_W = 2C, \tag{5}$$

where C is the surfactant concentration per unit volume of the system, which was kept constant in the experiments [4,5] which we attempt to interpret.

Eliminating C_O between Eq. (3) and Eq. (5), yields

$$C_W = \frac{2C}{1 + e^{\Delta\mu_0/kT}}. \tag{6}$$

The change of γ with h is given by the expression

$$\frac{d\gamma}{dh} = -\Gamma\frac{d\mu}{dh}, \tag{7}$$

in which, because

$$\frac{\mathrm{d}\mu}{\mathrm{d}h} = \frac{\mathrm{d}\mu_W^0}{\mathrm{d}h} + \frac{kT}{C_W}\frac{\mathrm{d}C_W}{\mathrm{d}h}$$

and

$$\frac{\mathrm{d}C_W}{\mathrm{d}h} = -\frac{C_W}{kT}\frac{1}{1+e^{-\Delta\mu^0/kT}}\frac{\mathrm{d}(\mu_W^0-\mu_O^0)}{\mathrm{d}h},$$

$$\frac{\mathrm{d}\mu}{\mathrm{d}h} = \frac{1}{1+e^{\Delta\mu_0/kT}}\frac{\mathrm{d}\mu_W^0}{\mathrm{d}h} + \frac{1}{1+e^{-\Delta\mu_0/kT}}\frac{\mathrm{d}\mu_O^0}{\mathrm{d}h}. \tag{8}$$

The standard chemical potential in water decreases with increasing hydrophilicity because of increasing favorable interactions between surfactant and water, and that in oil increases. Consequently,

$$\frac{\mathrm{d}\mu_W^0}{\mathrm{d}h} < 0 \quad \text{and} \quad \frac{\mathrm{d}\mu_O^0}{\mathrm{d}h} > 0. \tag{9}$$

For low values of h, $\Delta\mu_0$ is positive and large, because μ_W^0 has its largest and μ_O^0 its smallest value, and Eq. (8) becomes

$$\frac{\mathrm{d}\mu}{\mathrm{d}h} \simeq \frac{\mathrm{d}\mu_O^0}{\mathrm{d}h} > 0. \quad (\text{small } h) \tag{10a}$$

For large values of h, $\Delta\mu_0$ is negative and large in absolute value and one can similarly conclude that

$$\frac{\mathrm{d}\mu}{\mathrm{d}h} \approx \frac{\mathrm{d}\mu_W^0}{\mathrm{d}h} < 0. \quad (\text{large } h) \tag{10b}$$

The surface excess Γ is a positive quantity even when the interactions between surfactant and water become as favorable as those between surfactant and oil. In the latter case $\Delta\mu_0 = 0$ and the concentrations in the two phases become equal. Γ does not vanish in that case, because the interactions between water and head group and between oil and hydrocarbon chain are stronger than those between water and hydrocarbon chain and oil and head group, respectively. It is, however, small because the interactions at the interface are not much stronger than those in the bulk.

Consequently,

$$\frac{\mathrm{d}\gamma}{\mathrm{d}h} < 0, \quad (\text{small } h) \tag{11a}$$

$$\frac{\mathrm{d}\gamma}{\mathrm{d}h} > 0 \quad (\text{large } h) \tag{11b}$$

and it is clear that γ passes through a minimum. The position of the minimum is given by $d\mu/dh = 0$, hence by the equation

$$\frac{1}{1+e^{\Delta\mu_0/kT}}\frac{d\mu_W^0}{dh}+\frac{1}{1+e^{-\Delta\mu_0/kT}}\frac{d\mu_O^0}{dh}=0. \tag{12}$$

Because the derivatives $d\mu_W^0/dh$ and $d\mu_O^0/dh$ have opposite signs, there is a value h_c of h for which γ is minimum. If near the minimum, the curves μ_W^0 and μ_O^0 are symmetrical, then $\Delta\mu_0 = 0$ at the minimum and the concentrations of surfactant in oil and water become equal. Even when the curves are not completely symmetrical, $d\gamma/dh \approx 0$ for $\Delta\mu_0 = 0$, because Γ is small for $h = h_c$ and the factor which multiplies Γ in Eq. (7) is also small. A qualitative plot of γ against h is presented in Fig. 1.

If the concentration in water surpasses the CMC, aggregates are formed and the above equations are no longer valid. A conclusion can be, however, reached by combining Eq. (1) and Eq. (2) to obtain an expression [Eq. (13)] in terms of the oil phase where no important aggregation takes place:

$$\frac{d\gamma}{dh}=-\Gamma\left(\frac{d\mu_O^0}{dh}+\frac{kT}{C_O}\frac{dC_O}{dh}\right). \tag{13}$$

With increasing hydrophilicity, hence increasing h, μ_O^0 increases and C_O decreases. Consequently,

$$\frac{d\mu_O^0}{dh}>0 \quad \text{and} \quad \frac{dC_O}{dh}<0 \tag{14}$$

and γ can pass through an extremum. At low values of h, μ_O^0 is expected to increase more rapidly than at high values; in contrast, the concentration C_O is

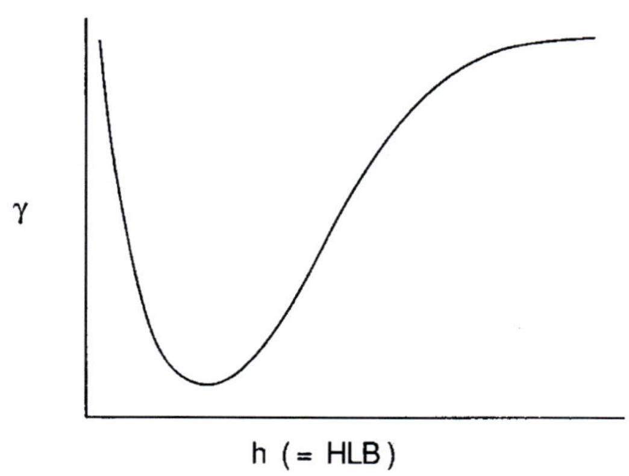

Fig. 1. Qualitative representation of the interfacial tension at a planar oil–water interface against HLB (h).

expected to vary more rapidly at high values of h, because of the aggregation of the surfactant as micelles in the aqueous phase. As a result,

$$\frac{\mathrm{d}\mu}{\mathrm{d}h} \approx \frac{\mathrm{d}\mu_{\mathrm{O}}^{0}}{\mathrm{d}h} > 0, \quad (\text{small } h) \tag{15a}$$

$$\frac{\mathrm{d}\mu}{\mathrm{d}h} \approx \frac{kT}{C_{\mathrm{O}}} \frac{\mathrm{d}C_{\mathrm{O}}}{\mathrm{d}h} < 0 \quad (\text{large } h) \tag{15b}$$

and γ passes through a minimum. A similar conclusion can be reached by assuming that the concentration of the nonaggregated surfactant is equal to the CMC $= C_{\mathrm{W1}}$, hence that

$$\mu = \mu_{\mathrm{W}}^{0} + kT \ln C_{\mathrm{W1}}. \tag{16a}$$

Combining Eq. (1) and Eq. (16a) yields

$$\frac{\mathrm{d}\gamma}{\mathrm{d}h} = -\Gamma\left(\frac{\mathrm{d}\mu_{\mathrm{W}}^{0}}{\mathrm{d}h} + \frac{kT}{C_{\mathrm{W1}}} \frac{\mathrm{d}C_{\mathrm{W1}}}{\mathrm{d}h}\right). \tag{16b}$$

With increasing hydrophilicity, μ_{W}^{0} decreases and C_{W1} increases, hence

$$\frac{\mathrm{d}\mu_{\mathrm{W}}^{0}}{\mathrm{d}h} < 0 \quad \text{and} \quad \frac{\mathrm{d}C_{\mathrm{W1}}}{\mathrm{d}h} > 0. \tag{17}$$

At low values of h, μ_{W}^{0} is expected to decrease less rapidly than at high values; in contrast, C_{W1} is expected to increase more rapidly at low values of h. Consequently,

$$\frac{\mathrm{d}\mu}{\mathrm{d}h} \approx \frac{kT}{C_{\mathrm{W1}}} \frac{\mathrm{d}C_{\mathrm{W1}}}{\mathrm{d}h}, \quad (\text{small } h) \tag{18a}$$

$$\frac{\mathrm{d}\mu}{\mathrm{d}h} \approx \frac{\mathrm{d}\mu_{\mathrm{W}}^{0}}{\mathrm{d}h} \quad (\text{large } h) \tag{18b}$$

and γ passes through a minimum.

It is clear from the above considerations that there is no parallelism between the stability of emulsions, which passes through two maxima and a minimum between them, and the interfacial tension γ which passes through a minimum. However, a parallelism is expected to occur between the amount of surfactant adsorbed per unit area and the stability of emulsion, because when the former is larger, the latter is higher due to greater steric repulsion.

2.2. Surface excess against HLB

The discussion which follows implies that the mixing process by which the emulsions are generated is the same, because in order to compare the stabilities, the surface areas generated by mixing should be nearly the same. Γ is expected to have low values for low and large values of h, as well as for that moderate value of h at which the concentrations of the surfactant in the two phases become equal ($\Delta\mu_0 = 0$). At low values of h, the surfactant is mostly distributed in oil and the strong interactions between oil and surfactant make the adsorbability of the latter low. At large values of h, the surfactant is mostly distributed in water and the strong interactions between water and surfactant lead again to a low adsorbability. Finally, at a moderate value h_0 of h, the concentrations of surfactant in oil and water become equal ($\Delta\mu_0 = 0$) and the surface excess low because the adsorbability is moderate.

Between the low value of h and h_0, it is convenient to write

$$\Gamma = \Gamma(C_O, h), \tag{19}$$

where $C_O = C_O(h)$. Consequently,

$$\frac{d\Gamma}{dh} = \left(\frac{\partial \Gamma}{\partial C_O}\right)_h \frac{dC_O}{dh} + \left(\frac{\partial \Gamma}{\partial h}\right)_{C_O} \tag{20}$$

An increase in the concentration C_O at constant hydrophilicity increases the surface excess, because more surfactant is adsorbed on the interface; similarly an increase in hydrophilicity at constant C_O increases the amount adsorbed, because of increased favorable interactions with water. However, C_O decreases with h because an increasing amount of surfactant is distributed in the water phase. Consequently,

$$\left(\frac{\partial \Gamma}{\partial C_O}\right)_h > 0, \quad \left(\frac{\partial \Gamma}{\partial h}\right)_{C_O} > 0 \quad \text{and} \quad \frac{dC_O}{dh} < 0. \tag{21}$$

On the basis of these inequalities, one can conclude that Γ passes through an extremum. The extremum is a maximum because: (i) Γ is small for small values of h, due to the low adsorbability caused by the strong interactions between surfactant and oil; (ii) it increases with h, due to the increasing favorable interactions between the head group of the surfactant and water; (iii) passes through a maximum when the decrease of C_O with increasing h compensates for the ratio of increases of Γ with h and C_O; it becomes again small for $h = h_0$.

Between $h = h_0$ and large values of h, it is convenient to write

$$\Gamma = \Gamma(C_W, h)$$

and considerations similar to the preceding ones lead to the conclusion that Γ passes through a maximum. Above the CMC, C_W should be replaced by C_{W1}.

The region between low values of h and h_0 corresponds to W/O emulsions and that between h_0 and large values to O/W emulsion. The highest stabilities occur at the maxima of Γ. The phase transition occurs at $h = h_0$, and the emulsions are very unstable at low and high values of h, as well as around $h = h_0$. Because the

E. Ruckenstein / Adv. Colloids Interface Sci. 79 (1999) 59–76 67

concentrations for $h = h_0$ are equal, $h_0 = h_c$, where h_c is the value of HLB at which the interfacial free energy exhibits a minimum. Fig. 2 provides a qualitative representation of Γ against h.

It should also be noted that the presence of micelles intensifies the coalescence of the globules of oil of an O/W emulsion because of the depletion effect [10].

3. Phase inversion temperature of macro and microemulsions [11]

3.1. The interfacial tension against temperature

This time the temperature is the field variable and the Gibbs adsorption equation has the form

$$d\gamma = -\Gamma d\mu - s dT, \tag{22}$$

where s is the surface excess entropy per unit area.

At sufficiently low temperatures, the hydrophilic surfactants have strong hydrogen bonding interactions with the water molecules and are mainly distributed in the water phase. (There are cases in which the surfactant is distributed in the oil phase over the entire range of temperature. Such cases will be discussed later in the paper.) The increase in temperature weakens the hydrogen bonding, thus increasing the hydrophobicity of the surfactant molecules. In other words, the surfactant is increasingly distributed in the oil phase as the temperature increases and the following inequalities can be written:

$$d\mu_W^0/dT > 0, \quad d\mu_O^0/dT < 0 \tag{23}$$

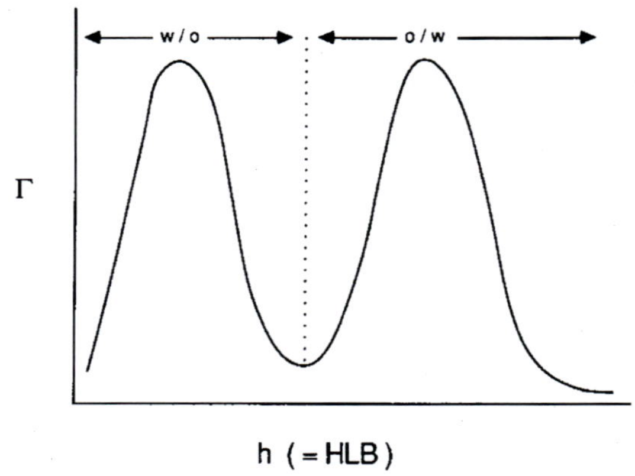

Fig. 2. Qualitative representation of the surface excess Γ against HLB (h).

and

$$dC_W/dT < 0, \quad dC_O/dT > 0. \tag{24}$$

One can define a temperature T_0 at which the concentrations in the two phases become equal. This temperature is provided by the equation:

$$\mu_W^0(T_0) = \mu_O^0(T_0). \tag{25}$$

Considering equal volumes of oil and water and concentrations in water below the critical micelle concentration (CMC), Eqs. (2), (5) and (22) lead to:

$$\frac{d\gamma}{dT} = -\Gamma\left\{k\ln C_W + \frac{1}{1+e^{\Delta\mu_0/kT}}\frac{d\mu_W^0}{dT} + \frac{1}{1+e^{-\Delta\mu_0/kT}}\frac{d\mu_O^0}{dT} + \frac{1}{1+e^{-\Delta\mu_0/kT}}\frac{\Delta\mu_0}{T}\right\} - s. \tag{26}$$

At low temperatures, μ_W^0 is small, because of the strong interactions between the surfactant and water molecules, and μ_O^0 large. Consequently, $\Delta\mu_0$ is negative and large in absolute value,

$$C_W \approx 2C \quad \text{and} \quad C_O \approx 0$$

and Eq. (26) leads to

$$\frac{d\gamma}{dT} = -\Gamma\left(k\ln 2C + \frac{d\mu_W^0}{dT}\right) - s \tag{27a}$$

At high temperatures, μ_W^0 is large and μ_O^0 small, hence $\Delta\mu_0$ is positive and large,

$$C_W = 0 \quad \text{and} \quad C_O \approx 2C$$

and Eq. (26) becomes

$$\frac{d\gamma}{dT} = -\Gamma\left(k\ln 2C + \frac{d\mu_O^0}{dT}\right) - s. \tag{27b}$$

Because the surfactants considered here are very sensitive to temperature, $d\mu_W^0/dT$ and $|d\mu_O^0/dT|$ are expected to be large compared to $|k\ln C|$. Also, because the freedom of motion in the bulk is higher than on the interface

$$\left|\Gamma\frac{d\mu_W^0}{dT}\right| \gg s \text{ and } \left|\Gamma\frac{d\mu_O^0}{dT}\right| \gg s$$

and the entropy s can be neglected in Eqs. (27a),(27b). As a result,

$$\begin{aligned} &\mathrm{d}\gamma/\mathrm{d}T < 0, \quad \text{(low temperatures)} \\ &\mathrm{d}\gamma/\mathrm{d}T > 0, \quad \text{(high temperatures)} \end{aligned} \tag{28}$$

and γ passes through a minimum. The minimum is expected to be located near the temperature T_0, because at this temperature Γ is small (see the next subsection) and the factor multiplying Γ in Eq. (26) is also small. The latter factor is small, because $\Delta\mu_0 = 0$ and $\mathrm{d}\mu_\mathrm{W}^0/\mathrm{d}T$ and $\mathrm{d}\mu_\mathrm{O}^0/\mathrm{d}T$ have opposite signs.

If the concentration in the water phase becomes greater than the CMC, one can make the approximate assumption that the concentration of the nonaggregated surfactant remains constant and equal to the critical micelle concentration, C_{W1}. One thus obtains

$$\frac{\mathrm{d}\gamma}{\mathrm{d}T} = -\Gamma\left(\frac{\mathrm{d}\mu_\mathrm{W}^0}{\mathrm{d}T} + k\ln C_{\mathrm{W1}} + \frac{kT}{C_{\mathrm{W1}}}\frac{\mathrm{d}C_{\mathrm{W1}}}{\mathrm{d}T}\right) - s. \tag{29}$$

The hydrogen bonding between the head group of the surfactant and water is very sensitive to temperature; for this reason $\mathrm{d}\mu_\mathrm{W}^0/\mathrm{d}T$ is expected to be larger at low temperatures. Because the hydrophobicity of the surfactant increases with temperature, $\mathrm{d}C_{\mathrm{W1}}/\mathrm{d}T < 0$. In reality, the critical micelle concentration passes through a minimum with increasing temperature [20]. The minimum is, however, located around 55°C for nonionic surfactants. As a result, the interfacial tension is expected to pass through a minimum.

3.2. Macroemulsions and microemulsions

When the minimum against temperature of the interfacial tension between the two phases separated by a planar surface has a positive value (Fig. 3), their mixing will lead to an emulsion whose stability depends upon the magnitude of the surface excess. When the surfactant is sufficiently hydrophilic, it will be mostly distributed in water at low temperatures, and the surface excess will be small because of the high compatibility of the surfactant with the water phase. At high temperatures, the surfactant will be mostly distributed in the oil phase and the surface excess will be small because of the high compatibility between the surfactant and oil. At the moderate temperature T_0 at which the concentrations in the two phases become equal, the surface excess is again small.

For the region between the low temperature and T_0 it is convenient to write that $\Gamma = \Gamma(C_\mathrm{W},T)$ where C_W is the total concentration below the CMC and the concentration of the nonaggregated surfactant (assumed equal to the CMC) above the CMC. Consequently,

$$\frac{\mathrm{d}\Gamma}{\mathrm{d}T} = \left(\frac{\partial\Gamma}{\partial C_\mathrm{W}}\right)_T \frac{\mathrm{d}C_\mathrm{W}}{\mathrm{d}T} + \left(\frac{\partial\Gamma}{\partial T}\right)_{C_\mathrm{W}} \tag{30}$$

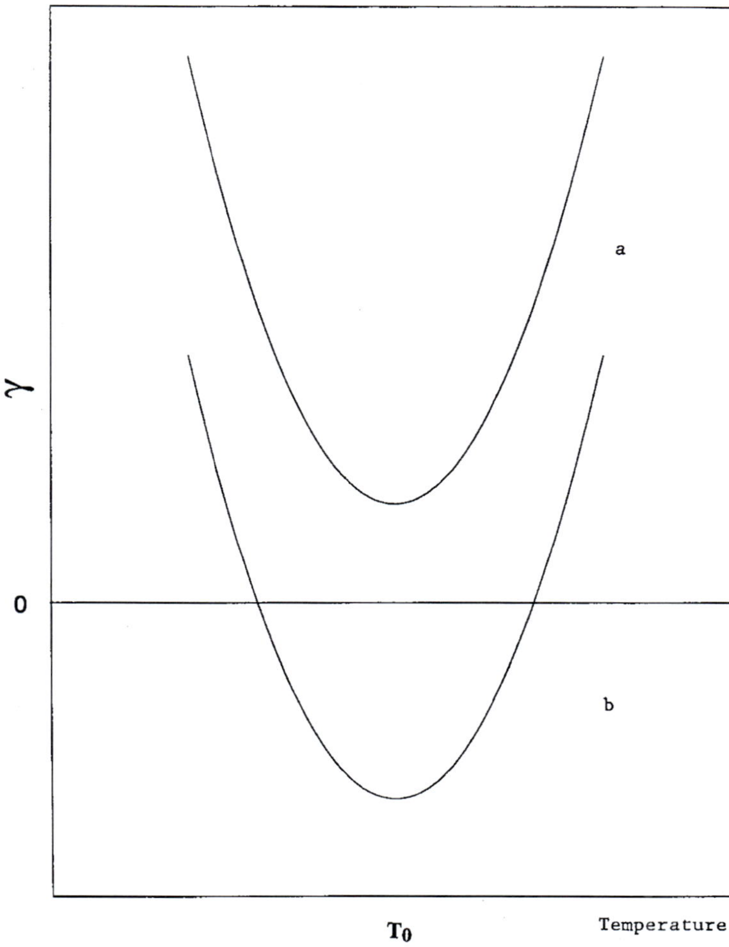

Fig. 3. Qualitative representation of the interfacial tension γ against temperature, (a) emulsions; (b) microemulsions.

At constant temperature, Γ increases with C_W, while at constant C_W, it increases with T, because of progressively stronger interactions between surfactant and oil. Hence

$$\left(\frac{\partial \Gamma}{\partial C_W}\right)_T > 0 \quad \text{and} \quad \left(\frac{\partial \Gamma}{\partial T}\right)_{C_W} > 0.$$

Because $(dC_W/dT) < 0$, Γ passes through an extremum. The surface excess Γ is small at low temperatures, increases with temperature, because of increasingly weaker interactions between the surfactant and water, passes through a maximum,

E. Ruckenstein / Adv. Colloids Interface Sci. 79 (1999) 59–76 71

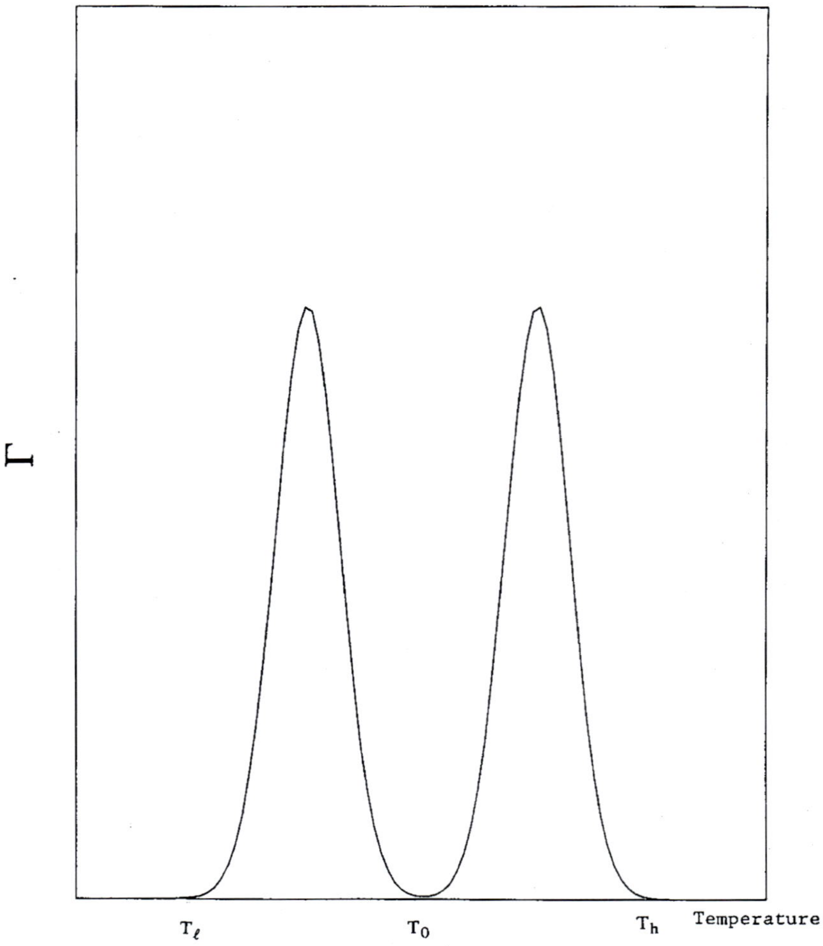

Fig. 4. Qualitative representation of the surface excess Γ against temperature.

when the decrease of C_W with temperature compensates for the ratio of increases of Γ with T and C_W, and finally becomes small when T reaches T_0.

Similarly, between T_0 and high temperatures it is convenient to write $\Gamma = \Gamma(C_O,T)$. Considerations similar to the previous ones indicate that Γ passes through a maximum between the two temperatures. Fig. 4 presents a qualitative picture regarding the dependence of Γ on T.

When the minimum of γ with respect to T is negative (Fig. 3), the interfacial tension is negative in the range of temperatures in which the curve $\gamma = \gamma(T)$ intersects the abscissa. A negative γ makes the water–oil interface unstable to thermal and mechanical perturbations and a spontaneous emulsification with the formation of globules of oil in water and water in oil takes place. At low temperatures, because of coalescence, only a fraction of the globules of oil survive after they attain the equilibrium radius; an O/W microemulsion in equilibrium with excess oil free of water is thus generated. At high temperatures, a W/O microemulsion in equilibrium with excess water free of oil is formed, and, at intermediate temperatures, a middle phase microemulsion in equilibrium with both excess phases is produced. The large interfacial area created increases the amount of surfactant adsorbed and decreases its concentrations in the bulk, thus increasing the interfacial tension at the surface of the globules and between the microemulsion and excess phase(s) to small positive values. At low temperatures, a microemulsion of the O/W type coexists with excess oil, while at high temperatures, a W/O microemulsion coexists with excess water, for reasons which will be discussed in Section 4. A phase inversion occurs at a temperature corresponding to the minimum interfacial tension, which is located in the middle of the middle phase microemulsion range.

The interfacial tension γ at the planar interface has a minimum near the temperature T_0. Indeed, at the latter temperature Γ is small, $\Delta\mu_0 = 0$ and because $\mathrm{d}\mu_{\mathrm{W}}^0/\mathrm{d}T$ and $\mathrm{d}\mu_{\mathrm{O}}^0/\mathrm{d}T$ have opposite signs and s is also small (because Γ is small), $\mathrm{d}\gamma/\mathrm{d}T \approx 0$. The temperature T_0 is provided by Eq. (25) and is independent of the concentration of surfactant. In other words, the two minima of Fig. 4 which correspond to the phase inversion temperatures of a macroemulsion (the curve with a positive minimum) and microemulsion (the curve with a negative minimum) are the same. When emulsions are generated from a microemulsion and its excess phase, the emulsion is of the same kind as the microemulsion, the phase inversion temperature is obviously located in the middle of the middle phase microemulsion range and the above conclusion remains valid. The above results explain the observation of Shinoda and Saito [6,7] that the phase inversion temperature (PIT) of emulsions can be provided by the ternary equilibrium phase diagram.

The sufficiently strong hydrophobic surfactants are preferentially distributed in the oil phase over the entire range of temperatures. One such surfactant, $C_{12}E_5$ (a nonionic surfactant of the polyoxyethylene glycol ether type), was investigated by Binks [21], and its behavior will be explained in Section 4. As in the previous cases, the inequalities (23) and (24) remain valid; however, $\Delta\mu_0$ does not change sign but is always positive. For concentrations below the CMC, the inversion temperature T_{in} can be obtained on the basis of Eq. (26), from the minimum of the interfacial tension. Neglecting the entropy s and $k\ln C_{\mathrm{W}}$ compared to the other terms, one obtains

$$\frac{1}{1+\mathrm{e}^{\Delta\mu_0(T_{\mathrm{in}})/kT_{\mathrm{in}}}}\left(\frac{\mathrm{d}\mu_{\mathrm{W}}^0}{\mathrm{d}T}\right)_{T_{\mathrm{in}}} + \frac{1}{1+\mathrm{e}^{-\Delta\mu_0(T_{\mathrm{in}})/kT_{\mathrm{in}}}}\left(\frac{\mathrm{d}\mu_{\mathrm{O}}^0}{\mathrm{d}T}\right)_{T=T_{\mathrm{in}}} + \frac{1}{1+\mathrm{e}^{-\Delta\mu_0(T_{\mathrm{in}})/kT_{\mathrm{in}}}}\frac{\Delta\mu_0(T_{\mathrm{in}})}{T_{\mathrm{in}}} = 0 \tag{31a}$$

If the derivatives $\mathrm{d}\mu_{\mathrm{W}}^0/\mathrm{d}T$ and $|\mathrm{d}\mu_{\mathrm{O}}^0/\mathrm{d}T|$ are sufficiently large for $k\ln C_{\mathrm{W}}$ to be negligible, T_{in} is independent of the surfactant concentration, and the phase inversion temperatures are expected to be the same for macro and micro-emulsions. It is likely that, because of the interactions involved, the adsorbability is such that Γ against temperature can be again represented qualitatively as in Fig. 4. For concentrations greater than the CMC, Eq. (29) leads, after neglecting the entropy s, to the following expression for the inversion temperature:

$$\left(\frac{\mathrm{d}\mu_{\mathrm{W}}^{0}}{\mathrm{d}T}\right)_{T=T_{\mathrm{in}}} + k\ln C_{\mathrm{W}1}(T_{\mathrm{in}}) + \frac{kT_{\mathrm{in}}}{C_{\mathrm{W}1}(T_{\mathrm{in}})}\left(\frac{\mathrm{d}C_{\mathrm{W}1}}{\mathrm{d}T}\right)_{T=T_{\mathrm{in}}} = 0 \qquad (31b)$$

It should be noted that an inversion temperature may not exist for a too-hydrophobic, or a too-hydrophilic surfactant.

4. The Bancroft rule

In most cases, in both micro- and macroemulsions, the phase in which the surfactant is most soluble constitutes the continuous phase. This rule was formulated by Bancroft [8,9] in 1913 for macroemulsions in the form 'a hydrophile colloid will tend to make water the dispersing phase while a hydrophobe colloid will tend to make water the dispersed phase'. Winsor [22] extended this rule to microemulsions. There are, however, some violations of the rule. Smith and Lim [23] investigated the morphologies and inversions of emulsions formed by pairs of conjugate liquids in the system $C_4H_9OC_2H_4OH$-n-$C_{10}H_{22}$-aqueous NaCl solution as a function of volume fraction of oil or water. They noted that the emulsions obeyed the Ostwald [24,25] stereometric hard sphere model and violated the Bancroft rule, because they inverted at a volume fraction of the dispersed phase of about 0.6. Ross and Kornbrekke [26], mixing two immiscible conjugate solutions of benzene–ethanol–water, observed that, for any volume ratio of the two liquids, only a probability of obtaining one or the other type of emulsion can be determined.

In the above cases, relatively weak surface active compounds were employed and the mixing process, hence hydrodynamics, was responsible for the outcome. In contrast, Binks [21] investigated a system containing a nonionic surfactant of the polyoxyethylene glycol ether type, namely $C_{12}E_5$, which is a stronger surface active agent. Regarding the phase behavior, he noted an O/W microemulsion coexisting with excess oil at low temperatures, a W/O microemulsion in equilibrium with excess water at high temperatures, and a middle phase microemulsion in equilibrium with both excess phases at intermediate temperatures. This is as expected. However, the Bancroft rule was violated at low temperatures, because the continuous phase was water and the surfactant was preferentially distributed in the oil phase. Regarding the macroemulsions, Binks noted that at low surfactant concentrations the Bancroft rule was violated since the higher concentration of surfactant was in the oil phase, which was the dispersed and not the continuous phase.

Obviously, the Bancroft rule is valid for microemulsions and macroemulsions for different reasons. Thermodynamics is responsible for it in the former and kinetics in the latter case. For macroemulsions three regimes were identified [12]. At low surfactant concentrations, or for weak surface active agents, the type of emulsion is controlled by the mixing process (hydrodynamics); at sufficiently high concentrations it is controlled by the mass transfer (only in this case the Bancroft rule is valid), and at even higher concentrations by hole nucleation in black films. Some of the violations of the rule noted by Binks for macroemulsions were at low surfactant concentrations (the mixing controlled regime for which the rule is not valid). The macroemulsion formed by mixing a microemulsion with its excess phase was of the same kind as the microemulsion because, for the surfactants and concentrations employed, the mass transfer is responsible for the outcome, and the Bancroft rule should be obeyed. Indeed, it is obeyed because the microemulsion contains much more surfactant than the excess phase and the rule indicates that it should become the continuous phase.

Regarding the phase behavior, it is clear that the state with the lowest free energy will be achieved. Detailed calculations are however complicated and for the time being a simple, intuitive approach will be employed. Winsor [22] explained qualitatively the phase behavior in terms of the R_1 ratio, which compares the surfactant–oil interactions to those between surfactant and water. Two ratios are, however, needed to describe the behavior of a system [12]. One of them is the ratio R_1

$$R_1 = A_{so}/A_{sw}, \tag{32}$$

where A_{so} is a measure of the strength of the surfactant–oil interactions and A_{sw} of the surfactant–water interactions in the bulk phases. Obviously, the surfactant will be preferentially distributed in the water phase when $R_1 < 1$ and in the oil phase when $R_1 > 1$.

The second ratio R_2 indicates the preferred curvature and can be obtained on the basis of an approximate expression for the free energy of formation of a microemulsion. The packaging of the surfactant molecules at the surface of the globules is expected to provide a major contribution to the free energy of the system. Because the globules are small ($\sim$ 10 nm), the thickness of the surfactant layer cannot be neglected compared to the radius of the globules and the interfacial free energy of a globule can be evaluated as

$$F_{int} = A_{hw}\gamma_{hw} + A_{co}\gamma_{co}, \tag{33}$$

where A_{hw} and A_{co} are the surface areas of the globule between the head groups and water, and between hydrocarbon chains and oil, respectively, and γ_{hw} and γ_{co} are the corresponding interfacial tensions, which, because the globules are small, are curvature dependent. At low temperatures, the interactions between the head groups of the surfactant and water are strong because of hydrogen bonding, and hence $\gamma_{hw} < \gamma_{co}$. In this case, F_{int} is smaller when the head groups are located on the external larger area and an O/W microemulsion in equilibrium with excess oil will be generated (the curvature of the globules is positive). At temperatures sufficiently high for the hydrogen bonding to be sufficiently weakened, $\gamma_{hw} > \gamma_{co}$

and a W/O microemulsion in equilibrium with excess water will be formed (the curvature of the globule is negative). At moderate temperatures, $\gamma_{hw} \approx \gamma_{co}$, and a middle phase microemulsion (zero curvature) will be generated. The ratio R_2 between the interactions between the hydrocarbon chains of the surfactant molecules of the interfacial layer and oil and those between the head groups in the same layer and water determine the kind of microemulsion and is expected to vary in the same direction as γ_{hw}/γ_{co}. The ratio R_1 determines the distribution of the surfactant between water and oil, while the ratio R_2 determines the type of microemulsion. When both ratios are smaller or larger than unity, Bancroft's rule is obeyed, while when one of them is larger and the other one smaller than unity, it is violated. In Bink's experiments, water was the continuous phase even though the surfactant was preferentially distributed in oil, because $\gamma_{hw} < \gamma_{co}$ ($R_2 < 1$), while $A_{so} > A_{sw}$ ($R_1 > 1$) [12].

5. Conclusion

In conclusion, thermodynamics provides some information about the overall behavior of macroemulsions, suggests to correlate the stability of macroemulsions to the surface excess, relates the phase inversion temperature or the phase inversion HLB to the minimum of the interfacial tension at a planar interface between oil and water and explains why the phase inversion temperature is the same for macroemulsions and the corresponding microemulsions.

References

[1] M. Bourrel, A. Graciaa, R.S. Schechter, W.H. Wade, J. Colloid Interface Sci. 72 (1979) 161.
[2] H.L. Rosano, D. Jon, J. Oil Am. Chem. Soc. 59 (1982) 360.
[3] A. Graciaa, Y. Barakat, R.S. Schechter, W.H. Wade, S.J. Yiv, J. Colloid Interface Sci. 89 (1982) 217.
[4] P.D. Berger, C.A. Hsu, J.P. Arendell, SPE (Society Petroleum Engineers) 16285, San Antonio, TX, 1987.
[5] J. Boyd, C. Parkinson, P. Sherman, J. Colloid Interface Sci. 41 (1972) 359.
[6] K. Shinoda, H. Saito, J. Colloid Interface Sci. 26 (1968) 70.
[7] K. Shinoda, H. Saito, J. Colloid Interface Sci. 30 (1969) 258.
[8] W.D. Bancroft, J. Phys. Chem. 17 (1913) 501.
[9] W.D. Bancroft, J. Phys. Chem. 19 (1915) 275.
[10] E. Ruckenstein, Langmuir 4 (1988) 1318.
[11] E. Ruckenstein, Langmuir 13 (1997) 2494.
[12] E. Ruckenstein, Langmuir 12 (1996) 6351.
[13] W.C. Griffin, J. Soc. Cosmet. Chem. 1 (1949) 311.
[14] W.C. Griffin, J. Soc. Cosmet. Chem. 5 (1954) 249.
[15] J.T. Davis, Proceedings 2nd International Congress on Surface Activity, London, 1957, p. 417.
[16] J.T. Davis, E.K. Rideal, Interfacial Phenomena, Academic Press, NY, 1963, p. 129.
[17] J.T. Davis, in: J.F. Danielli, K.G.A. Parkhurst, A.C. Riddford (Eds.), Progress in Surface Science, vol. 2, Academic Press, NY, 1964, p. 129.
[18] A.S. Kertes, H. Gutmann, in: E. Matijevic (Ed.), Surface and Colloids Science, Interscience, NY, 1975, p. 8.

[19] E. Ruckenstein, R. Nagarajan, J. Phys. Chem. 84 (1980) 1349.
[20] P. Becher, in: M.J. Schick (Ed.), Nonionic Surfactants, M. Dekker, NY, 1967, p. 478.
[21] B.P. Binks, Langmuir 9 (1993) 25.
[22] P.A. Winsor, Chem. Rev. 68 (1968) 1.
[23] D.H. Smith, K.H. Lim, J. Phys. Chem. 94 (1990) 3746.
[24] W. Ostwald, Kolloid Z. 6 (1910) 103.
[25] W. Ostwald, Kolloid Z. 7 (1910) 64.
[26] S. Ross, R.E. Kornbrekke, J. Colloid Interface Sci. 81 (1981) 58.

Colloids and Surfaces, 69 (1993) 271–275
Elsevier Science Publishers B.V., Amsterdam

On the stability of concentrated, non-aqueous dispersions

E. Ruckenstein
Department of Chemical Engineering, State University of New York at Buffalo, Buffalo, NY 14260, USA

(Received 22 June 1992; accepted 18 September 1992)

Abstract

It is demonstrated that the depth of the potential well between two colloidal particles which interact via van der Waals interactions can be minimized if they are covered with a shell of adsorbed surfactant molecules whose Hamaker constant is near to that of the liquid medium. This result is used to explain some experimental observations (I. Sushumna, R.P. Gupta and E. Ruckenstein, J. Mater. Res., 7 (1992) 2884) which indicate that long-chain fatty acids with an alkyl side chain of 5–10 carbon atoms located far from the head group of the surfactant lead to lower paste viscosities than those without side chains or with longer side alkyl chains. It is also suggested that, in concentrated suspensions, the collective interactions when the pair interaction is attractive can lead to an effective repulsion component between two neighboring particles, and that this effect may contribute to the stability of the concentrated dispersions.

Keywords: Concentrated dispersions; dispersion stability; non-aqueous dispersions.

Introduction

Low molecular weight molecules such as oleic acid, stearic acid or palm oil affect in a dramatic way the rheology of ceramic slurries in liquids of low dielectric constant [1]. It was also shown recently [2] that the viscosity of high solid loading (above or equal to 60 vol.%) suspensions of oxides, carbides or nitrides in paraffinic oils can be reduced if, instead of the above molecules, long-chain fatty acid molecules with an alkyl side chain of 5 to 10 carbon atoms, or a benzenic ring, located far from the head group, are employed. Dispersant molecules with long alkyl side chains as well as those without side chains lead to higher viscosities. In other words, there is an optimum side chain size for which the viscosity of the concentrated slurry is minimum.

Correspondence to: E. Ruckenstein, Dept. of Chemical Engineering, State University of New York at Buffalo, Buffalo, NY 14260, USA.

The stability of ceramic slurries in media with low dielectric constants is attributed either to steric stabilization, caused by polymer molecules adsorbed upon the surface of the particles, and/or to electrostatic stabilization. Albers and Overbeek [3] concluded that for concentrated slurries in low dielectric media, the electrostatic stabilization is not as effective as it is in aqueous systems. They use the argument that the low ionic strength in these media generates a large Debye length which implies a small variation of the electric potential with distance and hence a decrease of the height of the potential barrier. Fowkes et al. [4] disagree, arguing that in concentrated slurries in low dielectric media the concentration of the counter-ions is large enough for the thickness of the double layer to be comparable to that in water.

Concerning the stabilization of slurries in low dielectric media by small molecules, such as long-chain carboxylic acids, several experimental observations are available: Rehbinder et al. [5] noted

0166-6622/93/$06.00

that they could generate stable dispersions in benzene; Verwey and de Boer [6] observed that oleic acid can ensure the stability of powders in low dielectric media; van der Waarden [7] demonstrated that the stability of carbon black increased with increasing alkyl chain length and that molecules with two alkyl tails are more effective than those with one. Koelmans and Overbeek [8] carried out an extensive study of a large variety of powders dispersed in xylene and stabilized with aliphatic acids. They concluded that these systems are not truly stable and that their slight stability increases with increasing chain length. The greater stability caused by oleic acid is attributed by them to some long-chain impurities which can ensure steric stabilization. In contrast, van der Minne and Hermanie [9] made electrophoretic measurements on suspensions of carbon black in benzene, and demonstrated that dispersants stabilize dispersions only when appreciable zeta potentials resulted, suggesting that electrostatic charging in hydrocarbon solutions is responsible for stability.

More recently, highly concentrated dispersions have been investigated. Johnson and Morrison [1] reported that high-loading suspensions of alumina (up to 50 vol.%) in *p*-xylene are quite stable and exhibit lower viscosities with petroleum sulfonate (Petrosul H70) as dispersant than with oleic acid. Pugh et al. [10] reported that poly(isobutene succinimide) (OLOA 1200) is an effective dispersant for carbon black dispersions (about 35 vol.%) in mineral oil. Sushumna et al. [2] noted that fatty acids with chains greater than C_8 possessing either alkyl side chains with 5–10 carbon atoms or a benzenic molecule located far from the head group provide stable pastes of alumina (more than 60 vol.%) in mineral oil with lower viscosities than those without or longer side chains.

Johnson and Morrison [1] expressed the opinion that partial stability is provided in certain slurries by semi-steric stabilizers, i.e. surfactants that adsorb on the surface of the particles but do not have long enough carbon chains to ensure true steric stability. These semi-steric stabilizers decrease the van der Waals interactions between two particles because the shell of adsorbed molecules increases the distance between them and modifies the overall Hamaker constant. Assuming that the adsorbed molecules of surfactant form a shell surrounding the particles, Johnson and Morrison used the Vold [11] equation to calculate the van der Waals interactions between the particles. Their numerical calculations have shown that these interactions exhibit a minimum when the Hamaker constant of the shell of surfactant molecules is near (but not exactly equal to) that of the liquid medium.

The first goal of the present work is to establish an analytical expression for the optimum Hamaker constant of the shell. It will be also shown that the existence of an optimum size of the side chain for which the viscosity of the paste is minimum can be explained on the basis of this optimum Hamaker constant. Finally, it will be noted that, for concentrated dispersions, the collective behavior of particles interacting via a pairwise attractive potential can lead to an effective repulsion component which may play a role in the stability of the concentrated dispersions.

Optimum Hamaker constant for the shell of adsorbed surfactant molecules

Vold [11] has established Eqn (1) for the attractive van der Waals interactions between two spherical particles of radius R covered with a shell of a different material of thickness δ, by assuming pair-wise additivity of molecular interactions.

$$U = -\frac{1}{12}[(A_{mm} - 2A_{ms} + A_{ss})H_s + 2(A_{ps} - A_{pm} - A_{ss} + A_{ms})H_{ps} + (A_{pp} - 2A_{ps} + A_{ss})H_p] \quad (1)$$

where

$$H(z, y) = y/(z^2 + zy + z) + y/(z^2 + zy + z + y) + 2 \ln[(z^2 + zy + z)/(z^2 + zy + z + y)] \quad (2)$$

In Eqn(2) $z = x/2(R + \delta)$ and $y = 1$ for H_s;

$z = (x + \delta)/2R$ and $y = (R + \delta)/R$ for H_{ps}; and $z = (x + 2\delta)/2R$ and $y = 1$ for H_p, where x is the distance of closest approach of two covered spheres. In Eqn (1), A_{mm} is the Hamaker constant of the medium, A_{pp} is the Hamaker constant of the core of the particles, A_{ss} is the Hamaker constant of the shell of adsorbed molecules and A_{ms}, A_{ps} and A_{pm} are the Hamaker constants for the binary interactions between m and s, p and s, and p and m. The mixed Hamaker constants A_{ij} are evaluated using the geometric mean approximation

$$A_{ij} = (A_{ii}A_{jj})^{1/2} \tag{3}$$

In the cases of practical interest, $x/R \ll 1$ and $\delta/R \ll 1$. Under these conditions, Eqn (1) simplifies to

$$\begin{aligned} U = -\frac{R}{12}\{&[(A_s^{1/2} - A_m^{1/2})^2/x] \\ &+ [(A_p^{1/2} - A_s^{1/2})(A_s^{1/2} - A_m^{1/2})/(x+\delta)] \\ &+ [(A_s^{1/2} - A_m^{1/2})(A_p^{1/2} - A_s^{1/2})/(x+\delta)] \\ &+ [(A_p^{1/2} - A_s^{1/2})^2/(x+2\delta)]\} \end{aligned} \tag{4}$$

where instead of the double subscripts, ss, mm and pp, the single subscripts, s, m and p are used. Equation (4) can be derived more directly by employing the Derjaguin approximation.

For given values of A_m, A_p, δ and x, U is smallest when

$$\left(\frac{A_s}{A_m}\right)^{1/2} = 1 + \frac{x}{2\delta}\left[1 - \left(\frac{A_p}{A_m}\right)^{1/2}\right] \tag{5}$$

In order to calculate the potential well, Born repulsion should be included in the calculations. Instead of using an explicit expression for the Born forces, the interaction potential U is cut at a distance of about 4 Å. Considering δ of the order of 24 Å, $x = 4$ Å, $A_p = 15 \cdot 10^{-20}$ J and $A_m = 7.5 \cdot 10^{-20}$ J, one obtains that the optimum A_s is given by the expression

$$A_s = 1.06A_m$$

The optimum value of A_s is indeed near the value of A_m.

Calculations carried out by Johnson and Morrison [1] for alumina particles of 0.25 µm radius covered by 22.5 Å shells show that the potential energy of interaction is reduced from about $-175\,kT$ to about $-25\,kT$. The major effect of the shell is to eliminate the short-range attraction by separating the particles. Differences due to the Hamaker constant of the shell are also significant, the difference between the shallowest and deepest well being $20\,kT$.

Why an optimum side chain exists for which the viscosity of the paste is minimum

Experiments carried out in this laboratory [2] have shown that long-chain fatty acids possessing side chains far from the head group can be more effective dispersants when the side chain is benzenic or is an alkyl chain with 5–10 carbon atoms than when they are without side chains or with longer side chains. For these optimum side chains, the viscosity of the pastes is the lowest. The optimum Hamaker constant for the shell may explain this behavior. Indeed, the surfactant molecules are adsorbed with the head groups on the surface of the particles. If these adsorbed molecules possess side chains far from the head groups, they will keep the particles apart and simultaneously will generate a free space among the surfactant chains which will be filled with the liquid of the medium (Fig. 1). As a result, the Hamaker constant of the shell acquires a value near that of the liquid medium. This means that the attractive interactions between two particles become smaller, and less agglomeration among the particles will occur. If the side chain is paraffinic, the chain is flexible and if it is too long, entanglements among the side chains present on neighboring particles and in the liquid can occur, and this increases both the agglomeration of the particles and the viscosity of the system. The optimum size side chain is the outcome of the above two opposite effects.

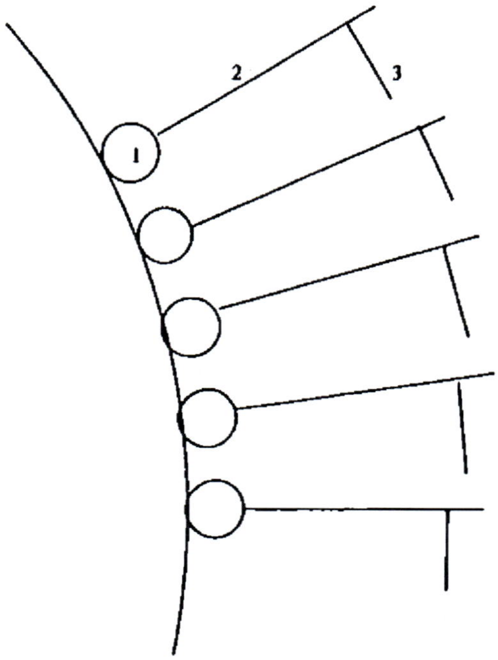

Fig. 1. Surfactant molecules with side chains adsorbed on a particle. The head group, hydrophobic chain and side chain are indicated by 1, 2 and 3, respectively.

Stability of concentrated dispersions

The electrostatic stabilization theory was developed for dilute colloidal systems and involves attractive van der Waals interactions and repulsive double layer interactions between two particles. They may lead to a potential barrier, an overall repulsion and/or to a minimum similar to that generated by steric stabilization. Johnson and Morrison [1] suggest that the stability in non-aqueous dispersions when the stabilizers are surfactant molecules, which are relatively small, is due to semi-steric stabilization, hence to a smaller van der Waals attraction between two particles caused by the adsorbed shell of surfactant molecules. The fact that such systems are quite stable suggests, however, that some repulsion is also present. In fact, it was demonstrated on the basis of electrophoretic measurements that a surface charge originates on solid particles suspended in aprotic liquids even in the absence of traces of water [12–14]. The liquids considered have both basicity, which can be expressed in terms of the Gutmann [15] donor-ability scale, and acidity, which can be expressed in terms of the Gutmann acceptor number [15]. The balance between the acceptor and donor properties determines the charge of the particles in organic media. Consequently, both surface charging and double layers of counterions can be generated in organic liquids which lead to repulsive forces. It is not yet clear, however, how many of these counterions form ion pairs with some of the charges on the solid particles and how many are free in the liquid. Is the number of the latter large enough for the Debye length to become comparable to that in aqueous solutions? Is it not more appropriate to treat the electrostatic interactions as dipole or dipole plus charge interactions, hence in a different way than for aqueous media?

In addition, if one considers a concentrated suspension, one cannot describe its behavior in terms of the interactions between two particles only, but in terms of the collective interactions among all the particles. Let us consider an ensemble of particles in which each pair interacts via the attractive interaction potential U, and within that ensemble, consider two adjacent particles 1 and 2, the first located on the left and the second on the right. The attraction which particle 1 experiences from the other particles except particle 2 can be dominated by the particles on its left. As a result, particle 1 can be pushed to the left. For the same reason, particle 2 can be pushed to the right. These attractions to the left of particle 1 and to the right of particle 2 generate an effective repulsion between particles 1 and 2 which may contribute to the stability of the system. Of course, this effect should be superimposed on the attraction between the two particles. Similarly, if the interaction potential is repulsive, an effective attraction can be generated superimposed on the repulsion between the two particles. These complexities must be accounted for in a more satisfactory treatment of concentrated dispersions.

References

1 R.E. Johnson, Jr., and W.H. Morrison, Jr., Advances in ceramics, Ceramic Powder Science, Vol. 21, American Ceramic Society, 1987, p. 323.
2 I. Sushumna, R.P. Gupta and E. Ruckenstein, J. Mater. Res., 7 (1992) 2884.
3 W. Albers and J.Th.G. Overbeek, J. Colloid Interface Sci., 14 (1959) 510.
4 F.M. Fowkes, H. Jinnai, M.A. Mostafa, F.W. Anderson and R.J. Moore, in M. Hair and M. Goucher (Eds), Colloids and Surfaces in Reprographic Technology, ACS Symp. Ser. 200, American Chemical Society, Washington, DC, p. 307.
5 P. Rehbinder, L. Lagutina and W. Wenstrom, Z. Phys. Chem., Abt. A, 146 (1930) 63.
6 E.J. Verwey and J.H. de Boer, Rec. Trav. Chim. Pays-Bas, 57 (1938) 383.
7 M. van der Waarden, J. Colloid Sci., 5 (1950) 317; 6 (1951) 443.
8 H. Koelmans and J.Th.G. Overbeek, Discuss. Faraday Soc., 18 (1954) 52.
9 J.L. van der Minne and P.H. Hermanie, J. Colloid Sci., 7 (1952) 600; 8 (1953) 38.
10 R.J. Pugh, T. Matsunaga and F.M. Fowkes, Colloids Surfaces, 7 (1983) 183.
11 M.J. Vold, J. Colloid Sci., 16 (1961) 1.
12 M.E. Labib and R. Williams, J. Colloid Interface Sci., 115 (1987) 330.
13 M.E. Labib and R. Williams, Colloid Polym. Sci., 264 (1986) 533.
14 M.E. Labib, Colloids Surfaces, 29 (1988) 293.
15 V. Gutmann, The Donor-Acceptor Approach to Molecular Interactions, Plenum Press, New York, 1978.

Introduction to CHAPTER 3
Stability of dispersions

3.1 E. Ruckenstein, R. Nagarajan: "*Critical micelle concentration. A transition point for micellar size distribution,*" JOURNAL OF PHYSICAL CHEMISTRY 79 (1975) 2622–2626.

3.2 E. Ruckenstein, R. Nagarajan: "*Critical micelle concentration and the transition point for micellar size distribution,*" JOURNAL OF PHYSICAL CHEMISTRY 85 (1981) 3010.

3.3 I.V. Rao and E. Ruckenstein: "*Phase behaviour of mixtures of sterically stabilized colloidal dispersions and free polymer,*" JOURNAL OF COLLOID AND INTERFACE SCIENCE 108 (1985) 389–402.

3.4 E. Ruckenstein and I.V. Rao: "*Effect of solvent on the stability of mixtures of sterically stabilized dispersions and free polymers,*" COLLOIDS AND SURFACES 17 (1986) 185–205.

3.5 E. Ruckenstein: "*Thermodynamics of microemulsions: relation between their structure and phase behaviour,*" FLUID PHASE EQUILIBRIA 20 (1985) 189–206.

3.6 E. Ruckenstein: "*Thermodynamics of microemulsion II,*" Handbook of Microemulsion Science and Technology, edited by P. Kumar and K. L. Mittal, Marcel Dekker, New York, 1999, 45–58.

3.7 R. Nagarajan, E. Ruckenstein: "*Molecular theory of microemulsions,*" LANGMUIR 16: 16 (2000) 6400–6415.

3.8 E. Ruckenstein: "*Phase transitions in insoluble monolayers of surfactants,*" Encyclopedia of Surface and Colloid Science (2007) Taylor & Francis.

3.9 E. Ruckenstein, M. Manciu: "*On the Stability of Lyotropic Lamellar Liquid Crystals and the Thicknesses of Their Lamellae,*" LANGMUIR 17 (2001) 5464–5475.

The critical concentration, C_{crit}, is defined as the concentration at which there is a change in the size distribution of the surfactant aggregates from a monotonically decreasing curve (below C_{crit}), to one that possesses a maximum and a minimum above the critical concentration. The transition in shape occurs at a size distribution that displays an horizontal inflexion point. The critical micelle concentration (CMC) has a value close to the critical value C_{crit} [3.1] and [3.2].

The effect of addition of a free polymer to a sterically stabilized non-aqueous colloidal dispersion is investigated using a second-order perturbation approach, which allows one to determine the conditions for phase separation. It is shown that the separation into a disordered dilute dispersion and an ordered concentrated dispersion occurs at a limiting value of the free polymer concentration, which decreases with increasing molecular weight [3.3]. The effect of the solvent on the stability of sterically stabilized dispersions containing free polymers is also investigated, and it is shown that the limiting concentration of the free polymer (at which separation occurs) decreases with increasing molecular weight of the polymer; with increasing particle size; with increasing temperature; with thinner steric layers; and with better solvents for the free polymers, in agreement with experiment [3.4].

In contrast to macroemulsions, which are kinetically stable, microemulsions are dispersions of oil in water or water in oil, stabilized by a surfactant, which are thermodynamically stable. They often contain a short chain alcohol (referred to as cosurfactant) and a salt. The microemulsions can be of the droplet type, either with spherical oil droplets dispersed in a continuous water phase (O/W microemulsions) or with spherical water droplets dispersed in a continuous oil phase (W/O microemulsions). The droplet type microemulsion can be either a single phase or part of a two phase system wherein the microemulsion coexists with an excess of dispersed phase (an upper phase of excess oil in the case of O/W and a lower phase of excess water in the case of W/O microemulsions).There are also non-droplet type microemulsions characterized by randomly distributed oil and water micro-domains and exhibiting bicontinuity in both oil and water. Such a system is known as a middle phase microemulsion, that occurs as part of a three phase system with the middle phase coexisting with an upper oil phase and a lower water phase. A novel thermodynamic treatment that explains the thermodynamic stability and phase behaviour of microemulsions ([3.5] and [3.6]) is suggested. Thermodynamic equations are derived regarding the single phase microemulsions, the transition to a microemulsion in equilibrium with an excess dispersed phase

E. Ruckenstein, M. Manciu, *Nanodispersions*, DOI 10.1007/978-1-4419-1415-6_3, © Springer Science+Business Media, LLC 2010

and the transition from a two phase system to one in which a middle phase microemulsion coexists with both excess phases.

A predictive molecular thermodynamics approach is developed for microemulsions, to determine their structural and compositional characteristics [3.7]. The theory is built upon a molecular level model for the free energy change. For illustrative purposes, numerical calculations are performed for the system water, cyclohexane, sodium dodecyl sulfate as surfactant, pentanol as cosurfactant and NaCl as electrolyte. The droplet radius, the thickness of the surfactant layer at the interface, the number of molecules of various species in the droplets, and the distribution of the components between droplets and the continuous phase are calculated. The theory also predicts the transition from a microemulsion to one in equilibrium with an excess phase as well as the transition from a two-phase droplet-type microemulsion to a three-phase system involving a bicontinuous microemulsion. In addition, the persistence length of the bicontinuous microemulsion, the surface tension between phases, and the curvature elasticity are also calculated [3.7].

A thermodynamic treatment, similar to that used for microemulsions, as well as an approximate statistical mechanical one, are developed to explain the phase transition in monolayers of insoluble surfactants [3.8]. A similar thermodynamic approach is applied to multilamellar liquid crystals, and it is shown that, for a given set of interactions and bending moduli, only narrow ranges of the thicknesses of the water and oil layers are allowed [3.9].

Critical Micelle Concentration. A Transition Point for Micellar Size Distribution

E. Ruckenstein* and R. Nagarajan

Faculty of Engineering and Applied Sciences, State University of New York at Buffalo, Buffalo, New York 14214 *(Received July 21, 1975)*

Publication costs assisted by the State University of New York at Buffalo

A critical concentration is defined which separates two kinds of behavior of the size distribution of micellar aggregates of surfactant molecules. Below this concentration, the size distribution is a monotonic decreasing function of size; above this concentration, the size distribution is a function exhibiting two extrema and the contribution of larger aggregates becomes important. Quantitative results are obtained from the condition of thermodynamic equilibrium, decomposing the standard chemical potential per amphiphile of various aggregates into size-dependent and size-independent terms, and using for these terms the empirical expressions given by Tanford.

Introduction

Micellization has been treated either as a stepwise association phenomenon or as a phase transition process.[1,2] In the first approach the micellar aggregates and the single molecules of surfactant are assumed to be in association–dissociation equilibrium and the law of mass action is applied. The critical micelle concentration (cmc) is defined as the concentration above which any added surfactant molecules appear with high probability as micellar aggregates. In the latter approach, micellization is regarded as a phase separation starting at cmc which now represents the saturation concentration of the phase containing single molecules of surfactant. Although both approaches can explain some features of the micelle formation, the available experimental evidence seems to be in agreement with the first of these. In the region close to the cmc, the physico-chemical properties do indeed change rapidly, but continuously, and the concentration of the single surfactant molecules increases slowly. The addition of surfactant above the cmc not only leads to an increase in the number of aggregates but also gives rise to larger average sizes of the aggregates. It is therefore reasonable to assume that for all concentrations of surfactant there are micellar aggregates of various sizes. At low concentrations, however, the number and size of these aggregates is small. As the concentration increases, both the number and the average size of the aggregates increase. The particular kind of behavior occurring at cmc is caused by an essential change in the shape of the size distribution of the aggregates; above the cmc the participation of the larger sizes is more important.

The model suggested here is similar to that provided by the first approach. The nature of the transition taking place at cmc is, however, better identified. In the present paper it is shown that at low concentrations of surfactant the size distribution of the aggregates is a monotonic decreasing function of the size. As the concentration increases, the size distribution of the aggregates changes from a monotonic decreasing function to one which has two extrema, a minimum and a maximum.

From this description a well-defined critical concentration emerges as that total amphiphile concentration corresponding to a transition from the monotonic decreasing size distribution function to a size distribution function exhibiting two extrema. This critical concentration corresponds to a surfactant solution with no appreciable amount of aggregates, whereas the cmc, as defined in the earlier papers based upon the first approach, corresponds to systems with about 5–10% of amphiphiles in micellar form. Hence, the critical concentration defined here as the transition point separating the two different behaviors of the aggregate size distribution function constitutes a close lower bound on the observed value of the cmc.

In the present paper the equilibrium condition defined by the minimum of the total free energy determines the size distribution of aggregates for a given concentration of the single surfactant molecules. The standard chemical potential per amphiphile of aggregates is decomposed into size-independent and size-dependent parts for which the empirical expressions given by Tanford[3] are used. The value of the critical concentration is computed as the concentration for which the size distribution curve has an inflection point. More refined computations, based on statistical mechanics, following the methods of Poland and Sheraga[4] and of Hoeve and Benson,[5] will be published in another paper.

Thermodynamic Formulation

The standard chemical potential of a single molecule of the solvent is denoted by μ_S and that of a single molecule of amphiphile in the aqueous medium by μ_A. The number of molecules of the solvent and the monomeric amphiphile are denoted by N_S and N_A, respectively, and the number of aggregates containing g molecules each by N_g. The standard chemical potential of the micelle per amphiphile is separated into two components: one is independent of the size of the aggregate and the second is size dependent. The size-independent part of the standard chemical potential for a single amphiphile, μ_B, accounts for the liquidlike behavior of the micellar interior made up of hydrocarbon tails of amphiphiles. The size of the aggregate and the compactness of the packing of the amphiphiles affect the repulsive interactions between the head groups, as well as the magnitude of the hydrocarbon surface area exposed to the aqueous medium which does not take part in the hydrophobic bonding. The size-dependent part of the standard chemical potential accounts for these contributions and is denoted by μ_g for an aggregate of size g.

Assuming ideal behavior of the solution, the free energy of the system is given by

$$\Phi = N_S\mu_S + N_A\mu_A + \sum_{g=2}^{\infty} N_g(\mu_B g + \mu_g) + kT\left[N_S \ln\frac{N_S}{F} + N_A \ln\frac{N_A}{F} + \sum_{g=2}^{\infty} N_g \ln\frac{N_g}{F}\right] \quad (1)$$

where

$$F = N_S + N_A + \sum_{g=2}^{\infty} N_g \quad (2)$$

k is the Boltzmann constant and T the absolute temperature. The total number of amphiphiles present as monomer and as aggregates is

$$N = N_A + \sum_{g=2}^{\infty} gN_g = \text{constant} \quad (3)$$

The equilibrium condition leads to the equation

$$-g\mu_A + g\mu_B + \mu_g + kT\left(-g \ln\frac{N_A}{F} + \ln\frac{N_g}{F}\right) = 0 \quad (4)$$

which can be rewritten as

$$\frac{N_g}{F} = \left(\frac{N_A}{F}\right)^g \exp\left(-\left[\frac{1}{kT}[(\mu_B - \mu_A)g + \mu_g]\right]\right) \quad (5)$$

Denoting by ξ the quantity

$$\xi = \left(\frac{N_A}{F}\right) \exp\left[-\left(\frac{\mu_B - \mu_A}{kT}\right)\right] \quad (6)$$

the size distribution function is expressed in the form

$$\frac{N_g}{F} = \xi^g \exp\left[-\left(\frac{\mu_g}{kT}\right)\right] \quad (7)$$

The decomposition of the standard chemical potentials per amphiphile of the aggregates permits the size distribution function to be written as a product of exp $[-(\mu_g/kT)]$, which is a decreasing function of g, and another factor containing the size independent quantity ξ. If ξ is small, the size distribution is a monotonic decreasing function of g. If, however, ξ is sufficiently large compared to unity, the size distribution can have a maximum. Therefore, a critical value of ξ exists, separating the two kinds of behavior. To compute this critical value as well as the corresponding critical concentration of amphiphile, explicit expressions for the terms of the standard chemical potentials are needed.

Explicit Expressions for the Free Energy Terms

The size-independent free-energy change $\mu_B - \mu_A$ corresponds to the change in the standard free energy in transferring an amphiphile from the dilute aqueous phase to the hydrocarbon phase of the micellar core. The transfer of an amphiphile into a micelle differs from the transfer into a bulk hydrocarbon phase because of the orientation of the amphiphile within the micelle with the head groups in contact with the aqueous medium. This free-energy change has significantly different values for ionic and nonionic micelles and also depends on the specificity of the hydrophilic head groups.

For illustrative purposes, the expression given by Tanford[3b] is used for the size-independent part $\mu_B - \mu_A$ of the standard free-energy change.

In general, for an alkyl chain with n_c carbon atoms

$$\frac{\mu_B - \mu_A}{kT} = \frac{S + I(n_c - 1)}{RT} \quad (8)$$

where R has a value of 2 cal/mol K. The subtraction of unity from n_c here accounts for the fact that the CH_2 group closest to the hydrophilic head group lies in its hydration sphere and, hence, does not contribute to the hydrophobic bonding energy. The term I depends upon the character of the head group. For nonionic head groups, it is virtually constant at a value of −700 cal/mol while for ionic head groups, it may be taken as having a value of approximately −420 cal/mol. The term S depends upon the specificity of the head group, particularly for nonionic micelles, but may be assumed to be of the order of −2000 cal/mol.

The size-dependent term μ_g/kT includes (a) the decrease in attractive hydrophobic bonding between hydrocarbon tails of amphiphiles, due to their partial exposure to the aqueous medium, and (b) the repulsive interaction between the hydrophilic head groups of the amphiphiles. This repulsive interaction is caused by steric repulsion between head groups in nonionic micelles and by electrical repulsion between the ionic head groups in ionic micelles.

The positive standard free-energy contribution due to contact between hydrocarbon tails of amphiphiles and the aqueous medium is given by Tanford[3b] as

$$\frac{\mu_g^{I}}{kT} = \frac{25(A - 21)g}{RT} \quad (9)$$

where A is the surface area in (ångstroms)^2 per amphiphile in an aggregate of size g. For a spherical aggregate of radius r_0

$$A = \frac{1}{g} 4\pi r_0^2 \quad (10)$$

The radius r_0 of the aggregate, calculated from the volume v_0 of the amphiphile, is

$$r_0 = \left(\frac{3v_0}{4\pi}\right)^{1/3} g^{1/3} = \beta g^{1/3} \quad (11)$$

The volume of the hydrocarbon tail of an amphiphile is determined from[3b]

$$v_0 = 27.4 + 26.9 n_c \text{ Å}^3 \quad (12)$$

The assumption of a spherical shape for the micelle is not a restrictive one and is used here for the sake of simplicity only. Any other shape of the aggregate or even changes in the shape of the aggregates with the growth in size can be incorporated in the formulation.

The second contribution to the size-dependent standard free-energy term arises from the repulsive head group interactions. The magnitude of this interaction depends on the separation between the head groups; the available area per head group is used as a measure of this separation. Tanford has shown[3b] that an expression of the form

$$\frac{\mu_g^{II}}{kT} = \frac{\alpha}{ART} g \quad (13)$$

can be used to represent this interaction, where α is a constant independent of g. The constant α has different values for ionic and nonionic micelles and depends, among other parameters, on the nature of the hydrophilic head group and the ionic strength. The size-dependent part of the free energy can now be written in the form

$$\frac{\mu_g}{kT} = \frac{25(A - 21)g}{RT} + \frac{\alpha g}{ART} \quad (14)$$

Expression for Critical Concentration

The value of the parameter ξ at which the size distribution function N_g/F has a point of inflection is denoted as ξ_{crit} and the corresponding value of the aggregation number is denoted as g_{crit}. The values of ξ_{crit} and g_{crit} are obtained from the solution of the system of equations

$$\frac{d}{dg}\left(\ln\frac{N_g}{F}\right) = \ln\xi - \frac{d}{dg}\left(\frac{\mu_g}{kT}\right) = 0 \quad \text{at } g = g_{crit} \quad (15a)$$

and

$$\frac{d^2}{dg^2}\left(\ln\frac{N_g}{F}\right) = -\frac{d^2}{dg^2}\left(\frac{\mu_g}{kT}\right) = 0 \quad \text{at } g = g_{crit} \quad (15b)$$

Substituting expression 14 for the size-dependent free-energy term μ_g and eq 10 and 11 for the surface area A per amphiphile in eq 15a and 15b yields for the inflection point

$$\left(\ln\xi + \frac{525}{RT}\right) - \left(\frac{25}{RT}\right)\left(\frac{2}{3}\right)4\pi\beta^2 g^{-1/3} - \left(\frac{\alpha}{RT}\right)\left(\frac{4}{3}\right)\frac{1}{4\pi\beta^2}g^{1/3} = 0 \quad (16)$$

and

$$\left(\frac{25}{RT}\right)\left(\frac{2}{9}\right)4\pi\beta^2 g^{-4/3} - \left(\frac{\alpha}{RT}\right)\left(\frac{4}{9}\right)\frac{1}{4\pi\beta^2}g^{-2/3} = 0 \quad (17)$$

The aggregation number at the inflection point is obtained from eq 17 and is given by

$$g_{crit}^{1/3} = 4\pi\beta^2\left(\frac{25}{2\alpha}\right)^{1/2} \quad (18)$$

The value of ξ_{crit} at the inflection point, obtained from eq 18 and eq 16, is

$$\ln\xi_{crit} = \left(\left(\frac{800\alpha}{9}\right)^{1/2} - 525\right)/RT \quad (19)$$

The total amphiphile concentration at this value of $\xi = \xi_{crit}$ gives the critical concentration C_{crit}.

$$C_{crit} = \left.\frac{N_A}{F}\right|_{\xi_{crit}} + \sum_{g=2}^{\infty} g\left.\frac{N_g}{F}\right|_{\xi_{crit}} \quad (20)$$

Replacing the summation by an integral, eq 20 becomes

$$C_{crit} = \xi_{crit}\exp\left(\frac{\mu_B - \mu_A}{kT}\right) + \int_2^{\infty} g\,\xi_{crit}^g \exp\left(-\frac{\mu_g}{kT}\right) dg \quad (21)$$

There is some analogy between the critical concentration as defined here and the critical temperature predicted by the van der Waals equation of state, since each of them separates two kinds of behavior of the size distribution function and pressure–volume relationship, respectively.

Phase Separation

There is a saturation value of the concentration of nonaggregated amphiphile at which phase separation occurs. The nonaggregated amphiphile and aggregates in solution are in this case in thermodynamic equilibrium with a bulk amphiphilar phase. The corresponding thermodynamic equilibrium condition introduces an upper limit, ξ_{max}, for ξ. The equilibrium condition leads to

$$\frac{N_A}{F} = \exp\left(\frac{\mu_B' - \mu_A}{kT}\right) = \xi_{max}\exp\left(\frac{\mu_B - \mu_A}{kT}\right) \quad (22)$$

and

$$\frac{N_g}{F} = \exp\left[\frac{g(\mu_B' - \mu_B)}{kT}\right]\exp\left(-\frac{\mu_g}{kT}\right) = \xi_{max}^g \exp\left(-\frac{\mu_g}{kT}\right) \quad (23)$$

where μ_B' is the chemical potential per molecule of the bulk amphiphilar phase. The value of ξ corresponding to this phase transition can be calculated from

$$\ln\xi_{max} = \frac{\mu_B' - \mu_B}{kT} = \left(\frac{\mu_B' - \mu_A}{kT}\right) - \left(\frac{\mu_B - \mu_A}{kT}\right) \quad (24)$$

The term $\mu_B' - \mu_A$ is the standard free-energy change for the transfer of an amphiphile molecule from the aqueous medium to the bulk amphiphilar phase. Since no information is available for this term at the present time, it is set equal to the standard free-energy change for transferring a hydrocarbon chain from an aqueous solution of monomer to a pure hydrocarbon phase.[3a] Consequently

$$\frac{\mu_B' - \mu_A}{kT} = \frac{(400 \text{ to } 2000) - 800n_c}{RT} \quad (25)$$

Results and Discussion

The size distribution of micellar aggregates N_g/F is plotted against the aggregation number, g, for an amphiphile with an octyl hydrocarbon tail and for $\alpha = 2 \times 10^4$ cal Å^2/mol[3b] (Figure 1). Equation 19 leads to $\xi_{crit} = 3.88$. For $\xi < \xi_{crit}$, the size distribution is a monotonic decreasing function of g. At $\xi = \xi_{crit}$, the size distribution function has an inflection point. At $\xi > \xi_{crit}$, the size distribution function has two extrema. It can be seen that if ξ increases both the number and the average size of the micellar aggregates increase.

In deriving eq 16–19, the surface area A is taken as the surface area of the micellar core since it is based on the volume of the hydrophobic core. In reality the surface area should account for the surface roughness due to the presence of the head groups. The actual surface area can be calculated by increasing the radius of the micellar core by a value δ.[3b] A is calculated using eq 10 where r_0 is now given by

$$r_0 = \left(\frac{3v_0}{4\pi}\right)^{1/3} g^{1/3} + \delta = \beta g^{1/3} + \delta \quad (26)$$

The introduction of this correction makes the algebra leading to the equations for g_{crit} and ξ_{crit} more complex. For this case, the value of ξ_{crit} can be computed only by numerical methods. Figure 2 represents the size distribution function for $\delta = 3$ Å and $\alpha = 8 \times 10^4$ cal Å^2/mol for an amphiphile with an octyl hydrocarbon tail.

The mole fractions of amphiphiles present both in nonaggregated form and as aggregates for various values of ξ are presented in Table I for the same parameters as in Figure 2. Table I shows that below $\xi_{crit} = 27.82$, the concentration of micellar aggregates remains extremely small. At $\xi = \xi_{crit}$, the concentration of micellar aggregates begins to increase even though their contribution to the total amphiphile concentration remains negligible. However as the value of ξ increases beyond ξ_{crit}, the number of aggregates increases significantly and their contribution to the total amphiphile concentration becomes important. The values of cmc normally reported should correspond to the total amphiphile concentration for values of ξ somewhat higher than ξ_{crit}.

The values of ξ_{crit} and the corresponding values of the

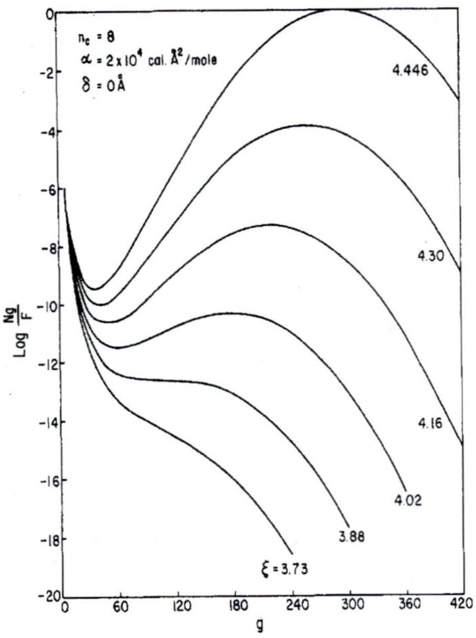

Figure 1. Variation in the aggregate size distribution function with nonaggregated amphiphile concentration for $\delta = 0$ Å, $\alpha = 2 \times 10^4$ cal Å^2/mol, $n_c = 8$.

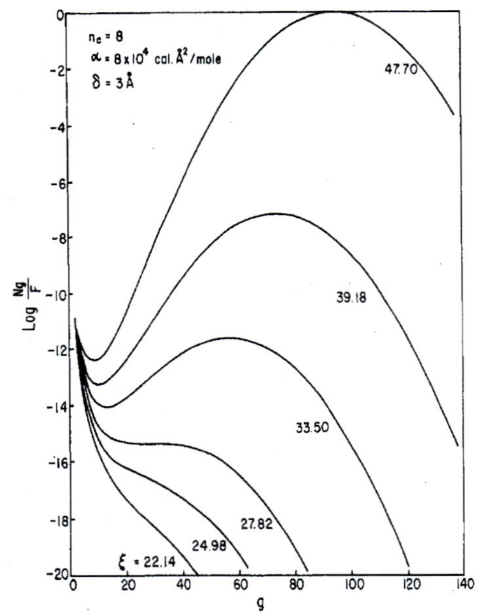

Figure 2. Variation in the aggregate size distribution function with nonaggregated amphiphile concentration for $\delta = 3$ Å, $\alpha = 8 \times 10^4$ cal Å^2/mol, $n_c = 8$.

critical concentration are computed for $\delta = 3$ Å, $\alpha = 8 \times 10^4$ cal Å^2/mol and for amphiphiles with various hydrocarbon chain lengths (Table II). Also shown in the table are the experimental cmc values[3a] of amphiphiles with hexoxyethylene glycol monoether head groups and various hydrocarbon chain lengths. It should be mentioned that the value of α used here is chosen for illustrative purposes and has not been computed on theoretical grounds for this amphiphile. It can be seen how closely the critical concentration C_{crit}, defined here predicts the cmc values normally reported.

For $\delta = 0$ the value of ξ_{crit} is solely determined by the repulsive interaction term α and remains a constant for amphiphiles with different hydrocarbon chain lengths. When $\delta > 0$, the value of ξ_{crit} is determined not only by α but also by n_c and δ. However, the value of ξ_{crit} for $\delta = 0$ is a good first approximation even for $\delta > 0$.

TABLE I: Variation of the Mole Fraction of the Aggregates as a Function of the Mole Fraction of the Nonaggregated Amphiphile[a]

ξ	N_A/F[b]	$\sum_{g=2}^{\infty} g(N_g/F)$[c]
22.14	2.09×10^{-4}	1.53×10^{-11}
24.98	2.34×10^{-4}	1.97×10^{-11}
27.82	2.61×10^{-4}	2.54×10^{-11}
33.50	3.14×10^{-4}	3.61×10^{-9}
39.18	3.67×10^{-4}	1.28×10^{-4}
44.86	4.21×10^{-4}	8.59[d]

[a] $n_c = 8$, $\alpha = 8 \times 10^4$ cal Å^2/mol, $\delta = 3$ Å. [b] Mole fraction of nonaggregated amphiphile. [c] Mole fraction of the aggregates. [d] See the explanation in the text.

TABLE II: Variation of the Critical Concentration as a Function of Hydrocarbon Chain Length[a]

n_c	ξ_{crit}	Calcd C_{crit}[b]	Exptl cmc[b,c]
6	27	2.64×10^{-3}	2.76×10^{-3}
8	27.8	2.61×10^{-4}	2.59×10^{-4}
10	29.0	2.58×10^{-5}	2.44×10^{-5}
12	29.5	2.52×10^{-6}	2.30×10^{-6}
14	30	2.49×10^{-7}	2.16×10^{-7}
16	31	2.45×10^{-8}	2.04×10^{-8}

[a] $\delta = 3$ Å, $\alpha = 8 \times 10^4$ cal Å^2/mol. [b] In mole fraction units. [c] Reference 3a.

Table I contains an aggregate molar fraction larger than unity because the corresponding value of ξ is larger than ξ_{max} which for this particular case is equal to 41. (The first constant in eq 25 was taken as 1750 cal/mol.) Of course, an upper bound of ξ_{max} can be obtained from the condition

$$N_A + \sum_{2}^{\infty} N_g + N_S = F$$

It has already been mentioned that the representation of aggregates as spheres is not essential to the definition of the critical concentration given here. Hoeve and Benson[5] suggest a spherical shape for micelles at low aggregation numbers and an oblate spherocylindrical shape at higher aggregation numbers. Tanford[3] suggests an ellipsoidal shape for the aggregates. It is possible to incorporate nonspherical shapes of aggregates as well as the variation in shapes accompanying growth in aggregate size in the present treatment. The quantitative results in terms of the size distribution function and critical concentrations will depend on these assumptions. However, the qualitative change in the pattern of the size distribution function from that of a monotonic decreasing function at low monomer concentration to one exhibiting extrema at higher monomer concentration is always present.

The explicit expressions used here for the various free-energy terms are of an empirical nature. Of course, it is possible to obtain expressions for the free energy on theoretical grounds following the approaches of Hoeve and Benson[5] or Poland and Scheraga[4] using the formalism of statistical thermodynamics.

Conclusions

A well-defined transition point in the aggregation process emerges as that separating two different types of behavior of the aggregate size distribution function. A critical concentration corresponding to this transition point is defined which is a close lower bound on the cmc values usually reported.

Acknowledgment. This work was supported by NSF.

References and Notes

(1) (a) P. Mukerjee, *Adv. Colloid Interface Sci.*, **1**, 241–275 (1967); (b) J. M. Corkill and J. F. Goodman, *ibid.*, **2**, 297–330 (1969).
(2) K. Shinoda, T. Nakagawa, B. Tamamushi, and T. Isemura, "Colloidal Surfactants", Academic Press, New York, N.Y., 1963.
(3) (a) C. Tanford, "The Hydrophobic Effect", Wiley, New York, N.Y., 1973; (b) C. Tanford, *J. Phys. Chem.*, **78**, 2469–2479 (1974).
(4) (a) D. C. Poland and H. A. Scheraga, *J. Phys. Chem.*, **69**, 2431–2442 (1965); (b) D. C. Poland and H. A. Scheraga, *J. Colloid Interface Sci.*, **21**, 273–283 (1966).
(5) C. A. J. Hoeve and G. C. Benson, *J. Phys. Chem.*, **61**, 1149–1158 (1957).

3010 J. Phys. Chem. 1981, 85, 3010–3014

Critical Micelle Concentration and the Transition Point for Micellar Size Distribution

E. Ruckenstein*

Department of Chemical Engineering, State University of New York at Buffalo, Amherst, New York 14260

and R. Nagarajan

Department of Chemical Engineering, Pennsylvania State University, University Park, Pennsylvania 16802 *(Received: February 18, 1981; In Final Form: May 22, 1981)*

The equation assumed by Ben-Naim and Stillinger for the free energy of micellization is used to demonstrate that, contrary to their assertion, the critical concentration C_{crit} corresponding to a transition in the shape of the micellar size distribution has a value close to the conventional cmc. The ratio cmc/C_{crit} is shown to decrease sharply from the large value of 20 obtained by Ben-Naim and Stillinger, to a value of 1.55, if one of the parameters (σ) is assigned a physically more plausible value. This value accounts better for the cooperative self-association of the surfactant molecules.

I. Introduction

In an earlier paper discussing the size distribution of micelles,[1] we defined a critical concentration C_{crit} as the total surfactant concentration at which the micellar size distribution displays a horizontal inflection point. Below the C_{crit}, this function decreases monotonically with size (Figure 1) indicating a small amount of aggregation. Above C_{crit}, it exhibits both a minimum and a maximum, and the concentration of the aggregates becomes significant. Far above C_{crit}, the size distribution goes through a minimum followed by a monotonic increase implying that a phase separation occurs. We anticipated that C_{crit} can be related to the conventional cmc because, at both of these concentrations, the number of micelles is small and only above them does it begin to increase appreciably. The connection between the two was tested by a number of illustrative calculations[1-3] which showed that the transition in the shape of the size distribution is closely accompanied by abrupt changes in properties like osmotic pressure, light scattering, dye solubilization, etc. The results obtained for a homologous series of alkylhexoxyethylene glycol monoethers by using Tanford's expression for the free energy of micellization,[4,5] and those obtained for homologous series of alkyl glucosides and sodium alkyl sulfates by using a statistical thermodynamic model,[2] have shown[3] good agreement among the calculated C_{crit}, the calculated cmc, and the experimental cmc values.

However, in a recent paper, Ben-Naim and Stillinger,[6] using an alternative equation for the free energy of micellization, arrived at the conclusion that the conventional cmc is about 20 times larger than C_{crit} over the range of parameter values they considered. Although they observe a parallelism between the conventional cmc and the C_{crit},

(1) E. Ruckenstein and R. Nagarajan, *J. Phys. Chem.*, **79**, 2622 (1975).
(2) R. Nagarajan and E. Ruckenstein, *J. Colloid Interface Sci.*, **60**, 221 (1977).
(3) E. Ruckenstein and R. Nagarajan, *J. Colloid Interface Sci.*, **57**, 388 (1976).
(4) C. Tanford, "The Hydrophobic Effect", Wiley, New York, 1973.
(5) C. Tanford, *J. Phys. Chem.*, **78**, 2469 (1971).
(6) A. Ben-Naim and F. H. Stillinger, *J. Phys. Chem.*, **84**, 2872 (1980).

0022-3654/81/2085-3010$01.25/0

The Journal of Physical Chemistry, Vol. 85, No. 20, 1981 3011

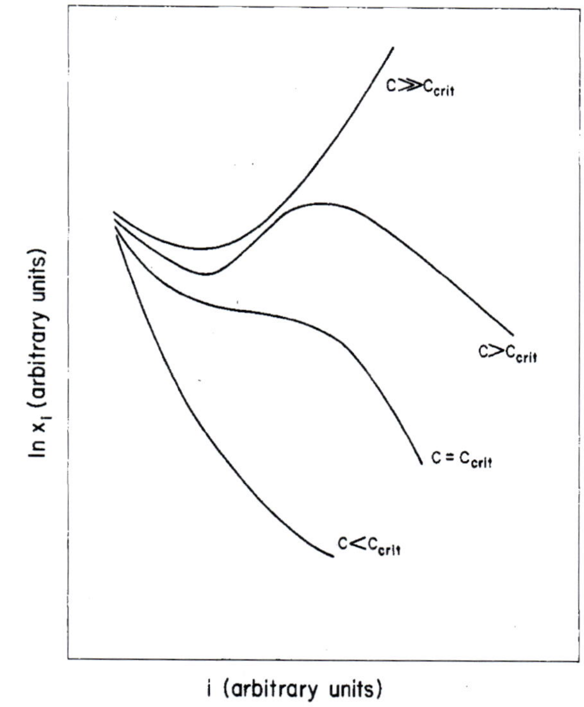

Figure 1. Qualitative micellar size distribution. The curves represent size distributions when the total surfactant concentration is less than C_{crit}, equal to C_{crit}, larger than C_{crit}, and far above C_{crit}.

they nevertheless conclude that "if the conventional cmc is to be related to any specific attribute of the aggregate size distribution, then it must be more subtle than the mere occurrence of a horizontal inflection point".

It is the objective of this note to show that, contrary to the conclusions of ref 6, a small change in the value of one of the parameters, σ, which appears in their expression for the free energy, can significantly reduce the ratio between cmc and C_{crit} to values near unity. For example, a change in σ from 3 to 4.5 reduces the ratio cmc/C_{crit} from about 20 to 2.5. A further increase of σ to 5.0 makes the above ratio as small as 1.55. We explain below that a value of 5 for σ is physically more plausible than the value of 3 used in ref 6, because it accounts better for the initial cooperative self-association of the surfactant molecules. Therefore, C_{crit} has values close to the conventional cmc, even for the free-energy expression assumed in ref 6, if a physically more plausible value is ascribed to σ.

In the following section, the micellization model and the free-energy expression used in ref 6 are described. In section III, the cmc and C_{crit} are calculated numerically for various values of σ. Because the numerical calculations for a complete parametric study would be too lengthy, approximate, analytical relations for the two concentrations are developed in section IV. Computations based on these relations show that, of all the parameters, σ is the only one which drastically affects the ratio between cmc and C_{crit}. The last section emphasizes again that C_{crit} has a value close to the conventional cmc as long as the expression used for the free energy is a physically acceptable one.

II. Micellization Model and Ben-Naim–Stillinger Free-Energy Expression

The micellization process can be described by the stepwise association equilibrium

$$A_{j-1} + A_1 \overset{k_j}{\rightleftharpoons} A_j \qquad j \geq 2 \tag{2.1}$$

where A refers to the amphiphilar aggregates, subscript j is the aggregation number, and k_j is the equilibrium constant for the stepwise association. The overall equilibrium constant K_i for i singly dispersed amphiphiles aggregating to form a micelle of size i is related to the stepwise association constants k_j by

$$K_i = \prod_{j=2}^{i} k_j = \prod_{j=2}^{i} (\rho_j/\rho_{j-1}\rho_1) \tag{2.2}$$

where ρ_j is the concentration of aggregates of size j.

A uniform stepwise association constant k_j cannot produce abrupt changes in the physical properties such as those accompanying micellization. This requires a cooperative self-association, i.e., k_j should increase with j in the initial stages.[7]

Ben-Naim and Stillinger[6] have taken

$$K_i = k^{i-1}F(i) \tag{2.3}$$

where k is a constant

$$F(i) = 1 \qquad i = 1$$

$$F(i) = 1 + \exp\left\{\Delta - \left(\frac{i - n^*}{\sigma}\right)^2\right\} \qquad 1 < i \leq n_{max} \tag{2.4}$$

$$F(i) = 0 \qquad i > n_{max}$$

and Δ, σ, n^*, and n_{max} are parameters. The factor k^{i-1} corresponds to a uniform association with all the stepwise association constants equal. The factor $F(i)$ is a Gaussian function which introduces nonuniformity in the overall equilibrium constant K_i to account for the cooperative association of amphiphiles. The stepwise association constant k_j can be calculated from

$$k_j = K_j/K_{j-1} = kF(j)/F(j-1) \tag{2.5}$$

and the micellar size distribution is given by

$$\rho_i = K_i\rho_1{}^i \tag{2.6}$$

Introducing the quantities

$$x_i = k\rho_i \qquad y = \sum_{i=1}^{n_{max}} ik\rho_i \qquad f = k\pi/RT \tag{2.7}$$

where π is the osmotic pressure, x_i is the reduced concentration of aggregates of size i, y is the reduced total concentration of surfactant, f is the reduced osmotic pressure, R is the universal gas constant, and T is the temperature in K, one can write

$$x_i = F(i)x_1{}^i \tag{2.8}$$

$$y = \sum_{i=1}^{n_{max}} iF(i)x_1{}^i \tag{2.9}$$

$$f = \sum_{i=1}^{n_{max}} F(i)x_1{}^i \tag{2.10}$$

In ref 6, y(cmc) is evaluated from a plot of f vs. y as the point at which an abrupt change in f occurs. y(crit) (denoted in ref 6 as y(RN)) is computed as the value of y at which ln x_i vs. i shows a horizontal inflection point.

While the expression 2.3 appears to be reasonable from a qualitative point of view, our computations seem to show that it cannot satisfactorily reproduce the results of previous expressions, such as that of Tanford, which have been obtained on the basis of somewhat more detailed physical considerations.

(7) P. Mukerjee in "Micellization, Solubilization and Microemulsions", Vol. 1, K. L. Mittal, Ed., Plenum, New York, 1977, p 171.

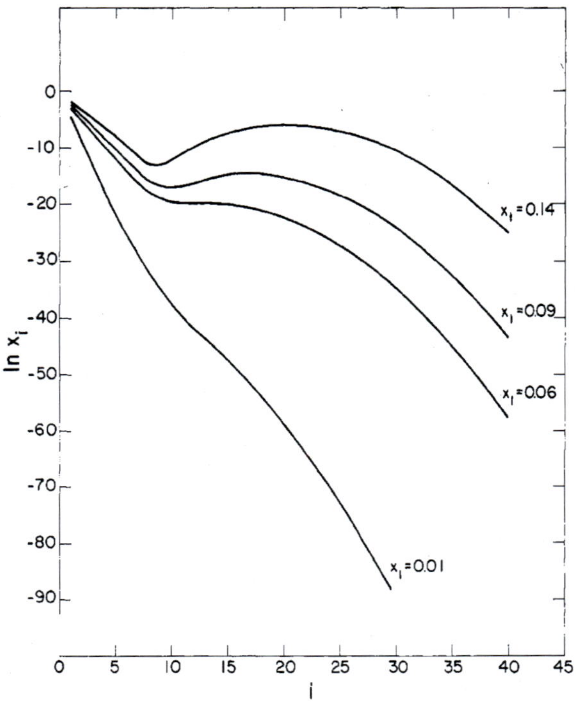

Figure 2. Aggregate size distributions for $\Delta = 50$, $\sigma = 4.5$, $n_{max} = 50$, and $n^* = 40$. The curves are labeled for various values of the reduced single amphiphile concentration x_1.

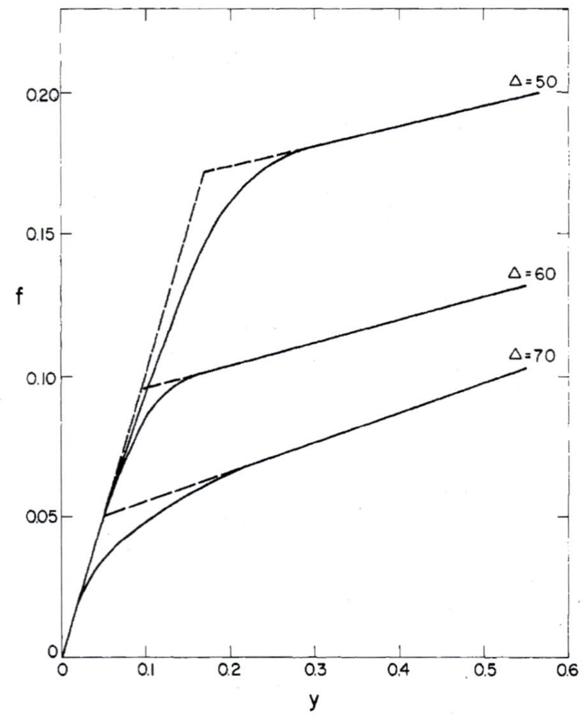

Figure 3. Reduced osmotic pressure f vs. reduced total amphiphilar concentration y. The curves are labeled for various values of the parameter Δ. The values of the other parameters are $\sigma = 4.5$, $n^* = 40$, and $n_{max} = 50$.

III. Estimation of y(cmc)/y(crit) for Various Values of σ

Numerical computations of the reduced variables x_i, y, and f have been carried out by using the Ben-Naim–Stillinger expression for the free energy. In Figure 2 the size distribution is plotted for $\sigma = 4.5$, $\Delta = 50$, $n^* = 40$, and $n_{max} = 50$. In this figure, all the parameters except σ have the values used in ref 6. The reduced osmotic pressure f is plotted in Figure 3 against the reduced total concentration y for the above values of the parameters and also for $\Delta = 60$ and $\Delta = 70$. Table I summarizes numerical results extracted from Figures 2 and 3. These results show that the difference between the cmc and C_{crit} is quite small when $\sigma = 4.5$ and is quite insensitive to changes in Δ. Additional computations have shown that if σ is increased further to a value of 5, the ratio cmc/C_{crit} decreases to 1.55. We explain later that a value of 5 for σ is more plausible from a physical point of view than the value 3 taken in ref 6.

In the next section, using approximate analytical relations for the cmc and C_{crit}, we present a parametric study which shows that the ratio cmc/C_{crit} is very sensitive to σ and less sensitive to all the other parameters.

IV. Relations between C_{crit}, cmc, and the Free-Energy Function

The transition in shape of the aggregate size distribution occurs when the size distribution displays a horizontal inflexion point, i.e., when

$$\mathrm{d}\ln x_i/\mathrm{d}i = \mathrm{d}^2(\ln x_i)/\mathrm{d}i^2 = 0 \qquad \text{at } i = i_{crit} \quad (4.1)$$

Combining eq 2.8 with the above expressions, we obtain

$$\mathrm{d}^2[\ln F(i)]/\mathrm{d}i^2 = 0 \qquad \text{at } i = i_{crit} \quad (4.2)$$

$$\ln x_1(\text{crit}) = -\mathrm{d}[\ln F(i)]/\mathrm{d}i \qquad \text{at } i = i_{crit} \quad (4.3)$$

TABLE I: y(cmc) and y(crit) for Various Values of σ Obtained via Numerical Computations

	$\sigma = 3.0$	$\sigma = 4.5$	$\sigma = 5.0$
	$\Delta = 50$, $n^* = 40$, $n_{max} = 50$		
y(crit)	0.013	0.064	0.078
y(cmc)	0.285	0.166	0.121
y(cmc)/y(crit)	21.9	2.59	1.55
	$\Delta = 60$, $n^* = 40$, $n_{max} = 50$		
y(crit)	0.008	0.042	
y(cmc)	0.180	0.095	
y(cmc)/y(crit)	22.5	2.26	
	$\Delta = 70$, $n^* = 40$, $n_{max} = 50$		
y(crit)	0.005	0.019	
y(cmc)	0.115	0.050	
y(cmc)/y(crit)	23.0	2.63	

When the Ben-Naim–Stillinger expression for the free energy is used, the above equations lead to

$$\exp\left\{\Delta - \left(\frac{i_{crit} - n^*}{\sigma}\right)^2\right\} = \frac{2}{\sigma^2}(i_{crit} - n^*)^2 - 1 \quad (4.4)$$

$$\ln x_1(\text{crit}) = \left[\frac{2}{\sigma^2}(i_{crit} - n^*)^2 - 1\right] \Big/ (i_{crit} - n^*) \quad (4.5)$$

It should be noted that the calculated x_1(crit) is almost equal to y(crit) (for the latter quantity, no simple analytical relation could be derived).

The conventional cmc can be approximately calculated by using the equation[7-9]

$$\ln y(\text{cmc}) = -\ln F(i)/i \qquad \text{at } i = i_M \quad (4.6)$$

where i_M is the aggregation number corresponding to the

(8) P. Mukerjee, *J. Pharm. Sci.*, **63**, 972 (1974).
(9) P. Mukerjee, *J. Phys. Chem.*, **73**, 2054 (1969).

Transition Point for Micellar Size Distribution

The Journal of Physical Chemistry, Vol. 85, No. 20, 1981 3013

TABLE II: Comparison of y(cmc)/y(crit) Estimated from Analytical Expressions with Those Obtained from Numerical Computations[a]

Δ	50	60	70	80
From Analytical Equations				
i_{crit}	19.7	17.4	15.5	13.7
i_M	33.9	32.6	31.1	29.7
y(crit)	0.0133	0.0084	0.0055	0.0037
y(cmc)	0.259	0.194	0.139	0.102
y(cmc)/y(crit)	19.5	23.1	25.3	27.6
From Numerical Computations (Ref 6)				
i_{crit}	19	17	15	13
i_M	32	30	28	26
y(crit)	0.013	0.008	0.005	0.003
y(cmc)	0.285	0.180	0.115	0.075
y(cmc)/y(crit)	21.9	22.5	23.0	25.0

[a] In both analytical and numerical computations, $\sigma = 3$, $n^* = 40$, and $n_{max} = 50$.

maximum in the size distribution curve. (Equations 4.6 and 4.7 involve setting $(i-1)/i \simeq 1$. Our numerical calculations have shown that this approximation is permissible.)

An upper bound on i_M is provided[7] by the value of i corresponding to an extremum of the free energy per amphiphile:

$$\frac{d}{di}\left[\frac{1}{i}\ln F(i)\right] = 0 \qquad \text{at } i = i_M \tag{4.7}$$

Near the cmc, the aggregation number corresponding to the maximum in the size distribution curve is lower than the value given by the above equation, but as the total surfactant concentration increases it approaches that value. Using the Ben-Naim–Stillinger expression for the free energy, we obtain

$$i_M = (n^{*2} - \sigma^2\Delta)^{1/2} \tag{4.8}$$

$$\ln y(\text{cmc}) = \frac{2}{\sigma^2}(i_M - n^*) \tag{4.9}$$

In deriving the above expressions, we assumed that

$$\exp\left\{\Delta - \left(\frac{i_M - n^*}{\sigma}\right)^2\right\} >> 1 \tag{4.10}$$

an inequality which is valid for the entire range of parameters considered in ref 6 and in this paper.

The results obtained from these relations are compared in Table II with those obtained from the numerical computations carried out by Ben-Naim and Stillinger[6] in order to verify the validity of the above approximate analytical expressions. The agreement between the analytical and the numerical results is very good. The somewhat higher values of the analytical y(cmc) compared to the numerical ones are due to the slight overestimation of the value of i_M. Therefore, the actual difference between y(cmc) and y(crit) is even smaller than that estimated by use of the above analytical relations.

The analytical expressions used to examine the effect of the model parameters n^*, n_{max}, Δ, and σ on y(crit) and y(cmc) are given in Table III. The results of the computations can be summarized as follows: y(crit) is insensitive to changes in n^*, whereas an increase in n^* from 40 to 50 (keeping $\Delta = 50$, $\sigma = 3$) results in a small increase in y(cmc) (by a factor of 1.36). A change in n_{max} has no influence on y(cmc) or y(crit) as long as n_{max} is chosen such that it accounts for all the aggregates which make a meaningful contribution to the total concentration. Even though an increase in the value of Δ causes a decrease in

TABLE III: Dependence of y(cmc) and y(crit) on the Parameters Appearing in the Free-Energy Expression (Eq 2.4)

parameter	d[ln y(crit)]/d[parameter]	d[ln y(cmc)]/d[parameter]
n^*	0	$\frac{2}{\sigma^2}[\frac{n^*}{i_M} - 1]$
n_{max}	0	0
Δ	$\dfrac{\frac{2}{\sigma^2}[i_{crit} - n^*]^2 - 1}{\frac{2}{\sigma^2}(i_{crit} - n^*)^3}$	$-\frac{1}{i_M}$
σ	$-\dfrac{\frac{2}{\sigma^2}[i_{crit} - n^*]^2 - 1}{\sigma[i_{crit} - n^*]}$	$-\dfrac{4i_M(i_M - n^*) + 2\sigma^2\Delta}{\sigma^3 i_M}$

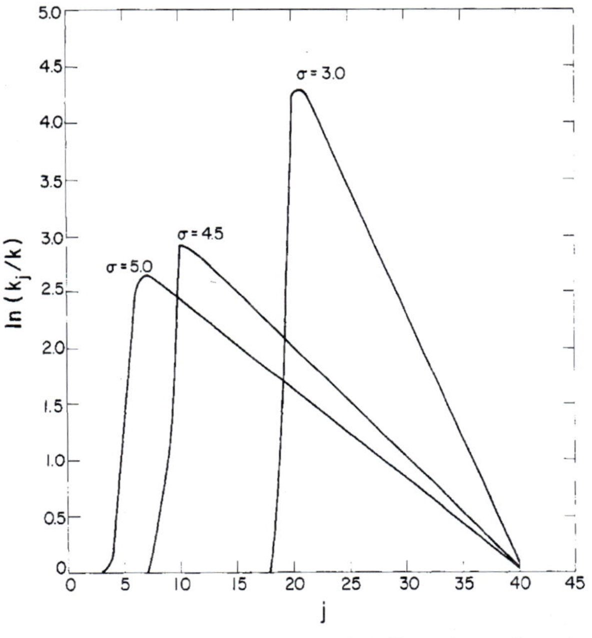

Figure 4. ln (k_j/k) vs. aggregation number j for various values of σ. k_j is the stepwise association constant and k is the uniform association constant appearing in eq 2.3. The values of the other parameters are $\Delta = 50$, $n^* = 40$, and $n_{max} = 50$.

both y(crit) and y(cmc) (see Table II), their ratio does not change too much. An increase in the value of σ produces a significant increase in y(crit) while y(cmc) slightly decreases; consequently, the ratio y(cmc)/y(crit) decreases sharply. For example, keeping $\Delta = 50$, $n^* = 40$, a change in σ from 3 to 4.5 increases y(crit) from 0.0133 to 0.064 while decreasing y(cmc) from 0.259 to 0.166. Therefore, the ratio y(cmc)/y(crit) is reduced from the value of 19.5 to a comparatively small value of 2.6. If σ is increased further to $\sigma = 5$, the ratio y(cmc)/y(crit) decreases further to 1.55. The parametric study thus reveals that σ is the only parameter which has a significant influence on the value of the ratio y(cmc)/y(crit).

V. Discussions and Conclusion

We have examined the sensitivity of the ratio cmc/C_{crit} to the parameters of the Ben-Naim and Stillinger free-energy expression in order to show that this ratio is largely insensitive to variations in the parameters Δ, n^*, and n_{max},

3014 J. Phys. Chem. 1981, 85, 3014–3018

while decreasing sharply when the parameter σ is increased. The main conclusion is that the concentration C_{crit} has values near those of the cmc if a sufficiently large value for σ, namely, $\sigma = 5$, is chosen.

Let us now emphasize that the value $\sigma = 5$ is more plausible from a physical point of view than the lower value $\sigma = 3$ taken in ref 6. Figure 4 shows the size dependence of the stepwise association constant. For the values of the parameters used in ref 6, namely, $\Delta = 50$, $n^* = 40$, $n_{max} = 50$, and $\sigma = 3$, the stepwise association constant remains constant up to $i = 17$. This implies that no cooperative association occurs until the aggregates contain 17 molecules. There is, however, good physical reason to expect this cooperation to set in from the beginning. Indeed, in the initial stages of micellar growth, the increment in the attractive interactions for each newly included amphiphile increases with the aggregation number and exceeds the increment in the repulsive head group interactions. This happens because in these initial stages an increase in the aggregation number gives rise to a larger number of interactions among the atoms of the hydrocarbon chains and, in particular, to an increasingly compact micellar core. The latter leads to greater and greater van der Waals attractions between the hydrocarbon tails as well as to a greater shielding from water. Consequently, the free energy per molecule associated with the hydrophobic bonding becomes more negative and the stepwise association equilibrium constant can be expected to increase with the aggregation number, i.e., an initial cooperativity of self-association[7,8] is expected to occur. The choice of $\sigma = 3$, in ref 6, imposes an artificial constraint on the nature of the association process for $i < 17$. When the value of σ is increased to 4.5 or 5.0, the region of noncooperative uniform association is limited to $i < 7$ and $i \leq 3$, respectively.

As mentioned earlier, the expression taken by Ben-Naim and Stillinger even though meaningful qualitatively is not able to yield the results of expressions such as that of Tanford, which have a somewhat better physical foundation. For this reason detailed computations have been carried out by using all the expressions available in the literature for the free energy of micellization.[10] These computations have shown that the ratio cmc/C_{crit} seldom exceeds a value of 1.5.

(10) R. Nagarajan and E. Ruckenstein, submitted for publication.

0022-3654/81/2085-3014$01.25/0

Phase Behavior of Mixtures of Sterically Stabilized Colloidal Dispersions and Free Polymer

I. V. RAO AND E. RUCKENSTEIN*

Department of Chemical Engineering, State University of New York, Buffalo, New York 14260

Received November 19, 1984; accepted April 22, 1985

The effect of the addition of free polymer to a sterically stabilized nonaqueous colloidal dispersion is examined theoretically by considering that the interparticle potential is a sum of contributions from (i) the forces due to the presence of adsorbed polymers, (ii) the forces due to the presence of the free polymer, and (iii) the forces due to the van der Waals attraction. After suitable division of the pair potential into reference and perturbation parts, the second-order perturbation theory is used to determine the conditions for phase separation. Separation into a disordered dilute dispersion and an ordered concentrated dispersion occurs when the concentration of the free polymer exceeds a certain limiting value. This limiting free polymer concentration is shown to decrease with increasing molecular weight of the free polymer, and with increasing particle concentration. This is in agreement with experimental results. The calculation of the forces due to the presence of the free polymer in solution has been carried out for two limiting cases; one in which the free polymer can penetrate into the adsorbed layer and the second in which the polymer sheath around the particle is impenetrable to the free polymer. Calculations for the low-molecular-weight free polymer indicate that the actual situation might correspond to one intermediate between the two limiting cases.

I. INTRODUCTION

Stability in mixtures of colloidal particles and polymer molecules, dispersed in a solvent, has been the subject of experimental and theoretical investigations for a long time and it has applications in diverse fields such as paint technology, wastewater treatment, emulsion polymerization, biology etc. It has now been well recognized that polymer molecules can be used to induce either stabilization or flocculation (phase separation) in colloidal dispersions. It is important to distinguish between polymers which are adsorbed on the particle surface and those that are free in solution because the two situations usually lead to qualitatively different effects. Stability imparted by adsorbed polymers is known as steric stabilization and the flocculation or phase separation caused by the free polymer is due to the so-called depletion effect, also known as the volume restriction effect.

Recent experimental studies (1–3), on systems of sterically stabilized colloidal particles that are dispersed in polymer solutions, have highlighted the role played by the free polymer molecules. These experiments are particularly relevant because the systems chosen are "model" dispersions in which the particles can be well approximated as monodisperse hard spheres. This simplifies the interpretation of the data and leads to a better understanding of the interparticle forces. DeHek and Vrij (1, 2) have added polystyrene molecules to sterically stabilized silica particles dispersed in cyclohexane and observed the separation of the mixtures into two phases—a silica-rich phase and a polystyrene-rich phase—when the concentration of the free polymer exceeds a certain limiting value. These experimental results indicate that the limiting polymer concentration decreases with increasing molecular weight of

* To whom correspondence should be addressed.

0021-9797/85 $3.00

the free polymer and with increasing particle diameter. Clarke and Vincent (3) also studied the effect of added polymer on sterically stabilized nonaqueous silica dispersions. These authors have observed the formation of a floc phase and a dispersed phase at equilibrium when the concentration of the free polymer exceeds the value corresponding to the onset of the semidilute regime. The molecular weight and the particle size dependence of the limiting polymer concentration for the onset of flocculation was found to be qualitatively the same as that observed by DeHek and Vrij. In addition to equilibrium state studies, Clarke and Vincent have also measured the initial rates of flocculation and observed that they increase with increasing free polymer concentration.

The experiments with aqueous dispersions (4) indicated a somewhat different behavior, both with respect to the initial rate of flocculation and the equilibrium state. The initial rate of flocculation increased from zero to a maximum value and decreased to zero as the free polymer concentration was further increased, indicating restabilization at higher concentrations. The amount of floc phase formed has also been observed to go through a maximum with increasing free polymer concentration. The restabilization phenomenon at higher polymer concentrations could not be detected in nonaqueous dispersions (1–3).

The possibility of occurrence of instability of colloidal dispersions in the presence of free polymer was first predicted by Asakura and Oosawa (5), who have shown that the exclusion of the free polymer molecules from the interparticle space generates an attractive force between particles. DeHek and Vrij (1) have developed a model in which the particles and the polymer molecules are treated as hard spheres and rederived in a simple and illuminating way the interaction potential proposed by Asakura and Oosawa. Using this potential, they calculated the second virial coefficient for the particles as a function of the free polymer concentration and have shown that phase separation results when this virial coefficient becomes sufficiently negative. Since no provision was made in the model for the steric stabilization of the particles, the agreement between the calculated and experimental limiting polymer concentrations for particles with steric layers was mainly qualitative. Vincent *et al.* (4) have proposed an alternate theory involving the displacement of the free polymer molecules from the interparticle space when the free polymer concentration exceeds the value corresponding to the onset of the semidilute regime. They have shown that the interparticle interaction energy has a minimum at a surface-to-surface separation distance of the order of twice the thickness of the adsorbed layer. The depth of this minimum energy is shown to increase with increasing concentration of the free polymer. The flocculation observed in the experiments has been attributed to this attractive energy between two particles. Feigin and Napper (6) have used the rotational isomeric state scheme and the Monte Carlo technique to calculate the segment density profile in the region between two flat, uncoated plates and extended it to the case of two interacting spheres. According to this theory, the free energy of interaction between two particles exhibits a minimum at very small surface-to-surface separation and a maximum when the distance of separation is of the order of the diameter of the polymer chains in solution. The increase in the height of the maximum and of the depth of the minimum with increasing concentration of the free polymer was considered to explain the flocculation and the restabilization phenomena observed in aqueous dispersions (7). Recently, Fleer *et al.* (8) have applied the polymer adsorption theory developed earlier by Scheutjens and Fleer (9), to calculate the segment density profile near an interface for a nonadsorbing polymer. They have shown that the depth of the depletion attraction potential passes through a maximum as a function of the free polymer concentration and that the restabilization at higher free polymer concentration is not due to the presence of a barrier in the interaction potential but is a thermodynamic phenomenon. When flocculation is caused by the free polymer alone, they show that the calculated value of the limiting polymer concentration agrees with the experimental data.

Using a theoretical model for the interparticle potential, the equilibrium properties of dispersions and the conditions for the onset of flocculation or phase separation can be predicted. van Megen and Snook (10) have used statistical mechanical techniques to locate the disorder–order transitions in electrostatically stabilized colloid particles dispersed in aqueous electrolyte solutions. The compositions of the ordered and of the disordered phases at equilibrium were predicted by using the cell model for the ordered phase and perturbation theory for the disordered phase. The results of this approximate technique have been shown to agree with the more exact computer simulation obtained by them earlier (11). With the availability of highly accurate expressions for the properties of hard-sphere solids, Gast *et al.* (12) have used the second-order perturbation theory to predict fluid–solid phase transitions in colloidal dispersions that have free polymer molecules in solution. They have presented phase diagrams that can be used to predict the compositions and the relative amounts of phases formed at equilibrium. However, many of the experiments have been carried out by starting with sterically stabilized colloidal dispersions. In such dispersions, the adsorbed and free polymers may be chemically similar or different. Since the sign and the magnitude of the interparticle forces are strongly dependent on the values of the polymer–solvent interaction parameters, it is possible to choose a solvent and a pair of polymers to control the stability of those dispersions. The particle size, the molecular weights of the adsorbed and free polymers, the concentration of the particles etc., are also expected to affect the behavior of such dispersions. For this reason, we consider in the present work the colloidal particles to be covered by a layer of adsorbed polymer in deriving the interparticle potential. The interaction potential between two particles is evaluated for two limiting cases. In one of them, the potential due to the steric repulsion is added to that due to the free polymer with the assumption that the free polymer does not penetrate into the steric layer. This restriction is relaxed in the second case. The second-order perturbation theory (13) is used to calculate the equilibrium properties of the dilute disordered (fluid) phase and of the ordered concentrated (solid) phase and to locate the fluid–solid phase transition boundaries. Phase diagrams relating the concentration of the particles in both phases to the free polymer concentration are presented and are further used to study the effects of particle concentration and molecular weight of the free polymer on the stability of dispersions. The theoretical predictions are compared with the experimental data of DeHek and Vrij.

Section II describes the model used for deriving the expressions for the interparticle potential. A brief account of the statistical mechanical approach to locate the phase transition boundaries is given in Section III and the results of the calculation are presented in Section IV.

II. INTERACTION POTENTIAL

Here a mixture of sterically stabilized colloidal particles, solvent, and free polymer molecules in solution is considered. When two particles approach one another during a Brownian collision, the interaction potential between the two depends not only on the distance of separation between them, but also on various parameters, such as the thickness and the segment density distribution of the adsorbed layer, the concentration and the molecular weight of the free polymer. The various types of forces that are expected to contribute to the interaction potential are (i) forces due to the presence of the adsorbed polymer, (ii) forces due to the presence of the free polymer, and (iii) van der Waals forces. It is assumed here that there are no electrostatic forces. A brief account of the nature of these forces as well as the expressions needed to estimate them are given below.

II.1. Forces Due to the Presence of Adsorbed Polymer

Napper (14) has derived analytical expressions for the free energy of interpenetration and compression of the adsorbed polymer chains as a function of the distance of separation between two particles. The region of close approach was divided into two domains:

(a) the interpenetration domain, $\delta < \Delta < 2\delta$, and

(b) the interpenetration plus compression domain, $\Delta < \delta$.

Here Δ is the surface to surface separation between the bare particles and δ is the thickness of the adsorbed layer. For distances of separation $\Delta \geqslant 2\delta$, the free energy of mixing of the chains is zero. Assuming constant segment density distribution in the adsorbed layers, Evans and Napper (15) derived the following expression for the free energy in the interpenetration domain, which is due only to the mixing of the chains:

$$\frac{V_s}{kT} = \frac{4\pi a\delta^2}{v_1}\left[v^2(\tfrac{1}{2} - \chi_1) + \frac{3v^3}{2}(\tfrac{1}{3} - \chi_2)\right] \times \left(1 - \frac{\Delta}{2\delta}\right)^2, \quad [1]$$

where a is the radius of the particle, v_1 is the molecular volume of the solvent, v is the volume fraction of the polymer in the adsorbed layer, k is the Boltzmann constant, and T is the absolute temperature. χ_1 and χ_2 are the constants appearing in the expression accounting for the concentration dependence of the Flory polymer–solvent interaction parameter χ ($\chi = \chi_1 + \chi_2 v$).

In the region $\Delta < \delta$, there are contributions from both the mixing V^{M} and compression of the adsorbed chains V^{El}. The expression for the free energy incorporating both effects, again for a constant segment density model, is given by

$$V_s = V^{\mathrm{M}} + V^{\mathrm{El}}, \quad [2]$$

where

$$\frac{V^{\mathrm{M}}}{kT} = \frac{4\pi a\delta^2}{v_1}\left[v^2(\tfrac{1}{2} - \chi_1)\left(\frac{\Delta}{2\delta} - \tfrac{1}{4} - \ln\frac{\Delta}{\delta}\right) + \frac{v^3}{2}(\tfrac{1}{3} - \chi_2)\left(1 + \frac{\Delta}{2\delta} + \frac{2\delta}{\Delta} - \tfrac{11}{4}\right)\right] \quad [3]$$

and

$$\frac{V^{\mathrm{El}}}{kT} = \frac{2\pi a\delta^2 v}{v_p}\left\{\frac{\Delta}{\delta}\ln\left[\frac{\Delta}{\delta}\left(\frac{3 - \frac{\Delta}{\delta}}{2}\right)^2\right] - 6\ln\left(\frac{3 - \frac{\Delta}{\delta}}{2}\right) + 3\left(1 - \frac{\Delta}{\delta}\right)\right\}. \quad [4]$$

Here v_p is the molecular volume of the polymer and the superscripts M and El refer to the mixing and elastic contributions, respectively.

II.2. Forces Due to the Presence of the Free Polymer

In the basic model, put forward by Asakura and Oosawa (5), the hard spherical particles immersed in a solution of macromolecules are considered to be surrounded by depletion layers from which the polymer molecules are excluded. When two particles are far apart with no overlap of the depletion zones, the thermal force acting over the entire particle surface is uniform. However, when the particles come closer, such that their depletion zones begin to overlap, there is a region in which the polymer concentration is zero and the force exerted over the surfaces facing this region is smaller compared to that exerted over the rest of the surface. This gives rise to an attractive force between the two particles which is proportional to the osmotic pressure of the polymer solution.

If the particles are covered by adsorbed polymer, it is not easy to identify the depletion zones. However, two limiting cases can be considered. In one of them (Fig. 1a), the free polymer molecules are allowed to penetrate

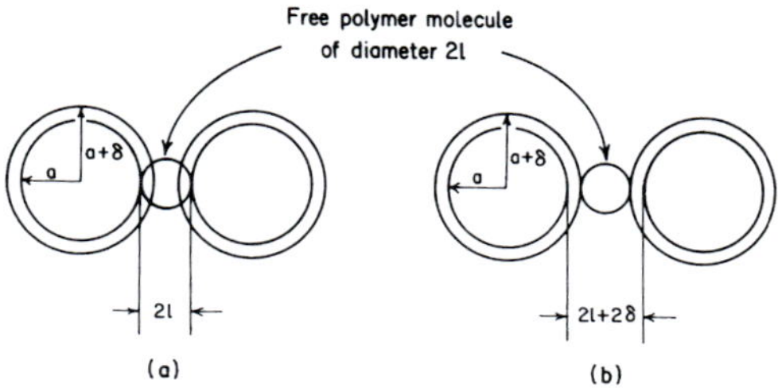

FIG. 1. Schematic diagram describing the close approach of two colloidal particles (radius a) covered with adsorbed polymer layer of thickness δ: (a) complete penetration of free polymer into the adsorbed layer; (b) polymer sheath is impenetrable to the free polymer.

the adsorbed layer. This means that attractive forces between the particles operate when the distance of separation Δ between their surfaces becomes less than the diameter $2l$ of the free polymer molecules. In the other limiting case (Fig. 1b), the polymer sheath around the particle is assumed to be impenetrable to the free polymer molecule and the range over which the attractive force is effective becomes now $2l + 2\delta$, which is larger than in the previous case.

In one of the limiting cases, the free polymer is allowed to penetrate the adsorbed layer around the particles. One may note that when the free polymer and the adsorbed polymer are both present in the steric layer around the particle, the interactions between the two must be taken into account while evaluating the interparticle forces. However, in the absence of a detailed knowledge of the structure of the adsorbed layer, it is difficult to evaluate this contribution to the interaction potential. Then, the situation is similar to the one considered by Asakura and Oosawa (16), and the force of attraction between two bare particles of radius a in the presence of free polymer molecules of radius l can be expressed as the product of the osmotic pressure P_{osm} and the area of the intersection of the two overlapping volumes:

$$F = -P_{osm}\frac{\pi}{4}[(2a + 2l)^2 - (2a + \Delta)^2]. \quad [5]$$

The potential energy of the attractive force is therefore given by

$$V_d = -\frac{2\pi}{3}P_{osm}a^3\left[2\left(1+\frac{l}{a}\right)^3 + \left(1+\frac{\Delta}{2a}\right)^3 - 3\left(1+\frac{\Delta}{2a}\right)\left(1+\frac{l}{a}\right)^2\right]. \quad [6]$$

For dilute polymer solutions, the osmotic pressure can be approximated by the limiting van't Hoff expression. A more general expression for the osmotic pressure of polymer solutions is given by (17)

$$P_{osm} = -\frac{kT}{v_1} \times \left[\phi\left(1-\frac{1}{x}\right) + \ln(1-\phi) + \chi\phi^2\right], \quad [7]$$

where ϕ is the volume fraction of the polymer, v_1 is the molecular volume of the solvent, χ is the polymer–solvent interaction parameter, and x is the ratio of the molar volumes of the polymer and solvent.

If the adsorbed polymer sheath is considered as an impenetrable barrier, the free polymer molecules are excluded from a larger area and the attractive force for this second limiting case now becomes

$$F = -P_{osm}\frac{\pi}{4}[(2a + 2l + 2\delta)^2 - (2a + \Delta)^2] \quad [8]$$

and the corresponding expression for the potential energy is given by

$$V_d = -\frac{2\pi}{3}P_{osm}a^3\left[2\left(1+\frac{(l+\delta)}{a}\right)^3 + \left(1+\frac{\Delta}{2a}\right)^3 - 3\left(1+\frac{\Delta}{2a}\right)\left(1+\frac{(l+\delta)}{a}\right)^2\right]. \quad [9]$$

It is relevant to examine the role played by the relative sizes of the steric layer and that of the free polymer molecules. If the molecular weight of the adsorbed polymer is small compared to that of the free polymer, the thickness of the adsorbed layers will also be small compared to the diameter of the free polymer and the depletion effects are expected to play the dominant role. In the opposite situation of thick steric layers and small-molecular-weight free polymer, the forces due to the presence of adsorbed layers should dominate the stability of the system. The contribution of steric stabilization and depletion flocculation would be comparable when the molecular weights of the adsorbed and of the free polymer are of the same order.

II.3. The van der Waals Forces

The van der Waals forces between surfaces covered with adsorbed layers can be calculated using the effective Hamaker constant A given by (18)

$$A = (A_1^{1/2} - A_2^{1/2})^2, \qquad [10]$$

where A_1 is the Hamaker constant for the particle and A_2 refers to the adsorbed polymer layer, including the solvent in this layer. If the adsorbed amount of polymer is small, A_2 can be approximated by the Hamaker constant of the pure solvent. The van der Waals attractive potential energy between two particles of radius a is given by

$$V_a = -\frac{A}{12}\left[\frac{1}{z^2+2z} + \frac{1}{z^2+2z+1} + 2\ln\frac{z^2+2z}{z^2+2z+1}\right], \qquad [11]$$

where z is the ratio of the surface-to-surface separation, Δ, to the particle diameter, $2a$.

The total potential energy of interaction between two particles, u, is taken to be equal to the sum of the three contributions V_s, V_d, and V_a. With the availability of expressions for the total interaction potential, the thermodynamic properties of the system and the conditions for phase transition can be predicted.

III. PERTURBATION THEORY APPROACH

The basic idea of the perturbation theory is that the behavior of the real systems at high densities closely resembles that of a system of hard spheres. Experimental evidence that the structure of a liquid is primarily determined by the hard-core repulsive forces, with the attractive forces merely acting to hold the molecules together, resulted in the use of the perturbation theory to predict the thermodynamic properties of liquids and the melting transitions in molecular systems (13, 19). Encouraged by the success of the technique for molecular systems, attempts have been made (12, 20–22) to apply the method to systems of colloidal particles.

In the application of the perturbation theory, the pair potential is expressed as the sum of two terms:

$$u = u_0 + u_1, \qquad [12]$$

where u_0 is the potential energy of a reference system and u_1 that due to the perturbation. Assuming pairwise additivity of the potentials, the total potential energy of a system of N particles is given by

$$U = U_0 + U_1, \qquad [13]$$

where

$$U_0 = \sum_{j>i=1}^{N} u_0(r_{ij}), \qquad [14]$$

and

$$U = \sum_{j>i=1}^{N} u(r_{ij}), \qquad [15]$$

r_{ij} being the center-to-center separation. With this division of the total interaction potential, the thermodynamic properties can be expressed as the sum of the contributions from the reference potential and from the perturbation potential. The obvious choice for the reference system is that of hard spheres, the properties of which are accurately known for both fluid and solid phases. The division of the pair potential into reference and perturbation terms is straightforward for a model potential such as the square-well potential, since the repulsive part of the potential is always infinite. The reference system in this case is indeed a system of hard spheres whose diameter is equal to the diameter of the particles. But for more realistic potentials such as the Lennard–Jones potential or the DLVO potential for colloid particles, the repulsive part is a continuous varying function from infinity to the minimum of the potential. In such cases, the division of the potential into reference and perturbation terms is done in such a way that the reference system can be equated to a system of hypothetical hard spheres.

The two most commonly used procedures for the division of the pair potentials into the reference and the perturbation parts have been proposed by Barker and Henderson (BH) (19) and Chandler, Weeks, and Anderson (CWA) (23). Both procedures have been found to be equally satisfactory and the BH scheme is used here because of its simplicity. The division of the pair potential into repulsive and attractive parts, according to the BH scheme, is

$$u_0(r) = u(r) \quad r < \sigma$$
$$= 0 \quad r \geq \sigma \qquad [16]$$

and

$$u_1(r) = u(r) \quad r \geq \sigma$$
$$= 0 \quad r < \sigma, \qquad [17]$$

where r is the center-to-center separation between two particles and σ is chosen such that $u(\sigma) = 0$. Usually, the repulsive part of the potential given by Eq. [16] is not always infinite and is therefore replaced by the following system of hypothetical hard spheres of diameter d_B:

$$u_0(r) = \infty \quad r < d_B$$
$$= 0 \quad r \geq d_B. \qquad [18]$$

Now a choice has to be made for d_B. Barker and Henderson have shown that the free energy of the reference system reduces to that of a system of hard spheres if d_B is chosen as

$$d_B = \int_0^{\sigma} [1 - \exp(-u_0/kT)]dr. \qquad [19]$$

For potentials which have a primary minimum, a barrier, and a secondary minimum as in the case of the DLVO potentials, the Barker–Henderson diameter d_B can be evaluated as (24)

$$S = \frac{d_B}{d} = R_m + \int_{R_m}^{\sigma/d} [1 - \exp(-u_0/kT)]dR. \qquad [20]$$

Here $R = r/d$ where d is the diameter of the particle and R_m is a reduced effective hard-sphere diameter, chosen such that $\exp(-u(R_m)/kT)$ is negligible (less than 10^{-6}). The particles cannot come closer than R_m because of the large repulsion at such separations.

The Helmholtz free energy F of the system of N particles of diameter d, in a volume V can now be expressed to second order in the perturbation as (13)

$$\frac{F}{NkT} = \frac{F_{HS}}{NkT} + \tfrac{1}{2}\rho \int_{\sigma}^{\infty} \frac{u_1(r)}{kT} g_0(r) \cdot 4\pi r^2 dr - \tfrac{1}{4}\rho kT\left(\frac{\partial\rho}{\partial p}\right)_{HS} \int_{\sigma}^{\infty} \left[\frac{u_1(r)}{kT}\right]^2 g_0(r) \cdot 4\pi r^2 dr, \qquad [21]$$

where ρ is the number density of the particles and F_{HS}, g_0, and $(\partial\rho/\partial p)_{HS}$ are the free energy, the radial distribution function, and the macroscopic compressibility of a system of hard spheres of diameter d_B, respectively. The second and third terms on the right-hand side of Eq. [21] represent the first- and second-order contributions due to the perturbing potential $u_1(r)$. The quantity $kT(\partial\rho/\partial p)_{HS}$ can be evaluated from the equation of state for hard spheres. The second-order contribution can be evaluated more accurately by using the "local" compressibility approximation instead of the "macroscopic" compressibility approximation employed in Eq. [21]. However, our calculations indicate that the second-order contribution is always a very small fraction (~1%) of the first-order term. Thus it is sufficient to evaluate the free energy using either one of the mentioned approximations. The above expression for the free energy can be expressed in terms of the volume fraction of the particles η which is related to the number density by the relation $\eta = (\pi/6)\rho d^3$, where d is the diameter of the particles. Then the Helmholtz free energy F_{HS} of the reference system is a function of an effective volume fraction η' expressed, in terms of the Barker–Henderson diameter d_B, as $\eta' = (\pi/6)\rho d_B^3$. The free energy is written as

$$\frac{F(\eta)}{NkT} = \frac{F_{HS}(\eta')}{NkT} + 12\eta \int_{\sigma/d}^{\infty} \left(\frac{u_1(R)}{kT}\right) \times g_0(R, \eta')R^2 dR - 6\eta kT\left(\frac{\partial\rho}{\partial p}\right)_{HS} \times \int_{\sigma/d}^{\infty} \left[\frac{u_1(R)}{kT}\right]^2 g_0(R, \eta')R^2 dR, \quad [22]$$

where the free energy and the radial distribution function are functions of the variables written in the brackets. For very steep repulsive potentials, the difference in the numerical values between η' and η is very small.

Once the Helmholtz free energy is calculated, other thermodynamic properties of the system can be obtained using the basic relationships between various quantities. The pressure P, given by

$$P = -\left(\frac{\partial F}{\partial V}\right)_{T,N} \quad [23]$$

can be expressed in an alternate way as

$$Z = \frac{PV}{NkT} = \left(\frac{PV}{NkT}\right)_{HS} + \eta \frac{\partial}{\partial\eta}\left(\frac{F - F_{HS}}{NkT}\right). \quad [24]$$

The Gibbs free energy is obtained from the relation

$$G = F + PV. \quad [25]$$

The conditions for the phase transition are provided by the equality of the pressures and of the chemical potentials of the particles in the two phases. In order to use the above expressions to locate the phase transitions from the dilute disordered phase to the concentrated ordered one, we need information about the radial distribution function and the equation of state for hard spheres in both the fluid and solid states.

III.1. Hard-Sphere Reference System

Molecular dynamics calculations of Hoover and Ree (25) have indicated that a fluid–solid transition occurs in a system of hard spheres even in the absence of attractive forces. The fluid exists for particle volume fractions up to a value $\eta = 0.49$ and at this point, a solid phase with $\eta = 0.55$ is predicted to coexist in equilibrium with the fluid phase. When the particle volume fraction lies in the range $0.55 < \eta < 0.74$, the solid phase is stable. The upper limit for η corresponds to the density at closest packing for a face-centered-cubic (fcc) arrangement of the particles.

III.1.1. The equation of state. Carnahan and Starling (26) have proposed a simple and accurate equation of state for hard-sphere fluids as a function of the volume fraction η, which is given by

$$Z_{HS} = \left(\frac{P}{\rho kT}\right)_{HS} = \frac{1 + \eta + \eta^2 - \eta^3}{(1 - \eta)^3}, \quad [26]$$

where ρ is the number density of the particles. The above equation is in good agreement with computer simulation studies in the density range over which the fluid phase is stable.

For the hard-sphere solid, Hall (27) proposed the following equation of state, that was found to be accurate in the range of solid densities:

$$Z_{HS} = 2.557696 + 0.1253077\beta + 0.176239\beta^2 - 1.053308\beta^3 + 2.818621\beta^4 - 2.921934\beta^5 + 1.118413\beta^6 + (12 - 3\beta)/\beta, \quad [27]$$

where

$$\beta = 4(1 - \eta/\eta_0) \quad \text{and} \quad \eta_0 = 0.74.$$

III.1.2. The radial distribution function. The radial distribution function is defined by the expression (28)

$$g(r) = \frac{1}{\rho^2}\frac{N!}{(N-2)!} \times \frac{\int \cdots \int e^{-U/kT} d\vec{r}_3 \cdots d\vec{r}_N}{Z_N}, \quad [28]$$

where ρ is the number density, N is the number of particles, and $U = \sum_{j>i=1}^{N} u(r_{ij})$. Z_N is the configurational integral for the N particles. Analytical expressions for the radial distribution function are generally difficult to obtain even for ideal potentials such as the hard-sphere potential. Smith and Henderson (29) have proposed an expression for the radial distribution function for a hard-sphere fluid that agrees closely with the Monte Carlo simulations. Verlet and Weis's (30) improved expression brings the calculated values of the radial distribution function to an even closer agreement with the computer simulations. For the hard-sphere solid, Kincaid and Weis (31) have suggested the following analytical expression for the radial distribution function

$$g(R, \eta) = \frac{\alpha}{R}\exp[-W_1^2(R - R_1)^2 - W_2^4(R - R_1)^4] + \frac{W}{24\eta\sqrt{\pi}} \times \sum_{i=2}^{\infty} \frac{n_i}{RR_i}\exp[-W^2(R - R_i)^2], \quad [29]$$

where n_i is the number of ith neighbors, R_i is the distance to the ith neighbor in a face-centered-cubic arrangement, and R_1 refers to the position of the first peak of the radial distribution function. The parameter α is chosen such that the function g satisfies the following consequence of the virial theorem:

$$g(1, \eta) = (Z_{HS} - 1)/4\eta, \quad [30]$$

where η is the volume fraction of the particles and Z_{HS} is calculated from the equation of state given by Eq. [27]. The parameters W, W_1, and W_2 are given by the expressions

$$W = [1 - 10.5896(\eta_0 - \eta)^{2.543}]/[0.694(\eta_0 - \eta)]^{1.072}, \quad [31]$$

$$W_1 = \frac{1.5523}{(\eta_0 - \eta)} - 2.0303\exp[5.8331(\eta_0 - \eta)] + 74.8732(\eta_0 - \eta)^2 \quad [32]$$

and

$$W_2 = [0.95596 - 5.8550(\eta_0 - \eta) + 39.7466(\eta_0 - \eta)^2 - 109.6264(\eta_0 - \eta)^3]/(\eta_0 - \eta), \quad [33]$$

where η_0 is the volume fraction at the closest packing ($\eta_0 = 0.74$).

III.2. Procedure

After determining the Barker–Henderson diameter for the potential, for a fixed molecular weight and concentration of the free polymer, the Helmholtz free energy is calculated, by numerical integration, according to Eq. [22] for various values of the volume fraction of the particles, η. Differentiation of the expression for the free energy, according to Eq. [24], gives the pressure. For the fluid phase, the differentiation was performed numerically. Values of PV/NkT for the solid phase were obtained by finding analytically the derivative of the solid-phase radial distribution function. The Gibbs free energy is calculated using the relation

$$\frac{G - G^\circ}{NkT} = \frac{F - F^\circ}{NkT} + (Z - 1), \quad [34]$$

where F° and G° are the values of F and G for an ideal gas at the same density and Z is given by Eq. [24]. The values of $P/\rho kT$ can be converted to the reduced pressures $P/\rho_0 kT$ by using the identity

$$\frac{P}{\rho_0 kT} = \left(\frac{P}{\rho kT}\right)\left(\frac{\rho}{\rho_0}\right) = (Z)\left(\frac{\eta}{0.74}\right). \quad [35]$$

The intersection of the plots of the chemical potential (given by Eq. [34]) versus the reduced pressure (given by Eq. [35]) for both phases provides the coexistence pressure. The densities of the coexisting phases can then be determined with the help of the isotherms, plotted as $P/\rho_0 kT$ versus ρ/ρ_0, of each of the phases. Now the concentration of the free polymer is varied and the above procedure is repeated to locate the corresponding coexisting pressures and densities. This maps out the phase diagram for a given molecular weight of the free polymer.

IV. RESULTS AND DISCUSSION

In order to apply the above procedure to determine the conditions of phase separation, we have chosen the system of polyisobutene-stabilized silica particles with polystyrene as the free polymer dissolved in cyclohexane. The system temperature is chosen to be the θ temperature for the polystyrene–cyclohexane system (34.5°C), corresponding to the experimental conditions of deHek and Vrij (1). The pertinent parameters required for the calculation of the contribution of the adsorbed layers to the total interaction potential are $a = 48$ nm, $v_1 = 0.18$ nm^3, $\delta = 5$ nm, $\chi_1 = 0.47(32)$, $\chi_2 = 0.10(32)$, $v = 0.10$, and $v_p = 2.36$ nm^3. It can be seen from Fig. 2 that these forces are repulsive, with very large positive values for the potential energy at small distances of separation and falling off to zero at separation distances of the order of 2δ, where δ is the thickness of the adsorbed layer. At the distance of separation δ, the expressions for the interpenetration domain and the interpenetration plus compression domain give the same value for the free energy, indicating a continuous transition from one domain to the other.

For fixed values of the molecular weight M and volume fraction of polystyrene in cyclohexane ϕ, the depletion contribution to the total potential is calculated according to Eq. [6], which involves complete penetration of the free polymer into the adsorbed layer, and is shown in Fig. 2. The range of this contribution can be seen to be of the order of polystyrene molecule diameter. Figure 3 shows how the range of the interaction increases with the increase in the molecular weight of the free polymer. In the calculation of the van der Waals forces, the effective Hamaker constant used in this work is 4.54 kT. The relative magnitude of this attraction is indicated in Fig. 2, where the total potential energy for a pair of particles is also depicted. The minimum value in the potential is a strong function of the volume fraction of polymer in solution.

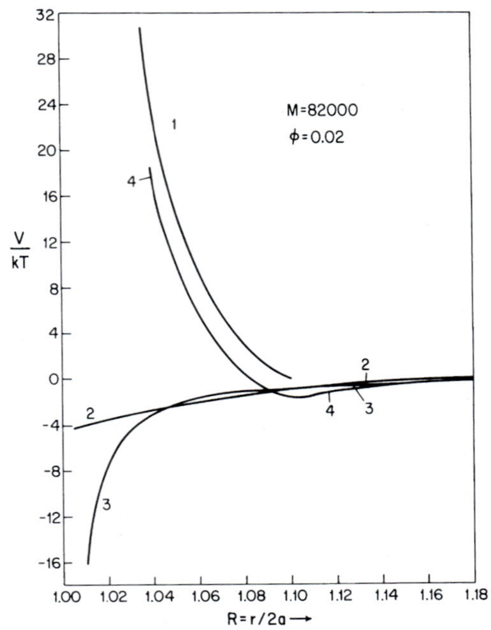

FIG. 2. Potential energy dependence on $R = r/2a$ where r is the center-to-center separation and a is the radius of the particle. The total potential (curve 4) is taken to be the sum of the contributions due to the adsorbed layer (curve 1), the free polymer assuming complete penetration (curve 2), and the van der Waals attraction (curve 3). M is the molecular weight of the free polymer and ϕ is the volume fraction of the free polymer.

The division of the potential into reference and perturbation terms is carried out according to the procedure described earlier and the values of the Barker–Henderson diameters are given in Table I. The Helmholtz free energy of the system is obtained by using Eq. [22], in which the upper limit of the integration is so chosen that raising it further does not change

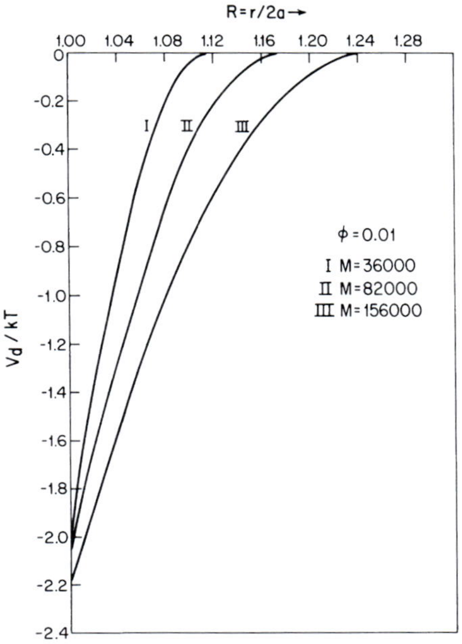

FIG. 3. Attractive potential energy due to the free polymer for different molecular weights of the sample. (I) 36,000; (II) 82,000; (III) 156,000. The volume fraction of the polymer is 0.01.

TABLE I

Barker–Henderson Diameters

Mol. wt. of polymer in solution, M	Volume fraction of polymer in solution, (ϕ)	Reduced BH diameter, S
	0.010	1.0836
	0.025	1.0819
36,000	0.050	1.0773
	0.075	1.0672
	0.100	1.0421
	0.015	1.0795
82,000	0.020	1.0775
	0.025	1.0754
	0.010	1.0793
156,000	0.025	1.0700
	0.050	1.0442

TABLE II

Thermodynamic Properties of Colloidal Dispersions in the Presence of Free Polymer

η'	$\frac{F - F_{HS}}{NkT}$	$\frac{P}{\rho kT}$	$\frac{\rho}{\rho_0}$	$\frac{P}{\rho_0 kT}$	$\frac{G - G^\circ}{NkT}$
		M = 82000, ϕ = 0.02			
		Fluid			
0.05	−0.098	1.137	0.054	0.061	0.259
0.15	−0.330	1.615	0.162	0.262	1.035
0.25	−0.655	2.491	0.271	0.673	2.236
0.35	−1.124	4.171	0.378	1.577	4.517
0.45	−1.777	7.701	0.486	3.743	8.884
		Solid			
0.60	−3.562	2.104	0.648	1.364	4.492
0.63	−4.247	3.256	0.681	2.216	5.509
0.64	−4.520	3.825	0.691	2.644	5.975
0.65	−4.830	4.282	0.702	3.006	6.302
		M = 82,000, ϕ = 0.025			
		Fluid			
0.15	−0.380	1.579	0.163	0.257	0.949
0.25	−0.749	2.422	0.272	0.658	2.073
0.35	−1.287	4.041	0.380	1.537	4.224
0.45	−2.043	7.469	0.489	3.652	8.386
		Solid			
0.6550	−5.848	1.330	0.711	0.946	2.432
0.6575	−5.954	2.363	0.714	1.688	3.409
0.6600	−6.036	3.992	0.717	2.863	5.006
0.6625	−6.124	4.762	0.719	3.428	5.738

the value of the integral appreciably (indicating the convergence of the integral). The function $(F - F_{HS})/NkT$, which constitutes the contribution to the free energy due to the perturbation is shown in Fig. 4. Table II gives the values of the reduced pressures and of the Gibbs free energy at different particle volume fractions. As discussed earlier, the coexistence pressure at the phase transition is determined by the intersection of the chemical potential versus reduced pressure plots for both phases, as shown in Fig. 5. The densities of the coexisting phases at equilibrium are determined with the help of the isotherms shown in Fig. 6.

The phase diagrams shown in Fig. 7 indicate the homogeneous regions and their boundaries with the two-phase region. For a fixed particle concentration, the increase of the free polymer concentration produces no phase change in the system, which remains as a fluid, until a limiting concentration of free polymer is reached. At this point, a further increase of the polymer concentration brings the system

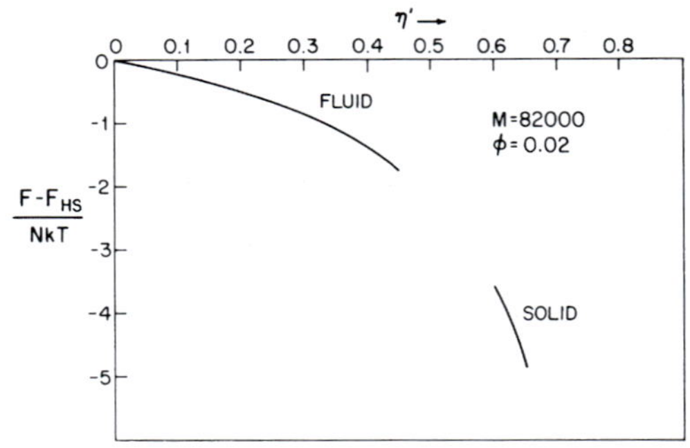

FIG. 4. The contribution of the first- and second-order perturbation terms to the Helmholtz free energy. η' is the volume fraction of the particles of diameter d_B.

into the two-phase region and phase separation into two phases takes place.

The effect of the molecular weight of the free polymer can be seen from Fig. 7. Our calculations confirm the experimental observation that the limiting polymer concentration for the onset of phase separation at a fixed particle concentration, decreases with increasing molecular weight of the free polymer. Figure 7 also includes the conditions for phase separation when there is no free polymer in solution, i.e., only steric repulsion and van der

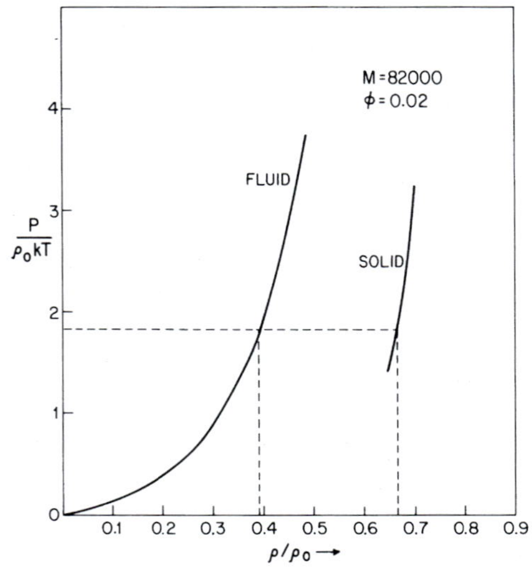

FIG. 6. Fluid- and solid-phase isotherms. The dashed lines indicate the densities of the coexisting phases at equilibrium.

Waals forces are acting. Since the attractive energy term is smaller in this case, the equilibrium particle concentrations in the two phases at equilibrium do not differ as much as in the presence of free polymer.

In the experiments of DeHek and Vrij, the highest particle concentration used was of the

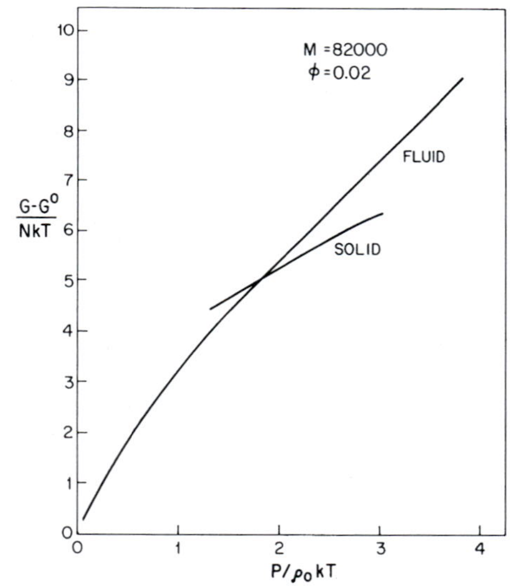

FIG. 5. Plot of the chemical potential versus the reduced pressure for both fluid and solid phases.

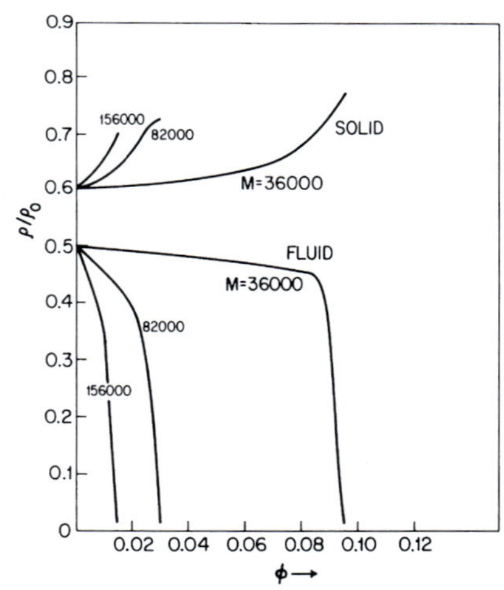

FIG. 7. Phase diagram relating the concentration of the particles in both phases to the free polymer concentration. The parameters of the curves are the molecular weights of the free polymer.

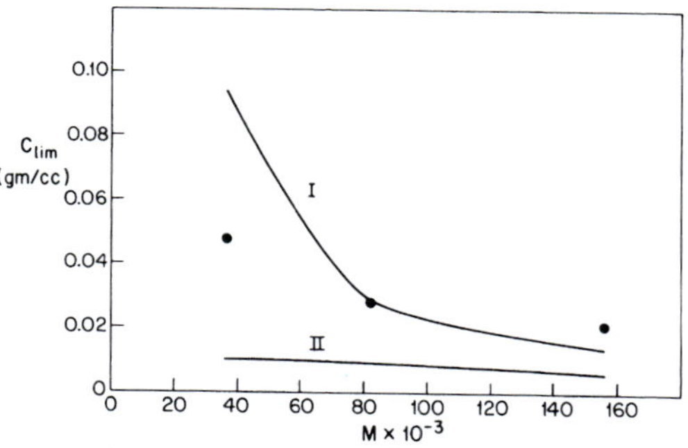

FIG. 8. Comparison of the experimental and the theoretical values of the limiting free polymer concentration C_{lim} for the onset of phase separation. (·) Experimental data (Ref. (1)). (I) Complete penetration of the free polymer, (II) the adsorbed layer is impenetrable.

order of 5 g per 100 cm^3. In Fig. 8, we have compared the calculated values of the limiting polymer concentration with the experimental data. For the same values of the concentration and the molecular weight of the free polymer, the depletion effect generates a stronger attraction when the polymer sheath is considered as an impenetrable barrier compared to the other extreme case in which it allows complete penetration by the free polymer. This is due to the fact that the area of intersection of the overlapping depletion zones, over which the imbalance of thermal forces exists, is larger and results in lower values of the limiting polymer concentration for the onset of phase separation in the former case. However, in actual situations, the adsorbed layer neither completely prevents nor completely allows the penetration by the free polymer. When the radius of the free polymer molecule and the thickness of the adsorbed layer are of the same order (in the present work this happens when the molecular weight of the free polymer is 36,000), the actual situation will be an intermediate between the two extreme cases. The experimental value is indeed between the two extreme cases as can be seen from Fig. 8. As the size of the free polymer increases in comparison to the thickness of the adsorbed layer, the adsorbed layer and the extent of penetration into it will become less important and in the limit of sufficiently large molecular weight of the free polymer, both extreme cases lead to the same result.

V. CONCLUSIONS

The presence of free polymer in sterically stabilized colloidal dispersions is shown to be responsible for the occurrence of phase separation/flocculation in dilute dispersions. The theoretical calculations show that the limiting polymer concentration for the onset of phase separation decreases with increasing molecular weight of the free polymer and with increasing particle concentration, as observed experimentally. The two limiting cases of no penetration and total penetration of the free polymer into the adsorbed layers represent extreme ends with the real situation being somewhat intermediary between the extreme cases.

REFERENCES

1. DeHek, H., and Vrij, A., *J. Colloid Interface Sci.* **84,** 409 (1981).
2. Pathmamanoharan, C., DeHek, H., and Vrij, A., *Colloid Polym. Sci.* **259,** 769 (1981).
3. Clarke, J., and Vincent, B., *J. Colloid Interface Sci.* **82,** 208 (1981).
4. Vincent, B., Luckham, P. F., and Waite, F. A., *J. Colloid Interface Sci.* **73,** 508 (1980).
5. Asakura, S., and Oosawa, F., *J. Chem. Phys.* **22,** 1255 (1954).
6. Feigin, R. I., and Napper, D. H., *J. Colloid Interface Sci.* **75,** 525 (1980).
7. Li-in-on, F. K. R., Vincent, B., and Waite, F. A., *J. Chem. Soc. Faraday Trans. 1* **74,** 337 (1978).
8. Fleer, G. J., Scheutjens, J. M. H. M., and Vincent, B., *in* "Polymer Adsorption and Dispersion Stability" (E. D. Goddard and B. Vincent, Eds.), ACS Symposium Series 240, p. 245. Amer. Chem. Soc., Washington, D. C., 1984.
9. Scheutjens, J. M. H. M., and Fleer, G. J., *J. Phys. Chem.* **83,** 1619 (1979); **84,** 178 (1980).
10. van Megen, W., and Snook, I., *J. Colloid Interface Sci.* **57,** 40 (1976).
11. Snook, I., and van Megen, W., *J. Chem. Soc. Faraday Trans. 2* **72,** 216 (1976).
12. Gast, A. P., Hall, C. K., and Russel, W. B., *J. Colloid Interface Sci.* **96,** 251 (1983).
13. Barker, J. A., and Henderson, D., *J. Chem. Phys.* **47,** 2856 (1967).

14. Napper, D. H., *J. Colloid Interface Sci.* **58,** 390 (1977).
15. Evans, R., and Napper, D. H., *J. Chem. Soc. Faraday Trans. 1* **73,** 1377 (1977).
16. Asakura, S., and Oosawa, F., *J. Polym. Sci.* **33,** 183 (1958).
17. Flory, P. J., "Principles of Polymer Chemistry." Cornell Univ. Press, Ithaca, N. Y., 1953.
18. Vincent, B., *J. Colloid Interface Sci.* **42,** 270 (1973).
19. Barker, J. A., and Henderson, D., *J. Chem. Phys.* **47,** 4714 (1967).
20. Victor, J. M., and Hansen, J. P., *J. Phys. Lett.* **45,** 307 (1984).
21. van Megen W., and Snook, I., *Faraday Discuss. Chem. Soc.* **65,** 92 (1978).
22. van Megen, W., and Snook, I., *J. Colloid Interface Sci.* **100,** 359 (1984).
23. Weeks, J. D., Chandler, D., and Anderson, H., *J. Chem. Phys.* **54,** 5237 (1971).
24. van Megen, W., and Snook, I., *Chem. Phys. Lett.* **35,** 399 (1975).
25. Hoover, W. G., and Ree, F. H., *J. Chem. Phys.* **49,** 3609 (1968).
26. Carnahan, N. F., and Starling, K. E., *J. Chem. Phys.* **51,** 635 (1969).
27. Hall, K. R., *J. Chem. Phys.* **57,** 2252 (1971).
28. McQuarrie, D. A., "Statistical Mechanics." Harper & Row, New York, 1976.
29. Smith, W. R., and Henderson, D., *Mol. Phys.* **19,** 411 (1970).
30. Verlet, L., and Weis, J. J., *Phys. Rev. A* **5,** 939 (1972).
31. Kincaid, J. M., and Weis, J. J., *Mol. Phys.* **34,** 931 (1977).
32. Flory, P. J., *Faraday Discuss. Chem. Soc.* **49,** 7 (1970).

Colloids and Surfaces, 17 (1986) 185—205 185
Elsevier Science Publishers B.V., Amsterdam — Printed in The Netherlands

EFFECT OF SOLVENT ON THE STABILITY OF MIXTURES OF STERICALLY STABILIZED DISPERSIONS AND FREE POLYMERS

E. RUCKENSTEIN and I.V. RAO

Department of Chemical Engineering, State University of New York at Buffalo, Buffalo, NY 14260 (U.S.A.)

(Received 11 March 1985; accepted in final form 5 August 1985)

ABSTRACT

The stability of mixtures of sterically stabilized colloidal particles and free polymer molecules in solution has been examined theoretically by assuming that the pair potential is the sum of contributions from (i) steric repulsion due to the adsorbed polymer layers, (ii) the attraction due to the exclusion of the free polymer molecules from the inter-particle space, and (iii) van der Waals attraction forces. Using the hard-sphere perturbation theory, the phase boundaries of the transition from a dilute, disordered (fluid) state to a concentrated, ordered (solid) state have been located. The systems chosen correspond to the situation in which the adsorbed polymer and the free polyner are chemically different and the molecular weight of the free polymer is much greater than that of the adsorbed polymer. The effect of varying the nature of the solvent from being good for both the adsorbed and free polymers to being good only for the adsorbed polymers is examined. If stable dispersions are desirable, it is advisable to have a solvent which is good for the adsorbed polymer but poor for the free polymer. The effect of other variables such as the molecular weight of the free polymer, the temperature, the thickness of the adsorbed layer and the particle size on the stability of the dispersion has also been examined. The limiting free polymer concentration, above which the disorder—order transition occurs, is found to decrease with increasing molecular weight of the free polymer, with increasing particle size, with increasing temperature, with thinner steric layers and with better solvents for the free polymer, in accordance with the trends observed experimentally. The case of pure steric stabilization is examined by setting the free polymer concentration equal to zero, and one concludes that destabilization is possible under better-than-theta conditions. It is also shown that for pure steric stabilization, the critical particle concentration above which the disorder—order transition occurs decreases with increasing particle size and increases with increasing thickness of the adsorbed layer, again as observed experimentally.

INTRODUCTION

Colloidal dispersions, in general, are rendered stable either by electrostatic stabilization or by steric stabilization. In the former case, the repulsive electrical double layer forces between two particles counteract the attractive van der Waals forces and generate a potential barrier between the primary and secondary minima. If the potential barrier is sufficiently higher than the

0166-6622/86/$03.50

average thermal energy of the particles, only a very small fraction of the particles can overcome the barrier and reach the primary minimum, leading to coagulation. Systems exhibiting high potential energy barriers can be considered as kinetically stable (thermodynamically metastable) and the DLVO theory, incorporating the electrical double layer forces and the van der Waals forces, has been found to be satisfactory for many charge-stabilized systems. Steric stabilization requires the presence of adsorbed polymer layers on the particle surface. When two such particles approach one another, such that the adsorbed layers begin to interpenetrate, a repulsive force is generated when the dispersion medium is a good solvent for the adsorbed polymer. However, when two particles come still closer, such that the surface-to-surface distance between them becomes less than or equal to the thickness of the adsorbed layer, the interparticle forces are always repulsive, irrespective of whether the solvent is good or poor for the adsorbed polymer. Various theoretical models that have been proposed to evaluate the repulsive forces due to the adsorbed polymer do not indicate a potential barrier in the interaction potential, suggesting that these systems might be considered as thermodynamically stable.

It is equally important to be aware of the conditions under which instabilities begin to appear in colloidal dispersions. While the polymer as an adsorbed species has a stabilizing effect on a dispersion, the free polymer in solution has been shown to have the opposite effect. The role of the free polymer in the continuous phase of a dispersion of bare colloidal particles was examined long ago by Asakura and Oosawa [1]. They have shown that, when two particles approach to a distance of separation less than twice the radius of gyration of the free polymer chain in solution (the diameter of the polymer), the interparticle space is occupied by pure solvent. As a result of the exclusion of the polymer molecules from the interparticle space, an attractive force, which is proportional to the osmotic pressure of the polymer solution, is generated between the particles. Experimental evidence for this attraction has been provided by the early work of Sieglaff [2] and more recently by Sperry et al. [3]. These experiments demonstrate that, when the concentration of the free polymer exceeds a certain limiting value, destabilization (either as flocculation into small aggregates or as phase separation) is observed. This limiting concentration is found to decrease with increasing molecular weight of the polymer.

The question of whether the free polymer molecules in solution can give rise to instabilities even when the particles are sterically stabilized, depends on the interplay between the steric repulsive forces and the attractive forces due to the presence of the free polymer and the van der Waals forces. Recent experimental data [4—6] on sterically stabilized systems having free polymer in solution clearly indicate that the limiting polymer concentration, above which instability sets in, decreases (i) with increasing molecular weight of the free polymer, (ii) with increasing particle size, (iii) with decreasing molecular weight of the adsorbed polymer, and (iv) with better solvents for the free polymer.

Various theoretical models have been presented to predict the onset of depletion flocculation caused by the free polymer in solution. Vrij [7] re-derived in an illuminating way the interparticle potential of Asakura and Oosawa. This potential was utilized by de Hek and Vrij [4] to calculate the second virial coefficient for bare particles. Instability, leading to phase separation, was shown to be the result when this virial coefficient becomes sufficiently negative. However, the assumption of hard-sphere behavior does not hold when the particles are 'soft', i.e. covered by an adsorbed polymer layer. An alternate theory proposed by Vincent et al. [6] accounts for the interpenetration of the free polymer chains in solution with the adsorbed polymer layer around the particles as well as for the steric repulsion. When the free polymer concentration exceeds the critical value corresponding to the beginning of the semi-dilute region, exclusion of the free polymer molecules from the interparticle space into the bulk occurs. This happens because of more favourable interactions for the polymer in the bulk and this leads to attraction between the particles. However, in reality the attraction occurs at concentrations even lower than the above critical value because of the exclusion of free polymer molecules from the interparticle space when the surface-to-surface separation between two particles becomes less than the diameter of the free polymer molecule. Feigin and Napper [8] have used the rotational isomeric scheme and Monte Carlo technique to calculate the segment density profile of the free polymer between two flat, uncoated plates and have shown that the free energy of interaction between the plates exhibits a minimum at very small separations and a maximum when the surface-to-surface separation is of the order of the diameter of the polymer chain in solution. Recently, Fleer et al. [9] considered two parallel surfaces separated by polymer solution and calculated the interaction free energy as a function of the separation distance. The depletion layer thickness around each particle, which remains almost constant for sufficiently low concentrations of the free polymer (below the onset of semi-dilute regime), is shown to decrease with increasing polymer concentration in the semi-dilute regime. As a result, the depletion attraction passes through a maximum. On this basis and using additional thermodynamic considerations for the entire dispersed system, they have shown that restabilization at higher free polymer concentrations is not due to the presence of the barriers in the interaction potential, as previously suggested [8], but is a thermodynamic phenomenon.

Using an interparticle potential, the characterization of the equilibrium state is possible by thermodynamic analysis. Van Megen and Snook [10,11] have adopted the statistical approach to predict the disorder—order phase transitions in concentrated dispersions that are stabilized electrostatically. Using the perturbation theory for the disordered phase and the cell model for the ordered phase, they have estimated the particle concentrations in the two coexisting phases when an electrostatically stabilized dispersion undergoes phase separation. Recently, Gast et al. [12] have used a similar approach to construct phase diagrams for colloidal dispersions that have free polymer molecules in solution. Using the interaction potential of Asakura and Oosawa as the perturbation term and the hard-sphere model as the reference system, they conclude that the limiting polymer concentration for the onset of phase separation decreases with increasing molecular weight of the free polymer. Their calculations provide the particle concentrations and the relative amounts of the two coexisting phases.

In our earlier work [13], the stability of mixtures of sterically stabilized colloidal particles and free polymer molecules in solution was examined theoretically by considering the total interparticle potential as the sum of contributions due to (i) the presence of the adsorbed polymer, (ii) the presence of free polymer in solution, and (iii) van der Waals attraction.

In sterically stabilized dispersions that have free polymer molecules in the continuous phase, the chemical nature of the adsorbed and free polymer may be the same or different. The quality of the solvent is expected to play a major role in the stability of such dispersions, since the sign and magnitude of the interaction forces between the particles are determined by the values of the Flory interaction parameters for the polymer—solvent pair. If the adsorbed and free polymer are chemically similar, the solvent may be uniformly good or poor for both of them. A suitable solvent in such cases, while increasing the steric repulsive forces, also increases the attractive forces due to the free polymer. However, it is desirable, from a stability point of view, to maximize the repulsive forces and reduce the magnitude of attractive forces. Such flexibility is possible in systems where the adsorbed and free polymer are chemically different. A solvent may, then, be chosen such that it is good for the adsorbed polymer but poor for the free polymer. Hence, in the present work, the role of the solvent in dispersions with chemically different polymers is examined. Since the quality of the solvent is dependent on temperature, the effect of temperature on the stability of such dispersions is also dwelt upon. In addition, the thickness of the adsorbed layer and the particle size are expected to be important variables affecting the behavior of the dispersion. The analysis is also applied to the case of pure steric stabilization by setting the value of the free polymer concentration equal to zero.

A brief description of the model used for the pair potential as well as of the perturbation theory used in the calculations is given in the next two sections, followed by the results of the calculations and a discussion of the effect of various variables on stability.

MODEL FOR THE PAIR POTENTIAL

The phase behavior of mixtures of sterically stabilized dispersions and free polymer molecules is determined by the interparticle potential. In such systems, the interplay between the repulsive steric forces and the attractive forces due to the presence of free polymer molecules and van der Waals interaction between the particles determines the onset of instability.

Assuming that the total potential is the sum of the individual contributions, the interaction potential, u, between two identical sterically stabilized particles can be written as:

$$u = V_s + V_d + V_a \tag{1}$$

where V_s is the free energy of interaction between two particles having attached polymer, V_d is the depletion contribution due to the exclusion of the free polymer from the interparticle space, and V_a is the van der Waals attraction between two particles. The free energy of interaction V_s between two particles with adsorbed layers is expressed separately for two different domains of particle approach distances. In one of them, the surface-to-surface separation distance between two particles, Δ, lies between δ and 2δ where δ is the thickness of the adsorbed layer. For this domain, known as the interpenetration domain, the free energy of mixing V_s is given by [14]:

$$\frac{V_s}{kT} = \frac{4\pi a \delta^2}{v_1}\left[v^2\left(\tfrac{1}{2} - \chi_1\right) + \frac{3v^3}{2}\left(\tfrac{1}{3} - \chi_2\right)\right]\left(1 - \frac{\Delta}{2\delta}\right)^2 \tag{2}$$

where a is the radius of the particle, v_1 is the molecular volume of the solvent, v is the volume fraction of the polymer in the adsorbed layer, k is the Boltzmann constant, and T is the absolute temperature. χ_1 and χ_2 are the constants appearing in the expression for the concentration dependence of the Flory interaction parameter χ ($\chi = \chi_1 + \chi_2 v$). While the interaction energy between two sterically stabilized particles is generally assumed to be the sum of osmotic and volume restriction effects, the above expression is only due to the osmotic effect. However, the contribution of the volume restriction effect (elastic term), in the interpenetration domain, is shown to be negligible [15], since the volume fraction of the polymer molecules is quite small. The second domain, known as the interpenetration plus compression domain, is characterized by $\Delta < \delta$. For these smaller distances of separation, the free energy of interaction is due to both osmotic and volume restriction effects and is given by [14]:

$$\frac{V_s}{kT} = \frac{V^{M}}{kT} + \frac{V^{El}}{kT} \tag{3a}$$

where

$$\frac{V^{M}}{kT} = \frac{4\pi a \delta^2}{v_1}\left[v^2\left(\tfrac{1}{2} - \chi_1\right)\left(\frac{\Delta}{2\delta} - \tfrac{1}{4} - \ln\frac{\Delta}{\delta}\right) + \frac{v^3}{2}\left(\tfrac{1}{3} - \chi_2\right)\left(1 + \frac{\Delta}{2\delta} + \frac{2\delta}{\Delta} - \tfrac{11}{4}\right)\right] \tag{3b}$$

and

$$\frac{V^{El}}{kT} = \frac{2\pi a \delta^2 v}{v_p}\left[\frac{\Delta}{\delta}\ln\left\{\frac{\Delta}{\delta}\left(\frac{3 - \Delta/\delta}{2}\right)^2\right\} - 6\ln\left(\frac{3 - \Delta/\delta}{2}\right) + 3(1 - \Delta/\delta)\right. \tag{3c}$$

where v_p is the volume of the polymer molecule and the other symbols have already been defined after Eqn (2).

The attractive potential energy between two particles due to the exclusion of the free polymer molecules from the interparticle space has been shown to be proportional to the product of the osmotic pressure of the polymer solution and the volume of the overlap of depletion layers around the particle [1]. In an earlier paper [13], two limiting cases were considered in the calculation of the attractive forces due to the free polymer molecules. In one limiting case, the adsorbed layer was considered impenetrable to the free polymer. As a result of this assumption, two colloidal particles experience an attractive force, due to the exclusion of the free polymer, when their surface-to-surface separation becomes less than or equal to $2\delta + 2l$, where δ is the thickness of the adsorbed layer and l is the radius of gyration of the free polymer chain in solution. In the second limiting case, the free polymer is allowed to penetrate the adsorbed layer completely and the attractive forces due to the free polymer begin to operate when the surface-to-surface distance between two particles becomes less than the diameter of the free polymer molecule. The potential energy of attraction is smaller in the second limiting case, because of the decrease in the volume of the interparticle space from which the free polymer molecules are excluded. The distinction between the two limiting cases is important only when the thickness of the adsorbed layer and the size of the free polymer are of the same order of magnitude but it is not important when the molecular weight of the free polymer is much larger than that of the adsorbed polymer. In the present work, the thickness of the adsorbed layer is small compared with the diameter of the free polymer molecules and, therefore, the effect of the thickness of the adsorbed layer on the depletion contribution can be neglected. In this case the expression for V_d can be written as:

$$\frac{V_d}{kT} = -\frac{2}{3}\pi \frac{P_{osm}}{kT} a^3 \left[2\left(1+\frac{l}{a}\right)^3 + \left(1+\frac{\Delta}{2a}\right)^3 - 3\left(1+\frac{\Delta}{2a}\right)\left(1+\frac{l}{a}\right)^2 \right] \tag{4}$$

where $2l$ is the diameter of the free polymer and P_{osm} is the osmotic pressure of the polymer solution. Equation (4) shows that V_d has the largest negative value when $\Delta = 0$ and approaches zero as Δ approaches the value $2l$. For infinitely dilute solutions, the osmotic pressure can be approximated by the limiting van 't Hoff expression, but a more general expression for the osmotic pressure which takes into account the nature of the solvent is given by [16]:

$$\frac{P_{osm}}{kT} = -\frac{1}{v_1}\left[\phi\left(1-\frac{1}{x}\right) + \ln(1-\phi) + \chi\phi^2\right] \tag{5}$$

where ϕ is the volume fraction of the free polymer, χ is the polymer—solvent interaction parameter, and x is the ratio of the molar volumes of the polymer and solvent.

The expression for the van der Waals attractive potential between two identical colloidal particles with adsorbed layers can be written in terms of a composite Hamaker constant A [17] as

$$\frac{V_a}{kT} = -\frac{A}{12kT}\left[\frac{1}{z^2+2z} + \frac{1}{z^2+2z+1} + 2\ln\frac{z^2+2z}{z^2+2z+1}\right] \quad (6a)$$

where z is the ratio of Δ to $2a$. The composite Hamaker constant A in Eqn (6a) depends on the Hamaker constants of the particle material, the

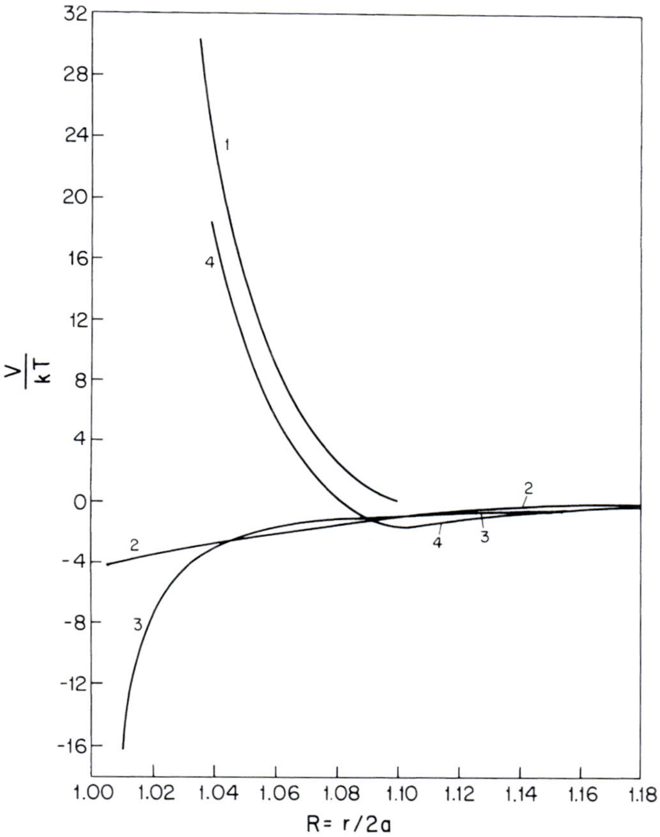

Fig. 1. Interaction potential between two colloidal particles as a function of the reduced centre-to-centre separation $R = r/2a$, where a is the radius of the particles. Curve 1, steric repulsion due to the adsorbed layer (V_s); curve 2, attraction due to the free polymer (V_d); curve 3, van der Waals attraction (V_a); curve 4, sum of the contributions given by curves 1—3. System: polyisobutene-stabilized silica particles and polystyrene (free polymer) in cyclohexane at 308 K. Molecular weight of the free polymer = 82,000, volume fraction of polystyrene, $\phi = 0.02$, $a = 48$ nm, thickness of the adsorbed layer δ = 5 nm, $\chi = 0.5$ for polystyrene—cyclohexane, $\chi_1 = 0.47$ and $\chi_2 = 0.10$ for polyisobutene—cyclohexane, $A/kT = 4.54$ and $v = 0.10$.

solvent and the adsorbed polymer through the following relationships:

$$A = (A_1^{1/2} - A_2^{1/2})^2 \tag{6b}$$

where A_1 is the Hamaker constant of the particle material and A_2 can be expressed in terms of the composition of the adsorbed layer as

$$A_2 = (vA_p^{1/2} + (1 - v)A_s^{1/2})^2 \tag{6c}$$

where A_p and A_s are the Hamaker constants of the polymer and solvent, respectively.

A typical interparticle potential energy diagram showing the individual contributions mentioned above is depicted in Fig. 1.

THE PERTURBATION THEORY

It is now well known that the structure of a liquid is primarily determined by the hard-core repulsive forces, while the attractive forces merely act to hold the molecules together. This led to the use of the hard-sphere perturbation theory to predict the thermodynamic properties of liquids and to locate the fluid—solid phase transitions in molecular systems. Recently, this theory has also been used [10,12,13,18] to predict the equilibrium properties of systems of colloidal particles. In the perturbation theory, the pair potential is divided into reference and perturbation terms. The thermodynamic properties can then be expressed as the sum of contributions from the reference and the perturbation potentials. A convenient choice for the reference system is that of the hard spheres, the properties of which are accurately known. The two most commonly used procedures for breaking the pair potential into reference and perturbation parts have been proposed by Barker and Henderson (BH) [19] and by Chandler, Weeks and Anderson (CWA) [20]. According to the BH scheme,

$$u = u_0 + u_1 \tag{7}$$

where the reference potential u_0 is defined by:

$$\begin{aligned} u_0(r) &= u(r) \quad r < \sigma \\ &= 0 \quad\quad\; r \geqslant \sigma \end{aligned} \tag{8}$$

where σ is chosen such that $u(\sigma) = 0$ and r is the centre-to-centre separation distance between two particles. The perturbation term u_1 is defined by

$$\begin{aligned} u_1(r) &= u(r) \quad r \geqslant \sigma \\ &= 0 \quad\quad\; r < \sigma \end{aligned} \tag{9}$$

For the potentials for which the repulsive part is not infinitely steep, the reference potential can be reduced to a system of hard spheres by introducing an effective diameter, the Barker—Henderson diameter d_B. This effective diameter is obtained from the repulsive part of the actual potential using the relation:

$$d_B = \int_0^\sigma [1 - \exp(-u_0/kT)]\, dr \tag{10}$$

With this choice for d_B, the reference system potential can be written as:

$$\begin{aligned} u_0(r) &= \infty \qquad r < d_B \\ &= 0 \qquad r \geqslant d_B \end{aligned} \tag{11}$$

Assuming pair-wise additivity of the potential, the Helmholtz free energy F of a system of N particles of diameter d in a volume V can be expressed, to second order, as [21]:

$$\frac{F}{NkT} = \frac{F_{HS}}{NkT} + 12\eta \int_{\sigma/d}^{\infty} \frac{u_1(R)}{kT}\, g_{HS}(R, \eta')R^2\, dR$$

$$- 6\eta kT\left(\frac{\partial \rho}{\partial p}\right)_{HS} \int_{\sigma/d}^{\infty} \left[\frac{u_1(R)}{kT}\right]^2 g_{HS}(R, \eta')R^2\, dR \tag{12}$$

where F_{HS}, g_{HS}, $\rho = N/V$ and $(\partial\rho/\partial p)_{HS}$ are the free energy, the radial distribution function, the number density of the particles and the macroscopic compressibility of a system of hard spheres (HS) of diameter d_B, respectively. R is the reduced centre-to-centre distance of separation between two particles, p is the pressure of the hard-sphere system or the osmotic pressure of the colloidal dispersion, η' is the volume fraction of the particles expressed in terms of d_B ($\eta' = (\pi/6)\rho d_B^3$) and η is the actual volume fraction of the particles of diameter d ($\eta = (\pi/6)\rho d^3$).

Once the free energy is calculated, other thermodynamic properties can easily be determined using the basic thermodynamic relations. For example, pressure p can be obtained by differentiating the Helmholtz free energy F:

$$Z = \frac{p}{\rho kT} = \left(\frac{p}{\rho kT}\right)_{HS} + \eta\, \frac{\partial}{\partial \eta}\left(\frac{F - F_{HS}}{NkT}\right) \tag{13}$$

where the subscript HS refers to the hard-sphere reference system.

In order to use the above expressions for calculating the thermodynamic properties, appropriate expressions for the radial distribution function and for the equation of state for the hard-sphere reference system are required which are given in Appendix A. Fortunately, accurate information for the hard-sphere fluid as well as for the hard-sphere solid is available and this enables the determination of the properties of the coexisting dilute and concentrated phases of colloidal dispersions.

The procedure to locate the phase transition boundaries from a dilute disordered (fluid) phase to a concentrated ordered (solid) phase involves the determination of the chemical potential of the particles in both phases. The condition of the equality of the pressures and of the chemical potentials of the phases locates the phase transition boundaries. The chemical potential of the particles in each phase is obtained using the following expression, separately for each phase:

$$\frac{G - G^0}{NkT} = \frac{F - F^0}{NkT} + (Z - 1) \tag{14}$$

where G is the Gibbs free energy and G^0 and F^0 are the values of G and F for an ideal gas of the same density. In order to equate the pressures, it is convenient to define the reduced pressure $p/\rho_0 kT$ for each phase, which is related to Z by

$$\frac{p}{\rho_0 kT} = \frac{Z\eta}{0.74} \tag{15}$$

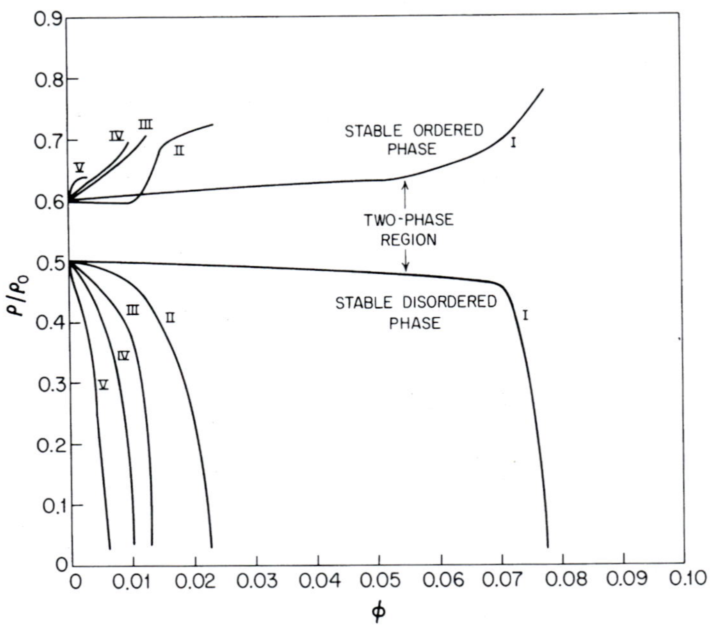

Fig. 2. Phase diagram at 340 K, showing the particle concentration ρ/ρ_0 in the coexisting ordered and disordered phases as a function of the volume fraction of the free polymer ϕ for different molecular weights of the free polymer. (I), 36,000; (II), 82,000; (III), 122,000; (IV), 176,000; (V), 490,000. System: polyisobutene-stabilized silica particles with polystyrene as the free polymer in cyclohexane, a = 48 nm, δ = 5 nm. The lower curves represent the transition boundary from a stable disordered phase to the two-phase region, and the upper curves the two-phase region to the stable ordered phase boundary. χ_1 = 0.47 and χ_2 = 0.10 for polyisobutene—cyclohexane, A/kT = 4.54 and v = 0.10. The value of χ for polystyrene—cyclohexane is calculated according to Eqn (16).

where ρ_0 is the density corresponding to the closest packing of the particles (face-centred-cubic lattice arrangement). The intersection of the plots of the chemical potential versus reduced pressure for each of the phases determines the coexistence pressure $p/\rho_0 kT$. The densities of the phases in equilibrium can then be obtained from the density isotherms, plotted as $p/\rho_0 kT$ versus ρ/ρ_0, for each of the phases. Now the concentration of the free polymer can be varied and the corresponding pair of coexisting phase densities can be found by repeating the above procedure. Thus, a phase diagram, relating the equilibrium particle concentrations of the coexisting phases as a function of the free polymer concentration, for a fixed molecular weight of the free polymer, can be constructed. Finally, such phase diagrams can be mapped for different molecular weights of the free polymer, for different particle sizes, and for different temperatures and solvents. A typical phase diagram is presented in Fig. 2.

RESULTS AND DISCUSSION

Comparison with experiment

In order to test the model used here, calculated values of the limiting free polymer concentration ϕ^* at which phase separation occurs are compared with the experimental data [6] on the aqueous dispersions of polystyrene latex particles with adsorbed polyethylene oxide and with polyethylene oxide as the free polymer. Since no information is available regarding the thickness of the adsorbed layer, the values used by Vincent et al. [6] in their theoretical calculations are adopted. Table 1 compares the experimental values of the limiting volume fraction of the free polymer with our calculated values for two different molecular weights of the free polymer. The simple model used here gives reasonably good agreement with the experimental values.

Effect of temperature

In our earlier work [13], we have calculated the limiting concentration of polystyrene in cyclohexane for silica particles, that have strongly anchored polyisobutene molecules, as a function of the molecular weight of polystyrene, where a temperature of 308 K was chosen (χ = 0.5), which is the theta temperature for polystyrene—cyclohexane. In this work, the temperature is varied from 290 to 390 K for the same system. The parameters used for the calculation of the pair potential are: a = 48 nm, v_1 = 0.18 nm^3, δ = 5 nm, v = 0.10 and v_p = 2.36 nm^3. The following values are taken from Ref. [17]: A (silica) = 35.76kT, A (polyisobutene) = 18.98kT, A (cyclohexane) = 14.35kT, hence A = 4.54kT, and these are considered temperature-independent. However, the polymer—solvent interaction parameter, χ, depends on temperature. The system polyisobutene—cyclohexane is known to be athermal [the reduced enthalpy of dilution χ_H defined as

196

TABLE 1

Comparison of calculated values (present work) of the limiting volume fraction of free polymer ϕ^*, at which instability sets in, with experimental values. System: Polyethylene oxide-stabilized polystyrene latex with polyethylene oxide as the free polymer at 298 K, with a = 85 nm, χ_1 = 0.466, χ_2 = 0.31 and δ = 5 nm

Mol. wt of the free polymer, M	ϕ^* exp.	ϕ^* calc.
20,000	0.04[a]	0.07
300,000	0.01[a]	0.008

[a] Experimental values from Ref. [6].

$-T\partial\chi/\partial T$ is found to be zero (Ref. [22])] and the following values of χ_1 and χ_2 reported for 298 K, can therefore be used at other temperatures: χ_1 = 0.470 and χ_2 = 0.100 [22]. The temperature dependence of χ for the system polystyrene—cyclohexane has been shown to be accurately represented by the following empirical relation [23]:

$$\chi = \alpha + \left(\beta_{00} + \frac{\beta_{01}}{T}\right)(1-\gamma)/(1-\gamma\phi)^2 \tag{16}$$

where ϕ is the volume fraction of polystyrene in cyclohexane. The values of the constants are α = −0.1597, β_{00} = 0.4987, β_{01} = 111.74 and γ = 0.2365. According to Eqn (16), the interaction parameter χ decreases with increasing temperature. For a fixed concentration of the free polymer, a reduced value of χ results in a higher value of the osmotic pressure of the polymer solution. The expression for P_{osm} given by Eqn (5) can be written for dilute solutions, after expanding the logarithmic term, as:

$$\frac{P_{osm} v_1}{kT} = \frac{\phi}{x} + \left(\tfrac{1}{2} - \chi\right)\phi^2 \tag{17}$$

At χ = 0.5, P_{osm} varies linearly with ϕ. But, if $\chi < 0.5$, the farther χ is away from 0.5, the higher the value of P_{osm}. For values of $\chi > 0.5$, the magnitude of the osmotic pressure decreases. Since the attractive energy due to the presence of the free polyner is proportional to the osmotic pressure of the polymer solution, higher temperatures are expected to result in increased attraction between particles. This should lead to increased values of the minimum in the total pair potential, thereby reducing the amount of polymer required to destabilize the dispersions (separation into dilute and concentrated coexisting phases). The calculated values of the limiting polymer concentration ϕ^* are plotted in Fig. 3 as a function of temperature and it can be seen that they fall off with increasing temperature. The same

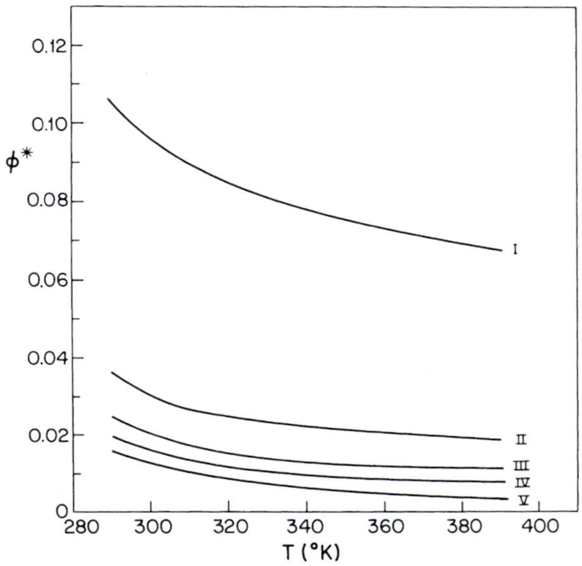

Fig. 3. Temperature dependence of the limiting volume fraction, ϕ^*, required for the onset of phase separation for different molecular weights of the free polymer. (I), 36,000; (II), 82,000; (III), 122,000; (IV), 176,000; (V), 490,000. Initial concentration of particles corresponds to ρ/ρ_0 = 0.05. System: polyisobutene-stabilized silica particles and polystyrene in cyclohexane. The values of the parameters are as in Fig. 2.

figure also indicates that for any given temperature in the range considered here, ϕ^* decreases with an increase in the molecular weight of the free polymer.

Direct experimental data providing the temperature dependence of ϕ^* are not available in the literature. However, as discussed earlier, the dependence of ϕ^* on the quality of the solvent (change in the values of the polymer—solvent interaction parameter) is expected to suggest the trend with temperature also. The experimental determination of ϕ^* values, for silica particles having polystyrene as the free polymer, indicated [5] that the amount of polymer required to produce phase separation decreased by a factor of three when the theta solvent cyclohexane (χ = 0.5) is replaced by the good solvent toluene (χ < 0.5). This implies that increased temperatures (reduced values for χ) should lead to lower values of the amounts of polymer required for phase separation. It can be safely concluded that the available experimental and theoretical information thus far, exhibits the trend of smaller values of the limiting polymer concentration at higher temperatures.

Effect of the solvent

In order to study the effect of changing the nature of the solvent, we have replaced cyclohexane with ethylbenzene which is a better solvent for polystyrene. With ethylbenzene as the solvent, the following values are used in the calculation: $\chi_1 = 0.477$ [16], $\chi_2 = (2/3)\chi_1 = 0.32$ [14] for the system polyisobutene—ethylbenzene and $\chi_1 = 0.466$ [24], $\chi_2 = (2/3)\chi_1 = 0.31$ for polystyrene—ethylbenzene and $v_1 = 0.203\ nm^3$. Replacing the theta solvent with a better solvent for the free polymer (polystyrene) is expected to increase the attractive energy contribution. Table 2 lists the limiting polymer volume fraction, ϕ^*, with cyclohexane and ethylbenzene as solvents. Our calculations indicate that, with ethylbenzene as the solvent, smaller amounts of polymer are required to destabilize the system.

TABLE 2

The limiting volume fraction of free polymer ϕ^* above which phase separation occurs for two different solvents. *M* is the molecular weight of the free polymer. System: polyisobutene-stabilized silica particles with polystyrene as the free polymer. $a = 48$ nm, $\delta = 5$ nm. Initial concentration of particles, $\rho/\rho_0 = 0.05$; temperature 308 K

M	ϕ^*	
	Cyclohexane	Ethylbenzene
36,000	0.094	0.053
82,000	0.030	0.024
176,000	0.012	0.011
490,000	0.007	0.006

Effect of the particle size

The effect of changing the particle diameter is reflected in all three contributions to the interparticle pair potential. An examination of the expressions used to evaluate the individual contributions indicates that each of them is directly proportional to the particle size. Hence, the depth of the potential well also increases with increasing particle size. It is natural, therefore, to expect that the larger particles make the dispersions less stable, i.e. require smaller amounts of free polymer for destabilization. Calculations in this work have been carried out for particle diameters ranging from 48 to 192 nm. Figure 4 exhibits the expected trend of lower values of the limiting free polymer concentration, ϕ^*, with increasing particle diameter. The experimental data on non-aqueous silica dispersions [4], with polystyrene as the free polymer, show indeed that larger particles are less stable than smaller particles.

Effect of the thickness of the adsorbed layer

The repulsive forces due to the presence of adsorbed polymer layers have a range of 2δ, where δ is the thickness of the adsorbed layer. The thickness of the adsorbed layer can be increased by increasing the molecular weight of the adsorbed polymer. Since the range and magnitude of the steric repulsion increase with δ, thicker layers result in the shifting of the minimum in the potential well to larger separations. In addition, higher values of δ also have the effect of reducing the depth of the potential well. This leads to larger

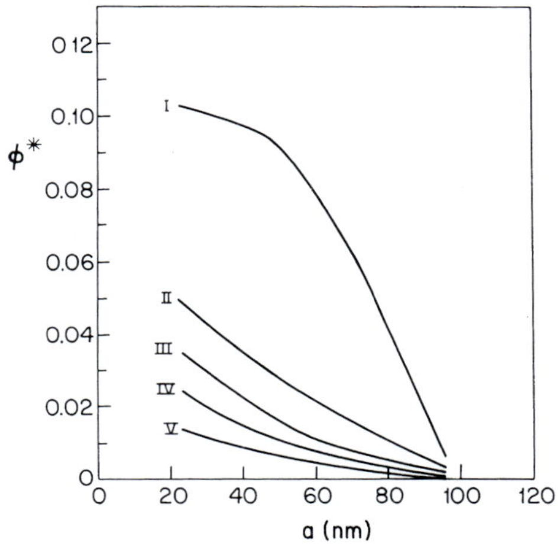

Fig. 4. The limiting volume fraction of the free polymer, ϕ^*, as a function of the particle radius a for different molecular weights of the free polymer. (I), 36,000; (II), 82,000; (III), 122,000; (IV), 176,000; (V), 490,000. Initial concentration of the particles ρ/ρ_0 = 0.05. System: polyisobutene-stabilized silica particles with polystyrene as the free polymer in cyclohexane at 308 K. The values of the other parameters are as in Fig. 1.

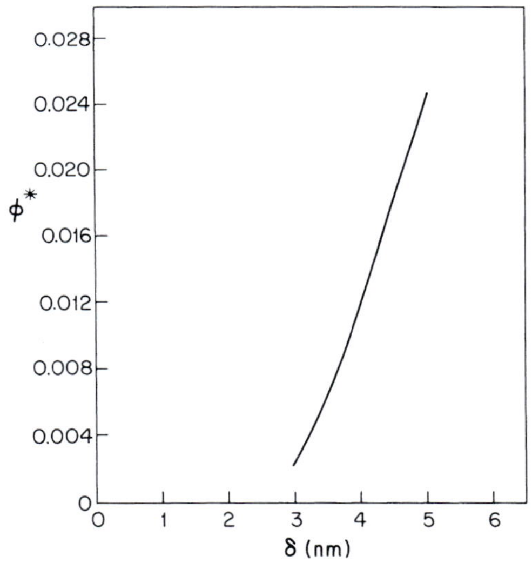

Fig. 5. The dependence of the limiting volume fraction of the free polymer, ϕ^*, on the thickness of the adsorbed layer, δ. System: polyisobutene-stabilized silica particles with polystyrene as the free polymer. Molecular weight of polystyrene = 82,000 and T = 308 K. Initial concentration of particles ρ/ρ_0 = 0.30. The values of the other parameters are as in Fig. 1.

values of the limiting concentrations of the free polymer. The experimental observation [4] that silica particles stabilized by polyisobutene are more stable than the particles coated with stearyl alcohol can be considered as a confirmation of the above theoretical arguments. Here, the thickness of the adsorbed layer is varied from 3 to 5 nm, and Fig. 5 gives the limiting polymer concentrations ϕ^* for polyisobutene-stabilized silica particles. It may be seen that, as δ increases, the system becomes more stable.

Effect of the amount of polymer adsorbed

Keeping the thickness of the adsorbed layer constant, the effect of varying the amount of polymer adsorbed was also examined. The volume fraction of the polymer in the adsorbed layer is varied from 0.10 to 0.30. The increase of this volume fraction again provides increased stability since it increases the magnitude of the repulsive steric forces. The calculated values shown in Table 3 exhibit the expected trend, although the variation is not as strong as that found with the thickness of the adsorbed layer.

TABLE 3

Dependence of ϕ^* on the volume fraction of polymer adsorbed on the particle surface, v. System: Polyisobutene-stabilized silica particles with polystyrene as free polymer in cyclohexane. M = 82,000, a = 48 nm, δ = 5 nm, temperature = 308 K, initial concentration of particles ρ/ρ_0 = 0.10, χ for polystyrene—cyclohexane = 0.5, χ_1 = 0.47, χ_2 = 0.10 for polyisobutene—cyclohexane and $A = 4.54kT$

Volume fraction of polymer adsorbed, v	Limiting free polymer volume fraction, ϕ^*
0.100	0.029
0.200	0.032
0.300	0.034

Steric stabilization

When the concentration of the free polymer is set equal to zero, the situation corresponds to pure steric stabilization. The free energy of interaction due to the interpenetration of the adsorbed polymer chains has a range of 2δ, where δ is the thickness of the adsorbed layer. This free energy is proportional to the quantity $(0.5 - \chi)$, where χ is the Flory interaction parameter for the polymer—solvent system. Thus, a repulsive potential is expected between two particles when $\chi < 0.5$; and this repulsion is absent when $\chi = 0.5$. For this reason, it was suggested [25] that instabilities in sterically stabilized dispersions occur for $\chi \geqslant 0.5$, hence for theta or worse-than-theta conditions. However, the correlation with the theta point only holds when the molecular weight of the added polymer is sufficiently high (resulting in higher values for δ) so that the effect of the van der Waals forces is negligible. In fact, dispersions sterically stabilized with low-molecular-weight polymer are known [26] to flocculate under better-than-theta conditions ($\chi < 0.5$) and this is due to the van der Waals forces between particles becoming important. The systems studied here correspond to better-than-theta solvents for the adsorbed polymer (polyisobutene) and are unstable above a critical particle concentration.

The dependence of the critical particle concentration of the disordered phase on the radius of the particles is shown in Fig. 6. When the radius is increased from 48 to 96 nm, the critical particle concentration decreases. This is due to the increase in the depth of the potential well with increasing particle diameter. One may note from Fig. 6 that in the range 24—48 nm for

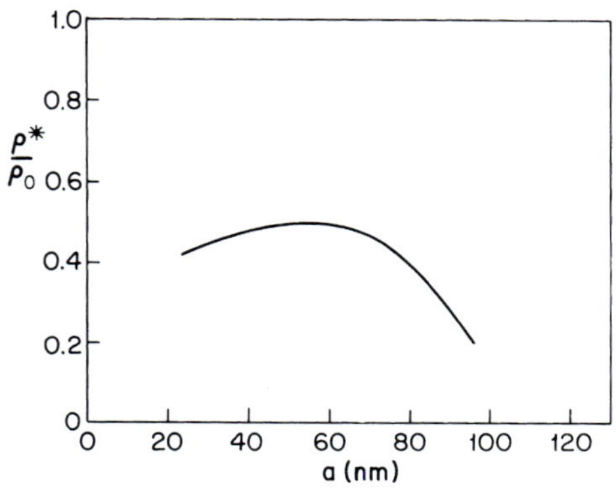

Fig. 6. Critical particle concentration, ρ^*/ρ_0, above which a dilute disordered phase and a concentrated ordered phase coexist in equilibrium for a sterically stabilized dispersion, as a function of the particle radius, *a*. System: polyisobutene-stabilized silica particles in cyclohexane, δ = 5 nm, T = 308 K, χ_1 = 0.47, χ_2 = 0.10, A = 4.54kT and v = 0.10.

TABLE 4

Dependence of the critical particle concentration ρ^*/ρ_0, above which disorder—order transition occurs for sterically stabilized dispersions, on the thickness of the adsorbed layer and on the nature of the solvent. System: polyisobutene-stabilized silica particles in cyclohexane at 308 K and in ethylbenzene at 298 K. a = 48 nm, A = 4.54kT, v = 0.10, χ_1 = 0.47, χ_2 = 0.10 for polyisobutene—cyclohexane, and χ_1 = 0.477, χ_2 = 0.32 for polyisobutene—ethylbenzene

Thickness of adsorbed layer, δ (nm)	ρ^*/ρ_0	Solvent[a]	ρ^*/ρ_0
3	0.40	Cyclohexane	0.500
4	0.50	Ethylbenzene	0.515
5	0.50		

[a]In studying the effect of the nature of the solvent, δ is kept constant at 5 nm, while studying the effect of δ, cyclohexane is the solvent.

the particle radius, there is an increase in the critical particle concentration even though the depth of the potential well increases with radius. This nonphysical behavior is due to the fact that very small particles have a long-range repulsive part for the pair potential and the perturbation theory based on the effective diameter can no longer be used. The perturbation theory is accurate only when the repulsive part of the potential is sufficiently steep. The effect of other variables on the critical particle concentration can be seen from Table 4. Higher stability is predicted with increasing thickness of the adsorbed layer in accordance with experimental observations [27].

CONCLUSIONS

In addition to the molecular weight of the free polymer, there are other variables, such as the nature of the solvent, particle size, temperature, and thickness of adsorbed layer which have a major influence on the amount of polymer required to cause destabilization in mixtures of sterically stabilized dispersions and free polymer in solution. Using the second-order perturbation theory and a simple model for the pair potential, phase diagrams relating the compositions of the disordered (dilute) and ordered (concentrated) phases to the concentration of the free polymer in solution have been presented which can be used for dilute as well as concentrated dispersions. Qualitative arguments show that, if the adsorbed and free polymer are chemically different, it is advisable to have a solvent which is good for the adsorbed polymer but is poor for the free polymer, for increased stability of such dispersions. Larger particles, higher temperatures, thinner steric layers and better solvents for the free polymer are shown to lead to decreased stability, i.e. require smaller amounts of free polymer for the onset of phase separation. These trends are in accordance with the experimental observations.

Analysis of the steric stabilization in the absence of free polymer in these dispersions indicates that destabilization is possible under better-than-theta conditions. It is shown that smaller particles and thicker layers of adsorbed polymer result in improved stability, i.e. the critical particle concentration, above which instability sets in, decreases with increasing particle size and with decreasing thickness of the adsorbed layer.

APPENDIX A

In order to determine the thermodynamic properties by means of the perturbation theory, the thermodynamic properties of the reference system are needed. Here, the expressions for the equation of state and the radial distribution function of a system of hard spheres are included for both the fluid and solid reference states. A face-centred-cubic arrangement of the particles at closest packing is assumed for the solid phase.

The hard-sphere solid

Equation of state

When the particle volume fraction η is in the range 0.55—0.74, the following equation of state, proposed by Hall [28], was found to be accurate:

$$Z_{HS} = \frac{pV}{NkT} = 2.557696 + 0.1253077\beta + 0.176239\,\beta^2$$

$$- 1.053308\,\beta^3 + 2.818621\,\beta^4 - 2.921934\,\beta^5$$

$$+ 1.118413\,\beta^6 + (12 - 3\beta)/\beta \qquad \text{(A1)}$$

where $\beta = 4(1 - \eta/\eta_0)$, $\eta_0 = 0.74$ and $\eta = (\pi/6)\rho d^3$. Here, d is the hard-sphere diameter and ρ is the number density.

The radial distribution function

Kincaid and Weis [29] have proposed the following analytical expression for the radial distribution function by fitting the Monte Carlo data:

$$g_{HS}(R,\eta) = \frac{\alpha}{R}\exp\left[-W_1^2(R - R_1)^2 - W_2^4(R - R_1)^4\right]$$

$$+ \frac{W}{24\eta\sqrt{\pi}}\sum_{i=2}^{\infty}\frac{n_i}{RR_i}\exp\left[-W^2(R - R_i)^2\right] \qquad \text{(A2)}$$

where n_i is the number of the i-th neighbors, R_i is the static lattice distance to the i-th neighbor (in units of d), R_1 refers to the position (in units of d) of the first peak of the radial distribution function, and R is the reduced centre-to-centre separation (in units of d). The parameter α is chosen such that it satisfies the following consequence of the virial theorem:

$$g(1,\eta) = (Z_{HS} - 1)/4\eta \qquad \text{(A3)}$$

where Z_{HS} is given by Eqn (A1). The parameters W, W_1 and W_2 are given by the following expressions:

$$W = [1 - 10.5896\,(\eta_0 - \eta)^{2.543}]/[0.694\,(\eta_0 - \eta)]^{1.072} \qquad \text{(A4)}$$

$$W_1 = \frac{1.5523}{(\eta_0 - \eta)} - 2.0303\exp[5.8331\,(\eta_0 - \eta)] + 74.8732\,(\eta_0 - \eta)^2 \qquad \text{(A5)}$$

and

$$W_2 = \frac{[0.95596 - 5.8550\,(\eta_0 - \eta) + 39.7466\,(\eta_0 - \eta)^2 - 109.6264\,(\eta_0 - \eta)^3]}{(\eta_0 - \eta)} \qquad \text{(A6)}$$

where $\eta_0 = 0.74$.

The hard-sphere fluid

Equation of state

The fluid phase for a hard-sphere system is stable up to $\eta = 0.49$, at which point a solid phase with $\eta = 0.55$ is predicted to coexist in equilibrium with the fluid phase. Carnahan and Starling [30] have proposed the following simple and accurate equation of state:

$$Z_{HS} = \frac{pV}{NkT} = \frac{1 + \eta + \eta^2 - \eta^3}{(1 - \eta)^3} \quad \text{(A7)}$$

The above equation predicts values which are in good agreement with the computer simulation studies for values of η up to 0.49.

The radial distribution function

Smith and Henderson [31] have derived an analytical expression for the radial distribution function of a hard-sphere fluid by solving the Percus–Yevick equation. The expression for the radial distribution function is given by:

$$g_{HS}(R) = \sum_{n=0}^{\infty} H(R - n) g_n(R) \quad \text{(A8)}$$

where $H(R)$ is a Heaviside step function and R is the reduced centre-to-centre separation. The various g_n are provided by the expressions:

$$g_0(R) = 0 \quad \text{(A9)}$$

$$R g_n(R) = \frac{(-12\eta)^{n-1}}{(n-1)!} \sum_{i=0}^{2} \lim_{t \to t_i} \frac{d^{n-1}}{dt^{n-1}} \left\{ (t - t_i)^n t \left[\frac{L(t)}{S(t)} \right]^n \exp\,[t(R - n)] \right\} \quad \text{(A10)}$$

where

$$S(t) = (1 - \eta)^2 t^3 + 6\eta(1 - \eta)t^2 + 18\eta^2 t - 12\eta(1 + 2\eta) \quad \text{(A11)}$$

$$L(t) = (1 + \tfrac{1}{2}\eta)t + 1 + 2\eta \quad \text{(A12)}$$

where $\eta = (\pi/6)\rho d^3$, t_i $(i = 0,1,2)$ are the zeros of $S(t)$, ρ is the number density N/V and d is the diameter of the hard sphere.

Better agreement between the values calculated using the above equations and the computer simulations is obtained by introducing the correction proposed by Verlet and Weis [32] which forces $g(1,\eta)$ to satisfy the consequence of the virial theorem given by Eqn (A3). For values of $R > 4$, the value of $g_{HS}(R)$ is set equal to unity.

REFERENCES

1 S. Asakura and F. Oosawa, J. Chem. Phys., 22 (1954) 1255.
2 C.L. Sieglaff, J. Polym. Sci., 41 (1959) 319.
3 P.R. Sperry, H.B. Hopfenberg and N.L. Thomas, J. Colloid Interface Sci., 82 (1981) 62.
4 H. de Hek and A. Vrij, J. Colloid Interface Sci., 84 (1981) 409.
5 C. Pathmamanoharan, H. de Hek and A. Vrij, Colloid Polym. Sci., 259 (1981) 769.
6 B. Vincent, P.F. Luckham and F.A. Waite, J. Colloid Interface Sci., 73 (1980) 508.
7 A. Vrij, Pure Appl. Chem., 48 (1976) 471.
8 R.I. Feigin and D.H. Napper, J. Colloid Interface Sci., 75 (1980) 525.
9 G.J. Fleer, J.M.H.M. Scheutjens and B. Vincent, ACS Symp. Ser., 240 (1984) 245.
10 W. van Megen and I. Snook, J. Colloid Interface Sci., 57 (1976) 40.
11 W. van Megen and I. Snook, J. Colloid Interface Sci., 100 (1984) 359.
12 A.P. Gast, C.K. Hall and W.B. Russel, J. Colloid Interface Sci., 96 (1983) 251.
13 I.V. Rao and E. Ruckenstein, J. Colloid Interface Sci., in press.
14 R. Evans and D.H. Napper, J. Chem. Soc. Faraday Trans. 1, 73 (1977) 1377.
15 J.B. Smitham, R. Evans and D.H. Napper, J. Chem. Soc. Faraday Trans. 1, 71 (1975) 285.
16 P.J. Flory, Principles of Polymer Chemistry, Cornell Univ. Press, Ithaca, NY, 1953.
17 B. Vincent, J. Colloid Interface Sci., 42 (1973) 270.
18 J.M. Victor and J.P. Hansen, J. Phys. Paris Lett., 45 (1984) 307.
19 J.A. Barker and D. Henderson, J. Chem. Phys., 47 (1967) 4714.
20 J.D. Weeks, D. Chandler and H. Anderson, J. Chem. Phys., 54 (1971) 5237.
21 J.A. Barker and D. Henderson, J. Chem. Phys., 47 (1967) 2856.
22 P.J. Flory, Discuss. Faraday Soc., 49 (1970) 7.
23 R. Koningsveld and L.A. Kleintjens, Macromolecules, 4 (1971) 637.
24 J. Pouchly, J. Biros, K. Solc and J. Vondrejsova, J. Polym. Sci. Part C, Polym. Symp., 16 (1967) 679.
25 D.H. Napper, J. Colloid Interface Sci., 58 (1977) 390.
26 M.J. Garvey, J. Colloid Interface Sci., 61 (1977) 194.
27 V.K. Dunn and R.D. Vold, J. Colloid Interface Sci., 54 (1976) 22.
28 K.R. Hall, J. Chem. Phys., 57 (1971) 2252.
29 J.M. Kincaid and J.J. Weis, Mol. Phys., 34 (1977) 931.
30 N.F. Carnahan and K.E. Starling, J. Chem. Phys., 51 (1969) 635.
31 W.R. Smith and D. Henderson, Mol. Phys., 19 (1970) 411.
32 L. Verlet and J.J. Weis, Phys. Rev. A, 5 (1972) 939.

Fluid Phase Equilibria, 20 (1985) 189—206 189
Elsevier Science Publishers B.V., Amsterdam — Printed in The Netherlands

THERMODYNAMICS OF MICROEMULSIONS: RELATION BETWEEN THEIR STRUCTURE AND PHASE BEHAVIOUR

E. RUCKENSTEIN

State University of New York at Buffalo, Department of Chemical Engineering, Buffalo, New York, 14260, USA

ABSTRACT

The treatment previously developed by the author to explain the thermodynamic stability and phase behaviour of microemulsions is presented and some of its physical consequences are emphasized. General thermodynamic equations are written concerning the single phase microemulsions, the transition to a microemulsion in equilibrium with an excess dispersed phase, and for the transition from the latter two phase system to one in which a microemulsion coexists with both excess phases (three phase system). On this basis, it is shown that the spherical shape of the globules of the dispersed phase becomes unstable to thermal perturbations, both in some single phase microemulsions as well as near the transition from the two to the three phase systems. While in the former case the radius of the globules at which the instability occurs is finite, in the latter case, it tends to infinity and the interfacial tension at the surface of the globules as well as their bending stress tend to zero proportionally to the square of the curvature. This explains the ill defined, fluctuating interfaces observed in the former case as well as the behaviour similar to that near a critical point observed in the latter case.

I. INTRODUCTION

Microemulsions can be prepared starting from a macroemulsion (of oil in water or water in oil) which is stabilized by the adsorption of surfactant molecules on the surface of the globules of the dispersed phase. The addition of a cosurfactant-a medium length alkyl alcohol-to the above system generates, sometimes spontaneously, an optically transparent dispersion of spherical globules (of oil in water or water in oil) of almost uniform size, having a diameter between 10^2 and 10^3Å (refs. 1-3). The small size of these globules has given rise to the name "microemulsion" to such a system. In contrast to the macroemulsions whose stability has a kinetic origin, microemulsions are thermodynamically stable. For electrically charged surfactants, the kinetic stability of the macroemulsions is determined by the potential barrier

0378-3812/85/$03.30

generated between two particles due to the competition between the attractive van der Waals interactions and the repulsive double layer forces (ref. 4). The thermodynamic stability of microemulsions can be explained as follows (ref. 5): The dispersion of one phase into the second is associated with an increase in the entropy of the system and results in the adsorption of the surfactant and cosurfactant (surfactants) on the large interfacial area thus created. The interfacial tension of the water-oil interface is therefore decreased from a value of the order of 50 dyne/cm, which is characteristic of a water-oil interface devoid of surfactants, to a very low value. In addition, the dilution of the continuous and dispersed media of the microemulsion in surfactants, as a result of adsorption, decreases the chemical potentials of the surfactants. The resulting decrease in the free energy of the system could be called the dilution effect. Thermodynamically stable dispersions can be formed when the negative free energy change due to the dilution effect and to the entropy of dispersion overcomes the positive product of the low interfacial tension and the large interfacial area generated by the dispersion of one phase into another. In contrast to molecular dispersions, the size of the dispersed phase globules is a result of the thermodynamic equilibrium condition, namely, the minimization of the free energy of the microemulsion with respect to the radius of the globules. Similar to molecular dispersions two or three phases can form, since a microemulsion phase can coexist in equilibrium with an excess phase or with both the excess dispersed and continuous phases (refs. 6-10).

In numerous single phase microemulsions, the dispersion occurs in the form of nearly spherical globules of a single size (refs. 2,3,11,12). However, Lindman, et al. (ref. 13) recently developed a novel Fourier transform NMR method to measure multicomponent self-diffusion coefficients in some single phase microemulsion systems. The large values obtained for the diffusion of water, hydrocarbon and alcohol, over a wide range of concentrations indicated that there are no well defined internal interfaces in the systems examined by them. In other words, the interfaces which separate the hydrophobic and hydrophilic regions open up and reform continuously, the spherical shape being unstable to thermal perturbations.

For microemulsions in equilibrium with an excess dispersed phase, the dispersion occurs in the form of nearly spherical globules of a single size (ref. 14) and there appears to be no instability of the spherical internal interface of the microemulsion. Such instabilities, however, arise again, near the transition point from the two phase system to one in which a (middle phase) microemulsion coexists with both the excess phases (ref. 14).

The aim of the present paper is to present the thermodynamic theory of microemulsions developed by the author (refs. 15,16) and to emphasize that it can explain the above features. While the basic theory is available in the above mentioned references, the present version contains a number of additional implications. It emphasizes a number of interfacial tensions of the internal interface of the microemulsion that can be defined and examines the relation between the interfacial tension at the surface of the globules and that between the microemulsion and the excess dispersed phase. In addition, relations between the interfacial tensions and the radius and volume fraction of the globules are obtained for some particular conditions. Similarly, the bending stress is also expressed in terms of the above two variables. On this basis one concludes that: (1) in single phase microemulsions there are conditions under which the spherical shape of the globules of the dispersed phase becomes unstable; (2) the spherical, internal surface of a microemulsion in equilibrium with an excess dispersed phase is always stable; (3) the transition from the latter two phase system to one in which a microemulsion coexists with both excess phases is associated with an infinite value of the radius of the globules, which causes their interfacial tension and bending stress to become equal to zero. This explains the large fluctuations and the similarity with the critical phenomena observed experimentally (ref. 14) near this transition point.

Single phase microemulsions are treated in the next section. Two general thermodynamic equations are derived from the condition that the free energy of the system should be a minimum with respect to both the radius r of the globules as well as the volume fraction ϕ of the dispersed phase. The first equation can be employed to calculate the radius while the second, a generalized Laplace equation, can be used to explain the instability of the spherical shape of the globules. The two and three phase systems are examined in Sections III and IV of the paper.

192

II. SINGLE PHASE MICROEMULSION

II, 1. The Free Energy

For the sake of simplicity, it will be assumed that the microemulsion contains spherical globules of a single size. Their dispersion in the continuous phase is accompanied by an increase in the entropy of the system and the corresponding free energy change per unit volume of microemulsion is denoted by Δf. The Helmholtz free energy f per unit volume of microemulsion is written as the sum

$$f = f_0 + \Delta f, \tag{1}$$

where Δf is separated from the remaining part f_0.

For curved interfaces, characterized by the principal curvatures c_1 and c_2, the Gibbs phenomenological theory leads to the following expression for df_0 (refs. 17-19):

$$df_0 = \gamma\, dA + C_1 dc_1 + C_2 dc_2 + \Sigma \mu_i dn_i - p_2 d\phi - p_1 d(1-\phi). \tag{2}$$

Here γ is a generalized interfacial tension, C_1 and C_2 are bending stresses associated with the curvatures c_1 and c_2, respectively; A is the internal interfacial area per unit volume of microemulsion; μ_i and n_i are the chemical potentials per molecule and the number of molecules of species i, respectively; ϕ is the volume fraction of the dispersed phase; and p_2 and p_1 are the pressures inside the globules and in the continuous phase in the space between the globules. Here the actual physical surface of the globule (to the extent to which it can be defined) of radius r is selected as the Gibbs dividing surface.

A few comments concerning γ, C_1 and C_2 and p_1 and p_2 are necessary. As emphasized previously (ref. 16), γ, C_1 and C_2 also include the effect of the interaction forces between the globules as well as the work done against these forces when the area A is changed at constant ϕ, T and n_i. Concerning the meaning of the pressures p_2 and p_1, it is clear from Eq. (2) that they are introduced on the basis of the free energy f_0 (by multiplying f_0 with the volume of the microemulsion and by performing the appropriate partial derivatives). This free energy does not include the free energy Δf due to the entropy of dispersion (an effect which manifests itself at a scale of the order of the microemulsion, which is large compared to the size of a globule). Therefore, the above pressures represent (micro)pressures on the scale of a single globule, p_2 being the pressure inside a globule and p_1 that outside, in its vicinity, in the continuous phase. The collective behaviour due to the entropy of dispersion generates an additional osmotic contribution to the thermodynamic pressure (the latter pressure being defined on the basis of f) which acts on the interface between the microemulsion and the environment. Of course, the mechanical equilibrium condition demands that the thermodynamic pressure be equal to the external pressure p.

For spherical globules $c_1 = c_2$ and $C_1 = C_2 = C/2$, and Eq. (1) becomes

$$df_0 = \gamma\, dA + Cd(1/r) + \Sigma\mu_i dn_i - p_2 d\phi - p_1 d\,(1 - \phi). \tag{3}$$

Note that γ and C are not independent quantities; their relationship is examined later in the paper.

Combining Eqs. (1) and (3), one obtains:

$$df = \gamma\, dA + Cd(1/r) + \Sigma\mu_i dn_i - p_2 d\phi - p_1 d\,(1 - \phi) + d\Delta f. \tag{4}$$

II, 2. The Equilibrium Radius and a Generalized Laplace Equation

The equilibrium state of the system is completely determined by n_i, the temperature T and the external pressure p. Therefore, the values of the radius r and of ϕ emerge from the condition that the microemulsion be in internal equilibrium, i.e., f be a minimum with respect to r and ϕ. This leads to the equations:

$$\gamma = -\left(\frac{\partial \Delta f}{\partial A}\right)_\phi - C\left(\frac{d(1/r)}{dA}\right)_\phi \tag{5}$$

and

$$p_2 - p_1 = \left(\frac{\partial \Delta f}{\partial \phi}\right)_A + C\left(\frac{d(1/r)}{d\phi}\right)_A. \tag{6}$$

Because

$$A = \frac{3\phi}{r}, \tag{7}$$

Eqs. (5) and (6) can be rewritten as (refs. 15,16):

$$\gamma = \frac{r^2}{3\phi}\left(\frac{\partial \Delta f}{\partial r}\right)_\phi - \frac{C}{3\phi} \tag{8}$$

and

$$p_2 - p_1 = \left(\frac{\partial \Delta f}{\partial \phi}\right)_r + \frac{r}{\phi}\left(\frac{\partial \Delta f}{\partial r}\right)_\phi - \frac{C}{r\phi}. \tag{9}$$

The mechanical equilibrium condition between the microemulsion and environment provides a second relationship between the micropressures p_2 and p_1. The variation of the total free energy F of the microemulsion of volume V can be written as:

$$dF = \gamma d(AV) + CVd(1/r) + \Sigma\mu_i dN_i - p_2 d(V\phi) - p_1 d(V(1 - \phi)) \quad (10)$$

$$+ d(V\Delta f),$$

where N_i is the number of molecules of species i in the entire microemulsion. The condition of mechanical equilibrium at constant T and N_i leads to

$$\gamma d(AV) + CVd(1/r) - p_2 d(V\phi) - p_1 d(V(1 - \phi)) + d(V\Delta f) - pdV_e = 0, \quad (11)$$

where p is the external pressure and dV_e (= $-dV$) is the variation of the volume of the environment. Combining Eq. (11) with Eqs. (8) and (9), one obtains:

$$\frac{3\phi}{r}\gamma - (p_2 - p_1)\phi + (p - p_1) + \Delta f = 0. \quad (12)$$

The system of equations formed by (9) and (12) leads to the following expressions for p_1 and p_2 (refs. 15,16):

$$p_2 - p = \Delta f + (1 - \phi)\left(\frac{\partial \Delta f}{\partial \phi}\right)_r - \frac{C}{\phi r} + \frac{r}{\phi}\left(\frac{\partial \Delta f}{\partial r}\right)_\phi \quad (13)$$

and

$$p_1 - p = \Delta f - \phi\left(\frac{\partial \Delta f}{\partial \phi}\right)_r. \quad (14)$$

Equation (8) constitutes the basic thermodynamic equation for the calculation of the radius of the globules. Of course, explicit expressions, in terms of the radius of the globules and volume fraction, are needed for γ, C and Δf before such a calculation can be carried out. Expressions for Δf will be provided in another section of the paper, but it is difficult to derive expressions for γ and C. One may, however, note that γ (and also C) depends on the radius for the following two reasons: (1) its value depends upon the concentrations of surfactant and cosurfactant in the bulk phases, which, because the system is closed, depend upon the amounts adsorbed on the area of the internal interface of the microemulsion; (2) in addition to the above mass balance effect, there is a curvature effect on γ (this point is examined later in the paper).

Even without explicit expressions for Δf and C, the generalized Laplace equation (9) is useful because it allows one to define various interfacial tensions at the surface of the globules and, more particularly, because it allows one to predict that there are conditions under which the spherical shape of the globules becomes unstable to perturbations. In this respect it is instructive to rewrite Eq. (9) in the form:

$$p_2 - p_1 = \frac{2\gamma}{r} - \frac{C}{3\phi r} + \frac{r}{3\phi}\left(\frac{\partial \Delta f}{\partial r}\right)_m , \qquad (15)$$

where m ($= \phi/\frac{4}{3}\pi r^3$) is the number of globules per unit volume of microemulsion and

$$\left(\frac{\partial \Delta f}{\partial r}\right)_m = \left(\frac{\partial \Delta f}{\partial r}\right)_\phi + \frac{3\phi}{r}\left(\frac{\partial \Delta f}{\partial \phi}\right)_r . \qquad (16)$$

A particular globule, which has the pressure p_2 inside, feels outside, in its vicinity, the micropressure p_1. Only a macroscopic part of the microemulsion, which of course should be large compared to the size of a globule (in order to contain a large number of globules), feels the thermodynamic pressure (which is equal to the external pressure p). The local condition of mechanical equilibrium of the interface of the globules demands $p_2 > p_1$. However, in contrast to the case of a liquid droplet surrounded by its vapors (ref. 18), this inequality does not require the difference $\frac{2\gamma}{r} - \frac{C}{3\phi r}$ to be a positive quantity, since the additional term, $\frac{r}{3\phi}\left(\frac{\partial \Delta f}{\partial r}\right)_m$ (in Eq. 15), is positive and can offset the negative effect. The latter term is positive because the increased volume exclusion (which arises when, at constant number m of globules, the radius r increases) decreases the disorder. As a result, the entropy of dispersion decreases with increasing radii, at constant m. In fact, it is this additional term in Eq. (15) which is responsible for some of the unusual characteristics of the microemulsions.

As noted above, in order to examine the mechanical equilibrium of the interface of the globules, it is not appropriate to compare the pressure p_2 with the external pressure p. In fact, in single phase microemulsions, p_2 will always be less than p. Indeed, as will be shown later (Eq. (24)), $p_2 = p$ when the chemical potentials in the dispersed phase of the microemulsion become equal to those in the excess dispersed phase. Since the chemical potentials in the dispersed phase of the single phase microemulsion are not sufficiently large for phase separation to take place, it follows that $p_2 < p$.

For the discussion which follows, it is convenient to introduce, in addition to the interfacial tension γ, the interfacial tension γ' defined by the expression:

$$p_2 - p_1 = \frac{2\gamma'}{r}. \qquad (17)$$

It is clear that the mechanical stability of the interface of the globules requires $\gamma' > 0$.

II,3. Instability of the Spherical Internal Interface in Single Phase Microemulsions

The spherical shape of the globules is stable to thermal perturbations as long as $p_2 > p_1$ and becomes unstable in the vicinity of the point where $p_2 = p_1$. Eqs. (9), (15), and (17) for $p_2 - p_1$ lead to the following equivalent expressions for the point near which this instability occurs:

$$\frac{2\gamma'}{r} \equiv \frac{2\gamma}{r} - \frac{C}{3\phi r} + \frac{r}{3\phi}\left(\frac{\partial \Delta f}{\partial r}\right)_m \equiv \left(\frac{\partial \Delta f}{\partial \phi}\right)_r + \frac{r}{\phi}\left(\frac{\partial \Delta f}{\partial r}\right)_\phi - \frac{C}{r\phi} = 0. \qquad (18)$$

Equations (8) and (18) could allow, at least in principle, the determination of the values of r and ϕ near which the spherical interface of the globules is no longer stable to thermal perturbations. Near such a point, fluctuations of the interface will occur and, as mentioned in the Introduction, this can explain the ill defined interfaces detected experimentally (ref. 13) by the Fourier transform NMR method. No expressions are as yet available for γ and C, to calculate the values of r and ϕ at which the spherical interface becomes unstable. However, inverting the problem, Eqs. (8) and (18) could provide some information about the values of γ and C which are needed for the instability of the spherical shape of the globules to arise. From the above equations one easily obtains:

$$\gamma = -\frac{r}{3}\left(\frac{\partial \Delta f}{\partial \phi}\right)_r \qquad (19)$$

and

$$C = r^2\left(\frac{\partial \Delta f}{\partial r}\right)_\phi + \phi r\left(\frac{\partial \Delta f}{\partial \phi}\right)_r. \qquad (20)$$

To relate γ and C to r and ϕ, expressions for Δf are, however, needed.

On the basis of a lattice model, upper and lower bounds have been established for the entropy of dispersion of spherical globules of radius r and volume fraction ϕ in the continuous phase (ref. 20). Here r and ϕ are the "actual" radius of the globules (including the adsorbed layer of surfactants) and the correspondin[g] volume fraction. The free energy corresponding to the upper bound of the entropy has the form:

$$\Delta f = -\frac{3\phi kT}{4\pi r^3}\{-\frac{1}{\phi}\ell n(1-\phi) + \ell n(\frac{4\pi r^3}{3v_c}[\frac{1-\phi}{\phi}])\} \qquad (21a)$$

and that corresponding to the lower bound has the form:

$$\Delta f = -\frac{3\phi kT}{4\pi r^3}\ell n\{\frac{4\pi r^3}{3v_c}[(\frac{0.74}{\phi})^{1/3} - 1]^3\}, \qquad (21b)$$

where k is as usual the Boltzmann constant and v_c is the molecular volume of the molecules of the continuous phase. The expression based on the Carnahan-Starling approximation for hard spheres (ref. 21) has the form:

$$\Delta f = -\frac{3\phi kT}{4\pi r^3}[1 - \ell n\,\phi - \phi\frac{4-3\phi}{(1-\phi)^2} + \ell n\frac{4\pi r^3}{3v_c}]. \qquad (21c)$$

Combining Eqs. (19) and (20) with each of the Eqs. (21), one obtains, in the same order:

$$\gamma = \frac{kT}{4\pi r^2}\ell n[\frac{4\pi r^3}{3v_c}\frac{(1-\phi)}{\phi}] \qquad (22a)$$

and

$$C = \frac{9\phi kT}{4\pi r^2}[-\frac{1}{\phi}\ell n(1-\phi) + \frac{2}{3}\ell n(\frac{4\pi r^3}{3v_c}\frac{(1-\phi)}{\phi}) - 1], \qquad (23a)$$

$$\gamma = \frac{kT}{4\pi r^2}[\ell n\{\frac{4\pi r^3}{3v_c}((\frac{0.74}{\phi})^{1/3} - 1)^3\} - \frac{(0.74/\phi)^{1/3}}{(0.74/\phi)^{1/3} - 1}] \qquad (22b)$$

and

$$C = \frac{3\phi kT}{4\pi r^2}[2\,\ell n\{\frac{4\pi r^3}{3v_c}((\frac{0.74}{\phi})^{1/3} - 1)^3\} + \frac{(0.74/\phi)^{1/3}}{(0.74/\phi)^{1/3} - 1} - 3], \qquad (23b)$$

$$\gamma = \frac{kT}{4\pi r^2}[-\ell n\phi + \ell n\frac{4\pi r^3}{3v_c} - \frac{\phi(8 - 9\phi + 3\phi^2)}{(1-\phi)^3}] \qquad (22c)$$

and

$$C = \frac{3kT\phi}{4\pi r^2}[2\ell n\frac{4\pi r^3}{3v_c} - 2\ell n\phi + \frac{12\phi^2 - 6\phi^3 - 4\phi}{(1-\phi)^3}]. \qquad (23c)$$

Table I provides some values of γ and C and shows that these quantities are small and become smaller with increasing radius. While γ and C are of the same order of magnitude, C is the larger of the two. Expression (21a) and (21c) for Δf lead to comparable results while expression (21b) leads to smaller values. It is of interest to note that the transition to unstable interfaces can

Table I

	r x 10^7 (cm)	γ (dyne/cm) Eq. (22a)	γ (dyne/cm) Eq. (22b)	γ (dyne/cm) Eq. (22c)	C (dyne/cm) Eq. (23a)	C (dyne/cm) Eq. (23b)	C (dyne/cm) Eq. (23c)
φ = 0.2		0.35	0.129	0.27	0.43	0.277	0.417
	3						
φ = 0.3		0.33	0.043	0.167	0.61	0.36	0.595
φ = 0.2		0.11	0.05	0.087	0.13	0.09	0.127
	6						
φ = 0.3		0.10	0.03	0.061	0.19	0.125	0.183
φ = 0.2		0.05	0.028	0.043	0.06	0.047	0.062
	9						
φ = 0.3		0.05	0.018	0.032	0.09	0.064	0.09
φ = 0.2		0.021	0.012	0.018	0.026	0.02	0.025
	15						
φ = 0.3		0.02	0.009	0.014	0.037	0.027	0.037
φ = 0.2		0.01	0.0056	0.008	0.011	0.0087	0.011
	24						
φ = 0.3		0.009	0.004	0.006	0.016	0.012	0.016
φ = 0.2		0.003	0.002	0.003	0.004	0.003	0.004
	42						
φ = 0.3		0.003	0.002	0.0023	0.006	0.005	0.006
φ = 0.2		0.0017	0.001	0.0015	0.002	0.0017	0.002
	60						
φ = 0.3		0.002	0.0009	0.0012	0.003	0.0024	0.003
φ = 0.2		0.001	0.0007	0.0009	0.0013	0.001	0.0013
	78						
φ = 0.3		0.001	0.00059	0.0008	0.002	0.0015	0.002

occur at finite values of r. This is in contrast to the instability which is associated with the formation of a three phase system that will be examined later in the paper. Using the above expression one can also verify that, at the point of transition to unstable interfaces, $p_2 < p$ in single phase microemulsions.

III. A MICROEMULSION IN EQUILIBRIUM WITH AN EXCESS DISPERSED PHASE

III,1. Equations for Phase Equilibrium

Equation (4) shows that the chemical potential, defined as the partial derivative of f with respect to n_i, at constant A, temperature, ϕ and n_j ($j \neq i$), equals the chemical potential μ_i defined on the basis of f_o. This arises because Δf depends only on ϕ and r and does not depend separately on n_i. A microemulsion is in equilibrium with an excess dispersed phase when the chemical potentials μ_i are equal to those in the excess dispersed phase. The composition of the excess dispersed phase is expected to be the same as that of the dispersed phase; therefore, the equality of the chemical potentials requires the equality of the corresponding pressures. The chemical potentials μ_i in the dispersed phase are expressed in Eq. (4) at the micropressure p_2. Since in the excess dispersed phase the pressure is equal to the external pressure p, the equilibrium condition leads to

$$p_2 = p, \tag{24}$$

which, when combined with Eq. (13), leads to (refs. 15,16):

$$\Delta f + (1 - \phi)\left(\frac{\partial \Delta f}{\partial \phi}\right)_r - \frac{C}{\phi r} + \frac{r}{\phi}\left(\frac{\partial \Delta f}{\partial r}\right)_\phi = 0. \tag{25}$$

Eqs. (8) and (25) (first derived in reference (22) without considering the bending stress) could provide the values of r and ϕ at the point at which an excess dispersed phase forms. Any addition to the system of excess dispersed phase, having the same composition as the dispersed phase, will not change r and ϕ as soon as the transition point is reached.

Appropriate expressions for γ and C, for use in Eqs. (8) and (25), to calculate r and ϕ are not yet available. Eqs. (8) and (25) combined with expressions (21) for Δf can be, however, employed to relate γ and C to r and ϕ. The expressions thus obtained are useful, not only because they provide some information about γ and C, but also in the discussion regarding the stability of the spherical interface.

For the three expressions of Δf given by Eqs. (21), one obtains, in the same order:

$$\gamma = \frac{kT}{4\pi r^2}\left[-\ln(1-\phi) + \ln\left(\frac{4\pi r^3}{3v_c}\frac{(1-\phi)}{\phi}\right)\right] \tag{26a}$$

and

$$C = \frac{3\phi kT}{4\pi r^2}\left[(1-(3/\phi))\ln(1-\phi) + 2\ln\left(\frac{4\pi r^3}{3v_c}\frac{(1-\phi)}{\phi}\right) - 3\right], \tag{27a}$$

$$\gamma = \frac{kT}{4\pi r^2}\left[\ln\left[\frac{4\pi r^3}{3v_c}\left\{\left(\frac{0.74}{\phi}\right)^{1/3} - 1\right\}^3\right] - (1-\phi)\frac{(0.74/\phi)^{1/3}}{(0.74/\phi)^{1/3} - 1}\right] \tag{26b}$$

and

$$C = \frac{3\phi kT}{4\pi r^2}\left[2\ln\left[\frac{4\pi r^3}{3v_c}\left\{\left(\frac{0.74}{\phi}\right)^{1/3} - 1\right\}^3\right] + (1-\phi)\frac{(0.74/\phi)^{1/3}}{(0.74/\phi)^{1/3} - 1} - 3\right] \tag{27b}$$

and

$$\gamma = \frac{kT}{4\pi r^2}\left[\ln\left(\frac{4\pi r^3}{3v_c\phi}\right) - \frac{8\phi - 5\phi^2}{(1-\phi)^2} + \phi\right] \tag{26c}$$

and

$$C = \frac{3\phi kT}{4\pi r^2}\left[-2\ln\phi + 2\ln\left(\frac{4\pi r^3}{3v_c}\right) + \frac{6\phi^2 - 5\phi - \phi^3}{(1-\phi)^2}\right]. \tag{27c}$$

Table II contains numerical values of γ and C as functions of r and ϕ. They show that C is larger than γ and both their values decrease tremendously with increasing r. These numerical values are useful in the qualitative discussion which follows.

III, 2. <u>The Stability of the Spherical Shape When a Microemulsion Coexists with an Excess Dispersed Phase</u>

When an excess dispersed phase coexists with a microemulsion, $p_2 - p = 0$. However, $p_2 - p_1$ is always positive. To prove this, it is convenient to use the interfacial tension γ' defined via Eq. (17). Using expressions (21) for Δf and the corresponding expressions for C provided by Eqs. 27, one obtains, in the same order:

$$\gamma' = -\frac{3kT}{8\pi r^2}\ln(1-\phi), \tag{28a}$$

$$\gamma' = \frac{3\phi kT}{8\pi r^2}\frac{(0.74/\phi)^{1/3}}{(0.74/\phi)^{1/3} - 1} \tag{28b}$$

and

$$\gamma' = \frac{3\phi kT}{8\pi r^2}(1+\phi+\phi^2-\phi^3)/(1-\phi)^3. \tag{28c}$$

The above equations show that $\gamma' > 0$ over the entire range of coexistence of a microemulsion with the excess dispersed phase; this ensures the stability of the internal spherical interface of the microemulsion over the above range.

It is of interest to note that the numerical calculations (Table II) show that expressions (21b) and (21c) lead to negative values for the difference $\frac{2\gamma}{r} - \frac{C}{3\phi r}$, while expression (21a) yields positive values for the same difference. As already noted, the derivative $\frac{r}{3\phi}\left(\frac{\partial \Delta f}{\partial r}\right)_m$ is, indeed, positive and compensates the negative value of $\frac{2\gamma}{r} - \frac{C}{3\phi r}$. This results in positive values for γ'.

III,3. <u>Relation between the interfacial tension at the surface of the globules and that between the microemulsion and the excess dispersed phase</u>

The interfacial tension γ and the bending stress C are not independent quantities, being related via the generalized Gibbs adsorption equation derived by Buff (refs. 18,16). It has the form:

$$d\gamma + \Sigma \Gamma_i d\mu_i = \frac{rC}{3\phi} d\,(1/r), \qquad (29)$$

where Γ_i is the surface excess of species i.

The chemical potentials of the surfactants are the same on the surface of the globules and the interface between the two phases. Therefore the interfacial tension γ_∞ between a microemulsion and an excess dispersed phase can be related to that at the surface of the globules γ, by integrating Eq. (29) at constant chemical potentials. One obtains:

$$\gamma_\infty - \gamma = \int_r^\infty \frac{rC}{3\phi} d\,(1/r) \equiv - \int_r^\infty \frac{C}{3\phi r} dr, \qquad (30)$$

where γ is given by Eqs. (26) and the integral has to be performed at constant chemical potentials. This integration involves the assumption that the planar interface does not have globules in its vicinity with which it interacts. Since the interfacial tension is largely a result of the adsorption of the surfactant and cosurfactant on the interface, the former effect is negligible. Using Eq. (8), Eq. (30) can be rewritten as:

$$\gamma_\infty = \frac{r^2}{3\phi}\left(\frac{\partial \Delta f}{\partial r}\right)_\phi - \frac{C}{3\phi} - \int_r^\infty \frac{C}{3\phi r} dr. \qquad (31)$$

Table II
$\phi = 0.2$

$r \times 10^7$ cm	C dyne/cm	γ dyne/cm	γ' dyne/cm	$\frac{2\gamma}{r} - \frac{C}{3\phi r}$ dyne/cm^2	$\frac{r}{3\phi}\left(\frac{\partial \Delta f}{\partial r}\right)_m$ dyne/cm^2	$\frac{\frac{2\gamma}{r} - \frac{C}{3\phi r}}{\frac{r}{3\phi}\left(\frac{\partial \Delta f}{\partial r}\right)_m}$	$\frac{\frac{2\gamma}{r} + \frac{r}{3\phi}\left(\frac{\partial \Delta f}{\partial r}\right)_m}{\left(\frac{C}{3\phi r}\right)}$	$\frac{C}{3\phi\gamma}$	$\frac{r^2}{3\phi}\left(\frac{\partial \Delta f}{\partial r}\right)_\phi$ dyne/cm
				Eq. (21a) for Δf					
3	0.425	0.36	0.012	39,325	42,360	0.93	1.034	1.97	1.07
6	0.130	0.11	0.003	4,915	5,295	0.93	1.028	1.973	0.324
9	0.063	0.053	0.001	1,456	1,568	0.93	1.026	1.975	0.16
15	0.025	0.021	0.00049	314	338	0.93	1.023	1.978	0.064
24	0.011	0.009	0.00019	76	82	0.93	1.021	1.98	0.027
				Eq. (21b) for Δf					
3	0.265	0.15	0.031	-470,000	678,000	-0.692	1.14	2.936	0.59
6	0.090	0.06	0.008	- 59,000	84,700	-0.692	1.10	2.62	0.21
9	0.046	0.03	0.004	- 17,000	25,000	-0.692	1.09	2.52	0.11
15	0.020	0.01	0.001	- 3,800	5,400	-0.692	1.08	2.43	0.045
24	0.008	0.006	0.00049	- 920	1,300	-0.692	1.07	2.37	0.02
				Eq. (21c) for Δf					
3	0.406	0.287	0.026	-339,000	515,000	-0.66	1.07	2.35	0.964
6	0.124	0.091	0.007	- 42,300	64,300	-0.66	1.064	2.28	0.298
9	0.061	0.045	0.003	- 12,500	19,100	-0.66	1.06	2.25	0.15
15	0.025	0.019	0.001	- 2,710	4,120	-0.66	1.05	2.22	0.06
24	0.011	0.008	0.0004	- 661	1,010	-0.66	1.046	2.19	0.026

Table II
$\phi = 0.3$

$r \times 10^7$ cm	C dyne/cm	γ dyne/cm	γ' dyne/cm	$\frac{2\gamma}{r} - \frac{C}{3\phi r}$ dyne/cm2	$\frac{r}{3\phi}\left(\frac{\partial \Delta f}{\partial r}\right)_m$ dyne/cm2	$\dfrac{\frac{2\gamma}{r} - \frac{C}{3\phi r}}{\frac{r}{3\phi}\left(\frac{\partial \Delta f}{\partial r}\right)_m}$	$\dfrac{\frac{2\gamma}{r} + \frac{r}{3\phi}\left(\frac{\partial \Delta f}{\partial r}\right)_m}{\left(\frac{C}{3\phi r}\right)}$	$\frac{C}{3\phi\gamma}$	$\frac{r^2}{3\phi}\left(\frac{\partial \Delta f}{\partial r}\right)_\phi$ dyne/cm
				Eq. (21a) for Δf					
3	0.605	0.346	0.02	61,410	69,156	0.9	1.06	1.95	1.02
6	0.188	0.11	0.005	7,676	8,644	0.9	1.05	1.96	0.312
9	0.09	0.052	0.002	2,274	2,561	0.9	1.043	1.96	0.153
15	0.037	0.021	0.00078	491	553	0.9	1.04	1.965	0.062
24	0.016	0.009	0.0003	119	135	0.9	1.03	1.97	0.027
				Eq. (21b) for Δf					
3	0.324	0.086	0.064	-630,000	1,050,000	-0.59	1.36	4.19	0.446
6	0.115	0.041	0.016	- 79,000	132,000	-0.59	1.25	3.16	0.168
9	0.060	0.023	0.007	- 23,000	39,000	-0.59	1.21	2.91	0.09
15	0.026	0.011	0.003	-5,030	8,400	-0.59	1.18	2.71	0.039
24	0.012	0.005	0.001	-1,230	2,060	-0.59	1.16	2.6	0.018
				Eq. (21c) for Δf					
3	0.556	0.211	0.065	-650,000	1,090,000	-0.6	1.21	2.92	0.83
6	0.173	0.072	0.016	- 81,500	136,000	-0.6	1.17	2.68	0.264
9	0.086	0.037	0.007	- 24,200	40,300	-0.6	1.15	2.60	0.132
15	0.035	0.016	0.003	- 5,200	8,710	-0.6	1.14	2.50	0.054
24	0.015	0.007	0.001	- 1,270	2,130	-0.6	1.12	2.45	0.024

204

Assuming (as suggested by the numerical calculations) that C is a positive quantity, Eq. (30) shows that $\gamma_\infty < \gamma$ and Eq. (31) indicates that the expression $\frac{r^2}{3\phi}\left(\frac{\partial \Delta f}{\partial r}\right)_\phi$ (suggested previously by the author (ref. 16) for the evaluation of γ_∞) provides only an upper bound. The integral in Eq. (31) cannot as yet be performed, because the dependence of C on r, when the chemical potentials remain constant, is not known. Eqs. (27) (which relate C to r and ϕ) cannot be used because they do not involve the restriction of constant chemical potentials. The numerical calculations of Table II show that $\frac{C}{3\phi\gamma}$ is about 2 or larger. If the effect of the integral is not taken into account, then γ_∞ is at least a factor of three smaller than the value provided by $\frac{r^2}{3\phi}\left(\frac{\partial \Delta f}{\partial r}\right)_\phi$. If C decays sufficiently rapidly to zero with increasing r, the above factor will not become larger than say 5, when the effect of the integral is also taken into account. Of course, γ given by Eqs. (26) constitute a closer upper bound of γ_∞, which is greater than γ_∞, perhaps, by a factor of about two.

Note that if C is not always positive in the integral, as assumed above, but changes sign, the relationship between γ_∞ and γ becomes more complex. However, if this is the case, γ may provide even a better evaluation of γ_∞.

IV. A MICROEMULSION IN EQUILIBRIUM WITH BOTH EXCESS PHASES

Considerations similar to those employed regarding the occurrence of an excess dispersed phase lead to the conclusion that an excess continuous phase emerges when

$$p_1 = p. \tag{32}$$

Combining Eq. (32) with Eq. (14), one obtains:

$$\Delta f - \phi\left(\frac{\partial \Delta f}{\partial \phi}\right)_r = 0. \tag{33}$$

One may observe that there cannot be a microemulsion in equilibrium with only an excess continuous phase. Indeed, since $p \geqslant p_2 \geqslant p_1$, the equality $p_1 = p$ implies $p_2 = p$, and hence the coexistence of the microemulsion with both excess phases. In addition, the equality $p_2 = p_1$ implies $\frac{2\gamma'}{r} = 0$, which, as each of the Eqs. (28) indicate happens only when $r \to \infty$. When $r \to \infty$, Eqs. (26) and (27) show that γ and $C \to 0$; γ and C tend to zero proportionally to the second power of the curvature. Consequently (as also suggested previously (ref. 16)) the equalities $p_2 = p_1 = p$ are not compatible with a finite curvature and a change in the structure of the microemulsion must take place. A planar structure of alternate oil and water layers will have a zero curvature, but it will not be stable. For such an ordered structure, the entropy of dispersion is expected to be much smaller than that of a globular dispersion. In fact, evaluating the entropy on the basis of the number of distinguishable configurations one obtains $\Delta f = 0$, since only one such configuration exists. Because Eqs. (5) and (6) have general validity, $\gamma = 0$ for this case. As a result of this the planar interfaces are extremely unstable. Consequently, a complex oscillatory behaviour takes place between more planar interfaces which are unstable and curved interfaces which are also unstable. Thus the interfaces have a turbulent behaviour and their motion is accompanied by breakup and coalescence. The peculiar behaviour at the transition point ($r \to \infty$ as well as γ and $C \to 0$) explains the large fluctuations observed experimentally (ref. 14) near this point and why this behaviour is similar to that existing near a critical point.

CONCLUSIONS

The equations established previously by the author in the framework of a thermodynamic theory of microemulsions enable one to arrive at the following conclusions: (1) An instability of the spherical shape of the globules occurs in single phase microemulsions at a finite value of the radius of the dispersed globules; (2) The nearly spherical shape of the globules is stable when a microemulsion coexists with an excess dispersed phase; (3) A microemulsion cannot be in equilibrium with an excess continuous phase alone, but either with an excess dispersed phase or with both excess phases; (4) The radius of the globules becomes infinity, and the interfacial tension at the surface of the globules and the bending stress tend to zero in direct proportion to the square of the curvature, at the transition from a microemulsion in equilibrium with an excess phase to one coexisting with both excess phases; (5) The interfacial tension γ at the surface of the globules (defined via Eq. (2)) is likely to be greater than that between the microemulsion and the excess dispersed phase, but of the same order of magnitude; (6) The expression for the excess pressure in the globules, as compared to that in the continuous phase nearby, contains (in addition to the terms which appear in the case of a droplet surrounded by its vapor) a positive term caused by the decrease of the entropy of dispersion with increasing radius at constant number of globules. As a result, the excess pressure can remain positive even when the traditional terms lead to negative values. This additional term is in fact responsible for some of the peculiarities associated with the microemulsion systems.

REFERENCES

1 T.P. Hoar and J.H. Schulman, Nature 152 (1943) 102.
2 J.H. Schulman and D.R. Riley, J. Colloid Sci. 3 (1948) 383.
3 W. Stoeckenius, J.H. Schulman and L.M. Prince, Kolloid Z. 169 (1960) 170.
4 E.J.W. Verwey and J.Th.G. Overbeek, Theory of the Stability of Lyophobic Colloids, Elsevier, Amsterdam, 1948.
5 E. Ruckenstein, Chem. Phys. Lett. 56 (1978) 518.
6 K. Shinoda and T. Ogawa, J. Colloid Interface Sci., 24 (1967) 56.
7 K. Kon-No and A.J. Kitahara, J. Colloid Interface Sci., 34 (1970) 22; 37 (1971) 469.
8 P.A. Winsor, Solvent Properties of Amphiphilic Compounds, Butterworth's Scientific Publications, London, 1954.
9 R.N. Healy, R.L. Reed and D.G. Stenmark, Soc. Pet. Eng. J. Trans., AIME, 261 (1976) 147.
10 M. Bourel, C. Koukounis, R. Schechter and W. Wade, J. Dispersion Sci. Tech., 1 (1980) 13.
11 J.E.L. Bowcott and J.H. Schulman, Z. Electrochem. 59 (1955) 283.
12 J.W. Falco, R.D. Walker, Jr. and D.O. Shah, AIChEJ., 20 (1974) 510.
13 B. Lindman, P. Stilbs and E. Moseley, J. Colloid Interface Sci., 83 (1981) 569.
14 A. M. Cazabat, D. Langevin, J. Meunier and A. Pouchelon, Adv. Colloid Interface Sci. 16 (1982) 175.
15 E. Ruckenstein, Chem. Phys. Letters 98 (1983) 573.
16 E. Ruckenstein, in K. L. Mittal (Ed.), Surfactants in Solution - Theoretical and Applied Aspects, Plenum Press, New York, 1983, Vol. 3, p. 1551.
17 J.W. Gibbs, Collected Works, Yale University Press, New Haven, CT, 1948, Vol. 1.
18 F.R. Buff, J. Chem. Phys. 19 (1951) 1591.
19. S. Ono and S. Kondo, In S. Flugge (Ed.), Handbuch der Physik, Springer, Berlin, 1960, Vol. X, p. 134.
20 E. Ruckenstein and J.C. Chi., J. Chem. Soc. Faraday Trans. II, 71 (1975) 1690.
21 J.Th.G. Overbeek, Faraday Discussions of the Chem. Soc. No. 65 (1978) 7.
22 E. Ruckenstein and R. Krishnan, J. Colloid Interface Sci. 71 (1979) 321.

3

Thermodynamics of Microemulsions II

Eli Ruckenstein
State University of New York at Buffalo, Buffalo, New York

I. INTRODUCTION

A. General Introduction

The goal of this chapter is to demonstrate that the thermodynamics of microemulsions presented by Ruckenstein in 1982 at the Lund meeting on surfactants in solution and published in Refs. 1 and 2, can provide insight with respect to some of the relevant experimental results gathered regarding microemulsions. The following experimental results are particularly meaningful in the present context:

1. Schulman and coworkers [3–7] observed that isotropic and optically transparent dispersions of oil in water or water in oil are formed spontaneously in the presence of a surfactant and a cosurfactant such as an alkyl alcohol. Using low-angle X-rays, light scattering, ultracentrifugation, electron microscopy, and viscosity measurements, they concluded that the dispersed phase consists of almost uniform spherical droplets with diameters ranging between 8 and 80 nm.

2. Winsor [8,9] found that in ternary nonionic surfactant–oil–water phase diagrams there are three types of systems. In Winsor's type I systems, an oil-in-water (O/W) microemulsion coexists with excess oil, while in Winsor II systems a water-in-oil (W/O) microemulsion coexists with excess water. In both types of systems, almost spherical globules are dispersed in the continuous phase. In Winsor III systems, a microemulsion coexists with both excess phases, and one can no longer identify dispersed and continuous phases. For given amounts of the components, the type I system is present at relatively low temperatures; at intermediate temperatures, the type III system is generated; and, finally, at relatively high temperatures, the type II system is formed. Similar observations have been made with more complex systems that contained oil, water, surfactant, cosurfactant, and salt [10,11]. Indeed, if the temperature, the cosurfactant content, or the ionic strength increases, the system passes from an O/W microemulsion in equilibrium with excess oil to a middle-phase microemulsion in equilibrium with both excess phases and finally to a W/O microemulsion in equilibrium with excess water.

Numerous attempts have been made to explain the above observations, and before presenting our 1982 thermodynamic treatment, it is appropriate to review them. Ruckenstein and Chi [12] suggested a drop model for the microemulsion and derived an expression for its free energy that contained the interfacial free energy of the droplets, the entropy of dispersion of the globules in the continuous phase and the double layer, and van der Waals interactions between globules. A lattice model was used to derive approximate expressions for the entropy of dispersion, in which the size difference between globules and the molecules of the continuous phase was taken into account. They could thus predict the existence of a radius of the globules for which the free energy is minimum.

Similar attempts were made by Likhtman et al. [13] and Reiss [14]. Reference 13 employed the ideal mixture expression for the entropy and Ref. 14 an expression derived previously by Reiss in his nucleation theory. These authors added the interfacial free energy contribution to the entropic contribution. However, the free energy expressions of Refs. 13 and 14 do not provide a radius for which the free energy is minimum. An improved thermodynamic treatment was developed by Ruckenstein [15,16] and Overbeek [17] that included the chemical potentials in the expression of the free energy, since those potentials depend on the distribution of the surfactant and cosurfactant among the continuous, dispersed, and interfacial regions of the microemulsion. Ruckenstein and Krishnan [18] could explain, on the basis of the treatment in Refs. 15 and 16, the phase behavior of a three-component oil–water–nonionic surfactant system reported by Shinoda and Saito [19].

A useful suggestion was made by Robins [20] and Miller and Neogi [21]. Robins explained qualitatively the behavior of microemulsions in terms of the bending stress at the interface of the globules, and Miller and Neogi included the bending stress in the thermodynamic treatment of Refs. 15–18. A different kind of droplet model was developed by Safran and his colleagues [22–25], one in which the bending free energy of the interface and the entropy of dispersion of the globules in the continuous phase were assumed to be the main contributions to the free energy. For the bending free energy, they used an expression suggested by Helfrich [26], while for the entropy they employed the Flory–Huggins expression. Of particular interest are their studies of the effect of fluctuations. Regarding the middle phase, Scriven [27] suggested a rigid bicontinuous representation, while Talmon and Prager [28] described it as a random interspersion of oil and water domains generated by a Voronoi tessellation and derived an expression for the entropy of the system. Using a somewhat simpler representation than that of Talmon and Prager and introducing additional terms in the free energy, Jouffroy et al. [29] examined the phase behavior.

Finally, one word about the lattice theories of microemulsions [30–36]. In these models the space is divided into cells in which either water or oil can be found. This reduces the problem to a kind of lattice gas, for which there is a rich literature in statistical mechanics that could be extended to microemulsions. A predictive treatment of both droplet and bicontinuous microemulsions was developed recently by Nagarajan and Ruckenstein [37], which, in contrast to the previous theoretical approaches, takes into account the molecular structures of the surfactant, cosurfactant, and hydrocarbon molecules. The treatment is similar to that employed by Nagarajan and Ruckenstein for solubilization [38].

The emphasis of the present chapter is on the droplet model, and the methodology used in the treatment is a thermodynamic one. While the information obtained is incomplete, it will be shown that thermodynamics can provide understanding regarding the phase behavior of microemulsions, with respect to both the equilibrium between a microemulsion and an excess dispersed phase and the transition to a middle-phase microemulsion in equilibrium with both excess phases.

B. Thermodynamic Description of a Microemulsion

From the point of view of traditional thermodynamics, a microemulsion is a multicomponent mixture formed of oil, water, surfactant, cosurfactant, and electrolyte. There is, however, a major difference between a conventional mixture and a microemulsion. In the former case, the components are mixed on a molecular scale, while in the latter, oil or water is dispersed as globules on the order of 10–100 nm in diameter in water or oil. The surfactant and cosurfactant are mostly located at the interface between the two phases but are also distributed at equilibrium between the two media. In conventional mixtures, the sizes of the component species are fixed. In the case of microemulsions, the sizes of the globules are not given but are provided by the condition of thermodynamic equilibrium.

The thermodynamics of microemulsions has to account for the entropy of dispersion of the globules in the continuous medium, for the free energy of formation of the interface between the two media of the microemulsion, for the interactions among globules, and for the distribution at equilibrium of the species between the two media of the microemulsion and their interface. It is convenient to split the Helmholtz free energy F of the entire system into two terms. One of them, F_0, is the free energy of a dispersion of fixed globules without interactions among them (called in what follows a frozen noninteracting system), and the other, ΔF, is the free energy due to the entropy of dispersion of the globules in the continuous phase and to their interactions. This decomposition allows us to employ the Gibbs thermodynamics for the first contribution and conventional statistical mechanics for the second contribution.

II. SINGLE-PHASE MICROEMULSIONS

A. Basic Equations

Let us consider that the microemulsion contains spherical globules of uniform size. Their dispersion in the continuous medium is accompanied by an increase in the entropy of the system. As noted in Sec. I, the Helmholtz free energy f per unit volume of microemulsion ($f=F/V$, where V is the volume of the microemulsion) will be written as the sum between a frozen noninteracting free energy f_0 and a free energy Δf due to the entropy of dispersion of the globules in the continuous phase and to their interactions:

$$f = f_0 + \Delta f \tag{1}$$

Gibbs thermodynamics provides the following expression for df_0:

$$df_0 = \gamma\, dA + C_1\, dc_1 + C_2\, dc_2 + \sum \mu_i\, dn_i - p_2\, d\phi - p_1\, d(1-\phi) \tag{2}$$

where γ is the interfacial tension; C_1 and C_2 are bending stresses associated with the curvatures c_1 and c_2, respectively; A is the interfacial area per unit volume between the two media of the microemulsion; μ_i and n_i are the chemical potential (at pressure p_1 for the species present in the continuous phase and at pressure p_2 for those present in the dispersed phase) and the number of molecules of species i per unit volume respectively; and ϕ is the volume fraction of the dispersed phase (which includes the surfactant and cosurfactant molecules present at the interface of the globules). It should be noted that p_1 represents the pressure in the continuous phase of the frozen noninteracting system, which differs from the pressure to which the continuous phase of the microemulsion is subjected.

For spherical globules, $c_1 = c_2 = 1/r$ and $C_1 = C_2 = C/2$, and Eqs. (1) and (2) lead to

$$df = \gamma\, dA + C\, d\left(\frac{1}{r}\right) + \sum \mu_i\, dn_i - p_2\, d\phi - p_1\, d(1-\phi) + d\Delta f \tag{3}$$

where r is the radius of the globule (which includes the surfactant and cosurfactant layer adsorbed on its surface).

The equilibrium state of a microemulsion is completely determined by n_i, the temperature T, and pressure p. Therefore, the values of r and ϕ emerge from the condition that f be a minimum with respect to r and ϕ. Since for spherical globules

$$A = 3\phi/r \tag{4}$$

one obtains

$$\gamma = \frac{r^2}{3\phi}\left(\frac{\partial \Delta f}{\partial r}\right)_\phi - \frac{C}{3\phi} \tag{5}$$

and

$$p_2 - p_1 = \left(\frac{\partial \Delta f}{\partial \phi}\right)_r + \frac{r}{\phi}\left(\frac{\partial \Delta f}{\partial r}\right)_\phi - \frac{C}{r\phi} \tag{6}$$

An additional relationship between p_2 and p_1 is provided by the mechanical equilibrium condition between microemulsion and environment. Considering a variation dV of the volume V of the microemulsion at constant $N_i(= n_i V)$ and T, one can write

$$\gamma\, d(AV) + VCd(1/r) - p_2\, d(V\phi) - p_1\, d(V(1-\phi)) + d(V\Delta f) - pdV_e = 0 \tag{7}$$

Since the variation in the volume of the environment dV_e is equal to $-dV$, combining Eqs. (5)–(7), one obtains

$$\frac{3\phi}{r}\gamma - (p_2 - p_1)\phi + (p - p_1) + \Delta f = 0 \tag{8}$$

Equations (6) and (8) yield

$$p_2 = p + \Delta f + (1-\phi)\left(\frac{\partial \Delta f}{\partial \phi}\right)_r - \frac{C}{\phi r} + \frac{r}{\phi}\left(\frac{\partial \Delta f}{\partial r}\right)_\phi \tag{9}$$

and

$$p_1 = p + \Delta f - \phi\left(\frac{\partial \Delta f}{\partial \phi}\right)_r \tag{10}$$

Equation (10) demonstrates that the pressure p_1 in the continuous medium of the frozen noninteracting system (the system of free energy f_0) differs from the pressure p, and that the pressure p is equal to the sum between the pressure p_1 and the osmotic contribution due to the free energy Δf. The pressure p acts both on the microemulsion and on the continuous medium of the microemulsion. A globule senses in its immediate vicinity the pressure p_1 of the frozen noninteracting system plus the osmotic pressure due to the free energy Δf. While occurring at a scale large compared to the size of a globule, the latter contribution is felt also at the scale of a globule because of the continuity of the continuous medium. My previous assertion [1,2], that the globules sense only the pressure p_1 must be revised.

It is also important to note that the conventional Laplace equation or the generalized one, which includes the bending stress, is no longer valid for a microemulsion. Equation (9) replaces the Laplace equation. It can be rewritten in a more revealing form by combining Eqs. (5) and (9):

$$p_2 - p = \frac{2\gamma}{r} - \frac{C}{3\phi r} + \Delta f + (1-\phi)\left(\frac{\partial \Delta f}{\partial \phi}\right)_r + \frac{r}{3\phi}\left(\frac{\partial \Delta f}{\partial r}\right)_\phi \tag{11}$$

In addition to the first two terms on the right-hand side, which are present in the case of a single droplet in a continuum, Eq. (11) contains terms that are due to the entropy of dispersion of the globules in the continuous medium and to the energy of interactions among globules.

B. An Alternative Derivation of the Basic Equations

From the point of view of traditional thermodynamics, a microemulsion is a multicomponent mixture in thermodynamic equilibrium. The change at constant temperature of the Helmholtz free energy of such a system can therefore be written as

$$dF = \sum \mu_i^* dN_i - p\, dV \tag{12}$$

where μ_i^* and N_i are the chemical potential in microemulsion and the number of moles of species i, respectively, and V is the volume of the microemulsion.

Let us now represent a microemulsion as a dispersion of globules of water in oil or of oil in water, with the surfactant and cosurfactant distributed at equilibrium among the dispersed and continuous media of the microemulsion and their interface. Assuming the globules to be spherical and of uniform radius, one can write the following expression for dF:

$$dF = \gamma\, d(AV) + CV\, d\left(\frac{1}{r}\right) + \sum \mu_i\, dN_i - p_2\, d(V\phi) - p_1\, d[V(1-\phi)] + d(V\Delta f) \tag{13}$$

Since Δf depends only on r and ϕ,

$$d\Delta f = \left(\frac{\partial \Delta f}{\partial r}\right)_\phi dr + \left(\frac{\partial \Delta f}{\partial \phi}\right)_r d\phi \tag{14}$$

and Eq. (13) can be rewritten in the form

$$dF = \left(\frac{3\phi\gamma}{r} - p_1(1-\phi) - p_2\phi + \Delta f\right) dV + \left[\frac{3\gamma V}{r} + p_1 V - p_2 V + V\left(\frac{\partial \Delta f}{\partial \phi}\right)_r\right] d\phi$$
$$+ \left[-\frac{3\phi\gamma V}{r^2} - \frac{CV}{r^2} + V\left(\frac{\partial \Delta f}{\partial r}\right)_\phi\right] dr + \sum \mu_i\, dN_i \tag{15}$$

Equation (15), which is based on a model, must, however, be equivalent to Eq. (12), which is based on the traditional thermodynamics of a multicomponent mixture. For the free energy changes given by Eqs. (12) and (15) to be the same for arbitrary changes in the independent variables V, ϕ, r, and N_i, the respective coefficients multiplying dV, $d\phi$, dr, and dN_i must be equal. It should be emphasized, however, that ϕ, r, V, and N_i are independent quantities only if the pressure p_1 is not imposed. Indeed ϕ depends on the distribution at equilibrium of the moles N_i of species i between the two media of the microemulsion and their interface, hence on N_is, V, the radius r, and the pressures p_1 and p_2. If pressure p_1 is imposed, then, because of the mechanical equilibrium condition [Eq. (16)], p_2 is a function of ϕ and r, and the N_i, V, ϕ, and r are not independent quantities. However, in the present case, p_1 is not imposed but is provided by the thermodynamic equilibrium condition (the minimum of the free energy F with respect to ϕ). Hence, ϕ, r, V, and N_i are independent variables. Consequently,

$$\frac{3\phi\gamma}{r} - p_1(1-\phi) - p_2\phi + \Delta f = -p \tag{16}$$

$$\frac{3\gamma}{r} + (p_1 - p_2) + \left(\frac{\partial \Delta f}{\partial \phi}\right)_r = 0 \tag{17}$$

$$\frac{-3\phi\gamma}{r^2} - \frac{C}{r^2} + \left(\frac{\partial \Delta f}{\partial r}\right)_\phi = 0 \tag{18}$$

and

$$\mu_i(\text{at } p_1 \text{ or } p_2) = \mu_i^*(p) \tag{19}$$

Equations (16), (17), and (18) can be rearranged in the form of Eqs. (5), (9), and (10), respectively. It is clear that the treatments in Secs. II.A and II.B lead to the same results. Equation (19) indicates that the chemical potential in the microemulsion at pressure p is equal to the chemical potential of the corresponding component in the frozen noninteracting system.

C. A Relation Between γ and C

Equation (5) can provide the radius of a microemulsion for a given ϕ, while Eqs. (9) and (10) allow us to calculate p_2 and p_1. However, to perform such calculations, expressions for Δf, γ, and C are needed. Traditional thermodynamics can provide only some information about C, which can be obtained as follows. Integrating Eq. (3) at constant $1/r$, γ, μ_i, p_2, and p_1 yields

$$f = A\gamma + \sum n_i\mu_i - p_2\phi - p_1(1-\phi) + \Delta f \tag{20}$$

which when differentiated leads to

$$df = A\,d\gamma + \gamma\,dA + \sum n_i\,d\mu_i + \sum \mu_i\,dn_i - p_2\,d\phi - p_1\,d(1-\phi) - \phi \times dp_2 - (1-\phi)\,dp_1 + d(\Delta f) \tag{21}$$

The Gibbs–Duhem equations for the dispersed and continuous phases have the forms

$$\sum n_i^d\,d\mu_i - \phi\,dp_2 = 0 \tag{22}$$

and

$$\sum n_i^c\,d\mu_i - (1-\phi)\,dp_1 = 0 \tag{23}$$

where the superscripts d and c stand for the dispersed and continuous media of the microemulsion, respectively.

Subtracting Eqs. (22) and (23) from Eq. (21) and using Eq. (3) yields

$$d\gamma + \sum \Gamma_i \, d\mu_i = \frac{rC}{3\phi} d\left(\frac{1}{r}\right) \equiv -\frac{C}{3\phi r} dr \tag{24}$$

where

$$\Gamma_i = \frac{n_i - (n_i^d + n_i^c)}{A}$$

is the surface excess of component i. Equation (24) leads to the expressions

$$C = \frac{3\phi}{r} \left(\frac{\partial \gamma}{\partial (1/r)}\right)_{\mu_i} \equiv -3\phi r \frac{\partial \gamma}{\partial r} \tag{25a}$$

and

$$\frac{\partial \Gamma_i}{\partial r} = \frac{1}{3\phi r} \frac{\partial C}{\partial \mu_i} \tag{25b}$$

The more compact arrangement of surfactant molecules on a surface of lower curvature suggests that $\partial \gamma / \partial r < 0$ and hence that C is a positive quantity. In addition, the surface excess is expected to increase with increasing radius, because the molecules can more easily pack on a surface of lower curvature. This means that $\partial C / \partial \mu_i$ is also a positive quantity.

III. MICROEMULSION COEXISTING WITH AN EXCESS DISPERSED PHASE

A. Phase Equilibria

The chemical potential of species i in the microemulsion can be defined as the partial derivative of F with respect to N_i at constant volume and temperature:

$$\mu_i^* = \left(\frac{\partial F}{\partial N_i}\right)_{V,T} \tag{26}$$

Equation (13), which is used to calculate μ_i^*, shows that F depends not only on N_i but also on r and ϕ. However, since $\partial F / \partial r = 0$ and $\partial F / \partial \phi = 0$, Eq. (26) can be replaced by

$$\mu_i^* = \left(\frac{\partial F}{\partial N_i}\right)_{V,T,r,\phi} \tag{27}$$

and Eq. (13) leads to

$$\mu_i^*(p) = \mu_i(\text{at } p_1 \text{ or } p_2) \tag{28}$$

Hence the chemical potential of any species of the microemulsion is equal to its chemical potential in the frozen noninteracting system, and this occurs because $\partial F / \partial r = 0$ and $\partial F / \partial \phi = 0$.

Since the composition of the excess dispersed phase is expected to be the same as that of the dispersed phase, the equality of their chemical potentials implies the equality of their pressures. The chemical potential in the dispersed phase is expressed in Eq. (3) at pressure p_2, while the pressure in the excess dispersed phase is equal to the external pressure p to which both the microemulsion and the excess phase are subjected. Consequently, at equilibrium,

$$p_2 = p \tag{29}$$

which, combined with Eq. (9), yields

$$\Delta f + (1-\phi)\left(\frac{\partial \Delta f}{\partial \phi}\right)_r - \frac{C}{\phi r} + \frac{r}{\phi}\left(\frac{\partial \Delta f}{\partial r}\right)_\phi = 0 \tag{30}$$

Equations (5) and (30) can provide the values of r and ϕ for a microemulsion in equilibrium with an excess dispersed phase. Of course, expressions for γ, C, and Δf are needed to carry out such calculations.

Δf is calculated in what follows by neglecting the interactions among globules. Expressions for the entropy of dispersion of the globules in the continuous phase were derived by Ruckenstein and Chi [12] on the basis of a lattice model, assuming (as is usually done in this kind of model) that the volume of a site is equal to the volume v_c of a molecule of the continuous phase. Only upper and lower bounds could be obtained for the entropy of dispersion of spherical globules of radius r and volume fraction ϕ in the continuous phase. For the free energy Δf corresponding to the upper bound of the entropy, it was found that

$$\Delta f = -\frac{3\phi kT}{4\pi r^3}\left\{-\frac{1}{\phi}\ln(1-\phi) + \ln\left[\frac{4\pi r^3}{3v_c}\left(\frac{1-\phi}{\phi}\right)\right]\right\} \tag{31}$$

and for that corresponding to the lower bound,

$$\Delta f = -\frac{3\phi kT}{4\pi r^3}\ln\left\{\frac{4\pi r^3}{3v_c}\left[\left(\frac{0.74}{\phi}\right)^{1/3} - 1\right]^3\right\} \tag{32}$$

where k is the Boltzmann constant.

The following empirical formula of Carnaham and Starling [39] for hard spheres, which is based on molecular dynamic simulations, will be used in this chapter:

$$\Delta f = -\frac{3\phi kT}{4\pi r^3}\left[1 - \ln\phi - \phi\frac{4-3\phi}{(1-\phi)^2} + \ln\left(\frac{4\pi r^3}{3v_c}\right)\right] \tag{33}$$

Equation (33) provides values between the above two bounds (nearer to the upper bound).

Combining Eqs. (5) and (30) with Eq. (33) yields

$$\gamma = \frac{kT}{4\pi r^2}\left[\ln\left(\frac{4\pi r^3}{3v_c\phi}\right) - \frac{8\phi - 5\phi^2}{(1-\phi)^2} + \phi\right] \tag{34}$$

and

$$C = \frac{3\phi kT}{4\pi r^2}\left[2\ln\left(\frac{4\pi r^3}{3v_c\phi}\right) + \frac{6\phi^2 - 5\phi - \phi^3}{(1-\phi)^2}\right] \tag{35}$$

Equations (34) and (35) provide some information about γ and C. They are plotted in Figs. 1 and 2, which show that γ is very small and that the ratio $C/6\phi\gamma$ [in Eq. (11), γ appears as $2\gamma/r$ and C as $C/3\phi r$] is on the order of unity. The conclusion is that 2γ cannot be neglected compared to $C/3\phi$. Figures 1 and 2 also show that both γ and C become very small with increasing values of r and tend to zero for $r \to \infty$.

It is of interest to notice that for values of ϕ greater than about 0.5, the interfacial tension has negative values, at least in some range of radii. For values greater than 0.6, it becomes negative over the entire range of radii. Figure 3 shows a clearer dependence of γ on ϕ and r than Fig. 1. Since, as shown later (see Sec. III.B), the interfacial tension γ_∞ between the microemulsion and dispersed excess phase is smaller than γ, γ_∞ will also be negative when the former is negative. For this reason, the interface between the

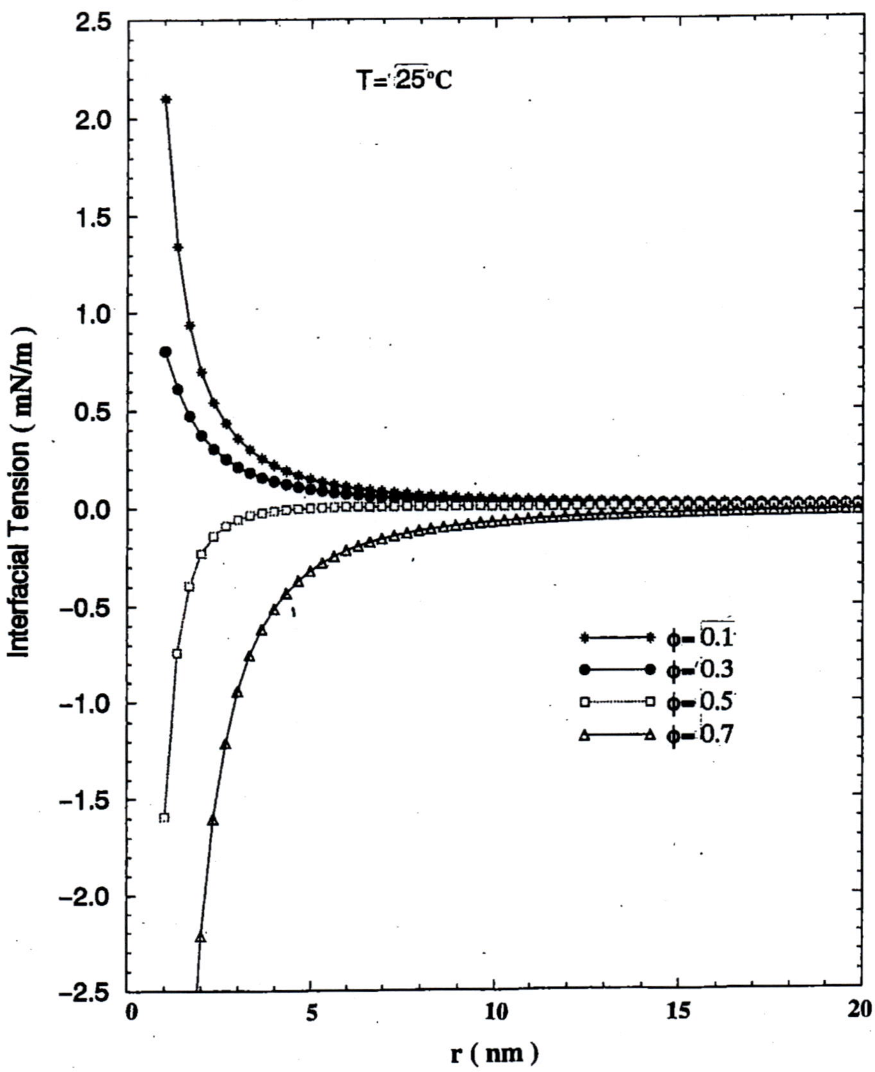

Figure 1 Interfacial tension γ versus radius r when a microemulsion coexists with an excess dispersed phase.

microemulsion and excess dispersed phase becomes unstable to thermal perturbations, spontaneous emulsification occurs, and the corresponding microemulsion cannot have physical reality. This may explain why for values of ϕ around 0.5 there is a transition to a middle-phase microemulsion that coexists with both excess phases.

The prediction of r and ϕ as a function of the natures of surfactant, cosurfactant, and hydrophobic phase requires the derivation of expressions for γ and C that should account for, among other things, their dependence on the curvature. Such expressions are not yet available.

It is of interest to note that, because at equilibrium $p_2 = p$, in a single-phase microemulsion the pressure p_2 in the globules is expected to be lower than that in the

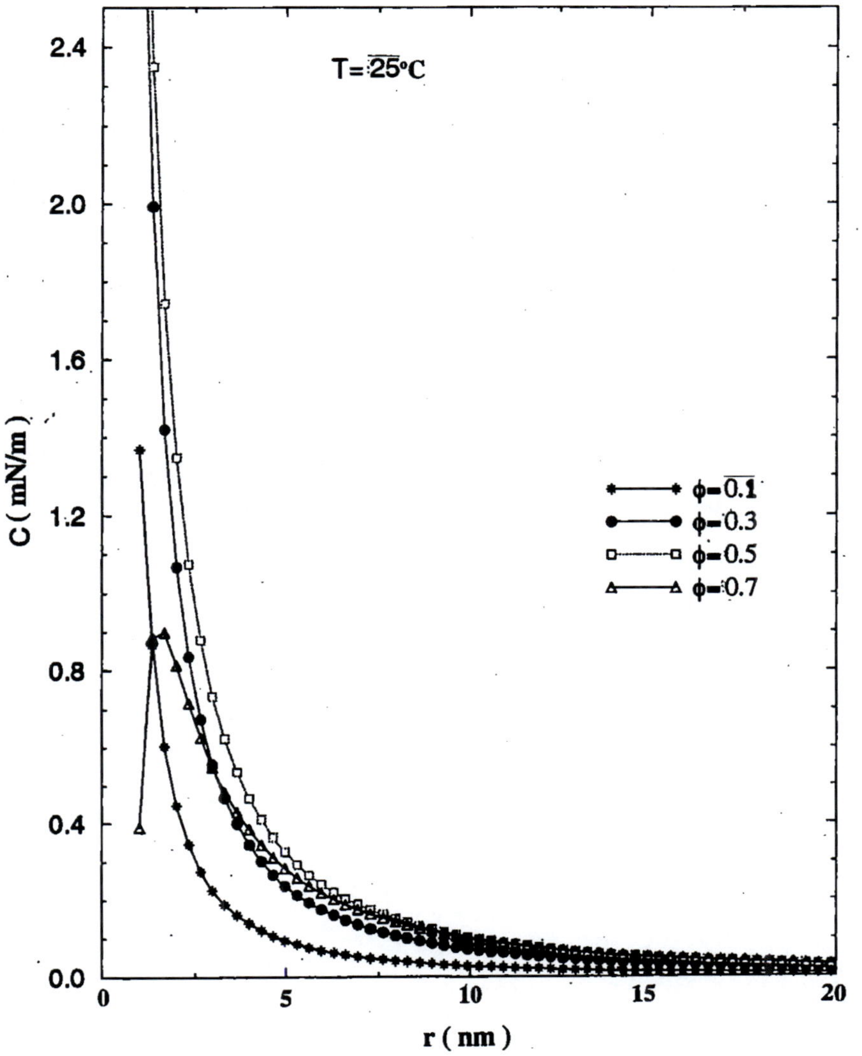

Figure 2 The bending stress C versus radius r when a microemulsion coexists with an excess dispersed phase.

continuous phase. Combining Eqs. (5) and (9) yields

$$p_2 - p = \frac{3\gamma}{r} + \Delta f + (1 - \phi)\left(\frac{\partial \Delta f}{\partial \phi}\right)_r$$

Because γ must be positive, $p_2 - p$ can become negative only if $\Delta f + (1 - \phi)\ \partial\Delta f/\partial\phi$ is sufficiently negative. The entropy of dispersion provides a negative contribution to Δf. When Δf is dominated by the entropy of dispersion, $\partial\Delta f/\partial\phi$ is expected to be negative for not too large values of ϕ, because the number of globules increases with increasing ϕ, and this increases the entropy of dispersion. At large values of ϕ, the excluded volume increases, which decreases the entropy. As long as the effect of the number of globules provides the main effect, $\partial\Delta f/\partial\phi < 0$.

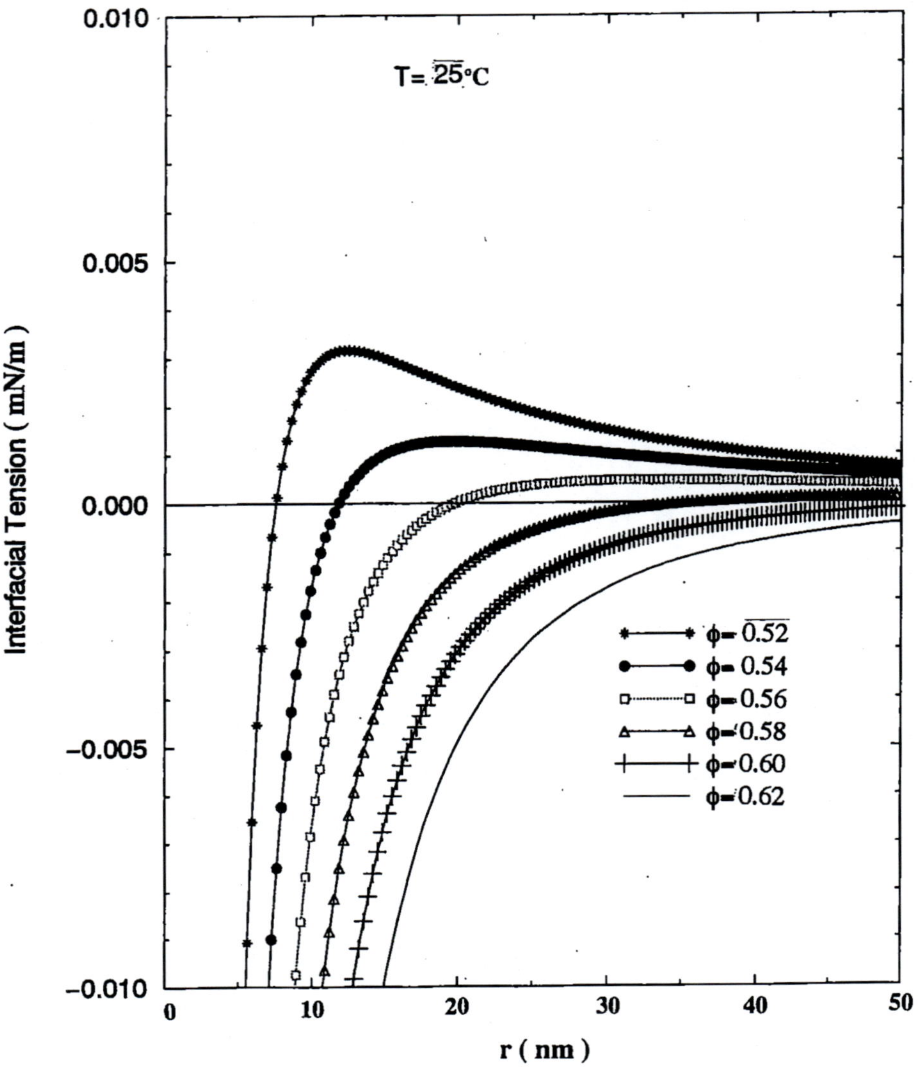

Figure 3 Interfacial tension γ versus radius r when a microemulsion coexists with an excess dispersed phase, for ϕ between 0.5 and 0.6.

B. Interfacial Tension Between Microemulsion and Excess Dispersed Phase

Integrating Eq. (24) at constant chemical potentials, one obtains

$$\gamma_\infty - \gamma = \int_r^\infty \frac{rC}{3\phi}\, d\left(\frac{1}{r}\right) \equiv -\int_r^\infty \frac{C}{3\phi r}\, dr \qquad (36)$$

where, because the chemical potentials of the surfactants are the same at the surface of the globules and the interface between microemulsion and excess phase, γ_∞ can be considered the interfacial tension between microemulsion and excess dispersed phase. As already noted, C is a positive quantity. For this reason, $\gamma > \gamma_\infty$.

In Eq. (36), the quantity γ represents the interfacial tension at the surface of globules of radius r and is given by Eq. (34). Equation (35) cannot, however, be used in Eq. (36) for the bending stress C, because it does not involve in its derivation the restriction of constant chemical potentials on which Eq. (36) is based.

IV. MICROEMULSION COEXISTING WITH BOTH EXCESS PHASES

A microemulsion can coexist with an excess continuous phase when the chemical potential of the continuous phase of the microemulsion becomes equal to the chemical potential of the excess continuous phase. Since the chemical potential in the continuous phase of the microemulsion is equal to that of the frozen noninteracting system, which is expressed at pressure p_1, and that of the excess continuous phase is expressed at the external pressure p, the equality of the chemical potentials implies that

$$p_1 = p \tag{37}$$

Combining Eqs. (1) and (37) yields

$$\Delta f - \phi\left(\frac{\partial \Delta f}{\partial \phi}\right)_r = 0 \tag{38}$$

Using Eq. (33) for Δf, one can conclude that the radius of the globules diverges when an excess continuous phase coexists with the microemulsion. In addition, using Eq. (33) for Δf and Eq. (35) for C, Eq. (6) leads to

$$p_2 - p_1 = \frac{3\phi kT}{4\pi r^3}\left(\frac{1+\phi+\phi^2-\phi^3}{(1-\phi)^3}\right) \tag{39}$$

Equation (39) shows that for $r \to \infty$,

$$p_2 - p_1 = 0 \tag{40}$$

In other words, the excess continuous phase cannot coexist alone with the microemulsion, since the equality $p_1 - p = 0$ implies that $p_2 = p_1 = p$, hence the coexistence of the microemulsion with both excess phases.

The preceding condition of thermodynamic equilibrium implies that the curvature of the dispersed phase becomes zero at the transition from two to three phases. In the middle-phase microemulsion, the pressures p_1 and p_2 fluctuate in time and space because of the instability of the interface between the two media (see below), and intuïtion suggests that Eqs. (37) and (40) be replaced with their average, hence that the condition of zero curvature be replaced with the condition of mean (with respect to time) average curvature.

Two structures are compatible with a zero average curvature. One of them is planar, and the other is bicontinuous [40,41]. A rigid bicontinuous structure for the middle phase was postulated by Scriven [27]. In contrast, in the present treatment, the zero average curvature is almost a natural result of the condition of thermodynamic equilibrium. Because of the extremely low interfacial tension, (Eq. (34) indeed suggests that the interfacial tension is extremely low in this case), the interface between the two media is extremely unstable to thermal perturbations; consequently, the interface cannot be rigid but probably has a complex oscillating, turbulent behavior accompanied by breakup and coalescence. The present considerations explain the large fluctuations observed experimentally [42] near this point and why this behavior is similar to that existing near a critical point.

V. CONCLUSIONS

1. General equations have been derived that can be used for single-phase microemulsions to calculate the dependence of the radius of the globules on the volume fraction of the dispersed phase, and for two-phase systems to calculate the values of both r and ϕ at equilibrium. Of course, thermodynamics provides only general equations among various quantities and is incomplete. Expressions for the interfacial tension and bending stress at the surface of the globules are needed to make predictions.
2. The phase equilibria between a microemulsion and the excess dispersed phase or both excess phases can be explained in terms of pressures.
3. The Laplace equation is no longer valid for a microemulsion. The new equation contains terms that are due to the entropy of dispersion of the globules in the continuous phase and to the energy of interaction among globules.
4. The curvature becomes zero at the transition to a microemulsion in equilibrium with both excess phases. This result is compatible with a planar or bicontinuous structure with a nonrigid interface because of its instability to thermal perturbations.

REFERENCES

1. E. Ruckenstein, Ann. N.Y. Acad. Sci. *404*:224 (1983).
2. E. Ruckenstein, Fluid Phase Equilib. *20*:189 (1985).
3. T. P. Hoar and J. H. Schulman, Nature *152*:102 (1943).
4. J. H. Schulman and D. R. Riley, J. Colloid Sci. *3*:383 (1948).
5. W. Stoeckenius, J. H. Schulman, and L. M. Prince, Kolloid Z. *169*:170 (1960).
6. J. E. L. Bowcott and J. H. Schulman, Z. Electrochem. *59*:283 (1955).
7. J. H. Schulman and J. A. Friend, J. Colloid Sci. *4*:1988 (1971).
8. P. A. Winsor, *Solvent Properties of Amphiphilic Compounds*, Butterworth, London, 1954.
9. P. A. Winsor, Chem. Rev. *68*:1 (1968).
10. R. N. Healy, R. L. Reed, and D. G. Stenmark, Soc. Petrol. Eng. J. Trans. AIME *261*:147 (1976).
11. M. Bourel, C. Konkounis, R. Schecter, and W. Wade, Dispersion Sci. Tech. *1*:13 (1980).
12. E. Ruckenstein and J. C. Chi, Faraday Trans. II *71*:1690 (1975).
13. V. I. Likhtman, E. D. Shchukin, and P. A. Rehbinder, *Physico-Chemical Mechanics of Metals*, Izdatelstvo Akad. Nauk SSSR, Moscow, p. 182. English translation by Israel Program for Scientific Translations, 1964.
14. H. Reiss, J. Colloid Interface Sci. *53*:61 (1975).
15. E. Ruckenstein, Chem. Phys. Lett. *56*:518 (1978).
16. E. Ruckenstein, Faraday Disc. Chem. Soc. *65*:141 (1978).
17. J. Th. G. Overbeek, Faraday Disc. Chem. Soc. *65*:7 (1978).
18. E. Ruckenstein and R. Krishman, J. Colloid Interface Sci. *71*:321 (1979).
19. K. Shinoda and H. Saito, J. Colloid Interface Sci. *26*:70 (1968); *30*:258 (1969).
20. M. Robbins, in *Micellization, Solubilization and Microemulsions*, Vol. 2 (K. L. Mittal, ed.), Plenum, New York, 1977, p. 713.
21. C. A. Miller and P. Neogi, AIChE J. *26*:212 (1980).
22. S. A. Safran, J. Chem. Phys. *78*:2073 (1981).
23. S. A. Safran and L. A. Turkevich, Phys. Rev. Lett. *50*:1930 (1983).
24. S. A. Safran, L. A. Turkevich, and P. A. Pincus, J. Phys. Lett. *45*:L69 (1984).
25. S. A. Safran, I. Webman, and G. S. Grest, Phys. Rev. A *32*:506 (1985).
26. W. Helfrich, Z. Naturforsch. *38*:6693 (1973).

27. L. E. Scriven, in *Micellization, Solubilization, and Microemulsions*, Vol. 2 (K. L. Mittal, ed.), Plenum, New York, 1977, p. 877.
28. Y. Talmon and S. Prager, J. Chem. Phys. *69*:2984 (1978).
29. J. Jouffroy, P. Levinson, and P. G. de Gennes, J. Phys. (Paris) *43*:1241 (1982).
30. B. Widom, J. Phys. Chem. *88*:6508 (1984).
31. B. Widom, J. Chem. Phys. *84*:6943 (1986).
32. M. Schick and W. H. Shih, Phys. Rev. *B34*:1797 (1986).
33. M. Schick and W. H. Shih, Phys. Rev. Lett. *59*:1205 (1987).
34. J. W. Halley and A. J. Kolan, J. Chem. Phys. *88*:3313 (1988).
35. R. G. Larson, J. Chem. Phys. *91*:2479 (1989).
36. G. Gompper and M. Schick, Phys. Rev. *A42*:2137 (1990).
37. R. Nagarajan and E. Ruckenstein, in *Equations of State for Fluids and Fluid Mixtures* (I.V. Sengers, ed.), (in press).
38. R. Nagarajan and E. Ruckenstein, Langmuir *7*:2934 (1991).
39. N. F. Carnaham and K. E. Starling, J. Chem. Phys. *51*:635 (1969).
40. E. R. Neovius, *Minimalflacken*, Frenkel, Helsingfors, Denmark, 1883.
41. M. A. Schwartz, *Gesammtte Mathematische Abhandlung*, Springer, Berlin, 1890.
42. A. M. Cazabat, D. Langevin, J. Meunier, and A. Pouchelon, Adv. Colloid Interface Sci. *16*:175 (1982).

6400 *Langmuir* **2000,** *16,* 6400–6415

Molecular Theory of Microemulsions

R. Nagarajan*

Department of Chemical Engineering, The Pennsylvania State University, University Park, Pennsylvania 16802

E. Ruckenstein

Department of Chemical Engineering, State University of New York at Buffalo, Amherst, New York 14260

Received December 2, 1999. In Final Form: May 25, 2000

In this paper, a molecular thermodynamic approach is developed to predict the structural and compositional characteristics of microemulsions. The theory can be applied not only to oil-in-water and water-in-oil droplet-type microemulsions but also to bicontinuous microemulsions. This treatment constitutes an extension of our earlier approaches to micelles, mixed micelles, and solubilization but also takes into account the self-association of alcohol in the oil phase and the excluded-volume interactions among the droplets. Illustrative results are presented for an anionic surfactant (SDS)–pentanol–cyclohexane–water–NaCl system. Microstructural features including the droplet radius, the thickness of the surfactant layer at the interface, the number of molecules of various species in a droplet, the size and composition dispersions of the droplets, and the distribution of the surfactant, oil, alcohol, and water molecules in the various microdomains are calculated. Further, the model allows the identification of the transition from a two-phase droplet-type microemulsion system to a three-phase microemulsion system involving a bicontinuous microemulsion. The persistence length of the bicontinuous microemulsion is also predicted by the model. Finally, the model permits the calculation of the interfacial tension between a microemulsion and the coexisting phase.

1. Introduction

Microemulsions are thermodynamically stable dispersions of oil and water stabilized by a surfactant and, in many cases, also a cosurfactant.[1–4] The microemulsions can be of the droplet type, either with spherical oil droplets dispersed in a continuous medium of water (oil-in-water microemulsions, O/W) or with spherical water droplets dispersed in a continuous medium of oil (water-in-oil microemulsions, W/O). The droplet-type microemulsions can be either a single-phase system or part of a two-phase system wherein the microemulsion phase coexists with an excess dispersed phase (an upper phase of excess oil in the case of O/W and a lower phase of excess water in the case of W/O microemulsions). There are also non-droplet-type microemulsions, referred to as middle-phase microemulsions. In this case, the microemulsion phase is part of a three-phase system with the microemulsion phase in the middle coexisting with an upper phase of excess oil and a lower phase of excess water. One possible structure of this middle-phase microemulsion, characterized by randomly distributed oil and water microdomains and bicontinuity in both oil and water domains, is known as the bicontinuous microemulsion. Numerous experimental studies have shown[1,2,4] that one can achieve a transition

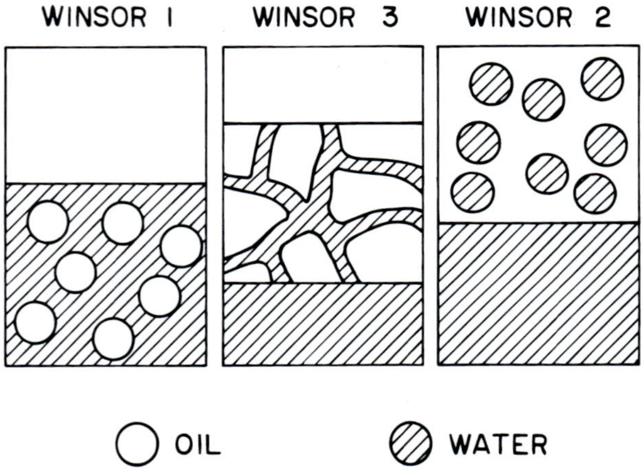

Figure 1. Schematic representation of the transition from a two-phase system with an oil-in-water droplet microemulsion to a three phase system with a bicontinuous microemulsion and to a two-phase system with a water-in-oil droplet microemulsion.

from a two-phase system involving an O/W droplet-type microemulsion to a two-phase system involving a W/O droplet-type microemulsion by passing through a three-phase system containing a bicontinuous microemulsion (Figure 1). Such phase transitions can be realized in many different ways, e.g., by increasing the concentration of added electrolytes when the surfactant employed is ionic, by increasing the temperature when the surfactant is nonionic, and by increasing the concentration of alcohol (typically butanol to hexanol) when the surfactant is ionic, nonionic, or zwitterionic.

Theoretical studies of microemulsions have focused on various features of the complex systems mentioned

* To whom correspondence should be addressed (rxn@psu.edu).

(1) Rosano, H. L., Clausse, M., Eds. *Microemulsion Systems*; Marcel Dekker: New York, 1987.
(2) Friberg, S. E.; Bothorel, P. *Microemulsions: Structure and Dynamics*; CRC Press: Boca Raton, FL, 1987.
(3) Ruckenstein, E.; Chi, J. *J. Chem. Soc., Faraday Trans. 2* **1975**, *71*, 1690. Ruckenstein, E. *Fluid Phase Equilib.* **1985**, *20*, 189; *J. Colloid Interface Sci.* **1998**, *204*, 142.
(4) Verhoeckx, G. J.; De Bruyn, P. L.; Overbeek, J. Th. G. *J. Colloid Interface Sci.* **1987**, *119*, 409. Overbeek, J. Th. G.; Verhoeckx, G. J.; De Bruyn, P. L.; Lekkerkerker, H. N. W. *J. Colloid Interface Sci.* **1987**, *119*, 422; De Bruyn, P. L.; Overbeek J. Th. G.; Verhoeckx, G. J. *J. Colloid Interface Sci.* **1989**, *127*, 244.

10.1021/la991578t CCC: $19.00
Published on Web 07/14/2000

above.[3–10] However, currently there exists no model which can predict the detailed phase behavior and the microstructural characteristics of the microemulsion systems starting from the composition and molecular structures of the constituent molecules. Such models were formulated in our earlier work for micelles, vesicles, and mixed micelles in aqueous media and for the solubilization of oil inside such aggregates.[11–15] These models are extended here to microemulsions. On this basis, a predictive approach for the various characteristics of microemulsions, such as the droplet radius, the number of molecules of various species in a droplet, and the size and composition dispersions of the droplets, is developed. The model also allows the prediction of the interfacial tension between the microemulsion and the coexisting phase. Further, the model allows one to examine the transition from a two-phase system involving a droplet-type microemulsion to a three-phase system involving a bicontinuous microemulsion. The persistence length in bicontinuous microemulsions is also predicted.

In developing a molecular treatment for microemulsions, we take into account the chain packing, which depends on the curvature of the aggregates, the self-association of alcohol in the oil, the penetration of the interfacial layer region of the microemulsion by the oil molecules, their effects on chain packing and interfacial energies, and the hard-core interactions among the droplets. Since the present treatment of microemulsions is an extension of our approach to micellization and solubilization, many of the above aspects of the model that were presented in detail in earlier papers are only briefly summarized here. Two additional features included in this paper are self-association of alcohol in oil domains and hard-sphere repulsive interactions among droplets.

To illustrate the present theory, calculations were carried out for a system consisting of an anionic surfactant, oil, water, alcohol, and electrolyte. Obviously, this model can be readily applied to simpler microemulsions free of alcohol, wherein the phase transitions are caused by varying the electrolyte concentration or the temperature.

2. Geometric Characteristics of Microemulsions

Both O/W and W/O droplet structures have a core region surrounded by an interfacial film, depicted schematically in Figure 2. The interfacial film excludes the headgroups of the surfactant and the alcohol. This region is analogous to the micelle core and, as confirmed by many experiments, is free of water molecules and is thus hydrophobic in character. Each droplet contains g_k molecules of kind k (with k = S for surfactant, A for alcohol, O for oil, and W for water), among which g_{kI} molecules are present in the interfacial layer and the remaining ones are present in the core containing mainly the dispersed phase. For the reason mentioned above, there are no water molecules in the interfacial layer (i.e., $g_{WI} = 0$). The radii R_O and R_W provide the boundaries of the interfacial layer. The volume of the aggregate V_g can be related to the radii R_O and R_W via

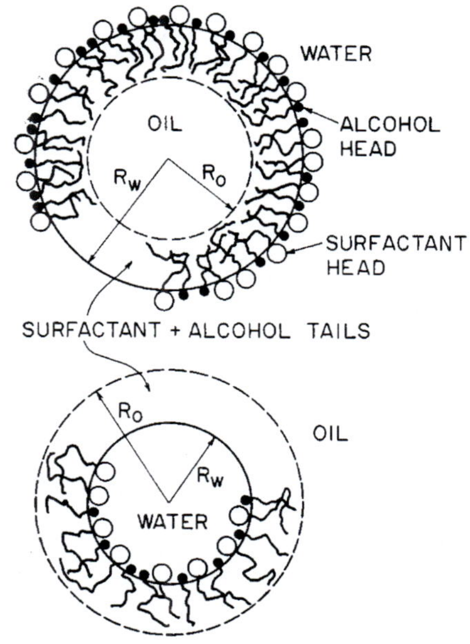

Figure 2. Geometrical variables R_W and R_O characterizing the size of O/W and W/O droplets in microemulsions.

$$V_g = \frac{4\pi R_O^3}{3} \tag{2.1}$$

for the water-in-oil droplets and

$$V_g = \frac{4\pi R_W^3}{3} + g_{SI}v_{SP} + g_{AI}v_{AP} \tag{2.2}$$

for the oil-in-water droplets, where v_{SP} and v_{AP} are the volumes of the polar headgroups of the surfactant and alcohol molecules, respectively. The volume V_I of the interfacial layer per surfactant molecule is given by

$$V_I = \frac{4\pi}{3g_{SI}}|R_O^3 - R_W^3| = \frac{g_{SI}v_{ST} + g_{AI}v_{AT} + g_{OI}v_O}{g_{SI}} \tag{2.3}$$

where v_{ST}, v_{AT}, and v_O are the volumes of surfactant tail, alcohol tail, and oil, respectively. The surface area of the droplet in contact with water, per surfactant molecule, is given by

$$a = \frac{4\pi R_W^2}{g_{SI}} \tag{2.4}$$

Combining eqs 2.3 and 2.4 yields

$$\frac{R_O}{R_W} = \left(\frac{3V_I}{a|R_O - R_W|} - \frac{3}{4}\right)^{1/2} - \frac{1}{2} \tag{2.5}$$

For the sake of simplicity, the thickness of the interfacial layer $|R_O - R_W|$ is taken to be the composition average of the surfactant tail length l_S and the alcohol tail length l_A:

(5) Scriven, L. In *Micellization, Solubilization and Microemulsions*; Mittal, K. L., Ed.; Plenum Press: New York, 1977; p 277.
(6) Talmon, Y.; Prager, S. *J. Chem. Phys.* **1978**, *69*, 2984.
(7) de Gennes, P. G.; Taupin, C. *J. Phys. Chem.* **1982**, *86*, 2284.
(8) Ruckenstein, E. In *Progress in Microemulsions*; Martellucci, S.; Chester, A. N., Eds.; Plenum Press: New York, 1989; pp 1 and 41.
(9) Peck, D. G.; Schechter, R. S.; Johnston, K. P. *J. Phys. Chem.* **1991**, *95*, 9541.
(10) Borkovac, M. *J. Chem. Phys.* **1989**, *91*, 6268; *Adv. Colloid Interface Sci.* **1992**, *37*, 195.
(11) Nagarajan, R.; Ruckenstein, E. *J. Colloid Interface Sci.* **1979**, *71*, 580; *Langmuir* **1991**, *7*, 2934.
(12) Nagarajan, R. *Langmuir* **1985**, *1*, 331; in *Mixed Surfactant Systems*; Holland, P. M., Rubingh, D. N., Eds.; ACS Symposium Series 501; American Chemical Society: Washington, DC, 1992; p 54.
(13) Rao, I. V., Ruckenstein, E. *J. Colloid Interface Sci.* **1986**, *113*, 375.
(14) Nagarajan, R.; Ruckenstein, E. *Sep. Sci. Technol.* **1981**, *16*, 1429; in *Surfactants in Solution. Theoretical and Applied Aspects*; Mittal, K. L., Lindman, B., Eds.; Plenum Press: New York, 1984; p 923. Nagarajan, R.; Chaiko, M. A.; Ruckenstein, E. *J. Phys. Chem.* **1984**, *88*, 2916.
(15) Ruckenstein, E. In *Progress in Microemulsions*; Martellucci, S., Chester, A. N., Eds.; Plenum Press: New York, 1989; p 31; *Colloids Surf.* **1992**, *65*, 95.

$$|R_O - R_W| = \frac{l_S + (g_{AI}/g_{SI})l_A}{1 + (g_{AI}/g_{SI})} \tag{2.6}$$

The determination of this thickness from free energy minimum considerations require a more detailed treatment of the interfacial film than justified by the available experimental information.

Consequently, a particular O/W or W/O droplet can be completely characterized by three geometrical variables. A set of convenient variables is (i) the surface area of the droplet in contact with water per surfactant molecule, a, (ii) the ratio of alcohol-to-surfactant molecules in the interfacial layer, g_{AI}/g_{SI}, and (iii) the ratio of oil-to-surfactant molecules in the interfacial layer, g_{OI}/g_{SI}. For a flat interfacial layer, denoting the layer thickness by $|R_O - R_W|_F$ and the area per surfactant molecule of the flat interface by a_F, one can write the following on the basis of molecular packing considerations:

$$a_F|R_O - R_W|_F = V_I \tag{2.7}$$

Therefore, for a flat interface, only two geometrical variables are needed for its complete specification, instead of the three needed for a spherical droplet.

3. Size Distribution of Droplet-Type Microemulsions

For a W/O microemulsion, the size and composition distribution of the aggregates can be written as for micelles, mixed micelles, and solubilization. However, whereas the earlier expressions were formulated for dilute solutions where interaggregate interactions are negligible, the size and composition distribution for microemulsions will take into account the hard-sphere repulsions among the droplets. For a single-phase microemulsion, the mole fraction X_{gO} of water-in-oil droplets is given by the following expression (see Appendix A for the derivation):

$$X_{gO}\gamma_{gO} = \Omega^{g_{SI}} f_{WO}^{\,g_W}$$

$$\Omega = (X_{SW}\gamma_{SW})(X_{AW}\gamma_{AW})^{g_{AI}/g_{SI}}(X_{OO}\gamma_{OO})^{g_{OI}/g_{SI}} \exp\left(-\frac{\Delta\mu_{gI}^{\circ}}{kT}\right)$$

$$\Delta\mu_{gI}^{\circ} = \frac{1}{g_{SI}}\mu_{gI}^{\circ} - \mu_{SW}^{\circ} - \frac{g_{AI}}{g_{SI}}\mu_{AW}^{\circ} - \frac{g_{OI}}{g_{SI}}\mu_{OO}^{\circ} \tag{3.1}$$

In the preceding equations, the subscript g refers to a droplet containing a total of g molecules ($g = g_S + g_A + g_O + g_W$), X_{gO} is the mole fraction of aggregates of size g in the continuous oil (O) phase, X_{SW} and X_{AW} are the mole fractions of singly dispersed surfactant and alcohol molecules, respectively, in the dispersed water phase, μ_{SW}° and μ_{AW}° are their standard chemical potentials, and γ_{SW} and γ_{AW} are their activity coefficients. X_{OO} denotes the mole fraction of oil in the continuous oil phase, μ_{OO}° is its standard chemical potential, and γ_{OO} is its activity coefficient.

It is important to emphasize that the alcohols used as cosurfactants have high solubility in the oil phase because they are present there as both single molecules and aggregates. Therefore, in calculating X_{OO}, one must consider the self-association of alcohol in the oil phase. The activity coefficients arise because of the hard-sphere interactions among the droplets. Therefore, in the dispersed phase, which is free of droplets, the activity coefficients are equal to unity, whereas, in the continuous phase, which contains the droplets, the activity coefficients differ from unity.

The fractional saturation of water in the oil phase is denoted f_{WO}. For a single-phase microemulsion, f_{WO} is less than unity. As the amount of water in the phase increases, f_{WO} progressively increases and reaches the constant value of unity when an excess aqueous phase coexisting with the microemulsion phase appears, creating a two-phase system.

The standard free energy difference $\Delta\mu_{gI}^{\circ}$ in eq 3.1 is due to the transfer of one surfactant molecule and (g_{AI}/g_{SI}) alcohol molecules from water and (g_{OI}/g_{SI}) oil molecules from pure oil to the interfacial layer of the microemulsion droplet. Expressions for $\Delta\mu_{gI}^{\circ}$ are provided in section 7 of the paper.

When eq 3.1 is used to calculate the size and composition distributions of droplets in a single-phase O/W microemulsion, X_{gO} and γ_{gO} should be replaced by X_{gW} and γ_{gW} and the factor $f_{WO}^{\,g_W}$ should be replaced by $f_{OW}^{\,g_O}$, where f_{OW} denotes the fractional saturation of oil in the water phase. As before, f_{OW} is less than unity for a single-phase microemulsion and reaches the constant value of unity when a two-phase system comes into existence with an excess oil phase coexisting with the microemulsion phase.

To predict the characteristics of a microemulsion on the basis of the size distribution relation (3.1), only two control variables, X_{SW} and X_{AW}, must be specified. This is equivalent to fixing the total amounts of surfactant and alcohol in the microemulsion. Detailed calculations of the size and composition distributions involved in eq 3.1 are not required. Instead, as shown in the following section and Appendix C, the size and composition dispersions can be obtained from calculations involving only the maximum term of the size distribution. By calculating how the size and composition corresponding to the most populous aggregates (i.e., the maximum in X_g) vary when the surfactant concentration is slightly perturbed, one can obtain the variance in the size and composition of aggregates with respect to their average values.

In eq 3.1, the activity coefficients appear as a result of the hard-sphere repulsions among the droplets. Since the calculations focus on the most populous aggregates, the hard-sphere repulsions will be expressed in terms of a single droplet size corresponding to the most populous aggregates. One can derive expressions for the activity coefficients γ_{kO} of a component k in the continuous phase O starting from an equation for the osmotic pressure of a hard-sphere fluid,[3,4] such as that based on the Carnahan–Starling equation of state (see Appendix B for the derivation):

$$\ln \gamma_{kO} = -\frac{v_k}{V_{gO}}\Phi_{gO}\frac{1 + \Phi_{gO} + \Phi_{gO}^2 - \Phi_{gO}^3}{(1 - \Phi_{gO})^3} \tag{3.2}$$

where V_{gO} is the volume of a droplet and Φ_{gO} is the volume fraction of the droplets in the microemulsion. Combining eq 3.2 with the Gibbs–Duhem equation, we obtain the following expression for the activity coefficient γ_{gO} of the droplet (Appendix B):

$$\ln \gamma_{gO} = \ln\left(\frac{\Phi_{gO}/V_{gO}}{X_{gO}\sum_{k\neq g}\Phi_{kO}/v_k}\right) + \frac{7\Phi_{gO} - 3\Phi_{gO}^2 - \Phi_{gO}^3}{(1 - \Phi_{gO})^2} \tag{3.3}$$

where Φ_{kO} is the volume fraction of component k in the continuous phase O.

Equation 3.1 is an implicit relation because the activity coefficients γ_{gO} and γ_{gW} depend on the volume fractions of the droplets and hence on the mole fractions X_{gO} and X_{gW}, respectively. The volume fractions of the droplets can be related to the mole fractions of various components and their respective volumes through the relations

$$\Phi_{gW} = \frac{X_{gW}V_{gW}}{X_{gW}V_{gW} + X_{SW}v_S + X_{WW}v_W + X_{AW}v_A}$$

$$\Phi_{gO} = \frac{X_{gO}V_{gO}}{X_{gO}V_{gO} + \sum jX_{jAO}v_A + X_{OO}v_O} \qquad (3.4)$$

In the first expression, X_{WW} can be replaced by $(1 - X_{gW} - X_{AW} - X_{SW})$ because the sum of all mole fractions should be unity. In the second expression, the total mole fraction $\sum jX_{jAO}$ of alcohol in the oil phase accounts for the alcohol present as both monomers and aggregates in the oil phase and is calculated on the basis of the continuous-association model[16] as described in Appendix D.

For given values of the control variables X_{SW} and X_{AW}, the maximum in X_{gO} (or X_{gW}) was found, using the IMSL subroutine ZXMWD, by solving the implicit eq 3.1 in combination with eq 3.4. As mentioned in section 2, the area per surfactant molecule a, the alcohol-to-surfactant ratio g_{AI}/g_{SI}, and the oil-to-surfactant ratio g_{OI}/g_{SI} in the interfacial layer were selected as the three independent variables with respect to which the maximization was carried out. The total volume fraction Φ_S of surfactant present in the microemulsion is given by

$$\Phi_S = \Phi_{gO}\frac{g_{SI}v_S}{V_{gO}} \text{ (for W/O)} \qquad \Phi_S = \Phi_{gW}\frac{g_{SI}v_S}{V_{gW}} \text{ (for O/W)} \qquad (3.5)$$

because the amount of singly dispersed surfactant molecules is relatively small and can be neglected. The total volume fraction of alcohol in the microemulsion, Φ_A, can be obtained from a species balance, since the concentrations of alcohol in the water domain, the oil domain, and the interfacial layer are known.

4. Size and Composition Dispersions of Droplets

The size and composition dispersions of droplets can be estimated by using the maximum-term method, without performing detailed distribution calculations. This is accomplished by adapting the approach used for micellar systems,[17,18] as described in Appendix C. The aggregate corresponding to the maximum of X_{gO} or X_{gW} is considered to provide the number-average size and composition of the equilibrium droplets. For each component k (=S, A, O, or W) present in the microemulsion, the mean-square deviation $\sigma_n^2(\mathrm{k})$ from the number-average number of molecules $g_n(\mathrm{k})$ can be shown to be

$$\frac{\sigma_n(\mathrm{k})}{g_n(\mathrm{k})} = \left(\frac{\Delta \ln g_n(\mathrm{k})}{\Delta \ln(g_n(\mathrm{k})X_{gM})}\right)^{1/2} \qquad (4.1)$$

where X_{gM} ($=X_{gO}$ or X_{gW}) is the mole fraction of the droplet corresponding to the maximum in the distribution, $g_n(\mathrm{k})$ are the numbers of molecules of kind k in that droplet, and Δ denotes the difference between two values. Consequently, by calculating the maximum X_{gM} and the corresponding numbers of molecules $g_n(\mathrm{k})$ for two slightly different conditions (for example, through marginal variations in X_{SW}), one can estimate $\sigma_n(\mathrm{k})$ for each of the four components k and thus provide information about the size and composition dispersions of the droplet.

The root-mean-square deviation in the droplet radius σ_R is obtained from purely geometric considerations using the calculated values of $\sigma_n(\mathrm{k})$. Equating the volume of the droplet to the volumes contributed by all the component molecules yields

$$\frac{4\pi R^3}{3} = \sum g_n(\mathrm{k})\, v_\mathrm{k}$$

$$\frac{4\pi(R + \sigma_R)^3}{3} - \frac{4\pi R^3}{3} = \sum \sigma_n(\mathrm{k})\, v_\mathrm{k} \qquad (4.2)$$

By rearranging eq 4.2, we obtain the following expression for the dispersion in the droplet radius:

$$\frac{\sigma_R}{R} = \left(1 + \frac{\sum \sigma_n(\mathrm{k})\, v_\mathrm{k}}{\sum g_n(\mathrm{k})\, v_\mathrm{k}}\right)^{1/3} - 1 \qquad (4.3)$$

5. Calculation of Interfacial Tension

The interfacial tension γ_F refers to that at the flat interface (denoted by subscript F) that separates the microemulsion from the coexisting excess dispersed phase. The flat interface can be treated as a planar infinite aggregate. The chemical potentials of the molecules in this interfacial layer should be equal to the chemical potentials of the molecules in the bulk phases and hence

$$\left(\frac{1}{g_{SI}}\right)_F \frac{(\mu_{gI}^\circ)_F}{kT} - \frac{\gamma_F a_F}{kT} = \frac{\mu_{SW}}{kT} + \left(\frac{g_{AI}}{g_{SI}}\right)_F \frac{\mu_{AW}}{kT} + \left(\frac{g_{OI}}{g_{SI}}\right)_F \frac{\mu_{OO}}{kT} \qquad (5.1)$$

In the preceding equation, the first term on the left-hand side is the reference state chemical potential per surfactant molecule of an infinite-size aggregate and the second term involving γ_F accounts for the interfacial tension to which this infinite aggregate is subjected. The terms on the right-hand side represent the chemical potentials of the surfactant, alcohol, and oil in the bulk phase in equilibrium with the molecules at the interface. Using eq 3.1, the above equation can be rewritten as

$$\frac{\gamma_F a_F}{kT} = \frac{(\Delta\mu_{gI}^\circ)_F}{kT} - \left[\ln \gamma_{SW}X_{SW} + \left(\frac{g_{AI}}{g_{SI}}\right)_F \ln g_{AW}X_{AW} + \left(\frac{g_{OI}}{g_{SI}}\right)_F \ln \gamma_{OO}X_{OO}\right] = -\ln \Omega_F \qquad (5.2)$$

where $(\Delta\mu_{gI}^\circ)_F$ is the standard free energy difference associated with the transfer of one surfactant molecule and $(g_{AI}/g_{SI})_F$ alcohol molecules from their infinitely dilute states in water and $(g_{OI}/g_{SI})_F$ oil molecules from a pure oil phase to the flat interfacial layer. For given chemical potentials of various components in the system, the composition of the flat interfacial film (namely, a_F, $(g_{AI}/g_{SI})_F$, and $(g_{OI}/g_{SI})_F$) is obviously different from the composition of the interfacial film of droplets in the microemulsion phase (namely, a, g_{AI}/g_{SI}, and g_{OI}/g_{SI}).

6. Bicontinuous Microemulsions

No well-defined structural characterization of a bicontinuous microemulsion is currently available. However, it has been suggested that oil and water domains are

(16) Prausnitz, J. M. *Molecular Thermodynamics of Fluid-Phase Equilibria*; Prentice Hall: Englewood Cliffs, NJ, 1969; Chapter 7. Acree, W. E., Jr., W. E. *Thermodynamic Properties of Non-Electrolyte Solutions*; Academic Press: New York, 1984; Chapter 8.

(17) Nagarajan, R. *Langmuir* **1994**, *10*, 2028.

(18) Nagarajan, R. In *Structure-Performance Relationships in Surfactants*; Esumi, K., Ueno, M., Eds.; Marcel Dekker: New York, 1997; pp 1–81.

randomly distributed within the bicontinuous microemulsion separated by locally planar interfaces.[3-8] A characteristic domain size or persistence length ξ_K has been defined[7] such that, for length scales smaller than ξ_K, the interface is essentially flat. Using this concept, one can establish conditions for the formation of a three-phase system containing a bicontinuous microemulsion phase coexisting with excess oil and water phases. One can also calculate, on the basis of the present molecular theory, the persistence length and the interfacial tensions between microemulsion and excess phases.

Considering a planar configuration, the size distribution of microemulsion domains in a bicontinuous microemulsion coexisting with two excess phases can be written, in a form similar to eq 3.1, as

$$X_g = \frac{1}{K}\Omega_F^{g_{SI}}$$

$$\Omega_F = X_{SW}X_{AW}^{(g_{AI}/g_{SI})_F}X_{OO}^{(g_{OI}/g_{SI})_F}\exp\left(-\frac{\Delta\mu^\circ_{gI}}{kT}\right)_F$$

$$(\Delta\mu^\circ_{gI})_F = \left(\frac{1}{g_{SI}}\right)_F(\mu^\circ_{gI})_F - \mu^\circ_{SW} - \left(\frac{g_{AI}}{g_{SI}}\right)_F\mu^\circ_{AW} - \left(\frac{g_{OI}}{g_{SI}}\right)_F\mu^\circ_{OO} \quad (6.1)$$

In these equations, the standard free energy terms are approximated as for a flat interface since the domains are considered locally flat and the activity coefficients have been taken as unity since there are no droplets in the water and oil domains. The factor K accounts for the fact that some molecules must remain at the corners and edges of the surfaces separating the oil and the water domains and not on the flat regions. Thus, K is a measure of the free energy cost of having molecules in such locations compared to those in the flat regions. While, for droplet microemulsions, an optimal value for g_{SI} could be obtained by maximizing X_g in eq 3.1, there is no maximum for X_g in eq 6.1, since Ω_F is independent of g_{SI}. The equilibrium characteristics of the microemulsion are obtained by maximizing Ω_F for given values of the control variables X_{SW} and X_{AW}.

One may note the analogy between the above equation and the equation $X_g = (1/K)Y^g$, where $Y = X_1 \exp(-\Delta\mu^\circ_{cyl}/kT)$, governing the formation of rodlike micelles.[11,18] The equilibrium structural characteristics of rodlike micelles are obtained from the maximum of Y or, equivalently, minimum of $\Delta\mu^\circ_{cyl}$, which is the standard state free energy difference between a molecule in the cylindrical part of the micelle and a singly dispersed molecule. K is the factor that accounts for the free energy cost of having molecules at the two globular ends of the cylinder rather than in the cylindrical part of the micelle and is thus analogous to the K of eq 6.1. Since Y is independent of g, the size distribution X_g does not have a maximum and rodlike micelles of all aggregation numbers are present in appreciable concentrations. Y is the analogue of Ω_F in eq 6.1.

The maximum of Ω_F is determined with respect to the two independent variables, the area per surfactant molecule of the interface, a_F, and the alcohol-to-surfactant ratio in the interfacial layer, $(g_{AI}/g_{SI})_F$. In contrast to that in the droplet-type microemulsion, the oil-to-surfactant ratio in the interfacial layer, $(g_{OI}/g_{SI})_F$, is not an independent variable but is determined by the packing constraint (eq 2.7) for flat layers.

The maximum value of Ω_F will be close to but smaller than unity, since otherwise the domain size distribution would diverge. Since the domain size g_{SI} can assume all possible values, the average domain size of the bicontinuous microemulsion is given by

$$(g_{SI})_{avg} = \frac{\sum g_{SI}X_g}{\sum X_g} = \frac{1}{1-\Omega_F} \quad (6.2)$$

where the last equality is obtained by performing the analytic summation of the two factors in the above expression on the basis of eq 6.1. The equilibrium area per surfactant molecule being a_F, the total surface area of the average domain is $a_F(g_{SI})_{avg}$, which can be used to obtain the following expression for the persistence length ξ_K:

$$6\xi_K^2 = (g_{SI})_{avg}a_F = \frac{a_F}{1-\Omega_F} \quad (6.3)$$

The numerical constant 6 in the preceding equation implies the cubic domain assumed by de Gennes and Taupin,[7] and the surface area $a_F(g_{SI})_{avg}$ is that of a cube of characteristic size ξ_K. For other domain geometries, such as the polyhedra considered by Talmon and Prager,[6] this numerical coefficient will be different. The relative proportions of the oil and water domains are related to the persistence length ξ_K and the volume fraction Φ_S of the surfactant present through the species balance relation

$$\xi_K = \frac{6\Phi(1-\Phi)v_S}{\Phi_S a_F} \quad (6.4)$$

In this equation, Φ stands for the volume fraction of the oil domain and includes the hydrophobic tails of the surfactant and alcohol in the interfacial layer, while $(1 - \Phi)$ stands for the volume fraction of the water domain and includes the headgroups of the surfactant and alcohol in the interfacial layer.

For given values of the control variables X_{SW} and X_{AW}, the maximum of Ω_F (eq 6.1) and the equilibrium characteristics of the flat layer can be calculated by employing the IMSL optimization subroutine ZXMWD. The concentration X_{SW} does not affect the equilibrium characteristics a_F, $(g_{AI}/g_{SI})_F$, and $(g_{OI}/g_{SI})_F$ of the interfacial layer but does influence the magnitude of Ω_F and the total amount of surfactant in the bicontinuous microemulsion. The persistence length can be calculated from eq 6.3 using the optimized value of Ω_F. For a specified total volume fraction Φ_S of surfactant in the microemulsion, the proportions of the oil and the water domains in the microemulsion can then be determined from eq 6.4. Finally, the volume fraction of the total alcohol Φ_A in the microemulsion can be calculated from a species balance, since the concentrations of alcohol in the water and the oil domains and in the interfacial layer are known.

To determine whether a droplet-type microemulsion coexists in equilibrium with an excess dispersed phase or a bicontinuous microemulsion coexists with both excess phases, one has to carry out the maximizations of X_g (eq 3.1) and Ω_F (eq 6.1) for given values of X_{AW} and X_{SW}. For low values of X_{AW} as well as for sufficiently large values of X_{AW}, a maximum in X_g occurs, indicating the formation of a two-phase droplet-type microemulsion system. However, for intermediate values of X_{AW}, no maximum of X_g exists and the radius of the equilibrium droplet diverges to infinity. In this region, a three-phase system involving a bicontinuous microemulsion is favored.

The interfacial tension at the flat interface is calculated using eq 5.2, but with the activity coefficients equated to unity. Hence

$$\frac{\gamma_F a_F}{kT} = \frac{(\Delta\mu^\circ_{gI})_F}{kT} - \left[\ln X_{SW} + \left(\frac{g_{AI}}{g_{SI}}\right)_F \ln X_{AW} + \left(\frac{g_{OI}}{g_{SI}}\right)_F \ln X_{OO}\right] = -\ln \Omega_F \quad (6.5)$$

Since for physical reasons Ω_F is close to unity, the logarithmic term in the above equation can be expanded in series, and after taking into account eq 6.3, one obtains

$$\gamma_F = \frac{kT}{a_F}(-\ln \Omega_F) = \frac{kT}{a_F}(1 - \Omega_F) = \frac{kT}{6\xi_K^{\ 2}} \quad (6.6)$$

This expression was suggested previously by de Gennes and Taupin[7] on the basis of scaling arguments. Experimentally, one can measure three different interfacial tensions in the three-phase system, namely, at the microemulsion−excess water phase boundary (γ_{MW}), at the microemulsion−excess oil phase boundary (γ_{MO}), and between the excess oil and excess water phases (γ_{OW}). On the basis of intuitive arguments, it has been suggested[8] that

$$\gamma_F = \gamma_{OW} \qquad \gamma_{MW} = \Phi\gamma_{OW} \qquad \gamma_{MO} = (1 - \Phi)\gamma_{OW} \quad (6.7)$$

7. Free Energy of Formation of the Interfacial Layer

The equations developed in previous sections can be used to calculate the structural features of microemulsions, provided explicit expressions for the standard free energies of transfer of surfactant and alcohol molecules from their infinitely dilute states in water and of oil molecules from the pure oil phase to the interfacial layer of the microemulsion droplets are available. Such expressions are given below for spherical layers of O/W droplets and W/O droplets and also for flat layers. The difference in the standard state free energy consists of a number of contributions:

$$\Delta\mu^\circ_{gI} = (\Delta\mu^\circ_{gI})_{tr} + (\Delta\mu^\circ_{gI})_{def} + (\Delta\mu^\circ_{gI})_{int} + (\Delta\mu^\circ_{gI})_{steric} + (\Delta\mu^\circ_{gI})_{dipole} + (\Delta\mu^\circ_{gI})_{ionic} + (\Delta\mu^\circ_{gI})_{mix} \quad (7.1)$$

which account respectively for the transfer of surfactant and alcohol tails from water to the interfacial film, the deformation of the tails to satisfy packing constraints, the free energy of formation of the oil−water interface, the steric, dipolar, and ionic interactions among the headgroups, and the free energy of mixing of surfactant, alcohol, and oil in the film region. These contributions were previously examined in our treatments for micelles, mixed micelles, and solubilization in micelles.[11,18] Consequently, only the modifications required for the treatment of microemulsions are emphasized below.

7.1. Transfer Free Energy of Surfactant and Alcohol Tails. For the interfacial layer that contains g_{SI} surfactant and g_{AI} alcohol molecules, the transfer free energy per surfactant molecule is obtained by adding the contributions of the surfactant and alcohol molecules:

$$\frac{(\Delta\mu^\circ_{gI})_{tr}}{kT} = \frac{(\Delta\mu^\circ_{gI})_{tr,S}}{kT} + \left(\frac{g_{AI}}{g_{SI}}\right)\frac{(\Delta\mu^\circ_{gI})_{tr,A}}{kT} \quad (7.2)$$

The expressions for the transfer free energy of methyl and methylene groups obtained previously[11] are used to calculate the transfer free energies of the surfactant and alcohol tails. The expression for a methylene group of an aliphatic tail at temperature T (in kelvins) has the form

$$(\Delta\mu^\circ_{gI})_{tr}/kT = 5.85 \ln T + 896/T - 36.15 - 0.0056T \quad (7.3)$$

while that for a methyl group in the aliphatic chain has the form

$$(\Delta\mu^\circ_{g})_{tr}/kT = 3.38 \ln T + 4064/T - 44.13 + 0.02595T \quad (7.4)$$

For the short-chain alcohols considered here, the methylene group adjacent to the hydroxyl group is considered to make no contribution to the transfer free energy because it is located close to the hydrophilic domain of the aggregate.

High concentrations of electrolyte affect the transfer free energy, and the extra contribution due to the presence of NaCl at 25 °C is given by[19]

$$(\Delta\mu^\circ_{gI})^{ex}_{tr}/kT = (-0.322 - 0.064n_C)C_{add} \quad (7.5)$$

where C_{add} is the molarity of NaCl in the aqueous domain of the microemulsion and n_C is the number of methyl and methylene groups of the surfactant or alcohol tails. Since no direct information on this extra contribution at other temperatures is available, the temperature dependence of the methyl and methylene group contributions in the presence of added electrolyte in the above equation will be taken as equal to that in salt-free systems.

7.2. Deformation Free Energy of Surfactant and Alcohol Tails. As in the treatment of mixed micelles,[12,18] this free energy contribution is calculated by adding the contributions from both surfactant and alcohol tails. Since the thickness of the interfacial layer $|R_O - R_W|$ is larger than the extended length l_A of the alcohol tail, a part of the interfacial film is not accessible to the alcohol tails. Such a packing problem for molecules of differing tail lengths was treated in an earlier paper,[12] giving

$$\frac{(\Delta\mu^\circ_{gI})_{def}}{kT} = \frac{10\pi^2}{80}\left(P\frac{|R_W - R_O|^2}{N_S L^2} + \frac{g_{AI}}{g_{SI}}\frac{l_A^{\ 2}}{N_A L^2}\right) \quad (7.6)$$

where N_S ($=l_S/L$) and N_A ($=l_A/L$) are the number of segments of the surfactant and alcohol tails, L (=4.6 Å) is the effective segment length (defined such that L^2 is the cross-sectional area of the tail), and P is the packing factor defined as

$$P = \frac{V_I}{a|R_W - R_O|} \quad (7.7)$$

For flat interfaces, $P = 1$, because of the packing constraint provided by eq 2.7.

7.3. Interfacial Free Energy of the Layer-Water Contact. This free energy contribution is calculated using the expression

$$\frac{(\Delta\mu^\circ_{gI})_{int}}{kT} = \frac{\sigma_{agg}}{kT}\left(a - a_{oS} - \frac{g_{AI}}{g_{SI}}a_{oA}\right) \quad (7.8)$$

where σ_{agg} is the macroscopic interfacial tension between the hydrophobic domain and water, a is the surface area per surfactant molecule of the interfacial layer, and a_{oS} and a_{oA} are the areas per molecule shielded from contact

(19) Tanford, C. *The Hydrophobic Effect*; Wiley: New York, 1973; 2nd ed., 1980.

6406 *Langmuir, Vol. 16, No. 16, 2000* *Nagarajan and Ruckenstein*

with water by the headgroups of surfactant and alcohol molecules, respectively. For surfactants, a_{oS} is equated to the smaller between a_{pS} and L^2, as described in refs 11 and 18, where a_{pS} is the cross-sectional area of the polar headgroup. For alcohols, the cross-sectional area of the hydroxyl headgroup a_{pA} is about 8 Å^2, which is smaller than L^2; hence, $a_{pA} = a_{oA} = 8$ Å^2.

The interfacial tension σ_{agg} depends on the composition of the hydrophobic domain of the layer and thus on the surfactant tail–water, alcohol tail–water, and oil–water interfacial tensions. Since the surfactant and alcohol tails considered here are aliphatic chains, their interfacial tensions against water are practically the same and are taken to be σ_{SW}, for which expressions are provided in refs 11 and 18. If the oil in the interfacial layer is a saturated hydrocarbon, the oil–water interfacial tension σ_{OW} is practically the same as the tail–water interfacial tension. However, if the oil is an aromatic hydrocarbon, its interfacial tension against water is appreciably different from that between the tails and water. Depending upon the oil–water interfacial tension, a preferential location of the oil molecules in contact with water will occur, and this affects the overall interfacial tension σ_{agg}. For a volume fraction η_O of the oil in the interfacial layer, the interfacial tension σ_{agg} can be estimated using the approach of Prigogine,[20,21] as in our earlier paper dealing with solubilization.[11] In that approach, the interfacial tension of a solution was obtained by first calculating the composition of the surface monolayer, thus accounting for the preferential location of any one component at this surface monolayer. It is not possible to identify a bulk region and surface monolayer region in the present case, since the interfacial layer itself consists only of a monolayer. Nevertheless, the formal application of the Prigogine approach provides a way to at least approximately take into account the specificity of the oil. Thus, the interfacial layer is considered to be a solution with a uniform oil volume fraction of η_O. Corresponding to this composition, the volume fraction of the oil in the surface monolayer η_O^S is determined using the equation

$$\ln\left[\frac{(\eta_O^S/\eta_O)^{v_{ST}/v_O}}{(1-\eta_O^S)/(1-\eta_O)}\right] = \frac{\sigma_{SW}-\sigma_{OW}}{kT}v_S^{2/3} + \chi_{OS}\left[\frac{3}{4}((1-\eta_O)-\eta_O)\right] - \chi_{OS}\left[\frac{1}{2}((1-\eta_O^S)-\eta_O^S)\right] \quad (7.9)$$

where the parameter χ_{OS} denotes the interaction parameter for the oil and the surfactant and alcohol tails. After eq 7.9 is solved for η_O^S, the interfacial tension σ_{agg} is calculated using the equation

$$\frac{\sigma_{agg}-\sigma_{SW}}{kT}v_S^{2/3} = \ln\left(\frac{1-\eta_O^S}{1-\eta_O}\right) + \left(1-\frac{v_{ST}}{v_O}\right)(\eta_O^S-\eta_O) + \chi_{OS}\left[\frac{1}{2}(\eta_O^S)^2 - \frac{3}{4}(\eta_O)^2\right] \quad (7.10)$$

The parameter for the interaction between the oil and the tails χ_{OS} is expressed in terms of the solubility parameters δ_S^H, δ_A^H, and δ_O^H of the surfactant tail, the alcohol tail and the oil, respectively, using the expression

$$\chi_{OS} = \frac{v_O(\delta_O^H - \delta_{mix}^H)^2}{kT} \qquad \delta_{mix}^H = \eta_S\delta_S^H + \eta_A\delta_A^H + \eta_O\delta_O^H \quad (7.11)$$

where η_S, η_A, and η_O are the volume fractions of the surfactant tail, alcohol tail, and oil, respectively, in the interfacial layer. The volume fractions are readily calculated from the respective molecular volumes and the composition of the interfacial layer. The solubility parameters[22,23] are estimated to be 16.76 MPa$^{1/2}$ for the surfactant tail, 9.1 MPa$^{1/2}$ for the pentanol tail, and 16.8 MPa$^{1/2}$ for cyclohexane (selected as the oil for illustrative purposes).

7.4. Headgroup Steric Interactions. This contribution is expressed as that for mixed surfactants. Because, in the present case, the free energy is expressed per surfactant molecule (not surfactant + alcohol molecule), one can write

$$\frac{(\Delta\mu_{gI}^\circ)_{steric}}{kT} = -\left(1+\frac{g_{AI}}{g_{SI}}\right)\ln\left(1-\frac{a_{pS}}{a}-\frac{g_{AI}}{g_{SI}}\frac{a_{pA}}{a}\right) \quad (7.13)$$

where, as mentioned before, a_{pS} is the cross-sectional area of the polar headgroup of the surfactant and a_{pA} is the cross-sectional area of the hydroxyl headgroup.

7.5. Headgroup Dipole Interactions. For zwitterionic surfactants, the dipole–dipole interactions are evaluated using the expression

$$\frac{(\Delta\mu_{gI}^\circ)_{dipole}}{kT} = \frac{2\pi e^2 d}{\epsilon a_\delta kT}\frac{1}{H} \quad (7.14)$$

where d is the dipole length, δ is the distance from the oil–water contact surface at which the dipole is located, a_δ is the surface area per surfactant molecule evaluated at the distance δ, and H is a curvature correction given by one of the following expressions

$$H = 1 + (d+\delta)/R_W, \quad a_\delta = a(1+\delta/R_W)^2 \quad \text{for O/W droplets}$$

$$H = 1 - (d+\delta)/R_W, \quad a_\delta = a(1-\delta/R_W)^2 \quad \text{for W/O droplets}$$

$$H = 1, \quad a_\delta = a \quad \text{for a flat film} \quad (7.15)$$

The dielectric constant ϵ is a function of temperature[24] and is evaluated using the relation

$$\epsilon = 87.74\exp[-0.0046(T-273)] \quad (7.16)$$

7.6. Headgroup Ionic Interactions. Ionic interactions arise at the interfacial layer if the surfactant has a charged headgroup and are calculated using the expression

$$\left(\frac{\Delta\mu_{gI}^\circ}{kT}\right)_{ionic} = 2\left\{\ln\left[\frac{S}{2}+\left(1+\left(\frac{S}{2}\right)^2\right)^{1/2}\right] - \frac{2}{S}\left[\left(1+\left(\frac{S}{2}\right)^2\right)^{1/2}-1\right] - \frac{2C_g}{\kappa S}\ln\left[\frac{1}{2}+\frac{1}{2}\left(1+\left(\frac{S}{2}\right)^2\right)^{1/2}\right]\right\} \quad (7.17)$$

where

(20) Defay, R.; Prigogine, I.; Bellemans, A.; Everett, D. H. *Surface Tension and Adsorption*; Wiley: New York, 1966.
(21) Siow, K. S.; Patterson, D. *J. Phys. Chem.* **1973**, *77*, 356.
(22) Hildebrand, J. H.; Prausnitz, J. M.; Scott, R. L. *Regular and Related Solutions*; Van Nostrand Reinhold: New York, 1970.
(23) Barton, A. F. M. *Handbook of Solubility Parameters and Other Cohesion Parameters*; CRC Press: Boca Raton, FL, 1983.
(24) *CRC Handbook of Chemistry and Physics*, 60th ed.; CRC Press: Boca Raton, FL, 1980.

$$S = \frac{4\pi e^2}{\epsilon\kappa a_\delta kT} \qquad \kappa = \left(\frac{8\pi n_o e^2}{\epsilon kT}\right)^{1/2}$$

$$n_o = \frac{C_{1\text{ion}} + C_{\text{add}}}{10^3} N_{\text{Av}} \quad (7.18)$$

The area per molecule a_δ which appears in the preceding equation is evaluated at a distance δ from the surface in contact with water and is curvature dependent. Expressions for a_δ are given by eq 7.15. The distance δ is estimated as the distance from the surface in contact with water to the surface where the center of the counterion is located. κ is the reciprocal Debye length, n_o is the number of counterions in solution per cubic centimeter, $C_{1\text{ion}}$ is the molar concentration of the singly dispersed ionic surfactant molecules in water, C_{add} is the molar concentration of the salt added to the surfactant solution, and N_{Av} is Avogadro's number. The last term in the right-hand side of eq 7.17 provides a curvature correction to the ionic interaction energy. For normal droplets, $C_g = 2/(R_W + \delta)$; for reverse droplets, $C_g = -2/(R_W - \delta)$; and for flat layers, $C_g = 0$.

7.7. Free Energy of Mixing Inside the Layer Region. This free energy contribution is estimated using the Flory–Huggins expression[25]

$$\frac{(\Delta\mu^\circ_{gI})_{\text{mix}}}{kT} = \left[\ln \eta_S + \frac{g_{AI}}{g_{SI}} \ln \eta_A + \frac{g_{OI}}{g_{SI}} \ln \eta_O\right] + \left[v_S \frac{(\delta^H_S - \delta^H_{\text{mix}})^2}{kT} + \frac{g_{AI}}{g_{SI}} v_A \frac{(\delta^H_A - \delta^H_{\text{mix}})^2}{kT} + \frac{g_{OI}}{g_{SI}} v_O \frac{(\delta^H_O - \delta^H_{\text{mix}})^2}{kT}\right] \quad (7.19)$$

8. Predictions for Droplet-Type Microemulsions

Calculations were carried out for a system consisting of the anionic surfactant sodium dodecyl sulfate, 1-pentanol (cosurfactant), cyclohexane, and water containing 0.3 M NaCl. As mentioned at the very beginning, the choice of this system was dictated by the possibility of identifying various types of phase behaviors for the same chemical components by merely changing the amount of added alcohol. In all calculations, we assumed the coexistence of an excess dispersed phase. This means that the droplet microemulsion phase is part of a two-phase system and that the amount of dispersed phase present in the droplet is the maximum achievable.

The total surfactant volume fraction Φ_S in the microemulsion is kept constant at 0.01 and the concentration X_{SW} of the singly dispersed surfactant is negligibly small compared to the total amount of surfactant in the microemulsion. The alcohol-to-surfactant ratio in the microemulsion and, correspondingly, the alcohol-to-surfactant ratio in the interfacial layer of the droplet increase with increasing X_{AW}. The radius R_W of the water–oil interface (see Figure 2 for the definition of R_W) predicted by the model is plotted as a function of the alcohol-to-surfactant volume ratio in the microemulsion in Figure 3. The calculations show that, for X_{AW} less than 0.00172 and, correspondingly, for Φ_A/Φ_S less than 2.0 and g_{AI}/g_{SI} less than 1.012, an O/W droplet-type microemulsion coexists with an excess oil phase. When X_{AW} is larger than 0.00184 (and, correspondingly, when Φ_A/Φ_S is larger than 3.15 and g_{AI}/g_{SI} is larger than 1.13), a W/O droplet-type microemulsion coexists with an excess water phase. For

(25) Flory, P. J. *Principles of Polymer Chemistry*; Cornell University Press: Ithaca, NY, 1953.

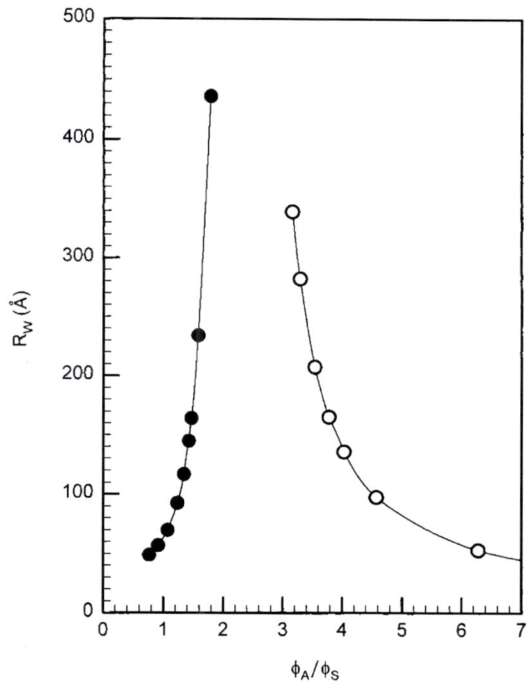

Figure 3. Predicted radii of microemulsion droplets as a function of the volume ratios of alcohol to surfactant in microemulsions. The predictions are for a system consisting of the surfactant SDS, 1-pentanol, 0.3 M NaCl, water, and cyclohexane. The total volume fraction of the surfactant in the microemulsions is 0.01. Filled circles denote O/W droplets, and open circles correspond to W/O droplets.

intermediate values of X_{AW} (corresponding to Φ_A/Φ_S in the range 2.0–3.15), no maximum in the aggregate size distribution occurs because the equilibrium aggregate radius diverges. For these conditions, a three-phase system involving a bicontinuous microemulsion comes into existence. The radius of the droplet increases significantly as the two-phase system (corresponding to either O/W or W/O droplet microemulsions) nears the transition to a three-phase system.

The number of surfactant molecules in the interfacial layer of a droplet is plotted in Figure 4 for the systems described in Figure 3. Detailed information regarding the composition of the microemulsions is provided by Figure 5. The volume fractions of oil, water, and alcohol plotted in the figures represent the total amounts in the microemulsion, and the volume fraction of surfactant Φ_S is 0.01 in all cases. As the alcohol-to-surfactant ratio increases, the volume fraction of oil increases while that of water decreases, as has been generally observed in numerous experiments.[1,3] The total volume fractions of the surfactant (Φ_S) and alcohol (Φ_A) determine the volume fraction of oil solubilized or water solubilized in the microemulsion. Figure 6 displays the predicted interfacial layer thickness and the area per surfactant molecule at the oil–water interface of the droplets as functions of the alcohol-to-surfactant ratio in the microemulsion. The calculations indicate that the area per surfactant molecule of the droplet is not a constant but increases with an increasing amount of alcohol. The extent of penetration of the oil into the interfacial layer is illustrated in Figure 7. The oil-to-surfactant ratio in the interfacial layer exhibits a monotonic increase as the alcohol-to-surfactant ratio in the microemulsion increases.

The volume fraction of droplets in the microemulsion phase predicted by the model is plotted in Figure 8, which

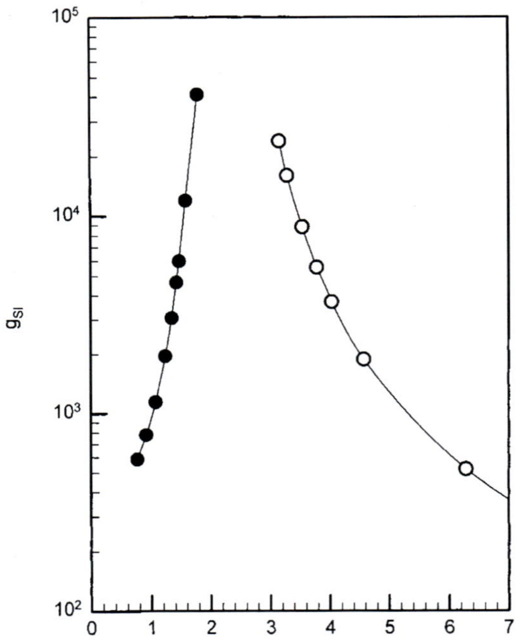

Figure 4. Numbers of surfactant molecules in the interfacial layers of droplets as a function of the volume ratios of alcohol to surfactant in microemulsions. All the system characteristics are identical to those described for Figure 3. Filled circles denote O/W droplets, and open circles correspond to W/O droplets.

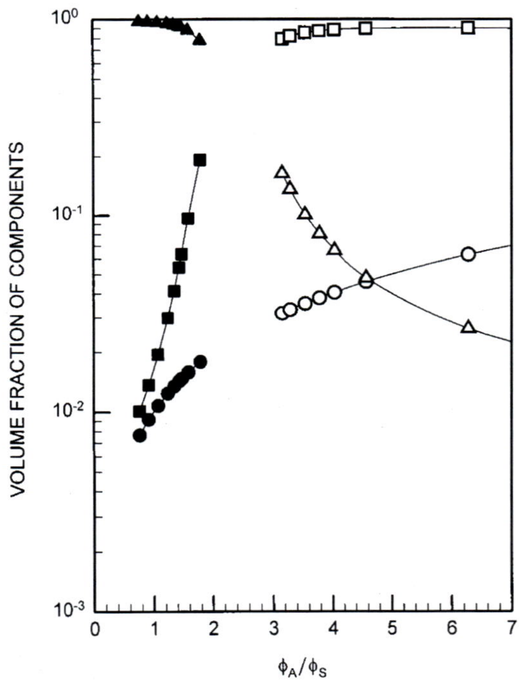

Figure 5. Volume fractions of oil (squares), water (triangles), and alcohol (circles) in microemulsions containing O/W and W/O droplets as a function of the volume ratios of alcohol to surfactant in the microemulsions. Filled symbols refer to O/W droplet microemulsions, and open symbols, to W/O droplet microemulsions. All the system characteristics are identical to those described for Figure 3. In all cases, the volume fraction of the surfactant in the microemulsions is 0.01.

shows that the volume fraction increases as one moves from the two-phase toward the three-phase system. The interfacial tension at the flat surface between the micro-

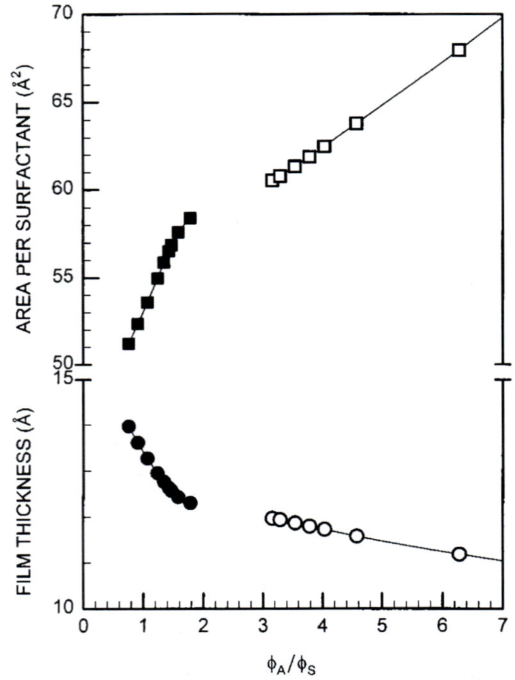

Figure 6. Thicknesses of the interfacial layers of droplets (circles) and areas per surfactant molecule of the droplets at an oil–water interface (squares) as functions of the volume ratios of alcohol to surfactant in microemulsions for systems containing O/W (filled symbols) and W/O (open symbols) droplets. All the system characteristics are identical to those described for Figure 3.

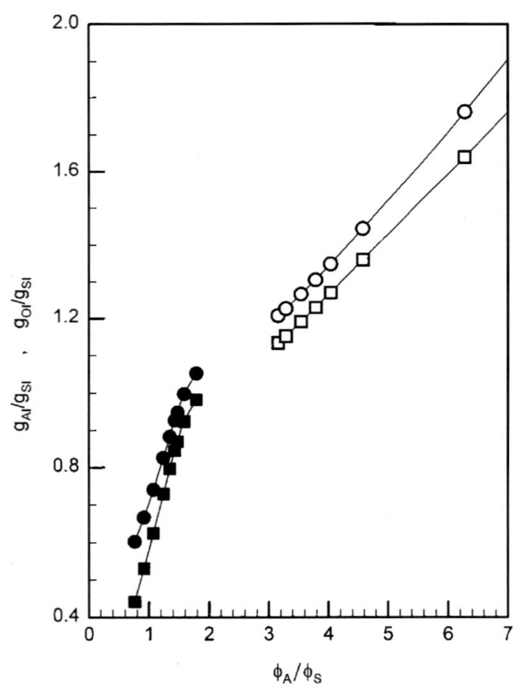

Figure 7. Alcohol-to-surfactant mole ratios and oil-to-surfactant mole ratios in the interfacial layers of O/W (filled symbols) and W/O (open symbols) droplets as functions of the volume ratios of alcohol to surfactant in microemulsions. All the system characteristics are identical to those described for Figure 3.

emulsion phase and the coexisting excess phase calculated with eq 5.2 is plotted in Figure 9, which reveals that the interfacial tension decreases as one moves from the two-phase to the three-phase system.

Molecular Theory of Microemulsions *Langmuir, Vol. 16, No. 16, 2000* 6409

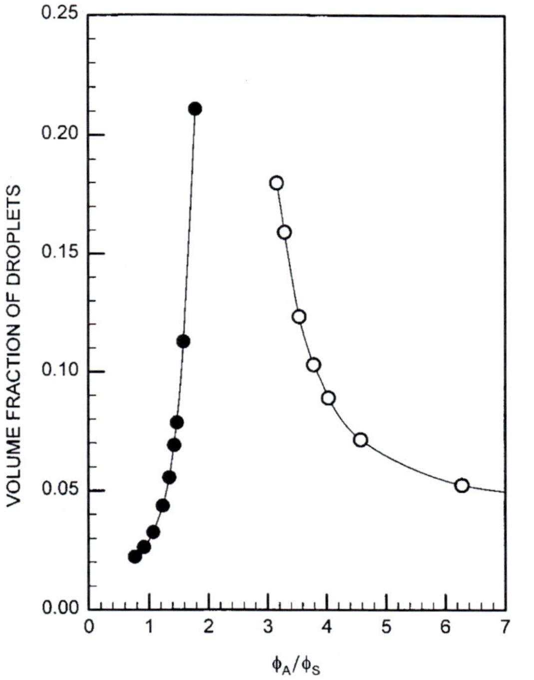

Figure 8. Volume fractions of droplets as a function of the volume ratios of alcohol to surfactant in microemulsions. All the system characteristics are identical to those described for Figure 3. Filled symbols are for O/W microemulsions, and open symbols refer to W/O microemulsions.

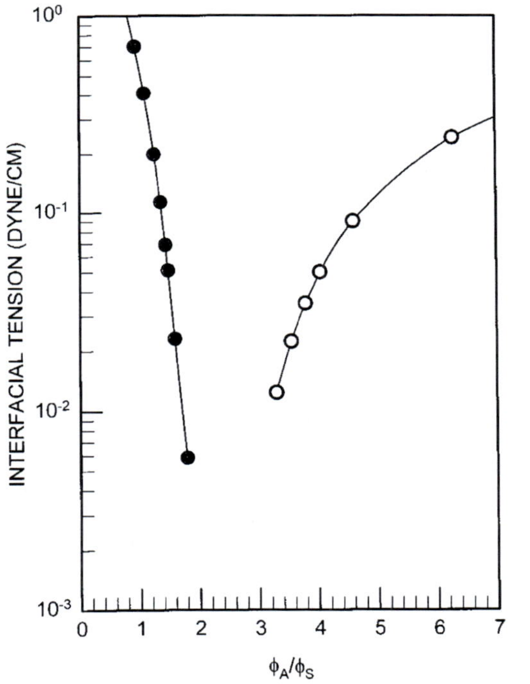

Figure 9. Oil–water interfacial tensions at the flat surfaces between microemulsion phases and excess dispersed phases as a function of the volume ratios of alcohol to surfactant in microemulsions. All the system characteristics are identical to those described for Figure 3. Filled symbols are for O/W microemulsions, and open symbols refer to W/O microemulsions.

Estimates of the size and composition dispersions of droplets in O/W and W/O droplet-type microemulsions, obtained using the maximum-term method described in Appendix C, are summarized in Table 1. The calculations for the O/W droplet-type microemulsions were carried out for X_{AW} = 0.001 62, while the calculations for the W/O droplet-type microemulsion were for X_{AW} = 0.00185. In all cases, the volume fraction of the surfactant in the microemulsion Φ_S was kept constant at 0.01. The average radii of the droplets, the average numbers of surfactant, alcohol, and oil molecules in the interfacial layers of the droplets, the overall compositions of the microemulsions, and their dispersions are listed in Table 1. The values of σ_n/g_n for all components are generally small, implying relatively narrow composition dispersions of the droplets. Consequently, the standard deviations σ_R of the droplet radii are also small.

Table 1. Predicted Size and Composition Dispersions of Microemulsion Droplets

variable	O/W microemulsion	W/O microemulsion
$\langle R_W \rangle$, Å	233.9	282
$\langle g_{SI} \rangle$	11 940	16 460
$\langle g_{AI}/g_{SI} \rangle$	0.9228	1.152
$\langle g_{OI}/g_{SI} \rangle$	0.9965	1.226
Φ_S	0.01	0.01
Φ_A	0.0158	0.0329
Φ_O	0.0963	0.8206
Φ_W	0.8779	0.1365
$\sigma_n(S)/g_n(S)$	0.219	0.266
$\sigma_n(A)/g_n(A)$	0.241	0.278
$\sigma_n(O)/g_n(O)$	0.274	0.260
$\sigma_n(W)/g_n(W)$		0.326
$\sigma_R/\langle R_W \rangle$	0.0822	0.096

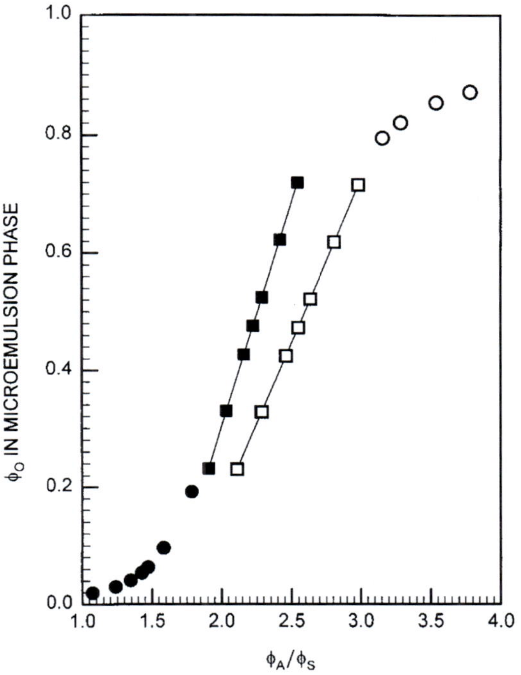

Figure 10. Volume fraction of oil in a bicontinuous microemulsion as a function of the volume ratio of alcohol to surfactant in the microemulsion. The values of X_{AW} and g_{AI}/g_{SI} are respectively 0.001 72 and 1.036 (filled squares) and 0.001 84 and 1.074 (open squares). In all cases, the volume fraction of surfactant is 0.01 and the volume fraction of water is found by subtraction from unity. The system consists of SDS, 1-pentanol, cyclohexane, water, and 0.3 M NaCl. Also shown are the corresponding results for O/W droplet microemulsions (filled circles) and for W/O droplet microemulsions (open circles).

9. Predictions for Bicontinuous Microemulsions

As mentioned previously, for alcohol concentrations X_{AW} in the water domain in the range 0.001 72–0.00 184 (corresponding to a Φ_A/Φ_S in the microemulsion between

6410 *Langmuir, Vol. 16, No. 16, 2000* *Nagarajan and Ruckenstein*

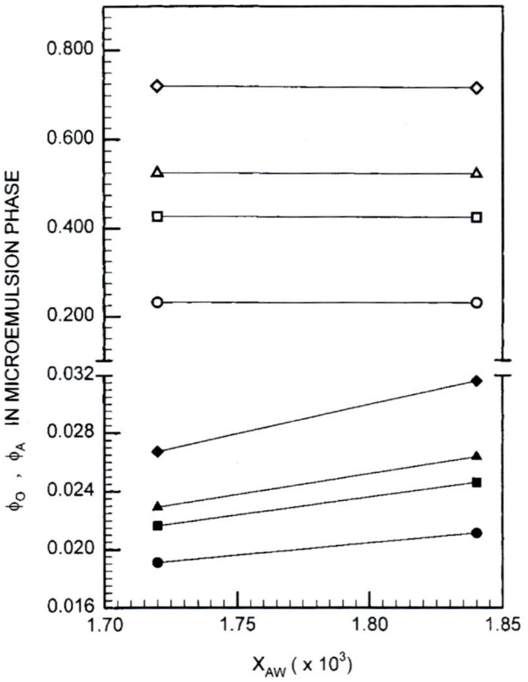

Figure 11. Dependence of the volume fractions of oil (open symbols) and alcohol (filled symbols) in a bicontinuous microemulsion as a function of the mole fraction of alcohol in the water phase. The different symbols refer to different volume fractions of oil and water domains in the microemulsion: Φ = 0.25 (circles), 0.45 (squares), 0.55 (triangles), and 0.75 (diamonds). The system consists of SDS, 1-pentanol, cyclohexane, water, and 0.3M NaCl.

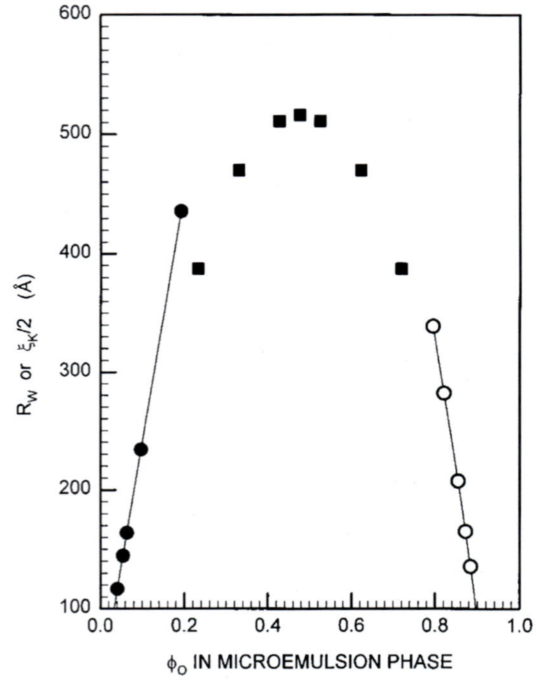

Figure 12. Dependence of the droplet radii in two-phase systems containing O/W (filled circles) or W/O microemulsions (open circles) and of persistence length/2 of a bicontinuous microemulsion (filled squares) on the volume fraction of oil in the microemulsion. In all cases, the volume fraction of the surfactant is 0.01. The system consists of SDS, 1-pentanol, cyclohexane, water, and 0.3 M NaCl.

2.0 and 3.25), three-phase systems involving a bicontinuous microemulsion are preferred thermodynamically. The overall composition limits within which a bicontinuous microemulsion can be generated are presented in Figure 10. The two lines correspond to X_{AW} = 0.001 72 and 0.001 84. For overall phase compositions between these two lines and for $\Phi_S = 0.01$, a three-phase system is favored thermodynamically. As X_{AW} increases from 0.00172 to 0.00184, the area per surfactant molecule a_F increases from 59.06 to 59.88 Å^2, while the alcohol-to-surfactant ratio in the interfacial layer $(g_{AI}/g_{SI})_F$ increases from 1.036 to 1.074. All the bicontinuous microemulsions marked by points on a line are characterized by the same values of X_{AW}, a_F, and $(g_{AI}/g_{SI})_F$.

One can observe from Figure 10 that, for a given Φ_A/Φ_S, there is a unique value of Φ_O for an O/W or W/O droplet microemulsion, but a range of Φ_O values lying between the two lines is possible for a bicontinuous microemulsion. The volume fraction of oil Φ_O in the bicontinuous microemulsion, for a given value of Φ_A/Φ_S, is determined by the concentration X_{AW} of the alcohol in the coexisting water phase. The volume fractions of oil Φ_O (unfilled symbols) and alcohol Φ_A (filled symbols) in the bicontinuous microemulsion are plotted in Figure 11 as a function of X_{AW}, with the volume fraction of surfactant Φ_S kept constant at 0.01. Because the sum of all volume fractions should be unity, the volume fraction of water Φ_W can be determined. Each symbol in Figure 11 corresponds to a given volume fraction of oil in the bicontinuous microemulsion: $\Phi = 0.25$ (circle), 0.45 (square), 0.55 (triangle), and 0.75 (diamond). For a given value of Φ_O, as X_{AW} increases, Φ_A increases, implying a corresponding decrease in Φ_W (because all volume fractions sum to unity). This predicted increase in the oil-to-water ratio in a bicontinuous microemulsion system with increasing alcohol content constitutes a well-known experimental result.

The persistence length of the bicontinuous microemulsion predicted by the model is plotted in Figure 12 as a function of the volume fraction of oil in the microemulsion for $X_{AW} = 0.001$ 72. The calculated persistence lengths for X_{AW} = 0.001 84 practically coincide with the former, because the areas per surfactant molecule in the interfacial layer a_F are close to one another for the two values of X_{AW}. The persistence length has a maximum when the oil and water domains have equal volume fractions. The radii of the O/W droplets increase with an increasing amount of alcohol in the low alcohol content range, while the radii of the W/O droplets decrease with an increasing amount of alcohol in the high alcohol content range. The interfacial tension calculated using eq 6.6 is plotted in Figure 13 as a function of the volume fraction of oil in the bicontinuous microemulsion. The calculated interfacial tensions for the two values of X_{AW} = 0.001 72 and 0.001 84 practically coincide (as expected, since the persistence lengths also practically coincide). The interfacial tension has a minimum when the oil and water domains have equal volume fractions.

10. Conclusions

In this paper, a predictive molecular thermodynamic approach is developed to calculate the structural and compositional characteristics of microemulsions. The theory applies not only to oil-in-water and water-in-oil droplet-type microemulsions but also to bicontinuous microemulsions. The treatment is an extension of our earlier theories for micelles, mixed micelles, and solubilization but also takes into account the self-association of alcohol in oil and the volume-excluded interactions among

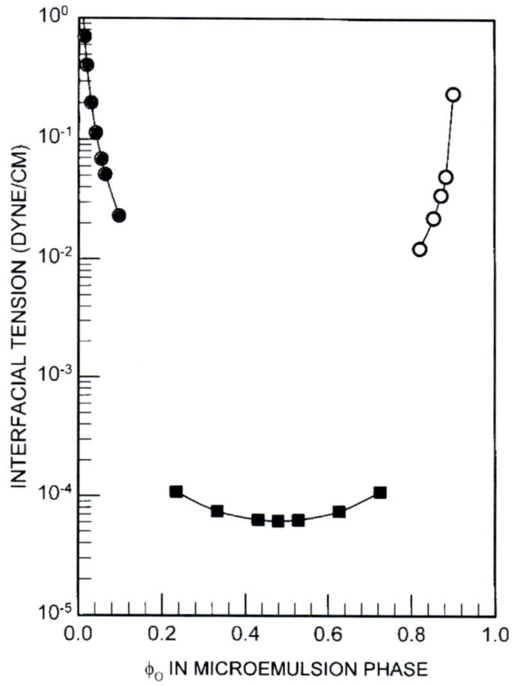

Figure 13. Interfacial tension at the oil–water interface in a bicontinuous microemulsion system (filled squares) as a function of the volume fraction of oil in the microemulsion. In all cases, the volume fraction of surfactant is 0.01. The system consists of SDS, 1-pentanol, cyclohexane, water, and 0.3 M NaCl. Also shown are the interfacial tensions at the flat surface between the O/W droplet microemulsion phase and the excess oil phase (filled circles) and between the W/O droplet microemulsion phase and excess water phase (open circles) in two-phase systems.

the droplets. Illustrative numerical results are provided for an anionic surfactant (SDS)–pentanol–cyclohexane–water–NaCl system. Microstructural features including the droplet radius, the thickness of the interfacial layer, the numbers of molecules of various species in a droplet, the size and composition dispersions of the droplet, and the distributions of the surfactant, oil, alcohol, and water molecules in the various microdomains have been calculated. Further, the model allows the identification of the transition from a two-phase to a three-phase system. The characteristic persistence length of the bicontinuous microemulsion is also predicted. Finally, the model permits the calculation of the interfacial tension between the microemulsion and the coexisting phase.

Appendix A: Equilibrium Size Distribution of Droplets

A.1. A Water-in-Oil Droplet Microemulsion. (a) Description of the Phase. The microemulsion system is composed of N_{gO} droplets of various sizes g (each consisting of surfactant, alcohol, oil and water molecules) and outside the droplets, N_{OO} oil molecules, N_{SO} surfactant molecules, N_{AO} alcohol molecules, and N_{WO} water molecules. The subscript g for a droplet denotes the total number of molecules of different kinds present in it (i.e., $g = g_S + g_A + g_O + g_W$). The N_{AO} alcohol molecules outside the droplets are present both as singly dispersed molecules and as aggregates. The number of alcohol aggregates containing j alcohol molecules is denoted by N_{jAO}.

(b) Free Energy of the Phase. The total Gibbs energy G of the microemulsion can be written as

$$G = N_{OO}\mu_{OO} + N_{SO}\mu_{SO} + \sum N_{jAO}\mu_{jAO} + N_{WO}\mu_{WO} + \sum N_{gO}\mu_{gO} \quad (A.1)$$

where μ refers to the chemical potentials of given kinds of species. The last subscript on N and μ stands for the nature of the continuous phase, with O denoting the oil. The subscripts O, W, S, A, and jA refer to molecules of oil, water, surfactant, alcohol, and alcohol aggregates containing j alcohol molecules, respectively. The summations over j extend from 1 to ∞ in all the equations in this paper. A summation over g in the last term of eq A.1 accounts for the size and the composition distribution of droplets.

(c) Equilibrium Condition. The chemical potentials of the singly dispersed alcohol molecules and alcohol aggregates are related via the expression

$$\mu_{jAO} = j\mu_{1AO} \quad (A.2)$$

which, combined with eq A.1, yields

$$G = N_{OO}\mu_{OO} + N_{SO}\mu_{SO} + N_{AO}\mu_{1AO} + N_{WO}\mu_{WO} + \sum N_{gO}\mu_{gO} \quad (A.3)$$

where

$$N_{AO} = \sum j N_{jAO} \quad (A.4)$$

For a given composition of the system

$$N_{kO} + \sum g_k N_{gO} = \text{constant} \qquad k = O, S, A, W \quad (A.5)$$

The condition $\delta G = 0$, subjected to the mass constraints of eq A.5, leads to

$$\mu_{gO} = g_O\mu_{OO} + g_S\mu_{SO} + g_A\mu_{1AO} + g_W\mu_{WO} \quad (A.6)$$

The chemical potentials of the components in the oil phase other than the droplets are expressed in the form

$$\mu_{kO} = \mu_{kO}^{\circ} + kT\ln(\gamma_{kO}X_{kO}) \qquad k = S, 1A, W, O \quad (A.7)$$

The standard chemical potential μ_{OO}° corresponds to the pure oil phase, the standard chemical potential μ_{1AO}° refers to the standard state of the monomeric alcohol in pure alcohol, and the standard chemical potentials μ_{SO}° and μ_{WO}° of the surfactant and water, respectively, correspond to their infinitely dilute solution states in the oil phase. For the droplets, the chemical potential is written as

$$\mu_{gO} = \mu_{gO}^{\circ} + kT\ln(\gamma_{gO}X_{gO}) \quad (A.8)$$

where μ_{gO}° is the standard chemical potential of the aggregate defined as that of an isolated aggregate in the continuous oil phase and γ_{gO} is the activity coefficient arising from the hard-sphere interactions.

The standard chemical potential μ_{gO}° of the aggregate can be split into a part due to the interfacial layer (denoted with the subscript I) and another part due to the dispersed water phase of the droplet (denoted with the subscript W). Thus, one can write

$$\mu_{gO}^{\circ} = \mu_{gI}^{\circ} + g_{OW}\mu_{OW} + g_{SW}\mu_{SW} + g_{AW}\mu_{AW} + g_{WW}\mu_{WW} \quad (A.9)$$

where μ_{gI}° is the standard chemical potential of the interfacial layer of the droplet, while μ_{kW} are the chemical potentials of the components k present in the dispersed water phase. The number of molecules g_{OW}, g_{SW}, g_{AW}, and g_{WW}, which refer to the oil, surfactant, alcohol, and water molecules in the dispersed water phase, are related to g_{OI}, g_{SI}, g_{AI}, and g_{WI}, the number of molecules in the interfacial layer of the droplet, via

$$g_{kI} = g_k - g_{kW} \qquad k = S, A, O, W \qquad (A.10)$$

(d) Relation between Chemical Potentials. The chemical potentials μ_{kO} of all the components k in the oil phase are equal to the chemical potentials μ_{kW} of the same components in the dispersed water phase. For oil, surfactant, and alcohol, one can write

$$\mu_{OO} = \mu_{OO}^{\circ} + kT\ln(\gamma_{OO}X_{OO}) = \mu_{OW} = \mu_{OW}^{\circ} + kT\ln(\gamma_{OW}X_{OW}) \qquad (A.11)$$

$$\mu_{SO} = \mu_{SO}^{\circ} + kT\ln(\gamma_{SO}X_{SO}) = \mu_{SW} = \mu_{SW}^{\circ} + kT\ln(\gamma_{SW}X_{SW}) \qquad (A.12)$$

$$\mu_{1AO} = \mu_{1AO}^{\circ} + kT\ln(\gamma_{1AO}X_{1AO}) = \mu_{AW} = \mu_{AW}^{\circ} + kT\ln(\gamma_{AW}X_{AW}) \qquad (A.13)$$

where X_{SW}, X_{AW}, and X_{OW} are the concentrations of the singly dispersed surfactant, alcohol, and water, respectively, in the dispersed water phase. μ_{SW}°, μ_{AW}°, and μ_{OW}° are their standard chemical potentials for infinitely dilute solutions in water, while γ_{SW}, γ_{AW}, and γ_{OW} are the activity coefficients in the dispersed water phase. For the fourth component, water, the concentration of singly dispersed water X_{WO} will be smaller than the value at saturation X_{WO}^{sat} if the microemulsion is a single-phase system and becomes equal to X_{WO}^{sat} when an excess water phase is present. Therefore

$$\mu_{WO} = \mu_{WO}^{\circ} + kT\ln(\gamma_{WO}X_{WO}^{sat}f_{WO}) = \mu_{WW} = \mu_{WW}^{\circ} + kT\ln(\gamma_{WW}X_{WW}) \qquad (A.14)$$

where

$$f_{WO} = X_{WO}/X_{WO}^{sat} \qquad (A.15)$$

(e) Size and Composition Distribution. One can now introduce into the equilibrium relation (A.6) the relations between the various chemical potentials, eqs A.7–A.15, to obtain the size and the composition distribution function of the droplets. For a single-phase microemulsion

$$X_{gO}\gamma_{gO} = \Omega^{g_{SI}} f_{WO}{}^{g_W} \qquad (A.16)$$

In eq A.16

$$\Omega = X_{SW}\gamma_{SW}(X_{AW}\gamma_{AW})^{g_{AI}/g_{SI}}(X_{OO}\gamma_{OO})^{g_{OI}/g_{SI}} \exp(-\Delta\mu_{gI}^{\circ}/kT) \qquad (A.17)$$

and

$$\Delta\mu_{gI}^{\circ} = \frac{1}{g_{SI}}\mu_{gI}^{\circ} - \mu_{SW}^{\circ} - \frac{g_{AI}}{g_{SI}}\mu_{AW}^{\circ} - \frac{g_{OI}}{g_{SI}}\mu_{OO}^{\circ} \qquad (A.18)$$

The standard free energy difference $\Delta\mu_{gI}^{\circ}$ is due to the transfer of one surfactant molecule and g_{AI}/g_{SI} alcohol molecules from water and g_{OI}/g_{SI} oil molecules from pure oil to the interfacial layer of the droplet. Equations A.16–A.18 imply that the number of water molecules in the interfacial layer (g_{WI}) is zero.

A.2. An Oil-in-Water Droplet Microemulsion. Similarly, for the oil-in-water droplet-type microemulsion, the size and composition distribution of droplets is given by

$$X_{gW}\gamma_{gW} = \Omega^{g_{SI}} f_{OW}{}^{g_O} \qquad (A.19)$$

where

$$\Omega = X_{SW}\gamma_{SW}(X_{AW}\gamma_{AW})^{g_{AI}/g_{SI}}(X_{OO}\gamma_{OO})^{g_{OI}/g_{SI}} \exp(-\Delta\mu_{gI}^{\circ}/kT) \qquad (A.20)$$

and

$$\Delta\mu_{gI}^{\circ} = \frac{1}{g_{SI}}\mu_{gI}^{\circ} - \mu_{SW}^{\circ} - \frac{g_{AI}}{g_{SI}}\mu_{AW}^{\circ} - \frac{g_{OI}}{g_{SI}}\mu_{OO}^{\circ} \qquad (A.21)$$

As in the previous case, f_{OW} is less than unity for a single-phase microemulsion but becomes equal to unity when an excess oil phase coexists. The activity coefficients γ_{SW} and γ_{SA} of the singly dispersed surfactant and alcohol molecules and the activity coefficient γ_{gW} of the aggregates differ from unity, since the water phase contains the droplets which contribute to solution nonideality through the hard-sphere interactions. For the same reason, the activity coefficient γ_{OO} of the oil in the dispersed oil phase is unity since no droplets are present in the oil phase.

Appendix B: Calculation of Activity Coefficients for Multicomponent Mixtures Containing Hard Spheres

Let us consider a solution composed of hard-sphere droplets of a single size g present in a multicomponent solvent. The expression for the osmotic pressure due to the hard spheres allows us to calculate the chemical potentials of the components in the mixed solvent. Subsequently, the Gibbs–Duhem equation is used to calculate the chemical potentials of the hard-sphere droplets.[26] The mole fraction X_i of the components and their volume fraction Φ_i are related via the expressions

$$X_i = \frac{\Phi_i/v_i}{\sum \Phi_i/v_i} \qquad \Phi_i = \frac{X_i v_i}{\sum X_i v_i} \qquad (B.1)$$

where v_i are the volumes of the various components including the hard-sphere aggregates. For all components other than $i = g$ (which refers to the hard spheres), considering their partial molar volumes equal to the molar volumes of the pure components, one can write

$$\left(\frac{\partial \mu_i}{\partial p}\right)_{T,N_j} = v_i \qquad \mu_i(\Phi_g = 0) - \mu_i(\Phi_g) = \Pi v_i \qquad (B.2)$$

where μ_i is the chemical potential of component i in the solution, while Π is the osmotic pressure due to the hard-sphere droplets of volume fraction Φ_g. The Carnahan–Starling equation of state leads to

$$\Pi = kT\frac{\Phi_g}{V_g}\frac{1 + \Phi_g + \Phi_g{}^2 - \Phi_g{}^3}{(1 - \Phi_g)^3} \qquad (B.3)$$

where V_g is the volume of a single hard-sphere droplet. Therefore, the chemical potential of component i is given by

(26) Reed, T. M.; Gubbins, K. E. *Applied Statistical Mechanics*; Butterworth-Heinemann: Boston, MA, 1973. Overbeek, J. Th. G. *Faraday Discuss. Chem. Soc.* **1978**, *65*, 7.

$$\mu_i = \mu_i^\circ + kT \ln X_i - kT\frac{v_i}{V_g}\frac{1 + \Phi_g + \Phi_g^2 - \Phi_g^3}{(1 - \Phi_g)^3} \quad (B.4)$$

The first term on the right-hand side μ_i° is the standard chemical potential of component i, the second term is the ideal mixture contribution, and the third term is the nonideality caused by the hard-sphere interactions. The chemical potential of the hard-sphere droplet (component g) is obtained using the Gibbs–Duhem equation

$$\sum X_i \, d\mu_i = 0 \qquad d\mu_g = -\frac{\sum_{i\neq g} X_i \, d\mu_i}{1 - \sum_{i\neq g} X_i} \quad (B.5)$$

and has the form

$$\mu_g = \mu_g^\circ + kT \ln X_g + kT\left[\ln\left(\frac{\Phi_g/V_g}{X_g \sum_{i\neq g} \Phi_i/v_i}\right) + \Phi_g\frac{7 - 3\Phi_g - \Phi_g^2}{(1 - \Phi_g)^2}\right] \quad (B.6)$$

In obtaining the expression for the activity coefficient part of the chemical potential, we have considered droplets of a single size represented by the most populous size (corresponding to the maximum in the size distribution). A more formal equation allowing for droplets of various sizes can be written according to the Mansoori–Carnahan–Starling equation of state for mixtures of hard spheres.[26] The results based on such an expression are not expected to be essentially different from those obtained on the basis of a single droplet size.

Appendix C: Approximate Method for Obtaining Size and Composition Polydispersity of Droplets Using the Maximum-Term Method

The size and composition distribution of droplets will be first rewritten in a general form applicable to both oil-in-water and water-in-oil microemulsions (by deleting the second subscripts):

$$X_g = X_S^{g_S} X_A^{g_A} X_O^{g_O} X_W^{g_W} Z(g_S, g_A, g_O, g_W) \quad (C.1)$$

where Z represents all the other factors that appear in the expression for the size distribution. The number- and weight-average numbers of the various constituents in the droplet are denoted by $g_n(k)$ and $g_w(k)$, respectively and are defined as

$$g_n(k) = \frac{\sum g_k X_g}{\sum X_g} \qquad g_w(k) = \frac{\sum g_k^2 X_g}{\sum g_k X_g} \quad (C.2)$$

where k stands for the four components present in a droplet (S, A, O, and W). In the above equation, as elsewhere in this paper, we employ the single-sum notation to represent the summation over all sizes and compositions of the droplets. One can write the following relations involving the derivatives by combining eqs C.1 and C.2:

$$\frac{\partial \ln \sum X_g}{\partial \ln X_k} = \frac{\sum g_k X_g}{\sum X_g} = g_n(k)$$

$$\frac{\partial \ln \sum g_k X_g}{\partial \ln X_k} = \frac{\sum g_k^2 X_g}{\sum g_k X_g} = g_w(k) \quad (C.3)$$

Further, from the definition of the number average (eq C.2), one can write the following derivative, which combined with eq C.3, leads to

$$\frac{\partial \ln g_n(k)}{\partial \ln \sum g_k X_g} = \frac{\partial \ln \sum g_k X_g}{\partial \ln \sum g_k X_g} - \frac{\partial \ln \sum X_g}{\partial \ln \sum g_k X_g} = 1 - \frac{g_n(k)}{g_w(k)} \quad (C.4)$$

The root-mean-square deviation of g_k from its number-average value $g_n(k)$ is given by

$$\sigma_n^2(k) = \frac{\sum [g_k - g_n(k)]^2 X_g}{\sum X_g} = g_n(k)\, g_w(k) - [g_n(k)]^2 \quad (C.5)$$

which, combined with eqs C.4 and C.5, leads to

$$\frac{\sigma_n(k)}{g_n(k)} = \left[\left(1 - \frac{\partial \ln g_n(k)}{\partial \ln \sum g_k X_g}\right)^{-1} - 1\right]^{1/2} \approx \left[\frac{\partial \ln g_n(k)}{\partial \ln \sum g_k X_g}\right]^{1/2} \quad (C.6)$$

Replacing the derivatives in eq C.6 by difference terms and restricting ourselves to the maximum term, we obtain eq 4.1 of section 4.

Appendix D: Model for Alcohol Self-Association in Oil

The partitioning of alcohol into the oil, water, and interfacial layer domains of a microemulsion controls whether a two-phase or a three-phase microemulsion system is formed, as well as the microscopic characteristics of the microemulsion phases. For the typical alcohols used, the amount of alcohol present in the oil domain can be large and comparable to the amount present in the interfacial layer. This is in contrast to the behavior of the surfactant, most of which remains at the interfacial layer and only a negligibly small amount of which are partitioned into the oil and the water domains. Therefore, the accurate accounting of the partitioning of alcohol into the oil domain is a necessary part of any quantitative theory of microemulsions. Such a theory must account for the facts that the alcohol is present in the oil phase as both monomers and aggregates and that the self-association of alcohol in the oil is responsible for its appreciable presence in the oil domain.

The thermodynamic treatments of the self-association of alcohols in hydrocarbon solvents[16] involve various levels of complexity and detail, depending on which of the following physical features are accounted for: (a) The alcohol molecules exist as self-associated aggregates. (b) The association can be represented by linear polymers, cyclic polymers, or aggregates of one or more distinct sizes. (c) The alcohol molecules can associate with the oil molecules to form oil–alcohol complexes. (d) The size differences between the various aggregates are appreciable and contribute to the solution properties. (e) In addition to the chemical association phenomenon, physical interactions between the various aggregates and the oil molecules constituting the solvent medium are also important. To keep the molecular theory of microemulsions at the simplest level possible, we treat the alcohol–oil systems by placing the emphasis on the most important source of nonideality, namely, the self-association of alcohol.[16]

D.1. Formulation of the Self-Association Model. Consider a mixture of N_{OO} oil molecules and N_{AO} alcohol molecules constituting a binary system from the stoichiometric point of view. The alcohol molecules are present as N_{jAO} aggregates of size j ($j = 1$, for the monomer), making the mixture a multicomponent system from the viewpoint of the true chemical species present. The true mole fractions of the aggregates of size j and of the oil are

$$X_{jAO} = \frac{N_{jAO}}{N_{OO} + \sum N_{jAO}} \qquad X_{OO} = \frac{N_{OO}}{N_{OO} + \sum N_{jAO}} \qquad \text{(D.1)}$$

while the stoichiometric mole fractions of alcohol and oil are

$$X_A = \frac{N_{AO}}{N_{AO} + N_{OO}} = \frac{\sum jN_{jAO}}{N_{OO} + \sum jN_{jAO}} \qquad X_O = 1 - X_A \qquad \text{(D.2)}$$

The self-association of alcohol in oil is represented by the association equilibria

$$A_1 + A_j \rightleftarrows A_{j+1} \qquad j = 1, 2, 3, ... \qquad \text{(D.3)}$$

where the association equilibrium constants are considered to be the same for all association steps. If we denote the stepwise association equilibrium constant by K_X, the mole fraction of alcohol aggregates of size j is

$$X_{jAO} = K_X^{j-1} X_{1AO}^{j} \qquad \text{(D.4)}$$

Because $K_X X_{1AO}$ must be less than unity to ensure that all mole fractions remain finite

$$\sum X_{jAO} = \frac{X_{1AO}}{1 - K_X X_{1AO}} \qquad \sum jX_{jAO} = \frac{X_{1AO}}{(1 - K_X X_{1AO})^2} \qquad \text{(D.5)}$$

$$X_{OO} = 1 - \sum_j X_{jAO} = \frac{1 - K_X X_{1AO} - X_{1AO}}{1 - K_X X_{1AO}} \qquad \text{(D.6)}$$

The average aggregation number of alcohol aggregates is therefore given by

$$j_{avg} = \frac{\sum jX_{jAO}}{\sum X_{jAO}} = \frac{1}{1 - K_X X_{1AO}} \qquad \text{(D.7)}$$

and the ratio of the total number of alcohol molecules to those in monomeric form is

$$\frac{N_{AO}}{N_{1AO}} = \frac{\sum jX_{jAO}}{X_{1AO}} = \frac{1}{(1 - K_X X_{1AO})^2} \qquad \text{(D.8)}$$

Finally, the stoichiometric mole fraction X_A can be used to calculate the true mole fraction X_{1AO} of the monomeric species using the expression

$$X_{1AO} = \frac{1 + 2K_X X_A - (1 + 4K_X X_A - 4K_X X_A^{\,2})^{1/2}}{2K_X X_A(1 + K_X)} \qquad \text{(D.9)}$$

The alcohol-partitioning calculations allow us to calculate X_{OO}, provided the self-association equilibrium constant K_X is known.

D.2. Determination of K_X from Thermodynamic Data for Alcohol–Hydrocarbon Mixtures. Using the available excess Gibbs energy and excess enthalpy data for alcohol–hydrocarbon mixtures,[16] the value of K_X can be estimated as described below. The Gibbs energy of the solution is given by

$$G = N_{OO}\mu_{OO} + \sum N_{jAO}\mu_{jAO} = N_{OO}\mu_{OO} + N_{AO}\mu_{1AO} \qquad \text{(D.10)}$$

where $\mu_{jAO} = j\mu_{1AO}$. The standard chemical potential of oil μ°_{OO} is taken to correspond to the pure oil, while the standard chemical potential of the monomeric alcohol μ°_{1AO} is taken to be that of the monomers in pure alcohol. One can find μ°_{1AO} by equating the chemical potential μ^{*}_{1AO} of the pure alcohol to the chemical potential of the monomers defined in terms of their standard state and their concentration X^{*}_{1AO} in pure alcohol:

$$\mu^{*}_{1AO} = \mu^{\circ}_{1AO} + kT \ln X^{*}_{1AO} \qquad \text{(D.11)}$$

The monomeric mole fraction X^{*}_{1AO} in the pure alcohol is obtained from eq D.9 by setting the stoichiometric alcohol mole fraction X_A equal to unity:

$$X^{*}_{1AO} = \frac{1}{1 + K_X} \qquad \text{(D.12)}$$

The excess Gibbs energy is thus provided by the expression

$$G^E = G - [N_{OO}\,\mu^{\circ}_{OO} + N_{AO}\,\mu^{*}_{1AO} + kT(N_{OO} \ln X_O + N_{AO} \ln X_A)] \qquad \text{(D.13)}$$

and the molar excess Gibbs energy by

$$\frac{G^E}{N_{OO} + N_{AO}} = kT\left[X_A \ln\left(\frac{X_{1AO}}{X^{*}_{1AO}}\right) + X_O \ln\left(\frac{X_{OO}}{X_O}\right)\right] \qquad \text{(D.14)}$$

The excess enthalpy can be related to the number of alcohol–alcohol hydrogen bonds formed in the alcohol–hydrocarbon solution compared to those formed in pure alcohol. Considering the enthalpy of formation of a hydrogen bond h_A to be independent of the degree of aggregation, the excess enthalpy is given by

$$H^E = h_A[\sum (j - 1)N_{jAO} - \sum (j - 1)N^{*}_{jAO}] \qquad \text{(D.15)}$$

where N^{*}_{jAO} denotes the number of aggregates with a degree of association j in pure alcohol. Because $\sum jN^{*}_{jAO} = N_{AO}$ and $\sum jN_{jAO} = N_{AO}$, eq D.7 can be written as

$$\frac{N_{AO}}{\sum N_{jAO}} = \frac{\sum jN_{jAO}}{\sum N_{jAO}} = \frac{\sum jX_{jAO}}{\sum X_{jAO}} = \frac{1}{1 - K_X X_{1AO}} \qquad \text{(D.16)}$$

Therefore

$$\sum N_{jAO} = N_{AO}(1 - K_X X_{1AO})$$
$$\sum N^{*}_{jAO} = N_{AO}(1 - K_X X^{*}_{1AO}) \qquad \text{(D.17)}$$

and, consequently, the molar excess enthalpy is given by the relation

$$\frac{H^E}{N_{OO} + N_{AO}} = h_A X_A K_X (X_{1AO} - X^{*}_{1AO}) \qquad \text{(D.18)}$$

The thermodynamic models available in the literature[16] employed a constant value of −6000 cal/mol for the hydrogen-bonding energies h_A of various alcohols. Therefore

$$\frac{d \ln K_X}{d(1/T)} = -\frac{h_A}{R} \tag{D.19}$$

where R is the gas constant.

By fitting the available excess Gibbs energy data to eq D.14, the excess enthalpy data to eq D.18, the average aggregation number of the alcohol to eq D.7, and the ratio between the total and the monomeric alcohol to eq D.8 and incorporating the temperature dependence provided by eq D.19, we have developed the following expression for an alcohol with n_{cA} carbon atoms:

$$K_X = 92 n_{cA}^{-0.47} \exp\left[3000\left(\frac{1}{T} - \frac{1}{323}\right)\right] \tag{D.20}$$

The preceding equation is valid for aliphatic alcohols ranging from propanol to decanol. For example, at 25 °C, the association equilibrium constants are 104.6, 94.2, 86.4, and 80.4, respectively, for butanol, pentanol, hexanol, and heptanol.

D.3. Use of the Alcohol Self-Association Model. In performing the computations for microemulsions, one first estimates K_X using eq D.20 for a given chain length of alcohol and a given temperature. By equating the chemical potential of the alcohol in the oil phase to the chemical potential of the alcohol in the water phase, one obtains

$$\frac{X_{AW}}{X_{AW}^S} = \frac{X_{1AO}}{X_{1AO}^*} = (1 + K_X) X_{1AO} \tag{D.21}$$

where X_{AW}^S is the saturation concentration of alcohol in water when the water phase coexists with a pure alcohol phase. At 25 °C, the solubility of aliphatic alcohols in water is given[19] by the expression

$$\ln X_{AW}^S = 1.40 - 1.38 n_{cA} \tag{D.22}$$

To estimate X_{AW}^S at any other temperature, the temperature dependence of alcohol solubility is assumed comparable to that of aliphatic hydrocarbons and the relations provided for the methyl and methylene groups by eqs 7.3 and 7.4, respectively, are used to obtain this temperature dependence. For a given value of X_{AW}, the mole fraction X_{1AO} of the monomeric alcohol in the oil phase is calculated by combining eqs D.21 and D.22. Finally, the true mole fraction of oil X_{OO} is calculated using eq D.6.

LA991578T

Phase Transitions in Insoluble Monolayers

Eli Ruckenstein
Department of Chemical and Biological Engineering, State University of New York at Buffalo, Buffalo, New York, U.S.A.

Abstract
A general thermodynamic theory and a statistical thermodynamic approach are presented, which describe the phase transitions in insoluble monolayers, particularly the inclined transition from a liquid-expanded to a liquid-condensed phase.

INTRODUCTION

The first goal of this article is to determine the conditions under which the transition from a liquid-expanded (LE) to a liquid-condensed (LC) phase is horizontal or inclined. A simple criterion involving the ranges of attractive and repulsive interactions is suggested, which can explain qualitatively the nature of the transition. Then, a thermodynamic approach for systems that exhibit inclined transitions is presented, followed by the derivation on the basis of statistical thermodynamics of a two-dimensional equation of state for the surface pressure against the surface area per molecule.

BACKGROUND

A large number of papers are concerned with the phase transitions in insoluble monolayers. Some of them are Refs.[1–48]. The insoluble surfactants, such as the fatty acids and the phospholipids, spread on the surface of water generating monolayers between the latter and air.[1,2] At low surface pressures, the monolayers are in a "gaseous" state, but the increase in surface pressure leads to a transition to a LE phase, followed by a transition from a LE to a LC phase. In contrast to the first phase transition, for which the isotherm in the transition region is horizontal, in the second, the isotherm in the transition region can be horizontal or inclined. For phospholipids at sufficiently low pH values, fluorescence microscopy[3–6] and Brewster angle microscopy[7–10] indicated that large islands of condensed phase of almost the same radius, of the order of 10–10^2 μm, which were organized in a superlattice, were dispersed in a LE phase. In those cases, the transition was inclined. For fatty acids at sufficiently low pH values, fluorescence microscopy experiments[2] have revealed the presence of very large islands of non-uniform size, distributed non-uniformly in a continuous phase and the transition was horizontal.

COMPARISON BETWEEN THE ATTRACTIVE AND REPULSIVE INTERACTION RANGES

Depending on the relative scale lengths of the attractive and repulsive interactions among molecules, the LE → LC transition can be horizontal or inclined. When the attractive interaction is of long range and the repulsive one of short range, the molecules will have the tendency to form a single large aggregate if the decrease in energy thus generated overcomes the loss of entropy. In such cases, the new phase will prefer to separate as a bulk phase from the initial one and the transition will be horizontal, hence of first order. When the repulsive interaction, due either to dipole–dipole and/or double layer interactions is of long range and the attractive interaction is of short range, the new phase will prefer to be distributed in the old one as islands, because the free energy due to repulsion is then smaller than when the phases are completely separated as bulk phases. The dipole moments of the headgroups of the phospholipid molecules are in most cases large and the dipole–dipole interactions of long range. The organization of the islands in a superlattice observed experimentally[12] additionally minimizes the free energy. At low pHs, the phospholipids are little dissociated, but their dipole–dipole interactions are of long range. At high pH, the phospholipids are highly dissociated and the repulsive interactions are again of long range. Consequently, for these molecules, the repulsive interactions appear to be of long range at any pH. It is shown later in the chapter that in such cases, the surface pressure increases as the surface fraction ϕ occupied by the LC phase increases, and, hence, that the transition is inclined.

Because the fatty acids have low dipole moments, it is most likely that when their charge is low (sufficiently low pH values), their repulsive interactions are smaller than the attractive interaction. Consequently, the LE → LC transition will be of first order, even though large islands of non-uniform size of condensed phase were identified via fluorescence spectroscopy.[2] These islands were, however, so large that their size could

Encyclopedia of Surface and Colloid Science DOI: 10.1081/E-ESCS-120023357

not affect the phase behavior. The slow coalescence of the islands was probably responsible for their incomplete aggregation. Their non-uniform sizes and distribution constitute an additional indication that the transition is of first order. Indeed, in the inclined transitions, the islands had almost the same radius and were organized in a superlattice. The horizontal transitions observed by Pallas and Pethica[13,14] for the pentadecanoic and hexadecanoic acids at low pH values, when they have been carefully purified, provide support for the above considerations. The occurrence of inclined transitions for some fatty acids might be, indeed, caused by the presence of impurities, as suggested by Pallas and Pethica. At high pH values, the fatty acids are charged and the double layer interaction generated can become of long range when compared with the attractive interactions and hence the transition inclined. Of course, the length of the hydrocarbon chain (or chains for phospholipids) and their tilting can affect the relative values of the ranges of attraction and repulsion.

THE THERMODYNAMICS OF THE INCLINED EXPANDED TO CONDENSED PHASE TRANSITION

Islands of uniform radius R of condensed phase are assumed to be distributed in a continuous LE phase. Of course, this assumption is no longer valid for $\phi > 0.6$, because transitions to stripe configurations and further to islands of LE in LC phases occur in such cases.[12]

It is convenient to write the Helmholtz free energy as the sum

$$F = F_0 + F_1 \tag{1}$$

where F_0 is the free energy of a fictitious system free of interactions among islands and of the entropy of dispersion of the islands into the continuous phase, and F_1 accounts for the interactions among islands and the entropy of dispersion. Such a decomposition of the free energy has the advantage that one can apply the Gibbs thermodynamics to F_0.

Denoting by ϕ the surface fraction occupied by the condensed phase, the perimeter P per unit area is given by

$$P = \frac{2\phi}{R}.$$

Consequently, at constant temperature,

$$dF_0 = \gamma_1 d\left(\frac{2\phi A}{R}\right) + CAd(1/R) + \mu_c dN_c + \mu_e dN_e - \pi_e d[A(1-\phi)] - \pi_c d(A\phi) \tag{2}$$

where A is the total area, γ_1 is the line tension, C is the bending line stress, μ_i is the chemical potential at the corresponding surface pressure π_i, N_i is the number of molecules, and the subscript i is c and e for the condensed and expanded phases, respectively. Further, denoting

$$f = F_1/A \tag{3}$$

and taking into account that the interaction among islands depends on the radius R and surface fraction ϕ,

$$df = \frac{\partial f}{\partial R}dR + \frac{\partial f}{\partial \phi}d\phi \tag{4}$$

Combining Eqs. (1–4) for $N_c + N_e = \text{const}$, one obtains

$$\begin{aligned} dF = {} & \left(2\gamma_1\frac{A}{R} - \pi_c A + \pi_e A + A\frac{\partial f}{\partial \phi}\right)d\phi \\ & + \left[2\gamma_1\frac{\phi}{R} - \pi_c\phi - \pi_e(1-\phi) + f\right]dA \\ & + \left(-\frac{2\gamma_1\phi A}{R^2} - \frac{CA}{R^2} + A\frac{\partial f}{\partial R}\right)dR + (\mu_c - \mu_e)dN_c \end{aligned} \tag{5}$$

However, at thermodynamic equilibrium, constant temperature and constant $N_e + N_c$,

$$dF = -\pi dA \tag{6}$$

where π is the surface pressure to which the system is subjected.

Because Eqs. (5) and (6) must be equivalent, the factor that multiplies dA in Eq. (5) must be equal to $-\pi$ and the factors that multiply $d\phi$, dR and dN_c must be equal to zero. After simple algebra, one obtains

$$\gamma_1 = \frac{R^2}{2\phi}\frac{\partial f}{\partial R} - \frac{C}{2\phi} \tag{7a}$$

$$\pi_e = \pi + f - \phi\frac{\partial f}{\partial \phi} \tag{7b}$$

$$\begin{aligned} \pi_c &= \pi + f + (1-\phi)\frac{\partial f}{\partial \phi} + \frac{R}{\phi}\frac{\partial f}{\partial R} - \frac{C}{R\phi} \\ &\equiv \pi + f + (1-\phi)\frac{\partial f}{\partial \phi} + \frac{2\gamma_1}{R} \end{aligned} \tag{7c}$$

$$\mu_c(\pi_c) = \mu_e(\pi_e) \tag{7d}$$

For a strip configuration, similar thermodynamic equations can be written.

One should note that (i) the real surface pressure in the continuous, expanded phase is π, (ii) π_e is the pressure in the continuous phase corresponding to the free energy F_0, and (iii) the conventional Laplace equation is no longer valid and is replaced by Eq. (7c).

For a given surface pressure π, Eq. (7a) and Eq. (7d) allow one to calculate R and ϕ [after Eq. (7b) and Eq. (7c) are employed to eliminate the surface pressures π_c and π_e from Eq. (7d)]. Similarly, for a given ϕ, Eq. (7a) and Eq. (7d) allow one to calculate R and π.

Of course, to perform the calculations, one must provide expressions for γ_1, C, f, and the chemical potentials μ_c and μ_e. The above equations allow one, however, to explain the nature of the phase transition from a LE to a condensed phase without such detailed calculations. Indeed, if this phase transition were a first order transition, two bulk phases would remain in contact as ϕ is varied and, as a result, $\pi_e = \pi_c = \pi =$ const. The LE/LC transition becomes inclined when the new phase remains dispersed as islands in the continuous phase, and Eqs. (7a), (7b), (7c), and (7d) show that in such cases $\pi_e \neq \pi_c \neq \pi$. This occurs because the free energy of the system becomes smaller when the new phase is dispersed as islands in the continuous phase, than when it forms a separate bulk phase.

Let us determine the conditions under which π increases as ϕ increases. Eq. (7b) can be useful in this respect. The surface pressure π_c (which is defined on the basis of the free energy F_0) is expected to vary slowly as ϕ increases. The repulsive free energy f is a positive quantity, which increases as ϕ increases at constant radius. If the increase is such that the difference

$$\phi \frac{\partial f}{\partial \phi} - f$$

is positive and increases with increasing ϕ, Eq. (7b) suggests that π is likely to increase as ϕ increases.

An expression for f for long-range dipole–dipole repulsions can be obtained using an equation proposed by McConnell.[15] Considering a hexagonal array of circular islands of radius R, one obtains

$$f = \frac{2\phi}{R}\left[(\Delta\mu)^2 \nu g \phi^{3/2}\right]$$

where $(\Delta\mu)$ is the difference in molecular dipole moment densities between the LC and LE phases, $\nu = (3\pi/2)\left(\sqrt{3}/2\pi\right)^{3/2}$ is a constant and, neglecting higher order terms in ϕ, g is given by[16]

$$g = 3.25 + 1.15\phi$$

Consequently,

$$f = \frac{2\nu(\Delta\mu)^2}{R}(3.25 + 1.15\phi)\phi^{5/2} \tag{8}$$

and

$$\phi \frac{\partial f}{\partial \phi} - f = \frac{2\nu(\Delta\mu)^2}{R}\left(\frac{9.75}{2}\phi^{5/2} + \frac{5.75}{2}\phi^{7/2}\right) \tag{9}$$

Using for R the following expression derived by McConnell[15]

$$R = R_0 \exp\left[\nu(3.25 + 1.15\phi)\phi^{3/2}\right] \tag{10}$$

where R_0 is a constant, one can conclude that $\phi \frac{\partial f}{\partial \phi} - f$ is a positive quantity, which increases as ϕ increases until ϕ becomes about 0.7, where it has a maximum. However, for $\phi > 0.6$, transitions to stripe configurations and further to islands of LE in a continuum of LC occur,[12] and the above expressions are no longer valid. A similar calculation for stripe phases, based on McConnell equations[15] leads to

$$\phi \frac{\partial f}{\partial \phi} - f = \alpha(\sin \pi\phi - \pi\phi \cos \pi\phi) \tag{11}$$

where α is a positive quantity independent of ϕ. The above equation shows that, in the range of interest ($\phi > 0.6$), $\phi \frac{\partial f}{\partial \phi} - f$ is positive and increases with increasing ϕ.

Of course, if f is sufficiently small, then the non-horizontal phase transition can be approximated by a horizontal one.

In conclusion, the LE/LC transition becomes inclined when, because of the lower repulsion than in a separated condensed bulk phase, the condensed phase prefers to be distributed as islands in the expanded one. This can occur only if the repulsive interaction is of long range and the attractive one is of short range. Whereas thermodynamics provides some insight into the phase transitions in monolayers, it cannot provide detailed predictions.

A STATISTICAL THERMODYNAMIC MODEL

In this section, a theoretical model[16] is proposed on the basis of which one can predict both the large sizes of the islands as well as the non-horizontal nature of the LE/LC phase transition. The present calculations involve not only the long-range dipole–dipole interactions, but also the van der Waals interactions among the hydrocarbon chains as well as the conformational free energy of the chains. In this manner, some of the molecular details are taken into account, and one can explain the dependence of the isotherms on the hydrocarbon chain length and the nature of the headgroup.

To describe the LE—LC phase transition, the insoluble monolayer containing N surfactant molecules at the air/water interface is treated as a two-dimensional mixture made up of N_d disordered molecules in the LE state, N_c islands of equal size with molecules in the LC state, and N_0 free sites, unoccupied by surfactant molecules. The free sites are introduced to take into account the translational freedom of the islands and of the disordered molecules at the surface. A two-dimensional lattice model is employed in which each lattice site occupies a constant surface area A_0 equal to the minimum surface area of a surfactant molecule (for phospholipid surfactants,[24] $A_0 = 40\,\text{Å}^2$). The free sites are considered to have a two-dimensional size only and to be located at the interface (called for convenience internal surface) between the headgroup region and the tail region of the monolayer, each occupying one site. An island is assumed to contain n surfactant molecules in the LC state, each having a constant surface area A_c and, consequently, occupying γ_c (= A_c/A_0) sites. Each surfactant molecule in the LE state is considered to occupy a surface area A_d in the cross section of the hydrocarbon chains, but an area A_0 at the internal surface where there are free sites. A_d varies with the surface pressure. The hydrocarbon chains in both the LC and LE state are considered to be compact because the LE/LC phase transition occurs at high surface densities and air is an extremely poor solvent for the hydrocarbon chains. They are, however, more extended in the LE state. Consequently, the insoluble monolayer behaves like a gas at the internal surface, because free sites are present, and like a liquid or even a solid in the region of the hydrocarbon chains, since the latter are compact. For the number of surfactant molecules N and the total surface area A_t, one can write

$$N = N_d + nN_c \tag{12}$$

and

$$\begin{aligned} A_t &= (N_0 + N_d)A_0 + nN_cA_c \\ &= N_dA_d + nN_cA_c = AN \end{aligned} \tag{13}$$

where

$$A_d = \left(1 + \frac{N_0}{N_d}\right)A_0 \tag{14}$$

and A is the average surface area of a surfactant molecule. Eq. (12) represents the number balance of the surfactant molecules and Eq. (13) their area balance. The quantities N_d, N_c, n, and N_0 are predicted by the theory.

Fig. 1 presents the formation of a two-dimensional mixture containing N surfactant molecules and N_0 free sites. First, N_d surfactant molecules change their state from an LC crystallized solid to a disordered liquid by their mixing with N_0 free sites. Thus, an LE monolayer is formed with an average cross sectional area A_d

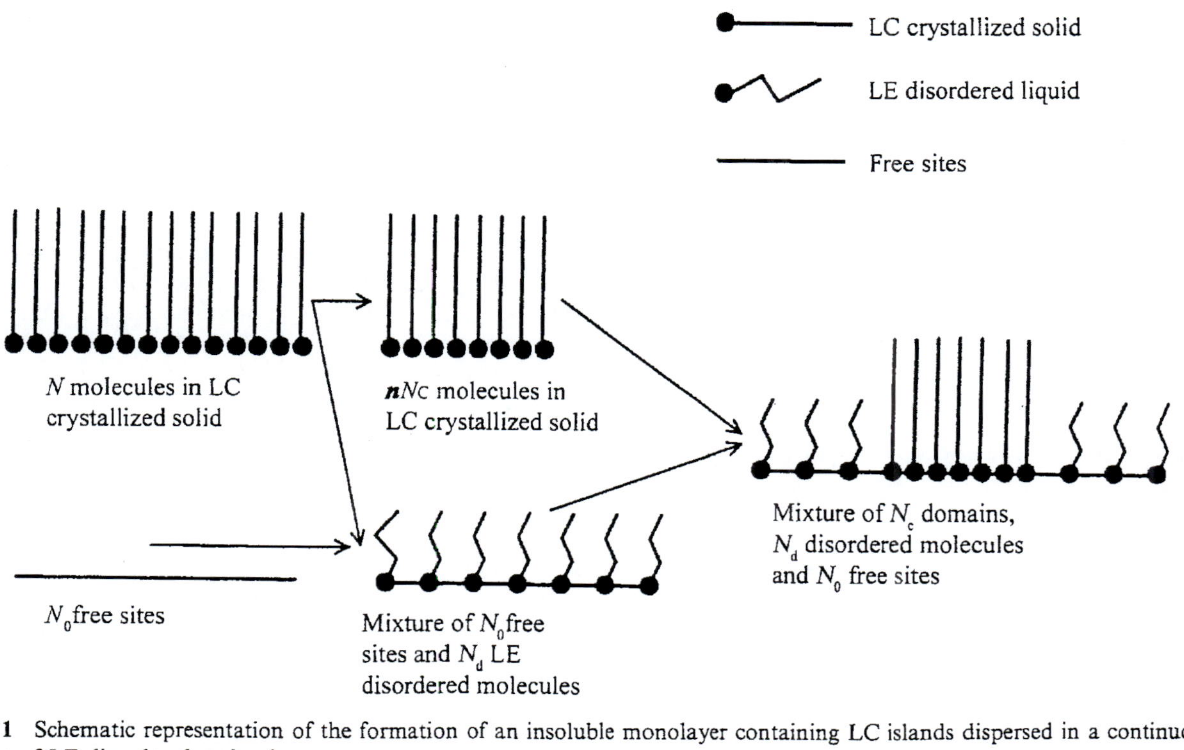

Fig. 1 Schematic representation of the formation of an insoluble monolayer containing LC islands dispersed in a continuous phase of LE disordered molecules.

per molecule. Then, a two-dimensional mixture is obtained by mixing the LE monolayer with the N_c domains obtained by the splitting of an LC monolayer containing nN_c surfactant molecules. The total Helmholtz free energy (F) of the two-dimensional mixture can be written as

$$F = N_0 f_0^0 + N f_c^0 + N_d \Delta f_{S/L}^0 + \Delta U_{mix} - T\Delta S_{mix} \quad (15)$$

where f_0^0 is the Helmholtz free energy of a free site, which is equal to $A_0\sigma_0$, with σ_0 denoting the pure water/air interfacial tension, f_c^0 is the Helmholtz free energy of a surfactant molecule in the LC monolayer, $\Delta f_{S/L}^0$ is the Helmholtz free energy change of a surfactant molecule from an LC crystallized solid state to a disordered LE liquid, and ΔU_{mix} and ΔS_{mix} are the two-dimensional mixing energy and entropy of the two-dimensional mixture.

The quantity $\Delta f_{S/L}^0$ does not involve the translational free energy change of a surfactant molecule since the latter was included in the mixing entropy contribution. The change from an LC crystallized solid to a disordered liquid is considered to occur in two steps: the first is from the LC crystallized solid (called below LC solid) to a liquid (called LC liquid), in which the molecules become somewhat disordered but occupy the same surface area A_c and the second step is from the LC liquid to the LE liquid, in which the molecular surface area varies from A_c to A_d. The former is a kind of melting process and has a melting free energy $\left(\Delta f_{L/L}^0\right)$. Hence,

$$\Delta f_{S/L}^0 = \Delta f_m^0 + \Delta f_{L/L}^0 \quad (16)$$

To calculate the two-dimensional mixing entropy, the following generalized Flory equation, proposed by Donohue and Prausnitz[17] is employed:

$$\Delta S_{mix} = -k \sum_i N_i \ln(y_i) \quad (17)$$

where N_i ($i = 0$, c, or d) is the number of molecules of species i and y_i is given by the following expression:

$$y_i = \frac{N_i \gamma_i^{p_i}}{\sum_j N_j \gamma_j^{p_j}} = \frac{x_i \gamma_i^{p_i}}{\sum_j x_j \gamma_j^{p_j}} = \frac{\Phi_i \gamma_i^{p_i - 1}}{\sum_j \Phi_j \gamma_j^{p_j - 1}} \quad (18)$$

In the above equation, γ_i is the number of sites occupied by species i at the internal surface, p_i is a parameter that has values between zero and unity, and x_i and Φ_i are the mole fraction and surface area fraction of species i, respectively, given by

$$x_i = \frac{N_i}{\sum_j N_j}, \quad \Phi_i = \frac{N_i \gamma_i}{\sum_j N_j \gamma_j} \quad (19)$$

The exponent p_i accounts for the effect of the size and shape of species i. When $p_i = 0$, $y_i = x_i$ and Eq. (17) reduces to the ideal mixing entropy, while when $p_i = 1$, $y_i = \Phi_i$ and Eq. (17) reduces to the Flory–Huggins equation. The former is suitable for mixtures of molecules of equal size and provides a lower limit of the mixing entropy, while the latter is suitable for mixtures of chain-like molecules of different sizes and provides an upper limit. The true mixing entropy lies in between, and consequently p_i has a value between zero and unity.

Because a free site and a disordered surfactant molecule are assumed to occupy a single site, each on the internal surface, $\gamma_0 = \gamma_d = 1$ and Eq. (18) becomes

$$y_i = \frac{\Phi_i}{1 + \Phi_c\left(\gamma_c^{p-1} - 1\right)} \quad i = 0 \text{ or d} \quad (20)$$

and

$$y_c = \frac{\Phi_c \gamma_c^{p-1}}{1 + \Phi_c\left(\gamma_c^{p-1} - 1\right)} \quad (21)$$

In Eqs. (20) and (21), $p = p_c$; that is, only one parameter is required in the present treatment for the entropy contribution. It will be shown later that p can be considered constant for different surfactants and different temperatures.

The two-dimensional mixing energy is the energy required to divide an LC monolayer into islands and to disperse them in the LE monolayer, and can be calculated using the equation derived by McConnell et al.[19–21] According to their treatment, the mixing energy involves two contributions: one is due to the dipole–dipole repulsion and another one arises from the line tension between the LC islands and the LE disordered molecules. The former tends to decrease and the latter increase the size of the islands. Considering a hexagonal array of circular islands of the same radius (R) dispersed in the LE phase, the mixing energy is given by[21]

$$\Delta U_{mix} = 2\pi R N_c \left[\lambda - (\Delta\mu)^2 \ln\left(\frac{4R}{e^2\delta}\right) + (\Delta\mu)^2 \upsilon g \Phi_c^{3/2}\right] \quad (22)$$

where λ is the line tension, $\Delta\mu$ is the difference in molecular dipole moment densities between the LC and the LE phases, δ is the nearest neighbor intermolecular dipole–dipole distance, $e = 2.718$, $v = (3\pi/2)\left(\sqrt{3/2\pi}\right)^{3/2}$, and g is given by

$$g = g_0 + 0.5168 g_1 \Phi_c \quad (23)$$

where g_0 and g_1 are constants equal to 3.25 and 2.23, respectively.

Inserting Eqs. (17) and (22) into Eq. (15) yields

$$\frac{F}{kT} = \frac{N_0 f_0^0 + N f_c^0 + N_d \Delta f_{S/L}^0}{kT} + \frac{2\pi R N_c (\Delta\mu)^2}{kT}\left[\frac{\lambda}{(\Delta\mu)^2} - \ln\left(\frac{4R}{e^2\delta}\right) + v g \Phi_c^{3/2}\right] + \sum_i N_i \ln(y_i) \quad (24)$$

THE EQUILIBRIUM RADIUS AND EQUILIBRIUM AREA FRACTION OF DOMAINS

Using the Helmholtz free energy given by Eq. (24), the equilibrium radius of the islands and the equilibrium area fraction occupied by the islands can be calculated via its minimization with respect to R and Φ_c, that is, using the equations $(\partial F/\partial R)_{N,A_t,\Phi_c} = 0$ and $(\partial F/\partial \Phi_c)_{N,A_t,R} = 0$, respectively. For the equilibrium radius, one obtains

$$R = R_0 \exp[u(\Phi_c)] \exp\left\{\frac{kT}{\pi R(\Delta\mu)^2}[\ln(y_c) + (1 - p)(1 - y_c)(1 - y_c^p)]\right\} \quad (25)$$

where

$$R_0 = \frac{e^3\delta}{4} \exp\left[\frac{\lambda}{(\Delta\mu)^2}\right] \quad (26)$$

and

$$u(\Phi_c) = v g \Phi_c^{3/2} = v(g_0 + 0.5168 g_1 \Phi_c)\Phi_c^{3/2} \quad (27)$$

in Eq. (25), R_0 is the equilibrium radius of an island in the absence of other islands, the factor $\exp[u(\Phi_c)]$ accounts for the effect of the dipole–dipole repulsion among the islands on the equilibrium radius R of the latter and shows that the radius increases exponentially with the area fraction of the domains, and the last factor is due to the mixing entropy of the two-dimensional mixture. When the last factor is taken as unity, Eq. (25) reduces to that derived by McConnell et al.[19–21] Because $kT/\left(\pi R(\Delta\mu)^2\right)$ is on the order of 10^{-5} for an island radius of 5 μm, that factor is very close to unity, indicating a negligible effect of the two-dimensional mixing entropy on the equilibrium radius. Consequently, the last factor in Eq. (25) was taken as unity in the present calculations.

For the equilibrium area fraction Φ_c of the islands, the equation $(\partial F/\partial \Phi_c)_{N,A_t,R} = 0$ leads to

$$y_c^{1/n} = y_d y_0^{(A_c/A_0 - 1)} \exp\left[\frac{A_c y_c}{A_0 \Phi_c}(1 - y_c^{-p})\right] K_m K \quad (28)$$

where K_m accounts for the effect of the melting free energy and K for the effects of the order–disorder free energy change from the LC liquid to the LE liquid and the dipole–dipole interactions among islands and between islands and disordered molecules. They are given by the expressions

$$K_m = \exp\left(\frac{\Delta f_m^0}{kT}\right) \quad (29)$$

and

$$K = \exp\left\{P - (A_d - A_c)P' + \frac{2A_c(\Delta\mu)^2}{kTR}[1 - u'\Phi_c] + \frac{(A_c - A_0)\sigma_0}{kT}\right\} \quad (30)$$

where

$$P = \Delta f_{L/L}^0/kT, \quad P' = \partial P/\partial A_d$$

and

$$u' = \frac{du(\Phi_c)}{d\Phi_c} = v\left(\frac{5}{2}g - g_0\right)\Phi_c^{1/2}$$

since no expression is available to calculate $\Delta f_m^0/kT$, K_m is determined from the molecular surface area A_{tr} at the onset of the LE/LC phase transition as follows: First, it should be noted that the term $(y_c)^{1/n}$ in Eq. (28) is always very near unity, because the number n of molecules in an island is on the order of 10^6–10^8. Taking $(y_c)^{1/n}$ equal to unity and Φ_c equal to zero and replacing A_d by A_{tr}, one obtains

$$K_m = \left(\frac{A_{tr}}{A_0}\right)\left(\frac{A_{tr}}{A_{tr} - A_0}\right)^{(A_c/A_0 - 1)} \times \exp\left[\frac{A_c}{\gamma_{c,tr} A_0}(1 - \gamma_{c,tr}^p)\right] K_{tr}^{-1} \quad (31)$$

where K_{tr} and $\gamma_{c,tr}$ are the values of K and γ_c for $A_d = A_{tr} = A_t/N$.

THE SURFACE EQUATION OF STATE

Using Eq. (24) for the Helmholtz free energy and the thermodynamic definition of the surface tension σ,

$$\sigma = \left(\frac{\partial F}{\partial A_t}\right)_{T,N} \tag{32}$$

one obtains

$$\frac{\pi A_0}{kT} = -\ln(y_0) - P'A_0 + \frac{2A_0(\Delta\mu)^2}{kTR}u'\Phi_c^2 + y_c(\gamma_c^{-p} - 1) + \frac{2u'\Phi_c \ln(y_c)}{\gamma_c} \times \left(\frac{\partial \Phi_c}{\partial A}A\right) \tag{33}$$

where $\pi = \sigma_0 - \sigma$ is the surface pressure. Differentiating Eq. (28), one obtains

$$A\frac{\partial \Phi_c}{\partial A} = \frac{A_0}{A}\left(\frac{A_d}{A_c} - 1\right) \times \left[\frac{\Phi_d}{\Phi_0} + P''A_0^2\left(\frac{A_d}{A_0}\right)\right]/(S_A + U_A) \tag{34}$$

where

$$S_A = \frac{2u'\Phi_d}{\gamma_c}\ln(y_c) - \left(\frac{A_0}{A_c} + \frac{\Phi_d}{\gamma_c \Phi_c}\right) + y_d(\gamma_c^{p-1} - 1)\left[\gamma_c^{-1} - 1 + \frac{y_c}{\Phi_c}(\gamma_c^{-p} - 1)\right]$$

$$U_A = \frac{2y_d P u'}{\gamma_c} - (P''A_0^2)\left(\frac{A_d}{A_c} - 1\right)^2 + \frac{2A_0(\Delta\mu)^2}{kTR}\Phi_d\left[(u')^2\Phi_c - 2u' - u''\Phi_c\right]$$

with

$$P'' = \frac{\partial^2}{\partial A_d^2}\left(\frac{\Delta f^0_{L/L}}{kT}\right)$$

and

$$u'' = \frac{d^2u(\Phi_c)}{d\Phi_c^2} = \frac{3}{2}v\left(\frac{5}{2}g - 2g_0\right)\Phi_c^{-1/2}$$

In Eq. (33), the first term accounts for the translational freedom of the islands and of the disordered molecules in the monolayer, the second accounts for the change in the state of the surfactant molecules from an LC liquid to an LE liquid, the third is due to the dipole–dipole repulsion among the islands, the fourth is due to the entropy, and the last one is due to the formation of new islands.

Eq. (33) reduces to a simple form in the absence of islands, since in this case, the two-dimensional mixture contains only two species: the free sites and the disordered molecules, each species occupying the same internal surface area A_0. Consequently, $y_c = \Phi_c = 0$, $A_d = A$, $y_0 = 1 - A_0/A$, and

$$\frac{\pi A_0}{kT} = -\ln\left(\frac{A - A_0}{A}\right) - P'A_0 \tag{35}$$

Using a low-density approximation for the first term of the right-hand side, Eq. (35) becomes

$$(\pi + kTP')(A - A_0) = kT \tag{36}$$

where $kTP' = \partial\left(\Delta f^0_{L/L}\right)/\partial A$ represents the surface cohesion pressure, which will be calculated later. The above equation constitutes a two-dimensional equation of state for a gaseous monolayer.

HELMHOLTZ FREE ENERGY CHANGE ($\Delta f^0_{l/l}$) FROM THE LC TO THE LE LIQUID

The Helmholtz free energy difference $\Delta f^0_{L/L}$ is considered to involve three contributions: a dipole–dipole interaction free energy change $\left(\Delta f^0_{\text{dipole}}\right)$, a chain conformational free energy change $\left(\Delta f^0_{\text{con}}\right)$, and a chain van der Waals interaction (volume-dependent) free energy change $\left(\Delta f^0_{\text{vdW}}\right)$,

$$\Delta f^0_{L/L} = \Delta f^0_{\text{dipole}} + \Delta f^0_{\text{con}} + \Delta f^0_{\text{vdW}} \tag{37}$$

the three kinds of free energy changes are calculated below.

The Dipole–Dipole Interaction Free Energy Change ($\Delta f^0_{\text{dipole}}$)

Using a capacitor as a model for a two-dimensional array of effective dipoles considered perpendicular to the air/water interface in a monolayer, the dipole–dipole interaction free energy can be approximated by[22]

$$f^0_{\text{dipole}} = \frac{e^2 d_e}{2\varepsilon_0 \varepsilon A} = \frac{e\mu}{2\varepsilon_0 \varepsilon} \tag{38}$$

where d_e is the effective charge separation of the dipole moment, e is the protonic charge (1.602×10^{-19}C), ε_0 is the permittivity of vacuum ($8.854 \times 10^{-12}\,C^2 J^{-1} m^{-1}$), and ε is the dielectric constant around the headgroup. The product $d_e e$ stands for the effective dipole moment, and consequently, the quantity $d_e e/A = \mu$ for the dipole moment density.

For a given surfactant, the effective dipole moment can be considered to be constant within the range of surface densities of concern here, but the dipole moment density (μ) varies with the two-dimensional molecular density. Indeed, for the 1,2-dilauroyl (DLPE) and 1,2-dimyristoylphosphatidylethanolamine (DMPE) monolayers in the LE state, the surface potential measurements indicated[23] that only the molecular density, and not the dipole orientation, was changed during the compression of the monolayer; hence, the normal dipole moment remained constant. We assume that the same dipole moment can be used for the LC liquid, in which the molecules are less extended than in the LE liquid, which is in a more disordered state. No equation is available for the calculation of the effective dipole moment. In the present treatment, the unknown effective dipole moment and the dielectric constant are included into a single parameter, which was selected to fit the experimental $\pi - A$ isotherms in the LE range. Thus, the dipole–dipole interaction free energy change is given by the expression

$$\frac{\Delta f^0_{\text{dipole}}}{kT} = \alpha_d \left(\frac{T_0}{T}\right)\left(\frac{A_0}{A_d} - \frac{A_0}{A_c}\right) \tag{39}$$

where

$$\alpha_d = \frac{e^2 d_e}{2kT_0\varepsilon_0\varepsilon A_0} = \frac{e\mu_0}{2kT_0\varepsilon_0\varepsilon}$$

with $T_0 = 298\,\text{K}$ as a reference temperature and $\mu_0 = ed_e/A_0$ as the dipole moment density in the limiting case in which the molecules become completely ordered.

The conformational free energy change (Δf^0_{con})

The hydrocarbon chains of the surfactant molecules are constrained to remain with the polar headgroups in contact with water and therefore have a conformational free energy relative to the unconstrained chains. The conformational free energy per chain can be calculated using the following expression:[22,24]

$$\frac{f^0_{\text{con}}}{kT} = \frac{\pi^2}{8}\frac{N_s L^4}{A^2} \tag{40}$$

where N_s is the number of segments of a chain located on a cubic lattice whose sites have a size L equal to the chain width (4.6 Å), and is given by $N_s = (n_c + 1)/3.6$ with n_c representing the number of carbon atoms of the chain. Eq. (40) was derived for polymers; it can also be used for the short chains of concern here, since our evaluations have shown that it provides results near to those obtained for short chains when the lattice statistical mechanical approach of Cantor and Dill[25] is used.

Using Eq. (40), the conformational free energy for a surfactant molecule with a double chain is given by

$$\frac{f^0_{\text{con}}}{kT} = \frac{\pi^2 N_s L^4}{A^2} \tag{41}$$

By increasing the molecular surface area from A_c to A_d, the packing constraint to a hydrocarbon chain is increasingly reduced, and this results in a conformational free energy decrease. Using Eq. (41), the conformational free energy change from an LC to an LE liquid is given by

$$\frac{\Delta f^0_{\text{con}}}{kT} = \alpha_c \left[\left(\frac{A_0}{A_d}\right)^2 - \left(\frac{A_0}{A_c}\right)^2\right] \tag{42}$$

where

$$\alpha_c = (\pi^2 N_s L^4)/(A_0^2)$$

The chain van der Waals interaction free energy change (Δf^0_{vdW})

With the increase of the two-dimensional molecular density, the chains become increasingly aligned and the anisotropic short-range steric repulsive interactions are reduced due to a better packing of the chains. This leads to a dependence of the chain van der Waals interaction free energy on the two-dimensional molecular density or the surface area occupied by a molecule. For the van der Waals interaction free energy among the chains in a monolayer, the following equation was derived:[25]

$$F_v(X) = F_v^\infty - \frac{15}{2}\frac{E_0}{\tilde{V}}(n_c + n_e)(1 - X) \tag{43}$$

where the quantity F_v^∞ represents the free energy per mole for the completely disordered chains, $\tilde{V} = V/V^* = 1.23$ is the reduced volume of a completely disordered chain, with V and V^* the volume and the hard-core volume of the chain, E_0 (= 1.57 kcal/mol) is the energy at $T = 0\,\text{K}$, n_c is the number of carbon atoms in the chain, and n_e is a constant of the order of unity. X accounts for the extent of reduction of the repulsive component of the chain pair potential with increased chain alignment. For completely disordered chains $X = 1$, while for total alignment (highly ordered chains) $X = X_r < 1$. Assuming that X is a linear function of an average order parameter (S^*) and considering that S^* depends on the molecular surface area, the van der Waals interaction free energy could be expressed in terms of the molecular surface area.[25] We make the same assumption regarding X, but calculate differently the average order parameter S^*.

Generally, the average order parameter S^* is given by

$$S^* = \frac{3}{2}P^* - \frac{1}{2}$$

where P^* is the average probability for a bond to be parallel to the normal to the interface. Denoting by n_0 the number of vertical bonds in a chain containing $N_b - 1$ bonds, one can write $P^* = n_0/(N_b - 1)$. Neglecting the bond's backtracks and considering that all the chains terminate at the same height (which are plausible assumptions for short-chain compact monolayers), one can write $n_0(N_b - 1) = A_0/A$. Consequently, the average order parameter S^* can be expressed in terms of the molecular surface area as

$$S^* = \frac{3}{2}\frac{A_0}{A} - \frac{1}{2} \tag{44}$$

The above equation provides reasonable limiting values. When $A = A_0$, $S^* = 1$, and when $A = 3A_0$ ($= 120\,\text{Å}^2$ for the phospholipid surfactants), which is a very large molecular surface area, $S^* = 0$, indicating complete disorder. In the LE monolayer and the LE/LC transition region, A is between A_0 and $3A_0$, and consequently, S^* varies between 0 and 1.

Assuming a linear relationship between X and S^*, one can write

$$X = 1 + (X_r - 1)\left(\frac{3}{2}\frac{A_0}{A} - \frac{1}{2}\right) \tag{45}$$

where X_r is the value of X for a highly ordered solid phase with $A = A_0$. Cantor and Dill[25] calculated X_r as a linear combination of contributions from ethylene and terminal methyl groups. In the present calculations,

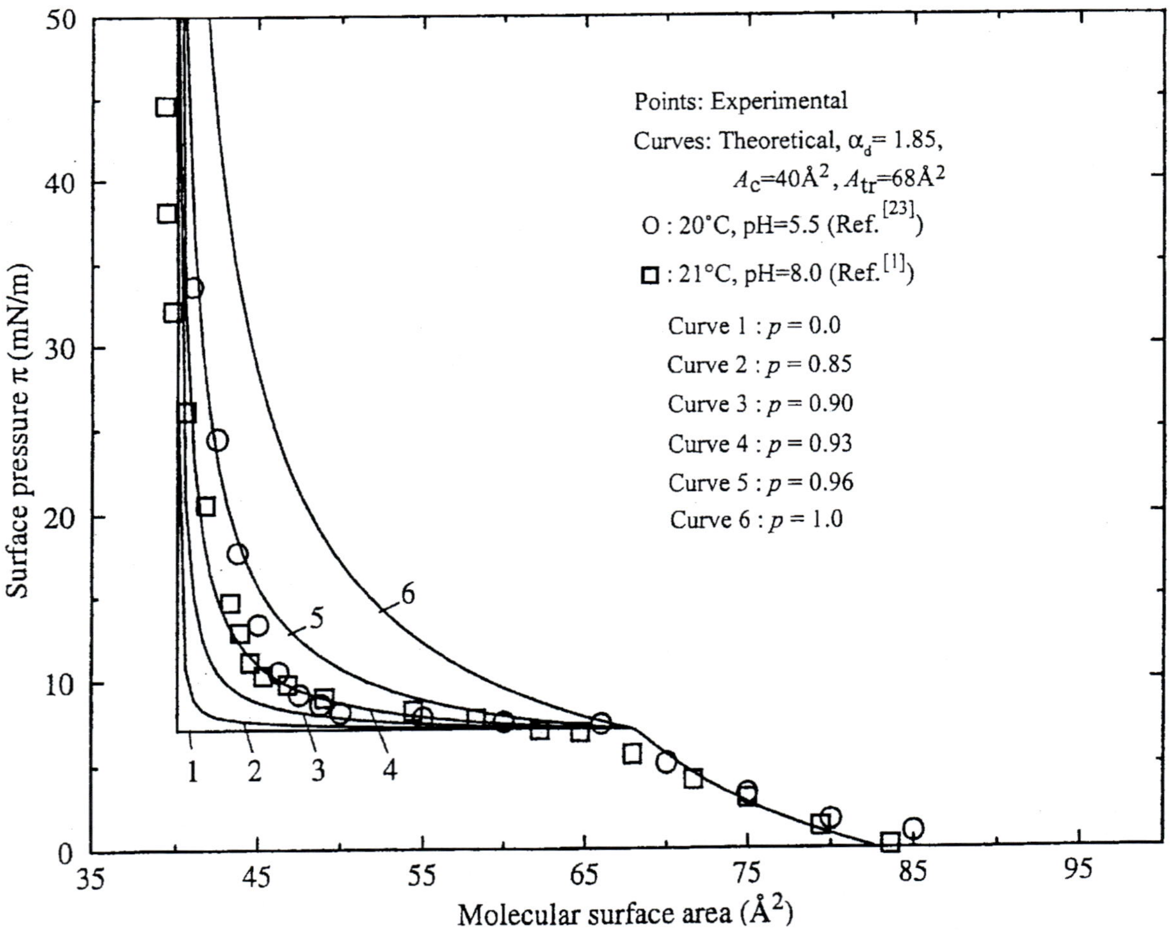

Fig. 2 Effect of the parameter p of the mixing entropy model on the theoretical $\pi - A$ isotherm of DMPE at 20.5°C.

the value 0.9765 suggested by them for the methylene group is used, since the number of methylene groups is large and the values of X_r for methyl and methylene groups are near to one another.

Substituting Eq. (45) into Eq. (43), one obtains the following expression for the van der Waals interaction free energy per chain:

$$\frac{f^0_{\text{volume}}}{kT} = \frac{F^{\infty}_{v}}{RT} + \alpha_v\left(\frac{T_0}{T}\right)\left(\frac{A_0}{A} - \frac{1}{3}\right) \quad (46)$$

where

$$\alpha_v = -\frac{45E_0}{4\widetilde{V}}(n_c + n_e)(1 - X_r)$$

Consequently, the van der Waals interaction free energy change from the LC to the LE liquid can be calculated using the expression

$$\frac{\Delta f^0_{\text{volume}}}{kT} = \alpha_v\left(\frac{T_0}{T}\right)\left(\frac{A_0}{A_d} - \frac{A_0}{A_c}\right) \quad (47)$$

Introducing the values of E_0, $\widetilde{V}$, and X_r and neglectin the small constant n_c, one obtains $\alpha_v = -0.57n_c$ pe hydrocarbon chain. For phospholipid surfactant with two identical chains, α_v is given by the expressio $\alpha_v = -1.14n_c$.

For an LE monolayer, Eq. (33) in which A_0/A wa replaced by the molecular surface density ρ, becomes

$$\frac{\pi A_0}{kT} = -\ln(1 - \rho) + (\alpha_d + \alpha_v)\left(\frac{T_0}{T}\right)\rho^2 + 2\alpha_c\rho$$

Expanding the first term on the right-hand side of th above equation up to the third order in densit the following two-dimensional virial equation of stat is obtained:

$$\frac{\pi A_0}{kT} = \rho + B_2\rho^2 + B_3\rho^3 \quad (48$$

where $B_2 = (\alpha_d + \alpha_v)(T_0/T) + 1/2$ and $B_3 = 2\alpha_c$ - 1/3 represent the second and third virial coefficient respectively. The dipole–dipole interactions and th

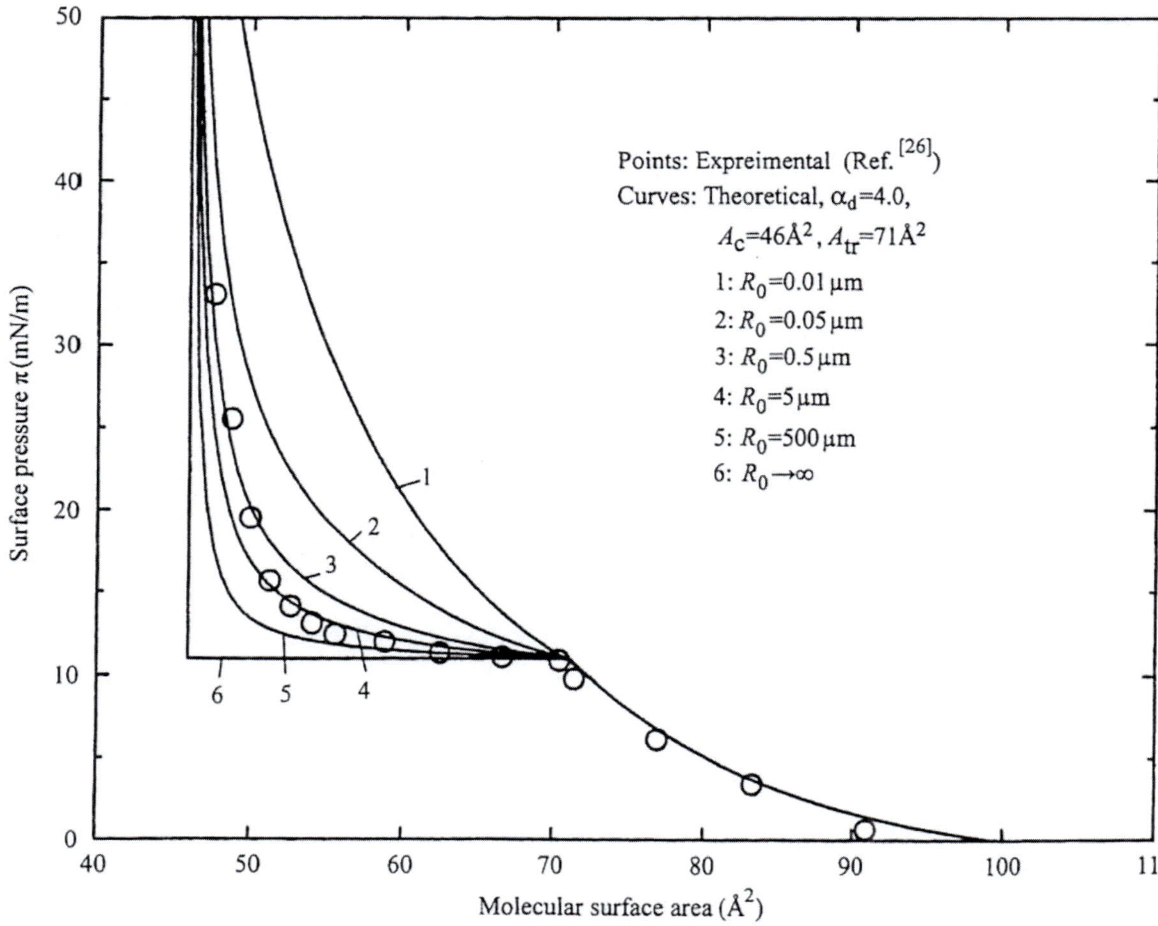

Fig. 3 Effect of the island size on the theoretical $\pi - A$ isotherm of 1,3-DPPC at 26°C.

chain van der Waals interactions affect the second virial coefficient, while the chain conformation affects the third virial coefficient. Since the parameter α_d depends on the nature of the headgroup and the parameters α_υ and α_c upon the number of carbons in the hydrocarbon chain, the surface equations of state 33 and 48 account for both the nature of the headgroup and the length of the chain.

RESULTS AND DISCUSSION

In this section, the derived surface equation of state [Eq. (33)] is compared with the experimental $\pi - A$ isotherms involving the LE/LC phase transition for the following six phospholipid surfactants: 1,2-dimyristoylphosphatidylcholine (DMPC), 1,2-dipalmitoylphosphatidylcholine (1,2-DPPC), 1,3-dipalmitoylphosphatidylcholine (1,3-DPPC), 1,2-dimyristoylphosphatidic acid (DMPA), 1,2-dilauroylcephalin (DLPE), and 1,2-dimyristoylphosphatidylethanolamine (DMPE). The general formula of the phospholipid surfactants can be written as R–CO_2–(h)–O_2C–R, where R stands for the hydrocarbon chain, which has n_c carbon atoms, and h stands for the connecting chains and the headgroup.

Fig. 2 presents the effect of the two-dimensional mixing entropy on the theoretical $\pi - A$ isotherm of DMPE at 20°C. Since the mixing entropy is characterized by the parameter p [Eq. (17)], six isotherms for different values of p from zero to unity are presented in the figure. First, one can observe that with increasing value of p, the slope of the isotherm in the transition region increases. When $p = 0$, the mixing entropy is given by the ideal-solution expression and the isotherm in the transition region becomes horizontal (curve 1). In this case, the islands provide a negligible contribution to the two-dimensional mixing entropy because the mole fraction of the islands, which are large, is very small. With increasing p ($p = 0.85$, curve 2), the isotherm in the transition region starts to deviate from the horizontal line, and a smoothly rising continuous curve is obtained at the high surface density end of the phase transition. When $p = 1$ (curve 6), the two-dimensional mixing entropy is given by the Flory–Huggins expression and the isotherm in the transition

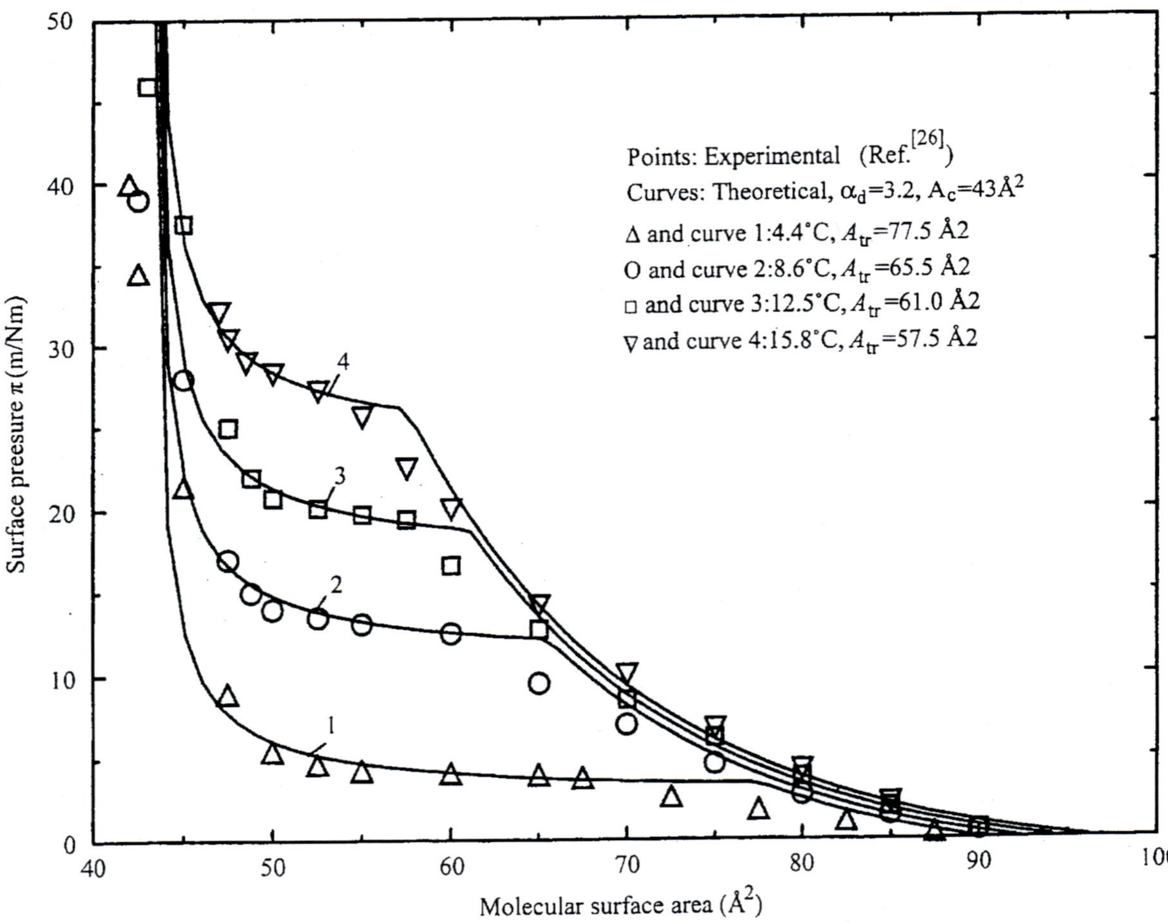

Fig. 4 $\pi - A$ isotherms of 1,2-dimyristoylphosphatidylcholine at four temperatures.

region exhibits the largest slope. For the DMPE surfactant, the theoretical $\pi - A$ isotherm is in good agreement with experiment when $p = 0.93$. For all the systems and at different temperatures, $p = 0.93$ results in good agreement with experiment. (In some cases, a somewhat better fitting can be reached by slightly changing the value of p). Consequently, the parameter p is taken equal to 0.93 in all calculations.

The islands can have various shapes. The present model considers only circular islands for the calculation of the mixing energy, but no restriction regarding the shape is made in the calculation of the mixing entropy. The parameter p in the entropy expression depends on the shape of the islands. In a three-dimensional solution, the molecular shape can be characterized by the ratio q/γ,[17] where γ represents the number of lattice sites occupied by the molecule (or the number of molecular segments) and qz (z is the lattice coordination number) provides the number of nearest neighboring sites of the molecule. As γ becomes very large, $q/\gamma \rightarrow 2/3$ for a linear chain and $q/\gamma \rightarrow 0$ for a sphere. For a globular molecule, q/γ lies between zero and unity. Similarly, for the present two-dimensional solution, the shape of an island can be characterized by the ratio L/nA_c between the perimeter (L) and the surface area (nA_c) of the island. The analogy between the two-dimensional and the three-dimensional mixing entropy suggests that the parameter p should increase with increasing L/nA_c. Consequently, the large circular islands require a smaller value for p than the long striped islands with the same surface area. Thus, the large circular islands are expected to lead to a lower slope of the $\pi - A$ isotherm in the transition region than the long striped ones.

According to Eq. (25), the equilibrium radius R of the island is the product of two factors: one of them is the equilibrium radius in the absence of other islands ($\Phi_c = 0$), and the other one increases exponentially with the area fraction of the islands. R_0 is determined by the line tension (λ) and the difference $\Delta\mu$ between the two-dimensional dipole moment densities of the LC islands and the LE molecules. Both λ and $\Delta\mu$ depend on the molecular surface area difference ($A_d - A_c$) between the LC and LE phases. Since A_c is assumed to be constant and A_d is considered to

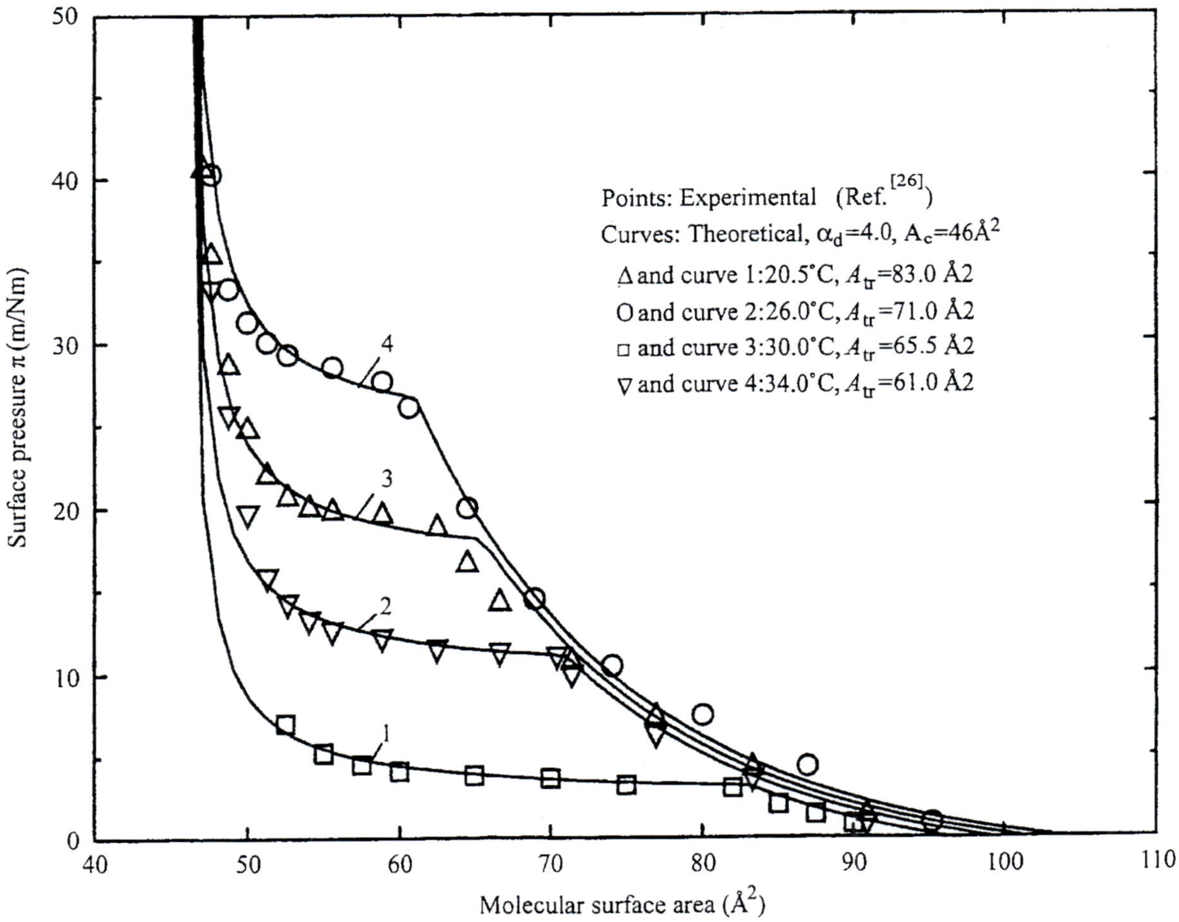

Fig. 5 $\pi - A$ isotherms of 1,2-dipalmitoylphosphatidylcholine at four temperatures.

vary with the surface pressure, λ and $\Delta\mu$ depend on the surface pressure (or on the area fraction of the islands). However, in the present treatment, the dependence (expected to be small) of R_0 on the area fraction of the islands is neglected.

Fig. 3 presents the effect of R_0 on the theoretical $\pi - A$ isotherm of the 1,2-DPPC surfactant at 26°C. As expected, a smaller island size results in a larger slope in the LE/LC transition region, while a larger one, in a smaller slope. For the range in which the calculations have been carried out, the theoretical isotherm has the largest slope for $R_0 = 0.01\,\mu m$, which corresponds to an island containing about 680 molecules. For $R_0 \to \infty$, the theoretical isotherm exhibits a horizontal line in the transition region. In the latter case, the two-dimensional mixture contains only a single LC island, and the surface area of a disordered molecule A_d becomes independent of the surface pressure. The experimental $\pi - A$ data in the transition region lie between the theoretical isotherms for $R_0 = 0.5\,\mu m$ and $R_0 = 500\,\mu m$. Using an intermediate value ($5\,\mu m$) for R_0, good agreement between the theoretical and experimental results is obtained. In addition, one can see that increasing R_0 by two orders of magnitude from $5\,\mu m$ to $500\,\mu m$, the theoretical isotherm does not vary too much. For this reason, in the present calculation, R_0 was taken equal to $5\,\mu m$ for all surfactants and all temperatures. It is important to emphasize that the radius R given by Eq. (25) increases exponentially with Φ_c. Taking $R_0 = 5\,\mu m$, R increases to 13, 36, and $105\,\mu m$ for $\Phi_c = 0.5$, 0.75, and 0.95, respectively, values that are in the range found experimentally.

Comparison with Experiment

A comparison between the theoretical and experimental isotherms is presented in Figs. 4–8. For all the surfactants, the exponent p in the mixing entropy model was taken equal to 0.93 and R_0 was taken $5\,\mu m$. The selected values of the parameters α_d, A_c, and A_{tr} are listed in the figures. For all the systems, the theoretical and experimental isotherms exhibit a

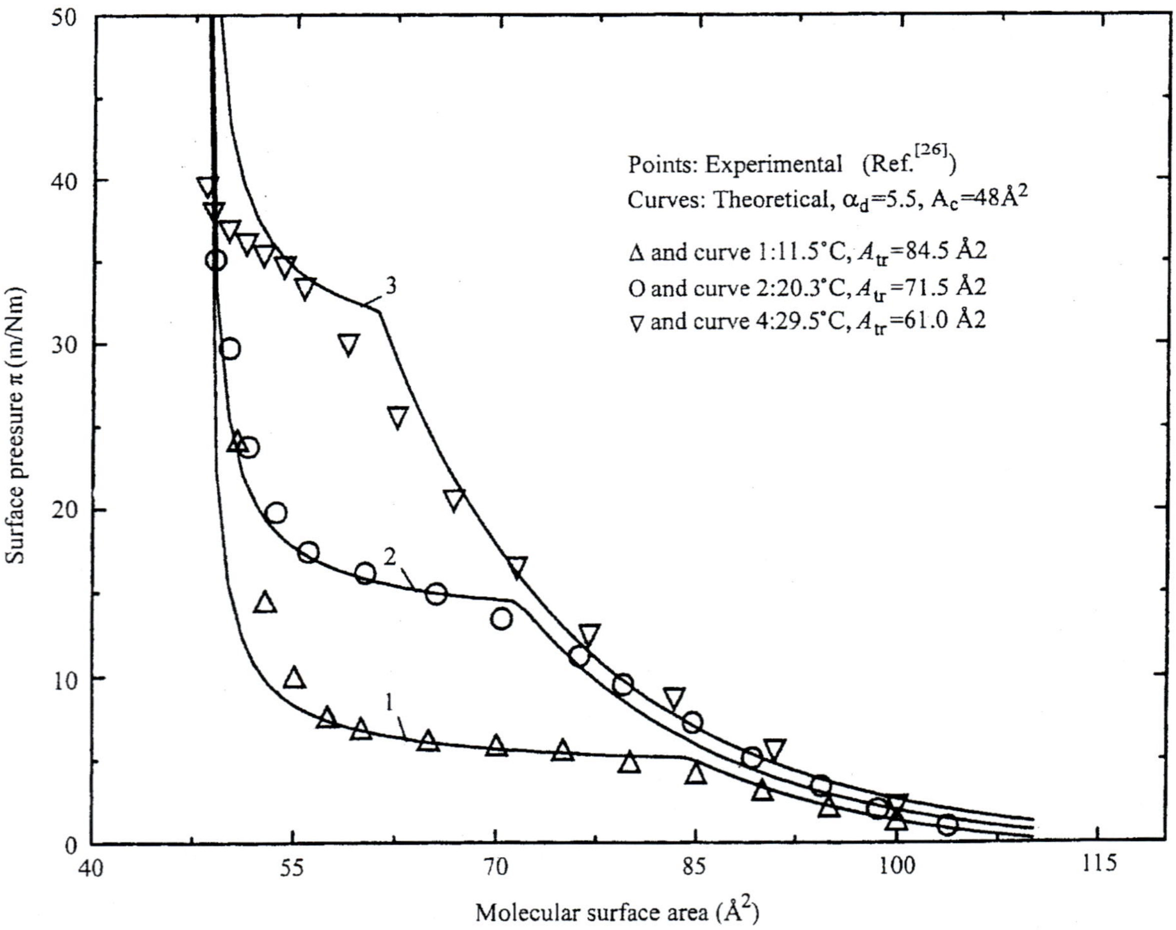

Fig. 6 $\pi - A$ isotherms of 1,3-dipalmitoylphosphatidylcholine at three temperatures.

non-horizontal line in the LE/LC transition region and a smoothly rising continuous curve at the high surface density end of the transition. For most of the 29 isotherms, very good agreement between the theoretical and experimental isotherms is obtained over the entire range of molecular surface densities.

The value of α_d, which accounts for the dipole–dipole interaction, was obtained by fitting the model to a single experimental point of the isotherm in the LE region. The determined value of α_d allowed one to estimate the effective dipole moment perpendicular to the air/water interface. Figs. 4–6 and 8 show that using the values of α_d determined from single points, the model can predict well the dependence of the isotherms on temperature in the LE region. For the surfactants with the same headgroup, the same value of α_d is expected. Indeed, the value $\alpha_d = 1.85$ could be used for both DMPE and DLPE surfactants (Fig. 7). However, for the pair of surfactants DMPC and 1,2-DPPC, which have the same headgroup, two different values ($\alpha_d = 3.2$ and 4.0, respectively) were required to fit the experimental data in the LE region. This probably occurred because the hydrocarbon chain of a surfactant molecule affects the dipole–dipole interactions. For the pair of surfactants, 1,2-DPPC and 1,3-DPPC, which possess the same headgroup and the same hydrocarbon chains, but different connections between the hydrocarbon chains and the headgroup, two very different values ($\alpha_d = 4.0$ and 5.5, respectively) were required to fit the experimental data in the LE region. Finally, the surfactant DMPA, which has a headgroup with a very small dipole moment, requires as expected a small value for α_d ($=0.25$).

In this model, the surface area of disordered molecules (A_d) is considered to vary with the surface pressure. In the LE region, A_d is equal to the average molecular surface area A, while in the transition region, A_d is larger than A and tends to become constant with decreasing slope of the isotherm. The molecular surface area in the LC state (A_c) is taken constant for a given surfactant, although experiment shows that it varies in a small range.[27] Among the factors that affect the value of A_c, the structure and size of the headgroup are expected to be the most

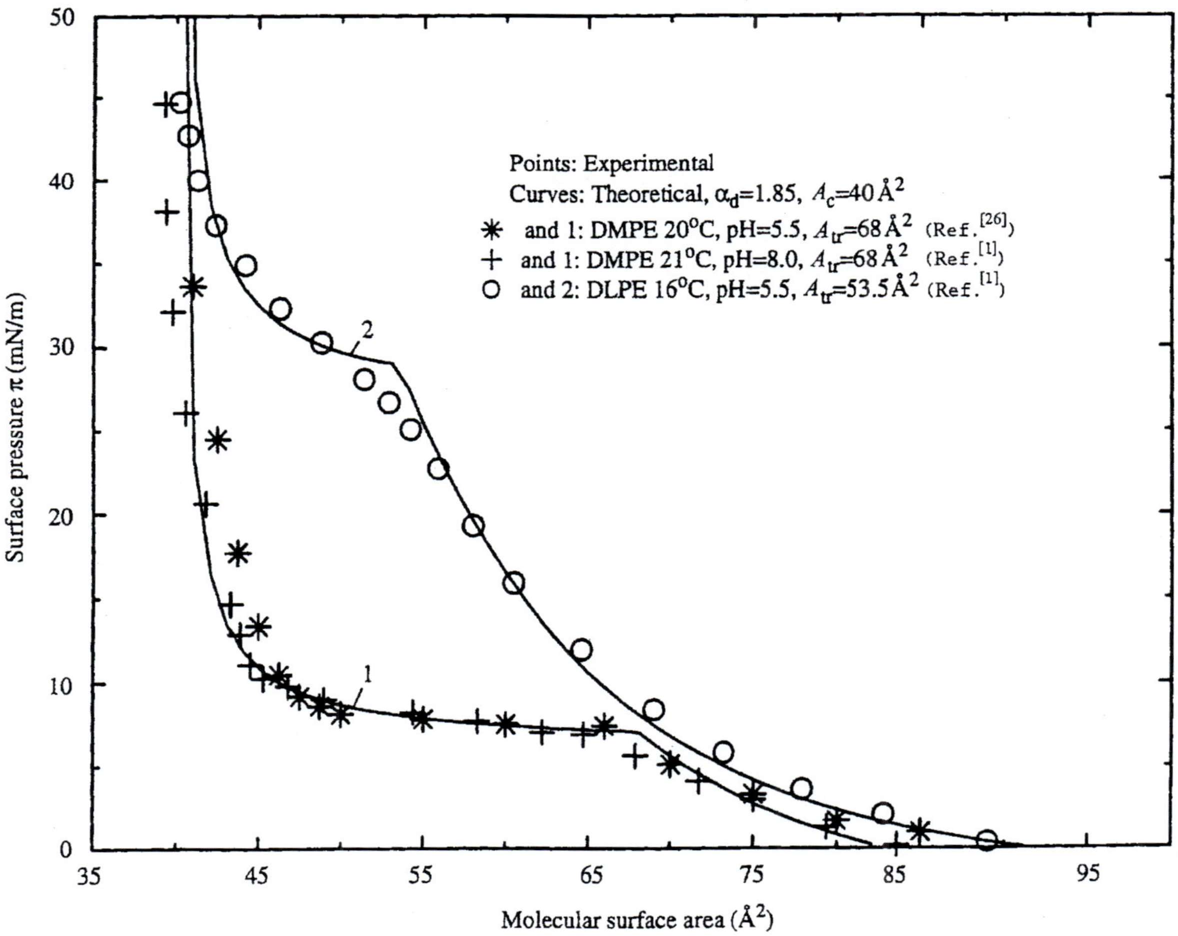

Fig. 7 $\pi - A$ isotherms of 1,2-dilauroylcephalin and 1,2-dimyristoylphosphatidyl-ethanolamine.

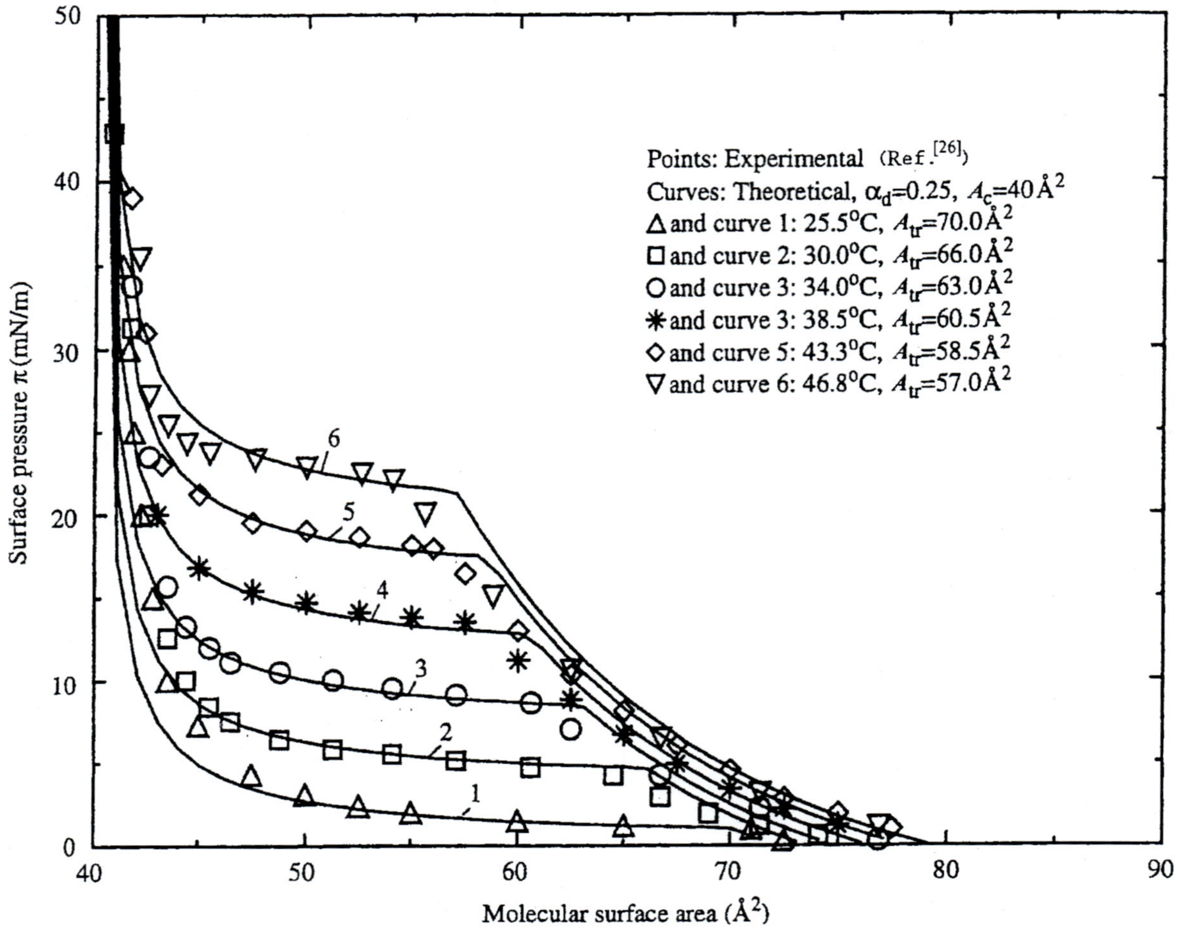

Fig. 8 $\pi - A$ isotherms of 1,2-dimyristoylphosphatidic acid at six temperatures.

important. Obviously, a large headgroup should result in a larger value of A_c. For the three surfactants, DMPC, 1,2-DPPC, and 1,3-DPPC (Figs. 4–6), with a larger headgroup (–PC), larger values of A_c (=43 Å^2 for DMPC, 46 Å^2 for 1,2-DPPC, and 48 Å^2 for 1,3-DPPC), whereas for the three surfactants, DMPE, DLPE, and DMPA (Figs. 7 and 8), with smaller headgroups (–PE or –PA), a smaller value of A_c (=40 Å^2) was required to obtain agreement with the experimental isotherms.

CONCLUSIONS

A thermodynamic treatment is first suggested on the basis of which one can explain the inclined phase transition that occurs in monolayers of insoluble surfactants. By minimizing the Helmholtz free energy of the monolayer, the equilibrium radius and the equilibrium area fraction of the LC islands are obtained as functions of the average molecular surface area A. The mixing entropy provides a negligible effect on the size of the islands, but is very important in the calculation of the equilibrium area fraction of the islands. Further, a surface equation of state is derived which leads to a $\pi - A$ isotherm that exhibits a nonhorizontal line in the phase transition region. The $\pi - A$ isotherm in the phase transition region depends on the two-dimensional mixing entropy and on the size of the islands. The theoretical $\pi - A$ isotherms are in good agreement with the experimental ones for six phospholipid surfactants.

REFERENCES

1. Mingotaud, A.F.; Mingotaud, C.; Patterson, L.K. *Handbook of Monolayers*; Academic Press: New York, 1993.
2. Knobler, C.M. Recent developments in the study of monolayers at the air-water interface. Adv. Chem. Phys. **1990**, *77*, 397.
3. von Tscharner, V.; McConnell, H.M. An alternative view of phospholipid phase behavior at the air-water interface. Microscope and film balance studies. Biophys. J. **1981**, *36*, 409.
4. Lösche, M.; Sackmann, E.; Möhwald, H. A fluorescence microscopic study concerning the phase diagram of phospholipids. Bunsenges. Phys. Chem. **1983**, *87*, 848.
5. Peters, R.; Beck, K. Translational diffusion in phospholipid monolayers measured by fluorescence microphotolysis. Proc. Natl. Acad. Sci. USA **1983**, *80*, 7183.
6. McConnell, H.M.; Tamm, L.K.; Weis, R.M. Periodic structures in lipid monolayer phase transitions. Proc. Natl. Acad. Sci. USA **1984**, *81*, 3249.
7. Henon, S.; Meunier, J. Microscope at the Brewster angle: direct observation of first-order, phase transitions in monolayers. Rev. Sci. Instrum. **1991**, *62*, 936.
8. Hönig, D.; Möbius, D. Reflectometry at the Brewster angle and Brewster angle microscopy at the air-water interface. Thin Solid Films **1992**, *210/211*, 64.
9. Henon, S.; Meunier, J. Phase transitions in Gibbs films: star texture defects in tilted mesophases. J. Chem. Phys. **1993**, *98*, 9148.
10. Overbeek, G.A.; Hönig, D.; Wolthaus, L.; Gnabe, M.; Möbius, D. Observation of bond orientational order in floating and transferred monolayers with Brewster angle microscopy. Thin Solid Films **1994**, *242*, 26.
11. Ruckenstein, E. Nonconventional first order transitions in insoluble monolayers of surfactants. Colloids and Surfaces A **2001**, *183*, 423.
12. McConnell, H.M. Structures and transitions in lipid monolayers at the air-water interface. Ann. Rev. Phys. Chem. **1991**, *42*, 171.
13. Pallas, N.R. Ph.D. thesis, Clarkson College of Technology: Potsdam, NY, 1983.
14. Pallas, N.R.; Pethica, B.A. Liquid-expanded to liquid-condensed transition in lipid monolayers at the air/water interface. Langmuir **1985**, *1*, 509.
15. McConnell, H.M. Theory of hexagonal and stripe phase in monolayers. Proc. Natl. Acad. Sci. USA **1989**, *86*, 3452.
16. Ruckenstein, E.; Li, B. Surface equation of state for insoluble surfactant monolayers at the air/water interface. J. Phys. Chem. **1998**, *102*, 981.
17. Donohue, M.D.; Prausnitz, J.M. Combinatorial entropy of mixing molecules that differ in size and shape: simple approximation for binary and multicomponent mixture. Can. J. Chem. **1975**, *53*, 1586.
18. Schwartz, D.K.; Knobler, M. Direct observation of transitions between condensed Langmuir monolayer phases by polarized fluorescence microscopy. J. Phys. Chem. **1993**, *97*, 8849.
19. Keller, D.; Korb, J.P.; McConnell, H.M. Theory of shape transitions in two-dimensional phospholipid domains. J. Phys. Chem. **1987**, *91*, 6417.
20. McConnell, H.M.; Moy, V.T. Shapes of finite two dimensional lipid domains. J. Phys. Chem. **1988**, *92*, 4520.
21. McConnell, H.M. Theory of hexagonal and stripe phase in monolayers. Proc. Natl. Acad. Sci. USA **1989**, *86*, 3452.
22. Nagarajan, R.; Ruckenstein, E. Theory of surfactant self-assembly: a predictive molecular thermodynamics. Langmuir **1991**, *7*, 2934.
23. Möhwald, H. Phospholipid and phospholipid-protein monolayers at the air/water interface. Annu. Rev. Phys. Chem. **1990**, *41*, 441.
24. Semenov, A.N. Contribution to the theory of microphase layering in block-copolymer melts. Sov. Phys.-JETP (Engl. Transl.) **1985**, *61*, 733.
25. Cantor, R.S.; Dill, K.A. Theory for the equation of state of phospholipid monolayers. Langmuir **1986**, *2*, 331.
26. Albrecht, O.; Gruler, H.; Sackmann, E. Polymorphism of phospholipid monolayers. J. Phys. **1978**, *39*, 301.
27. Thoma, M.; Möhwald, H. Phospholipid monolayers at hydrocarbon/water interfaces. J. Colloid Interface Sci. **1994**, *162*, 340.
28. Dill, K.A.; Flory, P.J. Interphases of chain molecules: monolayers and lipid bilayer membranes. Proc. Natl. Acad. Sci. USA **1980**, *77* (6), 3115.
29. Dill, K.A.; Flory, P.J. Molecular organization in micelles and vesicles. Proc. Natl. Acad. Sci. USA **1981**, *78* (2), 676.
30. Dill, K.A.; Cantor, R.S. Statistical thermodynamics of short-chain molecule interphases: 1. Theory. Macromolecules **1984**, *17*, 380.
31. Cantor, R.S.; Dill, K.A. Statistical thermodynamics of short-chain molecule interphase: 2. Configurational properties of amphiphilic aggregates. Macromolecules **1984**, *17*, 384.
32. Cantor, R.S.; Dill, K.A. Statistical thermodynamic theory for the melting of n-alkanes from their rotator phases. Macromolecules **1985**, *18*, 1875.
33. Cantor, R.S.; Melroy, R.M. Statistical thermodynamics of flexible-chain surfactant monolayer films. I. Theory of fluid phases. J. Chem. Phys. **1989**, *90* (8), 4423.
34. Cantor, R.S.; Melroy, R.M. Statistical thermodynamics of flexible-chain surfactants in monolayer films. II. Calculations for a modified cubic lattice model. J. Chem. Phys. **1989**, *90* (8), 4431.
35. Cantor, R.S. Statistical thermodynamics of flexible-chain surfactants in monolayer films: mixtures of different lengths. J. Chem. Phys. **1989**, *91* (10), 6306.
36. Shin, S.; Wang, Z.G.; Rice, S.A. An interpretation of the multiple fluid-fluid transitions in liquid supported amphiphile monolayers. J. Chem. Phys. **1990**, *92* (2), 1427.

37. Knobler, C.M. Seeing phenomena in flatland: studies of monolayers by fluorescence microscopy. Science **1990**, *249*, 870.
38. Honig, D.; Mobius, D. Direct visualization of monolayers at the air-water interface by Brewster angle microscopy. J. Phys. Chem. **1991**, *95*, 4590.
39. Overbeck, G.A.; Honig, D.; Mobius, D. Visualization of first and second order phase transitions in eicosanol monolayers using Brewster angle microscopy. Langmuir **1993**, *9*, 555.
40. Ruckenstein, E.; Bhakta, A. Clustering and its effects on adsorption. Langmuir **1994**, *10*, 2694.
41. Ruckenstein, E.; Li, B. A surface equation of state based on clustering of surfactant molecules of insoluble monolayers. Langmuir **1995**, *11*, 3510.
42. Riviere, S.; Henon, S.; Meunier, J. Electrostatic pressure and line tension in a Langmuir monolayer. Phys. Rev. Lett. **1995**, *75* (13), 2506.
43. Fainerman, V.B.; Vollhardt, D.; Melzer, V. Equation of state for insoluble monolayers of aggregating amphiphilic molecules. J. Phys. Chem. **1996**, *100*, 15478.
44. Ruckenstein, E.; Li, B. A simple surface equation of state for the phase transition in phospholipid monolayers. Langmuir **1996**, *12*, 2308.
45. McConlogue, C.W.; Vanderlick, T.K. A close look at domain formation in DPPC monolayers. Langmuir **1997**, *13*, 7158.
46. Kagner, V.M.; Mohwald, H.; Dutta, P. Structure and phase transitions in Langmuir monolayers. Rev. Modern Phys. **1999**, *71* (3), 779.
47. Knobler, C.M.; Schwartz, D.K. Langmuir and self-assembled monolayers. Curr. Opin. Colloid Interface Sci. **1999**, *4*, 46.
48. Cordero, S.R.; Weston, K.D.; Buratto, S.K. Near-field microscopy of collapsed Langmuir-Blodgett films. Thin Solid Films **2000**, *360*, 139.

5464 *Langmuir* **2001,** *17,* 5464–5475

On the Stability of Lyotropic Lamellar Liquid Crystals and the Thicknesses of Their Lamellae

Eli Ruckenstein* and Marian Manciu

Department of Chemical Engineering, State University of New York at Buffalo, Buffalo, New York 14260

Received March 13, 2001. In Final Form: June 19, 2001

The stability of lamellar lyotropic liquid crystals was investigated using a thermodynamic formalism. The interfacial tension γ was expressed in terms of the thicknesses δ_1 and δ_2 of the two kinds of lamellae. For given thicknesses, the system might be stable if $\gamma \geq 0$. In adition, the derivatives $\partial f/\partial \delta_i$ of the excess free energy f, due to van der Waals, hydration, steric, electrostatic, and undulation interactions, must be negative for the lamellar structure to be stable. The thermodynamic conditions employed show that the domain of existence of the lamellar phase is restricted to narrow ranges of layer thicknesses and parameters of interaction. The effect of the bending modulus and of the ionic strength on the stability was examined. The thicknesses of the two kinds of lamellae were calculated using equilibrium and mass balance equations. It was shown that the nonlinearity of these equations could be responsible for the observed deviations from the "ideal dilution law". A comparison between the calculated and experimental results was made.

Introduction

Lyotropic liquid crystals are composed of a surfactant, a cosurfactant (typically a medium chain length alcohol), water, and oil. It is well-known that, in relatively narrow ranges of composition and temperature, a lamellar structure that exhibits a long-range translational order along one direction (L_α) can be formed.[1] It was also shown that the lamellar phase could preserve its orientational order upon its dilution with water.[1] This was initially supposed to be a consequence of the repulsion between the charged interfaces formed by ionic surfactants. However, it was subsequently found that the lamellar phase can be hyperswollen by adding large amounts of an oil–cosurfactant mixture, a system for which the electrostatic interactions do not play a role;[2] a hyperswelling in water was also observed for nonionic surfactants.[3]

It was first suggested by Helfrich[4] that the confinement of the thermal undulations of uncharged phospholipid bilayers by the neighboring bilayers, in a lipid/water lamellar liquid crystal, leads to a long-range repulsive force, which can be responsible for the extensive swelling of the lipid bilayers in water.[5] However, for most bilayers, their high bending modulus K_C provides a small contribution of the entropic confinement (which is inversely proportional to K_C) to the free energy; hence only a limited swelling can occur.[6] It was shown by Safinya et al.[7] that the addition of pentanol to an uncharged lipid (dimyristoylphosphotidylcholine)/water multilayer decreases the rigidity of the bilayers and consequently the equilibrium separation distance increases from $d \approx 40$ to $d \geq 200$ Å.

It is then tempting to attribute the high swelling (distances over 5000 Å)[2] of uncharged lyotropic liquid crystals to a low value of the interface bending modulus, which provides a large entropic confinement repulsion. However, the matter is more complex in this case than in that of lipid bilayers. The lipid bilayers are generally formed before their swelling in water, and as a first approximation one can consider that the total area and the thickness of the bilayers remain constant during the formation of the lamellar liquid crystal for any water dilution. In this case, even for strong repulsions between the layers, the system remains stable. Indeed, Palmer and Schmitt[8] observed that charged lipid bilayers swelled indefinitely in distilled water but collapsed steadily upon exposure to solutions of progressively higher ionic strength.

In contrast, the lyotropic liquid crystals are self-assembled after the mixture of the constituents is allowed to equilibrate for a long time (typically, days), and there are additional contributions to the free energy due to the formation of the lamellae. In this case, a strong repulsion between layers will increase the free energy of the lamellar phase and as a result a nonlamellar phase may be formed. It was observed experimentally that the addition of salt, which screens the electrostatic interactions, and hence reduces repulsion, to a dilute ionic surfactant/cosurfactant/water system, can stabilize the lamellar phase.[9] Dilute lamellar phases typically occur in relatively narrow ranges of temperature and composition, and the addition of a cosurfactant or the increase of temperature (which by decreasing the bending modulus enhances the long-range undulating repulsion between layers, thus increasing the free energy) can lead to a phase transition.

The domains of existence of the lamellar liquid crystals can be in principle identified by calculating the free energies of the various possible phases. Because of the difficulty in performing such a calculation, a first purpose of this paper is to employ the thermodynamic formalism developed by Ruckenstein[10] to extract some information about the domain of stability of the lamellar phase.

A second goal is to determine the widths of the water and oil lamellae and to explain the deviation from the so-called "ideal dilution law". In a crude model, one can

* To whom correspondence may be addressed: e-mail, feaeliru@acsu.buffalo.edu; phone, (716) 645 2911/2214; fax, (716) 645 3822

(1) Ekwall, P. In *Advances in Liquid Crystals*; Brown, G. H., Ed.; Academic: New York, 1975.
(2) Larche, F. C.; Appell, J.; Porte, G.; Bassereau, P.; Marignan, J. *Phys. Rev. Lett.* **1986**, *56*, 1700.
(3) Strey, R.; Schomacker, R.; Roux, D.; Nallet, F.; Olsson, V. *J. Chem. Soc., Faraday Trans.* **1990**, *86*, 2253.
(4) Helfrich, W. Z. *Naturforschung* **1978**, *33a*, 305.
(5) Harbich, W.; Helfrich, W. *Chem. Phys. Lipids* **1984**, *36*, 39.
(6) Petrache, H. I.; Gouliaev, N.; Tristram-Nagle, S.; Zhang, R.; Sutter, R. M.; Nagle, J. F. *Phys. Rev. E* **1998**, *57*, 7014.
(7) Safinya, C. R.; Sirota, E. B.; Roux, D.; Smith, G. S. *Phys. Rev. Lett.* **1989**, *62*, 1134.
(8) Palmer, K. J.; Schmitt, F. O. J. *Cell. Comput. Physiol.* **1941**, *17*, 385.
(9) Guerin, C.; Bellocq, A. M. *J. Phys. Chem.* **1988**, *92*, 2550.
(10) Ruckenstein, E. *J. Chem. Phys.* **1998**, *109*, 6995.

10.1021/la010385e CCC: $20.00
Published on Web 08/09/2001

assume that the entire surfactant and cosurfactant is adsorbed on the interface and hence the total area of the interface, S, for a system of volume V is proportional to the volume fraction of surfactant and cosurfactant, ϕ. Because for planar interfaces $V = Sd/2$, the repeat distance d (the sum of the thicknesses of the hydrophobic and hydrophilic layers) becomes inversely proportional to ϕ. Consequently, the product between d and ϕ should be independent of ϕ. This behavior is sometimes called the "ideal dilution law". However, it was well established that deviations from the ideal dilution law occur frequently,[1] the area per headgroup of surfactant being variable during swelling, and dependent on the surfactant/cosurfactant ratio.

These deviations were accounted by Strey et al.,[3] who carried out experiments with the binary water$-C_{12}E_5$ system, by noting that the amplitude of the thermal undulations increased with the repeat distance d. If one considers the total area of the interface a constant S which depends only on the number of surfactant molecules, the projection S_0 of the total area on a plane perpendicular to d will decrease with increasing d. Hence, the apparent area per surfactant molecule, which is defined as the ratio between S_0 and the total number of surfactant molecules, decreases with increasing d, while the ideal dilution law implies that the apparent area per surfactant molecule is a constant. The excess area, defined as $\Delta S = S - S_0$, was related to the bending modulus of the interface,[3] and the experimental results for the deviations from the ideal dilution law were used to determine K_C.[3,11] However, it should be noted that there are binary systems for which the deviations from the ideal dilution law occur in the opposite directions. For instance, in the binary systems of fatty acid alkali soaps/water, the apparent area per headgroup increases with water dilution, because of the incorporation of water in the interface.[1]

In the second part of the paper, it will be shown that the partition of the alcohol between the phases can explain the deviations from the ideal dilution law for a water/oil/surfactant/cosurfactant lyotropic lamellar liquid crystal.

I. Thermodynamic Stability of Lyotropic Liquid Crystals

I.1. Thermodynamic Equations. We briefly describe the thermodynamic formalism of Ruckenstein.[10] Let us consider a liquid crystal of volume V per unit area, in equilibrium, and denote by δ_1 the average thickness of the water layer and by δ_2 the average thickness of the oil layer. The Helmholtz free energy of the system can be written as the sum of a free energy F_0 of a hypothetical system in which the lamellae are treated as bulk, planar, phases and a free energy F_1, which accounts for the smallness of the lamellae, the interactions between them, and their thermal undulations. One can write

$$dF_0(\delta_1,\delta_2,V,N_i) = 2\gamma d\left(\frac{V}{\delta_1+\delta_2}\right) + \sum \mu_i dN_i - p_1 d\left(\frac{V\delta_1}{\delta_1+\delta_2}\right) - p_2 d\left(\frac{V\delta_2}{\delta_1+\delta_2}\right) \quad (1a)$$

where γ is the interfacial free energy between the two kinds of lamellae, μ_i is the chemical potential of species i (at pressure p_1 in the layers of thickness δ_1 and pressure p_2 in the layers of thickness δ_2), and N_i is the number of molecules of species i in the volume V, and

$$dF_1(\delta_1,\delta_2,V) = d\left(f(\delta_1,\delta_2)\frac{V}{\delta_1+\delta_2}\right) \quad (1b)$$

where f is the free energy per unit area of one pair of lamellae 1 and 2, due to their thinness, interaction between them, and their thermal undulation ($V/(\delta_1+\delta_2)$ is the number of pairs of lamellae). It should be emphasized that the pressures p_1 and p_2 are not equal to the external pressure because they are defined on the basis of the free energy F_0.

However, at thermodynamic equilibrium one can write

$$dF_0 + dF_1 \equiv dF(N_i,V) = \sum \mu_i^* \, dN_i - p \, dV \quad (2)$$

where p is the external pressure and μ_i^* is the chemical potential of species i at the pressure p. By equating the coefficients of dV, dN_i, $d\delta_1$, and $d\delta_2$ from eqs 1 and 2, after some algebra the following equations are obtained[10]

$$p_1 = p + \frac{\partial f}{\partial \delta_1} \quad (3a)$$

$$p_2 = p + \frac{\partial f}{\partial \delta_2} \quad (3b)$$

$$2\gamma = -f + \delta_1\frac{\partial f}{\partial \delta_1} + \delta_2\frac{\partial f}{\partial \delta_2} \quad (3c)$$

$$\mu_i^*(p) = \mu_i(p_j) \quad (3d)$$

The free energy per unit area for a pair of successive lamellae, which includes the effect of thermal undulations, is computed as follows. Let us denote by $U(x_1,x_2)$ the interaction potential between planar water and oil lamellae of thicknesses x_1 and x_2, which include the van der Waals, hydration, steric, and electrostatic interactions

$$U(x_1,x_2) = f_{vdw} + f_{hydration} + f_{steric} + f_e \quad (4)$$

The first quantitative method to account for the undulation repulsion between membranes, due their mutual confinement, was developed by Helfrich,[4] which showed that the entropic contribution to the free energy, due to the confinement of the thermal fluctuations by the neighboring lamellae, for membranes interacting via a hard-wall potential, is given by $\Delta F \propto (kT)^2/K_C\sigma^2$, where k is the Boltzmann constant, T is the absolute temperature, K_C is the bending modulus, and σ is the root-mean-square fluctuation of the position of the membrane. There have been a number of theoretical attempts to extend the Helfrich theory to bilayers interacting via arbitrary potentials.[12−15] Here we will employ a variational procedure developed recently by the authors.[15] The free energy is composed of the energy of interaction between lamellae, averaged over the distance distributions, and an entropic term, due to the confinement of one interface by its neighbors. Denoting with $P_1(x_1)$ and $P_2(x_2)$ the distribution of distances through the water and oil lamellae, respectively, and considering that the free energy due to the entropic confinement of the membranes is inversely

(11) Roux, D.; Nallet, F.; Freyssingeas, E.; Porte, G.; Bassereau, P.; Skouri, M.; Marignan, J. *Europhys. Lett.* **1992**, *17*, 575.

(12) Sornette, D.; Ostrowsky, N. *J. Chem. Phys.* **1986**, *84*, 4062.

(13) Evans, E. A.; Parsegian, V. A. *Proc. Natl. Acad. Sci. U.S.A.* **1986**, *83*, 7132.

(14) Podgornik, R.; Parsegian, V. A. *Langmuir* **1992**, *8*, 557.

(15) Manciu, M.; Ruckenstein, E. *Langmuir* **2001**, *17*, 2455.

5466 *Langmuir, Vol. 17, No. 18, 2001* *Ruckenstein and Manciu*

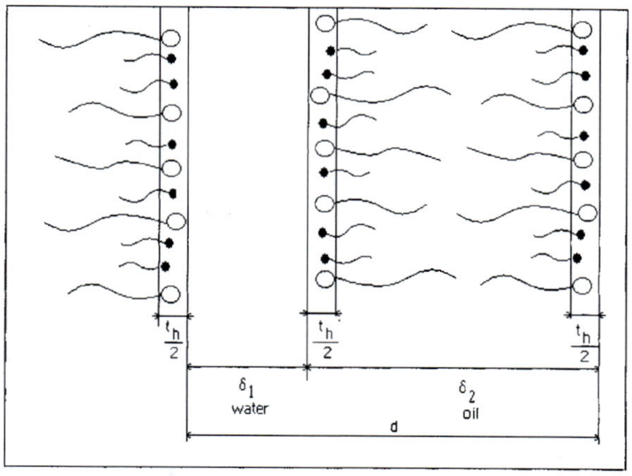

Figure 1. A sketch of a lyotropic lamellar liquid crystal, composed of water, oil, a surfactant (open circles), and cosurfactant (full circles). The distribution of the surfactant and cosurfactant between the water and oil phases and interface is discussed in section II.

proportional to the root-mean-square fluctuations of the distribution of the distances, $(\sigma_1)^2$ and $(\sigma_2)^2$, through water and oil, respectively, one can write, using Derjaguin approximation, that

$$f(\delta_1,\delta_2,\sigma_1,\sigma_2) = \int dx_1\, dx_2\, P_1(x_1)P_2(x_2)U(x_1,x_2) + \frac{(kT)^2}{64K_C{\sigma_1}^2} + \frac{(kT)^2}{64K_C{\sigma_2}^2} \quad (5)$$

where x_1 and x_2 are the fluctuating thicknesses of the water and oil layers, respectively.

In a first approximation one might consider that σ_1 and σ_2 are independent of each other. The energy is increased by distributions with large values of σ, because in this case a larger part of the interface is in a region of stronger repulsive interactions; on the other hand the entropic term decreases when the σ values are large. Consequently, there are optimum values of σ_1 and σ_2 for which the free energy is minimum. The following asymmetric truncated Gaussian function will be employed for the distance distributions ($i = 1, 2$)[15]

$$P_i(x_i) = 0 \qquad x_i < 0$$

$$P_i(x_i) = \frac{1}{C_a}\exp\left(-\frac{(x_i - \delta_i')^2}{2{\sigma_i}^2}\right) \qquad 0 < x_i < \delta_i'$$

$$P_i(x_i) = \frac{1}{C_a}\exp\left(-\frac{(x_i - \delta_i')^2}{2(\alpha\sigma_i)^2}\right) \qquad x_i > \delta_i' \quad (6)$$

where α is an asymmetry coefficient of the distribution; in the calculations, the value $\alpha = 1.4$, suggested by Monte Carlo simulations,[16] will be employed. The constant C_a and the position δ_i' of the maximum of the distribution can be calculated as a function of the average distance δ_i, through normalization and distance averaging, respectively. Then, the minimization of $f(\delta_1,\delta_2,\sigma_1,\sigma_2)$ of eq 5 with respect to σ_1 and σ_2 provides the free energy f for any values of δ_1 and δ_2.

I.2. The Interaction Potential $U(x_1,x_2)$. Let us consider the contributions to the interaction potential $U(x_1,x_2)$, for fixed x_1 and x_2 thicknesses of the planar water and oil layers, respectively (see Figure 1). In the hypothesis of pairwise additivity, the van der Waals excess interaction for a pair of lamellae in a periodic multilayer is given by (see Appendix A)

$$f_{vdW} = -\frac{A_{eff}}{12\pi}\left\{\frac{1}{(x_2 - t_h)^2} + \sum_{n=1}^{\infty}\left(\frac{1}{(n(x_1 + x_2) - x_2 + t_h)^2} + \frac{1}{(n(x_1 + x_2) + x_2 - t_h)^2} - \frac{2}{(n(x_1 + x_2))^2}\right)\right\} \equiv -\frac{A_{eff}}{12\pi}\left\{\frac{1}{(x_1 + t_h)^2} + \sum_{n=1}^{\infty}\left(\frac{1}{(n(x_1 + x_2) - x_1 - t_h)^2} + \frac{1}{(n(x_1 + x_2) + x_1 + t_h)^2} - \frac{2}{(n(x_1 + x_2))^2}\right)\right\} \quad (7)$$

where $A_{eff} = A_{11} + A_{22} - 2A_{12}$ is an effective Hamaker constant. In the calculation of the van der Waals attraction, the headgroups of thickness t_h of the surfactant will be considered as belonging to the water layer, since their dielectric properties are closer to water than to hydrocarbon. The van der Waals energy in eq 7 is symmetric in the distances $(x_1 + t_h)$ and $(x_2 - t_h)$ and can be interpreted as an interaction between lamellae (the terms under the sum) plus a contribution, the first term, due to the thinness of the lamellae; this term will be called effective self-energy (the true van der Waals self-energy is given by eq A.1 in Appendix A). When one of the thicknesses x_1 or $x_2 \to \infty$, the corresponding sum vanishes; however, the effective self-energy of the finite layer provides a negative contribution to the free energy. It will be shown later that this contribution is essential for the stability of the lamellar phase.

The second contribution to the energy comes from the hydration repulsion between the interfaces of the water layers, which is given by

$$f_{hydration} = A_H \exp\left(-\frac{x_1}{\lambda_H}\right) \quad (8)$$

The interaction parameters A_H and λ_H were determined for various lipid bilayers;[17] λ_H is typically between 1 and 3 Å, while A_H was found within a rather broad range, around 10^{-1} J/m^2.

The third contribution to the free energy comes from the steric repulsion in the oil layers between the hydrocarbon tails of the surfactant adsorbed on the interface. Extending the simple formula suggested by de Gennes[18] for the steric force between long-chain polymer-covered surfaces, to relatively short chains, one can write

$$f_{steric} = C\frac{2L(kT)}{s^3}\left[\frac{4}{5}\left(\left(\frac{2L}{x_2 - t_h}\right)^{5/4} - 1\right) + \frac{4}{7}\left(\left(\frac{x_2 - t_h}{2L}\right)^{7/4} - 1\right)\right] \quad (9)$$

(16) Gouliaev, N.; Nagle, J. F. *Phys. Rev. Lett.* **1998**, *81*, 2610.

(17) Rand, R. P.; Parsegian, V. A. *Biochim. Biophys. Acta* **1989**, *988*, 351.

(18) de Gennes, P. G. *Adv. Colloids Interface Sci.* **1987**, *27*, 189.

for $(x_2 - t_h) < 2L$ and

$$f_{steric} = 0$$

for $(x_2 - t_h) \geq 2L$, where $x_2 - t_h$ represents the actual thickness of the hydrocarbon lamella, L is the maximum length of the surfactant tail, and s is the separation distance between the headgroups of the surfactant along the interface. The proportionality coefficient C can be determined from a comparison with experimental data and was shown to be of the order of unity;[19] here it will be taken as unity.

The electrostatic contribution to the energy, in the presence of an electrolyte, depends on the behavior of the surface charge density, σ_e, when two surfaces approach each other. Assuming that the surface charge density remains constant, an excellent approximation for the electrostatic contribution to the energy is given by[20]

$$f_e(x_1) = \frac{2nkT}{\kappa}\left\{2y_0 \ln\left(\frac{B + y_0 \coth\left(\frac{\kappa x_1}{2}\right)}{1 + y_0}\right) - \ln(y_0^2 + \cosh(\kappa x_1) + B\sinh(\kappa x_1)) + \kappa x_1\right\} \quad (10)$$

where $y_0 = \tanh(e\psi_0/kT)$ and $B = (1 + y_0^2\,\mathrm{cosech}^2(\kappa x_1/2))^{1/2}$.

In the above expressions, n is the concentration of the 1:1 electrolyte, $\kappa = (2e^2n/\epsilon\epsilon_0 kT)^{1/2}$ is the reciprocal Debye length, e is the protonic charge, ϵ_0 is the permittivity constant, ϵ is the dielectric constant, and ψ_0 is the surface potential when the plates are infinitely far apart, which is related to the surface charge density via the expression $\sigma_e = (4ne/\kappa)\sinh(e\psi_0/2kT)$.

The electrostatic contribution to the energy when the only ions present in the solution are those of the counterions of the surfactant molecules, hence in the absence of an added electrolyte, is calculated by integrating the electrostatic pressure from infinity to the distance x_1. Denoting the surface charge $\sigma_e = a_d e/A$, where a_d is the degree of dissociation and A is the area per surfactant molecule adsorbed on the interface, the electrostatic energy per unit area is given by (see Appendix B)

$$f_e(x_1) = -\frac{2\epsilon\epsilon_0(kT)^2}{e^2}\int_{\infty}^{x_1} K^2(x)\,dx \quad (11a)$$

where

$$\frac{Kx}{2}\tan\left(\frac{Kx}{2}\right) = \frac{a_d xe^2}{4A\epsilon\epsilon_0 kT} \quad (11b)$$

The case $a_d = 0$ corresponds to a noncharged interface ($f_e = 0$).

I.3. Necessary Conditions for Stability. In a system with a fixed number of layers, such as the phospholipid bilayers, the equilibrium position (corresponding to the minimum of the free energy, F, of the whole system) is obtained when the free energy per unit area for the pair water/bilayer, f, is a minimum. This is no longer true when the number of pairs of layers is variable. In this case, at thermodynamic equilibrium one should use eq 3c. From this equation, if the interactions between lamellae are known, one can calculate the surface tension γ as a function of δ_1 and δ_2. While γ can be very small ($\sim 10^{-5}$ N/m), it cannot be negative, because then the interface will become unstable and a different phase (microemulsion, cubic, hexagonal, etc.) will be generated. Consequently, a necessary condition for the stability of the lamellar phase is

$$2\gamma = -f + \delta_1\frac{\partial f}{\partial \delta_1} + \delta_2\frac{\partial f}{\partial \delta_2} \geq 0 \quad (12a)$$

When the lamellar liquid crystal is in equilibrium with an excess phase i (either water or oil) and one assumes that the concentrations of the surfactant and cosurfactant are the same in the excess phase and in the lamellae of the same kind, the equality of the chemical potentials leads (because of eq 3d) to $p = p_i$, which when used in eqs3a or 3b provides the condition for equilibrium with an excess phase i, $\partial f(\delta_1,\delta_2)/\partial\delta_i = 0$. After the lamellar phase reaches the equilibrium with an excess phase, δ_i can no longer increase, since this will always increase the total free energy of the system. Consequently, the lamellar phase might exist only in the regions where

$$\partial f(\delta_1,\delta_2)/\partial\delta_i \leq 0 \qquad i = 1, 2 \quad (12b)$$

The domains in which $\partial f(\delta_1,\delta_2)/\partial\delta_i > 0$ correspond to unstable lamellar phases; in such cases, an excess phase will separate until thicknesses of the lamellae, which satisfy the conditions $\partial f(\delta_1,\delta_2)/\partial\delta_i = 0$, will be reached.

We will present now detailed calculations for a model lyotropic lamellar liquid crystal, using for the interaction parameters typical values from the literature: the thickness of the headgroups $t_h = 8$ Å;[6] the effective Hamaker constant $A_{eff} = 1.0 \times 10^{-20}$ J.[6,15] A recent comparison with experiment of the force per unit area and of the root-mean-square fluctuation of the separation distance between two neutral phospholipid bilayers as a function of the average separation distance provided the following values for the parameters involved in the hydration force: the decay length $\lambda_H \cong 1.5$ Å and the preexponential factor $A_H \cong 0.035$ J/m^2.[15] There are no experimental results for the lamellar lyotropic liquid crystals; however one can expect a comparable value for the decay length but a smaller one for the preexponential factor, because the former is due to the structure of water and the latter to the headgroup of the surfactant and the presence of cosurfactant. Here we will employ $\lambda_H = 1.5$ Å and $A_H = 0.02$ J/m^2. A value of $L = 8$ Å for the length of the surfactant tails and a value of $s = 7$ Å for the distance between the headgroups along the interface will be used. First, the surface charge will be set equal to zero, which corresponds to nonionic surfactants.

The elastic bending modulus K_C for lipid bilayers was found to be of the order of 10×10^{-20} J.[6] For the lyotropic lamellar liquid crystals, the additional presence of a cosurfactant at the interface should lead to smaller values. Various experimental measurements[21] provided values ranging between 0.08 and 5×10^{-20} J.

Figure 2 presents ($-f$) (the negative of the excess free energy per pair of layers), as a function of the average thicknesses of the layers, δ_1 and δ_2, for $K_C = 5 \times 10^{-20}$ J. From conditions (12a) and (12b), one can see that the lamellar phase cannot exist for positive values of f. To simplify the figure, we set $f = 0$ in Figure 2 when f was positive. In the case considered, the free energy has a minimum for $\delta_1 = 22$ Å and $\delta_2 = 39$ Å. Since both

(19) Ruckenstein, E.; Li, B. Q. *J. Chem. Phys.* **1997**, *107*, 932.
(20) Gregory, J. *J. Chem. Soc., Faraday Trans.* **1973**, *69*, 1723.
(21) Quist, P. O. *Langmuir* **1995**, *11*, 2201.

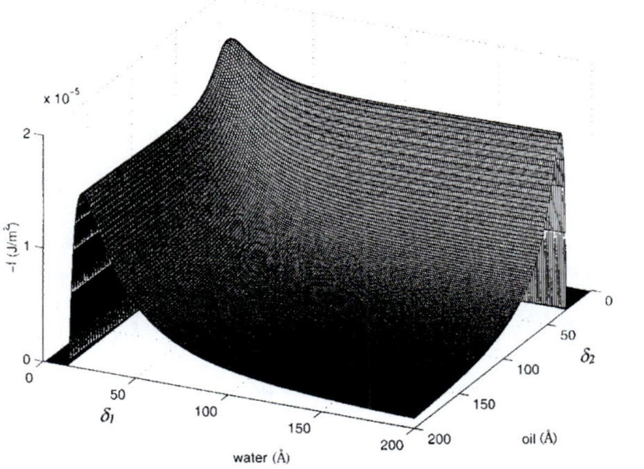

Figure 2. The negative of the excess free energy ($-f$) vs the thicknesses of water (δ_1) and oil (δ_2) layers. The interaction parameters are given in the text.

derivatives are vanishing for these values of δ_1 and δ_2, this point represents a triple point, where the lamellar phase is in thermodynamic equilibrium with both water and oil excess phases. Since eq 3c assigns a unique value to the interfacial tension ($2\gamma = -f$), the triple point can be reached only for certain concentrations of the constituents. There are however points in the vicinity of this minimum, where both derivatives of f are nonpositive and hence the lamellar phase might be stable for lower values of the interfacial tension than for the value at the triple point.

In Figure 3, the derivatives of the free energy with respect to δ_1 (Figure 3a) and δ_2 (Figure 3b) are plotted as functions of the distances δ_1 and δ_2 for the same parameters as in Figure 2. The lamellar phase might exist only in the regions where the derivatives are negative, which are the flat regions in parts a and b of Figure 3 (to simplify the figures, we set the values of the derivatives equal to zero in the regions where they are negative). One can see from the two-dimensional projections (Figure 3c) that condition (12b) is very restrictive, since most of the δ_1, δ_2 plane is inaccessible to the lamellar phase.

As already noted, when $\partial f/\partial \delta_i$ vanishes, the system is in equilibrium with an excess phase of kind i, and consequently the thickness of layer i cannot increase any longer. Figure 3c reveals that the above vanishing does not occur if at least one of the thicknesses is sufficiently small. This can be easily understood in terms of the interaction between lamellae of the same kind. While the repulsive interaction depends only on the separation between lamellae, the van der Waals attraction increases with their thickness (see eq 9). Therefore, thick lamellae of one kind are bound, because of increased van der Waals attraction, and the system cannot be swollen any longer in the other kind of liquid, while a system with thin lamellae of one kind might be unbound and hence be swollen in the other kind of liquid.

In a lipid/water system, the thickness of the bilayers is constant, and the unbinding transition can occur, in principle, by varying the Hamaker constant or temperature.[22] For lyotropic lamellar liquid crystals, hyperswelling in a liquid of one kind might also occur if the lamellae of the other kind are thin enough and hence unbound. However, thin lamellae might lead to a positive contribution to the free energy (since the repulsive forces overcome at short distances the van der Waals attraction), and hence the lamellar phase can become unstable.

(22) Lipowsky, R.; Leibler, S. *Phys. Rev. Lett.* **1986**, *56*, 2541.

To complete the analysis, we present in Figure 4 the domains in which both conditions (12a) and (12b) are fulfilled for various values of K_C, namely, (10, 5, 3, and 2) $\times\ 10^{-20}$ J. The figure indicates that outside the marked region the lamellar phase is certainly unstable; however, inside the marked region, the lamellar phase might compete for stability with another unknown phase, which might have a lower free energy.

Figure 4 indicates that the domains that might be accessible to the lamellar phase in the δ_1,δ_2 plane are rather narrow. For high values of K_C, hence low undulation repulsion, it is possible to generate lamellar liquid crystals when both kinds of lamellae are thin. This occurs because, if the undulation repulsion is negligible, there is always an absolute minimum of f at small thicknesses, since both the hydration repulsion through water and steric repulsion through oil have shorter ranges than the van der Waals attraction. Conditions (12a) and (12b) limit the existence of a lamellar phase only to a narrow region in the vicinity of this minimum, and further addition of water or oil will lead to the separation of an excess phase. However, there are other regions which are in principle accessible to the lamellar phase; they correspond to systems with thin lamellae of one kind and thick lamellae of the other kind, the effective self-van der Waals energy of the thin lamellae ensuring the thermodynamic stability of the system.

When K_C decreases, the undulation repulsion increases, and this has two consequences. On one hand, to maintain stability, the thickness of the thin lamellae should increase in order to diminish the effect of repulsion. This in turn increases their attraction and, hence, stimulates their binding. On the other hand, the repulsion between the thin lamellae increases, and this favors swelling. The results presented in Figure 4 show that, in a relatively broad range for K_C, it is possible to obtain a lamellar phase, if the lamellae of one kind are thin enough to be unbound. However, the contribution to the free energy due to their thinness (the self-energy) should remain sufficiently negative, to counteract the positive free energy due to the repulsion between them. When K_C further decreases, the undulation repulsion might become so large that no thin layers of one kind can provide enough effective self-energy for the formation of a stable lamellar liquid crystal. Indeed, for $K_C = 1 \times 10^{-20}$ J, the undulation repulsion between layers becomes so strong that the lamellar phase was found unstable in all the domains of the δ_1,δ_2 plane investigated ($\delta_1 \leq 200$ Å, $\delta_2 \leq 200$ Å).

Let us now consider the same system in the presence of an 1:1 electrolyte, with $K_C = 10 \times 10^{-20}$ J and the surface charge density, at large separation distance δ_1, $\sigma_e = 1e/200A^2 = 0.08C/m^2$. A higher bending modulus was selected to take into account the effect of the charge on the rigidity of the interfaces. In Figure 5 we present the region of stability of the lamellar phase, for various electrolyte concentrations. For $n = 0.5$ M, it is possible to obtain a lamellar phase with thin oil lamellae ($\delta_2 \approx 30$ Å) and thick water layers ($\delta_1 > 40$ Å), and also with thin water layers ($\delta_1 \approx 40$ Å) and thick oil layers ($\delta_2 > 170$ Å). In the latter case, the oil layers were thick enough for their attraction to overcome the electrostatic repulsion. When the electrolyte concentration was decreased, the electrostatic repulsion increased, and no lamellar phase containing thin water lamellae could be formed. However, a lamellar phase with thick water lamellae could be generated, when the thickness of the oil layers was sufficiently small to provide, via their self-energy, the required negative contribution to the free energy. Our calculations indicated that the thickness of the (thin) oil lamellae was almost

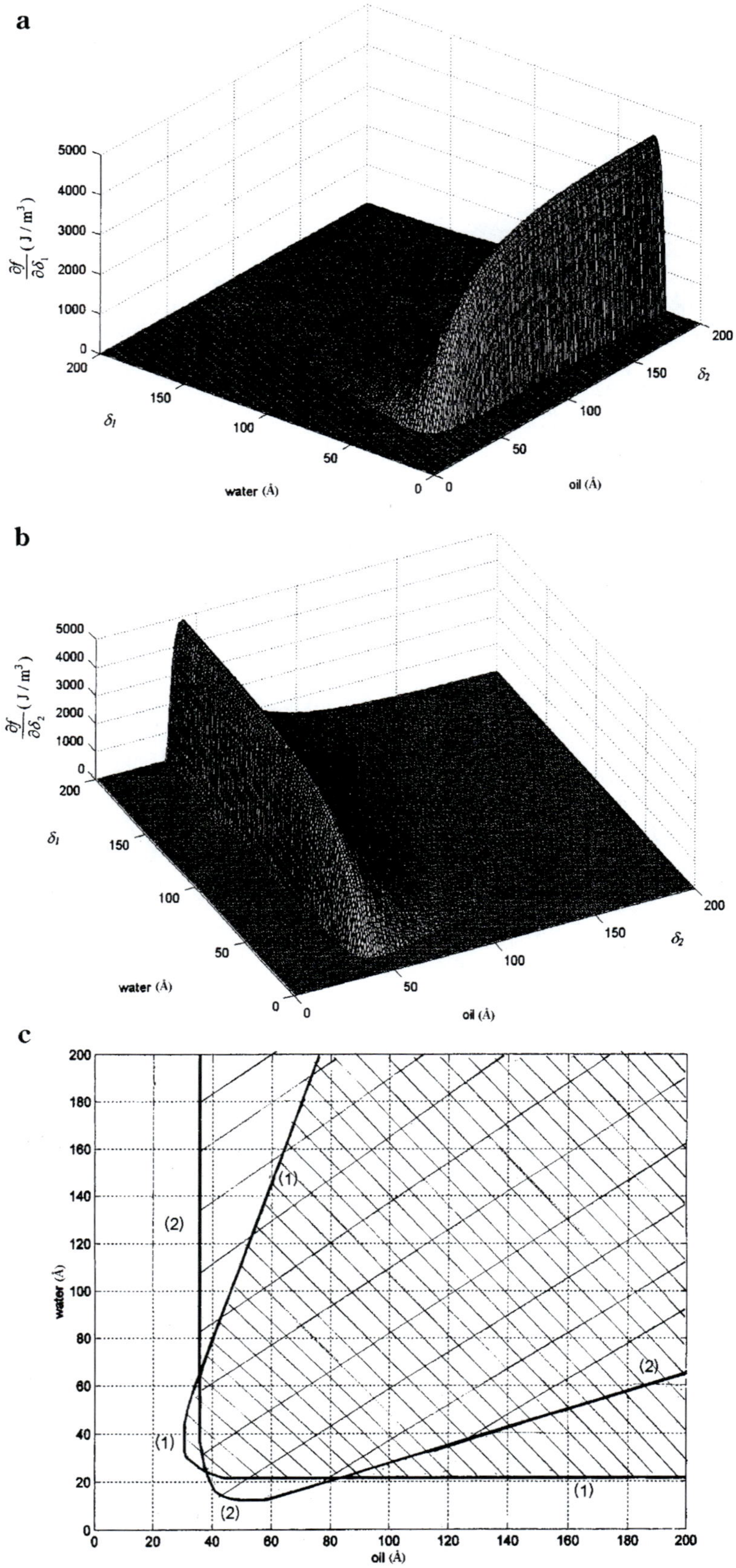

Figure 3. (a) The derivative of *f* with respect to δ_1. The lamellar phase cannot exist in the regions where the derivative is positive. (b) The derivative of *f* with respect to δ_2. The lamellar phase cannot exist in the regions where the derivative is positive. (c) The contour plot of the derivatives of *f* with respect to δ_1 and δ_2, respectively. At line 1, $\partial f/\partial \delta_1 = 0$ and at line 2, $\partial f/\partial \delta_2 = 0$ (inside each of the marked domains delimited by lines 1 and 2, an excess phase separates). The intersections of the line 1 and 2 are triple points, where the lamellar phase is in equilibrium with both excess phases.

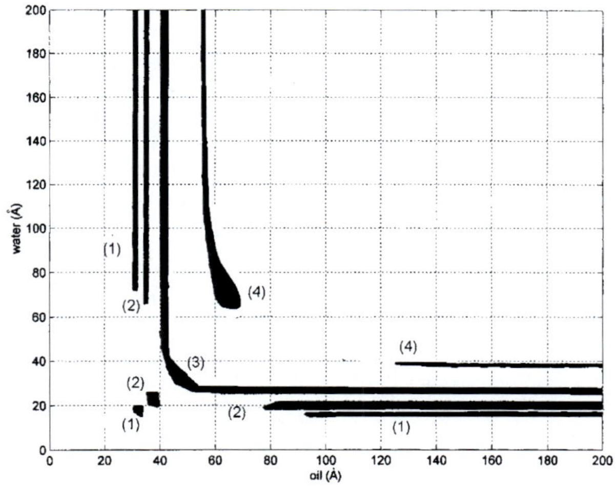

Figure 4. Domains in which the lamellar phase fulfills the conditions (12a) and (12b) and might be stable, for (1) $K_C = 10 \times 10^{-20}$ J, (2) $K_C = 5.0 \times 10^{-20}$ J, (3) $K_C = 3.0 \times 10^{-20}$ J, and (4) $K_C = 2.0 \times 10^{-20}$ J. For $K_C < 1.0 \times 10^{-20}$ J, no domain of stability was found in the region investigated in the δ_1,δ_2 plane ($\delta_1 \leq 200$ Å, $\delta_2 \leq 200$ Å).

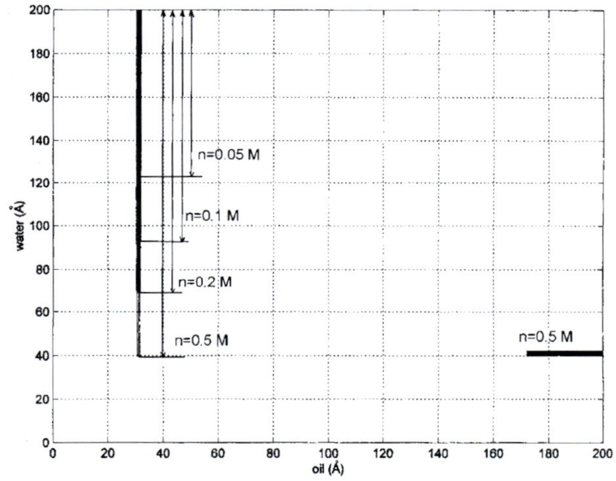

Figure 5. Possible domains of stability of the lamellar phase for charged lyotropic liquid crystals, in the presence of electrolyte: (1) $n = 0.5$ M; (2) $n = 0.2$ M; (3) $n = 0.1$ M; (4) $n = 0.05$ M. No domain of stability for the lamellar phase was found in the region investigated for $n < 0.01$ M.

independent of the electrolyte concentration n ($\delta_2 \approx 30$ Å), while the minimum thickness of the water lamellae was strongly dependent on n ($\delta_1 > 40$ Å for $n = 0.5$ M, $\delta_1 > 70$ Å for $n = 0.2$ M, $\delta_1 > 90$ Å for $n = 0.1$ M and $\delta_1 > 120$ Å for $n = 0.05$ M). A further decrease of the electrolyte concentration increased the free energy and provided negative values for the interfacial tension calculated with eq 3c and, hence, made the system unstable. For $n = 0.01$ M, no region of stability was found in the δ_1,δ_2 plane investigated (200 Å × 200 Å).

Most challenging are the systems that have ionic surfactants and are free of electrolyte. In this case it is possible to obtain a swollen phase, by the addition of a mixture of cosurfactant and oil.[2,23,24] The thickness of the water lamellae, evaluated from the composition of the liquid crystal and the repeat distance, is about 17 Å.[2] If one assumes that the surface charge is $\sigma_e = 1e/200A^2$, then the osmotic pressure due to counterions exceeds by 1 order of magnitude the van der Waals attraction, and the system cannot be stable. However, at the distances involved, the dielectric constant of water should be much smaller than that of the bulk water, since the water molecules are polarized by the headgroups of the surfactant adsorbed on the interface. In addition, while the dissociation increases the entropy and thus decreases the free energy, the repulsion between the surfaces generates a positive free energy, which can dominate. For this reason, the degree of dissociation which optimizes the free energy is expected to depend on the distance between the interfaces. When the latter distance is small, the dissociation can be low, because the dissociation will generate too strong a repulsion. A theory for the interaction between layers at low separations will be presented in a separate publication. Here a comparison between the results of the present calculations and experiment will be used to evaluate the degree of dissociation.

Stable lamellar liquid crystals can be obtained for $K_C = 10 \times 10^{-20}$ J and a surface charge $\sigma_e = 1e/50000A^2$. In this case, $a_d \approx 10^{-3}$ and the domain in which the lamellar phase might be stable corresponds to a thickness between 17 and 19 Å of the water lamellae; the system, however, can swell indefinitely in oil, the self-energy of the water lamellae providing the necessary negative contribution to the free energy.

In summary, information about the interactions between layers allowed one to identify the domains where the lamellar phase is unstable. A positive derivative of f with respect to δ_1 or δ_2 implies that the region is inaccessible to a lamellar phase. In this case, a water or oil phase will separate until the allowed values for δ_1 and δ_2 will be reached. In addition, a negative surface tension indicates that another phase (cubic,hexagonal, microemulsion, etc.) is stable. It should be, however, emphasized that when $\gamma > 0$, a phase other than the lamellar one may be the thermodynamically stable one.

It is generally believed that an extreme swelling is a consequence of a strong repulsion between layers and hence that a low bending modulus favors swelling. This is not, however, always accurate, because a high repulsion can lead to a large positive contribution to the free energy that makes the lamellar phase unstable.

II. The Distribution of the Components in the Lamellar Liquid Crystal

II.1. Uncharged Lyotropic Lamellar Liquid Crystals. A common procedure to evaluate the thicknesses of the water and oil layers was to assume that all the molecules of surfactant and cosurfactant are adsorbed on the interface, where they occupy constant areas. While very simple, this procedure is not always accurate, because the cosurfactant is also present in the water and particularly in the oil phase.[23] Larche et al. explained the deviation from the ideal dilution law by taking into account the distribution of the cosurfactant (pentanol) between the interface and the oil (decane) phase.[23]

A more complete procedure to compute the average thicknesses of the water and oil layers will be presented below. It is based on eq 3c, mass balances of components, and phase equilibrium equations. The calculations indicated that the interfacial tension of lamellar liquid crystals is very low, of the order of 10^{-5} N/m. It will be shown that $\gamma = 0$ is always an excellent approximation of eq 3c.

(23) Larche, F. C.; El Qebbaj, S.; Marignan, J. *J. Phys. Chem.* **1986**, *90*, 707.
(24) Safinya, C. R.; Roux, D.; Smith, G. S.; Sinha, S. K.; Dimon, P.; Clark, N. A.; Bellocq, A. M. *Phys. Rev. Lett.* **1986**, *57*, 2718.

For an uncharged lyotropic lamellar liquid crystal, the equilibrium concentrations are given by the expressions[25]

$$X_i^S = X_i^W \frac{f_i^W}{f_i^S} e^{\gamma \tau_i a/kT} e^{-(\mu_i^{0,S}-\mu_i^{0,W})/kT} \tag{13}$$

where a is the area occupied at the interface by a molecule of water, $\tau_i a$ is the area occupied at the interface by a molecule of species i, X_i^S is the surface fraction occupied by species i ($X_i^S = \tau_i N_i^S/N$, where N_i^S is the number of molecules of species i at the interface and N the total number of molecules of water which completely cover the interface, in the absence of surfactant and cosurfactant), X_i^W is the mole fraction of species i in the aqueous phase, μ_i^W and μ_i^S are the standard chemical potentials in the water phase and at the interface, respectively, and f_i^W and f_i^S are the activity coefficients in the water phase and at the interface, respectively. At room temperature (T = 300 K), and for $\gamma \approx 10^{-5}$ N/m and a = 10 Å^2, one obtains $\gamma a/kT \approx 2 \times 10^{-4}$ and the exponential involving γ can be approximated by unity. We will assume that the activity coefficients are unity in water, and the activity coefficients at the interface are computed using the expression

$$\ln f_i^S = \left(\frac{\partial \left(\frac{\Delta G^E}{kT} \right)}{\partial N_i^S} \right)_{T,P,N_{j\neq i}^S} \tag{14}$$

where for the excess free energy ΔG^E the following expression for ternary mixtures was employed[26]

$$\frac{\Delta G^E}{kT} = H_{12}^S N_1^S X_2^S + H_{13}^S N_1^S X_3^S + H_{23}^S N_2^S X_3^S \tag{15}$$

where H_{ij}^S is the interaction parameter for the pair $i-j$ at the interface.

We assume that no surfactant is present in the oil phase, and the main problem to be solved is to determine the relation between the concentrations of alcohol in the water and oil phases. A theory for the partition of the alcohol between oil and water, at chemical equilibrium, which takes into account the self-association of alcohol in the oil phase, is detailed in Appendix C. The equation that relates the mole fraction of the alcohol in the oil phase to the mole fraction of the alcohol in the water phase has the form

$$X_A^O = \frac{N_A^O}{N_O^O + N_A^O} = \frac{\frac{X_A^W}{(X_A^W)_{SAT}}}{1 + K_X\left(1 - \frac{X_A^W}{(X_A^W)_{SAT}}\right)^2} \tag{16}$$

where the subscripts stand for the type of molecule (A for alcohol and O for oil), the superscripts stand for the phase (O for oil and W for water), K_X is the self-association equilibrium constant, and $(X_A^W)_{SAT}$ is the saturation concentration of alcohol in water (see Appendix C).

Equations 13–16 with $\gamma = 0$ and mass balances can be used to determine the distribution of molecules between the interface and the water and oil phases. In what follows, we will examine the swelling of the quaternary mixture $C_{12}E_6$/pentanol/water/decane. The changes in the standard chemical potentials of the components between the water phase and interface, $\Delta\mu_i = \mu_i^{0,W} - \mu_i^{0,S}$, and the values of the pair interaction parameters H_{ij} can be determined in principle from the fit of eq 13 to experimental data regarding the oil/water interfacial tension at low concentrations of surfactant and cosurfactant. The earlier calculations of Ruckenstein and Rao[25] provided the values $\Delta\mu_S$ = 6500 cal/mol for $C_{12}E_6$ and $\Delta\mu_W$ = −718 cal/mol for water. For pentanol we selected the value $\Delta\mu_A$ = 3000 cal/mol, which is compatible with the experimental data for the lowering of the oil/water interfacial tension by alcohol.[27] The surface interaction parameters H_{ij}^S for different pair of molecules were taken as 0.5,[25] the value K_X = 94.2 was selected for the association constant of pentanol in oil,[28] and $(X_A^W)_{SAT}$ was taken as 4.1×10^{-3}.[28]

We denote by N_S, N_A, N_W, and N_O the total number of molecules of surfactant, cosurfactant, water, and oil, and by V_S, V_A, V_W, and V_O their volumes per molecule. The volumes and the area occupied on the interface were estimated from the density of alkanes and the earlier calculations of Nagarajan and Ruckenstein.[29] The values employed were V_W = 30 Å^3, V_O = 320 Å^3, V_S = 650 Å^3, V_A = 180 Å^3, a = 10 Å^2, τ_S = 3.4 and τ_A = 1.0. By denoting N_S^S, N_A^S, and N_W^S the number of molecules of surfactant, cosurfactant, and water, respectively, adsorbed on the interface, and with N_A^O the number of cosurfactant molecules in the oil phase, the system of equations to be solved is

$$X_S^S = \frac{(N_S - N_S^S)}{(N_W + N_S + N_A - N_W^S - N_S^S - N_A^S - N_A^O)} \times \frac{f_S^W}{f_S^S} e^{-(\mu_S^{0,S}-\mu_S^{0,W})/kT} \tag{17a}$$

$$X_A^S = \frac{(N_A - N_A^S - N_A^O)}{(N_W + N_S + N_A - N_W^S - N_S^S - N_A^S - N_A^O)} \times \frac{f_A^W}{f_A^S} e^{-(\mu_A^{0,S}-\mu_A^{0,W})/kT} \tag{17b}$$

$$X_W^S = \frac{(N_W - N_W^S)}{(N_W + N_S + N_A - N_W^S - N_S^S - N_A^S - N_A^O)} \times \frac{f_W^W}{f_W^S} e^{-(\mu_W^{0,S}-\mu_W^{0,W})/kT} \tag{17c}$$

$$\frac{N_A^O}{N_O + N_A^O} = \frac{\frac{1}{(X_A^W)_{SAT}} \frac{(N_A - N_A^S - N_A^O)}{(N_W + N_S + N_A - N_W^S - N_S^S - N_A^S - N_A^O)}}{1 + K_X\left(1 - \frac{1}{(X_A^W)_{SAT}} \frac{(N_A - N_A^S - N_A^O)}{(N_W + N_S + N_A - N_W^S - N_S^S - N_A^S - N_A^O)}\right)^2} \tag{17d}$$

(25) Ruckenstein, E.; Rao, I. V. *J. Colloid Interface Sci.* **1987**, *117*, 104.
(26) Reid, R. C.; Prausnitz, J. M.; Poling, B. E. *The properties of gases and liquids*; McGraw-Hill: New York, 1987.
(27) Aveyard, R.; Briscoe, B. J. *J. Chem. Soc., Faraday Trans.* **1972**, *68*, 478.
(28) Nagarajan, R.; Ruckenstein, E. *Langmuir* **2000**, *16*, 6400.
(29) Nagarajan, R.; Ruckenstein, E. *J. Colloid Interface Sci.* **1979**, *71*, 580.

5472 *Langmuir, Vol. 17, No. 18, 2001* *Ruckenstein and Manciu*

where the surface fractions are given by

$$X_S^{\ S} = \frac{\tau_S N_S^{\ S}}{\tau_S N_S^{\ S} + \tau_A N_A^{\ S} + N_W^{\ S}} \tag{17e}$$

$$X_A^{\ S} = \frac{\tau_A N_A^{\ S}}{\tau_S N_S^{\ S} + \tau_A N_A^{\ S} + N_W^{\ S}} \tag{17f}$$

$$X_W^{\ S} = \frac{N_W^{\ S}}{\tau_S N_S^{\ S} + \tau_A N_A^{\ S} + N_W^{\ S}} \tag{17g}$$

The activity coefficients on the interface are calculated using the expressions

$$f_S^{\ S} = \exp\left(H_{SA}^{\ \ S}(X_A^{\ S} - X_A^{\ S} X_S^{\ S}) + H_{SW}^{\ \ S}(X_W^{\ S} - X_W^{\ S} X_S^{\ S}) - H_{AW}^{\ \ S}(X_A^{\ S} X_S^{\ S})\frac{\tau_S}{\tau_W}\right) \tag{17h}$$

$$f_A^{\ S} = \exp\left(H_{SA}^{\ \ S}(X_S^{\ S} - X_S^{\ S} X_A^{\ S})\frac{\tau_A}{\tau_S} + H_{AW}^{\ \ S}(X_W^{\ S} - X_W^{\ S} X_A^{\ S}) - H_{SW}^{\ \ S}(X_W^{\ S} X_S^{\ S})\frac{\tau_A}{\tau_S}\right) \tag{17i}$$

$$f_W^{\ S} = \exp\left(H_{SW}^{\ \ S}(X_S^{\ S} - X_S^{\ S} X_W^{\ S})\frac{1}{\tau_S} + H_{AW}^{\ \ S}(X_A^{\ S} - X_A^{\ S} X_W^{\ S})\frac{1}{\tau_A} - H_{SA}^{\ \ S}(X_S^{\ S} X_A^{\ S})\frac{1}{\tau_S}\right) \tag{17j}$$

The following mass balance relations complete the system of equations

$$N_S = N_S^{\ S} + N_S^{\ W} \tag{17k}$$

$$N_A = N_A^{\ S} + N_A^{\ W} + N_A^{\ O} \tag{17l}$$

$$N_W = N_W^{\ S} + N_W^{\ W} \tag{17m}$$

where it was assumed that no oil is present at the interface or in the water phase and also that no water or surfactant is present in the oil phase.

The repeat distance d and the average thicknesses of the layers, δ_1(water) and δ_2(oil) can be obtained from the relation between the total interface area and total volume

$$d = \frac{2(N_S V_S + N_A V_A + N_W V_W + N_O V_O)}{a(\tau_S N_S^{\ S} + \tau_A N_A^{\ S} + N_W^{\ S})} \tag{18a}$$

$$\delta_1 = d\,\frac{N_S^{\ W} V_S + N_A^{\ W} V_A + N_W^{\ W} V_W}{N_S V_S + N_A V_A + N_W V_W + N_O V_O} \tag{18b}$$

$$\delta_2 = d\,\frac{N_O^{\ O} V_O + N_A^{\ O} V_A + N_S^{\ S} V_S + N_A^{\ S} V_A + N_W^{\ S} V_W}{N_S V_S + N_A V_A + N_W V_W + N_O V_O} \tag{18c}$$

where it was considered that the interface (including the polar headgroups) belongs to the oil layer.

In Figure 6 the repeat distance (Figure 6a) and the thickness of the oil lamellae (Figure 6b) are plotted as functions of ϕ (the ratio between the volume of surfactant + cosurfactant and the total volume), for fixed volume ratios, r, between surfactant and cosurfactant. The initial volumes of water and oil were taken each as 10% of the total volume of the system; further, the system was diluted by addition of water and the new equilibrium values were calculated, using eqs 17a–m and 18a–c.

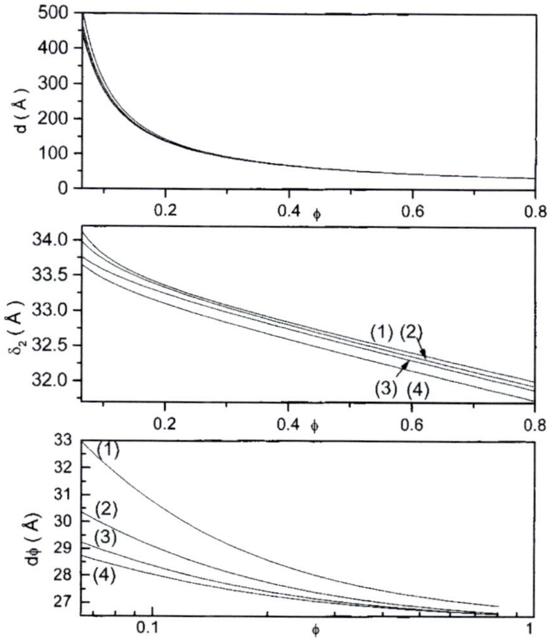

Figure 6. The repeat distance d (a), the thickness of the oil layer δ_2 (b), and the product $d\phi$ (c) vs ϕ (volume ratio of surfactant and cosurfactant) for various volume ratios r between surfactant and cosurfactant, for the quaternary system $C_{12}E_6$/pentanol/water/decane described in text: (1) $r = 1.0$; (2) $r = 2.0$; (3) $r = 3.0$; (4) $r = 4.0$.

One can see that the thickness of the oil layers does not remain constant upon dilution. This is a consequence of the nonlinearity of the system of eqs 17, which takes into account the partition of the alcohol between oil, water, and interface and the dependence of the surface activity coefficients on the surface concentrations of all the components. Figure 6c, in which the product between the repeat distance d and ϕ as a function of ϕ is plotted, clearly shows that there are deviations from the ideal dilution law; during the water dilution, alcohol and surfactant molecules leave the interface, and this shrinks its area.

II.2. Charged Lyotropic Lamellar Liquid Crystals. In the case of a charged system, two changes in the system of eqs 17 must be made.[30,31] First γ should be replaced in eq 17a, for the surfactant, by $\gamma + \int_{\psi=0}^{\psi_0} \sigma\, d\psi$. Second, the mole fraction of the surfactant in water, $X_S^{\ W}$, should be replaced in the same equation by the mole fraction of the surfactant in water, in the vicinity of the interface $(X_S^{\ W})^*$. Assuming that the surfactant in the aqueous phase is totally dissociated, the concentration of surfactant near the interface $(X_S^{\ W})^*$ can be related to the average surfactant concentration in the water phase, $X_S^{\ W}$, by using the equilibrium and mass balance relations

$$kT \ln(X_S^{\ W})^* + e\psi_0 = kT \ln(X_S(x)) + e\psi(x) \tag{19a}$$

with

(30) Ruckenstein, E.; Krishnan, R. *J. Colloid Interface Sci.* **1980**, *76*, 201.
(31) Hunter, R. J. *Foundations of Colloid Science*; Clarendon Press: Oxford, 1987.

$$\frac{1}{\delta_1}\int_{-\delta_1/2}^{\delta_1/2} X_S(x)\,dx = X_S^{W} \tag{19b}$$

where $X_S(x)$ is the local concentration of the surfactant in the water phase.

One thus obtains

$$(X_S^{W})^* = (X_S^{W})\frac{\delta_1 \exp\left(-\dfrac{e\psi_0}{kT}\right)}{\int_{-\delta_1/2}^{\delta_1/2} dx \exp\left(-\dfrac{e\psi(x)}{kT}\right)} \tag{20}$$

In the absence of electrolyte, both integrals $\int_{\psi=0}^{\psi_0} \sigma\, d\psi$ and $\int_{-\delta_1/2}^{\delta_1/2} dx \exp(-e\psi(x)/kT)$ can be carried out analytically (see Appendix B), and eq 17a is replaced by

$$X_S^{S} = A' \exp\left(\frac{aB'}{kT}\right) \frac{(N_S - N_S^{S})}{(N_W + N_S + N_A - N_W^{S} - N_S^{S} - N_A^{S} - N_A^{O})} \frac{f_S^{W}}{f_S^{S}} e^{-(\mu_S^{O,S} - \mu_S^{O,W})/kT} \tag{17a'}$$

where

$$A' = \frac{K\delta_1 \cos^2\left(\dfrac{K\delta_1}{2}\right)}{\left(\cos\left(\dfrac{K\delta_1}{2}\right)\sin\left(\dfrac{K\delta_1}{2}\right) + \dfrac{K\delta_1}{2}\right)} \tag{21a}$$

and

$$B' = \frac{8\epsilon\epsilon_0 (kT)^2}{e^2\delta_1}\left(\frac{K\delta_1}{2}\tan\left(\frac{K\delta_1}{2}\right) - \frac{1}{2}\ln\left(1 + \tan^2\left(\frac{K\delta_1}{2}\right)\right) - \frac{K^2\delta_1^{2}}{8}\right) \tag{21b}$$

with K the solution of

$$\frac{K\delta_1}{2}\tan\left(\frac{K\delta_1}{2}\right) = \frac{\sigma_e \delta_1 e}{4\epsilon\epsilon_0 kT} \tag{21c}$$

The calculated results in the absence of electrolyte will be now compared with the experimental results obtained regarding a lamellar lyotropic liquid crystal SDS (sodium dodecyl sulfate)/pentanol/water/dodecane swollen in a mixture of dodecane and pentanol.[24] The weight fraction water/surfactant was 1.552; from the dilution line in the phase diagram, we calculated that the initial concentration of pentanol in the oil-free system was 29 wt % and the concentration of pentanol in the dodecane-based diluant was 8 wt %. The experimental values for the repeat distance were obtained from the X-ray diffraction spectrum (Figure 2 in ref 24) for various dodecane concentrations.

In the calculations, we used $\Delta\mu_S = 6960$ cal/mol for SDS[25] and, as in section II.1, $\Delta\mu_W = -718$ cal/mol for water and $\Delta\mu_A = 3000$ cal/mol for pentanol. The area occupied on the surface was considered $\tau_S a = 34$ Å^2 for SDS and $\tau_A a = 10$ Å^2 for pentanol.[25] The values $V_S = 480$ Å^3 and $V_O = 380$ Å^3 were employed.[29] A very low surface charge, $\sigma_e = 1e/50000A^2$, as suggested by the stability analysis of section I, was employed. In Figure 7 the calculated repeat distance (curve 1) is compared with the experimental result of Safinya et al.[26] Curve 2 represents the repeat distance

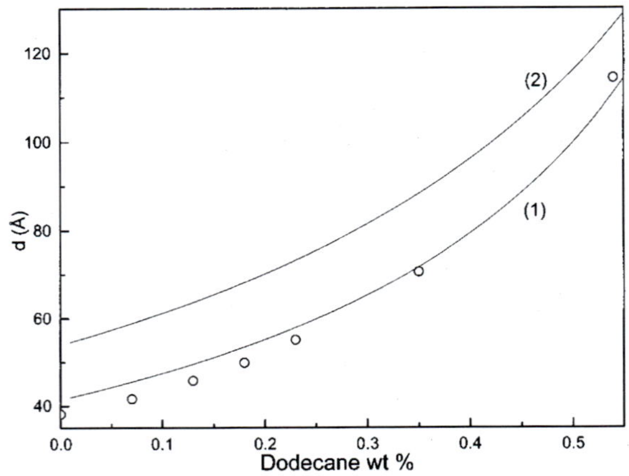

Figure 7. The repeat distance, as a function of dodecane weight fraction, for a quaternary system SDS/pentanol/water/dodecane: curve 1, computed using eqs 17 and 18; curve 2, computed by assuming that the entire surfactant and cosurfactant are adsorbed on the interface. The circles represent the experimental result of Safinya et al.[24]

calculated using a very simple model, in which the entire surfactant and cosurfactant was assumed to be adsorbed on the interface, and the total area of the interface was calculated as $N_S\tau_S a + N_A\tau_A a$. It is of interest to note that this simple evaluation provides a smaller value for the total area (larger repeat distance) than the more involved calculation in which the partition of surfactant and alcohol between phases was taken into account. The apparent paradox (less adsorbed surfactant/alcohol molecules and larger area in the latter case) is explained by the fact that eqs 17 predict that about 25% of the sites of the interface are occupied by water molecules.

III. Conclusions

In the first part of the paper, a thermodynamic formalism developed earlier[10] was used to obtain information about the domains of stability of the lamellar phase. It was shown that, for a set of interaction parameters between layers and bending modulus of the interface, only certain thicknesses are allowed for the water and oil layers.

In the second part, using the equilibrium relations for the various components present in the water and oil phases and on the interface and coupling them with mass balances, one could calculate the thicknesses of the water and oil layers as functions of the component concentrations. It was shown that the deviation from the ideal dilution law can be accounted for by the partition of the components between phases. A comparison between the calculated and experimental results was made.

Appendix A: Van Der Waals Free Energy for a Multilayer, Formed by Alternating Water and Oil Planar Lamellae

A complete expression for the van der Waals interaction energy can be obtained, in the hypothesis of pairwise additivity, by adding the self-energy of the planar layers

$$f_{I,1} = -\frac{A_{11}}{12\pi}\frac{1}{X_1^{2}}$$

$$f_{I,2} = -\frac{A_{22}}{12\pi}\frac{1}{X_2^{2}} \tag{A.1}$$

(where A_{11} and A_{22} are Hamaker constants and X_1 and X_2 are the thicknesses of the layers) to the interaction energy between all the pair of layers, computed as for layers separated by vacuum. The van der Waals interaction between two planar layers of thickness X_1 and X_2, separated by a distance d, is given by[31]

$$f_{12} = -\frac{A_{12}}{12\pi}\left(\frac{1}{d^2} + \frac{1}{(d + X_1 + X_2)^2} - \frac{1}{(d + X_1)^2} - \frac{1}{(d + X_2)^2}\right) \quad (A.2)$$

The total van der Waals energy (per pair of successive water and oil layers and unit area) is computed by adding all the interactions between each of the layers of the pair and all the layers in the system to their interaction and their self-energies. When two layers are in contact, it was considered that a small (but finite) distance e, which is related to the closest approach between molecules, separates the layers.

By rearranging the terms and taking the limit $e \to 0$, the excess van der Waals energy, dependent on the thicknesses X_1 and X_2, acquires the form

$$f_{\mathrm{vdW}} = -\frac{A_{\mathrm{eff}}}{12\pi}\left\{\frac{1}{(X_2)^2} + \sum_{n=1}^{\infty}\left(\frac{1}{(n(X_1+X_2) - X_2)^2} + \frac{1}{(n(X_1+X_2) + X_2)^2} - \frac{2}{(n(X_1+X_2))^2}\right)\right\}$$
$$\equiv -\frac{A_{\mathrm{eff}}}{12\pi}\left\{\frac{1}{(X_1)^2} + \sum_{n=1}^{\infty}\left(\frac{1}{(n(X_1+X_2) - X_1)^2} + \frac{1}{(n(X_1+X_2) + X_1)^2} - \frac{2}{(n(X_1+X_2))^2}\right)\right\} \quad (A.3)$$

where $A_{\mathrm{eff}} = A_{11} + A_{22} - 2A_{12}$ is an effective Hamaker constant.

Appendix B. Electrostatic Interaction in a Electrolyte-free Lamellar Liquid Crystal

In this Appendix, equations will be derived for the double layer interaction between two charged, planar surfaces in an electrolyte-free system. We assume that the potential $\psi(x)$ obeys the Poisson–Boltzmann equation

$$\frac{\partial^2\psi(x)}{\partial x^2} = -\frac{e\rho_0}{\epsilon\epsilon_0}\exp\left(-\frac{e\psi(x)}{kT}\right) \quad (B.1)$$

where ρ_0 is the concentration of counterions at the middle distance between the planes, x is the position coordinate, with $x = 0$ at the middle distance, ϵ is the dielectric constant of water, and ϵ_0 is the permittivity constant. The solution of this equation, for the boundary conditions

$$\psi(0) = 0, \quad \left.\frac{\partial\psi(x)}{\partial x}\right|_{x=0} = 0$$

is[32]

$$\psi(x) = \frac{2kT}{e}\ln(\cos(Kx)) \quad (B.2)$$

where K is given by

$$\frac{K\delta_1}{2}\tan\left(\frac{K\delta_1}{2}\right) = \frac{\sigma_e\delta_1 e}{4\epsilon\epsilon_0 kT} \quad (B.3)$$

δ_1 being the distance between the planes and σ_e the surface charge density. From (B.1) and (B.2), the following relation between ρ_0 and K is obtained

$$\rho_0 = \frac{2\epsilon\epsilon_0 kT}{e^2}K^2 \quad (B.4)$$

The osmotic pressure between the charged plates is given by[32]

$$p = \rho_0 kT = \frac{2\epsilon\epsilon_0(kT)^2}{e^2}K^2 \quad (B.5)$$

The integral $\int_{\psi=0}^{\psi_0}\sigma_e\,d\psi$ can be obtained by combining (B.2) and (B.3)

$$\int_{\psi=0}^{\psi_0}\sigma_e\,d\psi = \int_{K=0}^{K}\sigma_e\frac{d\psi}{dK}\,dK = \frac{8\epsilon\epsilon_0(kT)^2}{e^2\delta_1}\left(\frac{K\delta_1}{2}\tan\left(\frac{K\delta_1}{2}\right) - \frac{1}{2}\ln\left(1 + \tan^2\left(\frac{K\delta_1}{2}\right)\right) - \frac{K^2\delta_1^{\,2}}{8}\right) \quad (B.6)$$

where K is the solution of eq B.3 for the surface charge density, σ_e, and ψ_0 is the surface potential (at $x = \delta_1/2$).

It will be assumed that the surfactant molecules in the aqueous phase are completely dissociated and that their presence in the water phase does not affect the potential given by eq B.2, since their concentration is very low. The condition of equilibrium and the mass balance of the surfactant in the water phase lead to the following expression for the concentration $(X_S^{\,W})^*$ of surfactant ions near the interface, as a function of the average concentration of the surfactant in the water phase

$$(X_S^{\,W})^* = X_S^{\,W}\frac{K\delta_1\cos^2\left(\frac{K\delta_1}{2}\right)}{\left(\cos\left(\frac{K\delta_1}{2}\right)\sin\left(\frac{K\delta_1}{2}\right) + \frac{K\delta_1}{2}\right)} \quad (B.7)$$

Appendix C: Distribution of Alcohol between Water and Oil Phases

In what follows, we provide the derivation of an equation, which relates the concentration of alcohol in water and the oil phases, on the basis of a model developed earlier.[28] Let us assume that $N_O^{\,O}$ molecules of oil and $N_A^{\,O}$ molecules of alcohol are present in the oil phase. Denoting by $N_{jA}^{\,O}$ the number of aggregates containing j molecules of alcohol, the mole fraction of the aggregates of size j is given by

$$X_{jA}^{\,O} = \frac{N_{jA}^{\,O}}{N_O^{\,O} + \sum_j N_{jA}^{\,O}} \quad (C.1)$$

The association equilibrium for alcohol aggregates is written

(32) Cowley, A. C.; Fuller, N. L.; Rand, R. P.; Parsegian, V. A. *Biochemistry* **1978**, *1*, 7, 3163.

$$A_1 + A_j \leftrightarrow A_{j+1} \tag{C.2}$$

where the association equilibrium constant is considered to be independent of j. Denoting by K_x the stepwise association equilibrium constant, the mole fraction of alcohol aggregates of size j is given by

$$X_{jA}{}^{O} = \frac{(K_x X_{1A}{}^{O})^j}{K_x} \tag{C.3}$$

Since the summation over all mole fractions should remain finite, $K_x X_{1A}{}^{O} < 1$ and

$$\sum_j X_{jA}{}^{O} = \frac{X_{1A}{}^{O}}{(1 - K_x X_{1A}{}^{O})}$$

$$\sum_j jX_{jA}{}^{O} = \frac{X_{1A}{}^{O}}{(1 - K_x X_{1A}{}^{O})^2} \tag{C.4}$$

The true mole fraction of oil is then

$$X_O{}^{O} = \frac{N_O{}^{O}}{N_O{}^{O} + \sum_j N_{jA}{}^{O}} = 1 - \frac{\sum_j N_{jA}{}^{O}}{N_O{}^{O} + \sum_j N_{jA}{}^{O}} =$$

$$1 - \sum_j X_{jA}{}^{O} = \frac{(1 - K_x X_{1A}{}^{O} - X_{1A}{}^{O})}{(1 - K_x X_{1A}{}^{O})} \tag{C.5}$$

from where one obtains

$$\sum_j N_{jA}{}^{O} = N_O{}^{O} \frac{X_{1A}{}^{O}}{(1 - K_x X_{1A}{}^{O} - X_{1A}{}^{O})} \tag{C.6}$$

and further

$$N_{1A}{}^{O} = X_{1A}{}^{O}(N_O{}^{O} + \sum_j N_{jA}{}^{O}) =$$

$$N_O{}^{O} X_{1A}{}^{O} \frac{1 - K_x X_{1A}{}^{O}}{1 - K_x X_{1A}{}^{O} - X_{1A}{}^{O}} \tag{C.7}$$

The ratio between the total number of alcohol molecules and the number of alcohol monomers is given by

$$\frac{N_A{}^{O}}{N_{1A}{}^{O}} = \frac{\sum_j jX_{jA}{}^{O}}{X_{1A}{}^{O}} = \frac{1}{(1 - K_x X_{1A}{}^{O})^2} \tag{C.8}$$

Using eqs C.7 and C.8, the mole fraction of alcohol can be calculated as a function of the mole fraction of monomers

$$X_A{}^{O} = \frac{N_A{}^{O}}{N_O{}^{O} + N_A{}^{O}} = \frac{X_{1A}{}^{O}}{1 - 2K_x X_{1A}{}^{O} + K_x(K_x + 1)(X_{1A}{}^{O})^2} \tag{C.9}$$

The latter equation can be inverted as

$$X_{1A}{}^{O} = \frac{1 + 2K_x X_A{}^{O} - \sqrt{1 + 4K_x X_A{}^{O} - 4K_X (X_A{}^{O})^2}}{2K_x X_A{}^{O}(1 + K_x)} \tag{C.10}$$

The standard chemical potential of the monomeric alcohol in oil is considered equal to that of the monomers of alcohol in alcohol and is related to the chemical potential of the pure alcohol, μ_A, and the concentration of monomers in pure alcohol via

$$\mu_A = \mu_{1A}{}^{O} + kT \ln(X_{1A}{}^{O})^* \tag{C.11}$$

The monomeric mole fraction in pure alcohol is obtained from eq C.10 in the limit $X_A{}^{O} \to 1$

$$(X_{1A}{}^{O})^* = \frac{1}{1 + K_x} \tag{C.12}$$

By equating the chemical potential of alcohol in the oil phase to the chemical potential of alcohol in the water phase, one obtains

$$\frac{X_A{}^{W}}{(X_A{}^{W})_{SAT}} = \frac{X_{1A}{}^{O}}{(X_{1A}{}^{O})^*} = (1 + K_x)X_{1A}{}^{O} \tag{C.13}$$

where $(X_A{}^{W})_{SAT}$ is the saturation concentration of alcohol in water when the water phase coexists with a pure alcohol phase. At 25 °C, the solubility of alcohol in water is given by the expression[33]

$$\ln(X_A{}^{W})_{SAT} = 1.40 - 1.38 n_C \tag{C.14}$$

where n_C is the number of carbon atoms. The stepwise association equilibrium constant K_x for aliphatic alcohols ranging from propanol to decanol can be computed using the empirical expression[28]

$$K_x = 92(n_C)^{-0.47} e^{3000((1/T)-(1/323))} \tag{C.15}$$

Because K_x and $(X_A{}^{W})_{SAT}$ are known, a relation between the molar fraction of alcohol in water and oil, $X_A{}^{W}$ and $X_A{}^{O}$, respectively, can be established. Using eqs C.13 and C.9, one obtains

$$X_A{}^{O} = \frac{N_A{}^{O}}{N_O{}^{O} + N_A{}^{O}} = \frac{\dfrac{X_A{}^{W}}{(X_A{}^{W})_{SAT}}}{1 + K_x\left(1 - \dfrac{X_A{}^{W}}{(X_A{}^{W})_{SAT}}\right)^2} \tag{C.16}$$

LA010385E

(33) Tanford, C. *The hydrophobic effect*; John Wiley & Sons: New York, 1980.

Introduction to CHAPTER 4

Non-DLVO colloidal interactions: excluded volumes, undulation interactions, depletion forces and many-body effects

4.1 E. Ruckenstein, D. Schiby: "*Effect of the Excluded Volume of the Hydrated Ions on Double-Layer Forces,*" LANGMUIR 1 (1985) 612–615.

4.2 M. Manciu, E. Ruckenstein: "*Lattice Site Exclusion Effect on the Double Layer Interaction,*" LANGMUIR 18 (2002) 5178–5185.

4.3 M. Manciu, E. Ruckenstein: "*Free Energy and Thermal Fluctuations of Neutral Lipid Bilayers,*" LANGMUIR 17 (2001) 2455–2463.

4.4 M. Manciu, E. Ruckenstein: "*The Interaction between Two Fluctuating Phospholipid Bilayers,*" LANGMUIR 18 (2002) 4179–4182.

4.5 M. Manciu, E. Ruckenstein: "*Specific Ion Effects in Common Black Films: The Role of the Thermal Undulation of Surfaces,*" LANGMUIR 20 (2004) 1775–1780.

4.6 H. Huang, E. Ruckenstein: "*Interaction Force between Two Charged Plates Immersed in a Solution of Charged Particles. Coupling between Double Layer and Depletion Forces,*" LANGMUIR 20 (2004) 5412–5417.

4.7 H. Huang, E. Ruckenstein: "*Thermodynamically stable dispersions induced by depletion interactions*" JOURNAL OF COLLOID AND INTERFACE SCIENCE 290 (2005) 336–342.

4.8 J. Feng and E. Ruckenstein: "*Attractive interactions in dispersions of identical charged colloidal particles: a Monte-Carlo simulation,*" JOURNAL OF COLLOID AND INTERFACE SCIENCE 272 (2004) 430–437.

A long time ago, Stern noted that the traditional assumption that the ions interact only with a "mean" electrical field (the Poisson - Boltzmann approach) leads to an ionic density in the vicinity of the interface that exceeds the available volume. A simple way to avoid this difficulty is to consider that the ions are hydrated, and therefore there are fewer positions available to them in the vicinity of charged surfaces [4.1]. When compared to the traditional Poisson-Boltzmann result, this correction leads to an increase in the repulsive force at short separations [4.1]. This approach is extended to account for the ion competition for hydrating water molecules [4.2]. A position in the water lattice is considered available to an ion only when it is "free" and has a certain number of "free" neighboring water molecules (i.e., water molecules which do not participate in the hydration of other ions). Whereas at high surface potentials, this "excluded" volume effect increases the repulsion as compared to the traditional double layer force, at low surface potentials it can either increase or decrease the repulsion (the latter case occurs when the hydrated coion is larger than the hydrated counterion). The strong dependence of this effect on the hydration number of ions provides a possible explanation for ion-specific effects, which will be examined in the next chapter.

Other interactions that cannot be explained in the framework of the Poisson-Boltzmann approach include the repulsion between thermally undulating interfaces (the Helfrich force, which is partially responsible for the stability of neutral lipid bilayers [Helfrich, W. Z. *Naturforsch.* **1978**, *33a*, 305]). The Helfrich theory predicts that this repulsion is inversely proportional to the third power of the separation distance, while experimental evidence indicates an exponential dependence at small separations. The free energy of entropic confinement can be calculated exactly for harmonic interactions. Assuming that the same expression for the free energy remains valid when the interactions are slightly non-linear, the minimization of the total free energy leads, for the thermal undulation force, to an exponential behavior at short separations (as observed experimentally) and a power law dependence at large separations (as also predicted by Helfrich) [4.3]. Another approach considers that the undulating membrane is composed of independent pieces of a suitable area, which is selected in such a manner that, for an harmonic potential, the entropic confinement pressure of the separate pieces to be equal to the undulating repulsion. A harmonic approximation of any non-harmonic interaction potential allows one to calculate the area of the independent pieces and the corresponding repulsion force [4.4].

E. Ruckenstein, M. Manciu, *Nanodispersions*, DOI 10.1007/978-1-4419-1415-6_4, © Springer Science+Business Media, LLC 2010

It is well known that the traditional double layer theory is not accurate for small separations and high ionic strengths. However, experiment revealed that, for common black films, it is inaccurate even for moderate ionic strengths and large separation distances. Furthermore, strong specific ionic effects are exhibited by black films at large separation distances, again in contrast with the traditional theory. It is shown that by taking into account the thermal undulations of the black films and the traditional forces (double layer, hydration and van der Waals) and considering that different ions provide different bending moduli to the membrane, one can explain the dependence of the thickness of the films on the external pressure for various uni-univalent electrolytes [4.5]. An expression is derived for the force generated between two charged plates immersed in an electrolyte solution that contains small charged particles by coupling the double layer and depletion forces. The effects of particle charge, particle charge sign, particle size and particle volume fraction are examined [4.6]. When the depletion interactions are taken into account, together with double layer, van der Waals interactions and entropic effects, an increase in the volume fraction of the small particles leads to a more negative free energy of the system, which stabilizes a dispersion of large particles. This effect can explain the formation of gels observed experimentally in concentrated emulsions [4.7].

Iridescence was observed in a lattice-structured dispersion for distances between particles of the same order of magnitude as the wavelength of light, distances at which the van der Waals interactions are negligible. The collective interactions between identically charged monodisperse particles have been examined via Monte-Carlo simulations, using the Debye-Hückel pair potential. When the number of charges per particle and the particle volume fraction become sufficiently large, the pair long-range electrostatic repulsion generates an effective attractive interaction, because of many-body effects [4.8]. Small charges per particle and small particle volume fractions lead to disordered structures, while larger ones lead to ordered structures. Depending on these parameters, disordered liquid-like structures, ordered structures dispersed in disordered ones, disordered structures dispersed into ordered ones, and ordered crystal-like structures are observed [4.8].

612 *Langmuir* **1985**, *1*, 612–615

Effect of the Excluded Volume of the Hydrated Ions on Double-Layer Forces

E. Ruckenstein* and D. Schiby

Department of Chemical Engineering, State University of New York at Buffalo, Buffalo, New York 14260

Received April 30, 1985. In Final Form: June 24, 1985

A modified theory of the double layer is proposed in which the Boltzmann distribution is replaced with a distribution which accounts for the volume exclusion of the hydrated ions. The double-layer repulsion force between two parallel plates thus calculated can be much greater than that predicted by the traditional theory, when the distance between the plates is sufficiently small. This may at least partly explain the excess repulsive force that has been measured experimentally between two mica plates, when the distance between the plates is sufficiently small.

Introduction

In a number of experiments,[1] the measured repulsive force between two mica plates was greater than that predicted on the basis of the Poisson–Boltzmann equation, when the distance between them was below about 40 Å. This higher repulsive force arises only for sufficiently large ionic strengths and has a decay length in the range of 3–10 Å. In contrast, the repulsive force between two parallel phospholipid bylayers falls off with a decay length between 2 and 3 Å and is largely independent of the ionic strength.[2] Concerning the origin of the latter force, it was suggested[3,4] that the zwitterionic head groups of the phospholipid bilayers polarize the nearby water molecules, which in turn act as a self-consistent field at the location of the nearby dipoles. The polarization decays rapidly away from the wall. When the polarization layers of the two phospholipid bilayers overlap, the local average dipole moments of the solvent molecules decrease. This increases the free energy of the system, thus giving rise to a repulsive force.[4] The value of the decay length predicted[4] was between 2 and 3 Å, which is in good agreement with the experimental value. Additionally, since the ionic strength does not affect the force, the above polarization mechanism appears to offer a reasonable explanation of the observed phenomenon. However, in the experiments regarding mica, the ionic strength plays a part and the decay length is much larger. While a polarization mechanism can still explain the origin of the excess repulsive force,[5] the fact that this force cannot be predicted in the framework of the traditional double-layer theory does not necessarily mean that its origin is outside the electrical double layer. Indeed, the traditional theory, being based on the Poisson–Boltzmann equation, involves a number of assumptions that are not valid, particularly when the distance between the plates becomes sufficiently small. One important simplifying assumption employed by the traditional theory is that the size of the hydrated ions is negligible. As a result of this assumption, the charge density near the surface can exceed the maximum density allowed by the size of the ions. In order to avoid this difficulty a compact layer of hydrated ions (Stern layer) was assumed to cover the solid surface. This layer being assumed bound to the surface, the corresponding ions are no longer osmotically active and therefore they do not contribute to the repulsive force, which thus becomes smaller than the real one. This effect is expected to become important particularly for small distances between the interacting surfaces. Therefore, the introduction of the Stern layer is at least partially artificial and an approach that accounts for both the finite size of the hydrated ions as well as for the fact that they are osmotically active should be developed. The scope of the present paper is to suggest such a treatment. The basic equations are derived in the next section largely on the basis of thermodynamic arguments and rederived somewhat more rigorously in the third section of the paper. The repulsive force is calculated in the fourth section and the discussion in the fifth section concludes the paper.

Many authors, among whom we cite only a few,[6–9] have suggested corrections to the Poisson–Boltzmann equation and have dealt with the effect of the ion size, with the dependence of the dielectric constant on the electric field and ionic concentrations, and other effects. The suggested corrections have generally resulted in small changes in the repulsive force, which tended to be somewhat smaller or somewhat larger than in the traditional theory, depending upon the treatment involved. An interesting conclusion drawn by Levine and Bell[7] is that there is a partial cancellation of the various small corrections to the Poisson–Boltzmann equation, so that at least for 1:1 electrolyte concentration <0.1 M the Poisson–Boltzmann equation seems to be an adequate approximation. In addition, these authors emphasize that their treatment of certain corrections, e.g., the ionic size effect (which is based upon the approximate Flory–Huggins formula for the entropy of mixing), probably should be modified very close to the plate wall. The effect of the volume exclusion is treated in the present paper by a quite different approach to demonstrate that when the distance between the plates becomes sufficiently small, the repulsive force can increase by large factors.

Derivation of the Basic Equations

Let us consider two parallel plates immersed in an electrolyte solution. As is well-known the ions form a diffuse electric layer, bearing a charge opposite in sign to that of the surface, sandwiched between the two plates. Because of the strong interactions between the ions and water molecules, the ions are hydrated and the simplifying assumption is made that their configuration is the same everwhere in the system. Of course, if the ion–surface interactions favor adsorption of dehydrated counterions,

(1) Pashley, R. M. *J. Colloid Interface Sci.* **1981**, *83*, 531.
(2) Parsegian, V. A.; Fuller, N.; Rand, R. P. *Proc. Natl. Acad. Sci. U.S.A.* **1979**, *76*, 2750.
(3) Marcelja, S.; Radic, N. *Chem. Phys. Lett.* **1976**, *42*, 129.
(4) Schiby, D.; Ruckenstein, E. *Chem. Phys. Lett.* **1983**, *95*, 435.
(5) Ruckenstein, E.; Schiby, D. *Chem. Phys. Lett.* **1983**, *95*, 439.
(6) Sparnaay, M. J. *Recl. Trav. Chim. Pays-Bas* **1958**, *77*, 872; **1962**, *81*, 395.
(7) Levine, S.; Bell, G. M. *Discuss. Faraday Soc.*, **1966**, *42*, 69.
(8) Bell, G. M.; Levine, S. *Z. Phys. Chem. (Leipzig)* **1966**, *231*, 289.
(9) Sparnaay, M. J. *Discuss. Faraday Soc.* **1966**, *42*, 94.

0743-7463/85/2401-0612$01.50/0

this will weaken the repulsive force because the charge of the surface is decreased and also because these ions are no longer osmotically active. This effect will be discussed later in the paper but will not be included explicitly in the calculations. Specific adsorption of hydrated ions is also not considered in the calculations.

Denoting by $n_i(x)$ and q_i the concentration of the ions of species i at a distance x from one of the plates and their charge, respectively, the charge density at that point is given by

$$\rho = \sum q_j n_j \tag{1}$$

where q_i is positive for cations and negative for anions.

The Poisson equation has, in the unidimensional case considered here, the form

$$\frac{d^2\psi}{dx^2} = -\frac{4\pi\rho}{\epsilon} \tag{2}$$

where ψ is the electrical potential and ϵ is the dielectric constant. In the Boltzmann approximation

$$n_i(x) = n_i e^{-q_i\psi/kT} \tag{3}$$

where n_i is the concentration of ions of species i far from the surface, where $\psi = 0$. The above approximation involves, however, the important assumption that the ions are dimensionless points. The main goal of this paper is to replace expression 3 with one that accounts for the hydration and the finite sizes of the hydrated ions.

Equation 3 is derived from the equality of the electrochemical potentials at a point located at distance x from the plate, $\mu_i(x)$, and at large distances (where $\psi = 0$), μ_i. Since the ions are subject to the field $\psi(x)$ at distance x, the electrochemical potential $\mu_i(x)$ can be written as

$$\mu_i(x) = \mu_i^\circ + kT \ln \gamma_i(x) + kT \ln n_i(x) + q_i\psi(x) \tag{4}$$

where μ_i° is the standard chemical potential and $\gamma_i(x)$ is the activity coefficient. Similarly, the chemical potential at infinite distance, μ_i, is written as

$$\mu_i = \mu_i^\circ + kT \ln \gamma_i + kT \ln n_i \tag{5}$$

where γ_i is the activity coefficient at large distance from the surface. Equation 3 implies the additional assumption that $\gamma_i(x) = \gamma_i$. This assumption is relaxed in what follows, since the effect of the volume exclusion of the hydrated ions on $\gamma_i(x)$ is partially taken into account and this effect depends, via the concentrations of the ions, on x.

Let us consider the volume of a water molecule as the volume of a site and denote by τ_i the number of sites occupied by a hydrated ion of species i. The total number of sites per unit volume, $n(x)$, is given by

$$n(x) = n_w(x) + \sum_{j=1}^{2} \tau_j n_j(x) \tag{6}$$

where $n_w(x)$ is the concentration of the free water molecules at distance x.

A new ion introduced into the system associates with a few water molecules because of the strong interactions between them, thus generating a hydration sheath. This new hydrated ion cannot, however, occupy the sites already occupied by the other hydrated ions. Since the fraction of sites occupied by the hydrated ions is $\sum n_j(x)\tau_j/n(x)$, the fraction unoccupied by them is $1 - (\sum n_j(x)\tau_j/n(x))$ and the probability $P_i(x)$ of finding τ_i neighboring sites needed to introduce an ion of species i to generate the corresponding hydrated ion is (if a condition mentioned below is satisfied) given by

$$P_i(x) = [1 - (\sum n_j(x)\tau_j/n(x))]^{\tau_i} \tag{7}$$

Indeed, the probability of finding a site unoccupied by hydrated ions is $1 - (\sum n_j(x)\tau_j/n(x))$ and the probability of finding two such neighboring sites is $[1 - \sum \eta_j(x)\tau_j/\eta(x)]^2$. Intuition suggests that the latter expression can be generalized to τ_i sites if, as above, they are (first) neighbors among themselves. If the structure of a hydrated ion is approximated by one consisting of τ_i elements which are first neighbors among themselves, an additional ion of species i can therefore occupy $\eta(x)P_i(x)$ positions. In the traditional treatment based on the Poisson–Boltzmann equation, $P_i(x)$ is taken as unity and $n(x)$ is taken to be constant and equal to the number of molecules of water n'. Because the number of positions available to an ion is smaller in the present case, a decrease in the entropy of an ion Δs occurs in comparison to the traditional treatment, which is given by the expression

$$\Delta s = k \ln \frac{P_i(x)n(x)}{n'}$$

Assuming that this constitutes the only change in the activity coefficient in comparison to the traditional approach (this point is discussed later in the paper), the equality of the electrochemical potentials $\mu_i(x)$ and μ_i leads to

$$kT \ln n_i(x) - kT \ln P_i(x)n(x) + q_i\psi = kT \ln n_i - kT \ln P_i n \tag{8}$$

where P_i, n_i, and n are the values of $P_i(x)$, $n_i(x)$, and $n(x)$ far from the plate. Equation 8 can be rewritten in the form

$$\frac{n_i(x)/n(x)}{\left\{1 - \frac{\sum \tau_j n_j(x)}{n(x)}\right\}^{\tau_i}} = \frac{n_i/n}{\left\{1 - \frac{\sum \tau_j n_j}{n}\right\}^{\tau_i}} e^{-q_i\psi/kT} \tag{9}$$

Statistical Mechanical Approach

In the following derivation, the electrochemical potential is assumed again as given by

$$\mu_i(x) = \mu'_i(x) + q_i\psi$$

where $\mu'_i(x)$ is the chemical part of the electrochemical potential $\mu_i(x)$. In other words, the charge of the surface as well as the presence of the other ions is felt in a given point only via ψ, and $\mu'_i(x)$ can be calculated as for a uniform system having the concentrations of the particular point considered. For this reason, it is sufficient to derive an expression for the chemical potential at a large distance from the plate since a similar expression can be extended to any point in the system. The system is composed of ions and water molecules, and one assumes that the interactions are electrostatic and the water molecules constitute a continuous medium of uniform dielectric constant. The partition function of such a system is given by the expression

$$Q = Q_0 \sum_{\Omega} e^{-(u_{12} + u_{13} + \ldots)/kT - (u_{23} + u_{24} + \ldots)/kT\ldots} \tag{10}$$

where Q_0 is the partition function of the single noninteracting particles, Ω is the set of all the distinct configurations and u_{pj} is the electrostatic interaction energy between ions p and j. Denoting by q_1 the charge of particle 1 (assumed of species i), the electrical potential ϕ_1 felt by this particle is

$$q_1\phi_1 = \sum_j u_{1j} \qquad (j \neq 1) \tag{11}$$

Equation 10 can therefore be rewritten as

$$Q(\Omega) = \sum_{\Omega} Q_1 e^{-q_1\phi_1/kT} Q(\Omega') \tag{12}$$

where Ω' is the set of distinct configurations of the system after excluding particle 1 and Q_1 is the partition function of the noninteracting particle 1. Since the free energy F is given by the expression

$$F = -kT \ln Q$$

one obtains

$$\frac{\partial F}{\partial q_1} = \frac{\sum_{\Omega} \phi_1 e^{-q_1\phi_1/kT} Q(\Omega')}{\sum_{\Omega} e^{-q_1\phi_1/kT} Q(\Omega')} \equiv \langle \phi_1 \rangle \tag{13}$$

which, by integration, yields

$$F = \int_0^{q_1} \langle \phi_1(q') \rangle \, dq' + F(q_1 = 0) \tag{14}$$

Equation 14 can be rewritten as

$$Q = e^{-\alpha_{1i}/kT} Q(q_1 = 0) \tag{15}$$

where

$$\alpha_{1i} = \int_0^{q_1} \langle \phi_1(q') \rangle \, dq' \tag{16}$$

is the work of charging of an ion of species i and $Q(q_1 = 0)$ is the partition function if particle 1 is uncharged. The subscript i was included to emphasize that particle 1 is of species i.

Since the number of positions available to a hydrated ion of species i is

$$nP_i = n\left(1 - \frac{\sum \tau_j n_j}{n}\right)^{\tau_i} \tag{17}$$

eq 12 can be rewritten as

$$Q(n_i, q_1 = 0) = \frac{n}{n_i}\left\{1 - \frac{\sum \tau_j n_j}{n}\right\}^{\tau_i} Q_1 Q(n_i - 1) \tag{18}$$

where n_i in the denominator of the right-hand side accounts for the fact that any one of the n_i ions could be chosen as particle 1 without leading to distinct configurations. Combining eq 15 and 18 yields

$$\frac{Q(n_i)}{Q(n_i - 1)} = e^{-\alpha_{1i}/kT} Q_1 \frac{n}{n_i}\left(1 - \frac{\sum \tau_j n_j}{n}\right)^{\tau_i} \tag{19}$$

Since the chemical potential μ_i can be defined as

$$\mu_i = \frac{\partial F}{\partial n_i} = -kT \ln \frac{Q(n_i)}{Q(n_i - 1)} \quad \text{(constant volume)} \tag{20}$$

one finally obtains

$$\mu_i = \alpha_{1i} + kT \ln \frac{n_i}{n} - kT \ln Q_1 - \tau_i kT \ln \left(1 - \frac{\sum \tau_j n_j}{n}\right) \tag{21}$$

A similar approach can be used to derive an expression for the chemical part $\mu'_i(x)$ of the electrochemical potential at a point located at the distance x from the plate, to obtain

$$\mu'_i(x) = \alpha_{1i}(x) + kT \ln \frac{n_i(x)}{n(x)} - kT \ln Q_1 - \tau_i kT \ln \left(1 - \frac{\sum \tau_j n_j(x)}{n(x)}\right) \tag{22}$$

where $\alpha_{1i}(x)$ is the work of charging of an ion 1 of species i located at the distance x from the plate. Instead of the simplifying assumption $\gamma_i(x) = \gamma_i$ involved in the traditional treatment, the less severe assumption that $\alpha_{1i}(x) = \alpha_{1i}$ is employed in the present approach. With this simplifying assumption, eq 9 is again obtained.

Calculation of the Repulsive Force

At constant surface charge, the part of the free energy of the double layer per unit surface area that enters in the calculation of the force is given by the expression[10,11]

$$f = \int_0^{\sigma} \psi_0(\sigma') \, d\sigma' \tag{23}$$

where ψ_0 is the surface potential and σ is the surface charge density. The repulsive force per unit area is at constant surface charge density provided by the derivative

$$\mathcal{F} = \int_0^{\sigma} \frac{\partial \psi_0}{\partial h} d\sigma' \tag{24}$$

where $2h$ is the distance between the plates.

It is instructive to eliminate ψ_0 between eq 24 and 9 to obtain

$$\mathcal{F} = -\frac{kT}{q_i}\left\{\int_0^{\sigma} \frac{\partial \ln (n_i(0)/n(0))}{\partial h} d\sigma' - \tau_i \int_0^{\sigma} \frac{\partial \ln \left(1 - \frac{\sum \tau_j n_j(0)}{n(0)}\right)}{\partial h} d\sigma'\right\} \tag{25}$$

where $n_i(0)$ and $n(0)$ are the values of $n_i(x)$ and $n(x)$ for $x = 0$.

As the distance h decreases, the concentration $n_i(0)$ of the counterions at the surface increases in order to maintain the electroneutrality between the charged surface and the decreased volume. For this reason,

$$\frac{\partial \ln n_i(0)}{\partial h} < 0$$

and

$$\frac{\partial \ln \left(1 - \frac{\sum \tau_j n_j(0)}{n(0)}\right)}{\partial h} > 0$$

The Poisson equation, (2), together with eq 9 have been solved numerically, using the boundary conditions

$$\sigma = -\frac{\epsilon}{4\pi}\left(\frac{d\psi}{dx}\right)_{x=0} = \text{constant} \tag{26}$$

and

$$\left(\frac{d\psi}{dx}\right)_{x=h} = 0 \tag{27}$$

The results of the numerical calculations are plotted in Figure 1 for two negative surface charge densities $\sigma = 2$

(10) Verwey, E. J. W.; Overbeek, J. Th. G. "Theory of Stability of Lyophobic Colloids"; Elsevier: Amsterdam, 1948.
(11) Ruckenstein, E. *J. Colloid Interface Sci.* 1981, *82*, 490.

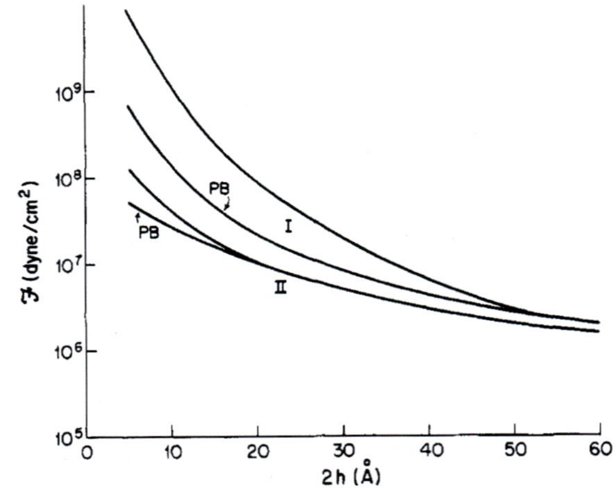

Figure 1. Repulsive force against distance. Concentration of uni-univalent electrolyte = 10^{-2} M; $\tau_{cation} = 5$, $\tau_{anion} = 1$. (I) $\sigma = 2 \times 10^{14}$ charges/cm^2; (II) $\sigma = 4 \times 10^{13}$ charges/cm^2. The curves marked by PB are based on the Poisson–Boltzmann equation.

$\times\ 10^{14}$ charges/cm^2 and $\sigma = 4 \times 10^{13}$ charges/cm^2 and a uni-univalent electrolyte with a reservoir concentration of 10^{-2} M. The values of τ_i have been taken as $\tau_1 = 5$ for the cation and $\tau_2 = 1$ for the anion.

As expected, the deviation from the approach based on the Poisson–Boltzmann equation is larger for the greater charge. The experimental results[1] have a shape similar to that predicted here.

Discussion

In order to avoid concentrations at the surface greater than those compatible with the volume occupied by the hydrated ions, a special kind of layer, the Stern layer, was frequently assumed to exist near the surface. If, instead of the Boltzmann distribution, a distribution that accounts for the volume exclusion of the hydrated ions is used, the above difficulty can be avoided in a more natural way. The repulsive force thus predicted is larger than that provided by the traditional theory, when the distance between the plates becomes sufficiently small. Of course, this effect is important if one assumes that the ions do not lose their hydration sheaths. In reality, they can lose their hydration layer, or at least the hydrated ions can have near the surface of the plate a configuration different from that in the bulk. This decreases the surface charge and binds the ions to the surface. The excess repulsive force is thus decreased for two reasons: (i) charge decrease and (ii) the bound ions are no longer osmotically active.

Experiment[1] has shown that the lower the pH, the smaller is the excess repulsive force. For instance, for 6×10^{-2} M LiCl, the excess repulsive force arises at pH 5.4 but does not occur at pH 4.5. For NaCl, the extra force appears at the lower concentration of 5×10^{-4} M if the pH is 8.9, while at the lower pH of 5.7, such forces appear at the higher concentration of 10^{-2} M and are much weaker. These experiments suggest that the surface adsorbs dehydrated H^+ ions which reduce its charge. At higher values of the pH, smaller amounts of dehydrated H^+ ions are adsorbed and the surface charge density is less decreased. If the hydrogen ions could remain hydrated near the surface (this can occur when their free energy of hydration is more negative than that due to the interaction with the surface), they will also contribute to the excess repulsive force. A minimum concentration of electrolyte is needed to generate the excess repulsive force, because, only above that value, the concentration of hydrated ions becomes sufficiently large near the mica surface. The large hydrated ions, such as Li, give rise to a higher volume exclusion than the smaller ones, such as Na, K, and Cs. As a result, a greater concentration in the bulk is needed for Li to achieve a sufficiently large concentration near the mica surface.

Acknowledgment is made to the donors of the Petroleum Research Fund, administrated by the American Chemical Society, for support of this research.

5178 *Langmuir* **2002,** *18,* 5178–5185

Lattice Site Exclusion Effect on the Double Layer Interaction

Marian Manciu and Eli Ruckenstein*

Department of Chemical Engineering, State University of New York at Buffalo, Buffalo, New York 14260

Received February 21, 2002. In Final Form: April 22, 2002

A lattice model for an electrolyte solution is proposed, which assumes that the hydrated ion occupies τ_i $(i = 1, 2)$ sites on a water lattice. A lattice site is available to an ion "i" only if it is "free" (it is occupied by a water molecule, which does not hydrate an ion) and has also at least $(\tau_i - 1)$ first-neighbors free. The model accounts for the correlations between the probabilities of occupancy of adjacent sites and is used to calculate the "excluded volume" (lattice site exclusion) effect on the double layer interactions. It is shown that at high surface potentials the thickness of the double layer generated near a charged surface is increased, when compared to that predicted by the Poisson–Boltzmann treatment. However, at low surface potentials, the diffuse double layer can be slightly compressed, if the hydrated co-ions are larger than the hydrated counterions. The finite sizes of the ions can lead to either an increase or even a small decrease of the double layer repulsion. The effect can be strongly dependent on the hydration numbers of the two species of ions.

I. Introduction

The distribution of ions near a charged surface immersed in an electrolyte solution was first calculated on the basis of the Poisson–Boltzmann equation, assuming that the ions are point charges subjected to a mean electrostatic field. One of the first corrections of the theory was a result of the observation that, at high charges, the Poisson–Boltzmann equation predicts a too high density of counterions in the vicinity of the surface, whose volume exceeds the available volume. For this reason, Stern[1] suggested to consider in the vicinity of the surface a layer (the Stern layer) that contains bound counterions and is not available to the thermally moving counterions, and Bikerman[2,3] proposed a modified Boltzmann expression which accounted for the finite volume of ions.

Ruckenstein and Schiby derived[4] an expression for the electrochemical potential, which accounted for the hydration of ions and their finite volume. The modified Poisson–Boltzmann equation thus obtained was used to calculate the force between charged surfaces immersed in an electrolyte. It was shown that at low separation distances and high surface charges, the modified equation predicts an additional repulsion in excess to the traditional double layer theory of Derjaguin–Landau–Verwey–Overbeek.

The volume exclusion effect has received recently renewed attention.[5–9] Paunov et al.[5] argued that the excluded volume should be taken as eight times the real volume of the ions (this corresponds to the first-order correction in the virial expansion for a hard-sphere fluid) and calculated on this basis the repulsive force. While this approximation is reasonable at very low concentrations, it breaks down when the volume of the spheres becomes about 10% of the total volume (since it predicts a maximum compaction of 12.5%). For hydrated ions with a large hydration radius, such as Na, this corresponds to a maximum electrolyte concentration of about 1 M.

In principle, the model can be improved by including in the "excluded volume", in addition to the term proportional to the ion density, higher-order corrections proportional to powers of the ion density, which are related to the virial expansion for hard-sphere fluids. However, only the first eight virial coefficients have been calculated,[10] and the power series converges slowly at high densities. To avoid this, one can account for the volume exclusion effect for hard-sphere fluids starting from an equation of state. The latter can be obtained with reasonable accuracy from the interpolation of the virial expansion, either using Padé approximants,[11] or other ad hoc analytical continuations of the series, a procedure first employed by Carnahan and Starling.[12] The method was employed by Lue et al.[9] to calculate the counterions distribution starting from the corresponding modified Poisson–Boltzmann equation. The advantage of this method consists of its easy application to ions of different sizes, since good approximations of the equation of state (and hence of the excess chemical potential with respect to an ideal solution) are known for mixtures of hard spheres of different sizes.[13]

The main question is whether the hydrated ions behave as hard spheres; while this seems plausible for ions much larger than the water molecules, it is probably not entirely applicable to small ions, whose hydration shells continuously change. Marcelja calculated recently the double layer interaction[8] using the anisotropic hypernetted chain method and potentials of mean force between pairs of ions in water, provided by Monte Carlo simulations. This

* Corresponding author: e-mail address, feaeliru@acsu.buffalo.edu; phone, (716) 645 2911/2214; fax, (716) 645 3822.

(1) Stern, O. Z. *Electrochem.* **1924**, *30*, 508.
(2) Bikerman, J. *J. Philos. Mag.* **1942**, *33*, 384.
(3) Rowlinson, J. S. In *Fundamentals of Inhomogeneous Fluids*; Henderson, D., Ed.; Marcel Dekker: New York, 1992.
(4) Ruckenstein, E.; Schiby, D. *Langmuir* **1985**, *1*, 612.
(5) Paunov, V. N.; Dimova, R. I.; Kralchevsky, P. A.; Broze, G.; Mehreteab, A. *J. Colloid Interface Sci.* **1996**, *182*, 239.
(6) Kralj-Iglic, V.; Iglic, A. *J. Phys. II* **1996**, *6*, 477.
(7) Borukhov, I.; Andelman, D.; Orland, H. *Phys. Rev. Lett.* **1997**, *79*, 435.
(8) Marcelja S. *Nature* **1997**, *385*, 689.
(9) Lue, L.; Zoeller N.; Blankschtein, D. *Langmuir* **1999**, *15*, 3726.
(10) van Rensburg, E. J. J. *J. Phys. A: Math. Gen.* **1993**, *26*, 4805.
(11) Ree, F. H.; Hoover, W. G. *J. Chem. Phys.* **1967**, *46*, 4181.
(12) Carnahan, N. F.; Starling, K. E. *J. Chem. Phys.* **1969**, *51*, 635.
(13) Boublik, T. J. *Chem. Phys.* **1970**, *53*, 471. Masoori, G. A.; Carnahan, N. F.; Starling, K. E.; Leland, T. W. *J. Chem. Phys.* **1971**, *54*, 1523.

10.1021/la020194r CCC: $22.00
Published on Web 06/18/2002

method implicitly accounted for the excluded volume and also for other correlations between ions; however, it required accurate information about the interactions between pairs of ions in water. It was shown[14] that the method is very sensitive to the details of the potential of the mean force between pairs of ions and that different models can lead to even qualitatively different results.

Another approach that accounted for the volume exclusion effect was based on lattice models.[4] Recently, Kralj-Iglic and Iglic[6] and, independently, Borukhov et al.[7] derived an expression for the counterions distributions, starting from a modified Poisson–Boltzmann equation, for a lattice whose cell size was that of a hydrated ion (both kind of ions were considered of the same size). The lattice cell can be either occupied or not by a hydrated ion. This type of lattice allows for a maximum compaction of unity. For this reason, in this case the excess chemical potential (with respect to an ideal solution) is, in general, important either for very large ions at sufficiently large concentrations or for small ions at high ionic strengths and high surface potentials. It is not yet clear how this model[6,7] can be extended to electrolytes with hydrated ions of different sizes.

A modality to overcome these difficulties was proposed earlier by Ruckenstein and Schiby,[4] who considered that the ions of different kinds ($i = 1, 2$) occupy τ_i sites on a water lattice. Since the size of the cell of the water lattice is independent of the size of the ions, the model can be easily extended to any number of ions of different kinds. To compare the prediction of various models, let us first consider the case in which the ions of different kinds have the same volume ($\tau_1 = \tau_2 = \tau$).

For n hard spherical particles of radius R and volume τv (v being the volume of a cell of the lattice) and for large values of τ (the continuum limit), the volume V' available to the particles can be calculated as follows.[15] The center of a new particle cannot be located within a distance $2R$ from an existing particle; hence each existing particle can be thought of as carrying a "forbidden region" of radius $2R$ (volume $8\tau v$), where a new particle cannot be placed. However, by subtracting $8n\tau v$ from the total volume V of the system, one obtains an underestimate of the available volume, since the "forbidden regions" of the existing particles overlap (and hence the overlapping regions are subtracted twice). Assuming that the particles are uniformly distributed in the available space, the overlapping volume per particle is[15] $34\tau v\, n\tau v/V' \cong 34\,(\tau v) \times (\tau\eta)$ (where $\eta = nv/V = n/N$ is the ratio between the total volume occupied by the ions and the total volume of the system, $\tau\eta$ is the ratio between the total volume of the hydrated ions and the total volume of the systems, and N is the total number of sites (n ions + $(\tau - 1)n$ water molecules hydrating the ions + free water molecules). Since the second correction should be counted only once per each pair of particles, the available volume is[15]

$$V' = V(1 - 8\tau\eta + 17(\tau\eta)^2 - ...) \qquad (1)$$

The first three terms of the series are overestimating the available volume, since they do not account for the overlapping of the "forbidden regions" for triplets, and a new correction (proportional to the third power of density) should be included, and so on. The calculation of the remaining terms of the (slowly) converging series is however increasingly difficult. The first two terms of the series (1) constitute the result proposed recently for the volume available to a hard sphere fluid by Paunov et al.[5]

On the other hand, if the hydrated ions are totally deformable (hence, they can occupy a site if the site is free and other $(\tau - 1)$ sites are free, regardless of their positions), the available volume is given by[15]

$$V' = V(1 - \tau\eta) \qquad (2)$$

which is the result of Kralj-Iglic and Iglic[6] or Borukhov et al.[7] However, the hydrated ions are probably not totally rigid (since the hydration shell changes its shape and its water molecules, which are bound to the ion in a transient manner only. To account for this, Ruckenstein and Schiby[4] considered that the ion of volume τv can occupy a site of the water lattice only if the site and its $(\tau - 1)$ neighbors are free (i.e., free of ions and hydration molecules). The probability for a randomly chosen site to be free is $(1 - \tau\eta)$. Considering that the probabilities of occupancy of neighboring sites are independent, the available volume is[4]

$$V' = V(1 - \tau\eta)^{\tau} \qquad (3a)$$

The result can be extended to ions of different sizes τ_i

$$V_i' = (1 - \tau_1\eta_1 - \tau_2\eta_2)^{\tau_i} \quad (i = 1, 2) \qquad (3b)$$

where $\eta_i = n_i/N$, n_i being the number of ions of kind "i". In this model, the "available volume" is different for each kind of ion.

For $\tau = 1$, eq 3a becomes identical to eq 2 and for not too large values of τ ($\tau < 8$) predicts an available volume between those of eqs 1 and 2 and, hence, can be interpreted as corresponding to a partially deformable hydrated ion. However, for large values of τ, the probability of occupation of one site becomes strongly affected by the occupation of their adjacent sites.

The purpose of this paper is to calculate the electrochemical potential and the double layer repulsion using a lattice model, applicable to hydrated ions of different sizes, that accounts for the correlation between the probabilities of occupancy of adjacent sites. As the other lattice models,[4–7] this model accounts only for the steric, excluded volume effects due to ionic hydration. In fact, short-ranged electrostatic interactions between the ions and the dipoles of the water molecules, as well as the van der Waals interactions between the ions and the water molecules, are responsible for the formation of the hydrated ions. The long-ranged interactions between charges are taken into account through an electrostatic (mean field) potential. The correlation between ions is expected to be negligible for sufficiently low ionic concentrations.

II. The Lattice Model

In the model of Ruckenstein and Schiby, the ion can replace a water molecule on the lattice of water only if the site and $(\tau - 1)$ of its first neighbor sites are "free" (i.e., occupied by water molecules which do not hydrate other ions); the probability of occupation of any of the τ sites was taken to be the same.[4] Here we will account to some extent for the correlation between the probabilities of occupancy of adjacent sites, assuming a coordination number w in the water lattice.

The number of sites available to an ion is (in the large N limit) the total number of sites, N, times the probability that a randomly chosen site can be occupied by an ion. This implies that the site and at least $(\tau - 1)$ of its first

(14) Otto, F.; Patey, G. N. *J. Chem. Phys.* **2000**, *113*, 2851.
(15) Ruckenstein, E.; Chi, J. C. *J. Chem. Soc., Faraday Trans. 2* **1975**, *71*, 1690.

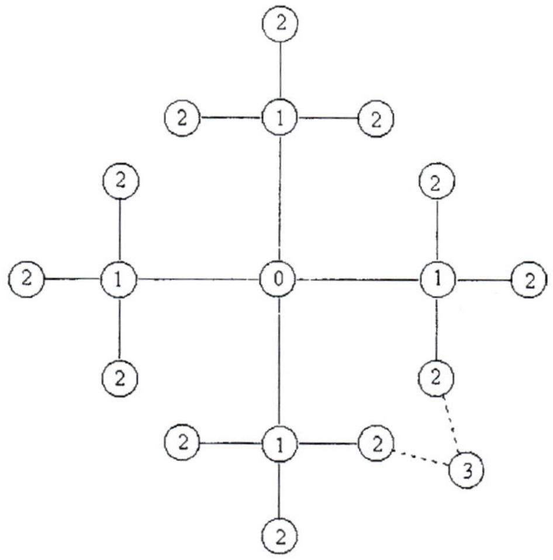

Figure 1. A schematic planar projection of the direct bonding of a water molecule in an ice lattice. A central site (0) has four first-neighbors (1), which are adjacent to the central site and to three other molecules (2). Each of the 12 second neighbors (2) of the central site (0) is adjacent to only one first-neighbor (1) of the central site. Some of the third-order neighbors (3) of the central site (molecules separated through three successive bondings by the central site) are, however, adjacent to two second neighbors. Only one of the third neighbors is drawn.

neighbors are "free" water molecules. It will be assumed that the probability of occupation of any first neighbor (of a central site) is independent of each other. This is obviously not valid for a normal liquid, which has a large number of nearest neighbors. Indeed, at close compaction, each of the 12 first neighbors of a selected site has as first neighbors (is adjacent to) other 3 first neighbors of the same selected site, and hence their occupation probabilities are not independent, because the ions that may occupy the first neighbors compete for hydration molecules. However, for sparse lattices, it is a reasonable approximation. Indeed, for ice (with a coordination number of 4), all the first and second neighbors of the central site are distinct (i.e., none of the first neighbors is adjacent to other first neighbors of the same central site and each of the second neighbors of the central site is adjacent to only one of the first neighbors of the same site; see Figure 1). Each of the third neighbors of the central site is, however, adjacent to two second neighbors of the central site; this leads to a small correlation between the occupancy of the second neighbors, which will be, however, neglected. A molecule of liquid water has between 4.4 and 4.9 first neighbors,[16] which is much closer to the coordination in ice (4) than to that in a normal liquid (~12). We will therefore assume that for a water lattice the probability of occupation of one of the w sites first-neighboring a central site is independent of the occupation of the other $(w - 1)$ sites.

In what follows an expression will be established for the probability for a random site to be available to a hydrated ion, by assuming first that a single kind of ion is present. The chance for a site, chosen randomly (which will be called central site), to be occupied by an ion is η, and the probability for the site to be "free" is

$$p_{f,0} = 1 - \tau\eta \tag{4a}$$

Let us now obtain an expression for the probability for an ion to occupy one site adjacent to a free central one.

(16) Eisenberg, D.; Kauzmann, W. *The structure and properties of water*; Oxford University Press: New York, 1969.

An ion surrounded by w water molecules can be hydrated by $(\tau - 1)$ molecules in $C_w^{\tau-1}$ distinct configurations, where $C_w^{\tau-1} = w!/[(w - \tau + 1)!(\tau - 1)!]$ is the number of combinations in which $(\tau - 1)$ objects (water molecules hydrating the ion) can be selected out of w objects (available water molecules). However, because one of the adjacent molecules (the central site) is "free", there are only $C_{w-1}^{\tau-1}$ configurations possible for the hydrated ion to sit on the lattice (the number of combinations in which $(\tau - 1)$ hydrating water molecules can be chosen out of the remaining $(w - 1)$ available water molecules).

It will be assumed that all the possible configurations of the ion hydrated by the $(\tau - 1)$ water molecules are equally probable and that the number of hydrated configurations of an ion is not affected by its concentration. The last approximation is valid only for small ion concentrations and is reasonable because the density of ions is in general much smaller than unity; only the density of hydrated ions $(\tau\eta)$ can become comparable to unity.

The probability for an ion to occupy a site adjacent to a free central one $(p_{oi,1})$ is therefore

$$p_{oi,1} = \eta \frac{C_{w-1}^{\tau-1}}{C_w^{\tau-1}} = \eta \frac{w - \tau + 1}{w} \tag{4b}$$

and consequently, the probability for the site to be not occupied by an ion $(p_{fi,1})$ is given by

$$p_{fi,1} = 1 - \eta \frac{w - \tau + 1}{w} \tag{4c}$$

We will now calculate the probability for a water molecule, first-neighboring a free central site, not to participate in the hydration of an ion. Since one of its neighbors (the central site) is a free water molecule, it can hydrate only an ion occupying one of the remaining $(w - 1)$ adjacent sites (which are second-neighbors of the central site). Even if an ion occupies one of these sites, it can be hydrated by $(\tau - 1)$ out of the other $(w - 1)$ of its first-neighbors, which are third-order-neighbors of the central site (the third-order-neighbors are connected to the central site through three successive bonds, regardless of their mutual angles). Since there are $C_{w-1}^{\tau-1}$ configurations in which the ion can be hydrated by the water molecules located in the third-neighboring sites of the central site, out of a total of $C_w^{\tau-1}$ configurations of possible hydration of an ion, the probability for an ion to occupy a site second neighbor to a central site, and not to be hydrated by the molecule occupying the first-neighboring site, is

$$p_{oi,2} = \eta \frac{C_{w-1}^{\tau-1}}{C_w^{\tau-1}} = \eta \frac{w - \tau + 1}{w} \tag{4d}$$

Therefore, a water molecule first-neighboring a free central site does not hydrate an adjacent ion, located on a second-neighboring site of the central site, in the following two circumstances: (1) there are no ions occupying any of the $(w - 1)$ sites, and (2) there is one or more ions occupying some or all of the $(w - 1)$ positions, ions which are hydrated by water molecules located on the third-order-neighboring sites of the central site. Since the probabilities of occupation of the first neighbors of each site are assumed independent of each other, the probability for none of the $(w - 1)$ sites to be occupied by ions is $(1 - \eta)^{(w-1)}$ and the probability for an ion to occupy one site and the other $(w - 2)$ sites to be free of ions is $\eta(1 - \eta)^{(w-2)}$. Because there are $(w - 1)$ modes in which an ion can occupy one of the $(w - 1)$ sites, the probability for an ion to occupy one of the $(w-1)$ sites and to be not hydrated by the water molecules first neighboring the central site is $(w-1)\eta(1-\eta)^{w-2}[(w-\tau+1)/w]$. Adding the probabilities for ions to occupy two, three, ..., $(w-1)$ sites, none of them being hydrated by the water molecules adjacent to the central free site, one obtains the following expression for the probability for a water molecule, first neighboring the central free site, not to hydrate any ion

$$p_{\mathrm{fh},1} = (1-\eta)^{w-1} + (w-1)\eta(1-\eta)^{w-2}\left(\frac{w-\tau+1}{w}\right) + \ldots + \eta^{w-1}\left(\frac{w-\tau+1}{w}\right)^{w-1} = \sum_{k=0}^{w-1} C_{w-1}^{\ k}\eta^{k}(1-\eta)^{w-1-k}\left(\frac{w-\tau+1}{w}\right)^{k} \quad (4e)$$

Consequently, the probability for one neighbor of the central site neither to be occupied by an ion nor to participate in the hydration of another ion is given by

$$p_{\mathrm{f},1} = p_{\mathrm{fi},1}p_{\mathrm{fh},1} \quad (4f)$$

Hence, the probability for all w first neighbors of the central site to be free is $(p_{\mathrm{f},1})^{w}$ and the probability for $(w-1)$ to be free and one not free (i.e., either occupied by an ion or hydrating another ion) is $w(p_{\mathrm{f},1})^{w-1}(1-p_{\mathrm{f},1})$ (there are w places that can be occupied by a nonfree water molecule). The last term of the sum is the probability for $(\tau-1)$ water molecules to be free and $(w-\tau+1)$ to be occupied, which is $C_w^{\ \tau-1}p_{\mathrm{f},1}^{\ \tau-1}(1-p_{\mathrm{f},1})^{w-\tau+1}$. Therefore, the probability for a central site (chosen randomly) and at least $(\tau-1)$ of its first neighbors to be "free" is given by

$$P = p_{\mathrm{f},0}(p_{\mathrm{f},1}^{\ w} + wp_{\mathrm{f},1}^{\ w-1}(1-p_{\mathrm{f},1}) + \ldots + C_w^{\ \tau-1}p_{\mathrm{f},1}^{\ \tau-1}(1-p_{\mathrm{f},1})^{w-\tau+1}) \quad (4g)$$

When two ionic species, of hydrated sizes τ_1 and τ_2 are present, the probabilities for a randomly chosen central site to be free is given by

$$p_{\mathrm{f},0} = 1 - \tau_1\eta_1 - \tau_2\eta_2 \quad (5a)$$

the probability for a first-neighboring site not to be occupied by an ion by

$$p_{\mathrm{fi},1} = 1 - \eta_1\frac{w-\tau_1+1}{w} - \eta_2\frac{w-\tau_2+1}{w} \quad (5b)$$

and the probability for a water molecule from a first-neighboring site of a central free one not to participate in the hydration of an ion by

$$p_{\mathrm{fh},1} = (1-\eta_1-\eta_2)^{w-1} + (w-1)(1-\eta_1-\eta_2)^{w-2} \times \left[\eta_1\left(\frac{w-\tau_1+1}{w}\right) + \eta_2\left(\frac{w-\tau_2+1}{w}\right)\right] + \ldots \quad (5c)$$

With $p_{\mathrm{f},1} = p_{\mathrm{fi},1}p_{\mathrm{fh},1}$, the probability for a random central site to be available to an ion is

$$P_i = p_{\mathrm{f},0}(p_{\mathrm{f},1}^{\ w} + wp_{\mathrm{f},1}^{\ w-1}(1-p_{\mathrm{f},1}) + \ldots + C_w^{\ \tau_i-1}p_{\mathrm{f},1}^{\ \tau_i-1}(1-p_{\mathrm{f},1})^{w-\tau_i+1}) \quad (5d)$$

$$(i = 1, 2)$$

where the volumes available to each species, $V_i' = P_iV$, are different $(\tau_1 \neq \tau_2)$.

III. The Excess Chemical Potential

The chemical potential for a fluid of n identical particles interacting with a hard-core potential (which is zero at separations larger than the distance of closest approach and infinity for smaller separations) can be calculated using the equation

$$\mu = -kT\left(\frac{\partial \ln Z}{\partial n}\right)_{V,T} \quad (6a)$$

where k is the Boltzmann constant, T the absolute temperature, and Z the partition function

$$Z = \langle z\rangle^{n}/n! \quad (6b)$$

In eq 6b, $\langle z\rangle$ represents the individual partition function, averaged over all possible configurations

$$\langle z\rangle = \frac{1}{h^3}z'\int\int\int \exp\left(-\frac{p^2}{2mkT}\right)\mathrm{d}^3p \times \left\langle\int\int_V\int \exp\left(-\frac{U}{kT}\right)\mathrm{d}^3x\right\rangle = \frac{z'V'}{\Lambda^3} \quad (6c)$$

where z' accounts for the internal degree of freedom of the particle, h is the Planck constant, and Λ is the thermal de Broglie wavelength. The individual partition function is proportional to the available volume V', which is the volume available to a particle, out of the total volume V, when the other $(n-1)$ particles occupy nonoverlapping positions. Using eqs 6a–c, one obtains for the chemical potential the expression

$$\mu = -kT\ln\left(\frac{z'}{\Lambda^3}\right) + kT\ln\frac{n}{V'} = \mu_0 + kT\ln\left(\frac{n}{V'}\right) = \mu_{\mathrm{id}} + \mu_{\mathrm{excees}} \quad (6d)$$

where

$$\mu_{\mathrm{id}} = -kT\ln\left(\frac{z'}{\Lambda^3}\right) + kT\ln\frac{n}{V} \quad (6e)$$

is the chemical potential in an ideal mixture and

$$\mu_{\mathrm{excess}} = kT\ln(V/V') \quad (6f)$$

is the excess chemical potential, due to the hard-core interactions. Since for a lattice the available volume is $V' = PV$, where P is the probability for a randomly chosen site to be occupied by an ion, eq 6f becomes

$$\mu_{\mathrm{excess}} = -kT\ln P \quad (6g)$$

The excess chemical potential for a dilute hard-sphere fluid in the Paunov et al. approximation (eqs 1 and 6f) is given by[5]

$$\mu_{\mathrm{excess}}^{\ \mathrm{P}} = -kT\ln(1-8\tau\eta) \quad (7a)$$

for a lattice model in which each ion occupies one site (Iglic–Iglic model) (eqs 2 and 6f) by[6,7]

$$\mu_{\mathrm{excess}}^{\ \mathrm{II}} = -kT\ln(1-\tau\eta) \quad (7b)$$

for the Ruckenstein and Schiby lattice model (eqs 3 and 6f) by[4]

$$\mu_{\text{excess}}^{\text{RS}} = -\tau kT \ln(1 - \tau\eta) \tag{7c}$$

and for the present lattice model (eqs 4e and 6f) by

$$\mu_{\text{excess}}^{\text{MR}} = -kT \ln[p_{f,0}(p_{f,1}{}^{w} + wp_{f,1}{}^{w-1}(1 - p_{f,1}) + ... + C_w^{\tau-1} p_{f,1}{}^{\tau-1}(1 - p_{f,1})^{w-\tau+1})] \tag{7d}$$

The excess chemical potential of a hard-sphere fluid can be calculated on the basis of an equation of state using the expressions[17]

$$A_{\text{excess}} = nkT \int_0^{\tau\eta} \left(\frac{\frac{pV}{nkT} - 1}{\tau\eta} \right) d(\tau\eta)$$

$$\mu_{\text{excess}} = \frac{1}{n} \left(A_{\text{excess}} + kT\tau\eta \frac{\partial A_{\text{excess}}}{\partial \tau\eta} \right) \tag{8}$$

where p is the pressure, A_{excess} represents the excess Helmholtz free energy, and $\tau\eta$ is the fraction of the volume occupied by the hydrated ions. Using the Carnahan–Starling equation of state[12]

$$\frac{pV}{nkT} = \frac{1 + \tau\eta + (\tau\eta)^2 - (\tau\eta)^3}{(1 - \tau\eta)^3} \tag{9}$$

the excess chemical potential becomes

$$\mu_{\text{excess}}^{\text{CS}} = kT \frac{\tau\eta(8 - 9\tau\eta + 3(\tau\eta)^2)}{(1 - \tau\eta)^3} \tag{7e}$$

In Figure 2 the excess chemical potentials provided by the models listed above (eqs 7a–e) are compared. We used $w = 5$, which is the closest integer to the estimate of the number of first neighbors in water (4.4–4.9),[16] and (a) $\tau = 6$, (b) $\tau = 5$, and (c) $\tau = 4$. A more realistic model (but more tedious to calculate) could have been obtained by the statistical averaging of systems with different values of w and τ.

When the ion is hydrated by all the adjacent molecules ($w = \tau - 1$), there is a strong lattice-site exclusion effect at high ionic concentrations, since the central site can be available to an ion only when both its first and second neighbors are free of ions (a total of $1 + w + w(w - 1) = w^2 + 1$ sites). On the other hand, when the hydration number ($\tau - 1$) is low, almost any "free" site can be occupied by an ion, since in this case there is a high chance to find at least ($\tau - 1$) free water molecules around a selected site. In the next section it will be shown that there is a strong dependence of the site-exclusion effect on the hydration number and, hence, that specific ion effects can be important in the double layer interactions.

IV. Excluded-Site Effect on the Double Layer Repulsion

IV.1. General Equations. At thermodynamic equilibrium, the electrochemical potential, (eq 6d) is constant in the system

$$\mu_i^0 + kT \ln(n_i(x)/V) - kT \ln P_i(n_1(x), n_2(x), \psi(x)) + q_i\psi(x) = \text{constant} \tag{10}$$

$$(i = 1, 2)$$

where μ_i^0 is the standard chemical potential of species i, q_i is the charge of an ion of species i ($q_i = (-1)^{i+1}e$, with e the protonic charge), ψ is the mean potential, and x

(17) Khoshbarchi, M. K.; Vera, J. H. *Fluid Phase Equilib.* **1998**, *142*, 131.

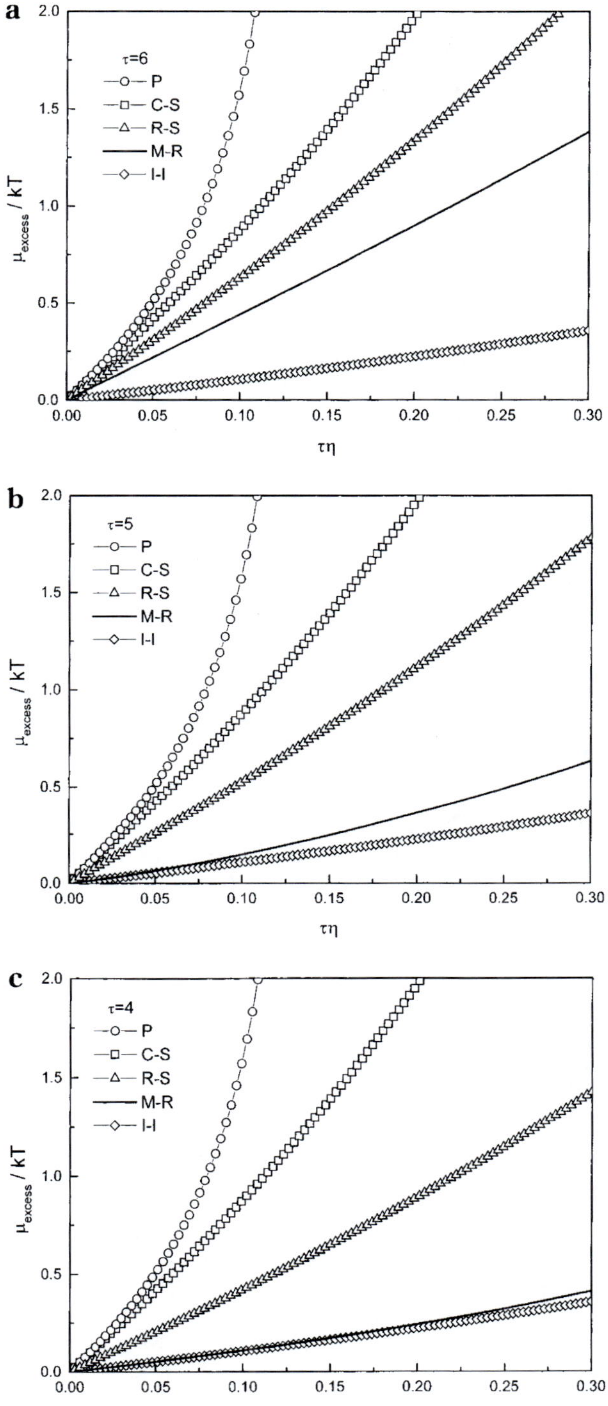

Figure 2. The excess chemical potential relative to that of an ideal solution, for the models discussed in the text (eqs 7a–e) and (a) $\tau = 6$, (b) $\tau = 5$, and (c) $\tau = 4$.

denotes the distance from one of the two charged, planar surfaces separated by a distance l. It will be assumed that the surface charge and potential are negative; hence the subscript 1 denotes the counterions and the subscript 2 is for co-ions. Consequently, the concentration of each species of ions for a 1:1 electrolyte is given by

$$n_i(x) = n\frac{P_i(x)}{P_i(x=\infty)}\exp\left(-\frac{q_i\psi(x)}{kT}\right) \qquad (i = 1, 2) \tag{11a}$$

where $n_1 = n_2 = n$ for $x = \infty$.

A modified Poisson–Boltzmann equation is obtained by replacing in the Poisson equation for parallel plates

$$\frac{\partial^2\psi(x)}{\partial x^2} = \frac{e(n_1(x) - n_2(x))}{\epsilon\epsilon_0 V} \tag{11b}$$

n_1 and n_2 by their expressions (11a), $\epsilon\epsilon_0$ being the dielectric constant of water. The above equation has to satisfy the boundary conditions $\partial\psi(x)/\partial x|_{x=l/2} \equiv 0$ at the middle distance between the plates, and either $\psi_S = \psi(x=0) =$ const (for constant surface potential) or $\partial\psi(x)/\partial x|_{x=0} =$ const (for constant surface charge density), where ψ_S represents the surface potential.

Caution should be taken when calculating the double-layer force between two parallel plates. It is clear that the force is not proportional to the excess concentration of ions at the middle distance (with respect to the concentration of ions at infinity), since this Langmuir equation involved the assumption of ions of negligible sizes. We will use instead the procedure introduced by Verwey and Overbeek,[18] which is based on general thermodynamic principles, and does not imply the Boltzmann distribution of ions.[19] The force, per unit area, between two parallel plates separated by a distance l is given by

$$p_\psi(l) = 2\frac{\partial}{\partial l}\int_0^{\psi_S}\sigma(\psi)\,d\psi \tag{12a}$$

at constant surface potential ψ_S and

$$p_\sigma(l) = -2\frac{\partial}{\partial l}\int_0^{\sigma_S}\psi(\sigma)\,d\sigma \tag{12b}$$

at constant surface charge density σ_S, where the surface charge density is related to the surface potential through the expression

$$\frac{\partial\psi(x)}{\partial x}\Big|_{x=0} = -\frac{\sigma}{\epsilon\epsilon_0} \tag{13}$$

The relation between the surface charge density and surface potential, as a function of the distance l, is obtained using the additional boundary condition required by symmetry $\partial\psi(x)/\partial x|_{x=l/2} = 0$. Consequently, eqs 12 can be integrated numerically to provide the value of the double layer force.

IV.2. Distribution of Ions near a Charged Surface. The distribution of counterions in the vicinity of a single charged surface, at a high surface potential ($\psi_s = 0.1$, Figure 3a) is represented as a function of the distance x from the surface for $w = 5$, $\tau_1 = 6$, and $\tau_2 = 2$ (large counterions) and for $\tau_1 = 2$, $\tau_2 = 6$ (small counterions). In both cases, the densities of counterions and co-ions in the vicinity of the surface are smaller than those predicted by the Poisson–Boltzmann equation, since the agglomeration

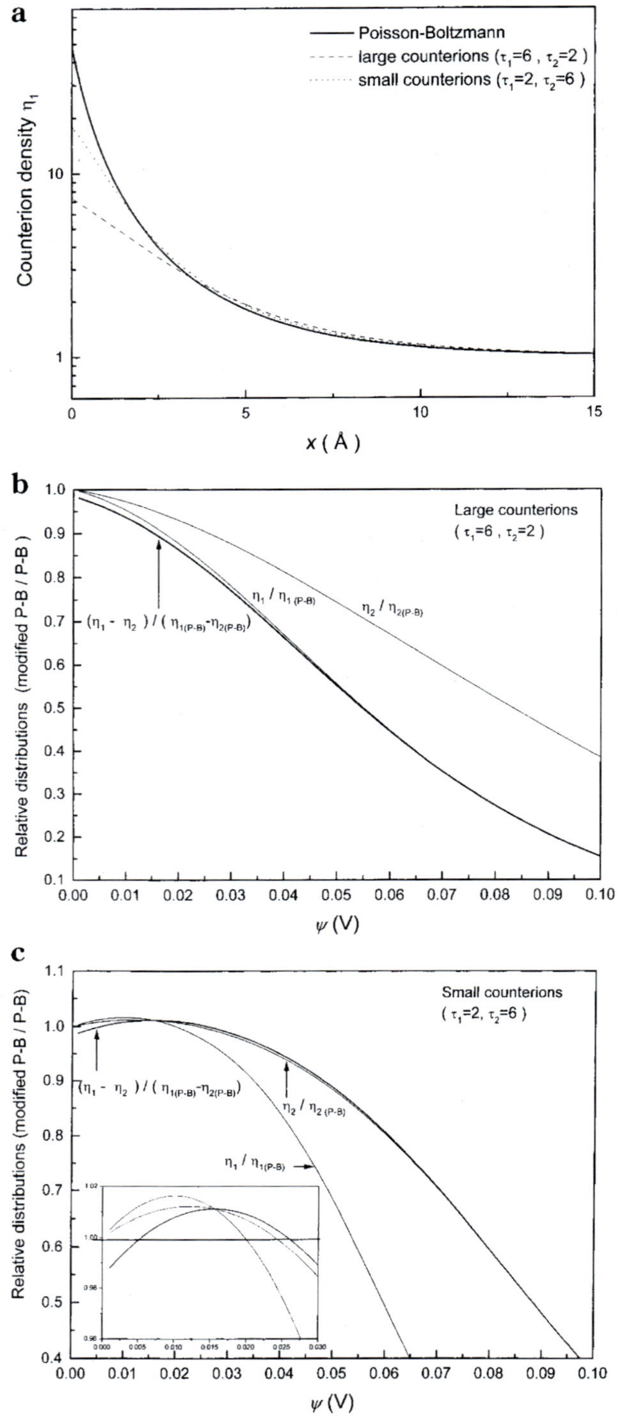

Figure 3. (a) Ion distributions in the vicinity of a single charged surface at high potentials ($\psi_S = 0.1$ V), for large ($\tau_1 = 6$, $\tau_2 = 2$) and small ($\tau_1 = 2$, $\tau_2 = 6$) counterions: $n/V = 1.0$ M and $T = 300$ K. (b) Ion and charge distributions (relative to those predicted by the Boltzmann distribution) as functions of the mean potential. Large counterions ($\tau_1 = 6$, $\tau_2 = 2$): $n/V = 1.0$ M and $T = 300$ K. (c) Ion and charge distributions (relative to those predicted by the Boltzmann distribution) as functions of the mean potential. Small counterions ($\tau_1 = 2$, $\tau_2 = 6$): $n/V = 1.0$ M and $T = 300$ K.

(18) Verwey, E. J. W.; Overbeek, J. Th. G. *Theory of stability of lyophobic colloids*; Elsevier: Amsterdam, 1948.

(19) For a detailed discussion, see: Ruckenstein, E. *Adv. Colloid Interface Sci.* **1998,** *75*, 169.

of the finite-size counterions near the surface reduces the available volume.

Let us examine now the effect of the excluded volume at low surface potentials. In the linear approximation of the Poisson–Boltzmann expression, the increase in the number of counterions in the vicinity of the interface equals the decrease in the number of co-ions. If the co-ions have a larger size, one expects the available volume near the surface to be larger than that in the bulk. As a result, a concentration of ions in excess to that predicted by the Poisson–Boltzmann equation is expected to occur in the vicinity of the surface, when the volume exclusion is taken into account.

Of course, when the potential of the surface becomes sufficiently large, the linear approximation fails, and the (exponential) increase in the counterions density in the vicinity of the surface predicted by the Poisson–Boltzmann equation largely exceeds the depletion of co-ions and the available volume is expected to become smaller than that in the bulk.

To illustrate this, let us use the simple approximation suggested by Bikerman[2]

$$n_i(\psi) = n\,\frac{\left(1 - \dfrac{n_1(\psi)v_1 + n_2(\psi)v_2}{V}\right)}{\left(1 - n\dfrac{v_1 + v_2}{V}\right)}\exp\left(-\frac{q_i\psi}{kT}\right) \quad (14)$$

(where v_i is the volume of the hydrated ions), which combined with Poisson equation becomes, in the linear approximation

$$\frac{\partial^2\psi}{\partial x^2} = \frac{2ne^2}{\epsilon\epsilon_0 kT}\psi\left(\frac{1}{1 + \dfrac{n(v_1 - v_2)}{V}\dfrac{e\psi}{kT}}\right) = \frac{1}{\lambda^2}\left(\frac{1}{1 + \dfrac{n(v_1 - v_2)}{V}\dfrac{e\psi}{kT}}\right)\psi = \frac{1}{\lambda'^2}\psi \quad (15)$$

When $v_1 > v_2$, the effective Debye–Hückel length λ' (which now depends on $\psi(x)$) is larger than that obtained for the Poisson–Boltzmann equation. Consequently, the diffuse double layer is larger in the vicinity of a charged surface, as predicted earlier.[4–7,9] However, when $v_2 > v_1$ (small counterions), $\lambda' < \lambda$ and the diffuse double layer is compressed. The effect is proportional to the ionic strength and is, in general, small for typical electrolyte concentrations, since $|n(v_1 - v_2)/V| \ll 1$ and $|e\psi/kT| < 1$.

Let us now calculate the dependence of the ions and charge densities on the mean potential in the present lattice model. For large counterions (Figure 3b) the ratios between counterion and co-ion densities and those calculated with the Poisson–Boltzmann equation are monotonic decreasing functions of the mean potential ψ. For small counterions, however, at low potentials (Figure 3c), the two ratios slightly exceed unity, because the available volume (as compared to V' at $\psi = 0$) increases. The ratio between the local charge density (which is proportional to $\eta_1 - \eta_2$) and that calculated from the Poisson–Boltzmann equation also exceeds slightly unity in a range of potentials. The subunit ratios at very low potentials occur because the available volumes are different for the two types of ions and depend differently on the potential ψ; one of them grows initially slowly and decreases later slowly, while the other grows initially quicker but decreases later also quicker.

IV.3. Modification of the Double Layer Repulsion Due to the Finite Volume of Ions. In the linear approximation of the Poisson–Boltzmann equation, the potential between two surfaces, separated by the distance l is given by

$$\psi(x) = \psi_S\frac{\cosh\left(\dfrac{x}{\lambda} - \dfrac{l}{2\lambda}\right)}{\cosh\left(\dfrac{l}{2\lambda}\right)} \quad (16)$$

and eq 13 yields

$$\sigma_S = \frac{\epsilon\epsilon_0\psi_S}{\lambda}\tanh\left(\frac{l}{2\lambda}\right) \quad (17)$$

The double layer forces obtained from eqs 12, 16, and 17 are

$$p_\psi(l) = \frac{\epsilon\epsilon_0{\psi_S}^2}{2\lambda^2\cosh^2\left(\dfrac{l}{2\lambda}\right)} \quad (18a)$$

at constant surface potential and

$$p_\sigma(l) = \frac{{\sigma_S}^2}{2\epsilon\epsilon_0\sinh^2\left(\dfrac{l}{2\lambda}\right)} \quad (18b)$$

at constant surface charge.

At constant surface charge, an increase in the Debye length implies an increase in the repulsion at any distance. However, this is not true for constant surface potential, since the function $(l/2\lambda)^2\cosh^{-2}(l/2\lambda)$ (for fixed l) has a maximum at $l/2\lambda \cong 1.2$. Consequently, an increase in the effective Debye length corresponds to an increase in repulsion only at large separations ($l > 2.4\lambda$) but to a decrease in repulsion at smaller separations.

At sufficiently high surface charges or potentials, the thickness of the double layer (the effective Debye length) increases when the finite volumes of the ions are taken into account. The replacement of λ by a larger value in eq 18b leads to a higher repulsion at any separation distance. This suggests that the accounting of the ion sizes will lead to a higher repulsion at a high constant surface charge. Figure 4a ($n/V = 0.01$ M, σ_S = constant = 0.32 C/m^2) confirms this expectation. The replacement of λ with a larger value in eq 18a suggests that, at high constant surface potentials, the double layer interaction will be decreased at short separation distances by the finite sizes of the ions. This expectation is confirmed by Figure 4b ($n/V = 0.01$ M, ψ_S = constant = 0.1 V). The modification of the double layer repulsion due to the sizes of the ions is, however, much smaller when the counterions are smaller and is strongly dependent on the hydration number, when the latter has a value near to w.

At low surface charges, the effective Debye length can either increase or decrease, depending on the relative sizes of the two species of ions. Consequently, the double layer repulsion might become smaller than that predicted by the Poisson–Boltzmann equation not only at constant surface potential but also at constant surface charge. In Figure 5, the double layer repulsion (relative to that predicted by Poisson–Boltzmann equation) is presented for (a) a small constant surface potential $\psi_S = 0.02$ V and (b) a small constant surface charge density $\sigma_S = 0.032$ C/m^2. Even at the relatively high ionic strength selected ($n/V = 1.0$ M), the decrease of the repulsion is small.

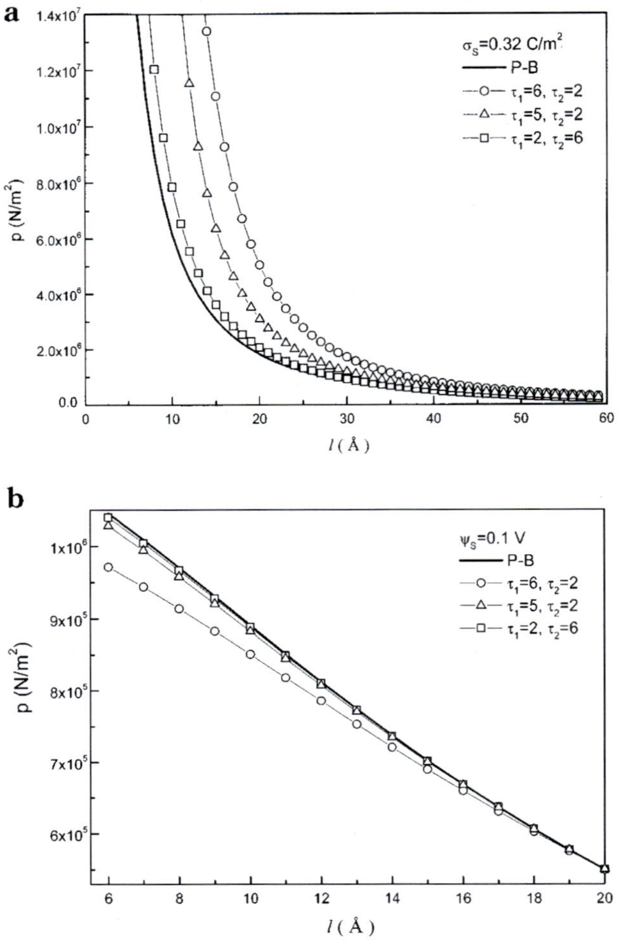

Figure 4. (a) Double layer force per unit area for constant surface charge density ($\sigma_S = 0.32$ C/m^2), $n/V = 0.01$ M and $T = 300$ K. (b) Double layer force per unit area for constant surface potential ($\psi_S = 0.1$ V), $n/V = 0.01$ M and $T = 300$ K.

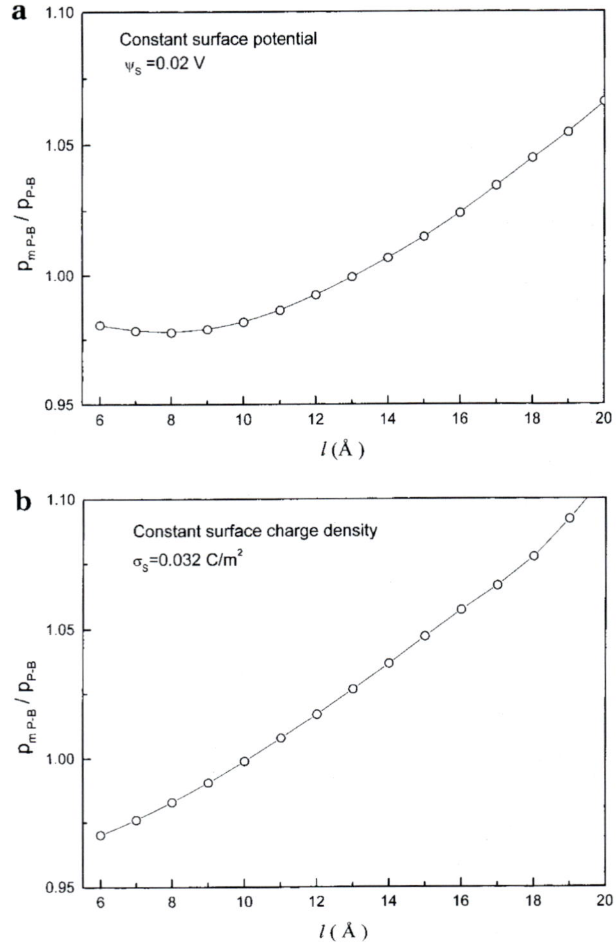

Figure 5. Ratio between the double layer force with site exclusion (modified Poisson–Boltzmann) and the double layer force provided by the Poisson–Boltzmann equation: (a) constant (small) surface potential ($\psi_S = 0.02$ V); (b) constant (small) surface charge density ($\sigma_S = 0.032$ C/m^2), $n/V = 1.0$ M and $T = 300$ K.

V. Conclusions

The lattice model of Ruckenstein and Schiby,[4] which accounted for the sizes of the ions, was extended to account for the correlations between the probabilities of occupancy of neighboring sites. In the present model, the excess chemical potential depends strongly on the number of water molecules that hydrate an ion; hence slightly different hydration numbers can lead (at high potentials) to large differences in the double layer repulsion. It was also shown that, at high surface potentials, the thickness of the double layer near a charged surface is increased (compared to that predicted by the Poisson–Boltzmann equation). However, at low surface potentials the diffuse layer might be slightly compressed, when the co-ions are larger than the counterions. The increase in the effective Debye length (thickness of the double layer) leads to an increase in the double layer repulsion at constant surface charge at any distance but to a slight decrease of the double layer repulsion at constant surface potential and small separation distances and an increase at large separations.

LA020194R

Langmuir **2001,** *17,* 2455–2463 2455

Free Energy and Thermal Fluctuations of Neutral Lipid Bilayers

Marian Manciu and Eli Ruckenstein*

Department of Chemical Engineering, State University of New York at Buffalo, Buffalo, New York 14260

Received November 27, 2000. In Final Form: January 31, 2001

A new method is proposed to calculate the free energy of lamellar liquid crystals lipid bilayers/water. The root-mean-square fluctuation of the distance between two neighboring bilayers is calculated for a number of distributions by minimizing the total free energy. Analytical solutions for the free energy are derived for a Gaussian distribution of distances, which are compared with numerical solutions obtained for more realistic distributions, which account for the correlation between the fluctuations of neighboring bilayers. Calculations are performed for typical values of the interaction parameters, and the comparison with experiment provides a more than qualitative agreement. The calculations also show that at infinite separation distance there is a minimum of zero free energy and that a local minimum can occur at a finite distance, which can be stable or unstable. The two minima are separated by a potential barrier. An unbinding transition occurs when the free energies of the two minima are equal to zero.

I. Introduction

It is well-known that the lipid molecules placed in water form multilamellar vesicles, which can be regarded locally as smectic liquid crystals, with stacks of bilayers aligned at constant separation, along a director axis. An important problem is to calculate the free energy of the system, from which one could derive the equilibrium separation distance and the corresponding binding free energy. It was recently noted[1,2] that the experimental data for both the interaction force per unit area (pressure) and the root-mean-square fluctuation of the distance between two neighboring bilayers as functions of the average separation between them are not satisfactorily explained by the existing theories.[3–5]

For neutral bilayers, there are no long-range double-layer forces which, coupled with the van der Waals attraction, could explain the stability of the lamellar structure. At small separations, the required repulsion is provided by the hydration force, which was investigated both experimentally[6–8] and theoretically.[9,10] However, it was experimentally observed that the lipid bilayers could be swollen in water up to very large interlayer distances,[11] where the short-range exponential hydration repulsion becomes negligible.

Helfrich was the first to suggest[3] that the entropic confinement (due to neighboring layers) of thermally excited out-of-plane undulations of the interfaces of the lipid bilayers (lamellae) leads to a long-range repulsion force, inverse proportional to the third power of the distance, which could be responsible for the hyperswelling of the bilayers. His original method was to assume a linear interaction between lamellae, dependent on an unknown elastic constant B. Because of the linearity, the oscillation modes are not coupled, and this allowed him to compute the free energy as a sum over the free energies of the individual modes. He showed that the unknown elastic constant could be obtained in a self-consistent manner, either from the bulk properties of the multilayer system or from the root-mean-square fluctuation σ_b of the bilayer positions.[3]

Since at large distances the van der Waals forces are inversely proportional to the fifth power of the distance, Helfrich showed that the entropic confinement repulsion would always dominate at large enough separations[3] and, consequently, that the separation between the bilayers can extend at infinity. Experiments appeared to confirm this conclusion.[11]

Some other experiments, however, indicated that the separation between membranes remained finite even in the presence of excess water.[1,12] A possible explanation could be that the Helfrich theory, which involves in the calculation of the entropic confinement only a hard-wall (steric) repulsion, does not account for other forces between membranes.

The role of undulation on the equilibrium of lipid bilayers was also examined by Lipowsky and Leibler,[13] who used a nonlinear functional renormalization group approach, and by Sornette,[14] who employed a linear functional renormalization approach. It was theoretically predicted that a critical unbinding transition (corresponding to a transition from a finite to an infinite swelling) can occur by varying either the temperature or the Hamaker constant. However, the renormalization group procedures do not offer quantitative information about the systems, when they are not in the close vicinity of this critical point.

* To whom correspondence may be addressed. E-mail address: feaeliru@acsu.buffalo.edu. Phone: (716) 645 2911/2214. Fax: (716) 645 3822.

(1) Petrache, H. I.; Gouliaev, N.; Tristram-Nagle, S.; Zhang, R.; Sutter, R. M.; Nagle, J. F. *Phys. Rev. E* **1998**, *57*, 7014.
(2) Gouliaev, N.; Nagle, J. F. *Phys. Rev. E* **1998**, *58*, 881.
(3) Helfrich, W. Z. *Naturforsch.* **1978**, *33a*, 305.
(4) Evans, E. A.; Parsegian, V. A. *Proc. Natl. Acad. Sci. U.S.A.* **1986**, *83*, 7132.
(5) Podgornik, R.; Parsegian, V. A. *Langmuir* **1992**, *8*, 557.
(6) Le Neveau, D. M.; Rand, R. P.; Parsegian, V. A.; Gingell, D. *Biophys. J.* **1977**, *18*, 209.
(7) Parsegian, V. A.; Fuller N.; Rand, R. P. *Proc. Natl. Acad. Sci. U.S.A.* **1979**, *76*, 2750.
(8) Rand, R. P.; Parsegian, V. A. *Biochim. Biophys. Acta* **1989**, *988*, 351.
(9) Marcelyia, S.; Radic, N. *Chem. Phys. Lett.* **1976**, *42*, 129.
(10) Schiby, D.; Ruckenstein, E. *Chem. Phys. Lett.* **1983**, *95*, 435.
(11) Harbich, W.; Helfrich, W. *Chem. Phys. Lipids* **1984**, *36*, 39.
(12) Lis, L. J.; McAlister, M.; Fuller N.; Rand, R. P.; Parsegian, V. A. *Biophys. J.* **1982**, *37*, 657.
(13) Lipowsky, R.; Leibler, S. *Phys. Rev. Lett.* **1986**, *56*, 2541.
(14) Sornette, D. *J. Phys. C: Solid State Phys.* **1987**, *20*, 4695.

10.1021/la0016266 CCC: $20.00
Published on Web 03/21/2001

The effect of the nonlinear interaction forces between the lipid bilayers on the undulation was investigated by several groups.[4,15] An important advance was made in the perturbative theory of "soft confinement" of Podgornik and Parsegian.[5] In their work, the nonlinear interactions between bilayers (due to DLVO forces plus hydration repulsion) was considered as a small perturbation to a hard wall (steric) confinement. Using a variational method due to Feynman,[16] they obtained an equation for the unknown interaction constant B of an hypothetical harmonic potential, as a function of the separation distance a, by minimizing the free energy with respect to both σ_b and B. The quantity σ_b represents the root-mean-square fluctuation of the distribution of the positions of membranes, $\rho(u)$, u being the displacement of a point of the membrane surface from the average position, located at half-distance between its two neighboring bilayers, assumed rigid. A Gaussian distribution of membrane positions was employed for averaging the interaction energy in the partition function, and then σ_b^2 was calculated by assuming that $\rho(u)$ should verify a diffusion-type equation, with the boundary conditions $\rho(u) = 0$ at $u = \pm a$. The latter distribution was employed because the long tails of the Gaussian will lead to a divergent energy for the hard (rigid wall) confinement, which is their unperturbed state.

The purpose of this article is to present a variational analysis which no longer considers that the fluctuating bilayer is confined by two rigid walls. A Gaussian, a truncated Gaussian, or an asymmetrical Gaussian distribution is assumed for the fluctuating distances between two neighboring bilayers.

In the case of a linear interaction between neighboring lipid bilayers, Helfrich has demonstrated that the repulsive free energy due to confinement is inversely proportional to σ_b^2. While this result is strictly valid for a harmonic interaction potential (linear force), we assume that it can be extended to any interaction. We will examine later under what conditions this approximation is accurate.

The free energy of a bilayer in a given potential is composed of the energy of interaction with the other membranes and an entropic term, due to the bilayer confinement, which is inverse proportional to σ_b^2. The energy is increased by a distribution which has a large value of σ_b, because then a large part of the bilayer is in a region of stronger repulsive interactions; on the other hand the entropic term decreases with increasing σ_b. Consequently, there is an optimum σ_b for which the free energy is minimum.

Instead of computing via optimization the hypothetical elastic constant B corresponding to the interaction potential of the problem, as in the work of Podgornik and Parsegian,[5] we express B in terms of σ_b via the equipartition principle and calculate σ_b which minimizes the free energy. In addition, whereas in the Podgornik and Parsegian work, the calculation involved a small perturbation of a rigid wall confinement, in the present approach the hard wall confinement is not used as a base state. However, while the confining bilayers are not considered rigid, the exclusion of permeation between two neighboring membranes provides indirectly such a confinement. Most importantly, the present approach can employ an asymmetrical distance distribution for the fluctuating distances between two neighboring bilayers,[17] which cannot be used when one assumes that a bilayer is confined by two rigid walls. The former distribution is expected, because the fluctuations of neighboring membranes are increasingly correlated, as the distance between them becomes smaller.

II. Free Energy of Entropic Confinement for a Linear Interaction

Following Helfrich,[3,18] one can compute the free energy of confinement, starting from the Hamiltonian of the confined bilayer

$$H = \int \mathrm{d}x\, \mathrm{d}y \left(\frac{1}{2} K_c \left(\frac{\partial^2 u(x,y)}{\partial x^2} + \frac{\partial^2 u(x,y)}{\partial y^2} \right)^2 + \frac{1}{2} B u^2(x,y) \right) \tag{1}$$

where K_c is the bending modulus of a bilayer, B is an unknown constant of the assumed linear interaction force between bilayers, and $u(x,y)$ denotes the displacement along z from the average position $z = 0$ of a point of a bilayer of coordinates x and y. Denoting by $\tilde{u}(q_x,q_y)$ the Fourier transform of $u(x,y)$, the partition function acquires the form

$$Z = \int \prod \mathrm{d}\tilde{u} \left(\exp\left(-\beta \sum_{q_x,q_y} \frac{1}{2} \tilde{u}^2 (K_c \vec{q}^4 + B) \right) \right) = \prod_{q_x,q_y} \left(\frac{2\pi kT}{K_c q^4 + B} \right)^{1/2} \tag{2}$$

where $q^2 = q_x^2 + q_y^2$.

The difference in the free energies of the free bilayer (which formally corresponds to $B \rightarrow 0$) and the confined one, per unit area, can be computed by replacing the summation over q_x,q_y with an integral

$$\Delta f = kT \frac{1}{(2\pi)^2} \int_0^\infty \mathrm{d}q\, 2\pi q \ln \left(\frac{K_c q^4 + B}{K_c q^4} \right)^{1/2} = \frac{kT}{8} \left(\frac{B}{K_c} \right)^{1/2} \tag{3}$$

The free energy due to the entropic confinement alone is obtained from eq 3 by subtracting the interaction energy per unit area

$$f_C = \frac{kT}{8} \left(\frac{B}{K_C} \right)^{1/2} - \frac{1}{2} B \langle u^2 \rangle \tag{4}$$

The unknown coupling constant B can be related to the mean-square fluctuation of the bilayer position $\sigma_b^2 = \langle u(x,y)^2 \rangle$ by using the equipartition principle

$$\tilde{u}^2 \left(\frac{1}{2} K_c q^4 + \frac{1}{2} B \right) = \frac{1}{2} kT \tag{5}$$

which leads to the following relation between σ_b^2 and B:

$$\sigma_b^2 = \langle u^2 \rangle = \sum_{q_x,q_y} \tilde{u}^2 = \frac{kT}{(2\pi)^2} \int_0^\infty \frac{2\pi q\, \mathrm{d}q}{K_c q^4 + B} = \frac{kT}{8 (K_c B)^{1/2}} \tag{6}$$

Introducing eq 6 in eq 4, yields:[5,18]

(15) Sornette, D.; Ostrovski, N. *J. Chem. Phys.* **1986**, *84*, 4062.
(16) Feynman, R. *Statistical Mechanics*; Benjamin: Reading, MA, 1972.
(17) Gouliaev N.; Nagle, J. F. *Phys. Rev. Lett.* **1998**, *81*, 2610.
(18) Sornette, D.; Ostrovsky, N. *Micelles, Membranes, Microemulsions and Monolayers*; Gelbart, W. M., Ben-Shaul, A., Roux, D., Eds.; Springer-Verlag: 1994.

$$f_c = \frac{(kT)^2}{128 K_c \sigma_b^2} \qquad (7)$$

a result which is essentially due to Helfrich.[3]

The Helfrich formula for the free energy (eq 7) is strictly valid only for a linear interaction. However, eq 7 constitutes a good approximation, even for nonlinear interactions, as long as the interaction energy between layers is much smaller than the elastic energy of the modes corresponding to bending, hence if $B \ll K_c q^4$. In this case, the coupling between modes is minimal and eq 5 is approximately valid. Using the expression of B from eq 6, the above condition reduces to an inequality for the wavelength λ_{xy} in the xy plane

$$\frac{\lambda_{xy}}{\sigma} \ll 2\pi \left(\frac{8K_c}{kT}\right)^{1/2} \qquad \left(\lambda_{xy} = \frac{2\pi}{q}\right) \qquad (8)$$

which indicates that eq 7 can be used for bilayers that possess a high bending modulus. For lipid bilayers, a typical value of the bending modulus is $K_c = 1.0 \times 10^{-19}$ J,[1] and the right-hand quantity becomes $\approx 10^2$.

III. Thermal Fluctuations for Soft Confinement

As already stated, the free energy due to the entropic confinement remains of the form of eq 7 even for nonlinear interactions, if the inequality (8) is satisfied. To compute the total free energy, one should calculate the pair distribution function of the distances between neighboring fluctuating bilayers. If the distribution of bilayer positions $\rho(u)$ is Gaussian with the root-mean-square σ_b and the undulations of neighboring bilayers are totally uncorrelated, then the distribution of distances z between two neighboring bilayers (see Figure 1) is the convolution of two Gaussians, which is again a Gaussian

$$P(z) = \int_{-\infty}^{\infty}\int_{-\infty}^{\infty} du_1\, du_2 \frac{1}{2\pi\sigma_b^2} e^{-u_1^2/2\sigma_b^2} e^{-u_2^2/2\sigma_b^2} \times \delta(z - (a - (u_1 - u_2))) = \frac{1}{(2\pi)^{1/2}\sigma} e^{-(z-a)^2/2\sigma^2} \qquad (9)$$

where $\sigma = 2^{1/2}\sigma_b$.

Let us denote by V the interaction potential for planar bilayers. Assuming that the fluctuations of adjacent bilayers are totally uncorrelated, and using locally the planar approximation, the free energy per unit area can be written in terms of the distance distribution function $P(z)$

$$f(a,\sigma) = \int dz\, P(z)V(z) + \frac{(kT)^2}{64 K_c \sigma^2} \qquad (10)$$

where $a = \langle z \rangle$ is the average separation between neighboring bilayers and $\sigma^2 \equiv \langle (z - \langle z \rangle)^2 \rangle$ is the root-mean-square fluctuation of the distribution $P(z)$.

The minimum of the free energy with respect to σ for any interaction potential $V(z)$ and all separation distances

$$\partial f(a,\sigma)/\partial\sigma = 0 \qquad (11)$$

provides an expression for σ.

However, the assumption that the fluctuations of neighboring bilayers are totally uncorrelated leads to the unphysical result that they can permeate each other ($P(z) > 0$ for $z < 0$). One can approximately account for the correlation of the motions of the neighboring bilayers as follows.

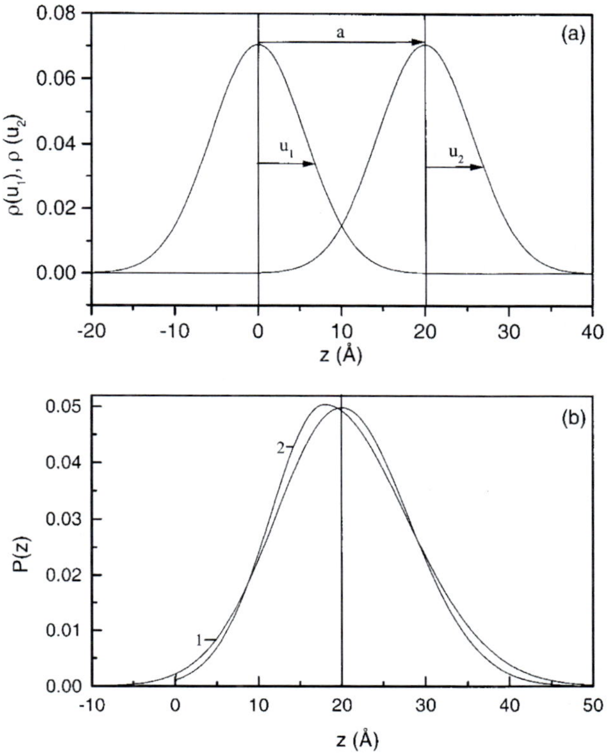

Figure 1. (a) Gaussian distributions of displacements from equilibrium positions, $\rho(u_1)$ and $\rho(u_2)$, for two neighboring bilayers, situated at the average distance a apart. (b) Distributions of distances between bilayers. (1) a Gaussian and a truncated Gaussian (the same distribution as Gaussian, except $P(z) = 0$ for $z < 0$) and (2) an asymmetric distribution ($\alpha = 1.4$).

First, we will assume that the fluctuations of neighboring bilayers are independent of each other, as long as the distance between them is positive, but forbid those fluctuations which involve their interpenetration. In this case, the distance distribution function is a truncated Gaussian (see Figure 1b). This is equivalent to superimposing an additional hard-wall potential to the interaction potential, and we will show that, in the asymptotic limit, the results are in qualitative agreement with those of Podgornik and Parsegian.[5]

Second, we will consider that the degree of correlation between the fluctuating bilayers depends on their separation distance, with total correlation as $z \to 0$ and no correlation as $z \to \infty$. This provides an asymmetric distribution (Figure 1b), which will be described using an asymmetry coefficient α in a truncated Gaussian distribution

$$P(z) = \frac{1}{C} e^{-(z-a')^2/2\sigma'^2}, \qquad 0 < z \le a'$$

$$P(z) = \frac{1}{C} e^{-(z-a')^2/2(\alpha\sigma')^2}, \qquad a' < z < \infty \qquad (12)$$

where a' is the position of the maximum and the constant C is determined through normalization. The average separation distance $a = \langle z \rangle$ and the variance of the distribution (the mean-square fluctuation) $\sigma^2 = \langle z^2 \rangle - \langle z \rangle^2$ can be computed as functions of the parameters a', α, and σ'.

Recent Monte Carlo simulations of bilayer fluctuations[17] revealed that the distance distribution function is asymmetric. While this result is expected, since it implies that the correlation between fluctuating bilayers decreases

2458 *Langmuir, Vol. 17, No. 8, 2001* *Manciu and Ruckenstein*

with their separation, there are no quantitative analytic theories to predict the degree of asymmetry from the bilayer interaction parameters. However, the simulations show that the asymmetry is stronger only when a few bilayers are present. For a large number of bilayers, the distribution function tends to a limiting distribution which can be approximated by eq 12, with $\alpha \approx 1.4$.

Let us now present explicit calculations for each distribution.

III.1. Gaussian Distribution. The Gaussian distribution allows one to obtain analytical expressions for σ and for the free energy, when only the hydration force is present. It will be shown later that, for small distances, these expressions provide also good approximations for the truncated and asymmetric Gaussian distributions and that the van der Waals attraction does not modify substantially the expression of σ.

The hydration interaction energy between two neighboring planar bilayers,[6-10] separated by a distance z, is given by

$$V_H(z) = A_H e^{-z/\lambda} \tag{13}$$

The free energy per unit area $f(a,\sigma)$ acquires the form

$$f(a,\sigma) = \int_{-\infty}^{\infty} \frac{dz}{(2\pi)^{1/2}\sigma} e^{-(a-z)^2/2\sigma^2} A_H e^{-z/\lambda} + \frac{(kT)^2}{64K_c\sigma^2} = A_H e^{-a/\lambda} e^{\sigma^2/2\lambda^2} + \frac{(kT)^2}{64K_c\sigma^2} \tag{14}$$

and the minimum with respect to σ leads to

$$4 \ln\left(\frac{\sigma}{\lambda}\right) + \frac{\sigma^2}{2\lambda^2} - \ln\left(\frac{(kT)^2}{32A_H K_c \lambda^2}\right) = \frac{a}{\lambda} \tag{15}$$

The solution of the above equation is plotted in Figure 2, for $T = 300$ K and the following typical values of the parameters:[1,8,19] $K_c = 1 \times 10^{-19}$ J, $A_H = 0.035$ J/m^2, and $\lambda = 1.5$ Å. One can see from Figure 2a that σ remains finite even when $a = 0$ (i.e., the neighboring bilayers touch each other). In previous theories, which include the hard wall confinement,[3,5] the free energy diverges for $a \to 0$; in the present case, it remains finite because the repulsion is provided by the hydration force, which is finite at any distance.

One can see from Figure 2a that for separation distances less than 10 Å, the quadratic term can be neglected compared to the logarithms, and eq 15 becomes

$$\sigma^2 = \frac{\lambda kT}{4(2A_H K_c)^{1/2}} e^{a/2\lambda} \tag{16}$$

A recent improvement of the Caille theory of X-ray scattering by smectic liquid crystals[20,21] opens the possibility to obtain $\sigma = \sigma(a)$ from the diffraction spectrum. Extensive experimental measurements on lipid bilayers[1] indicated that, at low separations, there is an exponential dependence as predicted by eq 16. However, to obtain reliable interaction parameters from the fit of the experimental data, one should identify the validity domain of eq 16. As shown in Figure 2a, eq 16 is valid only for small values of a.

(19) McIntosh, T. J.; Simon, S. A. *Biochemistry* **1993**, *32*, 8374.
(20) Caille, A. *C. R. Seances Acad. Sci., Ser. B* **1972**, *174*, 891.
(21) Zhang, R.; Sutter, M.; Nagle, J. F. *Phys. Rev. E* **1994**, *50*, 5047.

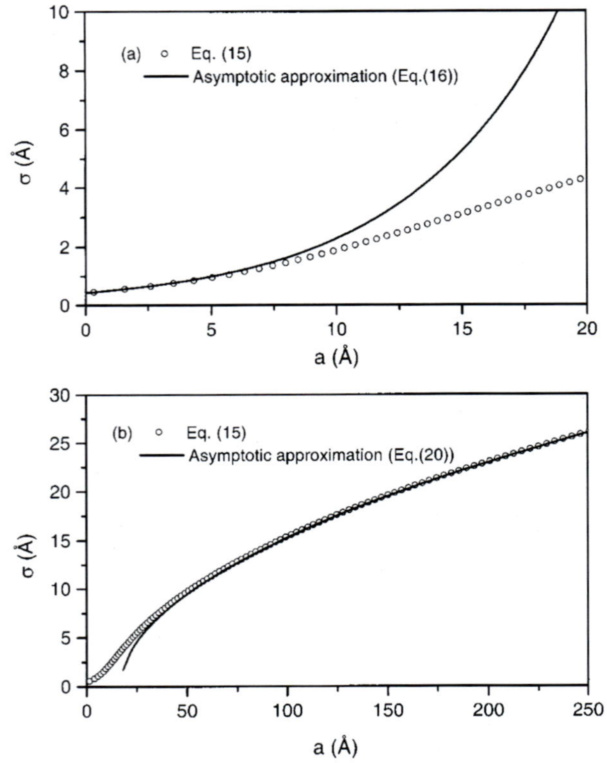

Figure 2. Equation 15 (circles): (a) asymptotic approximation for small values of a (eq 16), continuous line; (b) asymptotic approximation for large values of a (eq 20), continuous line.

Let us now calculate the free energy for small values of a. Equations 14 and 16 lead to

$$f(a) = A_H e^{-a/\lambda} e^{\sigma^2/2\lambda^2} + \frac{kT}{16(2^{1/2})}\left(\frac{A_H}{K_c\lambda^2}\right)^{1/2} e^{-a/2\lambda} \tag{17}$$

an equation which contains two repulsive contributions to the free energy. The first is due to hydration, which is enhanced by an exponential factor $e^{\sigma^2/2\lambda^2}$ due to undulation; this enhancement was also suggested by Sornette and Ostrovsky.[15] The other term is due to the entropic confinement. Both terms have exponential decays, but with decay lengths of λ and 2λ, respectively. The exponential decay in the second term is due to the exponential dependence of σ on the separation a.

A similar functional form for the free energy, with two exponential decays, was proposed by Evans and Parsegian[4,22]

$$f_{EP}(a) = A_H e^{-a/\lambda} + \frac{\pi kT}{16}\left(\frac{A_H}{K_c\lambda^2}\right)^{1/2} e^{-a/2\lambda} \tag{18}$$

The differences consist in the absence of the undulation enhancement in the first term of eq 18 and the numerical constant in the second term.

The variational treatment of hydration confinement by Podgornik and Parsegian[5] provided an expression similar to eq 18 in the limit of small a; however the coefficient multiplying the exponential of the second term was dependent on a.

(22) Evans, E.; Needham, D. *J. Phys. Chem.* **1987**, *91*, 4219.

For large values of a, the quadratic term dominates the left-hand term of eq 15; neglecting the other terms, one can write

$$\sigma = (2a\lambda)^{1/2} \tag{19}$$

A better approximation is obtained by evaluating the neglected terms using eq 19. This leads to

$$\sigma = \left(2a\lambda + 2\lambda^2 \ln\left(\frac{(kT)^2}{128A_H K_c a^2}\right)\right)^{1/2} \tag{20}$$

Figure 2b shows that the asymptotic value of σ provided by eq 20 is in very good agreement with eq 15 for $a > 30$ Å.

Using eqs 14 and 20, the free energy acquires the form

$$f(a) = \frac{(kT)^2}{128K_c a^2} + \frac{(kT)^2}{64K_c\left(2a\lambda + 2\lambda^2 \ln\frac{(kT)^2}{128A_H K_c a^2}\right)} \tag{21}$$

For large values of a, the logarithm in the second term becomes negligible, and the free energy becomes proportional to a^{-1}

$$f(a) \approx \frac{(kT)^2}{128K_c a\lambda} \tag{22}$$

The derived equations involve the assumption that the bilayers can permeate each other. It will be shown later that, when the permeation is forbidden, σ becomes proportional to a for large values of a, as predicted by the existing hard wall confinement theories.[3,5] However, for small values of a, eq 15 constitutes a very good approximation.

III.2. Truncated Gaussian Distribution. As already noted, to avoid permeation, the integration range should start from zero and, when only the hydration forces are taken into account, eq 10 becomes

$$f(a,\sigma) = \int_0^{\infty} \frac{dz}{C}\, e^{-(a-z)^2/2\sigma'^2} A_H e^{-z/\lambda} + \frac{(kT)^2}{64K_c\sigma^2} = A_H e^{-a/\lambda} e^{\sigma'^2/2\lambda^2} \int_0^{\infty} \frac{dz}{C}\, e^{-(a'-z)^2/2\sigma'^2} + \frac{(kT)^2}{64K_c\sigma^2} \tag{23}$$

with $a' = a - (\sigma'^2/\lambda)$, where σ' is the root-mean-square fluctuation of the Gaussian distribution, which is different from σ, the root-mean-square fluctuation of the truncated Gaussian distribution However, when σ' is at least a few times smaller than a

$$\int_{-\infty}^{0} dz\, e^{-(a-z)^2/2\sigma'^2} \ll \int_0^{\infty} dz\, e^{-(a-z)^2/2\sigma'^2}$$

and one can write that $\sigma \cong \sigma'$ and $C \cong \sigma(2\pi)^{1/2}$. When, in addition, σ' is a few times smaller than a', the integral of the first term of eq 23 is close to unity, and the results are well approximated by eq 15. However, σ' is not always smaller than a', and then numerical methods are necessary to obtain the values of σ which minimize the free energy in eq 23. This happens either when $a \to 0$ or, at large separations, when σ^2/λ becomes comparable or even larger than a. In these cases, the results differ from those obtained in the previous section.

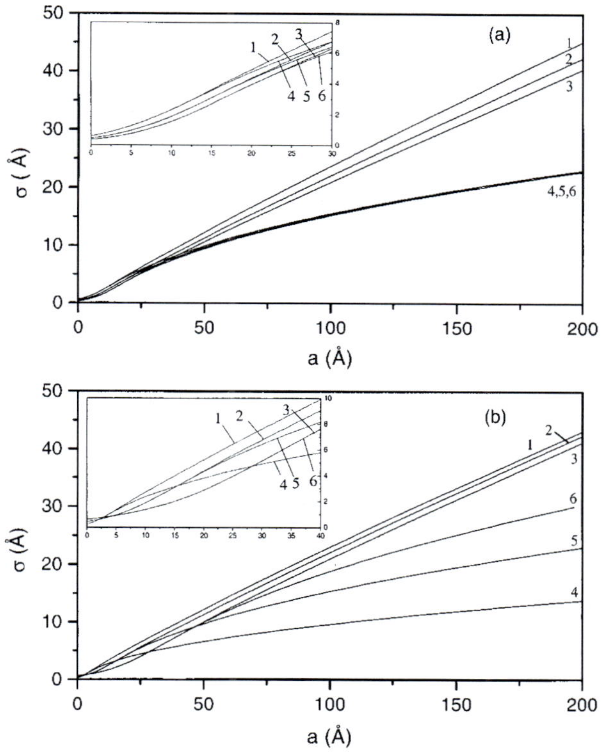

Figure 3. Root-mean-square fluctuation σ vs average separation distance a, for a Gaussian and a truncated Gaussian distribution. (a) $K_c = 1 \times 10^{-19}$ J and $\lambda = 1.5$ Å: truncated Gaussian, (1) $A_H = 0.01$ J/m^2, (2) $A_H = 0.035$ J/m^2, (3) $A_H = 0.1$ J/m^2; Gaussian, (4) $A_H = 0.01$ J/m^2, (5) $A_H = 0.035$ J/m^2, (6) $A_H = 0.1$ J/m^2. (b) $K_c = 1 \times 10^{-19}$ J and $A_H = 0.035$ J/m^2: truncated Gaussian, (1) $\lambda = 0.5$ Å, (2) $\lambda = 1.5$ Å, (3) $\lambda = 3$ Å; Gaussian (4) $\lambda = 0.5$ Å, (5) $\lambda = 1.5$ Å, (6) $\lambda = 3$ Å. In the inserts, the same functions are presented for small values of a at a larger scale.

The results of the numerical computations are compared with the analytical ones provided by eq 15 in Figure 3, for various values of the hydration interaction constant A_H (Figure 3a) and of the decay length λ. In all the cases investigated, a linear asymptotic behavior was obtained for σ as a function of a, as suggested by Helfrich. As already noted, this occurs because the truncation is equivalent to a hard wall repulsion, which limits the fluctuations of the bilayers. However, the proportionality coefficient $\mu = \sigma^2/a^2$ differs from those obtained from the theories for hard wall confinement[3,5] and is dependent on the parameter A_H of the hydration force. It is of interest to note that the asymptotic behavior of $\sigma(a)$ is almost independent of λ for the truncated Gaussian distribution, but almost independent of A_H for the Gaussian distribution (see eq 20).

III.3. Asymmetric Distribution. When in eq 12 σ' is at least a few times smaller than a', one can neglect the integral from $-\infty$ to 0, and the following expressions are obtained for C, a', and $\sigma^2 = \langle z^2\rangle - \langle z\rangle^2$:

$$C = \left(\frac{\pi}{2}\right)^{1/2} \sigma'(\alpha + 1)$$

$$a' = a - \sigma'\left(\frac{2}{\pi}\right)^{1/2} (\alpha - 1) \tag{24}$$

$$\sigma^2 = \sigma'^2\left(\alpha^2 - \alpha + 1 - \frac{2}{\pi}(\alpha - 1)^2\right)$$

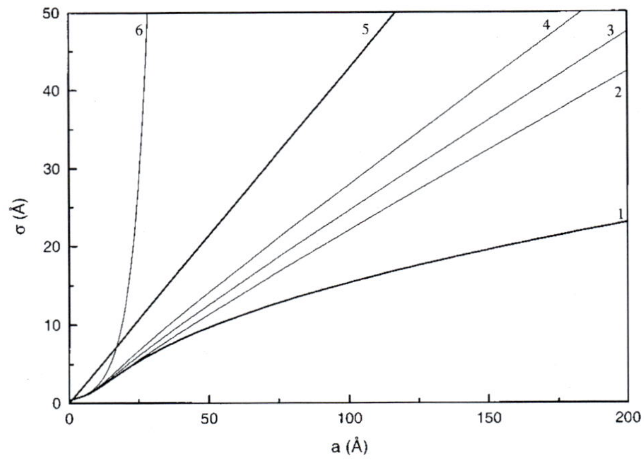

Figure 4. Root-mean-square fluctuation σ vs the separation distance a, for (1) eq 15 and (2–4) σ calculated for truncated Gaussian distribution with various asymmetry coefficients: (2) $\alpha = 1$; (3) $\alpha = 1.4$; (4) $\alpha = 2.0$; (5) Helfrich proportionality relation, $\sigma^2 = \mu a^2$ ($\mu = 0.183$); (6) approximate solution for small distances, eq 16.

The following approximate expression can be written for the free energy:

$$f(a,\sigma') = \int_0^\infty dz\, P(z) A_H e^{-z/\lambda} + \frac{(kT)^2}{64K_c\left(\alpha^2 - \alpha + 1 - \frac{2}{\pi}(\alpha - 1)^2\right)\sigma'^2} \approx \frac{2A_H e^{-a/\lambda} e^{\sigma'^2/2\lambda^2}}{(\alpha + 1)} \int_0^\infty \frac{dz}{(2\pi)^{1/2}\sigma'} e^{-(a''-z)^2/2\sigma'^2} + \frac{(kT)^2}{64K_c\left(\alpha^2 - \alpha + 1 - \frac{2}{\pi}(\alpha - 1)^2\right)\sigma'^2} \tag{25}$$

where

$$a'' = a - \sigma'\left(\frac{2}{\pi}\right)^{1/2}(\alpha - 1) - \frac{\sigma'^2}{\lambda}$$

It should be emphasized that eqs 24 and 25 are only approximate. In the calculation, we used eqs 10 and 12 for the numerical minimization of $f(a,\sigma')$, with respect to σ', for various values of the asymmetry coefficient α. The results are presented in Figure 4. For a smaller than about 20 Å, eq 15, and for a smaller than about 10 Å, eq 16, constitute accurate approximations. The figure indicates that again σ is proportional to a in the asymptotic limit $a \to \infty$, with a proportionality coefficient dependent on the degree of asymmetry α. The value $\mu = 0.183$ was derived for the hard wall confinement.[5] Recent Monte Carlo simulations of soft confinement of bilayers (interacting via van der Waals and hydration forces) provided lower values for the proportionality coefficient $\mu = 0.06{-}0.12$.[17] The results in Figure 3 show that, for a truncated Gaussian distributions, involving only hydration interactions, even lower values are obtained. However, using asymmetric Gaussian distributions, as in Figure 4, the values of μ, which are in the range 0.06–0.08, becomes compatible with those obtained from Monte Carlo simulations.

IV. The Inclusion of the van der Waals Interaction

Let us now complete the calculation by adding the van der Waals interaction potential to the hydration interaction. The bilayers are composed of hydrophilic headgroups, which are hydrated by water, and of hydrophobic hydrocarbon chains. A complete theory should take into account that their dielectric properties are different, by using different Hamaker constants.[22] The problem can be simplified by considering that, for the purpose of van der Waals interaction, the polar headgroups, being hydrophilic, are part of the water lamellae. Denoting by t_h the total thickness of the headgroups in the lamellae and by t_c the thickness of the layer of hydrocarbon chains, the van der Waals attraction between two bilayers separated by a distance z, measured in water, becomes (we include only the interactions between the nearest neighbors)

$$V_{vdW}(z) = -\frac{H}{12\pi}\left(\frac{1}{(z + t_h)^2} + \frac{1}{(z + t_h + 2t_c)^2} - \frac{2}{(z + t_h + t_c)^2}\right) \tag{26}$$

where H is the Hamaker constant.

Typical values for the Hamaker constant were estimated to range from 10^{-21} to 10^{-20} J.[1,19,22,23] The values of the bending modulus are either not known for some lipids or uncertain by a factor of 4, around 10^{-19} J, for other lipids.[1] The values of the parameters of the hydration force were estimated from the fit of experimental data[1,19] with a formula derived by Evans and Parsegian[4] (reproduced here as eq 18). The values lie in the rather large range of $\lambda = 1.3{-}2.1$ Å and $A_H = 6.0 \times 10^{-3}$ to 2.5×10^{-2} J/m^2. In addition to depending upon the expression employed in the fitting of the data, large differences in A_H, by orders of magnitude, can be obtained whether the distance z in the hydration force is measured from the boundary between the hydrocarbon chains and headgroups or between the headgroups and water.[1]

In this section we present the results of the calculations performed for σ and the free energy for $T = 300$ K and the following values of the parameters: $H = 1.0 \times 10^{-20}$ J, $t_c = 37.8$ Å, $t_h = 7.6$ Å, $K_c = 1.0 \times 10^{-19}$ J, $\lambda = 1.5$ Å, and $A_H = 0.035$ J/m^2.

The root-mean-square fluctuation of the distance distributions, σ, will be computed for the following cases: (1) using the analytical formula (15), which involves only the hydration force ($H = 0$); (2) for a truncated Gaussian distribution, with only the hydration force present ($H = 0$); (3) for a truncated Gaussian distribution, with hydration force and van der Waals interactions with a Hamaker constant $H = 1.0 \times 10^{-20}$ J; (4) for an asymmetric Gaussian distribution with $\alpha = 1.4$ and $H = 0$; (5) for an asymmetric Gaussian distribution with $\alpha = 1.4$ and $H = 1.0 \times 10^{-20}$ J; (6) for an asymmetric Gaussian distribution with $\alpha = 2.0$ and $H = 0$; (7) for an asymmetric Gaussian distributions with $\alpha = 2.0$ and $H = 1.0 \times 10^{-20}$ J; (8) using Helfrich's proportionality relation, with $\mu = 0.183$.

The results of these computations are presented in Figure 5, which shows that (i) σ is only slightly affected by the van der Waals interactions, but depends on the distribution considered, and (ii) for $a < 20$ Å, the results for all the distributions employed are well approximated by the simple expression, eq 15, which can be used to extract some of the interaction parameters from the experimental values of $\sigma(a)$ for small values of a.

The free energy calculations are presented in Figure 6 as a function of the average distance a, for the following cases:

(23) Petrache, H. I.; Tristram-Nagle, S.; Nagle, J. F. *Chem. Phys. Lipids* **1998**, *95*, 83.

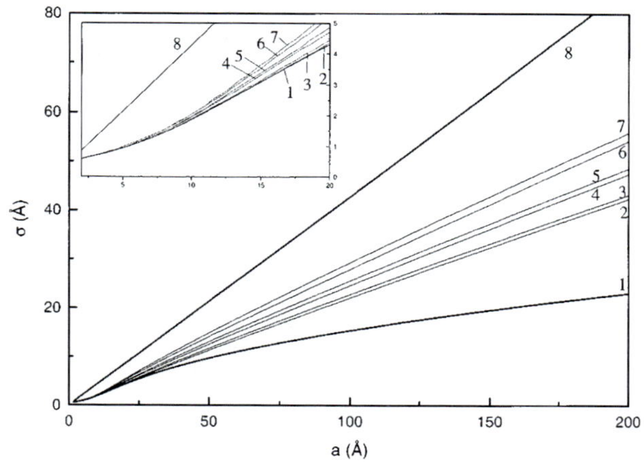

Figure 5. Root-mean-square fluctuation σ vs separation distance a, for $K_c = 1 \times 10^{-19}$ J, $t_h = 7.6$ Å, $t_c = 37.8$ Å, $A_H = 0.035$ J/m^2, $\lambda = 1.5$ Å for: (1) eq 15 ($H = 0$); (2) truncated Gaussian distribution, $H = 0$; (3) truncated Gaussian distribution, $H = 1.0 \times 10^{-20}$ J; (4) asymmetric Gaussian distribution, $\alpha = 1.4$, $H = 0$; (5) asymmetric Gaussian distribution, $\alpha = 1.4$, $H = 1.0 \times 10^{-20}$ J; (6) asymmetric Gaussian distribution, $\alpha = 2.0$, $H = 0$; (7) asymmetric Gaussian distribution, $\alpha = 2.0$, $H = 1.0 \times 10^{-20}$ J (8) Helfrich proportionality relation ($\mu = 0.183$). In the inserts, the same functions are presented for small values of a at a larger scale.

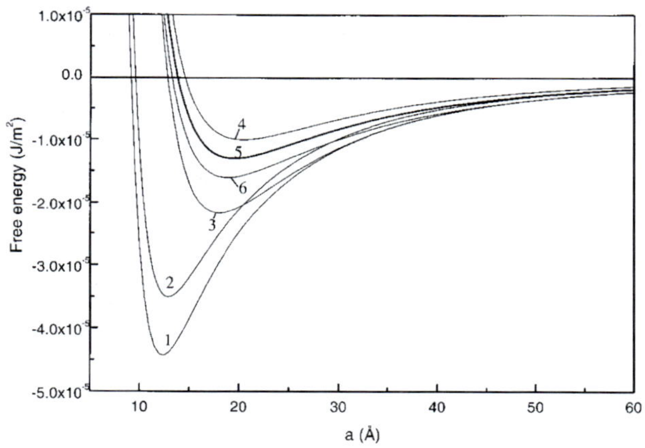

Figure 6. Free energy per unit area of bilayer, for $H = 1.0 \times 10^{-20}$ J, $K_c = 1 \times 10^{-19}$ J, $t_h = 7.6$ Å, $t_c = 37.8$ Å, $A_H = 0.035$ J/m^2, $\lambda = 1.5$ Å: (1) $K_c \to \infty$ (energy of planar bilayers); (2) calculated by adding the entropic term (eq 7 with $\sigma^2 = \mu a^2$) to the energy of planar bilayers; (3) eq 18; (4) truncated Gaussian distribution; (5, 6) asymmetric Gaussian distribution with (5) $\alpha = 1.4$ and (6) 2.0.

(1) $K_c \to \infty$. In this case, the entropic term vanishishes and the free energy is provided by the energy of planar (nonundulating) bilayers.

(2) The free energy is obtained by adding to the energy obtained under (1) the entropic term, using for σ the relation $\sigma^2 = \mu a^2$ suggested by Helfrich,[3] with the value $\mu = 0.183$.[5] This procedure neglects both the fluctuation enhancement of the energy and the more involved dependence of σ on a, obtained through the minimization of the free energy with respect to σ.

(3) The free energy was computed by adding to the Evans–Parsegian formula (eq 18) the van der Waals attraction between nonundulating bilayers. This kind of calculation, which was often used to fit the experimental data, offers results in qualitative agreement with our approach only for small values of a. However, even for small a, the fluctuation enhancement of the energy (as in eq 17) is neglected. It should be noted that the free energy computed in this manner would always have a stable minimum at finite distance, since the exponential has a shorter range than the van der Waals attraction.

(4) The free energy is computed via minimization, by using the truncated Gaussian distribution in eq 10.

(5, 6) The free energy is computed by using the asymmetric Gaussian distribution, with $\alpha = 1.4$ and 2.0, respectively, in eq 10.

The addition of the hard wall entropic confinement free energy to the interaction energy, as under (2), only raises somewhat the minimum, but almost does not modify its position. In contrast, in all the other cases, which take into account the other interactions in the calculation of entropy, considerable shifts of the equilibrium distance and of the value of the free energy at the minimum occur. This observation indicates that a common procedure, to add to the conventional interactions, calculated as for planar layers, the free energy due to the hard wall entropic confinement is inaccurate.

The asymptotic value of the free energy is zero for $a \to \infty$. Since the entropic free energy has a slower decay, for all distance distributions employed (4–6), than the van der Waals energy, the free energy will be positive at sufficiently large distances. This does not mean, however, that an unbinding of the bilayers will occur, since a negative free energy minimum (hence an absolute minimum) can exist at a finite distance. Reciprocally, the existence of a local minimum for a given set of parameters does not necessarily imply that the bilayers are bound, since the local minimum can be positive and hence unstable. However, this unstable minimum can be separated from the absolute minimum (which is zero and located at infinity) by a potential barrier and, if the latter is sufficiently high, the metastable state can survive for a long time. Similarly, if the minimum is negative but small in absolute value, a perturbation can cause the bilayers to extend to the unstable state at infinity. These results can explain the contradictory experimental observations[11,12] for the same phospholipid bilayers. The first authors[11] observed that the bilayers extend at very large separations, while the latter[12] authors observed that they swell up to small distances.

In Figure 7 we present the free energy for an asymmetric Gaussian distribution ($\alpha = 1.4$) as a function of distance for various values of the Hamaker constant (with all the other parameters unchanged). For $H > 3.825 \times 10^{-21}$ J, a stable minimum is obtained at a finite distance. For $H < 3.825 \times 10^{-21}$ J, the stable minimum is at infinite distance; however, for $3.825 \times 10^{-21} J > H > 3.45 \times 10^{-21}$ J, a local (unstable) minimum is still obtained at finite distance. For $H = 3.825 \times 10^{-21}$ J, a critical unbinding transition occurs, since the minima at finite and infinite distances become equal. However, these two minima are separated by a potential barrier, with a maximum height of 1.68×10^{-7} J/m^2, located at a separation distance of 90 Å. The results remained qualitatively the same for any combination of the interaction parameters.

V. Comparison with Experimental Data

A common procedure to determine the values of the interaction parameters between bilayers is to fit the experimental data with a functional form predicted by theory. It was shown[1,19] that the equilibrium separation distance between bilayers as a function of the osmotic pressure applied can be well fitted by expression (18). A recent improvement of Caille theory[20,21] of X-ray scattering by smectic liquid crystals allowed the simultaneous measurement of the average separation distance and the

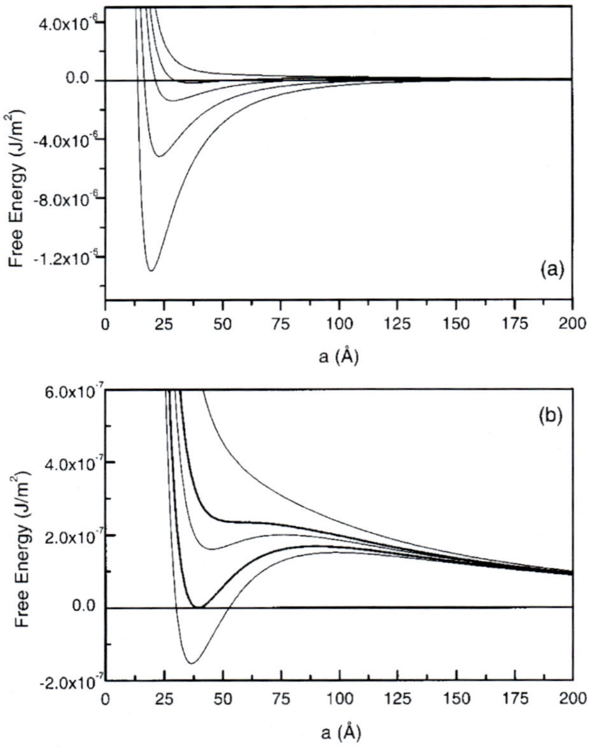

Figure 7. Free energy per unit area, computed for an asymmetric Gaussian distribution, for $K_c = 1 \times 10^{-19}$ J/m², $t_h = 7.6$ Å, $t_c = 37.8$ Å, $A_H = 0.035$ J/m², $\lambda = 1.5$ Å, $\alpha = 1.4$, and various values of H: (a) from bottom to top, $H = 10.0$, 7.0, 5.0, 4.0, and 3.0×10^{-21} J; (b) from bottom to top, $H = 4.0$, 3.825, 3.6, 3.45, and 3.0×10^{-21} J. For $H > 3.825 \times 10^{-21}$ J, the minimum at finite distance is stable. For $H < 3.825 \times 10^{-21}$ J, the minimum at infinity is the stable one. For $H < 3.45 \times 10^{-21}$, the minimum at infinity is the only one.

root-mean-square fluctuation σ as a function of the applied osmotic pressure, and the authors[1,2] noted disagreement between the existing theories and experiment.

Figure 8 reproduces the experimental data of ref 1 for EPC (egg phosphatidylcholine) (circles) and DMPC (1,2-dimyristoyl-*sn*-glycero-3-phosphatidylcholine) (squares). Figure 8a presents $p = p(a)$ and Figure 8b presents $\sigma = \sigma(a)$. Petrache et al.[1] have shown that agreement can be obtained by fitting eq 18 to the pressure data, for various values of H ((1.65−7.13) × 10^{-21} J) and K_c ((0.5−2.0) × 10^{-19} J). In all the cases, the fitting provided $\lambda = 2.0 \pm 0.3$ Å. The experimental data for $\sigma = \sigma(a)$ are well described by the exponential function $\sigma^2 \propto e^{a/\lambda_{fl}}$ for distances between 5 and 20 Å, with $\lambda_{fl} \approx 6$ Å.[1] Combining the Helfrich expression (eq 7) with that of Evans and Parsegian (eq 18), one obtains $\sigma^2 \propto e^{a/2\lambda}$, which is in disagreement with experiment because $\lambda_{fl} \neq 2\lambda$.

Now σ and the pressure (computed as the derivative of the free energy per unit area) will be calculated, using the procedure outlined in this article. For the Hamaker constant and the bending modulus, typical values from literature, namely, 1.0×10^{-20} and 1.0×10^{-19} J, respectively, will be used. For t_h and t_c we employed the values obtained from X-ray data,[1] $t_h = 7.6$ Å and $t_c = 37.8$ Å for EPC and $t_h = 7.6$ Å and $t_c = 36.4$ Å for DMPC, respectively. Because the hydrocarbon thicknesses t_c of EPC and DMPC produced almost no difference (see eq 27), in the following only the results for the EPC are presented ($t_c = 37.8$ Å). For the degree of asymmetry α the value of 1.4, which is in agreement with the Monte Carlo simulations,[17] was selected.

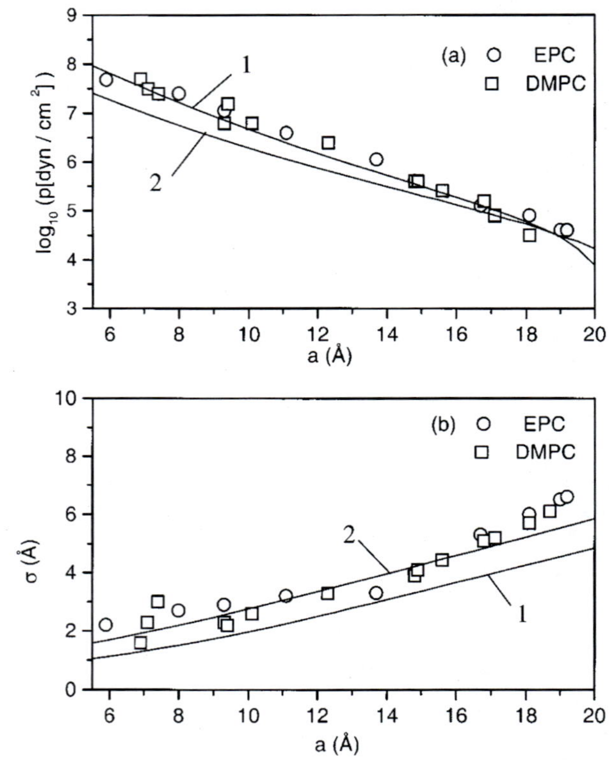

Figure 8. Experimental data from ref 1 for EPC (circles) and DMPC (squares) bilayers compared with predictions obtained using some typical values for the interaction parameters: (1) $K_c = 1.0 \times 10^{-19}$ J, $H = 1.0 \times 10^{-20}$, $t_h = 37.8$ Å, $t_c = 7.6$ Å, $\alpha = 1.4$, $A_H = 0.035$ J/m², $\lambda = 1.5$ Å; (2) $K_c = 0.5 \times 10^{-19}$ J, $A_H = 0.01$ J/m², and the other parameters as in (1).

It is clear that one cannot use the values of the parameters obtained through the fitting of a different expression to experimental data, to compare the present calculations to the same experimental data. One can proceed in two different ways: either to determine all the parameters from fitting the present equations to the experimental data or to use the experimental data for one of the physical quantities (p or σ) to determine some parameters and to verify if the second (σ or p) is in agreement.

Because the number of parameters is large, it is obvious that one can obtain "agreement" with experiment by fitting the present equations. It is, however, more meaningful to employ the second procedure.

The values of $\lambda = 1.5$ Å and $A_H = 0.035$ J/m² provided agreement with the $p = p(a)$ data (curve 1 of Figure 8a). Because for these values, the σ experiments were underestimated, a lower value for K_c was chosen. Selecting $K_c = 0.5 \times 10^{-19}$ J required a value of A_H of 0.01 J/m² to preserve the position of the minimum of the free energy. Under these conditions (all the other parameters remaining unchanged), curve 2 was obtained. While in better agreement with $\sigma(a)$, the agreement with $p(a)$ was somewhat deteriorated. There is no doubt that better agreement can be further achieved by fitting the parameters; however, even using some typical values from literature for the parameters, a more than qualitative agreement with experimental data was obtained.

VI. Conclusions

A new variational analysis of the interactions of fluctuating bilayers is proposed, in which the entropic confinement energy is considered inverse proportional to the root-mean-square fluctuation of the distribution function of the distance between two successive bilayers. This approximation can be used, even for nonlinear interactions, when the bending elastic energy of the bilayer provides the dominating contribution to the mode energy. For a given functional form of the distribution, σ was obtained by minimizing the free energy. Analytical results were derived for a Gaussian distribution and compared with numerical results obtained for truncated and asymmetric Gaussian distributions.

It is shown that, in the asymptotic limit, the present approach recovers the results of other existing theories, but important differences occur at relatively short distances. As reported in the literature,[1,2] there are no theories or simulations, which are in agreement with the experimental data for both the pressure and the root-mean-square fluctuations of the distance between two neighboring bilayers, as a function of the average distance, for the same set of interaction parameters. Comparison with recent experimental data shows that, with the present approach, a more than qualitative agreement can be obtained.

The present approach recovers the well-known result that a transition from a bounded to an unbounded state can be achieved by varying the Hamaker constant. However, the stable and the unstable states (which can be either at infinity or at finite distance) are separated, in the vicinity of the transition point, by a potential barrier. This might explain the contradictory experimental observations on infinite or finite swelling of some lipid bilayers.[11,12]

LA0016266

Langmuir
The ACS Journal of Surfaces and Colloids

MAY 28, 2002
VOLUME 18, NUMBER 11

Letters

The Interaction between Two Fluctuating Phospholipid Bilayers

Marian Manciu and Eli Ruckenstein*

Department of Chemical Engineering, State University of New York at Buffalo, Buffalo, New York 14260

Received February 12, 2002. In Final Form: April 5, 2002

The dependence of the interaction force between two undulating phospholipid bilayers and of the root-mean-square fluctuation of their separation distances on the average separation can be determined once the distribution of the intermembrane separation is known as a function of the applied pressure. However, most of the present theories for interacting membranes start by assuming that the distance distribution is symmetric, a hypothesis invalidated by Monte Carlo simulations. Here we present an approach to calculate the distribution of the intermembrane separation for any arbitrary interaction potential and applied pressure. The procedure is applied to a realistic interaction potential between neutral lipid bilayers in water, involving the hydration repulsion and van der Waals attraction. A comparison with existing experiments is provided.

The role of thermal undulations on the interaction between elastic phospholipid bilayers has received considerable attention because of its relevance to biological processes. Helfrich suggested that,[1] for membranes interacting via a rigid wall potential, the attenuation of the thermal undulation of a membrane by the neighboring membranes generates an entropic repulsion force (fluctuation pressure), $p_{\mathrm{fl}} = \alpha(kT)^2/K_C d^3$, where k is the Boltzmann constant, T the absolute temperature, K_C the bending modulus of the elastic membrane, and d the average distance between membranes. Helfrich also provided the first estimates of α, ranging from $\alpha = 0.84$[1] to $\alpha = 0.0242$;[2] more accurate values were obtained later, either assuming that the distribution of the membrane positions satisfies a diffusion equation ($\alpha = 0.0854$)[3] or via a diagrammatic expansion ($\alpha = 0.0797$),[4] in good agreement with Monte Carlo simulations ($\alpha = 0.079$).[5]

The role of thermal fluctuations for membranes interacting via arbitrary potentials, which constitutes a problem of general interest, is however still unsolved. Earlier treatments [6,7] coupled the fluctuations and the interaction potential and revealed that the fluctuation pressure has a different functional dependence on the intermembrane separation than that predicted by Helfrich for rigid-wall interactions. The calculations were refined later by using variational methods.[3,8] The first of them employed a symmetric functional form for the distribution of the membrane positions as the solution of a diffusion equation in an infinite well.[3] However, recent Monte Carlo simulations of stacks of lipid bilayers interacting via realistic potentials indicated that the distribution of the intermembrane distances is asymmetric;[9] the root-mean-square fluctuations obtained from experiment were also shown to be in disagreement with this theory.[10]

* Corresponding author: e-mail address, feaeliru@acsu.buffalo.edu; phone, (716) 645 2911/2214; fax, (716) 645 3822.

(1) Helfrich, W. *Z. Naturforsch.* **1978**, *33a*, 305.
(2) Helfrich, W.; Servuss, R.-M. *Nuovo Cimento* **1984**, *3D*, 137.
(3) Podgornik, R.; Parsegian, V. A. *Langmuir* **1992**, *8*, 557.
(4) Bachmann, M.; Kleinert, H.; Pelster, A. *Phys. Lett. A* **1999**, *261*, 127.
(5) Janke W.; Kleinert, H. *Phys. Lett. A* **1986**, *117*, 353.
(6) Sornette D.; Ostrowsky, N. *J. Chem. Phys.* **1986**, *84*, 4062.
(7) Evans E. A.; Parsegian, V. A. *Proc. Natl. Acad. Sci. U.S.A.* **1986**, *83*, 7132.
(8) Manciu M.; Ruckenstein, E. *Langmuir* **2001**, *17*, 2455.
(9) Gouliaev N.; Nagle, J. F. *Phys. Rev. Lett.* **1998**, *81*, 2610.

10.1021/la0201568 CCC: $22.00
Published on Web 05/03/2002

The variational method proposed earlier by the authors relied on Monte Carlo simulations to select an intermembrane distance distribution function.[8] The purpose of this paper is to present a new approach, in which the interaction between two membranes, in the presence of thermal fluctuations, is calculated directly by employing a suitable approximate partition function. Thus, the asymmetry of the distance distributions results in a natural manner from the calculation.

First, let us summarize the results obtained for a fluctuating membrane placed in a harmonic potential.[1,11] Assuming that the total area of the membrane is constant, the Hamiltonian is given by

$$\mathcal{H} = \int \mathrm{d}x\, \mathrm{d}y \left(\frac{1}{2} K_\mathrm{C}\left(\frac{\partial^2 u(x,y)}{\partial x^2} + \frac{\partial^2 u(x,y)}{\partial y^2}\right)^2 + \frac{1}{2} B u^2(x,y)\right) \quad (1)$$

where K_C is the curvature elastic modulus of a bilayer, B is the spring constant (per unit area of the bilayer) of the harmonic potential, and $u(x,y)$ denotes the displacement along the normal to the $x-y$ plane from the average position u_0 of a point of the membrane of coordinates x and y. Denoting by $\tilde{u}(q_x,q_y)$ the Fourier transform of $u(x,y)$, the average energy of each mode for a membrane of unit area was obtained from the equipartition principle as[1,11]

$$\langle \tilde{u}^2 \rangle \left(\frac{1}{2} K_\mathrm{C} q^4 + \frac{1}{2} B\right) = \frac{1}{2} kT \quad (2)$$

Hence the mean square fluctuation is given by

$$\sigma^2 = \langle (u - u_0)^2 \rangle = \sum_{q_x,q_y} \langle \tilde{u}^2 \rangle = \frac{kT}{(2\pi)^2} \int_0^\infty \frac{2\pi q\, \mathrm{d}q}{K_\mathrm{C} q^4 + B} = \frac{kT}{8(K_\mathrm{C}B)^{1/2}} \quad (3)$$

and the distribution of the positions of the membrane is Gaussian

$$\rho(u) = \frac{1}{(2\pi)^{1/2}\sigma} \exp\left(-\frac{(u - u_0)^2}{2\sigma^2}\right) \quad (4)$$

This distribution is formally identical with the normal (Boltzmann) distribution of small independent, planar membranes of area S in the potential $(1/2)B(u - u_0)^2$

$$\rho(u) = \left(\frac{SB}{2\pi kT}\right)^{1/2} \exp\left(-\frac{SB(u - u_0)^2}{2kT}\right) \quad (5)$$

if the area S of the independent, nonundulating small membranes is given by

$$S = \frac{kT}{B\sigma^2} = 8\left(\frac{K_\mathrm{C}}{B}\right)^{1/2} \quad (6)$$

where eq 3 was used to derive the last equality.

The treatment of fluctuating membranes as small planar pieces moving independently was employed by Helfrich and Servuss,[2] who showed that the kinetic (ballistic) pressure exerted by an 1D gas of planar pieces in a rigid-wall confinement, is of the form $p_\mathrm{fl} = \alpha(kT)^2/K_\mathrm{C}d^3$, for a suitable choice of the area of the pieces (different from eq 6). However, Monte Carlo simulations for membranes interacting via realistic potentials[12] indicated that in these cases the ballistic term, due to the collisions between a membrane and the rigid walls, provides only a negligible contribution. In what follows, the ballistic contribution to the fluctuation pressure will be neglected.

When two fluctuating bilayers interact, the intermembrane distance distribution is still given by eq 5, since the interaction depends only on the distance z between the bilayers. For an arbitrary interaction (per unit area) $U(z)$ and a constant applied pressure p, the Boltzmann distribution of distances between the small independent surfaces of area S_0 can be calculated using the enthalpy (per unit area) $H(z) = U(z) + pz$ instead of the energy

$$\rho(z) = \frac{1}{N} \exp\left(-\frac{S_0(U(z) + pz)}{kT}\right) \quad (7)$$

where the constant N can be obtained by normalization.

The last task is to calculate the area of the independent pieces S_0, which is provided by eq 6 only for a harmonic interaction. To accomplish this, we will seek a harmonic potential $H_0(z) = (1/2)B_0(z - z_0')^2 + C$ (with B_0, z_0', and C independent of z), which best approximates $H(z)$ and use $S_0 = 8(K_\mathrm{C}/B_0)^{1/2}$.

A simple procedure consists of using the parabolic approximation of the enthalpy in the vicinity of its minimum at z_m

$$B_0 = \left.\frac{\mathrm{d}^2 H(z)}{\mathrm{d}z^2}\right|_{z=z_\mathrm{m}} \quad (8a)$$

where $z_\mathrm{m} = z_0'$ is the solution of

$$\mathrm{d}H(z)/\mathrm{d}z = 0 \quad (8b)$$

In this case

$$C = H(z_\mathrm{m}) \quad (8c)$$

A more rigorous procedure to calculate the harmonic approximate of the enthalpy is to seek the minimum of the expression

$$W(B_0,z_0',C) = \frac{1}{N} \int \exp\left(-\frac{S_0 H_0(z)}{kT}\right)(H(z) - H_0(z))^2\, \mathrm{d}z \quad (9)$$

with respect to B_0, z_0', and C. It will be shown later that for not too small pressures almost identical results are obtained by both procedures for usual interaction potentials.

For illustration purposes, we will apply the method described above to a potential typical for the interaction between two neutral, planar, lipid bilayers in water

$$U(z) = A \exp\left(-\frac{z}{\lambda}\right) - \frac{A_\mathrm{H}}{12\pi}\left(\frac{1}{(z + t_\mathrm{h})^2} + \frac{1}{(z + t_\mathrm{h} + 2t_\mathrm{c})^2} - \frac{2}{(z + t_\mathrm{h} + t_\mathrm{c})^2}\right) \quad (10)$$

where z represents the thickness of the water layer, λ is the decay length of the hydration force, A is a hydration

(10) Petrache, H. I.; Gouliaev, N.; Tristram-Nagle, S.; Zhang, R.; Suter, R. M.; Nagle, J. F. *Phys. Rev. E* **1998**, *57*, 7014.
(11) Sornette D.; Ostrowsky, N. In *Micelles, Membranes, Microemulsions and Monolayers*; Gelbart, W. M., Ben-Shaul, A., Roux, D. Eds.; Springer-Verlag: Berlin, 1994.
(12) Gouliaev N.; Nagle, J. F. *Phys. Rev. E* **1998**, *58*, 881.

parameter, A_H is the Hamaker constant, and the thicknesses t_h and t_c are defined below.

An accurate theory for the van der Waals interaction between bilayers should account for the variation of the dielectric constant inside the bilayer. Here, instead of considering (as usual) the whole bilayer as a hydrocarbon, we assume that the dielectric properties of a part of the dipolar headgroups are closer to those of water and the dielectric properties of the remaining part closer to those of the hydrocarbon chains. As in our previous paper,[8] the region between the phosphate groups, of thickness t_c (denoted D_{HH} in ref 13), is treated as a hydrocarbon and the rest of the bilayer, of thickness $t_h = t - t_c$, as water (where t is the thickness of the bilayer).

Using eqs 6 and 8a, one obtains

$$S_0 = 8\left[K_C/\left\{\frac{p}{\lambda} + \frac{A_H}{2\pi}\left[\frac{1}{(z_m + t_h)^3}\left(\frac{1}{3\lambda} - \frac{1}{z_m + t_h}\right) + \frac{1}{(z_m + t_h + 2t_c)^3}\left(\frac{1}{3\lambda} - \frac{1}{z_m + t_h + 2t_c}\right) - \frac{2}{(z_m + t_h + t_c)^3} \times \left(\frac{1}{3\lambda} - \frac{1}{z_m + t_h + t_c}\right)\right]\right\}\right]^{1/2} \quad (11)$$

where z_m is the solution of eq 8b and the distribution of the intermembrane distances is given by eq 7. Once the distribution is known, the average position $\langle z\rangle$ and the root-mean-square fluctuation σ can be calculated from

$$\langle z\rangle = \int_0^\infty z\rho(z)\,dz = \frac{\int_0^\infty z \exp\left(-\frac{S_0(U(z) + pz)}{kT}\right) dz}{\int_0^\infty \exp\left(-\frac{S_0(U(z) + pz)}{kT}\right) dz} \quad (12a)$$

$$\sigma = \left(\frac{\int_0^\infty (z - \langle z\rangle)^2 \exp\left(-\frac{S_0(U(z) + pz)}{kT}\right) dz}{\int_0^\infty \exp\left(-\frac{S_0(U(z) + pz)}{kT}\right) dz}\right)^{1/2} \quad (12b)$$

The asymmetry of the distribution can be defined as the ratio r

$$r = \left(\frac{\int_{z_m}^\infty (z - z_m)^2 \exp\left(-\frac{S_0(U(z) + pz)}{kT}\right) dz}{\int_0^{z_m} (z - z_m)^2 \exp\left(-\frac{S_0(U(z) + pz)}{kT}\right) dz}\right)^{1/2} \quad (12c)$$

which is unity in the symmetrical case.

Figure 1 presents the enthalpy and the distributions ρ_1 and ρ_2, calculated with the two procedures described (eqs 8 and 9, respectively), for typical values of the interaction parameters, namely, $A = 0.05$ J/m^2, $\lambda = 1.5$ Å, $A_H = 1.0 \times 10^{-20}$ J, $t_h = 5.0$ Å, $t_c = 40$ Å, $K_C = 1.0 \times 10^{-19}$ J, $T = 300$ K, and (a) $p = 1 \times 10^4$ N/m^2 and (b) $p = 1 \times 10^6$ N/m^2. For large pressures, both procedures lead to almost identical results, because most of the intermembrane distances are distributed in such cases in the vicinity of the minimum of the enthalpy. The asymmetry of the distribution function follows the asymmetry of the enthalpy, and, most importantly, is dependent on the pressure, being larger at low pressures.

In Figure 2, the mean square fluctuation σ (Figure 2a), the asymmetry ratio r (Figure 2b), and the pressure p (Figure 2c) are plotted versus the average distance $\langle z\rangle$, for various values of K_C, using the first method. Figure 2d

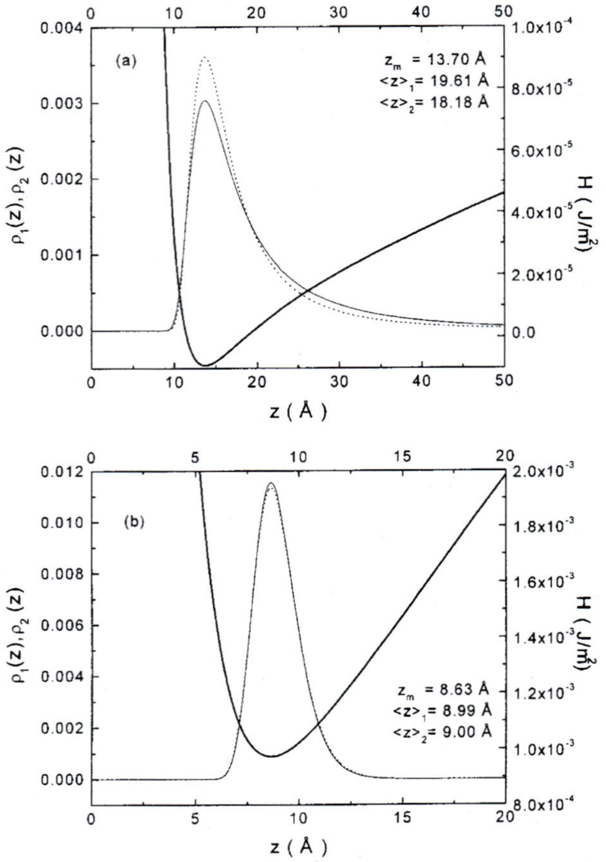

Figure 1. The enthalpy H (thick line) and the distributions of the intermembrane separation ρ_1 (continuous thin line) and ρ_2 (dotted line) (calculated from eqs 8 and 9, respectively) vs the distance z for the interaction potential (eq 10) with $A = 0.05$ J/m^2, $\lambda = 1.5$ Å, $A_H = 1.0 \times 10^{-20}$ J, $t_h = 5.0$ Å, $t_c = 40$ Å, $K_C = 1.0 \times 10^{-19}$ J, and $T = 300$ K, the applied pressure being: (a) $p = 1.0 \times 10^4$ N/m^2; (b) $p = 1.0 \times 10^6$ N/m^2. The subscripts 1 and 2 refer to quantities calculated using eqs 8 and 9, respectively.

represents the fluctuation pressure p_{fl}, defined as

$$p_{fl}(\langle z\rangle) = p(\langle z\rangle) - p(\langle z\rangle)|_{K_C=\infty} \quad (13)$$

vs the average separation $\langle z\rangle$. At low separation distances, $p_{fl} \sim \exp(-z/2\lambda)$, which is consistent with previous predictions.[3,7,8]

The average separation distance and its root-mean-square fluctuation were recently experimentally determined as functions of the applied pressure, for lipid bilayers/water multilayers,[10] and a disagreement between the existing theories and experiment was noted.[12] No comparable experimental results are yet available for two lipid bilayers, and it is difficult to extend the present theory to multilayers. For this reason, we will employ the present procedure to estimate the interaction parameters of eq 10 from the experimental results for multilayers.[10]

Figure 3 represents the root-mean-square fluctuations (Figure 3a) and the pressure (Figure 3b) as functions of the average separation distance, for (i) EPC (egg phosphatidylcholine, circles) and (ii) DMPC (1,2-dimyristoyl-*sn*-glycero-3-phosphatidylcholine, squares)/water multilayers. The values $A = 0.044$ J/m^2, $\lambda = 1.54$ Å, $A_H = 8.13 \times 10^{-21}$ J, and $K_C = 2.4 \times 10^{-19}$ J (EPC, continuous line)

(13) Petrache, H. I.; Tristram-Nagle S.; Nagle, J. F. *Chem. Phys. Lipids* **1998**, *95*, 83.

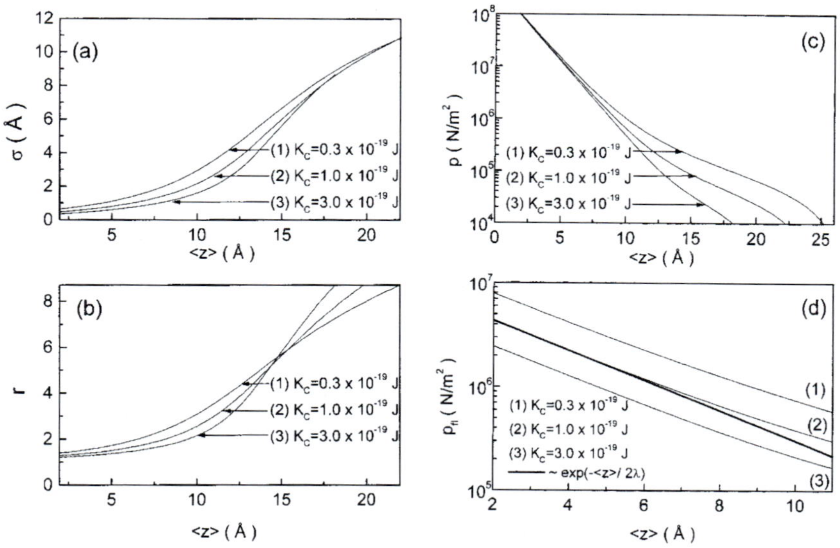

Figure 2. (a) The root-mean-square fluctuation σ, (b) the asymmetry ratio r, (c) the applied pressure p, and (d) the fluctuation pressure p_{fl} vs average thickness $\langle z \rangle$ for (1) $K_C = 0.3 \times 10^{-19}$ J, (2) $K_C = 1.0 \times 10^{-19}$ J, and (3) $K_C = 3.0 \times 10^{-19}$ J ($A = 0.05$ J/m^2, $\lambda = 1.5$ Å, $A_H = 1.0 \times 10^{-20}$ J, $t_h = 5.0$ Å, $t_c = 40$ Å, and $T = 300$ K).

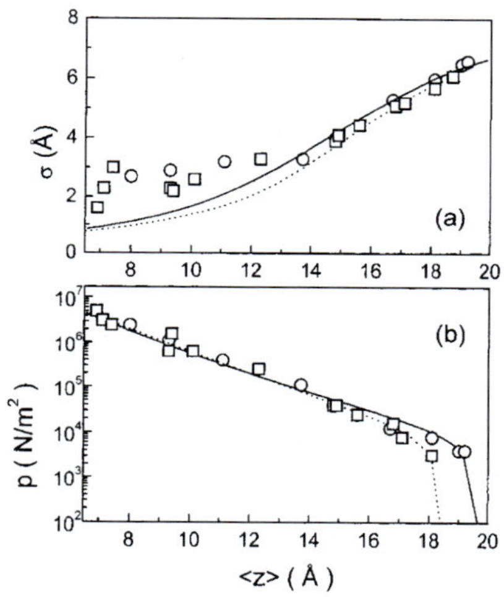

Figure 3. Fit of the experimental data of ref 10 for (a) the root-mean-square fluctuation σ and (b) the applied pressure p as functions of the average separation distance $\langle z \rangle$, for EPC/water (circles/continuous line) and DMPC/water (squares/dotted line) multilayers.

and $A = 0.043$ J/m^2, $\lambda = 1.62$ Å, $A_H = 1.04 \times 10^{-20}$ J, and $K_C = 3.4 \times 10^{-19}$ J (DMPC, dotted line) were obtained from a simultaneous least-squares fit of the experimental data of Petrache et al.[10] for both the pressure and the root-mean-square fluctuation. For the other parameters, we employed the values provided by Petrache et al.:[10] $t_h = 7.6$ Å, $t_c = 45.4$ Å for EPC; $t_h = 7.6$ Å, $t_c = 44.0$ Å for DMPC. The agreement is good except for the root-mean-square fluctuation at low separation distances. There are several reasons for this deterioration of the fit. First, the experiments were carried out for stacks of lipid bilayers, where the correlation between the fluctuations of neighboring membranes is expected to play a role, especially at low separation distances. Second, we assumed that both K_C and t_c were independent of the applied pressure, which is not accurate at large pressures.[13] Third, while X-ray diffraction allowed an extremely precise determination of the total periodicity distance (one bilayer plus one water layer), the average thicknesses of the water and hydrocarbon layers were determined with an error of the order of 1 Å,[10,13] which is clearly relevant at large applied pressures, for which the separation distance is small. It should be noted that the values obtained for the parameters are in the ranges characteristic for phospholipid bilayers.

Since the model was developed for two bilayers in water and was compared to experimental results for lipid bilayer/water multilayers, no complete accuracy is expected for the values obtained by fitting. Particularly, the values obtained for the bending moduli are somewhat larger than those provided by literature. This might be a consequence of the increased rigidity of a multilayer, when compared to that of two bilayers, due to the correlation between the fluctuations of adjacent bilayers.

To conclude, we presented a new method to account for the effect of the thermal fluctuations on the interactions between elastic membranes, based on a predicted intermembrane separation distribution. It was shown that for a typical potential, the distribution function is asymmetric, with an asymmetry dependent on the applied pressure and on the interaction potential between membranes. Equations for the pressure, root-mean-square fluctuation, and asymmetry as functions of the average distance (and the parameters of the interacting membranes) were derived. While no experimental data are available for two interacting lipid bilayers, a comparison with experimental data for multilayers of lipid bilayer/water was provided. The values of the parameters, determined from the fit of experimental data, were found within the ranges determined from other experiments.

LA0201568

Langmuir **2004,** *20,* 1775–1780 1775

Specific Ion Effects in Common Black Films: The Role of the Thermal Undulation of Surfaces

Marian Manciu and Eli Ruckenstein*

Department of Chemical Engineering, State University of New York at Buffalo, Buffalo, New York 14260

Received September 5, 2003. In Final Form: November 26, 2003

It has long been demonstrated that the traditional double layer theory does not provide an accurate description for the repulsion between colloidal particles at small separations and high ionic strengths. Recent experiments on black films revealed that the traditional theory seems to be inadequate even for large separation distances and moderate ionic strengths. These results raise doubts about the validity of the DLVO approximations. It is argued here that the results can be, however, understood within the framework of the traditional theory if one accounts for the thermal undulation of the film interfaces, and a new treatment to account for the thermal undulations is suggested.

1. Introduction

When two charged particles immersed in an electrolyte approach each other, the overlap of their ionic atmospheres (the double layers) generates a repulsive force. The traditional Derjaguin–Landau–Verwey–Overbeek (DLVO) theory assumes that the stability of charged colloids is a consequence of a balance between this double layer repulsion and the attractive van der Waals interactions.[1]

The first theoretical description of the double layers assumed that the ions interact via a mean potential, which obeys the Poisson equation.[2] Such a simple theory is clearly only approximate and sometimes predicts ionic concentrations in the vicinity of the surface that exceed the available volume.[3] There were a number of attempts to improve the model, by accounting for the variation of the dielectric constant in the vicinity of the surface,[4] for the volume-exclusion effects of the ions,[5] or for additional interactions between ions and surfaces, due to the screened image force potential,[6] to the van der Waals interactions of the ions[7] with the system, or to the change in hydration energy when an ion approaches the interface.[8]

It is clear that the DLVO theory is incomplete, a simple example being the stability of neutral lipid bilayers,[9] or of the water films involving nonionic surfactants,[10] where there is no double layer to provide the required repulsion. Another example is provided by the specific ion effects, namely, the different behaviors of systems immersed in different electrolytes of the same valence. Various electrolytes have been classified long ago by Hofmeister in an empirical series, based on their efficiency in protein precipitation.[11] More recently, it was shown that, in a certain pH range, amphoteric latex particles coagulated when the concentration of $CsNO_3$ exceeded about 0.3 M but remained stable even at 3.0 M of KNO_3.[12] Since the DLVO theory of double layer forces accounts only for the ion valency and concentration, it cannot explain why different results were obtained when Cs was replaced by K.

Experiments regarding the force between mica surfaces revealed that the traditional DLVO theory provides good agreement at large separations but fails at low separations and high electrolyte concentrations.[13] For some electrolytes, even at relatively low ionic strengths (10^{-3} M) there is an additional strong repulsion at low separations between surfaces.

All of the above experiments can be understood qualitatively if one assumes an additional short-range repulsion between surfaces. It was suggested that such a repulsion, the hydration force, is provided by the organization of water in the vicinity of the surface (the "hydration" of the surfaces).[14] However, there is no agreement regarding the microscopic origin of the hydration repulsion. Some models assume that the hydrogen bonding of water near a surface is responsible for this repulsion and that the ions do not play any role.[15] Another model considers that, when two surfaces approach each other, the electrolyte ions are losing their hydrating molecules and this unfavorable free energy process generates a repulsion, hence that the hydrated electrolyte ions are vital to the hydration force.[12,16] The local interaction between dipoles (surface–water and water–water dipoles) was also proposed to explain the hydration repulsion.[17] Since the dipole interactions are electrostatic, this hydration and the double layer should be coupled into a single force.[18] In this case,

* Corresponding author. E-mail address: feaeliru@acsu.buffalo.edu. Phone: (716) 645-2911 ext. 2214. Fax: (716) 645-3822.

(1) Deryagin, B. V.; Landau, L. *Acta Physicochim. URSS* **1941**, *14*, 633. Verwey, E. J.; Overbeek, J. Th. G. *Theory of stability of lyophobic colloids*; Elsevier: Amsterdam, 1948.
(2) Gouy, G. J. *Phys. Radium* **1910**, *9*, 457. Chapman, D. L. *Phyl. Mag.* **1913**, *25*, 475.
(3) Stern, O. *Z. Eletrochem.* **1924**, *30*, 508.
(4) Spaarnay, M. J. *Recl. Trav. Chim. Pays-Bas* **1958**, *77*, 872.
(5) Ruckenstein, E.; Schiby, D. *Langmuir* **1985**, *1*, 612.
(6) Jonsson, B.; Wennerstrom, H. *J. Chem. Soc., Faraday Trans. 2* **1983**, *79*, 19.
(7) Ninham, B. W.; Yaminsky, V. *Langmuir* **1997**, *13*, 2097.
(8) Manciu, M.; Ruckenstein, E. *Adv. Colloid Interface Sci.* **2003**, *105*, 63.
(9) Rand, R. P.; Parsegian, V. A. *Biochim. Biophys. Acta* **1989**, *988*, 351.
(10) Strey, R.; Schomacker, R.; Roux, D.; Nallet, F.; Olsson, U. *J. Chem. Soc., Faraday Trans.* **1990**, *86*, 2253.
(11) Hofmeister, F. *Naunin-Schmiedebergs Arch. Exp. Pathol. Pharmakol.* **1888**, *24*, 247.
(12) Healy, T. W.; Homola, A.; James, R. O.; Hunter, R. J. *Faraday Discuss. Chem. Soc.* **1978**, *65*, 156.
(13) Pashley, R. M. *J. Colloid Interface Sci.* **1981**, *83*, 531.
(14) Derjaguin, B. V.; Churaev, N. V.; Muller, V. M. *Suface Forces*; Plenum: New York, 1987.
(15) Attard, P.; Batchelor M. T. *Chem. Phys. Lett.* **1988**, *149*, 206.
(16) Paunov, V. N.; Kaler, E. W.; Sandler, S. I.; Petsev, D. N. *J. Colloid Interface Sci.* **2001**, *240*, 640.
(17) Schiby, D.; Ruckenstein E. *Chem. Phys. Lett.* **1983**, *95*, 435. Manciu M.; Ruckenstein, E. *Langmuir* **2001**, *17*, 7061.
(18) Ruckenstein, E.; Manciu, M. *Langmuir* **2002**, *18*, 7584.

10.1021/la0356598 CCC: $27.50
Published on Web 01/20/2004

while a repulsion is present even in the absence of electrolyte ions, the force can depend on the electrolyte concentration via the Poisson equation,[18] and particularly because of the charge recombination on the surface, which, while decreasing the surface charge, generates new surface dipoles.[17]

A common feature of the experiments discussed above is that the hydration repulsion is short ranged. In contrast, recent experiments with AOT-based black films (thin water films with surfactant monolayers adsorbed at the water/air interfaces) showed that the traditional double layer theory is not accurate even at large distances, of the order of 100 Å.[19] Indeed, different cations (Cs and Li) led to different dependencies of the thickness of the film on the applied pressure, for electrolyte concentrations exceeding 0.05 M. For concentrations between 0.05 and 0.2 M CsCl, the results could be explained in terms of the DLVO interactions, when a surface potential of about 0.1 V was employed in the calculations. However, this implies that the surface charge density doubles when the electrolyte concentration increases from 0.05 to 0.2, whereas a decrease of the surface charge density because of charge recombination is expected to occur. When Li replaced Cs, the equilibrium thickness of the film increased by as much as 10 Å, and the slope of pressure vs distance plot decreased, suggesting an increase in the Debye decay length. Sentenac and Benattar[19] pointed out that the difference is too large to be explained by different Stern layers.[3] The difference is also too large to be accounted for by additional interactions between ions and surfaces, such as ion-dispersion[7] or ion-hydration interactions.[8] It was suggested[19] that the hydration of ions might be responsible for these long-range effects, the Li ion, which is much strongly hydrated than Cs, losing some of its ability to screen the electrostatic repulsion. However, this effect should have been observed in the double layer repulsion between rigid surfaces, such as those reported by Pashley,[13] and not only in the black films.

In the present paper, we will argue that the experimental results of Sentenac and Benattar,[19] which apparently contradict sharply the DLVO theory, can be explained in terms of the traditional theory, when the thermal undulations of the film interfaces are also taken into account. The interface will be treated as a membrane subjected to thermal undulations, which can take place at either constant or nonconstant surface area. In the first case the fluctuations generate bending, while in the second they can produce both bending and change of area.

The bending free energy of a normal mode of wavelength λ, $u(x,y) = \alpha \sin(2\pi(x/\lambda))$, where $u(x,y)$ denotes the displacement of a point of coordinates (x,y) of the membrane, along the normal to the (x,y) plane and α the amplitude, is given by[20]

$$U_{\text{Bending}} = \int_0^{\lambda} dx \int dy \left(\frac{1}{2} K_C \left(\frac{\partial^2 u(x,y)}{\partial x^2} + \frac{\partial^2 u(x,y)}{\partial^2}\right)^2\right) = \int dy \frac{K_C \alpha^2}{4\lambda^4} \quad (1a)$$

where K_C is the bending modulus, while the free energy due to surface tension is[20]

$$U_{\text{ST}} = \int_0^{\lambda} dx \int dy \left(\frac{1}{2}\gamma\left[\left(\frac{\partial u(x,y)}{\partial x}\right)^2 + \left(\frac{\partial u(x,y)}{\partial y}\right)^2\right]\right) = \int dy \frac{\gamma\alpha^2}{4\lambda^2} \quad (1b)$$

where γ is the surface tension.

While the exact value for the bending modulus of an AOT layer between air and water is not known, experiment provided a value of the order of 1 kT for AOT monolayers between water and oil.[21] These values are not very accurate; indeed for water/decane, the neutron spin–echo experiments provided 5.0 kT, while the small-angle neutron scattering led to 0.5 kT.[21] Assuming that $K_C = 1$ kT and $\gamma = 0.025$ N/m,[19] the bending energy (1a) and the surface tension energy (1b) become equal for a mode with $\lambda = (K_C/\gamma)^{1/2} \sim 4$ Å. At smaller wavelengths, the free energy due to the change in area is smaller than that due to bending. However, for much larger wavelengths, the bending free energy becomes much smaller. Consequently, for $\lambda \gg 4$ Å, the bending at constant area becomes favorable thermodynamically. Since the wavelengths of the thermal undulation of AOT interface are in general much larger than 4 Å (the distance between two neighboring AOT molecules on the surface being about 7–8 Å),[22] in what follows it will be assumed that the interfaces bends at constant area.

It was shown that when the interfaces interact through a harmonic potential, the entropic contribution to the free energy due to the finite confinement is inversely proportional to the mean square fluctuation, σ^2, of the distribution of distances between interfaces.[23] In a seminal paper, Helfrich conjectured that for a hard-wall interaction between membranes, the entropic confinement contribution has the same functional form as for a harmonic potential and that σ is proportional to the average intersurface distance $\langle d \rangle$.[23] This conjecture leads to a supplementary pressure, due to the confinement of the thermal undulations, $\Pi_u \propto \langle d \rangle^{-3}$. This law is valid at large separations, for elastic membranes interacting via short-range potentials, such as the neutral lipid bilayers;[24] however, it is not accurate at small separations, where experiment[25] and Monte Carlo simulations[26] showed that σ depends exponentially on $\langle d \rangle$.

The approach proposed here combines a Boltzmann-like distribution of the film thicknesses[27] with the minimization of the total Gibbs free energy.[28] One of the undulating interfaces is assumed flat and rigid (denoted 1 in Figure 1), while the other (2 in Figure 1) is considered to be composed of many flat pieces of finite area A, which behave independently of each other and are Boltzmannian distributed in the potential created by the rigid interface. As shown below, the intersurface distance distribution based on this model coincides, for a harmonic interaction, with the exact result obtained from the partition function of two undulating membranes. The model is extended to cases involving nonharmonic interactions for which analytic solutions are not available.

(19) Sentenac, D.; Benattar, J. J. *Phys. Rev. Lett.* **1998**, *81*, 160.
(20) Sornette D.; Ostrowsky, N. In *Micelles, Membranes, Microemulsions and Monolayers*; Gelbart, W. M., Ben-Shaul, A., Roux, D., Eds.; Springer-Verlag: Berlin, 1994.
(21) Kellay, H.; Binks, B. P.; Hendrikx, Y.; Lee, L. T.; Meunier, J. *Adv. Colloid Interface Sci.* **1994**, *49*, 85.
(22) Bulavchenko, A. I.; Batischev, A. F.; Batischeva, E. K.; Torgov, V. G. *J. Phys. Chem. B* **2002**, *106*, 6381.
(23) Helfrich, W. *Z. Naturforsch.* **1978**, *33a*, 305.
(24) Larche, F. C.; Appell, J.; Porte, G.; Bassereau, P.; Marignan, J. *Phys. Rev. Lett.* **1986**, *56*, 1700.
(25) Petrache, H. I.; Gouliaev, N.; Tristram-Nagle, S.; Zhang, R.; Suter, R. M.; Nagle, J. F. *Phys. Rev. E* **1998**, *57*, 7014.
(26) Gouliaev N.; Nagle, J. F. *Phys. Rev. Lett.* **1998**, *81*, 2610.
(27) Manciu M.; Ruckenstein, E. *Langmuir* **2002**, *18*, 4179.
(28) Manciu M.; Ruckenstein, E. *Langmuir* **2001**, *17*, 2455.

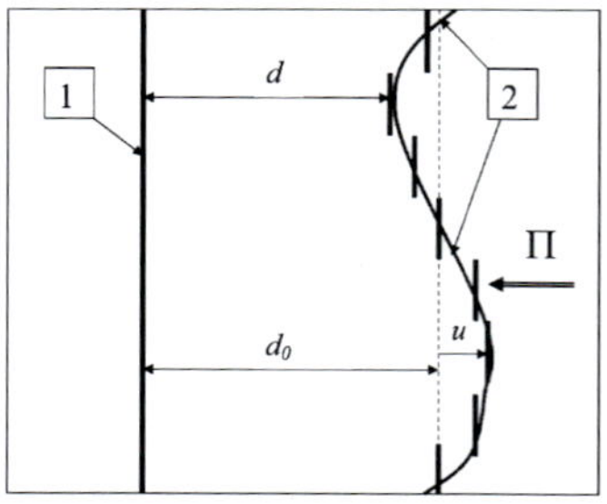

Figure 1. The distribution of distances between an undulating membrane and a rigid surface interacting via a harmonic potential is the same as the Boltzmann distribution of distances between a rigid surface and independent planar pieces of a suitable chosen area A, interacting via the same potential.

At constant external pressure the enthalpy is the relevant thermodynamic potential for the Boltzmann distribution. A large area A implies that the distribution of the thicknesses is confined near the minimum of the enthalpy, while a small value of A corresponds to large thickness fluctuations. The first case corresponds to a low enthalpy, but a large entropy, whereas the second to a large enthalpy, but a low entropy. The real distribution is provided by the minimization of the Gibbs free energy with respect to A at constant external pressure Π.

2. Elastic Interfaces Interacting via a Nonlinear Potential

Let us first summarize the results obtained for a thermal undulating membrane interacting with a rigid, flat surface via a harmonic potential, $U_h(u) = (1/2)Bu^2$. The total energy, per unit area, is given by[20]

$$U_h = \int dx\, dy \left(\frac{1}{2}K_C\left(\frac{\partial^2 u(x,y)}{\partial x^2} + \frac{\partial^2 u(x,y)}{\partial y^2}\right)^2 + \frac{1}{2}Bu^2(x,y)\right) \quad (2)$$

where $u(x,y)$ denotes the displacement of a point of coordinates (x,y) of the membrane, along the normal to the (x,y) plane, from the position of the potential minimum, $u_0 \equiv 0$, K_C is the bending modulus of the membrane and B is the constant of the harmonic potential of interaction between the undulating membrane and the rigid surface. Denoting by $\tilde{u}(q_x,q_y)$ the Fourier transform of $u(x,y)$, the average energy of each undulation mode obtained from the equipartition principle is given by[20,23]

$$\langle \tilde{u}^2 \rangle \left(\frac{1}{2}K_C q^4 + \frac{1}{2}B\right) = \frac{1}{2}kT \quad (3)$$

Therefore for the mean square fluctuations one obtains

$$\sigma^2 = \langle u^2 \rangle = \sum_{q_x,q_y} \langle \tilde{u}^2 \rangle = \frac{kT}{(2\pi)^2}\int_0^\infty \frac{2\pi q\, dq}{K_C q^4 + B} = \frac{kT}{8(K_C B)^{1/2}} \quad (4)$$

and the distribution of distances is Gaussian

$$\rho(u) = \frac{1}{(2\pi)^{1/2}\sigma}\exp\left(-\frac{u^2}{2\sigma^2}\right) \quad (5)$$

This distribution of thicknesses is formally identical with the Boltzmannian distribution of distances between the fixed, rigid membrane (1 in Figure 1) and a collection of planar pieces of membranes of area A, independent of each other, but free to move in the potential $(1/2)Bu^2$ (see Figure 1)

$$\rho(u) = \left(\frac{AB}{2\pi kT}\right)^{1/2} \exp\left(-\frac{ABu^2}{2kT}\right) \quad (6)$$

if the area A is selected as[27]

$$A = \frac{kT}{B\sigma^2} = 8\left(\frac{K_C}{B}\right)^{1/2} \quad (7)$$

An alternate procedure to calculate the distribution of an elastic membrane in a harmonic potential has as starting point the direct integration of the partition function of the canonical ensemble in Fourier space[21]

$$Z = \int \prod d\tilde{u} \left(\exp\left(-\beta \sum_{q_x,q_y} \frac{1}{2}\tilde{u}^2(K_C q^4 + B)\right)\right) = \prod_{q_x,q_y}\left(\frac{2\pi kT}{K_c q^4 + B}\right)^{1/2} \quad (8)$$

The difference in the free energies of a free membrane (which formally corresponds to $B \to 0$) and that interacting with the rigid wall via the harmonic potential $(1/2)Bu^2$ is given by

$$\Delta F = kT\frac{1}{(2\pi)^2}\int_0^\infty dq\; 2\pi q \ln\left(\frac{K_C q^4 + B}{K_C q^4}\right)^{1/2} = \frac{kT}{8}\left(\frac{B}{K_C}\right)^{1/2} \quad (9)$$

The entropic term of the free energy (per unit area) due to the confinement is obtained by subtracting the interaction energy per unit area from eq 9, a result which is essentially due to Helfrich[23]

$$\Delta F_{ent} = \frac{kT}{8}\left(\frac{B}{K_C}\right)^{1/2} - \frac{1}{2}B\langle u^2 \rangle = \frac{(kT)^2}{128 K_C \sigma^2} \quad (10)$$

Let us now assume that the functional form for the entropic contribution, eq 10, holds even when the membrane is subjected to an arbitrary potential, $U(u)$, with a minimum at $u = 0$. Therefore, the Gibbs free energy of the membrane per unit area at a constant external pressure Π is given by

$$G(A) = \int_u \rho(u)(U(u) + \Pi u)\, du + \frac{(kT)^2}{128 K_C \sigma^2} \quad (11a)$$

where

$$\rho(u) = \frac{1}{N}\exp\left(-\frac{A(U + \Pi u)}{kT}\right) \quad (11b)$$

with the normalization factor N given by

$$N = \int_u \rho(u)\, du \quad (11c)$$

The average distance between the membrane and the rigid surface is obtained from

$$\langle d \rangle = d_0 + \frac{1}{N}\int_u u\rho(u)\, du \quad (11d)$$

where d_0 is the distance between the position of the potential minimum ($u = 0$) and the rigid surface. The mean square fluctuation of the membrane is given by

$$\sigma^2 = \frac{1}{N}\int_u (u - (\langle d\rangle - d_0))^2 \rho(u)\, du \tag{11e}$$

The above integrations are performed over the entire range allowed for u.

By employing eqs 11b–e, the Gibbs free energy eq 11a becomes a function of A alone for any value of the external pressure Π, and consequently the value of A can be obtained from the minimization of G. For the harmonic potential $U_h(u) = (1/2)Bu^2$, eq 11a can be integrated

$$G(A) = \int_{-\infty}^{+\infty} \rho(u)\left(\frac{1}{2}Bu^2 + \Pi u\right) du + \frac{(kT)^2}{128K_C \int_{-\infty}^{+\infty} u^2\rho(u)\, du} = \frac{kT}{2A} + \frac{kTBA}{128K_C} \tag{12}$$

where the Boltzmannian distribution of the independent pieces of area A (eq 5) was taken into account. Minimization of G with respect to A recovers the relation between the area A and the elastic parameters of the membrane, eq 7, obtained by employing the equipartition principle (eq 4).

The main assumption in this approach is that the functional form of the entropic term is given by eq 10. This constitutes a very good approximation when most of the membrane is confined near the minimum of the enthalpy (the equilibrium position). This assumption is probably acceptable even when parts of the membrane wander far from the equilibrium position, since the Helfrich law (based on eq 10) seems to be obeyed at large separations.[10,24]

3. Long-Range Interactions in Black Films

Let us now return to the long-range interactions between AOT-based black films at moderately high electrolyte concentrations (0.05–0.2 M CsCl and LiCl).[19] It has been long known that the DLVO theory is not accurate for uni-univalent electrolytes above about 0.05 M.[14] It is also clear that the differences between the hydration behavior of the Li and Cs ions should lead to different interactions of the ions with the interfaces and to different rates for their recombination with the surface charges. Furthermore, the different polarizabilities of the ions should lead to different van der Waals interactions between ions and the rest of the system. In addition, the excluded volume effects are more important for the much strongly hydrated Li^+ than for Cs^+. There is no doubt that all of the above corrections to the DLVO theory, as well as many others, should be included in a complete theory. However, the purpose of this article is not to provide a set of parameters which lead to an excellent agreement with experiment but to show that the traditional DLVO theory can still provide a reasonable accurate description of the experiments, when the thermal undulations of the interfaces are taken into account. Therefore, the specific ion effects will be taken into account in what follows only via the ability of different ions to modify the bending modulus of the interface. There are no reliable data for the bending modulus of AOT interfaces between air and water, and it is not known how the bending modulus depends on the electrolyte concentration. However, there is experimental evidence that the addition of salt (NaCl) can decrease drastically the bending modulus of the AOT film, possibly because of the decrease of the surface charge density at high ionic strength.[29]

(29) Alexandridis, P.; Holzwarth J. F.; Hatton, T. A. *Langmuir* **1993**, *9*, 2045.

The concentration of AOT in all experiments was 2.5×10^{-3} M, which roughly corresponds to its critical micelle concentration (cmc) in the absence of an electrolyte. When an electrolyte is added, the cmc concentration is lowered. For the ionic strengths employed in the experiments of Sentenac and Bennatar,[19] the surface tension reached saturation well before 2.5×10^{-3} M.[19] It will be therefore assumed that in all those experiments, the amount of AOT adsorbed did not change and that each surfactant molecule occupies in average an area of 60 Å^2, which is in agreement with the values (40–70 Å^2) reported in the literature.[22]

The concentration of Na cation, resulting from the dissociation of AOT, will be neglected, since it is at least a few times lower than the concentration of the electrolyte cation. The reassociation of the AOT adsorbed on the surface with the electrolytes cations is taken into account via the association–dissociation equilibrium, which leads to a surface charge density σ_S given by[30]

$$\sigma_S = -\frac{e\Gamma}{1 + \dfrac{C_E}{K_D}\exp\left(-\dfrac{e\psi_S}{kT}\right)} \tag{13}$$

where e is the protonic charge, Γ is the surfactant surface density (1 molecule/60 Å$^2 \equiv 2.77 \times 10^{-6}$ mol/m^2), C_E is the bulk electrolyte concentration (assumed completely dissociated), K_D is the dissociation constant, and ψ_s the surface potential. While specific ion effects can be introduced by employing different dissociation constants for AOT–Li and AOT–Cs, here the same value will be used for both, $K_D = 0.5$ M, implying that the most important contribution to specificity is due to the different bending moduli of the interfaces.

Let us first calculate the free energy of interactions between two planar, nonundulating interfaces. The potential obeys the Poisson–Boltzmann equation

$$\nabla^2\psi = \frac{2eC_E}{\epsilon\epsilon_0}\sinh\left(\frac{e\psi}{kT}\right) \tag{14a}$$

with the boundary conditions

$$\left.\frac{d\psi}{dz}\right|_{z=d/2} = -\frac{\sigma_S}{\epsilon\epsilon_0} \tag{14b}$$

and

$$\left.\frac{d\psi}{dz}\right|_{z=0} = 0 \tag{14c}$$

where z is measured from the middle distance between the planar (nonundulating) interfaces separated by a distance d, ϵ is the dielectric constant, and ϵ_0 is the vacuum permittivity.

After the solution of the Poisson–Boltzmann equation is obtained, the total double layer free energy per unit area is obtained by adding the electrostatic

$$F_{el}(d) = \frac{1}{2}\int_{-d/2}^{d/2} \epsilon\epsilon_0(\nabla\psi)^2\, dz \tag{15a}$$

entropic

(30) Prieve, D. C.; Ruckenstein, E. *J. Theor. Biol.* **1976**, *56*, 205.

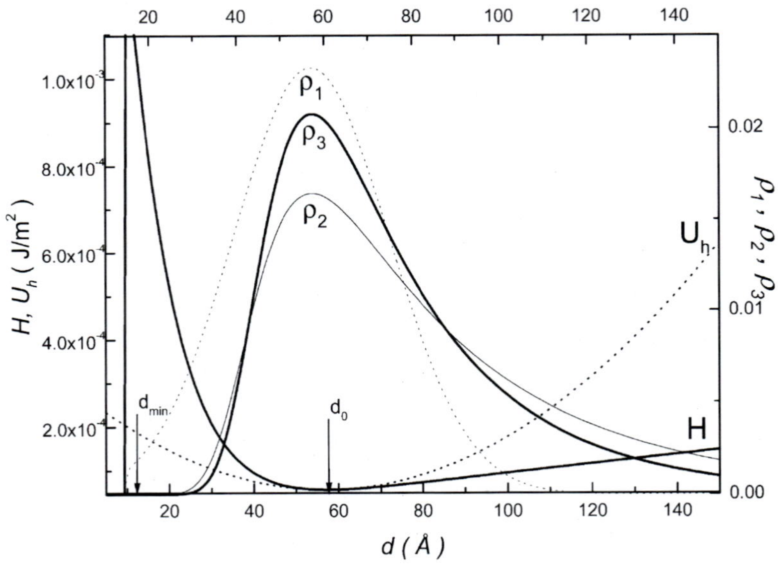

Figure 2. The enthalpy H and its harmonic approximation U_h in the vicinity of the minimum, for planar, rigid surfaces separated by a distance d ($C_E = 0.1$ M, $\Pi = 1 \times 10^4$ N/m², $\Gamma = 3.3 \times 10^{-6}$ mol/m², $K_D = 0.5$ M, $b_1 = -3.08 \times 10^{-22}$ J, $b_2 = -6.28 \times 10^{-14}$ J/m, $b_3 = 8.28 \times 10^7$ J/m, $b_4 = 6.13 \times 10^{15}$ J/m², $b_5 = -9.00 \times 10^{-23}$ J, and $T = 300$ K). ρ_1 is the distribution of the intersurface distances for interfaces with bending modulus $K_C = 2\,kT$ interacting via the potential U_h. This distribution coincides with the Boltzmann distribution of finite pieces of area A. ρ_2 is the Boltzmann distribution of the pieces of area A, but now the enthalpy H and not its harmonic approximation U_h is the thermodynamic potential. ρ_3 is calculated using for A the value obtained from the minimization of the Gibbs free energy (eq 11a).

$$F_{\text{ent,ions}}(d) = 2C_E kT \int_{-d/2}^{d/2} \left[\left(\frac{e\psi}{kT}\right)\sinh\left(\frac{e\psi}{kT}\right) + 1 - \cosh\left(\frac{e\psi}{kT}\right)\right] dz \quad (15b)$$

and chemical contributions[31]

$$\Delta F_{\text{ch}}(d) = -2\int_{\sigma_S(\infty)}^{\sigma_S(d)} \psi_S(\sigma_S)\, d\sigma_S \quad (15c)$$

where ψ_S is the surface potential and $\sigma_S(\infty)$ is the surface charge at infinite separation.

For the van der Waals interaction between nonundulating AOT interfaces, whose hydrocarbon chains are about 7 Å long,[22] the Pade aproximant, calculated on the basis of the Lifshitz theory by Donners et al., will be used[32]

$$U_{\text{vdW}}(d) = \frac{1}{d^2}\left(\frac{b_1 + b_2 d}{1 + b_3 d + b_4 d^2} + b_5\right) \quad (16)$$

where d is the thickness of the water layer and the empirical parameters b_i are assumed equal to those provided by Donners et al. for dodecane films with thickness of 9 Å[32] (in SI units): $b_1 = -3.08 \times 10^{-22}$, $b_2 = -6.28 \times 10^{-14}$, $b_3 = 8.28 \times 10^7$, $b_4 = 6.13 \times 10^{15}$, and $b_5 = -9.00 \times 10^{-23}$.

The enthalpy per unit area, H, between the surface 1 and one of the flat surfaces of area A is given by

$$H(u) = [F_{\text{el}}(d) - F_{\text{el}}(\infty) + F_{\text{ent,ions}}(d) - F_{\text{ent,ions}}(\infty) + \Delta F_{\text{ch}}(d) + U_{\text{vdw}}(d)] + \Pi u \quad (17)$$

where $d = d_0 + u$, d_0 being the distance between the rigid membrane and the position of the local minimum of the enthalpy. The enthalpy defined by eq 17 has an absolute minimum ($H = -\infty$) for $d = 0$, due to the divergence of the van der Waals attraction. However, the black film (before rupture) is in a metastable state, with the intersurface distances distributed around the local minimum of the enthalpy and all the intersurface distances larger than $d_{\min}$ (see Figure 2). The integration over u in eqs 11 are therefore performed in the range corresponding to $d_{\min} < d < \infty$ ($d_{\min} - d_0 < u < \infty$). When, because of thermal undulations, one region of one of the interfaces is separated from the corresponding region of the other interface by distances smaller than $d_{\min}$, the black film ruptures. Details of this process are provided elsewhere.[33]

In Figure 2 the enthalpy H of the system, per unit area, for a nonundulating interface is plotted as a function of the separation distance d for $C_E = 0.1$ M and $\Pi = 1 \times 10^4$ N/m², the other parameters having the values given above. When both interfaces are rigid, their separation distance d_0 corresponds to the local minimum of the enthalpy. If the bending modulus is large, the separation between the interfaces is distributed in the vicinity of this minimum, where the enthalpy can be well approximated by the harmonic potential U_h. The distribution of the separation distances ρ_1 was calculated using this harmonic approximation and eq 5, for a bending modulus $K_C = 2\,kT$ ($T = 300$ K). As already noted, the distribution is identical to a Boltzmann distribution of independent pieces of interfaces of area A, given by eq 7. A further improvement in calculating the distribution of the intersurface distances, proposed previously,[27] consisted in using the value of A provided by the harmonic approximant of the enthalpy, U_h, but employing the real enthalpy H for the Boltzmann distribution (ρ_2 in Figure 2). A new better approximation is to consider the value of A as a parameter which minimizes the Gibbs free energy, eq 11a (curve ρ_3 in Figure 2). The differences between ρ_2 and ρ_3 are in general significant only for small values of K_C (flexible interfaces) and large separation distances.

The bending moduli of the AOT films in the presence of various electrolytes are not known. Experiment showed that[19] at low AOT concentrations, the surface tension

(31) Manciu, M.; Ruckenstein, E. *Langmuir* **2003**, *19*, 1114.
(32) Donners, W. A. B.; Rijnbout, J. B.; Vrij, A. *J. Colloids Interface Sci.* **1977**, *60*, 540.
(33) Ruckenstein, E.; Manciu, M. *Langmuir* **2002**, *18*, 2727.

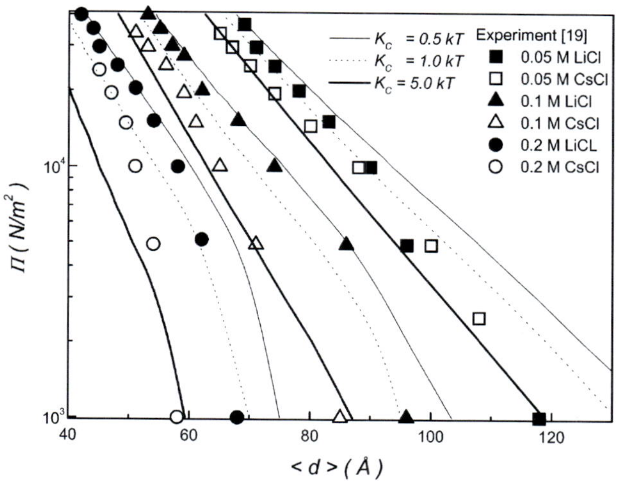

Figure 3. The experimental data for the external pressure Π vs average thickness ⟨*d*⟩ are compared to the results of the present theory, which accounts for the undulation of the interfaces combined with the traditional DLVO treatment of interactions. Different bending moduli affect not only the strength of the interactions but also its slope (traditionally related to the Debye screening length).

increased by about 5×10^{-3} N/m when Cs was replaced by Li. This indicates that the more strongly hydrated Li ion somehow diminishes the adsorption of AOT on the interface and also that the (negative) surface adsorption of Li is larger (in absolute value) than the (negative) surface adsorption of Cs. Since the Li ions cannot approach the water/AOT interface (a simple theory for surface tension showed that the distance of closest approach might be a few times larger than the ion's diameter),[8] the water in the vicinity of the surface is not bound to the hydrated ions; hence the interface can more easily undulate in the "free" water. On the other hand, the Cs ions can approach closer to the interface binding of some water molecules. The presence of these more rigid structures (hydrated ions) in the vicinity of the interface attenuates the undulations. On this ground, one expects the AOT interfaces to be more flexible in the presence of Li ions. Another possibility is that different recombination constants for Cs and Li lead to different surface charging and hence to different bending moduli. However, in this article the same dissociation constant was employed in both cases, to emphasize the importance of membrane flexibility.

In Figure 3, the experimental results of ref 19 are compared with the calculations involving the traditional DLVO theory for the interactions between parallel (nonundulating) interfaces, combined with the present approach to account for the thermal undulations. The only parameter modified was the unknown bending modulus, whose values were changed from 0.5 to 5 kT, which are of the right order of magnitude for the water/oil AOT interfaces.[21] A reasonable agreement with experiment was obtained for K_C in the range 0.5–5 kT even though the other corrections to the traditional DLVO interactions (which are clearly non-negligible) were completely disregarded. It is of interest to note that the calculations at low ionic strengths seems to be better explained by a larger bending modulus, which is expected on intuitive grounds because the higher charged interface should be also more rigid. Consequently, the specific ion effects could be explained in terms of a change in interface flexibility within the traditional DLVO theory.

4. Conclusions

It is well-known that the traditional interactions of the DLVO theory are not accurate at high electrolyte interactions, particularly at small separation distances. Recent experiments on AOT-based black films apparently suggested that the traditional theory is also not valid at moderate ionic strengths and large separations. In this paper, it was argued that the experiments can be understood in terms of the traditional theory, when the thermal undulations of the interfaces are taken into account.

A new approach to account for the thermal undulations of elastic interfaces interacting via arbitrary potentials was suggested. For interfaces interacting via a harmonic potential, the distribution of distances between interfaces is formally identical to a Boltzmann distribution of many independent pieces of area A, where A is related to the bending modulus and to the constant of the interaction potential. Assuming that the distribution remains Boltzmannian in a nonharmonic potential, and that the entropic term due to undulation confinement remains proportional to σ^{-2}, as for a harmonic potential, the value of A can be obtained via the minimization of the Gibbs free energy of the system.

On the basis of this approach, it was shown that the experiments on AOT-based black films, which apparently disagree drastically with the DLVO theory, can still be understood in the traditional framework, if the bending modulus of the AOT-based interfaces depends on the kind of electrolyte used Li or Cs. One of the reasons for this explanation is that such a long-range departure from the classical theory was observed in thin films with flexible charged interfaces[34] and not between rigid interfaces.[13]

It should be emphasized that, without any doubt, there are many differences between the double layers formed in LiCl and CsCl electrolytes, particularly at high ionic strengths, which are completely disregarded here. The dissociation constant of Li–AOT is probably different from that of Cs–AOT. In addition, the dependence of the bending modulus on the electrolyte concentration for the two types of electrolytes is unknown. The purpose of this paper was not to provide a set of parameters, which can describe the experiment, but merely to suggest that the DLVO theory, the traditional workhorse of colloid science, still provides reasonable results, when the undulations of the interfaces are taken into account.

LA0356598

(34) Kolarov, T.; Cohen, R.; Exerowa, D. *Colloids Surf.* **1989**, *42*, 49.

5412 *Langmuir* **2004,** *20,* 5412−5417

Interaction Force between Two Charged Plates Immersed in a Solution of Charged Particles. Coupling between Double Layer and Depletion Forces

Haohao Huang and Eli Ruckenstein*

Department of Chemical and Biological Engineering, State University of New York at Buffalo, Amherst, New York 14260

Received February 3, 2004. In Final Form: April 9, 2004

When two parallel plates are immersed in a solution of small charged particles, the center of the particles is excluded from a region of thickness $D/2$ near the plate, where D is their diameter. The approach which Langmuir developed for the double layer repulsion in the presence of an electrolyte with ions of negligible size is extended to the case in which one of the "ions" is a charged particle of finite, relatively small size. A general expression for the force generated between the two charged plates immersed in an electrolyte solution containing relatively small charged particles is derived. In this expression, only the electrical potential at the middle distance between the plates is required to calculate the force. A Poisson−Boltzmann equation which accounts for the volume exclusion of the charged particles in the vicinity of the surface is solved to obtain the electrical potential at the middle between the two plates. Starting from this expression, some results obtained previously for the depletion force acting between two plates or two spheres are rederived. For charged plates immersed in a solution of an electrolyte and charged small particles, the effects of the particle charge, particle charge sign, particle size, and volume fraction of the particles on the force acting between the two plates are examined.

1. Introduction

The presence of nonadsorbing polymer particles in a colloidal dispersion affects the colloidal stability.[1] The depletion force has been the subject of many experimental and theoretical studies. A depletion force between two colloidal particles immersed in a solution of small particles, such as polymer particles or micelles, arises when the gap between the two particles becomes smaller than the diameter of the small particles. At small gaps, the small particles are expelled from the gap and an attraction between the two large particles is generated. Asakura and Oosawa were the first to provide an explanation for this attractive force.[2,3] They considered the cases of two plates or two relatively large particles immersed in a solution containing rigid, spherical, relatively small macromolecules. When the gap between the two large particles becomes smaller than the diameter of the macromolecules, the depletion of macromolecules in the gap leads to a concentration difference of the macromolecules between the bulk and the gap. As a result, an attraction between the two large particles is generated due to the difference between the osmotic pressures in the gap and in the bulk.

The interactions between two parallel plates or spheres in a solution of nonadsorbing polymers assumed to be spherical have been extensively investigated experimentally,[4−6] theoretically,[7−13] and by simulations.[14,15] The second virial coefficients of the polymer particles were introduced into the theory by Sieglaff and Vrij.[16−18] Walz and Sharma calculated the force acting between charged plates or particles immersed in an electrolyte solution also containing small but finite size particles using a "force balance" approach.[19] The effect of the polydispersity of the small particles was examined by Chu et al.[20] and by Walz.[21]

In this paper, the approach suggested by Langmuir[22] for the double layer repulsion is employed to calculate the force between two charged parallel plates immersed in an electrolyte solution containing also relatively small charged particles. Langmuir's approach has the advantage that only the electrical potential at the middle between the two plates appears in the expression for the force. Consequently, the force can be directly obtained from the electrical potential distribution provided by a Poisson−Boltzmann equation which accounts for the excluded volume due to the finite size of the small charged particles. This paper is organized as follows: first, a general

* To whom correspondence should be addressed. Phone: (716) 645-2911 ext. 2214. Fax: (716) 645-3822. E-mail: feaeliru@acsu.buffalo.edu.

(1) Napper, D. H. *Polymeric stabilization of colloidal dispersions*; Academic Press: New York, 1983.
(2) Asakura, S.; Oosawa, F. *J. Chem. Phys.* **1954**, *22*, 1255.
(3) Asakura, S.; Oosawa, F. *J. Polym. Sci.* **1958**, *33*, 183.
(4) Verma, R.; Crocker, J. C.; Lubensky, T. C.; Yodh, A. G. *Phys. Rev. Lett.* **1998**, *81*, 4004.
(5) Liang, W.; Tadros, T. F. *J. Colloid Interface Sci.* **1993**, *158*, 152.
(6) Ohshima, Y. N.; Sakagami, H.; Okumoto, K.; Tokoyada, A.; Igarashi, T.; Shintaku, K. B.; Toride, S.; Sekino, H.; Kabuto, K.; Nishio, I. *Phys. Rev. Lett.* **1997**, *78*, 3963.
(7) Vrji, A. *Pure Appl. Chem.* **1976**, *48*, 471.
(8) Joanny, J. F.; Leibler, L.; de Gennes, P. G. *J. Polym. Sci., Polym. Phys. Ed.* **1979**, *17*, 1073.
(9) Attard, P. *J. Chem. Phys.* **1989**, *91*, 3083.
(10) Henderson, D. *J. Colloid Interface Sci.* **1988**, *121*, 486.
(11) Gast, A. P.; Hall, C. K.; Russel, W. B. *J. Colloid Interface Sci.* **1983**, *96*, 251.
(12) Rao, I. V.; Ruckenstein, E. *J. Colloid Interface Sci.* **1985**, *108*, 389.
(13) Ruckenstein, E.; Rao, I. V. *Colloids Surf.* **1986**, *17*, 185.
(14) Dickman, R.; Attard, P.; Simonian, V. *J. Chem. Phys.* **1997**, *107*, 205.
(15) Feigin, R. I.; Napper, D. H. *J. Colloid Interface Sci.* **1980**, *75*, 525.
(16) Sieglaff, C. L. *J. Polym. Sci.* **1959**, *41*, 319.
(17) Vrij, A. *Pure Appl. Chem.* **1976**, *48*, 471.
(18) de Hek, H.; Vrij, A. *J. Colloid Interface Sci.* **1981**, *84*, 409.
(19) Walz, J. Y.; Sharma, A. *J. Colloid Interface Sci.* **1994**, *168*, 485.
(20) Chu, X. L.; Nikolov, A. D.; Wasan, D. T. *Langmuir* **1996**, *12*, 5004.
(21) Walz, J. Y. *J. Colloid Interface Sci.* **1996**, *178*, 505.
(22) Langmuir, I. *J. Chem. Phys.* **1938**, *6*, 893.

10.1021/la049703a CCC: $27.50
Published on Web 05/27/2004

expression for the force acting between two parallel plates immersed in an electrolyte solution containing small charged particles of finite size is derived; second, the obtained expression is used to derive in a consistent manner expressions for the depletion force previously established; finally, the effects of the charge of the small finite size particles, their sign, their size, and the volume fraction as well as the electrolyte concentration on the force between the two charged plates are examined.

2. Theory

Langmuir considered that the force acting per unit area between two charged plates is equal to the difference between the osmotic pressures at the middle between the plates and the bulk. On the basis of this idea, Langmuir derived the following expression for the force acting between two charged parallel plates (p) immersed in a symmetric v/v electrolyte solution[22,23]

$$p = 2nkT\left[\cosh\left(\frac{ve\Psi_0}{kT} - 1\right)\right] \quad (1)$$

where n is the number density of the electrolyte molecules in the bulk, e is the protonic charge, Ψ_0 is the electrical potential at the midway between the two plates, v is the valency of the ions, k is the Boltzmann constant, and T is the absolute temperature. This equation implies that the ion distributions are Boltzmannian and the solution is dilute.

If small finite size charged particles are also present in the electrolyte solution, Langmuir's approach can be generalized by writing the following expression for the force acting between the plates:

$$p = \Pi_0 - \Pi_b = \sum_i n_{i,0}kT - \sum_i n_{i,b}kT \quad (2)$$

where Π_0 and Π_b are the osmotic pressures at the midway between the two plates and in the bulk, respectively, and $n_{i,0}$ and $n_{i,b}$ are the number densities of ion i, including the small finite size charged particles, at the midway between the plates and in the bulk, respectively. The second equality in the above equation implies that the solution satisfies the van't Hoff equation.

Let us consider two parallel plates immersed in a solution of a 1:1 electrolyte and of small finite size particles which can dissociate monovalent counterions, thus acquiring a charge of ze per particle. The system contains two ions of the electrolyte which are assumed to be of negligible size, one kind of charged particles which have a finite size, and their counterions which are of negligible size. Assuming Boltzmann distributions for the ions and charged particles, eq 2 becomes

$$p = kT\left\{2n\left[\cosh\left(\frac{e\Psi_0}{kT}\right) - 1\right] + |z|n_m\left[\exp\left(\mathrm{sgn}(z)\frac{e\Psi_0}{kT}\right) - 1\right] + n_m\left[\exp\left(-\frac{ze\Psi_0}{kT}\right) - 1\right]\right\} \quad (3)$$

for distances L between the plates that are larger than the diameter D of the charged small particles and

$$p = kT\left\{2n\left[\cosh\left(\frac{e\Psi_0}{kT}\right) - 1\right] + |z|n_m\left[\exp\left(\mathrm{sgn}(z)\frac{e\Psi_0}{kT}\right) - 1\right] - n_m\right\} \quad (4)$$

for distances L between the plates that are smaller than the diameter D, because the gap is in this case depleted of particles. In eqs 3 and 4, ze is the charge of the particles, sgn is the sign function, and n_m is the number density of the charged particles in the bulk.

The main assumption in eq 3 is that the small particles can be treated in the framework of the Poisson–Boltzmann approximation. However, being of finite size and having a relatively large charge, the interactions among particles as well as their van der Waals interactions with the plates should also be involved in the distribution of their concentration. The latter treatment[24] will allow identification of the conditions under which the present simple treatment is valid.

The electrical potential distribution can be calculated using the Poisson equation

$$\frac{d^2\Psi}{dx^2} = -\frac{\rho}{\epsilon\epsilon_0} \quad (5)$$

where ρ is the charge density in solution, ϵ is the dielectric constant, ϵ_0 is the vacuum permittivity, and x is the distance measured from the middle between the plates.

Let us consider that the distance L between the plates is larger than D. The center of the charged particle cannot be located in a layer of thickness $D/2$ near the plate. Consequently, the charge density in region I of thickness $D/2$ close to the plate is given by

$$\rho = -2ne\sinh\left(\frac{e\Psi}{kT}\right) - zn_m e\exp\left(\mathrm{sgn}(z)\frac{e\Psi}{kT}\right) \quad -L/2 < x < -(L-D)/2,\ L > D \quad (6)$$

and the charge density outside this region, region II, is given by

$$\rho = -2ne\sinh\left(\frac{e\Psi}{kT}\right) - zn_m e\exp\left(\mathrm{sgn}(z)\frac{e\Psi}{kT}\right) + zn_m e\exp\left(-\frac{ze\Psi}{kT}\right) \quad -(L-D)/2 < x < 0,\ L > D \quad (7)$$

When the distance between the two plates becomes smaller than the diameter D of the small particles, eq 6 for the charge density can be employed in the whole gap between the plates.

To obtain the electrical potential at the midway between the plates, boundary conditions are required. At the plate surface with a charge density of σ,

$$\frac{d\Psi}{dx} = -\frac{\sigma}{\epsilon\epsilon_0} \quad \text{for} \quad x = -L/2 \quad (8)$$

(23) Verwey, E. J.; Overbeek, J. Th. G. *Theory of the Stability of Lyophobic Colloids*; Elsevier: Amsterdam, The Netherlands, 1948.

(24) Huang, H.; Ruckenstein, E. Manuscript in preparation.

At the boundary between the two regions, I and II, the electrical potential and its derivative with respect to x should be continuous.

$$\Psi|^{\mathrm{I}}_{x=(D-L)/2} = \Psi|^{\mathrm{II}}_{x=(D-L)/2} \tag{9}$$

$$\left.\frac{\mathrm{d}\Psi}{\mathrm{d}x}\right|^{\mathrm{I}}_{x=(D-L)/2} = \left.\frac{\mathrm{d}\Psi}{\mathrm{d}x}\right|^{\mathrm{II}}_{x=(D-L)/2} \tag{10}$$

At the midway between the plates, the symmetry of the system provides another boundary condition.

$$\frac{\mathrm{d}\Psi}{\mathrm{d}x} = 0 \qquad (x = 0) \tag{11}$$

Combined with the boundary conditions (eqs 8–11) and the charge density distributions (eqs 6 and 7), eq 5 can be solved numerically to obtain the electrical potential distribution. Introducing the value obtained for the electrical potential at the middle between the plates in eq 3 or 4, one can obtain the force that is acting between the two plates.

The force can also be calculated starting from an expression for the free energy. The free energy (F) of a colloidal system composed of two charged parallel plates immersed in an electrolyte solution containing charged particles is composed of three contributions: electrical (F_{ele}), entropic (F_{ent}), and chemical (F_{che}).[25,26]

$$F = F_{\mathrm{ele}} + F_{\mathrm{ent}} + F_{\mathrm{che}} \tag{12}$$

The electrical contribution, the electrostatic energy of the system, can be expressed as

$$F_{\mathrm{ele}} = \frac{1}{2}\epsilon\epsilon_0 \int_{-L/2}^{L/2} \left(\frac{\mathrm{d}\Psi}{\mathrm{d}x}\right)^2 \mathrm{d}x \tag{13}$$

The entropic contribution with respect to that in the bulk is given by

$$F_{\mathrm{ent}} = kT\int_{-L/2}^{L/2} [\sum(n_i \ln x_i - n_{i,\mathrm{b}} \ln x_{i,\mathrm{b}}) + n_{\mathrm{w}} \ln x_{\mathrm{w}} - n_{\mathrm{w,b}} \ln x_{\mathrm{w,b}}]\, \mathrm{d}x \tag{14}$$

where n_{w} is the number density of the water molecules located in the gap, x_i and x_{w} are the mole fractions of ion i and the water molecules located in the gap, respectively, and the subscript b indicates the bulk.

The chemical free energy with respect to the bulk can be calculated using the expression[26]

$$\Delta F_{\mathrm{che}} = -2\int_{\sigma(\infty)}^{\sigma(L)} \Psi_{\mathrm{s}}\, \mathrm{d}\sigma \tag{15}$$

where $\sigma(L)$ is the surface charge density for a distance between the plates equal to L and Ψ_{s} is the surface potential.

The force is provided by the derivative of the free energy with respect to the distance L

$$p = -\frac{\mathrm{d}F}{\mathrm{d}L} \tag{16}$$

3. Depletion Force Acting between Uncharged Plates or Spheres

If the two plates are uncharged, the electrical potential in the middle is 0, and the force acting between the plates is the depletion force. By the substitution of $\Psi_0 = 0$ into eqs 3 and 4, the equations acquire the form

$$p = \begin{cases} -n_{\mathrm{m}}kT & \text{for } L < D \\ 0 & \text{for } L \geq D \end{cases} \tag{17}$$

where the negative sign indicates an attractive force. Consequently, the depletion free energy (E) is given by

$$E = -\int_{\infty}^{L} p\, \mathrm{d}x = \begin{cases} n_{\mathrm{m}}kT(D-L) & \text{for } L < D \\ 0 & \text{for } L \geq D \end{cases} \tag{18}$$

Equation 12 for the free energy also leads to the above expression for the depletion force. Indeed, because there is no charge on the plates, the electrical and chemical contributions become 0 and only the entropic contribution remains. For distances smaller than the particle diameter, no particle is located between the plates and the entropic contribution is provided by the difference of the mixing entropies of the water molecules in the gap and the bulk. Because the number density of the small particles (n_{m}) of diameter D in the bulk is much smaller than that of the water molecules ($n_{\mathrm{w,b}}$), the free energy can be approximated by

$$F = 2kT\int_{-L/2}^{0}(0 - n_{\mathrm{w,b}} \ln x_{\mathrm{w,b}})\, \mathrm{d}x = -n_{\mathrm{w,b}}kTL \ln x_{\mathrm{w,b}} \cong n_{\mathrm{w}}kTLx_{\mathrm{m}} \cong n_{\mathrm{m}}kTL \tag{19}$$

Therefore, the depletion force for a distance of $L < D$ is

$$p = -\frac{\mathrm{d}F}{\mathrm{d}L} = -n_{\mathrm{m}}kT \tag{20}$$

For distances larger than the diameter D,

$$F = 2kT\int_{-L/2}^{-(L-D)/2}(0 - n_{\mathrm{w,b}} \ln x_{\mathrm{w,b}})\, \mathrm{d}x = -n_{\mathrm{w,b}}kTD \ln x_{\mathrm{w,b}} \cong n_{\mathrm{w}}kTDx_{\mathrm{m}} \cong n_{\mathrm{m}}kTD \tag{21}$$

and the entropic contribution becomes independent of the distance L. Therefore, the force is 0 for $L > D$.

Using Derjaguin's approximation, the force between two uncharged spheres (p_{s}) of diameter a separated by a center-to-center distance of l can be calculated from the force between two parallel plates (p) using the expression[23]

$$p_{\mathrm{s}} = \int_0^{\infty} 2\pi h p(h)\, \mathrm{d}h \tag{22}$$

where h is defined in Figure 1.

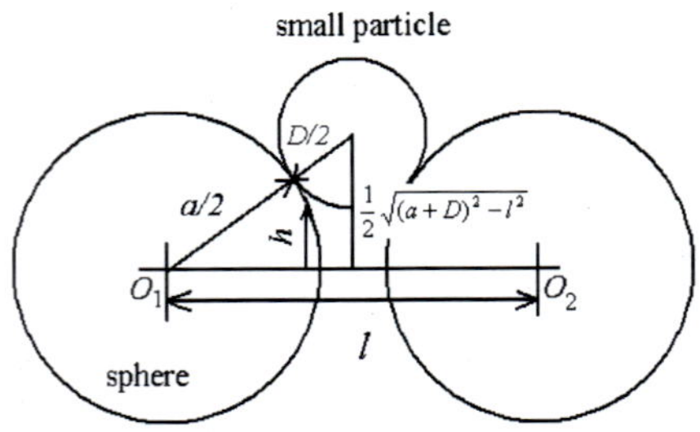

Figure 1. Schematic of two large spheres immersed in a solution of small particles. When the center-to-center distance between the two spheres of diameter a is l, the distance from the connecting line of the centers of the large particles to the center of the particle of diameter D, which contacts the two spheres, is $\frac{1}{2}\sqrt{(a+D)^2-l^2}$.

(25) Overbeek, J. Th. G. *Colloids Surf.* **1990**, *51*, 61.
(26) Manciu, M.; Ruckenstein, E. *Langmuir* **2003**, *19*, 1114.

Since, for $D > l - a$, the center of the small particles cannot be located in the region where $h < {}^1\!/_2\sqrt{(a+D)^2-l^2}$ (see Figure 1), eq 17 becomes

$$p = \begin{cases} -n_{\mathrm{m}}kT & \text{for } h < (\sqrt{(a+D)^2 - l^2})/2 \\ 0 & \text{for } h \geq (\sqrt{(a+D)^2 - l^2})/2 \end{cases} \tag{23}$$

For $D \leq l - a$, $p = 0$.

Equations 22 and 23 lead to

$$p_{\mathrm{s}} = \begin{cases} -\dfrac{1}{4}n_{\mathrm{m}}\pi kT[(a+D)^2 - l^2] & \text{for } a < l < a + D \\ 0 & \text{for } l \geq a + D \end{cases} \tag{24}$$

Equations 17, 18, and 24 have been obtained previously by different methods.[2,19]

4. Effect of Polydispersity of the Particle Size on the Depletion Force

Let us now consider the case of two uncharged plates immersed in a solution of small polydisperse particles. An expression for the force between the plates can be obtained directly from eq 2. Only the particles with sizes smaller than the distance between the plates can be present in the gap. However, the particles with larger sizes cannot be located in the gap. Therefore, the depletion force can be expressed as

$$p = \int_0^L n_{\mathrm{m}}kTf(D)\,\mathrm{d}D - \int_0^\infty n_{\mathrm{m}}kTf(D)\,\mathrm{d}D = -n_{\mathrm{m}}kT\int_L^\infty f(D)\,\mathrm{d}D \tag{25}$$

where $f(D)$ is the probability density of the particle size distribution. If the size distribution of the particles is Gaussian, $f(D)$ has the form

$$f(D) = \frac{1}{\alpha\sqrt{2\pi}}\exp\left[-\frac{1}{2}\left(\frac{D - \bar{D}}{\alpha}\right)^2\right] \tag{26}$$

where $\bar{D}$ is the mean particle diameter and α is the standard deviation. Substituting eq 26 into eq 25, one can readily obtain for the depletion force between two plates immersed in a solution of normally distributed particles the expression

$$p = -\frac{1}{2}n_{\mathrm{m}}kT\,\mathrm{erfc}\left(\frac{L - \bar{D}}{\alpha}\right) = \frac{1}{2}n_{\mathrm{m}}kT\left[\mathrm{erf}\left(\frac{L - \bar{D}}{\alpha}\right) - 1\right] \tag{27}$$

where erfc and erf are the complementary and the error functions, respectively.

The depletion energy is

$$E = \frac{n_{\mathrm{m}}kT\alpha}{2}\left\{\frac{L - \bar{D}}{\alpha}\,\mathrm{erfc}\left(\frac{L - \bar{D}}{\alpha}\right) - \frac{1}{\sqrt{\pi}}\exp\left[-\left(\frac{L - \bar{D}}{\alpha}\right)^2\right]\right\} \tag{28}$$

The force changes from $-n_{\mathrm{m}}kT$ to 0 as the distance increases. For a standard deviation equal to 0, the force reduces to eq 7.

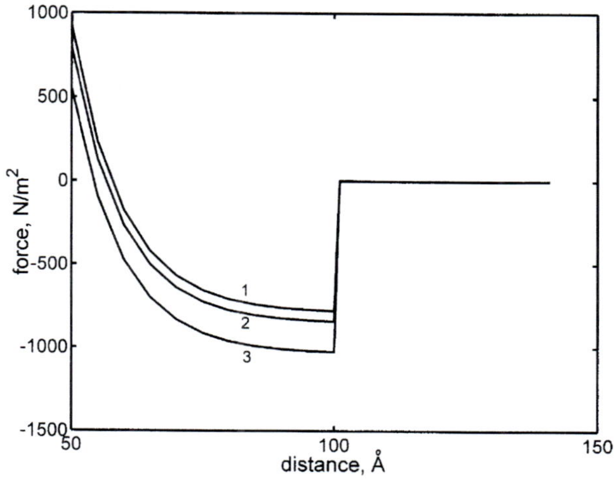

Figure 2. Effect of the charge of the small particles on the force. The force was calculated using eqs 3 and 4. The parameters used in the calculation are $\sigma = -0.01$ C/m², electrolyte concentration $c = 0.1$ M, volume fraction of the small particles = 0.01, and $D = 100$ Å: (1) $z = 0$; (2) $z = -10$; (3) $z = -20$.

Using eq 27 and Derjaguin's approximation, one obtains for the force between two spherical particles immersed in a solution of particles of polydisperse sizes the expression

$$p_{\mathrm{s}} = \frac{an_{\mathrm{m}}kT\pi\alpha}{4}\left\{\frac{L - \bar{D}}{\alpha}\,\mathrm{erfc}\left(\frac{L - \bar{D}}{\alpha}\right) - \frac{1}{\sqrt{\pi}}\exp\left[-\left(\frac{L - \bar{D}}{\alpha}\right)^2\right]\right\} \tag{29}$$

Equations 27 and 29 were first derived by Walz who employed a force balance procedure.[21]

5. Force between Two Charged Plates Immersed in an Electrolyte Solution Containing Small Charged Particles

When the two plates are charged, an electrical potential is generated between them and the ions and the charged small particles are no longer uniformly distributed between the plates.

5.1. Effect of the Charge of the Small Particles on the Force. Let us assume that both the plates and the particles are negatively charged. As the charge of the small particles becomes more negative, the concentration of the small particles between the two plates decreases because they have the same sign of the charge as the plates. Therefore, the contribution of the small particles to the osmotic pressure at the middle distance also decreases. Consequently, the force decreases (becomes less positive) as the charge of the small particles becomes more negative (Figure 2).

The interaction between the two plates is repulsive at small distances and becomes attractive at large distances. At small distances, the electrostatic interactions between the two plates dominate and the net force is repulsive. As the distance increases but remains smaller than the diameter of the particles, the electrostatic repulsion between the plates decreases but, because the attractive depletion force remains almost constant, the net force can become attractive. As soon as the distance between the two plates becomes larger than the particle diameter, the electrostatic and depletion contributions become comparable and the net force becomes small. The above considerations are valid when D is sufficiently large for the double layer repulsion at the distance $L = D$ to become

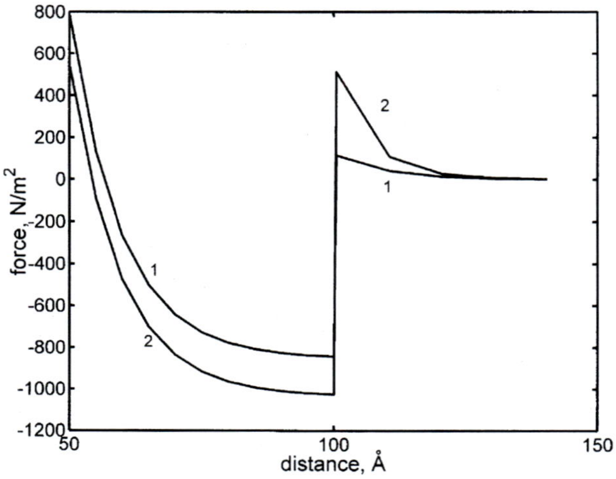

Figure 3. Effect of the sign of the charge of the small particles on the force. The force was calculated using eqs 3 and 4. The charge of the small particles is positive. The parameters used in the calculation are $\sigma = -0.01$ C/m^2, electrolyte concentration $c = 0.1$ M, volume fraction of the small particles = 0.01, and $D = 100$ Å: (1) $z = 10$; (2) $z = 20$.

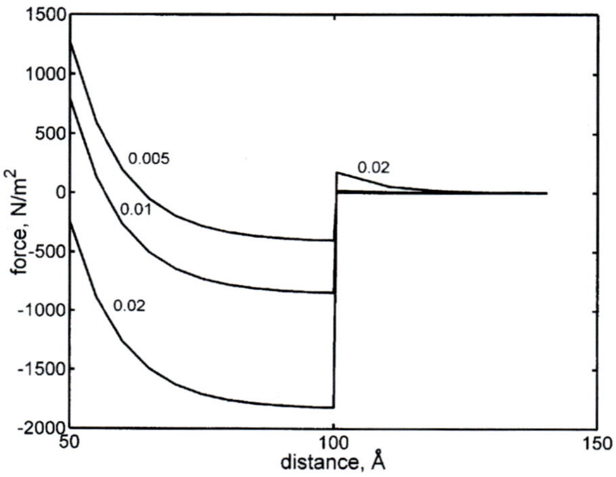

Figure 4. Effect of the concentration of the small particles on the force. The force was calculated using eqs 3 and 4. The parameters used in the calculation are $\sigma = -0.01$ C/m^2, electrolyte concentration $c = 0.1$ M, $D = 100$ Å, and $z = -10$. The volume fractions of the small particles with respect to the entire volume are noted in the figure.

sufficiently small. Cases in which D is small are examined later in the paper.

5.2. Effect of the Sign of the Charge of the Particle on the Force. If the particle possesses a charge with a sign opposite to that of the plates (the latter are assumed negatively charged in all calculations), a larger number of particles will be located between the plates. Therefore, the contribution of the positively charged particles to the osmotic pressure is larger than that of the negatively charged particles. As a result, the net force decreases (becomes less positive) as the charge of the particles becomes more positive (Figure 3).

As the distance increases, the force decreases, being repulsive at short distances and attractive at large ones, until the distance L becomes equal to the diameter D. For values of $L > D$, the force becomes again repulsive and tends to 0 for large distances.

5.3. Effect of the Concentration of the Small Particles on the Force. Figure 4 shows that, for the same sign of charges of the plates and the particles, the force decreases as the volume fraction of the particles in the solution increases. Obviously, for distances between plates smaller than D, the particle concentration difference between the bulk and the gap is larger for higher volume fractions. Therefore, the net force is smaller (less positive) for higher volume fractions.

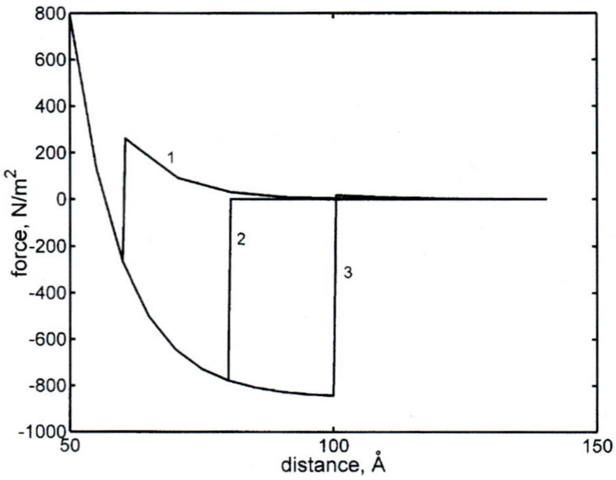

Figure 5. Effect of the size of the small particles on the force. The force was calculated using eqs 3 and 4. The parameters used in the calculation are $\sigma = -0.01$ C/m^2, electrolyte concentration $c = 0.1$ M, volume fraction of the small particles = 0.01, and $z = -10$: (1) $D = 60$ Å; (2) $D = 80$ Å; (3) $D = 100$ Å.

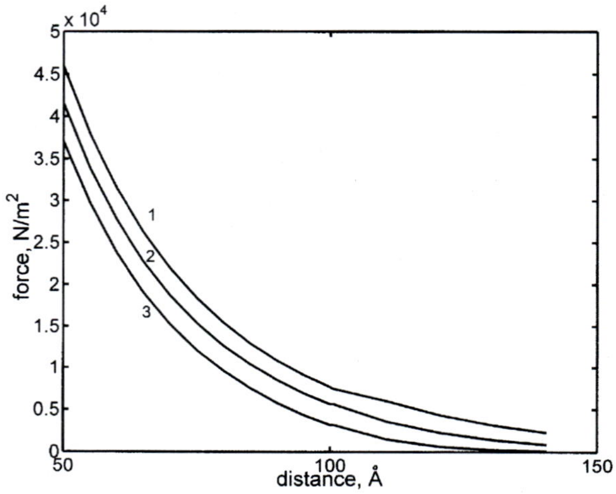

Figure 6. Effect of the charge of the small particles on the force for an electrolyte concentration of $c = 0.01$ M. The force was calculated using eqs 3 and 4. The parameters used in the calculation are $\sigma = -0.01$ C/m^2, volume fraction of the small particles = 0.01, $D = 100$ Å: (1) $z = 0$; (2) $z = -10$; (3) $z = -20$.

5.4. Effect of the Size of the Small Particles on the Force. Three values for the particle diameter are considered in Figure 5, and the force was calculated for negative charges of the plates and the particles. The force is the same for all three sizes for distances between the plates smaller than the smallest particle diameter. However, when the distance between the plates becomes equal to the diameter of the smallest particle, the force acting on the plates first increases and then becomes positive. At this distance, the electrostatic interaction is dominant and the net force repulsive. For the other two sizes, the force increases as soon as the distance between the plates becomes equal to their diameters. These two sizes of particles can enter in the gap between the plates for relatively large distances between the plates at which the electrostatic repulsion is relatively weak. For this reason, the net forces for particles of these two sizes are smaller than that for the particle of the smallest size.

The calculation also showed that, as expected, the net force for very small charged particles approached the double layer repulsion between two plates immersed in a solution of two electrolytes.

5.5. Effect of Electrolyte Concentration on the Force. As the electrolyte concentration increases, the double layer interaction decreases. Comparing Figures 2 and 6, one can see that the net force for the 0.01 M electrolyte solution (Figure 6) is much larger than that for the 0.1 M electrolyte solution (Figure 2). For the 0.1 M electrolyte solution, there is a sharp increase from attraction to small repulsion when the distance between the plates becomes equal to the diameter of the small particle. However, for the 0.01 M electrolyte solution, no such increase occurs. Therefore, in the latter case, the depletion force was negligible compared to the strong electrostatic interactions.

6. Conclusions

A general expression for the net force acting between two charged plates immersed in an electrolyte solution containing charged particles was obtained. The derivation of equations for the force based on this expression is simpler than those available in the literature. The same results as those in previous papers for the depletion force acting between two plates or two spheres were obtained. An expression for the interaction between two plates or two spheres immersed in a solution of particles of polydisperse size could also be derived. In addition, the present approach could be used to treat the case of the interaction between two charged plates immersed in an electrolyte solution containing charged particles of finite size, a case which has not been previously treated.

LA049703A

Available online at www.sciencedirect.com
SCIENCE DIRECT
Journal of Colloid and Interface Science 290 (2005) 336–342
ELSEVIER
Journal of Colloid and Interface Science
www.elsevier.com/locate/jcis

Thermodynamically stable dispersions induced by depletion interactions

Haohao Huang, Eli Ruckenstein *

Department of Chemical and Biological Engineering, State University of New York at Buffalo, Buffalo, NY 14260, USA

Received 16 December 2004; accepted 20 April 2005
Available online 28 June 2005

Abstract

When small particles are added to a colloidal dispersion of large particles, a depletion interaction between large particles occurs because the small ones are depleted from the gaps between the former particles. In the present paper, a cell model is employed to examine the behavior of a dispersion of large particles immersed in an electrolyte solution containing small particles. In this model, each cell consists of one large particle in its center and an associated atmosphere. Double-layer, van der Waals, and depletion interactions, as well as entropic effects, have been taken into account. When the change of the free energy with respect to that of the electrolyte solution is negative (and this happens in most cases), the dispersions of large particles are stable from a thermodynamic point of view. With increasing volume fraction of the small particles, the free energy change becomes more negative. The formation of gels observed experimentally in concentrated emulsions is explained through the formation of a thermodynamically stable dispersion.

Keywords: Depletion force; Electrostatic interaction; Van der Waals interactions; Thermodynamic stability

1. Introduction

During recent years, the behavior of colloidal particles immersed in an electrolyte solution containing small particles has been extensively studied experimentally [1–3], theoretically [4–8], and by simulations [9–11]. In addition to van der Waals and double-layer interactions, a depletion interaction was present in such systems. By changing the volume fraction of the small particles and the ratio of the diameters of the small and large ones, one can control the strength of the depletion interactions and hence the behavior of the system. Experiment has shown that colloidal dispersions of large particles can form a gel or a glass in the presence of depletion interactions caused by polymer molecules or surfactant micelles present in the dispersion. It was observed that with increasing small particle (polymer) concentration a solidlike gel was generated when the ratio of the radius of gyration of the polymer to the large particle radius became sufficiently small [3].

A first explanation for the origin of the depletion interaction between two large particles present in a solution containing small particles was suggested by Asakura and Oosawa [12]. In their theory, the hard small particles are excluded from the gaps between the large particles, and for this reason, there is a difference between the osmotic pressures in the gap and the bulk solution. For uncharged small particles, this interaction is caused by an entropic change due to the depletion of small particles. However, if the small particles that induce the depletion force are charged, the latter macroions affect the electrical potential distribution in the colloidal system and hence the depletion and the double-layer interactions [13].

The phase behavior of nonaqueous colloidal suspensions containing nonadsorbing polymer was investigated by Gast et al. [3] on the basis of statistical mechanics. In their theory, a second-order perturbation approach was used to calculate the free energy. Rao and Ruckenstein [4,5] examined the phase behavior of systems involving steric, depletion, and van der Waals interactions.

If a depletion interaction is present in a colloidal system, it generates a negative free energy change, and, as a result,

* Corresponding author. Fax: +1 (716) 645 3822.
E-mail address: feaeliru@acsu.buffalo.edu (E. Ruckenstein).

0021-9797/$ – see front matter
doi:10.1016/j.jcis.2005.04.059

the system can become thermodynamically stable. If the system contains a high volume fraction of dispersed phase, it can become a gel.

The goal of the present paper is to investigate the behavior of a dispersion of large particles in an electrolyte solution containing small uncharged or charged particles. A cell model is employed to calculate the free energy of the system. In this model, the charged small particles of radius R_S are depleted from layers of thickness R_S around the large particles of radius R_L due to hard-sphere and electrostatic interactions. The small particles also introduce a charge into the Poisson–Boltzmann equation, which affects the electrical potential distribution. In addition to the above interactions, the van der Waals interactions between the large particles and an entropic contribution are also taken into account.

2. The model

The colloidal system is divided into a number of Wigner–Seitz cells. A large particle of radius R_L is located at the center of each cell, and a liquid atmosphere is associated with the particle. The atmosphere contains an electrolyte solution and small particles of radius R_S (Fig. 1). Each of the small particles has a charge ze, and the large particles have a surface electrical potential Ψ_0, which is assumed to be negative in the calculations.

In the present model, the Wigner–Seitz cell is approximated by a sphere. The radius of the sphere R_C can be obtained from the number density n_L of the large particles,

$$R_C^3 = \frac{3}{4\pi n_L}. \tag{1}$$

The small particles cannot be located in a layer of thickness R_S around the large particles due to hard-sphere interactions. Outside this region, the distribution of the small particle is assumed to be Boltzmannian.

To calculate the free energy of the system, one must solve the Poisson–Boltzmann equation to obtain the electrical potential Ψ and the ionic distribution. Because the Wigner–Seitz cell is approximated by a sphere, the electrical potential depends only on the radial coordinate r,

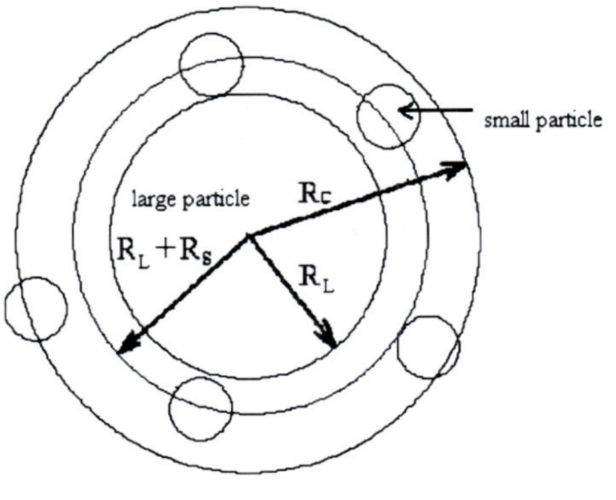

Fig. 1. The cell model. The small particles are depleted from the layer between the radii R_L and $R_L + R_S$.

$$\nabla^2\Psi = -\frac{\rho}{\varepsilon\varepsilon_0}, \tag{2}$$

where ρ is the charge density in the solution (which includes the ions of the electrolyte, the charged small particles, and the counterions of the large and small particles), ε is the dielectric constant, and ε_0 is the vacuum permittivity.

The concentrations of ions can be expressed in the form

$$c_+ = c_{+0}\exp\left[-\frac{e(\Psi - \Psi_0)}{kT}\right], \tag{3}$$

$$c_- = c_{-0}\exp\left[\frac{e(\Psi - \Psi_0)}{kT}\right], \tag{4}$$

$$c_s = \begin{cases} c_{sc}\exp\left[\frac{-ze(\Psi-\Psi_c)}{kT}\right] & \text{for } R_L + R_S < r < R_C, \\ 0 & \text{for } R_L < r < R_L + R_S, \end{cases} \tag{5}$$

where k is the Boltzmann constant, T is the absolute temperature in K, c_+ and c_- are the concentrations of cations and anions, respectively, c_{+0} and c_{-0} are the corresponding concentrations at the surface of the large particles, c_s is the concentration of small particles, and c_{sc} and Ψ_c are the concentration of small particles and the electrical potential at the periphery of the spherical cell, respectively.

The average electrolyte and small particle concentrations in each cell are equal to their overall concentrations in the whole system,

$$\bar{c}_+ = 3\int_{R_L}^{R_C} c_+ r^2\,dr \big/ \left(R_C^3 - R_L^3\right) = c + c_e, \tag{6}$$

$$\bar{c}_- = 3\int_{R_L}^{R_C} c_- r^2\,dr \big/ \left(R_C^3 - R_L^3\right) = c, \tag{7}$$

and

$$\bar{c}_s = 3\int_{R_L+R_S}^{R_C} c_s r^2\,dr \big/ \left(R_C^3 - (R_L + R_S)^3\right), \tag{8}$$

where c is the electrolyte concentration and c_e is the concentration of the counterions of the particles. The overall concentration of cations also includes the counterions dissociated from all particles, which are assumed to be negatively charged.

To solve the Poisson–Boltzmann equation, boundary conditions are required. For a constant surface potential Ψ_0 on the surface of the large particles,

$$\Psi(r = R_L) = \Psi_0. \tag{9}$$

Due to the periodicity of the cells, one can write that at the periphery of the spherical cells,

$$\frac{d\Psi}{dr} = 0 \quad \text{for } r = R_C. \tag{10}$$

338 H. Huang, E. Ruckenstein / Journal of Colloid and Interface Science 290 (2005) 336–342

Equation (2) and the boundary conditions (9) and (10) can be solved numerically to obtain the electrical potential distribution for selected overall concentrations of electrolyte and small particles.

The change of the free energy F_{cell} of each cell with respect to that of the electrolyte solution involves three contributions: electrical F_{ele}, entropic F_{ent}, and chemical F_{che} [14,15],

$$F_{\mathrm{cell}} = F_{\mathrm{ele}} + F_{\mathrm{ent}} + F_{\mathrm{che}}. \tag{11}$$

The electrical contribution, the electrostatic free energy of the system, can be calculated using the expression

$$F_{\mathrm{ele}} = \frac{1}{2}\varepsilon\varepsilon_0 \int_{R_{\mathrm{L}}}^{R_{\mathrm{C}}} 4\pi r^2 \left(\frac{d\Psi}{dr}\right)^2 dr. \tag{12}$$

The entropic contribution is given by

$$F_{\mathrm{ent}} = kT \int_{R_{\mathrm{L}}}^{R_{\mathrm{C}}} 4\pi r^2 \left[\left(\sum_i c_i \ln x_i + n_{\mathrm{s}} \ln x_{\mathrm{s}} + c_{\mathrm{w}} \ln x_{\mathrm{w}}\right) - \left(\sum_i c_{i\mathrm{b}} \ln x_{i\mathrm{b}} + c_{\mathrm{wb}} \ln x_{\mathrm{wb}}\right)\right] dr, \tag{13}$$

where c_i is the concentration of ion i in the cell, c_{w} is the concentration of water in the cell, x is the molar fraction, subscript i refers to ions i, subscript s to the small particles, subscript b indicates the overall average concentration of the ion, and subscript w refers to water. Obviously, the concentration of small particles is zero for $R_{\mathrm{L}} < r < R_{\mathrm{L}} + R_{\mathrm{S}}$. The calculated free energy change is relative to that of a uniform electrolyte solution.

The chemical contribution F_{che} is 0 at the constant surface charge density of the large particles and $-\Psi_0\sigma$ at the constant surface potential Ψ_0 of the large particles, where σ is the surface charge per large particle.

The van der Waals interaction among the large particles of the system was calculated on the basis of a model that was previously employed to examine the stability of emulsions [16] and microemulsions [17]. In this model, a particular particle is considered to be surrounded by 12 nearest neighbor particles randomly located on a sphere of radius R_1, and all the other particles of the system are assumed to be homogeneously distributed outside a sphere of radius R_2 (Fig. 2). The radius R_1 is selected as the average distance between large particles, and is therefore given by

$$\varphi_1 = 0.74 \frac{R_{\mathrm{L}}^3}{(R_1/2)^3}, \tag{14}$$

where φ_1 is the volume fraction of the large particles, and the constant 0.74 represents the volume fraction of the most compact arrangement of large particles. The number of large particles in the sphere of radius R_2 is 13. Assuming that they are arranged in a compact manner, one can write

$$\frac{4\pi}{3} R_2^3 = \frac{13}{0.74} \frac{4\pi}{3} (R_1/2)^3. \tag{15}$$

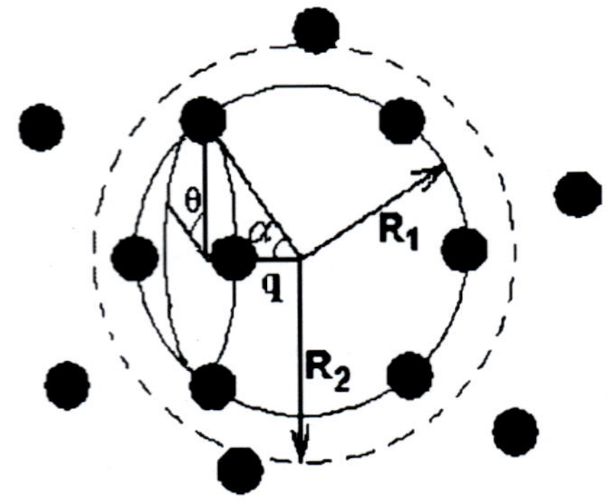

Fig. 2. Interaction between one particle and the other particles. The neighboring 12 particles are located on a sphere of radius R_1 and the other are located outside a sphere of radius R_2.

The probability of finding one particle per unit area of the surface of radius R_1 is $12/4\pi R_1^2$. Therefore, the van der Waals interaction between one particle located at a distance q from the center of the sphere of radius R_1 and the particles on the surface element $R_1^2 \sin\alpha\, d\alpha\, d\theta$ is given by

$$dV = \frac{3}{\pi} U \sin\alpha\, d\alpha\, d\theta, \tag{16}$$

where U, the van der Waals interaction between two identical particles with the center-to-center distance $x = \sqrt{R_1^2 + q^2 - 2qR_1\cos\alpha}$, is provided by the expression

$$U = -\frac{A_{\mathrm{H}}}{12} \frac{R_{\mathrm{L}}}{(x - 2R_{\mathrm{L}})^3} = -\frac{A_{\mathrm{H}}}{12} \frac{R_{\mathrm{L}}}{(-2R_{\mathrm{L}} + \sqrt{R_1^2 + q^2 - 2qR_1\cos\alpha})^3}, \tag{17}$$

where A_{H} is the Hamaker constant.

The overall van der Waals interaction between one particle located at the distance q from the center of the sphere of radius R_1 and the other particles can be expressed in the form

$$V = \frac{3}{\pi} \iint\limits_{0<\alpha<\pi,\ 0<\theta<2\pi} U \sin\alpha\, d\alpha\, d\theta = -\frac{A_{\mathrm{H}} R_{\mathrm{L}}}{q R_1} \left(R_{\mathrm{L}} \ln \frac{R_1 + q - 2R_{\mathrm{L}}}{R_1 - q - 2R_{\mathrm{L}}} + q\right). \tag{18}$$

If one assumes that the particle can be located in a sphere of radius $R_1 - 2R_{\mathrm{L}}$ at any place with equal probability, the following average van der Waals interaction between a particle and the neighboring 12 is obtained:

$$\bar{V} = \int_0^{R_1 - 2R_{\mathrm{L}}} \frac{4\pi q^2 V\, dq}{4\pi (R_1 - 2R_{\mathrm{L}})^3/3} = -\frac{A_{\mathrm{H}} R_{\mathrm{L}} (R_1 + R_{\mathrm{L}})}{R_1 (R_1 - 2R_{\mathrm{L}})}. \tag{19}$$

Because the van der Waals interaction is a short-range one, the van der Waals interaction between one large particle and the large particles other than the neighboring ones can be neglected.

The free energy change as calculated above refers to one cell. However, the total free energy change F_{sys} of the system also contains an entropic contribution due to the mixing of the large particles with the dispersing medium containing an electrolyte and small particles.

The entropic contribution for the dispersion of the large particles in a medium containing small particles is given in Appendix A. An upper bound for the entropy of mixing is given by

$$\Delta S_u = k\left\{n_L \ln\frac{n_l}{n_L} + \frac{N_t - n_2 n_S}{n_l}\ln\frac{N_t - n_2 n_S}{n_l} - \left(\frac{N_t - n_2 n_S}{n_l} - n_L\right)\ln\left(\frac{N_t - n_2 n_S}{n_l} - n_L\right)\right\}, \quad (20)$$

where N_t denotes the total number of sites per unit volume of the system (each site is considered to have the volume of a molecule of the dispersing medium, a water molecule for instance), n_l the number of sites occupied by each large particle, and n_2 is the number of sites occupied by each small particle. Although Eq. (20) provides an upper bound for the entropic contribution, the calculations revealed that this contribution to the free energy is negligible.

Adding the entropy of mixing to the double-layer depletion and the van der Waals interactions, one obtains the following expression for the free energy change of the system,

$$F_{sys} = n_L(F_{cell} + V/2) - T\Delta S_u, \quad (21)$$

when the upper bound of the entropy of mixing is employed.

3. Numerical calculations

3.1. The free energy change of the system for uncharged small particles

The small particles are depleted from the gaps between the large particles when $R_C < R_L + R_S$. As a result, an attractive force is induced between the large particles. The effect of the volume fraction φ_2 of the small particles on the free energy change is presented in Fig. 3. The dispersion is thermodynamically stable because the free energy change with respect to that of the electrolyte solution is negative. As expected, the free energy change becomes more negative as the concentration of uncharged small particles increases. This happens because with the introduction of a larger number of small particles, the entropic contribution provided by Eq. (13) increases and the free energy change becomes more negative.

As the size of the small particles increases, the thickness of the depletion zone becomes larger, and at constant φ_2, the free energy change becomes less negative (Fig. 4). The free energy change becomes less negative with increasing φ_1 and the van der Waals interaction between the large particles becomes significant at large φ_1.

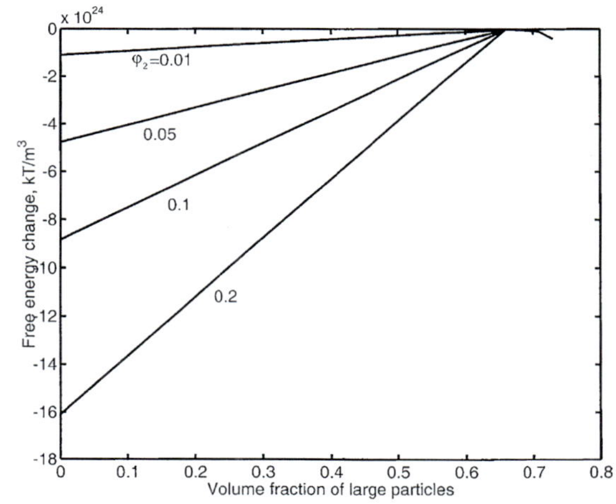

Fig. 3. Effect of the volume fraction of uncharged small particles. $R_L = 200$ Å, $R_S = 10$ Å, $c = 0.01$ M, $\Psi_0 = -0.05$ V, $A_H = 0.1 \times 10^{-20}$ J.

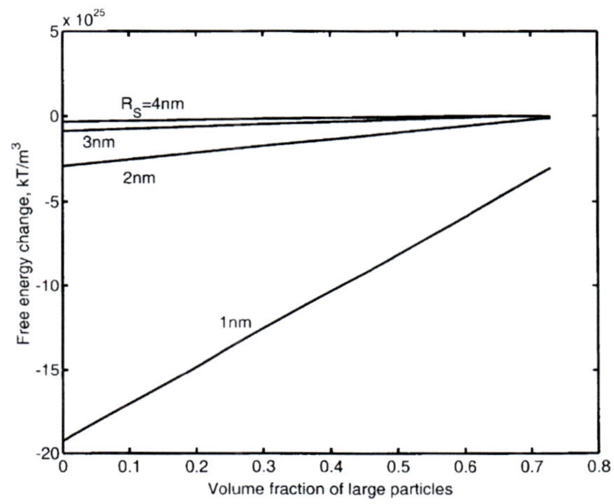

Fig. 4. Effect of the size of the uncharged small particles. $R_L = 200$ Å, $c = 0.01$ M, $\Psi_0 = -0.05$ V, $\varphi_2 = 0.1$, $A_H = 0.1 \times 10^{-20}$ J.

As the electrolyte concentration increases at constant φ_1 and φ_2, the free energy change becomes less negative (Fig. 5). Two effects play a role in this case: (i) the electrical free energy (Eq. (12)), which becomes less positive with increasing electrolyte solution, and (ii) the entropic contribution (Eq. (13)), which becomes more negative with increasing electrolyte concentration. It is clear that under the conditions examined the entropic contribution is dominant (Fig. 6) and the free energy change increases as the electrolyte concentration increases because of the entropic contribution.

3.2. Distributions of electrical potential and concentrations of ions in the system with charged small particles

Fig. 7 shows that the electrical potential first increases with the distance from the surface of a large particle and then decreases. The magnitude of the change of the electrical potential is, however, small, less than 10% of the surface electrical potential in the case examined.

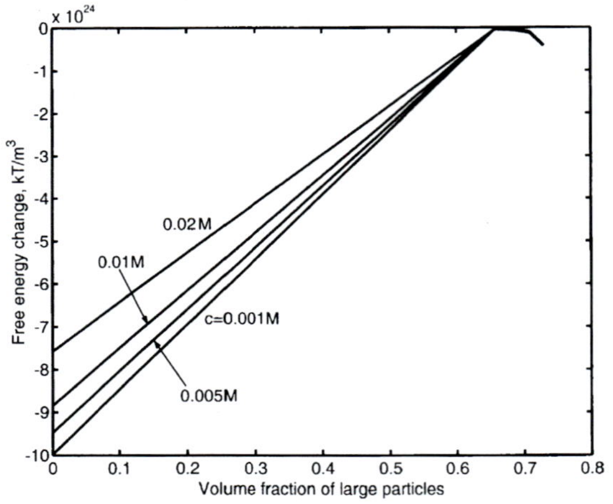

Fig. 5. Effect of the electrolyte concentration. $R_L = 200$ Å, $R_S = 30$ Å, $\Psi_0 = -0.05$ V, $\varphi_2 = 0.1$, $A_H = 0.1 \times 10^{-20}$ J.

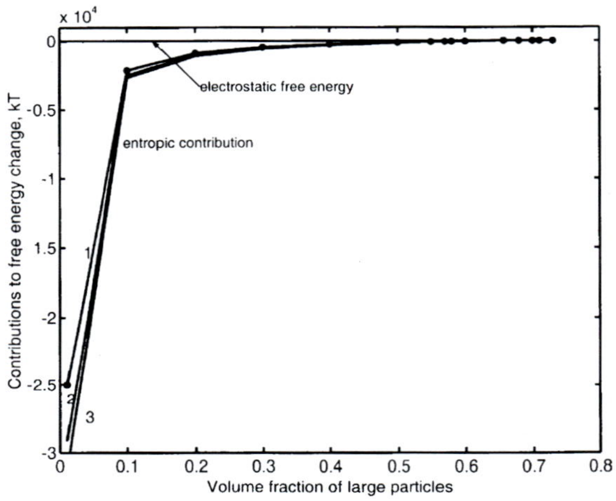

Fig. 6. Contributions to the free energy change: (1) $c = 0.02$ M; (2) $c = 0.01$ M; (3) $c = 0.005$ M. The values of the other parameters are as in Fig. 5.

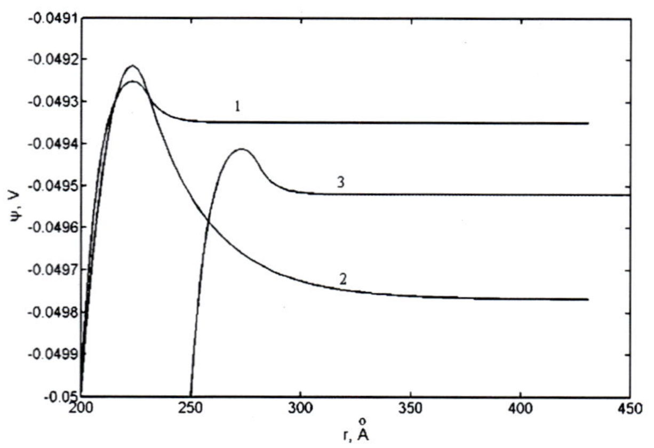

Fig. 7. Electrical potential distribution in the radial direction. $\Psi_0 = -0.05$ V, $z = -3$, $\varphi_1 = 0.1$, $\varphi_2 = 0.05$, $R_S = 30$ Å. (1) $c = 0.15$ M, $R_L = 200$ Å; (2) $c = 0.005$ M, $R_L = 200$ Å; (3) $c = 0.15$ M, $R_L = 250$ Å.

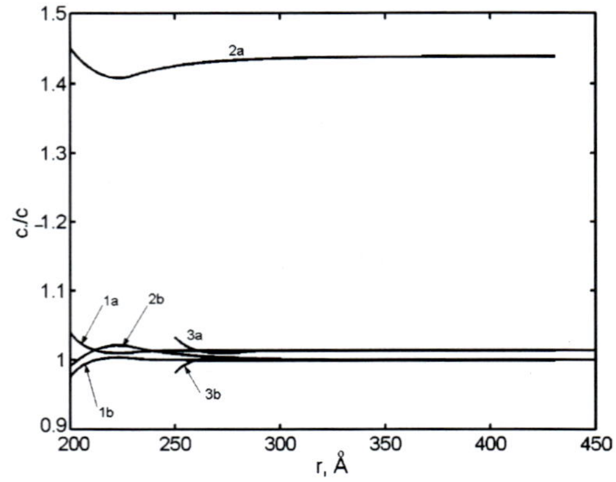

Fig. 8. Distributions of cation and anion concentrations in the radial direction. (1) $c = 0.15$ M, $R_L = 200$ Å; (2) $c = 0.005$ M, $R_L = 200$ Å; (3) $c = 0.15$ M, $R_L = 250$ Å. (a) Anion; (b) cation. The other parameter values are as in Fig. 7.

The concentrations of the cations and the anions are plotted in Fig. 8. Because both the large and the small particles are negatively charged, their counterions make the average cation concentration higher than the average anion concentration. With increasing distance from the surface of a large particle, the cation concentration first decreases because the electrical potential increases. However, at a distance corresponding to the maximum in the electrical potential, the cation concentration begins to increase until it reaches the boundary of the cell. The highest difference between the concentrations of the cations and anions is at the surface of the particles. The concentration difference at the surface is 8.7 mM for an electrolyte concentration of 150 mM and 2.3 mM for an electrolyte concentration of 5 mM, both for $\varphi_1 = 0.1$ and $\varphi_2 = 0.05$.

3.3. The free energy change of the system containing charged small particles

If the small particles are charged, they affect the electrical potential and the depletion interactions. The small particles are repelled by the large ones if they have the same sign of charge and the depletion interaction becomes stronger. As the volume fraction of small particles increases, the free energy change becomes more negative (Fig. 9).

As for the uncharged small particles, the free energy change of the system containing charged small particles becomes less negative at high electrolyte concentrations (Fig. 10).

As the charge of the small particles increases, the free energy change becomes more negative (Fig. 11). This occurs because (i) the ionic strength increases very much and this decreases the repulsion between the large particles; (ii) the depletion interaction is stronger for the charged small particles than the uncharged small particles.

The free energy change of the system becomes less negative with increasing volume fraction of large particles, be-

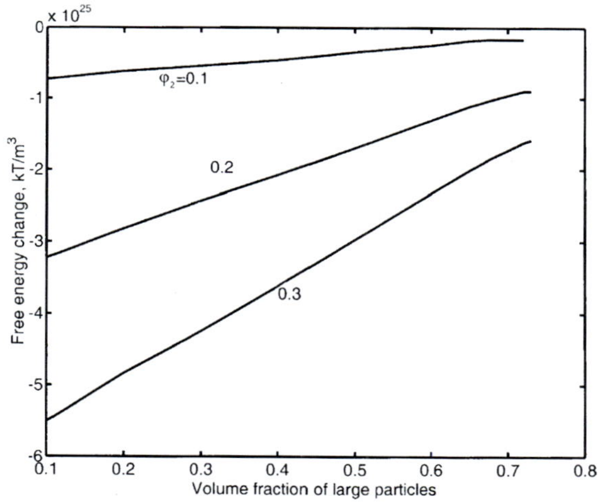

Fig. 9. Effect of the volume fraction of small particles. $R_L = 200$ Å, $\Psi_0 = -0.05$ V, $R_S = 30$ Å, $z = -3$, $A_H = 1 \times 10^{-21}$ J, $c = 0.01$ M.

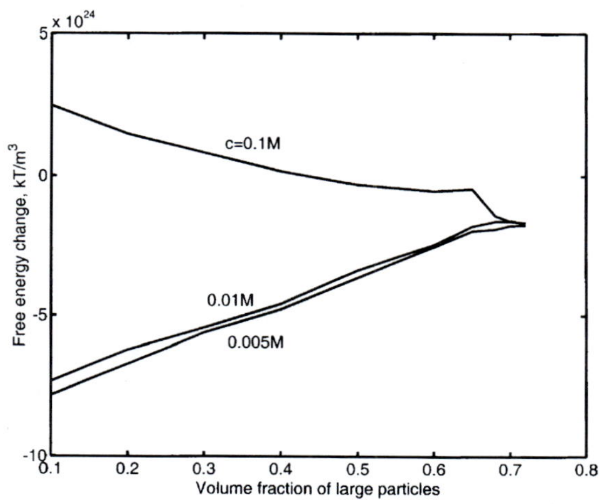

Fig. 10. Effect of salt concentration. $R_L = 200$ Å, $\Psi_0 = -0.05$ V, $R_S = 30$ Å, $z = -3$, $\varphi_2 = 0.1$, $A_H = 1 \times 10^{-21}$ J.

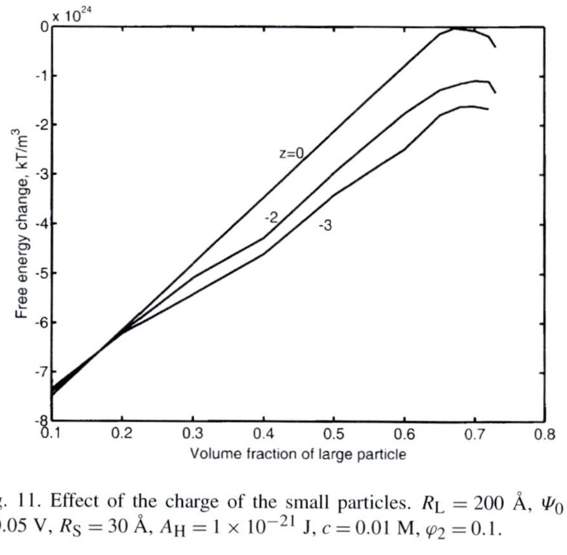

Fig. 11. Effect of the charge of the small particles. $R_L = 200$ Å, $\Psi_0 = -0.05$ V, $R_S = 30$ Å, $A_H = 1 \times 10^{-21}$ J, $c = 0.01$ M, $\varphi_2 = 0.1$.

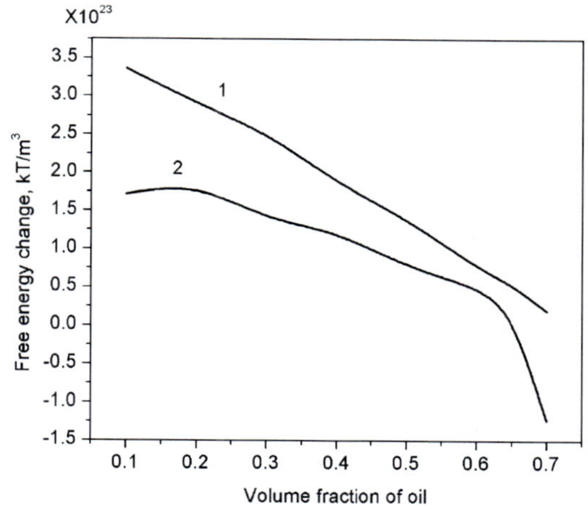

Fig. 12. Prediction of the behavior of an experimental system. $\Psi_0 = -40$ mV. $T = 293$ K, $R_L = 180$ nm, $R_S = 10$ nm, $z = -5$, $A_H = 1 \times 10^{-21}$ J, and $\varepsilon = 68.9$. (1) $\varphi_2 = 0.0336$; (2) $\varphi_2 = 0.0206$.

cause the separation between the large particles becomes smaller and the repulsion between them stronger.

3.4. Prediction of the phase behavior of an experimental system

Berli et al. have investigated the rheological behavior of protein-stabilized oil-in-water emulsions [18]. They measured the viscosity of an oil-in-water emulsion as a function of the shear stress at various concentrations of sodium caseinate. The sodium caseinate molecules aggregate in water-generating micelles, which can be considered as the small particles. The oil droplets are considered as the large particles. For large values of φ_1, as the concentration of sodium caseinate increased, the viscosity of the emulsion exhibited a sharp increase because of the formation of a gel. The gel was generated because of the depletion interactions induced by the large number of sodium caseinate micelles.

Using the values $\Psi_0 = -40$ mV, $T = 293$ K, $R_L = 180$ nm, and $R_S = 10$ nm provided in Ref. [18], and selecting the charge of the small particle $z = -5$, $A_H = 1 \times 10^{-20}$ J, $\varepsilon = 68.9$ for the mixture of water and alcohol, the free energy of the system is plotted in Fig. 12. As the volume fraction of oil increases, the free energy decreases. For a volume fraction $\varphi_2 = 0.0206$, the free energy is positive for $\varphi_1 < 0.60$, but becomes negative for $\varphi_1 > 0.60$. A thermodynamically stable gel is formed at large volume fractions of the oil droplets and this explains the sharp increase in the viscosity of the system for $\varphi_1 \approx 0.60$.

4. Conclusions

A spherical cell model was employed to describe the behavior of large particles immersed in a solution of electrolyte containing small particles. An attractive depletion force was included in the free energy change of the system. The free energy change becomes more negative as the volume fraction of small uncharged particles increases because of a stronger depletion force. As the size of the small uncharged particles increases, the free energy change becomes less negative. As the charge of the small particles with the same sign of charge as the large particles increases, the free energy change becomes more negative than for uncharged small particles because of a stronger depletion force induced by the charged small particles. Some experimental facts regarding the behavior of the viscosity of an oil-in-water emulsion with increasing concentration of a surfactant are explained via the formation of a gel induced by depletion forces.

Appendix A

Let us derive an expression for the entropy of mixing of the large particles with a fluid containing small particles. The procedure employed is similar to that of Ref. [17]. The entropy of mixing is given by the Boltzmann relation

$$\Delta S = k \ln \Omega, \tag{A.1}$$

where Ω is the number of configurations of a mixture formed from N_1 molecules of liquid, n_L identical large particles, and n_S identical small particles, all per unit volume.

If N_t is the total number of sites per unit volume and the number of sites occupied by each small particle is n_2, the center of the first large particle can be located at any of the $N_t - n_S n_2$ remaining sites, the center of the second large particle at any of the $N_t - n_S n_2 - \alpha_2$ sites, the center of the jth large particle at any of $N_t - n_S n_2 - \alpha_j$, and so on. Therefore, the number of configurations Ω can be calculated using the following expression:

$$\Omega = (N_t - n_S n_2)(N_t - n_S n_2 - \alpha_2) \times (N_t - n_S n_2 - \alpha_3)\dots(N_t - n_S n_2 - \alpha_{n_L}) \times n_L!^{-1}. \tag{A.2}$$

An upper bound of Ω can be obtained if α_j is selected in the form

$$\alpha_j = (j-1)n_1, \tag{A.3}$$

where n_1 is the number of sites occupied by each large particle. Consequently,

$$\Omega = (N_t - n_S n_2)(N_t - n_S n_2 - \alpha_2) \times (N_t - n_S n_2 - \alpha_3)\dots(N_t - n_S n_2 - \alpha_{n_L}) \times n_L!^{-1} = \frac{n_1^{n_L}[(N_t - n_S n_2)/n_1]!}{[((N_t - n_S n_2)/n_1) - n_L]!n_L!}. \tag{A.4}$$

Substituting Eq. (A.4) into the Boltzmann relation, and using the Stirling approximation for the factorials, one obtains

$$\Delta S_u = k\left\{n_L \ln\frac{n_1}{n_L} + \frac{N_t - n_2 n_S}{n_1}\ln\frac{N_t - n_2 n_S}{n_1} - \left(\frac{N_t - n_2 n_S}{n_1} - n_L\right)\ln\left(\frac{N_t - n_2 n_S}{n_1} - n_L\right)\right\}. \tag{A.5}$$

One can show that the contribution of the entropy of mixing to the total system free energy change is in the present case negligible.

References

[1] E.K. Hobbie, Phys. Rev. Lett. 81 (1998) 3996.
[2] E.H.A. de Hoog, W.K. Kegel, A. van Blaaderen, H.N.W. Lekkerkerker, Phys. Rev. E 64 (2001) 021407.
[3] A.P. Gast, C.K. Hall, W.B. Russel, J. Colloid Interface Sci. 96 (1983) 251.
[4] I.V. Rao, E. Ruckenstein, J. Colloid Interface Sci. 108 (1985) 389.
[5] E. Ruckenstein, I.V. Rao, Colloids Surf. 17 (1986) 185.
[6] S. Ramakrishnan, M. Fuchs, K.S. Schweizer, C.F. Zukoski, J. Chem. Phys. 116 (2002) 2201.
[7] Y.L. Chen, K.S. Schweizer, M. Fuchs, J. Chem. Phys. 118 (2003) 3880.
[8] Y.L. Chen, K.S. Schweizer, J. Chem. Phys. 120 (2004) 7212.
[9] J.J. Cerda, T. Sintes, C.M. Sorensen, A. Chakrabarti, Phys. Rev. E 70 (2004) 011405.
[10] K.G. Soga, J.R. Melrose, R.C. Ball, J. Chem. Phys. 110 (1999) 2280.
[11] D. Costa, P. Ballone, C. Caccamo, J. Chem. Phys. 116 (2002) 3327.
[12] S. Asakura, F. Oosawa, J. Chem. Phys. 22 (1954) 1255.
[13] H.H. Huang, E. Ruckenstein, Langmuir 20 (2004) 5412.
[14] J.Th.G. Overbeek, Colloids Surf. 51 (1990) 61.
[15] M. Manciu, E. Ruckenstein, Langmuir 19 (2003) 1114.
[16] W. Albers, J.Th.G. Overbeek, J. Colloid Sci. 14 (1959) 519.
[17] E. Ruckenstein, J.C. Chi, J. Chem. Soc. Faraday Trans. II 71 (1975) 1690.
[18] C.L. Berli, D. Quemada, A. Parker, Colloids Surf. A 215 (2003) 201.

Available online at www.sciencedirect.com
SCIENCE DIRECT
Journal of Colloid and Interface Science 272 (2004) 430–437
Journal of Colloid and Interface Science
www.elsevier.com/locate/jcis

Attractive interactions in dispersions of identical charged colloidal particles: a Monte Carlo simulation

Jie Feng and Eli Ruckenstein *

Department of Chemical Engineering, State University of New York at Buffalo, Buffalo, NY 14260-4200, USA

Received 30 June 2003; accepted 12 November 2003

Abstract

Using the Debye–Hückel pair potential, the collective interactions between identical charged particles were examined via Monte Carlo simulations. The results have shown that when the number of charges per particle and the particle volume fraction were sufficiently large, the pair long-range electrostatic repulsion generated an effective attractive interaction between identical charged particles because of many-body effects. Disordered liquid-like structures, ordered crystal-like structures, ordered structures dispersed in disordered ones, and disordered structures dispersed in ordered ones have been found. The structures are dependent on the volume fraction and charge of the particles, as well as on the electrolyte concentration.

Keywords: Effective attractive interaction; Colloidal dispersions; Monte Carlo simulation

1. Introduction

Experiment has revealed that concentrated dispersions of monodispersed latices exhibit iridescence [1,2]. This phenomenon occurs because of Bragg diffraction of the visible light by ordered arrangements of spheres and the equality of their lattice parameter with the optical wavelength [3,4]. Because of the large distances involved, the van der Waals interactions are negligible and the only interactions that can explain the ordered lattice arrangement of the colloidal particles are the electrostatic ones. Microscopy experiments carried out by Ise and co-workers [5–7] with colloidal dispersions of latex particles indicated that the distance between particles for low particle charges was approximately equal to that corresponding to a uniform distribution. However, for highly charged particles, the distance between some of them became smaller than that for a uniform distribution. To explain this observation, Ise and co-workers suggested that the smaller distances occur in ordered regions generated by the long-range attractive interactions caused by the electrostatic interactions between identical highly charged particles. A theory was formulated by Sogami and Ise [8,9] that starting from the Poisson–Boltzmann equation, predicted that the electrostatic interaction between two identical charged colloidal particles immersed in an electrolyte solution is repulsive at short distances but becomes attractive at long ones. This conclusion contradicts the DLVO theory [10], where the electrostatic interactions between two colloidal particles are repulsive at all distances. This issue has been discussed in several papers, some in favor of the Sogami–Ise theory [10–12] but most against [13–18]. Thermodynamics rigorously demonstrates that in the framework of the Poisson–Boltzmann equation, the interaction between two identical charged particles is repulsive at all distances [18].

The fact that interactions between ions of the same kind can become attractive was noted a long time ago. Using the integral equation approach and considering higher order correlations between ions [19], Kirkwood predicted attractive interactions between identical charged ions. Outhwaite and Hutson arrived at a similar conclusion, starting from the symmetrized Poisson–Boltzmann equation [20]. Schmidt and Ruckenstein [21,22], considering the colloidal particles a component of a particles–electrolyte–liquid mixture, and assuming coulombic interactions among the colloidal particles, anions and cations, found conditions under which electrostatic interaction generated attraction between particles. They attributed this phenomenon to a collective Coulomb interaction among all the charged species. The Ornstein–Zernike equation in its zero-order approximation,

* Corresponding author.
E-mail address: feaeliru@acsu.buffalo.edu (E. Ruckenstein).

0021-9797/$ – see front matter © 2003 Elsevier Inc. All rights reserved.
doi:10.1016/j.jcis.2003.11.059

which constitutes a symmetrized Poisson–Boltzmann equation, was employed. The calculations revealed that the pair correlation function of the particles became larger than unity for particle concentrations larger than a critical value. This result indicates the presence of a potential of mean force representing an effective attraction between particles. Chu and Wasan [23] solved the Ornstein–Zernike equation with the mean spherical approximation. They found an oscillatory attractive effective interaction potential between two identical charged particles at high particle charges and concentrations and a repulsive interaction at low particle charges and concentrations.

The scope of this paper is to demonstrate via Monte Carlo simulations that the conventional Debye–Hückel screened pair potential, which is based on the DLVO theory and is repulsive at all distances, can explain the experimental observations of Ise et al. Consequently, the pair potential does not have to become attractive at large distances to explain the Ise et al. observations. The simulations have been carried out for a wide range of charges and volume fractions of the particles as well as a few electrolyte concentrations. The paper is organized as follows: In Section 2, the model and the Monte Carlo algorithm will be described. In Section 3, the main results will be presented. Section 4 will emphasize the conclusions.

2. Model and algorithm

N identical charged hard spherical particles with a radius of 670 Å located in a three-dimensional cubic box were considered in the simulations. The pair interaction between two particles was represented by the Debye–Hückel screened potential ($E_{\mathrm{D-H}}(r_{ij})$),

$$E_{\mathrm{D-H}}(r_{ij}) = \begin{cases} \infty, & \text{for } r_{ij} < 2r, \\ \frac{(Ze)^2}{4\pi\varepsilon_0\varepsilon_d r_{ij}} \frac{\exp(-\kappa(r_{ij}-2r))}{(1+\kappa r)^2}, & \text{for } r_{ij} \geqslant 2r, \end{cases} \quad (1)$$

where r_{ij} is the distance between the centers of two particles, r is the radius of the particles, e is the protonic charge, ε_0 is the vacuum permittivity, ε_d is the dielectric constant, Z represents the number of charges per particle, and κ is the reciprocal Debye length, which is given by the expression

$$\kappa^2 = \frac{4\pi e^2}{\varepsilon_0\varepsilon_d k_B T}\left(n_c z_c^2 + \sum_i n_i z_i^2\right). \quad (2a)$$

Because of electroneutrality, $n_p Z = n_c z_c$ and Eq. (2a) becomes

$$\kappa^2 = \frac{4\pi e^2}{\varepsilon_0\varepsilon_d k_B T}\left(n_p Z z_c + \sum_i n_i z_i^2\right). \quad (2b)$$

In Eqs. (2), n_p is the particle concentration, n_i is the concentration of ions of species i and valence z_i, n_c is the concentration of counterions of the particles, z_c is the valence of counterions of the particles, k_B is the Boltzmann constant, and T is the temperature in K. In what follows z_i and z_c will be considered equal to unity.

The effective potential U was calculated from the radial distribution function ($g(r)$) using the expression

$$U = -k_B T \ln g(r). \quad (3)$$

The effects of the number of charges Z per particle and of the volume fraction f of the particles on the effective interaction potential between identical charged particles were examined. The number of charges per particle was varied between 150 and 1200 and the particle volume fraction f was varied between 0.005 and 0.12. The electrolyte concentration was assumed to be 10^{-5} or 10^{-4} M.

In the simulations, the traditional algorithm was applied to an off-lattice at each step, and periodic boundary conditions were employed. The interactions were truncated using the minimum image convention, and the information was recorded every 50 cycles, after the system reached equilibrium; 1000 samples were averaged in each simulation.

3. Results and discussion

As shown in Fig. 1, for the low particle volume fraction f of 0.005 and an electrolyte concentration of 10^{-5} M, there was little attraction between identical charged particles and very small oscillations occurred when the number of charges per particle Z was 300. With number of charges per particle Z increasing to 600 and 1200, the curves of the effective interaction between identical charged particles versus distance between particles became oscillatory, but the amplitudes of oscillations were relatively small because of the low particle volume fraction. However, it is clear that the larger the charge per particle the larger the oscillations. Fig. 2 presents the structure generated at a large charge per particle ($Z = 1200$) for $f = 0.005$.

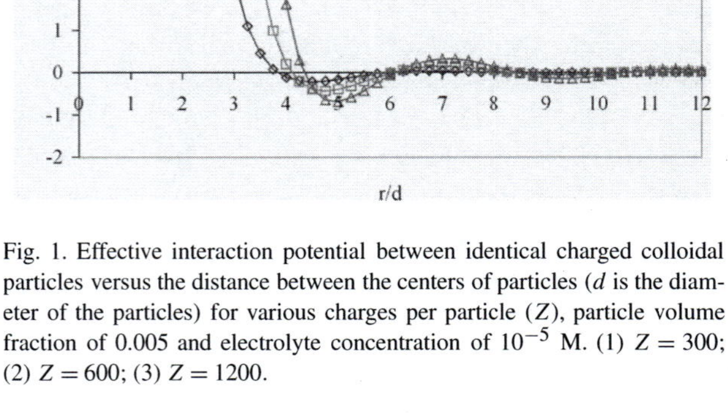

Fig. 1. Effective interaction potential between identical charged colloidal particles versus the distance between the centers of particles (d is the diameter of the particles) for various charges per particle (Z), particle volume fraction of 0.005 and electrolyte concentration of 10^{-5} M. (1) $Z = 300$; (2) $Z = 600$; (3) $Z = 1200$.

432 J. Feng, E. Ruckenstein / Journal of Colloid and Interface Science 272 (2004) 430–437

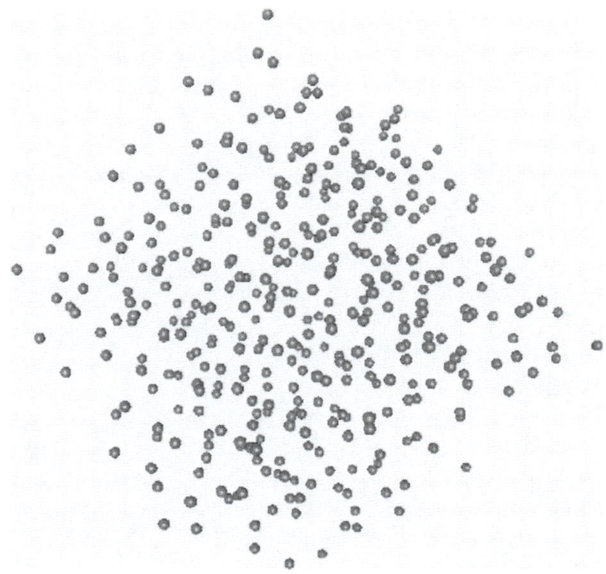

Fig. 2. Structure of charged colloidal particles with $Z = 1200$, particle volume fraction of 0.005, and electrolyte concentration of 10^{-5} M.

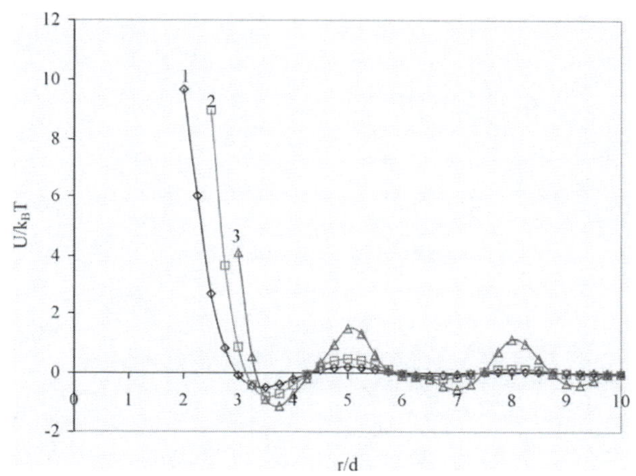

Fig. 3. Effective interaction potential between identical charged colloidal particles versus the distance between the centers of particles (d is the diameter of the particles) for various charges per particle (Z), particle volume fraction of 0.015, and electrolyte concentration of 10^{-5} M. (1) $Z = 300$; (2) $Z = 600$; (3) $Z = 1200$.

When the particle volume fraction f was increased to 0.015, the oscillations of the effective interaction between identical charged particles became larger than those for $f = 0.005$ at an electrolyte concentration of 10^{-5} M (Fig. 3). The effective interaction between identical charged particles versus the distance between particles is plotted in Fig. 3 for various values of Z ($Z = 300$, $Z = 600$, $Z = 1200$). As shown in Fig. 4, the colloidal dispersion has a disordered liquid-like structure for $Z = 600$, but a more ordered structure for $Z = 1200$. When the electrolyte concentration was increased to 10^{-4} M, the interaction between identical charged particles became completely screened. As shown in Fig. 5, no oscillations of the effective interaction potential were present for $Z = 300, 600$, and 1200.

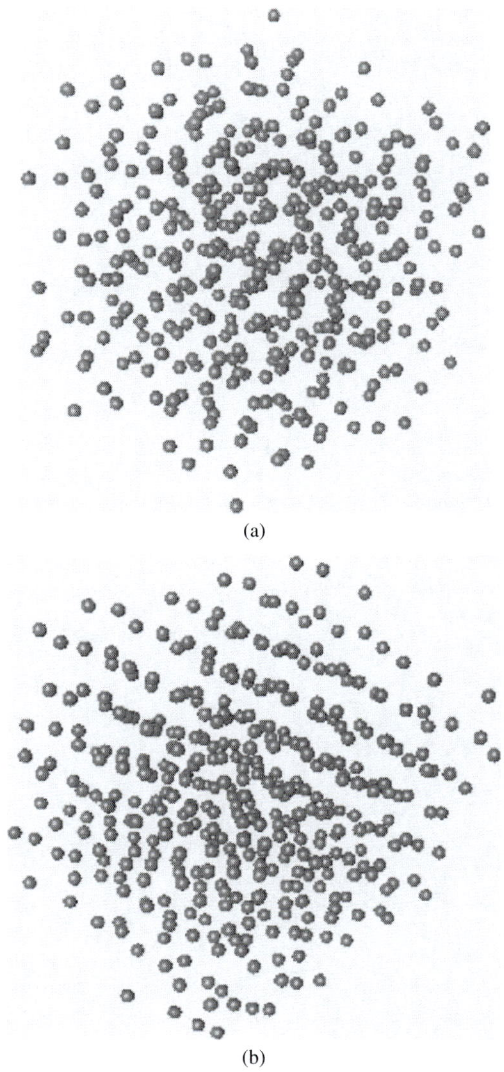

Fig. 4. Structure of charged colloidal particles with particle volume fraction of 0.015 and electrolyte concentration of 10^{-5} M. (a) $Z = 600$; (b) $Z = 1200$.

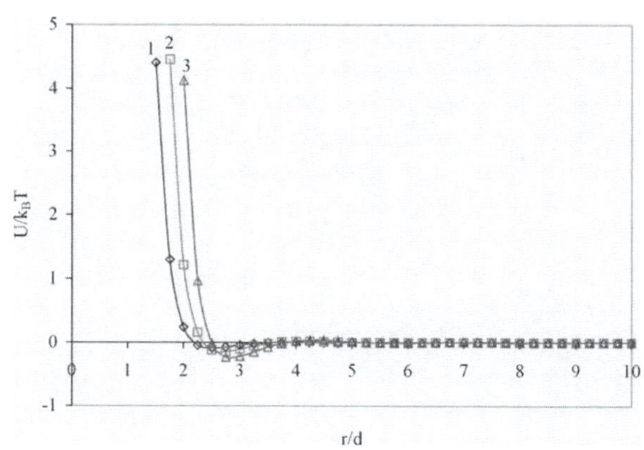

Fig. 5. Effective interaction potential between identical charged colloidal particles versus the distance between the centers of particles (d is the diameter of the particles) for various charges per particle (Z), particle volume fraction of 0.015, and electrolyte concentration of 10^{-4} M. (1) $Z = 300$; (2) $Z = 600$; (3) $Z = 1200$.

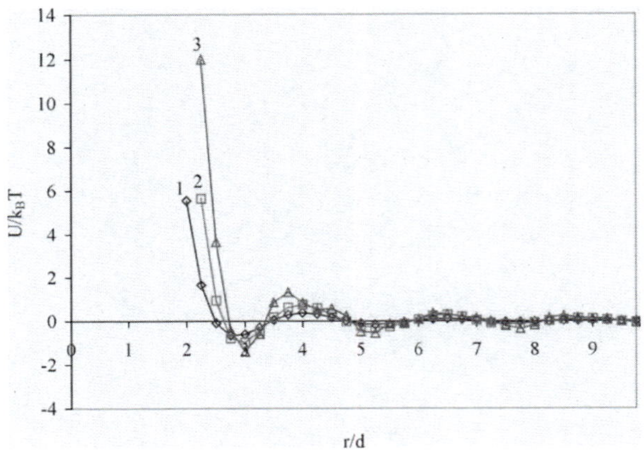

Fig. 6. Effective interaction potential between identical charged colloidal particles versus the distance between the centers of particles (d is the diameter of the particles) for various charges per particle (Z), particle volume fraction of 0.03 and electrolyte concentration of 10^{-5} M. (1) $Z = 300$; (2) $Z = 600$; (3) $Z = 1200$.

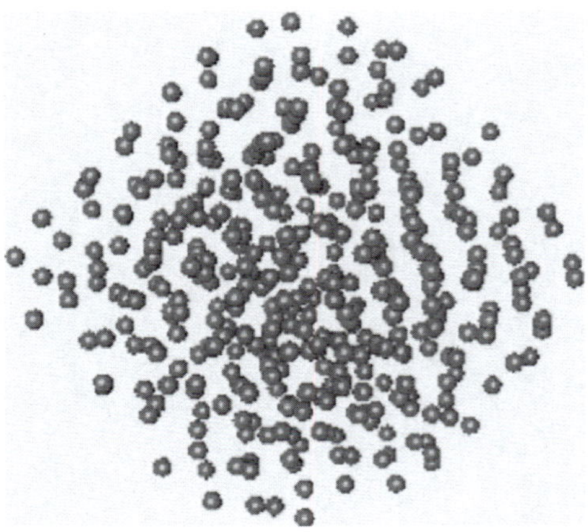

(a)

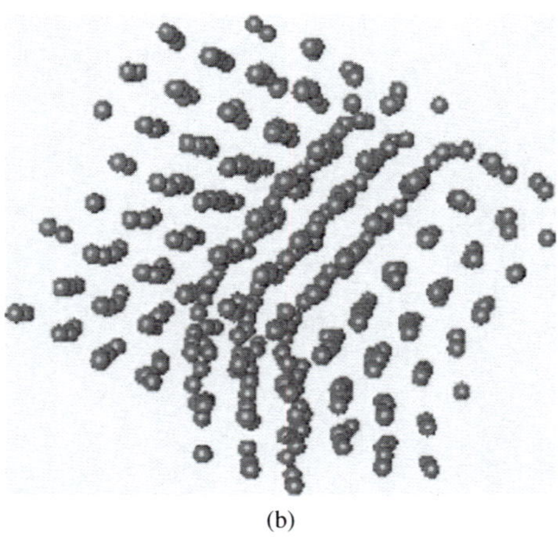

(b)

Fig. 7. Structure of charged colloidal particles with particle volume fraction of 0.03 and electrolyte concentration of 10^{-5} M. (a) $Z = 600$; (b) $Z = 1200$.

When the particle volume fraction f was increased to 0.03 (Fig. 6), oscillations could be observed at the electrolyte concentration of 10^{-5} M. Fig. 7a shows that a disordered structure surrounded by an ordered one occurred for

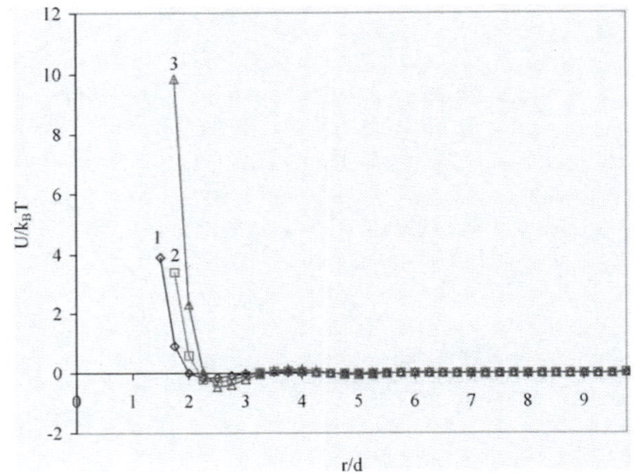

Fig. 8. Effective interaction potential between identical charged colloidal particles versus the distance between the centers of particles (d is the diameter of the particles) for various charges per particle (Z), particle volume fraction of 0.03, and electrolyte concentration of 10^{-4} M. (1) $Z = 300$; (2) $Z = 600$; (3) $Z = 1200$.

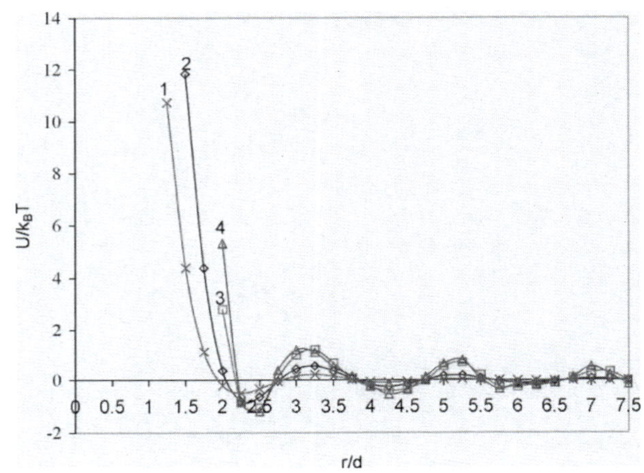

Fig. 9. Effective interaction potential between identical charged colloidal particles versus the distance between the centers of particles (d is the diameter of the particles) for various charges per particle (Z), particle volume fraction of 0.06, and electrolyte concentration of 10^{-5} M. (1) $Z = 150$; (2) $Z = 300$; (3) $Z = 600$; (4) $Z = 1200$.

$Z = 600$. For $Z = 1200$ (Fig. 7b), the structure became more ordered. When the electrolyte concentration was increased to 10^{-4} M, the effective interaction became almost completely screened and no oscillations could be found (Fig. 8).

When the particle volume fraction f was further increased to 0.06, oscillations of the effective interaction potential between identical charged particles occurred under a wide range of charges per particle (from $Z = 150$ to 1200) at an electrolyte concentration of 10^{-5} M (Fig. 9). The effective interaction potential curves between identical charged particles versus distance between particles can be divided into two groups: the curves for $Z = 600$ and $Z = 1200$, and those for $Z = 150$ and $Z = 300$. Disordered structures characterize the second group and very ordered structures the first one (Fig. 10). When the electrolyte concentration was

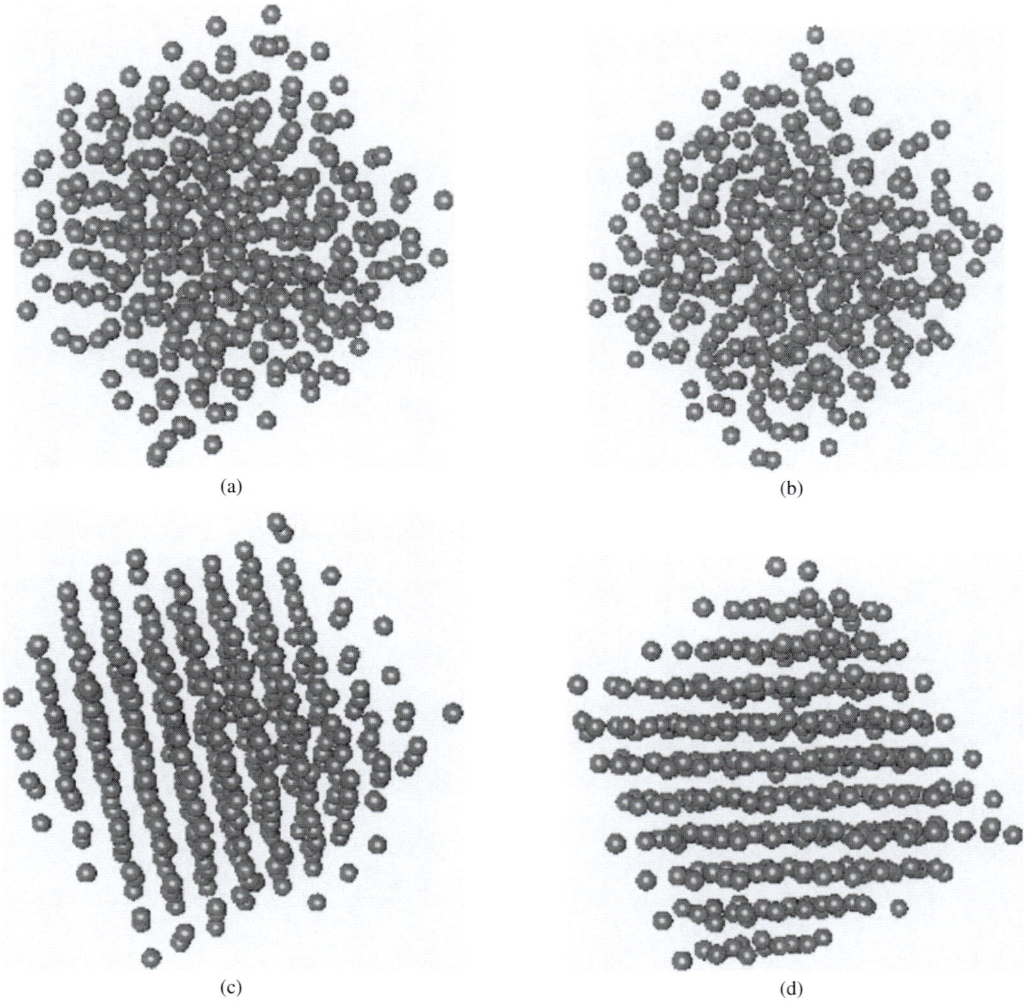

Fig. 10. Structure of charged colloidal particles with particle volume fraction of 0.06 and electrolyte concentration of 10^{-5} M. (a) $Z = 150$; (b) $Z = 300$; (c) $Z = 600$; (d) $Z = 1200$.

increased to 10^{-4} M, the oscillations of the effective interactions remained, but because of increased screening, their amplitudes were smaller than for an electrolyte concentration of 10^{-5} M (Fig. 11). Fig. 12 shows that the structure was no longer ordered even for $Z = 1200$.

As shown in Fig. 13, when the particle volume fraction f was further increased to 0.12 and the electrolyte concentration was 10^{-5} M, oscillations of the effective interaction occurred from $Z = 150$ to $Z = 1200$. As shown in Fig. 14, for $Z = 150$ the system had a disordered structure, but for $Z = 300$, 600, and 1200 the systems became very ordered. When the electrolyte concentration was increased to 10^{-4} M, the effective interaction exhibited large oscillations for $Z = 1200$, but the oscillations for $Z = 150$, 300, and 600 became much smaller (Fig. 15). As shown in Fig. 16, the structure was very ordered for $Z = 1200$, but disordered for $Z = 150$, 300, and 600.

The effective interaction potential is a result of collective behavior. Let us select two neighboring particles immersed in a dispersion of particles of the same kind. Two kinds of

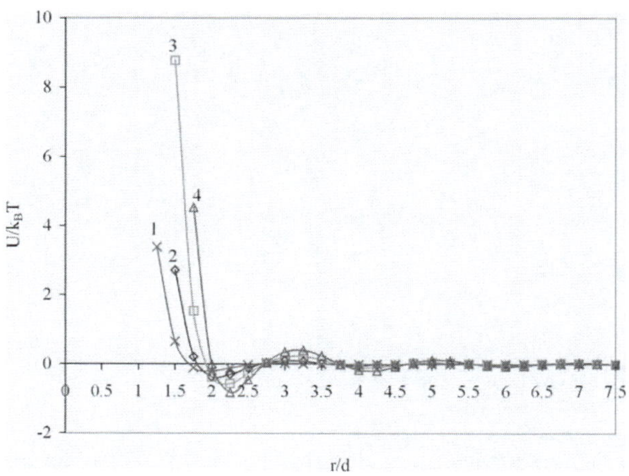

Fig. 11. Effective interaction potential between identical charged colloidal particles versus the distance between the centers of particles (d is the diameter of the particles) for various charges per particle (Z), particle volume fraction of 0.06, and electrolyte concentration of 10^{-4} M. (1) $Z = 150$; (2) $Z = 300$; (3) $Z = 600$; (4) $Z = 1200$.

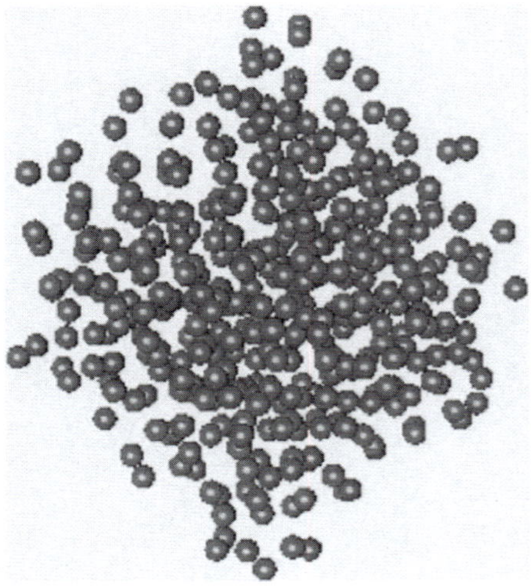

Fig. 12. Structure of charged colloidal particles with particle volume fraction of 0.06, electrolyte concentration of 10^{-4} M, and charge per particle $Z = 1200$.

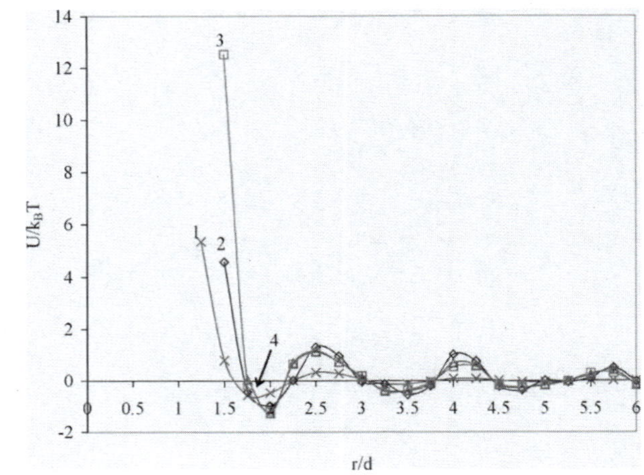

Fig. 13. Effective interaction potential between identical charged colloidal particles versus the distance between the centers of particles (d is the diameter of the particles) for various charges per particle (Z), particle volume fraction of 0.12, and electrolyte concentration of 10^{-5} M. (1) $Z = 150$; (2) $Z = 300$; (3) $Z = 600$; (4) $Z = 1200$.

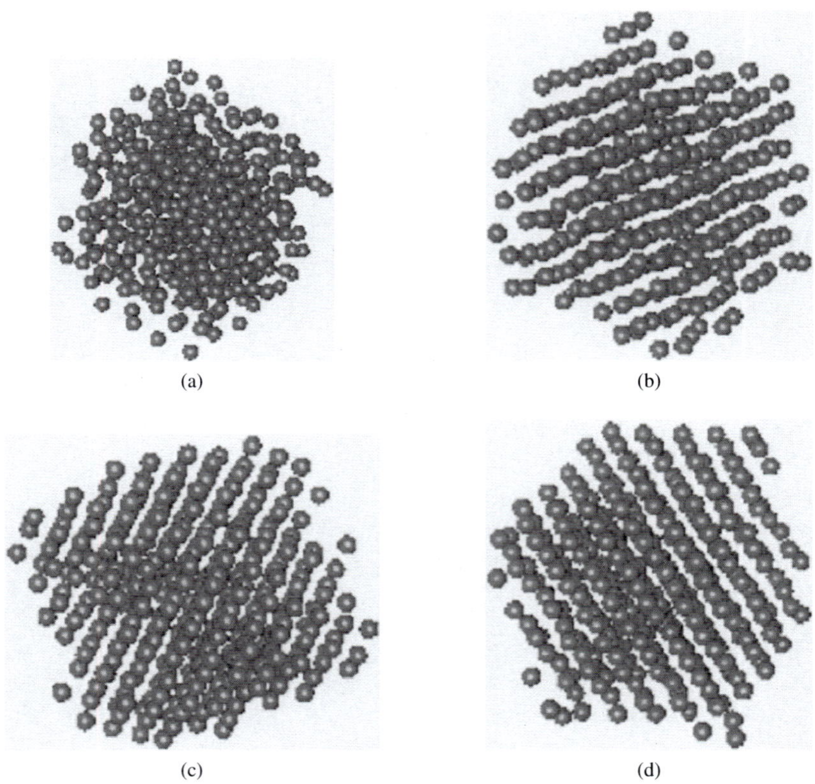

Fig. 14. Structure of charged colloidal particles with particle volume fraction of 0.12 and electrolyte concentration of 10^{-5} M. (a) $Z = 150$; (b) $Z = 300$; (c) $Z = 600$; (d) $Z = 1200$.

interactions are involved: (1) a repulsive interaction between the two selected particles, and (2) a collective interaction to which each of the two selected particles is subjected by all the other particles of the system. If the latter interaction is a repulsive one which can overcome the former, then the two selected particles will be subjected to an effective attraction. At equilibrium, the repulsion between the two neighboring particles becomes equal to the repulsion to which they are subjected by all the other particles of the system.

As shown in Fig. 17, for $Z = 300$ and an electrolyte concentration of 10^{-5} M, the oscillations of the effective interaction potential were weak for low particle-volume frac-

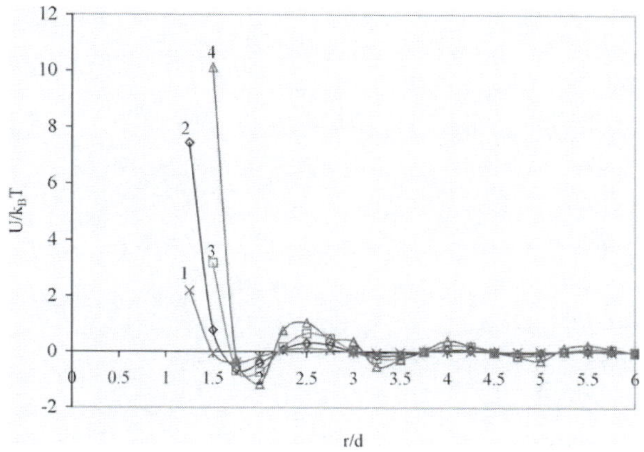

Fig. 15. Effective interaction potential between identical charged colloidal particles versus the distance between the centers of particles (d is the diameter of the particles) for various charges per particle (Z), particle volume fraction of 0.12, and electrolyte concentration of 10^{-4} M. (1) $Z = 150$; (2) $Z = 300$; (3) $Z = 600$; (4) $Z = 1200$.

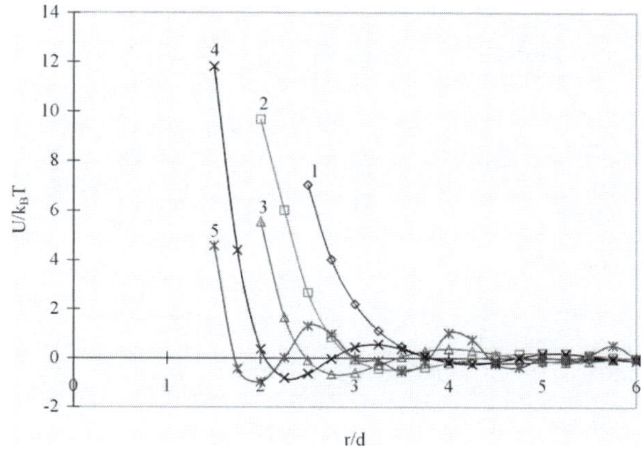

Fig. 17. Effective interaction potential between identical charged colloidal particles versus the distance between the centers of particles (d is the diameter of the particles) for various particle volume fractions (f), for charges per particle $Z = 300$ and electrolyte concentration 10^{-5} M. (1) $f = 0.005$; (2) $f = 0.015$; (3) $f = 0.03$; (4) $f = 0.06$; (5) $f = 0.12$.

tions, but became increasingly large as the volume fraction increased. The strengthening of the oscillations is caused by the increased collective interaction. As shown in Fig. 18, for $Z = 1200$ and an electrolyte concentration of 10^{-5} M, large oscillations of the effective interaction were present for most particle volume fractions because of the large number of charges per particle.

4. Conclusion

Using Monte Carlo simulations and a Debye–Hückel pair repulsive potential, effective interaction potentials between identical charged particles were calculated and the phase structures were obtained for various charges per particle and various particle volume fractions. The simulation re-

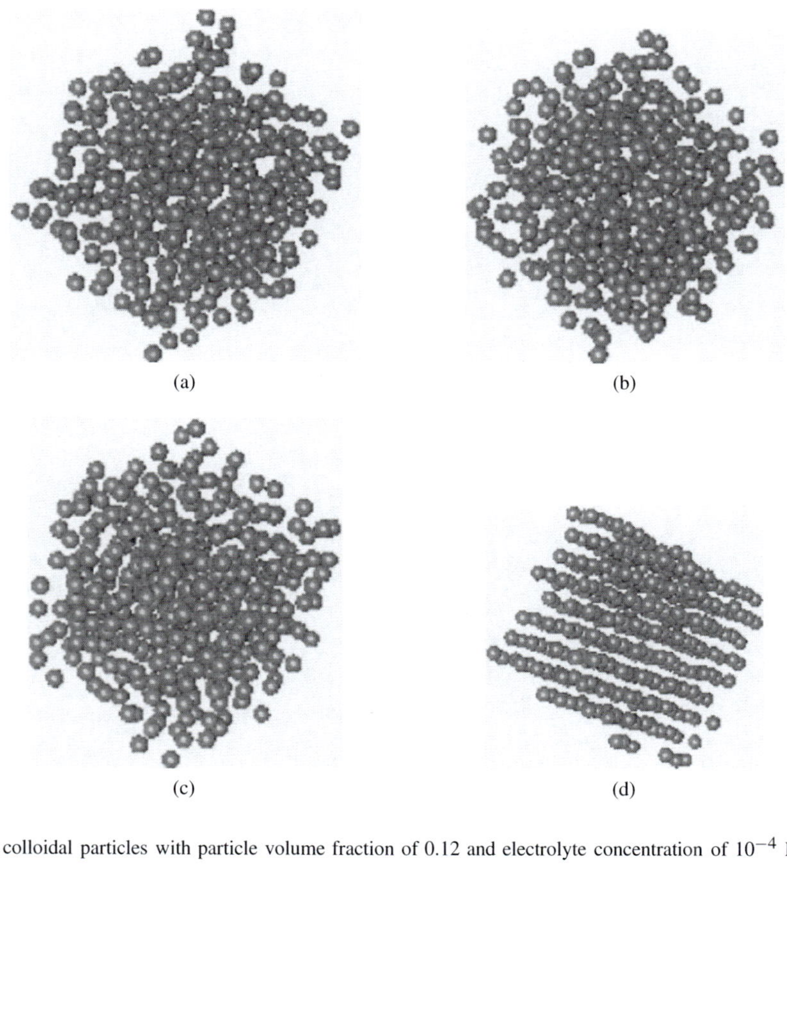

Fig. 16. Structure of charged colloidal particles with particle volume fraction of 0.12 and electrolyte concentration of 10^{-4} M. (a) $Z = 150$; (b) $Z = 300$; (c) $Z = 600$; (d) $Z = 1200$.

Table 1
The effective interaction and the phase structure for identical charged particles

	$Z = 150$	$Z = 300$	$Z = 600$	$Z = 1200$
	Diameter of particle $d = 670$ Å and electrolyte concentration $n_i = 10^{-5}$ M			
$f = 0.005$		Weak oscillations; disordered structure	Weak oscillations; disordered structure	Weak oscillations; disordered structure
$f = 0.015$		Oscillations; disordered structure	Oscillations; disordered structure	Strong oscillations; partially ordered structure
$f = 0.030$		Oscillations; disordered structure	Oscillations; local ordered structures	Oscillations; local ordered structures
$f = 0.060$	Oscillations; disordered structure	Oscillations; disordered structure	Large oscillations; ordered structure	Large oscillations; ordered structure
$f = 0.1200$	Oscillations; disordered structure	Large oscillations; ordered structure	Large oscillations; ordered structure	Large oscillations; ordered structure
	Diameter of particle $d = 670$ Å and electrolyte concentration $n_i = 10^{-4}$ M			
$f = 0.015$		No oscillations	No oscillations	No oscillations
$f = 0.030$		No oscillations	No oscillations	No oscillations
$f = 0.060$	Oscillations; disordered structure	Oscillations; disordered structure	Oscillations; disordered structure	Oscillations; disordered structure
$f = 0.1200$	Oscillations; disordered structure	Oscillations; disordered structure	Oscillations; disordered structure	Large oscillations; ordered structure

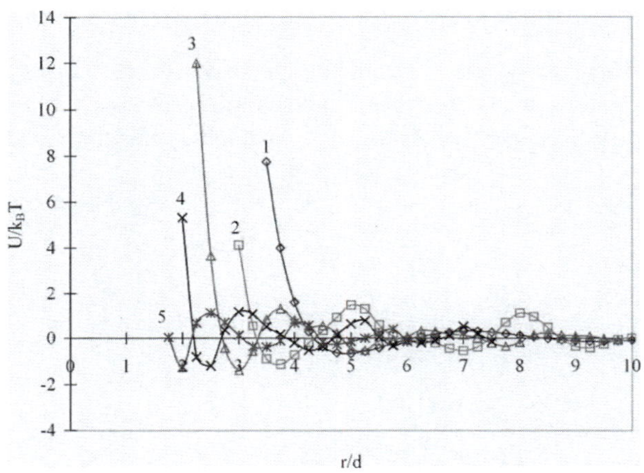

Fig. 18. Effective interaction potential between identical charged colloidal particles versus the distance between the centers of particles (d is the diameter of the particles) for various particle volume fractions (f), for charges per particle $Z = 1200$ and electrolyte concentration 10^{-5} M. (1) $f = 0.005$; (2) $f = 0.015$; (3) $f = 0.03$; (4) $f = 0.06$; (5) $f = 0.12$.

sults have shown that an effective attractive interaction can be generated even though there is no attractive component in the pair potential employed. This effective attraction is a result of many-body effects. The general equilibrium properties of the identical charged particles systems are summarized in Table 1. Generally, when the number of charges per particle is small and/or the particle volume fraction is low, no oscillatory effective interactions between identical charged particles are present. As the number of charges per particle and the particle volume fraction increased, the effective interaction between identical charged particles became oscillatory. Large charges per particle and particle volume fractions lead to very ordered structures, while smaller ones lead to disordered structures.

References

[1] T. Alfrey Jr., E.B. Bradford, J.W. Vanderhoff, G. Oster, J. Opt. Soc. Am. 44 (1954) 603.
[2] W. Luck, M. Klier, H. Wesslau, Ber. Bunsenges. Phys. Chem. 67 (1963) 75.
[3] S. Hachisu, Y. Kobayashi, A. Kose, J. Colloid Interface Sci. 42 (1973) 342.
[4] A. Kose, M. Ozaki, K. Takano, K. Kobayashi, S. Hashisu, J. Colloid Interface Sci. 44 (1973) 330.
[5] N. Ise, T. Okubo, M. Sugimura, K. Ito, H.J. Nolte, J. Chem. Phys. 78 (1982) 536.
[6] Y. Ishii, H. Matusuoka, N. Ise, Ber. Bunsenges. Phys. Chem. 90 (1986) 50.
[7] N. Ise, Angew. Chem. Int. Ed. 25 (1986) 323.
[8] I. Sogami, N. Ise, Phys. Lett. A 96 (1983) 199.
[9] I. Sogami, N. Ise, J. Chem. Phys. 81 (1984) 6320.
[10] M.V. Smalley, Mol. Phys. 71 (1990) 1251.
[11] I.S. Sogami, T. Shinohara, M.V. Smalley, Mol. Phys. 74 (1991) 599.
[12] M.V. Smalley, Langmuir 11 (1995) 1813.
[13] J.T.G. Overbeek, J. Chem. Phys. 87 (1987) 4406.
[14] C.E. Woodward, J. Chem. Phys. 89 (1988) 5140.
[15] S. Levine, D.G. Hall, Langmuir 8 (1992) 1090.
[16] R. Ettelare, Langmuir 9 (1993) 1888.
[17] J.T.G. Overbeek, Mol. Phys. 80 (1993) 685.
[18] E. Ruckenstein, Adv. Colloid Interface Sci. 75 (1998) 169.
[19] J.G. Kirkwood, J. Chem. Phys. 3 (1935) 300.
[20] C.W. Outhwaite, V.C.L. Hutson, Mol. Phys. 29 (1975) 1521.
[21] A.B. Schmidt, E. Ruckenstein, J. Colloid Interface Sci. 150 (1992) 169.
[22] E. Ruckenstein, A.B. Schmidt, J. Chem. Soc. Faraday Trans. 88 (1992) 2365.
[23] X. Chu, D.T. Wasan, J. Colloid Interface Sci. 184 (1996) 268.

Introduction to CHAPTER 5

Non-DLVO colloidal interactions: specific ion effects explained by ion-hydration forces

5.1 M. Manciu, E. Ruckenstein: "*Specific ion effects via ion hydration: I. Surface tension*" ADVANCES IN COLLOID AND INTERFACE SCIENCE 105 (2003) 63–101.

5.2 E. Ruckenstein, M. Manciu: "*Specific ion effects via ion hydration: II. Double layer interaction*" ADVANCES IN COLLOID AND INTERFACE SCIENCE 105 (2003) 177–200.

5.3 M. Manciu, E. Ruckenstein: *Comments on the "Osmotic Coefficients and Surface Tensions of Aqueous Electrolyte Solutions: Role of the Dispersion Forces*" JOURNAL OF PHYSICAL CHEMISTRY B 108 (2004) 20479–20481.

5.4 M. Manciu, E. Ruckenstein: "*On the Interactions of Ions with the Air/Water Interface,*" LANGMUIR 21(2005) 11312–11319.

5.5 M. Manciu, E. Ruckenstein: "*Ions at the air/water interface,*" JOURNAL OF COLLOID AND INTERFACE SCIENCE 304 (2006) 541–544.

One of the main shortcomings of the traditional DLVO theory is its inability to predict specific ionic effects. These effects are sometimes much stronger than what the excluded volume effects (examined in the previous chapter) can explain. It was recently suggested by Ninham and his group that the van der Waals interactions between ions and the rest of the system might be responsible for the ionic specificity [M. Boström, D.M.R. Williams, B.W. Ninham, Phys. Rev. Lett. 87 (2001) 168103]. However, these ion-dispersion forces lead to qualitatively incorrect results, namely that the cations are attracted toward the air/water interface and the anions are repelled by that interface. Experiment shows that the opposite is actually true.

A simple model that accounts for the large specific interactions between ions and surfaces is based on the long-known role of ions in structuring the water around them; the structure-making ions interact preferentially with the bulk water, hence they are effectively repelled by the surface, while the structure-breaking ions disturb the order of bulk water and therefore they are expelled toward the interface [5.1]. This simple Structure Making/Structure Breaking (SM/SB) model can explain quantitatively many effects related to ion specificity, such as the dependence of the surface potential and surface tension on *pH* and type and concentration of electrolyte; the existence of a minimum (dependent on the nature of the ions) of the surface tension at low electrolyte concentrations; and an almost linear dependence (with a positive or sometimes negative slope, again dependent on the nature of the ions) of the surface tension on the electrolyte concentration at sufficiently high ionic strengths [5.1]. The SM/SB model for ion-hydration forces is used to calculate the force between two parallel plates, and it is shown that specific ion effects emerge through changes in the surface charge and potential induced by ion hydration forces [5.2].

There are difficulties in accounting for the ion-dispersion forces at short separation distances, particularly when other strong interactions (such as ion-hydration or Born hard-sphere repulsion) are also present. It is demonstrated that the dispersion force between ions is negligible at distances larger than about 5 Å. At smaller distances it is not appropriate to base its calculation on a continuous dielectric model, because a small change in the cut-off parameter changes the value of the dispersion interaction from negligible into one that exceeds the mutual electrostatic repulsion between ions of the same kind. If real, the latter value can lead to the clustering of ions of the same kind, an interesting effect which was not yet observed [5.3].

The SM / SB model for the ion hydration force is compared with the traditional model for ion hydration (based on the Born energy of ions) and with a recent model due to Bostrom, Kunz and Ninham [Bostrom, M.; Kunz, W.; Ninham, B. W. *Langmuir* **2005**, *21*, 2619], which accounts for the decrease in hydration energy caused by the water density decrease in the vicinity of the air/water interface. Even when the ion-dispersion force is included in the latter model, it leads to an interfacial depletion of both

E. Ruckenstein, M. Manciu, *Nanodispersions*, DOI 10.1007/978-1-4419-1415-6_5, © Springer Science+Business Media, LLC 2010

anions and cations, which contradicts the Molecular Dynamics simulations results of the Jungwirth and Tobias group [Mucha, M.; Frigato, T.; Levering, L. M.; Allen, H. C.; Tobias, D.J.; Dang, L. X.; Jungwirth, P. *J. Phys. Chem. B* **2005**, *109*, 7617]. In contrast, the SM / SB model reproduces more accurately the results of the simulations [5.4].

There are an increasingly large number of experiments which demonstrate that the cations (except H^+) are repelled by the air-water interface whereas the anions (except F^-) accumulate there. The simple SM/SB model can account, both qualitatively and quantitatively, for these experimental results [5.5].

ELSEVIER
Advances in Colloid and Interface Science
105 (2003) 63–101
ADVANCES IN COLLOID AND INTERFACE SCIENCE
www.elsevier.com/locate/cis

Specific ion effects via ion hydration: I. Surface tension

Marian Manciu, Eli Ruckenstein*

Department of Chemical Engineering, State University of New York at Buffalo, Buffalo, NY 14260, USA

Abstract

A simple modality is suggested to include, in the framework of a modified Poisson–Boltzmann approach, specific ion effects via the change in the ion hydration between the bulk and the vicinity of the surface. This approach can account for both the depletion of the interfacial region of structure-making ions as well as for the accumulation of structure-breaking ions near the interface. Expressions for the change in interfacial tension as a function of electrolyte concentrations are derived. On the basis of this theory, one explains the dependence of the surface potential on pH and electrolyte concentration, the existence of a minimum in the surface tension at low electrolyte concentrations and the linear dependence, with a positive or sometimes negative slope, of the surface tension on the electrolyte concentration at sufficiently high ionic strengths.

Keywords: Specific ion effect; Ion hydration; Structure-making ion; Structure-breaking ion

Contents

*Corresponding author. Tel.: +1-716-645-2911/2214; fax: +1-716-645-3822.
E-mail address: feaeliru@acsu.buffalo.edu (E. Ruckenstein).

0001-8686/03/$ - see front matter
doi:10.1016/S0001-8686(03)00018-6

1. Introduction

The distribution of electrolyte ions around a charged surface was first calculated by Gouy [1] and Chapman [2]. Their theory assumed that the ions are point charges embedded in a continuum of constant dielectric constant, which are distributed according to Boltzmann statistics and interact with all the other charges via a mean field obeying the Poisson equation. The limitations of a theory with such drastic approximations was obvious from the beginning, and Stern pointed out [3] that this Poisson–Boltzmann approach can predict, in the vicinity of the surface, an ion concentration larger than that corresponding to close compaction. Since then, the theory was corrected in many ways to account for image forces [4], finite sizes of the hydrated ions [5], ion correlations [6,7], dependence of dielectric constant on the field [8] and electrolyte concentration [9] and many other effects.

The Poisson–Boltzmann approach [1,2] has the advantage of simplicity and is surprisingly accurate, at least for univalent ions in a certain range of electrolyte concentrations (1.0×10^{-3}–5×10^{-2} M) and not too close to the interface. It was later employed [10] to explain the repulsion due to the overlap of two double layers, and the stability of colloids thereafter [11,12].

It is instructive to compare the Poisson–Boltzmann approach as applied to colloid stability to a related example regarding the surface tension of electrolyte solutions. For most electrolytes, the surface tension of the water/air interface increases with increasing ionic strength. This increase is caused (via the Gibbs equation) by the negative adsorption of electrolyte ions on the surface [13]. The depletions of such ions near the interface were attributed for a long time to a more favorable free energy when the ion is fully hydrated [13], which occurs when the ion is sufficiently far from the interface. However, the Poisson–Boltzmann approach always predicts, for a charged interface and a symmetric electrolyte, an increase in the total number of ions in the vicinity of the interface, as compared to the bulk, because the counterions increase always exceeds the coions depletion. Hence, to calculate the distribution of ions in the vicinity of a surface and their surface adsorption, additional interactions have to be included in the Poisson–Boltzmann treatment. A charge embedded in a dielectric is repelled by the charge density it induces on the dielectric boundary (the 'image force'). The use of the image force to calculate the (negative) surface adsorption of ions led to a diverging increase in surface tension [14]. However, Wagner [14], inspired by the then-recent Debye–Hückel theory of electrolytes, realized that the other ions screen the image force. The screening length around an ion depends on the local concentration of ions and is, therefore, a function of the distance from the surface. Onsager and Samaras [15] simplified the calculations of Wagner, by assuming a constant screening length through electrolyte, and obtained an analytic 'limiting law' for the dependence of the surface tension on electrolyte concentration. However, the theory failed at high electrolyte concentrations (above ≈ 0.2 M), and the authors pointed out various sources of error and suggested various modalities to account for them into a more complete theory.

In contrast, the bare-bone Poisson–Boltzmann approach has had an unusual success in explaining quantitatively the double layer repulsion and the colloid stability. A few reasons for this success might be the partial cancellation of some effects decorating the bare-bone equations [5], or the fact that the corrections are important particularly within a few molecular diameters from the interface. The model inaccuracies could be reasonably well accounted for by a simple theoretical construction (the Stern layer) or by using the surface charge, surface potential or the dissociation constant of surface groups as a fitting parameter. The series of decorations of the Poisson–Boltzmann equation is long, and in spite of the rich literature available in the field, the calculations of the interactions are still mostly performed in the traditional DLVO manner (the bare-bone Poisson–Boltzmann approach).

It was recently shown that such a decoration of the theory with van der Waals interactions between individual ions and the interface might account in a simple manner for the inaccuracies of the Poisson–Boltzmann approach regarding the ion specific effects [16] and also for those of the Wagner–Onsager–Samaras theory [17,18]. In this article, we will first review the general framework of the Poisson–Boltzmann approach and will discuss the various kinds of approximations it involves. Some of the decorations will be briefly reviewed. The emphasis will be, however, on the recent van der Waals correction and arguments will be brought that the specific ion effects can be explained through ion hydration. A simple manner to account quantitatively for ion hydration and its relevance to ion specific effects will be suggested, and a comparison with the approach involving the van der Waals interactions made.

2. The mean-field formalism (the Poisson–Boltzmann approach)

A fundamental hypothesis of the Gouy–Chapman theory is that the interaction of an ion with all the other charges can be described by a 'mean potential' $\psi(x)$, where x is the distance from the surface. The chemical potential of an ion of species 'i' with charge q_i in the liquid is then given by:

$$\mu_i(x) = \mu_{0i} + kT\ln(c_i(x)) + q_i\psi(x) + W_i(x), \quad (1)$$

where μ_{0i} represents the standard chemical potential of the ion of species 'i', of concentration $c_i(x)$ and charge q_i. The interaction free energy $W_i(x)$ includes all the other interactions of the ion with the medium (not accounted for by the mean field). Some examples are examined below.

The distribution of ions of species 'i', $c_i(x)$, is related to the bulk concentrations c_{0i} of ions by the Boltzmann equation:

$$c_i = c_{0i}\exp\left(-\frac{q_i\psi(x) + \Delta W_i(x)}{kT}\right), \tag{2}$$

where $\Delta W_i(x)$ is calculated with respect to the bulk and can depend also on ψ, c_i and other parameters, k is the Boltzmann constant and T is the absolute temperature. The mean field is assumed to satisfy the Poisson equation:

$$\nabla^2\psi(x) = -\frac{\sum_i c_i(x) q_i}{\varepsilon_0} + \frac{\partial P(x)}{\varepsilon_0 \partial x}, \tag{3}$$

where the polarization $P(x)$ is typically assumed proportional to the macroscopic electric field:

$$P(x) = -\varepsilon_0(\varepsilon - 1)\nabla\psi(x), \tag{4}$$

ε_0 being the vacuum permittivity and ε the dielectric constant of the medium.

2.1. The Poisson–Boltzmann equation

The most simple approximation is $\Delta W_i(x) = 0$; for an uni-univalent electrolyte of concentration c_E the potential in the vicinity of a planar surface can be calculated from Eqs. (2)–(4), which lead to,

$$\frac{d^2\psi(x)}{dx^2} = \frac{2ec_E}{\varepsilon\varepsilon_0}\sinh\left(\frac{e\psi(x)}{kT}\right) \tag{5}$$

the well-known Poisson–Boltzmann equation, where e is the proton charge. It can be solved for suitable boundary conditions, such as constant surface charge density, constant surface potential, dissociation or adsorption equilibrium at the surface. Once the ion distribution is known, the other characteristics of the system, such as the free energy, can be calculated.

Several kinds of approximations are involved in the derivation of Eq. (5). First, it implies that the interactions between charges can be described by an 'average' potential, which is a mean field type of approximation. A theory devoid of this assumption can be developed in the general framework of statistical mechanics, using pair correlation functions. One of the successful approaches, based on the inhomogeneous hypernetted chain approximation, was presented by Kjellander and Marcelja [7]. However, the results provided by this method depend strongly on the ion pair potentials employed and it was shown [19] that minor changes in the ion-pair potentials can lead to qualitatively different results. It is not yet clear if the ion-correlation effects can be described to a good approximation by a suitable-chosen function $\Delta W_i(x)$.

Another limitation of the Poisson–Boltzmann approach is that the interaction between two surfaces immersed in water might not be exclusively due to the electrolyte ions. For instance, water has a different structure in the vicinity of the surface than in the bulk and the overlapping of such structures generates a repulsion even in the absence of electrolyte [20]. In this traditional picture, the 'hydration' repulsion is not related to ion hydration; actually it is not related at all to electrolyte ions. However, as recently suggested [21], this 'hydration' interaction can still be accounted for within the Poisson–Boltzmann framework, assuming that the polarization is not proportional to the macroscopic electric field, but depends also on the field generated by the neighboring water dipoles and by the surface dipoles.

There are also other effects, which cannot be incorporated easily in the Poisson–Boltzmann formalism, such as the non-planarity of the interface, its thermal fluctuations, the charge fluctuations and so on.

The other kinds of approximations are due to the oversimplification $\Delta W_i = 0$. Many effects can be accounted for by decorating the Poisson–Boltzmann equation with more suitable expressions for $\Delta W_i(x)$. The rest of the article is devoted to the examination of a few possible choices.

2.2. *The image forces* (*Wagner–Onsager–Samaras theory*)

As already noted, the Poisson–Boltzmann equation fails to describe the ion depletion in the vicinity of the surface, which is basically due to the preference of the ions for the bulk liquid, where they can be fully hydrated.

One of the first corrections to the formalism was the addition of the interaction between the ions of charge q_i and the surface via the (screened) 'image force' [14]:

$$\Delta W_{\mathrm{W},i}(x) = \left(\frac{\varepsilon - \varepsilon'}{\varepsilon + \varepsilon'}\right)\frac{q_i^2}{16\pi\varepsilon_0\varepsilon x}\exp\left(-\frac{2x}{\lambda(x)}\right), \tag{6a}$$

where ε' is the dielectric constant of the medium which has an interface with water and $\lambda(x)$ is the screening length. This interaction is the same for both kinds of ions of a symmetric electrolyte, hence the ion depletions are in this case the same for both ions and, in the absence of surface charges or other interactions, the electric potential is vanishing everywhere. To account for different ion sizes (and hence to obtain, to some extent, ion-specific effects), Onsager and Samaras [15] proposed an interaction potential modified in a Debye–Hückel spirit:

$$\Delta W_{\mathrm{OS},i}(x)=\left(\frac{\varepsilon-\varepsilon'}{\varepsilon+\varepsilon'}\right)\frac{\exp\left(\frac{2a_i}{\lambda}\right)}{1+\frac{2a_i}{\lambda}}\frac{q_i^2}{16\pi\varepsilon_0\varepsilon x}\exp\left(-\frac{2x}{\lambda}\right), \tag{6b}$$

where a_i are the ion radii and $\lambda=\sqrt{(\varepsilon\varepsilon_0 kT/\Sigma_i c_{0i}q_i^2)}$ is the Debye–Hückel screening length.

The Onsager–Samaras theory fails at electrolyte concentrations above approximately 0.2 M. One of the reasons is the assumption that the screening length λ does not depend on the distance to the interface. In reality, the screening in the ion-depleted region near the surface is smaller than in the bulk. Only when λ is large compared to the thickness of the depleted layer, the above approximation is accurate. Other reasons for the inaccuracy (additional interactions and surface charge) will be examined later in the paper.

The image force formalism was intended primarily to relate, via the Gibbs adsorption equation, the increase of the surface tension of water to the negative adsorption of the electrolyte ions. Its effect on the double layer interactions was examined by Jönsson and Wennerström [4].

2.3. Ion hydration effects via the dielectric constant

Another modality to include the ion hydration effects in the Poisson–Boltzmann formalism is via the dielectric constant. If the surface charge is sufficiently large, then there is a large electric field and a high concentration of counterions near the surface. Both of these factors lead to a decrease of the dielectric constant. The main contribution to the polarizability of a water molecule is a result of the alignment of its dipole by the electric field, the polarization being proportional with the field when the latter is sufficiently low, but becoming saturated at high fields. In addition, the presence of an electrolyte hinders the polarizability of the water molecules, thus decreasing the dielectric constant. To account for these effects, one should first account for the dependence of the dielectric constant on the magnitude of the field [9] and on the counterions concentration [8] in Eq. (4). Second, one should account, via the Born energy, for the change of the ion hydration free energy due to the change in the dielectric constant of the medium in Eq. (2).

The Born hydration energy is the electrostatic energy of an ion in a dielectric and depends on the field generated by the ion in the entire space. When there is a boundary between two media with different dielectric constants, the change in the electrostatic energy when the ion approaches the boundary generates an image force, which induces a diverging negative surface adsorption. One can avoid this behavior by accounting for the screening of the interaction by the electrolyte [14]. This treatment does not include the non-uniformity of the dielectric constant.

Another possible treatment [9] is to define the Born energy in term of a local dielectric constant $\varepsilon(x)$:

$$\Delta W_{\mathrm{B},i}(x) = \frac{q_i^2}{8\pi\varepsilon_0 a_i}\left(\frac{1}{\varepsilon(x)} - \frac{1}{\varepsilon}\right), \tag{7}$$

where $\varepsilon = \varepsilon(\infty)$ is the bulk dielectric constant. In this case, the ions are repelled by the interface only because the dielectric constant of water in its vicinity, $\varepsilon(x)$, is lower than in the bulk. If the dielectric constant would be larger near the interface, the ions would be attracted there.

When the interface is charged, the Poisson–Boltzmann equation predicts stronger fields and higher ionic concentrations in the vicinity of the surface, therefore a dielectric constant smaller than in the bulk, and the 'local' Born energy indicates that the ions are repelled by the interface. However, a neutral surface depleted of ions should have a dielectric constant larger than the bulk and the ions should be attracted, not repelled by such an interface. Therefore, other interactions should be included to explain the ion depletion near a neutral surface.

2.4. Ion size effects

It has been long accepted that the Poisson–Boltzmann equation can lead to ion concentrations that exceed the volume available in the vicinity of a surface. Stern included the effect of the ion size on its distribution around a charged surface by assuming the existence of a layer in the vicinity of the surface containing only condensed counterions [3]. The effect of the ion size on the double layer interaction was later incorporated in the modified Poisson–Boltzmann formalism in a manner similar to that described here [22]. The difference in the free energy of ions between bulk and interface can be written as a function of ion concentrations, either in a lattice model [22], in terms of the occupation probabilities or in a continuum treatment [23], by assuming that the hydrated ions behave as rigid spheres and employing the excess chemical potential with respect to an ideal gas.

By accounting for the size of the ions one can explain the depletion of ions in the vicinity of the surface and obtain in most cases an increase in the double layer repulsion as compared to the Poisson–Boltzmann equation (in some special cases, it might lead to a decrease of the repulsion [24]). It also can account, at least partially, for some specific ion effects. However, the differences in hydrated radii between various ions are small and the specific effects become important in this treatment only at high ionic strengths and high surface potentials. While the hydrated ion radius hinders the ions to approach to close the interface, the 'depletion region', which will be estimated later in this paper from the change in the interfacial tension, might either exceed or be much smaller (sometimes, even negative) than the hydrated radius of the ions.

2.5. The van der Waals interactions between ions and media

Onsager and Samaras were interested in a qualitative explanation of the experimentally determined increase in surface tension with increasing electrolyte concentrations. The screened image force provided a good agreement with experiment, up to approximately 0.2 M. They suggested that better agreement could be obtained if higher multipolar correction would be included in the treatment. Onsager and Samaras also remarked that these interactions should be important only in the close vicinity of the surface, since they should decrease with at least the third power of the distance.

It was later conjectured by Sienko [25] that the multipolar corrections to the Onsager–Samaras theory might account for specific ion effects, because of different ion polarizabilities. A quantitative attempt to investigate this conjecture was made recently by Stairs [25], by including the interaction between a dipole and its image, using the (screened) Onsager–Samaras potential. However, because of screening, the effect is small at high ionic strengths.

Ninham and Yaminsky [26], and Karraker and Radke [18] realized that the van der Waals interactions between the ions and interface are not screened by the electrolyte and hence might become more important than the image force, at large electrolyte concentrations. Recognizing that the hydration of ions might also play a role, Boström et al. [16,17] showed that the van der Waals interactions alone (with suitable values selected for the interaction parameters) might account for the ion specific effects.

The van der Waals interaction between an ion and the interface between two dielectric media, calculated in the Lifshitz approach, has as leading term [26]:

$$\Delta W_{\mathrm{NY}}(x) = +\frac{B}{x^3}. \tag{8}$$

Let us first address some quantitative issues, using the more simple Hamaker approach. If one assumes pairwise additivity, the interaction between a sphere of kind (1) immersed in a fluid of kind (3) and a planar half-space of kind (2) is given by [27]:

$$\Delta W_{\mathrm{H}}(x) = -\frac{A_{132}a}{6(x-a)}\left[1+\frac{x-a}{x+a}+\frac{x-a}{a}\ln\left(\frac{x-a}{x+a}\right)\right] = -\frac{2A_{132}a^3}{9x^3}-\frac{4A_{132}a^5}{15x^5}-\ldots, \tag{9}$$

where x is the distance from the interface to the center of the ion of radius a and A_{132} is the effective Hamaker constant between ion (1) embedded in a fluid (3) and a medium (2) on the other side of the interface, given by [27]:

$$A_{132} = A_{12} + A_{33} - A_{13} - A_{23}, \tag{10}$$

where A_{ij} are the Hamaker constants of the interactions between the media 'i' and 'j' in vacuum. Eq. (8), with $B = -(2A_{132}a^3)/9$, is a good approximation of Eq. (9) for $a \ll x$ (the difference is less than 10% for $x > 4a$).

Since three media are involved in these van der Waals interactions, the effective Hamaker constant can become negative. For the water/air interface, $A_{12} = A_{23} = 0$, and hence:

$$A_{132} = +A_{33} - A_{13}. \tag{11}$$

If the Hamaker constant between ion and water A_{13} (in vacuum) is larger than that between water and water A_{33}, the effective Hamaker constant is negative and the ions are repelled by the interface. On the other hand, if $A_{13} < A_{33}$, the ions are attracted by the interface.

Boström et al. [16,17] showed that the positive ions are less polarizable and the negative ions more polarizable than water. Therefore, the former should be attracted to and the latter repelled by the interface. The values of B_i have been, however, obtained by fitting the experimental interfacial tensions.

Karraker and Radke [18], accounting only for the interactions between the individual ions and the water and air continua (thus ignoring the term A_{33} in Eq. (11)), concluded that the dispersion interactions always repel the ions from the water/air interface, with the exception of the non-polarizable H^+, for which $B_H = 0$.

By assuming a radius $a = 1.8$ Å for the Cl^- ion embedded in a water continuum, one obtains the value $B_{Cl} = 31 \times 10^{-50}$ J m^3 employed by Boström et al. [16,17] if $A_{132} = 23 \times 10^{-20}$ J, which is typical for metal–metal interactions through vacuum ($10–30 \times 10^{-20}$ J [27]). If the 'Hamaker constants' A_{13} of ion–water interactions would be of this order of magnitude, then the water–water interaction ($A_{33} = 3.7 \times 10^{-20}$ J) can be neglected in Eq. (11). However, the derivation of Eq. (10) is based on the (macroscopic) hypothesis that the 'ion sphere' replaces a 'water sphere' of the same size. If the ions are thought as occupying interstitial places in a (rigid) water lattice, the displacement of ions is not followed by that of water. In such a case, the effect of the 'intervening medium' has to be discarded and the Hamaker constants calculated (in the hypothesis of pairwise additivity) as through vacuum.

Furthermore, these van der Waals interactions are important only near the interface, where it is unlikely that either Lifshitz or Hamaker approaches are accurate for spheres of molecular sizes. For example, the magnitude of the interaction for Na^+ ions at 5 Å from the interface is only approximately $0.02kT$ (the values of B used in the calculation, $B_{Na} = -1 \times 10^{-50}$ J m^3 was obtained from fit by Boström et al. [17] and $B_{Na} = +0.8 \times 10^{-50}$ J m^3 was calculated by Karraker and Radke [18]). Eq. (8) might provide a convenient way to account for the interfacial interactions, if suitable values for B_i (not related to the 'macroscopic' Hamaker constants) would be selected.

There are, however, some more important qualitative issues, which should be discussed. If the cations would be attracted by the interface, they will condense there (since this attraction diverges for $x \rightarrow 0$) and the accumulation of ions on the interface will lead to an infinite decrease of the interfacial tension. Boström et al. [17] avoided this 'cation catastrophe' by selecting a cut-off of 2 Å in the calculations and obtained good agreement with experiment by fitting the values of *B*. As will be shown in the next section, if a cut-off of approximately 4 Å is used, good agreement with experiment can be obtained by completely disregarding the van der Waals interactions of ions.

More dramatic are the long-range consequences of the van der Waals attraction, even when a cut-off is included. As shown by Boström et al. [16], the double layer interactions in electrolytes of physiological concentrations become in this case attractive at distances less than 30 Å. This unexpected theoretical prediction has yet to be verified experimentally.

Recent molecular dynamics simulations [28] showed that the 'commonly held opinion' that the interfacial region is depleted of electrolyte ions might not be true, and that some ions might approach the interface. Their accumulation on the interface can be predicted by a suitable negative value for the parameter B_i in the van der Waals interactions of the ions (coupled with an interaction cut-off). However, there are the large negative ions (Cl^-, Br^-) which prefer to accumulate in the vicinity of the interface [28], and not the less-polarizable small and positive ions, as expected from the van der Waals interactions.

In the treatment of Karraker and Radke [18], the interaction coefficients B_i are always positive, and the cation catastrophe is avoided. However, the van der Waals repulsions are still larger for the negative ions than for the positive ones; therefore, the accounting for the van der Waals interaction alone would imply that all air/electrolyte interfaces should have a positive surface potential. Boström et al. [17] argued that this might be indeed the case, and that the negative values of the potentials typically obtained experimentally might be caused by the double layers generated around the electrodes used in the experiments, and do not represent intrinsic properties of the water/air interface. Karraker and Radke [18] suggested that the negative potentials of the surfaces are due to the preference of OH^- for the surface, thus including another interaction (in their treatment, the van der Waals interactions still repel the OH^- ions from the interface, stronger than they repel the cations).

The increase of the surface tension of water by the addition of an electrolyte was traditionally related (via Gibbs adsorption equation) to the negative adsorption of ions on the interface. However, some electrolytes decrease the interfacial tension [29], hence should be positively adsorbed. Therefore, if the van der Waals interactions would repel all the ions from the interface, some additional interactions have to be included to explain the positive adsorption.

The accounting for the van der Waals interactions of ions has the advantage of simplicity, since the determination of the values of the B_i coefficients for all the ions could lead to a quantitative treatment of ion specific effects within the Poisson–Boltzmann framework. However, as mentioned above, this treatment provides the debatable prediction that the cations would approach the interface closer than the anions.

The change in ion hydration between bulk and interface was suspected since a long time [13] to be responsible for the distribution of ions in the vicinity of the interface; whereas this effect can be easily understood qualitatively, a consistent quantitative theory has not yet been developed. The reason is that accurate information about the structure of water in the vicinity of an interface and about the corresponding changes in the ion hydration with the distance are yet beyond our reach. In the next section, a simple modality to account in a semi-quantitative manner for the ion hydration effects will be suggested. The effects of the van der Waals interactions and image forces will be compared with those of hydration.

3. A simple treatment of ion hydration. General behavior

It is well known that, in general, there is a strong correlation between the standard enthalpy of hydration of an electrolyte and the modification of the surface tension by that electrolyte [30]. The effect of ion hydration on interfacial tension can be easily understood qualitatively. For the structure-making ions, their approach to the interface is unfavorable, because they can better organize the water dipoles in bulk water than at the interface. The opposite is true for the structure-breaking ions; in such cases, the total free energy of the system is minimized by pushing the structure-breaking ions toward the interface, because the bulk water molecules can better organize their hydrogen bonding network without these ions.

The surface tension can either increase or decrease with electrolyte concentration, depending whether the depletion of the structure-making ions in the vicinity of the surface or the adsorption of the structure-breaking ions dominates the process. The distribution of ions in the vicinity of the interface could be calculated if accurate expressions for $\Delta W_i(x)$ would be available.

There is, however, a simple manner to obtain reasonable estimations of ion distribution, even when the expressions $\Delta W_i(x)$ are not known. The hydration free energy per ion, in bulk water, is of the order of -10^2kT [31]. An increase in the total free energy by $5kT$ implies (via the Boltzmann distribution) a concentration decrease by more than two orders of magnitude. Therefore, if at a particular distance d_i from the interface, the hydration free energy of a structure-making ion is increased by only a few percent from the bulk hydration free energy, one can consider, for calculation purposes, that the region from the interface to the distance d_i is totally depleted of those ions.

Hence, one can describe the hydration interaction of a structure-making ion by a parameter, representing a distance of closest approach to the surface. It is plausible to consider d_i related to the hydrated radii of the ions of species 'i'; however, they are not the same because the structure of water is different in the vicinity of the interface from that in the bulk.

This approach might seem to be rather crude; however, it should be noted that recent molecular simulations of simple electrolytes interfaces [28] show that the variation of the electrolyte concentrations is about as steep as the variation of water density at the interface. Therefore, if one assumes a sharp water/air interface, one can also assume step distributions for the ions of the electrolyte.

A similar simple treatment is less accurate for the structure-breaking ions. The change in the total free energy between surface and bulk cannot exceed a few kT, because otherwise the ions would be trapped at the interface and generate a large surface potential, possibility ruled out by experiment. Since $\Delta W_i(x)$ is not very large, it cannot be well approximated, in general, by a step function. However, in the spirit of the above approximation and in the absence of more accurate information, we will assume that it can be described by a potential well and will examine the consequences.

In this section, we will investigate the effect of the changes in the free energies of ions between bulk and interfacial region, on the ion distributions in the vicinity of the interface, and consequently on the surface tension, for an uni-univalent electrolyte in the following cases:

1. Both kinds of ions are structure-making, and $d_1 = d_2$;
2. Both kinds of ions are structure-making, but the cations (2) are repelled more strongly than the anions (1): $d_2 > d_1$;
3. The cations are structure-making ($d_2 > 0$) but the anions are structure-breaking, hence positively adsorbed on the interface.

3.1. $d_1 = d_2 = d_E$

In this case (in the absence of a surface charge) the interface is neutral. The dependence of the interfacial tension γ on the uni-univalent electrolyte concentration c_E can be calculated using the Gibbs adsorption equation:

$$d\gamma = -\sum_{i=1,2} \Gamma_i d\mu_i, \tag{12}$$

where μ_i is the (total) electrochemical potential and Γ_i is the surface excess of ion i:

$$\Gamma_i(c_E) = \int_0^{\infty} (c_i(x) - c_E) dx \tag{13}$$

with the water/air interface located at $x = 0$. The total electrochemical potential, which is independent of the distance from the interface, can be related to the bulk electrolyte concentration $c_E = c_i(\infty)$ via:

$$d\mu_i = d(kT \ln(c_i(\infty) f_i)), \tag{14}$$

where f_i is the activity coefficient of the electrolyte in the bulk.

Using Eqs. (12)–(14) and assuming $f_i \equiv 1$, the change in the interfacial tension due to the addition of an electrolyte is given by:

$$\frac{1}{kT} \frac{\partial \gamma}{\partial c_E} = -\frac{\Gamma_1 + \Gamma_2}{c_E} \equiv 2\Delta, \tag{15}$$

where the ratio $-(\Gamma_1+\Gamma_2)/2c_E$ has the dimension of a length which is denoted by Δ, the depletion thickness.

The surface tension as a function of electrolyte concentration can be obtained by integrating Eq. (15):

$$\gamma(c_E)-\gamma(0)=\int_{c=0}^{c_E}\frac{\partial\gamma}{\partial c}\mathrm{d}c=-kT\int_0^{c_E}\frac{\Gamma_1(c)+\Gamma_2(c)}{c}\mathrm{d}c. \tag{16}$$

In the present case, there is no surface charge and $d_1=d_2=d_E$, therefore $c_i(x)=c$ for $x>d_E$ and $c_i(x)=0$ for $0<x<d_E$, which provides $\Gamma_i(c_i)\equiv -c\Delta=-cd_E$ and

$$\Delta\gamma(c_E)=(2kTd_E)c_E\,\alpha\,c_E. \tag{17}$$

This expression was employed before to account for the dependence of γ on the electrolyte concentration [30], which is roughly linear up to large electrolyte concentrations. It should be noted that this simple model is in better agreement with experiment up to large ionic concentrations [29] than the Onsager–Samaras theory [15], which assumes that the image forces are screened by the electrolyte and hence are small at large electrolyte concentrations. However, d_E has to be regarded as a parameter, which depends on both kinds of ions of the electrolyte; therefore, the 'specific ion effects' are transformed into 'specific electrolyte effects'. For some electrolytes, the value of d_E determined through fitting is reasonable and seems to be close to the hydrated ion radius, for example $d_E=3.9$ Å for NaCl and $d_E=3.8$ Å for KCl [30]. For these electrolytes, the selection of a cut-off distance equal to $d_E(\approx 4$ Å) in the Poisson–Boltzmann equation would lead to an excellent agreement with surface tension experiments, even if the van der Waals interactions and other interactions that decorate the Poisson–Boltzmann approach are disregarded [30].

It is instructive to compare how the ion-interface interactions (image force, van der Waals interactions, ion hydration) affect the depletion thickness at various electrolyte concentrations. To estimate the magnitudes of these effects, it will be assumed that the interactions are non ion-specific (which implies to use Eq. (6a) for the potential of the image force, and $B_1=B_2=B$ in Eq. (8) for the van der Waals interactions). Consequently, the interface is neutral and the surface adsorption can be calculated from the simple expression:

$$\frac{\Gamma_1}{c_E}=\frac{\Gamma_2}{c_E}=-\Delta=\int_0^{\infty}\left(\exp\left(-\frac{\Delta W(x)}{kT}\right)-1\right)\mathrm{d}x. \tag{18}$$

As that generated by ion hydration for $d_1=d_2$, the depletion thickness generated by the van der Waals interactions, when $B_1=B_2$, does not depend on concentration. When only the van der Waals interactions are accounted for, Eqs. (8) and (18) lead

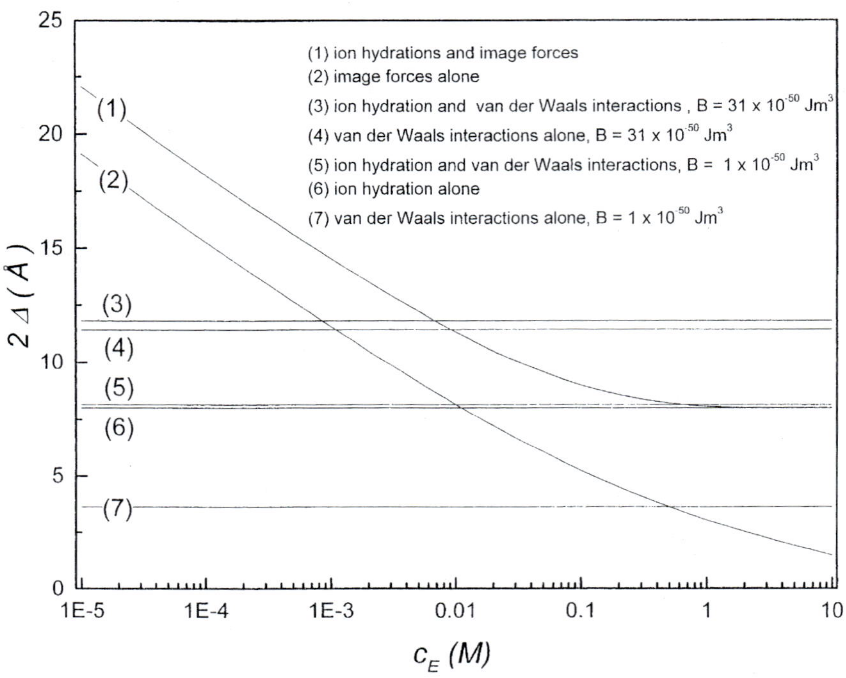

Fig. 1. Comparison between the depletion distances generated by several symmetric (the same for both ions) interactions: the image forces (Eq. (6a), $\varepsilon=80$, $\varepsilon'=1$, $T=300$ K), the van der Waals interaction (Eq. (8), $B_1=B_2=31\times10^{-50}$ or 1×10^{-50} J m^3) and ion hydration ($d_1=d_2\equiv d_E=4$ Å). The interactions accounted for on each curve are: (1) the ion hydration and the image forces; (2) the image forces alone; (3) the ion hydration and the van der Waals interactions, $B=31\times10^{-50}$ J m^3; (4) the van der Waals interactions alone, $B=31\times10^{-50}$ J m^3; (5) the ion hydration and the van der Waals interactions, $B=1\times10^{-50}$ J m^3; (6) the ion hydration alone; (7) the van der Waals interactions alone, $B=1\times10^{-50}$ J m^3.

to the following depletion distance [17]:

$$\Delta\equiv\int_0^\infty\left(1-\exp\left(-\frac{B}{kTx^3}\right)\right)\mathrm{d}x=\left(\frac{B}{kT}\right)^{1/3}\int_0^\infty\left(1-\exp\left(-\frac{1}{y^3}\right)\right)\mathrm{d}y\cong1.354\left(\frac{B}{kT}\right)^{1/3}. \tag{19a}$$

while when both the ion hydration and van der Waals are accounted, the effect of the van der Waals interactions is to increase (decrease) the cut-off distance:

$$\Delta\equiv\int_0^{d_E}\mathrm{d}x+\int_{d_E}^\infty\left(1-\exp\left(-\frac{B}{kTx^3}\right)\right)\mathrm{d}x\cong d_E+\frac{B}{2kTd_E^2}. \tag{19b}$$

The correction of the cut-off distance in Eq. (19b) is significant only for large values of B.

There is no reason to assume that $B_1 = B_2$ or that $d_1 = d_2$. However, when the surface potential is low (hence in the absence of a surface charge and at sufficiently low electrolyte concentrations), Eq. (18) can be used to calculate separately the depletion thicknesses for each kind of ions.

For comparison purposes, the depletion thickness is plotted in Fig. 1 as a function of electrolyte concentration ($\varepsilon = 80$, $\varepsilon' = 1$, $T = 300$ K) for a number of cases listed there. The value of the cut-off distance $d_E = 4$ Å was selected to obtain agreement with the surface tension data for the aqueous solutions of some common electrolytes (NaCl, KCl) at high ionic strengths. The effects of the van der Waals interactions or of ion hydration on the depletion thickness are qualitatively the same. When B is large ($B = 31 \times 10^{-50}$ Jm^3), the additional accounting of ion hydration provides only a small correction to the result obtained with van der Waals interactions alone. When B is small, the additional accounting of the van der Waals interactions does not affect much the result obtained with ion hydration interactions alone. It should be noted than neither B nor d_E are precisely known; what can be inferred from the data on surface tension at high ionic strengths is that both effects together should lead to a depletion thickness of approximately 4 Å. Therefore, the rather large value $B_{Cl} = 31 \times 10^{-50}$ J m^3, evaluated from fitting in Ref. [17], might have been obtained because the ion hydration was neglected.

Assuming an electrolyte with a depletion thickness of approximately 4 Å at high ionic strengths, Fig. 1 suggests that the image forces play a dominant role at very low electrolyte concentrations ($c_E < 1.0 \times 10^{-4}$ M), but are negligible at relatively large concentrations ($c_E > 0.1$ M).

For some electrolytes, the 'characteristic distance' d_E required to fit the experimental data on surface tension with Eq. (17) is smaller than the hydrated radius of the ions and might even become negative (e.g. HCl, HNO_3, $HClO_4$ [29]). This means that at least one component of the electrolyte is preferentially adsorbed on the interface. Therefore, d_E can be regarded only as a parameter in an extremely simplified theory.

3.2. $d_1 < d_2$

The differences in the hydration energies of the ions and in their hydrated radii are expected to lead to $d_1 \neq d_2$. As indicated by simulations [28], the negative ions are less repelled, therefore they can approach the interface closer than the positive ions. The asymmetric ion depletions generate a surface potential, even in the absence of an 'external' surface charge. The charge that generates the diffuse double layer is located a few Ångstroms from the interface, and is due to the anions.

The corresponding modified Poisson–Boltzmann equations for the planar interface located at $x = 0$ are:

$$\frac{d^2\psi}{dx^2} = 0 \quad 0 < x < d_1, \tag{20a}$$

$$\frac{d^2\psi}{dx^2}=\frac{ec_E}{\varepsilon\varepsilon_0}\exp\left(\frac{e\psi}{kT}\right)\quad d_1<x<d_2, \tag{20b}$$

$$\frac{d^2\psi}{dx^2}=\frac{2ec_E}{\varepsilon\varepsilon_0}\sin h\left(\frac{e\psi}{kT}\right)\quad d_2<x<\infty, \tag{20c}$$

where e is the protonic charge. Eqs. (20a), (20b) and (20c) can be solved for the boundary conditions:

$$\frac{d\psi(x)}{dx}=0\quad \text{(no surface charge density)}\quad x=0, \tag{21a}$$

$$\psi(x)\to 0\quad x\to\infty. \tag{21b}$$

The surface potential generated by the asymmetric distribution of the ions in the vicinity of the surface is small when the electrolyte concentration is sufficiently low. It is, therefore, worthwhile to obtain an analytical solution in the linear approximation. The general solution of the linear approximation of Eqs. (20a), (20b) and (20c) is:

$$\psi_I(x)=C_1x+C_2\quad 0<x<d_1, \tag{22a}$$

$$\psi_{II}(x)=-\frac{kT}{e}+C_3\exp\left(\frac{x-d_1}{\sqrt{2}\lambda}\right)+C_4\exp\left(-\frac{x-d_1}{\sqrt{2}\lambda}\right)\quad d_1<x<d_2, \tag{22b}$$

$$\psi_{III}(x)=C_5\exp\left(\frac{x-d_2}{\lambda}\right)+C_6\exp\left(-\frac{x-d_2}{\lambda}\right)\quad d_2<x<\infty, \tag{22c}$$

where λ is the Debye–Hückel length, $\lambda=\sqrt{(\varepsilon\varepsilon_0 kT/2e^2c_E)}$.

The boundary condition Eq. (21a) leads to $C_1=0$ and the boundary condition Eq. (21b) to $C_5=0$. The continuity of ψ and $d\psi/dx$ at $x=d_1$ and $x=d_2$ leads finally to the solution:

$$\psi_I(x)=-\frac{kT}{e}\left(1-\frac{1}{\cos h\left(\frac{d_2-d_1}{\lambda\sqrt{2}}\right)+\frac{1}{\sqrt{2}}\sin h\left(\frac{d_2-d_1}{\lambda\sqrt{2}}\right)}\right)\quad 0<x<d_1, \tag{23a}$$

$$\psi_{II}(x)=-\frac{kT}{e}\left(1-\frac{\cos h\left(\frac{x-d_1}{\lambda\sqrt{2}}\right)}{\cos h\left(\frac{d_2-d_1}{\lambda\sqrt{2}}\right)+\frac{1}{\sqrt{2}}\sin h\left(\frac{d_2-d_1}{\lambda\sqrt{2}}\right)}\right)\quad d_1<x<d_2, \tag{23b}$$

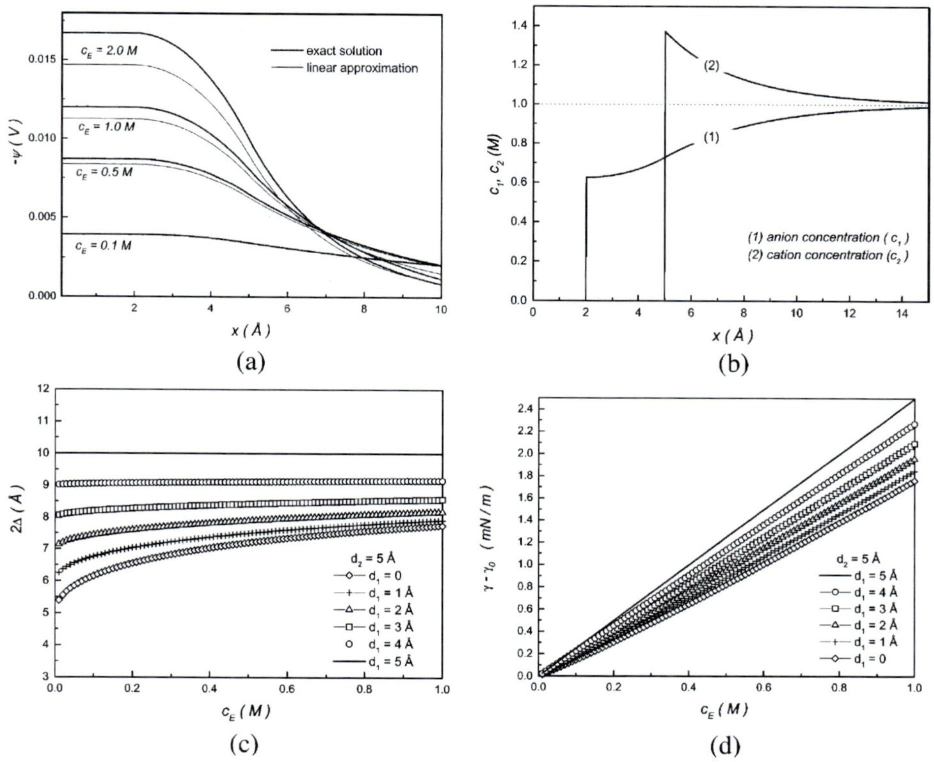

Fig. 2. (a) The potential generated by the asymmetric distribution of the electrolyte ions ($d_1 = 2$ Å, $d_2 = 5$ Å, $\varepsilon = 80$, $T = 300$ K) provided by Eq. (23a), Eq. (23b) and Eq. (23c) is compared with the numerical solution of Eqs. (20a), (20b) and (20c) for various electrolyte concentrations. (b) The distribution of electrolyte ions in the vicinity of the interface ($d_1 = 2$ Å, $d_2 = 5$ Å, $c_E = 1$ M, $\varepsilon = 80$, $T = 300$ K). (c) The depletion thickness for $d_2 = 5$ Å and $d_1 = 0$, 1, 2, 3, 4 and 5 Å, as a function of electrolyte concentration ($\varepsilon = 80$, $T = 300$ K). (d) The change in surface tension as a function of electrolyte concentration, for $d_2 = 5$ Å and $d_1 = 0$, 1, 2, 3, 4 and 5 Å. The value γ_0 corresponds to $c_E \equiv 0$.

$$\psi_{\mathrm{III}}(x) = -\frac{kT}{e}\frac{1}{1+\sqrt{2}\left(\tan h\left(\dfrac{d_2-d_1}{\lambda\sqrt{2}}\right)\right)^{-1}}\exp\left(-\frac{x-d_2}{\lambda}\right) \quad d_2 < x < \infty. \tag{23c}$$

The magnitude of the potential is larger for larger values of $(d_2 - d_1)/\lambda$, but is everywhere smaller than $kT/e = 0.026$ V. The potential is generated by the asymmetric distributions of the two kinds of ions (it vanishes for $d_1 = d_2$) and increases with increasing c_E. At sufficiently low electrolyte concentrations, the linear approximation is always accurate; however, at sufficiently high ionic strengths, the linear approximation is no longer valid and the magnitude of the surface potential can exceed kT/e. The potential obtained via the numerical integration of Eqs. (20a), (20b) and (20c) is compared with Eqs. (23a), (23b) and (23c) in Fig. 2a, for $d_1 =$

2 Å, $d_2=5$ Å and various electrolyte concentrations. Up to $c_E=1$ M ($\lambda=3$ Å), the numerical integration of the nonlinear Eqs. (20a), (20b) and (20c) provides a solution $\psi(x)$ within a few percents from the analytical solution of the linear approximation. The distributions of electrolyte ions for $c_E=1.0$ M are plotted in Fig. 2b.

The change in the interfacial tension due to the electrolyte ions can be calculated using the Gibbs adsorption equation (Eqs. (12)–(14)). The surface adsorptions are given by:

$$\Gamma_1=\int_0^\infty (c_1(x)-c_E)\mathrm{d}x=\int_0^{d_1} -c_E\mathrm{d}x+\int_{d_1}^{d_2}(c_1(x)-c_E)\mathrm{d}x+\int_{d_2}^\infty (c_1(x)-c_E)\mathrm{d}x$$
$$=-c_E d_2+\int_{d_2}^\infty (c_1(x)-c_E)\mathrm{d}x+\int_{d_1}^{d_2} c_1(x)\mathrm{d}x \qquad (24a)$$

and

$$\Gamma_2=\int_0^\infty (c_2(x)-c_E)\mathrm{d}x=-c_E d_2+\int_{d_2}^\infty (c_2(x)-c_E)\mathrm{d}x. \qquad (24b)$$

For small values of ψ_{II}, the last integral in Eq. (24a) can be evaluated using Eq. (23b):

$$\int_{d_1}^{d_2} c_E \exp\left(\frac{e\psi_{II}(x)}{kT}\right)\cong c_E\int_{d_1}^{d_2}\frac{\cos h\left(\frac{x-d_1}{\lambda\sqrt{2}}\right)}{\cos h\left(\frac{d_2-d_1}{\lambda\sqrt{2}}\right)+\frac{1}{\sqrt{2}}\sin h\left(\frac{d_2-d_1}{\lambda\sqrt{2}}\right)}\mathrm{d}x$$
$$=2c_E\lambda\frac{1}{1+\sqrt{2}\tan h^{-1}\left(\frac{d_2-d_1}{\lambda\sqrt{2}}\right)}. \qquad (25)$$

Since in the linear approximation the second integrals in Eqs. (24a) and (24b) cancel each other, Eq. (15) becomes:

$$\frac{1}{kT}\frac{\partial\gamma}{\partial c_E}=-\frac{\Gamma_1+\Gamma_2}{c_E}\cong 2\Delta=2\left(d_2-\frac{\lambda}{1+\sqrt{2}\tan h^{-1}\left(\frac{d_2-d_1}{\lambda\sqrt{2}}\right)}\right). \qquad (26)$$

The slope of the surface tension as a function of electrolyte concentration is no longer constant, since λ depends on c_E. For low electrolyte concentrations,

$$2\Delta=(d_1+d_2)+\frac{(d_2-d_1)^2}{2\lambda^2}-\frac{(d_2-d_1)^3}{12\lambda^3}+\cdots \tag{27a}$$

For $c_E\to 0$, the surface potential generated by the asymmetry of the ion concentrations vanishes, the anions are uniformly distributed from d_1 to infinity, and the total depletion thickness is given by the sum of the depletion thicknesses of each kind of ions, calculated independently.

For sufficiently high ionic strengths, but low enough for the linear approximation to remain still valid, Eq. (26) leads to:

$$2\Delta\cong 2d_2-\frac{\lambda}{1+\sqrt{2}}. \tag{27b}$$

The latter equation shows that the cation hydration (described by the cut-off distance d_2) dominates the process. Because of the increase in potential with electrolyte concentration, the region between d_1 and d_2 is practically depleted of anions.

The depletion distance 2Δ (given by Eq. (26)) is plotted in Fig. 2c as a function of electrolyte concentrations for $d_2=5$ Å and various values of d_1. The slope increases with ionic strength because the surface potential increases. The change in interfacial tension (obtained by integrating Eq. (26)) is represented for the same cases in Fig. 2d. The dependence is well approximated in all cases by a straight line, with a slope dependent on both d_2 and d_1.

As already noted in Section 3.1, for neutral systems with $d_1=d_2$ and $B_1=B_2$, the accounting of ion hydration through a cut-off distance provides similar results as the ion-surface van der Waals interactions. The surface potential generated by the asymmetric distribution of ions, due to van der Waals interaction alone (with $B_1\neq B_2$) is plotted in Fig. 3a. We employed for curve (1) the values $B_1=+15\times10^{-50}$ J m^3 and $B_2=+0.8\times10^{-50}$ J m^3, which were calculated for Cl^- and Na^+ in Ref. [18]. The values $B_1=+31\times10^{-50}$ J m^3 and $B_2=-1.0\times10^{-50}$ J m^3, employed for the curves (2–4), were obtained in Ref. [17] by fitting the experimental surface tension data. In the latter case, it was necessary to use a cut-off, $\Delta_{\text{cut-off}}$, in the cation interactions, to avoid their infinite accumulation on the interface. The value $\Delta_{\text{cut-off}}=2$ Å was suggested in Ref. [17]; as shown in Fig. 3, the results are strongly dependent upon this choice. The corresponding dependence of surface tension on electrolyte concentration is presented in Fig. 3b.

The only significant difference between the two models (one accounting for hydration alone and the other for van der Waals interactions alone) is that the surface potential is always positive in the latter method, because always $B_1>B_2$. At sufficiently large electrolyte concentrations, both models predict an almost linear increase of the surface tension, and agreement with experiment can be obtained by using suitable choices for the parameters (d_i or B_i).

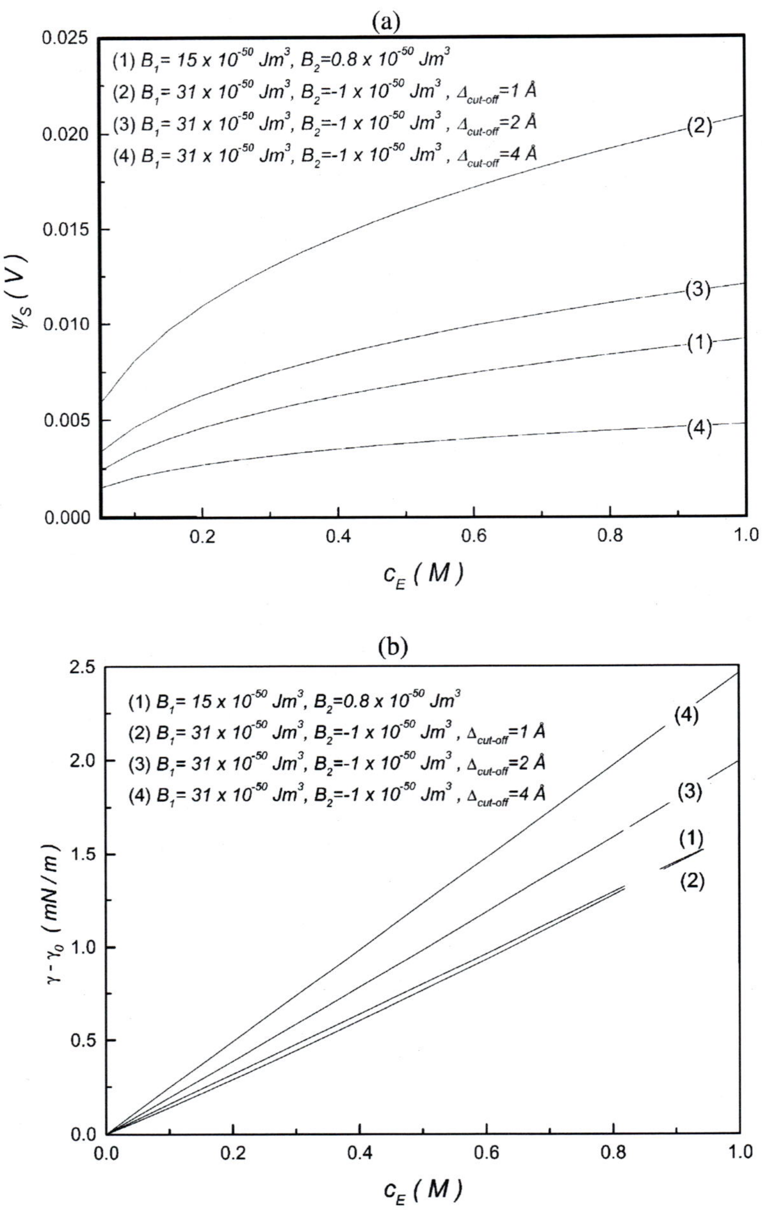

Fig. 3. (a) The surface potential and (b) the surface tension vs. electrolyte concentration, calculated by accounting for the van der Waals interactions alone in the Poisson–Boltzmann formalism ($\varepsilon=80$, $T=300$ K), for (1) $B_1=15\times10^{-50}$ J m^3 and $B_2=0.8\times10^{-50}$ J m^3; (2) $B_1=31\times10^{-50}$ J m^3, $B_2=-1\times10^{-50}$ J m^3 and $\Delta_{\text{cut-off}}=1$ Å; (3) $B_1=31\times10^{-50}$ J m^3, $B_2=-1\times10^{-50}$ J m^3 and $\Delta_{\text{cut-off}}=2$ Å; (4) $B_1=31\times10^{-50}$ J m^3, $B_2=-1\times10^{-50}$ J m^3 and $\Delta_{\text{cut-off}}=4$ Å.

3.3. The anions are adsorbed on and the cations are repelled by the interface

The change in hydration energy for the cations will be assumed, as before, given by:

$$\Delta W_2(x) = \infty \quad 0 < x < d_2 \tag{28a}$$

$$\Delta W_2(x) = 0 \quad d_2 < x < \infty \tag{28b}$$

The interactions between anions and the interface will be, however, approximated by a potential of the type:

$$\Delta W_1(x) = -W_1 \quad 0 < x < d_1 \tag{29a}$$

$$\Delta W_1(x) = 0 \quad d_1 < x < \infty \tag{29b}$$

In this case, two parameters (W_1, d_1) are used to describe the interfacial interactions of the anions. The modified Poisson–Boltzmann equation is (for $d_1 < d_2$):

$$\frac{d^2\psi_I}{dx^2} = \frac{ec_E}{\varepsilon\varepsilon_0}\exp\left(\frac{W_1}{kT}\right)\exp\left(\frac{e\psi_I}{kT}\right) \quad 0 < x < d_1, \tag{30a}$$

$$\frac{d^2\psi_{II}}{dx^2} = \frac{ec_E}{\varepsilon\varepsilon_0}\exp\left(\frac{e\psi_{II}}{kT}\right) \quad d_1 < x < d_2, \tag{30b}$$

$$\frac{d^2\psi_{III}}{dx^2} = \frac{2ec_E}{\varepsilon\varepsilon_0}\sin h\left(\frac{e\psi_{III}}{kT}\right) \quad d_2 < x < \infty \tag{30c}$$

with the same boundary conditions as before:

$$\frac{d\psi}{dx} = 0 \quad x = 0, \tag{31a}$$

$$\psi \to 0 \quad x \to \infty. \tag{31b}$$

To evaluate the effect of the potential well on the ion distributions, it will be assumed that $d_1 = d_2 = d_E$.

3.3.1. $d_1 = d_2 = d_E$. Linear approximation

For sufficiently small electrolyte concentrations, the surface potential generated is small and the linear approximation is valid. The solution of Eqs. (30a), (30b) and (30c) for the boundary conditions Eqs. (31a) and (31b), is given in this

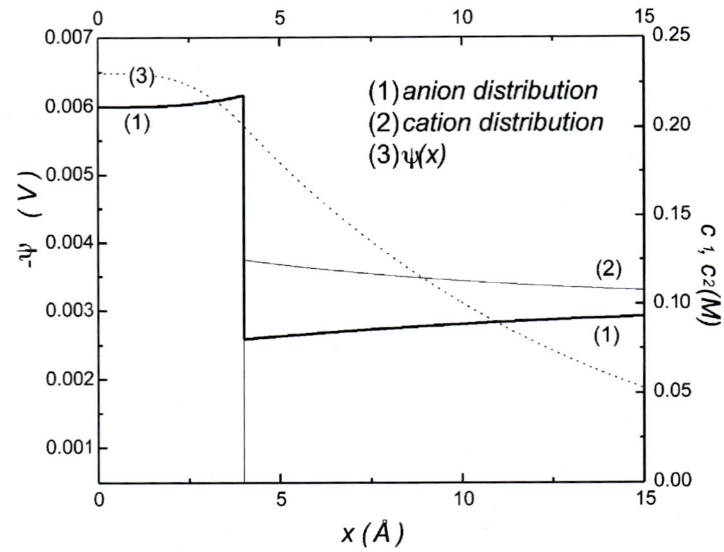

Fig. 4. The anion (1) and cation (2) distributions and the potential (3) in the vicinity of an interface, $d_1=d_2=4$ Å, $W_1=1kT$, $c_E=0.1$ M, $\varepsilon=80$, $T=300$ K.

approximation by:

$$\psi_I(x)=-\frac{kT}{e}\left(1-\frac{\cos h\left(\frac{x}{\lambda}\sqrt{\frac{A}{2}}\right)}{\cos h\left(\frac{d_E}{\lambda}\sqrt{\frac{A}{2}}\right)+\sqrt{\frac{A}{2}}\sin h\left(\frac{d_E}{\lambda}\sqrt{\frac{A}{2}}\right)}\right)\quad 0<x<d_E, \tag{32a}$$

$$\psi_{II}(x)=-\frac{kT}{e}\frac{1}{1+\sqrt{\frac{2}{A}}\tan h^{-1}\left(\frac{d_E}{\lambda}\sqrt{\frac{A}{2}}\right)}\exp\left(-\frac{x}{\lambda}\right)\quad d_E<x<\infty, \tag{32b}$$

where $A\equiv\exp(W_1/kT)$. The potential in the diffuse layer and the corresponding ion distributions are presented in Fig. 4 for $d_E=4$ Å, $c_E=0.1$ M and $W_1=1kT$.

For the depletion thickness, one obtains in the linear approximation:

$$\begin{aligned}-\frac{\Gamma_1+\Gamma_2}{c_E}&\equiv 2\Delta=2d_E-\exp\left(\frac{W_1}{kT}\right)\int_0^{d_E}\exp\left(\frac{e\psi_I}{kT}\right)\mathrm{d}x\\&=2d_E-\frac{2\lambda}{1+\sqrt{\frac{2}{A}}\tan h^{-1}\left(\frac{d_E}{\lambda}\sqrt{\frac{A}{2}}\right)}\cong d_E\left(2-\frac{A}{1+\frac{Ad_E}{2\lambda}}\right).\end{aligned} \tag{33}$$

The last approximate expression shows that, at low electrolyte concentrations ($\lambda \to \infty$), the total adsorption is positive when the potential well is sufficiently deep ($A \equiv \exp(W_1/kT) > 2$). In this case, the surface tension decreases almost linearly with increasing electrolyte concentration. However, the slope does not remain constant, its magnitude decreasing with increasing electrolyte concentration, because of the surface potential generated by the asymmetric distribution of electrolyte ions. The slope $\partial\gamma/\partial c$ becomes positive at concentrations higher than a critical electrolyte concentration, obtained from:

$$2 - \frac{A}{1 + \frac{A d_E}{2\lambda_c}} = 0 \Rightarrow \lambda_c = \frac{d_E}{2} \frac{1}{\frac{1}{2} - \frac{1}{A}}. \tag{34}$$

3.3.2. $d_1 = d_2 = d_E$. Approximate solution of the nonlinear equation

At high electrolyte concentrations, the linear approximation fails and Eqs. (31a), (31b), (32a), (32b) and (33) are no longer valid. A simple solution, even approximate, of the non-linear Poisson–Boltzmann equation is more difficult to obtain; however, the general behavior of the system can be understood from the following semi-quantitative analysis.

The diffuse double layer is generated by the anions trapped in the potential well. The 'surface charge density', due to their distribution between 0 and d_E, is given by:

$$\sigma = \int_0^{d_E} \rho(x)\mathrm{d}x = -ec_E \exp\left(\frac{W_1}{kT}\right) \int_0^{d_E} \exp\left(\frac{e\psi(x)}{kT}\right) \mathrm{d}x \equiv -ec_E d_E \exp\left(\frac{W_1 + e\bar{\psi}}{kT}\right), \tag{35}$$

where $\bar{\psi}$ is a suitable average value between $\psi(0)$ and $\psi(d_E)$.

Starting from Poisson equation, one can show that, because of the electroneutrality, the potential gradient at distance d_E is related to the 'surface charge density' via:

$$\left.\frac{\mathrm{d}\psi}{\mathrm{d}x}\right|_{x=d_E} = -\int_{d_E}^{\infty} \frac{\mathrm{d}^2\psi}{\mathrm{d}x^2} \mathrm{d}x = \int_{d_E}^{\infty} \frac{\rho(x)}{\varepsilon\varepsilon_0} \mathrm{d}x = -\int_0^{d_E} \frac{\rho(x)}{\varepsilon\varepsilon_0} \mathrm{d}x \equiv -\frac{\sigma}{\varepsilon\varepsilon_0}. \tag{36}$$

Outside the well, the integration of the Poisson–Boltzmann equation (with the boundary condition Eq. (31a)) leads to [27]:

$$\frac{\mathrm{d}\psi(x)}{\mathrm{d}x} = \frac{2kT}{e\lambda} \sin h\left(-\frac{e\psi(x)}{2kT}\right) \quad d_E < x < \infty. \tag{37}$$

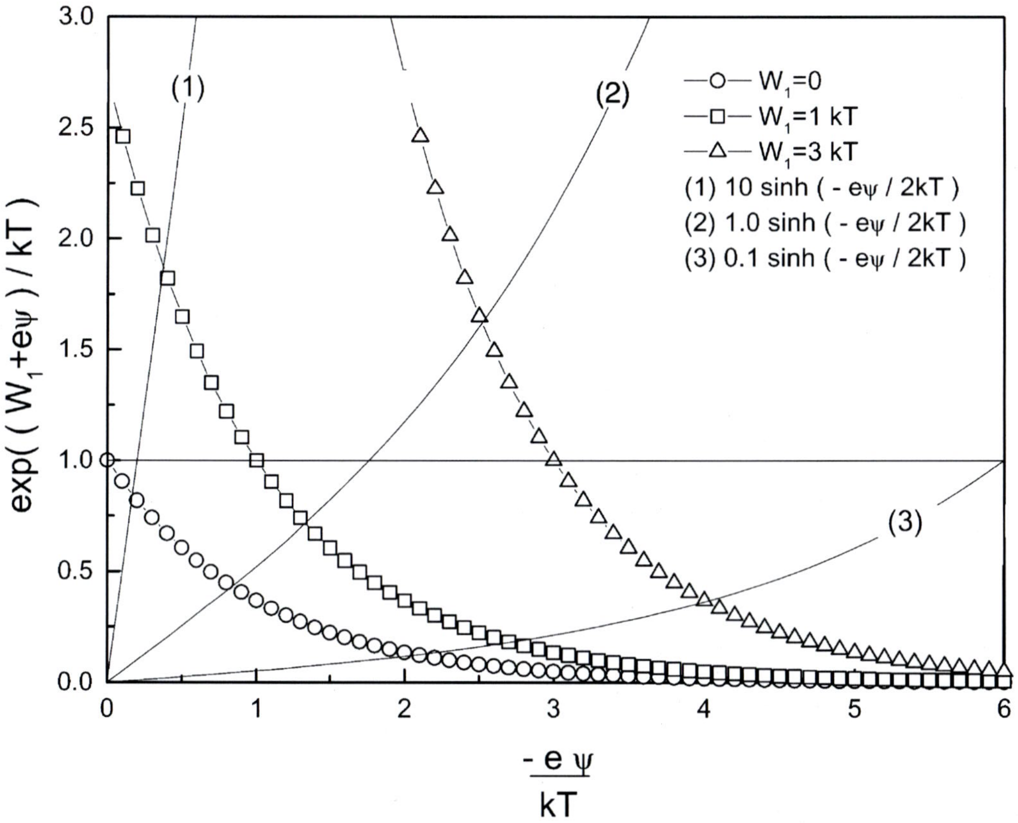

Fig. 5. Graphical solution of Eq. (38). If $W_1 > 0$, there is a positive adsorption of anions in the potential well if the electrolyte concentration is sufficiently low. The increase of the electrolyte concentration displaces $|\bar{\psi}|$ toward larger values. If the electrolyte concentration is sufficiently high, the anion adsorption in the well becomes negative $\left(\exp\left((W_1 + e\bar{\psi})/kT\right) < 1\right)$.

Combining Eqs. (35)–(37), yields:

$$\exp\left(\frac{W_{\mathrm{I}}+e\bar{\psi}}{kT}\right)=\frac{4\lambda}{d_{\mathrm{E}}}\sin h\left(-\frac{e\psi(d_{\mathrm{E}})}{2kT}\right). \tag{38}$$

If the well is not very broad and the potential is sufficiently large, then the potential varies slowly between 0 and d_{E} and one can approximate $\bar{\psi}\approx\psi(d_{\mathrm{E}})$. Under such conditions, Eq. (38) allows to calculate $\psi(d_{\mathrm{E}})$ at any electrolyte concentration. In Fig. 5, the left-hand-side term of Eq. (38) is plotted against $-e\bar{\psi}/kT$ ($\bar{\psi}$ is negative) for $W_{\mathrm{I}}=0$, $1kT$ and $3kT$. The right-hand-side term is plotted for $4\lambda/d_{\mathrm{E}}=$ 0.1, 1, 10. The intersection of the two curves provides $\bar{\psi}$ for the value selected for $4\lambda/d_{\mathrm{E}}$.

The cations cannot approach the interface closer than d_{E}. When $W_{\mathrm{I}}=0$, the region $0<x<d_{\mathrm{E}}$, which is depleted of cations, is also partially depleted of anions, since $\exp\left((W_{\mathrm{I}}+e\bar{\psi})/kT\right)<1$, because of the potential generated by the asymmetric distributions of ions. At low electrolyte concentrations (curve (1) in Fig. 5), the potential is small and the anion depletion negligible. The increase of the electrolyte concentration displaces the intersection toward larger magnitudes of $\bar{\psi}$, and at high electrolyte concentrations (curve (3) in Fig. 5) the potential well can be considered as being completely depleted of anions, because $\exp\left((W_{\mathrm{I}}+e\bar{\psi})/kT\right)\approx 0$. When $W_{\mathrm{I}}>0$, there is a positive adsorption of anions at the interface if the electrolyte concentration is sufficiently low, but the unbounded increase of the surface potential with increasing ionic strength generates a negative adsorption of the anions in the potential well at large electrolyte concentrations. Consequently, the surface tension can first decrease (if the potential well is deep enough) and then increase with increasing electrolyte concentration.

When the surface potential is large, the hyperbolic sinus in the right term can be approximated by its largest exponential term and Eq. (38) becomes:

$$\bar{\psi}=-\frac{2}{3}\left(\frac{W_{\mathrm{I}}}{e}\right)\ln\left(\frac{d_{\mathrm{E}}}{2\lambda}\right), \tag{39}$$

which shows that the surface potential increases without bound with the electrolyte concentration. Therefore, the potential well will be always depleted by the anions at sufficiently high electrolyte concentrations, in spite of their favorable interaction with the interface.

To illustrate the above discussions, the surface potential and the surface tension are plotted vs. electrolyte concentration in Fig. 6a and b, respectively, for $d_1=d_2=d_{\mathrm{E}}=5$ Å and various values for W_{I}. The Poisson–Boltzmann Eq. (30b) was integrated numerically and its solution was employed in Eq. (15) to calculate the surface tension. If the potential well for anions is shallow, the total depletion thickness is positive at any electrolyte concentration and the surface tension always

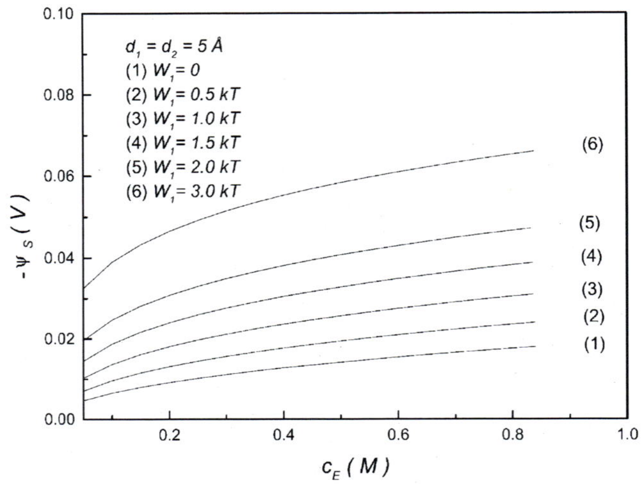

(a)

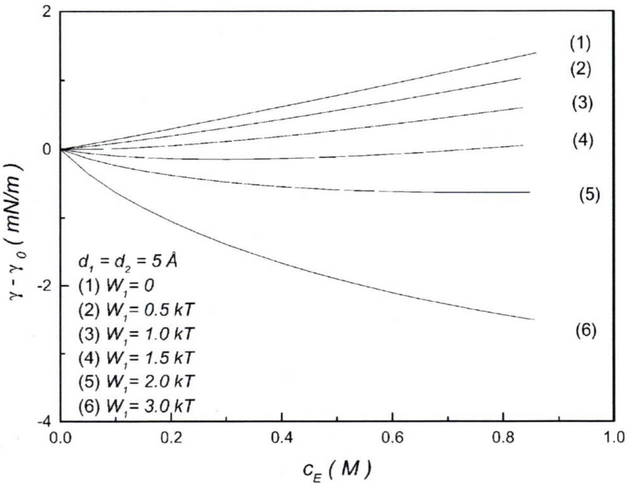

(b)

Fig. 6. (a) The surface potential and (b) the surface tension vs. electrolyte concentration, calculated for a potential well, with $d_1 = d_2 = 5$ Å, $\varepsilon = 80$, $T = 300$ K.

increases with increasing ionic strength. If the potential well is sufficiently deep ($A \equiv \exp(W_1/kT) > 2$), the total depletion thickness is negative at low ionic strength (because of the anion adsorption in the potential well) and the surface tension is decreased by the addition of electrolyte. However, the increase in surface potential with ionic strength expels the anions from the potential well and, at large electrolyte concentrations, the depletion thickness becomes positive. Consequently, the surface tension passes through a minimum. For $W_1 = 2kT$, the minimum is at approximately

$c_E=0.8$ M; the position of the minimum depends on the parameters of the interactions and increases with the depth of the well.

4. Applications

4.1. Zeta potential

As noted above, the surface potential generated by the asymmetric distributions of electrolyte ions increases with increasing ionic strength. In contrast, electrophoretic experiments on air bubbles indicated that the surface potential decreases with the addition of electrolyte and depends strongly on the pH of the solution. For this reason, the effect of pH was included in the model, by assuming that the OH^- ions are adsorbed on the interface. The adsorption of the H^+ will be neglected in comparison to that of OH^-.

Assuming an equilibrium constant K_{OH} for the adsorption of OH^- on the interface sites, one can write:

$$K_{OH}=\frac{(1-x)[OH^-]_S}{x}, \tag{40}$$

where $(1-x)$ is the fraction of free sites and $[OH^-]_S$ the concentration of OH^- in the vicinity of the interface sites. Consequently, the surface charge density is given by:

$$\sigma_{OH}=-exN=-\frac{eN}{1+\dfrac{K_{OH}}{[OH^-]_S}} \tag{41a}$$

where N is the number of adsorption sites per unit area of the interface.

It will be assumed that the pH value was attained by adding an acid (or a base) with the same anion (cation) as the electrolyte. The salt and the acid or base will be assumed totally dissociated. The surface charge is therefore screened by monovalent ions with a total bulk concentration $c'=c_E+c_A+c_B$, where $c_A(c_B)$ is the concentration of the added acid (base).

The concentration of OH^- ions in the vicinity of the surface, $[OH^-]_S$, is related to the bulk concentration $[OH^-]$ through:

$$[OH^-]_S=[OH^-]\exp\left(-\frac{\Delta W_{OH}(0)}{kT}\right)\exp\left(\frac{e\psi_S}{kT}\right), \tag{42}$$

where $\psi_S=\psi(x=0)$ is the surface potential, which is negative, $[OH^-]$ is the bulk concentration, and ΔW_{OH} includes all the ion interactions which are not contained in the mean field ψ. The latter interactions can be accounted for by using a new

90 M. Manciu, E. Ruckenstein / Advances in Colloid and Interface Science 105 (2003) 63–101

apparent dissociation constant K'_{OH}. With this choice, Eq. (41a) becomes:

$$\sigma_{OH} = -\frac{eN}{1+\frac{K'_{OH}}{[OH^-]}\exp\left(-\frac{e\psi_S}{kT}\right)}, \tag{42b}$$

A more complete treatment should include in the Poisson–Boltzmann formalism all the interactions between the electrolyte ions and the interface, summarized in Section 2. Here it will be shown that, by accounting only for the hydration interaction of the electrolyte ions, one can capture the experimental dependence of the surface potential on pH and electrolyte concentration.

It will be assumed that the cations are structure-making, with a cut-off distance $d_2 = d_E$ and the anions structure-breaking, with interactions given by Eq. (29a) and Eq. (29b). For simplicity, it will be assumed that $d_1 = d_2 \equiv d_E$, and that the interactions of H^+ and OH^- with the interface ($\Delta W_H(x)$, $\Delta W_{OH}(x)$) are negligible. The latter approximation affects the results only at very low or very high pH values.

Consequently, the potential satisfies the equations:

$$\frac{d^2\psi}{dx^2} = \left((c_E + c_A)\exp\left(\frac{W_1}{kT}\right) + c_B\right)\frac{e}{\varepsilon\varepsilon_0}\exp\left(\frac{e\psi}{kT}\right) + c_A\frac{e}{\varepsilon\varepsilon_0}\exp\left(-\frac{e\psi}{kT}\right) \quad 0 < x < d_E \tag{43a}$$

$$\frac{d^2\psi}{dx^2} = \frac{2e(c_E + c_A + c_B)}{\varepsilon\varepsilon_0}\sin h\left(\frac{e\psi}{kT}\right) \quad d_E < x < \infty \tag{43b}$$

where c_E is the electrolyte concentration and $c_A(c_B)$ is the concentration of the added acid (base). The boundary conditions are:

$$\left.\frac{d\psi}{dx}\right|_{x=0} = -\frac{\sigma_{OH}}{\varepsilon\varepsilon_0} = \frac{eN}{\varepsilon\varepsilon_0}\frac{1}{1+\frac{K'_{OH}}{[OH^-]}\exp\left(-\frac{e\psi_S}{kT}\right)} \tag{44a}$$

and:

$$\psi(x)|_{x\to\infty} = 0 \tag{44b}$$

A boundary condition, which is equivalent to Eq. (44b), but is more convenient for calculations, is provided by Eq. (37), which is valid because for $x > d_E$ the potential satisfies the traditional Poisson–Boltzmann equation. In this case, it is enough to solve Eq. (43a) for $0 < x < d_E$, with the boundary conditions Eq. (44a) and Eq. (37).

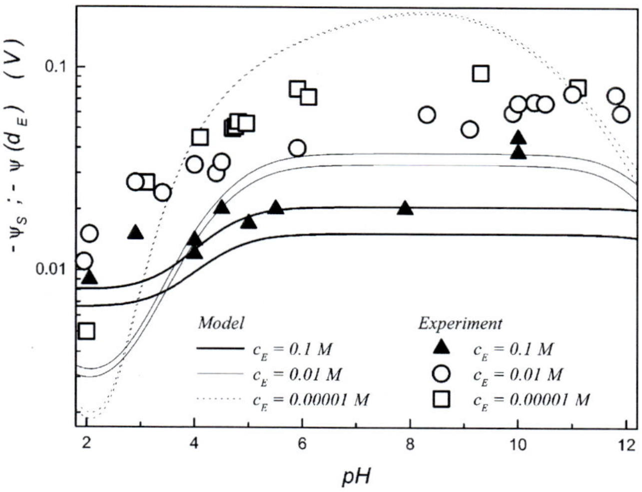

Fig. 7. The experimental values of the zeta potential for water/air interface in presence of NaCl, as a function of pH, from Ref. [32] (triangles: 0.1 M; circles: 0.01 M; squares: 0.00001 M) are compared with the potential values (ψ_S and ψ_d) predicted by the present model: $d_1 = d_2 = 4$ Å, $W_1 = 0.5kT$, $K'_{OH} = 10^{-10}$ M, $N = 5.0\times10^{16}$ sites/m^2, $\varepsilon = 80$ and $T = 300$ K.

The numerical solution for the surface potential as a function of pH is compared in Fig. 7, for various NaCl concentrations, with the experimental results provided by Li and Somasundaran [32]. The potentials $-\psi_S = -\psi(0)$ and $-\psi_d = -\psi(d_E)$ are plotted as functions of distance, since the zeta potential determined by electrophoresis is not defined at the surface, but at an unknown location, the 'plane of slip'. The magnitude of ψ_S is always larger than that of ψ_d, since the potential decays with the distance. The value $d_E = 4$ Å, which is provided by the dependence of the surface tension of water on the NaCl concentration at high ionic strengths was employed. For the equilibrium constant, the value $K'_{OH} = 10^{-10}$ M, which is consistent with the experimental data for pH values between 3 and 6, was selected. A reasonable agreement with the data (which have a rather large error) was obtained by selecting $N = 5.0\times10^{16}$ sites/m^2 and $W_1 = 0.5kT$.

One important consequence of the present hydration model is that the predicted surface potential does not vanish at high ionic strength, because of the cut-off distance for cations. Since there is no screening in the cation depletion region, the surface potential at large electrolyte concentrations is roughly given by:

$$\psi_S\big|_{c_E\to\infty} \cong \frac{\sigma d_E}{\varepsilon\varepsilon_0} \tag{45}$$

Another consequence is that the asymmetric distributions of electrolyte ions, due to their different hydration interactions, generate themselves a surface potential, which increases with increasing electrolyte concentration. At sufficiently high pH values (larger than ≈ 3) the surface charge generated by the adsorption of OH^- on the interface provides the dominant contribution to the surface potential. In this case, the increase of the electrolyte concentration enhances the screening of the surface charge and hence the absolute value of the surface potential decreases. However, at low pH values, the adsorption of the OH^- ion is negligible and the surface potential is generated by the asymmetric distributions of the two kinds of ions. In this case, the absolute value of the surface potential is increased by the addition of electrolyte. The decrease of the absolute value of the surface potential with increasing electrolyte concentration at most pH values and its increase at very low pH values (Fig. 7) appears to be supported by experiment [32].

Karraker and Radke [18] compared the same experimental data with a model in which the surface potential was generated by the charge of the $[OH^-]$ adsorbed on the interface, which is screened by the ions of the electrolyte. While good agreement for the surface potential was obtained for $c_E = 1 \times 10^{-5}$ M, the predicted potential was much lower than the calculated one for 10^{-2} and particularly for 10^{-1} M. They have taken into account the image and van der Waals forces acting on the ions while we accounted only for the ion hydration instead. As noted in Section 3.2, the two approaches can lead to similar results, for suitable choices of the values of the parameters. However, for the values of B_1 and B_2 calculated from the polarizabilities of ions [18], the surface potential is positive al low pH values, and not negative, as indicated by experiment. Its magnitude at 0.1 M, ($\approx$0.003 V, Fig. 3a) is also smaller than that provided by experiment. At large pH values, the adsorption of the OH^- ions on the interface is responsible for the negative surface potential.

A possibility to generate large surface potentials, at low pH values, would be to employ values for B_1 and B_2 of different signs. If one would select $B_1 < 0$ and $B_2 > 0$ and a suitable cut-off distance (to avoid the divergent accumulation of anions at the interface), results similar to those obtained here would be obtained. However, this implies that the van der Waals interactions attract the anions toward and repel the cations from the interface, which contradicts the general theory of the van der Waals interactions.

4.2. Surface tension at low ionic strength. Jones–Ray effect

The dependence of interfacial tension on electrolyte concentration can be related, via the Gibbs adsorption equation, to the depletion of the interfacial region of ions. At relatively large concentrations (between 0.1 and 1 M) of simple electrolytes (NaCl, KCl), the increase in surface tension with increasing ionic strength has an almost constant slope [29]. In the absence of accurate experimental data, it was reasonable to expect that this trend can be extended to low ionic strengths. The Wagner–Onsager–Samaras theory supported this expectation by predicting a positive slope at all electrolyte concentrations, which decreases with increasing ionic strength.

However, a few years after the Onsager–Samaras theory, experiments performed by Jones and Ray [33] indicated that the surface tension at the water/air interface is decreased by the addition of small amounts of KCl and exhibits a minimum at approximately $c_E = 0.001$ M. Because the relative changes in surface tension are of the order of 0.01% at that electrolyte concentration, hence very susceptible to experimental error, the existence of this minimum was doubted. The negative slope of the surface tension at low ionic strength contradicted the Onsager–Samaras theory, which was considered valid in that range. Dole and Swartout [34] confirmed, however, the experimental results of Jones and Ray.

Dole [35] explained the Jones–Ray effect using a modified Langmuir adsorption equation. He assumed that the interface contains 'active spots' which attract anions and that at low electrolyte concentration, the adsorption of the negative ions on these spots provides the main contribution in the Gibbs adsorption equation. By adding electrolyte, all the 'active spots' become occupied, no further adsorption of negative ions occurs and the change in surface tension is controlled by the overall depletion of ions due to image forces.

Karraker and Radke [18] proposed another explanation of this minimum. At low electrolyte concentrations, the positive adsorption of OH^- leads to a decrease of surface tension, while at sufficiently high ionic strengths, the depletion of Cl^-, repelled by the negative surface charge as well as by the van der Waals and image forces, becomes dominant and the surface tension increases.

A simpler mechanism is suggested here. The adsorption of OH^- ions does not affect, in a first approximation, the surface tension via the Gibbs adsorption equation. However, it generates a surface charge, which is responsible for the accumulation of cations in the vicinity of the surface. Consequently, the total surface adsorption of ions is positive and the slope of the surface tension, as a function of electrolyte concentration, is negative. The surface potential is, however, decreased by the addition of electrolyte, and the accumulation of cations is compensated by the anion depletion. Hence, by adding electrolyte, the ions depletions of the interfacial region, due to other interactions (ion hydration, image or van der Waals forces) dominate the process, and the slope becomes positive.

The surface adsorption of OH^- is obviously not negligible. However, the chemical potential of OH^- in the bulk is given by:

$$\mu_{OH} = \mu_{OH}^0 + kT\ln([OH^-]f_{OH}) \tag{46}$$

and at constant pH, μ_{OH} depends on c_E only via the activity coefficient, which was taken unity. Therefore, Eq. (15) yields:

$$\frac{1}{kT}\frac{\partial\gamma}{\partial c_E} = -\frac{\Gamma_{Cl^-} + \Gamma_{K^+}}{c_E}. \tag{47}$$

It should be noted that the same result can be obtained by using an activity coefficient of 1 in Eq. (11) of Ref. [18]. At concentrations of approximately 0.001 M of uni-univalent electrolytes, (relevant for the Jones–Ray effect) $f = 1$ is, in general, a good approximation.

Let us evaluate how much the surface potential affects the surface adsorption of electrolyte ions. When only the short-range hydration interactions of ions with the interface are taken into account, the depletion thickness is given by:

$$-\frac{\Gamma_{Cl^-}+\Gamma_{K^+}}{c_E}=\int_0^{\infty}\left(\frac{2c_E-c_1(x)-c_2(x)}{c_E}\right)dx$$
$$=\int_0^{d_E}\left(2-\frac{c_1(x)}{c_E}-\frac{c_2(x)}{c_E}\right)dx-2\int_{d_E}^{\infty}\left(\cos h\left(\frac{e\psi(x)}{kT}\right)-1\right)dx, \quad (48)$$

where the potential satisfies Eqs. (43a) and (43b). The first term of the right-hand-side is less than $2d_E$, which is approximately 10 Å. At $c_E=0.001$ M and assuming $-e\psi(d_E)=2kT$, the negative depletion thickness (the adsorption thickness) provided by the second term of the right-hand-side of Eq. (48) is approximately 200 Å, largely exceeding the magnitude of the depletion thickness due to the cut-off distance. This value is so large, that the total surface adsorption remains positive and large, even when other long-range repulsions (such as the image or the van der Waals force) are taken into account. The inclusion of the image potential decreases the adsorption thickness by approximately 10 Å, and a similar effect is obtained by including the van der Waals interactions.

When the electrolyte concentration is increased, the range of the double layer decreases dramatically (the Debye–Hückel length decreases) and the magnitude of the surface potential also decreases. In the linear approximation, $\psi(x)=\psi(d_E)\exp(-(x-d_E)/\lambda)$ (for $x>d_E$) and the second right-hand-side term of Eq. (48) becomes:

$$2\int_{d_E}^{\infty}\left(\cos h\left(\frac{e\psi(x)}{kT}\right)-1\right)dx\cong\int_{d_E}^{\infty}\left(\frac{e\psi(x)}{kT}\right)^2dx=\left(\frac{e\psi(d_E)}{kT}\right)^2\frac{\lambda}{2} \quad (49)$$

In summary, at very low electrolyte concentrations (below ≈0.001 M) and in the absence of a surface charge, the image forces provide the most important contribution to the (negative) ion adsorptions, and the other repulsions (ion hydration, van der Waals interactions) can be neglected. However, if there is a surface charge, the overall accumulation of ions in the vicinity of the surface (the positive overall adsorption) due to the mean field ψ (the Poisson–Boltzmann effect) becomes important. Furthermore, at electrolyte concentrations smaller than 0.001 M and if $e\psi(d_E)$ is at least of the order of kT, all the other interactions, including the image forces, become negligible. The overall ion adsorption, which is positive, is well approximated in this case by the traditional (bare-bone) Poisson–Boltzmann equation. At very large electrolyte concentrations, the range of the mean field Poisson–Boltzmann interactions as well as the image forces decrease drastically, since their decay lengths are roughly proportional to λ and $\lambda/2$, respectively. In this case, other interactions, such as ion hydration or van der Waals forces, become dominant.

In Fig. 8, curve (1) represents the dependence of surface tension on electrolyte concentration, based on the present model. It is assumed that OH^- ions are adsorbed on the surface, with an apparent equilibrium constant $K'_{OH}=1\times10^{-10}$ M, and that there are a cut-off $d_E=4$ Å for the cations, and a potential well of depth $W_1=0.5kT$

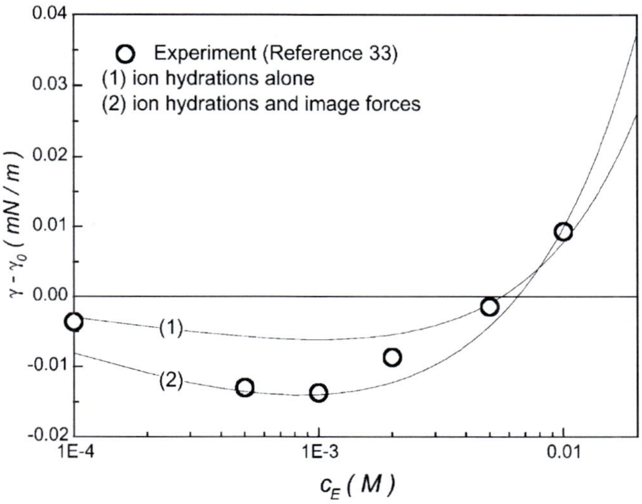

Fig. 8. The change in surface tension at low electrolyte concentrations (the Jones–Ray effect). Circles: Experimental data from Ref. [33]. Curve (1): predicted surface tension for the simple model which accounts for OH^- adsorption and only ion hydration effects for electrolyte ions, within the Poisson–Boltzmann approach. Curve (2): predicted surface tension when image forces are also included in the model. As noted in Section 3, the image forces cannot be neglected at concentration lower than 0.01 M.

and width $d_E=4$ Å for the anions. These values of the parameters are those used in the previous section, which was concerned with the zeta potential of air bubbles determined from electrophoretic experiments. However, to recover the Jones–Ray minimum approximately 0.001 M (for pH 7), a much smaller value for the number of adsorption sites on the interface N than the one used there had to be employed, namely $N=3\times10^{15}$ sites/m^2.

At the minimum of the surface tension, the positive adsorption due to the mean field (the Poisson–Boltzmann effect) is compensated by the negative adsorption due to the ion hydration, image forces and van der Waals interactions. As already noted (Fig. 1), at the low ionic strengths where the Jones–Ray minimum occurs ($\approx$0.001 M), the image force contribution to the depletion thickness cannot be neglected. By including the image potential in the calculation of the surface adsorption, good agreement with experiment was obtained for $N=6\times10^{15}$ sites/m^2 (Fig. 8, curve (2)).

From the experimental values of the zeta potentials for air bubbles [32], one could infer that the surface potential should be approximately 0.060 V for 0.001 M NaCl at pH 7. However, this value is much too large to be in agreement with the experiments of Jones and Ray [33]. The depletion thickness vanishes at the minimum of the surface tension, and at this surface potential, the adsorption thickness, due to the Poisson–Boltzmann effect, would be over 200 Å and could not be compensated by the other interactions discussed here. The value which had to be selected for N

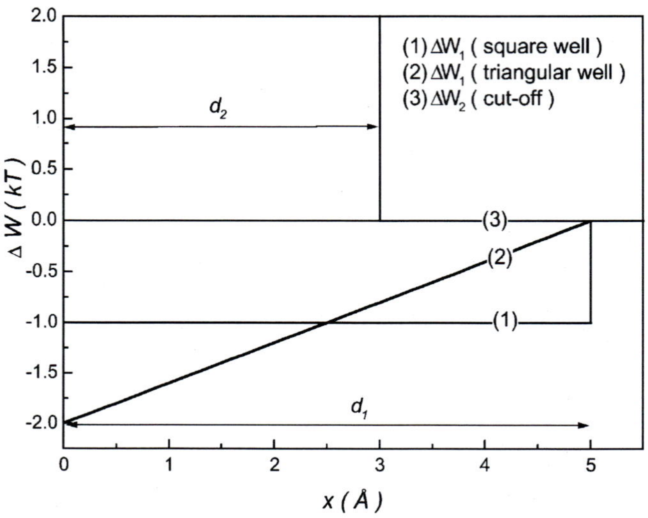

Fig. 9. Approximation of the interactions between ions and the interface: (1) square well for anions; (2) triangular well for anions; (3) hard wall for cations.

($N=6\times10^{15}$ sites/m^2) to provide a good agreement with Jones–Ray minimum, led to a surface potential of only approximately 0.014 V.

4.3. Surface tension at high ionic strength

At high ionic strength, the adsorption of OH^- no longer plays a central role. Because the potential at the surface and its decay length are both small, the behavior of the surface tension is determined by the ion distributions within the first few Ångstroms from the interface.

The almost linear increase of the surface tension with the concentration of the common uni-univalent electrolytes can be easily described by employing a cut-off distance for the cations. However, experiment [29] indicated that for other electrolytes (HCl, HNO_3, $HClO_4$) the surface tension decreases with increasing electrolyte concentration up to 1 M. The adsorption thickness due to the Poisson–Boltzmann effect, which is responsible for the Jones–Ray effect at low ionic strength, cannot explain the linear decrease of surface tension at high ionic strength, since it decays too strongly with the electrolyte concentration. The image forces and the repulsive van der Waals interactions ($B_i>0$) generate a depletion thickness (a negative adsorption), and, therefore, also cannot explain this behavior.

Can the change in cation hydration, between bulk and interface explain this effect? If the cations (such as Na^+ or K^+) cannot approach the interface, and the potential well for anions is not very deep, the overall adsorption of ions is negative at any electrolyte concentration. However, if the cations (such as H^+) can penetrate at least a part of the interfacial potential well of the anions ($d_2<d_1$), they are

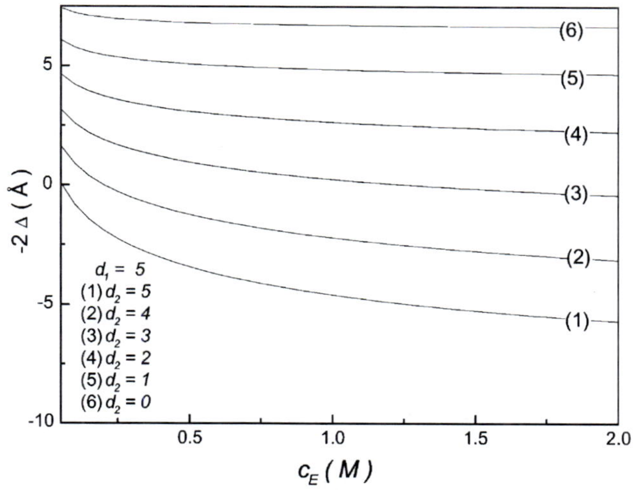

(a)

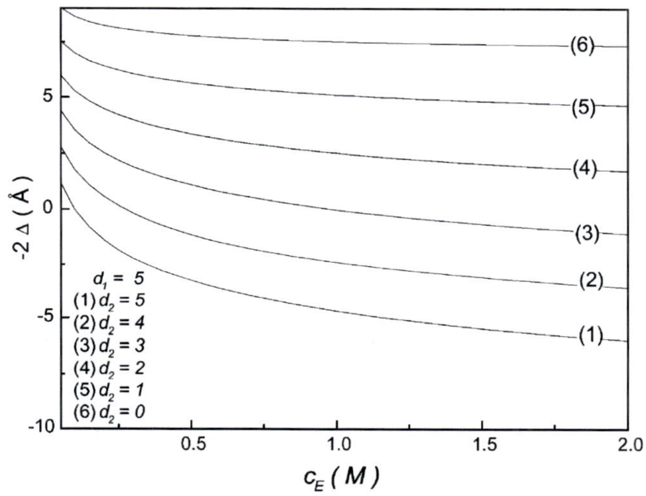

(b)

Fig. 10. The depletion thickness as a function of electrolyte concentration for (a) a square well interaction for anions, with $d_1=5$ Å and $W_1=1kT$; (b) a triangular well interaction for anions, with $d_1=5$ Å and $\Delta W(0)=-2kT$; and a cut-off distance for cations $d_2=0$, 1, 2, 3, 4 and 5 Å; $\varepsilon=80$ and $T=300$ K.

attracted there by the accumulated anions and the overall adsorption can become positive.

Two cases have been investigated. First, the solution of the Poisson–Boltzmann equation was obtained for an interfacial square-well potential for the anions, with $W_1=1kT$ and $d_1=5$ Å. Second, a triangular well was employed for the potential (the interaction potential varying linearly form $\Delta W_1=-2kT$ at $x=0$ to $\Delta W_1=0$ at

$d_1 = 5$ Å) (Fig. 9). For cations, the distance of closest approach to the interface was selected $d_1 = 5$, 4, 3, 2, 1 and 0 Å. The dependence of the slope of the interfacial tension, $(1/kT)(\partial\gamma/\partial c_E) = -(\Gamma_1 + \Gamma_2)/c_E$ on the electrolyte concentration is presented in Fig. 10a (for a square-well) and Fig. 10b (for a triangular well). The results are qualitatively the same: if the cations are allowed to penetrate most of the well, the surface tension decreases with increasing electrolyte concentration. If the cations are not allowed to penetrate the well, the potential is large in the well and the total surface adsorption is negative. At intermediate situations, the total surface adsorption is positive at low electrolyte concentrations (corresponding to a decrease of the surface tension with increasing electrolyte concentration) and negative at high ionic strengths (corresponding to an increase of the surface tension with increasing electrolyte concentration). The corresponding changes in surface tension as a function of electrolyte concentration are presented in Fig. 11a and b, for the square and the triangular well, respectively. While in the intermediate cases the surface tension passes through a minimum, at the extreme cases the dependence of the surface tension on electrolyte concentration is almost linear.

It should be noted that a decrease of the interfacial tension with increasing electrolyte concentration was observed [29] only when the cation was H^+. The major differences in the hydration of H^+ as compared to those of Na^+ or K^+, are probably responsible for the fact that the H^+ can approach closer the interface than any other cation. Consequently, the simple model suggested here can predict a negative slope for the surface tension when H^+ is the cation and a positive slope for the other cations.

5. Conclusions

It has been long known that the over-simplified Poisson–Boltzmann equation is accurate in predicting the double layer interaction only in a relatively narrow range of electrolyte concentrations. One obvious weakness of the treatment is the prediction that the ions of the same valence produce the same results, regardless of their nature. In contrast, experiment shows marked differences when different kinds of ions are used. The ion-specific effects can be typically ordered in series (the Hofmeister series [36]), and the placement of ions in this series correlates well with the hydration properties of the ions in bulk water.

It was suggested recently [18,26] that the van der Waals interactions between ions and two media separated by an interface can account, at least partially, for the order in the Hofmeister series and therefore can explain ion specific effects in the distribution of electrolyte ions in the vicinity of an interface. However, there is no consensus regarding the values of the interaction constants which should be used in the van der Waals interactions. It was expected that, due to the van der Waals interactions, the negative ions should be more strongly repelled by the water/air interface than the positive ions. This seems to be, however, contradicted by experiment and simulations.

In this paper, the old idea that the ion hydration should be mainly responsible for the ion-specific effects is reiterated. After a brief review of the Poisson–Boltzmann

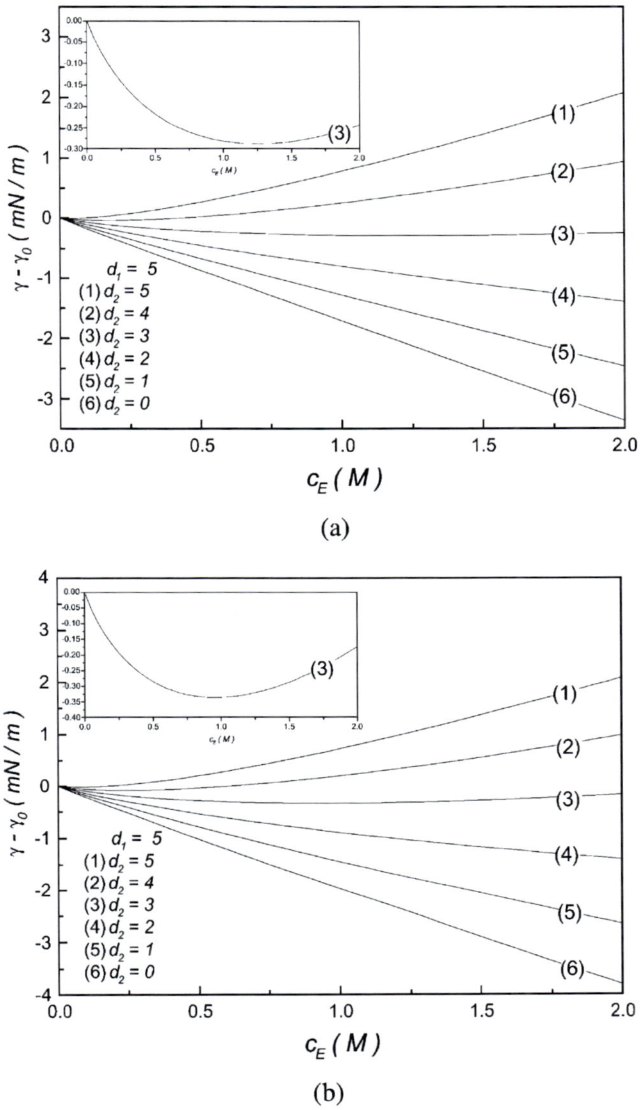

Fig. 11. The change in surface tension as a function of electrolyte concentration for (a) a square well interaction for anions, with $d_1 = 5$ Å and $W_1 = 1kT$; (b) a triangular well interaction for anions, with $d_1 = 5$ Å and $\Delta W(0) = -2kT$; and a cut-off distance for cations $d_2 = 0$, 1, 2, 3, 4 and 5 Å; $\varepsilon = 80$ and $T = 300$ K.

formalism and of a few approaches that describe ion-specific effects, a simple approach is proposed to account for ion hydration. Assuming that most of the hydration free energy is due to the interaction between ion and its water neighbors, the approach of a structure-making ion to the interface is unfavorable, because it cannot be fully hydrated there. Since the hydration energy is large, only a minute percentage change of its value can lead to a complete depletion of ions. Therefore, a cut-off distance from the interface can describe well the interaction between ion and the interface. The long-range Wagner–Onsager–Samaras image force is important for neutral interfaces at very low electrolyte concentration, but is negligible when the interface is charged or above the physiological concentrations ($\approx$0.1 M). The van der Waals interactions can be approximately taken into account by a small change in the cut-off distance.

More difficult is to treat the case of structure-breaking ions, which are pushed toward the interface, because they have there more favorable interactions with water. The consequence of a potential well for the anions in the vicinity of the interface is therefore investigated. However, the depth of the well should not be larger than a few kT, otherwise huge surface potentials would be generated at high ionic strengths, and this was not observed experimentally.

It is not yet clear whether the total interaction of ions with the interface can be accounted by a simple function of distance; however, the simple treatment presented here leads to qualitatively appealing predictions. The assumptions that $d_1 = d_2$ or the simple functional form chosen for the anion interaction with the interface were made only to simplify the calculations. Better agreement with experiment can be obtained if other details are included in the model (dependence of the activity coefficient on the electrolyte concentration, additional interactions).

The dependence of the surface potential on the electrolyte concentration and on the pH, determined experimentally, was reasonably well described by the traditional Poisson–Boltzmann equation, using a model involving the OH^- adsorption on the interface and the treatment suggested for ion hydration.

The same simple model can also explain the Jones–Ray effect (a minimum of the surface tension at a concentration of approximately 1.0×10^{-3} KCl [33]). At low electrolyte concentrations, the accumulation of the electrolyte ions due to the surface charge (as predicted by the Poisson–Boltzmann equation) overwhelms all the other contribution to the surface adsorptions. Consequently, the interfacial tension first decreases with increasing electrolyte concentration. However, the Poisson–Boltzmann adsorption thickness decays strongly with increasing electrolyte concentration, and the repulsion of ions by the interface (due to image forces, change in ion hydration and van der Waals interactions) becomes dominant, and the slope of the interfacial tension becomes positive.

The behavior of surface tension at high ionic strength also can be understood on the basis of changes in ion hydration changes between bulk and interface. The tendency of the structure-breaking ions to accumulate at the surface can lead to a positive surface adsorption. However, if the cations cannot approach the interface, the asymmetry of the ion distributions generates a potential, which repels the anions from the interface, and the total adsorption becomes negative. Consequently, the surface tension increases with electrolyte concentration; this occurs for simple salts (NaCl, KCl). If the cations can approach the interface, the accumulation of anions in the vicinity of the interface is also followed by an accumulation of cations, without generating large surface potentials. Consequently, the total adsorption becomes positive and the surface tension decreases with electrolyte concentrations, up to high ionic strengths. This occurs, for example, for acids (HCl, HNO_3, $HClO_4$), and can be attributed to the ability of the H^+ ion to approach the interface.

References

[1] G. Gouy, J. Phys. Radium 9 (1910) 457.
[2] D.L. Chapman, Phyl. Mag. 25 (1913) 475.
[3] O. Stern, Z. Eletrochem. 30 (1924) 508.
[4] B. Jönsson, H. Wennerström, J. Chem. Soc. Faraday Trans. II 79 (1983) 19.
[5] S. Levine, G.M. Bell, Discuss. Faraday Soc. 42 (1966) 69.
[6] H. Wennerström, B. Jönsson, P. Linse, J. Chem. Phys. 76 (1982) 4665.
[7] R. Kjellander, S. Marcelja, J. Chem. Phys. 82 (1985) 2122.
[8] M.J. Spaarnay, Recl. Trav. Chim. Pays-Bas 77 (1958) 872.
[9] Y. Gur, I. Ravina, A.J. Babchin, J. Colloid Interface Sci. 64 (1978) 333.
[10] I. Langmuir, J. Chem. Phys. 6 (1938) 893.
[11] B.V. Deryagin, L. Landau, Acta Physicochim. USSR 14 (1941) 633.
[12] E.J. Verwey, J.Th.G. Overbeek, Theory of Stability of Lyophobic Colloids, Elsevier, Amsterdam, 1948.
[13] N.K. Adam, Physics and Chemistry of Surfaces, Oxford University Press, London, 1941.
[14] C. Wagner, Phys. Zeit. 25 (1924) 474.
[15] L. Onsager, N.N.T. Samaras, J. Chem. Phys. 2 (1934) 628.
[16] M. Boström, D.M.R. Williams, B.W. Ninham, Phys. Rev. Lett. 87 (2001) 168103.
[17] M. Boström, D.M.R. Williams, B.W. Ninham, Langmuir 17 (2001) 4475.
[18] K.A. Karraker, C.J. Radke, Adv. Colloid Interface Sci. 96 (2002) 231.
[19] F. Otto, G.N. Patey, J. Chem. Phys. 113 (2000) 2851.
[20] B.V. Deryagin, N.V. Churaev, V.M. Muller, Surface Forces, Plenum, New York, 1987.
[21] E. Ruckenstein, M. Manciu, Langmuir 18 (2002) 7584.
[22] E. Ruckenstein, D. Schiby, Langmuir 1 (1985) 612.
[23] L. Lue, N. Zoeller, D. Blankschtein, Langmuir 15 (1999) 3726.
[24] M. Manciu, E. Ruckenstein, Langmuir 18 (2002) 5178.
[25] R.A. Stairs, Can. J. Chem. 73 (1995) 781.
[26] B.W. Ninham, V. Yaminsky, Langmuir 13 (1997) 2097.
[27] R.J. Hunter, Foundations of Colloid Science, Oxford Science Publications, 1987.
[28] P. Jungwirth, D.J. Tobias, J. Phys. Chem. 105 (2001) 10468.
[29] P.K. Weissenborn, R.J. Pugh, J. Colloid Interface Sci. 184 (1996) 550.
[30] K. Johansson, J.C. Eriksson, J. Colloid Interface Sci. 49 (1974) 469.
[31] B.E. Conway, Ionic Hydration in Chemistry and Biophysics, Elsevier, Amsterdam, 1981.
[32] C. Li, P. Somasundaran, J. Colloid Interface Sci. 146 (1991) 215.
[33] G. Jones, W.A. Ray, J. Am. Chem. Soc. 59 (1937) 187.
[34] M. Dole, J.A. Swartout, J. Am. Chem. Soc. 62 (1940) 3039.
[35] M. Dole, J. Am. Chem. Soc. 60 (1939) 904.
[36] K.D. Collinns, M.W. Washabaugh, Q. Rev. Biophys. 18 (1985) 323.

ELSEVIER

Advances in Colloid and Interface Science
105 (2003) 177–200

ADVANCES IN COLLOID AND INTERFACE SCIENCE

www.elsevier.com/locate/cis

Specific ion effects via ion hydration: II. Double layer interaction

Eli Ruckenstein*, Marian Manciu

Department of Chemical Engineering, State University of New York at Buffalo, Buffalo, NY 14260, USA

Abstract

A simple modified Poisson–Boltzmann formalism, which accounts also for those interactions between electrolyte ions and colloidal particles not included in the mean potential, is used to calculate the force between two parallel plates. It is shown that the short-range interactions between ions and plates, such as those due to the change in the hydration free energy of a structure-making/breaking ion that approaches the interface, affect the double layer interaction at large separations through the modification of the surface potential and surface charge density. While at short separations (below the range of the short-range ion-hydration forces) the interaction can be attractive, at larger separations the interaction is always repulsive, as in the traditional theory. When the long-range van der Waals interactions between the ions and the system (ion-dispersion interactions) are accounted for in the modified Poisson–Boltzmann approach, an attractive force between plates can be generated. At sufficiently large separations, this attraction can become even stronger than the traditional van der Waals attraction between plates of finite thickness, thus generating a dominant long-range 'double layer attraction'. At small plate separations, the attraction generated by the ion-dispersion forces combined with the electrostatic repulsion due to the double layers overlap can lead to a variety of interactions, from a weak attraction (which is typically by at least one order of magnitude smaller than the traditional van der Waals attraction between plates) to a strong double layer repulsion (for sufficiently large surface charges). Both types of ion interactions (long-range van der Waals or short-range ionic hydration) strongly affect the magnitude of the double layer interaction, and can account for the specific ion effects observed experimentally. However, they do not affect qualitatively the traditional theory of the colloid stability, which predicts that the colloid is stable when there is a sufficiently large charge on the surface, and coagulates when the van der Waals interactions between two colloidal particles dominate. The only qualitative difference found when the ion-dispersion

*Corresponding author. Tel.: +1-716-645-2911, ext. 2214; fax: +1-716-645-3822.
E-mail address: feaeliru@acsu.buffalo.edu (E. Ruckenstein).

doi:10.1016/S0001-8686(03)00068-X

178 E. Ruckenstein, M. Manciu / Advances in Colloid and Interface Science 105 (2003) 177–200

interactions were incorporated into the traditional double layer theory was the emergence of a 'double layer attraction' at very large separations, which, however, does not affect much the stability of colloids.

Keywords: Colloidal particles; Double layer interaction; Ion-hydration forces

Contents

1. Introduction

Langmuir demonstrated long ago [1] that the overlap of the ionic atmospheres generated near two charged parallel plates immersed in an electrolyte, calculated within the Poisson–Boltzmann formalism [2,3], leads always to a repulsive force between the plates. This repulsion (the double layer force), combined with the van der Waals attraction, was later successfully employed to explain the stability of colloids (the DLVO theory) [4,5]. The original Poisson–Boltzmann treatment (which assumes that the ions obey the Boltzmann equation and interact only via a 'mean potential', which satisfies the Poisson equation) is clearly a drastic simplification. However, it provided in many cases a good qualitative agreement with experimental data, and sometimes—in a certain range of electrolyte concentrations—even a good quantitative agreement.

The limitations of the traditional Poisson–Boltzmann treatment are numerous. A particularly important one refers to specific ion effects. The amphoteric latex particles coagulate when the concentration of $CsNO_3$ is approximately 1 M, but remain stable at high pH values even when the concentration of KNO_3 exceeds 3 M [6]. Nevertheless, in the traditional Poisson–Boltzmann approach the double layer repulsions for Cs and K are the same. A number of attempts were made to include additional interactions in the formalism, some of them being briefly discussed in the first part of this article [7]. With these corrections, many of them dependent on unknown parameters, the 'modified Poisson–Boltzmann approach' regained its explanatory power, and in general most of the experiments could be accounted for by using suitable values for the unknown interaction parameters. However, both the simplicity and the predictability of the traditional treatment were lost, since in general, the choice of the parameter values was dependent on the particular problem investigated.

In contrast, the ‘specific ion effects’ seems to be rather well organized in a ‘lyotropic series’, first obtained empirically by Hofmeister from data on protein salting out [8]. The ions seems to obey rather well the lyotropic series sequence in many different experiments (an exhaustive compilation being provided in Ref. [9]), and the specific ionic effects relevant in biophysics are particularly strong for anions at moderate ionic strengths (0.01–1 M [10]). It was also noted that different experiments might lead to slight alterations of the series [9]. An example is provided by the surface force experiments of Pashley [11] between crossed mica cylinders in various electrolytes, which indicated the occurrence of an additional repulsion in the vicinity of the surface in excess to the double layer interaction predicted by the traditional Poisson–Boltzmann equation. Such an interaction was observed at low ionic strengths ($\sim 10^{-4}$–10^{-2} M) for NaCl, KCl and CsCl, but only at much larger electrolyte concentrations ($\sim 6\times10^{-2}$ M) for LiCl. Therefore, the ‘specific ionic effect’ of Li seems to be quite different from those of Na, K and Cs. In contrast, the amphoteric latex colloids [6] were stable when the cation was K or Li, but coagulated for Cs.

The relative regularity of the Hofmeister series suggested that the specific ion effects might be associated to the properties of the ions in the solvent, and it was noted long ago that the Hofmeister series is related to the hydration of ions [12] and to the ability of the ions to organize the neighboring water molecule [10]. More recently, it was proposed than the sequence of ions in the lyotropic series could be explained by the van der Waals interaction between the ions and the rest of the system [13]. When suitable constants for the van der Waals interactions were selected, the accounting for these ion-dispersion forces provided a quantitative agreement with the experiments regarding the surface tension of aqueous electrolytes [14,15].

More dramatic results were obtained when the ion-dispersion forces were accounted in the double layer interactions [16]. At low electrolyte concentrations (below 0.01 M), the specific ionic effects due to the van der Waals interactions of ions were negligible. However, under physiological conditions ($c_E = 0.1$ M NaCl, surface potential $\psi_S \approx 0.010$ V) the double layer repulsion between parallel plates became attractive for separation distances smaller than approximately 20 Å. The maximum interaction free energy due to the double layer repulsion calculated in Ref. [16] was only approximately 5×10^{-5} kT/Å^2, which is one order of magnitude smaller than a typical van der Waals attraction between proteins at that separation distance. Therefore, the net interaction (namely, the double layer repulsion, which includes the van der Waals interactions of the electrolyte ions with the system, plus the van der Waals interactions between two parallel plates) is in general attractive and all the ‘physiological colloids’ should coagulate.

It should be noted that there is an extensive literature about the existence of a ‘double layer attraction’ [17], and there is no general agreement yet about its nature.

However, such an attraction occurs at large separation distances, and does not affect much the interactions at low separations, hence the coagulation of colloids. It is also well accepted that the classic Poisson–Boltzmann treatment is inaccurate at concentrations higher than approximately 0.05 M, and in general either additional interactions or unknown fitting parameters had to be employed for quantitative agreement.

One of the major successes of the inclusion of the ion-dispersion forces into the Poisson–Boltzmann formalism was the explanation of the dependence of the surface tension of electrolytes on ionic strength [14,15] and pH [15]. In a previous article, it was shown that this dependence could be also explained by a simple accounting of the interactions between the ion and surface, through the changes in the ion hydration free energy between bulk and interface [7]. The latter treatment avoided some difficulties that occurred when only the van der Waals interactions of the ions with the system were taken into account (the accumulation of cations near the interface and the positive surface potential generated, in contrast to the accumulation of most anions and the negative surface potential provided by experiment and molecular simulations [18]). It also showed that the rather high values of the coefficients of the ion-dispersion interactions, obtained from the fit of the experimental data [14], might have been caused by the fact that the ion-hydration interactions were neglected. In this article, a general formalism to calculate the interactions between colloidal particles, when additional interactions are included in the Poisson–Boltzmann formalism is presented. The effects of ion-hydration and ion-dispersion interaction potentials on the force between parallel plates are investigated. While quantitative effects strongly dependent on the ion specificity are obtained, the calculations indicate that the DLVO theory remains essentially valid, namely that the colloids are stable if their surface is sufficiently charged. The values of the surface charge density and surface potential, which provide stability, are of course different if additional interactions are accounted in the Poisson–Boltzmann formalism. It will be also shown that the ion-dispersion interactions might lead, at large separation distances, to an attraction between plates. This 'double layer attraction' might become at large separations, even stronger than the traditional van der Waals attraction between parallel plates of finite thickness.

2. Ion distribution and interaction free energy: general formalism

As in the previous article [7], the equality of the electrochemical potential through the system leads to the following equation for the distribution of the ions of an uni-univalent electrolyte of concentration c_{E}:

$$c_i = c_{\mathrm{E}} \exp\left(-\frac{(-1)^i e\psi + \Delta W_i}{kT} \right) \tag{1}$$

where 'i' stands for the kind of ions (1 for anions and 2 for cations), e is the protonic charge, k is the Boltzmann constant, T is the absolute temperature, ΔW_i is the change in the interaction energy of an ion 'i' (not included in the mean potential ψ) with respect to the bulk. The potential ψ obeys the Poisson equation:

$$\nabla^2\psi = \frac{e(c_1 - c_2)}{\varepsilon\varepsilon_0} \tag{2}$$

where ε_0 is the vacuum permittivity and ε is the dielectric constant of the medium, which here is assumed to be constant through the solvent. The distribution of electrolyte ions can be calculated, for suitable boundary conditions, from Eqs. (1) and (2).

The total free energy of the system is composed of several contributions: an 'electrostatic energy' due to the distribution of charges, an energy due to the 'additional interactions of ions' (not included in the mean potential), an 'entropic' term due to the mobile ions, and a 'chemical free energy' which is responsible for the formation of the double layer.

The electrostatic energy is given by:

$$F_{\mathrm{el}} = \frac{1}{2}\int_V \varepsilon\varepsilon_0(\nabla\psi)^2 \mathrm{d}V, \tag{3}$$

and the entropic contribution can be calculated from [19]:

$$F_{\mathrm{ent}} = -T\Delta S = kT\sum_i \int_V \left(c_i \ln\left(\frac{c_i}{c_{\mathrm{E}}}\right) - c_i + c_{\mathrm{E}}\right)\mathrm{d}V \tag{4}$$

where both integrals are calculated over the volume V of the solvent. The interaction free chemical energy (calculated with respect to the infinite separation of the components of the system) is given by [20]:

$$\Delta F_{\mathrm{chem}} = \int_A \left(-\int_{\sigma(\infty)}^{\sigma} \psi_{\mathrm{S}}(\sigma)\mathrm{d}\sigma\right)\mathrm{d}A \tag{5}$$

where σ is the surface charge density, ψ_{S} the surface potential, the surface integral is performed over the entire area A of the system (the interface between the solvent and the interacting surfaces), and the other integral is performed along the trajectory of the system from infinite separation distance between the surfaces to the final configuration of the system.

To these components of the traditional double layer free energy one should add an energetic contribution due to the supplementary ion interactions:

$$F_{\mathrm{int}} = \int_V (c_1\Delta W_1 + c_2\Delta W_2)\mathrm{d}V. \tag{6}$$

Consequently, the interaction free energy of the system (with respect to infinite separation distance between components) is given by:

$$\Delta F = \Delta F_{\mathrm{el}} + \Delta F_{\mathrm{ent}} + \Delta F_{\mathrm{chem}} + \Delta F_{\mathrm{int}} \tag{7}$$

3. Interaction between parallel plates immersed in a gas of neutral particles

3.1. Free energy of the system and the Gibbs adsorption equation

The formalism described in Section 2 can be applied to systems of any geometry. To simplify the calculations, it will be applied here to the interactions between two parallel plates. Let us first assume that two large parallel plates are immersed into a reservoir containing a neutral gas of concentration c_{N}. If the gas interacts with the inner side of the plates (the interactions with the other sides are neglected), the free energy of the system (the plates and the adsorbed gas between them) per unit area is given by:

$$\begin{aligned} F_{\mathrm{N}}(2d) &= F_{\mathrm{int}}(2d) + F_{\mathrm{ent}}(2d) = \int_{-d}^{d} c(x)\Delta W(x)\mathrm{d}x \\ &\quad + kT\int_{-d}^{d}\left(c(x)\ln\left(\frac{c(x)}{c_{\mathrm{N}}}\right) - c(x) + c_{\mathrm{N}}\right)\mathrm{d}x \\ &= c_{\mathrm{N}}kT\int_{-d}^{d}\left(1-\exp\left(-\frac{\Delta W(x)}{kT}\right)\right)\mathrm{d}x \end{aligned} \tag{8}$$

where $2d$ is the distance between plates, $\Delta W(x)$ is the change in the energy of a particle located at a distance x from the middle between the plates with respect to that at infinite distance and the gas particles are assumed to obey a Boltzmann distribution. The same result can be obtained using a Gibbs adsorption procedure, introduced by van der Waals and Makor [21] and generalized by Ash et al. [22]. The positive or negative surface excess adsorption of gas particles, Γ_{N}, between plates is related, via the Gibbs adsorption equation, to the change in the surface tension γ:

$$\begin{aligned} \mathrm{d}\gamma &= -\Gamma_{\mathrm{N}}\mathrm{d}\mu_{\mathrm{N}} = -\left(\int_{0}^{d}(c(x)-c_{\mathrm{N}})\mathrm{d}x\right)\mathrm{d}\mu_{\mathrm{N}} \\ &= -c_{\mathrm{N}}\left(\int_{0}^{d}\left(\exp\left(-\frac{\Delta W(x)}{kT}\right)-1\right)\mathrm{d}x\right)\mathrm{d}\mu_{\mathrm{N}} \end{aligned} \tag{9}$$

where μ_{N} is the chemical potential of the neutral particles, which is constant through the system and can be related to the concentration of the neutral particles in the reservoir, for a perfect gas, through:

$$d\mu_N = d(kT \ln(c_N)). \tag{10}$$

Integrating Eqs. (9) and (10) for a concentration of particles varying from 0 to c_N, one obtains:

$$\gamma(c_N) - \gamma(0) = -kT\int_0^{c_N} \frac{\Gamma_N(c)}{c} dc = -c_N kT \int_0^d \left(\exp\left(-\frac{\Delta W(x)}{kT}\right) - 1\right) dx \tag{11}$$

and, as expected, the force between plates [21],

$$\Pi(2d) = -\frac{\partial \gamma}{\partial d} \tag{12}$$

is identical to that obtained from the derivative of the free energy, Eq. (8), with respect to the separation distance $2d$.

For a neutral gas, the adsorption (which is proportional to the free energy of the system, see Eq. (8)) can be calculated easily. For charged particles one should account in the Gibbs adsorption equation for the adsorption of all particles of the system (including those responsible for the charging of the surface [23]). Therefore, one should first identify the mechanism of formation of the double layer. In this case,

$$\gamma(c_E) - \gamma(0) = -kT \sum_i \int_{-\infty}^{\mu(c_G)} \Gamma_i d\mu_i \tag{11a}$$

where i is taken over all the ionic species present in the system and Γ_i is the adsorption of ion i. The result is incomplete when one accounts only for the distributions of electrolyte ions between the plates. Indeed, for an uni-univalent electrolyte of concentration c_E, the adsorption Γ' of the two species of ions between two plates with constant surface charges σ, is given at low surface potentials by:

$$\begin{aligned}\frac{2\Gamma'(c_E)}{c_E} &= 2\int_{-d}^{d}\left(\cosh\left(\frac{e\psi(x)}{kT}\right) - 1\right)dx \cong \left(\frac{e}{kT}\right)^2 \int_{-d}^{d} (\psi(x))^2 dx \\ &\cong \left(\frac{e\sigma\lambda}{kT\varepsilon\varepsilon_0}\right)^2 \int_{-d}^{d} \frac{\cosh^2\left(\frac{x}{\lambda}\right)}{\sinh^2\left(\frac{d}{\lambda}\right)} dx = \left(\frac{e\sigma\lambda}{kT\varepsilon\varepsilon_0}\right)^2 \frac{\frac{\lambda}{2}\sinh\left(\frac{2d}{\lambda}\right) + d}{\sinh^2\left(\frac{d}{\lambda}\right)}\end{aligned} \tag{13}$$

where λ is the Debye length. Since $(\partial/\partial(2d))(\Gamma'(c_E)/c_E) < 0$ for any c_E, using

Eqs. (9)–(12) would predict an attraction between the plates ($\Pi<0$), which is incorrect.

In what follows, the direct calculation of the free interaction energy by adding the energetic and entropic contributions (Eq. (7)) will be employed.

3.2. Long-range interactions between neutral particles and plates

To have an idea about how important are the dispersion interactions of electrolyte ions, we will consider in this section the force between plates generated by a neutral gas, whose particles interact with the plates by long-range dispersion forces. An interaction of the type [16]:

$$\Delta W_{vdW}(x)=\frac{B}{(d-x)^3}+\frac{B}{(d+x)^3} \tag{14}$$

will be assumed, where $2d$ is the separation distance between plates.

For a molecule of gas, B is negative (the particles are attracted by the plates). However, if the particles are immersed in a medium, it is possible to have positive B values, depending on the Hamaker constants of the plates, particles and medium. Assuming a cut-off distance Δ for the interaction, the free energy per unit area (Eq. (8)) becomes, in the linear approximation (which is accurate if B is small or if the particle is sufficiently far from any interface):

$$F_N(2d)\cong c_N kT\int_{-d}^{d}\left(\frac{\Delta W_{vdW}(x)}{kT}\right)dx=Bc_N\frac{4d(d-\Delta)}{\Delta^2(2d-\Delta)^2} \tag{15}$$

and the corresponding interaction free energy, per unit area, is:

$$F(2d)-F(\infty)=-\frac{Bc_N}{(2d-\Delta)^2} \tag{16}$$

This interaction, due to the interaction between particles and plates, has the same functional form at large separations ($d\gg\Delta$) as the well-known van der Waals attraction between thick parallel plates [24]:

$$F_{vdW,\mathrm{plates}}=-\frac{A_H}{12\pi(2d)^2} \tag{17}$$

where A_H is the Hamaker constant.

When the particles are attracted by the plates through dispersion forces ($B<0$), the net interaction between plates is repulsive, while when the interactions between particles and plates are repulsive ($B>0$), the plates are attracted to each other.

The magnitude of the interaction generated by the dispersion forces between the neutral particles and the plates (Eq. (16)) is proportional to the concentration of

E. Ruckenstein, M. Manciu / Advances in Colloid and Interface Science 105 (2003) 177–200 185

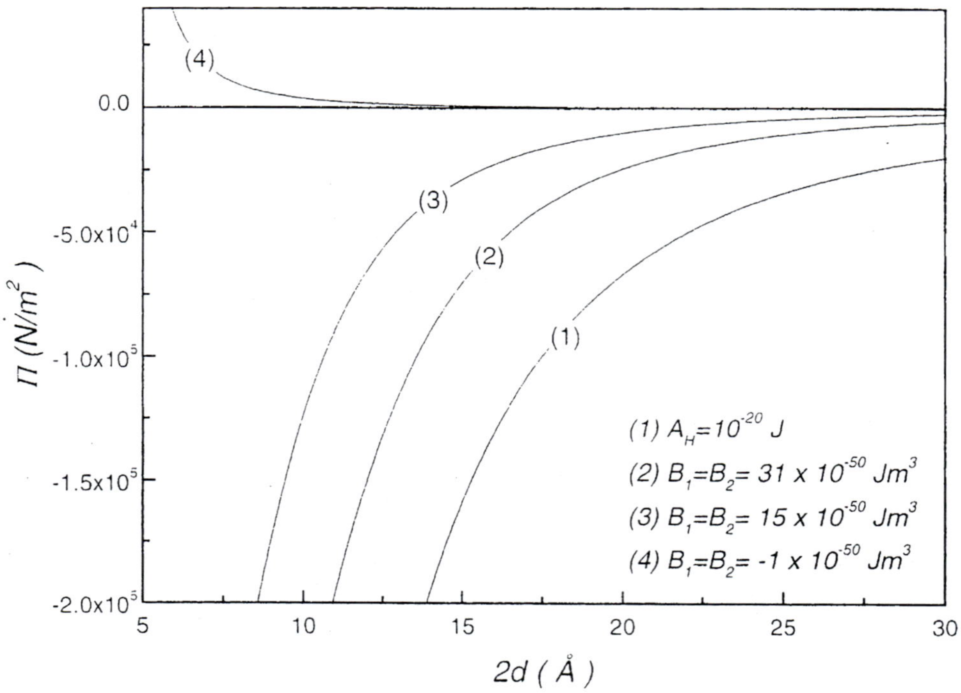

Fig. 1. The force between parallel plates for a neutral system, generated by ion-dispersion forces at c_E= 0.1 M, for $B_1=B_2=31\times10^{-50}$ J m^3 (curve (2)); $B_1=B_2=15\times10^{-50}$ J m^3 (curve (3)); $B_1=B_2=-1\times10^{-50}$ J m^3 (curve (4)) are compared to the traditional van der Waals attraction between plates for $A_H=10^{-20}$ J (curve (1)).

particles in the reservoir (c_N) and is typically much smaller than the traditional van der Waals attraction between plates. The two interactions (Eqs. (16) and (17)) become comparable at large separation distances for:

$$c_N \cong \frac{A_H}{12\pi B}. \tag{18}$$

Since the Hamaker constant for proteins in a physiological solution is of the order of 10^{-20} J, Eq. (18) provides a concentration of approximately 2 M even for the large value of $B=31\times10^{-50}$ J m^3 corresponding to Cl^- ions [14]. This corresponds to an ideal gas pressure of approximately 50 atm. Consequently, the long-range repulsion or attraction between plates due to the van der Waals interactions between a gas of neutral particles and the plates is expected to be negligible in most cases.

The interactions between neutral plates immersed in an electrolyte can be calculated from Eqs. (11) and (12), providing that the dispersion forces are the same for both kinds of ions and hence the potential vanishes everywhere. In Fig. 1, the forces (per unit area) between parallel plates, calculated numerically for $c_1 =$ 0.1 M, $T = 300$ K, a cut-off distance $\Delta = 2$ Å and various values $B_1 = B_2$ (curves (2–4)) are compared with the traditional van der Waals attraction due to the polarizability of the plates, for a typical Hamaker constant of $A_H = 1 \times 10^{-20}$ J. The attraction generated by the ion-dispersion forces at this concentration is in general much smaller than the traditional van der Waals attraction (curve 1). When $B_1 \equiv B_2 < 0$ (curve 4), the interaction generated by ion-dispersion forces is a repulsion.

In general, $B_1 \neq B_2$ and a double layer is generated by the asymmetric distribution of the electrolyte ions even if there is no charge on the plate surface. A treatment based on the modified Poisson–Boltzmann formalism will be presented in Section 4.2.

3.3. Short-range interactions between neutral particles and plates

In a previous article [7], it was suggested that the changes in ion hydration between bulk and the vicinity of the surface might be approximated by a simple step function. In what follows the interaction between parallel plates will be investigated for an interaction between particles and plates of the type:

$$\Delta W(\xi) = W_1 \quad 0 < \xi < d_1 \tag{19a}$$

$$\Delta W(\xi) = 0 \quad d_1 < \xi < \infty \tag{19b}$$

where ξ is the distance to one of the plates and W_1 can be negative (corresponding to an attraction between particles and plates) or positive (repulsion). It is also assumed that the interactions between particles and plates are additive (Fig. 2):

$$\Delta W = \Delta W(\xi) + \Delta W(2d - \xi), \tag{19c}$$

$2d$ being the separation between plates.

Due to the short-range of this interaction, the force between plates vanishes for $d > d_1$. For $(d_1/2) < d < d_1$, the free energy per unit area of the system (Eq. (8)) is given by:

$$\begin{aligned} F(2d) &= c_N kT \int_{-d}^{d} \left(1 - \exp\left(-\frac{\Delta W(x)}{kT}\right)\right) \mathrm{d}x \\ &= 2c_N kT \left[(d_1 - d)\left(1 - \exp\left(\frac{2W_1}{kT}\right)\right) - (2d - d_1)\left(1 - \exp\left(\frac{W_1}{kT}\right)\right)\right] \\ &= 2c_N kT \left[d\left(\exp\left(\frac{W_1}{kT}\right) - 1\right)^2 + d_1\left(\exp\left(\frac{W_1}{kT}\right) - \exp\left(\frac{2W_1}{kT}\right)\right)\right] \end{aligned} \tag{20}$$

E. Ruckenstein, M. Manciu / Advances in Colloid and Interface Science 105 (2003) 177–200 187

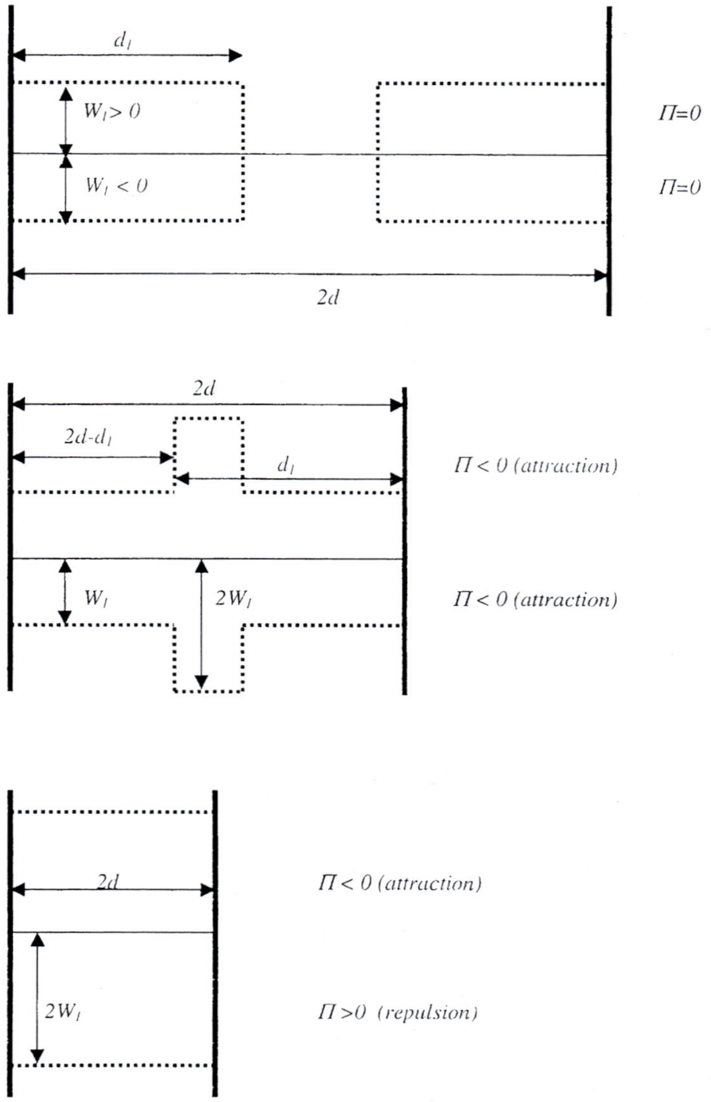

Fig. 2. An additive step-like interfacial interaction between neutral particles and plates. There is no force between plates for $d > d_1$, attraction when $(d_1/2) < d < d_1$ and either attraction or repulsion for $0 < d < (d_1/2)$.

and the force per unit area between plates by

$$\Pi(2d) = -\frac{\partial F(2d)}{\partial(2d)} = -c_N kT\left(\exp\left(\frac{W_1}{kT}\right) - 1\right)^2 < 0 \quad \frac{d_1}{2} < d < d_1 \tag{21}$$

is always attractive in this region ($\Pi < 0$), regardless of whether the particles are attracted toward ($W_1 < 0$) or repelled ($W_1 > 0$) by the plates.

For the range $0 < d < (d_1/2)$, the force per unit area is given by:

$$\Pi(2d) = c_N kT\left(\exp\left(\frac{W_1}{kT}\right) - 1\right) \quad 0 < d < \frac{d_1}{2} \tag{22}$$

which leads to a net repulsion between plates if the neutral particles are attracted by the plates and to an attraction when the neutral particles are repelled by the plates.

In summary, short-ranged step-like interactions for ions generate only short-range interactions between plates, which are always attractive for $(d_1/2) < d < d_1$ and become either attractive or repulsive, depending on the sign of W_1, for $0 < d < (d_1/2)$.

4. The interaction between parallel plates immersed in an electrolyte

4.1. Short-range interactions (ion-hydration forces)

The change in the ion hydration energy between the bulk and the water–air interface for structure-making ions is so large compared to kT, that these ions practically do not approach the interface. The change is not so steep when a solid interface is immersed in water, particularly when the surface has sites which can bound structure-making ions. In what follows, it will be assumed that the anions are structure breaking and hence interact with the surface via an attractive potential of the type ($W_1 > 0$, see Fig. 3a):

$$\Delta W_1(\xi) = -W_1 \quad 0 < \xi < d_1 \tag{23a}$$

$$\Delta W_1(\xi) = 0 \quad d_1 < \xi < \infty \tag{23b}$$

and the cations are structure making and interact with the surface via a repulsive potential ($W_2 > 0$):

$$\Delta W_2(\xi) = W_2 \quad 0 < \xi < d_2 \tag{24a}$$

$$\Delta W_2(\xi) = 0 \quad d_2 < \xi < \infty \tag{24b}$$

The modified Poisson–Boltzmann equation for a uni-univalent electrolyte confined between parallel plates is:

$$\frac{d^2\psi}{d^2x} = \frac{ec_E}{\varepsilon\varepsilon_0}\left(\exp\left(-\frac{\Delta W_1}{kT}\right)\exp\left(\frac{e\psi}{kT}\right) - \exp\left(-\frac{\Delta W_2}{kT}\right)\exp\left(-\frac{e\psi}{kT}\right)\right) \tag{25}$$

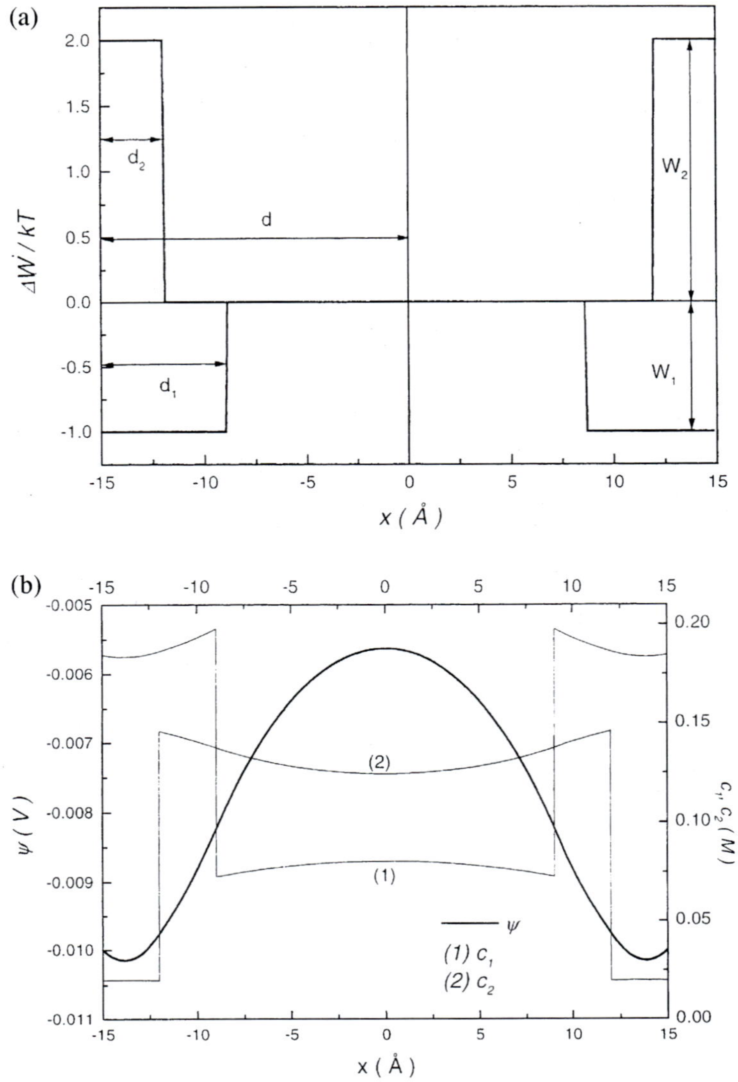

Fig. 3. (a) A typical ion-hydration potential: the anions are attracted toward the interface ($W_1 = 1$ kT, $d_1 = 6$ Å) while the cations are repelled by the surface ($W_2 = 2$ kT, $d_2 = 3$ Å). (b) The potential and the corresponding ion distributions, for $W_1 = 1$ kT, $d_1 = 6$ Å, $W_2 = 2$ kT, $d_2 = 3$ Å, $d = 15$ Å, $c_E = 0.1$ M, $\psi_S = -0.010$ V, $\varepsilon = 80$, $T = 300$ K.

which has to satisfy two boundary conditions, one due to the symmetry:

$$\left.\frac{d\psi}{dx}\right|_{x=0} = 0 \tag{25a}$$

and the other providing the surface charge density, the surface potential or a relation between them.

Since the selected $\Delta W_i(\xi)$ is independent of ξ on each of the intervals $0<\xi<d_2$, $d_2<\xi<d_1$, $d_1<\xi<d$ (Fig. 3a), the solution in the linear approximation (valid at low potentials) in the interval 'j' is of the type:

$$\psi(\xi) = -\frac{kT}{e}\frac{A_j - D_j}{A_j + D_j} + C_{2j}\exp\left(\sqrt{\frac{A_j+D_j}{2}}\frac{\xi}{\lambda}\right) + C_{2j+1}\exp\left(-\sqrt{\frac{A_j+D_j}{2}}\frac{\xi}{\lambda}\right) \tag{26}$$

where $A_j = \exp(-(\Delta W_1)_j/(kT))$, $D_j = \exp(-(\Delta W_2)_j/(kT))$, $\lambda = \sqrt{((\varepsilon\varepsilon_0 kT)/(2e^2 c_E))}$ is the Debye length and in each interval j, C_{2j}, C_{2j+1} are constants to be determined from the boundary conditions and the continuity of the electric potential and electric field. A typical result for the potential and the corresponding ion distribution is presented in Fig. 3b, for: $W_1 = 1$ kT, $d_1 = 6$ Å, $W_2 = 2$ kT, $d_2 = 3$ Å, $d = 15$ Å, $c_E =$ 0.1 M, $\psi_S = -0.010$ V, $\varepsilon = 80$, $T = 300$ K. Once the potential has been determined, the interaction free energy of the system can be calculated using Eq. (7).

For separation distances $d > d_1$, the Poisson–Boltzmann equation is obeyed within $d_1 < \xi < 2d - d_1$. The system is completely equivalent with a traditional double layer, but formed between the surfaces $S'(\xi = d_1)$ and not between the real surfaces $S(\xi = 0)$. The corresponding 'surface potential' is the potential at S', $\psi_S \equiv \psi(\xi = d_1)$, and the corresponding surface charge density, σ', is given by:

$$\sigma' = \sigma + \int_0^{d_1} \rho(\xi)\mathrm{d}\xi = -\varepsilon\varepsilon_0 \left.\frac{\mathrm{d}\psi}{\mathrm{d}\xi}\right|_{\xi = d_1}. \tag{27}$$

Therefore, the short-range interactions due to ion hydration do not modify qualitatively the double layer interaction, which remains repulsive at distances larger than the range of the interfacial interaction of the ions. The magnitude of the repulsion is, however, modified, since the relation between the 'true' surface charge density σ and the 'apparent' surface charge density σ' depends on ΔW_i and hence on the specificity of the ions.

To illustrate this, the apparent charge density σ' is calculated in Fig. 4a and b as a function of the separation distance, for $d_I \equiv d_2 = 5$ Å and constant surface charge density. The apparent charge density is proportional to the derivative of the electric potential at the distance $\xi = d_1 \equiv d_2$. When the surface is negatively charged ($\sigma = -0.010$ C/m^2, Fig. 4a), both the accumulation of anions ($\Delta W_1 < 0$) and the cation depletion ($\Delta W_2 > 0$) lead to an increase of the charge in the vicinity of the surface, therefore to an increase in the double layer repulsion. When the surface is positively charged ($\sigma = 0.010$ C/m^2, Fig. 4b), the ion–hydration interaction can lead to a decrease in the apparent surface charge density (curve (4)), which can even change sign.

The apparent charge density is plotted in Fig. 4c as a function of electrolyte concentration, for $2d = 20$ Å and $\sigma = +0.010$ C/m^2. Depending on the potentials of

E. Ruckenstein, M. Manciu / Advances in Colloid and Interface Science 105 (2003) 177–200 191

ion hydrations, σ', which is positive at sufficiently low concentrations, can become negative when the ionic strength is sufficiently high. The electrolyte concentration which correspond to $\sigma'=0$ and hence, roughly, to coagulation depends strongly, via W_1 and W_2, on the type of electrolyte. Therefore, short-range interfacial interactions of ions, of the order of kT, can account for specific ion effects in the double layer interactions.

The force between neutral plates ($\sigma=0$) immersed in an electrolyte of concentration $c_E=0.1$ M, calculated from the derivative of the free energy of the system (Eq. (7)) is plotted in Fig. 5a. The double layer, which is generated by the different distributions of the positive and negative ions of the electrolyte alone (it vanishes when $\Delta W_1=\Delta W_2$), is always repulsive for $d>d_1\equiv d_2$, and can be either attractive or repulsive for $d<d_1$.

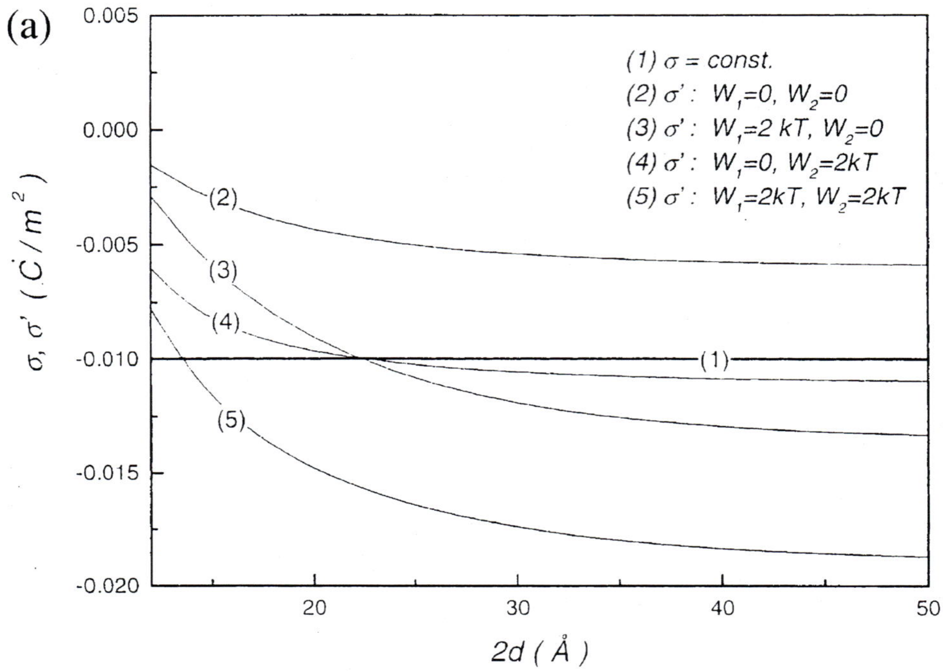

Fig. 4. (a) The apparent surface charge density σ' for an interaction at 'constant surface charge density' ($\sigma=-0.01$ C/m^2, $c_E=0.1$ M) as a function of the separation between plates. The long-range double layer interaction is controlled by σ', which in turn depends on the short-range potential of interaction between ions and surface. (b) The apparent surface charge density σ' for an interaction at a positive 'constant surface charge density' ($\sigma=+0.01$ C/m^2, $c_E=0.1$ M) as a function of the separation between plates. The apparent surface charge density can be decreased or even change its sign when the short-range interaction between ions and surface is accounted for. (c) The apparent surface charge density σ' for $\sigma=+0.01$ C/m^2 and $2d=20$ Å as a function of the electrolyte concentration. While the double layer interaction is always repulsive at sufficiently large concentrations, the electrolyte concentration corresponding to $\sigma'=0$ (hence, to a vanishing double layer repulsion) depends strongly on the short-range interactions between ions and surface.

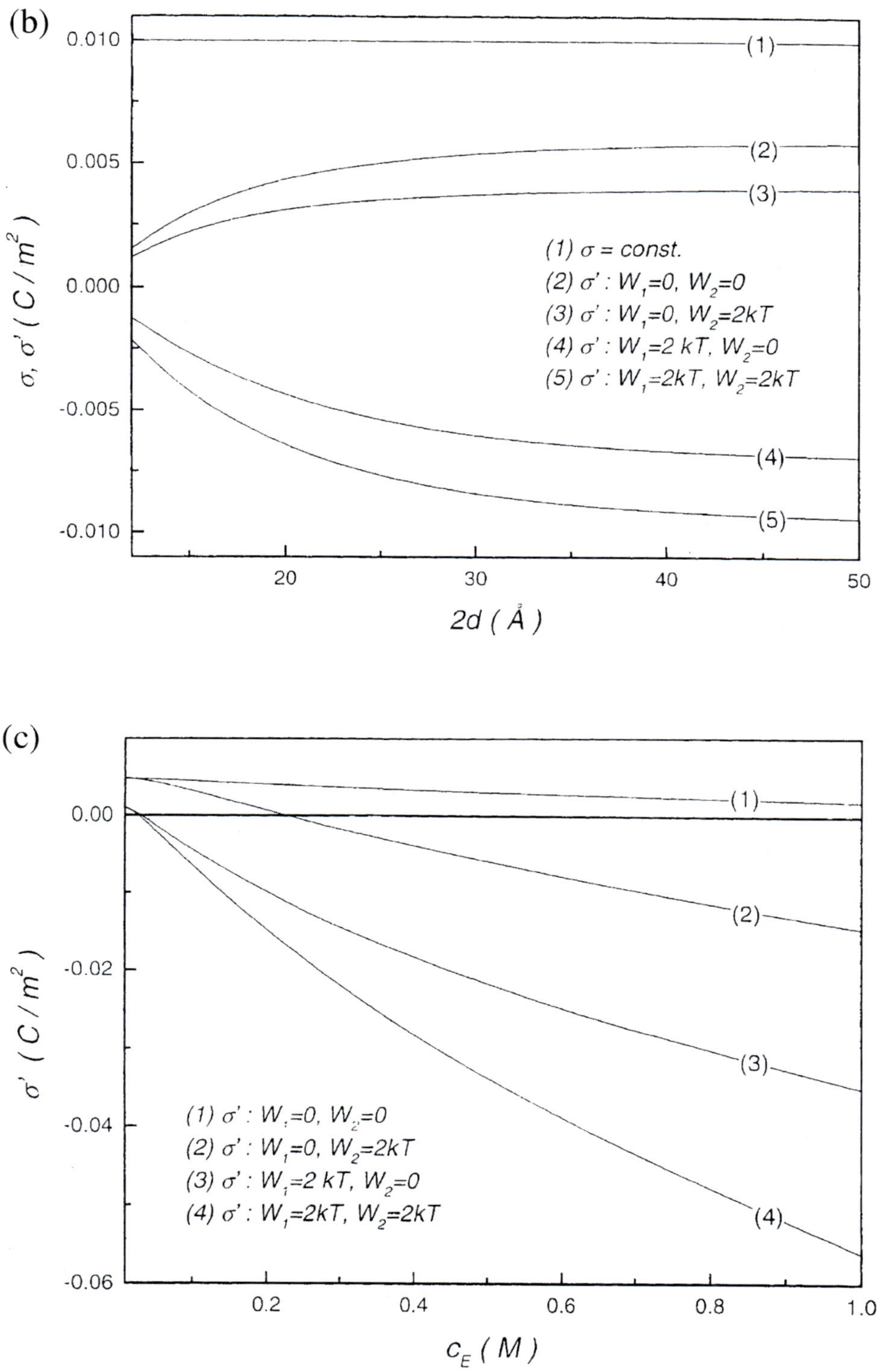

Fig. 4 (Continued).

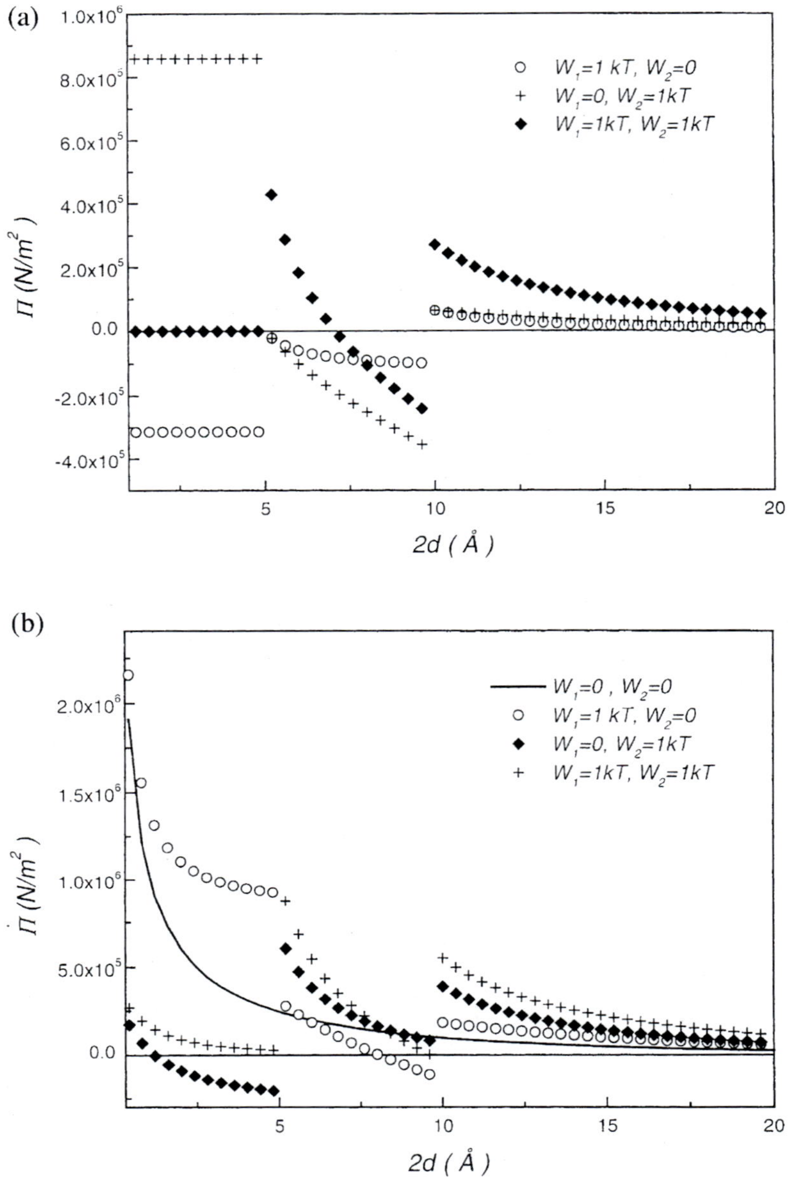

Fig. 5. (a) The force per unit area between neutral parallel plates immersed in an electrolyte with $c_E = 0.1$ M. The (asymmetric) short-range interactions between ions and interface lead to an asymmetric distribution of electrolyte ions, which in turn generates a long-range double layer repulsion. In the vicinity of the surface ($2d < 10$ Å), the force can become attractive. (b) The force per unit area between negatively charged parallel plates immersed in an electrolyte with $c_E = 0.1$ M. The interactions between ions and surfaces increase in this case the apparent surface charge density, hence increase the repulsion at large separation distances. The force can become attractive for $2d < 10$ Å.

A more realistic calculation is presented in Fig. 5b, which implies that there is a surface charge density generated by the dissociation of surface groups. Assuming that the cations can be bound to N negative sites per unit area, the dissociation equilibrium provides the expression:

$$K_D = \frac{(\eta N)c_E^+}{(1-\eta)N} = \frac{\eta c_E^+}{(1-\eta)} = \frac{\eta}{1-\eta} c_E \exp\left(-\frac{e\psi_S}{kT} - \frac{W_2}{kT}\right) \tag{28}$$

where K_D is the dissociation constant, c_E^+ denotes the cation concentration in the liquid in the vicinity of the surface and η is the fraction of dissociated sites.

The surface charge density is therefore given by:

$$\sigma = -e\eta N = -\frac{eN}{1 + \frac{c_E}{K_D}\exp\left(-\frac{e\psi_S}{kT} - \frac{W_2}{kT}\right)} \tag{29}$$

The ion–hydration interactions lead to an accumulation of anions and a depletion of cations near the interface, thus increasing the magnitude of the apparent surface charge density σ'. Consequently, the double layer repulsion is enhanced at large distances. When the surface charge density is positive at low electrolyte concentrations (e.g. when the charge is generated by the dissociation of basic surface groups), the ion hydration can decrease the magnitude of the repulsion at large distances, but cannot transform it there into an attraction.

The apparent charge density, which is ultimately responsible for the double layer interaction, depends strongly on the hydration potential of ions. Although the double layer interaction is always repulsive, for some electrolytes σ' can become sufficiently large to prevent coagulation, while for other electrolytes sufficiently low to allow coagulation. Therefore, the ion–hydration interaction can provide a suitable explanation for the specific ion effects observed in double layer interactions.

4.2. *Long-range interactions (ion-dispersion forces)*

In this section, the effect of the ion-dispersion forces on the double layer interactions will be investigated, by assuming that they provide the only correction to the traditional Poisson–Boltzmann equation.

When the dispersion interaction coefficients are different for anions and cations, the different distributions of the two kinds of ions generate a potential even when the plates are neutral. The potential is obtained from the solution of the correspondingly modified Poisson–Boltzmann equation (Eqs. (1) and (2)) and the interaction free energy can be calculated via the numerical integration of Eq. (7).

The force between neutral plates immersed in an uni-univalent electrolyte of concentration $c_E = 0.1$ M, when $B_1 \neq B_2$, is presented in Fig. 6(a), for a cut-off distance $\Delta = 2$ Å. While the force was always attractive when $B_1 = B_2 > 0$ (Fig. 1), the double layer generated by the asymmetric distributions of electrolyte ions, when

E. Ruckenstein, M. Manciu / Advances in Colloid and Interface Science 105 (2003) 177–200 195

$B_1 \neq B_2$, can dominate the interaction (which becomes repulsive) at intermediate separations ($B_1 = 31 \times 10^{-50}$ J m^3, $B_2 = 1 \times 10^{-50}$ J m^3, curve (1) in Fig. 6a). If one type of the ions (the cation) is attracted by the interface ($B_2 = -1 \times 10^{-50}$ J m^3, curve (2)), the repulsion induced by the ion-dispersion forces becomes, as expected, even stronger.

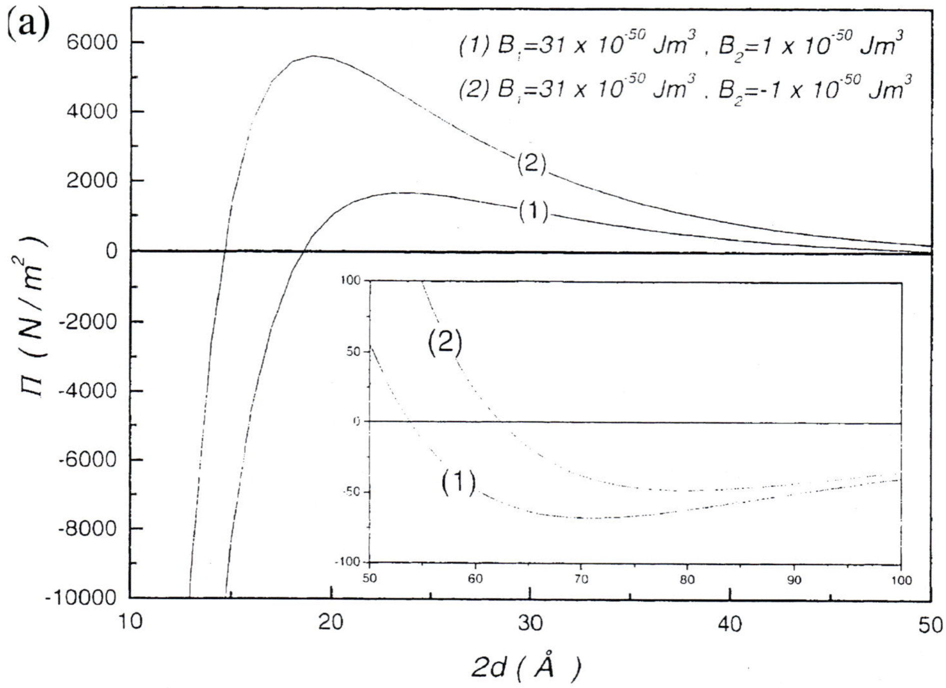

Fig. 6. (a) The force between neutral plates immersed in an electrolyte with $c_E = 0.1$ M, due to the ion-dispersion forces. At intermediate separations, the double layer generated by the asymmetric distribution of electrolyte ions leads to a net repulsion, although both types of ions are repelled by the interface ($B_1 = 31 \times 10^{-50}$ J m^3, $B_2 = 1 \times 10^{-50}$ J m^3). When the ions of one kind are attracted by the interface ($B_1 = -1 \times 10^{-50}$ J m^3), the repulsion increases. The interactions at large separation distances (inset) are attractive. (b) The force between negatively charged plates immersed in an electrolyte with $c_E = 0.1$ M, (1) $N = 3 \times 10^{16}$ sites/m^2, $K_D = 0.1$ M, $B_1 = B_2 = 0$; (2) $N = 3 \times 10^{16}$ sites/m^2, $K_D = 0.1$ M, $B_1 = 31 \times 10^{-50}$ J m^3 and $B_2 = -1 \times 10^{-50}$ J m^3; (3) $\sigma = 0$, $B_1 = 31 \times 10^{-50}$ J m^3 and $B_2 = -1 \times 10^{-50}$ J m^3; (4) $N = 1 \times 10^{17}$ sites/m^2, $K_D = 0.1$ M, $B_1 = 31 \times 10^{-50}$ J m^3 and $B_2 = -1 \times 10^{-50}$ J m^3; the 'double layer attraction' (inset) occurs at large separations, even in the presence of large surface charges. (c) The force between parallel plates, for a moderate surface charge generated by the dissociation of surface groups, for $N = 1 \times 10^{17}$ sites/m^2, $K_D = 0.1$ M, $\varepsilon = 80$, $T = 300$ K and various sets of ion-dispersion coefficients: (1) $B_1 = 0$, $B_2 = 0$; (2) $B_1 = 15 \times 10^{-50}$ J m^3, $B_2 = 3 \times 10^{-50}$ J m^3; (3) $B_1 = 15 \times 10^{-50}$ J m^3, $B_2 = 1 \times 10^{-50}$ J m^3; (4) $B_1 = 15 \times 10^{-50}$ J m^3, $B_2 = -1 \times 10^{-50}$ J m^3; (5) $B_1 = 15 \times 10^{-50}$ J m^3, $B_2 = -3 \times 10^{-50}$ J m^3; (6) $B_1 = 31 \times 10^{-50}$ J m^3, $B_2 = 3 \times 10^{-50}$ J m^3; (7) $B_1 = 31 \times 10^{-50}$ J m^3, $B_2 = 1 \times 10^{-50}$ J m^3; (8) $B_1 = 31 \times 10^{-50}$ J m^3, $B_2 = -1 \times 10^{-50}$ J m^3; (9) $B_1 = 31 \times 10^{-50}$ J m^3, $B_2 = -3 \times 10^{-50}$ J m^3; In the inset, the calculations are repeated for a strong double layer ($N = 2 \times 10^{17}$ sites/m^2, $K_D = 1$ M).

196 *E. Ruckenstein, M. Manciu / Advances in Colloid and Interface Science 105 (2003) 177–200*

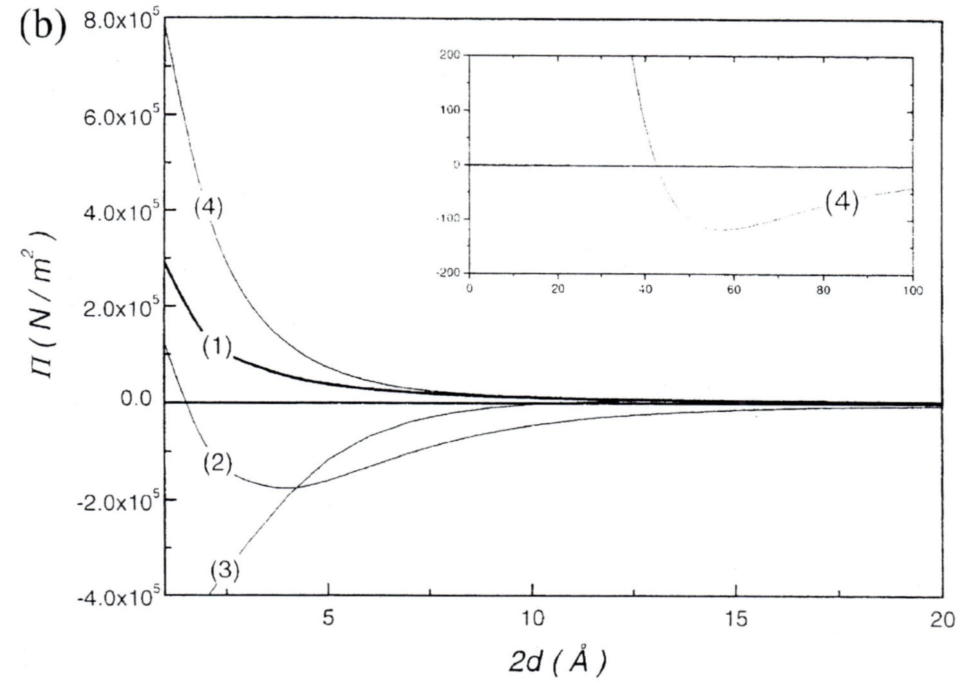

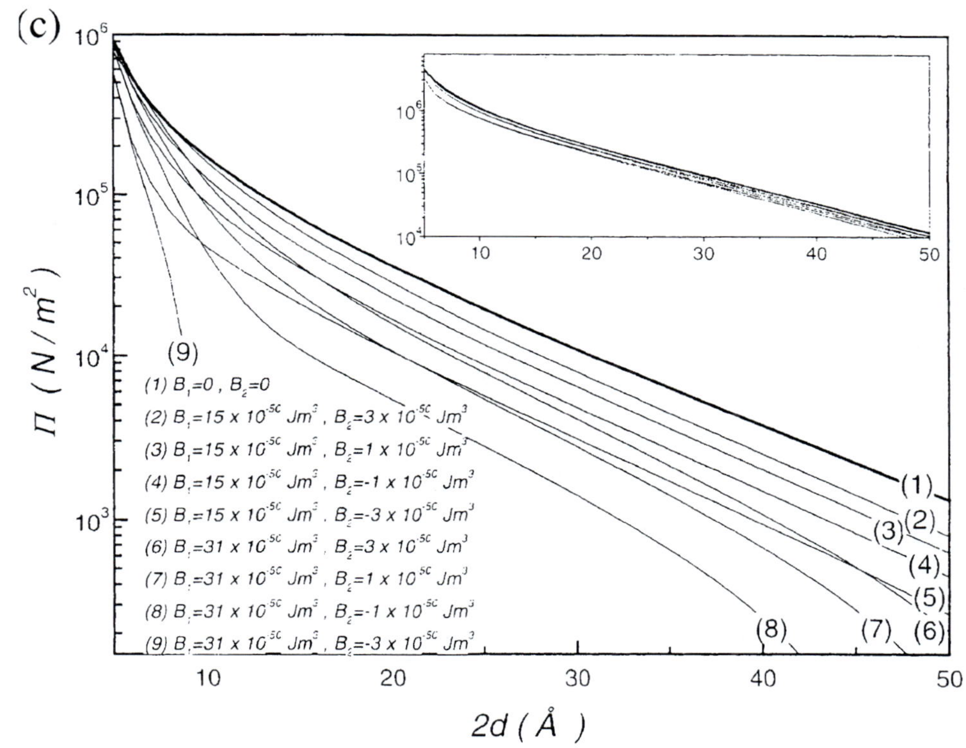

Fig. 6 (Continued).

The maximum repulsion generated by the asymmetric distribution of ions is however, small when the surface charge is zero (approximately one order of magnitude smaller than a typical van der Waals attraction) and therefore the neutral physiological colloids will coagulate, in agreement with the DLVO theory.

More interesting are the effects of ion-dispersion forces when the plates are charged. Let us first assume that a small negative surface charge is generated by the dissociation of surface groups (Eqs. (28) and (29), with $c_E=0.1$ M, $N=3\times10^{16}$ sites/m^2 and $K_D=0.1$ M). In the absence of the van der Waals interactions of ions ($B_1=B_2=0$), a weak double layer repulsion is provided by the traditional theory (curve (1), Fig. 6b). By accounting in the modified Poisson–Boltzmann equation for the van der Waals interactions of ions with $B_1=31\times10^{-50}$ J m^3, $B_2=-1\times10^{-50}$ J m^3 and $\Delta=2$ Å, but assuming $\sigma=0$, curve (3) was obtained. When both the van der Waals interaction and the surface charge density were accounted for (curve (2)), the interaction became attractive at most distances.

However, if the surface charge density is sufficiently large (curve (4) in Fig. 6b, with $N=1\times10^{17}$ sites/m^2) then the interaction in the vicinity of the surface is repulsive. Therefore, a strong double layer interaction is not modified much at small separations by ion-dispersion forces.

The ion-dispersion forces (as the ion-hydration forces, discussed previously) might constitute a suitable candidate to account for the ion specific effects within the Poisson–Boltzmann formalism, provided that suitable values for the interaction parameters are chosen. In Fig. 6c, the force between plates was calculated for a moderate double layer, generated by the dissociation of surface groups ($N=1\times10^{17}$ sites/m^2, $K_D=0.1$ M), and for $c_E=0.1$ M, $\varepsilon=80$ and $T=300$ K. In the absence of ion–dispersion interactions ($B_1=B_2=0$), the surface potential is approximately -0.020 V. When the ion–dispersion interactions were included, the repulsion in the vicinity of the surface has always decreased. The decrease of the repulsion seems to be particularly affected by B_1, which typically has larger magnitudes than B_2 [14,15] ($B_1=15\times10^{-50}$ J m^3 for the curves (2)–(5) and $B_1=31\times10^{-50}$ J m^3 for the curves (6)–(9)). The change of B_2 from $B_2=1\times10^{-50}$ to $B_2=3\times10^{-50}$ J m^3 (curves (2)–(3) and (6)–(7)) had a much smaller effect. The specific ionic effects of cations become more important when $B_2<0$, because the accumulation of positive charges at the surface compensates for the negative charge generated through the dissociation and this attenuates the double layer.

In the inset of Fig. 6c, the calculations were repeated for the same values of the parameters, but a stronger double layer ($N=2\times10^{17}$ sites/m^2, $K_D=1.0$ M). The ion-dispersion forces have in this case only a minor effect, and the interaction can be well approximated by the traditional Poisson–Boltzmann approach, with slightly modified parameters (density of sites, dissociation constant).

There is, however, a range in which the ion-dispersion forces generate qualitatively new results. The double layer between two plates immersed in an electrolyte decays exponentially with the distance and becomes vanishingly small at large separations. The interactions generated by the van der Waals interactions of the ions with the system decay slower and become dominant at those separations. In the inset of Fig.

6b, the 'double layer force' between plates becomes attractive at distances larger than approximately 40 Å, despite the relatively large surface charge density, which generates a repulsion in the vicinity of the surface.

The 'equivalent' Hamaker constant of such an attraction $\approx 12\pi c_E(B_1+B_2)$ is small, for usual electrolyte concentrations ($c_E<1$ M), when compared to the Hamaker constants of proteins ($A_H\approx 10^{-20}$ J). However, if the plates have a finite thickness l, their van der Waals interactions at large separations is given by [24]:

$$F_{vdW,\text{plates}}=-\frac{A_H}{12\pi}\left(\frac{1}{(2d)^2}+\frac{1}{(2d+2l)^2}-\frac{2}{(2d+l)^2}\right)\approx-\frac{A_H l^2}{2\pi(2d)^4}\alpha-\frac{1}{(2d)^4} \quad (30)$$

At very large separations ($d\gg l$), the attraction induced by the van der Waals interactions of ions, which decays only with the second power of the distance (Eq. (16)), becomes much larger than the traditional van der Waals attraction between plates, and its magnitude is proportional to the electrolyte concentration.

5. Conclusions

The traditional Poisson–Boltzmann treatment of the electrolyte ions, while simple and relatively accurate, is incomplete, since the interactions are treated only via a mean field. It predicts that the double layer interactions depend only on the valence of the electrolyte ions, whereas it is well documented that different electrolytes of the same valence lead to qualitatively different results.

It was recently suggested that the van der Waals interactions of the ions with the system can explain the ion specific effects [13], and it was shown that by accounting for this interactions one can successfully predict the dependence of surface tension of aqueous solutions of electrolytes on their concentration [14,15]. In the first part of this article [7], it was shown that the ion–hydration interaction can also explain the surface tension behaviour.

In this article, the effect of the interfacial interaction of ions (ion-hydration and ion-dispersion forces) on the double layer interaction was investigated. When only short-ranged interactions were taken into account (such as the ion–hydration interactions), the magnitude of the repulsion between plates at large separations became dependent on the magnitude of an 'apparent charge density', which is the sum between the real surface charge density and the charge due to the accumulation/depletion of electrolyte ions in the vicinity of the interface. The accounting of the ion–hydration interactions can increase or decrease the apparent surface charge density, hence increases or decreases the interaction, which remained, however, always repulsive at large separations. The long-range double layer interactions are strongly dependent on the short-range interfacial interaction of ions. Consequently, specific ion effects can be accounted for in this manner.

When the ion-dispersion forces are included in the Poisson–Boltzmann treatment, an attraction was obtained at most distances, in the absence of a surface charge density. The magnitude of the attraction was in general smaller than the traditional van der Waals attraction between plates. When a small (negative) charge was added on the surface, the attraction was even increased. However, when a large charge density was present on the interface, the interaction became repulsive (and strong) at small separations.

When either ion–hydration interaction or ion-dispersion forces were included in the treatment, the results were qualitatively identical to the traditional DLVO theory, which roughly predicts that 'strongly charged' colloidal particles are stable, and ' weakly charged' particles coagulate. Significant quantitative differences, which can account for specific ion effects, could be introduced by either mechanism, when suitable interaction parameters were selected.

There are some qualitative difficulties when the specific ion effects are explained via the dispersion forces of the ions. Particularly the anions, for which the dispersion coefficients B_1 are large, affect the double layer interactions. However, experiments on colloid stability [6] or colloidal forces [11] revealed strong specific ion effects especially for cations. Furthermore, the ions which affect most strongly the solvating properties of the proteins are those from their vicinity, since they perturb mostly the structure of water near the proteins. However, the van der Waals interactions of ions predict that the cations remain in the vicinity of an interface, and the anions are strongly repelled, while Hofmeister concluded that anions are mainly responsible for the salting out of proteins.

Accounting for ion hydration is more appealing qualitatively. The structure-making ions (mostly cations) are repelled by the surface, which therefore becomes more negatively charged. Consequently, the cations can generate via this mechanism long-range specific effects. Furthermore, there are the chaotropes (structure-breaking) anions, which are expelled from the bulk water toward the interface, therefore they modify more strongly the structure of water near the proteins. The structure-making cations prefer the bulk of the electrolyte, hence, they are expected to affect less the solvation properties of the proteins.

It is noteworthy that the ion–dispersion interactions induce an attraction between plates, which prevails at sufficiently large separations, regardless of their surface charge density. At separations much larger than the thickness of the plates, this attraction can even exceed the traditional van der Waals attraction between plates.

References

[1] I. Langmuir, J. Chem. Phys. 6 (1938) 893.
[2] G. Gouy, J. Phys. Radium. 9 (1910) 457.
[3] D.L. Chapman, Phyl. Mag. 25 (1913) 475.
[4] B.V. Deryagin, L. Landau, Acta. Physicochim. URSS 14 (1941) 633.
[5] E.J. Verwey, J.Th.G. Overbeek, Theory of Stability of Lyophobic Colloids, Elsevier, Amsterdam, 1948.
[6] T.W. Healy, A. Homola, R.O. James, R.J. Hunter, Farady Discuss. Chem. Soc. 65 (1978) 156.
[7] M. Manciu, E. Ruckenstein, Adv. Colloid Inter. Sci. 105 (2003) 63.
[8] F. Hofmeister, Naunin-Schmiedebergs Archiv für Experimentelle Pathologie und Pharmakologie (Leipzig) 24 (1888) 247.
[9] K.D. Collinns, M.W. Washabaugh, Q. Rev. Biophys. 18 (1985) 323.

[10] M.G. Cacace, E.M. Landau, J.J. Ramsden, Q. Rev. Biophys. 30 (1997) 241.
[11] R.M. Pashley, J. Colloid. Interf. Sci. 83 (1981) 531.
[12] N.K. Adam, The Physics and Chemistry of Surfaces, Oxford University Press, London, 1941.
[13] B.W. Ninham, V. Yaminsky, Langmuir 13 (1997) 2097.
[14] M. Boström, D.M.R. Williams, B.W. Ninham, Langmuir 17 (2001) 4475.
[15] K.A. Karraker, C.J. Radke, Adv. Colloid. Interf. Sci. 96 (2002) 231.
[16] M. Boström, D.M.R. Williams, B.W. Ninham, Phys. Rev. Lett. 87 (2001) 168103.
[17] N. Ise, T. Okubo, M. Sugimura, K. Ito, H.J. Nolte, J. Chem. Phys. 78 (1983) 1474.
[18] P. Jungwirth, D.J. Tobias, J. Phys. Chem. 105 (2001) 10468.
[19] J.Th.G. Overbeek, Colloid. Surf. 51 (1990) 61.
[20] M. Manciu, E. Ruckenstein, Langmuir 19 (2003) 1114.
[21] E.L. Makor, J.H. van der Waals, J. Colloid. Sci. 7 (1952) 535.
[22] S.G. Ash, D.H. Everett, C.J. Radke, J. Chem. Soc. Faraday Trans. 69 (1973) 1256.
[23] E. Ruckenstein, J. Colloid. Interf. Sci. 82 (1981) 490.
[24] R.J. Hunter, Foundations of Colloid Science, Oxford Science Publications, 1987.

J. Phys. Chem. B **2004,** *108,* 20479−20481 **20479**

COMMENTS

Comments on the "Osmotic Coefficients and Surface Tensions of Aqueous Electrolyte Solutions: Role of the Dispersion Forces"

Marian Manciu† and Eli Ruckenstein*

Department of Chemical and Biological Engineering, State University of New York at Buffalo, Buffalo, New York 14260

Received: May 12, 2004

In the past decade or so, a series of papers have been published suggesting that the van der Waals interactions acting on ions near an interface can explain a plethora of phenomena, such as the ion specificity (the Hofmeister series),[1] the failure of the DLVO theory in biological systems[2] or the surface tension of electrolytes.[3] More recently, the same idea was extended to the ion specificity in bulk solutions.[4]

However, the dispersion interactions (between ions and the whole system) generate repulsive forces between the water/air interface and the highly polarizable ions (Cl^-, Br^-, I^-) and attractive forces between the interface and the less polarizable ions (Na^+, Li^+, K^+).[3] Some recent experimental results[5] also challenged the traditional Langmuir picture of a surface layer depleted of ions; they revealed, however, the opposite, namely that the more polarizable anions are positively adsorbed on the interface.[5] The conclusion of Hu et al.[5] was supported by the numerical simulations of Jungwirth and Tobias,[6] which demonstrated that the polarizability of halogen anions (Cl^-, Br^-, I^-) is directly related to their propensity for the surface. The less-polarizable ions (Na^+ and F^-), however, preferred the bulk water.[6] These results are opposite to the predictions of the ion dispersion theories.[3]

The failure of the ion dispersion forces to predict even the qualitative behavior of ions near an interface is probably due to the existence of other short-ranged forces that dominate the interactions in the vicinity of the interface. We suggested that such a role might be played by the interactions of the ions with their neighboring water molecules.[7] The approach to the water/air interface is unfavorable for the structure-making ions, because they have to lose a part of their hydration shell; however, it is favorable for the structure-breaking ions because the bulk water can organize better in their absence.[7] A simple model based on short range ion-hydration forces explained fairly accurately the experiments on surface tension (namely, its increase for salt solutions but decrease for acidic solutions with increasing electrolyte concentration, its pH dependence, and the Jones-Ray effect).[7] Kunz et al. reached a similar conclusion,[4] namely, that the difficulties of the previous treatments based on ion dispersion forces were due to the fact that the incomplete hydration was not taken into account.[4]

The focus of the present Comment is on ion−ion dispersion forces, which in the opinion of Kunz et al.[4] are important in bulk electrolyte solutions. One of the questions raised by them is why these forces have been neglected in previous theories. They answered by suggesting that the general belief is that the Coulomb interactions dominate the van der Waals ones, and hence the latter can be neglected. Kunz et al.[4] also noted that the van der Waals interactions might be negligible at low ionic strength ($c_E < 0.02$ M), but not at high ionic strength ($c_E \sim 1$ M), for which the Coulomb interactions are screened. As a simple calculation shows, at large separations the van der Waals interactions, though larger than the screened Coulomb interactions, are much smaller than kT, and hence negligible.

To understand why the ion−ion dispersion interactions have been ignored by most theories of electrolytes, let us perform some simple calculations. The London theory provides for the ion−ion dispersion interactions:[8,9]

$$U_L(r) \cong -\frac{3}{4}\frac{h}{2\pi}\omega_0\left(\frac{r_0}{r}\right)^6 \equiv -\frac{C_{AB}}{r^6} \qquad (1)$$

where r is the distance between the centers of two ions, h is the Planck constant, and C_{AB} is a constant dependent on ω_0 and r_0, which are given by

$$\omega_0 = \frac{4}{\pi\alpha_A^0\alpha_B^0}\int_0^\infty \alpha_A(i\xi)\,\alpha_B(i\xi)\,d\xi \qquad (2a)$$

and

$$r_0 = \left(\frac{\alpha_A^0\alpha_B^0}{(4\pi\epsilon_0^{\,2})}\right)^{\frac{1}{6}} \qquad (2b)$$

where α_A is the polarizability of atom "A", α_A^0 is its zero frequency polarizability, and ϵ_0 is the vacuum permittivity. Typical values for r_0 and ω_0 are 2 Å and 10^{16} rad/s, respectively,[9] which provide a typical value of about $C_{AB} \approx 12\,000\ kT$ Å^6.

The screened Coulomb force is given by

$$U_C(r) \cong \frac{1}{4\pi\epsilon\epsilon_0}\frac{e^2}{r}\exp\left(-\frac{r}{\lambda}\right) \qquad (3)$$

where e is the protonic charge, ϵ is the dielectric constant in water, and λ is the Debye length. The van der Waals interactions (eq 1) dominate the screened Coulomb interactions (eq 3) at both very small and very large separations. In the latter case, for a concentration of 1 M of a uni-univalent electrolyte, $\lambda \cong 3$ Å, and by employing $\epsilon = 80$, one finds that the ion−ion dispersion forces dominate the interactions only for distances larger than about 30 Å (see the inset in Figure 1). However, both interactions are negligible at those separations, being of the order of 10^{-5} kT. Because the typical value provided in ref 9, $C_{AB} \approx 12\,000\ kT$ Å^6, is rather large, it is likely that most ion−ion dispersion forces are in general negligible at large separations.

The dispersion forces play an important role only for separations less than about 5 Å. The problem is that in this region, they seem to play an enormous role (curve 1 in Figure 1). A hard-sphere repulsion (Born interaction) is typically

* Corresponding author. E-mail address: feaeliru@acsu.buffalo.edu. Phone: (716) 645 2911/2214. Fax: (716) 645 3822.
† E-mail address: manciu@acsu.buffalo.edu.

10.1021/jp047959j CCC: $27.50
Published on Web 12/02/2004

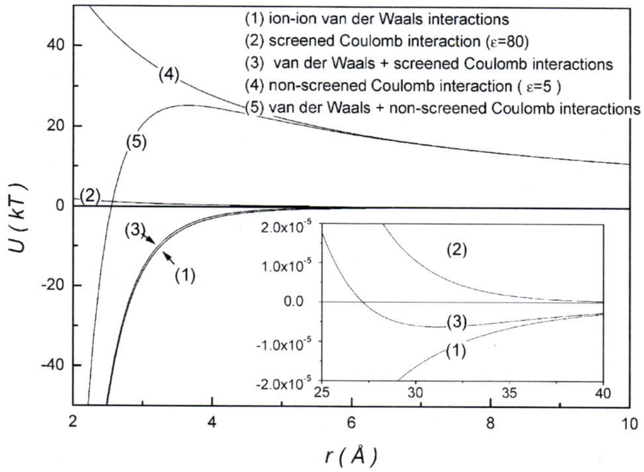

Figure 1. Ion–ion dispersion forces (curve 1) dominate the screened Coulomb interactions (curve 2, $\epsilon = 80$, $\lambda = 3$ Å) for separations smaller than about 7 Å and larger than about 30 Å (see the inset). However, the nonscreened Coulomb interactions (curve 4, $\epsilon = 5$, $\lambda = \infty$) are much larger than the ion–ion dispersion forces (1) for distances larger than about *3.5* Å.

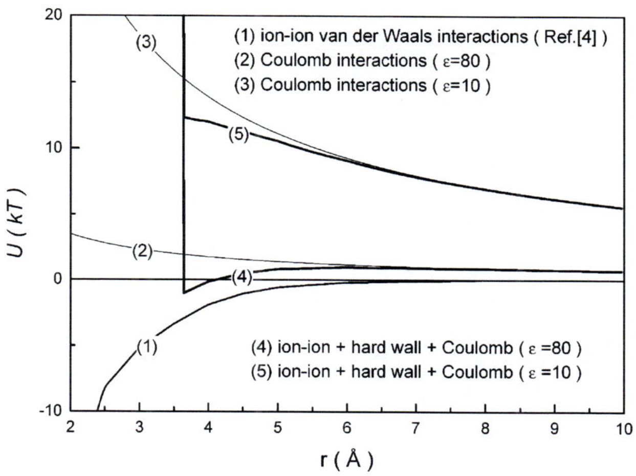

Figure 2. Ion–ion dispersion interactions (curve 1) calculated using the Lifshitz approach in ref 4 dominate at small separations the water-screened Coulomb interactions (curve 2, $\epsilon = 80$, $\lambda = \infty$); however, they are negligible compared to the nonscreened Coulomb interactions ($\epsilon = 10$, $\lambda = \infty$) (curve 3), in the presence of a hard wall repulsion at $r = 3.6$ Å.

considered to be responsible for preventing the ion–ion collapse due to their (divergently large) van der Waals attractions. If the hard sphere repulsion would start at $r_{HS} = 2$ Å, the depth of the energy minimum would be about 200 *kT*, and all the ions would cluster, regardless of their charge. On the other hand, if the hard sphere repulsion would start at $r_{HS} = 5$ Å, the depth of the energy minimum would be less that 1 *kT* and the effect of ion–ion dispersion would be much less noticeable. The effect of the ion–ion interactions (from huge to negligible) is strongly dependent on the position selected for the hard sphere repulsion. Kirkwood considered long ago[10] that a square well with a depth of the order of about 1 *kT* and with the width of about $(1/4)r_0$ can represent the van der Waals interactions between molecules. This choice probably influenced the conclusion that the ion–ion dispersion is not important in electrolytes, drawn by Kirkwood in his theory of strong electrolytes.[11] Later, Kirkwood and Poirier indicated a manner to include the short-ranged interactions (hard core repulsions and van der Waals attractions) and the long-ranged interactions (Coulomb force) into a unitary theory of strong electrolytes.[12] They ignored, however, the van der Waals interactions at large separations, probably for the reasons outlined above.

One might argue that the London theory for van der Waals interactions is inaccurate and the dispersion forces should be calculated using the Lifshitz approach (a macroscopic continuum theory) extrapolated to the limit of atomic distances.[1,4] Though such a theory would not predict a divergence at $r \rightarrow 0$,[4] the attractive minimum suggested by Figure 2 of ref 4 would be probably much larger than 20 *kT*. Even for the ion–ion dispersions potential plotted in Figure 2 of ref 4, (reproduced here for convenience in Figure 2, curve 1), the location of the hard-sphere potential strongly affects the result. Na^+ has a relatively low polarizability, and because $r_{HS} = 4.10$ Å, it seems that the ion–ion dispersions are not important for the Na^+–Na^+ interactions.[4] For NO_3^-–NO_3^-, $r_{HS} = 3.6$ Å, and the depth of about 2 *kT* of the attractive minimum generates a significant clustering of these ions.[4] In fact, the probability of the formation of NO_3^-–NO_3^- pairs is even larger that the probability of the formation of Na^+–NO_3^- pairs (see Figure 3 of ref 4), a surprising result. Whether the hard sphere repulsion would start at $r_{HS} \cong 2$ Å instead of $r_{HS} \cong 3.6$ Å (the r_{HS} were adjusted in ref 4 to fit the osmotic coefficients data), an attractive minimum of over 10 *kT* would predict a considerable clustering of ions of the same kind. This effect (if real) would be certainly impressive.

There is, however, a more important issue. At separations of a few Ångstroms, the Coulomb interactions are hardly screened either by other ions or by the water molecules. The large dielectric constant of bulk water is due to the ability of the water molecules to reorient their permanent dipole in an electric field. However, the molecules near ions are already oriented by the strong electric fields generated by ions; therefore the local dielectric constant of water (in the vicinity of the ions) is decreased. Using a lower value $\epsilon = 5$ for the dielectric constant in the vicinity of ions, one finds only repulsive interactions between ions of the same type (curve 5 in Figure 1) for distances *r* larger than about 3.5 Å. Though at shorter distances divergently large attractions can be found, even in the absence of a hard-wall repulsion it is unlikely for the ions to be able to escalate a repulsive potential barrier larger than 20 *kT* to collapse on the attractive minimum (for $\epsilon = 1$, the repulsive barrier would exceed 150 *kT*). For the ion–ion dispersion forces calculated via the Lifshitz approach,[4] employing a dielectric constant $\epsilon \cong 10$ for separations distances of a few Ångstroms (curve 5 in Figure 2) instead of the bulk value $\epsilon = 78.2$ in the Coulomb interactions (curve 2 in Figure 2) would completely remove the attractive well induced by the ion–ion dispersion forces, the latter being much less important than the change in the local dielectric constant.

In conclusion, we suggest that the ion dispersion forces were ignored by most (but by no means all) electrolyte theories mainly because they are important only for separations between ions smaller than about *5* Å, and the interactions at these distances are not well-known. It is hard to believe that at these distances the interactions can be accurately described by a sum between a hard-wall repulsion, a Coulomb interaction, and a London attraction. Even if the latter would be true, a correction in the local dielectric constant (because of incomplete screening by water molecules) would render again the van der Waals interactions negligible, up to distances of the order of ion diameters.

It is true that, by adjusting the parameters of such interactions, a plethora of interesting phenomena can be predicted, such as the collapse of all the ions of the same kind of an electrolyte. We completely agree with Kunz et al.[4] that the short-range

Comments *J. Phys. Chem. B, Vol. 108, No. 52, 2004* **20481**

interactions between ions are important for the behavior of the system; this was also emphasized in the work of Jungwirth and Tobias.[6] In fact, the latter work showed that the local interactions between ions and neighboring water molecules are so important, that they control the behavior of ions near interfaces. Whereas the highly polarizable ions are drawn toward the interface in their discrete, microscopic treatment,[6,7] a macroscopic, continuum van der Waals approach would predict exactly the opposite.[3]

Whether or not two ions of the same kind have an attractive minimum, possibly of many *kT*, when they are located at a few Ångstroms apart in water it is quite an interesting issue. But this issue has probably to be resolved by ab initio microscopic theories that include all the interactions (Coulomb, dispersion, hydrogen bonding, etc.) of a representative system. Today's computational capability (for up to about 20 water molecules) might be sufficient for such ab initio calculations.

References and Notes

(1) Ninham, B. W.; Yaminsky, V. *Langmuir* **1997**, *13*, 2097.
(2) Boström, M.; Williams, D. M. R.; Ninham, B. W. *Phys. Rev. Lett.* **2001**, *87*, 168103.
(3) Boström, M.; Williams, D. M. R.; Ninham, B. W. *Langmuir* **2001**, *17*, 4475.
(4) Kunz, W.; Belloni, L.; Bernard, O.; Ninham, B. W. *J. Phys. Chem. B* **2004**, *108*, 2398.
(5) Hu, J. H.; Shi, Q.; Davidovits, P.; Worsnop, D. R.; Zahniser, M. S.; Kolb, C. E. *J. Phys. Chem.* **1995**, *99*, 8768.
(6) Jungwirth, P.; Tobias, D. J. *J. Phys. Chem.* **2001**, *105*, 10468.
(7) Manciu, M.; Ruckenstein, E. *Adv. Colloid Interface Sci.* **2003**, *105*, 63.
(8) London, F. *Z. Phys.* **1930**, *63*, 245.
(9) Hunter, R. J. *Foundations of Colloid Science*; Clarendon Press: Oxford, U.K., 1987.
(10) Kirkwood, J. G. *J. Chem. Phys.* **1935**, *3*, 300.
(11) Kirkwood, J. G. *J. Chem. Phys.* **1934**, *2*, 767.
(12) Kirkwood, J. G.; Poirier, J. C. *J. Phys. Chem.* **1954**, *58*, 591.

11312 *Langmuir* **2005,** *21,* 11312–11319

On the Interactions of Ions with the Air/Water Interface

Marian Manciu*

Department of Physics, University of Texas at El Paso, El Paso, Texas 79968

Eli Ruckenstein*

Department of Chemical and Biological Engineering, State University of New York at Buffalo, Buffalo, New York 14260

Received July 21, 2005. In Final Form: September 15, 2005

In the vicinity of a charged interface, the Poisson–Boltzmann approach considers that the ions obey Boltzmann distributions in a mean electrical field that satisfies the Poisson equation. However, the boundary between two dielectrics generates additional interactions between ions and the interface. The traditional models of ion hydration interactions, that assume that water is a homogeneous dielectric, predict that these interactions are repulsive for all kinds of ions, since all ions should prefer the medium with a larger dielectric constant, where they are better hydrated. In reality, the interactions between the ions and the neighboring water molecules can generate additional short-range ion-hydration interactions, which are either repulsive (for structure-making ions) or attractive (for structure-breaking ions). In the present paper, various models for the ion-hydration forces are examined and compared with the results of molecular dynamics simulations. At large ionic strengths, the latter results could be reproduced qualitatively only when short-ranged attractions between the structure-breaking ions and the interface were taken into account.

1. Introduction

It has long been known that the addition of a salt increases the surface tension of the air/water interface, an experimental result that indicates, via the Gibbs adsorption equation, that the ions are repelled by the air/water interface. The latter result can be intuitively understood in terms of simple electrostatics; the ions, being charged, prefer the medium with larger dielectric constant, in which their electrostatic energy is lower. When a charge approaches the interface between two dielectrics, the field generated by the charge induces a net charging of the interface. Because the interaction between the charge and the induced surface charge is mathematically equivalent to that between the charge and an imaginary charge placed on the other side of the interface (the image charge), the corresponding electrostatic force is commonly named the "image force".

Because of the long-range nature of Coulomb forces, the repelling of ions from the air/water interface due to the "image forces", calculated as free of ion screening, is so large that the Gibbs adsorption equation predicts a divergently large surface tension. This difficulty has been solved by Wagner, who noted that the image force acting on an ion is screened by the cloud formed by the other ions.[1] By assuming that the Debye–Hückel screening length λ_{DH} of the electrolyte solution does not depend on the distance from the interface, Onsager and Samaras obtained an analytical expression for the surface tension of dilute electrolyte solutions.[2]

At high salt concentrations, the surface tension of aqueous electrolyte solutions increases almost linearly with the salt concentration,[3] and the Onsager-Samaras law, which is valid only for low electrolyte concentrations, is not applicable. The qualitative behavior of the surface tension at high ionic strengths can be, however, understood by using a simple model suggested by Langmuir, who noted that it could be explained by the complete depletion of ions from the first water layer of the interface, of a thickness of about 4 Å.[4] Langmuir suggested that the specific ion effects (different slopes of the surface tension of salt aqueous solutions as functions of ionic strength) might be explained by different dissociation constants of the salts.[4] By selecting suitable "depletion lengths" for all kinds of salts, the behavior of surface tension at large electrolyte concentrations could be explained even quantitatively.[5]

Since then, the depletion of ions from the air/water interface became the commonly accepted representation of the surface tension of electrolyte solutions. However, this common picture is not always true; near a charged interface, there is an accumulation of counterions, predicted by the Poisson–Boltzmann treatment, which at very low electrolyte concentrations (less than 10^{-3} M) can dominate the interfacial depletion of ions due to ion hydration forces,[6] and consequently, the surface tension of aqueous electrolytes can decrease with increasing ionic strength (The Jones-Ray effect).[7] Even more significant, the surface tension of acids decreases with increasing

* Corresponding authors. Phone: (915) 747 7531. Fax: (915) 747 5447. E-mail: mmanciu@utep.edu. (Marian Manciu); Phone: (716)6452911/2214. Fax: (716)6453822. E-mail: feaeliru@acsu.buffalo.edu (Eli Ruckenstein).

(1) Wagner, C. *Phys. Zeit.* **1924**, *25*, 474.
(2) Onsager, L.; Samaras, N. N. T. *J. Chem. Phys.* **1934**, *2*, 628.
(3) Weissenborn, P. K.; Pugh, R. J. *J. Colloid Interface Sci.* **1996**, *184*, 550.
(4) Langmuir, I *J. Am. Chem. Soc.* **1917**, *39*, 1848.
(5) Aveyard, R.; Saleem, S. N. *J. Chem Soc. Faraday I* **1976**, *72*, 1609.
(6) Manciu, M.; Ruckenstein, E. *Adv. Colloid Interface Sci.* **2003**, *105*, 63.
(7) Jones, G.; Ray, G. W. A. *J. Am. Chem. Soc.* **1937**, *59*, 187.

10.1021/la051979a CCC: $30.25
Published on Web 10/27/2005

electrolyte concentration,[3] which clearly points to an overall accumulation of ions at the water/air interface.

The possible accumulation of negative ions at the air/water interface was first predicted by Perera and Berkowitz,[8] who found out from molecular dynamics simulations, surprisingly, that the large anions (Cl^-, Br^-, and I^-) are expelled from water clusters to their interface. Their predictions are supported by the recent large-scale molecular dynamics simulations for the air/water interface of various electrolyte solutions, which reveal that, when the polarization of ions and water molecules is explicitly taken into account, the large anions are accumulated near the interface.[9]

The traditional models, such as the Wagner-Onsager-Samaras "screened image force" or the Langmuir "interfacial depletion length" imply that the ions prefer the medium with a higher dielectric constant, where they are better hydrated (surrounded by water molecules that screen their electric field). It is clear that these models cannot explain the accumulation of ions at the air interface of aqueous acid solutions, and the main question is what kind of ion interactions might be responsible for this effect. It was suggested that the ion-dispersion forces might explain the interfacial distribution of ions.[10] However, they predict that the cations are attracted by the interface and that the anions are repelled by the interface,[6] in disagreement with experiment[11] and simulations.[8,9]

Recently, two new models have been proposed to explain the interactions between ions and the air–water interface, based on changes of the hydration energy of ions with the distance from the interfaces.[6,12] In the first one,[6] the discrete nature of water molecule has been taken explicitly into account; when a structure-making ion approaches the interface, it loses some of its hydrating water molecules. Because the energetic penalty is much larger than kT, the interfacial region is practically forbidden to the structure-making ions, and the "Langmuir depletion length" approach can describe well the ion distribution. Because each ion has its particular hydrated structure, there is a particular "Langmuir depletion length" that can be associated with each kind of ion. In contrast, the structure-breaking ions are expelled from the bulk water toward the interface, and this implies an attraction between them and the interface, an interaction that can be roughly approximated by a square well described by two parameters. The other model[12] suggests that, because there is a lower density of water in the interfacial region than in bulk water, the dielectric constant is lower than in the bulk, and this leads to repulsions between all kinds of ions and the air/water interface.

The purpose of this article is to examine the effect of the ion-hydration forces, suggested by these models, on the distribution of ions near an interface, and to compare their predictions with those obtained from molecular dynamics simulations.[9]

2. Modified Poisson–Boltzmann Approach

Under the assumption that the ions interact with a "mean field" ψ generated by all of the other charges of the system, the Boltzmann distribution of ions in the vicinity of an interface is provided by the expression

$$c_i(x) = c_E \exp\left(-\frac{q_i\psi(x) + \Delta W_i(x)}{kT}\right) \tag{1a}$$

where c_i is the concentration of ions of kind "i", x is the distance from the interface (assumed planar), c_E is the electrolyte concentration in a reservoir in contact with the system, q_i is the charge of ions of kind "i", k is the Boltzmann constant, T is the absolute temperature, and $\Delta W_i(x)$ is an additional free energy of interaction between the ion and surface, not included in the mean field ψ (note that $\Delta W_i(x) = W_i(x) - W_i(\infty)$ is calculated with respect to a reservoir far away from the surface[6]). The mean field "ψ" is assumed to obey the Poisson equation

$$\Delta\psi(x) = -\frac{\sum_i q_i c_i(x)}{\epsilon\epsilon_0} \tag{1b}$$

where ϵ is the dielectric constant and ϵ_0 the vacuum permittivity.

Once the additional interactions are known, the distribution of the concentration of ions near the interface can be obtained from eqs 1a and 1b (for the appropriate boundary conditions), and consequently, the surface tension of electrolytes can be easily calculated, via the Gibbs adsorption equation.

However, the additional interactions ΔW_i are not known, and there is no consensus about their microscopic origin. One natural candidate for this "additional interaction" is the screened image force (Wagner-Onsager-Samaras)[2]

$$\Delta W_{OS,i}(x) = \left(\frac{\epsilon - \epsilon'}{\epsilon + \epsilon'}\right)\frac{\exp\left(\frac{a_i}{\lambda_{DH}}\right)}{1 + \frac{a_i}{\lambda_{DH}}}\frac{q_i^2}{16\pi\epsilon_0\epsilon x}\exp\left(-\frac{2x}{\lambda_{DH}}\right) \tag{2}$$

where a_i is the radius of ion "i", ϵ is the bulk dielectric constant of water, $\epsilon' = 1$ is the dielectric constant of air and

$$\lambda_{DH} = \sqrt{\frac{\epsilon\epsilon_0 kT}{\sum_i c_E q_i^2}}$$

is the Debye–Hückel screening length. Although the screened image forces are important at low electrolyte concentrations, simple calculations show that they become too weak at large ionic strengths and therefore cannot explain the surface tension of electrolytes at those concentrations.[6]

A simple treatment, which circumvents this difficulty, uses an alternative approximate treatment of the image forces. Because the Born energy of ions in bulk water is given by

$$W_{Born,i}(x) = -\frac{q_i^2}{8\pi\epsilon_0 a_i}\frac{1}{\epsilon} \tag{3}$$

one can suggest a distance-dependent interaction, based

(8) Perera, L.; Berkowitz, M. L. *J. Chem. Phys.* **1993**, *99*, 4222.
(9) Mucha, M.; Frigato, T.; Levering, L. M.; Allen, H. C.; Tobias, D. J.; Dang, L. X.; Jungwirth, P. *J. Phys. Chem. B* **2005**, *109*, 7617.
(10) Ninham, B. W.; Yaminsky, V. *Langmuir* **1997**, *13*, 2097.
(11) Knipping, E. M.; Lakin M. J.; Foster K. L.; Jungwirth, P.; Tobias, D. J.; Gerber, R. B.; Dabdub, D.; Finlayson-Pitts, B. J. *Science* **2000**, *288*, 301.
(12) Bostrom, M.; Kunz, W.; Ninham, B. W. *Langmuir* **2005**, *21*, 2619.

on a local approximation of the Born energy[13]

$$\Delta W_{\mathrm{Born},i}(x) = \frac{q_i^2}{8\pi\epsilon_0 a_i}\left(\frac{1}{\epsilon(x)} - \frac{1}{\epsilon}\right) \qquad (4)$$

where $\epsilon(x)$ is the variable dielectric constant near the interface.

In the vicinity of the air/water interface, the dielectric constant is expected to be lower that in the bulk, because the rotation of the water molecules is there somewhat obstructed. As a result, eq 4 indicates a repulsion between ions and the interface. This approximate treatment of electrostatic interactions has the advantages that it does not lead to a diverging surface tension, and that the dependence of the dielectric constant on distance can be selected "a posteriori", such as to match the results provided by the experiment. However, the local approximation of the Born energy has two shortcomings. First, it predicts that all ions are repelled by the interface, which is not true. Second, eq 4 is not an accurate approximation: the correct Born energy has to be calculated by integrating the energy of the electric field over the entire space, including both dielectrics (note that eq 3 implies only one infinite dielectric). Consequently, the derivative of the Born energy with respect to the distance between the charge and the interface should provide rigorously the "image force". It is clear that the derivative of the approximate expression, eq 4, does not lead to the image force. However, it should be noted that both the "screened image force" and the "local approximation of the Born energy" are approximations of the same physical process, namely the change in the electrostatic energy of the system when a charge approaches a boundary between two continuous dielectrics. Therefore, they do not represent different contributions to the free energy.

Both the "image force" and the "local approximation for the Born energy" approaches account for the ion hydration, through the macroscopic dielectric constant of water, assumed a continuous dielectric. Because none of them could explain satisfactorily the distribution of ions near a water–air interface, it was suggested that another effect plays the main role, namely the van der Waals interactions between ions and the rest of the system. These ion-dispersion forces can be described by an equation of the type[10]

$$\Delta W_{\mathrm{vdW},i} = \frac{B_i}{x^3} \qquad (5)$$

The theory based on ion-dispersion forces predicts, however, that the cations are attracted toward the interface and that the anions are repelled toward the bulk,[6] in contrast with the experimental[11] and computational[8,9] findings, which indicate the opposite. It was therefore suggested[6] that another short-range ion-hydration force, which involves changes in the hydration energy when an ion approaches the interface, due to interactions between the ion and the neighboring water molecules, should be employed to calculate the distribution of ions in the vicinity of the interfaces. In what follows, the predictions of the two models for the short-range ion-hydration force of refs 6 and 12 will be compared with the results of molecular dynamics simulations provided in ref 9 for various electrolyte solutions (NaCl, NaOH, HCl and HBr).

(13) Spaarnay, M. J. *Recl. Trav. Chim. Pays-Bas* **1958**, *77*, 872.

3. Distribution of Simple Ions near the Water/Air Interface

3.1. Bostrom–Kunz–Ninham (BKN) Model for Ion-Solvation Forces. A new model for the short-ranged ion-solvation forces has been proposed recently.[12] It is based on the observation that the air/water interface is not sharp, the water density increasing gradually, over a distance of a few angstroms, from zero (in air) to the density of bulk water. The water density profile was obtained by fitting the results of the molecular dynamics simulations with the empirical expression[12]

$$p(x) = 1 - 1.0302\exp(3.5 - x)/(1 + \exp(3.5 - x)) \qquad (6a)$$

where $p(x) = \rho(x)/\rho_0$, $\rho(x)$ being the density of water, ρ_0 the density of bulk water, and x is measured in angstroms.

In the region where the water density is lower than that of the bulk, the dielectric constant is expected to be smaller than in the bulk, and a Spaarnay-like local approximation of the Born energy of ions is given by[12]

$$\Delta W_{\mathrm{solvation}}(x) = \Delta G_{\mathrm{solvation}} \frac{\dfrac{1}{1 + p(x)(\epsilon - 1)} - \dfrac{1}{\epsilon}}{1 - \dfrac{1}{\epsilon}} \qquad (7)$$

where $\Delta G_{\mathrm{solvation}}$ is the free energy of solvation in the bulk. The total additional interaction was considered to be the sum of the potentials of a screened image force (eq 2), an ion-dispersion force (eq 5), and an ion-solvation force (eq 7)[12]

$$\Delta W_i(x) = \Delta W_{\mathrm{OS},i} + \Delta W_{\mathrm{vdW},i} + \Delta W_{\mathrm{solvation},i} \qquad (8)$$

In Figure 1a, the interaction potentials at $c_E = 1$ M are plotted for Cl^-, using in calculations the values $\Delta G_{\mathrm{solvation,Cl}} = 361.79$ kJ/mol and $B_{Cl} = 13.6 \times 10^{-50}$ Jm^3, provided in ref 12. The ion-solvation contribution clearly dominates the interactions in the vicinity of the interface, for $x > 1$ Å. At lower distances ($x < 1$ Å), the total additional energy is larger than 20 kT, and practically no ions are allowed in that region; therefore, the fact that neither eqs 2 nor 5 are accurate for $x < 1$ Å is not relevant. It should also be noted that both the screened-image and ion-dispersion forces are longer-ranged than the ion-solvation force, and they dominate at large distances (see the inset of Figure 1a). However, at large distances and large ionic strengths, they are small compared to kT and hence negligible. In contrast, at low ionic strengths, the screened image force provides an important contribution at large distances, even when the short-range ion-solvation forces are strong: see the inset of Figure 1b, in which the additional interactions are calculated for Na^+ at $c_E = 0.001$ M, using $\Delta G_{\mathrm{solvation,Na}} = 407.65$ kJ / mol and $B_{Na} = 1.4 \times 10^{-50}$ Jm^3.[12] From Figure 1b it is clear than the ion-dispersion forces are in this case negligible in the vicinity of the air/water interface. At large distances from the interface, they dominate the interactions (because of their $1/x^3$ dependence), but are too small compared to kT to affect the ion distribution.

In summary, the recent BKN model[12] supports the previous predictions[6] that the long-ranged screened image forces govern the interfacial distribution of ions at low ionic strengths, that the short-range ion-hydration forces govern their distribution at high ionic strengths, and that the ion-dispersion forces provides only minor corrections to the ion distributions.

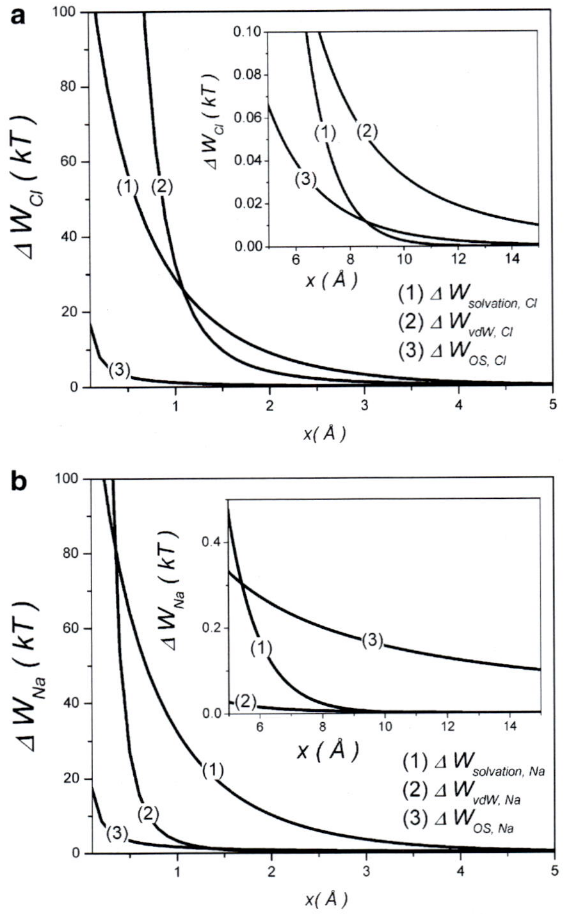

Figure 1. Contributions of the screened image, ion dispersion and ion-solvation to the additional energy $\Delta W_i(x)$ predicted by BKN model for (a) Cl^-, at $c_E = 1$ M and (b) Na^+ at $c_E = 0.001$ M. In the inset are represented the long-range behavior of the energetic contributions.

Because the solvation energy of any kind of ion in the BKN model is extremely large near the air/water interface, the interface is practically depleted of any kind of ions. Let us assume, for the sake of simplicity, that the mean field electrical potential is negligible (because of the large ionic strength considered); therefore, the distribution of ions in the vicinity of the air/water interface is well approximated by

$$c_i(x) \cong c_E \exp\left(-\frac{\Delta W_{\text{solvation},i}(x) + \Delta W_{\text{OS},i} + \Delta W_{\text{vdW},i}(x)}{kT} \right) \quad (9)$$

The distributions of the ions Na^+, Cl^-, and I^-, for which both the solvation energies and the ion-dispersion constants B_i have been provided in ref 12 ($\Delta G_{\text{solvation},i} = 407.65$, 361.79, and 209.49 kJ/mol; $B_i = 1.4$, 13.6, and 15.1×10^{-50} J m^3, respectively), are plotted in Figure 2, together with the density profile of the air/water interface. The ion distributions are also compared in Figure 2 with those predicted by a simple model, in which "Langmuir depletion lengths" (e.g., $\Delta W_i = \infty$ for $x < d_{\text{cut-off}}$ and $\Delta W_i = 0$, $x > d_{\text{cut-off}}$), with $d_{\text{cut-off}} = 1$ and 2 Å, have been arbitrarily selected, and all of the other interactions have been ignored. The calculations for the latter distributions take into account the interfacial profile of the water using a statistical distribution of sharp interfaces, a procedure which is detailed in the next section.

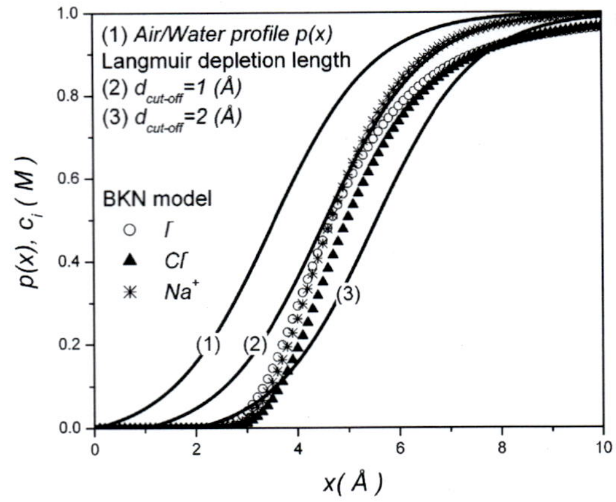

Figure 2. Distribution of ions predicted by BKN ion-hydration model, for $\psi(x) \cong 0$. The first curve is the water profile obtained from fitting MD simulations. The circles, triangles and stars represent the distribution of I^-, Cl^-, and Na^+, respectively. Note that the interfacial region is depleted of all kind of ions, and therefore, the distribution of ions can be well approximated by a simple model using suitable chosen Langmuir depletion lengths (the distributions corresponding to $d_{\text{cut-off}} = 1$ Å and $d_{\text{cut-off}} = 2$ Å are represented in the figure).

Although providing comparable results with the simple model based on "Langmuir depletion lengths", the main advantage of the BKN model is that it allows the direct calculation of the ion distributions without any fitting parameters, once the profile of the interface eq 6, the van der Waals parameters B_i in eq 5, and the solvation energies in eq 7 are known. There are, however, some difficulties associated with the model. The most important is that it predicts that all of the ions are repelled from the interface, in contrast with the experimental data[11] and simulations,[8,9] which indicate that some anions (e.g., Cl^-, Br^-, I^-, but not F^-) are attracted by the interface.

The second difficulty is that it predicts very small changes in the interfacial forces for different ions and, therefore, can hardly account for ion specific effects (Figure 2 shows only a minute difference between the distributions of Na^+ and I^- ions). In Figure 3a, the predictions of the BKN model for $c_E \cong 1.2$ M are compared with the distributions of Na^+ and Cl^- ions near the air/water interface, obtained via molecular dynamics simulations in ref 9; note that in the molecular dynamics simulations the water molecules occupy the region from $x = 0$ to the water/air interface, located at $x_0 = 13.9$ Å; therefore, the expression

$$p(x) = 1 - 1.0302 \exp(x - 13.9)/(1 + \exp(x - 13.9)) \quad (6b)$$

has been employed instead of eq 6a.

The distributions of Na^+ and Cl^- predicted by the BKN model are very close to each other, both kinds of ions being repelled from the air/water interface. Furthermore, the Na^+ cations more closely approach the interface than the Cl^- anions, in qualitative disagreement with the simulations.

Whereas the average interfacial depletion of ions might be close to the value required to explain the behavior of the surface tension of NaCl solution, the distributions of ions does not reproduce even qualitatively the Molecular Dynamics simulations. The situation is particularly dramatic for acids: the hydrogen ion has a Gibbs free

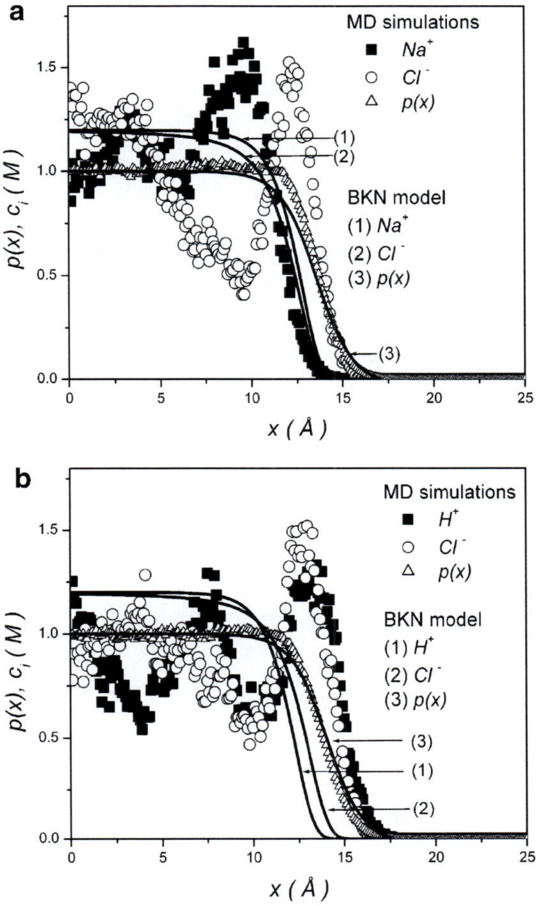

Figure 3. (a) Distributions of Na^+ and Cl^- ions near the air/water interface of a NaCl solution, predicted by the BKN model (note that the Na^+ ions approaches the interface closer than the Cl^- ions), are compared with the molecular dynamics simulations of ref 9. (b) The distributions of H^+ and Cl^- ions near the air/water interface of a HCl solution, predicted by the BKN model (note that the interface is depleted of both H^+ and Cl^- ions), are compared with the molecular dynamics simulations of ref 9, that show an accumulation of both H^+ and Cl^- ions at the interface.

energy of hydration of about $\Delta G_{\text{solvation,H}} \cong 1100$ kJ/mol[14] and consequently should be repelled from the interface by ion-solvation forces much stronger than the simple cations (Na^+, K^+, and Li^+). Assuming a negligible ion-dispersion interaction for H^+, the ions distributions in the vicinity of the air/water interface of the HCl solution at $c_E = 1.2$ M, predicted by the BKN model, are compared to molecular dynamics simulations[9] in Figure 3b. In this case, the BKN model leads to an average interfacial depletion of ions larger than that predicted for the NaCl solution (Figure 3a), which implies that the surface tension of the HCl solution should increase with ionic strength stronger than the surface tension of the NaCl solution. This is not true; in reality, the surface tension of acid solutions is not even increasing but decreases with the ionic strength.[3] The reason for this behavior can be clearly seen from the results of the molecular dynamics simulations[9] reproduced in Figure 3b: both H^+ and Cl^- prefer the air/water interface, where they accumulate.

3.2. Structure Making/Structure Breaking (SM/SB) Model for the Short-Range Ion-Hydration Forces. A simple model has been suggested recently for the ion-hydration forces,[6] based on the old common intuition that some ions are able to structure the water (the structure making ions), and for this reason they prefer the bulk, while other ions generate disorder in water (the structure-breaking ions), and for this reason, they are expelled toward the interface. When a structure making ion approaches an interface, it loses some of its hydration molecules, a process which is thermodynamically unfavorable. The hydration energies are of the order of 10^2 kT, and even a change of a few percents of this energy affects drastically the ion distribution; an increase in the total free energy by 5 kT implies (via the Boltzmann distribution) a decrease in concentration by more than 2 orders of magnitude. For this reason, one can assume that this region is practically depleted of structure-making ions, in agreement with the Langmuir suggestion.[4]

The main difference between the traditional ion-hydration models and the SM/SB model is that the previous ones assumed that water is a continuous dielectric, characterized by a macroscopic dielectric constant, whereas the latter takes into account that, microscopically, water is constituted of discrete molecules. The change in hydration occurs when the hydrated ion loses some of its hydrating water molecules, a process that starts from a certain distance $d_{\text{cut-off}}$ from the interface. Therefore, for $x > d_{\text{cut-off}}$, there are no "ion-hydration" interactions, whereas the region with $x < d_{\text{cut-off}}$ can be considered completely depleted of ions. Furthermore, because different structure-making ions have different cutoff thicknesses in which they lose hydrating water molecules, $d_{\text{cut-off}}$ is a parameter that describes the specificity of the structure-making ions (in contrast to the suggestion of Aveyard and Saleem, who suggested that a $d_{\text{cut-off}}$ should be associated with the specificity of salts, not of the corresponding ions).[5]

For structure-breaking ions, it was suggested that their attraction toward the interface is governed by a simple surface potential well, described by two adjustable parameters (the width and the depth of the surface potential well).[6] When this simple model was employed for the air/water interface, it could explain the dependence of the surface potential on the electrolyte concentration and on pH, the behavior of surface tension of salt and acids at high ionic strengths and the Jones-Ray effect (the existence of a minimum in the surface tension of salts at a small electrolyte concentration).[7]

To compare the results with the existing simulations,[9] one should note that the simulations were performed at a fixed number of ions, and not for a system in contact with a reservoir. Therefore, instead of eq 1 one should use

$$c_i(x) = c_{0,i} \exp\left(- \frac{q_i \psi(x) + \Delta W_i(x)}{kT} \right) \tag{10a}$$

with the constant $c_{0,i}$ provided by the normalization condition

$$\iint_A \mathrm{d}y\mathrm{d}z \int_0^{x_{\max}} c_i(x)\, \mathrm{d}x = \frac{N}{N_A} \tag{10b}$$

where the integral in the interface plane yz is on an area 30×30 Å^2, the integral along the distance x, perpendicular to the interface, is up to $x_{\max} = 25$ Å, $N = 18$ represents the number of ions (either cations or anions) from the box and N_A is the Avogadro number.

The total additional interaction ΔW_i is the sum of contributions from a short-range ion-hydration force, a long-range ion-hydration force (the screened image force)

(14) Tissandier, M. D.; Cowen, K. A.; Feng, W. Y.; Gundlach, E.; Cohen, M. H.; Earhart, A. D.; Coe, J. V.; Tuttle, T. R. *J. Phys. Chem. A* **1998**, *102*, 7787.

and a ion-dispersion force. The long-range ion hydration forces are due to the behavior of water as a homogeneous dielectric, characterized by a macroscopic dielectric constant, whereas the short-range ion-hydration forces are due to the local, discrete structure of water. At low electrolyte concentrations, the screened image forces are dominant (leading to the Onsager–Samaras law),[2] whereas at large ionic strengths, they are negligible and the behavior of ions near interfaces is determined by the short-ranged ion-hydration forces.[6] As noted earlier, the inclusion of the ion-dispersion interactions is equivalent to a slight modification of the parameters of the short-range ion-hydration forces.[6]

Explicit expressions for the additional interactions between ions and interfaces, $\Delta W_i(x)$, allow us to calculate the distribution of ions $c_i(x)$ (for a sharp water/air interface) eqs 10a and 10b, for the appropriate boundary conditions

$$\frac{d\psi}{dx}\Big|_{x=0} = -\frac{\sigma_s}{\epsilon\epsilon_0} \tag{11a}$$

and

$$\psi(x)|_{x=x_{\max}} \cong 0 \tag{11b}$$

where σ_s is the surface charge density (assumed $\sigma_s = 0$ in what follows) and with the last approximation holding because $x_{\max}$ of the numerical simulations is at about 5 Debye-Hückel lengths from the air/water interface.

The traditional Modified Poisson–Boltzmann (MPB) approach (eqs 1a and 1b) assumes a sharp air/water interface. It is more reasonable to consider that our system is constituted from a statistical ensemble of sharp interfaces, with a statistical distribution

$$\varphi(\xi) = \frac{1}{\sqrt{2\pi}\sigma}\exp\left(-\frac{(\xi - x_0)^2}{2\sigma^2}\right) \tag{12a}$$

around the average position $x=x_0$ of the interface, the variable ξ denoting the position of the sharp air/water interface in one component of the statistical ensemble.

Once the ion distributions for sharp air/water interfaces, $c'_i(x)$, are calculated, the concentrations of ions for the statistical ensemble, $c_i(x)$, are obtained by summing up the corresponding contributions to concentrations provided by each element of the ensemble. For an element with a sharp interface located at ξ, the concentration of ions at the distance $(x - \xi)$ from the sharp interface represents the contribution to the concentration of ions at location x in the statistical ensemble (because $\xi + (x - \xi) = x$). Adding up all of these contributions (from the sharp interfaces located at all possible ξ values) multiplied with the probability of a sharp interface at the distance ξ in the statistical ensemble (eq 12a), one obtains the convolution

$$c_i(x) = \int_{\xi=-\infty}^{\infty} \varphi(\xi)c'_i(x - \xi)\, d\xi \tag{12b}$$

where the prime denotes the concentration calculated for a sharp air/water interface. The water density profile for a sharp interface located at ξ is described by the Heaviside function $\Theta(x - \xi) = 0$ for $x - \xi < 0$, $\Theta(x - \xi) = 1$ for $x - \xi > 0$; therefore, the water profile for the statistical ensemble is given by

$$p(x) = \int_{\xi=-\infty}^{\infty} \varphi(\xi)\Theta(x - \xi)\, d\xi = \frac{1 + erf\left(\frac{x - x_0}{\sigma}\right)}{2} \tag{12c}$$

The distributions of ions and the water density profile for a water/air interface (water molecules located in the region $x < x_0$), as considered in the molecular dynamics simulations[9] (instead of an air/water interface) are simply obtained by the transformation $(x - \xi) \rightarrow (\xi - x)$ for $c_i'(x - \xi)$ in eq 12b and $\Theta(x - \xi)$ in eq 12c.

This treatment does not generate large repelling forces between ions and interfaces, as the BKN model does, because the structure-breaking ions expelled at the interface are still hydrated in each subsystem of the statistical ensemble. Intuitively, this means that the structure-breaking ions arrive at the interface surrounded by their hydration water molecules. Whereas, on average, the water density is lower near the air/water interface, so there are sufficient water molecules locally to hydrate each interfacial ion. In contrast, the BKN model suggests that the (average) low density of water leads to a low dielectric constant near the air/water interface, and consequently, no ion will approach the interface.

The results of the numerical calculations for the distributions of ions for NaCl are presented in Figure 4a, in which we employed the following values for $\Delta W_i(x)$

$$\Delta W_{Na} = \infty \quad \text{for } 0 < x < 3 \text{ Å} \tag{13a}$$

and

$$\Delta W_{Na} = 0 \quad \text{for x} > 3 \text{ Å} \tag{13b}$$

which involves a total depletion of ions in the first $d_{cut-off}$ = 3 Å of water and

$$\Delta W_{Cl} = -1.7kT \quad \text{for } 0 < x < 2.5 \text{ Å} \tag{14a}$$

and

$$\Delta W_{Cl} = 0 \quad \text{for x} > 2.5 \text{ Å} \tag{14b}$$

which involves a simple square potential near the interface, with a width of 2.5 Å and a depth of 1.7 kT (the parameters selected in eqs 13 and 14 are assumed to include also the ion-dispersion interactions). The parameters of statistical distribution of the sharp air/water interfaces have been selected $\sigma = 1.4$ Å and $x_0 = 13.9$ Å, in order to fit the profile of the interface provided by the molecular dynamics simulations.[9]

The simple SM/SB model clearly cannot explain the behavior of ions at distances larger than 10 Å from the interface, but it is not yet clear whether longer-ranged forces are involved or whether the strong variation of the ions distributions in that region are artifacts of the limited size of the system employed in simulations.

To reproduce the simulations for the ion distributions in the aqueous solution of HCl (see Figure 1b), ΔW_{Cl} has been selected as above (eqs 14a and 14b) and an attractive interaction between hydrogen ions and the air/water interface has been employed

$$\Delta W_H = -1kT \quad \text{for } 0 < x < 1 \text{ Å} \tag{15a}$$

and

$$\Delta W_H = 0 \quad \text{for } x > 1 \text{ Å} \tag{15b}$$

Such an interaction has been suggested by experiment regarding the surface tension of aqueous acid solutions.[3] The accumulation of H^+ at the interface has been recently

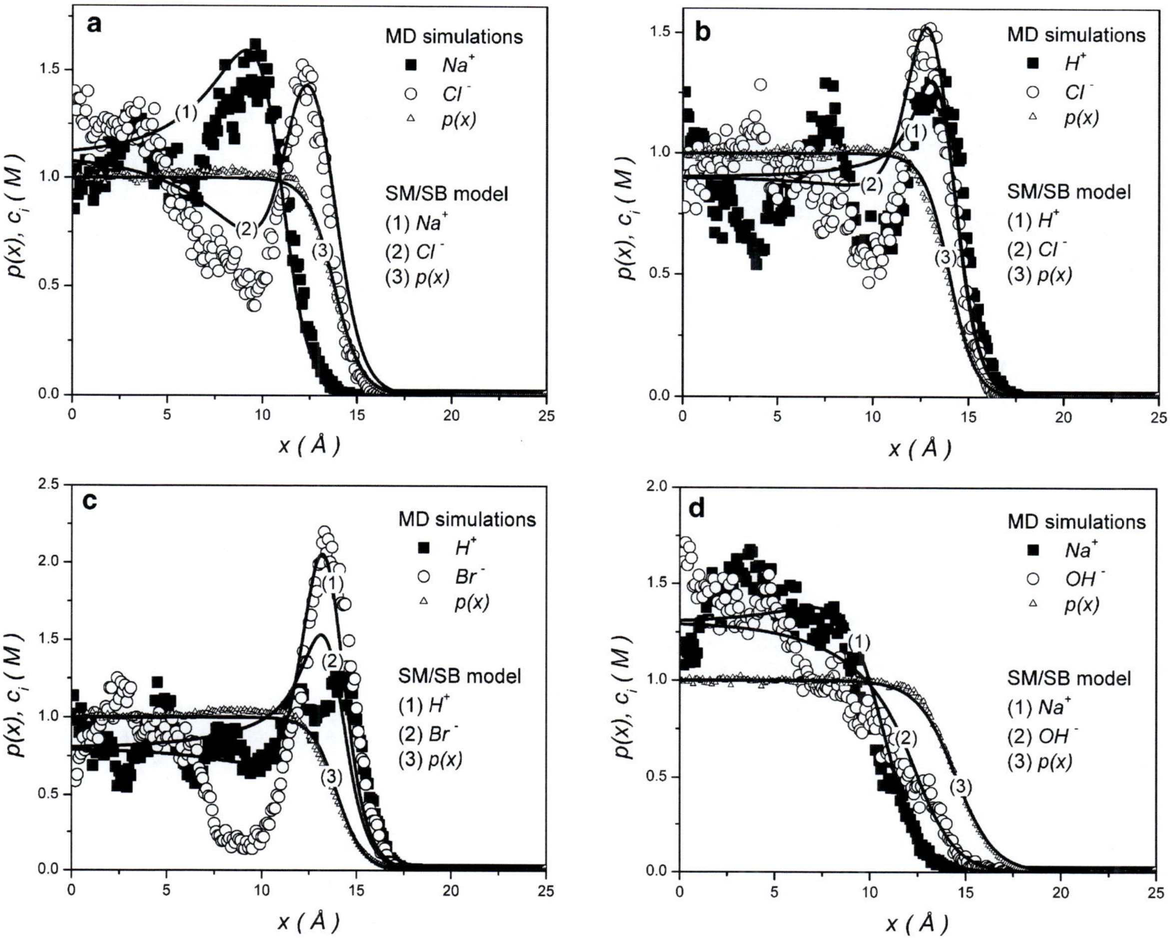

Figure 4. (a–d) Distribution of cations (1), anions (2), and water (3) at the air/water interface. The continuous curves are calculations based on the SM/SB model for short-range ion-hydration forces. The squares (cations), circles (anions) and triangles (water) are the molecular dynamics results of ref 9. (a) NaCl; (b) HCl; (c) HBr; (d) NaOH.

confirmed directly by experiment.[15] One should note that there is a minimum in the concentrations of both anions and cations around 10 Å. This minimum cannot be explained either by the simple potentials selected for the ion-hydration forces or by the traditional electrostatic mean field interactions (for which the depletion of one kind of ions is associated with the accumulation of ions of the other kind). Furthermore, the Poisson–Boltzmann treatment predicts a monotonic variation of charge, whereas in Figure 4b successive inversions of charges with the distance are apparent. One possible reason for this behavior might be another interaction between ion and the surface, perhaps due to the structuring of the water near the interface.

The behavior of ions is similar for other aqueous acid solutions (HBr, Figure 4c) for which an attractive potential has been selected for the interaction between Br^- and the interface, whereas for H^+ the same potential as above (eqs 15a,b) was employed

$$\Delta W_{Br} = -2.0kT \quad \text{for } 0 < x < 1.5 \text{ Å} \qquad (16a)$$

and

$$\Delta W_{Br} = 0 \quad \text{for x} > 1 \text{ Å} \qquad (16b)$$

It is of interest to note that, for a basic solution (NaOH, Figure 4d), both the cations and the anions are repelled by the interface (in contrast with the simple acids, for which both H^+ and Cl^- or Br^- are accumulated on the interface). To obtain a reasonable agreement with experiment, we used a repulsive potential for OH^- with a cut off $d_{OH} = 2.0$ Å, whereas for Na^+, we had to use $d_{cut-cff} = 3.5$ Å, which is slightly different from the value selected in eqs 13.

The agreement between the numerical results and those obtained via the simple SM/SB model for the short-range ion-hydration forces is satisfactory, if one takes into account that the interactions between ions and interfaces have been assumed to have a very simplified form, and the possible charging of the air/water interface by other mechanisms, other than the accumulation/depletion of ions, was ignored.

4. Conclusions

The distributions of ions in the vicinity of an air/water interface are due not only to the electrostatic interactions among ions, but also to the additional interactions between ions and the interface. Whereas it is plausible to consider that the ions prefer the medium with a larger dielectric constant and hence that the corresponding interactions should repel the ions from the interface, the traditional treatments of the hydration of ions via screened image forces[1,2] or local approximations of the Born energy,[12,13] which consider water as a continuous dielectric, failed to reproduce the experiment at high ionic strengths. This

(15) Petersen, P. B.; Saykally, R. J. *J. Phys. Chem. B* **2005**, *109*, 7976.

failure led to a search for additional forces, such as the ion-dispersion forces, due to van der Waals interactions between ions and the rest of the system.[10] Such interactions predicted, however, qualitatively wrong ions distributions (an accumulation of cations and a depletion of anions at the air/water interface, instead of the opposite).

We proposed recently a simple model (SM/SB) for the short-range interactions between ions and interface, based on the fact that the water surrounding the ion is formed of discrete molecules.[6] Consequently, the repelling ion-hydration forces, acting on structure-making ions could be described by one parameter (a cutoff length), whereas the attractive ion-hydration forces, acting on structure-breaking ions, could be described by two parameters (a simple potential well at the interface). It is clear that the ion-hydration forces are more complex, but there is no theory yet that can provide an "ab initio" prediction of these interactions. The reason for this failure consists of the large number of molecules involved in these interactions. A natural way to calculate the interactions would be to simulate the air/water interface of an aqueous solution and to calculate "a posteriori" the potential that matches the distribution of ions. However, only systems with a limited number of molecules (for the time being, about 1000) could be treated in a reasonable time.[9] It is apparent that in these cases, boundary effects are influencing the distribution of ions in all figures (Figure 4a–d), because the systems investigated were limited only to a slab of water of a thickness of about 15 Å from the air/water interface.

We showed previously that a simple model for the ion-hydration interactions, which separates the ion-hydration forces in a long-range term due to the behavior of water as a continuous dielectric (the screened image force) and a short-range term due to the discreetness of the water molecules (SM/SB), can explain almost quantitatively a number of phenomena related to the electrolyte interfaces.[6] In this article, we examined the limitations of the model in predicting the distributions of ions near the air/water interface, by comparison with molecular dynamics simulations. It is clear that the real ion-hydration forces are more complicated than the simple model employed here; however, the interfacial phenomena (including specific ionic effects) can be understood, at least qualitatively, in terms of this simple approach.

The situation is similar to the success of the traditional Poisson–Boltzmann approach: its ability in describing, at least qualitatively, and many times even quantitatively, the behavior of most colloidal systems probably resides in the use of at least one adjustable parameter (surface charge, surface potential, recombination constant and so on) in the fitting of the experimental results. If that parameter could be accurately measured, one would have to address the inaccuracies generated by the mean field treatment itself.

Acknowledgment. We are indebted to Dr. Pavel Jungwirth for kindly providing us the files of the molecular dynamics simulations, reproduced in Figures 3a–b and 4a–d.

LA051979A

Available online at www.sciencedirect.com
ScienceDirect
Journal of Colloid and Interface Science 304 (2006) 541–544
JOURNAL OF Colloid and Interface Science
www.elsevier.com/locate/jcis

Note

Ions at the air/water interface

Marian Manciu [a,*], Eli Ruckenstein [b]

[a] *Physics Department, University of Texas at El Paso, USA*
[b] *Chemical and Biological Engineering Department, State University of New York at Buffalo, USA*

Received 26 July 2006; accepted 15 September 2006
Available online 20 September 2006

Abstract

In a recent review of this topic [B.C. Garett, Science 303 (2004) 1146] the emphasis was on some recent experiments, in which it was found that some anions accumulate at the air/water interface and not in the bulk, as usually happens to the cations, and on some simulations which explained those positive surface adsorption excesses. Because a large number of these experiments could be explained on the basis of some simple physical models proposed by the authors for the interaction between the ions and the air/water interface [M. Manciu, E. Ruckenstein, Adv. Colloid Interface Sci. 105 (2003) 63; Adv. Colloid Interface Sci. 112 (2004) 109; Langmuir 21 (2005) 11312], those models are reviewed in the present note, the goal being to draw attention to them.

Keywords: Interface; Ions; Double layer; Ion-hydration forces

Garrett [1] has briefly reviewed the recent progress in the understanding of the behavior of ions at the water/air interface, by emphasizing both some experiments [2,3], which have shown that some anions can accumulate at the interface and not in the bulk as usually considered, as well as some numerical simulations [4], which can explain such a behavior. His review stimulated us to draw attention to some physical models, which can provide additional clarifications and which have been considered recently by the authors [5–7].

First let us note that experiment revealed long ago that not all ions prefer the bulk to the interface [8]. Gibbs adsorption equation predicts that the surface tension increases with the electrolyte concentration when the total surface excess of ions is negative. The conventional picture, that the ions prefer the bulk, is probably due to Langmuir, who noted that the increase in the surface tension of aqueous solutions of simple salts with increasing concentration can be explained by assuming a surface layer of pure solvent with a thickness of about 4 Å [9]. However, because the aqueous solutions of some simple acids (such as HCl) possess surface tensions smaller than that of pure water [8], Gibbs adsorption equation indicates a positive total surface excess adsorption of ions, hence at least one of the ion species must accumulate in this case at the interface. Therefore, experimental evidence for accumulation of at least one species of ions on the interface has been around as long as reliable measurements for the surface tension of acid solutions have been available.

The first explanation for the negative ion adsorption was based on the "image force," i.e., the electrostatic interactions between ions and the boundary between two dielectric media. However, the long range Coulomb force leads to a diverging large negative adsorption. The paradox was solved by Wagner [10] who, inspired by the then-developed Debye–Huckel theory of electrolytes, suggested that a screened image force could explain the surface tension behavior of salt solutions. A simplifying assumptions about the screening length provided later an analytic solution for the surface tension of electrolytes, the celebrated Onsager–Samaras theory [11]. Subsequently, the theory was challenged by the surface tension measurements of Jones and Ray [12], who observed an initial decrease, followed by a minimum at low concentrations, and a further increase with increasing ionic strength, which has been attributed to the charging of the interface [13]. The important conclusion was that there are other interactions between ions and interfaces in addition to image forces.

* Corresponding author. Fax: +1 915 747 5447.
E-mail address: mmanciu@utep.edu (M. Manciu).

0021-9797/$ – see front matter
doi:10.1016/j.jcis.2006.09.034

542 M. Manciu, E. Ruckenstein / Journal of Colloid and Interface Science 304 (2006) 541–544

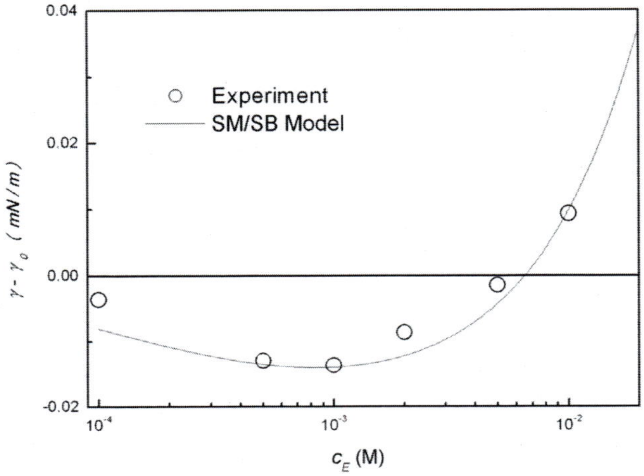

Fig. 1. Change in surface tension at low electrolyte concentrations (the Jones–Ray effect). Circles: experimental data from Ref. [12]; line: predicted surface tension by the SM/SB model which accounts for OH^- adsorption on the surface, ion hydration and image forces within the Poisson–Boltzmann approach (Ref. [5]).

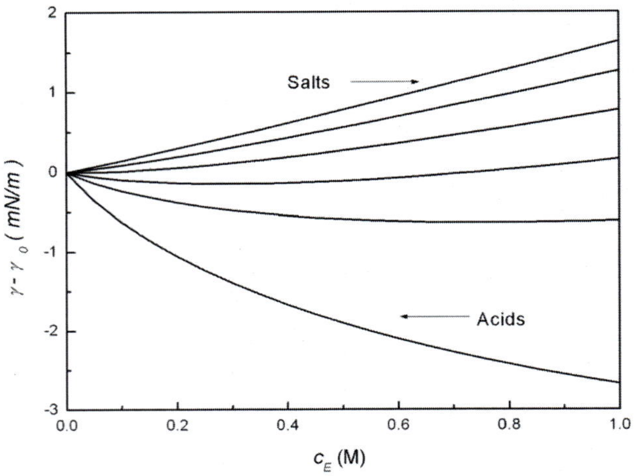

Fig. 2. Possible dependence of the surface tension on electrolyte concentration, for various ion-hydration potentials for the anions (Ref. [5]). When the surface excess of the SB ions exceeds the surface depletion of the SM ions, the surface tension decreases with ionic strength.

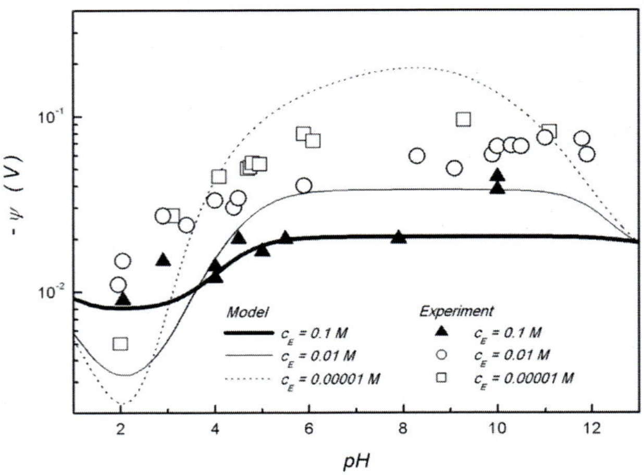

Fig. 3. Experimental values of the zeta potential for the air/water interface in the presence of NaCl, as a function of pH, from Ref. [15] (triangles: 0.1 M; circles: 0.01 M; squares: 0.00001 M) are compared with the potential values predicted by the SM/SB model.

In a mean field theory, the forces acting on each ion because of all the other species in the system can be described by a potential U_i, which can be separated into an electrical part due to the interaction between the charge q_i of each ion "i" and the mean electrical potential ψ, and an additional term ΔW_i, which accounts for all the other interactions between the ion and the rest of the system, not included in ψ,

$$U_i = q_i \psi + \Delta W_i. \tag{1}$$

Assuming a planar surface (hence, that the mean field depends only on the distance z from the surface) the chemical equilibrium for each species of ions implies that their concentration $c_i(z)$ is related to their bulk concentration c_{0i} by

$$c_i(z) = c_{0i} \exp\left(-\frac{q_i \psi(z) + \Delta W_i(z)}{kT}\right), \tag{2}$$

where k is the Boltzmann constant, T is the absolute temperature, and z is the distance to the interface. Equation (2) also implies $\Delta W_i(\infty) \equiv 0$ and that the electric potential obeys the Poisson equation:

$$\frac{d^2\psi}{dz^2} = -\frac{q_A c_A + q_C c_C}{\varepsilon \varepsilon_0}, \tag{3}$$

where ε is the dielectric constant of water, ε_0 is the vacuum permittivity, and the subscripts A and C stand for anion and cations, respectively.

The main difficulty consists in the determination of ΔW_i, which should include (but is not limited to) image forces. It was recently suggested by Ninham and co-workers [14] that the different van der Waals interactions of anions and cations (due to their different polarizabilities) might explain the accumulation of ions near the interface (and not only the depletion provided by the image force alone). However, as emphasized in Ref. [5], the calculations based on ion-dispersion forces predicted that the cations should accumulate and the anions should be repelled by the interface, in contradiction with both experiment [2,3] and simulations [4], which indicate the opposite.

Another possible explanation might be related to the ability of some ions to organize (structure-making ions) and others to disorganize (structure-breaking ions) the water structure. The approach to the interface of the structure-making ions is unfavorable, because they can better organize the water dipoles in bulk water than at the interface, thus decreasing the free energy of the system. The opposite is true for the structure-breaking ions: the total free energy of the system is minimized by pushing these ions toward the interface, because bulk water molecules can better organize their hydrogen bonding network without them. A recent attempt [5] to account for this structure-making/structure-breaking (SM/SB) interactions (ion-hydration forces) within the traditional mean field framework has shown that the surface excesses of ions can be either

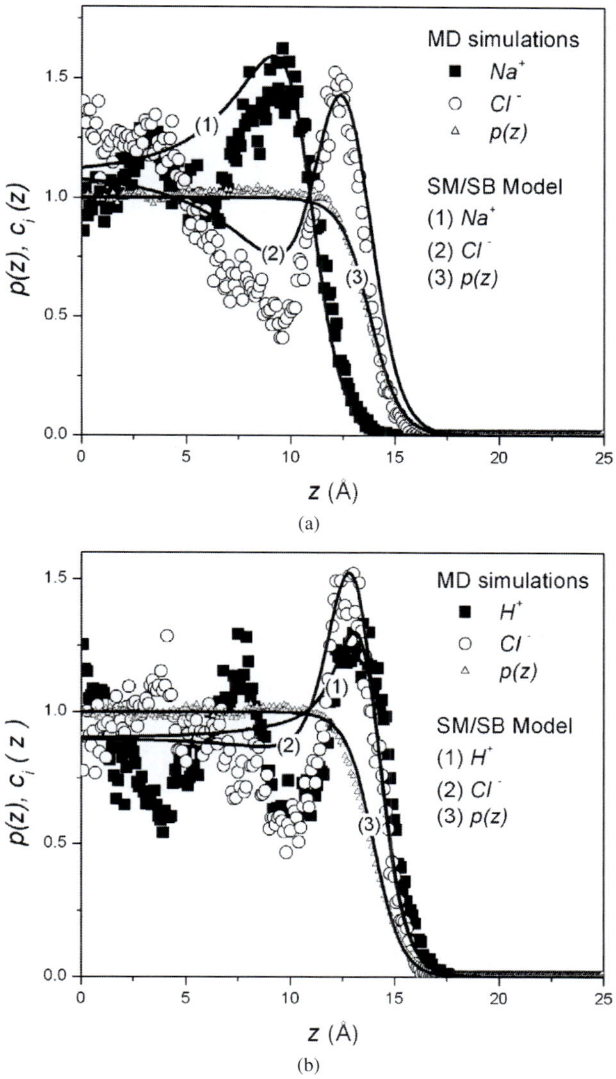

Fig. 4. Distribution of cations (1), anions (2), and water (3) at the air/water interface. The continuous curves are calculations based on the short-range ion-hydration forces (the SM/SB model) (Ref. [7]). The squares (cations), circles (anions), and triangles (water) are the Molecular Dynamics results of Ref. [4]. (a) NaCl; (b) HCl.

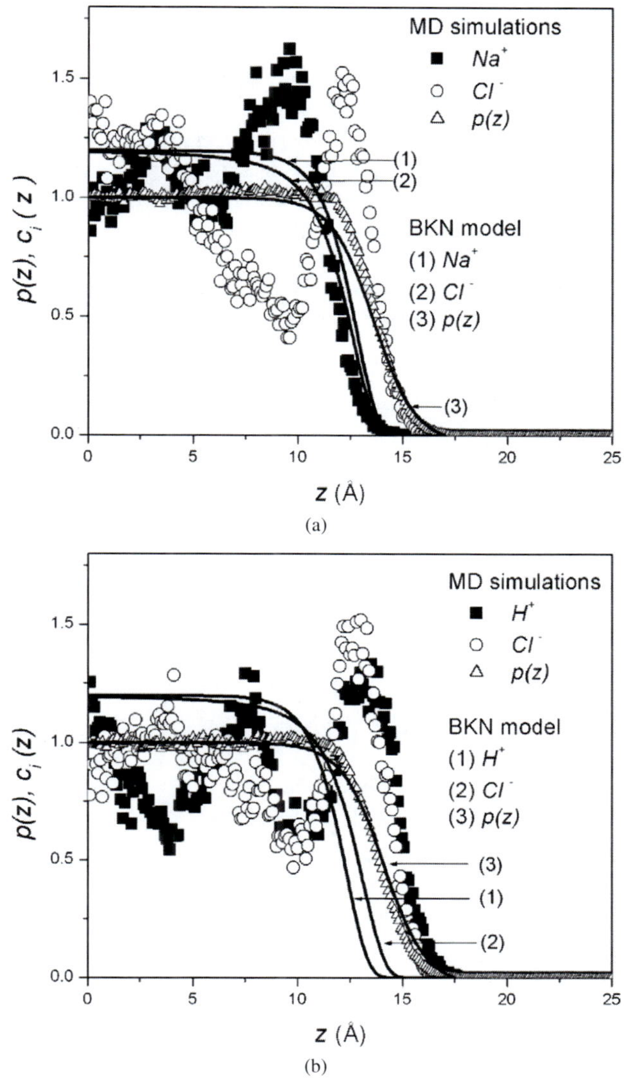

Fig. 5. Distribution of cations (1), anions (2), and water (3) at the air/water interface. The continuous curves are calculations based on ion dispersion and ion-solvation forces (the BKN model, Ref. [16]). The squares (cations), circles (anions), and triangles (water) are the Molecular Dynamics results of Ref. [4]. (a) NaCl; (b) HCl.

positive or negative. In this model the potentials ΔW_i have been taken as potential wells with depths of the order of kT and widths of a few Å for the structure breaking ions, and as generating depletion lengths of a few Å for the structure-making ions. It was shown that this simple SM/SB model for the "ion-hydration forces" can explain the Jones–Ray effect (Fig. 1) [5], the increase of the surface tension of an aqueous salt solution with concentration and its decrease for aqueous acid solutions (Fig. 2) [5], as well as the dependence of the zeta potential on pH and electrolyte concentration (Fig. 3) [5]. The agreement is not surprising since the above simple structure-making/breaking potentials can reproduce reasonably well the distributions of ions around the air/water interface, obtained by MD simulations [4], for various kinds of simple electrolytes, for an interfacial water profile assumed to be given by a statistical distribution of sharp interfaces [7] (see Figs. 4a and 4b). While the interactions potentials ΔW_i used in the SM/SB model are qualitatively understandable [5], it is not yet possible to predict them a priori. It was recently suggested [16] that the short ranged ion-hydration interactions (denoted ion-solvation forces) might be calculated using the expression:

$$\Delta W_{i,\mathrm{solvation}}(z) = \Delta G_{i,\mathrm{solvation}} \frac{\frac{1}{1+p(z)(\varepsilon-1)} - \frac{1}{\varepsilon}}{1 - \frac{1}{\varepsilon}}, \tag{4}$$

where $\Delta G_{i,\mathrm{solvation}}$ is the free energy of solvation of an ion "i" in the bulk [17], and $p(z)$ is the profile of the water/air interface, which is reasonable well approximated by the empirical equation [16]:

$$p(z) = 1 - 1.0302 \exp(3.5 - z)/\big(1 + \exp(3.5 - z)\big). \tag{5}$$

The latter treatment has the advantage of simplicity; however it predicts that the interface is depleted by all the kinds of ions, prediction contradicted by both experiment and simulations. In contrast, the earlier accounting of ion-dispersion forces [14] predicted that the cations are accumulated at the surface and the anions are depleted, whereas in reality the opposite occurs. In addition, the Bostrom–Kunz–Ninham (BKN) treatment of ion-solvation forces [16] predicts that the surface tension of the simple acid solutions increases with ionic strength more rapidly than the surface tension of simple salts (since the free energy of solvation of H^+ is larger than those of the other cations, such as Na^+ [17]). Actually, the former surface tension decreases with ionic strength, because both H^+ and the anions are accumulated on the surface. As shown in Ref. [7], the inclusion of both ion-dispersion and ion-solvation forces (provided by the BKN model) in the interaction potential between ions and the air/water interface (see Figs. 5a and 5b) does not lead to good agreement with simulations. In contrast, the use of the ion interactions potentials employed in the SM/SB model [5] in conjunction with a statistical model for the profile of the air/water interface [7] provides a reasonable agreement with the simulated distribution of ions near the air/water interface [4]. In summary, we can emphasize that our simple models provide insights regarding both the experimental and simulation results on the behavior of ions near the air/water interface.

References

[1] B.C. Garett, Science 303 (2004) 1146.
[2] E.A. Raymond, G.L. Raymond, J. Phys. Chem. B 108 (2004) 5051.
[3] D. Liu, G. Ma, L.M. Levering, H.C. Allen, J. Phys. Chem. B 108 (2004) 2252.
[4] P. Jungwirth, D.J. Tobias, J. Phys. Chem. B 105 (2001) 10468.
[5] M. Manciu, E. Ruckenstein, Adv. Colloid Interface Sci. 105 (2003) 63.
[6] M. Manciu, E. Ruckenstein, Adv. Colloid Interface Sci. 112 (2004) 109.
[7] M. Manciu, E. Ruckenstein, Langmuir 21 (2005) 11312.
[8] J.W. Belton, Trans. Faraday Soc. 31 (1935) 1413.
[9] I. Langmuir, J. Am. Chem. Soc. 39 (1917) 1848.
[10] C. Wagner, Phys. Zeit. 25 (1924) 474.
[11] L. Onsager, N.N.T. Samaras, J. Chem. Phys. 2 (1934) 628.
[12] G. Jones, W.A. Ray, J. Am. Chem. Soc. 59 (1939) 187.
[13] M. Dole, J. Am. Chem. Soc. 60 (1940) 904.
[14] M. Bostrom, D.M.R. Williams, B.W. Ninham, Langmuir 17 (2001) 4475.
[15] C. Li, P. Somasundaran, J. Colloid Interface Sci. 146 (1991) 215.
[16] M. Bostrom, W. Kunz, B.W. Ninham, Langmuir 21 (2005) 2619.
[17] B.E. Conway, Ionic Hydration in Chemistry and Biophysics, Elsevier, Amsterdam, 1981.

Introduction to CHAPTER 6

Polarization Model: a unified framework for hydration and double layer interactions

6.1 D. Schiby, E. Ruckenstein: "*The role of the polarization layers in hydration forces,*" CHEMICAL PHYSICS LETTERS 95 (1983) 435–438.
6.2 E. Ruckenstein, D. Schiby: "*On the origin of the repulsive hydration forces between mica plates,*" CHEMICAL PHYSICS LETTERS 95 (1983) 439–443.
6.3 D. Schiby, E. Ruckenstein: "*On the coupling between the double-layer and the solvent polarization-fields,*" CHEMICAL PHYSICS LETTERS 100 (1983) 277–281.
6.4 M. Manciu, E. Ruckenstein: "*Oscillatory and Monotonic Polarization. The Polarization Contribution to the Hydration Force,*" LANGMUIR 17 (2001) 7582–7592.
6.5 M. Manciu, E. Ruckenstein: "*Polarization of Water near Dipolar Surfaces: A Simple Model for Anomalous Dielectric Behavior,*" LANGMUIR 21(2005) 11749–11756.
6.6 E. Ruckenstein M. Manciu: "*The Coupling between the Hydration and Double Layer Interactions,*" LANGMUIR 18 (2002) 7584–7593.
6.7 M. Manciu, E. Ruckenstein: "*On the Chemical Free Energy of the Electrical Double Layer,*" LANGMUIR 19 (2003) 1114–1120.

Because the double layer force vanishes in the absence of surface charges, one expects the attractive van der Waals force to cause the coagulation of all neutral (or even weekly charged) colloids. The absence of such a behavior has been explained by the existence of an additional (non-DLVO) force, the hydration interaction, which is due to the structuring of water in the vicinity of hydrophilic surfaces. This chapter is devoted to the identification of the microscopic origin of the hydration force, and to the presentation of a unified treatment of the double layer and hydration forces, the Polarization Model.

The basic idea is that the dipoles of the surface are responsible for the partial alignment of the dipoles of the neighboring water molecules via electrostatic dipole-dipole interactions. The dipoles of the partially aligned water molecules induce partial alignment in the next layer of water molecules and so on; consequently a decaying polarization field is generated in the vicinity of a dipolar surface even in the absence of a surface charge [6.1]. When two such surfaces are brought close to each other, the overlap of the polarization fields (which have opposite orientations) increases the free energy of the system, thus generating repulsion. This repulsive force is similar (both in magnitude and distance dependence) to the experimentally determined hydration force [6.2]. Since both hydration and double layer forces are based on electrostatic interactions (dipole-dipole and charge-electric field, respectively), a unified theory for hydration and double layer force is proposed [6.3].

Whereas in the traditional electrodynamics a uniform dipolar field on a surface does not induce an electric field outside the surface, it is suggested that the discrete nature of the water molecules (which cannot be ignored at molecular distances) is responsible for a local, non vanishing electric field. This field is calculated by assuming that water is organized in ice-like layers in the vicinity of the interface, and leads to interactions compatible with the experimentally observed hydration force [6.4]. Two consequences emerge from this model: (i) in water, the polarization is no longer proportional to the electric field in the vicinity of dipolar surfaces (hence, an anomalous dielectric constant) , and (ii) the polarization oscillates near interfaces (consequently the corresponding hydration force is non-monotonic). The experimentally observed monotonic hydration force is most likely a result of the statistical roughness of the interacting surfaces: the oscillations are averaged out and thus a monotonically decreasing repulsion results [6.4].

Experiments to measure the electric field and water polarization within 10 Å of the surface are difficult to perform. However, recent Molecular Dynamics simulations carried out by Faraudo and Bresme for water between two sodium dodecyl sulfate layers revealed oscillatory behaviors for both the polarization and the electric fields near the surface, and non-proportionality between them [Faraudo, J.; Bresme, F. *Phys. Rev. Lett.* **2004**, *92*, 236102]. Our polari-

E. Ruckenstein, M. Manciu, *Nanodispersions*, DOI 10.1007/978-1-4419-1415-6_6, © Springer Science+Business Media, LLC 2010

zation model is able to account for such unexpected behaviors [6.5]

When all the electrical interactions (between surface charges and dipoles, ions and water molecule dipoles are taken into account, the electric "mean field" potential and the polarization become coupled via two nonlinear differential equations of second order (in which the polarization is no longer proportional to the electric field). The same constitutive equations are also derived using a variational approach [6.6]. In the linear approximation (valid for small potentials), the equations can be solved analytically and the interaction is the result of the sum of two free energy terms with two decay lengths, which (in the limit of vanishing small electrolyte concentrations) are related to the large Debye-Hückel length, and to a small hydration decay length of about 2 Å which is independent of ionic strength [6.6]. The latter decay length is comparable to that measured experimentally for neutral lipid bilayers. Therefore, if the "coupling" between the interactions is small, they can be perceived as two different interactions with two different decay lengths - the double layer and hydration forces [6.6].

In order to extend the polarization model to systems of arbitrary geometries and arbitrary surface conditions, the corresponding chemical part of the free energy has to be accounted for. It is shown that the change in the chemical free energy depends not only on the surface charge density and potential in the final state, but also on their values at each distance between infinity and the final state. An exact expression for the chemical free energy has been derived and a simple approximation suggested [6.7].

Volume 95, number 4,5 CHEMICAL PHYSICS LETTERS 11 March 1983

THE ROLE OF THE POLARIZATION LAYERS IN HYDRATION FORCES

D. SCHIBY and E. RUCKENSTEIN
Department of Chemical Engineering, State University of New York at Buffalo, Buffalo, New York 14260, USA

Received 5 November 1982; in final form 9 February 1983

Hydration forces between two hydrophilic surfaces are generated by the overlap between the two polarization layers which are formed near each of them. The differential equation that describes the polarization field as well as the corresponding boundary conditions are derived on the basis of electrostatics. Expressions for the decay length and for the magnitude of the force have been derived and compared with experiment.

1. Introduction

The DLVO theory has been frequently used to explain the effect of electrolyte concentration on the stability of colloid suspensions and emulsions. The double layer forces which are responsible in this theory for stability become weaker as the electrolyte concentration increases. However, recent experiments [1] with emulsions of oil in water, in which a surfactant is adsorbed on the surface of the globules, have shown that at high electrolyte concentrations flocculation occurs even though the DLVO theory predicts coagulation. Similarly, recent measurements of the repulsive force between two mica plates, in a dilute electrolyte solution, reveal that for plate separations smaller than ≈40 Å, the force is larger than that resulting from the DLVO theory [2,3]. These experiments clearly prove the existence of short-range repulsive forces different in nature from the double layer forces. For reasons which will become clear later, they are called hydration forces, even though the name solvation forces would be more appropriate, because they can also occur in polar liquids other than water.

In addition to the above experiments on colloidal systems, it is important to mention the pioneering experiments on phospholipid bilayers published by Parsegian and co-workers starting from 1976 [4,5]. These authors regard the strong repulsive force they measured between parallel bilayers as a hydration force. In contrast to the colloidal systems, in this case the extra repulsion is insensitive to the concentration of electrolyte. In the framework of the theory presented below, this can be explained by observing that, in the latter case, the polar head groups are zwitterionic and are by themselves able to polarize sufficiently strong the neighboring water molecules. In the former case, however, a sufficiently strong polarization is generated only after the adsorption of a sufficient number of hydrated ions, which together with the charges of the surface generate a bilayer of charge on the surface [6].

The object of this note is to show that the organization of water in the vicinity of the surface propagates into the liquid because of the interactions between the water molecules. The overlap of the polarization layers of the two surfaces decreases the average dipole moment. This increases the free energy of the molecules, thus generating a repulsive force. The idea that the propagation into the liquid of a polarization generated by the surface is responsible for hydration forces was first formulated, in a pioneering paper [7], by Marčelja and Radić. They used the Landau phenomenological expansion for the free energy to predict the polarization field between two parallel plates, in terms of a parameter – the decay length – assuming as a boundary condition that the polarization is constant on the plates. The present

Volume 95, number 4,5 CHEMICAL PHYSICS LETTERS 11 March 1983

treatment of the problem differs from theirs, firstly because it is based on electrostatics and, secondly, because the boundary condition at the plate, which is derived in the same manner as the differential equation for the polarization field, is more realistic. As a result of this boundary condition, the polarization at the surface depends upon the distance h between the plates. This result is quite plausible physically because for small values of h the orientation of the water molecules adsorbed on one plate is expected to be affected by the other plate. In addition, the present approach, based on electrostatics, is able to provide improved descriptions if more adequate relations between the average dipole moment and the electric field are used. When the field is strong, the above relation is non-linear and the differential equation describing the polarization field is also non-linear. The earlier phenomenological approach can also lead to a non-linear equation but at the price of additional empirical parameters. Lastly we mention that, while in the phenomenological treatment, the decay length is a parameter to be determined from experiment, in the present theory an expression is derived for this quantity in terms of known physical quantities.

2. The physical model

Points on the surface occupied by charges or dipoles induce a polarization on the water molecules adjacent to them. The latter induce a polarization in the neighboring water molecules and so on. In what follows we assume that the polarization on the surface is uniform and hence that the polarization is a function only of the distance z to one of the plates. The field E that acts upon a molecule at z contains two components:

$$E(z) = E^0(z) + E^i(z), \tag{1}$$

where $E^0(z)$ is the external field and $E^i(z)$ is the field generated by the neighboring molecules. Here only the nearest neighbors are taken into account. The average dipole moment $\bar{\mu}(z)$ of the molecule at z is related to the polarization P via the equation

$$P = \bar{\mu}/a^2\delta, \tag{2}$$

where δ is the distance along the z axis between two molecules and a is the distance between two molecules in the plane normal to z ($a \approx \delta$). Further, the average dipole moment $\bar{\mu}$ has to be related to the electric field E. For sufficiently small values of E,

$$\bar{\mu} = \gamma E, \tag{3}$$

where the polarizability γ can be evaluated from the Clausius–Mosotti equation:

$$\gamma = (3/4\pi n)[(\epsilon - 1)/(\epsilon + 2)], \tag{4}$$

ϵ being the dielectric constant and n the number of molecules per unit volume. When E is very large one can expect the saturation behavior:

$$\bar{\mu} = \mu. \tag{5}$$

The non-linear case will be examined later in a more extended paper; here, for illustrative purposes, the linear case [eq. (3)] will be considered.

Assuming that the field $E^{(i)}$ acting on a water molecule is generated by the nearest neighbors only, one can write

$$E^{(i)}(z) = E'(z + \delta) + E'(z) + E'(z - \delta), \tag{6}$$

where $E'(z \pm \delta)$ is the field due to the nearest neighbors located on the plane $z \pm \delta$ normal to the direction of polarization and $E'(z)$ is that due to the nearest neighbors situated on the plane z. These fields are given by the expressions

$$E'(z \pm \delta) = \Lambda(\pm\delta)\bar{\mu}(z \pm \delta), \quad E'(z) = \Lambda(0)\bar{\mu}(z), \tag{7}$$

where

$$\Lambda(\pm\delta) = \frac{2}{\delta^3} + \frac{6}{(a^2 + \delta^2)^{3/2}} \frac{2\delta^2 - a^2}{\delta^2 + a^2} \tag{8a}$$

and

$$\Lambda(0) = -6/a^3. \tag{8b}$$

In order to calculate Λ we assumed that the molecules of water are located in planes parallel to the surface which can be superimposed by translation. In each plane the molecules are arranged in an hexagonal (close packed) configuration

If $\bar{\mu}(z \pm \delta)$ is expanded in series, eq. (6) leads to

$$E^{(i)}(z) = L\bar{\mu}(z) + \delta^2 M \mathrm{d}^2\bar{\mu}/\mathrm{d}z^2, \tag{9}$$

where

$$L = 2\Lambda(\delta) + \Lambda(0), \quad M = \Lambda(\delta). \tag{10}$$

Volume 95, number 4,5 CHEMICAL PHYSICS LETTERS 11 March 1983

Combining eqs. (1), (9), and (3), one obtains

$$d^2\bar{\mu}/dz^2 = \alpha^2\bar{\mu} + E^{(0)}/\delta^2 M\gamma, \tag{11}$$

where

$$\alpha^2 = (1 - L\gamma)/\gamma M\delta^2. \tag{12}$$

3. The polarization profile

Eq. (11) needs two boundary conditions for its solution. They can be derived in a simple manner. The first,

$$\bar{\mu}(z) = 0 \quad \text{at the midplane} \quad (z = \tfrac{1}{2}h), \tag{13}$$

is a consequence of the mirror symmetry of the system about the midplane between the plates (assumed identical). The second boundary condition can be derived by observing that $\bar{\mu}(z = 0)$ is a function of $E(z = 0)$. When $E(z = 0)$ is sufficiently small, the dependence between $\bar{\mu}(z = 0)$ and $E(z = 0)$ is linear:

$$\bar{\mu}(0) = \gamma E(0) \equiv \gamma[\Lambda(\delta)\bar{\mu}(\delta) + \Lambda(0)\bar{\mu}(0) + E^0(0)]. \tag{14}$$

The external field $E^{(0)}$, which in the present case is generated only by the surface, is considered non-zero only in the first layer of molecules, i.e. it represents the effect of an adsorption potential for the water molecules. Of course, in general $E^{(0)}$ is not zero in the liquid, for instance because of the occurrence of electrical double layers. We assume, however, that for small values of h the double layer field is small compared with the polarization field. Experiment [3] as well as more detailed calculations presented in the more extended version of this paper show that, indeed, when the hydration forces are important, the effect of double layers can be neglected.

The solution of eq. (11) for the boundary conditions (13) and (14) has the form

$$\bar{\mu} = \frac{\gamma E^0(0)(e^{-\alpha z} - e^{-\alpha(h-z)})}{A' e^{-\alpha h} + B'}, \tag{15}$$

with

$$A' = -1 + \gamma[\Lambda(\delta)e^{\alpha\delta} + \Lambda(0)] \tag{16a}$$

and

$$B' = 1 - \gamma[\Lambda(\delta)e^{-\alpha\delta} + \Lambda(0)]. \tag{16b}$$

It is of interest to observe that $\bar{\mu}(0)$ is a function of the distance h between the plates which tends to zero when $h \to 0$.

When, however, the field $E^0(0)$ is extremely large, but $E^{(i)}(z)$ remains sufficiently small, eq. (11) remains valid, but the boundary condition (14) has to be replaced with the saturation value of $\bar{\mu}$,

$$\bar{\mu}(0) = \mu. \tag{17}$$

4. The hydration force

The hydration force acting per unit area equals the rate of change of the free energy of the liquid per unit area, W, as h decreases,

$$F = -dW/dh. \tag{18}$$

The free energy density of molecules at the distance z is given by

$$f = -nE^{(0)}(z)\bar{\mu}(z) - \tfrac{1}{2}nE^{(i)}(z)\bar{\mu}(z) \tag{19}$$

and

$$W = \int_0^h f\,dz = -2nE^0(0)\bar{\mu}(0)\delta - \tfrac{1}{2}n\int_0^h E^{(i)}(z)\bar{\mu}(z)\,dz. \tag{20}$$

Using eqs. (15), (18) and (20), the following expression is obtained for the repulsive force F:

$$F = \frac{P_1(e^{-2\alpha h} - e^{-\alpha h} + \alpha h\, e^{-\alpha h})}{(A' e^{-\alpha h} + B')^2} + \frac{P_1 A'(e^{-\alpha h} - e^{-3\alpha h} - 2\alpha h\, e^{-2\alpha h})}{(A' e^{-\alpha h} + B')^3} + \frac{P_2 e^{-\alpha h}}{(A' e^{-\alpha h} + B')} + \frac{P_2 A'(e^{-\alpha h} - e^{-2\alpha h})}{(A' e^{-\alpha h} + B')^2}, \tag{21}$$

where

$$P_1 = n\gamma[E^0(0)]^2, \quad P_2 = 2P_1\alpha\delta. \tag{22}$$

The first two terms in eq. (21) represent the contribu-

Volume 95, number 4,5 CHEMICAL PHYSICS LETTERS 11 March 1983

tion of the diffuse layer of centers of polarization because they contain the interactions among the water molecules themselves, while the last two terms represent the contribution of the surface since they contain the interactions between the field of the surface and the water molecules. With decreasing values of h, the contribution of the diffuse layer diminishes, while that of the surface increases, the latter taking its maximum value when $h = 0$.

5. Results and discussion

As mentioned earlier the present theory provides an expression for the decay length l which is equal to α^{-1}. Because for water $\mu = 1.8 \times 10^{-18}$ esu, $\gamma \approx 6 \times 10^{-24}$ esu and $\delta \approx a = 2.8$ Å, eq. (12) leads to

$l = 2.4$ Å.

This value is of the same order as that obtained experimentally by Parsegian et al. [5]. Using the above value for the decay length and taking for $E^0(0)$ various values, we have calculated the dependence of the force F on the distance h. The results of the calculations are plotted in fig. 1, where, for comparison, we have introduced also a curve calculated on the basis of the boundary condition $\bar{\mu}(0) = \mu$. Fig. 1 shows that the experimental results of Parsegian et al. are, for $E^0(0) = 1.75 \times 10^5$ esu, near to those calculated by us. One may note that $E^0(0)$ can be calculated from the interaction of the bilayer with a water molecule. Preliminary evaluations lead to a value of 10^5 esu. Details will be given in the expanded version of this paper.

The force generated by the overlap of the polarization fields can be, for certain values of $E^0(0)$, of the same order of magnitude as that found experimentally between two mica plates by Pashley [3]. However, in that case, the force depends also on the detailed structure of the layer of hydrated ions adsorbed on the surface. In particular, their deformation produced by the interactions with the other adsorbed hydrated ions and with the polarization field

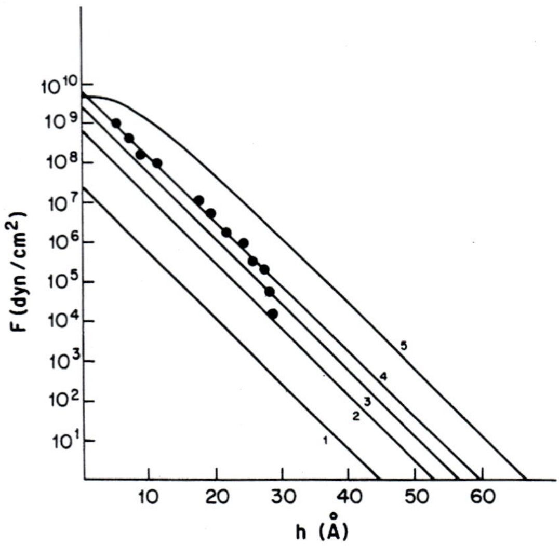

Fig. 1. The hydrated repulsive force against the distance h between the plates. The other values of the parameter are given in the text. Boundary condition eq. (14): curve 1, $E^0(0) = 10^4$ esu; curve 2, $E^0(0) = 5 \times 10^4$ esu; curve 3, $E^0(0) = 10^5$ esu; curve 4, $E^0(0) = 1.75 \times 10^5$ esu, (•) experimental data of ref. [5]; curve 5, $\bar{\mu}(0) = \mu$. The deviations of the experimental data from the straight line which occur for $h > 27$ Å are due, as explained in ref. [5], to the van der Waals forces.

of the free water molecules plays a major role. A detailed examination of this problem will be published elsewhere [6].

References

[1] H.M. Princen, M.P. Aronson and G.C. Moser, J. Colloid Interface Sci. 75 (1980) 246.
[2] J.N. Israelachvili, Faraday Discussions 65 (1978) 20.
[3] R.M. Pashley, J. Colloid Interface Sci. 83 (1981) 531.
[4] D.M. LeNeveu, R.P. Rand and V.A. Parsegian, Nature 259 (1976) 601.
[5] V.A. Parsegian, N. Fuller and R.P. Rand, Proc. Natl. Acad. Sci. US 76 (1979) 2750.
[6] E. Ruckenstein and D. Schiby, Chem. Phys. Letters 95 (1983) 439.
[7] S. Marčelja and N. Radić, Chem. Phys. Letters 42 (1976) 129.

Volume 95, number 4,5 CHEMICAL PHYSICS LETTERS 11 March 1983

ON THE ORIGIN OF REPULSIVE HYDRATION FORCES BETWEEN TWO MICA PLATES

E. RUCKENSTEIN and D. SCHIBY
Department of Chemical Engineering, State University of New York at Buffalo, Buffalo, New York 14260, USA

Received 31 August 1982; in final form 14 December 1982

Recent experiments on forces between two mica plates, in relatively dilute electrolyte solutions, have identified, for separations below ≈40 Å, a repulsive force that is larger than that predicted by the DLVO theory. It is suggested here that the origin of the additional force has to be sought in the bilayer of charge formed by the adsorption of hydrated ions on the charged mica surface. The bilayer generates a polarization in the neighboring water molecules which propagates into the liquid. The repulsive force is generated by the overlap of the polarization layers of the two plates. The boundary condition imposed by the bilayer of charge on the polarization field and an expression for the additional repulsive force are derived.

1. Introduction

Experiments on the interaction forces between two phospholipid bilayers have revealed the existence of a repulsive force that is greater than that predicted by the DLVO theory [1,2]. More recently, similar results have been obtained for the repulsive force between two mica plates, in relatively dilute electrolyte solutions, for plate separations smaller than ≈40 Å [3]. In the preceding paper [4], a theory was developed to explain the additional repulsive force which occurs between two phospholipid bilayers. The basic assumption of that theory is that the zwitterionic head groups of the phospholipid molecules generate a sufficiently strong polarization of the neighboring water molecules. This polarization then propagates further into the liquid. The overlap of the polarization layers of the two surfaces decreases the average dipole moment; this increases the free energy of the molecules, thus generating a repulsive force. While an initial theoretical approach had already been given by Marčelja and Radić [5], the theory developed in ref. [4] is predictive, yielding a value for the decay length which had to be supplied in the earlier treatment. In addition, the equation for the polarization profile and the boundary condition have been formulated in a more appropriate way.

The goal of the present paper is to extend the previous theory to deal with the repulsive force between two mica plates. In contrast to the case of the two phospholipid bilayers considered previously, experiment shows that in this case the electrolyte plays an important part. Indeed, measurements referred to above indicate that the additional repulsive force, which is not predictable within the DLVO theory, occurs only when the bulk concentration of electrolyte is sufficiently large. A possible explanation, which constitutes the basis of the present theoretical treatment, is that hydrated ions are adsorbed on the negatively charged surface of mica, thus generating a bilayer of charge. This bilayer of charge generates a sufficiently strong polarization of the neighboring water molecules which propagates further into the liquid as a result of the interactions among the water molecules. We remark that the bilayer of charge here plays the same role as do the zwitterionic head groups in the case of the phospholipids. It is also important to emphasize that it is natural to expect that the double charge of the zwitterionic head groups and of the bilayer of charge both have a stronger directional effect than a single charged surface has. However the two cases differ in one important respect: the adsorbed hydrated ions are expected to deform when the separation distance between the plates decreases,

Volume 95, number 4,5 CHEMICAL PHYSICS LETTERS 11 March 1983

while the zwitterions do not; as shown later in the paper, this deformation has a major effect on the repulsive force.

2. The physical model

Hydrated ions are assumed adsorbed on the negatively charged surface of mica. A hydrated ion is considered as a composite unit in which the water molecules are in contact with the ion with their dipoles oriented radially. This assumption is justified in view of the enormous electric field ($\approx 10^6$ esu) generated by the ion. In a bulk liquid the symmetry of a hydrated ion is spherical; a hydrated ion adsorbed on a surface, however, has symmetry only around an axis that passes through the ion and is normal to the surface. Capacitance measurements [6] indicate that the small ions are separated from the surface by water molecules. If the hydration number is four, one may therefore expect one water molecule to be aligned with the axis of symmetry and to lie between the negative charge of the surface and the positive charge of the adsorbed hydrated ion, while the other three are distributed symmetrically around the axis of symmetry. The molecule below the horizontal plane through the center of the ions is called an S molecule, while those above are called N molecules (fig. 1). The water molecules of the hydration sheath of the ion interact with the neighboring water molecules which are located outside the bilayer of charge. Now, the average dipole moment of the latter is affected by the other mica surface because of the overlap of the polarization fields. As a consequence, the angle θ

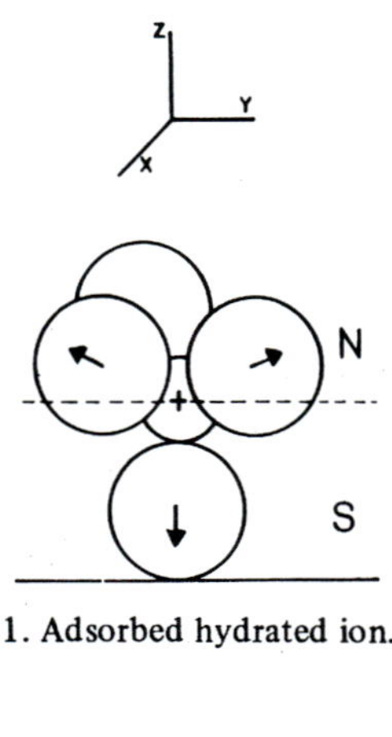

Fig. 1. Adsorbed hydrated ion.

between the dipole of each N molecule and the symmetry axis of the ion depends upon the distance between the two surfaces. In turn the angle θ affects the polarization of the nearby water molecules which are located outside the bilayer of charge, i.e. it will affect the boundary condition. The theory formulated below begins with the differential equation for the polarization field derived in the previous paper [4]; the essential contribution of the present work is the derivation of a boundary condition which takes into account not only the polarization generated by the bilayer of charge in the neighboring water molecules, but also the effect of the latter on the structure of hydrated ions. One of the basic assumptions is that the bilayer of charge contains a sufficiently large number of centers of polarization to generate a polarization in all the water molecules of the first neighboring layer. If the above condition is not satisfied (see section 6), only some molecules of the first layer are polarized. In such cases, the number of centers of polarization increases from layer to layer, till after a few layers all the water molecules are polarized.

3. The polarization field

The polarization P satisfies the following equation [4]

$$\mathrm{d}^2P/\mathrm{d}z^2 = \alpha^2 P, \qquad (1)$$

where z is the distance to one of the plates and α^{-1} is the decay length. P is related to the average dipole moment $\bar{\mu}$ via the expression

$$P = \bar{\mu}/a^2\delta, \qquad (2)$$

where δ is the distance between two water molecules along the z axis and a is the distance between two centers of polarization in a plane parallel to the charged surface. As already mentioned, the calculations have been carried out in the present paper for the case in which $a \approx \delta$. In this case $\alpha^{-1} \approx 2.4$ Å [4].

One obvious boundary condition is

$$P(0) = -P(h), \qquad (3)$$

where h is the distance between the layers of adsorbed ions (which in the present treatment are considered fixed). The solution of eq. (1) for the boundary

Volume 95, number 4,5 CHEMICAL PHYSICS LETTERS 11 March 1983

condition (3) is

$$P(z) = A(e^{-\alpha z} - e^{-\alpha(h-z)}), \tag{4}$$

where A is a constant. To obtain an expression for A an additional boundary condition has to be derived which takes into account the interactions between the layer of adsorbed hydrated ions and the neighboring water molecules.

4. The boundary condition due to the interactions between the adsorbed hydrated ions and the neighboring water molecules

It is important to emphasize that in the present model one assumes that the interaction of the hydration sheaths of the adsorbed ions with the neighboring free water molecules outside the bilayer of charge produces changes in the N molecules but no change in the S water molecules. Denoting by μ_a the z component of the total dipole moment of the three water molecules of the N region and by n_a the number of hydrated ions adsorbed per unit area, we can write that:

$$P(0) = (n_a/\delta)\,\mu_a. \tag{5}$$

The dipole moment μ_a can be calculated if the angle θ between the axis of a water molecule of the N region and the axis of symmetry of the hydrated ion is known. Indeed,

$$\mu_a = 3\mu \cos\theta, \tag{6}$$

where μ is the dipole moment of water.

The angle θ is a result of various forces acting on each of the N water molecules and can be obtained from the condition that the tangential field acting on such a molecule of water be zero. It is convenient to decompose the total field into three parts: $\boldsymbol{E}_m$ generated by the monopoles (ions, surface charges), $\boldsymbol{E}_d$ due to the dipoles of the hydrated sheaths of the adsorbed ions (including the other water molecules belonging to the hydration sheath of the same ion), and $\boldsymbol{E}_b$ generated by the neighboring free water molecules located outside the bilayer of charge. The condition for the tangential component to be zero leads to

$$E_{m,t} + E_{d,t} + E_{b,t} = 0, \tag{7}$$

where the subscript t denotes the tangential component. The dependence of the fields $E_{m,t}$ and $E_{d,t}$ on θ has been calculated numerically for various values of n_a, assuming a hexagonal configuration for the adsorbed ions using the expressions:

$$(E_m^{(n)})_i = \sum_{n'} \frac{(r_i^{(n)} - r_i^{(n')})e}{|r^{(n)} - r^{(n')}|^3} \tag{8}$$

and

$$(E_d^{(n)})_i = \sum_{n''} - \frac{\delta_{ij}\mu_j^{(n'')}}{|r^{(n)} - r^{(n'')}|^3} + \frac{3(r_i^{(n)} - r_i^{(n'')})(r_j^{(n)} - r_j^{(n'')})\mu_j^{(n'')}}{|r^{(n)} - r^{(n'')}|^5}. \tag{9}$$

Here $\boldsymbol{r}$ is a position vector with components r_j, μ_j are the components of the dipole moment and the superscripts denote dipoles (n, n'') or ions (n'), the latter having charge e. As usual, δ_{ij} is the Kronecker δ and repetition of the same subscript implies summation with respect to that subscript. To perform the sum in eq. (9), the following comments are useful: When the distance between two neighboring adsorbed hydrated ions is sufficiently large, each of them can have a large number of configurations obtained through rotation around the axis of symmetry; in this case, one can consider that each hydrated ion sees the total dipole moment of the N region of the other hydrated ion distributed uniformly over the lateral surface of a cone of angle 2θ. When the distance is small, however, the hydration sheaths of the two are strongly correlated and a single configuration exists.

Because $E_{b,t} = E_{b,z} \sin\theta$, where $E_{b,z}$ is the component in the z direction of the field generated by the free water molecules located outside the bilayer of charge, eq. (7), together with (8) and (9), provides a relation between $E_{b,z}$ and θ:

$$E_{b,z} = f(\theta), \tag{10}$$

which by taking into account eq. (6) can be transformed into a relation between μ_a and $E_{b,z}$. However, according to ref. [4],

$$E_{b,z} = \Lambda(\delta)\delta a^2 P(\delta), \tag{11}$$

where

Volume 95, number 4,5 CHEMICAL PHYSICS LETTERS 11 March 1983

$$\Lambda(\delta) = \frac{2}{\delta^3} + \frac{6}{(a^2+\delta^2)^{3/2}} \frac{2\delta^2 - a^2}{\delta^2 + a^2} \qquad (12)$$

and has approximately the value $3/\delta^3$ for $a \approx \delta$.

Eqs. (5), (6), (10), and (11) provide the necessary boundary condition. It is, however, convenient to proceed as follows:

If the value of θ resulting from eq. (7) when $E_{b,t}$ is taken zero is θ_0, then the interactions with the neighboring water molecules change the angle θ by $\Delta\theta = \theta - \theta_0$, where

$$\Delta\theta = f_1(E_{b,z}, \theta) \equiv g_1(\theta)E_{b,z} + g_2(\theta)E^2_{b,z} + \dots \qquad (13)$$

Because the change $\Delta\mu_a$ produced by $\Delta\theta$ is given by $-3\mu \sin\theta\, \Delta\theta$, eq. (13) leads to

$$\Delta\mu_a \approx \gamma_1(\theta)E_{b,z} + \gamma_2(\theta)E^2_{b,z}. \qquad (14)$$

The numerical calculations are long but simple and show that for θ between 60 and 120°, γ_1 is a weak function of θ and that the higher-order terms can be neglected. In the range of angles mentioned, γ_1 varies between 3 and 5×10^{-23} esu, the values being dependent on the surface density n_a of adsorbed hydrated ions ($\gamma_1 = 3 \times 10^{-23}$ for $n_a = 2 \times 10^{14}$ ions/cm^2 and $\gamma_1 = 5 \times 10^{-23}$ for $n_a = 5 \times 10^{13}$ ions/cm^2). Hence

$$\mu_a = \mu_{a,0} + \Delta\mu_a, \qquad (15)$$

where $\mu_{a,0}$ is the value of μ_a for $\theta = \theta_0$.

Combining eqs. (4), (5), (15), (14), (11), and (12) (for $a \approx \delta$), one obtains

$$A = \frac{n_a\mu_{a,0}}{\delta[(1-e^{-\alpha h}) - (3n_a\gamma_1/\delta)(e^{-\alpha\delta} - e^{-\alpha(h-\delta)})]}. \qquad (16)$$

One may observe that for $h \to 0$, $P(0) \to 0$ and the three water molecules of the N region tend to become arranged in a plane parallel to the surface of mica. For $h \to \infty$, $P(0)$ is larger than $n_a\mu_{a,0}/\delta$ because of the effect of the bulk of the liquid.

5. The repulsive force per unit area between the two surfaces

The free energy G per unit area of the system is equal to the sum contributions G_b from the bulk and G_a from the adsorbed layers of hydrated ions. The repulsive force per unit area is provided by the derivative:

$$F = -\partial G/\partial h. \qquad (17)$$

The free energy G_b is given by the expression:

$$G_b = -\tfrac{1}{2}n \int_0^h E(z)\bar{\mu}(z)\,dz, \qquad (18)$$

where n is the molecular density of water and $E(z)$ is the field. Since $\bar{\mu}(z) = \gamma E(z)$, where γ is the polarizability of the bulk, given by the Clausius–Mosotti equation, eq. (18) becomes:

$$G_b = -\frac{1}{2n\gamma}\int_0^h P^2(z)\,dz. \qquad (19)$$

The free energy per unit area of the layer of adsorbed hydrated ions is given by

$$G_a = -2n_a\left(\frac{1}{2}\sum_{m'=1}^{m} \mathbf{E}^{(m')}\cdot\boldsymbol{\mu}^{(m')} + (m-1)\mathbf{E}_m\cdot\boldsymbol{\mu}^{(N)} + (m-1)\mathbf{E}_b\cdot\boldsymbol{\mu}^{(N)}\right), \qquad (20)$$

where $\boldsymbol{\mu}^{(m')}$ is the dipole moment of the water molecule m' in a hydration sheath, m is the hydration number, $\mathbf{E}^{(m')}$ is the field acting on molecule m' that is generated by the other water molecules in the same

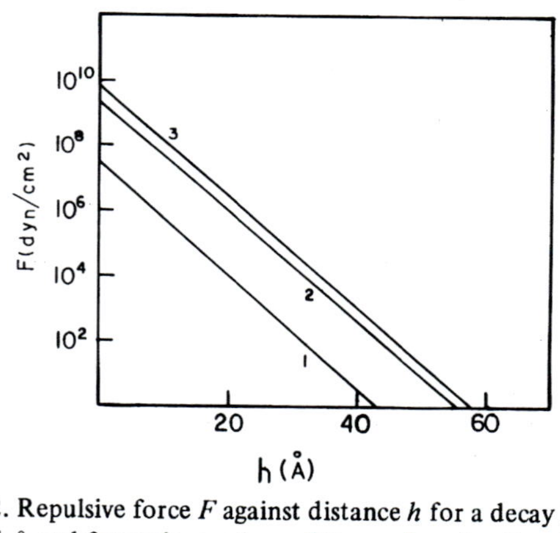

Fig. 2. Repulsive force F against distance h for a decay length of 2.4 Å and for various values of the surface density of hydrated ions, n_a [(1) $n_a = 5 \times 10^{13}$; (2) $n_a = 10^{14}$; (3) $n_a = 2 \times 10^{14}$ ions/cm^2].

Volume 95, number 4,5 CHEMICAL PHYSICS LETTERS 11 March 1983

sheath as well as by those in the sheaths of the other adsorbed ions, E_m is the field due to the monopoles, E_b that generated by the bulk, and $\boldsymbol{\mu}^{(N)}$ the dipole moment of each of the water molecules of the N region. By using eq. (19) and evaluating G_a numerically from eq. (20), the results plotted in fig. 2 have been obtained. The computations show that at sufficiently large values of h the repulsive force is dominated by the overlapping polarization fields, whereas for sufficiently small values of h the layers of adsorbed hydrated ions and their deformation dominate the repulsive force.

6. Discussion

Experiment [3] shows that the decay length has values between ≈3 and 12 Å. For small ions, such as Li^+ and Na^+, the decay length is ≈9 Å for a surface coverage of $\approx 2 \times 10^{14}$ ions/cm^2. For large ions, such as Cs^+ or K^+, the decay length is ≈3 Å at a surface coverage of $\approx 10^{14}$ ions/cm^2. However, when the surface density of K^+ is 3.5×10^{13} ions/cm^2, the decay length was near to 6 Å, whereas for a surface density of $\approx 2 \times 10^{14}$ ions/cm^2, it increases to 12 Å. As mentioned previously in the paper, the theory predicts a decay length of ≈2.4 Å when the first layer of water molecules at the boundary of the adsorbed hydrated ions is completely polarized. For this to occur, there must be a sufficient density of N-type water molecules, each of them having an appreciable dipole moment in the z direction. In this case $a = \delta$. When the surface concentration of hydrated ions is sufficiently low, or when their surface concentration is large, but the configuration of the hydrated ions is such that fewer N molecules have an appreciable dipole moment in the z direction, only some of the water molecules of the first layer will be polarized. In such cases $a > \delta$ and the equation for α in ref. [4] predicts that the decay length is larger than 2.4 Å. It becomes equal to 2.4 Å after a few layers. Because the small ions generate a strong electric field, their hydration sheaths probably have a higher rigidity than the larger ions do. Therefore, the hydrated ions adsorbed on the solid surface keep in some measure the spherical symmetry they have in the bulk. Since they do not deform appreciably, they influence a smaller area of the layer of neighboring water molecules. In contrast, larger ions, such as Cs and K, have hydration sheaths that are not rigid, bur rearrange during adsorption, generating polarization over a larger area. This explains why the decay length is, in general, greater in the first case than in the second. At very high surface concentration of K^+ ions, the overlapping of the hydration sheaths probably leads to a change of their configuration which, in turn, leads to a larger decay length.

Let us compare the experimental results of ref. [3] with our theoretical results. The Derjaguin approximation allows us to write for the hydration force $\mathcal{F}$ between two curved surfaces of radius R the expression

$$\mathcal{F}/R = 2\pi G,$$

where G represents the free energy per unit surface area between two parallel plates. Because the decay length predicted by the theory (2.4 Å) is in reasonable agreement with experiment, it is sufficient to compare the experimental values of the above ratio with theory for $h = 0$. The theoretical value of $2\pi G$ is $\approx 4 \times 10^3$ dyn/cm. The experimental value of $\mathcal{F}/R$ for $h = 0$ (and $\alpha^{-1} = 3$) is between 10^3 and 10^4 dyn/cm.

References

[1] D.M. LeNeveu, R.P. Rand and V.A. Parsegian, Nature 259 (1976) 601.
[2] V.A. Parsegian, N. Fuller and R.P. Rand, Proc. Natl. Acad. Sci. US 76 (1979) 2750.
[3] R.M. Pashley, J. Colloid Interface Sci. 83 (1981) 531.
[4] D. Schiby and E. Ruckenstein, Chem. Phys. Letters 95 (1983) 435.
[5] S. Marčelja and N. Radić, Chem. Phys. Letters 42 (1976) 129.
[6] J.O'M. Bockris and A.K.N. Reddy, Modern electrochemistry, Vol. 2 (Plenum Press, New York, 1970).

Volume 100, number 3 CHEMICAL PHYSICS LETTERS 9 September 1983

ON THE COUPLING BETWEEN THE DOUBLE LAYER AND THE SOLVENT POLARIZATION FIELDS

D. SCHIBY and E. RUCKENSTEIN
Department of Chemical Engineering, State University of New York at Buffalo, Buffalo, New York 14260, USA

Received 23 May 1983

An approach is developed to incorporate in the standard theory of the double layer the effect of the solvent polarization. This polarization is due to a source of polarization on the surface and to the charges in the diffuse layer. The new equations lead to the conclusion that a diffuse layer can be generated even in the absence of a surface charge as long as there is some source of polarization on the surface, such as a layer of dipoles. The solvent polarization also changes the surface potential.

1. Introduction

Recent experiments by Parsegian et al. [1,2] have revealed strong repulsive forces between two parallel phospholipid bilayers immersed in an electrolyte solution, at distances below ≈40 Å; these forces are insensitive to the concentration of electrolyte. Later experiments [3,4] concerned with the interaction force between two mica plates immersed in an electrolyte solution have found a repulsive force much stronger than that predicted by DLVO theory; as in the previous case the additional repulsion occurs for distances below ≈40 Å, but only when the concentration of electrolyte is sufficiently large. Both kinds of experiments clearly demonstrate the occurrence of a short-range repulsive force different in nature from double-layer forces. They have been called hydration forces even though the name solvent polarization forces is more adequate. A first treatment of these forces based on the Landau phenomenological expansion for the free energy was provided by Marčelja and Radić [5] and further developed by Cevc et al. [6]. A very different approach based on electrostatics, was developed by Schiby and Ruckenstein [7,8]. In contrast to previous work, this provides an expression for the decay length of this force in terms of known physical quantities.

The origin of the above repulsive forces can be explained qualitatively as follows:

The zwitterionic headgroups of the phospholipid bilayers polarize the water molecules nearby, which in turn act as a self-consistent field at the location of the nearby dipoles. The effect decays rapidly away from the wall. When the polarization layers thus produced by the two phospholipid bilayers overlap, the local average dipole moment of the solvent molecules decreases, thus increasing their free energy. This generates a repulsive force. Similarly, the adsorption of hydrated ions on a negatively charged mica surface generates a bilayer of charge which plays the same role as do the zwitterionic headgroups in phospholipid bilayers. In the case of mica surfaces the concentration of electrolyte plays a role since it controls the adsorption of hydrated ions. In the previous papers [7,8], the polarization profiles in the solvent, produced by the zwitterionic headgroups in the case of phospholipid bilayers and by the bilayer of charge in the case of mica, have been calculated ignoring the effect of the double layers. A natural question is: in what way does the polarization of the solvent affect (interact with) the double-layer field. The main difficulty in considering this problem is the formulation of the basic equations which account for these interactions. This constitutes the main goal of the present note. It will be shown that because of the polarization of the solvent one can have a diffuse layer of ions even in the absence of a surface charge and that there is an excess surface potential due to the polarization of the solvent.

Volume 100, number 3 CHEMICAL PHYSICS LETTERS 9 September 1983

2. The equations

The traditional theory of the double layer is based on a combination of the Poisson equation and Boltzmann distribution. While this involves the approximation that the potential of mean force used in the Boltzmann expression equals the mean value of the electrical potential [9], the results thus obtained are satisfactory at least for 1 : 1 electrolytes. The equations proposed in the present paper use the approximations inherent in the Poisson–Boltzmann equation, but also include the effect of the polarization field of the solvent which is caused by a polarization source assumed uniformly distributed on the surface and by the double layer itself.

The Poisson equation

$$dE/dz = 4\pi(\rho - dP/dz) \quad (1)$$

constitutes the starting point of the treatment. Here E is the macroscopic field acting at a distance z from the plate, ρ is the charge density of free charges (ions) and P is the polarization. The usual constitutive equation which is used in the case of double layers has the form

$$P = \chi E, \quad (2)$$

where χ is the electric susceptibility. However, this expression is valid when the field E is uniform or when the gradients of E (or P) are sufficiently small. Near the solid surface the gradients of P are very large (see below) and for this reason the above expression is no longer valid. For the present situation therefore eq. (2) must be replaced by another which we now proceed to establish. Each water molecule is treated as a dipole. Its moment is taken to be equal to the average dipole moment along the direction of the average electric field and is induced by the self-consistent field generated by the surrounding water molecules (dipoles), in combination with the diffuse layer field and any other field present. The field E_{mic} that acts on a molecule situated at a distance z from the plate contains two components:

$$E_{mic}(z) = E^0(z) + E^i(z), \quad (3)$$

where $E^0(z)$ is the field due to the double layer and to any other field present, and $E^i(z)$ is the field generated by the neighboring water molecules. Here we consider that $E^0(z)$ is due only to the double layer.

Further, the polarization P can be related to the field E_{mic} via the expression

$$P = \gamma n E_{mic}, \quad (4)$$

where γ is the molecular polarizability and n is the number of water molecules per unit volume. Assuming as in ref. [7] that the field $E^{(i)}$ acting on a water molecule is generated by the dipole interactions with the nearest neighbours only, one obtains [7]:

$$nE^{(i)}(z) = LP + \delta^2 M\, d^2P/dz^2, \quad (5)$$

where

$$L = \frac{4}{\delta^3} + \frac{12}{(a^2+\delta^2)^{3/2}}\,\frac{2\delta^2 - a^2}{\delta^2 + a^2} - \frac{6}{a^3} \quad (6a)$$

and

$$M = \frac{2}{\delta^3} + \frac{6}{(a^2+\delta^2)^{3/2}}\,\frac{2\delta^2 - a^2}{\delta^2 + a^2}\,. \quad (6b)$$

Here, a is the shortest distance between two neighboring water molecules in the plane normal to z and δ ($\approx a$) the shortest distance between two such molecules along the z distance. Since $\delta \approx a$ it results that $L \approx 0$. $L = 0$ is equivalent to assuming that the mean arrangement of water molecules that are oriented by a uniform field has sufficient symmetry for the dipole fields at a central molecule due to its neighbors to cancel. Combining eqs. (3), (4) and (5), one obtains*

$$d^2P/dz^2 = \alpha^2 P - nE^0(z)/\delta^2 M \equiv \alpha^2[P - n\gamma E^0(z)], \quad (7)$$

where

$$\alpha^2 = 1/\gamma M \delta^2. \quad (8)$$

Because for water $\gamma \approx 6 \times 10^{-24}$ esu and $\delta \approx a = 2.8$ Å, eq. (8) leads to $\alpha^{-1} \approx 2.4$ Å. Because α^2 is a very large quantity, $P \neq n\gamma E^0(z)$ only in the region in which d^2P/dz^2 is also sufficiently large. Outside that region, $P = n\gamma E^0(z)$. In eq. (7) the term $n\gamma E^0(z)$ depends on the charge density ρ and remains constant if the latter quantity is kept constant. Let us now consider a fictitious process in which $\alpha^2 \to \infty$ while ρ is kept constant. In this limit the polarization P tends to a value P' which, from eq. (7) is given by

* There is an error in eq. (11) of ref. [7]. The second term on the right-hand side should read $-E^{(0)}/\delta^2 M$ instead of $+E^{(0)}/\delta^2 M\gamma$.

$$P' = \gamma n E^{(0)}. \qquad (9)$$

Because in this case the decay length $\alpha^{-1} = 0$, the inhomogeneous field $E^{(i)}$ of the solvent dipoles disappears completely and the Poisson equation (1) reduces to that usually encountered in the double-layer theory:

$$dE'/dz = 4\pi(\rho - dP'/dz) \equiv 4\pi\rho/\epsilon, \qquad (10)$$

with

$$P' = \chi E', \qquad (11)$$

ϵ being the dielectric constant of the bulk liquid.

Consequently the state of the layer near the solid surface can be described by the system of equations

$$dE/dz = -d^2\psi/dz^2 = 4\pi(\rho - dP/dz), \qquad (12)$$

$$d^2P/dz^2 = \alpha^2(P - \chi E') \qquad (13)$$

and

$$dE'/dz = 4\pi\rho/\epsilon, \qquad (14)$$

where

$$\rho = \sum \nu_i e c_i \exp[-(\nu_i \psi e/kT)]. \qquad (15)$$

Here ψ is the electrical potential, ν_i is the valence of the ion (positive for cations), c_i is the number of ions of species i per unit volume in the reservoir with which the space between the plates is in contact, k is the Boltzmann constant, e the protonic charge and T the absolute temperature.

3. Two parallel plates immersed in an electrolyte solution

As is shown in fig. 1 a layer of hydrated ions is assumed adsorbed on the negatively charged solid surface. The plane SH, taken as the origin, is located a distance $\frac{1}{2}h$ from the midplane, and is determined by the (average position of) the centers of those water molecules in the hydration sheaths of the adsorbed ions that lie on the water side of the ions. The minimum distance δ_1 of the counterions in the diffuse layer from the plane SH is about the radius $\frac{1}{2}\delta'$ of the hydrated ion, while the minimum distance of the co-ion from the same plane, $\delta_2 \approx 0$. Expression (15)

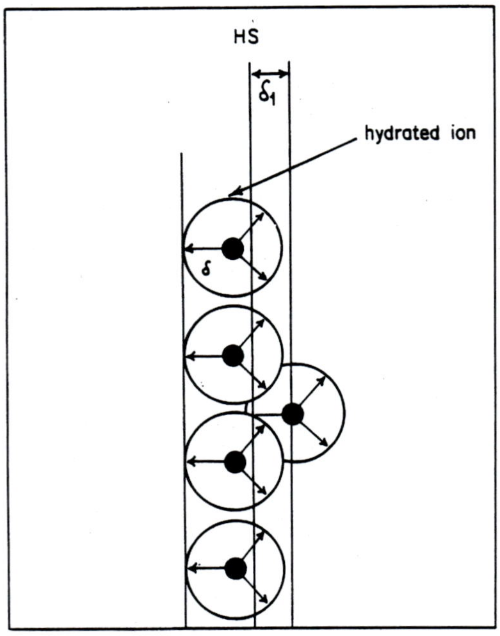

Fig. 1. Geometrical characteristics for the two parallel plates system.

is valid in the range $\delta_1 \leqslant z \leqslant \frac{1}{2}h$. In the range $z < \delta_1$, eq. (15) has to be replaced by

$$\rho = \nu_c e c_c \exp[-(\nu_c \psi e/kT)], \qquad (15')$$

where the subscript c refers to the co-ion.

The boundary conditions are

$$\epsilon E' = 4\pi\sigma, \quad \text{for } z = 0, \qquad (16)$$

$$P = P_0, \quad \text{for } z = 0, \qquad (17a)$$

$$P = 0, \quad \text{for } z = \tfrac{1}{2}h, \qquad (17b)$$

$$d\psi/dz = -4\pi(\sigma - P_0), \quad \text{for } z = 0, \qquad (18a)$$

$$d\psi/dz = 0, \quad \text{for } z = \tfrac{1}{2}h. \qquad (18b)$$

Here σ is the net surface charge density (the difference between the charge of the surface and the charge of hydrated ions adsorbed) and

$$P_0 = n_a \mu_a/\delta,$$

where μ_a is the mean dipole moment of that part of an adsorbed hydrated ion that faces the liquid, and n_a is the surface density of these ions. The above equations have been solved numerically for the boundary

Volume 100, number 3 CHEMICAL PHYSICS LETTERS 9 September 1983

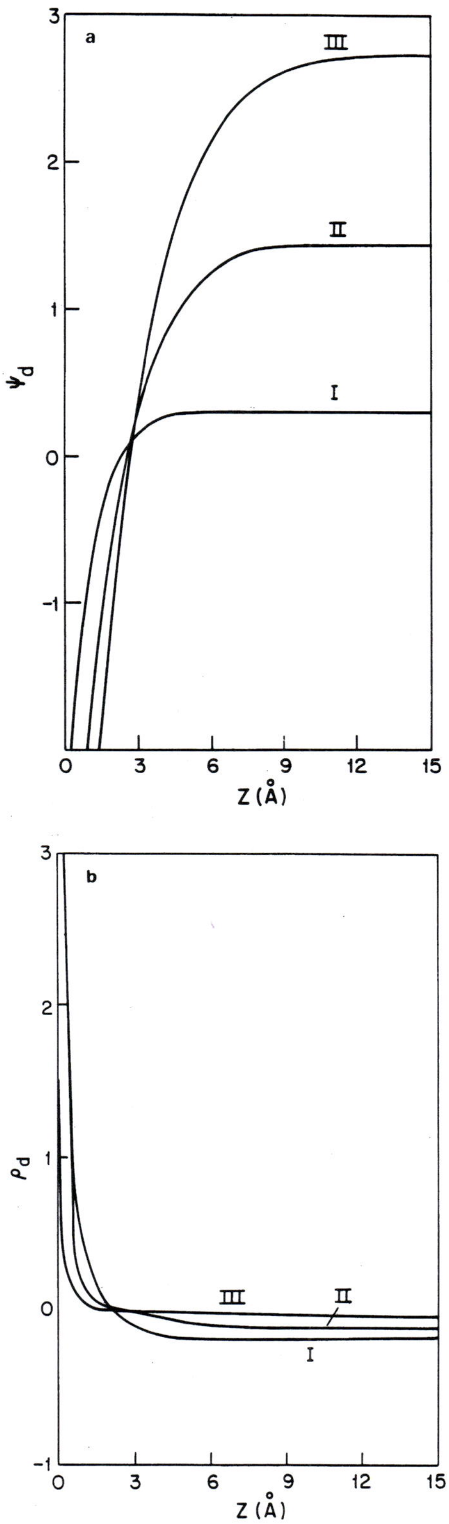

conditions (16)–(18) and some results have been plotted in figs. 2–4.

4. Results and discussions

In contrast to the DLVO theory, the present theory predicts a diffuse layer even for zero surface charge (figs. 2a and 2b). Since the total charge of this layer has to be zero, the charge density ρ will change sign at a certain distance from the wall. To explain the occurrence of this diffuse layer one may note that a positive polarization generates by itself a negative electric field which leads to the migration of the cations towards the surface. Thus, there is an accumulation of cations near the surface and of anions further from the surface. However, this accumulation leads to a positive field which tends to displace cations away from the surface. At equilibrium the flux caused by the resultant of these two fields equals the opposite flux due to the concentration gradient. Figs. 2a and 2b give for h = 30 Å and for bulk concentrations c_i = 0.1, 0.01 and 0.001 M the dimensionless potential $\psi_d = \psi e/kT$ and dimensionless charge density $\rho e\delta^2/kT$. Clearly the results of figs. 2a and 2b are at variance with the DLVO theory which would predict no diffuse layer at all. Sufficiently far from the wall, the deviations from the DLVO theory are small and increase with increasing c_i. It is of interest to note that the wall potential generated by the polarization P_0 can be (even for $\sigma = 0$) as high as 300 mV if $\delta e P_0/kT = 10$. For $\sigma \neq 0$ and sufficiently small distances between plates, the DLVO theory predicts almost constant electrical potential and charge density. In the present theory, when σ is negative and the surface polarization is positive a larger number of cations are accumulated near the surface than predicted by the DLVO theory. As in the previous case, the deviations from the DLVO theory are very large near the wall and become small sufficiently far from the wall if the distance between the plates is sufficiently large (see figs. 3 and 4).

◀ Fig. 2. (a) Dimensionless potential $\psi_d = \psi e/kT$ as a function of z (for $z \leqslant \frac{1}{2}h$). h = 30 Å, $\sigma = 0$, $P_{0,d} = P_0 e\delta/kT = 1$ and the ionic concentration (1 : 1 electrolyte) c_i = 0.1 M (curve I), 0.01 M (curve II) and 0.001 M (curve III). Here δ_1 is taken zero. (b) Dimensionless charge density $\rho_d = \rho\delta^2/kT$ as a function of z for the same conditions as in in (a).

Volume 100, number 3 CHEMICAL PHYSICS LETTERS 9 September 1983

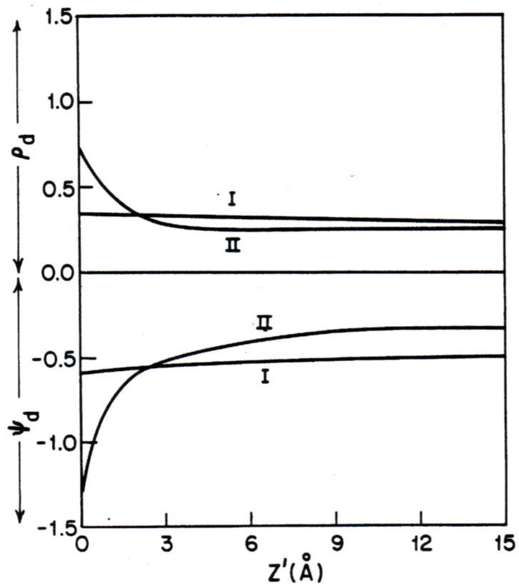

Fig. 3. ψ_d and ρ_d as a function of $z' = z - \delta_1$. The distance between the planes SH of the two plates is $h = 30 + 2\delta_1$ Å; $\delta_1 = 3.5$ Å, $\sigma_d = \sigma e\delta/kT = -0.75$, $c_i = 0.1$ M, $P_{0,d} = 0$ (curve I), $P_{0,d} = 1$ (curve II).

We have also calculated the total free energy of the system by adding the free energy W [7] due to the polarization field in the absence of free charges to that, resulting from a Debye charging procedure, which provides the contribution of the free charges. The calculations show that the repulsive force due to the polarization of the solvent is not affected by the double layer when the surface polarization is sufficiently strong. Details will be included in an expanded version of this paper.

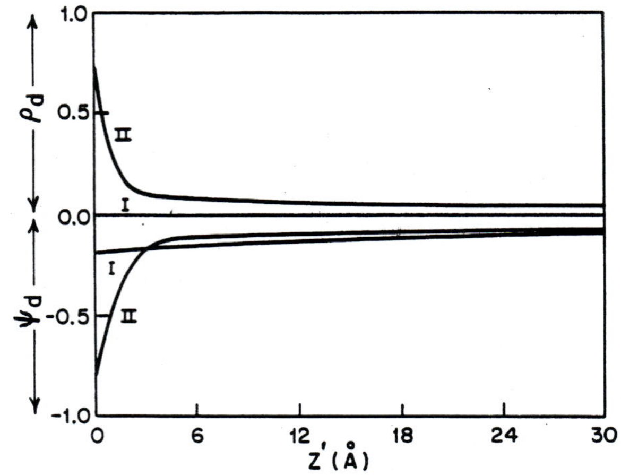

Fig. 4. ψ_d and ρ_d as functions of $z' = z - \delta_1$ for $h = 60 + 2\delta_1$ Å; the values of the other parameters are as in fig. 3.

References

[1] D.M. LeNeveu, R.P. Rand and V.A. Parsegian, Nature 258 (1976) 601.
[2] V.A. Parsegian, N. Fuller and R.P. Rand, Proc. Natl. Acad. Sci. US 76 (1979) 2750.
[3] J.N. Israelachvili, Faraday Discussions 65 (1978) 20.
[4] R.M. Pashley, J. Colloid Interface Sci. 83 (1981) 531.
[5] S. Marčelja and N. Radić, Chem. Phys. Letters 42 (1976) 129.
[6] G. Cevc, R. Podgornik and B. Žekš, Chem. Phys. Letters 91 (1982) 193.
[7] D. Schiby and E. Ruckenstein, Chem. Phys. Letters 95 (1983) 435.
[8] E. Ruckenstein and D. Schiby, Chem. Phys. Letters 95 (1983) 439.
[9] J.G. Kirkwood, J. Chem. Phys. 2 (1934) 767.

7582 *Langmuir* **2001,** *17,* 7582–7592

Oscillatory and Monotonic Polarization. The Polarization Contribution to the Hydration Force

Marian Manciu and Eli Ruckenstein*

Department of Chemical Engineering, State University of New York at Buffalo, Buffalo, New York 14260

Received June 25, 2001. In Final Form: September 20, 2001

The propagation of polarization in water, in the vicinity of a planar surface containing dipoles, is considered to be a consequence of the nonuniformity of the dielectric constant near the dipoles. Discrete finite difference equations have been derived by assuming a layered icelike structure. These equations predicted a polarization that oscillates with the distance from the surface. When the polarizing surface was considered rough or fluctuating, the oscillations were smoothed out and a monotonically decaying average polarization was obtained. Assuming that there are local icelike structures around any of the water molecules, a second-order differential equation for the polarization and expressions for its decay length were derived.

I. Introduction

The direct measurement of the interaction force between two mica surfaces[1] indicated a large repulsion at relatively short distances, which could not be accounted for by the DLVO theory. This force was associated with the structuring of water in the vicinity of the surface.[2] Theoretical work and computer simulations[3–5] indicated that, in the vicinity of a planar surface, the density of the liquid oscillates with the distance, with a periodicity of the order of molecular size. This reveals that, near the surface, the liquid is ordered in quasi-discrete layers. When two planar surfaces approach each other at sufficiently short distances, the molecules of the liquid reorder in discrete layers, generating an oscillatory force.[6]

When the surfaces were not rigid, as in the case of lipid bilayers, the oscillations of the force were smoothed out and the interactions became monotonic. The short-range repulsion between neutral[7] or weakly charged[8] bilayers, often called hydration force, was exhaustively investigated experimentally and was found to have an exponential decay, with a decay length between 1.5 and 3 Å, while the preexponential factor varied by more than an order of magnitude.

A phenomenological treatment of the hydration repulsion, based on a Landau expansion of the free energy density, was proposed by Marcelya and Radic.[9] They showed that, if the free energy density is a function of an "order parameter" that varies continuously from the surface, and if only the quadratic terms in this parameter and its derivative are nonnegligible, an exponential decay of the repulsion could be obtained for not too short distances. They also suggested that the correlation (decay) length of the order parameter should be of the order of the molecular size. However, the microscopic origin of the order parameter, and more importantly, the dependence of the free energy density on the order parameter, remained obscure.

The interaction was later associated with the propagation of the polarization in water, and two models were proposed. Schiby and Ruckenstein[10] suggested that the polarization propagates because of interaction at the molecular level (mutual orientation of neighboring dipoles); we will return later to this model. Gruen and Marcelja[11] introduced a continuum model, in which equations for both the polarization and the electric field, considered independent quantities, were derived via the minimization of the total free energy of the system. The spatial variation of the polarization was considered dependent on the concentration of the Bjerrum defects, which constitute the source of the polarization field, while the spatial variation of the electric field was considered dependent on the concentration of ions, which constitute the source of the electric field. They showed that the polarization varies smoothly from the interface and that the electrostatic double-layer repulsion is enhanced by a contribution due to the surface polarization. At low separation distances, the effect of polarization becomes dominant and the repulsion decays exponentially, with a decay length inversely proportional to the square root of the concentration of the Bjerrum defects.

A simplified version of the theory was presented by Cevc and Marsh.[12] They started from the Marcelja–Radic phenomenological treatment, assumed that the polarization constitutes the order parameter, and used the Gruen–Marcelja model to explain the various contributions to the free energy density.

The subsequent molecular dynamics simulations[13,14] indicated that the polarization in water does not vary smoothly near the surface, but oscillates, and hence is not

* Corresponding author. E-mail address: feaeliru@acsu.buffalo.edu. Phone: (716) 645-2911/2214. Fax: (716) 645-3822.

(1) Horn, R. G.; Israelachvili, J. N. *J. Chem. Phys.* **1981**, *75*, 1400.
(2) Derjaguin, B. V.; Churaev, N. V. *J. Colloid Interface Sci.* **1974**, *49*, 249.
(3) Nicholson, D.; Parsonage, N. G. *Computer Simulation and the Statistical Mechanics of Adsorption;* Academic Press: New York, 1982.
(4) Ciccotti, G.; Frenkel, D.; MacDonald, I. R. *Simulation of Liquids and Solids;* North-Holland: Amsterdam, 1987.
(5) Evans, R.; Parry, A. O. *J. Phys.: Condens. Matter* **1990**, *2*, SA15–SA32.
(6) Israelachvili, J. N. *Intermolecular and Surface Forces;* Academic Press: New York, 1992.
(7) Rand, R. P.; Parsegian, V. A. *Biochim. Biophys. Acta* **1989**, *988*, 351.
(8) Cowley, A. C.; Fuller, N. L.; Rand, R. P.; Parsegian, V. A. *Biochemistry* **1978**, *17*, 3163.
(9) Marcelja, S.; Radic, N. *Chem. Phys. Lett.* **1976**, *42*, 129.
(10) Schiby, D.; Ruckenstein, E. *Chem. Phys. Lett.* **1983**, *95*, 435.
(11) Gruen, D. W. R.; Marcelja, S. *Faraday Trans. 2* **1983**, 211 and 225.
(12) Cevc, G.; Marsh, D. *Biophys. J.* **1985**, *47*, 21.
(13) Kjellander, R.; Marcelja, S. *Chem. Scr.* **1985**, *25*, 73; *Chem. Phys. Lett.* **1985**, *120*, 393.
(14) Berkowitz, M. L.; Raghavan, K. *Langmuir* **1991**, *7*, 1042.

10.1021/la010979h CCC: $20.00
Published on Web 11/01/2001

accurately described by the Gruen–Marcelja theory.[11] It was therefore concluded that the polarization is not a suitable order parameter. It was also noted[15] that the exponential decay of the hydration force in the Gruen–Marcelja model was a consequence of an inconsistent choice for the functional form of the "non-local dielectric function".

The molecular dynamics simulation indicated that, unlike polarization (which oscillates), the disruption of hydrogen bonding varies smoothly in the vicinity of the surface and constitutes therefore a better choice as an order parameter. There were a number of computational approaches[13,14,16] and lattice model calculations[17] which related the hydration force to the disruption of the hydrogen bonding networks when two surfaces approach each other.

An essentially different point of view was suggested by Israelachvili and Wennerstrom, who consider that any interaction that may arise from water structuring effects should be either oscillatory or attractive but not repulsive, and hence that the microscopic origin of the hydration repulsion should be sought elsewhere than in hydrogen bonding or polarizability.[18] They proposed that the protrusion of the molecules belonging to the lipid bilayers could explain the hydration repulsion.[19] It was however argued that the interactions due to protrusion are much smaller than the experimentally measured hydration repulsion,[20] and experiment[21] indicated that the polymerization of the bilayers does not modify the hydration repulsion.

Another mechanism for the hydration repulsion between lipid bilayers was more recently proposed by Marcelja.[22] It is based on the fact that in water the ions are hydrated and hence occupy a larger volume. The volume exclusion effects of the ions are important corrections to the Poisson–Boltzmann equation and modify substantially the double-layer interaction at low separation distances. The same conclusion was reached earlier by Ruckenstein and Schiby,[23] and there is little doubt that the hydration of individual ions modifies the double-layer interaction, providing an excess repulsion force.[23] However, while the hydration of ions affects the double-layer interactions, the hydration repulsion is strong even in the absence of an electrolyte, or double-layer repulsion.

One can therefore conclude that there is no commonly accepted explanation for the microscopic origin of the hydration repulsion. The main purpose of this paper is to show that a suitable model for the polarization of water layers, based on the earlier work of Schiby and Ruckenstein,[10] is compatible with both simulations and experiments on hydration repulsion.

The treatment of polarization based on the assumption that water has a uniform dielectric constant involves a fundamental difficulty. Indeed, a uniform, continuous distribution of dipoles on a planar surface does not generate a field in a homogeneous medium and hence is not able to polarize the water. If the dipoles are distributed in the sites of a 2D planar square lattice, the field is oscillating above the lattice, with an amplitude decaying exponentially with the normal distance from the lattice. Hence, the average field is vanishing, and the polarization does not propagate in water. In the model of Gruen and Marcelja the spatial variation of polarization was associated with the existence of the Bjerrum defects. Consequently, the absence of such defects implies that the decay length is infinite and the polarization is constant in the medium, instead of being zero, as expected on the above grounds. Cevc and Marsh suggested that the polarization of water near the interface is proportional to the normal component of the dipolar density,[12] which is a plausible assumption. However, the proportionality constant was provided without explanation, and it is not clear why the surface dipoles polarize the continuous medium.

While one cannot rule out that there are contributions of different origins to the hydration repulsion, the polarization contribution might be the dominant one, at least for not too small separations, and this can explain, as shown later in the paper, the quadratic dependence, determined experimentally by Simon and McIntosh,[24] of the hydration repulsion on the surface dipolar potential.

The model proposed in the present paper for the polarization is based on a simple idea. In a real medium, the interactions between remote dipoles are screened by the intervening solvent molecules and hence are decreased by a factor equal to the inverse of the dielectric constant of the medium. On the other hand, the interactions between adjacent dipoles are much less screened and hence are more important. In a medium with homogeneous dielectric properties, a planar layer of dipoles produces a vanishing average electric field above it. However, because the medium is not homogeneous at the molecular level, the field of a nearby dipole is screened less and an electric field is generated above the plane.

To calculate the net field in water, when the dielectric constant is inhomogeneous, one should know the distribution of molecules around a given site. In the following, we will consider that a water molecule belongs locally to an icelike structure. This was suggested by Nemethy and Scheraga,[25] who treated the bulk water as an aggregate of icelike clusters of various sizes. In the vicinity of a surface, it is reasonable to assume that the clusters (which have a layered structure) are aligned by the surface. The alignment is expected to gradually decrease away from the surface.

It will be shown that local inhomogeneities in the dielectric constant of the medium are responsible for the propagation of polarization inside the medium. A discrete representation of the water layers leads to equations with finite differences whose solution allows one to relate the total electrostatic energy (due to polarization) to the density and location of the dipole moments on the surface. We will show that a long-range order in the water structure, normal to the interface, can lead to oscillations of the polarization and that in this case the repulsive force cannot be approximated by an exponentially decaying function. However, the roughness or the undulations of the polarizing surface can smooth out the oscillations and lead to an exponentially decaying repulsion.

II. Hydration Interaction

II.1. Net Field above a Plane of Dipoles. In a homogeneous dielectric described by a constant ϵ, the field

(15) Attard, P.; Wei, D. Q.; Patey, G. N. *Chem. Phys. Lett.* **1990**, *172*, 69.
(16) Besseling, N. A. M. *Langmuir* **1997**, *13*, 2113.
(17) Attard, P.; Batchelor, M. T. *Chem. Phys. Lett.* **1988**, *149*, 206.
(18) Israelachvili, J. N.; Wennerstrom, H. *Nature* **1996**, *379*, 219.
(19) Israelachvili, J. N.; Wennerstrom, H. *Langmuir* **1990**, *6*, 873.
(20) Parsegian, V. A.; Rand, R. P. *Langmuir* **1991**, *7*, 1299.
(21) Binder, H.; Dietrich, U.; Schalke, M.; Pfeiffer, H. *Langmuir* **1999**, *15*, 4857.
(22) Marcelja, S. *Nature* **1997**, *385*, 689.
(23) Ruckenstein, E.; Schiby, D. *Langmuir* **1985**, *1*, 612.
(24) Simon, S. A.; McIntosh, T. J. *Proc. Natl. Acad. Sci. U.S.A.* **1989**, *86*, 9263.
(25) Nemethy, G.; Scheraga, H. A. *J. Chem. Phys.* **1962**, *36*, 3382.

generated by an infinite planar, uniform distribution of dipoles vanishes. The dipole distribution can be thought of as two planar surface charge distributions, of equal magnitude and opposite sign, separated by the infinitesimal distance δ ($\delta \to 0$ while the surface dipole density $\delta\sigma$ = constant, where σ is the surface density of the charges on the two planes). Both planar charge densities generate constant electric fields, which cancel in the region outside the planes.

The field, normal to the surface, generated by the surface dipoles assumed to be distributed on the sites of a square lattice can be calculated as follows. The field along the z direction, normal to the surface, generated by the positive charges is given by[6]

$$E_z^{\,+} = \frac{\sigma}{2\epsilon\epsilon_0}\left(1 + 2\left(\cos\frac{2\pi x}{d} + \cos\frac{2\pi y}{d}\right)\exp\left(-\frac{2\pi z}{d}\right) + ...\right) \tag{1a}$$

where ϵ is the dielectric constant, ϵ_0 is the permittivity constant, x and y are the coordinates in the plane of the surface, and the charges are located at the positions $(x, y, z) = (nd, md, 0)$ with n and m integers and d the length of the unit cell of the lattice. The field generated by the negative charges located at $(x, y, z) = (nd, md, -\delta)$ is given by

$$E_z^{\,-} = -\frac{\sigma}{2\epsilon\epsilon_0}\left(1 + 2\left(\cos\frac{2\pi x}{d} + \cos\frac{2\pi y}{d}\right)\exp\left(-\frac{2\pi(z+\delta)}{d}\right) + ...\right) \tag{1b}$$

Hence, for the total field generated by the dipoles normal to the surface one obtains

$$E_z^{\,d} = E_z^{\,+} + E_z^{\,-} = \frac{2\pi\sigma\delta}{\epsilon\epsilon_0 d}\left(\cos\frac{2\pi x}{d} + \cos\frac{2\pi y}{d}\right)\exp\left(-\frac{2\pi z}{d}\right) \tag{2}$$

where the terms of higher order in δ were neglected. The field generated by the surface dipoles, averaged on a planar surface equal to a unit cell of the lattice, vanishes.

However, the medium is not homogeneous at molecular distances, and this can be taken into account by assuming that the field caused by the nearest dipole is screened by a different dielectric constant, $\epsilon_1 < \epsilon$. If one denotes by $E_{z,i}^{\,d}$ the field along the z direction, generated in a vacuum by the dipole i of the surface, the field above the lattice is given by

$$E_z^{\,d} = \frac{1}{\epsilon_1}E_{z,\text{nearest}}^{\,d} + \frac{1}{\epsilon}\sum_{\text{all others}} E_{z,i}^{\,d} = \left(\frac{1}{\epsilon_1} - \frac{1}{\epsilon}\right)E_{z,\text{nearest}}^{\,d} + \frac{1}{\epsilon}\sum_{\text{all}} E_{z,i}^{\,d} = \frac{(\epsilon - \epsilon_1)}{\epsilon\epsilon_1}E_{z,\text{nearest}}^{\,d} \equiv \frac{1}{\epsilon'}E_{z,\text{nearest}}^{\,d} \tag{3}$$

where the sum of the fields generated by all the dipoles in a medium of constant ϵ vanishes on average. Therefore, the net contribution to the field can be calculated by considering only the nearest dipole in a medium with an effective dielectric constant ϵ'.[26] A correction to eq 3 could be obtained if the dependence of ϵ on distance would be known.

While the effective dielectric constant ϵ' is expected to be small near the surface dipoles, it is expected to acquire the bulk value after a few molecular distances. When calculating the net field generated by the surface dipoles in water, it will be assumed that effective dielectric constants can be used only for the interactions with the water molecules close to the surface (the first water layer). The interactions with more distant water molecules will be considered as in a continuum medium; hence, the net field will vanish on average.

To calculate the net field produced by a surface dipole, we will assume that one dipole of the surface polarizes the water molecules of the first layer located nearby above it. For the sake of simplicity, the area polarized by one dipole will be consider circular of radius R and equal to the area corresponding to a surface dipole (the inverse of the number of dipoles per unit area). The surface dipole $\vec{p}$ has the components p_{xy} in the plane and p_z normal to the surface. Because of symmetry, the field produced by p_{xy} vanishes, and so does the electric field parallel to the surface. The electric field, normal to the surface, generated by a dipole in a medium with an effective constant ϵ', at a point whose position vector makes an angle θ with the z direction, is given by

$$E_z^{\,d} = \frac{p_z}{4\pi\epsilon'\epsilon_0\varsigma^3}(3\cos^2\theta - 1) \tag{4}$$

where $\varsigma = \Delta'/\cos\theta$, Δ' being the distance between the center of the dipole and the center of the first water layer. For the average field produced by a surface dipole one obtains[26]

$$\bar{E} = \frac{1}{\pi R^2}\int_0^R \frac{2\pi\rho d\rho}{4\pi\epsilon'\epsilon_0(\Delta'^2 + \rho^2)^{3/2}}\left\{p_z\left(3\frac{\Delta'^2}{(\Delta'^2 + \rho^2)} - 1\right)\right\} = \frac{p_z}{\epsilon'}\frac{1}{2\pi\epsilon_0(R^2 + \Delta'^2)^{3/2}} \tag{5}$$

II.2. Polarization of an Ordered, Icelike Water Structure. While the water molecules in the vicinity of the surface are polarized by the dipoles of the surface and by the neighboring water molecules, the other water molecules are polarized only by the field generated by the neighboring water molecules. To calculate this field, the structure of water between the external surfaces must be known.

Here we will assume that icelike clusters which, as suggested by Nemethy and Scheraga,[25] constitute the liquid water are aligned by the external surfaces and hence that a layered, icelike structure is formed between the surfaces. The water layers, parallel to the surface, are made of out-of-plane hexagonal rings of water molecules. Each layer is composed of two planar sublayers; a water molecule from a sublayer has three hydrogen bonds with three molecules of the other sublayer of the same layer, and one with a molecule belonging to a sublayer of one of the adjacent layers (a schematic picture is presented in Figure 1). All the molecules of a sublayer are assumed to have the same average dipole moment, and the local field, normal to the surface, acting on a site of a sublayer (I or II) of a layer j, can be formally written as

(26) Manciu, M.; Ruckenstein, E. *Langmuir* **2001**, *17*, 7061.

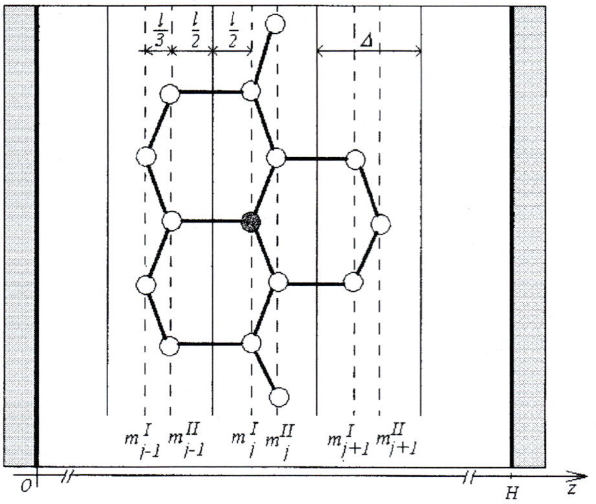

Figure 1. Sketch of an icelike cluster around a selected water molecule. In this structure, a molecule from a sublayer (dotted lines) of one layer (delimited with full lines) is connected via hydrogen bonds with three molecules of the other sublayer of the same layer (only two bonds are drawn in the picture; it should be noted that the molecules drawn are not in the same plane) and with one water molecule from an adjacent layer. The first four neighbors are located at the vertexes of a tetrahedron. The projection of the position of the water molecules from one layer in the plane of the surface (denoted in the text as xy) is a hexagonal network. Each icelike cluster involves 26 molecules around the selected molecule (marked); they are located in three water layers: a central and the two adjacent ones.

$$E_j^{I} = D_0^{I,I}m_j^{I} + D_0^{II,I}m_j^{II} + D_{-1}^{I,I}m_{j-1}^{I} + D_{-1}^{II,I}m_{j-1}^{II} + D_{+1}^{I,I}m_{j+1}^{I} + D_{+1}^{II,I}m_{j+1}^{II} + \ldots \quad (6a)$$

$$E_j^{II} = D_0^{I,II}m_j^{I} + D_0^{II,II}m_j^{II} + D_{-1}^{I,II}m_{j-1}^{I} + D_{-1}^{II,II}m_{j-1}^{II} + D_{+1}^{I,II}m_{j+1}^{I} + D_{+1}^{II,II}m_{j+1}^{II} + \ldots \quad (6b)$$

where m_j^{α} is the average dipole moment of a molecule from the sublayer α (α = I or II) of layer j and the interaction coefficients $D_{\pm k}^{\alpha,\beta}$ account for the contribution of the dipoles of the sublayer α of layer $j \pm k$ to the local field at a site of the sublayer β (α, β = I or II) of layer j. The estimation of the coefficients D on the basis of the icelike model used is provided in Appendix A. It will be assumed that only the molecules within a radius $2l$ from the given site (where l is the distance between adjacent molecules) generate a nonnegligible field at a given site and hence that $k = -1$, 0, and 1.

The average polarization of a molecule is proportional to the local field:

$$m_j^{\alpha} = \gamma E_j^{\alpha} \quad (7)$$

where the proportionality constant is the molecular polarizability.

Equations 6 and 7 constitute a system of $(2N - 4)$ linear equations with constant coefficients for the average polarizations m_j^{α}, where N is the number of water layers. The other four equations are provided by the boundary conditions. Seeking solutions of the form[27]

$$m_j^{I} = Ar^j \quad \text{and} \quad m_j^{II} = \bar{A}r^j \quad (8)$$

(27) Mickens R. E. *Difference Equations;* Van Nostrand Reinhold Company: New York, 1987.

where A, $\bar{A}$, and r are constants, eqs 6 and 7 lead to ($j = 2, \ldots, N - 1$)

$$\left(D_{-1}^{I,I} + \left(D_0^{I,I} - \frac{1}{\gamma}\right)r + D_{+1}^{I,I}r^2\right)m_{j-1}^{I} + (D_{-1}^{II,I} + D_0^{II,I}r + D_{+1}^{II,I}r^2)m_{j-1}^{II} = 0 \quad (9a)$$

$$(D_{-1}^{I,II} + D_0^{I,II}r + D_{+1}^{I,II}r^2)m_{j-1}^{I} + \left(D_{-1}^{II,II} + \left(D_0^{II,II} - \frac{1}{\gamma}\right)r + D_{+1}^{II,II}r^2\right)m_{j-1}^{II} = 0 \quad (9b)$$

The corresponding characteristic equation, which provides the values of r, has the form

$$\left(D_{-1}^{I,I} + \left(D_0^{I,I} - \frac{1}{\gamma}\right)r + D_{+1}^{I,I}r^2\right)\left(D_{-1}^{II,II} + \left(D_0^{II,II} - \frac{1}{\gamma}\right)r + D_{+1}^{II,II}r^2\right) - (D_{-1}^{II,I} + D_0^{II,I}r + D_{+1}^{II,I}r^2)(D_{-1}^{I,II} + D_0^{I,II}r + D_{+1}^{I,II}r^2) = 0 \quad (9c)$$

The general solution of the system in eqs 9a,b is

$$m_j^{I} = A_1r_1^j + A_2r_2^j + A_3r_3^j + A_4r_4^j \quad (10a)$$

$$m_j^{II} = \bar{A}_1r_1^j + \bar{A}_2r_2^j + \bar{A}_3r_3^j + \bar{A}_4r_4^j \quad (10b)$$

where the constants A_k ($k = 1, \ldots, 4$) have to be determined using the four boundary conditions given below. The relation between the constants A_k and their conjugate $\bar{A}_k$ is obtained by substituting eq 8 in eq 9a or b:

$$\bar{A}_k = -A_k\frac{D_{-1}^{I,I} + \left(D_0^{I,I} - \frac{1}{\gamma}\right)r_k + D_{+1}^{I,I}r_k^2}{D_{-1}^{II,I} + D_0^{II,I}r_k + D_{+1}^{II,I}r_k^2} = \eta_kA_k \quad (11)$$

The boundary conditions should take into account the effect of both the field produced by the surface dipoles and the field produced by the molecules of the first two layers.

$$m_1^{I} = \gamma(\bar{E}_{I} + D_0^{I,I}m_1^{I} + D_0^{II,I}m_1^{II} + D_{+1}^{I,I}m_2^{I} + D_{+1}^{II,I}m_2^{II}) \quad (12a)$$

$$m_1^{II} = \gamma(\bar{E}_{II} + D_0^{I,II}m_1^{I} + D_0^{II,II}m_1^{II} + D_{+1}^{I,II}m_2^{I} + D_{+1}^{II,II}m_2^{II}) \quad (12b)$$

$$m_N^{I} = \gamma(-\bar{E}_{II} + D_{-1}^{I,I}m_{N-1}^{I} + D_{-1}^{II,I}m_{N-1}^{II} + D_0^{I,I}m_N^{I} + D_0^{II,I}m_N^{II}) \quad (12c)$$

$$m_N^{II} = \gamma(-\bar{E}_{I} + D_{-1}^{I,II}m_{N-1}^{I} + D_{-1}^{II,II}m_{N-1}^{II} + D_0^{I,II}m_N^{I} + D_0^{II,II}m_N^{II}) \quad (12d)$$

The fields generated by the surface dipoles on the sublayers of the first layer, $\bar{E}_{I}$ and $\bar{E}_{II}$, are calculated using eq 5, for the corresponding distances from the center of the surface dipole to the centers of the sublayers, Δ_{I}' and Δ_{II}', respectively. The field in the first sublayer of the first water layer, generated by the surface dipoles, is calculated by considering an effective dielectric constant $\epsilon' < \epsilon$, while, for the field generated in the second sublayer of the same layer, a dielectric constant $\xi\epsilon'$ is employed ($\xi > 1$) to account for the dependence of the screening on the distance.

7586 *Langmuir, Vol. 17, No. 24, 2001* *Manciu and Ruckenstein*

Since the interaction coefficients D are invariant to the transformation I → II, $k \to -k$, and, additionally, $D_{-1}^{I,I} = D_{+1}^{I,I}$ (see Appendix A), the characteristic equation is invariant to the transformation $r \to 1/r$, which can be directly verified by substitution in eq 9c. Therefore, the solutions of the characteristic equation come in pairs,

$$r_1 = \rho_1,\quad r_2 = \frac{1}{\rho_1},\quad r_3 = \rho_2,\quad r_4 = \frac{1}{\rho_2}$$

Because for identical surfaces the polarization should be antisymmetric with respect to the middle distance between the two, one can write

$$m_j^{\mathrm{I}} = -m_{N+1-j}^{\mathrm{II}} \quad \text{for} \quad j = 1, N$$

Using eqs 10 and 11, the latter equation leads to

$$A_1\rho_1^{\,j} + A_2\rho_1^{\,-j} + A_3\rho_2^{\,j} + A_4\rho_2^{\,-j} = -(\eta_1 A_1\rho_1^{\,N+1-j} + \eta_2 A_2\rho_1^{\,-(N+1-j)} + \eta_3 A_3\rho_2^{\,N+1-j} + \eta_4 A_4\rho_2^{\,-(N+1-j)}) \quad (13)$$

which can be rearranged as

$$(A_1\rho_1^{\,n} + \eta_2 A_2\rho_1^{\,-n})\rho_1^{\,-k} + (A_2\rho_1^{\,-n} + \eta_1 A_1\rho_1^{\,n})\rho_1^{\,k} + (A_3\rho_2^{\,n} + \eta_4 A_4\rho_2^{\,-n})\rho_2^{\,-k} + (A_4\rho_2^{\,-n} + \eta_3 A_3\rho_2^{\,n})\rho_2^{\,k} = 0 \quad (14)$$

where $n = (N + 1)/2$ and $k = (N + 1 - 2j)/2$. Since, for a given value of N, the equations should be valid for any value of j, the coefficients in parentheses should vanish. The four equations thus obtained reduce to two equations, because $\eta_1\eta_2 = \eta_3\eta_4 \equiv 1$, and the constants A_2 and A_4 can be expressed in terms of A_1 and A_3, respectively. Consequently, the general solution has the form

$$m_j^{\mathrm{I}} = A_1(\rho_1^{\,j} - \eta_1\rho_1^{\,N+1-j}) + A_3(\rho_2^{\,j} - \eta_3\rho_2^{\,N+1-j}) \quad (15a)$$

$$m_j^{\mathrm{II}} = \eta_1 A_1(\rho_1^{\,j} - \eta_2\rho_1^{\,N+1-j}) + \eta_3 A_3(\rho_2^{\,j} - \eta_4\rho_2^{\,N+1-j}) \quad (15b)$$

For illustration purposes we will present a numerical example.

The interaction coefficients D are evaluated in Appendix A, selecting for the effective dielectric constant for the interaction between neighboring water molecules $\epsilon'' = 1$. This constitutes a lower bound for ϵ''. For the molecular polarizability, we will employ an expression[26] based on the hypothesis of hindered molecular rotations introduced by Debye. This provides a lower bound for γ, which is given by[26]

$$\gamma = \frac{\epsilon_0 v}{\dfrac{\epsilon + 2}{3(\epsilon - 1)} + \epsilon_0 v(D_{-1}^{\mathrm{I,I}} + D_{-1}^{\mathrm{II,I}} + D_0^{\mathrm{I,I}} + D_0^{\mathrm{II,I}} + D_{+1}^{\mathrm{I,I}} + D_{+1}^{\mathrm{II,I}})} \quad (16)$$

where ϵ is the dielectric constant of water, ϵ_0 is the permittivity constant, and v is the volume occupied by a water molecule.

The Onsager theory for the molecular polarizability[28] implies that the water molecules can rotate freely in the liquid water and that their permanent dipole moments are higher in the liquid than in the vapor phase (because of their interaction with a strongly polarizable medium, the liquid water). The value predicted for the molecular polarizability is about four times larger than that provided by eq 16; however, this does not affect the qualitative features of the discussion which follows.

For an icelike structure, the distance between the centers of two adjacent molecules is $l = 2.76$ Å and $v = (8/3\sqrt{3})l^3 = 32.37$ Å^3. Using $\epsilon = 80$, the characteristic equation acquires the form

$$r^4 + 0.575r^3 - 25.78r^2 + 0.575r + 1 = 0 \quad (17)$$

The solutions r_k are

$$r_1 = \rho_a,\quad r_2 = \frac{1}{\rho_a},\quad r_3 = -\rho_b,\quad r_4 = -\frac{1}{\rho_b} \quad (18)$$

with $\rho_a = 4.783$ and $\rho_b = 5.381$. Consequently, the general antisymmetric solutions are

$$m_j^{\mathrm{I}} = A_1(\rho_a^{\,j} - \eta_1\rho_a^{\,N+1-j}) + A_3(-1)^j(\rho_b^{\,j} - \eta_3(-1)^{N+1}\rho_b^{\,N+1-j}) \quad (19a)$$

$$m_j^{\mathrm{II}} = \eta_1 A_1(\rho_a^{\,j} - \eta_2\rho_a^{\,N+1-j}) + \eta_3 A_3(-1)^j(\rho_b^{\,j} - \eta_4(-1)^{N+1}\rho_b^{\,N+1-j}) \quad (19b)$$

which depend on two constants, A_1 and A_3 to be determined from the boundary conditions, eqs 12a and 12b. (The other two boundary conditions, eqs 12c and 12d, are equivalent to eqs 12b and 12a, respectively.)

The above equations are defined only for integer values of j; however, since both ρ_a and ρ_b are real and positive, the above equations can be analytically extended to noninteger values of j. A continuous function of z, which provides the same values as eqs 19 at the discrete positions of the sublayers and interpolates between them can be obtained using the equality

$$\rho^j = \exp(\ln(\rho^j)) = \exp(j\ln(\rho)) = \exp\left(\frac{z}{\Delta}\ln\rho\right) = \exp\left(\frac{z}{\lambda}\right)$$

where $\Delta = {}^4/_3 l$ is the distance between the centers of two adjacent water layers. In what follows, we will define the origin of z at one of the interfaces, located at the external boundary of the first water layer. The first sublayer of the first water layer is located at the distance $z = {}^1/_2 l = {}^3/_8\Delta$ and corresponds to $j = 1$ in eq 19a. Hence, in eq 19a, $j = z/\Delta + {}^5/_8$ and $N + 1 - j = (H - z)/\Delta + {}^3/_8$, where $H = N\Delta$ is the separation distance between the two surfaces. From eqs 19, one obtains

$$m^{\mathrm{I}}(z) = A_1\left(\exp\left(\frac{5\Delta}{8\lambda_a}\right)\exp\left(\frac{z}{\lambda_a}\right) - \eta_1\exp\left(\frac{3\Delta}{8\lambda_a}\right)\exp\left(\frac{H-z}{\lambda_a}\right)\right) + A_3\cos\left(\pi\left(\frac{z}{\Delta} + \frac{5}{8}\right)\right)\left(\exp\left(\frac{5\Delta}{8\lambda_b}\right)\exp\left(\frac{z}{\lambda_b}\right) - (-1)^{N+1}\eta_3\left(\frac{3\Delta}{8\lambda_b}\right)\exp\left(\frac{H-z}{\lambda_b}\right)\right) \quad (20a)$$

and

$$m^{\mathrm{II}}(z) = -m^{\mathrm{I}}(H - z) \quad (20b)$$

where $\lambda_a = \Delta/\ln\rho_a = 2.35$ Å and $\lambda_b = \Delta/\ln\rho_b = 2.18$ Å are the decay lengths of the amplitudes of the monotonic and oscillatory terms, respectively.

(28) Onsager, L. *J. Am. Chem. Soc.* **1936**, *58*, 261.

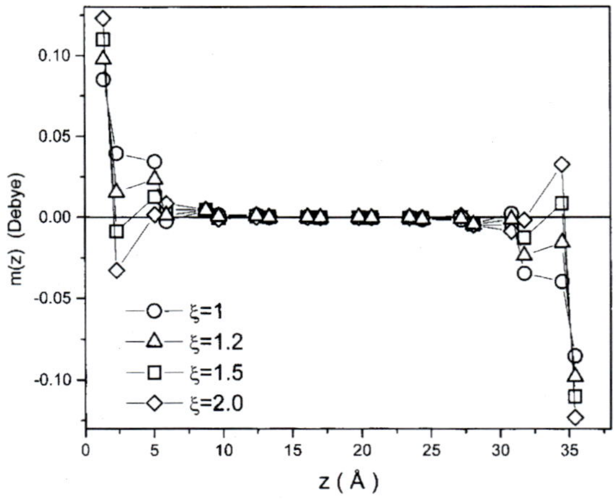

Figure 2. Dipole moment as a function of the distance from one surface, calculated in the discrete approach for $N = 10$ and various values of ξ. The values of the other parameters are $l = 2.76$ Å, $(p_z/\epsilon') = 1.0$ D, $\epsilon'' = 1$, $\epsilon = 80$, $\pi R^2 = 50$ Å^2, $\Delta'_{\rm I} = 2.88$ Å, and $\Delta'_{\rm II} = 3.8$ Å.

The numerical solutions for $N = 10$, $p_z/\epsilon' = 1.0$ D, $A = \pi R^2 = 50$ Å^2, $\Delta'_{\rm I} = 2.88$ Å, and $\Delta'_{\rm II} = 3.8$ Å are presented in Figure 2, for various values for ξ. The contributions of the monotonic and oscillatory parts of the polarizations depend strongly on the boundary conditions, which basically means the values of $\bar{E}_{\rm I}$ and $\bar{E}_{\rm II}$. Both the monotonic and oscillatory parts decay from the surface, but with different decay lengths.

The oscillations of the polarization are a consequence of the particular geometry of the system. The field generated by the average dipole moment of one sublayer tends to align in the same direction as the dipole moments of the molecules of the adjacent layers, but in the opposite direction as the dipole moments of the molecules of the other sublayer of the same layer. If the difference between $\bar{E}_{\rm I}$ and $\bar{E}_{\rm II}$ is large enough, this will produce a polarization which oscillates with the distance from the surface. The magnitude of the oscillations, which are attenuated with the distance, depends on the ratio $\bar{E}_{\rm II}/\bar{E}_{\rm I}$ and on the strength of the electrostatic interaction between adjacent layers. The total free energy per unit area, due to the polarization of the water molecules, can be calculated using the expression

$$F(N) = -\frac{N_{\rm w}}{2}\sum_{j=1}^{N}(m_j^{\rm I}E_j^{\rm I} + m_j^{\rm II}E_j^{\rm II}) = -\frac{N_{\rm w}}{2\gamma}\sum_{j=1}^{N}((m_j^{\rm I})^2 + (m_j^{\rm II})^2) \quad (21)$$

where $N_{\rm w}$ is the number of water molecules in a sublayer, per unit area, and the factor $^1/_2$ avoids double counting. The above expression can be computed only for integer values of N, that is, an integer number of water layers. The interaction free energy $F_{\rm int}(N)$, obtained by subtracting the free energy $F(N\to\infty)$, is represented in Figure 3 for various values of ξ. For higher values of ξ, the electrostatic interaction might become attractive for odd values of N (i.e. $F(N) < F(\infty)$).

II.3. Effect of Partial Disorder. The oscillations of the polarization are a consequence of the long-range order along the direction normal to the surface. For planar surfaces, some disorder is, however, induced by the gradual decrease of the cluster alignment away from the surface.

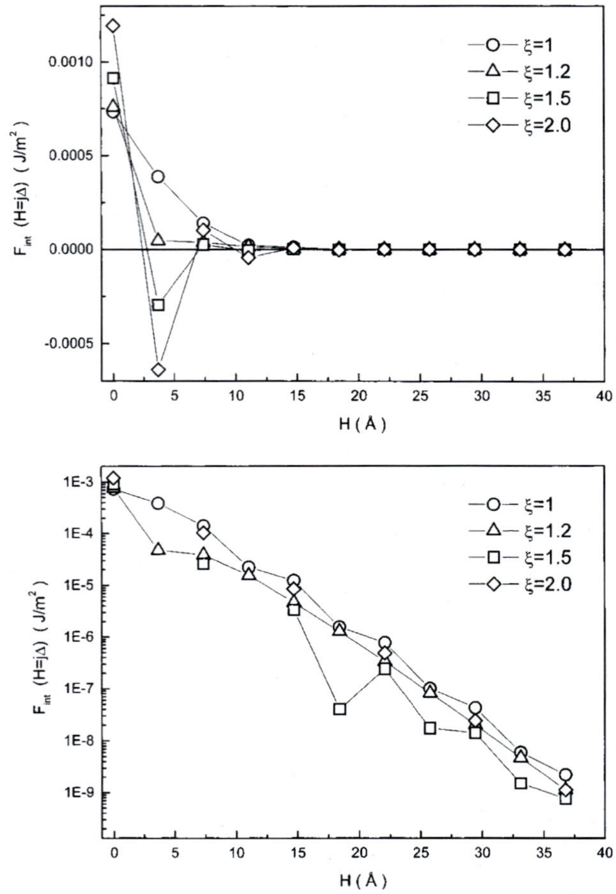

Figure 3. Interaction free energy in the discrete approach as a function of the separation distance between surfaces, calculated in the discrete model for various values of ξ: (a, top) linear scale (which shows better the oscillatory behavior near the surface; (b, bottom) logarithmic scale (which better reveals the behavior at large distances).

In addition, if the surfaces are not planar, but are sufficiently rough or fluctuating, the average polarization is no longer defined only for discrete values of z (the positions of the water sublayers with respect to one planar surface) but becomes a continuous function of distance, $m(z)$. The system with rough or undulating surfaces will be approximated here by a statistical ensemble of systems with planar surfaces in which the water molecules are organized in long-range icelike structures. Assuming that in the statistical ensemble the surfaces are distributed Gaussian with the root-mean-square fluctuation σ, the average m at a position z from the average surface is obtained from

$$m(z) = \frac{1}{\sqrt{2\pi}\sigma}\int_{-\infty}^{\infty}\exp\left(-\frac{s^2}{2\sigma^2}\right)m(z-s)\,{\rm d}s = \frac{1}{\sqrt{2\pi}\sigma}\sum_{\alpha={\rm I,II}}\sum_{j=1,N} m_j^{\alpha}\exp\left(-\frac{(z-z_j^{\alpha})^2}{2\sigma^2}\right) \quad (22)$$

where z_j^{α} is the position of the sublayer α of the layer j.

During the fluctuations, the number N of layers changes. However, the values of m_j^{α} in the layers near the surfaces are almost independent of N, for large N. Figure 4a presents the dependence of m on z for various values of

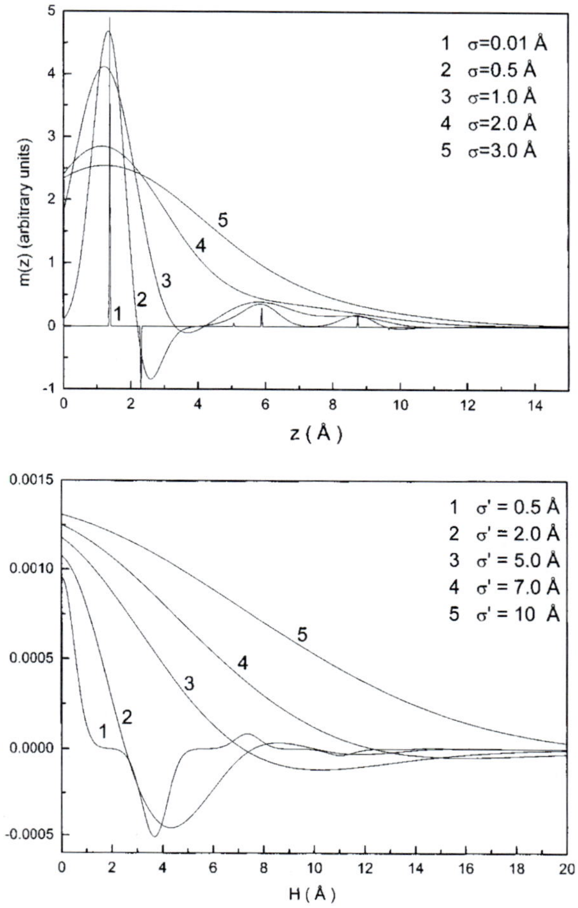

Figure 4. (a, top) Average polarization m as a function of the distance from one surface (the other surface is located at a large separation distance), for a rough or fluctuating surface, calculated using eq 22 for $\xi = 2$ and various values of σ. The functions are multiplied with arbitrary factors for easier comparison of their shapes. (The factors have the following values: (1) 1; (2) 50; (3) 100; (4) 150; (5) 200.) (b, bottom) Interaction free energy as a function of separation distance, for a rough or fluctuating surface, calculated using eq 23 for $\xi = 2$ and various values of σ'. The functions are multiplied with arbitrary factors for easier comparison of their shapes. (The factors have the following values: (1) 1; (2) 5; (3) 20; (4) 35; (5) 50.)

the root-mean-square fluctuation and a large value of N ($N = 20$). For $\sigma \to 0$, the average dipole moment is nonvanishing only around $z = z_j^{\alpha}$, while, for $\sigma > 5$ Å, the average dipole moment becomes a smooth, monotonically decreasing function of the distance from the surface. Such a behavior was observed in computer simulations. Molecular dynamics studies of water between rigid surfaces[13,14] indicated that the polarization is oscillating in the vicinity of the surface, with a spatial period of the order of molecular size. However, when the dynamic restrictions on bilayers' headgroups were removed, the polarization became a smooth function of z, monotonically decaying.[29]

A similar procedure can be used to calculate the interaction free energy between two undulating (or rough) surfaces. Assuming that the separation distance between surfaces is normally distributed around H with a dispersion σ', the interaction free energy is given by

$$F_{\text{int}}(H) = \frac{1}{\sqrt{2\pi}\sigma'}\sum_{j=0}^{\infty}(F(H_j) - F(\infty))\exp\left(-\frac{(H - H_j)^2}{2\sigma'^2}\right) \tag{23}$$

where $H_j = j\Delta$ is the distance between surfaces in each of the elements of the statistical ensemble. The interaction free energy for various values of σ' is presented in Figure 4b.

II.4. Monotonic Polarization. Let us assume that it is possible to define an average m as a monotonic function of distance to the average surface and that locally an icelike order is preserved in each of the clusters with radius $2l$, l being the distance between the centers of two adjacent water molecule, containing 26 water molecules around a selected water molecule (see Figure 1). If the cluster is aligned with its layers parallel to the external surface, and its center is located at the position z (measured from one of the surfaces), the distance between the centers of two adjacent water layers is $\Delta = {}^4/_3 l$, the positions of the sublayers of the central layer are $z \pm {}^1/_6 l$, and the positions of the other sublayers are $z \pm {}^7/_6 l$ and $z \pm {}^3/_2 l$, respectively. The equations which describe the polarizations of the two sublayers of the central layer are

$$\begin{aligned} m\left(z - \frac{l}{6}\right) = \gamma\Big(& D_{-1}^{\text{I,I}} m\left(z - \frac{3l}{2}\right) + D_{-1}^{\text{II,I}} m\left(z - \frac{7l}{6}\right) + \\ & D_0^{\text{I,I}} m\left(z - \frac{l}{6}\right) + D_0^{\text{II,I}} m\left(z + \frac{l}{6}\right) + D_{+1}^{\text{I,I}} m\left(z + \frac{7l}{6}\right) + \\ & D_{+1}^{\text{II,I}} m\left(z + \frac{3l}{2}\right)\Big) \end{aligned}$$

$$\begin{aligned} m\left(z + \frac{l}{6}\right) = \gamma\Big(& D_{+1}^{\text{II,I}} m\left(z - \frac{3l}{2}\right) + D_{+1}^{\text{I,I}} m\left(z - \frac{7l}{6}\right) + \\ & D_0^{\text{II,I}} m\left(z - \frac{l}{6}\right) + D_0^{\text{I,I}} m\left(z + \frac{l}{6}\right) + D_{-1}^{\text{II,I}} m\left(z + \frac{7l}{6}\right) + \\ & D_{-1}^{\text{I,I}} m\left(z + \frac{3l}{2}\right)\Big) \end{aligned} \tag{24a,b}$$

where because of the symmetry of the system to the transformation I ↔ II, $k \leftrightarrow -k$ only six of the interaction coefficients D were used (see Appendix A). When the cluster is not perfectly aligned parallel to the surface but makes a small angle α, the coefficients of interactions are slightly modified (see Appendix B). If the distribution of the tilt angles and their dependence on z would be known, it could be included in the calculations which follow. However, as shown later, the tilting of the local clusters by as much as 20° changes the decay length less than 10%. In addition, the tilting is expected to be lower near the surface. For these reasons, the solution with a constant decay length probably constitutes a good approximation.

We will employ the procedure used by Schiby and Ruckenstein[10] and will expand m around z. Adding the two eqs 24, one obtains (neglecting the terms of order 4 and higher)

$$\begin{aligned} 2m(z) + \frac{\partial^2 m(z)}{\partial z^2}\frac{l^2}{36} = \\ \gamma\Big\{& (D_{-1}^{\text{I,I}} + D_{+1}^{\text{II,I}})\left(2m(z) + \frac{\partial^2 m(z)}{\partial z^2}\frac{9l^2}{4}\right) + \\ & (D_0^{\text{I,I}} + D_0^{\text{II,I}})\left(2m(z) + \frac{\partial^2 m(z)}{\partial z^2}\frac{l^2}{36}\right) + \\ & (D_{+1}^{\text{I,I}} + D_{-1}^{\text{II,I}})\left(2m(z) + \frac{\partial^2 m(z)}{\partial z^2}\frac{49l^2}{36}\right)\Big\} \end{aligned} \tag{25}$$

(29) Perera, L.; Essmann, U.; Berkowitz, M. L. *Prog. Colloid Polym. Sci.* **1997**, *103*, 107.

which can be rewritten as

$$\frac{\partial^2 m(z)}{\partial z^2} - \frac{1}{\lambda_1^{\,2}} m(z) = 0 \tag{26}$$

with λ_1 given by

$$\lambda_1 = l\sqrt{\frac{\frac{9}{4}(D_{-1}^{\mathrm{I,I}} + D_{+1}^{\mathrm{II,I}}) + \frac{49}{36}(D_{-1}^{\mathrm{II,I}} + D_{+1}^{\mathrm{I,I}}) + \frac{1}{36}\left(D_0^{\mathrm{I,I}} + D_0^{\mathrm{II,I}} - \frac{1}{\gamma}\right)}{2\left(\frac{1}{\gamma} - (D_{-1}^{\mathrm{I,I}} + D_{-1}^{\mathrm{II,I}} + D_0^{\mathrm{I,I}} + D_0^{\mathrm{II,I}} + D_{+1}^{\mathrm{I,I}} + D_{+1}^{\mathrm{II,I}})\right)}} \tag{27}$$

A simpler procedure[26] is to consider that the average dipole moment of the water molecules of one layer is obtained by averaging over the two sublayers, $m_j = (m_j^{\mathrm{I}} + m_j^{\mathrm{II}})/2$. In this case, the quantities representing the interaction coefficients with the same (C_0) and adjacent (C_{+1} and C_{-1}) layers are given by

$$C_0 = D_0^{\mathrm{I,I}} + D_0^{\mathrm{II,I}},$$
$$C_1 = C_{-1} = C_{+1} = \frac{1}{2}(D_{-1}^{\mathrm{I,I}} + D_{-1}^{\mathrm{II,I}} + D_{+1}^{\mathrm{I,I}} + D_{+1}^{\mathrm{II,I}}) \tag{28}$$

The average dipole moment is related in this case to the average dipole moment of the molecules from the same and adjacent layers via

$$m(z) = \gamma(C_{-1}m(z-\Delta) + C_0 m(z) + C_1 m(z+\Delta)) \tag{29}$$

where $\Delta = 4l/3$ is the distance between adjacent layers. After a Taylor expansion and neglecting the terms of order 4 and higher, a differential equation of the same type as eq 26 is obtained, but with a decay length λ_2 given by[26]

$$\lambda_2 = \frac{4l}{3}\sqrt{\frac{C_1}{\frac{1}{\gamma} - (C_0 + 2C_1)}} \tag{30}$$

The solution of eq 26 for two planar, parallel interfaces at a distance H apart, if the average dipole moment of the first water layers (located at $z = \pm\Delta/2$) is $\pm m_1$ (hence, for the boundary condition $m(z{=}\Delta/2) = m_1$, $m(z{=}H{-}\Delta/2) = -m_1$) has the form

$$m(z) = -m_1\frac{\sinh\left(\frac{2z-H}{2\lambda}\right)}{\sinh\left(\frac{H-\Delta}{2\lambda}\right)} \tag{31}$$

where λ can be λ_1 or λ_2.

The value of m_1 is related to the local field, which can be obtained by adding to the field generated by the surface dipoles, $\bar{E}$, the field generated by the molecules of the first two layers from the surface

$$m_1 = \gamma E_{\mathrm{e}} = \gamma(\bar{E} + C_0 m_1 + C_1 m_2) \tag{32}$$

Using eq 31 for $m_2 = m(3\Delta/2)$, one obtains

$$m_1 = \frac{\gamma\bar{E}}{\left(1 - \gamma C_0 - \gamma C_1\frac{\sinh\left(\frac{H-3\Delta}{2\lambda}\right)}{\sinh\left(\frac{H-\Delta}{2\lambda}\right)}\right)} \tag{33}$$

The total electrostatic free energy per unit area, between the two parallel plates, due to the orientation of dipoles $m(z)$ in the local field $E = m(z)/\gamma$ is given by[26]

$$F(H) = -\frac{1}{2S}\sum_{\text{all molecules}}\frac{(m(z))^2}{\gamma} = -\frac{1}{2S}\int_{\text{volume}}\frac{(m(z))^2}{\gamma v}\,\mathrm{d}V = -\frac{1}{2\gamma v}\int_0^H (m(z))^2\,\mathrm{d}z = \frac{\gamma\bar{E}^2}{\left(1 - \gamma\left(C_0 + C_1\frac{\sinh\left(\frac{H-3\Delta}{2\lambda}\right)}{\sinh\left(\frac{H-\Delta}{2\lambda}\right)}\right)\right)^2}\,\frac{H - \lambda\sinh\left(\frac{H}{\lambda}\right)}{4v\sinh^2\left(\frac{H-\Delta}{2\lambda}\right)} \tag{34}$$

where $V = SH$ is the volume occupied by the water molecules and S is the area of the planar interfaces. The interaction free energy is obtained by subtracting $F(H{\to}\infty)$ from $F(H)$.

Because $\bar{E}$ is proportional to p_z/ϵ', the free energy is proportional to $(p_z/\epsilon')^2$. This result is in agreement with the experiments of Simon and McIntosh[24] and consistent with the theory of Schiby and Ruckenstein.[10]

To verify the accuracy of the continuum approximation, we compared the average dipole moments along the z axis calculated using both the discrete and continuous procedures for integer values of N.

The general solution of the finite differences equation (eq 29) is given by

$$m_j = A_1 r_1^{\,j} + A_2 r_2^{\,j} \tag{35}$$

where r_1 and r_2 are the solutions of the characteristic equation

$$r^2 + \frac{C_0 - \frac{1}{\gamma}}{C_1}r + 1 = 0 \tag{36}$$

and the constants A_k are obtained from the boundary conditions. For the conditions specified in section II.2, $C_0 < 0$ and $-(C_0 - 1/\gamma) > 2C_1$. Consequently, both solutions of the characteristic equation are real and positive:

$$r_1 = \rho,\quad r_2 = \frac{1}{\rho},\quad \rho = \frac{1-\gamma C_0}{2\gamma C_1} + \sqrt{\left(\frac{1-\gamma C_0}{2\gamma C_1}\right)^2 - 1} \tag{37}$$

For a system with identical surfaces, the average dipole moment should be antisymmetric with respect to the middle distance; hence, $m_j = -m_{N+1-j}$, $j = 1, N$:

$$A_1\rho^j + A_2\rho^{-j} = -(A_1\rho^{N+1-j} + A_2\rho^{-(N+1-j)}) \tag{38}$$

which can be rearranged as

$$(A_1\rho^{(N+1)/2} + A_2\rho^{-(N+1)/2})(\rho^{(N+1-2j)/2} + \rho^{-(N+1-2j)/2}) = 0 \tag{39}$$

and can be satisfied for any j (and a given N) only if $A_2 = -A_1\rho^{N+1}$. The corresponding (discrete) solution is

$$m_j = A_1(\rho^j - \rho^{N+1-j}) \tag{40}$$

Table 1

	$\alpha = 0°$	$\alpha = 5°$	$\alpha = 10°$	$\alpha = 15°$	$\alpha = 20°$
$4\pi\epsilon_0\epsilon'' l^3 D_{-1}{}^{I,I}$	0.688 92	0.681 07	0.657 76	0.619 70	0.568 04
$4\pi\epsilon_0\epsilon'' l^3 D_{-1}{}^{II,I}$	1.844 62	1.823 61	1.761 19	1.659 28	1.520 95
$4\pi\epsilon_0\epsilon'' l^3 D_0{}^{I,I}$	−1.377 84	−1.362 14	−1.315 52	−1.239 39	−1.136 07
$4\pi\epsilon_0\epsilon'' l^3 D_0{}^{II,I}$	−2.388 44	−2.361 22	−2.280 41	−2.148 44	−1.969 35
$4\pi\epsilon_0\epsilon'' l^3 D_{+1}{}^{I,I}$	0.688 92	0.681 07	0.657 76	0.619 70	0.568 04
$4\pi\epsilon_0\epsilon'' l^3 D_{+1}{}^{II,I}$	0.432 00	0.427 08	0.412 46	0.388 59	0.356 20
$4\pi\epsilon_0\epsilon'' l^3 C_0$	−3.766 28	−3.723 36	−3.595 93	−3.387 84	−3.105 42
$4\pi\epsilon_0\epsilon'' l^3 C_1$	1.827 23	1.806 41	1.744 58	1.643 63	1.506 61
λ_1 (Å)	2.770 4	2.755 0	2.708 6	2.630 8	2.520 8
λ_2 (Å)	2.964 9	2.948 6	2.899 6	2.817 5	2.701 6
λ_3 (Å)	3.137 7	3.122 2	3.075 7	2.997 8	2.888 1

Since z is measured from the external boundary of the first water layer, the center of the first water layer is located at $z = \Delta/2$ and corresponds to $j = 1$. Therefore $j = z/\Delta + 1/2$ and $N + 1 - j = (H - z)/\Delta + 1/2$, where $H = N\Delta$ is the distance between surfaces. The analytical extension for $m(z)$ (the continuous expression which provides the exact values at the discrete points and interpolates between them) is thus given by

$$m(z) = 2A_1 \exp\left(\frac{H+\Delta}{2\lambda_3}\right) \sinh\left(\frac{2z - H}{2\lambda_3}\right)$$

where A_1 is obtained from the boundary condition $m(z{=}\Delta/2) = m_1$; hence,

$$m(z) = -m_1 \frac{\sinh\left(\frac{2z-H}{2\lambda_3}\right)}{\sinh\left(\frac{H-\Delta}{2\lambda_3}\right)} \tag{41}$$

This solution is of the same type as eq 31, but with a decay length λ_3 given by

$$\lambda_3 = \frac{\Delta}{\ln \rho} = \Delta\left(\ln\left(\frac{1-\gamma C_0}{2\gamma C_1} + \sqrt{\left(\frac{1-\gamma C_0}{2\gamma C_1}\right)^2 - 1}\right)\right)^{-1} \tag{42}$$

Figure 5a presents the values of $m(z)$ calculated with the discrete and continuous approaches for $N = 10$. The interaction free energies derived by the two approaches are compared in Figure 5b; the agreement is very good for distances larger than about the thicknesses of two water layers (7.36 Å). However, when the separation distance between surfaces becomes comparable to the molecular size, one expects the free energy associated with the restructuring of water to become dominant. Therefore, in this region neither the discrete nor its continuum approximation is likely to represent an accurate description of the hydration interaction.

Let us now consider that the layers of the local clusters make an angle α with the surface. In this case, $\cos^2\theta$ in eq A.1 of Appendix A should be replaced by (Appendix B, eq B.7)

$$\cos^2\theta' = \cos^2\theta\cos^2\alpha + \frac{\sin^2\theta\sin^2\alpha}{2} \tag{B.7}$$

Assuming that the average dipole moment remains constant in a layer regardless of the tilt angle, one obtains for the interaction coefficients the values listed in Table 1. Additionally, we computed the decay lengths for the parameter values given in section II.4: λ_1 from the continuous approximation, eq 27; λ_2 from the continuous approximation which treats the polarization of one layer as an average over its two sublayers, eq 30; and λ_3 from the analytical extension of the solution of finite difference equation, eq 42.

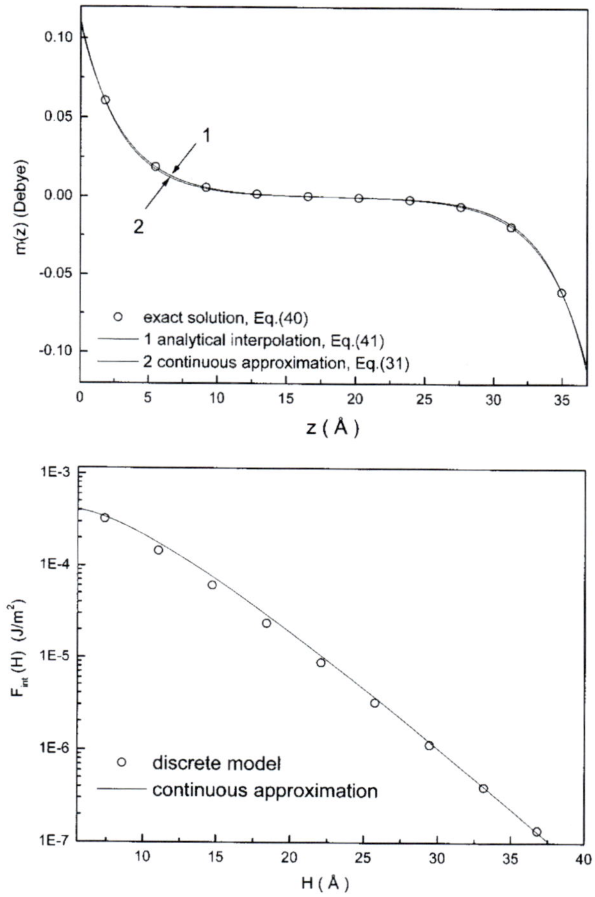

Figure 5. (a, top) Polarization as a function of the distance from one surface. The solution of the discrete approach (eq 40, circles) and its analytical interpolation (eq 41, line 1) are compared to the solution obtained via the continuous approximation (eq 31, line 2). (b, bottom) Interaction energy, as a function of separation distance, for the discrete approach and for the continuous approximation.

As already noted, if the average tilting angle, α, remains below 20°, the interaction coefficients and the decay lengths are only slightly modified.

III. Summary and Conclusions

We followed the model introduced by Schiby and Ruckenstein.[10] Arguments are brought that the local inhomogeneities of the dielectric constant are responsible for the propagation of polarization in water, in the vicinity of a planar surface which has a dipole moment density. If the water has a long-range icelike order along the direction normal to the surface, the system of finite difference equations, corresponding to the layered structure, leads to discrete solutions, which can oscillate with the distance. However, if some disorder is present, due, for instance, to the roughness or fluctuations of the surfaces, the average polarization becomes a monotonically decaying function of the average distance from the surface. A continuous equation is derived for $m(z)$, and expressions are obtained for the decay length.

Appendix A. Evaluation of the Coefficients $D_k^{\alpha,\beta}$

The coefficients $D_{\pm k}^{\alpha,\beta}$ account for the contributions of the dipoles of the sublayer α of layer $j \pm k$ to the local field at a site of the sublayer β (α, β = I or II) of layer j. While the field produced by remote dipoles can be treated as screened by a medium with a large dielectric constant ($\epsilon \cong 80$), the screening of the neighboring dipoles is much weaker. As noted in section II.1, the net contribution to the local field generated by a layer of dipoles is due to the neighboring dipoles, whose interactions take place in a medium with an effective dielectric constant smaller than ϵ. It is assumed that only the dipoles located within a radius $2l$ (where l denotes the distance between the centers of two adjacent water molecules) from the given site contribute to the local field and that the effective dielectric constant for them has a value ϵ'', smaller than ϵ. The electric field along z, caused by a neighboring molecule having an average dipole moment m along the z direction, is given by

$$E_z = \frac{m(3\cos^2\theta - 1)}{4\pi\epsilon''\epsilon_0\varsigma^3} \tag{A.1}$$

where ς is the distance between a neighboring molecule and the selected site and θ is the angle between $\vec{\varsigma}$ and z. We will consider that the cluster of radius $2l$ has an icelike structure with the layers perpendicular to z (the effect of a tilting angle is evaluated in Appendix B). In an ice-I structure of perfect tetrahedrons, the distances between the sublayers of the same layer is $l/3$ and the distance between adjacent layers is $4l/3$. The edges of the tetrahedron formed by the four first neighbors of a water molecule have the length $l(\sqrt{8/3})$, while the planar projection of the tilted hexagonal lattice has the side $l(\sqrt{8/9})$. The volume occupied in this structure by a water molecule is $v = [8/(3\sqrt{3})]l^3$.

Because the system is invariant at the transformation I → II, $+k \rightarrow -k$ (see Figure 1), only six interaction coefficients should be computed:

$$D_{-1}^{\mathrm{I,I}} = D_{+1}^{\mathrm{II,II}} = \frac{1}{4\pi\epsilon''\epsilon_0 l^3} 3\frac{3\left(\sqrt{\frac{2}{3}}\right)^2 - 1}{\left(\sqrt{\frac{8}{3}}\right)^3} = \frac{0.6889}{4\pi\epsilon''\epsilon_0 l^3} \tag{A.2a}$$

$$D_{-1}^{\mathrm{II,I}} = D_{+1}^{\mathrm{I,II}} = \frac{1}{4\pi\epsilon''\epsilon_0 l^3}\left(2 + 6\frac{3\left(\sqrt{\frac{3}{11}}\right)^2 - 1}{\left(\sqrt{\frac{11}{3}}\right)^3}\right) = \frac{1.8446}{4\pi\epsilon''\epsilon_0 l^3} \tag{A.2b}$$

$$D_0^{\mathrm{I,I}} = D_0^{\mathrm{II,II}} = \frac{1}{4\pi\epsilon_0\epsilon'' l^3}\left(6\frac{(-1)}{\left(\sqrt{\frac{8}{3}}\right)^3}\right) = -\frac{1.3778}{4\pi\epsilon_0\epsilon'' l^3} \tag{A.2c}$$

$$D_0^{\mathrm{II,I}} = D_0^{\mathrm{I,II}} = \frac{1}{4\pi\epsilon_0\epsilon'' l^3}\left(3\frac{\left(3\left(\frac{1}{3}\right)^2 - 1\right)}{1} + 3\frac{\left(3\left(\frac{1}{\sqrt{33}}\right)^2 - 1\right)}{\left(\sqrt{\frac{11}{3}}\right)^3}\right) = -\frac{2.3884}{4\pi\epsilon_0\epsilon'' l^3} \tag{A.2d}$$

$$D_{+1}^{\mathrm{I,I}} = D_{-1}^{\mathrm{II,II}} = \frac{1}{4\pi\epsilon''\epsilon_0 l^3} 3\frac{3\left(\sqrt{\frac{2}{3}}\right)^2 - 1}{\left(\sqrt{\frac{8}{3}}\right)^3} = \frac{0.6889}{4\pi\epsilon''\epsilon_0 l^3} \tag{A.2e}$$

$$D_{+1}^{\mathrm{II,I}} = D_{-1}^{\mathrm{I,II}} = \frac{1}{4\pi\epsilon''\epsilon_0 l^3}\frac{2}{\left(\frac{5}{3}\right)^3} = \frac{0.432}{4\pi\epsilon''\epsilon_0 l^3} \tag{A.2f}$$

The averaged coefficients of interaction C between layers are obtained from

$$C_0 = D_0^{\mathrm{I,I}} + D_0^{\mathrm{II,I}} = -\frac{3.76628}{4\pi\epsilon''\epsilon_0 l^3} \tag{A.3a}$$

$$C_1 = C_{-1} = \frac{1}{2}(D_{-1}^{\mathrm{I,I}} + D_{-1}^{\mathrm{II,I}} + D_{+1}^{\mathrm{I,I}} + D_{+1}^{\mathrm{II,I}}) = \frac{1.82723}{4\pi\epsilon''\epsilon_0 l^3} \tag{A.3b}$$

Appendix B. Effect of a Tilt Angle of the Cluster

The coefficients $D_k^{\alpha,\beta}$ were calculated in Appendix A by assuming that the layers of the cluster are perpendicular to z. When the cluster is tilted, the angles between z and the direction to the neighboring molecules is modified. To calculate the new angle, let us consider that the neighboring molecule, located at the point (x_1, y_1, z_1) is rotated with an angle α (the tilt angle) around an axis in the xy plane, which makes the angle φ with the x axis. The final coordinate z_3 of the molecule after rotation can be obtained by first calculating the coordinates (x_2, y_2, z_2) in a system $x'y'z$ whose axis x', located in the xy plane, makes the angle φ with x

$$\begin{pmatrix} x_2 \\ y_2 \\ z_2 \end{pmatrix} = \begin{pmatrix} \cos\varphi & \sin\varphi & 0 \\ -\sin\varphi & \cos\varphi & 0 \\ 0 & 0 & 1 \end{pmatrix} \begin{pmatrix} x_1 \\ y_1 \\ z_1 \end{pmatrix} \tag{B.1}$$

and then rotating the molecule around the x' axis with an angle α:

$$\begin{pmatrix} x_3 \\ y_3 \\ z_3 \end{pmatrix} = \begin{pmatrix} 1 & 0 & 0 \\ 0 & \cos\alpha & -\sin\alpha \\ 0 & \sin\alpha & \cos\alpha \end{pmatrix} \begin{pmatrix} x_2 \\ y_2 \\ z_2 \end{pmatrix} \tag{B.2}$$

From eqs B.1 and B.2 one obtains

$$z_3 = z_1\cos\alpha - x_1\sin\alpha\sin\varphi + y_1\sin\alpha\cos\varphi \tag{B.3}$$

The angles between the axis z and the direction of the molecule, θ and θ', before and after rotation are given by

$$\cos^2\theta = \frac{z_1^2}{x_1^2 + y_1^2 + z_1^2} \quad \text{and} \quad \cos^2\theta' = \frac{z_3^2}{x_3^2 + y_3^2 + z_3^2} \tag{B.4}$$

Since the distance is not modified by rotation, $x_1^2 + y_1^2 + z_1^2 = x_3^2 + y_3^2 + z_3^2$, and from eqs B.3 and B.4 one obtains

$$\cos^2\theta' = \frac{(z_1 \cos \alpha - x_1 \sin \alpha \sin \varphi + y_1 \sin \alpha \cos \varphi)^2}{x_1^2 + y_1^2 + z_1^2} \tag{B.5}$$

Averaging over the angle φ

$$\frac{1}{2\pi}\int_{\varphi=0}^{2\pi}(z_1 \cos \alpha - x_1 \sin \alpha \sin \varphi + y_1 \sin \alpha \cos \varphi)^2 \, d\varphi = z_1^2 \cos^2\alpha + (x_1^2 + y_2^2)\frac{\sin^2\alpha}{2} \tag{B.6}$$

one obtains from eq B.5 the result

$$\langle\cos^2\theta'\rangle = \cos^2\theta \cos^2\alpha + \frac{\sin^2\theta \sin^2\alpha}{2} \tag{B.7}$$

where $\langle\ \rangle$ denotes the average over the angle φ.

LA010979H

Langmuir **2005,** *21,* 11749–11756 11749

Polarization of Water near Dipolar Surfaces: A Simple Model for Anomalous Dielectric Behavior

Marian Manciu*

Department of Physics, University of Texas at El Paso, El Paso, Texas 79968

Eli Ruckenstein‡

Department of Chemical and Biological Engineering, State University of New York at Buffalo, Buffalo, New York 14260

Received July 5, 2005. In Final Form: September 17, 2005

A model for the electrostatic interactions in water in the vicinity of a surface is suggested, which accounts, within the Poisson–Boltzmann mean field approach, for the screening of the charges and for the coupling interactions between neighboring dipoles. When the water molecules near a solid surface are assumed to be organized in icelike layers, the polarization is not a continuous function but exists only at the discrete positions of the water molecules. The particular positions of the water molecules in the icelike structure govern the manner in which the average water dipoles align with each other. On the basis of this model, one could explain the nonmonotonic behavior of the polarization and the electrical potential as well as the anomalous dielectric response of water (the nonproportionality of the polarization and the macroscopic electric field), which were obtained recently via molecular dynamics simulations.

1. Introduction

The interactions between charged particles in water is described in the traditional Derjaguin–Landau–Verwey–Overbeek (DLVO) model by a van der Waals attraction between particles, coupled with a double layer repulsion due to the overlapping of the ion clouds formed near charged surfaces.[1] In the traditional approach, the ions are assumed to have Boltzmannian distributions in a mean electric field generated by charges, which in turn obeys the Poisson equation with a uniform dielectric constant (corresponding to bulk water).[2] These assumptions are oversimplified, and many corrections to the model have been proposed to improve its accuracy (by accounting for image forces, finite sizes of the hydrated ions, ion correlations, dependence of dielectric constant on the field and electrolyte concentration, ion hydration and ion dispersion forces, etc.; for a recent review, see ref 3).

Whereas the corrections to the traditional Poisson–Boltzmann approach could explain many experimental results, there are systems, such as the vesicles formed by neutral lipid bilayers in water, for which an additional force is required to explain their stability.[4] This force was related to the organization of water in the vicinity of hydrophilic surfaces; therefore it was called "hydration force".[5]

The first physical models for the hydration force related the interactions to the polarization of water near a surface, and two different approaches were proposed concomitantly.[6,7] One of these models[7] was based on the observation that the exponential decaying force, observed experimentally, is mathematically equivalent to a Landau order-parameter expansion if only the quadratic terms of the expansion are retained.[8] While the physical grounds for such an expansion remained obscure,[8] it was later conjectured that the Bjerrum defects might constitute the source of the polarization field in analogy with the ions, which provide the source of the electric field.[7] The polarization and the electric fields are no longer considered proportional to each other, and coupled equations for them were obtained through a variational procedure. A suitable concentration of Bjerrum defects had to be selected to match the experimental results for the hydration force.[7]

The Gruen–Marcelja model could relate the hydration force to the physical properties of the surfaces by assuming that the polarization of water near the interface is proportional to the surface dipole density.[9] This assumption led to the conclusion that the hydration force is proportional to the square of the surface dipolar potential of membranes (in agreement with the Schiby–Ruckenstein model),[6] a result that was confirmed by experiment.[10] However, subsequent molecular dynamics simulations revealed that the polarization of water oscillated in the vicinity of an interface, instead of being monotonic.[11] Because the Gruen–Marcelja model was particularly built to explain the exponential decay of the polarization, it was clearly invalidated by the latter simulations. Other conceptual difficulties of this model have been also reported.[12,13]

* Corresponding author. Phone: (915) 747-7531; fax: (915) 747-5447; e-mail: mmanciu@utep.edu.

‡ Phone: (716) 645-2911/2214; fax: (716) 645-3822; e-mail: feaeliru@acsu.buffalo.edu.

(1) Deryagin, B. V.; Landau, L. *Acta Physicochim. URSS* **1941**, *14*, 633; Verwey, E. J.; Overbeek, J. Th. G. *Theory of Stability of Lyophobic Colloids*; Elsevier: Amsterdam, 1948.

(2) Gouy, G. *J. Phys. Radium* **1910**, *9*, 457; Chapman, D. L. *Philos. Mag.* **1913**, *25*, 475.

(3) Manciu M.; Ruckenstein, E. *Adv. Colloid Interface Science* **2003**, *105*, 63.

(4) Rand, R. P.; Parsegian, V. A. *Biochim. Biophys. Acta* **1989**, *988*, 351.

(5) Derjaguin, B. V.; Churaev, N. V.; Muller, V. M. *Suface Forces*; Plenum Publishing: New York, 1987.

(6) Schiby, D.; Ruckenstein, E. *Chem. Phys. Lett.* **1983**, *95*, 435.

(7) Gruen, D. W. R.; Marcelja, S. *J. Chem. Soc., Faraday Trans. 2* **1983**, *79*, 211.

(8) Marcelja, S.; Radic, N. *Chem. Phys. Lett.* **1976**, *42*, 129.

(9) Cevc, G.; Marsh, D. *Biophys. J.* **1985**, *47*, 21.

(10) Simon, S. A.; McIntosh, T. J. *Proc. Natl. Acad. Sci. U.S.A.* **1989**, *86*, 9263.

(11) Berkowitz. M. L.; Raghavan, K. *Langmuir* **1991**, *7*, 1042.

(12) Attard, P.; Wei, D. Q.; Patey, G. N. *Chem. Phys. Lett.* **1990**, *172*, 69.

10.1021/la051802g CCC: $30.25
Published on Web 10/29/2005

Recent molecular dynamics simulations of water between two surfactant (sodium dodecyl sulfate) layers, reported by Faraudo and Bresme,[14] revealed oscillatory behaviors for both the polarization and the electric fields near a surface and that the two fields are not proportional to each other. While the nonmonotonic behavior again invalidated the Gruen–Marcelja model for the polarization, the nonproportionality suggested that a more complex dielectric response of water might be at the origin of the hydration force. The latter conclusion was also supported by recent molecular dynamics simulations of Faraudo and Bresme, who reported interactions between surfactant surfaces with a nonmonotonic dependence on distance.[15]

This type of oscillatory hydration was previously observed experimentally in interactions between mica surfaces in water,[16] and has been associated with the layering of water in the vicinity of a surface.[16,17] The discrete nature of the water molecules, considered hard spheres, was suggested to be responsible for these nonmonotonic interactions;[18,19] however, the high fluidity of the water confined in molecularly thin films[20] seems to be inconsistent with the "crystallization" of water predicted by the hard-sphere model.[18]

It is commonly accepted that nonassociative liquids have the tendency to crystallize near surfaces.[21] Whether or not water is layered near a surface, however, is still under debate. Experiments on phospholipid bilayers involving coherent anti-Stokes Raman scattering (CARS) microscopy[22] or electron spin resonance (ESR)[23] indicate that the water is highly ordered in the vicinity of a surface containing strong dipoles, with the permanent dipoles of the water molecules aligned parallel to the surface dipoles. The alignment is so strong that it weakens the hydrogen bonding in water near a strong dipolar interface.[22]

Even if one assumes that the water near a surface has the same structure as it does in bulk, the oscillations of the short-range interactions between surfaces could be explained by a nonlocal dielectric constant for water.[24] This model assumes that the dielectric displacement field ($D = \epsilon_0 E + P$) at a position r not only depends on the local electric field [$D(r) = \epsilon(r)E(r)$], but also depends on the electric field in the whole space: $D(r) = \int \epsilon(r,r')E(r')dr'$. In this model, the oscillations of the interactions are due to charge overscreening[25] and are analogous to the charge density waves in plasmas.[24]

Whereas in the Henderson and Lozada–Cassou model[18] the oscillatory hydration interactions are not affected much by the electrolyte concentration, in the Cheperanov model[24] the electrolyte concentration plays a major role (the oscillations increasing resonantly for concentrations larger than 0.01 M). Both models[18,24] predict many oscillations in the hydration interactions with a periodicity of ~2 Å, which is in excellent agreement with the experiment for the force between mica surfaces at 10^{-3} M. However, they are inconsistent with the polarization data obtained via molecular dynamics simulations of electrolyte solutions of higher ionic strengths confined between flat surfaces.[11,14] Because the above models do not relate the magnitude of the hydration force to the physical properties of the surfaces, they cannot explain why in most systems the hydration interactions are monotonic functions of the separation distance (such as for neutral lipid bilayers),[4] why their magnitude depends on the surface dipole density,[4,10] and why the hydration decay length itself depends on the nature of the surface (being ~2 Å for neutral lipid bilayers, but on the order of 10 Å for the interactions between mica surfaces).[26] It was recently shown that all of the above effects could be explained within the framework of Ruckenstein's model for water polarization.[27]

Let us now briefly review the basics of the model for water polarization proposed by Schiby and Ruckenstein.[6] The main idea was that the neighboring dipoles are able to orient each other, a hypothesis than seems to be validated by more recent experiments.[22,23] A planar surface with a homogeneous, constant surface dipole density cannot generate any electric field in a continuous medium. At any point, the field generated by the neighboring dipoles is exactly compensated by the opposing field of the more remote dipoles. Consequently, in a continuum theory (such as the Gruen–Marcelja model),[7] the surface dipoles are not able to polarize the nearby water molecules. However, at the molecular level water is hardly a continuum, and the interactions between neighboring dipoles are much stronger than the interactions between remote dipoles, which are screened by the intervening water molecules.[13] Therefore, one can expect on intuitive grounds that the electrical interactions between neighboring dipoles take place in a medium with a much lower dielectric constant than that of bulk water. Because of the change in dielectric screening with distance, the fields generated by the neighboring and the remote dipoles do not cancel each other any longer, and a net electric field is generated in the medium. This field is able to polarize the neighboring dipoles, which in turn generate electric fields in their vicinity, polarizing their neighboring dipoles and so on, which is in agreement with the recent CARS microscopy results for the orientation of water molecules in the vicinity of a phospholipid bilayer surface.[22]

To calculate the dipole correlations, Schiby and Ruckenstein assumed a homogeneous distribution of water molecules, which simplified the calculations considerably and led to an exponential decay of the polarization of water from a dipolar surface. The overlap of the polarized regions, when two surfaces approach one another, increases the free energy of the system, and this generates a repulsion with a roughly exponential behavior, which has a magnitude and a decay length that are very similar to the hydration forces measured in neutral lipid bilayers systems.[4]

The monotonic decay of the polarization in the Schiby–Ruckenstein model (the main critique of the polarization models) is a consequence of the assumption of the homogeneous distribution of water molecules in the vicinity of the surface. However, when the water was assumed to be structured in icelike layers in the vicinity of the surface, the polarization became an oscillatory function of the distance from the interface.[13] This result was due to the particular locations of the water molecules

(13) Manciu, M.; Ruckenstein, E. *Langmuir* **2001**, *17*, 7582.
(14) Faraudo, J.; Bresme, F. *Phys. Rev. Lett.* **2004**, *92*, 236102.
(15) Faraudo, J.; Bresme, F. *Phys. Rev. Lett.* **2005**, *94*, 077802.
(16) Israelachvili, J. N.; Pashley, R. M. *Nature* **1983**, *306*, 249.
(17) Cleveland, J. P.; Schaeffer, T. E.; Hansma, P. K. *Phys. Rev. B* **1995**, *52*, R8692.
(18) Henderson, D.; Lozada-Cassou, M. *J. Colloid Interface Sci.* **1986**, *114*, 180.
(19) Trokhymchuck, A.; Henderson, D.; Wasan, D. T. *J. Colloid Interface Sci.* **1999**, *210*, 320.
(20) Zhu, Y. X.; Granick, S. *Phys. Rev. Lett.* **2001**, *87*, 096104.
(21) Israelachvili, J. N.; McGuigan, P. M.; Homola, A. M. *Science* **1988**, *240*, 189.
(22) Cheng, J.-X.; Pautot, S.; Weitz, D. A.; Xie, X. S. *Proc. Natl. Acad. Sci. U.S.A.* **2003**, *100*, 9826.
(23) Ge, M.; Freed, J. H. *Biophys. J.* **2003**, *85*, 4023.
(24) Cherepanov, D. A. *Phys. Rev. Lett.* **2004**, *93*, 266104.
(25) Kornyshev, A. A.; Leikin, S.; Sutmann, G. *Electrochim. Acta* **1997**, *42*, 849.
(26) Pashely, R. M. *J. Colloid Interface Sci.* **1981**, *83*, 531.
(27) Manciu, M.; Ruckenstein, E. *Adv. Colloid Interface Sci.* **2004**, *112*, 109.

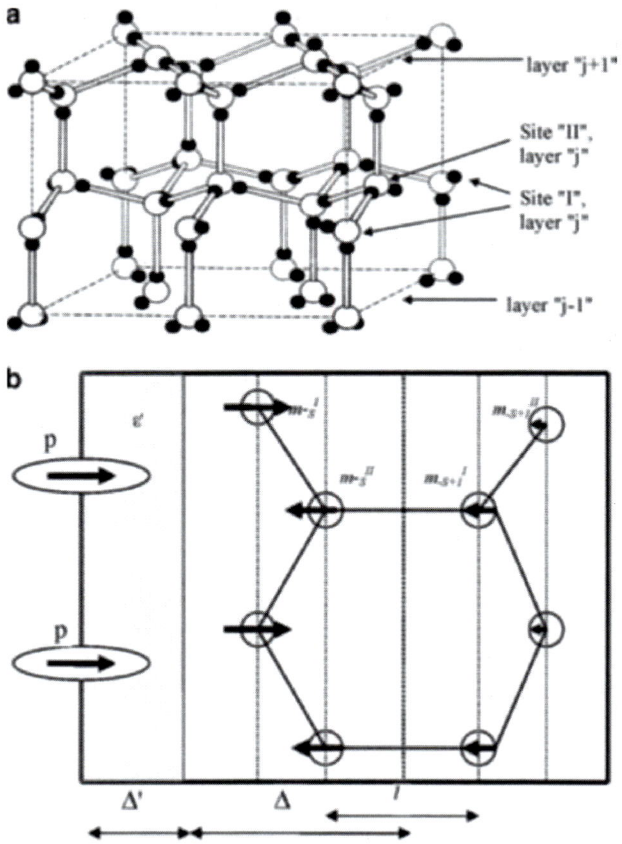

Figure 1. (a) The structure of ice-I. Each water molecule is connected by hydrogen bonding with three water molecules from sites of different kind of the same layer and one water molecule from one of the adjacent layers. The two different sites I and II of the same layer are marked on the Figure. (b) A schematic model: the dipoles p of the surface groups are embedded in a medium of dielectric constant ϵ', formed by the bound water molecules. The centers of the dipoles are at a distance Δ' from the first water layer (of thickness Δ). The polarization of the water molecules is due to the interactions with neighboring dipoles as well as with all the other charges of the system (surface charges and electrolyte ions).

in the icelike structure (see Figure 1a). In each icelike layer, the water molecules can occupy two distinct sites. As it will be shown later, the dipoles generate an electric field in the same direction in neighboring water molecules from adjacent layers; however, they generate a field in the opposite direction in the neighboring water molecules from the same layer, thus polarizing them in opposite directions. Therefore, it is possible for the dipoles of the water molecules belonging to different sites of the same layer to become oriented in antiparallel directions, leading to an oscillatory behavior in the polarization. This picture is the electrostatic equivalent of the antiferromagnetic materials, in which neighboring spins align with each other in opposite directions.

A schematic drawing is presented in Figure 1b in which the surface dipoles p align with the dipoles of the water molecules of sites "I" of the first layer "−S" in the same direction, generating the average dipolar moment m_S^I, whereas the average dipole moment m_S^{II} of the water molecules of sites "II" of the same layer are oriented in opposite direction. This picture is supported by the calculations presented in Section 3. It should be noted that molecular dynamics simulations of water between flat surfaces[11,14,15] showed an oscillatory behavior in the polarization; however, when the dynamic restrictions on the surface headgroups were removed, the statistical fluctuations of the surfaces render the polarization a monotonic function with the distance from the surface.[28] This explains the monotonic decay of the hydration force obtained in most experiments[13] and justifies the approximation employed in the Schiby−Ruckenstein model.[6] This "smearing out" of oscillations in disordered systems was suggested by Israelachvilli and Pashley[16] and was analyzed more quantitatively elsewhere.[13] It should also be noted that, if the surface dipoles are strong enough to orient all of the dipoles of the neighboring water molecules parallel to those of the surface by weakening the hydrogen bonding,[22] the icelike layering of the water near the interface will be disrupted. Consequently, no oscillations in the polarization would occur.

A simple model that illustrated the behavior of the polarization when the water molecules are organized in water layers between perfectly flat surfaces was previously suggested.[13] That model took into account the nearest-neighbor dipole interactions, but ignored the surface charges and the electrolyte ions. The model is extended here to cases in which an electrolyte as well as surface charges are also present. It will be shown that a treatment of all electrostatic interactions, in the assumption of an icelike structuring of water near interfaces, can predict an oscillatory behavior for both the polarization and the electric potential as well as a nonproportionality between the polarization and the electric fields.

It will be shown in what follows that the oscillations of the polarization are due to the structuring of water in a particular form, the coupling interactions between neighboring dipoles, the electrolyte concentration, and the boundary conditions (surface charge and surface dipole density).

2. Basic Equations

The average polarization of a water molecule is due to the electric field generated by all of the other charges and dipoles and obeys the Poisson equation, which in the vicinity of a planar surface has the form[29]

$$\frac{d^2\psi(x)}{dx^2} = -\frac{\sum_i c_i(x) q_i}{\epsilon_0} + \frac{1}{\epsilon_0}\frac{dP(x)}{dx} \tag{1}$$

in which ψ is the electrical "mean field" potential, considered to depend only on the distance x from the planar surface; q_i is the charge of an ion of kind i and concentration $c_i(x)$; $P(x)$ is the polarization of the medium; and ϵ_0 is the vacuum permittivity. The local concentrations of the ions in the "mean" electric field are assumed to have Boltzmannian distributions:

$$c_i = c_E \exp\left(-\frac{q_i\psi(x)}{kT}\right) \tag{2}$$

in which c_E represents the electrolyte concentration in the reservoir far from the plate. By assuming that the polarization is proportional to the macroscopic field, $P = -\epsilon_0(\epsilon - 1)d\psi/dx$, the Poisson−Boltzmann equation is obtained from eqs 1 and 2. However, the Schiby and Ruckenstein model suggested that the neighboring dipoles create a supplementary field, and the average polarization

(28) Perera, L.; Essmann, U.; Berkowitz, M. L. *Prog. Colloid Polym. Sci.* **1997**, *103*, 107.

(29) Ruckenstein, E.; Manciu, M. *Langmuir* **2002**, *18*, 7584.

11752 *Langmuir, Vol. 21, No. 25, 2005* *Manciu and Ruckenstein*

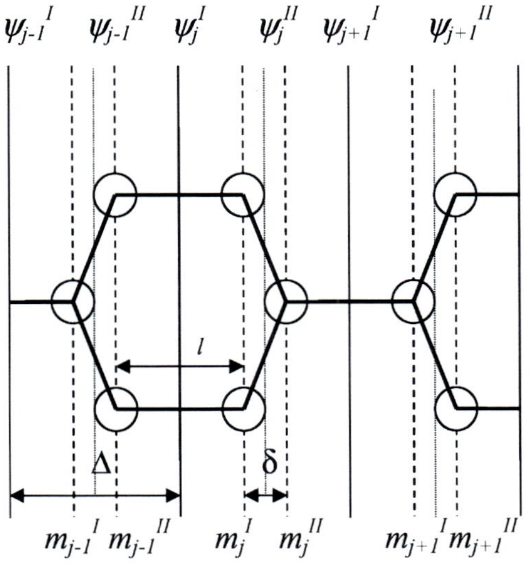

Figure 2. Sketch of the structuring of water in icelike layers in the vicinity of surfaces and the points at which the discrete variables are defined. The water molecules (circles) for the two sites of each icelike layer j have the average polarization m_j^{I} and m_j^{II}, respectively. The average value of the potential ψ_j^{I} is defined at the position between two adjacent layers $j-1$ and j, whereas ψ_j^{II} is defined at the middle of layer j.

of a water molecule is proportional to the total field acting at its location.

The molecular field acting at the location of a water molecule is composed of a macroscopic field, $E = -(\mathrm{d}\psi/\mathrm{d}x)$, a local field (the Lorentz field E_{L}) that occurs because each molecule is surrounded by a dielectric medium (constituted of all the other water molecules), $E_{\mathrm{L}} = P/3\epsilon_0$, plus an additional field, E_{P}, due to the neighboring dipoles.[13,27] Assuming that the additional field is generated only by the neighboring molecules from the same and adjacent icelike water layers (see Figure 2), at each of the sites (I or II) of layer "j" the field is provided by[13]

$$E_{\mathrm{P}}^{\mathrm{I},j} = D_0^{\mathrm{I,I}}\, m_j^{\mathrm{I}} + D_0^{\mathrm{II,I}}\, m_j^{\mathrm{II}} + D_{-1}^{\mathrm{I,I}}\, m_{j-1}^{\mathrm{I}} + D_{-1}^{\mathrm{II,I}}\, m_{j-1}^{\mathrm{II}} + D_{+1}^{\mathrm{I,I}}\, m_{j+1}^{\mathrm{I}} + D_{+1}^{\mathrm{II,I}}\, m_{j+1}^{\mathrm{II}} \quad (3a)$$

$$E_{\mathrm{P}}^{\mathrm{II},j} = D_0^{\mathrm{I,II}}\, m_j^{\mathrm{I}} + D_0^{\mathrm{II,II}}\, m_j^{\mathrm{II}} + D_{-1}^{\mathrm{I,II}}\, m_{j-1}^{\mathrm{I}} + D_{-1}^{\mathrm{II,II}}\, m_{j-1}^{\mathrm{II}} + D_{+1}^{\mathrm{I,II}}\, m_{j+1}^{\mathrm{I}} + D_{+1}^{\mathrm{II,II}}\, m_{j+1}^{\mathrm{II}} \quad (3b)$$

in which the m's and the D's are average dipole moments and coupling coefficients, respectively (m_j^{I} represents the average dipole moment of a molecule at site I of layer j, and the coupling coefficient $D_0^{\mathrm{I,II}}$ represents the ratio between the electric field generated by all of the molecules of sites I of layer j at the positions of sites II of layer j and the average dipole moment of the former molecules, m_j^{I}).

The molecular polarizability of a water molecule, γ, can be related to macroscopic quantities by taking into account that, in an uniformly polarized medium, the fields generated by the near neighbors cancel each other. Consequently, one obtains[29]

$$m = \gamma\left(E + \frac{m}{3\epsilon_0 v_0}\right) = P v_0 = \epsilon_0 v_0 (\epsilon - 1)E \quad (4)$$

in which v_0 represent the average volume occupied by a water molecule, which leads to the Claussius–Mossotti equation, $\gamma = 3\epsilon_0 v_0\, (\epsilon - 1)/(\epsilon + 2)$. The average dipole moments of the molecules in the two sites, I and II, of layer j are therefore provided by

$$m_j^{\mathrm{I}} = \gamma\left(E + E_{\mathrm{P}} + \frac{m_j^{\mathrm{I}}}{3\epsilon_0 v_0}\right) = \epsilon_0 v_0 (\epsilon - 1)\left[-\left(\frac{\mathrm{d}\psi}{\mathrm{d}x}\right)_j^{\mathrm{I}} + D_{-1}^{\mathrm{I,I}}\, m_{j-1}^{\mathrm{I}} + D_{-1}^{\mathrm{II,I}}\, m_{j-1}^{\mathrm{II}} + D_0^{\mathrm{I,I}}\, m_j^{\mathrm{I}} + D_0^{\mathrm{II,I}}\, m_j^{\mathrm{II}} + D_{+1}^{\mathrm{I,I}}\, m_{j+1}^{\mathrm{I}} + D_{+1}^{\mathrm{II,I}}\, m_{j+1}^{\mathrm{II}}\right] \quad (5a)$$

$$m_j^{\mathrm{II}} = \gamma\left(E + E_{\mathrm{P}} + \frac{m_j^{\mathrm{II}}}{3\epsilon_0 v_0}\right) = \epsilon_0 v_0 (\epsilon - 1)\left[-\left(\frac{\mathrm{d}\psi}{\mathrm{d}x}\right)_j^{\mathrm{II}} + D_{-1}^{\mathrm{I,II}}\, m_{j-1}^{\mathrm{I}} + D_{-1}^{\mathrm{II,II}}\, m_{j-1}^{\mathrm{II}} + D_0^{\mathrm{I,II}}\, m_j^{\mathrm{I}} + D_0^{\mathrm{II,II}}\, m_j^{\mathrm{II}} + D_{+1}^{\mathrm{I,II}}\, m_{j+1}^{\mathrm{I}} + D_{+1}^{\mathrm{II,II}}\, m_{j+1}^{\mathrm{II}}\right] \quad (5b)$$

in which the derivatives of the potentials of eqs 5a and 5b should be taken at the location of sites I and II, respectively, of layer j.

3. General Behavior of the Solutions of the System

Although the nonlinear system of equations can be solved numerically, here we focus on the linear approximation of the Poisson–Boltzmann equation (which is accurate for small values of the potentials ψ, $q\psi/kT \ll 1$). Because the average polarization of water is $P(x) = m(x)/v_0$, in this approximation eqs 1 and 2 become

$$\frac{\mathrm{d}^2\psi}{\mathrm{d}x^2} = \frac{\epsilon}{\lambda_{\mathrm{DH}}{}^2}\psi + \frac{1}{\epsilon_0 v_0}\frac{\mathrm{d}m(x)}{\mathrm{d}x} \quad (6)$$

in which ϵ is the dielectric constant of water, and $\lambda_{\mathrm{DH}} = (\epsilon\epsilon_0 kT/2e^2 c_{\mathrm{E}})^{1/2}$ is the Debye–Hückel length for a uni-univalent electrolyte of concentration c_{E}, where e is the elementary charge, k is the Boltzmann constant, and T is the absolute temperature.

The polarizations of the water molecules are discrete quantities, being defined only at the locations of the water molecules (sites I and II in an icelike layer). To combine the discrete eq 5a,b with the continuous Poisson–Boltzmann eq 6, the latter equation must also be discretized. To do this, we define for each layer j the average discrete potentials ψ_j^{I} and ψ_j^{II}, which are defined at the boundary and the middle of the layer, respectively (see Figure 2). These positions were selected because the potential undergoes a steep change at the crossing of a dipolar surface (at the locations of dipoles m_j^{I} and m_j^{II}), whereas, in the remainder of the space, it varies slowly (only because of the distribution of electrolyte ions). By integrating eq 6 between site II of layer $j-1$ and site I of layer j, and between site I of layer j and site II of the same layer j one obtains

$$\frac{\mathrm{d}\psi}{\mathrm{d}x}\Big|_j^{\mathrm{I}} - \frac{\mathrm{d}\psi}{\mathrm{d}x}\Big|_{j-1}^{\mathrm{II}} \cong \left(\frac{\psi_j^{\mathrm{II}} - \psi_j^{\mathrm{I}}}{\Delta/2}\right) - \left(\frac{\psi_j^{\mathrm{I}} - \psi_{j-1}^{\mathrm{II}}}{\Delta/2}\right) \equiv 2\frac{\psi_j^{\mathrm{II}} - 2\psi_j^{\mathrm{I}} + \psi_{j-1}^{\mathrm{II}}}{\Delta} = \frac{\epsilon}{\lambda_{\mathrm{DH}}{}^2}\psi_j^{\mathrm{I}}(\Delta - \delta) + \frac{1}{\epsilon_0 v_0}(m_j^{\mathrm{I}} - m_{j-1}^{\mathrm{II}}); \quad (7a)$$

$$\frac{d\psi}{dx}\Big|_j^{\rm II} - \frac{d\psi}{dx}\Big|_j^{\rm I} \cong \left(\frac{\psi_{j+1}^{\rm I} - \psi_j^{\rm II}}{\Delta/2}\right) - \left(\frac{\psi_j^{\rm II} - \psi_j^{\rm I}}{\Delta/2}\right) \equiv 2\frac{\psi_{j+1}^{\rm I} - 2\psi_j^{\rm II} + \psi_j^{\rm I}}{\Delta} = \frac{\epsilon}{\lambda_{\rm DH}{}^2}\psi_j^{\rm II}\delta + \frac{1}{\epsilon_0 v_0}(m_j^{\rm II} - m_j^{\rm I}) \quad (7b)$$

in which Δ is the distance between the centers of two adjacent icelike layers, δ is the distance between the sites I and II of the same layer, and $d\psi/dx\,|_j^{\rm I}$ represents the derivative of the potential with respect to distance, calculated at site I of layer j, which is approximated using finite differences. The same approximations in eqs 5a and 5b lead to

$$m_j^{\rm I} = \epsilon_0 v_0(\epsilon - 1)\left(2\frac{\psi_j^{\rm I} - \psi_j^{\rm II}}{\Delta} + D_{-1}^{\rm I,I} m_{j-1}^{\rm I} + D_{-1}^{\rm II,I} m_{j-1}^{\rm II} + D_0^{\rm I,I} m_j^{\rm I} + D_0^{\rm II,I} m_j^{\rm II} + D_{+1}^{\rm I,I} m_{j+1}^{\rm I} + D_{+1}^{\rm II,I} m_{j+1}^{\rm II}\right) \quad (7c)$$

$$m_j^{\rm II} = \epsilon_0 v_0(\epsilon - 1)\left(2\frac{\psi_j^{\rm II} - \psi_{j+1}^{\rm I}}{\Delta} + D_{-1}^{\rm I,II} m_{j-1}^{\rm I} + D_{-1}^{\rm II,II} m_{j-1}^{\rm II} + D_0^{\rm I,II} m_j^{\rm I} + D_0^{\rm II,II} m_j^{\rm II} + D_{+1}^{\rm I,II} m_{j+1}^{\rm I} + D_{+1}^{\rm II,II} m_{j+1}^{\rm II}\right) \quad (7d)$$

Consequently, there are 4 independent variables $m_j^{\rm I}$, $m_j^{\rm II}$, $\psi_j^{\rm I}$, and $\psi_j^{\rm II}$ for each icelike layer j, and a system of 4 linear equations with finite differences (eqs 7a–7d) is obtained. We try solutions of the type[30]

$$\psi_j^{\rm I} = ar^j$$
$$\psi_j^{\rm II} = br^j$$
$$m_j^{\rm I} = cr^j$$
$$m_j^{\rm II} = dr^j \quad (8)$$

which, when introduced in the system of eqs 7a–7d, lead to

$$\left[\frac{4}{\Delta} + \frac{\epsilon}{\lambda_{\rm DH}{}^2}(\Delta - \delta)\right]a - \frac{2}{\Delta}\left(1 + \frac{1}{r}\right)b + \frac{1}{\epsilon_0 v_0}c - \frac{1}{\epsilon_0 v_0}\frac{1}{r}d = 0 \quad (9a)$$

$$-\frac{2}{\Delta}(1 + r)a + \left[\frac{4}{\Delta} + \frac{\epsilon}{\lambda_{\rm DH}{}^2}(\Delta - \delta)\right]b - \frac{1}{\epsilon_0 v_0}c + \frac{1}{\epsilon_0 v_0}d = 0 \quad (9b)$$

$$m_j^{\rm I} = \frac{2}{\Delta}a - \frac{2}{\Delta}b + \left[\frac{D_{-1}^{\rm I,I}}{r} + D_0^{\rm I,I} + D_{+1}^{\rm I,I}r - \frac{1}{\epsilon_0 v_0(\epsilon - 1)}\right]c + \left(\frac{D_{-1}^{\rm II,I}}{r} + D_0^{\rm II,I} + D_{+1}^{\rm II,I}r\right)d = 0 \quad (9c)$$

$$m_j^{\rm II} = -\frac{2r}{\Delta}a + \frac{2}{\Delta}b + \left(\frac{D_{-1}^{\rm I,II}}{r} + D_0^{\rm I,II} + D_{+1}{}^{\rm I,II}r\right)c + \left[\frac{D_{-1}^{\rm II,II}}{r} + D_0^{\rm II,II} + D_{+1}^{\rm II,II}r - \frac{1}{\epsilon_0 v_0(\epsilon - 1)}\right]d = 0 \quad (9d)$$

The linear system of eqs 9a–9d has nontrivial solutions for the constants a, b, c, and d only if its determinant vanishes. This condition leads to an equation of 6 degrees in r, which provides 6 values for r. The physical symmetry of the system (the invariance at a simultaneous exchange of layers $j-1$ and $j+1$ and of sites I and II) requires that for each solution r_k, its reciprocal $1/r_k$ should also be a solution of the system; hence, only three values of r_k are independent.

There is a simple correspondence between the solutions r_k of a linear system with finite differences and the characteristic lengths of a linear system of differential equations. By denoting $x = j\Delta$ as a continuous position variable, one obtains

$$r_k^j = r_k^{x/\Delta} = \exp\left(\frac{x}{\Delta}\ln r_k\right) = \exp\frac{x}{\lambda_k} \quad (10a)$$

with

$$\lambda_k \equiv \frac{\Delta}{\ln(r_k)} \quad (10b)$$

as a characteristic length associated to the solution r_k. The associate solution $1/r_k$ corresponds to the characteristic length $-\lambda_k$. While a description in terms of characteristic lengths is more appealing intuitively, one should note that those solutions are defined only at the discrete points $x = j\Delta$ (with j integer). Because the determinant of the system (eq 9) vanishes for any r_k, three of the constants for each set of a_k, b_k, c_k, and d_k can be determined as functions of the forth one. This means that the solution depends on six constants, which have to be determined from the boundary conditions. When the surfaces are identical, the symmetry implies that there are only three independent boundary conditions (For an even number of layers, $\psi^{\rm I}$ is a symmetric function with respect to the middle distance between surfaces and $m_j^{\rm I} = m_{-j-1}^{\rm II}$, whereas, for an odd number of layers, $\psi^{\rm II}$ is symmetric with respect to the middle distance and $m_j^{\rm I} = m_{-j}^{\rm II}$). Assuming a surface charge density σ and a surface dipole density $p_\perp/A$, in which $p_\perp$ is the normal component of the dipole moment of a polar surface group and A is the average area occupied by each dipole located at a distance Δ' from the boundary of the first icelike water layer, the Poisson equation provides the first boundary condition:

$$\frac{d\psi_{-S}{}^{\rm I}}{dx} = -\frac{\sigma}{\epsilon_0\epsilon'} \quad (11a)$$

in which ϵ' is the dielectric constant of the medium formed by the surface polar groups and the disorganized water molecules between them (see Figure 1b). The remaining two boundary conditions are provided by the polarization of the water molecules located on the two sites of the first water layers (denoted by $j = \pm S$):

$$m_{-S}^{\rm I} = \epsilon_0 v_0(\epsilon - 1)\left(2\frac{\psi_{-S}^{\rm I} - \psi_{-S}^{\rm II}}{\Delta} + D_0^{\rm I,I}\, m_{-S}^{\rm I} + D_0^{\rm II,I}\, m_{-S}^{\rm II} + D_{+1}^{\rm I,I}\, m_{-S+1}^{\rm I} + D_{+1}^{\rm II,I} m_{-S+1}^{\rm II} + E_{-S}^{\rm I}\right) \quad (11b)$$

(30) Mickens, R. E. *Difference Equations*; Van Nostrand Reinhold Company: New York, 1987.

$$m^{\mathrm{II}}_{-S} = \epsilon_0 v_0(\epsilon - 1)\left(2\frac{\psi^{\mathrm{II}}_{-S} - \psi^{\mathrm{I}}_{-S+1}}{\Delta} + D_0^{\mathrm{I,II}} m^{\mathrm{I}}_{-S} + D_0^{\mathrm{II,II}} m^{\mathrm{II}}_{-S} + D_{+1}^{\mathrm{I,II}} m^{\mathrm{I}}_{-S+1} + D_{+1}^{\mathrm{II,II}} m^{\mathrm{II}}_{-S+1} + E^{\mathrm{II}}_{-S}\right) \quad (11c)$$

The main differences between eqs 11a–11c and eqs 5a and 5b exist in the absence of water dipoles in the layer $-S-1$ (because the layer $-S$ is the first icelike layer) and the presence of the fields E^{I}_{-S} and E^{II}_{-S}, generated at sites I and II of layer $-S$, respectively, by the surface dipoles. Because the dielectric constant for the latter interaction, ϵ', is much smaller than the dielectric constant of bulk water, only the nearest surface dipole has a significant contribution to these fields, which can be approximated by[13]

$$E^{\mathrm{I}}_{-S} = \frac{p_{\perp}}{\epsilon'}\frac{1}{2\pi\epsilon_0\left[\frac{A}{\pi} + \left(\Delta' + \frac{\Delta - \delta}{2}\right)^2\right]^{3/2}} \quad (12a)$$

$$E_{-S}^{\mathrm{II}} = \frac{p_{\perp}}{\epsilon'}\frac{1}{2\pi\epsilon_0\left[\frac{A}{\pi} + \left(\Delta' + \frac{\Delta + \delta}{2}\right)^2\right]^{3/2}} \quad (12b)$$

The linear system of eqs 9a–9d with finite differences, together with the three boundary conditions (for identical surfaces), can be solved analytically to provide solutions of the type[30]

$$\psi^{\mathrm{I}}_j = \sum_{k=1}^{6} a_k r_k^j$$

$$\psi^{\mathrm{II}}_j = \sum_{k=1}^{6} b_k r_k^j$$

$$m^{\mathrm{I}}_j = \sum_{k=1}^{6} c_k r_k^j$$

$$m^{\mathrm{II}}_j = \sum_{k=1}^{6} d_k r_k^j \quad (13)$$

In the following section, a model solution of the system is presented for surfaces separated by seven icelike layers (corresponding to a separation distance of ~25 Å). This will clarify the general behavior of the system and will demonstrate the nonproportionality between the polarization and the electric field.

4. A Model Solution for Polarization and Electric Potential

In what follows, we examine the general behavior of the solutions of the system of eqs 9a–9d for the boundary conditions given by eqs 11a–11c by assuming perfect icelike layers. It is taken into account that the distance between the centers of two adjacent water molecules is $l = 2.76$ Å, the distance between the centers of two adjacent icelike water layers is $\Delta = 4/3l = 3.68$ Å, the distance between the two distinct sites in the same layer is $\delta = l/3 = 0.92$ Å (see Figure 2), and the volume occupied by a water molecule is $v_0 = 8l^3/3\sqrt{3} = 32.37$ Å^3.

A legitimate first question is related to the magnitude of errors generated by the discretization of the Poisson–Boltzmann equation (the replacement of eq 6 by eqs 7a–7d). Considering that all the coupling coefficients D vanish (therefore, neglecting the coupling between neighboring dipoles), one should recover the traditional Poisson–Boltzmann equation. Indeed, the condition of vanishing of the determinant of the system (eqs 9a–9d) generates in this case only two solutions, r_1 and $1/r_1$, and the corresponding decay lengths (provided by eq 10b) are almost identical to $\pm\lambda_{\mathrm{DH}}$, at least for distances larger than Δ (see Figure 3a). To calculate the coupling coefficients D, we employ the approximations used in ref 13, which assume that only the dipoles of the neighboring water molecules, within a distance $2l$ from a particular point, generate a significant electric field at that point and that the fields generated by all the other dipoles are negligible. This means that the field E_p is generated at a site of a water molecule only by the dipoles of its first 26 neighbors, which are located either in the same icelike layer or in one of the adjacent layers. By assuming an effective dielectric constant, $\epsilon'' = 1$, for this interaction, one obtains[13]

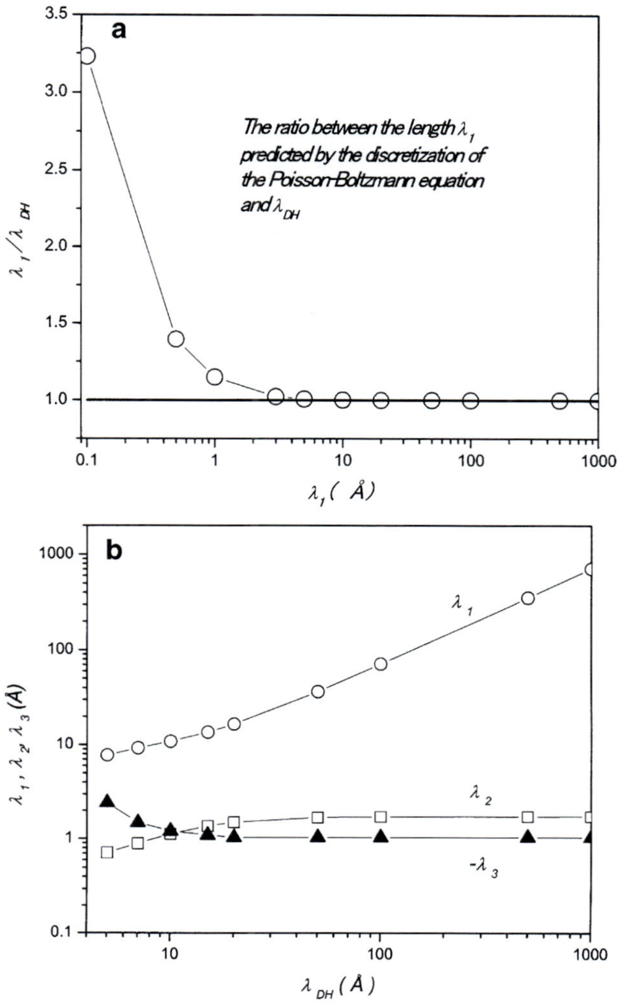

Figure 3. (a) The ratio between the Debye–Hückel length λ_{DH} and the characteristic length of the system (eqs 9a–9d), in the absence of the coupling interactions between dipoles (all D coefficients vanish), as a function of λ_{DH}. The results obtained via the discretization of the Poisson–Boltzmann equation are accurate, except for very small values of λ_{DH}. (b) The characteristic lengths of the linear system with finite differences as a function of the Debye–Hückel length, for the coupling coefficients given by eqs 14a–14f.

$$D_{-1}^{\mathrm{I,I}} \equiv D_{1}^{\mathrm{II,II}} \frac{0.689}{4\pi\epsilon_0 l^3} \quad (14a)$$

$$D_{-1}^{\mathrm{II,II}} \equiv D_{1}^{\mathrm{I,I}} \frac{0.689}{4\pi\epsilon_0 l^3} \quad (14b)$$

$$D_{-1}^{\mathrm{I,II}} = D_{1}^{\mathrm{II,I}} = \frac{0.432}{4\pi\epsilon_0 l^3} \quad (14c)$$

$$D_{-1}^{\mathrm{II,I}} = D_{1}^{\mathrm{I,II}} = \frac{1.845}{4\pi\epsilon_0 l^3} \quad (14d)$$

$$D_{0}^{\mathrm{I,I}} = D_{0}^{\mathrm{II,II}} = -\frac{1.378}{4\pi\epsilon_0 l^3} \quad (14e)$$

$$D_{0}^{\mathrm{I,II}} = D_{0}^{\mathrm{I,II}} = -\frac{2.388}{4\pi\epsilon_0 l^3} \quad (14f)$$

The condition of vanishing of the determinant provides three pairs of solutions $r_1, 1/r_1$; $r_2, 1/r_2$; and $r_3, 1/r_3$, with the last one being negative. The 3 decay lengths corresponding to the 6 solutions are plotted in Figure 3b as functions of the decay length λ_{DH}. The length λ_1 is reminiscent of the Debye–Hückel length (at least at low electrolyte concentrations), whereas λ_2 is comparable to the hydration decay length, which was determined experimentally for the interactions between neutral lipid bilayers.[4] The negative sign of r_3 implies that the terms proportional to r_3^j and $1/r_3^j$ have different signs in adjacent layers; hence, λ_3 corresponds to a decay length of an oscillatory contribution to the polarization and potential (with an oscillation length 2Δ). These types of oscillations are provided by the system of equations with finite differences (eqs 9a–9d) and are generated by the coupling coefficients between neighboring dipoles.

The physical origin of this behavior is related to the particular structure of the icelike layers: the field generated by a dipole $\vec{p}$, at a distance $\vec{r}$ from the dipole, is given by

$$\vec{E} = \frac{1}{4\pi\epsilon_0 \epsilon r^5}[3(\vec{p}\vec{r})\vec{r} - r^2\vec{p}] \quad (15a)$$

Because we are interested only in the polarization along the x axis, the field along this axis, at a position that makes an angle θ with the x axis, becomes

$$E_x = \frac{p_x}{4\pi\epsilon_0 \epsilon r^3}(3\cos^2\theta - 1) \quad (15b)$$

Consequently, the neighboring water molecules belonging to different layers (with small θ) tend to align with each other in the same direction, whereas the neighboring dipoles from the different sites of the same layer (for which the angle $\theta = \cos^{-1}(1/3) \cong 71°$) tend to align with each other in opposite directions (these interactions dominate the coupling coefficient $D_0^{\mathrm{I,II}}$, which is negative). The relatively low magnitude of the decay length of the "oscillatory" part of the polarization and electrical potential indicates that the oscillations are attenuated rapidly with the distance from the surface. In a continuous model, in which the water molecules are not constrained to particular positions corresponding to an icelike structure, the water dipoles tend to align themselves in the same directions. Under these conditions, the oscillations vanish,

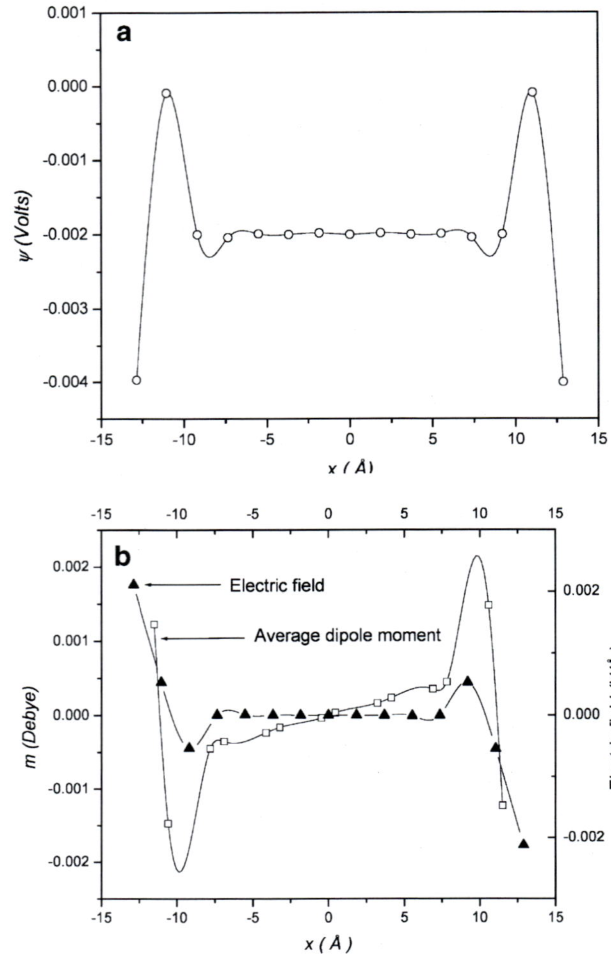

Figure 4. (a) The electric potential (circles) as a function of the distance from the surface for the system described in the text. The continuous line represents a Spline interpolation. (b) The average polarization of a water molecule (squares) as a function of the distance from surface for the system described in text; the macroscopic electric field (triangles) obtained through the numerical derivative of the potential is not proportional to the average polarization.

and only the characteristic lengths λ_1 and λ_2 remain.[29] The vanishing of the oscillations of polarization due to structural disorder (the absence of the icelike structuring of water near interfaces) has also been observed in molecular dynamics simulations. In the latter case, when surface disorder was generated by allowing the motion of the surface groups, the polarization decayed monotonically from the surface.[28]

Let us investigate the electrical potential and the polarization of a typical system. For two surfaces separated by 7 icelike water layers ($S = 3, j = 0$ corresponding to the middle layer), with $\sigma = 1.9 \times 10^{-4}$ C/m^2, $p_\perp/\epsilon' = 0.1$ D, $A = 50$ Å^2 and $\lambda_{\mathrm{DH}} = 50$ Å, the solutions of the system (eqs 9a–9d) with boundary conditions given in eqs 11a–11c for the potential and polarization as functions of the distance between surfaces are plotted in Figure 4, panels a and b, respectively. Although the solutions are defined only at particular positions in each layer (dots), for clarity they have been numerically interpolated using splines. Both the electrical potential (Figure 4a, circles) and the polarization (Figure 4b, squares) have an oscillatory behavior with respect to the distance. Furthermore, the numerical derivative of the electrical potential (Figure 4b, triangles) shows that the polarization is not proportional to the macroscopic electric field. The result is particularly clear in the middle region between the surfaces, where the polarization varies slowly (mainly because of the coupling between dipoles), whereas the electric potential is almost constant (because the electrolyte ions can attenuate the electric field). Near the surfaces, the rapid variation of ψ generates a strong electric field, which has a large contribution to the polarization of the water molecules. Accounting for the facts that, in a real system, the water molecules are not perfectly structured in icelike layers (the consequences of this disorder have been investigated elsewhere),[13] and that the charges and surface dipoles were selected here to be sufficiently small for the linear approximation to hold, the behavior of the calculated electrical potential and polarization are in excellent qualitative agreement with the results of molecular dynamics simulations (see Figures 1 and 2 of ref 14).

5. Conclusions

It was recently shown via molecular dynamics simulations[14] that, in the close vicinity of a surface, water molecules exhibit an anomalous dielectric response, in which the local polarization is not proportional to the local electric field. The recent findings are also in agreement with earlier molecular dynamics simulations, which showed that the polarization of water oscillates in the vicinity of a dipolar surface,[11,14] leading therefore to a nonmonotonic hydration force.[15] Previous models for oscillatory hydration forces, based either on volume-excluded effects,[18,19] or on a nonlocal dielectric constant,[24] predicted many oscillations with a periodicity of ~2 Å, which is inconsistent with these molecular dynamics simulations,[11,13,14] in which the polarization exhibits only a few oscillations in the vicinity of the surface, with a larger periodicity.

It is suggested here that this behavior of water in the vicinity of a surface is a result of a coupling between neighboring dipoles when the water is organized in an icelike layer in the vicinity of the surface. This coupling constituted the basis of a model for water polarization, which was proposed earlier to explain the monotonic hydration interactions[6] (either in the absence of water structuring, or when statistical disorder has been taken into account).[13,27] The interactions between neighboring dipoles are the most important because they are much less screened than those between more remote dipoles. Whereas in the latter case, there are many intervening water molecules that screen the interactions, in the former case there are no intervening water molecules. This behavior, due to the discrete (molecular) nature of water, is in sharp contrast with the classical results obtained by treating water as a continuum with a constant bulk dielectric constant. The continuum treatment assumes that a homogeneous surface dipolar density cannot polarize the water, whereas, in this model, the water is polarized by the surface dipoles.[6,13]

As shown earlier,[13] if one takes into account the coupling between neighboring dipoles and one also assumes a structuring of the water in icelike layers near the surfaces, one can explain the oscillation of the water polarization in the vicinity of the surfaces. The oscillations are essentially due to the particular positions of water molecules in an icelike structure. Whereas the water molecules from different icelike layers tend to align with each other in the same direction, the water molecules from two sites of the same layer tend to align with each other in opposite directions (other crystalline structures may not necessarily lead to an oscillatory polarization). That model, which involved a vanishing electrolyte concentration, was extended here to the general case that takes into account all of the electrostatic interactions between charges (electrolyte ions and surface charges) and dipoles (water as well as surface dipoles). A general formalism, which employs an analytical solution in the linear approximation, is suggested, and the general behavior of the system is investigated. It was shown that, in the linear approximation, the polarization is a superposition of three exponentially decreasing functions, two of them being monotonic and one oscillating with the periodicity of two icelike water layers (~7Å). The decay lengths of the functions depend on the electrolyte concentration and on the coupling interactions between neighboring dipoles.

A model calculation predicted an anomalous dielectric behavior of water in the vicinity of a surface and an oscillatory dependence on distance of the water polarization and the electric potential in the close vicinity of a surface, which are very similar to the results obtained via molecular dynamics simulations.[14]

LA051802G

7584 *Langmuir* **2002,** *18,* 7584−7593

The Coupling between the Hydration and Double Layer Interactions

Eli Ruckenstein* and Marian Manciu

Department of Chemical Engineering, State University of New York at Buffalo, Buffalo, New York 14260

Received May 8, 2002. In Final Form: July 31, 2002

The electric potential and the polarization between two charged, flat surfaces immersed in water are calculated without the usual assumption that the polarization is proportional to the electric field. The new constitutive equation accounts for an additional interaction, due to the orientational correlation of the water dipoles, which is a result of the mutual interaction between neighboring dipoles. The surfaces should be characterized not only by their charges or potentials but also by their dipole densities. Because both the double layer and the hydration forces are dependent in the present model on the polarization, the repulsion cannot be separated into two additive terms, one being the traditional "double layer" repulsion (DLVO theory) and the other the "structural" repulsion (hydration). In the absence of surface dipoles, the repulsion between charged surfaces becomes stronger than that predicted by the DLVO theory, particularly at high ionic strengths. The total repulsion can be increased or even decreased by the presence of dipoles on the surfaces, which contradicts the additivity of the repulsions. The repulsion between uncharged surfaces that possess dipoles was found to depend on the electrolyte concentration, and to be extended over a much longer distance than the conventional exponential decay, particularly at high ionic strengths. As a consequence of the coupling between the double layer and hydration, the decay length of the repulsive force becomes larger than those of the two conventional repulsions and at high ionic strength the difference becomes increasingly larger.

I. Introduction

Gouy[1] and Chapman,[2] who were the first to predict the distribution of electrolyte ions in water around a charged flat surface, demonstrated that the ions form a diffuse layer (the electric double layer) in the liquid near the interface. The interaction between two charged surfaces, due to the overlapping of the double layers, was calculated much later by Deryaguin and Landau[3] and Verwey and Overbeek.[4] The stability of the colloids was successfully explained by them in terms of a balance between the double layer and van der Waals interactions (the DLVO theory).[3,4]

However, it is well-known that the DLVO theory is reliable only for a limited range of electrolyte concentrations. Deryaguin himself has noted that this theory is valid between approximately 10^{-3} and 5.0×10^{-2} M. There have been numerous attempts to improve the DLVO theory, by accounting for the saturation of the molecular polarizability at high fields,[5] the image forces,[6] finite ion sizes,[7] or the correlation between ions,[8] to cite only a few. Experiments with uncharged lipid bilayers in water[9] and the restabilization of some colloids at high ionic strengths[10] indicated that another, non-DLVO repulsion is also present. This repulsion was related to the structuring of the solvent around the interface and is known (when water is the solvent) as the hydration force.

While the existence of the hydration force is undisputed, its origin is still a matter of debate. Marcelja and Radic showed that the exponential repulsion observed experimentally can be obtained if a suitable Landau free energy density, dependent on an unknown "order parameter", is associated with the correlation of the water molecules in the vicinity of the surface.[11] Later, Schiby and Ruckenstein[12] and Gruen and Marcelja[13] presented two different models, both involving the polarization of the water molecules.

Gruen and Marcelja considered that the electric and polarization fields are not proportional in the vicinity of a surface and that while the electric field has the ion concentrations as its source, the source of the polarization field is provided by the Bjerrum defects. The coupled equations for the electric and polarization fields were derived through a variational method. Attard et al.[14] contested the Gruen−Marcelja model because, to obtain an exponential decay of the repulsion, the nonlocal dielectric function was assumed to have a simple monotonic dependence upon the wavelength (eq 33 in ref 13). This was found to be inconsistent with the exact expression for multipolar models.[14] In addition, the characteristic decay length for polarization (denoted ξ in eq 18, ref 13) is inversely proportional to the square of the (unknown) concentration of Bjerrum defects in ice. While at large concentrations of Bjerrum defects the disordered ice becomes similar to water and the traditional Poisson−

* Corresponding author: e-mail address, feaeliru@acsu.buffalo.edu; phone, (716) 645-2911 (ext 2214); fax, (716) 645-3822.

(1) Gouy, G. *J. Phys. Radium* **1910**, *9*, 457.
(2) Chapman, D. L. *Philos. Mag.* **1913**, *25*, 475.
(3) Deryaguin, B. V.; Landau, L. *Acta Physicochim. URSS* **1941**, *14,* 633.
(4) Verwey, E. J.; Overbeek, J. Th. G. *Theory of stability of lyophobic colloids*; Elsevier: Amsterdam, 1948.
(5) Henderson, D.; Lozada-Cassou, M. *J. Colloid Interface Sci.* **1986**, *114,* 180.
(6) Jönnson, B.; Wennerström, H. *J. Chem. Soc., Faraday Trans.* **1983**, *79*, 19.
(7) Levine, S.; Bell, G. M. *Discuss. Faraday Soc.* **1966**, 42, 69. Ruckenstein, E.; Schiby, D. *Langmuir* **1985**, *1*, 612.
(8) Wennerström, H.; Jönnson, B.; Linse, P. *J. Chem. Phys.* **1982**, *76*, 4665.
(9) Rand, R. P.; Parsegian, V. A. *Biochim. Biophys. Acta* **1989**, *988*, 351.
(10) Healy, T. W.; Homola, A.; James, R. O.; Hunter, R. J. *Faraday Discuss. Chem. Soc.* **1978**, *65*, 156.
(11) Marcelja, S.; Radic, N. *Chem. Phys. Lett.* **1976**, *42*, 129.
(12) (a) Schiby, D.; Ruckenstein, E. *Chem. Phys. Lett.* **1983**, *95*, 435. (b) Schiby, D.; Ruckenstein, E. *Chem. Phys. Lett.* **1983**, *100*, 277.
(13) Gruen, D. W. R.; Marcelja, S. *J. Chem. Soc., Faraday Trans. 2* **1983**, *79*, 211, and 225.
(14) Attard, P.; Wei, D. Q.; Patey, G. N. *Chem. Phys. Lett.* **1990**, *172*, 69.

10.1021/la020435v CCC: $22.00
Published on Web 09/06/2002

Boltzmann equation is recovered, a low concentration of Bjerrum defects provided a much too large polarization decay length (ref 13 indicated as an example the value $\xi = 332$ Å), which is by more than 2 orders of magnitude larger than the decay length of the hydration force.

The model of Schiby and Ruckenstein[12a] predicted that the surface dipoles induced a polarization, even in the absence of a double layer, which decayed exponentially from the surface. The electrodynamics of continuous media predicts that the field generated by a planar surface with an uniform dipole density immersed in a fluid of uniform dielectric constant vanishes outside the surface. However, at a molecular scale, the fluid is not uniform. The interactions between remote charges or dipoles can be considered "screened" by the intervening medium which has a large dielectric constant ($\epsilon \sim 80$ for water); however, the interaction between adjacent dipoles is much less screened ($\epsilon = 1$ represents no screening).[16] Consequently, a net electric field is generated by the surface dipoles which polarizes the nearby water molecules. These dipoles generate in turn electric fields in the neighboring molecules and so on. The decay length for polarization, calculated from this model, was found in good agreement with the values determined from experiment on neutral lipid bilayers.[9] Schiby and Ruckenstein also suggested a new constitutive equation that related the polarization to a local electric field, which included the interaction between neighboring dipoles.[12b]

Both models predicted a monotonic decay of the polarization from the surface. Because the two models could not explain the oscillatory profile of the average polarization obtained by Monte Carlo simulations, it was suggested that the density of hydrogen bonding is a more suitable order parameter.[15] Recently, it was, however, shown[17] that the Schiby–Ruckenstein model can lead to an oscillatory polarization, if the water in the vicinity of a flat surface is assumed organized in icelike layers. The oscillations are smoothed out when the surface is rough (or fluctuating) or if some disorder is assumed to exist in the icelike structure. It was also shown that the continuum approximation, based on a second order differential equation, can describe well the average interaction.[17]

One cannot yet rule out that other interactions contribute to the hydration, such as the disruption of the hydrogen bond networks when two surfaces approach each other. However, at least a part of this disruption is already contained in the dipole–dipole interactions included in the polarization model. In addition, the polarization model of hydration can relate the magnitude of the hydration force to the density of dipoles on the surface. This can explain the dependence of the hydration repulsion on the surface dipolar potential[18] or the restabilization of some colloids at high ionic strength[16] observed experimentally.[10]

It is usually assumed that the total repulsion is the sum between a "double layer" repulsion, due to the charges on the interface, and a "hydration" repulsion, due to the structuring of water in the vicinity of the interface, and that the two effects are independent of each other. This is, however, not accurate when the hydration is induced by the orientational correlation of neighboring dipoles, because both forces depend on polarization.

The presence of dipoles on a surface free of charge generates an electric field and a polarization, dependent on the distance from the surface. The gradient of polarization affects, via the Poisson equation, the macroscopic electric field and, hence, the double layer, even in the absence of a surface charge. If ions are present in the system (because of the addition of an electrolyte), they interact with the electric field and hence affect both the electric field and the polarization. Therefore, the total repulsion should depend on the electrolyte concentration, even for uncharged surfaces.

On the other hand, in the absence of surface dipoles, the correlation between the water dipoles induced by a surface charge is expected to increase the repulsion as compared to the DLVO theory. This is due to the tendency of the water dipoles to orient in the same direction the adjacent water dipoles, thus increasing the decay length of the polarization.

Consequently, even when only a surface charge is present (and the surface dipole density is zero), the repulsion is not accurately described by the DLVO theory (particularly at high ionic strength, as will be shown later) because it disregards the orientational correlation of neighboring dipoles. In addition, even for uncharged surfaces, the repulsion induced by the surface dipoles is not accurately described by the traditional "hydration" force, independent of the electrolyte concentration, because the gradient of polarization generates an electric field which is affected by the ionic strength. The problem is further complicated when surface charges as well as surface dipoles are present. Can the total repulsive interaction still be described by a superposition between two independent interactions, the "double layer" and the "hydration" repulsions?

The purpose of this paper is to present a model that accounts in an unitary manner for both the double layer and hydration repulsions. The model is an extension of an earlier treatment of Schiby and Ruckenstein.[12] The equation coupling the polarization and the electric potential are derived here using an analysis based on the Lorentz model for the polarization and also a variational treatment (see Appendix). It will be shown that if the mutual interaction between neighboring water dipoles is taken into account, the double layer repulsion increases in the absence of surface dipoles, when compared to the DLVO theory, particularly at high ionic strengths. Similarly, in the absence of a surface charge, but the presence of a surface dipole density, the repulsion is increased by the addition of an electrolyte, an effect that is important particularly at high ionic strengths. For a surface whose charge is generated by the dissociation of a surfactant, and the surface dipoles are provided by the nondissociated surfactant molecules, it will be shown that the "hydration" and the "double layer" repulsions are not only nonadditive but that the presence of surface dipoles can decrease the repulsion.

II.1. The Basic Equations

We will employ here the polarization model developed in refs 12 and 17. The water molecules are considered dipoles localized at the site of a lattice; however, the fields generated by the adjacent dipoles are treated differently from those of the remote dipoles, because the latter are screened by the intervening water molecules. Starting from the dipoles of the surface, the polarization propagates through water because of the interactions between the neighboring dipoles. When two external surfaces approach each other, the overlap of the polarization layers decreases the dipole moments and hence increases the free energy of the system, thus generating a repulsion.

(15) Kjellander, R.; Marcelja, S. *Chem. Scr.* **1985**, *25*, 73. Attard, P.; Batchelor, M. T. *Chem. Phys. Lett.* **1988**, *149*, 206.
(16) Manciu, M.; Ruckenstein, E. *Langmuir* **2001**, *17*, 7061.
(17) Manciu, M.; Ruckenstein, E. *Langmuir* **2001**, *17*, 7582.
(18) Simon, S. A.; McIntosh, T. J. *Proc. Natl. Acad. Sci. U.S.A.* **1989**, *86*, 9263.

7586 *Langmuir, Vol. 18, No. 20, 2002* *Ruckenstein and Manciu*

The electric field E_P at a site of the layer "i" of a water lattice, situated at z_i, generated by all the other dipoles of magnitude m_i of the same layer and those of magnitude $m_{i\pm1}$ of the adjacent layers (the contribution of the other layers is neglected because of the screening) can be expressed as[17]

$$E_P(z_i) = C_1 m_{i-1} + C_0 m_i + C_1 m_{i+1} = C_0 m(z_i) + C_1(m(z_{i-1}) + m(z_{i+1})) \approx (C_0 + 2C_1)m(z_i) + C_1\Delta^2 \left.\frac{\partial^2 m(z)}{\partial z^2}\right|_{z=z_i} \approx C_1\Delta^2 \left.\frac{\partial^2 m(z)}{\partial z^2}\right|_{z=z_i} \quad (1)$$

where Δ is the distance between the water layers, $m(z)$ is a continuous function which is equal in every layer (located at $z_i = i\Delta$) to the average polarization of the water molecules in that layer ($m(z)|_{z=z_i} = m_i$), and C_j ($j = 0, 1$) are interaction coefficients (see ref 17)

$$C_0 = -\frac{3.766}{4\pi\epsilon_0\epsilon'' l^3} \qquad C_1 = \frac{1.827}{4\pi\epsilon_0\epsilon'' l^3} \quad (2)$$

with l the distance between the centers of two adjacent water molecules, ϵ_0 the vacuum permittivitty, and ϵ'' the dielectric constant for the interaction between neighboring molecules, which is expected to be nearer to unity than to the dielectric constant of water, $\epsilon = 80$.

In addition to the field generated by the adjacent dipoles, there is a macroscopic field E due to the presence of charges and of the "average" polarization P of the medium. In the Lorentz treatment of polarization, for a constant macroscopic field in a linear and homogeneous medium of dielectric constant ϵ (hence satisfying $P = \epsilon_0(\epsilon - 1)E$), the local field E_{local} at a site of a selected dipole is related to the macroscopic field E via[19]

$$E_{local} = E + \frac{P}{3\epsilon_0} \quad (3)$$

the macroscopic field being generated by all the charges and dipoles present, while the local field is generated by all the charges and dipoles, except the selected dipole.

Extending the Lorentz relation when the additional field E_P is present, one obtains

$$E_{local} = E + \frac{P}{3\epsilon_0} + E_P \quad (4)$$

The average dipole moment $m(z) = P(z)v_0$, where v_0 is the volume of a water molecule, is related to E_{local} via

$$m(z) = \gamma\left(E + E_P + \frac{m(z)}{3\epsilon_0 v_0}\right) \Rightarrow m(z) = \frac{\gamma}{1 - \dfrac{\gamma}{3\epsilon_0 v_0}}(E + E_P) \quad (5)$$

where γ is the molecular polarizability. At constant polarization, $E_P = (C_0 + 2C_1)m \cong 0$. Hence eq 5 becomes

$$m = \gamma\left(E + \frac{m}{3\epsilon_0 v_0}\right) \quad (6)$$

The macroscopic relation between the polarization and the electric field, $m = \epsilon_0 v_0(\epsilon - 1)E$ can be employed in eq 6 to determine the value of γ.

Assuming that the molecular polarizability remains the same in a nonconstant field, eqs 1 and 5 lead to

$$m(z) = \epsilon_0 v_0(\epsilon - 1)E(z) + \epsilon_0 v_0(\epsilon - 1)C_1\Delta^2 \frac{\partial^2 m(z)}{\partial z^2} \quad (7)$$

The Poisson equation

$$-\frac{\partial E(z)}{\partial z} \equiv \frac{\partial^2 \psi(z)}{\partial z^2} = -\frac{\rho}{\epsilon_0} + \frac{1}{\epsilon_0 v_0}\frac{\partial m(z)}{\partial z} \quad (8)$$

where ρ is the local charge density, given, as in the traditional theory, by

$$\rho(z) = -ec_E\left(\exp\left(\frac{e\psi(z)}{kT}\right) - \exp\left(-\frac{e\psi(z)}{kT}\right)\right) = -2ec_E \sinh\left(\frac{e\psi(z)}{kT}\right) \quad (9)$$

where c_E is the bulk electrolyte concentration, e is the protonic charge, k is the Boltzmann constant, and T is the absolute temperature, can be rewritten as

$$\frac{\partial^2 \psi(z)}{\partial z^2} = \frac{2ec_E}{\epsilon_0}\sinh\left(\frac{e\psi}{kT}\right) + \frac{1}{\epsilon_0 v_0}\frac{\partial m(z)}{\partial z} \quad (10)$$

Equations 7 and 10 constitute a complete system of equations for $\psi(z)$ and $m(z)$, which replaces the traditional equations of the double layer.

II.2. The Linear Approximation

For the sake of simplicity, in what follows it will be considered that the double layer potential is sufficiently small to allow the linearization of the Poisson–Boltzmann equation (the Debye–Hückel approximation). The extension to the nonlinear cases is (relatively) straightforward; however, it will turn out that the differences from the DLVO theory are particularly important at high electrolyte concentrations, when the potentials are small. In this approximation, the distribution of charge inside the double layer is given by

$$\rho = -2ec_E \sinh\left(\frac{e\psi}{kT}\right) \cong -\frac{2e^2 c_E}{kT}\psi \equiv -\frac{\epsilon\epsilon_0}{\lambda_D^2}\psi \quad (11)$$

where $\lambda_D = (\epsilon\epsilon_0 kT/2e^2 c_E)^{1/2}$ is the Debye–Hückel length.

In the linear approximation, the system of eqs 7 and 10 becomes

$$\frac{\partial^2 \psi(z)}{\partial z^2} = \frac{\epsilon}{\lambda_D^2}\psi + \frac{1}{\epsilon_0 v_0}\frac{\partial m(z)}{\partial z} \quad (12a)$$

$$\lambda_m^2 \frac{\partial^2 m(z)}{\partial z^2} = m(z) + \epsilon_0 v_0(\epsilon - 1)\frac{\partial \psi(z)}{\partial z} \quad (12b)$$

where $\lambda_m^2 \equiv \epsilon_0 v_0(\epsilon - 1)C_1\Delta^2$.

In the absence of an electrolyte ($c_E = 0$, $\lambda_D = \infty$), eq 12a leads to

$$m(z) = \epsilon_0 v_0 \frac{\partial \psi(z)}{\partial z} + \text{constant} \quad (13)$$

The constant is zero since the average polarization should

(19) Frankl, D. R. *Electromagnetic Theory*; Prentice Hall: Englewood Cliffs, NJ, 1986.

vanish when the electric potential is constant (this can be seen better from eq 12b, since a constant polarization and constant electric potential imply a vanishing polarization). Introducing the result (with constant = 0) in eq 12b yields

$$m(z) = \frac{\lambda_m^2}{\epsilon}\frac{\partial^2 m}{\partial z^2} \equiv \lambda_H^2 \frac{\partial^2 m(z)}{\partial z^2} \qquad (14)$$

which predicts a monotonic decay of the polarization from the surfaces, with a decay length $\lambda_H = (\epsilon_0 v_0 C_1 \Delta^2(\epsilon - 1)/\epsilon)^{1/2}$. The solution of eq 14, which accounts for the antisymmetry of the polarization ($m(-d) = m_0$, $m(d) = -m_0$), is

$$m(z) = -m_0 \frac{\sinh(z/\lambda_H)}{\sinh(d/\lambda_H)} \qquad (15)$$

where z is measured from the middle distance and $2d$ is the distance between the plates.

Employing the values $l = 2.76$ Å for the distance between the centers of two adjacent water molecules in an icelike structure, $\epsilon'' = 1$ for the dielectric constant for the interaction between neighboring molecules (selected as in refs 16 and 17), $\Delta = {}^4/_3 l = 3.68$ Å for the distance between two adjacent water layers in an icelike structure, $v_0 = 30$ Å^3 for the volume occupied by one water molecule, and $\epsilon = 80$, one obtains $\lambda_m = 14.9$ Å and $\lambda_H = \lambda_m/\epsilon^{1/2} = 1.67$ Å, which is in good agreement with the hydration length experimentally determined for neutral lipid bilayers.[9] This exponential dependence of the polarization was found also (in the continuum approximation) by earlier analysis.[12,16,17] It should also be noted that using eq 14 (which is valid only in the absence of electrolyte), the free energy of the system (discussed in detail in section II.5) reduces to the form employed in previous polarization-based treatments of hydration forces.[16,17] However, as will be shown below, the addition of electrolyte affects the hydration even in the absence of surface charges.

II.3. The Characteristic Decay Lengths

In the presence of an electrolyte ($\lambda_D \neq \infty$), we will seek solutions of the type $\psi = \psi_0 \exp(z/\lambda)$ and $m = m_0 \exp(z/\lambda)$ for the homogeneous system of the two linear equations (12a,b). The condition for existence of nontrivial solutions leads to the characteristic equation

$$\left(\frac{1}{\lambda^2} - \frac{\epsilon}{\lambda_D^2}\right)\left(\frac{\lambda_m^2}{\lambda^2} - 1\right) - \frac{\epsilon - 1}{\lambda^2} = 0 \Leftrightarrow \lambda^4 - (\lambda_D^2 + \lambda_m^2)\lambda^2 + \frac{\lambda_D^2\lambda_m^2}{\epsilon} = 0 \quad (16)$$

The solutions of eq 16, $\pm\lambda_1$ and $\pm\lambda_2$ are always real, since the dielectric constant of the medium is higher than the vacuum dielectric constant (the discriminant $\lambda_D^4 + \lambda_m^4 + 2\lambda_D^2\lambda_m^2(1 - 2/\epsilon) > 0$ for $\epsilon > 1$).

The dependence of the decay lengths λ_1 and λ_2 on the electrolyte concentration (λ_D) for $\epsilon = 80$ and $\lambda_m = 14.9$ Å are presented in Figure 1. At low electrolyte concentrations ($\lambda_D \gg \lambda_m$), the decay lengths λ_1 and λ_2 are well approximated by λ_D and $\lambda_H = \lambda_m/\epsilon^{1/2}$; however, when λ_D becomes comparable to λ_m, λ_1, and λ_2 differ markedly from λ_D and λ_H.

II.4. Boundary Conditions for Two Identical Surfaces Immersed in Water

The symmetry of the system implies that the potential is symmetric and the average polarization is antisymmetric with respect to the middle distance; hence, the general solutions of the system (12a,b) are

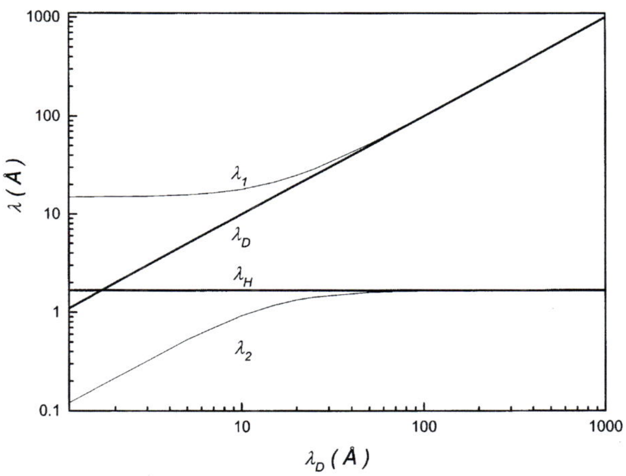

Figure 1. The dependence of the characteristic decay lengths λ_1, λ_2 of the system on the Debye–Hückel length λ_D ($\lambda_m = 14.9$ Å, $\lambda_H = \lambda_m/\epsilon^{1/2} = 1.67$ Å).

$$\psi(z) = a_1 \cosh\left(\frac{z}{\lambda}\right) + a_2 \cosh\left(\frac{z}{\lambda_2}\right) \qquad (17a)$$

$$m(z) = \bar{a}_1 \sinh\left(\frac{z}{\lambda_1}\right) + \bar{a}_2 \sinh\left(\frac{z}{\lambda_2}\right) \qquad (17b)$$

where the constants $\bar{a}_1$ and $\bar{a}_2$ are related to the constants a_1 and a_2 via eq 12a

$$\bar{a}_1 = a_1\epsilon_0 v_0 \lambda_1\left(\frac{1}{\lambda_1^2} - \frac{\epsilon}{\lambda_D^2}\right) \qquad (18a)$$

$$\bar{a}_2 = a_2\epsilon_0 v_0 \lambda_2\left(\frac{1}{\lambda_2^2} - \frac{\epsilon}{\lambda_D^2}\right) \qquad (18b)$$

The remaining two independent constants, a_1 and a_2 can be determined using the boundary conditions for the electrical potential and polarization at the surfaces.

For constant surface potential ψ_0, the boundary condition is

$$a_1 \cosh\left(\frac{d}{\lambda_1}\right) + a_2 \cosh\left(\frac{d}{\lambda_2}\right) = \psi_0 \qquad (19a)$$

For constant surface charge density σ, the condition of overall electroneutrality $2\sigma = -\int_{-d}^{d} \rho(z)\, dz$ leads to

$$a_1\lambda_1 \sinh\left(\frac{d}{\lambda_1}\right) + a_2\lambda_2 \sinh\left(\frac{d}{\lambda_2}\right) = \frac{\lambda_D^2}{\epsilon\epsilon_0}\sigma \qquad (19b)$$

The double layer charge distribution (eq 11) was employed in the derivation of the last equation.

When the value of the polarization at the boundary ($m_0 = m(-d)$) is known, the corresponding boundary condition is

$$\bar{a}_1 \sinh\left(-\frac{d}{\lambda_1}\right) + \bar{a}_2 \sinh\left(-\frac{d}{\lambda_2}\right) = m_0 \qquad (20a)$$

A more realistic approach[17] is to assume that the average polarization of the water molecules of the first water layer near the surface is proportional to the local field, generated by the surface charges, surface dipoles, and the water dipoles of the first two water layers.

7588 *Langmuir, Vol. 18, No. 20, 2002* *Ruckenstein and Manciu*

The average electric field generated by surface dipoles with a surface density of $1/A$ and with a dipole moment p normal to the surface, whose centers are located at a distance Δ' from the center of the first water layer, is given by[17]

$$E_S = \left(\frac{p}{\epsilon'}\right)\frac{1}{2\pi\epsilon_0}\frac{1}{\left(\frac{A}{\pi} + \Delta'^2\right)^{3/2}} \quad (21)$$

where ϵ' is the local dielectric constant for the field generated by the surface dipoles in the neighboring water molecules.

The macroscopic field at the interface is related to the electric potential, via the expression $E = -\partial\psi(z)/\partial z$, while the field generated by the water dipoles is[17]

$$E_{SP} = C_0 m_1 + C_1 m_2 \quad (22)$$

where m_1 and m_2 are the average polarization of the water molecules of the first and second water layer near the surface, respectively, and C_0 and C_1 are given by eq 2. Hence, the boundary condition for polarization has the form

$$m_1 = \gamma\left(E + \frac{m_1}{3\epsilon_0 v_0} + E_S + E_{SP}\right) = \epsilon_0 v_0(\epsilon - 1)(E + E_S + E_{SP}) \quad (23)$$

where eq 6 was employed to derive the last equality.

Assuming that the positions of the centers of the first water layers are at the distances $\pm d$, and using eqs 12a and 12b, eq 23 becomes

$$a_1\left[\lambda_1\epsilon_0 v_0\left(\frac{1}{\lambda_1^2} - \frac{\epsilon}{\lambda_d^2}\right)\left((1 - \epsilon_0 v_0(\epsilon - 1)C_0)\sinh\left(-\frac{d}{\lambda_1}\right) - \epsilon_0 v_0(\epsilon - 1)C_1\sinh\left(-\frac{d-\Delta}{\lambda_1}\right)\right) + \frac{\epsilon_0 v_0(\epsilon - 1)}{\lambda_1}\sinh\left(-\frac{d}{\lambda_1}\right)\right] + a_2\left[\lambda_2\epsilon_0 v_0\left(\frac{1}{\lambda_2^2} - \frac{\epsilon}{\lambda_d^2}\right)\left((1 - \epsilon_0 v_0(\epsilon - 1)C_0)\sinh\left(-\frac{d}{\lambda_2}\right) - \epsilon_0 v_0(\epsilon - 1)C_1\sinh\left(-\frac{d-\Delta}{\lambda_2}\right)\right) + \frac{\epsilon_0 v_0(\epsilon - 1)}{\lambda_2}\sinh\left(-\frac{d}{\lambda_2}\right)\right] = \left(\frac{p}{\epsilon'}\right)\frac{v_0(\epsilon - 1)}{2\pi}\frac{1}{\left(\frac{A}{\pi} + \Delta'^2\right)^{3/2}} \quad (20b)$$

II.5. The Free Energy of the System

In the DLVO framework, the free energy of a system of two overlapped double layers is composed of an electrostatic energy, an entropic contribution due to the ions in the double layer, and a chemical term, where applicable.[4]

The electrostatic energy per unit area of the double layer is provided by the familiar expression[19,20]

$$F_{el} = \frac{1}{2}\int_{-\infty}^{\infty}\rho'(z)\psi(z)\,dz = \frac{1}{2}\int_{-\infty}^{\infty}\nabla D(z)\psi(z)\,dz = \frac{1}{2}D(z)\psi(z)\Big|_{z=-\infty}^{z=\infty} - \frac{1}{2}\int_{-\infty}^{\infty}D(z)\nabla\psi(z)\,dz = \frac{1}{2}\int_{-\infty}^{\infty}D(z)E(z)\,dz = \frac{1}{2}\int_{-d}^{d}D(z)E(z)\,dz \quad (24)$$

where ρ' is the total charge density (which includes the charge distributed between surfaces, ρ, and the surface charge density, σ), ψ is the electric potential, E is the macroscopic electric field, z is the distance measured from the middle between the surfaces and $D = \epsilon_0 E + P$ is the displacement field. The above expression accounts for the fact that the field is nonvanishing only between the surfaces, located at $\pm d$; the Poisson equation $\nabla D = \rho'$ and the relation $E = -\nabla\psi$ were also employed.

The excess entropy contribution (with respect to the bulk) per unit area due to the ions of an electrolyte, calculated as for an ideal solution, is given by[21]

$$-T\Delta S = kT\sum_i\int_{-d}^{d}\left(c_i\ln\left(\frac{c_i}{c_{i0}}\right) - c_i + c_{i0}\right)dz \quad (25)$$

where c_i is the actual (double layer) concentration of ions of species "i", c_{i0} is the concentration at large distances, and the subscript "i" runs over all ion species. For an 1−1 electrolyte of concentration c_E, which obeys the Boltzmann distribution, the above expression becomes

$$-T\Delta S = -c_E kT\sum_{i=1,2}\int_{-d}^{d}\left(\frac{(-1)^i e\psi}{kT}\exp\left(\frac{(-1)^{i+1}e\psi}{kT}\right) + \exp\left(\frac{(-1)^{i+1}e\psi}{kT}\right) - 1\right)dz \quad (26)$$

The entropic contribution to the free energy (per unit area) becomes in the Debye−Hückel approximation

$$-T\Delta S = \frac{c_E e^2}{kT}\int_{-d}^{d}(\psi(z))^2\,dz \equiv \frac{\epsilon\epsilon_0}{2\lambda_D^2}\int_{-d}^{d}(\psi(z))^2\,dz \quad (27)$$

The chemical contribution to the free energy, per unit area, due to the adsorption of n molecules (per unit area) of charge q on each surface of potential ψ_S is[4]

$$F_{ch} = 2n\Delta\mu = -2nq\psi_S = -2\sigma\psi_S \quad (28)$$

at constant surface potential, $\Delta\mu$ being the change in the chemical part of the electrochemical potential of a molecule at its adsorption from the bulk on the surface, and σ the surface charge density. At equilibrium, the electrochemical potential is constant through the system ($\Delta\mu = -q\psi_S$). At constant surface charge the chemical contribution to the free energy is zero.[4]

In addition to these well-known free energy contributions, one has to consider another one, which accounts for the mutual interactions between neighboring dipoles

$$F_m = -\frac{1}{2}mE_P = -\frac{C_1\Delta^2}{2}m\frac{\partial^2 m}{\partial z^2} \quad (29)$$

where eq 1 was employed. It should be noted that the above expression contains both the energy and the entropy of the dipole, in the weak field (linear) approximation.

(20) Schwinger, J.; DeRaad, L. L., Jr.; Milton, K. A.; Tsai, W.-Y. *Classical Electrodynamics*; Perseus Books: Reading, MA, 1998.

(21) Overbeek, J. T. G. *Colloids Surf.* **1990**, *51*, 61.

Consequently, the total electrostatic free energy per unit area (accounting for all the interactions between charges and dipoles) is provided by the expression

$$F_{\mathrm{el}} = \int_{-d}^{d}\left[\frac{1}{2}E(\epsilon_0 E + P) - \frac{1}{2}PE_{\mathrm{P}}\right]\mathrm{d}z = \frac{1}{2}\int_{-d}^{d}\left[\left(\epsilon_0 E(z) + \frac{m(z)}{v_0}\right)E(z) - \frac{m(z)}{v_0}E_{\mathrm{P}}(z)\right]\mathrm{d}z \quad (30)$$

The free energy of the surface layer formed by surface dipoles and the water molecules between them is assumed to be independent of the distance $2d$.

At constant surface potential, the free energy per unit area of a system of two identical charged surfaces at distances $\pm d$ from the middle distance can be written as the sum between a surface term (the chemical energy) and an integral over the density of the entropy of the mobile ions and all the electrostatic interactions between charges and dipoles

$$F_{\psi} = -2\sigma\psi_{\mathrm{S}} + \frac{1}{2}\int_{-d}^{d}\left[\left(\epsilon_0\frac{\partial\psi}{\partial z} - \frac{m}{v_0}\right)\frac{\partial\psi}{\partial z} + \frac{\epsilon\epsilon_0}{\lambda_{\mathrm{D}}^2}(\psi)^2 - \frac{C_1\Delta^2}{v_0}m\frac{\partial^2 m}{\partial z^2}\right]\mathrm{d}z \quad (31a)$$

The same results (eqs 12a,b and 31) were obtained using a variational method. The details are given in Appendix. At constant surface charge, the free energy is given by

$$F_{\sigma} = \frac{1}{2}\int_{-d}^{d}\left[\left(\epsilon_0\frac{\partial\psi}{\partial z} - \frac{m}{v_0}\right)\frac{\partial\psi}{\partial z} + \frac{\epsilon\epsilon_0}{\lambda_{\mathrm{D}}^2}(\psi)^2 - \frac{C_1\Delta^2}{v_0}m\frac{\partial^2 m}{\partial z^2}\right]\mathrm{d}z \quad (31b)$$

III.1. The Influence of the Dipole–Dipole Interactions on the Double Layer Force

At low electrolyte concentrations, because the Debye–Hückel length is large, the polarization predicted by the DLVO theory is slowly varying in space. Therefore, when the dipole density on the surface is negligible, one expects the additional interaction, due to the mutual interaction between neighboring dipoles, to be also small, its density being proportional to $m(\partial^2 m/\partial z^2) \propto (m^2/\lambda_{\mathrm{D}}^2)$. Consequently, the contribution of the interaction between neighboring dipoles to the total free energy becomes negligible, and the DLVO theory is recovered. However, when λ_{D} is sufficiently small and becomes comparable to λ_{m}, a coupling between the two effects is expected to occur. As a consequence, a larger decay length ($\lambda_1 > \lambda_{\mathrm{D}}$) appears in the system (see Figure 1).

There are two main reasons for the departure of the present model from the DLVO theory. First, the constitutive equations, which relate the polarization to the electric potential, are different. Second, the boundary conditions are different, since the average polarization in the DLVO theory is directly related to the surface charge, while in the present treatment it depends also on the surface dipole density.

Let us first investigate the effect of the new equations alone, by using for both the DLVO theory and the present equations the same boundary conditions. For the surface charge density the constant value $\sigma = 5 \times 10^{-4}$ C/m^2 was employed (the value selected is low enough for the linear approximation to be accurate for all the electrolyte concentrations investigated here), while the polarization at the surface was considered induced by the surface charge only, as in the DLVO theory

$$m_0 = -\epsilon_0 v_0(\epsilon - 1)\left.\frac{\partial\psi(z)}{\partial z}\right|_{z=-d} \quad (32a)$$

In parts a and b of Figure 2, the electric potential and the polarization, respectively, calculated using eqs 12a,b are compared to those predicted by DLVO, with the boundary conditions (19b) and (32a), using for the parameters the values $v_0 = 30$ Å^3, $\epsilon = 80$, $T = 300$ K, $\lambda_{\mathrm{m}} \equiv (\epsilon_0 v_0(\epsilon - 1)C_1\Delta^2)^{1/2} = 14.9$ Å, and the values $d = 20$ Å and $\lambda_{\mathrm{D}} = 10$ Å.

The force per unit area between surfaces, $\Pi = -[\partial F(2d)]/[\partial(2d)]$, with $F(2d)$ given by eq 31, at constant surface charge $\sigma = 5 \times 10^{-4}$ C/m^2 and for a polarization m_0 given by eq 32a, is compared to that predicted by the DLVO theory in parts c and d of Figure 2, for various electrolyte concentrations ($\lambda_{\mathrm{D}} = 100$, 30, 10, and 3 Å). As expected, at low electrolyte concentrations, the interaction is well described by the DLVO theory. At large electrolyte concentrations, the repulsion is, however, markedly larger than that provided by the DLVO theory, because the interactions between neighboring dipoles attenuate the decay of the polarization.

To examine the effect of the surface dipole density, we consider that the polarization on the surface acquires the value

$$m_0' = -\epsilon_0 v_0(\epsilon - 1)\left.\frac{\partial\psi(z)}{\partial z}\right|_{z=-d} + \delta m \quad (32b)$$

where the change δm of the average dipole moment of the water molecules at the interface is generated by the surface dipoles. It should be noted that the boundary condition (32b) affects the value of both ψ and m on the surface and hence $m_0' \neq m_0 + \delta m$.

Let us consider that the surface charge arises via the dissociation of surfactant molecules adsorbed on the interfaces and that the surface dipoles are due to the undissociated surfactant molecules adsorbed. In this case, the electric field induced by the surface dipoles is opposite to that generated by the surface charge ($\delta m < 0$). Hence, the presence of surface dipoles actually decreases the repulsion. The effect is illustrated in parts c and d of Figure 2, for $\delta m = -0.2$ D. Therefore, in this case the total repulsion cannot be obtained (as usually assumed) by adding two independent terms, a "double layer force" due to the surface charges and a "hydration force" due to the surface dipoles.

The second important difference with the DLVO theory arises from the boundary condition for the polarization; while the classical theory ignores the interactions between neighboring dipoles, eq 20b takes them into account. The electrical potential and the polarization calculated with eqs 12a,b and the boundary conditions eqs 19b and 20b with $\sigma = 5 \times 10^{-4}$ C/m^2 and $p/\epsilon' = 0$ or $p/\epsilon' = -0.1$ D are compared in parts a and b of Figure 3 with the DLVO predictions (the value $A = 100$ Å^2 was selected for the area occupied by a surface dipole). The repulsion forces, per unit area, at constant surface charge density are presented in Figure 3c and Figure 3d. The differences from the DLVO theory become again important at large electrolyte concentrations, and the presence of surface dipoles decreases the repulsion. At high ionic strengths, the range of the interaction is much longer than that predicted by the DLVO theory. At large separation distances, the first term of eqs 17a and 17b, with the decay length λ_1 ($\lambda_1 \gg \lambda_2$), dominates both the electrical potential

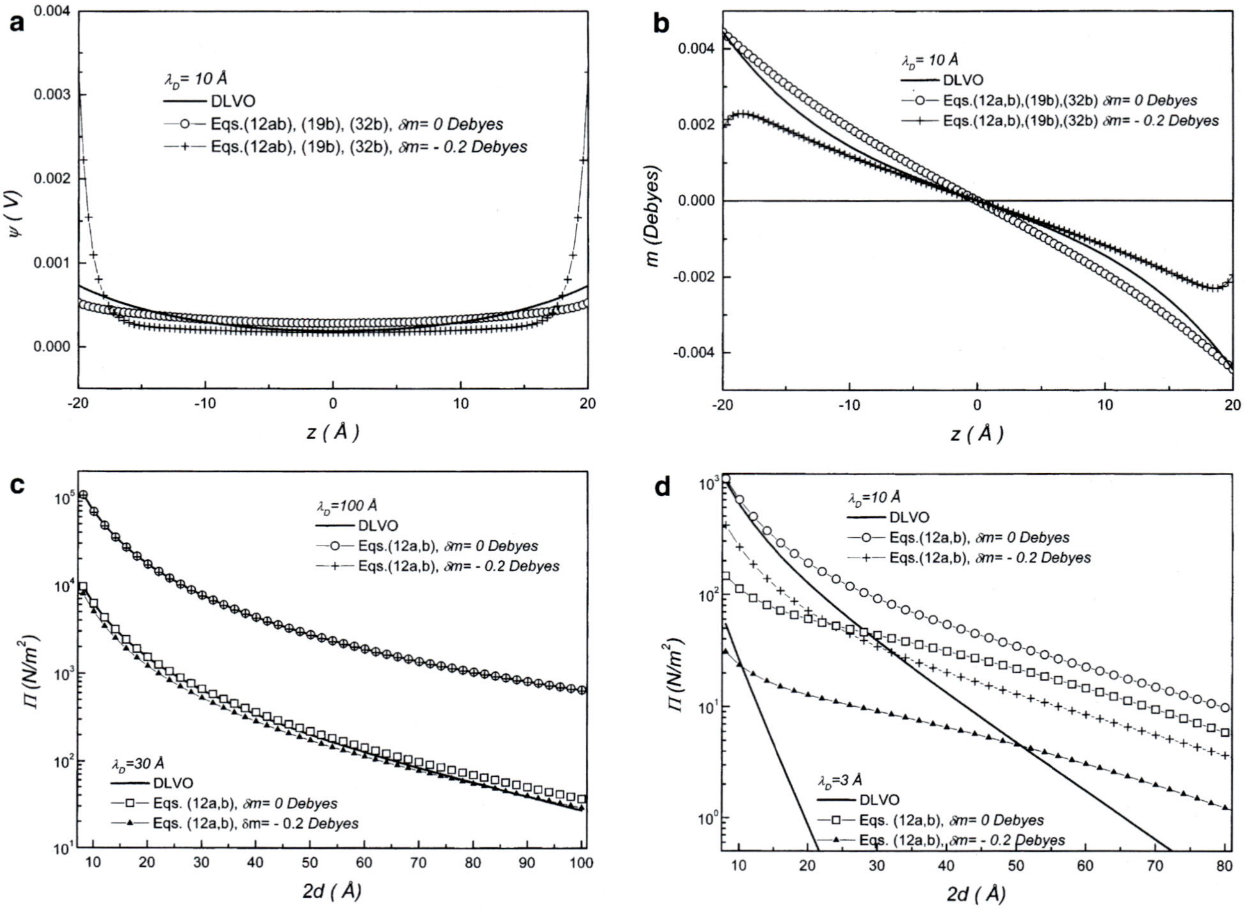

Figure 2. (a) The electric potential and (b) the average polarization of a water molecule between two surfaces with $\sigma = 5 \times 10^{-4}$ C/m^2, separated by a distance $2d = 40$ Å, as a function of the position from the middle distance. The average polarization of the water molecules from interface, m_0, is calculated from eq 32a. A perturbation $\delta m = -0.2$ D illustrates the effect of surface dipoles. (c, d) The interaction force calculated, at various electrolyte concentrations, from eqs 12a,b with the boundary conditions (19b) and (32a), versus the separation distance, compared to the DLVO predictions. A perturbation $\delta m = -0.2$ D illustrates the effect of surface dipoles.

and the polarization, respectively. Consequently, at large distances, the force depends mainly on the term containing the decay length λ_1. For $\lambda_D = 0$, one obtains from eq 16 the minimum of λ_1

$$\lambda_1(\lambda_D{=}0) = \lim_{\lambda_D \to 0}\left(\frac{1}{2}\left(\lambda_D^{\,2} + \lambda_m^{\,2} + \left(\lambda_D^{\,4} + \lambda_m^{\,4} + 2\lambda_D^{\,2}\,\lambda_m^{\,2}\left(1 - \frac{2}{\epsilon}\right)\right)^{1/2}\right)\right)^{1/2} \equiv \lambda_m \quad (33)$$

with $\lambda_m = 14.9$ Å for the values of the parameters employed here. Therefore, a long-ranged interaction, with a decay length larger than λ_m, is always present in the system, at any electrolyte concentration. The magnitude of this long-range interaction depends markedly on the ionic strength (see Figure 3d) and is much larger than the Debye and hydration decay lengths.

When the interaction due to the surface charge is large, the presence of a low dipole density on the surface (generated by surface charge association with counterions, as discussed above) decreases the repulsion, since the electric field generated by the surface dipoles is opposite to the electric field generated by the charges. However, a sufficiently high surface dipole density, which still induces an electric field opposite to that generated by the charge, can eventually lead to an increase in the repulsion (the regime when the "hydration" dominates). This effect is illustrated in Figure 3e, for weakly charged surfaces ($\sigma = -0.002$ C/m^2) and $\lambda_D = 3$ Å. In this case, the increase of the dipole strength decreases initially the repulsion (when $0 < p/\epsilon' < 0.7$ D); the repulsion is, however, increased for $p/\epsilon' > 0.7$ D. When $p/\epsilon' > 1.7$ D, the total repulsion becomes larger than that in the absence of surface dipoles.

III.3. The Repulsion between Uncharged Surfaces

The potential $\psi(z)$ and the dipole moment $m(z)$ provided by the system of eqs 12a,b are presented in parts a and b of Figure 4, for $\sigma = 0$, $p/\epsilon' = 1$ D and $\lambda_D = 3$, 10, 30, and 100 Å. The forces between surfaces at various electrolyte concentrations (with the other parameters unchanged) are plotted in Figure 4c. At large separation distances, the first term in eqs 17a and 17b, which decays much slower, becomes dominant, with $\lambda_1 \sim \lambda_D$ at low ionic strengths and $\lambda_1 \sim \lambda_m$ at high ionic strengths. At short distances and low ionic strengths, the first term is, however, small, and the force is well described by an exponential "hydration force", with constant preexponential factor and decay length ($\lambda_2 \cong \lambda_H = \lambda_m/\epsilon^{1/2}$). In the limiting cases $\sigma = 0$ and $\lambda_D \to \infty$, the free energy expression eq 31 leads (because of eq 13) to

$$F_H = -\frac{1}{2}\int_{-d}^{d}\left[\frac{C_1\Delta^2}{v_0}\, m\, \frac{\partial^2 m}{\partial z^2}\right] dz \quad (34)$$

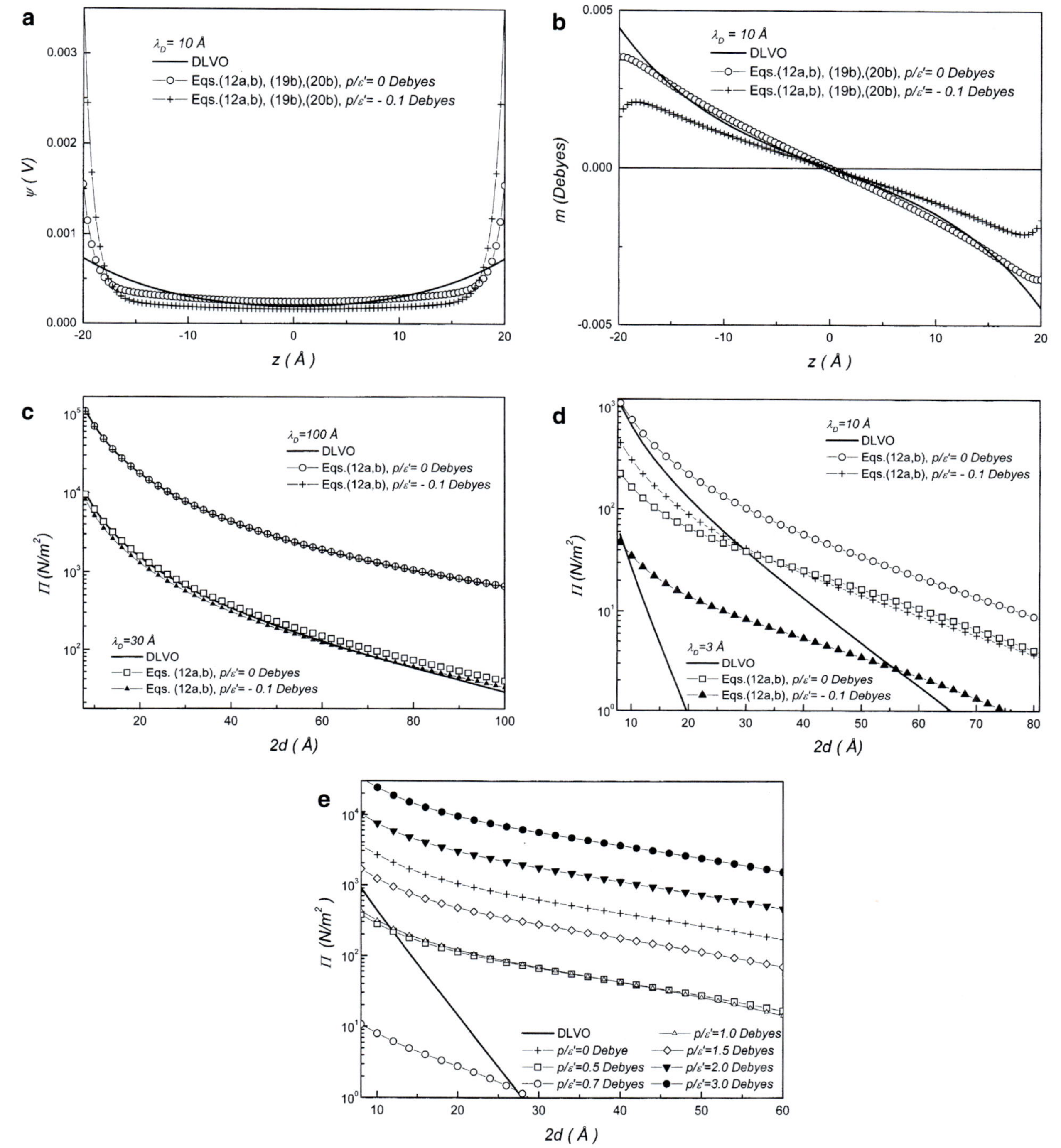

Figure 3. (a) The electric potential and (b) the average polarization of a water molecule between two surfaces with $\sigma = 5 \times 10^{-4}$ C/m^2, separated by a distance $2d = 40$ Å, as a function of the position from the middle distance. The average polarization of the water molecules from interface, m_1, was calculated using eq 20b, for $p/\epsilon' = 0$ and $p/\epsilon' = -0.1$ D, and $A = 100$ Å^2. (c, d) The interaction force calculated, at various electrolyte concentrations, from eqs 12a,b with the boundary conditions (19b) and (20b), versus separation distance, compared to the DLVO predictions. (e) The interaction force for $\lambda_D = 3$ Å and $\sigma = -0.002$ C/m^2 for various dipole strengths ($0 < p/\epsilon' < 3$ D, $A = 100$ Å^2). The repulsion initially decreases and then increases with the increasing strength of the surface dipoles.

where m is given by eq 15 and the hydration free energy from the previous work[16,17] is recovered.

At high ionic strength, both terms, with decay length λ_1 and λ_2, are important and the force is smaller at short separation distances and larger at long separation distances when compared to the force for $\lambda_D \rightarrow \infty$ (see Figure 4c).

The long-range repulsion induced by the electrolyte concentration (see Figure 4c) can be explained as follows. At zero surface charge, the total charge of the electrolyte ions between the surfaces (therefore the integral over potential (eq 11) between $-d$ and d) vanishes. This can occur only if the potential changes sign between the surfaces, hence if the coefficients a_1 and a_2 in eq 17a have opposite signs, with the ratio of their magnitude determined by electroneutrality. The rapidly varying, positive potential near the surfaces (we assumed a positive surface charge) is compensated by a slowly varying, negative potential in the middle range (see Figure 4a. This generates a polarization, which decays with a decay length

7592 *Langmuir, Vol. 18, No. 20, 2002* *Ruckenstein and Manciu*

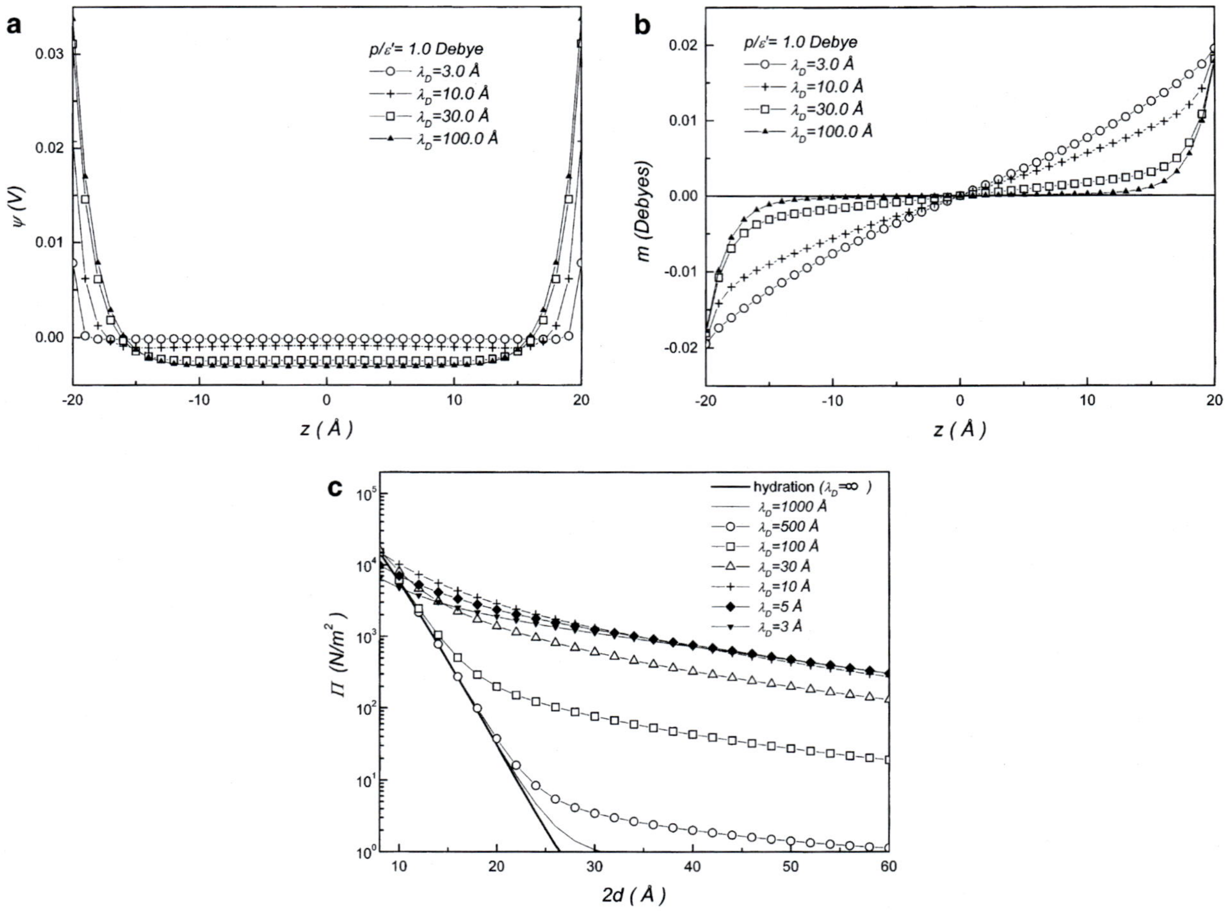

Figure 4. (a) The electric potential and (b) the average polarization of a water molecule between two neutral surfaces ($\sigma = 0$), separated by a distance $2d = 40$ Å and $\lambda_D = 3$, 10, 30, and 100 Å, as a function of the position from the middle distance, calculated using eqs 12a,b. The average polarization of the water molecules at the interface, m_1, was calculated from eq 20b, for $p/\epsilon' = 1$ D and $A = 100$ Å^2. (c) The interaction force between two neutral surfaces calculated, at various electrolyte concentrations, using eqs 12a,b with the boundary conditions (19b) and (20b) for $p/\epsilon' = 1$ D and $A = 100$ Å^2 versus the separation distance $2d$.

$>\lambda_m$, which is much larger than λ_H (see Figure 4b). It should be, however, noted that the magnitude of the long-ranged repulsive force is small at low ionic strengths (see Figure 4c).

IV. Conclusions

The interaction between two charged surfaces immersed in a liquid was traditionally described by assuming that the polarization of the liquid is proportional to the electric field. However, an additional interaction, caused by the structuring of the liquid in the vicinity of a surface, should be also taken into account. This can be included, at least partially, via the mutual interactions between the water dipoles. This induces a polarization that propagates, from the surfaces, through water, being generated by both the surface charge and surface dipoles. The polarization gradient produces an electric field, which interacts both with the dipole moments of the water molecules and with the charges of the electrolyte.

In this paper a model was presented, which allowed one to calculate both the electric potential and the polarization between two surfaces, without assuming, as in the traditional theory, that the polarization and the macroscopic electric field are proportional. An additional local field, due to the interaction between neighboring dipoles, was introduced in the constitutive equation which relates the polarization to the local field. The basic equations were also derived using a variational approach.

It was shown that the interaction between dipoles increases markedly the repulsion at high ionic strength and large separation distances, when compared to the DLVO theory. When both charges and dipoles are present on the surface, the repulsion is not provided by the sum of two independent repulsions, a "double layer" and a "hydration" repulsion. The presence of dipoles on the surface can even decrease the repulsion.

It was also shown that the presence of an electrolyte generates a long-range repulsion, even at zero surface charge, if the interfaces carry a surface dipole density. At low ionic strength, this repulsion can be described, in the vicinity of the surface, by an exponential with both the decay length and the preexponential factor almost independent of electrolyte concentration, as usually considered for the hydration forces. The long-ranged interactions between the neutral surfaces is in this case small. However, at high ionic strengths, the repulsion between neutral surfaces differs markedly from this description, the force being smaller at short distances and larger at large distances than that at zero electrolyte concentration.

Appendix. Derivation of the Equations for *m* and ψ from a Variational Principle

The Maxwell equation of electrostatics in a vacuum for a system with planar *xy* symmetry ($\partial\psi/\partial z = -E$, $\partial E/\partial z = \rho'/\epsilon_0$), ρ' being the total charge density present in the system, can be derived as extremals of the electrostatic free energy functional[20]

$$F_{el} = \int_V \left(\rho'\psi + \epsilon_0 E \frac{\partial\psi}{\partial_z} + \frac{\epsilon_0}{2}E^2\right) dV = \int_V \mathcal{J} dV \quad (A.1)$$

using the Euler–Lagrange equations

$$\frac{d}{dz}\left(\frac{\partial\mathcal{J}}{\partial\left(\frac{\partial\psi}{\partial z}\right)}\right) = \frac{\partial\mathcal{J}}{\partial\psi} \quad (A.2)$$

$$\frac{d}{dz}\left(\frac{\partial\mathcal{J}}{\partial\left(\frac{\partial E}{\partial z}\right)}\right) = \frac{\partial\mathcal{J}}{\partial E}$$

Let us try to find a free energy functional of a polarizable medium, which can be extended to any constitutive relation between E and P. Since the Poisson equation in a medium is $\nabla(\epsilon_0 E) = \rho' - \nabla P$, a natural choice for this functional is

$$F_{el} = \int_V \left(\left(\rho' - \frac{\partial P}{\partial z}\right)\psi + \epsilon_0 E\frac{\partial\psi}{\partial z} + \frac{\epsilon_0}{2}E^2 + f(P)\right) dV \quad (A.3)$$

The Euler–Lagrange equations of this functional, with respect to ψ and E, are $(\partial/\partial z)(\epsilon_0 E + P) = \rho'$ and $E = -\partial\psi/\partial z$, respectively, provided that ρ' and the arbitrary function $f(P)$ do not depend on either E or ψ.

Let us now obtain a functional, which represents the free energy density of a linear, homogeneous, and isotropic medium, that satisfies the constitutive equation $P = \epsilon_0(\epsilon - 1)E$. To obtain the classical result for the free energy density, $(1/2)E(\epsilon_0 E + P)$, the function $f(P)$ must acquire the form

$$f(P) = \frac{P^2}{2\epsilon_0(\epsilon - 1)} \quad (A.4)$$

In this case, the conditions of extremum of the functional given by eq A.3 with respect to ψ, E, and P, considered as independent functions (the Euler–Lagrange equations), lead to the Maxwell equations and the equation that relates the polarization to the field. It should be noted that the above equations imply that ρ' is independent of E and P. Of course, this assumption is not valid in the presence of an electrolyte.

For two overlapping double layers, the free energy per unit area F_{DLVO} can be written as the sum between the electrostatic energy, the chemical energy, and the entropic term of the electrolyte ions. Since the total charge density ρ' is composed of the surface charges, σ, and the charge density ρ distributed between the surfaces, the later obeying (in the linear approximation) eq 11, the electrostatic energy per unit area becomes

$$F_{el} = 2\sigma\psi_S + \int_{-d}^{d}\left(-\frac{\epsilon\epsilon_0}{\lambda_D^2}\psi^2 - \frac{\partial P}{\partial z}\psi + \epsilon_0 E\frac{\partial\psi}{\partial z} + \frac{\epsilon_0}{2}E^2 + \frac{P^2}{2\epsilon_0(\epsilon - 1)}\right) dz \quad (A.5)$$

where ψ_S is the surface potential. The chemical energy per unit area (eq 28) is given by

$$F_{ch} = -2\sigma\psi_S \quad (A.6)$$

at constant surface potential and the entropy contribution (per unit area, eq 27 is

$$-T\Delta S = \frac{\epsilon\epsilon_0}{2\lambda_D^2}\int_{-d}^{d}(\psi(z))^2\, dz \quad (A.7)$$

Adding eqs A.5, A.6, and A.7 and writing $E = -\partial\psi/\partial z$, one obtains

$$F_{DLVO,\psi} = \int_{-d}^{d}\left(-\frac{\epsilon_0}{2}\left(\frac{\partial\psi(z)}{\partial z}\right)^2 - \frac{\epsilon\epsilon_0}{2\lambda_D^2}(\psi(z))^2 - \psi\frac{\partial P(z)}{\partial z} + \frac{P^2(z)}{2\epsilon_0(\epsilon - 1)}\right) dz \quad (A.8)$$

The extremals of $\psi(z)$ and $P(z)$, obtained through the Euler–Lagrange equations are given by

$$\frac{\partial^2\psi(z)}{\partial z^2} = \frac{\epsilon}{\lambda_D^2}\psi + \frac{1}{\epsilon_0}\frac{\partial P(z)}{\partial z} \quad (A.9)$$

$$P(z) = -\epsilon_0(\epsilon - 1)\frac{\partial\psi(z)}{\partial z} \quad (A.10)$$

which represent the linear Poisson–Boltzmann equation and the proportionality relation between polarization and electric field, respectively.

Until now, the classical DLVO results have been recovered. Let us suppose that another interaction, whose free energy density is $-\alpha P[\partial^2 P(z)/\partial z^2]$ is also present. An integration by parts leads to

$$-\alpha\int_{-d}^{d} P(z)\frac{\partial^2 P(z)}{\partial z^2}\, dz = -\alpha P\left.\frac{\partial P}{\partial z}\right|_{z=-d}^{z=d} + \int_{-d}^{d}\alpha\left(\frac{\partial P(z)}{\partial z}\right)^2 dz \quad (A.11)$$

Hence, the free energy at constant potential F_ψ per unit area is given by

$$F_\psi = -\alpha P\left.\frac{\partial P}{\partial z}\right|_{z=-d}^{z=d} + \int_{-d}^{d}\left(-\frac{\epsilon_0}{2}\left(\frac{\partial\psi}{\partial z}\right)^2 - \frac{\epsilon\epsilon_0}{2\lambda_D^2}(\psi)^2 - \psi\frac{\partial P}{\partial z} + \frac{P^2}{2\epsilon_0(\epsilon - 1)} + \alpha\left(\frac{\partial P}{\partial z}\right)^2\right) dz \quad (A.12)$$

and the extremals function $\psi(z)$, $P(z)$ satisfy the Euler equations

$$\frac{\partial^2\psi(z)}{\partial z^2} = \frac{\epsilon}{\lambda_D^2}\psi + \frac{1}{\epsilon_0}\frac{\partial P(z)}{\partial z} \quad (A.13)$$

$$2\alpha\frac{\partial^2 P(z)}{\partial z^2} = \frac{P(z)}{\epsilon_0(\epsilon - 1)} + \frac{\partial\psi(z)}{\partial(z)} \quad (A.14)$$

Since $P(z) = m(z)/v_0$ and $\alpha = C_1\Delta^2/2$, the system of eqs A.13 and A.14 becomes identical to the system of eqs 12a,b, while the free energy eq A.12 coincides with eq 31a. Equation 31b, for constant surface charge, can be derived in a similar manner.

LA020435V

1114 *Langmuir* **2003,** *19,* 1114–1120

On the Chemical Free Energy of the Electrical Double Layer

Marian Manciu and Eli Ruckenstein*

Department of Chemical Engineering, State University of New York at Buffalo, Buffalo, New York 14260

Received September 26, 2002. In Final Form: December 16, 2002

The free energy of interaction between colloidal particles due to the overlap of their double layers was traditionally calculated either for systems of arbitrary geometry interacting at constant surface potential or at constant surface charge or for parallel plates interacting under arbitrary surface conditions. An expression is obtained for the change in the chemical contribution to the free energy of the double layers during a general interaction, which allows the calculation (within the Poisson–Boltzmann formalism) of the interaction free energy for systems of arbitrary geometry and surface conditions. The change in chemical free energy depends not only on the values of the surface charge and potential at the final state but also on their values at each distance between infinity and the final state. A simple approximate expression for the change in the chemical free energy contribution is also proposed, which involves only the states at infinite and final distances. Its accuracy is tested for planar and parallel surfaces, with charges generated via the dissociation of surface groups.

I. Introduction

Gouy[1] and Chapman[2] were the first to predict the distributions of electrolyte ions in the vicinity of a charged surface, by assuming that the ions obey Boltzmann distributions and interact with a mean potential, which satisfies the Poisson equation. The distribution of ions between two planar charged surfaces (the difference in ions concentrations between middle distance and infinity) was later related to the interaction force between the two surfaces by Langmuir.[3] The interaction free energy between parallel planar plates, separated by a distance l, could then be obtained by integrating the force from infinity to the distance l.

The Langmuir procedure cannot be applied directly to nonplanar surfaces. However, an approximate method, the Derjaguin approximation,[4] in which the interaction between curved surfaces is calculated in terms of the interactions between planar surfaces, could be employed. The method is accurate only when the radii of curvature of the surfaces are sufficiently large compared to the Debye–Hückel length and to the distance of closest approach. Major progress was achieved by Verwey and Overbeek,[5] who proposed a direct method to calculate the interaction free energy between particles of arbitrary shapes, when either the surface potentials or the surface charges remain constant during the interaction. Their approach involved an imaginary charging process, in which the relation between the surface potential and the surface charge density was obtained by integrating the Poisson–Boltzmann equation for one of the above constraints.

Another approach consisted of writing the free energy as the sum between the energy of the electrostatic field, an entropic term due the free ions and a chemical term due to the charge transfer to the surface.[6] However, expressions for the chemical free energy contribution have been derived only for interactions that occur either at constant surface charge density or at constant surface potential. For these cases, the latter method was shown to be equivalent to the imaginary charging approach.[6]

Whereas the charging approach could be applied to any geometry but only at constant surface charge or potential, the Langmuir expression could be employed for any surface conditions but only for parallel planar plates. The addition of electrostatic, entropic, and chemical contributions would allow the calculation of the free energy of interaction for systems of any shape and any surface conditions, if one could derive a general expression for the chemical free energy contribution.

There are colloidal systems for which the interactions can be well described by either constant surface charge density or potential. For example, when the surface charge is generated by the dissociation of surface groups—such as those of surfactants—and the dissociation constant is sufficiently large, the constraint of constant charge is a suitable one. In addition, in many cases the radii of curvature of the surfaces are large enough for the interactions to be accurately calculated by combining the Langmuir equation with the Derjaguin approximation.

There are, however, colloidal systems, for which the above requirements are not fulfilled. For example, the surfaces of the proteins can have small radii, and their acidic and basic groups are not completely dissociated. In this case, the Verwey–Overbeek approach cannot be employed and the Langmuir–Derjaguin approximation is not accurate.

The purpose of this article is to obtain a general expression for the chemical free energy contribution, when neither the surface potential nor the surface charge density are constant on the surfaces during the interaction. It will be shown that the change in the chemical free energy contribution depends on the trajectory $\psi_S = \psi_S(\sigma)$, where ψ_S is the surface potential and σ the surface charge density, during the interactions from an infinite to the final separation distance. Using this expression for the change

* To whom correspondence may be addressed: e-mail address, feaeliru@acsu.buffalo.edu; phone, (716) 645-2911 ext 2214; fax, (716) 645-3822

(1) Gouy, G. *J. Phys. Radium* **1910**, *9*, 457.
(2) Chapman, D. L. *Philos. Mag.* **1913**, *25*, 475.
(3) Langmuir, I. *J. Chem. Phys.* **1938**, *6*, 893.
(4) Derjaguin, B. V. *Kolloid-Z.* **1934**, *69*, 155.
(5) Verwey, E. J.; Overbeek, J. T. G. *Theory of the stability of lyophobic colloids*; Elsevier: Amsterdam, 1948.
(6) Overbeek, J. T. G. *Colloids Surf.* **1990**, *51*, 61.

10.1021/la0266132 CCC: $25.00
Published on Web 01/28/2003

in the chemical free energy, one can recover the results obtained via the imaginary charging process, for constant surface charge or potential conditions, or via the Langmuir procedure for parallel plates and arbitrary surface conditions. However, it also can be used without any of the restrictions involved in the traditional theories.

Furthermore, a simple approximate expression for the chemical free energy will be suggested, which is more efficient computationally. In that approximation, the free energy of the system does not depend on the trajectory $\psi_S = \psi_S(\sigma)$, but only on the values of σ and ψ_S at infinite and at the final separation distance. The approximate expression is exact in the two limiting cases, namely, interactions at constant surface potential and at constant surface charge density. Finally, the approximate free energy is compared with the exact one (obtained both from the present theory and the Langmuir procedure) for two planar surfaces, when the charge is generated via the dissociation of surface groups.

II. The Double Layer Free Energy

The total free energy of a double layer is the sum of an electrostatic free energy F_{el}, an entropic term due to the mobile ions F_{ent} and a chemical free energy.[6] The first one is given by

$$F_{el} = \frac{1}{2}\int_V \epsilon\epsilon_0(\nabla\psi)^2\, dV \quad (1)$$

the integral being carried out over the entire volume V of the system. In the above expression, ψ is the electric potential, ϵ_0 the vacuum permittivity, and ϵ the relative dielectric constant. Note that in eq 1 the charges are assumed fixed (hence their entropy is not included), but the entropy of the dipoles of the medium is accounted for through ϵ.

The entropic term due to the mobile ions of the electrolyte takes into account the differences between the ions distributions in the double layer and their distribution in the bulk[6]

$$F_{ent} = -T\Delta S = kT\sum_i \int_V \left(c_i \ln\left(\frac{c_i}{c_{i0}}\right) - c_i + c_{i0}\right) dV \quad (2)$$

where k is the Boltzmann constant, T the absolute temperature, c_i the concentration of ions of species "i" in the double layer, and c_{i0} its concentration in the bulk and the subscript "i" runs over all ionic species. For an 1−1 electrolyte of concentration c_E, whose ions obey Boltzmann distributions, the above expression becomes

$$-T\Delta S = 2c_E kT \int_V \left[\left(\frac{e\psi}{kT}\right)\sinh\left(\frac{e\psi}{kT}\right) + 1 - \cosh\left(\frac{e\psi}{kT}\right)\right] dV \quad (3)$$

where e is the protonic charge.

The above two contributions to the free energy of the double layer are both positive. The third term, the chemical free energy contribution, which is responsible for the spontaneous formation of the double layer, should be negative and larger than the sum of the previous two terms.

During interactions at constant surface potential ψ_S, the surface charge decreases when two surfaces approach each other and ion adsorption or molecular group dissociation is involved in the surface charging. The system (liquid and surfaces) is in thermodynamic equilibrium, and a transfer of charges from the liquid to the surfaces occurs to ensure the equality of the electrochemical potentials of the ions in the bulk and near the surface. Denoting by $\Delta\mu$ the change in the chemical part of the electrochemical potential of an ion of charge q brought from the bulk to the surface and by n the number of ions adsorbed per unit surface area, the chemical free energy (due to the adsorption of charges) per unit area, is given by[5]

$$F_{ch}|_{\psi_S=\text{const}} = n\Delta\mu \quad (4)$$

The thermodynamic equilibrium implies that

$$\mu_{bulk} = \mu_{surface} + q\psi_S \quad (5)$$

Since $\Delta\mu = \mu_{surface} - \mu_{bulk} = -q\psi_S$, eq 4 becomes

$$F_{ch}|_{\psi_S=\text{const}} = n\Delta\mu = -nq\psi_S = -\sigma\psi_S \quad (6a)$$

where σ is the surface charge density and ψ_S is the surface potential.

For an interaction at constant surface charge, the charge is fixed on the surface and it is not at thermodynamic equilibrium, and the change in the chemical free energy during the interaction is zero. Hence, the chemical free energy is a constant which was taken as zero[5]

$$F_{ch}|_{\sigma=\text{const}} = 0 \quad (6b)$$

As a result of this choice, the double layer free energy is in this case always positive. Actually this constant is negative, equal to the free energy required to charge the surfaces at infinite separation distance, and the double layer free energy is negative.

Let us now derive an expression for the chemical free energy for a general interaction (when both the surface charge density and surface potential are changing during the interaction). When two particles (of arbitrary geometry) approach each other from an infinite separation, the surface charge density and the surface potential, for an infinitesimal area δA on the surface of a particle, depend on the distance and follow a trajectory $\psi_S = \psi_S(\sigma)$ (see Figure 1). This trajectory can be thought of as composed of successive infinitesimal transformations at constant surface potential ($\sigma_i, \psi_{S,i} \rightarrow \sigma_{i+1}, \psi_{S,i}$) and constant surface charge density ($\sigma_{i+1}, \psi_{S,i} \rightarrow \sigma_{i+1}, \psi_{S,i+1}$). The change of the chemical free energy during the transformations at constant surface potential is given by $-\psi_{S,i}(\sigma_{i+1} - \sigma_i)\delta A$, and it is zero for the transformations at constant surface charge. Adding all the changes when two particles approach from infinity to a final separation distance l, one obtains for the change $\Delta F_{ch}(l)$ of the chemical contribution to the free energy

$$\Delta F_{ch}(l)\delta A = \left(-\int_{\sigma(\infty)}^{\sigma(l)} \psi_S(\sigma)\, d\sigma\right)\delta A \quad (7)$$

In the above integral, the Poisson−Boltzmann equation should be solved for each separation distance l and the appropriate surface boundary conditions (constant surface charge density, constant surface potential, dissociation of surface groups, or adsorption of ions). Consequently, ψ_S and σ are functions of l; eliminating l one obtains $\psi_S = \psi_S(\sigma)$. The function $\psi_S(\sigma)$ describes the unique trajectory of the system during the interaction, and the integral is performed along this trajectory (the path 1 → 2 in Figure 1).

The total free energy of the double layer is obtained by adding eqs 1, 3, and 7

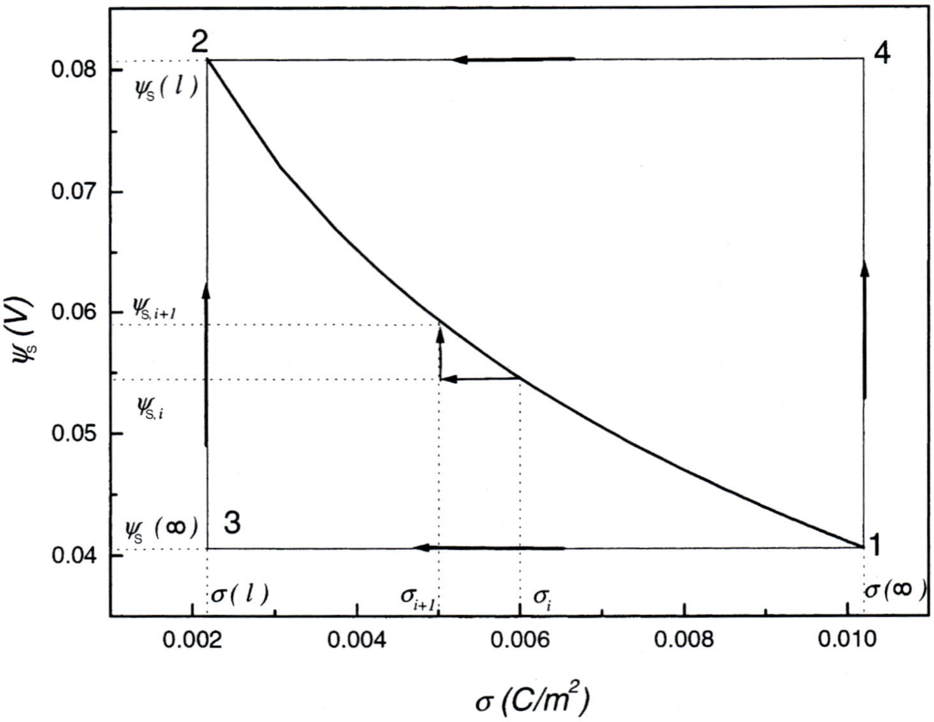

Figure 1. The surface potential ψ_S as a function of the surface charge density σ when two interacting particles approach from infinite separation to the final separation distance l.

$$F_{DL}(l) = \frac{1}{2}\int_V \epsilon\epsilon_0(\nabla\psi)^2\,dV + 2c_E kT\int_V \left[\left(\frac{e\psi}{kT}\right)\sinh\left(\frac{e\psi}{kT}\right) + 1 - \cosh\left(\frac{e\psi}{kT}\right)\right] dV + \int_A \left(C - \int_{\sigma(\infty)}^{\sigma(l)} \psi_S(\sigma)\,d\sigma\right) dA \quad (8)$$

the last integral being taken over the entire surface of the system, and the constant C (independent of l) representing the chemical free energy at infinite separation.

As shown by Overbeek, the first two terms of eq 8 can be transformed into surface integrals using Green's theorem[6]

$$\frac{1}{2}\int_V \epsilon\epsilon_0(\nabla\psi)^2\,dV + 2c_E kT\int_V \left[\left(\frac{e\psi}{kT}\right)\sinh\left(\frac{e\psi}{kT}\right) + 1 - \cosh\left(\frac{e\psi}{kT}\right)\right] dV = \int_A \left(\int_0^{\sigma(l)} \varphi_S(\xi,l)\,d\xi\right) dA \quad (9)$$

To perform the integral in the right-hand side of the above expression, the Poisson–Boltzmann equation should be solved for a fixed value of l and all the "imaginary surface charges" ξ between 0 and $\sigma(l)$ to obtain the "imaginary surface potentials" φ_S. The surface potential φ_S depends in this case on the values of ξ and l.

From eqs 8 and 9 one obtains

$$F_{DL}(l) = \int_A \left(\int_0^{\sigma(l)} \varphi_S(\xi,l)\,d\xi\right) dA + \int_A \left(C - \int_{\sigma(\infty)}^{\sigma(l)} \psi_S(\sigma)\,d\sigma\right) dA \quad (10)$$

Expression 10 is valid for a system of arbitrary shape and arbitrary surface conditions. It will be shown in what follows that one can recover the results obtained via the traditional procedures, within their domains of validity.

For interactions at constant surface charge, the integral over σ in the second term vanishes and the well-known expression due to Verwey and Overbeek, which assumed $C \equiv 0$, is recovered[5]

$$F_{DL}(l)|_{\sigma=\text{const}} = \int_A \left(\int_0^{\sigma} \varphi_S(\xi,l)\,d\xi\right) dA \quad (11a)$$

For interactions at constant surface potential, $\psi_S(l) = \psi_S(\infty) = \psi_S$ and using the corresponding Verwey–Overbeek choice $C \equiv -\sigma(\infty)\psi_S(\infty)$, one obtains, integrating by parts the first right-hand side term of eq 10, the expression

$$F_{DL}(l)|_{\psi=\text{const}} = \int_A \left(\int_0^{\sigma(l)} \varphi_S(\xi,l)\,d\xi\right) dA + \int_A \left(-\sigma(\infty)\psi_S(\infty) - \psi_S(\infty)\int_{\sigma(\infty)}^{\sigma(l)} d\sigma\right) dA = \int_A \left(\varphi_S\xi\Big|_{\xi=0,\varphi_S=0}^{\xi=\sigma(l),\varphi_S=\psi_S(l)} - \int_0^{\psi_S(l)} \xi(\varphi_S,l)\,d\varphi_S\right) dA - \int_A (\psi_S(\infty)\sigma(l))\,dA = -\int_A \left(\int_0^{\psi_S} \xi(\varphi_S,l)\,d\varphi_S\right) dA \quad (11b)$$

which is the Verwey–Overbeek expression for the free energy for interactions at constant surface potential.

The free energy of the system is always defined up to an arbitrary constant; the choice $C \equiv 0$ in the Verwey–Overbeek approach leads to positive values for the free energy of the system at constant surface charge and negative at constant surface potential. However, the interaction free energy (the difference between the free energy of the system at the final separation distance and the free energy of the system at infinite separation) is not affected by the choice of the constant C.

Let us now show that eq 10 leads to the same result as the Langmuir equation for the force between two identical planar surfaces and arbitrary surface conditions. Using eq 10, one obtains the following expression for the force per unit area, between two identical planar surfaces

$$\Pi_{DL}(l) = -\frac{1}{A}\frac{dF_{DL}(l)}{dl} = -\frac{d}{dl}2\int_0^{\sigma(l)} \varphi_S(\xi,l)\,d\xi + \frac{d}{dl}2\left(\int_{\sigma(\infty)}^{\sigma(l)} \psi_S(\sigma)\,d\sigma\right) \quad (12)$$

where the factor 2 accounts for the two identical surfaces.

The first term is the Verwey–Overbeek expression for the interactions at constant surface charge density and the second accounts for the variation of the surface charge with the separation distance.

The derivative of the first term of eq 12 is given by

$$\frac{\mathrm{d}}{\mathrm{d}l}\left(\int_0^{\sigma(l)} \varphi_S(\xi,l)\,\mathrm{d}\xi\right) = \int_0^{\sigma(l)} \frac{\partial\varphi_S(\xi,l)}{\partial l}\,\mathrm{d}\xi + \psi_S(\sigma)\frac{\mathrm{d}\sigma(l)}{\mathrm{d}l} \quad (13)$$

and the derivative of the second term by

$$\frac{\mathrm{d}}{\mathrm{d}l}\left(\int_{\sigma(\infty)}^{\sigma(l)} \psi_S(\sigma)\,\mathrm{d}\sigma\right) = \psi_S(\sigma)\frac{\mathrm{d}\sigma(l)}{\mathrm{d}l} \quad (14)$$

Consequently, from eqs 12–14, one obtains

$$\Pi_{DL}(l) = -2\int_0^{\sigma(l)} \frac{\partial\varphi_S(\xi,l)}{\partial l}\,\mathrm{d}\xi \quad (15)$$

It remains only to prove that eq 15 is equivalent to the Langmuir equation

$$\Pi_{DL,Langmuir}(l) = 2c_E kT\left(\cosh\left(\frac{e\psi_m(l)}{kT}\right) - 1\right) \quad (16)$$

where $\psi_m(l)$ is the potential at the middle distance between plates. This task was performed earlier.[7] For the completeness of the presentation, the details of the proof that eqs 15 and 16 are equivalent are reproduced in the Appendix.

Therefore, the present treatment of the double layer interaction leads to the same results for the interaction free energy as the imaginary charging approach for systems of arbitrary shapes and constant surface potential or constant charge density and to the same results as the Langmuir equation for parallel plates and arbitrary surface conditions. It can be, however, used for systems of any shape and any surface conditions, since it does not imply any of the above restrictions.

III. Approximate Expressions for the Change in Chemical Free Energy

A notable consequence of eq 7 is that the values of surface charge density and surface potential at a given separation distance l are not sufficient to calculate the free energy of interaction between surfaces. To calculate that free energy, one has to solve the Poisson–Boltzmann equation at each separation distance along a trajectory from infinity to the separation distance l. Here we will propose a simple approximation for the change of the chemical free energy, which simplifies significantly the problem.

As shown in Figure 1, the transformation from $1(\infty) \rightarrow 2(l)$ is bounded by two extremal processes, $1 \rightarrow 3 \rightarrow 2$ and $1 \rightarrow 4 \rightarrow 2$. For the first, the change in the chemical free energy, per unit area, calculated using eqs 6a and 6b, is

$$\Delta F_{132} = \Delta F_{13} + \Delta F_{32} = -\psi_S(\infty)(\sigma(l) - \sigma(\infty)) + 0 \quad (17a)$$

while for the second

$$\Delta F_{142} = \Delta F_{14} + \Delta F_{42} = 0 - \psi_S(l)(\sigma(l) - \sigma(\infty)) \quad (17b)$$

Therefore, the change in the chemical free energy is bounded by ΔF_{132} and ΔF_{142}. A simple approximation for the change in the chemical free energy, per unit area, is the arithmetic average of those two bounds

$$\Delta F_{ch}(l) = -\frac{(\psi_S(l) + \psi_S(\infty))(\sigma(l) - \sigma(\infty))}{2} \quad (18)$$

This expression is exact for both constant surface charge and constant surface potential interactions.

One can also construct approximate expressions for the change in the chemical free energy by extending those at constant surface charge (6b) or at constant surface potential (6a) to general processes. In the first case

$$\Delta F_{ch}(l) \approx 0 \quad (19)$$

which will be named in what follows "constant surface charge approximation". In the second case, since the chemical free energy is defined in terms of σ and ψ_S at each distance l, a natural approximation is

$$\Delta F_{ch}(l) \approx -(\sigma(l)\psi_S(l) - \sigma(\infty)\psi_S(\infty)) \quad (20a)$$

which, as our calculations have shown, is a very poor approximation when the surface potential is not constant. Since eq 6a implies that $\psi_S(l) \equiv \psi_S(\infty)$, other two legitimate choices for the approximate change in the chemical free energy would be

$$\Delta F_{ch}(l) = -(\sigma(l)\psi_S(\infty) - \sigma(\infty)\psi_S(\infty)) \quad (20b)$$

or

$$\Delta F_{ch}(l) = -(\sigma(l)\psi_S(l) - \sigma(\infty)\psi_S(l)) \quad (20c)$$

where the "constant surface potential" is assumed to be either at infinite separation distance or at the distance l. Expressions 20b and 20c are the two bounds proposed earlier (eqs 17a and 17b, respectively). A simple modality to improve the approximation is to use as "constant surface potential" in eq 6a an intermediate value between $\psi_S(\infty)$ and $\psi_S(l)$. By employing their average, one obtains

$$\Delta F_{ch}(l) = -(\sigma(l) - \sigma(\infty))\frac{(\psi_S(l) + \psi_S(\infty))}{2} \quad (20d)$$

which coincides with eq 18. This expression will be named in what follows "the constant surface potential approximation", although it is exact when either the surface charge or the surface potential is constant.

The accuracy of these approximations, when neither the surface potential nor the surface charge density are constants, will be investigated below, for two identical, parallel planar surfaces.

It will be assumed that the surface charge is due to the dissociation of surface groups (for instance dissociation of surfactant molecules).[8,9] In general, the electrolyte counterions are the most abundant ions in the vicinity of the surface and therefore they can control the surface charge via reassociation. Denoting by c_E the concentration (in the reservoir) of an 1:1 electrolyte, by N the number of sites per unit area, and by x the fraction of dissociated sites, the dissociation equilibrium provides the expression

$$K_D = \frac{(xN)c_E^{\,+}}{(1-x)N} = \frac{xc_E^{\,+}}{(1-x)} = \frac{x}{1-x}c_E \exp\left(-\frac{e\psi_S}{kT}\right) \quad (21)$$

where K_D is the dissociation constant, c_E^{+} denotes the cation concentration in the liquid in the vicinity of the surface, e is the protonic charge, and the Boltzmann distribution was assumed for the counterions (note that ψ_S is negative).

The surface charge density is therefore given by

(7) Ruckenstein, E. *J. Colloid Interface Sci.* **1981**, *82*, 490.
(8) Ninham, B. W.; Parsegian, V. A. *J. Theor. Biol.* **1971**, *31*, 405.
(9) Prieve, D. C.; Ruckenstein, E. *J. Theor. Biol.* **1976**, *56*, 205.

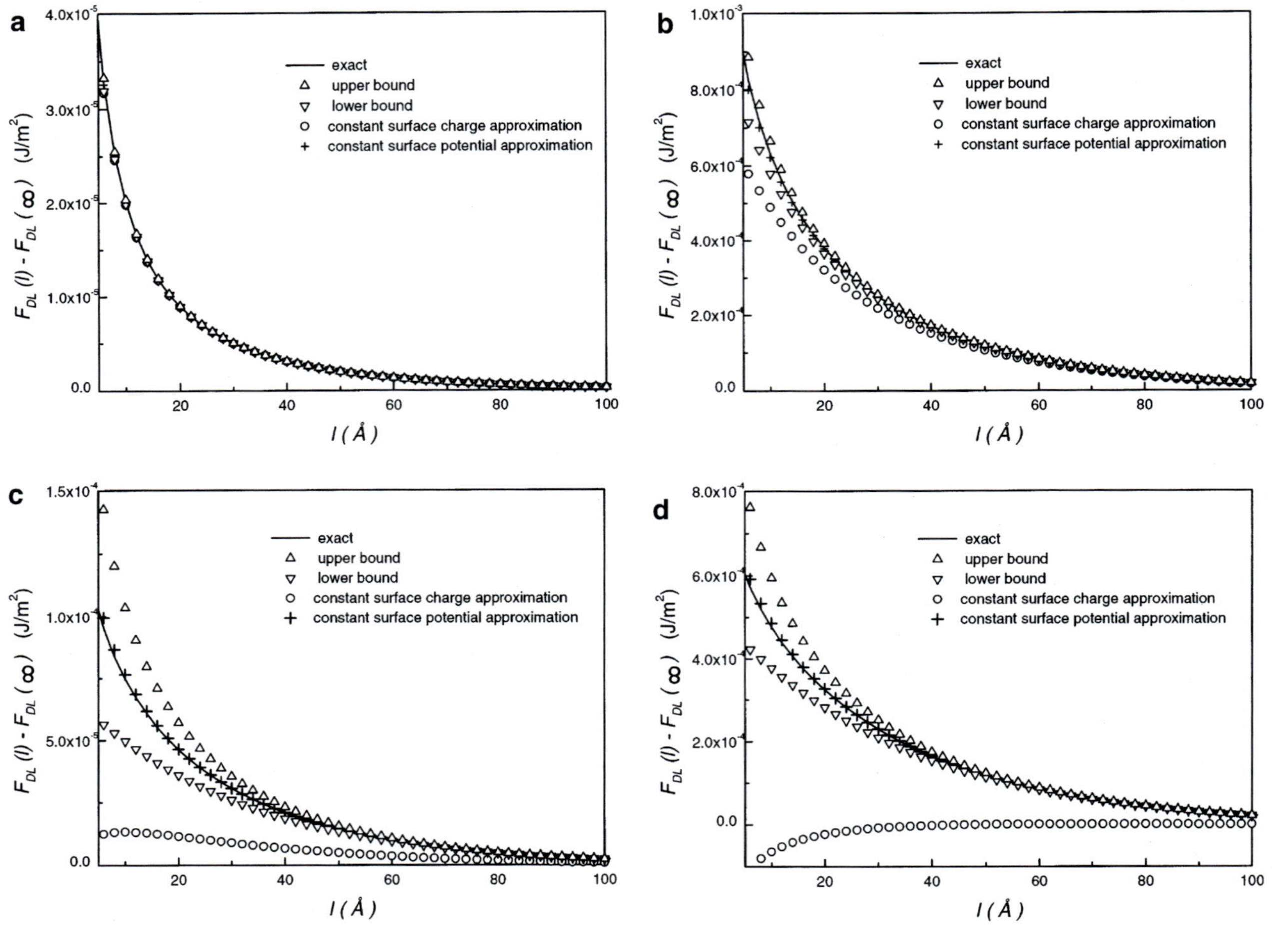

Figure 2. The interaction free energy (per unit area) as a function of the separation distance l for two identical plates, planar and parallel, at $T = 300$ K, $\epsilon = 80$, $c_E = 0.01$ M and: (a) $K_D = 1.0$ M, $N = 6.25 \times 10^{15}$ sites/m^2; (b) $K_D = 1.0$ M, $N = 6.25 \times 10^{16}$ sites/m^2; (c) $K_D = 0.001$ M, $N = 3.125 \times 10^{17}$ sites/m^2; (d) $K_D = 0.001$ M, $N = 3.125 \times 10^{18}$ sites/m^2. The continuous thick line represents the exact result; the up triangles represent the upper bound, the down triangles represent the lower bound, the circles represent the "constant surface charge approximation", and the crosses represent the "constant surface potential approximation".

$$\sigma = -exN = -\frac{eN}{1 + \frac{c_E}{K_D}\exp\left(-\frac{e\psi_S}{kT}\right)} \tag{22}$$

When the dissociation constant K_D is large, $(c_E/K_D)\exp(-e\psi_S/kT) \ll 1$ and $\sigma \cong -eN$. Then the interaction can be well approximated by that at constant surface charge density. However, for low values of K_D, the interaction can no longer be approximated by assuming constant surface charge or constant surface potential.

Another relation between the surface charge and surface potential is provided by the equation

$$\sigma = -\epsilon\epsilon_0\left(\frac{d\psi(z)}{dz}\right)_S \tag{23}$$

where the z direction is normal to the surface. Equation 23, combined with eq 22 provides one of the boundary conditions. The other boundary condition is provided by the symmetry of the system, which implies that the derivative of the potential at the middle distance vanishes

$$\frac{d\psi(z)}{dz}\Big|_{z=l/2} = 0 \tag{24}$$

Consequently, one has to solve the Poisson–Boltzmann equation, which for a uni-univalent electrolyte has the form

$$\frac{d^2\psi}{dz^2} = \frac{2ec_E}{\epsilon\epsilon_0}\sinh\left(\frac{e\psi}{kT}\right) \tag{25}$$

for the above boundary conditions.

The free energy of interaction between two identical plates with charges generated through dissociation will be calculated both exactly and approximately, and the results will be compared for various values of the dissociation constant.

For a given separation distance l, the solution of the Poisson–Boltzmann equation was obtained via numerical integration. From the value of the potential at the middle distance between the plates $\psi_m(l) \equiv \psi(l/2)$, the disjoining pressure was calculated using the Langmuir equation. The interaction free energy was subsequently obtained by integrating the disjoining pressure from infinite to the final separation distance l.

In a second method, the free energy was calculated by adding the electric (eq 1), entropic (eq 3), and chemical contributions, and the interaction free energy was obtained by subtracting the free energy for infinite separation. The integrals were calculated using the numerical solution of the Poisson–Boltzmann equation for $\psi(z,l)$ and $\sigma(l)$.

The change in the chemical free energy was calculated using several expressions, namely, eq 7 (the exact expression), eq 19 (the "constant surface charge approximation"), eq 20b (the lower bound), eq 20c (the upper bound), and eq 20d (the "constant surface potential approximation").

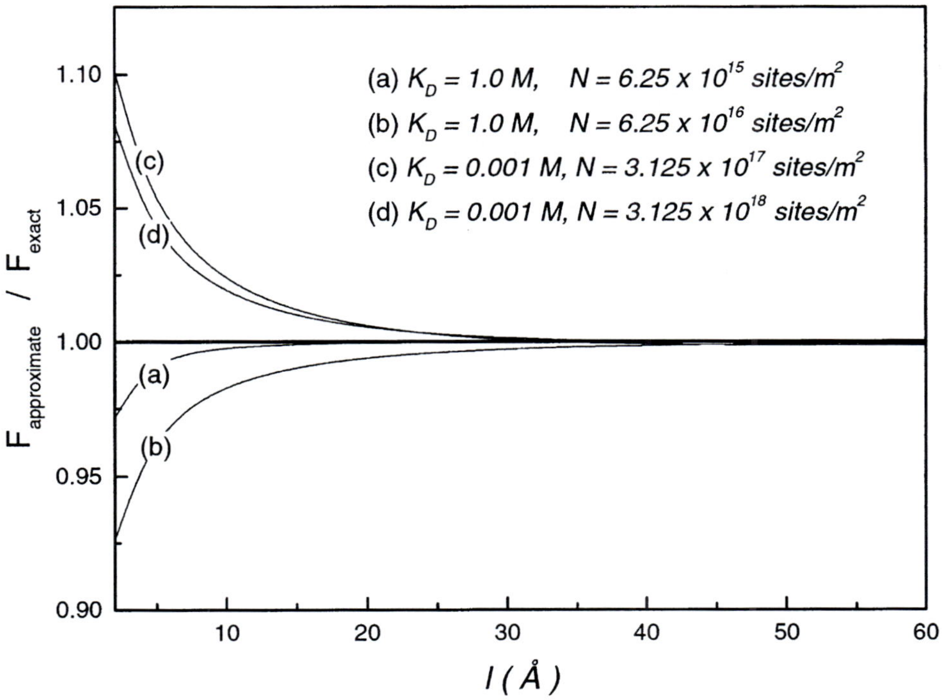

Figure 3. The ratio between the approximate interaction free energy, based on the change in chemical free energy contribution provided by eq 18 and the rigorous interaction free energy. The values of the parameters are the same as those given in Figure 2.

When the exact expression was employed, the results obtained always coincided (within numerical errors), as expected, with those provided by the Langmuir procedure.

For large dissociation constants (implying small c_E/K_D ratio), the interaction is well described, at low surface potentials, by the constant surface charge density approximation. In Figure 2a, the free energy is calculated using $c_E = 0.01$ M, $T = 300$ K, $\epsilon = 80$, $K_D = 1.0$ M, and $N = 6.25 \times 10^{15}$ sites/m^2. In this case, the calculations based on the expressions for the change in the chemical free energy (eq 19 and eqs 20b–d)) are all good approximations to the rigorous result obtained either using eq 7 for the change in the chemical energy or via the Langmuir procedure.

The increase of N to 6.25×10^{16} sites/m^2 (Figure 2b) displaces the $\sigma - \psi_S$ equilibrium toward larger surface potentials. The "constant surface potential approximation" (based on eq 20d) provides in this case a much better agreement with the exact result than the "constant surface charge approximation" (based on eq 19).

For the low dissociation constant $K_D = 0.001$ M ($c_E = 0.01$ M, $T = 300$ K, $\epsilon = 80$, $N = 3.125 \times 10^{17}$ sites/m^2, Figure 2c), the "constant surface charge approximation" is inaccurate even at low potentials ($\psi_S < 0.010$ V). When the site density was increased to $N = 3.125 \times 10^{18}$ sites/m^2 ($c_E = 0.01$ M, $T = 300$ K, $\epsilon = 80$, $K_D = 0.001$ M, Figure 2d), the "constant surface charge approximation" predicted attraction between the plates.

In Figure 3 the ratio between the interaction free energy calculated on the basis of the approximate (eq 20d) and the rigorous (eq 7) expressions for the change in the chemical free energy is plotted against the distance. In all the cases investigated, the approximate expression for the chemical free energy was accurate.

In summary, when the surface charge was generated by the dissociation equilibrium of surface groups, the surface charge density was almost constant for large dissociation constants and low surface potentials. In this case, the neglecting of the change in the chemical free energy (eq 19) was a good approximation. However, in the other cases (small dissociation constants or large surface potentials) it was, in general, inaccurate and sometimes even predicted double layer attraction.

In all the cases investigated, the surface potential has changed, sometimes drastically, during interactions. However, a suitable choice of the surface potential (the arithmetic average of the surface potentials at infinity and at distance l) in the "constant surface potential approximation" for the change in the chemical free energy provided always a good approximation. This approximation is exact when either the surface potential or the surface charge are constant.

IV. Conclusions

One can calculate the free energy of a double layer by adding to the free energy of the electric field, an entropic contribution due to mobile ions and a chemical free energy contribution due to the charge transfer between bulk and interface. However, expressions for the latter contribution are provided in the literature only for interactions at constant surface charge density or constant surface potential. In this paper, an expression for the change in the chemical free energy which is valid for interactions at any surface conditions was derived; it depends not only on the values of the surface charge and potential at the final separation distance but also on their values at all distances between infinite and the final one.

The present approach reduces to the traditional ones within their range of application (imaginary charging processes for double layer interactions between systems of arbitrary shape and interactions either at constant surface potential or at constant surface charge density, and the procedure based on Langmuir equation for interactions between planar, parallel plates and arbitrary surface conditions). It can be, however, employed to calculate the interaction free energy between systems of arbitrary shape and any surface conditions, for which the traditional approaches cannot be used.

Since the change in the chemical energy depends on all the values of the surface potential and surface charge density, when the particles approach from infinity to the final separation distance, the calculation of the interaction free energy involves the solution of the Poisson–Boltzmann equation at all the points along this trajectory. For systems with complicated shapes, this procedure is inconvenient. For this reason, an approximate expression for the change in the chemical free energy was also proposed, which depends only on the values of the surface potential and charge density at the infinite and final separation distances. It was shown that this approximation is accurate.

Appendix

Integrating eq 15 by parts, one obtains for the disjoining pressure between two identical, parallel and planar surfaces separated by the distance l

$$\Pi_{DL}(l) = -2\int_0^{\sigma(l)} \frac{\partial \varphi_S(\xi,l)}{\partial l}\,d\xi = -2\frac{\partial}{\partial l}\left(\int_0^{\sigma(l)} \varphi_S(\xi,l)\,d\xi\right) + 2\psi_S(\sigma)\frac{\partial \sigma(l)}{\partial l} = -2\frac{\partial}{\partial l}\left(\varphi_S\xi\Big|_{\xi=0,\varphi_S=0}^{\xi=\sigma(l),\varphi_S=\psi_S(l)} - \int_0^{\psi_S(l)} \xi(\varphi_S,l)\,d\varphi_S\right) + 2\psi_S(\sigma)\frac{\partial \sigma(l)}{\partial l} = 2\frac{\partial}{\partial l}\left(\int_0^{\psi_S(l)} \xi(\varphi_S,l)\,d\varphi_S\right) - 2\sigma(\psi_S)\frac{\partial \psi_S(l)}{\partial l} \tag{A.1}$$

For the integral in the last right-hand side of the equality, one can use an expression derived by Verwey and Overbeek (eq 37b in ref 5)[5]

$$-\int_0^{\psi_S} \xi(\varphi_S,l)\,d\varphi_S = -c_E kTl\left(\cosh\left(\frac{e\psi_m(l)}{kT}\right) - 1\right) + \left(\frac{2\epsilon\epsilon_0 c_E (kT)^3}{e^2}\right)^{1/2} \int_{e\psi_S(l)/kT}^{e\psi_m(l)/kT} \left(2\cosh(y) - 2\cosh\left(\frac{e\psi_m(l)}{kT}\right)\right)^{1/2} dy \tag{A.2}$$

the derivative of which with respect to the distance is given by

$$-\frac{\partial}{\partial l}\left(\int_0^{\psi_S} \xi(\varphi_S,l)\,d\varphi_S\right) = -c_E kT\left(\cosh\left(\frac{e\psi_m(l)}{kT}\right) - 1\right) - c_E kTl\sinh\left(\frac{e\psi_m}{kT}\right)\frac{\partial \psi_m}{\partial l} + \left(\frac{2\epsilon\epsilon_0 c_E (kT)^3}{e^2}\right)^{1/2} \times \left(2\cosh(y) - 2\cosh\left(\frac{e\psi_m(l)}{kT}\right)\right)^{1/2} \frac{\partial y}{\partial l}\Bigg|_{(e\psi_S(l)/kT)}^{(e\psi_m(l)/kT)} + \left(\frac{2\epsilon\epsilon_0 c_E (kT)^3}{e^2}\right)^{1/2} \sinh\left(\frac{e\psi_m}{kT}\right)\frac{\partial \psi_m}{\partial l}\int_{e\psi_S(l)/kT}^{e\psi_m(l)/kT} \times \frac{1}{\left(2\cosh(y) - 2\cosh\left(\frac{e\psi_m(l)}{kT}\right)\right)^{1/2}}\,dy \tag{A.3}$$

Integrating once the Poisson–Boltzmann equation, one obtains

$$\frac{\partial \psi(z,l)}{\partial z} = -\left(\frac{2c_E kT}{\epsilon\epsilon_0}\right)^{1/2}\left(2\cosh\left(\frac{e\psi(z,l)}{kT}\right) - 2\cosh\left(\frac{e\psi_m(l)}{kT}\right)\right)^{1/2} \tag{A.4}$$

which leads to the relation

$$\sigma(l) = -\epsilon\epsilon_0 \frac{\partial \psi(z,l)}{\partial z}\Big|_{z=0} = (2c_E\epsilon\epsilon_0 kT)^{1/2}\left(2\cosh\left(\frac{e\psi_S(l)}{kT}\right) - 2\cosh\left(\frac{e\psi_m(l)}{kT}\right)\right)^{1/2} \tag{A.5}$$

and by direct integration, from the surface to the middle distance to the expression[5]

$$\int_{e\psi_S(l)/kT}^{e\psi_m(l)/kT} \frac{dy}{\left(2\cosh(y) - 2\cosh\left(\frac{e\psi_m(l)}{kT}\right)\right)^{1/2}} = -\frac{l}{2}\left(\frac{2c_E e^2}{\epsilon\epsilon_0 kT}\right)^{1/2} \tag{A.6}$$

By employing eq A.6, the last right-hand-side term in expression A.3 becomes

$$\left(\frac{2\epsilon\epsilon_0 c_E (kT)^3}{e^2}\right)^{1/2} \sinh\left(\frac{e\psi_m}{kT}\right)\frac{\partial \psi_m}{\partial l}\int_{e\psi_S(l)/kT}^{e\psi_m(l)/kT} \times \frac{1}{\left(2\cosh(y) - 2\cosh\left(\frac{e\psi_m(l)}{kT}\right)\right)^{1/2}}\,dy = c_E kTl\sinh\left(\frac{e\psi_m}{kT}\right)\frac{\partial \psi_m}{\partial l} \tag{A.7}$$

and cancels the second term of the right-hand side of eq A.3. Using expression A.5, eq A.3 becomes

$$-\frac{\partial}{\partial l}\left(\int_0^{\psi_S} \xi(\varphi_S,l)\,d\varphi_S\right) = -c_E kT\left(\cosh\left(\frac{e\psi_m(l)}{kT}\right) - 1\right) - \left(\frac{2\epsilon\epsilon_0 c_E (kT)^3}{e^2}\right)^{1/2}\left(2\cosh\left(\frac{e\psi_S(l)}{kT}\right) - 2\cosh\left(\frac{e\psi_m(l)}{kT}\right)\right) \times \frac{e}{kT}\frac{\partial \psi_S(l)}{\partial l} = -c_E kT\left(\cosh\left(\frac{e\psi_m(l)}{kT}\right) - 1\right) - \sigma(\psi_S)\frac{\partial \psi_S(l)}{\partial l} \tag{A.8}$$

Finally, from eqs A.1 and A.8, one obtains

$$\Pi_{DL}(l) = 2\frac{\partial}{\partial l}\left(\int_0^{\psi_S(l)} \sigma(\varphi_S,l)\,d\varphi_S\right) - 2\sigma(\psi_S)\frac{\partial \psi_S(l)}{\partial l} = 2c_E kT\left(\cosh\left(\frac{e\psi_m(l)}{kT}\right) - 1\right) \tag{A.9}$$

which is the Langmuir equation, eq 16.

LA0266132

Introduction to CHAPTER 7
Polarization Model and ion specificity: applications

7.1 M. Manciu, E. Ruckenstein: "*Role of the Hydration Force in the Stability of Colloids at High Ionic Strengths,*" LANGMUIR 17 (2001) 7061–7070.
7.2 M. Manciu, E. Ruckenstein: "*Long Range Interactions between Apoferritin Molecules,*" LANGMUIR 18 (2002) 8910–8918.
7.3 E. Ruckenstein, M. Manciu: "*On the Stability of the Common and Newton Black Films,*" LANGMUIR 18 (2002) 2727–2736.
7.4 M. Manciu, E. Ruckenstein: "*On possible microscopic origins of the swelling of neutral lipid bilayers induced by simple salts*" JOURNAL OF COLLOID AND INTERFACE SCIENCE 309 (2007) 56–67.
7.5 H. Huang, M. Manciu, E. Ruckenstein: "*The effect of surface dipoles and of the field generated by a polarization gradient on the repulsive force,*" JOURNAL OF COLLOID AND INTERFACE SCIENCE 263 (2003) 156–161.
7.6 E. Ruckenstein, H. Huang: "*Colloid Restabilization at High Electrolyte Concentrations: Effect of Ion Valency,*" LANGMUIR 19 (2003) 3049–3055.
7.7 H. Huang, M. Manciu, E. Ruckenstein: "*On the Restabilization of Protein-Covered Latex Colloids at High Ionic Strengths,*" LANGMUIR 21 (2005) 94–99.
7.8 M. Manciu, E. Ruckenstein: "*The polarization model for hydration/double layer interactions: the role of the electrolyte ions*" ADVANCES IN COLLOID AND INTERFACE SCIENCE 112 (2004) 109–128.
7.9 M. Manciu, O. Calvo, E. Ruckenstein: "*Polarization model for poorly-organized interfacial water: Hydration forces between silica surfaces,*" ADVANCES IN COLLOID AND INTERFACE SCIENCE 127 (2006) 29–42.

The polarization model predicts that the interaction between nanoparticles depends not only on the surface charge density, but also on the surface dipole density. As the concentration of electrolyte increases, the surface charge density decreases, due to the recombination of ions with surface groups, but the density of surface dipoles increases. At relatively low salt concentrations, the repulsion due to the double layer is dominant and decreases with increasing ionic strength; however, at high electrolyte concentrations, the repulsive force due to surface dipoles becomes dominant and increases with increasing ionic strength. Consequently, there is a minimum in the repulsion between nanoparticles. This mechanism explains the restabilization of some colloids at high ionic strengths [7.1].

This mechanism also is employed to explain the light scattering phenomena observed for apoferritin molecules at various NaCH3COO concentrations. The second virial coefficient of apoferritin decreases to a value corresponding to hard spheres for an electrolyte concentration of 0.15 M but increases to a large value at a concentration of 0.25 M. In the traditional framework, these results can be explained either by considering that the apoferritin molecules acquire a huge charge at 0.25 M, or that the traditional hydration force increases by orders of magnitude at that ionic strength. Neither of these possibilities is, however, plausible. The polarization model can explain this behavior as being a result of the recombination of surface charges with ions and their replacement by dipoles. The increase of the surface dipole density with increasing electrolyte concentration generates a long-range repulsion at high ionic strengths [7.2].

The recombination of surface charges with ions and the corresponding increase in repulsion also has consequences regarding the stability of common and Newton black films. However, in the latter case, the repulsive thermal undulations of the films also play a role [7.3]. As revealed by experiment and intuitively expected, an increase in temperature (which in turn increases the thermal undulations) decreases the stability of the films. In contrast, the insight provided by the traditional approach suggests that the increase in thermal undulations should lead to a larger Helfrich repulsion, hence to a more stable system. However, the treatment of this repulsion by considering the membrane as composed of pieces of a suitable area implies that an increase in temperature decreases the area of the pieces, hence increases the probability of film rupture [7.3].

By combining the thermal undulations of membranes with the polarization model it is shown that an increase in

E. Ruckenstein, M. Manciu, *Nanodispersions*, DOI 10.1007/978-1-4419-1415-6_7, © Springer Science+Business Media, LLC 2010

electrolyte concentration can lead to an increase in the repulsion between neutral lipid bilayers, thus explaining their experimentally observed swelling at high electrolyte concentrations [7.4].

Whereas the coupled equations of the polarization model can be solved analytically in the linear approximation (which is valid only for small potentials), in the general case one must rely on numerical solutions [7.5]. The polarization model can explain the restabilization of protein-stabilized polymer latexes, for which the increase in the repulsive force generated by the surface dipoles more than compensates for the decrease in repulsion caused by the decrease in the surface charge and the increase in the screening of the electrostatic field by the increasing ionic strength [7.5].

The theory is extended to multivalent ions, in which case an inversion of the surface charge can occur with increasing electrolyte concentration. Depending on the sign of the surface charge, the adsorption of dipolar molecules on the surface can increase or decrease the repulsion [7.6]. When association equilibria are taken into account for all ions, including the acidic and basic sites of a protein, our model can explain, more than qualitatively, the experimental results of Lopez-Leon et al. [Lopez-Leon, T.; Gea-Jodar, P. M.; Bastos-Gonzalez, D.; Ortega-Vinuesa, J. L. *Langmuir* **2005**, *21*, 87-93], regarding the restabilization of protein-covered latex colloids at high ionic strengths [7.7].

The polarization model is extended to account for additional interactions, not included in the "mean" field, such as the ion-hydration forces [7.8].

In order to explain the interactions between silica surfaces, the polarization model is adapted to poorly-organized surfaces. To account for the disorder induced in water by the rough surfaces of silica, the dipole correlation length λm, which is the main parameter of the polarization model, is allowed to decrease from $\lambda m = 14.9Å$ obtained for water perfectly organized in ice-like layers in the vicinity of a surface to smaller values. For $\lambda m = 4Å$, good agreement with experiment is obtained for reasonable values of the parameters involved (such as surface dipole and charge densities) [7.9].

Langmuir **2001,** *17,* 7061–7070 7061

Role of the Hydration Force in the Stability of Colloids at High Ionic Strengths

Marian Manciu and Eli Ruckenstein*

Department of Chemical Engineering, State University of New York at Buffalo, Buffalo, New York 14260

Received May 18, 2001. In Final Form: August 28, 2001

It is shown that the repulsion between colloidal particles or emulsion droplets depends both on the surface charge density and on the surface dipole density, the latter being a result of the presence of ion pairs on the surface. For illustration purposes, one considers the case in which a surfactant, such as sodium dodecyl sulfate, is adsorbed on the surface of droplets or particles. As the concentration of electrolyte (NaCl) increases, the charge on the surface decreases, and the number of ion pairs increases, because of the association–dissociation equilibrium. At relatively low salt concentrations, the repulsion due to the double layer is dominant and decreases as the electrolyte concentration increases. At relatively high electrolyte concentrations, the hydration repulsive force due to the ion pairs present on the surface becomes dominant. Consequently, as the salt concentration increases, the total repulsion decreases and passes through a minimum, after which it increases. If the hydration repulsion is large, the emulsion or colloidal system will remain stable at any electrolyte concentration. If the hydration repulsion is small, the system will be stable only for sufficiently low electrolyte concentrations. At intermediate strength of the hydration repulsion, the stability depends on the size of the particles or droplets and the Hamaker constant. The rate of coagulation of particles of small radii and small Hamaker constants reaches a maximum and then decreases with increasing ionic strength. For particles of large radii, the increase of the maximum of the interaction energy with increasing electrolyte concentration can be so large (~30 kT) that it can forbid the coagulation at the primary minimum; however, the particles can aggregate at the secondary minimum, which is deep.

I. Introduction

The cornerstone of colloid science is the Derjaguin–Landau–Verwey–Overbeek (DLVO) theory, which explains the stability of colloids in terms of a repulsive (double-layer) force and an attractive (van der Waals) one. The double-layer repulsion, due to the charge on the surface of the particles, can be indefinitely diminished by increasing the electrolyte concentration. The combination of the two interactions typically generates a potential barrier, between a low (and hence stable) primary minimum and a weak secondary minimum. The latter, with a typical depth of the order of kT, generates a weak bonding between particles (flocculation), which can be reversed by stirring or shaking the system. Coagulation of colloids occurs when two neighboring particles acquire enough energy from the random Brownian motion to overcome the potential barrier. At low electrolyte concentrations, the potential barrier is high, and this ensures a long lifetime for the colloidal dispersion; however, by adding electrolyte, the double-layer repulsion is screened and the colloidal particles eventually coagulate.

While the DLVO theory has the advantage of simplicity and often offers "almost quantitative results",[1] it is well-known to be incomplete. A striking example is offered by some colloids (e.g., silica) which do not coagulate even at high electrolyte concentrations, when the double layer is expected to collapse.[1] Recently, it was also shown that the rate constant for aggregation of paraffin wax particles covered by a long-chain carboxylic acid (C_{22}) passes through a maximum when plotted against NaCl concentration.[2] This means that the repulsive force passes through a minimum.

The above behaviors can be explained by additional forces, which also exist between noncharged surfaces and are particularly strong at low separation distances. These forces are generically called non-DLVO forces. An important interaction, for colloids dispersed in water, was related to the organization of water in the vicinity of a surface.[3] This structural force, often called hydration force, was exhaustively investigated experimentally for neutral[4] or weakly charged[5] phospholipid bilayers and was reasonably well described by an exponential decay, with a decay length between 1.5 and 3 Å and a preexponential factor that varied in a rather large range. While the origin of this interaction is attributed to the changes in the local organization of water, there are still debates over the theoretical models which describe the interaction at the molecular level. The interaction was associated with the orientational correlation of water molecules,[6] with the mutual polarization of water layers,[7] disruption of hydrogen bonding,[8] protrusion interactions,[9] or truncation of ion atmosphere by walls.[10] A detailed examination of the above models is made in another paper dedicated to hydration forces.[11]

* Corresponding author. E-mail: feaeliru@acsu.bufalo.edu. Phone: (716) 645-2911/2214. Fax: (716) 645-3822.

(1) Hunter, R. J. *Foundations of Colloid Science*; Oxford Science Publications: New York, 1987.

(2) Alfridsson, M.; Ninham, B.; Wall, S. *Langmuir* **2000**, *16*, 10087.

(3) Derjaguin, B. V.; Churaev, N. V.; Muller, V. M. *Surface Forces*; Plenum: New York, 1987.

(4) Rand, R. P.; Parsegian, V. A. *Biochim. Biophys. Acta* **1989**, *988*, 351.

(5) Cowley, A. C.; Fuller, N. L.; Rand, R. P.; Parsegian, V. A. *Biochemistry* **1978**, *17*, 3163.

(6) Marcelja, S.; Radic, N. *Chem. Phys. Lett.* **1976**, *42*, 129.

(7) Schiby, D.; Ruckenstein, E. *Chem. Phys. Lett.* **1983**, *95*, 435.

(8) Attard, P.; Batchelor, M. T. *Chem. Phys. Lett.* **1988**, *149*, 206.

(9) Israelachvili, J. N.; Wennerstrom, H. *Langmuir* **1990**, *6*, 873.

(10) Marcelja, S. *Nature* **1997**, *385*, 689.

(11) Manciu, M.; Ruckenstein, E. *Langmuir*, in press.

10.1021/la010741t CCC: $20.00
Published on Web 10/04/2001

The hydration force is important for distances between plates less than a few nanometers. Since the DLVO potential barrier between spherical particles or droplets is located at separation distances of the order of the Debye length, it is clear that at least at high electrolyte concentrations the hydration force becomes relevant. The qualitative effect of hydration was earlier recognized, regarding the stability of silica colloids at high electrolyte concentrations[1] or the stability of amphoteric latex particles at high concentrations of some electrolytes.[12]

The main goal of this paper is to show that a variant of an earlier model of Schiby and Ruckenstein[7] can account for the stabilization of some colloids or emulsions at high electrolyte concentrations. In the present model, the hydration force has a strong dependence on the density of ion pairs on the surface of the particles or droplets, which is enhanced by the addition of electrolyte. This is due to the reassociation on the surface, which decreases the charge but generates additional ion pairs. The dipole moments of the ion pairs polarize the nearby water molecules and thus induce a hydration force. For large surface dipole densities, the hydration repulsion is strong enough to prevent (together with the double-layer repulsion) the coagulation at any electrolyte concentration. For low surface dipole density, the system coagulates for ionic strength higher than a critical value, as predicted by the DLVO theory. At intermediate surface dipole densities, the increase in the hydration repulsion with increasing electrolyte concentration can explain the restabilization of some colloidal and emulsion systems at high ionic strength.

In what follows, one considers for illustration purposes the case in which the charge is generated on the surface of colloidal particles or droplets by the adsorption of a surfactant, namely sodium dodecyl sulfate (SDS). We selected this case because information about the adsorption of SDS on an interface is available in the literature, and as it will become clearer later the number of parameters involved is smaller than in the case of silica. A more complex calculation about the silica and the amphoteric latex particles will be presented in a forthcoming paper. It involves several kinds of surface dipoles and equilibrium constants.

II. Theoretical Framework

In this section, the equations needed to calculate the interaction potential between two spherical particles will be presented. It will be assumed that the double-layer, hydration, and van der Waals interactions are independent of each other.

II.A. Surfactant Adsorption. The surface density of surfactant has a relevant role for both the double-layer (via the surface charge density) and the hydration interaction (via the dipole moment density of the ion pairs formed). It will be computed using the Frumkin adsorption isotherm:

$$b_1 C_{R^-} = \frac{\dfrac{\Gamma}{\Gamma_\infty}\exp\left(-2b_2\dfrac{\Gamma}{\Gamma_\infty}\right)}{1-\dfrac{\Gamma}{\Gamma_\infty}} \qquad (1)$$

where Γ is the surfactant surface density, Γ_∞ is the surface density at saturation, b_1 and b_2 are empirical parameters, and C_{R^-} is the concentration of the surfactant anions in the liquid in the vicinity of the interface. The surfactant in the bulk is assumed to be completely dissociated, and hence the concentration of surfactant in the vicinity of the interface is related to the bulk surfactant concentration through the expression

$$C_{R^-} = C_s \exp\left(\frac{e\psi_s}{kT}\right) \qquad (2)$$

where C_s is the bulk surfactant concentration; ψ_s is the surface potential, which is negative; e is the protonic charge; k is the Boltzmann constant; and T is the absolute temperature.

II.B. The Surface Charge. Denoting by α the dissociation constant, the surface charge density σ is given by

$$\sigma = -e\alpha\Gamma \qquad (3)$$

For simplicity, it will be assumed that the electrolyte is 1:1 and has the same type of cations as the surfactant (e.g., NaCl and SDS). The charge is generated through the dissociation

$$\mathrm{R{-}X}|_{\mathrm{SURFACE}} \leftrightarrow \mathrm{R}^-|_{\mathrm{SURFACE}} + \mathrm{X}^+|_{\mathrm{LIQUID\ AT\ SURFACE}}$$

where $\mathrm{R}^-|_{\mathrm{SURFACE}}$ denotes the surface surfactant anion group and X^+ denotes the cation in the liquid at the surface. At equilibrium, one can write

$$K_D = \frac{C_{R^-,S}C_{X^+}}{C_{R-X}} = \frac{\alpha\Gamma C_{X^+}}{(1-\alpha)\Gamma} = \frac{\alpha}{1-\alpha}C_{X^+} \qquad (4)$$

where $C_{R^-,S}$ is the surface density of the surfactant anions, C_{R-X} is the surface density of the nondissociated surfactants, C_{X^+} is the concentration of the cations in the liquid in the vicinity of the surface, and K_D is the equilibrium constant. Assuming that the electrolyte is completely dissociated, one obtains

$$C_{X^+} = (C_s + C_e)\exp\left(-\frac{e\psi_s}{kT}\right) \qquad (5)$$

where C_e is the bulk electrolyte concentration.

II.C. Double-Layer Interaction. The calculation of the double-layer interaction accounts for the association–dissociation at the interface. Assuming two parallel plates at a distance x apart, the potential ψ_m at the middle distance between plates can be very well approximated by the following equation, due to Ohshima and Kondo:[13]

$$\tanh\left(\frac{e\psi_m}{4kT}\right) = \gamma_0\frac{1}{\cosh\left(\frac{\kappa x}{2}\right)} - \gamma_0^3\frac{\left(\frac{\kappa x}{2}\right)\sinh\left(\frac{\kappa x}{2}\right)}{\cosh^4\left(\frac{\kappa x}{2}\right)} + \gamma_0^5\left\{\frac{\cosh^2\left(\frac{\kappa x}{2}\right) - 2\left(\frac{\kappa x}{2}\right)^2 - 1}{4\cosh^5\left(\frac{\kappa x}{2}\right)} - \frac{3\left(\frac{\kappa x}{2}\right)\sinh\left(\frac{\kappa x}{2}\right)}{4\cosh^6\left(\frac{\kappa x}{2}\right)}\left(1 - 4\left(\frac{\kappa x}{2}\right)\tanh\left(\frac{\kappa x}{2}\right)\right)\right\} \qquad (6)$$

where $\gamma_0 = \tanh(e\psi_s/4kT)$, ψ_s is the surface potential, $\kappa = \sqrt{2e^2C_e/\epsilon\epsilon_0 kT}$ is the reciprocal Debye length for a 1:1

(12) Healy, T. W.; Homola, A.; James, R. O.; Hunter, R. J. *Faraday Discuss. Chem. Soc.* **1978**, *65*, 156.

(13) Ohshima, H.; Kondo, T. *J. Colloid Interface Sci.* **1988**, *122*, 591.

electrolyte, ϵ_0 is the vacuum permittivity, and ϵ the dielectric constant of water. The surface charge density is related to the surface and midplane potentials via[1]

$$\sigma = -\frac{\kappa\epsilon\epsilon_0 kT}{e}\sqrt{2\left(\cosh\left(\frac{e\psi_s}{kT}\right) - \cosh\left(\frac{e\psi_m}{kT}\right)\right)} \quad (7)$$

Equations 1−7 can be solved simultaneously to obtain the surface charge density and the surface and midplane potentials. The double-layer force per unit area is given by the Langmuir equation,[1]

$$p_{DL} = 2C_e kT\left(\cosh\left(\frac{e\psi_m}{kT}\right) - 1\right) \quad (8)$$

and the double-layer interaction energy per unit area, between two parallel plates at a distance H apart, is obtained by integrating the double layer force:

$$U_{DL,p}(H) = \int_H^{\infty} p_{DL}(x)\,dx \quad (9)$$

For two identical spheres of radius a separated by the distance of closest approach z, the interaction energy is given, in the Derjaguin approximation, by[1]

$$U_{DL,s}(z) = \pi a\int_z^{\infty} U_{DL,p}(H)\,dH \quad (10)$$

II.D. The van der Waals Attraction. For the van der Waals attraction between identical spheres of radius a, separated by the distance of closest approach z, we will use the formula derived by Hamaker:[14]

$$U_{vdW} = -\frac{A_H}{6}\left\{\frac{2a^2}{(z+2a)^2-(2a)^2} + \frac{2a^2}{(z+2a)^2} + \ln\left(\frac{(z+2a)^2-(2a)^2}{(z+2a)^2}\right)\right\} \quad (11)$$

where A_H is the Hamaker constant.

The Lifshitz theory of dispersion forces, which does not imply pairwise additivity and takes into account retardation effects, shows that the Hamaker constant A_H is actually a function of the separation distance. However, for the stability calculations that follow, only the values of the attraction potential at distances less than a few nanometers are relevant, and in this range one can consider that A_H is constant.

II.E. Hydration Interaction. A molecule in liquid water has a first-neighbor coordination number between 4 and 5, which is closer to the coordination number in ice, 4, than to that in a normal liquid, 12.[15] This suggests that the tetrahedral coordination in ice, due to the hydrogen bonding, is almost preserved in the liquid state. To account for this characteristic, Nemethy and Scheraga proposed a model in which the liquid water is composed of icelike clusters of various sizes.[16] It is then reasonable to assume that in the vicinity of a surface the icelike clusters are reorganized in a layered structure, similar to that of ice I, with successive layers containing out-of-plane hexagonal rings of water molecules, parallel to the external surface, as depicted in Figure 1. Each layer is composed of two planar sublayers; a water molecule from a sublayer has three hydrogen bonds with three molecules of the other

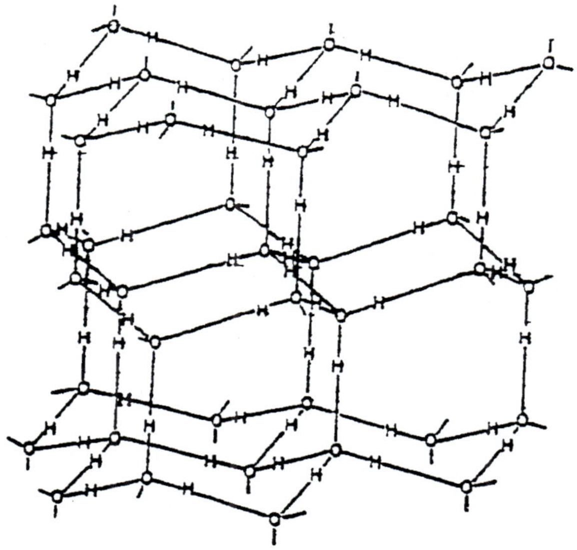

Figure 1. The structure of ice I (reproduced from ref 16). Each site is connected by hydrogen bonding with three sites from the same layer and one site from either one or the other of the neighboring layers. Copyright 1962 by the American Institute of Physics.

sublayer of the same layer and one with a molecule belonging to a sublayer of one of the adjacent layers. While the alignment gradually decreases with the distance from the surface, it is reasonable to consider that for not too large distances this decrease can be neglected.

The treatment of the hydration interaction follows the model employed earlier by Schiby and Ruckenstein,[7] based on the mutual interactions of dipoles. In this model, it is considered that the dipoles of the surface polarize the water molecules of the first layer of water and the polarization propagates from layer to layer. When two surfaces approach each other, the polarized layers will increasingly overlap. As a result, the local dipole moment of the water molecules will be decreased. This increases the free energy of the system and thus generates a repulsive force.

II.E.1. Mutual Polarization of the Water Molecules. Assuming that all the water molecules from the layer i have the same average polarization m_i, normal to the layer, the field E_i^{local}, acting on a site of layer i, due to all the other polarized molecules can be formally written as

$$E_i^{local} = (C_0 m_i + C_1(m_{i-1} + m_{i+1}) + ... + C_k(m_{i-k} + m_{i+k}) + ...) \quad (12)$$

where the coefficient C_k accounts for the contribution of the dipoles of layer $i \pm k$ to the local field at a site of layer i (C_0 corresponds to the field at a site of layer i generated by all the other dipoles of the same layer i). An evaluation of the coefficients C will be provided in section II.E.4.

If there is no external electric field and the molecules of the first water layer have an average dipole moment m_1 oriented perpendicular to the layer, caused by the surface dipoles, the average dipole moment of the molecules of the layer i situated at the distance $x = i\Delta$, $m_i = m(x)|_{x=i\Delta}$ is given by

$$m_i = m(x)|_{x=i\Delta} = \gamma E_i^{t} = \gamma E_i^{local} = \gamma(C_0 m(x) + C_1(m(x-\Delta) + m(x+\Delta)) + ...) \quad (13)$$

where Δ is the distance between the centers of two

(14) Hamaker, H. C. *Physica IV* **1937**, *10*, 1058.
(15) Eisenberg, D.; Kauzmann, W. *The Structure and Properties of Water*; Oxford University Press: New York, 1969.
(16) Nemethy, G.; Scheraga, H. A. *J. Chem. Phys.* **1962**, *36*, 3382.

7064 *Langmuir, Vol. 17, No. 22, 2001* *Manciu and Ruckenstein*

successive layers and γ is the molecular polarizability, as defined and calculated in the Appendix.

An approximate solution can be obtained by expanding $m(x\pm\Delta)$ in series:

$$m(x \pm \Delta) = m(x) \pm \frac{\mathrm{d}m(x)}{\mathrm{d}x}\Delta + \frac{1}{2}\frac{\mathrm{d}^2m(x)}{\mathrm{d}x^2}\Delta^2 \pm ...$$

Because of symmetry the odd derivatives cancel, and when the terms of order 4 and higher are neglected, eq 13 becomes

$$m(x)(1 - \gamma(C_0 + 2C_1 + 2C_2 + ...)) = \gamma\Delta^2(C_1 + 4C_2 + ...)\frac{\mathrm{d}^2m(x)}{\mathrm{d}x^2}$$

which is of the form

$$\frac{\mathrm{d}^2m(x)}{\mathrm{d}x^2} - \frac{1}{\lambda^2}m(x) = 0 \qquad (14)$$

with a decay length λ given by

$$\lambda = \sqrt{\frac{(C_1 + 4C_2 + 9C_3 + ...)\gamma\Delta^2}{1 - (C_0 + 2C_1 + 2C_2 + ...)\gamma}} \qquad (15)$$

II.E.2. Polarization of the First Water Layer. In a homogeneous medium with uniform dielectric properties, an infinite, planar, uniform and continuous distribution of dipoles produces a vanishing electric field. This is due to the long range of the electrostatic interactions, which leads to the cancellation of the strong fields, generated by the nearest dipoles, by the weakest fields produced by the infinite number of remote dipoles. A discrete distribution of planar dipoles generates an oscillating field above the surface, whose average vanishes.

However, the dielectric constant of the interactions of the water molecule from the first layer with the nearest dipole of the surface, denoted ϵ_1, is expected to be much smaller than the bulk dielectric constant of water, ϵ. Denoting as E^{d} the field generated (in a vacuum) by a surface dipole, the field generated by all the dipoles of the surface is

$$E = \frac{1}{\epsilon_1}E^{\mathrm{d}}_{\mathrm{nearest}} + \frac{1}{\epsilon}\sum_{\mathrm{all\ others}} E^{\mathrm{d}} = \frac{\epsilon - \epsilon_1}{\epsilon\epsilon_1}E^{\mathrm{d}}_{\mathrm{nearest}} + \frac{1}{\epsilon}\sum_{\mathrm{all\ dipoles}} E^{\mathrm{d}} = \frac{\epsilon - \epsilon_1}{\epsilon\epsilon_1}E^{\mathrm{d}}_{\mathrm{nearest}} \equiv \frac{1}{\epsilon'}E^{\mathrm{d}}_{\mathrm{nearest}} \qquad (16)$$

where it was assumed that the interaction with the nearest dipole is screened by a medium with the dielectric constant ϵ_1, while all the other interactions are screened by a medium with the dielectric constant ϵ. Equation 16 accounts for the fact that the average field of all the dipoles distributed on the interface, $(1/\epsilon)\sum_{\mathrm{all\ dipoles}} E^{\mathrm{d}}$, in a medium with constant ϵ vanishes. This implies that the net average field acting on the first layer can be calculated by considering only the field generated by the nearest dipole, in a medium of effective dielectric constant ϵ'.

After some distance from the surface, the fields of all dipoles of the interface are screened by the intervening water molecules. Hence, the spatial average of the field induced by the surface dipoles vanishes, since the depolarization field of the remote dipoles is no longer negligible compared to the polarization field of the nearest dipoles.

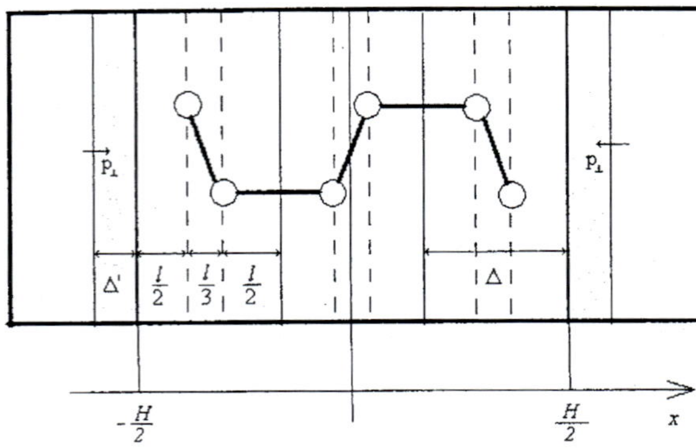

Figure 2. Sketch of three water layers between parallel plates. The dotted lines represent the two planar sublayers of the layers. Only two hydrogen bonds (thick lines) are shown for each water molecule (circles): one with a molecule from the other sublayer of the same layer and one with a molecule from an adjacent layer. The distance between two water molecules is denoted by l; Δ is the distance between two water layers, and Δ' is the distance between the center of the surface dipoles and the interface.

It will be assumed that the field induced by the surface dipoles becomes negligible after the first water layer.

The interfaces are located at the external boundaries of the first organized water layers, while the surface dipoles are situated at a distance Δ' below the interface (see Figure 2). There are also water molecules absorbed among the headgroups of the surface, which do not have the icelike structure but are organized differently, due to the strong interactions with the surfactant headgroups. The high local electric field generated by a dipole of the surface, as well as the small number of water molecules which are screening the water molecules of the first layer, make the local dielectric constant, ϵ_1, near the dipole small.

Let us approximate the area A per surface dipole, which is polarized by the corresponding dipole, by a disk of radius R, located at a distance Δ' from the dipole $\vec{p}$, which has the components $p_\parallel$ and $p_\perp$ parallel and normal, respectively, to the surface.

The field produced by $p_\parallel$ vanishes because of symmetry, and so does the electric field parallel to the surface. The electric field, normal to the surface, generated by the dipole, at a point whose position vector makes an angle θ with the normal, is given by

$$E^{\mathrm{d}}_\perp = \frac{1}{4\pi\epsilon'\epsilon_0 r^3}\{p_\perp(3\cos^2(\theta) - 1)\} \qquad (17)$$

where $r = \Delta'/\cos(\theta)$. For the average field produced by a surface dipole, one obtains

$$\bar{E} = \frac{1}{\pi R^2}\int_0^R \frac{2\pi\rho\,\mathrm{d}\rho}{4\pi\epsilon'\epsilon_0(\Delta'^2 + \rho^2)^{3/2}}\left\{p_\perp\left(3\frac{\Delta'^2}{(\Delta'^2 + \rho^2)} - 1\right)\right\} = \frac{p_\perp}{\epsilon'}\frac{1}{2\pi\epsilon_0(R^2 + \Delta'^2)^{3/2}} \qquad (18)$$

II.E.3. The Hydration Interaction. The solution of eq 14 for two planar, parallel surfaces separated by the distance H (hence for the boundary condition $m(-H/2) = m_1, m(H/2) = -m_1$) is

$$m(x) = -m_1\frac{\sinh\left(\frac{x}{\lambda}\right)}{\sinh\left(\frac{H}{2\lambda}\right)} \qquad (19)$$

where m_1 is the dipole moment of the water molecules at the interface, which is proportional to the total field $E^t = \bar{E} + E_1^{local}$:

$$m_1 = \gamma(\bar{E} + E_1^{\text{local}}) = \gamma(\bar{E} + m_1C_0 + m_2C_1 + ...) \quad (20)$$

Since from eq 19, $m_2 = m[-(H/2) + \Delta] = m_1 \sinh[(H - 2\Delta)/2\lambda]/\sinh(H/2\lambda)$, eq 20 becomes

$$m_1(H) = \frac{\gamma\bar{E}}{\left(1 - \gamma\left(C_0 + C_1\frac{\sinh\left(\frac{H-2\Delta}{2\lambda}\right)}{\sinh\left(\frac{H}{2\lambda}\right)} + ...\right)\right)} \quad (21)$$

The electrostatic energy of a water molecule with the polarizability γ, which acquires the polarization m in the field E_e, is

$$U = -\int_0^{E_e} m(E)\,\mathrm{d}E = -\int_0^{E_e}\gamma E\,\mathrm{d}E = -\frac{1}{2}\gamma E_e^{\,2} = -\frac{m^2}{2\gamma}$$

The electrostatic interaction energy per unit area, between the two parallel plates, is given by

$$U_{\mathrm{H,p}}(H) = -\frac{1}{2S}\sum_{\text{all molecules}}\frac{(m(x))^2}{\gamma} = -\frac{1}{2S}\int_{\text{Volume}}\frac{(m(x))^2}{\gamma v}\mathrm{d}V = -\frac{1}{2\gamma v}\int_{-H/2}^{H/2}(m(x))^2\,\mathrm{d}x = \frac{\gamma\bar{E}^2}{\left(1-\gamma\left(C_0 + C_1\frac{\sinh\left(\frac{H-2\Delta}{2\lambda}\right)}{\sinh\left(\frac{H}{2\lambda}\right)} + ...\right)\right)^2}\frac{H - \lambda\sinh\left(\frac{H}{\lambda}\right)}{4v\sinh^2\left(\frac{H}{2\lambda}\right)} \quad (22)$$

where V is the volume occupied by all the icelike organized water layers, v is the volume occupied by a water molecule, and S is the surface area of a plate. Here, we will assume that the electrostatic energy of the remaining water molecules, located among the headgroups of the surfactant molecules, does not depend on the separation distance H between plates and hence does not contribute to the hydration interaction.

The interaction energy is obtained by subtracting the electrostatic energy at large separation distances, $U_{\mathrm{H,p}}(\infty)$. Using the Derjaguin approximation, the hydration interaction energy between two spheres is given by

$$U_{\mathrm{H,s}}(z) = \pi a\int_z^{\infty}(U_{\mathrm{H,p}}(H) - U_{\mathrm{H,p}}(\infty))\,\mathrm{d}H \quad (23)$$

II.E.4. Evaluation of the Local Field. For a tetrahedral coordination of the water molecules, the distance between the two planar sublayers of the same layer of icelike structure is $(1/3)l$, l being the distance between the centers of two adjacent water molecules, while the distance between the centers of two adjacent layers is $\Delta = (4/3)l$. The vertexes of the tetrahedron formed by the four first neighbors of a water molecule have the length $l\sqrt{8/3}$, while the planar projection of the tilted hexagonal lattice has the side $l\sqrt{8/9}$. The volume occupied in this structure by a water molecule is $v = (8/3\sqrt{3})l^3$.

While the field produced by remote dipoles can be treated as screened by a medium with a large dielectric constant ($\epsilon \cong 80$), the screening of the neighboring dipoles is much weaker. In the present treatment, we will simply assume that E^{local} is produced only by the dipoles located within a radius $2l$ from the given site and that the dielectric constant for them is a constant ϵ''. The electric field caused by a neighboring molecule is given by eq 17 (with ϵ'' replacing ϵ'). It is important to emphasize that the local dielectric constant ϵ'' is smaller than ϵ, the bulk dielectric constant of water.

The electric field $E_{i,0}^{\text{local}}$ generated at a site of layer i by the other dipoles from the same layer is given by

$$E_{i,0}^{\text{local}} = C_0m_i = \left(3\frac{\left(3\left(\frac{1}{3}\right)^2 - 1\right)}{1} + 6\frac{(3(0)^2-1)}{\left(\sqrt{\frac{8}{3}}\right)^3} + 3\frac{\left(3\left(\frac{1}{\sqrt{33}}\right)^2 - 1\right)}{\left(\sqrt{\frac{11}{3}}\right)^3}\right)\frac{m_i}{4\pi\epsilon_0\epsilon'' l^3} = -\frac{3.7663}{4\pi\epsilon_0\epsilon'' l^3}m_i \quad (24)$$

The field generated by the neighbors (within a radius $2l$ from the given site) located in the adjacent water layers $i \pm 1$, averaged over the two sublayer sites of layer i, is given by

$$E_{i,\pm1}^{\text{local}} = C_1m_{i\pm1} = \frac{\left(2 + 3\frac{3\left(\sqrt{\frac{2}{3}}\right)^2 - 1}{\left(\sqrt{\frac{8}{3}}\right)^3} + 6\frac{3\left(\sqrt{\frac{3}{11}}\right)^2 - 1}{\left(\sqrt{\frac{11}{3}}\right)^3}\right) + \left(\frac{2}{\left(\frac{5}{3}\right)^3} + 3\frac{3\left(\sqrt{\frac{2}{3}}\right)^2 - 1}{\left(\sqrt{\frac{8}{3}}\right)^3}\right)}{2} \times \frac{m_{i\pm1}}{4\pi\epsilon''\epsilon_0 l^3} = \frac{1.8272}{4\pi\epsilon''\epsilon_0 l^3}m_{i\pm1} \quad (25)$$

Only the contributions from the first 26 neighbors (12 from the same layer and 14 from the two adjacent layers) were taken into account. The field produced by the more remote dipoles is neglected, because it is screened by the intervening water molecules.

III. Results

III.1. General Behavior of the Interaction Energy. We will try first to obtain some general information about the interaction energy between two identical spheres of radius a, separated by the distance of closest approach z, at high electrolyte concentrations, using some simple approximations. The following expressions will be used for the interaction energies:

$$U_{\mathrm{vdW,s}} = -\frac{A_{\mathrm{H}}a}{12z} \quad (26a)$$

$$U_{\mathrm{H,s}} = \pi a\lambda B\exp\left(-\frac{z}{\lambda}\right) \quad (26b)$$

which are rough approximations of the interaction energies obtained with the methods described above for the van der Waals and hydration interactions (we approximate the hydration interaction energy per unit area between two parallel plates at distance x apart by $U_{\mathrm{H,p}} = B\exp(-x/\lambda)$). When the double-layer interaction at high ionic strengths is neglected, the interaction energy has an

7066 *Langmuir, Vol. 17, No. 22, 2001* *Manciu and Ruckenstein*

absolute minimum at $z = 0$ (primary minimum) and a local minimum at some finite z (secondary minimum), separated by a potential barrier. The extrema of the interaction energy are obtained from the condition of vanishing of the first derivative,

$$\frac{dU(z)}{dz} = -\pi a B \exp\left(-\frac{z}{\lambda}\right) + \frac{A_H a}{12 z^2} = 0$$

which can be rewritten as

$$F(\xi) \equiv \xi^2 \exp(-\xi) = \frac{A_H}{12\pi B \lambda^2} \quad (27)$$

where $\xi = z/\lambda$. $F(\xi)$ has a maximum at $\xi = 2$, $F(2) = 2^2 \exp(-2)$. Hence, if $B < B^* = A_H/[48\pi\lambda^2 \exp(-2)]$, eq 27 has no solution and there is no potential barrier to prevent coagulation. When $B > B^*$, eq 27 has two solutions, the smaller one corresponding to the maximum and the larger one to the secondary minimum of the energy. Another useful value for the preexponential factor B of hydration is obtained from the condition of vanishing of the energy maximum,

$$U(z) = \pi a \lambda B \exp\left(-\frac{z}{\lambda}\right) - \frac{A_H a}{12 z} = 0 \quad (28)$$

which, together with eq 27, leads to

$$z_{max} = \lambda \quad (29a)$$

and

$$B = B_0 = \frac{A_H \exp(1)}{12\pi\lambda^2} \quad (29b)$$

Although in this case $U(z_{max}) - U(\infty) = 0$, the secondary minimum can be sufficiently deep for the potential barrier between the latter and the maximum to hinder the coagulation at the primary minimum. By substitution of eq 29b in eq 27, the positions of the two extrema satisfy the equation

$$\xi^2 \exp(-\xi) = \exp(-1) \quad (30)$$

with the solutions $\xi = \xi_1 = 1$ for the maximum (eq 29a) and $\xi = \xi_2 = 3.5129$ for the minimum. Therefore,

$$z_{min} = \xi_2 \lambda \quad (31a)$$

and

$$U(z_{min}) = -\frac{A_H a}{12\lambda}\left(\frac{\xi_2 - 1}{\xi_2^{\,2}}\right) \quad (31b)$$

By use of some typical values for the parameters, namely, $A_H = 1 \times 10^{-20}$ J, $a = 1000$ Å, and $\lambda = 2.96$ Å, for an absolute temperature $T = 300$ K, the depth of the minimum is about $-13.8\ kT$, and hence the transition from the secondary to the primary minimum is very slow. The depth of the minimum increases as the radius a and the Hamaker constant A_H increase.

We will now examine the consequences of a small increase in the hydration force, by assuming that the preexponential factor becomes $B = B_0(1 + \eta)$, with η small compared to 1. Equation 27 acquires the form

$$F(\xi) = \xi^2 \exp(-\xi) = \frac{\exp(-1)}{(1 + \eta)} \quad (32)$$

Because the perturbation η is small, $F(\xi)$ can be replaced around ξ_1 and ξ_2 by the linear approximations

$$F_1(\xi) = \frac{\xi}{\exp(1)} \quad (33a)$$

$$F_2(\xi) = \frac{\xi(2 - \xi_2) + \xi_2^{\,2} - \xi_2}{\xi_2 \exp(1)} \quad (33b)$$

For the maximum, the linear approximation (eq 33a) inserted in eq 32 provides

$$z_{max} = \frac{\lambda}{1 + \eta} \quad (34a)$$

and

$$U(z_{max}) = \frac{A_H a}{12\lambda}(\eta + \eta^2) \quad (34b)$$

For the values of the parameters employed above, the factor in the front of the parenthesis is about 68 kT, and consequently even a small η produces a large increase of the energy of the maximum.

For the minimum, eqs 33b and 32 lead to

$$z_{min} = \frac{\lambda\xi_2}{\xi_2 - 2}\left(\xi_2 - \frac{2 + \eta}{1 + \eta}\right) \quad (36a)$$

and

$$U(z_{min}) = \frac{A_H a}{12\lambda}\left(-\frac{\xi_2 - 1}{\xi_2^{\,2}} + \frac{\eta}{\xi_2^{\,2}} + O(\eta^2)\right) \quad (36b)$$

where the terms in η^2 and higher were neglected. The first term represents the depth of the minimum for $\eta = 0$. Because $\xi_2^2 = 12.34$, the increase in the energy of the secondary minimum, when the hydration increases, is about 1 order of magnitude smaller than the corresponding increase of the maximum energy. For small values of η, the increase is negligible compared with the depth of the unperturbed secondary minimum ($\xi_2 - 1 \cong 2.5129 \gg \eta$).

As general conclusions, one can notice the following: (1) if the preexponential factor B is smaller than the critical value B^*, there is no potential barrier which can prevent aggregation; (2) if $B \gg B_0$, the maximum of the interaction energy is large and no aggregation takes place at the primary minimum but can take place at the secondary minimum; and (3) small changes of B around B_0 dramatically affect the value of the maximum, but much less the minimum. Because B is proportional to $(p_\perp/\epsilon')^2$, the behavior of the system is expected to be very sensitive to changes in the ratio $p_\perp/\epsilon'$.

III.2. Numerical Calculations. In what follows, the equations from section II will be employed to calculate the effect of the hydration force on the stability of colloids or emulsions in water in the presence of NaCl and SDS, which represent a common electrolyte and surfactant, respectively.

For the adsorption isotherm, the following values of the parameters will be used: $b_1 = 881$ m^3/mol, $b_2 = -1.53$, and $\Gamma_\infty = 5 \times 10^{-6}$ mol/m^2. These coefficients were obtained from experimental data at high electrolyte concentrations.[17] While the experiments have been carried out for a water–air interface, the coefficients probably provide good approximations also for the present case, because the hydrophobic bonding (the disorganization of water by the presence of surfactant) constitutes the main driving force for adsorption. The equilibrium constant for the association–dissociation at the interface is not known. The only related relevant information is the dissociation constant of the ion pair $NaSO_4^-$, which was estimated to be $K_D = 10^{-0.7} = 0.1995$ mol/dm^3.[18] Because the repulsion among the headgroups increases as the dissociation increases (and this provides a positive contribution to the free energy), the dissociation constant is expected to be lower in this case.

A discussion regarding the molecular polarizability is provided in the Appendix. In the present paper, we will use for the molecular polarizability expression A.7, which provides a lower bound for γ, in the absence of saturation effects. When eq A.7 is inserted into eq 15, the decay length acquires the simple form ($C_k = 0$ for $k > 1$)

$$\lambda = \sqrt{\frac{3\epsilon_0 v(\epsilon - 1)C_1\Delta^2}{\epsilon + 2}} \qquad (37)$$

We will use in the calculations l = 2.76 Å (which corresponds to the distance between molecules in the structure of ice I, as compared to about 2.9 Å for molecules in water), and $\epsilon = 80$. For the local dielectric constant, we will assume $\epsilon'' = 1$, which constitutes a lower bound. In a perfect tetrahedral coordination, the average distance between two successive water layers is $\Delta = (4/3)l$, and the decay length of the hydration interaction calculated using eq 37 is $\lambda = 2.96$ Å. It should be noted that the latter value is in the range determined experimentally for the hydration force between phospholipid bilayers.[4] Lower values of λ can be obtained for higher ϵ''. For the distance between the center of the ion pair and the interface (located at the boundary of the first organized water layer), the value Δ' = 1.0 Å was selected.

The model employed for the hydration force implies a continuum, even for low separation distances between particles. It was also assumed that the hydration, double-layer, and van der Waals interactions are additive. The double layer affects the hydration force mainly because of the decrease of the molecular polarizability by large electric fields. However, in the present case, in the vicinity of the surface the electric field due to the double layer is smaller than $\bar{E}$ by an order of magnitude and hence its influence on the hydration force can be neglected at small separation distances.

The magnitude of the hydration force is proportional to the ratio $(p_\perp/\epsilon')^2$, which can vary in a quite large range. The dipole moments of the common salt molecules are of the order of 10 D (e.g., 8.5 D for NaCl and 10.4 D for CsCl);[19] the values are expected to be larger for the ion pairs on the particle surface, because the cations are hydrated. The value of ϵ' is expected to depend strongly on the particular organization of the water molecules absorbed among the headgroups of the surface.

To have some feelings about their order of magnitude, the ratio Γ/Γ_∞, the dissociation constant α, the area per ion pair A, and the average dipole moment of the first water layer m_1 are plotted in Figure 3a–d as functions of electrolyte concentration for $C_s = 0.001$ mol/dm^3, $(p_\perp/\epsilon') =$ 1.0 D, large distances between particles or droplets ($z \rightarrow \infty$), and various dissociation constants K_D, ranging between 0.05 and 0.5 mol/dm^3. The low dissociation of the surfactant headgroups is a consequence of the low value selected for the equilibrium dissociation constant K_D; however, the results are not significantly modified even when K_D varies with 1 order of magnitude. In what follows, the following values will be used for some of the parameters: $K_D = 0.1$ mol/dm^3, $b_1 = 881$ m^3/mol, $b_2 = -1.53$, $\Gamma_\infty = 5 \times 10^{-6}$ mol/m^2, $\lambda = 2.96$ Å, $\epsilon = 80$, $\Delta' = 1.0$ Å, and $C_s = 0.001$ mol/dm^3. The other parameters, namely, a, A_H, and the ratio $(p_\perp/\epsilon')$, will be varied.

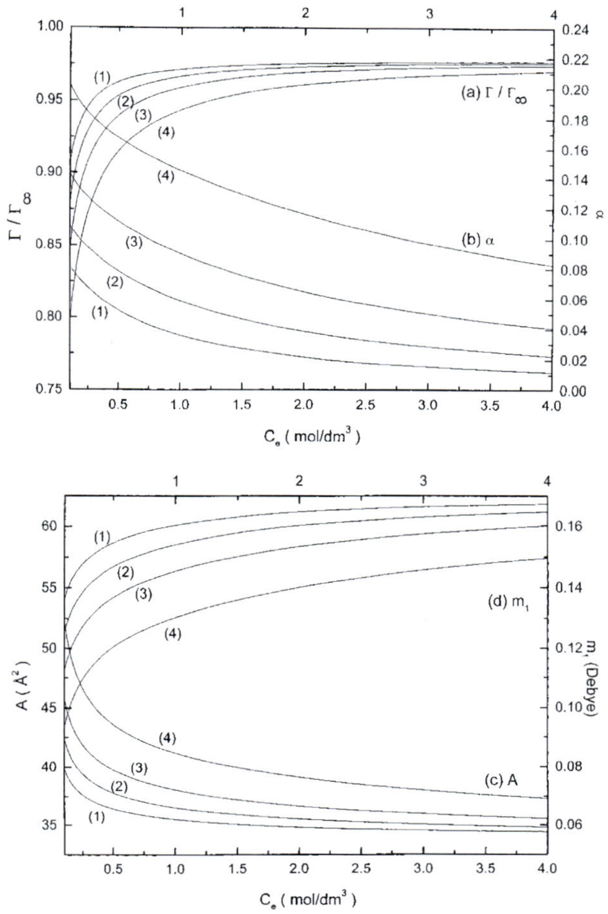

Figure 3. (a) The ratio Γ/Γ_∞, (b) the dissociation constant α, (c) the area per ion pair A, and (d) the polarization of the water molecules near the interface, m_1, as functions of the electrolyte concentration C_e for various values for the dissociation constant: (1) K_D = 0.05 mol/dm^3; (2) K_D = 0.1 mol/dm^3; (3) K_D = 0.1995 mol/dm^3; (4) K_D = 0.5 mol/dm^3.

The values of the interaction energy at the maximum and at the secondary minimum are proportional with the radius a of the particle or droplet and depend strongly on the Hamaker constant A_H and on the hydration repulsion. As shown in the previous section, small modifications in the ratio $(p_\perp/\epsilon')$ (and hence in the hydration repulsion) can lead to a large increase in the potential barrier between the primary and secondary minima, thus affecting the stability of the system.

First, calculations will be presented regarding the interaction energy between two identical spherical particles of radius a = 100 Å, Hamaker constant $A_H = 2.0 \times 10^{-20}$ J, and $p_\perp/\epsilon' = 2.6$ D. The interaction energy is plotted as a function of distance for electrolyte concentrations ranging from 0.1 to 4.0 mol/dm^3 (Figure 4a). At relatively low electrolyte concentrations, the height of the barrier

(17) Feinerman, V. B. *Colloids Surf.* **1991**, *57*, 249.
(18) Davies, C. W. *Ion Association*; Butterworth: London, 1962.
(19) Landolt-Börnstein, Springer, Heidelberg, 1985.

7068 *Langmuir, Vol. 17, No. 22, 2001* *Manciu and Ruckenstein*

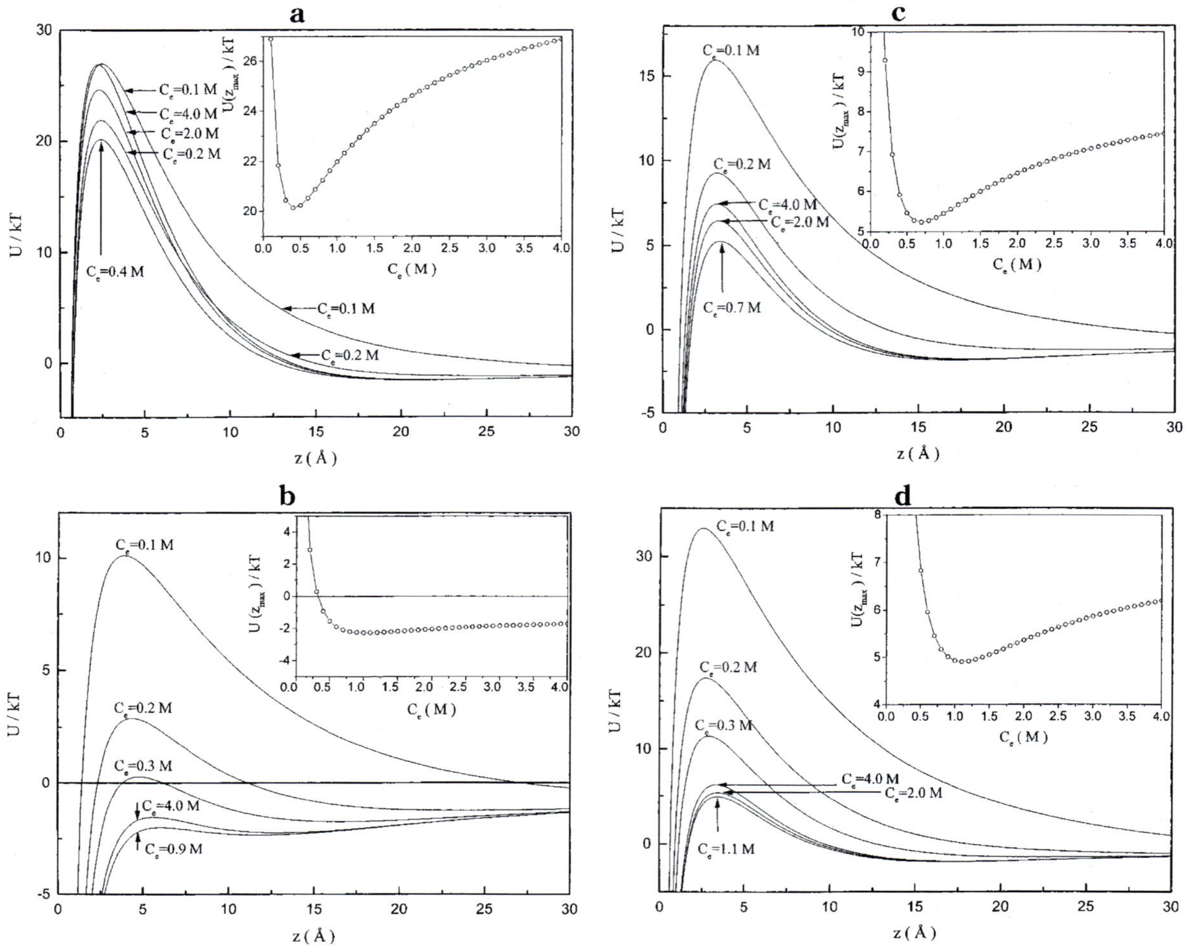

Figure 4. (a) The interaction energy as a function of separation distance, for a = 100 Å, $p_{\perp}/\epsilon'$ = 2.6 D, A_H = 2.0 × 10^{-20} J, and various electrolyte concentrations. In the inset, the maximum value of the interaction energy is plotted versus electrolyte concentration. (b) The interaction energy as a function of separation distance, for a = 100 Å, $p_{\perp}/\epsilon'$ = 1.3 D, A_H = 2.0 × 10^{-20} J, and various electrolyte concentrations. In the inset, the maximum value of the interaction energy is plotted versus electrolyte concentration. (c) The interaction energy as a function of separation distance, for a = 100 Å, $p_{\perp}/\epsilon'$ = 1.9 D, A_H = 2.0 × 10^{-20} J, and various electrolyte concentrations. In the inset, the maximum value of the interaction energy is plotted versus electrolyte concentration. (d) The interaction energy as a function of separation distance, for a = 200 Å, $p_{\perp}/\epsilon'$ = 1.3 D, A_H = 1.0 × 10^{-20} J, and various electrolyte concentrations. In the inset, the maximum value of the interaction energy is plotted versus electrolyte concentration.

ΔU, which is measured with respect to an infinite separation distance, $\Delta U = U(z_{max}) - U(\infty)$, decreases with increasing electrolyte concentration, as predicted by the DLVO theory. However, after reaching a minimum at C_e = 0.4 M, the barrier slowly increases with increasing ionic strength (the inset presents the maximum value of the energy as a function of electrolyte concentration). Since in this case the height of the barrier is always higher than 20 kT, the system is stable at any electrolyte concentration.

In Figure 4b, the interaction energy is calculated using $p_{\perp}/\epsilon'$ = 1.3 D. At low electrolyte concentrations, the height of the potential barrier decreases with increasing C_e, because of the screening of the double layer. The increase in the hydration force with increasing C_e, due to the reassociation, is not sufficient to generate a high enough potential barrier to prevent the rapid coagulation. For electrolyte concentrations larger than about 0.25 mol/dm^3, the system is unstable.

Figure 4c provides U/kT for $p_{\perp}/\epsilon'$ = 1.9 D. The maximum of the interaction energy decreases with increasing electrolyte concentration and reaches a minimum of 5.26 kT at C_e = 0.7 mol/dm^3. However, with a further increase of the ionic strength the maximum energy gradually increases and at C_e = 4.0 mol/dm^3 it becomes 7.45 kT. The rate of aggregation at 4.0 mol/dm^3 is hence decreased by a factor of the order of exp(7.45 − 5.26) = 8.93, as compared to the maximum aggregation rate, obtained at C_e = 0.7 mol/dm^3. There is therefore a restabilization of the system with increasing electrolyte concentration.

The calculations were repeated for a system with a = 200 Å, A_H = 1.0 × 10^{-20} J, and $p_{\perp}/\epsilon'$ = 1.5 D. As shown in Figure 4d, the minimum potential barrier is reached now at C_e = 1.1 mol/dm^3. The rate of aggregation is in this case about 3.7 times slower at C_e = 4.0 mol/dm^3 than at C_e = 1.1 mol/dm^3.

The matters become more complex for large particles and large Hamaker constants. Figure 5a presents the interaction energy for a = 1000 Å, A_H = 5.0 × 10^{-20} J, and $p_{\perp}/\epsilon'$ = 2.45 D (while keeping the values of the other parameters as before). The maximum of the interaction energy becomes negative at C_e ~ 0.3 mol/dm^3, reaches a minimum at C_e ~ 0.5 mol/dm^3, and then strongly increases, becomes again positive at C_e ~ 1.0 mol/dm^3, and exceeds 30 kT at C_e = 4.0 mol/dm^3. However, as anticipated in the section where the general behavior was examined, the secondary minimum is extremely deep in this case. The increase of the hydration force with increasing ionic strength has only a moderate effect on the depth of the

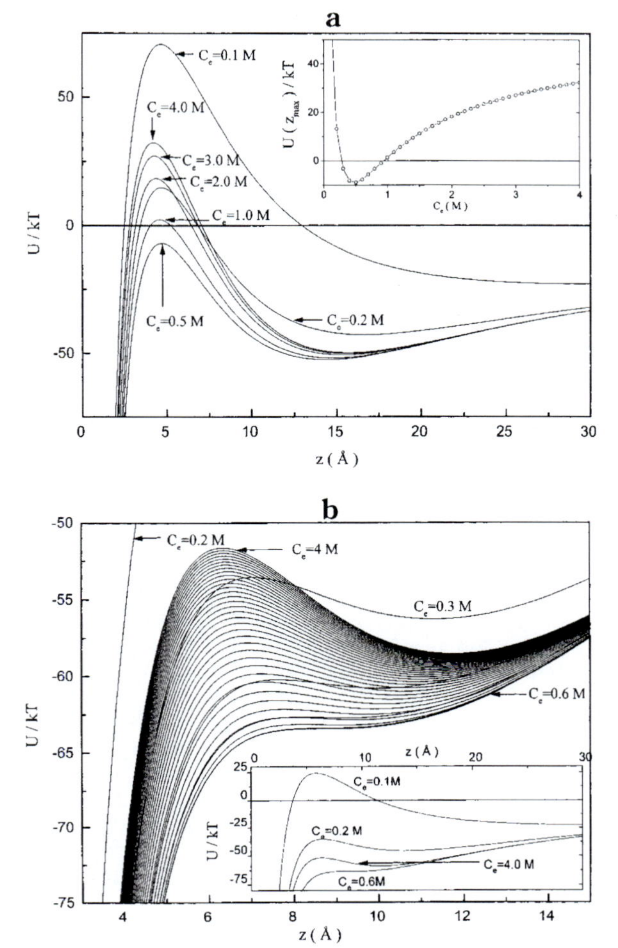

Figure 5. (a) The interaction energy as a function of separation distance, for $a = 1000$ Å, $p_{\perp}/\epsilon' = 2.45$ D, $A_H = 5.0 \times 10^{-20}$ J, and various electrolyte concentrations. In the inset, the maximum value of the interaction energy is plotted versus electrolyte concentration. (b) The interaction energy as a function of separation distance, for $a = 1000$ Å, $p_{\perp}/\epsilon' = 1.95$ D, and $A_H = 5.0 \times 10^{-20}$ J for various electrolyte concentrations, in the region where a potential barrier between the primary and the secondary minimum is formed at large electrolyte concentrations. The electrolyte concentration is varied in steps of 0.1 M. The interaction energy is plotted over larger ranges in the inset.

secondary minimum (see eq 36b and Figure 5a). Therefore, the particles will coagulate at the secondary minimum for all the electrolyte concentrations considered, with the exception of a narrow range around $C_e \sim 0.5$ mol/dm^3. At concentrations of the order of 0.5 mol/dm^3, the particles can overcome the barrier and aggregate at the primary minimum if the frictional losses are small enough.

If the frictional losses are large, an interesting situation occurs at a lower value of $p_{\perp}/\epsilon'$ than that used in Figure 5a, namely, $p_{\perp}/\epsilon' = 1.95$ D (see Figure 5b). When $C_e = 0.5$ mol/dm^3, there is no potential barrier and rapid coagulation at the primary minimum occurs. However, a relatively large potential barrier of the order of 10 kT between the primary and the secondary minima ($\Delta U = U(z_{max}) - U(z_{min}$)) is established at very large electrolyte concentrations ($C_e > 4$ mol/dm^3). In this case, the aggregation occurs at the secondary minimum.

There is some experimental evidence that for some colloidal systems, the rate of aggregation has a maximum and then decreases with increasing electrolyte concentration. In the experiment of Alfridsson et al.,[2] the rate constant of aggregation for a suspension of charged paraffin particles (in the presence of a long-chain carboxylic acid, C_{22}) increased with the concentration of NaCl at low ionic strength, reached a maximum at about 0.5 mol/dm^3 NaCl, and then decreased for higher electrolyte concentrations.

There are also colloidal systems (e.g., silica) whose stability at any electrolyte concentration can be explained by a large hydration repulsive force.[1] It is interesting to note that there are also systems whose stability depends on the nature of the electrolyte. Healy et al.[12] reported that amphoteric latex particles coagulated when the concentration of $CsNO_3$ exceeded 0.8 mol/dm^3. However, they remained stable even when concentrations up to 3.0 mol/dm^3 of KNO_3 or $LiNO_3$ were employed. The effect was qualitatively interpreted in terms of a "hydration barrier"; to allow close particle–particle aggregation, the cations adsorbed on the surface should be dehydrated. In the present paper, the hydration force is considered as originating from the ability of the ion pairs formed on the surface to polarize the water layers. The cation Cs^+ can be easily dehydrated because its heat of hydration is relatively low. In contrast, the cations Li^+ and K^+ have higher heats of hydration and it is difficult to dehydrate them. As a result, the dipole moment of Cs is smaller than those of Li and K, and hence the hydration repulsion is weaker for the former cation. As our calculations indicate, changes of the order of 10% in the ratio $p_{\perp}/\epsilon'$ might be sufficient to convert an unstable system into a stable one at high electrolyte concentrations. A detailed treatment of the problem will be presented in a forthcoming paper.

IV. Conclusions

The surface charge is responsible for the double-layer repulsion, which is attenuated by increasing the ionic strength. However, the ion pairs formed through the association–dissociation of the charges with the counterions induce a repulsive hydration force that increases with ionic strength. At the same surface density of ion pairs, the magnitude of the hydration force is related to the dipole moment, normal to the surface, of the surface ion pair, and to the local dielectric constant in the vicinity of the headgroups. A small value of $p_{\perp}/\epsilon'$ corresponds to a system which coagulates when the electrolyte concentration exceeds a critical value, as predicted by the DLVO theory. If $p_{\perp}/\epsilon'$ is large, the colloidal or emulsion system can remain stable at any ionic strength.

At intermediate values of $p_{\perp}/\epsilon'$ and for small particles or droplets and low van der Waals forces, the increase in the hydration force with electrolyte concentration can reduce the rate of aggregation by 1 order of magnitude; for large particles or droplets, the increase in hydration force with the ionic strength can increase the maximum interaction energy by more than 30 kT. In the latter case, the coagulation at the primary minimum can occur at intermediate electrolyte concentrations but at large electrolyte concentrations can be prevented by the increase of the hydration repulsion with increasing ionic strength. The particles can however aggregate at the secondary minimum, which is deep.

Appendix: Evaluation of the Molecular Polarizability

We considered two models for the evaluation of the molecular polarizability in the condensed phase.

One of the models follows the Lorentz–Debye theory, as summarized by Jackson[20] and Frenkel.[21] In a constant macroscopic field E_0, the macroscopic polarization P is

7070 *Langmuir, Vol. 17, No. 22, 2001* *Manciu and Ruckenstein*

given by

$$P = \epsilon_0(\epsilon - 1)E_0 \qquad (A.1)$$

where ϵ is the macroscopic dielectric constant.

The effective field E_e is the sum between the macroscopic field E_0 and the internal field E_{int}, due to the other molecules of the medium. The latter can be calculated by adding the individual contributions of the other molecules close to the selected one, E^{local}, and subtracting the contribution from the same molecules treated in an average continuum approximation described by the polarization P:

$$E_{int} = E^{local} - E_P = E^{local} + \frac{P}{3\epsilon_0} \qquad (A.2)$$

The field E^{local} at a site of an icelike lattice generated by the dipoles located at the other sites of the lattice, assuming that all the water molecules have the average dipole moment m, normal to the layer, is given by

$$E^{local} = m(C_0 + 2C_1 + ... + 2C_k + ...) \qquad (A.3)$$

where the coefficient C_k accounts for the contribution of the dipoles of layer $i \pm k$ to the E^{local} at a site of layer i. The average dipole moment m is proportional to the effective field; hence,

$$m = \gamma E_e = \gamma\left(E_0 + \frac{P}{3\epsilon_0} + m(C_0 + 2C_1 + 2C_2 + ...)\right) \qquad (A.4)$$

where γ is the molecular polarizability.

Since m is related to the macroscopic polarization P via $m = Pv$, where v is the volume of a water molecule, eqs A.1 and A.4 provide the following expression for the macroscopic dielectric constant:

$$\epsilon = \frac{1 - \gamma(C_0 + 2C_1 + 2C_2 + ...) + \dfrac{2\gamma}{3\epsilon_0 v}}{1 - \gamma(C_0 + 2C_1 + 2C_2 + ...) - \dfrac{\gamma}{3\epsilon_0 v}} \qquad (A.5)$$

(20) Jackson, J. D. *Classical Electrodynamics*; John Wiley & Sons: New York, 1975.

(21) Frenkel. J. *Kinetic theory of liquids*; Clarendon Press: Oxford, 1946.

The polarizability of a polar molecule has two components: an electronic polarizability, γ_e, due to the displacement of electrons, and an orientational polarizability, γ_d, due to the rotation of the permanent dipole in an electric field. In the vapor phase, assuming that the molecule can rotate freely, the orientational polarizability is given by the Langevin function. Therefore, for low effective electric fields ($pE_e \ll kT$)

$$\gamma = \gamma_e + \gamma_d \cong \gamma_e + \frac{p^2}{3kT} \qquad (A.6)$$

where p is the permanent dipole moment of the molecule.

For vapors, the local field vanishes because of the spherical symmetry and eqs A.5 and A.6 provide good agreement with experiment. However, for liquids one can no longer use eq A.6 for the polarizability in the Lorentz–Debye model. Indeed, for liquid water, eq A.5 diverges for values of γ about 4 times smaller than the value of γ for its vapor, which at 300 K is $\gamma = \gamma_e + \gamma_d = 32.3 \times 10^{-40}$ C^2 m^2 J^{-1}. One can regard eq A.5 as the definition of the molecular polarizability and calculate γ in terms of the macroscopic dielectric constant ϵ. The lower value of the polarizability in liquid than in vapor can be explained in the framework of the Lorentz–Debye model by the hindered rotation of the permanent dipole moment by the neighboring molecules in the condensed state.

In contrast, the Onsager–Kirkwood model provides a polarizability in polar liquids larger than that in vapors.[21] This is a result of the increase of the dipole moment by the strong electric field, which is generated when a molecule with a permanent dipole moment is introduced in a polarizable medium (Onsager), and the correlation between the orientations of neighboring molecules (Kirkwood).[21]

In the present paper, we will use for the molecular polarizability the following expression (obtained from eq A.5):

$$\gamma = \frac{\epsilon_0 v}{\dfrac{\epsilon + 2}{3(\epsilon - 1)} + (C_0 + 2C_1 + 2C_2 + ...)\epsilon_0 v} \qquad (A.7)$$

which provides a lower bound for γ in the absence of saturation. In a nonuniform field, we consider that $m = \gamma E^t$, where E^t is the total field acting on a molecule.

LA010741T

8910 *Langmuir* **2002,** *18,* 8910−8918

Long Range Interactions between Apoferritin Molecules

Marian Manciu and Eli Ruckenstein*

Department of Chemical Engineering, State University of New York at Buffalo, Buffalo, New York 14260

Received June 25, 2002. In Final Form: September 11, 2002

Recent light scattering experiments regarding the apoferritin molecules (the hollow shells of the iron-storage protein ferritin) indicated a surprising dependence of the repulsion between proteins on the electrolyte concentration ($NaCH_3COO$). The second virial coefficient decreased to a value close to that corresponding to hard spheres for 0.15 M but increased to a very large value at 0.25 M. The results are difficult to be interpreted in the classical framework through the addition of double layer and hydration repulsive forces. While the double layer theory can predict the behavior of the virial coefficient at low electrolyte concentrations, only an abnormally large charge can explain the values of the virial coefficient at high ionic strengths. Alternatively, the traditional hydration force should increase with orders of magnitude between 0.15 and 0.25 M, to be consistent with experiment, and this is unlikely to happen. In this paper, it is shown that a unitary treatment of the repulsion (the double layer and the polarization-based hydration repulsions) might explain the unexpected values of the second virial coefficient and the corresponding long-ranged repulsion.

I. Introduction

The interaction between charged particles immersed in an electrolyte solution is traditionally described by the Derjaguin−Landau−Verwey−Overbeek theory in terms of an attraction, due to the correlations between the instantaneous electronic dipoles of the particles (van der Waals force), and a screened Coulomb interaction, due to the charges on the particles' surfaces (the double layer force).[1] It is known that the theory is accurate only in a certain range of electrolyte concentrations (1.0×10^{-3} to 5×10^{-2} M), and a number of improvements were proposed for the calculation of the interactions, such as, for example, the accounting of the retardation[2] and the field-theoretical treatment for the van der Waals interactions[3] and the accounting of the dielectric saturation at high fields,[4] the image forces,[5] finite ion sizes,[6] and the correlation between ions[7] for the double layer interactions.

A long time ago,Voet prepared stable sols of various metals (Pt, Pd) and salts (sulfides, halides) in highly concentrated solutions of sulfuric acid, phosphoric acid, and calcium chloride in water.[8] Dilution with water induced their coagulation. More recent experiments revealed that amphoteric latex particles did not coagulate, even at high ionic strengths (above 1 M) of $LiNO_3$.[9] All the above experiments contradict the DLVO theory. The stability of some colloids at high electrolyte concentrations led to the conjecture that another repulsion (a non-DLVO one), due to the organization of the solvent around particles, should also be present. The existence of such a force was demonstrated both experimentally and theoretically for any solvent,[10] being particularly strong for the polar ones. The force (called hydration force when the solvent is water) is partly responsible for the stability of multilayers of neutral lipid bilayers/water, a system that was extensively investigated because of its relevance to biology.[11]

One of the first quantitative models[12] of the hydration force was based on the polarization of water molecules in the vicinity of a surface. Another model involving polarization, suggested by Gruen and Marcelja,[13] was shown later to have some inconsistencies.[14] Because Monte Carlo simulations[15] indicated that the average polarization of water molecules oscillates in the vicinity of a surface, both polarization models have been contested. However, it was recently shown[16] that the Schiby and Ruckenstein model[12] can lead to an oscillatory profile of the polarization, if the water is assumed to be organized in icelike layers in the vicinity of the surface. This model can relate the strength of the hydration force to the surface dipole density, a dependence which was observed experimentally,[17] and could also explain the restabilization of some colloids at high ionic strengths.[18]

The hydration repulsion is expected to be affected by the hydrogen bonding. When two surfaces approach each other, the increase of the free energy (due to the disruption of the hydrogen bonds) generates repulsion. Indeed, Monte

* Corresponding author. E-mail address: feaeliru@acsu.buffalo.edu. Phone: (716)645-2911/2214. Fax: (716)645-3822.

(1) Verwey, E. J. W.; Overbeek, J. Th. G. *Theory of stability of lyophobic colloids*; Elsevier: Amsterdam, 1948.
(2) Casimir, H. B.; Polder, D. *Phys. Rev.* **1948**, *73*, 360.
(3) Dzyaloshinskii, I. E.; Lifshitz, E. M.; Pitaevskii, L. P. *Adv. Phys.* **1961**, *10*, 165.
(4) Henderson, D.; Lozada-Casou, M. *J. Colloid Interface Sci.* **1986**, *114*, 180.
(5) Jönnson, B.; Wennerström, H. *J. Chem. Soc., Faraday Trans.* **1983**, *79*, 19.
(6) Levine, S.; Bell, G. M. *Discuss. Faraday Soc.* **1966**, *42*, 69. Ruckenstein, E.; Schiby, D. *Langmuir* **1985**, *1*, 612.
(7) Wennerström, H.; Jönnson, B.; Linse, P. *J. Chem. Phys.* **1982**, *76*, 4665. Kjellander, R.; Marcelja, S. *J. Chem. Phys.* **1985**, *82*, 2122.
(8) Voet, A. Thesis, Amsterdam, 1935. See also: Kruyt, H. R. *Colloid Science*; Elsevier: Amsterdam, 1952.
(9) Healy, T. W.; Homola, A.; James, R. O.; Hunter, R. J. *Faraday Discuss. Chem. Soc.* **1978**, *65*, 156.
(10) Derjaguin, B. V.; Churaev, N. V.; Muller, V. M. *Suface Forces*; Plenum: 1987.
(11) Rand, R. P.; Parsegian, V. A. *Biochim. Biophys. Acta* **1989**, *988*, 351.
(12) Schiby, D.; Ruckenstein, E. *Chem. Phys. Lett.* **1983**, *95*, 435.
(13) Gruen, D. W. R.; Marcelja, S. *Faraday Trans. 2* **1983**, 211 and 225.
(14) Attard, P.; Wei, D. Q.; Patey, G. N. *Chem. Phys. Lett.* **1990**, *172*, 69.
(15) Kjellander, R.; Marcelja, S. *Chem. Scr.* **1985**, *25*, 73; *Chem. Phys. Lett.* **1985**, *120*, 393.
(16) Manciu, M.; Ruckenstein, E. *Langmuir* **2001**, *17*, 7582.
(17) Simon, S. A.; McIntosh, T. J. *Proc. Natl. Acad. Sci. U.S.A.* **1989**, *86*, 9263.
(18) Manciu, M.; Ruckenstein, E. *Langmuir* **2001**, *17*, 7061.

10.1021/la026126m CCC: $22.00
Published on Web 10/15/2002

Carlo simulations[15] indicated that a disruption of the hydrogen bonding occurs when two surfaces approach each other, and lattice models for the calculations of such interactions have been proposed.[19] We consider, however, that an important contribution to the hydrogen bonding is already contained in the dipole–dipole interactions included in the polarization model.

The restabilization of colloids at high ionic strengths observed experimentally[20] (the coagulation rate having a maximum with increasing electrolyte concentration) could be explained in terms of the reassociation of charges on the interface and the formation of surface ion pairs (dipoles).[18] While, at low ionic strengths, the repulsion decreases with increasing electrolyte concentration because of the electrostatic screening, at high ionic strengths, the reassociation increases the density of ion pairs on the surface, and this generates a strong repulsion. A restabilization of some colloidal systems at high ionic strength could be explained if the magnitude of the hydration force increases moderately with the addition of electrolyte.[18] There are, however, some recent striking experimental results regarding the interaction between apoferritin molecules,[21] which are more difficult to explain.

Light scattering experiments[21] on solutions of apoferritin molecules (the hollow shells of ferritin, an iron-storage globular protein) in an acetate buffer at pH = 5.0 have indicated that the dimensionless second virial coefficient (the ratio between the second virial coefficient and the volume of the molecule) first decreases with increasing electrolyte concentration, becomes close to 4 (value corresponding to hard-core interactions alone) for an electrolyte concentration in the range 0.08–0.18 M, but increases markedly at higher ionic strengths, reaching a very large value (~13) at 0.25 M.

The dimensionless second virial coefficient is defined as

$$\tilde{B}_2 = \frac{3}{2a^3}\int_{r=0}^{\infty} r^2\left(1 - \exp\left(-\frac{F(r)}{kT}\right)\right) dr = 4 + \frac{3}{2a^3}\int_{r=2a}^{\infty} r^2\left(1 - \exp\left(-\frac{F(r)}{kT}\right)\right) dr \quad (1)$$

where a = 63.5 Å represents the radius of the spherical apoferritin molecules, r the distance between the centers of two particles, k the Boltzmann constant, T the absolute temperature, and $F(r)$ the interaction potential between particles, which was assumed to be the sum between a hard core repulsion ($F(r) = \infty$ for $r < 2a$) and the other interactions (double layer, hydration and van der Waals). The second virial coefficient is not very sensitive to the magnitude of the repulsion, as long as the interaction energy exceeds a few kT, but is extremely sensitive to the range of the interaction (the distance between the centers of the particles at which the interaction becomes comparable to kT).

At low electrolyte concentrations, the experimental results can be explained within the DLVO theory. A well-known approximation for the double layer interaction between weakly charged spheres, at constant surface charge, is[1]

$$F_{\rm DL}(r) = \frac{(ne)^2 \exp\left(-\frac{r - 2a}{\lambda_{\rm D}}\right)}{4\pi\epsilon_0\epsilon\left(1 + \frac{a}{\lambda_{\rm D}}\right)^2 r} \quad (2)$$

where n is the number of charges on each protein, e the protonic charge, ϵ_0 the vacuum permittivity, ϵ the dielectric constant, and $\lambda_{\rm D}$ the Debye–Hückel length ($\lambda_{\rm D} = (\epsilon\epsilon_0 kT/2e^2C_{\rm E})^{1/2}$, $C_{\rm E}$ being the concentration of the 1:1 electrolyte). Using eqs 1 and 2, the experimental virial coefficient for $C_{\rm E}$ = 0.01 M could be obtained (neglecting the van der Waals interactions) by assuming $n = -38$. This value is reasonable, since the isoelectric point of apoferritin is about 4.0 and the experiments were performed at pH = 5.0.[21] This is a consequence of the long-ranged double layer repulsion at low electrolyte concentrations. However, at 0.25 M the Debye–Hückel length is about 6.2 Å and the value $\tilde{B}_2 = 13$ can be obtained only if the charge of each molecule would be of the order of $|10^3 e|$, which is most unlikely.

Can the conventional hydration force be responsible for this large value of $\tilde{B}_2$? If one assumes an exponential form (a commonly used approximation)[11] for the hydration interaction between two planar surfaces separated by a distance D

$$F_{\rm H}^{\rm planar}(D) = A \exp\left(-\frac{D}{\lambda}\right) \quad (3a)$$

where λ is the hydration decay length and A is a preexponential constant, the interaction between two spheres of radius a becomes in the Derjaguin approximation

$$F_{\rm H}(r) = \pi a \int_{r-2a}^{\infty} F_{\rm H}^{\rm planar}(D)\, dD = \pi a \lambda A \exp\left(-\frac{r - 2a}{\lambda}\right) \quad (3b)$$

For a decay length of 14.4 Å (as selected in ref 21), a = 63.5 Å and T = 300 K, a value $\tilde{B}_2 \sim 13$ could be obtained for $A = 5 \times 10^{-3}$ J/m^2. These values of λ and A are, however, too large, since the values for the hydration decay length reported recently are closer to 2 Å.[11] Assuming that λ = 14.4, independent of the electrolyte concentration, the value of the parameter A has to decrease by 2 orders of magnitude to provide the low value $\tilde{B}_2 \sim 4.5$ (close to the value corresponding to the hard-core repulsion). As noted above, a small increase of A by a few percent with increasing electrolyte concentration is not unexpected (due to the charge recombination), but what mechanism could explain the 10^2 fold increase of A when the ionic strength changes from 0.15 to 0.25 M?

If one assumes a decay length λ = 2 Å and a preexponential factor A = 0.05 J/m^2 (which are typical values for lipid bilayers),[11] one obtains $\tilde{B}_2 \sim 4.9$. However, the preexponential factor must increase by more than 11 orders of magnitude to provide $\tilde{B}_2 \sim 13$, which is, of course, unreasonable.

To summarize the difficulties, if the hydration repulsion would be responsible for the long-ranged repulsion of apoferritin protein at high ionic strengths (0.25 M), a theory of the hydration force should be able to predict a large decay length at large separations and explain why such a repulsion was not observed in the interactions between two phospholipid bilayers. Second, it should also explain why a repulsion with a much shorter decay length (about 2 Å) has been typically determined in the latter experiments. Third (and more importantly), it should

(19) Attard, P.; Batchelor, M. T. *Chem. Phys. Lett.* **1988**, *149*, 206.
(20) Molina-Bolivar, J. A.; Galisteo-Gonzalez, F.; Hidalgo-Alvarez, R. *Phys. Rev. E* **1997**, *55*, 4522.
(21) Petsev, D. N.; Vekilov, P. G. *Phys. Rev. Lett.* **2000**, *84*, 1334.

explain why there is such a strong increase of the repulsion when the electrolyte concentration changes from 0.15 to 0.25 M.

As already noted, the restabilization of some colloids at high ionic strength found in previous experiments[20] can be explained in the traditional framework of the additivity between double layer and hydration forces, by a slight increase of the hydration repulsion caused by the increase in surface ion pair (dipole) density with electrolyte concentration. However, the increase in repulsion due to this mechanism is much too low to explain the strong increase of the second virial coefficient.

To have an idea about the range of the repulsion required to provide such a high virial coefficient, it should be noted that, if the hard-core repulsion, infinite in magnitude, is extended with 15 Å (above the $2a$ separation), $\tilde{B}_2$ increases from 4 to only 5.6. If the range of the hard-core repulsion is extended with 30 Å, $\tilde{B}_2$ increases to 7.55, while 60 Å leads to 12.8. From these simple estimations one can infer that the repulsion needed to explain the measured second virial coefficient for apoferritin molecules should have a much longer range than that typically observed for the traditional hydration force.

A new theory for the hydration force was proposed recently by Paunov et al. and used to explain the existence of a minimum of $\tilde{B}_2$ with increasing electrolyte concentration.[22] However, the predicted interactions vanish for distances that exceed two hydration diameters (about 14 Å), while the hydration force necessary to explain the high values of $\tilde{B}_2$ determined experimentally requires a much longer range.

The purpose of this article is to show that a recent model,[23] which provided a unitary treatment of the double layer and hydration forces, can satisfy all the above requirements and might explain the high values of the second virial coefficient of apoferritin at high ionic concentrations. The total repulsive force is described in this model by two equations for the electrical potential ψ and the average dipole moment m of a water molecule. The latter is no longer assumed to be proportional to the macroscopic electric field, as in the traditional theory, but depends also on the field generated by the neighboring dipoles. In the linear approximation, the solutions for both ψ and m are linear superpositions of functions with two distinct decay lengths. At low electrolyte concentrations, one of the decay lengths is close to the Debye–Hückel length, which is characteristic for the traditional double layer repulsion, and the other is close to a length characteristic for the traditional hydration repulsion. However, at high electrolyte concentrations, the two decay lengths characteristic of the system are markedly different from the latter two, with one of them being above 14.9 Å at any ionic strength and the other being small, below 1.67 Å. The large length is responsible for a long-ranged repulsion. At small separation distances, the terms with a shorter decay length become dominant, in agreement with the experiments on the hydration force between neutral lipid bilayers.

The most important feature of this model consists of the nonadditivity of the hydration and double layer repulsions. The "double layer-like" repulsion is generated by the charge on the surface, while the "hydration-like" repulsion is generated by the surface ion pairs (dipoles), which are formed through the reassociation of the surface charges with counterions. These dipoles orient the neighboring water molecules in a direction opposite to that produced by the electric field generated by the surface charge (due to the dissociated groups), because they expose an opposite charge to the water molecules. The increase of the electrolyte concentration in the acetate buffer increases the amount of Na^+ ions adsorbed on the acidic sites of the surface of apoferritin, and this lowers the surface charge but enhances the surface dipole density. At a particular concentration, the electric fields generated by the surface charge and the surface dipole densities in the neighboring water molecules compensate each other to a high extent and the repulsion passes through a deep minimum. At low ionic strengths, the charges dominate the interaction, while, at high ionic strengths, the surface dipoles dominate. It will be shown below that this approach can explain the strong variation of the second virial coefficient over a relatively narrow range of electrolyte concentrations.

II. Theoretical Framework

II.A. Surface Association–Dissociation Equilibria. As already noted, we suggest that the behavior of the second virial coefficient of the apoferritin in acetate buffer is due to the adsorption of Na^+ ions upon the negative sites of the protein surface, which depends on the concentration of the Na^+ ions in the liquid in the vicinity of the surface. In what follows, the adsorption of acetate ions upon the positive sites or of neutral $Na^+-CH_3COO^-$ pairs on the neutral sites of the protein surface will be neglected and it will be assumed that only the dipoles of the ion pairs formed through the association of Na^+ to the acidic sites of the surface polarize the neighboring water molecules.

The equilibrium constants for the association–dissociation of H^+, Na^+, and OH^- to the apoferritin acidic and basic groups, respectively, are not precisely known. We will try to estimate them as follows:

The apoferritin protein has $N_A = 624$ acidic and $N_B = 576$ basic amino acid residues on its surface and an isoelectric point of ~4.0.[24] The dissociation equilibria of the acidic and basic sites, at negligible electrolyte concentrations, are[25]

$$xN_A \rightleftarrows (1-x)N_A + [H^+]_S \tag{4a}$$

$$yN_B \rightleftarrows (1-y)N_B + [OH^-]_S \tag{4b}$$

where $(1-x)$ and $(1-y)$ represent the fractions of the dissociated acidic and basic sites and $[H^+]_S$ and $[OH^-]_S$ represent the concentrations of hydrogen and hydroxyl ions, respectively, in the liquid in the vicinity of the surface. At equilibrium one can write

$$K_H = \frac{(1-x)[H^+]_S}{x} \tag{5a}$$

and

$$K_{OH} = \frac{(1-y)[OH^-]_S}{y} \tag{5b}$$

where K_H and K_{OH} are the dissociation constants of the acidic and basic sites, respectively.

The total charge q on the surface, obtained using eqs 5a and b, is given by

(22) Paunov, V. N.; Kaler, E. W.; Sandler, S. I.; Petsev, D. N. *J. Colloid Interface Sci.* **2001**, *240*, 640.
(23) Ruckenstein, E.; Manciu, M. *Langmuir* **2002**, *18*, 7584.
(24) Petsev, D. N.; Thomas, B. R.; Yau, S.-T.; Vekilov, P. G. *Biophys. J.* **2000**, *78*, 2060.
(25) Prieve, D. C.; Ruckenstein, E. *J. Theor. Biol.* **1976**, *56*, 205.

$$q = -e((1-x)N_A - (1-y)N_B) = e\left(-\frac{N_A}{\frac{[H^+]_S}{K_H}+1} + \frac{N_B}{\frac{[OH^-]_S}{K_{OH}}+1}\right) \quad (6)$$

Since at the isoelectric point $q = 0$, the surface potential $\psi_S = 0$ and the concentrations of H^+ and OH^- in the liquid are the same in the vicinity of the surface and in the bulk. In this case, eq 6 leads to a relation between the dissociation constants:

$$K_{OH} = \frac{10^{-10}}{\frac{N_B}{N_A}\frac{10^{-4}}{K_H} - \frac{N_A - N_B}{N_A}} \text{ mol/L} \quad (7)$$

Because the dissociation constants must be positive, the condition $K_{OH} > 0$ in eq 7 provides an upper bound for K_H:

$$K_H < \frac{N_B}{N_A - N_B}10^{-4} = 1.2 \times 10^{-3} \text{ mol/L} \quad (8)$$

One can determine the dissociation constants if the values of the charge are known at two distinct pH values. Unfortunately, only the value $q = 0$ at pH = 4.0 (the isoelectric point) is known.

At low ionic strength, the association–dissociation equilibria of H^+ and OH^- are only slightly affected by the presence of the electrolyte. However, when the sodium acetate concentration increases, Na^+ ions compete with the H^+ ions for adsorption on the acidic sites (the adsorption of the acetate ion on the basic sites is neglected). The corresponding equilibria are

$$xN_A \rightleftarrows (1-x-w)N_A + [H^+]_S \quad (9a)$$

$$yN_B \rightleftarrows (1-y)N_B + [OH^-]_S \quad (9b)$$

$$wN_A \rightleftarrows (1-x-w)N_A + [Na^+]_S \quad (9c)$$

where wN_A is the fraction of acidic sites upon which Na atoms are adsorbed and xN_A is the fraction of associated acidic sites. After some algebra, one obtains[25]

$$q = e\left(-\frac{N_A}{\frac{[H^+]_S}{K_H} + \frac{[Na^+]_S}{K_{Na}} + 1} + \frac{N_B}{\frac{[OH^-]_S}{K_{OH}}+1}\right) \quad (10)$$

where K_{Na} is the dissociation constant of sodium. The concentrations of the ions in liquid near the surface, $[Na^+]_S$, $[H^+]_S$, and $[OH^-]_S$, are related to those in the bulk via

$$[Na^+]_S = [Na^+]\exp\left(-\frac{e\psi_S}{kT}\right) \quad (11a)$$

$$[H^+]_S = [H^+]\exp\left(-\frac{e\psi_S}{kT}\right) \quad (11b)$$

$$[OH^-]_S = [OH^-]\exp\left(\frac{e\psi_S}{kT}\right) \quad (11c)$$

Let us first investigate whether the traditional double layer theory can explain the behavior of the repulsion between apoferritin molecules. For low surface potentials, the potential around a single spherical particle of radius a in an electrolyte is given in the Debye–Hückel approximation by[26]

$$\psi(r') = \psi_S\frac{a}{r'}\exp\left(-\frac{r'-a}{\lambda_D}\right) \quad (12)$$

where $r' > a$ is the distance from the center of the particle and ψ_S is the surface potential. The surface potential is related to the surface charge density through the expression

$$\frac{q}{4\pi a^2\epsilon\epsilon_0} = -\frac{\partial\psi(r')}{\partial r}\Big|_{r'=a} = \psi_S\left(\frac{1}{a}+\frac{1}{\lambda_D}\right) \quad (13)$$

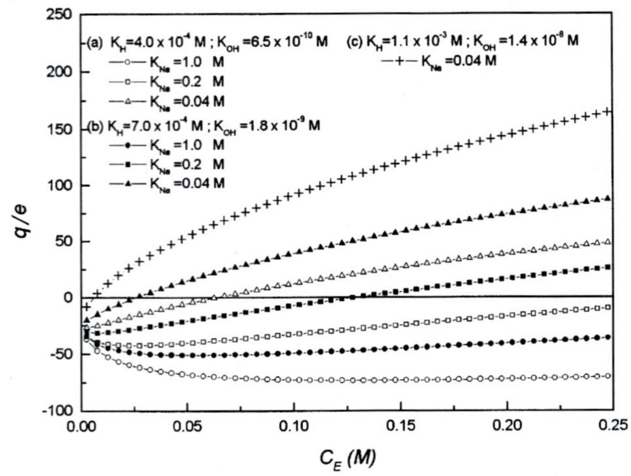

Figure 1. Total charge on apoferritin as a function of electrolyte concentration, for three pairs of equilibrium constants (K_H – K_{OH}) compatible with the isoelectric point of the apoferritin. The equilibrium constant K_{Na} was varied in the range 0.04–1.0 M.

Using eqs 7, 10, 11, and 13, one can determine the charge on an apoferritin molecule at any pH, provided that the values of the dissociation constants K_H and K_{Na} are known. Further, using eqs 1 and 2, one can evaluate the second virial coefficient for the interaction between two particles at constant surface charge.

The dissociation constant of sodium (from the acidic sites of apoferritin) is not accurately known; a reasonable value might be 0.2 M, which is compatible with the values for the Na^+ and an amino acid[25] (see Figure 3 of ref 25). The charge of apoferritin molecules is plotted in Figure 1 as a function of electrolyte concentration, for three pairs of values of K_H and K_{OH} compatible with the isoelectric point (eq 7) and for K_{Na} = 1.0, 0.2 and 0.04 M. In some cases, the molecule, which is negatively charged at very low electrolyte concentrations, becomes positively charged at higher ionic strength, because of the adsorption of Na^+ ions.

The change of the charge sign might constitute an appealing explanation for the abnormal behavior of the second virial coefficient. Indeed, the double layer repulsion first decreases with the electrolyte concentration, vanishes at a concentration corresponding to neutral molecules, and then increases when the charge becomes positive. However, the charge at 0.25 M is much smaller than that required to reach $\tilde{B}_2 = 13$. Also, a large final positive charge implies that the molecules become neutral at concentrations less than 0.01 M (see Figure 1) and not at about 0.10–0.15 M, the location of the minimum determined

(26) Hunter, R. J. *Foundations of Colloid Science*; Oxford Science Publications: 1987.

experimentally. We performed nondenaturating electrophoretic measurements of apoferritin in an acetate buffer, and the results indicated that the molecules do not become positively charged at any acetate concentration tested (up to 0.25 M). Consequently, the large repulsion at high ionic strength cannot be due to a large positive charge acquired by the apoferritin molecules.

II.B. Electrostatic Repulsion between Molecules. In the traditional double layer theory, the polarization is assumed to be proportional to the macroscopic electric field. However, this assumption is valid only when the field varies sufficiently slowly with the distance. When the variation is stronger, the interactions between neighboring dipoles become important and have to be accounted for. In addition, the dipoles present on a surface generate an electric field, which polarizes the neighboring water molecules, which in turn generate electric fields, and so on. The overlap of the polarization fields generates a repulsion when two surfaces approach each other. When both dipoles and charges are present on the surface, the traditional double layer and hydration repulsions are no longer independent, because both depend on the polarization. A unitary treatment of the interaction was presented recently[23] and will only be summarized here.

The average polarization of a water molecule, $m(z)$, between two identical, charged parallel plates separated by a distance $2d$ is related to the macroscopic electric field, $E(z)$, and to the local field produced by the neighboring dipoles via[23]

$$m(z) = \epsilon_0 v_0(\epsilon - 1)E(z) + \epsilon_0 v_0(\epsilon - 1)C_1\Delta^2\frac{\partial^2 m(z)}{\partial z^2} \tag{14}$$

where ϵ_0 is the vacuum permittivity, ϵ is the bulk dielectric constant, v_0 is the volume occupied by one water molecule, Δ is the distance between the centers of two adjacent icelike layers, and $C_1 = 1.827/4\pi\epsilon_0\epsilon'' l^3$ is an interaction coefficient, with ϵ'' denoting the dielectric constant for the interaction between neighboring water molecules and l the distance between the centers of two adjacent water molecules. Equation 14, coupled with the Poisson equation

$$\frac{\partial^2 \psi(z)}{\partial z^2} = \frac{2ec_E}{\epsilon_0}\sinh\left(\frac{e\psi}{kT}\right) + \frac{1}{\epsilon_0 v_0}\frac{\partial m(z)}{\partial z} \tag{15}$$

where the first term on the right side was obtained by assuming Boltzmann distributions for the ions, constitutes a system of differential equations for $m(z)$ and $\psi(z)$. For small surface potentials, hence in the linear approximation, the above system of equations becomes

$$\frac{\partial^2 \psi(z)}{\partial z^2} = \frac{\epsilon}{\lambda_D^2}\psi + \frac{1}{\epsilon_0 v_0}\frac{\partial m(z)}{\partial z} \tag{16a}$$

$$\lambda_m^2\frac{\partial^2 m(z)}{\partial z^2} = m(z) + \epsilon_0 v_0(\epsilon - 1)\frac{\partial \psi(z)}{\partial z} \tag{16b}$$

where $\lambda_m^2 \equiv \epsilon_0 v_0(\epsilon - 1)C_1\Delta^2$. The following solutions, which account for the symmetry of the system, can be written for ψ and m:

$$\psi(z) = a_1\cosh\left(\frac{z}{\lambda_1}\right) + a_2\cosh\left(\frac{z}{\lambda_2}\right) \tag{17a}$$

$$m(z) = \bar{a}_1\sinh\left(\frac{z}{\lambda_1}\right) + \bar{a}_2\sinh\left(\frac{z}{\lambda_2}\right) \tag{17b}$$

where z is measured from the middle distance between the parallel plates. The constants $\bar{a}_1$ and $\bar{a}_2$ are related to a_1 and a_2 through

$$\bar{a}_1 = a_1\epsilon_0 v_0\lambda_1\left(\frac{1}{\lambda_1^2} - \frac{\epsilon}{\lambda_D^2}\right) \tag{18a}$$

$$\bar{a}_2 = a_2\epsilon_0 v_0\lambda_2\left(\frac{1}{\lambda_2^2} - \frac{\epsilon}{\lambda_D^2}\right) \tag{18b}$$

and λ_1 and λ_2 are the characteristic lengths of the interaction, given by

$$\lambda_{1,2} = \left(\frac{\left(\lambda_D^2 + \lambda_m^2 \pm \left(\lambda_D^4 + \lambda_m^4 + 2\lambda_D^2\lambda_m^2\left(1 - \frac{2}{\epsilon}\right)\right)^{1/2}\right)}{2}\right)^{1/2} \tag{19}$$

At low ionic strengths, $\lambda_1 \cong \lambda_D$ and $\lambda_2 \cong \lambda_H = \lambda_m/\sqrt{\epsilon}$, which indicates that the double layer is not affected much by the local interaction due to neighboring dipoles. The presence of a surface dipole density can either increase or decrease the repulsion, depending on the dipole's orientation. Selecting $v_0 = 30$ Å^3, $\epsilon = 80$, and $\epsilon'' = 1$ for the dielectric constant for the interaction between neighboring water molecules, $l = 2.76$ Å for the distance between the centers of two adjacent water molecules, and $\Delta = 3.68$ Å,[16] one obtains $\lambda_H = 1.67$ Å,[23] which is in agreement with the value determined experimentally for neutral lipid bilayers in water.[11] However, at high ionic strength, λ_1 and λ_2 are markedly different from λ_D and λ_H, respectively. For $C_E \rightarrow \infty$ ($\lambda_D \rightarrow 0$), $\lambda_1 \rightarrow \lambda_m = 14.89$ Å, which indicates the presence in the system of an interaction with a much longer range than those of the traditional hydration or double layer.

To obtain the solution of the system of eqs 16, two boundary conditions are needed. The overall neutrality leads to the following relation between potential and surface charge density σ:

$$a_1\lambda_1\sinh\left(\frac{d}{\lambda_1}\right) + a_2\lambda_2\sinh\left(\frac{d}{\lambda_2}\right) = \frac{\lambda_D^2}{\epsilon\epsilon_0}\sigma \tag{20a}$$

while the average polarization of the first water layer from the surface is given by

$$\begin{aligned} a_1\Bigg[\lambda_1\epsilon_0 v_0\left(\frac{1}{\lambda_1^2} - \frac{\epsilon}{\lambda_D^2}\right)\Bigg((1 - \epsilon_0 v_0(\epsilon - 1)C_0) \times \\ \sinh\left(-\frac{d}{\lambda_1}\right) - \epsilon_0 v_0(\epsilon - 1)C_1\sinh\left(-\frac{d - \Delta}{\lambda_1}\right)\Bigg) + \\ \frac{\epsilon_0 v_0(\epsilon - 1)}{\lambda_1}\sinh\left(-\frac{d}{\lambda_1}\right)\Bigg] + a_2\Bigg[\lambda_2\epsilon_0 v_0\left(\frac{1}{\lambda_2^2} - \frac{\epsilon}{\lambda_D^2}\right) \times \\ \Bigg((1 - \epsilon_0 v_0(\epsilon - 1)C_0)\sinh\left(-\frac{d}{\lambda_2}\right) - \epsilon_0 v_0(\epsilon - 1)C_1 \times \\ \sinh\left(-\frac{d - \Delta}{\lambda_2}\right)\Bigg) + \frac{\epsilon_0 v_0(\epsilon - 1)}{\lambda_2}\sinh\left(-\frac{d}{\lambda_2}\right)\Bigg] = \\ \left(\frac{p}{\epsilon'}\right)\frac{v_0(\epsilon - 1)}{2\pi}\frac{1}{\left(\frac{A_{Na}}{\pi} + \Delta'^2\right)^{3/2}} \end{aligned} \tag{20b}$$

where p is the dipole moment normal to the surface of a surface dipole, A_{Na} is the area corresponding to one surface dipole, ϵ' is the dielectric constant for the interaction between a dipole and the adjacent water molecules, and

Δ' is the distance between the center of a dipole and the center of the first water layer. The interaction coefficients C_0 and C_1 are given by $C_0 = -3.766/4\pi\epsilon_0\epsilon'' l^3$ and $C_1 = 1.827/4\pi\epsilon_0\epsilon'' l^3$, where l is the distance between the centers of two adjacent water molecules in ice.[16]

We will assume, as before,[23] that the free energy of the surface layer formed by the surface dipoles and the water molecules between them is independent of the distance between the two plates. Consequently, the free interaction energy, per unit area, of two planar plates separated by a distance $D = 2d$ is composed of a chemical term $F_{\rm ch}$, an entropic term due to the mobile counterions, and a term due to the electrostatic fields:[23]

$$F_{\rm flat} = F_{\rm ch} + \frac{1}{2}\int_{-d}^{d}\left(\frac{\epsilon\epsilon_0}{\lambda_{\rm D}^{2}}(\psi)^2 + \left[\left(\epsilon_0\frac{\partial\psi}{\partial z} - \frac{m}{v_0}\right)\frac{\partial\psi}{\partial z} - \frac{C_1\Delta^2}{v_0}m\frac{\partial^2 m}{\partial z^2}\right]\right)\mathrm{d}z \quad (21)$$

For interaction at constant surface charge, $F_{\rm ch} \equiv 0$, since the (fixed) surface charge is not in thermodynamic equilibrium with the medium, while, at constant surface potential, $F_{\rm ch} = -2\sigma\psi_{\rm S}$, where $\psi_{\rm S}$ is the surface potential.[1] When neither σ nor $\psi_{\rm S}$ is constant, the change in chemical free energy, per unit area, from infinite separation to a distance $D = 2d$ between plates is given by

$$F_{\rm ch}(D) - F_{\rm ch}(\infty) = -2\int_{\infty}^{D}\psi_{\rm S}(\eta)\,\mathrm{d}\sigma(\eta) \quad (22)$$

The latter expression can be derived by decomposing the trajectory of $\psi_{\rm S}$ versus σ in a succession of small changes at constant σ followed by changes at constant $\psi_{\rm S}$. At constant σ, the change in chemical energy is zero, while, at constant surface potential, the change in chemical energy is $-2\psi_{\rm S}\,\mathrm{d}\sigma$.

The interaction between two identical spherical particles of radius a, separated by a distance r between their centers, is calculated using a variant of the Derjaguin approximation (see Appendix)

$$F_{\rm sph}(r) = \frac{\pi}{2}\int_{r-2a}^{r}(F(D) - F(\infty))(r - D)\,\mathrm{d}D \quad (23)$$

where the concentration of ions in liquid in the vicinity of the surface (eqs 11) is calculated using the renormalized surface potential $\psi_{\rm S}'$ (see Appendix)

$$\psi_{\rm S}' = \frac{\psi_{\rm S}}{\left(1 + \frac{\lambda_{\rm D}}{a}\right)} \quad (24)$$

with $\psi_{\rm S}$ being the surface potential for planar surfaces calculated from the system (eq 16) subjected to the boundary conditions (eq 20).

II.C. van der Waals Attraction. The van der Waals attraction between the hollow shells of the ferritin can be easily computed in the pairwise summation approximation. Let us consider a large sphere B made up of a small sphere b and a spherical shell S. The interaction free energy F_{BB} between two large spheres can be written (assuming pairwise interactions) as

$$F_{BB} = F_{(b+S)(b+S)} = F_{bb} + F_{bS} + F_{Sb} + F_{SS} \quad (25)$$

The interaction between a large sphere B and a small one b can be separated into

$$F_{Bb} = F_{(b+S)b} = F_{bb} + F_{bS} \quad (26)$$

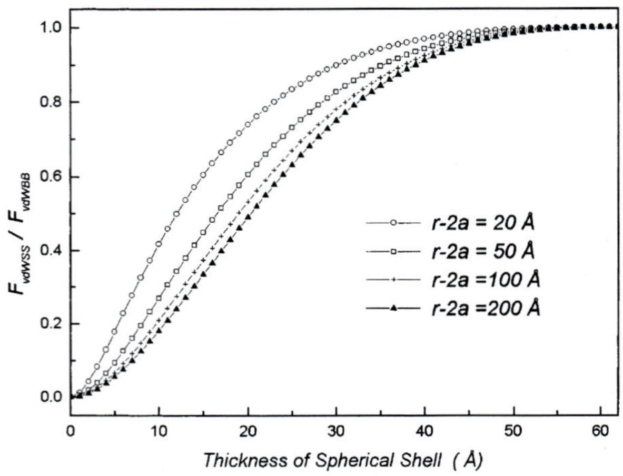

Figure 2. Ratio between van der Waals attraction of compact and hollow spheres of external radius a = 63.5 Å, at various distances, as a function of the thickness of the shell.

Using eqs 25 and 26, one can calculate the interaction between two spherical shells

$$F_{SS} = F_{BB} + F_{bb} - 2F_{Bb} \quad (27)$$

by taking into account that the van der Waals attraction between two spheres of radii a_1 and a_2, separated by a distance r between their centers, is given by[26]

$$F_{\rm vdW} = -\frac{Ha_1a_2}{6(r-(a_1+a_2))(a_1+a_2)}\left[\frac{1}{1+\frac{r-(a_1+a_2)}{2(a_1+a_2)}} + \frac{1}{1+\frac{2a_1a_2}{(r-(a_1+a_2))(a_1+a_2)} + \frac{r-(a_1+a_2)}{2(a_1+a_2)}} + \left(\frac{(r-(a_1+a_2))(a_1+a_2)}{a_1a_2}\right)\ln\left(\frac{(r-(a_1+a_2))(a_1+a_2)}{2a_1a_2} \times \left(\frac{1+\frac{r-(a_1+a_2)}{2(a_1+a_2)}}{1+\frac{(r-(a_1+a_2))(a_1+a_2)}{2a_1a_2}+\frac{(r-(a_1+a_2))^2}{4a_1a_2}}\right)\right)\right] \quad (28)$$

where H is the Hamaker constant. In Figure 2, the ratio between the van der Waals interactions of hollow and compact spheres of radius a = 63.5 Å is plotted, at various distances of closest approach ($r - 2a$), as a function of the shell thickness.

While the outer radius of the apoferritin molecule (a = 63.5 Å) can be accurately determined, for example by dynamic light scattering,[21] the inner radius (of the hollow shell) is more difficult to estimate. The iron cores of the ferritin have a maximum diameter of the order of 80 Å, while the six hydrophobic channels have a length of about 12 Å.[27] In what follows the value 20 Å will be employed for the average thickness of the shell. The Hamaker

(27) Ford, G. C.; Harrison, P. M.; Rice, D. W.; Smith, J. M. A.; Treffry, A.; White, J. L.; Yariv, J. *Philos. Trans. R. Soc. London* **1984**, *B304*, 551.

constant of the apoferritin is not known; the Hamaker constants reported for proteins cover a rather large range, from $0.06kT$ for α-crystallin to about $10kT$ for α-chymotrypsin and bovine serum albumin.[28] In the present calculations, the value $H = 5.0kT$ will be used.

III. Second Virial Coefficient of Apoferritin Molecules

In what follows, the model presented will be employed to calculate the dependence of the dimensionless second virial coefficient of the apoferritin molecules. Assuming that the surface dipolar density is generated only by the Na^+ ions adsorbed on the surface, the area corresponding to an ion is $A_{Na} = (4\pi a^2/wN_A)$. As already emphasized, one of the characteristic decay lengths of the system is larger than $\lambda_m = 14.9$ Å at any electrolyte concentration; hence, a relatively long-ranged interaction is always present in the system, even at high ionic strengths. The magnitude of this interaction per unit area depends on the boundary conditions and is generally small. Apoferritin is, however, a large protein, and since in the Derjaguin approximation the interaction is proportional to the radius, the interaction is of the order of kT at large separations. Another particularity of the apoferritin is its hollow nature, which implies a relatively small van der Waals attraction. For other proteins, the attraction might overcome the weak long-ranged repulsion, leading to a small second virial coefficient at high ionic strengths.

The electrostatic repulsion is calculated by combining eqs 10 and 11 for the charge (using the renormalized surface potential given by eq 24) with the system of eqs 16a and b solved with the boundary conditions (eqs 20a and b). For all separation distances $D = 2d$ a solution was obtained for planar surfaces by successive approximations and the corresponding free energy per unit area was calculated using eqs 21 and 22. The free energy of the repulsive interaction between two spherical apoferritin molecules was calculated using eq 23, and the van der Waals interaction was obtained from eqs 27 and 28.

Figure 3a presents the interaction free energy (in kT units) as a function of the closest approach distance $r - 2a$ for various electrolyte concentrations. The following values have been employed for the parameters: $K_H = 1 \times 10^{-4}$ M, $K_{OH} = 1.18 \times 10^{-10}$ M, $K_{Na} = 0.4$ M, $T = 300$ K, $\epsilon = 80$, $\epsilon'' = 1$, $(p/\epsilon') = 5$ D, $\Delta' = 1.5$ Å, $l = 2.76$ Å, $v_0 = 30$ Å^3, $H = 5kT$, and 20 Å for the thickness of the ferritin shell. The repulsion first decreases with increasing electrolyte concentration and then attains a minimum around 0.1 M, after which it increases. This behavior is a result of the adsorption of Na^+ ions on the acidic sites. While the total charge of the protein decreases with increasing ionic strength, the strong dipole moment of the adsorbed Na^+ orients the water molecules and generates a repulsion.

The repulsive free energy (without the van der Waals attraction) is plotted in Figure 3b for various separations against the electrolyte concentration. The repulsion has a strong minimum at an ionic strength of about 0.08 M. The reason for this unexpected behavior is the nonadditivity of the double layer and hydration repulsions.[23] When the effect of the charge is dominant, the formation of the surface dipoles decreases the free energy, because the surface dipoles and the surface charges orient the neighboring water molecules in opposite directions. Similarly, when the effect of the surface dipoles becomes dominant, the recombination of charges increases the repulsion, by

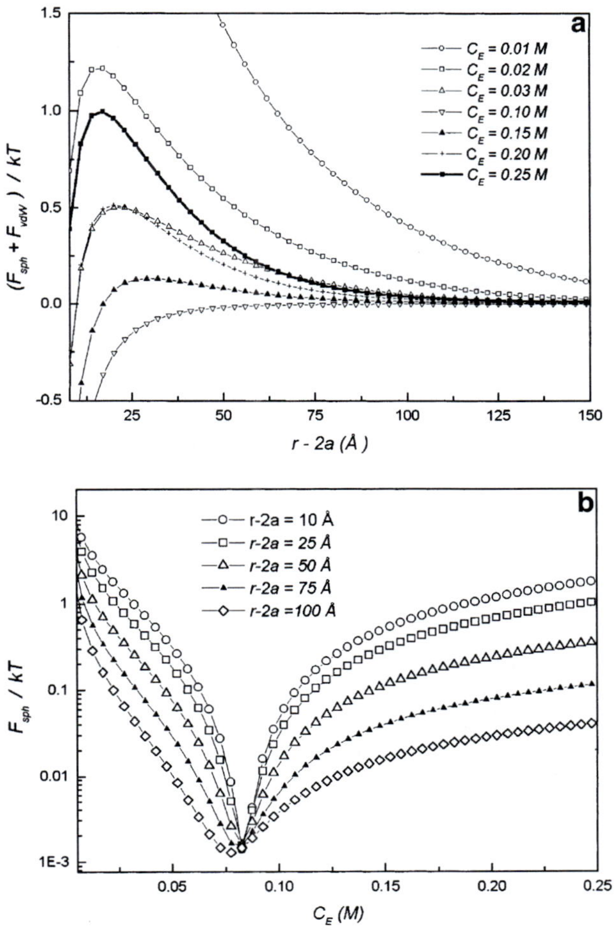

Figure 3. (a) Interaction energy between two apoferritin molecules (in kT units) as a function of the distance of closest approach, for various electrolyte concentrations. (b) Repulsive interaction as a function of electrolyte concentration, at various distances of closest approach between particles.

increasing the number of surface dipoles and decreasing the charge. For the values of the parameters employed, about 64% of the acidic sites are occupied by Na^+ ions at 0.25 M, a value which is compatible with the increase of the apoferritin mass determined by light scattering.[22]

For the values of the parameters employed (a relatively large Hamaker constant), the potential barrier is only a few kT or less; hence, the apoferritin should coagulate at almost all the concentrations studied. Since experiment shows that the proteins did not coagulate, another repulsion should be present, at least al low separation distances. This repulsion, while essential for the stability of the system, did not affect much, because of its short range, the behavior of the second virial coefficient. In the calculation of the second virial coefficient, it was assumed that the distance of closest approach between apoferritin proteins cannot be less than 8 Å. This value leads to a dimensionless second virial coefficient for the hard spheres repulsion of 4.8 instead of 4.

The dimensionless second virial coefficient calculated from eq 1 for various values of (p/ϵ') is compared in Figure 4 with the experimental results (circles) of ref 21. There is a minimum in the repulsion at about 0.1 M. At higher ionic strengths, the accumulation of Na^+ ions via association with the acidic sites of the surface increases the surface dipole density and decreases the surface charge, leading to a higher repulsion. The value of $(p/\epsilon') = 5.5$ D, which provides a good agreement between the theory and

(28) Broide, M. L.; Tominc, T. M.; Saxowsky, M. D. *Phys. Rev. E* **1996**, *53*, 6325.

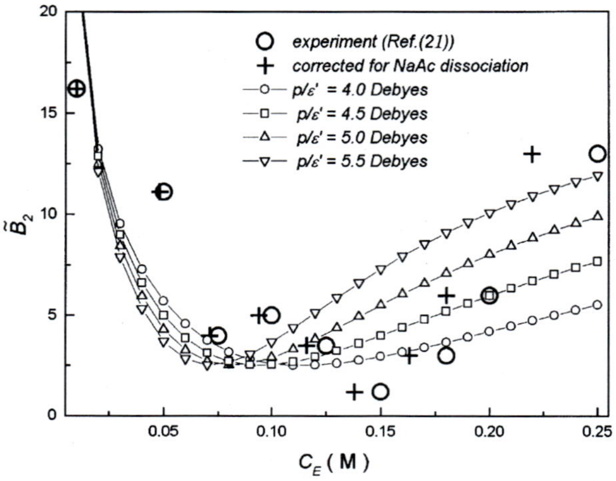

Figure 4. Dimensionless virial coefficient of the interaction, calculated as a function of the electrolyte concentration for various (p/ϵ') ratios, compared with the experimental results of ref 22 (circles). The crosses represent the experimental results when the partial dissociation of the sodium acetate is taken into account.

experiment, is reasonable, since the water dipole is about 1.85 D and the dipole formed by two elementary charges separated by 1 Å is about 5 D. Good agreement with experiment can be obtained when the equilibrium constants K_H and K_{OH} are varied in a rather large range (over 1 order of magnitude), if K_{Na} and (p/ϵ') are suitably chosen. Small changes in the values of the Hamaker constant, shell thickness, or cutoff distance for the virial integration do not affect essentially the behavior of the second virial coefficient. While the values of the parameters chosen are reasonable, the results presented in this article should be considered as qualitative only, for the reasons outlined below.

The results presented in ref 21 are, to our knowledge, the only experiments with 1:1 electrolytes in which such a strong increase of the repulsion with ionic strength was observed. If, for $C_E = 0.25$ M, one would disregard the last experimental point from the Debye plot, which provided the second virial coefficient (Figure 1b in ref 21), the corresponding $\tilde{B}_2$ would be about 6 instead of 13. In addition, the sodium acetate in the buffer is not completely dissociated. In water, sodium acetate forms both inner and outer sphere complexes. In the former, the ions are directly bound, while in the latter, Na^+ and CH_3COO^- ions are separated by one or more water molecules. Raman spectroscopy identifies only the inner complexes as associated species, while potentiometry identifies both inner and outer complexes as associated species.[29] Using the dissociation constants provided by ref 29, one can infer that, at room temperature, the inner sphere complexes are almost completely dissociated for the whole range of electrolyte concentrations (0.01−0.25 M) investigated here. However, some of the Na^+ and CH_3COO^- remain bound in neutral pairs (outer sphere complexes) and do not participate as free charges to the electrostatic screening; hence, the concentration of truly dissociated Na^+ (C_E) ions differs slightly from the $NaCH_3COO$ buffer concentration. The results of our estimations for the truly dissociated Na^+ are presented as crosses in Figure 4.

In addition, the large dipoles of the outer sphere complexes can affect the interactions, particularly when adsorbed on the surface of apoferritin. The simplified

(29) Fournier, P.; Oelkers, E. H.; Gout, R.; Pokrovski, G. *Chem. Geol.* **1998**, *151*, 69.

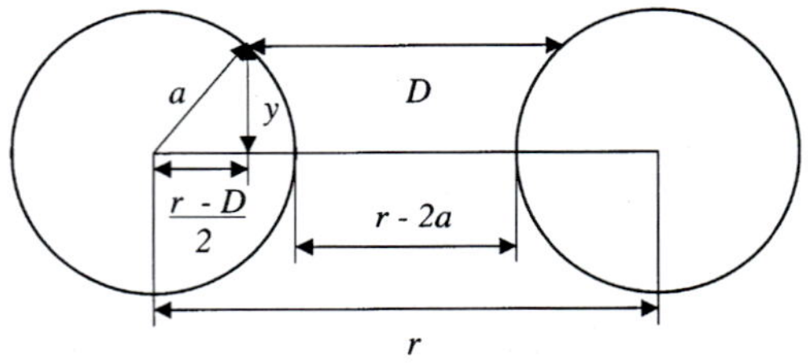

Figure 5. Geometry of two interacting spherical particles.

model presented here has neglected the influence of the dipoles of H^+ and OH^- ions associated with the surface and the effect of the sodium acetate neutral pairs adsorbed on the surface. It should be again emphasized that the Derjaguin approximation is not accurate at large separations, particularly at low electrolyte concentrations, when the Debye−Hückel length is not sufficiently small compared to the protein radius. At very low ionic strengths (0.01 M), the surface potential is large and the linear approximation employed here is not accurate.

The main point of the article is that the outlined theory can explain the strong increase in repulsion with increasing electrolyte concentration. This is a result of the nonadditivity of the repulsions generated by the surface charge and surface dipole densities.

IV. Conclusions

Recent light scattering experiments on apoferritin proteins in an acetate buffer at pH 5.0 revealed an unexpected behavior of the second virial coefficient, which passed through a minimum when the electrolyte concentration was about 0.15 M but increased to a high value at a concentration of 0.25 M. The results are incompatible with the traditional theoretical framework, which assumes that the total repulsion is the sum between a "double layer repulsion" due to the charges on the interface and a "hydration repulsion" due to the structuring of the solvent near the surface. The results could have been explained by the traditional double layer theory, if the apoferritin would have acquired an unexpectedly large positive charge at 0.25 M. This possibility was, however, ruled out by nondenaturating electrophoretic experiments. The result could also have been explained by postulating that the hydration provides a very long range exponential repulsion, with a decay length about 10 times larger than what is generally accepted. Even in this case, it would still have been difficult to explain why the repulsion increased by orders of magnitude from a concentration of 0.15 M to 0.25 M.

In this paper, we presented an alternate explanation for the unexpected behavior of the virial coefficient, based on our previous unitary treatment of the electrostatic repulsion between particles immersed in a polar solvent, which combined the double layer and the hydration forces into a single repulsive force.[23] The main features of the model are the existence of two characteristic decay lengths for the interaction (a small one, important at small separations, and a large one, which is relevant at large separations) and the nonadditivity of the effects of surface charges and surface dipoles, which in the present case tend to orient the water molecules neighboring the surface in opposite directions. The number of Na^+ ions adsorbed on the surface increases with the electrolyte concentration, decreasing the surface charge but increasing the surface dipole density. At low electrolyte concentrations, the charge effect is dominant and the interaction is well approximated by the traditional DLVO theory. At high ionic strength, the surface dipoles are mainly responsible for the repulsion. At intermediate concentrations, the opposite fields generated by the charges and the surface dipoles in the vicinity of the surface compensate each other to some extent, and consequently the repulsion has a strong minimum.

Acknowledgment. We are indebted to Prof. M. Ettinger for performing the nondenaturating electrophoretic measurements on apoferritin and to Prof. G. Frens for drawing our attention to the experimental results discussed in ref 8.

Appendix

The interactions between spherical particles can be calculated, in the Derjaguin approximation, in terms of the interactions between planar surfaces, by decomposing the spherical surface in small areas (which can be considered locally planar). The free energy of interaction, corresponding to each small piece, is assumed to be equal to the product between the area of the piece and the interaction free energy per unit area of some virtual infinite planar surfaces, located at the same separation distance D (see Figure 5). The total free energy of interaction between identical spherical particles of radius a is the integral over the surface of one particle:[26]

$$F_{\text{sph}}(r) = \int_0^a (F_{\text{flat}}(D) - F_{\text{flat}}(\infty))2\pi y \, dy \approx \pi a \int_{r-2a}^{\infty} (F_{\text{flat}}(D) - F_{\text{flat}}(\infty)) \, dD \quad \text{(A1)}$$

with r being the distance between the centers of the spheres. The approximation is accurate for large particle radii (compared to both the Debye–Hückel length and the distance of closest approach, $r - 2a$). Because (see Figure 5) $y^2 = a^2 - [(r - D)/2]^2 \Rightarrow 2y\,dy = ((r - D)/2)\,dD$, eq A.1 becomes

$$F_{\text{sph}}(r) = \frac{\pi}{2}\int_{r-2a}^{r} (F_{\text{flat}}(D) - F_{\text{flat}}(\infty))(r - D) \, dD \quad \text{(A2)}$$

Another difficulty in using the Derjaguin approximation arises when the charge is related to the surface potential via various ionic equilibria. Indeed, in the linear approximation, the surface potential is related to the surface charge density, σ, of a single planar surface by

$$\frac{\sigma}{\epsilon\epsilon_0} = -\frac{\partial \psi_{\text{flat}}(z)}{\partial z}\Big|_{z=0} = \frac{\psi_{\text{S,flat}}}{\lambda_D} \quad \text{(A3)}$$

while for a spherical particle (eq 13)

$$\frac{\sigma}{\epsilon\epsilon_0} = \psi_{\text{S,sph}}\left(\frac{1}{a} + \frac{1}{\lambda_D}\right) \quad \text{(A4)}$$

To account for the effect of the radius of the particles, an approximate renormalized surface potential ψ_S'

$$\psi_S' = \frac{\psi_{\text{S,flat}}}{\left(1 + \frac{\lambda_D}{a}\right)} \quad \text{(A5)}$$

will be used to calculate the ion density in the vicinity of the surface. Both corrections (in eqs A2 and A5) become negligible at high ionic strengths, since then $(\lambda_D/a) \ll 1$.

LA026126M

Langmuir **2002,** *18,* 2727–2736 2727

On the Stability of the Common and Newton Black Films

Eli Ruckenstein* and Marian Manciu

Department of Chemical Engineering, State University of New York at Buffalo, Buffalo, New York 14260

Received October 17, 2001. In Final Form: January 17, 2002

The effect of electrolyte concentration on the transition from common to Newton black films and the stability of both types of films are explained using a model in which the interaction energy for films with planar interfaces is obtained by adding to the classical DLVO forces the hydration force. The theory takes into account the reassociation of the charges of the interface with the counterions as the electrolyte concentration increases and their replacements by ion pairs. This affects both the double layer repulsion, because the charge on the interface is decreased, and the hydration repulsion, because the ion pair density is increased by increasing the ionic strength. The theory also accounts for the thermal fluctuations of the two interfaces. Each of the two interfaces is considered as formed of small planar surfaces with a Boltzmannian distribution of the interdistances across the liquid film. The area of the small planar surfaces is calculated on the basis of a harmonic approximation of the interaction potential. It is shown that the fluctuations decrease the stability of both kinds of black films.

I. Introduction

It is well-known that free films of water stabilized by surfactants can exist as somewhat thicker primary films, or common black films, and thinner secondary films, or Newton black films. The thickness of the former decreases sharply upon addition of electrolyte, and for this reason its stability was attributed to the balance between the electrostatic double-layer repulsion and the van der Waals attraction. A decrease in its stability leads either to film rupture or to an abrupt thinning to a Newton black film, which consists of two surfactant monolayers separated by a very thin layer of water. The thickness of the Newton black film is almost independent of the concentration of electrolyte; this suggests that another repulsive force than the double layer is involved in its stability. This repulsion is the result of the structuring of water in the vicinity of the surface. Extensive experimental measurements of the separation distance between neutral lipid bilayers in water as a function of applied pressure[1] indicated that the hydration force has an exponential behavior, with a decay length between 1.5 and 3 Å, and a preexponential factor that varies in a rather large range.

The interactions between the surfactant monolayers of the black film consist of a long-range van der Waals attraction, a short range repulsion due to the hydration force, and an electrostatic repulsion with a longer range than the hydration repulsion but shorter than the van der Waals attraction. Their combination can lead, in principle, to an interaction energy possessing three minima: one at a separation distance $d \rightarrow 0$, which implies the rupture of the film, another one at a relatively large distance, which corresponds to the common black film, and finally a third one at an intermediate separation distance, which corresponds to the Newton black film. Only the minimum at $d \rightarrow 0$ is stable; the other two, while metastable, can have relatively long lifetimes.

Experiments on the stability of water/surfactant films at various pressures were performed by Exerowa et al.[2,3] For a dilute aqueous solution of a nonionic surfactant,[3] tetraoxyethylene decyl ether ($D(EO)_4$, 5×10^{-4} mol/dm^3) or eicosaoxyethylene nonylphenol ether ($NP(EO)_{20}$, 1×10^{-5} mol/dm^3), and electrolyte (KCl), thick films (with thicknesses of the order of 100 Å) were observed at low electrolyte concentrations. With an increase of the electrolyte concentration, the film thickness first decreased, which suggests that the repulsion was caused by the double layer. This repulsive force was generated because of the different adsorptions of the two species of ions on the water/surfactant interface. At a critical electrolyte concentration, a black film was formed, and the further addition of electrolyte did not modify its thickness, which became almost independent of the external pressure, until a critical pressure was reached, at which it ruptured. While for $NP(EO)_{20}$ only one metastable equilibrium thickness was found at each electrolyte concentration, in the case of $D(EO)_4$ a hysteresis of the film thickness with increasing and decreasing pressure (i.e., two metastable minima) was observed in the range 5×10^{-4} to 3×10^{-3} mol/dm^3 KCl. The maximum pressure used in these experiments was relatively low, $\sim 5 \times 10^4$ N/m^2, and the Newton black films did not rupture in the range of pressures employed.

In the case of ionic surfactants (10^{-3} mol/dm^3 sodium lauryl sulfate, NaLS), abrupt transitions from common black films to Newton black films were observed, for NaCl concentrations in the range 0.165–0.31 mol/dm^3, by increasing the external pressure.[2] The Newton black films ruptured at a pressure of about 12×10^4 N/m^2.

The purpose of this article is to present a model and to calculate on its basis the metastable equilibrium thicknesses of the film as a function of the applied pressure. In section II, the interaction energy of the film was calculated, assuming planar interfaces free of thermal fluctuations. The double layer interaction was calculated by accounting for the charge recombination at the surface with increasing electrolyte concentration. An approximate

* To whom correspondence may be addressed. E-mail address: feaeliru@acsu.buffalo.edu. Phone: (716) 645-2911/2214. Fax: (716) 645-3822.

(1) Rand, R. P.; Parsegian, V. A. *Biochim. Biophys. Acta* **1989**, *988*, 351.

(2) Exerowa, D.; Kolarov, T.; Khristov, Khr. *Colloids Surf.* **1987**, *22*, 171.

(3) Kolarov, T.; Cohen, R.; Exerowa, D. *Colloids Surf.* **1989**, *42*, 49.

10.1021/la011569w CCC: $22.00
Published on Web 02/20/2002

expression derived by Donners et al.[4] on the basis of the Lifshitz theory was used to calculate the van der Waals interaction energy. For the hydration force, a relation derived recently[5,6] was used, which accounts for the increase in the hydration force with increasing electrolyte concentration, due to a higher ion-pair (dipole) density produced by the recombination of some charges on the interfaces with counterions. This enhancement of the repulsion at relatively high ionic concentrations might be at the origin of the stability of the Newton black films, as suggested recently;[7] however, in the present article, the field generated by an ion pair located at the interface is calculated more accurately and the theory is further developed to include the effect of thermal fluctuations.

II. Interaction Energy of Soap Films

Throughout this section, the energy of the film will be calculated by assuming planar, non-undulating, interfaces. Most of the equations of section IIA–E were taken from our previous paper.[5] We include them without detailed derivation for the completeness of the presentation.

II.A. The Surface Density of Surfactant. The surface density of surfactant is relevant for both the double layer (caused by the surface charge density generated via the dissociation of the surfactant adsorbed at the interface) and the hydration interaction (caused by the ion pair density of the non-dissociated surfactant molecules). The surface density of an anionic surfactant, Γ, will be related to the saturation surface density, Γ_∞, via the Frumkin adsorption isotherm

$$a_1 C_{R^-} = \frac{\frac{\Gamma}{\Gamma_\infty}\exp\left(-2a_2\frac{\Gamma}{\Gamma_\infty}\right)}{1-\frac{\Gamma}{\Gamma_\infty}} \tag{1}$$

where a_1 and a_2 are empirical parameters available from experiment and C_{R^-} is the concentration of the surfactant anions in the liquid in the vicinity of the interface. The surfactant in the bulk is considered completely dissociated, and hence the concentration of surfactant in the vicinity of the interface is related to the bulk surfactant concentration through the expression

$$C_{R^-} = C_s \exp\left(\frac{e\psi_s}{kT}\right) \tag{2}$$

where C_s is the bulk surfactant concentration, ψ_s is the surface potential, which is negative (hence $C_{R^-} < C_s$), e is the protonic charge, k is the Boltzmann constant, and T is the absolute temperature. The parameters a_1, a_2, and Γ_∞ are provided by experiment. In what follows the data obtained by Fainerman[8] at high electrolyte concentrations, namely, $\Gamma_\infty = 5 \times 10^{-6}$ mol/m², $a_1 = 881$ m³/mol, and $a_2 = -1.53$, will be used.

IIB. The Surface Charge. With α denoting the dissociation constant, the surface charge density σ is given by

$$\sigma = -e\alpha\Gamma \tag{3}$$

(4) Donners, W. A. B.; Rijnbout J. B.; Vrij, A. *J. Colloids Interface Sci.* **1977**, *60*, 540.
(5) Manciu, M.; Ruckenstein, E. *Langmuir* **2001**, *17*, 7061.
(6) Manciu, M.; Ruckenstein, E. *Langmuir* **2001**, *17*, 7582.
(7) Ruckenstein, E.; Bhakta, A. *Langmuir* **1996**, *12*, 4134.
(8) Fainerman, V. B. *Colloids Surf.* **1991**, *57*, 249.

For simplicity, it will be assumed that the electrolyte is 1:1 and that it has the same type of cation as the surfactant (e.g., NaCl and NaLS). The charge is generated through the following dissociation equilibrium

$$\text{R–X}|_{\text{SURFACE}} \leftrightarrow \text{R}^-|_{\text{SURFACE}} + \text{X}^+|_{\text{LIQUID-AT-INTERFACE}}$$

where R denotes the surfactant anion group and X the cation. At equilibrium, one can write

$$K_D = \frac{C_{R^-}{}^S C_{X^+}}{C_{R\text{-}X}} = \frac{\alpha\Gamma C_{X^+}}{(1-\alpha)\Gamma} = \frac{\alpha}{1-\alpha}C_{X^+} \tag{4}$$

where $C_{R^-}{}^S$ is the surface density of the dissociated surfactant anions, $C_{R\text{-}X}$ is the surface density of nondissociated (ion pair) surfactant molecules, C_{X^+} is the concentration of the cations in the liquid in the vicinity of the surface, and K_D is the equilibrium constant. Assuming that the surfactant in solution and the electrolyte are completely dissociated, one can write

$$C_{X^+} = (C_s + C_e)\exp\left(-\frac{e\psi_s}{kT}\right) \tag{5}$$

where C_e is the bulk electrolyte concentration.

II.C. The Double Layer Interaction Energy. The calculation of the double layer interaction is based on the accurate approximation of the solution of the Poisson–Boltzmann equation due to Ohshima and Kondo.[9] For two parallel plates at a distance z apart, at a surface potential ψ_s, and with a midplane potential ψ_m, they obtained the equation

$$\tanh\left(\frac{e\psi_m}{4kT}\right) = \gamma_0\frac{1}{\cosh\left(\frac{\kappa z}{2}\right)} - \gamma_0^3\frac{\left(\frac{\kappa z}{2}\right)\sinh\left(\frac{\kappa z}{2}\right)}{\cosh^4\left(\frac{\kappa z}{2}\right)} + \gamma_0^5\left\{\frac{\cosh^2\left(\frac{\kappa z}{2}\right) - 2\left(\frac{\kappa z}{2}\right)^2 - 1}{4\cosh^5\left(\frac{\kappa z}{2}\right)} - \frac{3\left(\frac{\kappa z}{2}\right)\sinh\left(\frac{\kappa z}{2}\right)}{4\cosh^6\left(\frac{\kappa z}{2}\right)} \times \left(1 - 4\left(\frac{\kappa z}{2}\right)\tanh\left(\frac{\kappa z}{2}\right)\right)\right\} \tag{6}$$

where $\gamma_0 = \tanh(e\psi_s/4kT)$, $\kappa = (2e^2C_e/\epsilon\epsilon_0 kT)^{1/2}$ is the Debye length for a 1:1 electrolyte and $\epsilon\epsilon_0$ is the dielectric constant of water. The surface charge density is related to the surface and midplane potentials via

$$\sigma = -\frac{\kappa\epsilon\epsilon_0 kT}{e}\left[2\left(\cosh\left(\frac{e\psi_s}{kT}\right) - \cosh\left(\frac{e\psi_m}{kT}\right)\right)\right]^{1/2} \tag{7}$$

The double layer force per unit area is given by the Langmuir equation

$$p_{DL}(z) = 2C_e kT\left(\cosh\left(\frac{e\psi_m(z)}{kT}\right) - 1\right) \tag{8a}$$

Equations 1–8a can be simultaneously solved to obtain the surfactant density, the degree of dissociation, and the double layer force. The double layer interaction energy

(9) Ohshima, H.; Kondo, T. *J. Colloids Interface Sci.* **1988**, *122*, 591.

per unit area, U_{DL}, is obtained by integrating the double layer force per unit area from d to ∞, where d is the thickness of the water layer

$$U_{DL}(d) = \int_d^{\infty} p_{DL}(z)\,dz \tag{8b}$$

II.D. The van der Waals Energy. The van der Waals energy per unit area for a triple layer system (hydrocarbon/water/hydrocarbon) in air was calculated on the basis of the Lifshitz theory by Donners et al.[4] and is well represented by the expression

$$U_{vdW}(d) = \frac{1}{d^2}\left(\frac{b_1 + b_2 d}{1 + b_3 d + b_4 d^2} + b_5\right) \tag{9}$$

where d is the thickness of the water layer and, for dodecane films with a thickness of 9 Å, the parameters are $b_1 = -3.08 \times 10^{-22}$ J, $b_2 = -6.28 \times 10^{-14}$ J/m, $b_3 = 8.28 \times 10^{7}$ m^{-1}, $b_4 = 6.13 \times 10^{15}$ m^{-2}, and $b_5 = -9.00 \times 10^{-23}$ J.

II.E. The Hydration Interaction Energy. The calculation of the hydration interaction is based on a model proposed by Schiby and Ruckenstein,[10] which considers that, in the vicinity of an interface, the water is organized in layers parallel to the interface with the structure of ice. The water molecules from the first water layer are polarized by the nearest dipoles of the headgroups of the surfactant molecules; these dipoles induce an electric field into the adjacent layer, whose induced dipoles generate in turn electric fields in both adjacent layers and so on. When two surfaces approach one another, the polarization layer will increasingly overlap. This decreases the average polarization of the water molecules and hence increases the electrostatic energy, thus generating a repulsion between surfaces. We will only outline the theory here; a detailed presentation was made recently.[5]

The field generated by the surface dipoles, which polarize the first water layer, is given by

$$\bar{E} = \frac{p_\perp}{\epsilon'}\,\frac{1}{2\pi\epsilon_0\left(\frac{A}{\pi} + \Delta'^2\right)^{3/2}} \tag{10}$$

where $p_\perp$ is the component, normal to the surface, of the dipole moment of an ion pair of the surface, ϵ_0 is the vacuum permittivity, $1/A$ is the ion pair density on the surface, ϵ' is an effective dielectric constant for the interaction between the water molecules of the first water layer and the nearest surface dipole, and Δ' is the distance between the first water layer and the center of the surface dipole. It is important to emphasize that ϵ' is smaller than the dielectric constant of water, ϵ, because of the lower screening of the electric field by the fewer intervening water molecules.

The average polarization of the water molecule is well described by

$$m(z) = m_1\,\frac{\sinh\left(\frac{d - 2z}{2\lambda}\right)}{\sinh\left(\frac{d}{2\lambda}\right)} \tag{11}$$

where z is the distance normal to the surface, measured from one surface (considered at the external boundary of the first water layer), d is the distance between surfaces, m_1 is the polarization of the first water layer, and λ is the decay length of the polarization. For a layered structure of water, λ can be calculated using the expression[5]

$$\lambda = \left(\frac{C_1\Delta^2}{\frac{1}{\gamma} - (C_0 + 2C_1)}\right)^{1/2} \tag{12}$$

where Δ is the distance between the centers of the adjacent layers, γ is the molecular polarizability, and C_k are interaction coefficients that account for the electric field generated by the dipoles of the layer $i + k$ at a site of layer i

$$E_i^{\,\mathrm{local}} = C_{-1}m_{i-1} + C_0 m_i + C_1 m_{i+1} \tag{13}$$

For an icelike structure of water, the values of the coefficients are[5,6]

$$C_0 = -\frac{3.7663}{4\pi\epsilon_0\epsilon'' l^3} \tag{14a}$$

and

$$C_1 = C_{-1} = \frac{1.8272}{4\pi\epsilon_0\epsilon'' l^3} \tag{14b}$$

where l is the distance between the centers of two adjacent water molecules and ϵ'' is an effective dielectric constant for the interaction between neighboring water molecules; again $\epsilon'' < \epsilon$.

The polarization of the first water layer is produced by the surface dipoles and the water dipoles of the first two layers, hence

$$m_1 = \frac{\gamma\bar{E}}{\left(1 - \gamma\left(C_0 + C_1\,\frac{\sinh\left(\frac{d - 2\Delta}{2\lambda}\right)}{\sinh\left(\frac{d}{2\lambda}\right)}\right)\right)} \tag{15}$$

and the total energy, per unit area, due to the polarization of water molecules by the surface dipoles, is given by

$$U_H^{\,\mathrm{total}}(d) = \frac{\gamma\bar{E}^2}{\left(1 - \gamma\left(C_0 + C_1\,\frac{\sinh\left(\frac{d - 2\Delta}{2\lambda}\right)}{\sinh\left(\frac{d}{2\lambda}\right)}\right)\right)^2} \times \frac{d - \lambda\sinh\left(\frac{d}{\lambda}\right)}{4v\sinh^2\left(\frac{d}{2\lambda}\right)} \tag{16}$$

where v is the volume occupied by a water molecule. The hydration interaction energy per unit area is given by

$$U_H(d) = U_H^{\,\mathrm{total}}(d) - U_H^{\,\mathrm{total}}(\infty) \tag{17}$$

II.F. A Simple Model for the Interaction Energy. We will try first to obtain some information about the interaction energy by using some simple approximations.

(10) Schiby, D.; Ruckenstein, E. *Chem. Phys. Lett.* **1983**, *95*, 435.

The following approximate expressions will be used for the interaction energies per unit area:

$$U'_{\mathrm{H}} = A_1 \exp\left(-\frac{d}{\lambda_1}\right) \quad (18a)$$

$$U'_{\mathrm{DL}} = \frac{64 C_e kT}{\kappa} \tanh^2\left(\frac{e\psi_s}{4kT}\right) \exp(-\kappa d) = A_2 \exp\left(-\frac{d}{\lambda_2}\right) \quad (18b)$$

$$U'_{\mathrm{vdW}} = -\frac{A_{\mathrm{H}}}{12\pi d^2} \quad (18c)$$

which are rough approximations of the interaction energies provided by the equations presented above (eqs 17, 8b, and 9, respectively).

Neglecting the double layer interaction (neutral interfaces), one obtains for the disjoining pressure (the negative of the derivative of the interaction energy) the expression

$$\Pi(d) = \frac{A_1}{\lambda_1} \exp\left(-\frac{d}{\lambda_1}\right) - \frac{A_{\mathrm{H}}}{6\pi d^3} \quad (19)$$

Newton black films, which are the only ones that can exist in the absence of double layer interactions, can be obtained only if the disjoining pressure has a positive value (otherwise, the film will collapse). The extrema of the disjoining pressure are obtained through the derivation of eq 19 with respect to d

$$\frac{d^4}{\lambda_1^4} \exp\left(-\frac{d}{\lambda_1}\right) = \xi^4 \exp(-\xi) = F_1(\xi) = \frac{A_{\mathrm{H}}}{2\pi A_1 \lambda_1^2} \quad (20)$$

where $\xi = d/\lambda_1$. The function $F_1(\xi)$ in eq 20 has a maximum at $\xi = 4$ (see Figure 1a); therefore, eq 20 has solutions only if the Hamaker constant $A_{\mathrm{H}} < 2\pi A_1\lambda_1^2 F_1(4) = 29.46 A_1\lambda_1^2$. In such cases, eq 20 has two solutions. The smaller corresponds to a maximum disjoining pressure and the other one to a local minimum. A useful lower bound is obtained when the maximum value of the disjoining pressure vanishes, $\Pi(d) = 0$. Using eqs 19 and 20, one obtains for the latter case

$$d^* = 3\lambda_1 \quad (21a)$$

$$A_1^* = \frac{A_{\mathrm{H}}}{162\pi\lambda_1^2 \exp(-3)} \quad (21b)$$

A stable film can be obtained only if the hydration repulsion is stronger than the above critical value, hence if $A_1 > A_1^*$, and the maximum disjoining pressure occurs at a distance $d < d^* = 3\lambda_1$.

When all three interactions are taken into account, the extrema of the disjoining pressure are given by

$$\xi^4(\exp(-\xi) + r_1^2 r_2 \exp(-r_1\xi)) = F_2(\xi) = \frac{A_{\mathrm{H}}}{2\pi A_1 \lambda_1^2} \quad (22)$$

where $r_1 = \lambda_1/\lambda_2$ and $r_2 = A_2/A_1$. The left-hand term of eq 22 is a positive function of ξ, which vanishes at $\xi = 0$ and $\xi \to \infty$. This function is represented in Figure 1 for various values of the ratios r_1 and r_2. If the right-hand term is sufficiently large to exceed the maximum value of $F_2(\xi)$ (strong attraction), eq 22 has no solution and the film collapses. For $r_2 = 0$ (Figure 1a), $F_2(\xi) \equiv F_1(\xi)$ and there are two solutions if $A_{\mathrm{H}} < 29.46 A_1\lambda_1^2$. For $r_2 = 0.05$ and

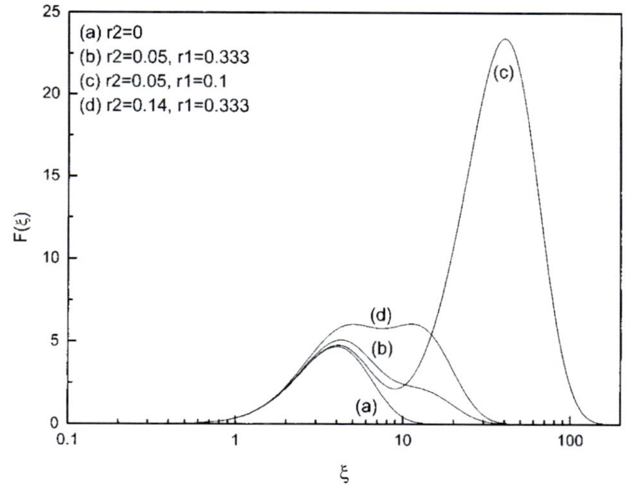

Figure 1. $F_2(\xi) = \xi^4(\exp(-\xi) + r_1^2 r_2 \exp(-r_1\xi))$ (eq 22) for (a) $r_2 = 0$, (b) $r_1 = 0.333$ and $r_2 = 0.05$, (c) $r_1 = 0.1$ and $r_2 = 0.05$, and (d) $r_1 = 0.333$ and $r_2 = 0.14$.

$r_1 = 0.333$ (Figure 1b), eq 22 still has only two solutions; hence there is only one maximum of the disjoining pressure. At metastable equilibrium, the disjoining pressure equals the external pressure. Starting with a thick film and increasing the external pressure, the thickness decreases continuously, until the film ruptures. However, if the ratio r_1 is sufficiently small ($r_1 = 0.1$ and $r_2 = 0.05$, Figure 1c), there is a range of Hamaker constants ($13.5 A_1\lambda_1^2 < A_{\mathrm{H}} < 29.9 A_1\lambda_1^2$), in which eq 22 has four solutions. In these cases there are two maxima in the disjoining pressure, located at distances d_1 and d_2. Let us first consider that the maximum at d_1 is higher (see Figure 2a). Starting with a thick film and increasing the external pressure, the thickness of the film decreases continuously up to the distance d_2, when the external pressure reaches the value Π_2. A further increase of the pressure produces a jump from the distance d_2 to the distance d', which corresponds to a stable Newton black film. On the other hand, if the maximum at d_2 is the higher one (Figure 2b), then the increase of the pressure above Π_2 produces the rupture of the film. However, the existence of a domain with four solutions for eq 22 does not necessarily imply that both Newton and common black films can be stable. While two maxima of the disjoining pressure exist for $r_1 = 0.333$ and $r_2 = 0.14$ (Figure 1d), the value of the second one is negative (Figure 2c) and hence only the Newton black film is stable.

II.G. Model Calculations. In what follows it will be shown that the general behavior discussed above is consistent with realistic calculations. An important parameter, which is however unknown, is the equilibrium constant of the association–dissociation equilibrium, K_{D}. The dissociation constant of the ion pair $NaSO_4^-$ was estimated to be $K_{\mathrm{D}} = 10^{-0.7}$ mol/dm^3 = 0.1995 mol/dm^3.[11] Because of the repulsion among the headgroups on the interface, the dissociation constant is expected to be lower in the present case. In what follows, we will use K_{D} = 0.050 mol/dm^3. For this value, the pressures at which the transitions from the common to the Newton black films occur are in agreement with the experiments of Exerowa et al.[2]

As noted in the previous section, to obtain a transition from the common to a Newton black film, the decay lengths of the hydration and double layer repulsions must be

(11) Davies, C. W. *Ion Association*; Butterworth: London, 1962.

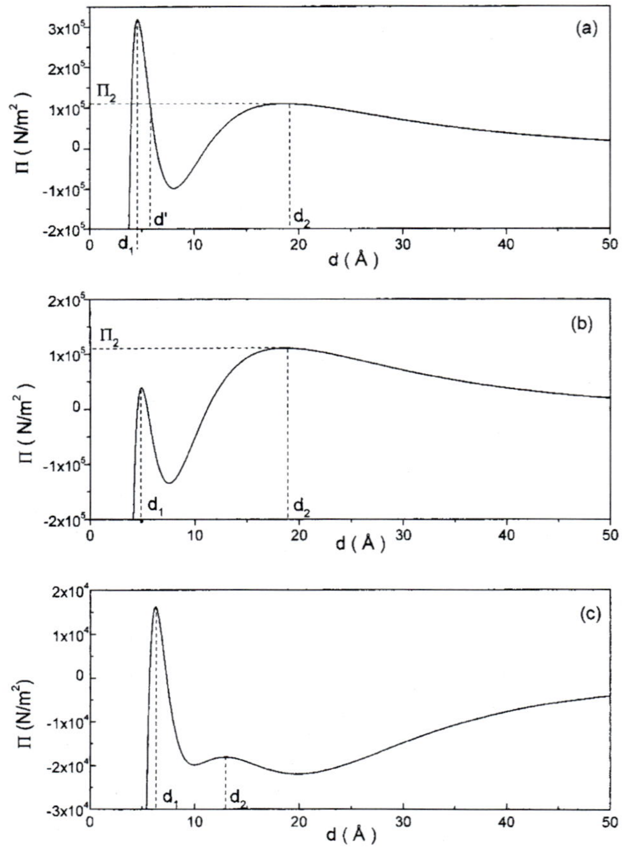

Figure 2. Disjoining pressure $\Pi(d)$ vs d: (a) $A_H = 10^{-20}$ J, $A_1 = 0.017$ J/m^2, $\lambda_1 = 1.5$ Å, $A_2 = 0.001$ J/m^2, $\lambda_2 = 15.0$ Å; (b) the same values for the parameters as in (a), except $A_1 = 0.016$ J/m^2; (c) $A_H = 10^{-20}$ J, $A_1 = 0.0121$ J/m^2, $\lambda_1 = 1.5$ Å, $A_2 = 6.05 \times 10^{-4}$ J/m^2, $\lambda_2 = 4.5$ Å (these values correspond to the domain of four solutions of eq 22 of curve d of Figure 1; hence $r_1 = 0.333$, $r_2 = 0.05$, $F_2(\xi) = 5.85$).

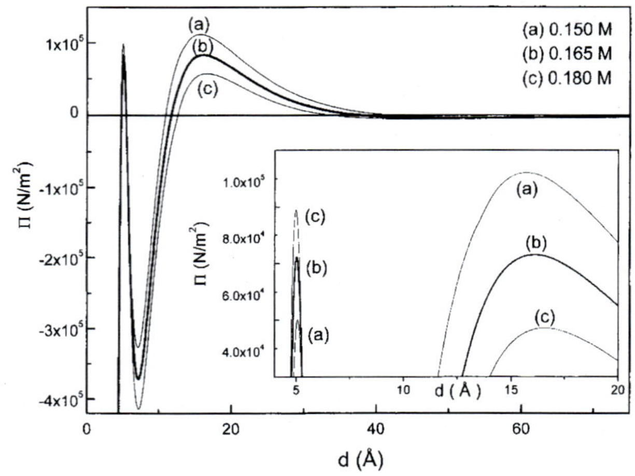

Figure 3. Disjoining pressure $\Pi(d)$ vs d for $\Gamma_\infty = 5 \times 10^{-6}$ mol/m^2, $a_1 = 881$ m^3/mol, $a_2 = -1.53$, $b_1 = -3.08 \times 10^{-22}$ J, $b_2 = -6.28 \times 10^{-14}$ J/m, $b_3 = 8.28 \times 10^7$ m^{-1}, $b_4 = 6.13 \times 10^{15}$ m^{-2}, $b_5 = -9.00 \times 10^{-23}$ J, $K_D = 0.050$ mol/dm^3, $C_s = 1 \times 10^{-3}$ mol/dm^3, $\lambda_1 = 1.0$ Å, $p_\perp/\epsilon' = 3.45$ D: (a) $C_e = 0.150$ mol/dm^3; (b) $C_e = 0.165$ mol/dm^3; (c) $C_e = 0.180$ mol/dm^3. In the inset, the peaks of the disjoining pressure are presented at a smaller scale.

sufficiently different. A transition from the common to the Newton black film was observed experimentally [2] to start when the electrolyte concentration exceeded 0.165 mol/dm^3, and transitions were observed to take place up to the highest concentration (0.31 mol/dm^3) used in experiment. For these two concentrations, the Debye lengths are $\lambda_2 = (\kappa)^{-1} = 7.5$ and 5.5 Å, respectively. To obtain a separation between the pressure peaks, the decay length for the hydration repulsion should be much smaller. For a molecular polarizability $\gamma = 8.6 \times 10^{-40}$ C^2 m^2/J, calculated using the Debye–Lorentz model, the upper bound for the hydration decay length, λ_1, calculated assuming $\epsilon'' = 1$, is 2.96 Å.[5] Lower values can be obtained when the effective dielectric constant ϵ'' for the interaction between molecules is larger than 1. The decay length is also decreased by the structural disorder of the water layers.[6] While recent experiments indicated that for lipid bilayers λ_1 is close to 2 Å, [12] there are no such experimental results for soap films. In what follows, we will use $\epsilon'' = 9$, which introduced in eqs 14 and using eq 12 provides $\lambda_1 = 1.0$ Å.

In Figure 3, the disjoining pressure, calculated as described above, is plotted for $C_s = 10^{-3}$ mol/dm^3 and $p_\perp/\epsilon' = 3.45$ D. For $C_e = 0.15$ mol/dm^3 (Figure 3a), the first maximum of the disjoining pressure is somewhat lower than the second maximum, which corresponds to the common black film. Therefore, the increase of the external pressure produces in this case the rupture of the common black film. The increase of the electrolyte concentration has two effects. First, it decreases the double layer repulsion leading to a lower height of the second peak, and secondly, the recombination of surface charges on the interfaces with counterions provides an increase in the hydration repulsion (via the increase of the ion pair density) and hence an increase of the height of the first peak. In Figure 3b ($C_e = 0.165$ mol/dm^3), there is a very narrow range of pressures in which the Newton black film is the only stable one (the value $p_\perp/\epsilon' = 3.45$ D was selected to provide about the same height for the two peaks of the disjoining pressure at $C_e = 0.165$ mol/dm^3). A further increase of the electrolyte concentration ($C_e = 0.18$ mol/dm^3, Figure 3c) lowers the transition pressure from the common to the Newton black film (because of the screening of the double layer) and also slightly increases the rupture pressure of the Newton black film (because of the increase in the hydration repulsion with increasing ionic strength). At high electrolyte concentrations, the Newton black films might be the only stable ones.

The calculations presented in this section show that the behavior of the black films can be understood in terms of the interaction energy between planar films. However, they cannot explain why, for the same electrolyte concentration, the transition from the common to the Newton black film occurs at various pressures (for example, for $C_e = 10^{-3}$ mol/dm^3, $p = (2.5 \div 9.8) \times 10^4$ N/m^2).[2] In addition, the thickness at the transition apparently does not depend on the electrolyte concentration (while the Debye length λ_2 does) and is larger than the upper bound $3\lambda_2$ (which is obtained, when only the double layer and van der Waals interactions are present, using the approach employed to derive eq 21). In the next section, it will be shown that by accounting for the thermal fluctuations of the interfaces one can provide answers to these questions.

III. The Role of Thermal Fluctuations of the Interfaces

The first quantitative theory, which accounted for the repulsion between phospholipid bilayers due to a steric

(12) Petrache, H. I.; Gouliaev, N.; Tristram-Nagle, S.; Zhang, R.; Suter, R. M.; Nagle, J. F. *Phys. Rev. E: Stat. Phys., Plasmas, Fluids, Relat. Interdiscip. Top.* **1998**, *57*, 7014.

2732 *Langmuir, Vol. 18, No. 7, 2002* *Ruckenstein and Manciu*

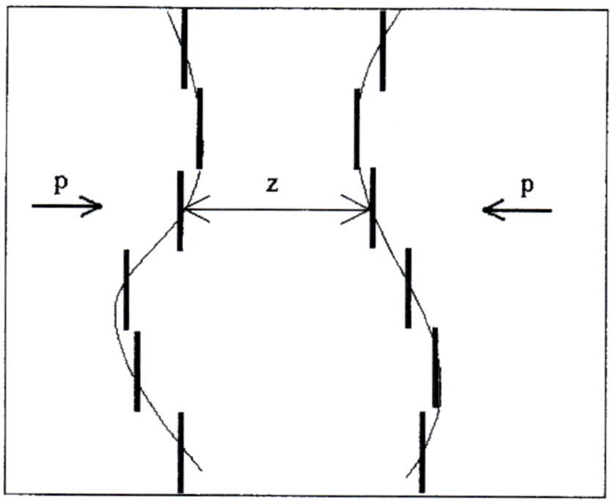

Figure 4. The fluctuating interfaces are approximated by independent small, planar surfaces of area S separated by the distance z.

(rigid wall) confinement of their undulation by the neighboring bilayers, was developed by Helfrich.[13] Since then a number of attempts were made to extend the Helfrich theory to arbitrary interaction potentials.[14–16] The last two approaches involved the observation that the partition function of an undulating bilayer in an external potential can be calculated exactly for a harmonic potential. However, in the present case, we are interested in the equilibrium position as a function of the applied pressure in a metastable state and the interaction potential (which possesses two local minima) cannot be approximated by a harmonic one. A new approach is therefore suggested below.

III.A. The Thermal Fluctuations of the Interfaces for Arbitrary Interactions. After the Helfrich initial theory,[13] Helfrich and Servuss[17] suggested an alternate derivation of the entropic repulsion due to the confinement of a membrane between rigid walls, by considering the lipid bilayer composed of many independent "pieces", whose area is related to the root mean square fluctuations of the positions of the undulating bilayer. As shown below, this representation can be extended to interfaces interacting via arbitrary potentials.

In what follows, the two fluctuating interfaces will be replaced by many small, independent surfaces of area S, separated by a distance z (see Figure 4). The (metastable) distribution of the distances between the surfaces, in an ensemble subjected to a constant pressure, will be assumed Boltzmannian. It will be also assumed that the fluctuating interfaces have constant total areas and an elastic bending modulus K_C. Let us first consider that the interfaces interact per unit area through a harmonic potential

$$U(z) = \frac{1}{2} B(z - z_0)^2 \tag{23}$$

where B is the spring constant, z is the distance between two small surfaces, and z_0 is the average thickness of the film. The Boltzmannian distribution of the thicknesses has the form

$$\rho(z) \propto \exp\left(-\frac{SU(z)}{kT}\right) \propto \exp\left(-\frac{SB(z - z_0)^2}{2kT}\right) \tag{24}$$

However, as demonstrated by Helfrich,[13] the distribution of thicknesses in a harmonic potential is given by

$$\rho(z) = \frac{1}{(2\pi)^{1/2}\sigma} \exp\left(-\frac{(z - z_0)^2}{2\sigma^2}\right) \tag{25}$$

where the root mean square fluctuation of the thickness, σ, obtained from the partition function has the form[13,18]

$$\sigma^2 = \frac{kT}{8(K_C B)^{1/2}} \tag{26}$$

Comparing eqs 24 and 25, and taking into account eq 26, one obtains

$$S = \frac{kT}{B\sigma^2} = 8\left(\frac{K_C}{B}\right)^{1/2} \tag{27}$$

For a constant applied pressure p, the distribution of distances between the small surfaces can be calculated using the enthalpy instead of the energy

$$\rho(z) \propto \exp\left(-\frac{S\left(\frac{1}{2}B(z - z_0)^2 + pz\right)}{kT}\right) \equiv \exp\left(-\frac{S\left(\frac{1}{2}B(z - z_0')^2 + C\right)}{kT}\right) \tag{28}$$

where z_0 is the average thickness at $p = 0$, $z_0' = z_0 - (p/B)$ is the average thickness at pressure p, and $C = z_0 p - (p^2/2B)$. The constant C is independent of z and hence does not affect the distribution $\rho(z)$, being eliminated through normalization. As expected, the new equilibrium distance is the thickness where the elastic force equilibrates the external pressure, $B(z_0 - z_0') = p$. Let us first note that in this case, there is no repulsion due to the entropic confinement, because the spring constant does not depend on the applied pressure, and therefore the intersurfaces distance distribution, $\rho(z)$, and the mean square fluctuation, $\sigma^2 = \langle(z - z_0')^2\rangle$, are independent of the applied pressure and, hence, on the average distance. Because the free energy due to the entropic confinement is given by $(kT)^2/(128K_C\sigma^2)$,[13,18] the undulation force is zero in this case.

The above derivation will be extended to a well-behaved interaction potential, $U(z)$, for which $U(0) \to \infty$, $U(\infty) \to 0$ and has a minimum at a distance z_0. In the vicinity of the minimum of the enthalpy (per unit area) $H(z) = U(z) + pz$, the potential will be approximated by a harmonic one, with the effective spring constant $B' = \partial^2 H(z)/\partial z^2|_{z=z_0'}$, where z_0' is the solution of $\partial H(z)/\partial z = 0$. It will be assumed that the areas of the small independent surfaces into which the interfaces are decomposed are still given by eq 27, but

(13) Helfrich, W. *Z. Naturforsch.* **1978**, *33a*, 305.
(14) Evans E. A.; Parsegian V. A. *Proc. Natl. Acad. Sci. U.S.A.* **1986**, *83*, 7132.
(15) Podgornik, R.; Parsegian, V. A. *Langmuir* **1992**, *8*, 557.
(16) Manciu, M.; Ruckenstein, E. *Langmuir* **2001**, *17*, 2455.
(17) Helfrich, W.; Servuss, R.-M. *Nuovo Cimento* **1984**, *3D*, 137.
(18) Sornette, D.; Ostrowsky, N. In *Micelles, Membranes, Microemulsions and Monolayers*; Gelbart, W. M., Ben-Shaul, A., Roux, D., Eds.; Springer-Verlag: Berlin, 1994.

with a new spring constant B'. In this case, the distribution of distances is given by

$$\rho(z) = \frac{1}{N} \exp\left(-8 \left(\frac{K_C}{\left.\frac{\partial^2 H(z)}{\partial z^2}\right|_{z=z_0'}} \right)^{1/2} \frac{H(z)}{kT} \right) \tag{29}$$

where N is obtained through normalization: $N = \int_0^\infty \rho(z)\, dz$.

The average thickness, $\langle z \rangle$, is obtained from

$$\langle z \rangle = \int_0^\infty z\rho(z)\, dz \tag{30}$$

while the mean square fluctuation is given by

$$\sigma^2 = \int_0^\infty (z - \langle z \rangle)^2 \rho(z)\, dz \tag{31}$$

For nonvanishing values of p, both integrals (30) and (31) converge.

Using eqs 29–31, one can calculate the average thickness and the root mean square fluctuation of the intersurfaces separation as a function of the applied pressure, once the interaction potential is known. For illustration purposes, let us first apply the method to the simple interaction potential

$$U'(z) = A_1 \exp\left(-\frac{z}{\lambda_1}\right) - \frac{A_H}{12\pi z^2} \tag{32}$$

with $A_1 = 0.07$ J/m^2, $\lambda_1 = 1.0$ Å, and $A_H = 10^{-20}$ J, which are some typical values of the parameters. The potential is not well-behaved (in the sense described above) because $U(0) \to -\infty$. The stable equilibrium position corresponds in this case to the trivial case $z = 0$ (the film ruptures). However, we are interested in the behavior of the film at the metastable equilibrium (the secondary minimum of the enthalpy, at z_0', see Figure 5). At metastable equilibrium, the film can have a thickness between z_1', corresponding to the maximum of the enthalpy, and $z = \infty$.

Figure 5 presents the enthalpy and the distribution $\rho(z)$ for various values of K_C and (a) $p = 1.0 \times 10^4$ N/m^2 and (b) $p = 1.0 \times 10^6$ N/m^2. The asymmetry of the distributions depends on the pressure applied and is a consequence of the anharmonicity of the interaction potential. The lifetime of the film (until it ruptures) depends on the value of $\rho(z)$ at $z = z_1'$. Curves 1–3 provide the ratios $\rho(z)/\rho(z_0')$, for various values of K_C. They are negligible at $z = z_1'$; hence the metastable state at the secondary minimum has a long lifetime.

In Figure 6, the pressure (a) and the root mean square fluctuation (b) are plotted as functions of the average thickness of the film, for $K_C = 10 \times 10^{-19}$ J and for (1) the anharmonic (eq 32) and (2) the harmonic (eq 23) interaction potentials. The spring constant for the second case was obtained from the harmonic approximation of eq 32 around its minimum, at $p = 0$.

We already noted that, for a harmonic interaction, σ_h = constant and $\langle z \rangle = z_0' = z_0 - (p/B)$ varies linearly with the applied pressure. However, for an anharmonic interaction, σ_a is a function of the applied pressure (or, equivalently, of the average thickness of the film $\langle z \rangle$, which differs from z_0' because of the asymmetry of the distribution). The functional dependence of the pressure on the average thickness differs in the anharmonic and harmonic cases. The pressure is higher in the former case, because of the

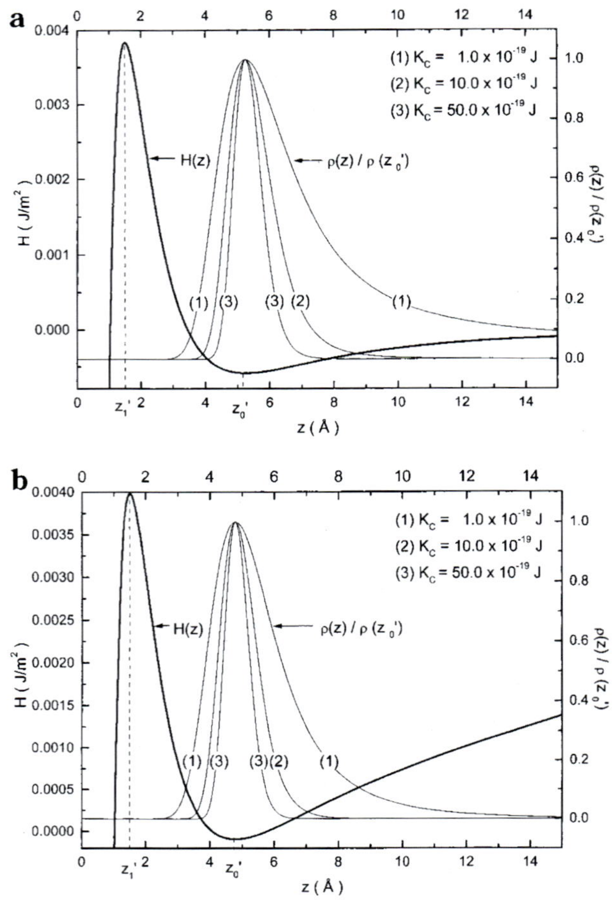

Figure 5. The enthalpy H (thick line) and the ratio $\rho(z)/\rho(z_0')$ vs the distance z for an anharmonic interaction potential (eq 32) with $A_H = 10^{-20}$ J, $A_1 = 0.07$ J/m^2, $\lambda_1 = 1.0$ Å: (1) $K_C = 1.0 \times 10^{-19}$ J; (2) $K_C = 10.0 \times 10^{-19}$ J; (3) $K_C = 50.0 \times 10^{-19}$ J. The applied pressures were (a) $p = 1.0 \times 10^4$ N/m^2; (b) $p = 1.0 \times 10^6$ N/m^2.

"*undulation repulsion*", which increases with decreasing bending modulus. The *undulation repulsion*, however, depends on the interaction potential and vanishes for a harmonic interaction.

The lifetime of the film depends on $\rho(z_1')$; hence it decreases for interfaces with low bending moduli (less rigid interfaces are likely to fluctuate with a larger amplitude and thus can reach easier the maximum of the enthalpy). Therefore, while the *undulation repulsion* increases the average thickness of the film at a given external pressure (Figure 6a), it decreases its stability (Figure 5).

III.B. The Role of Thermal Fluctuations on the Transition from Common Black Films to Newton Black Films. The method described in the previous section will be now applied to thin films with fluctuating interfaces, with the interaction energy calculated as in section II.G. For low values of the external pressure, the enthalpy has two metastable minima at z_N and z_C, and a stable one at $z \to 0$ (the former two correspond to the Newton and to the common black films, respectively, and the latter implies the rupture of the film), separated by two maxima located at z_1 and z_2 (see Figure 7a). At metastable equilibrium the distances between the surfaces are distributed between z_1 and z_2 for the Newton black film and between z_2 and $z \to \infty$ for the common black film. The stability of the metastable states depends on the chance for a small area S of the interface to reach the

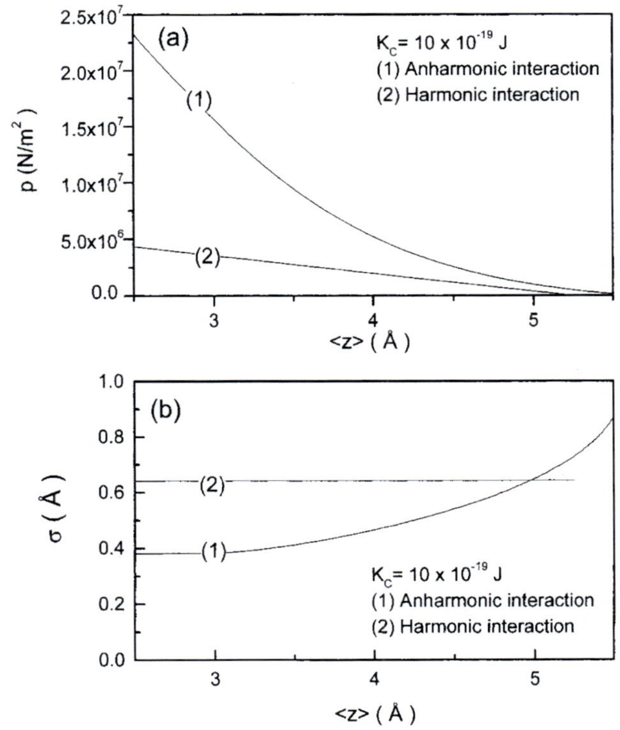

Figure 6. (a) The applied pressure p and (b) the root mean square fluctuation σ vs average thickness $\langle z \rangle$ for (1) anharmonic interaction potential (eq 32) and (2) its harmonic approximation (at $p = 0$).

limits of the domain (the heights of the adjacent peaks). The lifetime of a state in the potential well is proportional to the factor $\exp(S\Delta H/kT)$; hence the lifetime until the common black film transforms into a Newton black film, $t_{C\to N}$ is proportional to

$$t_{C\to N} \propto \exp\left(\frac{S_C(H(z_2) - H(z_C))}{kT}\right) \propto \frac{\rho_C(z_C)}{\rho_C(z_2)} \quad (33a)$$

while the lifetimes of a Newton black film until it ruptures, $t_{N\to R}$, or until its transition to a common black film, $t_{N\to C}$, are given by

$$t_{N\to R} \propto \exp\left(\frac{S_N(H(z_1) - H(z_N))}{kT}\right) \propto \frac{\rho_N(z_N)}{\rho_N(z_1)} \quad (33b)$$

and

$$t_{N\to C} \propto \exp\left(\frac{S_N(H(z_2) - H(z_N))}{kT}\right) \propto \frac{\rho_N(z_N)}{\rho_N(z_2)} \quad (33c)$$

respectively.

Both minima of the enthalpy, while metastable, might have a relatively long lifetime, and a system prepared in one state (common black film or Newton black film) might remain in that state during the time of the experiment, if the potential barriers are sufficiently high. However, because of the thermal fluctuations, it is possible to have, for the same experimental conditions, a transition in a range of pressures, and this explains one of the experimental results of Exerowa et al.[2]

When both the Newton and the common black films have lifetimes exceeding the duration of the experiment in a domain of applied pressures, the common black film transforms into a Newton one at the upper bound of the domain, while the Newton black film can have a transition to the common one at the lower bound of the pressure domain. This results in a hysteresis in thickness with increasing and decreasing pressure.

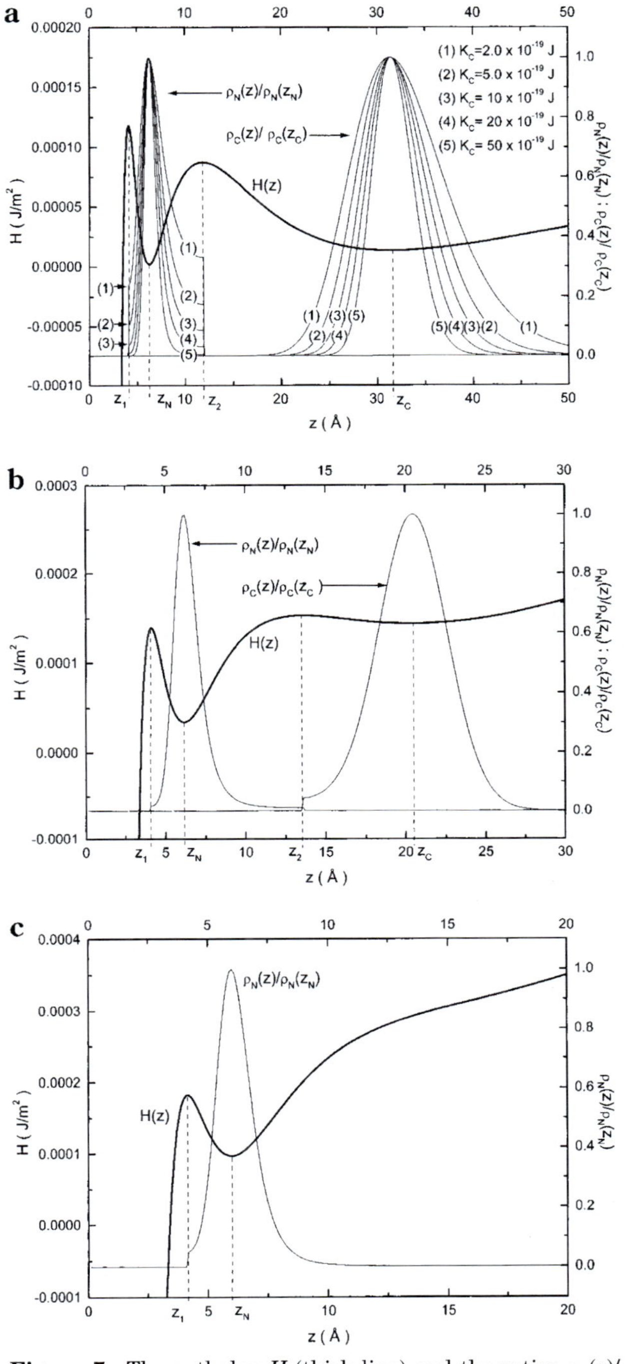

Figure 7. The enthalpy H (thick line) and the ratios $\rho_N(z)/\rho_N(z_N)$ and $\rho_C(z)/\rho_C(z_C)$ vs the distance z for the parameters $\Gamma_\infty = 5 \times 10^{-6}$ mol/m², $a_1 = 881$ m³/mol, $a_2 = -1.53$, $b_1 = -3.08 \times 10^{-22}$ J, $b_2 = -6.28 \times 10^{-14}$ J/m, $b_3 = 8.28 \times 10^7$ m^{-1}, $b_4 = 6.13 \times 10^{15}$ m^{-2}, $b_5 = -9.00 \times 10^{-23}$ J, $K_D = 0.050$ mol/dm³, $C_s = 1 \times 10^{-3}$ mol/dm³, $\lambda_1 = 1.0$ Å, $p_\perp/\epsilon' = 3.7$ D, $C_e = 0.165$ mol/dm³: (a) $p = 1.0 \times 10^4$ N/m², $K_C = (2.0, 5.0, 10.0, 20.0$, and $50.0) \times 10^{-19}$ J; (b) $p = 6.2 \times 10^4$ N/m², $K_C = 20.0 \times 10^{-19}$ J; (c) $p = 1.6 \times 10^5$ N/m², $K_C = 20.0 \times 10^{-19}$ J.

The bending modulus K_C of the interfaces of the soap films is unknown; however, one expects to be comparable to those of the lipid bilayers in water, which are of the order of 10^{-19} J.[12] In Figure 7a, the distributions of the thicknesses of the Newton and common black films, $\rho_N(z)/\rho_N(z_N)$ and $\rho_C(z)/\rho_C(z_C)$, respectively, are calculated for $C_e = 0.165$ mol/dm^3, $p_\perp/\epsilon' = 3.7$ D, $p = 1.0 \times 10^4$ N/m^2, and various values of K_C (2, 5,10, 20, and 50 $\times 10^{-19}$ J), the other parameters being as in section II.G. While the first metastable minimum (corresponding to a Newton black film) is somewhat deeper, the second one (corresponding to a common black film) is more stable to fluctuations, because $\rho_C(z_2)/\rho_C(z_C)$ is negligible, while $\rho_N(z_2)/\rho_N(z_N)$ is much larger. This is a result of the stronger spatial confinement of the undulation modes in the first minimum, which leads to a higher effective spring constant.

In what follows, it will be considered that the lifetime of a metastable state exceeds the duration of experiment if the potential barrier for a small surface of area S, $S\Delta H$, exceeds $3kT$. For the first minimum in Figure 7a, the potential barrier is higher than $3kT$ for $K_C > 20 \times 10^{-19}$ J; hence the Newton black films, once formed, remain stable in cases 4 and 5 during the experiment. In contrast, if the interfaces are more flexible (low K_C, cases 1, 2, and 3), the probability for an individual small surface to reach the height which separates two minima is much higher. In this case, the metastable equilibrium of the Newton black film has a shorter lifetime; while the film can either rupture or have a transition to a common black film, the second process has a higher chance. On the other hand, the common black films are more stable, because, while the minimum of the enthalpy is higher, the interfaces have more room to fluctuate.

The increase of the applied pressure decreases the stability of the common black film and the probability for a transition from a Newton to a common black film but also increases the probability of the rupture of the Newton black films. Figure 7b plots the enthalpy and the ratios $\rho_N(z)/\rho_N(z_N)$ and $\rho_C(z)/\rho_C(z_C)$, for $p = 6.2 \times 10^4$ N/m^2 ($K_C = 20 \times 10^{-19}$ J). The common black film becomes unstable (and a transition to a Newton black films occurs), while the Newton black film has a longer lifetime. At even a higher pressure (Figure 7c, $p = 1.6 \times 10^5$ N/m^2, $K_C = 20 \times 10^{-19}$ J) the Newton black film ruptures.

Assuming that a film is stable when the potential barrier for the individual surfaces in which the interface were decomposed is higher than a critical value (selected $3kT$ in the present paper), one can identify the regions of stability of the films. When the ratio of the distributions at the adjacent maximum and minimum of the enthalpy, $\rho(z_{max})/\rho(z_{min})$, exceeds $\exp(-3) = 0.05$, a transition to the next minimum is likely. In Figure 8a $\exp(-S\Delta H/kT)$ is plotted for $C_e = 0.165$ mol/dm^3 and $K_C = 20 \times 10^{-19}$ J against the applied pressure for Newton black films at the first ($\rho_N(z_1)/\rho_N(z_N)$, curve 1) and second maximum ($\rho_N(z_2)/\rho_N(z_N)$, curve 2) and for the common black film, at the second maximum ($\rho_C(z_2)/\rho_C(z_C)$, curve 3). Below $p = 6.0 \times 10^4$ N/m^2, both the Newton and the common black films have a long lifetime. When the external pressure exceeds 6.0×10^4 N/m^2 (marked C - N in Figure 8a) the common black film is likely to have a transition to a Newton black film, and the latter has a long lifetime. However, when the pressure exceeds 1.5×10^5 N/m^2 (marked N - R in Figure 8a), the Newton black film is likely to rupture. Therefore, both kinds of films have a long lifetime below 6.0×10^4 N/m^2, only the Newton black films exist for pressures in the range 6.0×10^4 to 1.5×10^5 N/m^2 and no black films exist at higher pressures.

It is clear that the stability of the films depends on the interaction between surfaces; in addition, the domains of stability of the black films are strongly dependent on their rigidity. The calculations presented in Figure 8b differ

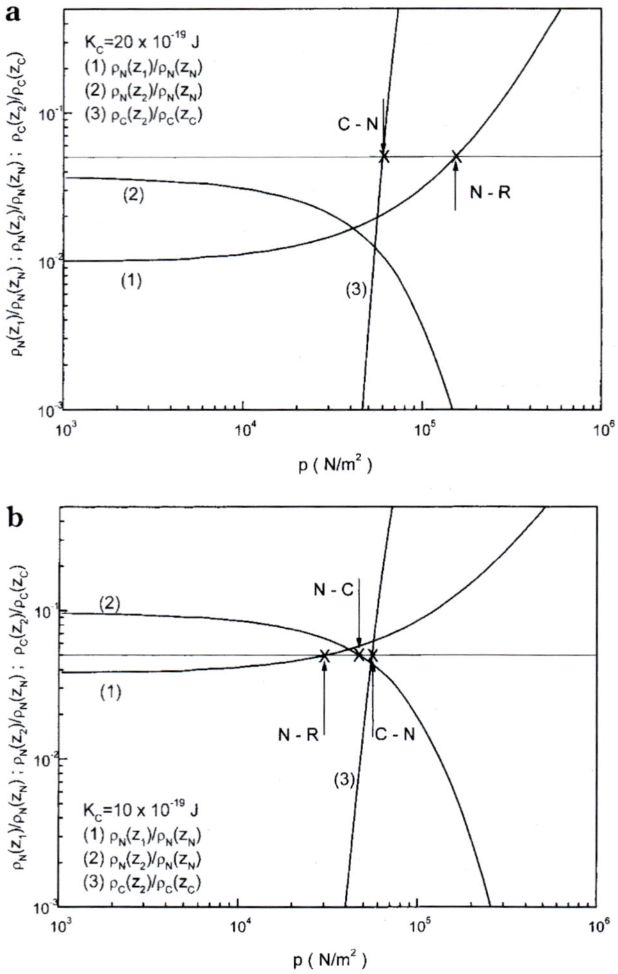

Figure 8. The ratios $\rho_N(z_1)/\rho_N(z_N)$, $\rho_N(z_2)/\rho_N(z_N)$, and $\rho_C(z_2)/\rho_C(z_N)$ as functions of the applied pressure for the same parameters as before ($\Gamma_\infty = 5 \times 10^{-6}$ mol/m^2, $a_1 = 881$ m^3/mol, $a_2 = -1.53$, $b_1 = -3.08 \times 10^{-22}$ J, $b_2 = -6.28 \times 10^{-14}$ J/m, $b_3 = 8.28 \times 10^7$ m^{-1}, $b_4 = 6.13 \times 10^{15}$ m^{-2}, $b_5 = -9.00 \times 10^{-23}$ J, $K_D = 0.050$ mol/dm^3, $C_s = 1 \times 10^{-3}$ mol/dm^3, $\lambda_1 = 1.0$ Å, $p_\perp/\epsilon' = 3.7$ D, $C_e = 0.165$ mol/dm^3): (a) $K_C = 20.0 \times 10^{-19}$ J; (b) $K_C = 10.0 \times 10^{-19}$ J.

from those of Figure 8a, only through a new value of the bending modulus, $K_C = 10 \times 10^{-19}$ J. However, the Newton black films are in this case unstable at any pressure; they are likely to have a transition to a common black film up to $p = 3.0 \times 10^4$ N/m^2, and either to rupture or to transform into common black films for pressures between 3.0×10^4 and 4.7×10^4 N/m^2. For higher pressures the Newton black films rupture immediately. The common black films are stable for pressures up to 5.5×10^4 N/m^2, after which they rupture. Therefore, in this case, the black films can exist only as common black films, at low applied pressures, but rupture at higher pressures.

The fluctuations increase the equilibrium thicknesses of the films (see Figures 5, 6a, and 7a), an effect which is particularly important for the common black films (see Figure 7a). This effect was found in the experiments of Exerowa et al.,[2] who noted that the DLVO theory (for planar surfaces) predicts a too small repulsion and suggested that the hydration force (which has a shorter range) cannot account for the discrepancy.

IV. Conclusions

In the first part of the paper, the interaction energy of thin soap films was calculated, assuming planar, parallel interfaces ($K_C \to \infty$). The interaction energy (and hence the disjoining pressure) could explain qualitatively the behavior of the black films and the existence of a transition from the common black films to the Newton black films. At low electrolyte concentrations, the disjoining pressure peak due to the double layer force (corresponding to the common black film) is high and might exceed the first peak, caused by the hydration repulsion (corresponding to the Newton black film). Hence, the increase of the pressure leads to the rupture of the common black film. The increase of electrolyte concentration has two opposite consequences: the double layer repulsion decreases and, hence, the height of the second peak decreases, while the charge recombination increases the ion-pair density on the surface and, hence, increases the hydration force, raising the height of the first peak.

However, the theory involving planar interfaces does not account completely for the transition from the common to the Newton black films and for the rupture of the common black films, which can occur in a range of pressures for the same remaining conditions. In the second part it is shown that the fluctuations of the thicknesses of the films lead to transitions which occur with different probabilities at an applied pressure.

The stability of the undulating films depends not only on the difference between the corresponding local minimum of the enthalpy and the adjacent maximum but also on the shape of the enthalpy in the vicinity of the minimum. The thermal fluctuations affect especially the stability of the Newton black films, because their spatial confinement is stronger. The confinement of the fluctuating interface can drive the Newton black films either to the common black film or to rupture. The present analysis indicates that the fluctuations of the interfaces, while decreasing the stability of the film, lead to larger thicknesses, in agreement with experiment.

LA011569W

Available online at www.sciencedirect.com
ScienceDirect
Journal of Colloid and Interface Science 309 (2007) 56–67
Journal of Colloid and Interface Science
www.elsevier.com/locate/jcis

On possible microscopic origins of the swelling of neutral lipid bilayers induced by simple salts

Marian Manciu [a], Eli Ruckenstein [b,*]

[a] Physics Department, University of Texas at El Paso, 500 W. University Ave, El Paso, TX 79968, USA
[b] Department of Chemical and Biological Engineering, State University of New York at Buffalo, Clifford C. Furnas Hall, Box 604200, Buffalo, NY 14260-4200, USA

Received 27 November 2006; accepted 1 February 2007
Available online 6 February 2007

Abstract

It was recently suggested that the swelling of neutral multilipid bilayers upon addition of a salt can be simply explained only by the electrolyte screening of the van der Waals attractions, while assuming that the hydration force and the repulsion due to thermal undulations of membranes are unaffected by the salt. While we agree that the screening of the van der Waals interactions plays a role, we suggest that the increase in the hydration force upon addition of a salt has also to be taken into account. In a statistical model, which accounts for the membrane undulations, parameters could be found to explain the multibilayer swelling even when the van der Waals attraction is considered unaffected by the electrolyte screening. These results point out that the decrease by a factor of three of the Hamaker constant upon addition of a salt, suggested recently to be responsible for the swelling of neutral multilipid bilayers, is perhaps too large, and a smaller decrease in Hamaker constant, coupled with the above mentioned effects might explain the swelling.

Keywords: Lipid bilayers; Swelling; Effect of salt

1. Introduction

The interest in lipid bilayers is due to their relevance to biological membranes [1]. They exhibit a richness of structures due to the interplay between many different inter- and intrabilayer forces. Among all the multilamellar bilayer structures, probably the most pertinent to biological membranes are the lamellar ones. Their equilibrium spacing is considered to be the result of a balance between attractive and repulsive forces. While the former forces are just the usual van der Waals interactions, the latter are composed of double layer forces (for charged bilayers) [2], hydration forces (due to the structuring of water near interfaces) [3] and repulsive forces generated by the thermal undulation of the membranes [4].

The hydration forces in neutral multilamellar lipid bilayers have been thoroughly investigated experimentally by subjecting them to high osmotic pressures [3] and it was observed that they can be well approximated by an exponential repulsion with a decay length λ_H of about 1–3 Å, which is almost independent of the electrolyte concentration [3]:

$$p_H = A_H \exp\left(-\frac{z}{\lambda_H}\right), \tag{1}$$

where z is the separation distance between bilayers in water and A_H is a constant. The unbinding transition of lamellar lipid bilayers (their swelling to very large repeat distances), observed experimentally [5], cannot be explained by an exponential repulsion (which is short-ranged), but requires another long range repulsion. It was suggested that this long-range repulsion is due to the confinement of the thermal undulations of the bilayers [4]:

$$p_{\text{und}} = \mu \frac{(kT)^2}{K_C z^3}, \tag{2}$$

where k is the Boltzmann constant, T the absolute temperature, K_C the bending rigidity and the Helfrich proportionality con-

* Corresponding author.
E-mail addresses: mmanciu@utep.edu (M. Manciu), feaeliru@acsu.buffalo.edu (E. Ruckenstein).

0021-9797/$ – see front matter
doi:10.1016/j.jcis.2007.02.001

stant $\mu = 0.115$. This force has a longer range than the van der Waals attraction, which at large separations becomes inverse proportional to the fifth power of the separation distance:

$$p_{\mathrm{vdW}}(z) = \frac{H}{6\pi}\left(\frac{1}{z^3} + \frac{1}{(z+2b)^3} - \frac{2}{(z+b)^3}\right) \approx \frac{2b^2 H}{\pi z^5} \quad (z \gg b), \tag{3}$$

where b is the bilayer thickness and H the Hamaker constant.

While the increase in temperature is expected to increase the undulation repulsion (Eq. (2)), and consequently the repeat distance between the lipid bilayers, it turned out that the repeat distance increases by lowering the temperature (anomalous swelling [6]). This behavior was attributed to a strong decrease of the bending rigidity K_C of the lipid bilayer as the temperature decreases [7,8] and hence to a drastic increase in the Helfrich repulsion at lower temperatures. It was also suggested that the unbinding of the lipid bilayers might be driven by a decrease in the van der Waals attraction [9]. Leibler and Lipowsky calculated a critical value for the Hamaker constant and suggested that it could be reached by modifying the polarizability of water, by the addition of a solute [9]. It was subsequently observed that the repeat distance increases upon addition of a carbohydrate solute [10] or some alkali-halide salts [11] to water, which can indeed decrease the Hamaker constant for lipid–lipid interactions through water.

It is natural to consider that the swelling of neutral lipid bilayers in water upon addition of a salt can be induced by a decrease of the van der Waals attraction (effect predicted long ago [12]), and not by the increase of the hydration or undulation forces, for the following reasons. If the increase of the Helfrich repulsion Eq. (2) would be responsible for the swelling, they will dominate the van der Waals attraction at any distance and the membranes will become completely unbound, since no equilibrium position can be generated by a long range repulsion coupled with a short range attraction. On the other hand, it is very unlikely that a hydration force of the form of Eq. (1) can be responsible for such a swelling. In order to double the equilibrium distance from 20 to about 40 Å, the pre-exponential factor A_H must increase about 1000 times, which is extremely unlikely to occur. Furthermore, the swelling of lipid bilayers upon addition of a salt remains large even for large ionic strengths; hence the charging of the membrane because of the salt cannot play a very important role. Indeed, for electrolyte concentrations larger than 1 M, the Debye–Hückel length is less than 3 Å, and therefore the double layer force has a short range effect comparable to the hydration provided by Eq. (1).

While Korreman and Posselt already suggested that the decrease in Hamaker constant, because of electrolyte screening, might be responsible for the swelling of neutral lipid bilayers by salts [11], an accurate test of this hypothesis has been carried out only recently by Petrache et al. [13,14]. Upon addition of 1 M of KCl or KBr, the repeat distance in DLPC (a 12 carbon-chain lipid, 1,2-dilauroyl-*sn*-glycero-3-phosphocholine) bilayers has increased, at 25 °C, from 58 Å, to 68 and 74 Å, respectively [13]. These increases correspond to increases in the water spacing from about 26 Å to about 35 and 41 Å, respectively [13]. Careful X-ray structural investigation showed that the membrane bending rigidity, K_C, is not affected much by the addition of salt (see Fig. 5A of Ref. [13]), and osmotic pressure experiments have shown that, at short separations, the interactions between bilayers are only weakly affected by the salt addition.

Petrache et al. also suggested that the different distances obtained for the multilamellar bilayer swellings by KCl and KBr can be explained by the preferential binding of Br^- (but not Cl^-) on the bilayers. By assuming that the "hydration force" (Eq. (1)) and the undulation repulsion (Eq. (2)) are not affected by the salt, and considering a binding constant $K = 0.22\ M^{-1}$ for Br^- (and $K = 0$ for Cl^-), Petrache et al. calculated the dependence of the Hamaker constant on the electrolyte concentration required to fit their experimental data [14]. Note that the suggestion of Petrache et al. [14], that Br^- binds to the surface whereas Cl^- does not, might explain the differences in swelling upon addition of KBr or KCl, but cannot explain the differences upon addition of KBr and NaBr, which have been revealed by the Korreman and Posselt experiments [11].

When plotted as functions of $\frac{z}{\lambda_{\mathrm{DH}}}$, where z is the separation distance through water and λ_{DH} is the Debye–Hückel length, the Hamaker constants H calculated for both kinds of salts collapsed on a single curve, that can be very well represented by [14]:

$$H = H_0\left(1 + 2\chi\frac{z}{\lambda_{\mathrm{DH}}}\right)\exp\left(-\chi\frac{2z}{\lambda_{\mathrm{DH}}}\right) + H', \tag{4}$$

where H_0 and H' represent the low-frequency and the high-frequency contributions, respectively, to the Hamaker constant and χ is an empirical correction parameter. For $\chi = 1$, the prediction of Mahanty and Ninham [12] could be recovered; however the value $\chi = 0.2$ was required to fit the experimental data.

The work of Petrache et al. [13,14] represents, in our opinion, an important contribution to the understanding of the stability of multilamellar lipid bilayers, and we agree that the decrease in the Hamaker constant via electrolyte screening constitutes an important effect that drives the swelling of lipid bilayers upon salt addition. The "universality" of the Hamaker constant dependence on the Debye–Hückel length, e.g., its independence of ion-specificity obtained experimentally, constitutes a strong argument in favor of their hypothesis. However, whether the "universality law" holds for other salts (than KBr and KCl) is not yet clear. Until more experiments will be carried out, we like to discuss some issues that are raised by the work of Petrache et al. [13,14] and we will show that by combining the recently proposed model for the origin of hydration forces [15] and the treatment of membrane undulations [16] might clarify some of these issues.

Firstly, the physical origin of the empirical parameter χ is not known. As suggested by Petrache et. al [13], it might simply represent a correction due to the interlamelar salt deficit (compared to the salt concentration in reservoir). However, since the Debye–Hückel length is inverse proportional to the square root of the ionic strength, to account for the value $\chi = 0.2$ the

58 M. Manciu, E. Ruckenstein / Journal of Colloid and Interface Science 309 (2007) 56–67

concentration of the interlamellar electrolyte must be about 25 times smaller than that in the reservoir. Such a salt deficit should provide a strong osmotic pressure, particularly at high ionic strengths, which would affect strongly the interactions between the lipid bilayers. Such an interaction has not been taken into account in the calculation of the Hamaker constant [14], where it was assumed that only the van der Waals interactions are affected by the salt addition.

The second issue is the extent of the decrease of the van der Waals interaction. Experiment and calculation of the van der Waals interactions between polystyrene latex beads and either a bare glass plane or a polystyrene coated glass plane [17] revealed that the Hamaker constant decreases only by about 25% at complete screening, while the experiments of Petrache et al. for neutral lipid bilayers require a decrease of about 75% (from $1.2kT$ to $0.4kT$). Such a strong decrease of the van der Waals interaction upon addition of salt would be expected to have strong consequences in the general theory of colloid stability, and not only in the stability of lipid bilayers.

The third issue is that, in determining the appropriate values of the Hamaker constant, it was supposed that both the exponential hydration (Eq. (1)) and the Helfrich repulsion (Eq. (3)) remain unchanged upon addition of a salt (and that the osmotic pressure due to the interlamellar salt deficit, discussed above, is negligible). While the exponential form for the hydration force has been determined at high osmotic pressures (corresponding to separations of around 10 Å) it is not clear that this behavior will remain the same at distances of the order of 40 Å. For a decay length of 2.2 Å [14], the magnitude of the hydration forces decreases about 10^6 times between 10 and 40 Å, and even a minuscule deviation from the exponential behavior will have drastic consequences in the evaluation of the Hamaker constant. The problems associated with a constant Helfrich repulsion (due to the thermal undulations of the bilayers) are even more complicated and will be addressed in detail in Section 3.

The forth issue is the increase in the repulsion between bilayers at short distances. In Fig. 1, the osmotic pressure is plotted as a function of separation distance (data from Ref. [13]) for no added salt, for 1 M KCl and for 1 M KBr. They reveal an increase in repulsion at short separation distances upon addition of salt. While the relatively small difference between 1 M KCl and 1 M KBr can be attributed to the charging of the neutral lipid bilayers by the binding of Br^- (but not Cl^-) [14], the relatively large difference between no salt and 1 M KCl is more difficult to explain. Even a zero value for the Hamaker constant (continuous line (2) in Fig. 1), in the 1 M KCl case, is not enough to explain the increase in repulsion, determined experimentally. The screening of the van der Waals interaction, at distances of the order of three Debye–Hückel lengths (about 10 Å) should lead, according to Petrache et al. calculations, to a decrease of only about 30% of the Hamaker constant (from $1.2kT$ to about $0.8kT$, see Fig. 5C of Ref. [14]). Therefore, an additional mechanism to increase the hydration repulsion or the undulation force (or both) upon addition of salt should exist to explain the experiments.

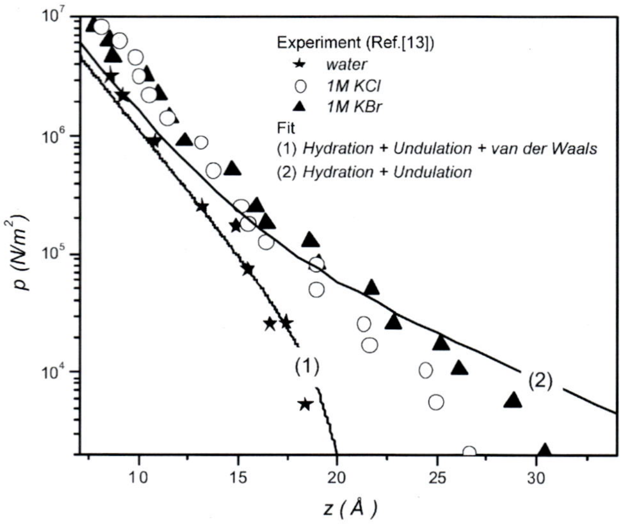

Fig. 1. Experimental values for the osmotic pressure as a function of separation distance from Ref. [13] (stars: water; circles: 1 M KCl; triangles: 1 M KBr) are compared with calculations based on the simple equations ((1), (3) and (9)), with parameters reported in Ref. [13] (note that in Ref. [13] it was suggested that hydration interaction increases upon addition of salt): $A_H = 1.6 \times 10^8$ N/m², $\lambda_H = 2.1$ Å, $H = 9.2 \times 10^{-21}$ J, $b = 39$ Å, $K_C = 5.8 \times 10^{-20}$ J, $A_{fl} = 1.06$ Å^{-2}, $\lambda_{fl} = 6.0$ Å (Line 1). Even for $H = 0$ and the rest of the parameters as before (Line 2), the repulsion at short separations is weaker than in the experiment. This points out that either hydration and/or undulation forces must increase upon addition of electrolyte.

2. Predictions of the polarization model for hydration forces between bilayers

We noted recently that the polarization model for hydration forces [18] can explain a modest increase in repulsion at short separations (mostly via the recombination of the charges of the surface with the ions of the electrolyte, as surface dipoles, which generate a polarization field in the adjacent water molecules). The polarization model predicted also an increase in repulsion at large separations, because of the correlations between neighboring dipoles and the coupling interactions between the electrolyte ions and the water dipoles [15]. The framework of the polarization model is briefly summarized in Appendix A. The coupling between the double layer and the polarization leads to the following relation between the average dipole of a water molecule m and the electric field $E = -\frac{d\psi}{dz}$:

$$\lambda_m^2 \frac{d^2 m(z)}{dz^2} = m(z) + \varepsilon_0 v_0 (\varepsilon - 1) \frac{d\psi(z)}{dz}, \tag{5}$$

where v_0 is the volume of a water molecule, ε_0 is the vacuum permittivity, ε the dielectric constant of water, and λ_m the correlation length of neighboring dipoles. When the water is organized in ice-like layers in the vicinity of the surface, the latter has a value of about $\lambda_m \sim 15$ Å [15], which is much larger than the decay lengths corresponding to both the traditional hydration force (~1–3 Å) and to the double layer interactions at high ionic strength ($\lambda_{DH} \sim 3$ Å at 1 M KCl).

The polarization model can be combined with ion-specific effects induced by the ions that approach the surface (with preference for the bulk water for structure making (SM) ions and preference for the interface for structure-breaking (SB) ones), which can be well described by simple potentials of interactions between the ions and the surface [19].

The chemical potential of an ion of species "i" and of charge q_i in the liquid is given by:

$$\mu_i(z) = \mu_{0i} + kT\ln\big(\gamma_i c_i(z)\big) + q_i\psi(z) + W_i(z), \tag{6}$$

where k is the Boltzmann constant, T the absolute temperature, and μ_{0i} represents the standard chemical potential of the ions of species "i," of charge q_i and concentration $c_i(z)$. The activity coefficient γ_i of the ions of species "i" will be assumed unity. The interaction free energy $W_i(z)$ should include all the other interactions of the ion with the medium, not accounted for by the mean field electrical potential ψ. Assuming that the system is in contact with an infinite reservoir in which the ion concentrations are c_{0i}, the total charge density is provided by:

$$\rho_{\mathrm{el}}(z) = \sum_i q_i c_{0i}\exp\left(-\frac{q_i\psi(z) + \Delta W_i(z)}{kT}\right), \tag{7}$$

where $\Delta W_i(z)$ is calculated with respect to the bulk.

The Poisson equation can be combined with Eq. (7) to yield:

$$\varepsilon_0\frac{\mathrm{d}^2\psi(z)}{\mathrm{d}z^2} = -\sum_i q_i c_{0i}\exp\left(-\frac{q_i\psi(z) + \Delta W_i(z)}{kT}\right) + \frac{\mathrm{d}m(z)}{v_0\,\mathrm{d}z}. \tag{8}$$

Equations (5) and (8) represent a system of coupled nonlinear equations, which can be solved for $\psi(z)$ and $m(z)$, under appropriate boundary conditions, which are discussed in Appendix A. The various contributions to the free energy are also calculated in Appendix A.

In Fig. 2a, the free energies f for the hydration interaction, per unit surface, plotted as functions of the distance $2d$ between two identical surfaces, are calculated for various concentrations of an 1:1 electrolyte, by assuming that $\frac{p_\perp}{\varepsilon'} = 2$ Debyes (where $p_\perp$ is the dipole moment of the surface dipoles and ε' is the dielectric constant in their neighborhood), the average area occupied by each dipole on the surface $A = 50$ Å^2, the surface charge density $\sigma = 0$, and neglecting the SM/SB interactions of the ions (e.g., $\Delta W_C \equiv \Delta W_A \equiv 0$, see Appendix A). While at large separations the hydration increases strongly upon addition of electrolyte, in the vicinity of the surface the repulsion actually decreases at high ionic strengths. However, this repulsion is smaller, at any distance, than the van der Waals attraction (continuous line in Fig. 2a, calculated with the values $H = 9.2 \times 10^{-21}$ J and $b = 39$ Å, employed in Refs. [13] and [14]). By assuming that for $d < 4$ Å, $\Delta W_C = 1kT$ and $\Delta W_A \equiv 0$ (Fig. 2b) or that $\Delta W_C = 1kT$ and $\Delta W_A = 1kT$ (Fig. 2c), the free energy of interaction becomes larger in the vicinity of the surface, but still remains smaller than the van der Waals attraction. This means that other forces must be present to explain the equilibrium of neutral lipid bilayers. Of course, one might assume that these additional forces are due to the confinement of the membranes [4]. Their discussion is postponed for the next section.

Let us also note that at large separations, the polarization model provides a very large repulsion, hence that the swelling upon addition of salt might be well over-predicted for concentrations of the order of 1 M. In addition, to represents the experiments at high osmotic pressures (see Fig. 1) an exponential hydration (Eq. (1)) with a decay length of about 2.2 Å is required [14], while the polarization model generates a repulsion with a much larger decay length. There is a simple explanation for this behavior. A dipole correlation length $\lambda_m = 14.9$ Å has been employed for water that is strongly organized in ice-like layers in the vicinity of a surface [15]. Since it is obvious that the lipid bilayers undulate thermally (otherwise, their anomalous swelling or their unbinding cannot be explained), it is impossible for the water molecules to remain organized in ice-like layers near the undulating bilayers. A model involving the assumption of a totally random orientation of water molecules, proposed recently [20], indicates that the correlation length might become in this case as low as $\lambda_m = \frac{6.27}{\varepsilon''}$ Å, where ε'' is the dielectric constant for the screening of the interactions between neighboring atoms (much lower than the bulk dielectric constant of water, $\varepsilon = 80$). A value $\lambda_m = 4$ Å has been shown to represent well the hydration forces between silica surfaces [20]. However, a low value for λ_m still does not provide a viable solution for the interactions between neutral lipid bilayers; indeed, for $\lambda_m = 4$ Å, the free energy decreases sufficiently rapid in the vicinity of the surface (Fig. 3a, $\frac{p_\perp}{\varepsilon'} = 2$ Debyes, $A = 50$ Å^2, $\sigma = 0$ and $\Delta W_C \equiv \Delta W_A \equiv 0$), but still remains much smaller that the van der Waals interaction (continuous line, Fig. 3a). This result can be easily understood. Indeed, if the coupling between dipoles is small, the energy stored in the dipole–dipole interactions is also small. To generate a hydration interaction comparable with the van der Waals attraction, huge dipoles (about 50 Debyes for each 60 Å^2 of surface) must reside on the surface, which is unrealistic. To form an idea about the magnitude of the repulsion required to explain the stability of neutral lipid bilayers, in Fig. 3b the exponential free energy of hydration derived in Ref. [13] to fit the experiments on lipid bilayers (namely, $f_{\mathrm{hyd}}(z) = 0.0336\exp(-\frac{z}{2.1})$ J/m^2 is compared with an exponential "hydration force" employed recently, $f_{\mathrm{hyd}}(z) = 0.00053\exp(-\frac{z}{5.4})$ J/m^2 [20] to fit the experiments on the interactions between silica surfaces [21]. The later surfaces are known for their strong hydration interactions; silica colloids typically do not coagulate, regardless of the ionic strength. In the vicinity of the surface it seems that the hydration repulsion in neutral lipid bilayers is about two orders of magnitude larger than those in silica, which is again unrealistic. It is therefore likely that the thermal undulations play a considerable role for lipid bilayers, particularly at low separations. Let us therefore address in what follows in some detail the issue of the thermal undulations of the membranes.

3. Thermal undulations of lipid bilayers

The role of thermal undulations of membranes in their stability [4] and particularly in the unbinding transition [5] has been early recognized. Helfrich derived the first analytical expression for the repulsion due to the entropic confinement of the membrane, by assuming hard-wall repulsions between membranes. When the membranes are far apart, they can undulate freely;

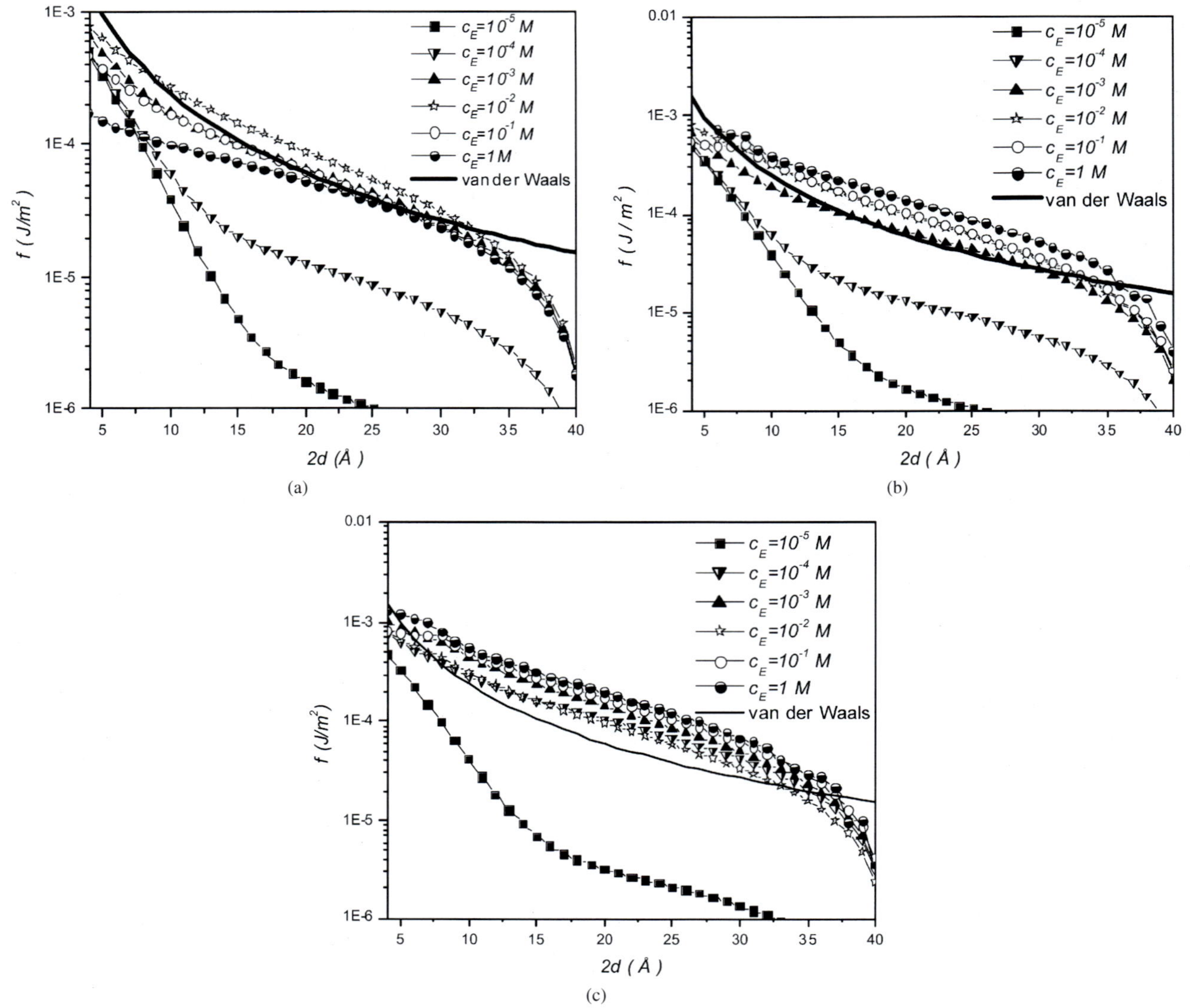

Fig. 2. Hydration interaction energy as a function of distance between surfaces for various electrolyte concentrations and $\frac{p_\perp}{\varepsilon'} = 2$ Debyes, $A = 50$ Å^2, $\sigma = 0$, $\lambda_m = 14.9$ Å, for: (a) $\Delta W_C \equiv \Delta W_A \equiv 0$; (b) $\Delta W_C = 1kT$ and $\Delta W_A = 0$ for $z < 4$ Å, $\Delta W_C \equiv \Delta W_A \equiv 0$, $z > 4$ Å; (c) $\Delta W_C = 1kT$ and $\Delta W_A = -1kT$ for $z < 4$ Å, $\Delta W_C \equiv \Delta W_A \equiv 0$, $z > 4$ Å. The continuous line represents the van der Waals interaction (Eq. (3)), for $H = 9.2 \times 10^{-21}$ J and $b = 39$ Å.

however, when two membranes approach each other, the hard wall repulsion between them attenuates the undulations. The entropic confinement of the membranes generates a repulsion and the corresponding free energy is inversely proportional to the square of the average separation between membranes [4].

The Helfrich theory is valid for hard-wall interaction between membranes. However, at small separation additional interactions have to be taken into account.

For typical membranes interacting via an exponential repulsion (hydration force) and a van der Waals interaction, it was shown that at small separations the undulation interaction has an exponential behavior [16,22,23]. However, the exponential repulsion cannot be valid at large separation, because it cannot predict the unbinding of the membranes (an exponential is always of shorter range that the van der Waals attraction). In general, the undulation repulsion is treated as an additive interaction, which is exponential at short separations [22]:

$$p_{\text{und}} = \left(\frac{kT}{2\pi}\right)^2 \frac{1}{K_C} \frac{A_{\text{fl}}}{\lambda_{\text{fl}}} \exp\left(-\frac{z}{\lambda_{\text{fl}}}\right), \quad z \text{ small}, \tag{9}$$

where the scale factor A_{fl} and the decay length λ_{fl} might be regarded as fitting parameters [13], while it has a power law behavior at large separations, as provided by Helfrich [4]:

$$p_{\text{und}} = \mu \frac{(kT)^2}{K_C z^3}, \quad z \text{ large}. \tag{2}$$

The above two different functions have been used by Petrache et al. to fit the interactions between bilayers at small (Ref. [13], Eq. (9)) and large (Ref. [14], Eq. (2)) separations. Since the ranges of validity of the above equations (Eq. (2) and Eq. (9)) are not known, it is clear that the calculation of the Hamaker constant based on either of them is not very accurate.

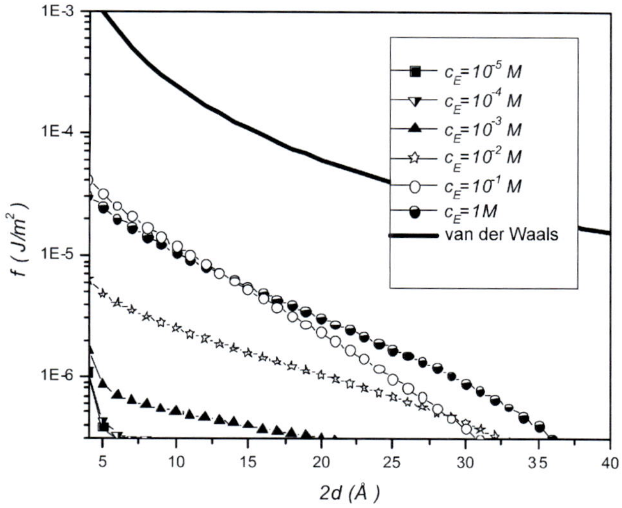

(a)

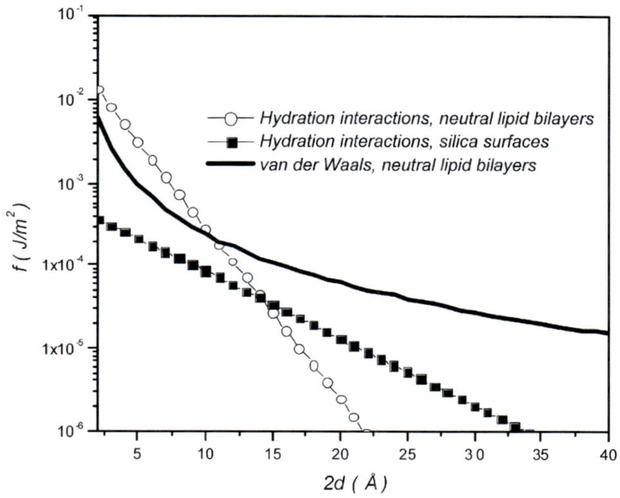

(b)

Fig. 3. (a) Hydration interaction energy as a function of distance between surfaces for various electrolyte concentrations and $\frac{p_\perp}{\varepsilon'} = 2$ Debyes, $A = 50$ Å^2, $\sigma = 0$, $\lambda_m = 4$ Å and $\Delta W_C \equiv \Delta W_A \equiv 0$. The thick line represents the van der Waals interaction (Eq. (3)), for $H = 9.2 \times 10^{-21}$ J and $b = 39$ Å. (b) The exponential approximation for the hydration interaction, required to explain the experiment for neutral lipid bilayers (Ref. [13]) is compared with the exponential approximation for the hydration interaction, required to explain the experiment for silica surfaces [21]. The thick line represents the van der Waals interaction (Eq. (3)), for neutral lipid bilayers ($H = 9.2 \times 10^{-21}$ J and $b = 39$ Å).

More importantly, the undulation force cannot be treated as an additive interaction [16].

Helfrich' original theory assumed that the membranes interact only with a hard-wall potential, but when interactions become longer range, they affect themself the undulation of the membranes, contributing to their confinement. In this case, there is a mutual interdependence between the thermal fluctuations and the interaction potentials, which cannot be any longer assumed independent of each other, hence they cannot be simply additive.

To account for the interdependence between thermal fluctuations and interaction potentials, we proposed recently an alternate approach [16,24,25], in which the elastic membranes were allowed to undulate thermally in a potential provided by all the interactions between membranes. Perfectly rigid membranes (i.e., $K_C = \infty$) will align themselves parallel to each other, at a separation distance corresponding to the minimum of the interaction potential. However, when $K_C \neq \infty$, the equilibrium of the membrane involves two contributions to the free energy, which generate a continuous distribution of distances between bilayers, $\rho_{\text{und}}(z)$ [16,24] (see Appendix B). Large thermal undulations of the membranes (wide distribution $\rho_{\text{und}}(z)$) increase the total energy, because parts of the membranes wander into regions of higher potentials; in contrast, the suppression of the undulations (narrow distribution $\rho_{\text{und}}(z)$, hence low values for the root mean square fluctuation σ_{und}) increases the entropy of the system. The equilibrium between membranes is provided by the minimum of the total free energy of the system, which is the sum between an energetic and an entropic contribution [16,24,25] (see Appendix B). By using an approximate equation for the entropic contribution [16]:

$$(-T\Delta S)_{\text{und}} = \frac{(kT)^2}{64K_C\sigma_{\text{und}}^2}, \tag{10}$$

where σ_{und} is the root mean square fluctuation of the interbilayer distance (see Appendix B), it was shown that the undulation repulsion can be approximated by an exponential at short separations but has a power law behavior at large separations [16].

When the system is subjected to an external pressure p_{ext}, the undulations of the membranes occur in such a manner to minimize the total Gibbs free energy of the system, which is the sum between the enthalpy of the system and the entropic contribution provided by Eq. (10) (see Appendix B):

$$G(s) = \int_u \rho_{\text{und}}(u)\big(f(u) + p_{\text{ext}}u\big)\,\mathrm{d}u + \frac{(kT)^2}{64K_C\sigma_{\text{und}}^2}, \tag{11}$$

where f is the free energy of interaction per unit area between planar surfaces. For each value of the external pressure, an average separation between membranes can be found. This treatment was presented in details elsewhere [16,24,25], and only a brief summary is reproduced in Appendix B. It should be noted that $G(s)$ represents the total free energy of the system, including the interaction potentials (hydration and van der Waals interactions) as well as the undulations entropy of the membranes, and s is a parameter which is used for the minimization of G. Equation (11) shows that the interactions and undulations are intermingled and that their effects are not additive.

Let us evaluate the effect of this treatment of the fluctuations on the interactions between bilayers. Let us first consider the simple interaction potential between membranes, employed by Petrache et al., with the hydration interaction corresponding to Eq. (1):

$$f_{\text{hyd}}(z) = A_H\lambda_H \exp\left(-\frac{z}{\lambda_H}\right), \tag{12}$$

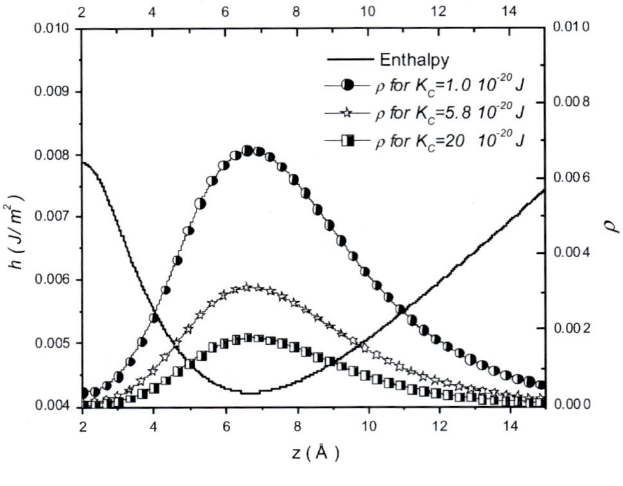

Fig. 4. The enthalpy as a function of distance between bilayers for $p_{ext} = 5 \times 10^6$ N/m², $A_H = 1.6 \times 10^8$ N/m², $\lambda_H = 2.1$ Å, $H = 9.2 \times 10^{-21}$ J and $b = 39$ Å and the corresponding distribution of membrane interdistances, $\rho(z)$, obtained from the minimization of the Gibbs free energy for $K_C = 1 \times 10^{-20}$ J, $K_C = 5.8 \times 10^{-20}$ J and $K_C = 20 \times 10^{-20}$ J.

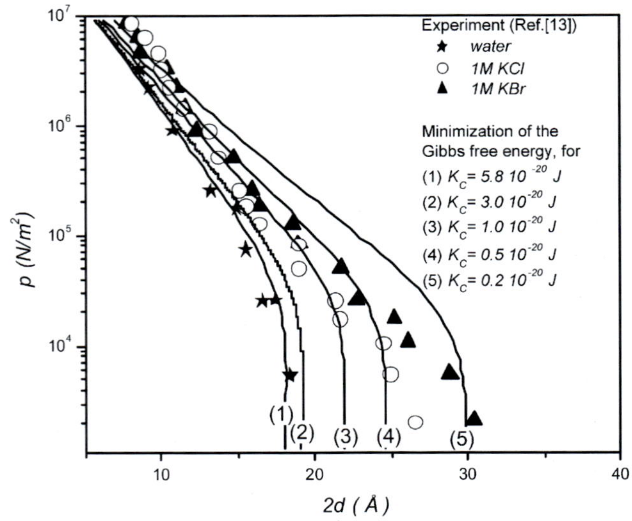

Fig. 5. Experimental values for the osmotic pressure as a function of separation distance from Ref. [13] (stars: water; circles: 1 M KCl; triangles: 1 M KBr) are compared with calculations for constant parameter values for hydration ($A_H = 1.6 \times 10^8$ N/m², $\lambda_H = 2.1$ Å) and van der Waals ($H = 9.2 \times 10^{-21}$ J and $b = 39$ Å) interactions, but various values for K_C.

where $A_H = 1.6 \times 10^8$ J/m³ and $\lambda_H = 2.1$ Å, and a van der Waals interaction:

$$f_{vdW}(z) = -\frac{H}{12\pi}\left(\frac{1}{z^2} + \frac{1}{(z+2b)^2} - \frac{2}{(z+b)^2}\right). \tag{13}$$

For an external pressure p_{ext}, the membranes undulations minimize the Gibbs free energy, based on the enthalpy:

$$h(z) = p_{ext}z + A_H\lambda_H \exp\left(-\frac{z}{\lambda_H}\right) + \frac{H}{12\pi}\left(\frac{1}{z^2} + \frac{1}{(z+2b)^2} - \frac{2}{(z+b)^2}\right) \tag{14}$$

and the entropic term provided by Eq. (10) (see Fig. 4).

Let us note that an additional, hard-wall like repulsion has to be taken into account at short separation distances, otherwise during their undulations, parts of the membranes will approach at distances less than about 1 Å, and therefore will be driven in the deep first minimum of the interaction. However, experiment shows that such a collapse does not occur, and therefore separations less than a cut-off distance $d_{cut\text{-}off} = 1$ Å should be considered inaccessible to the membranes. The distances between membranes will obey a Boltzmann distribution, with the prefactor determined in such a manner to minimize the total Gibbs free energy of the system (a large prefactor corresponds to membranes confined to their minimum enthalpy, but having a large entropy because σ_{und} is small; whereas a small prefactor corresponds to a small entropy, but large enthalpy, because the membranes with higher elasticity will undulate at larger distances from the minimum of the enthalpy, see Fig. 4)

For the interaction potentials provided by Eqs. (12) and (13) and various values of K_C, the average separation as a function of the external pressure is compared in Fig. 5 with experiment (Ref. [13]) for neutral lipid bilayers at different salt concentrations. Note that the force due to the confinement of the undulation is not simply additive to the other interactions (hydration and van der Waals) and that a change in the elasticity of the membrane can have a strong effect on the equilibrium distance, even if it does not affect much the interactions at short separations.

One interesting questions is why the membranes never collapses, even when the undulations might drive part of them at separations less than 1 Å. One possible explanation is provided by the strong, hard-wall like, Born repulsion between the surface dipoles and their bound water molecules. Another possible explanation is that the polar headgroups of the neutral lipid bilayers, being hydrophilic, are more closely related to water than to the hydrocarbon region of the bilayers. Therefore, the van der Waals interaction between neutral bilayers might be better described by [16]:

$$f_{vdW}(z) = \frac{H}{12\pi}\left(\frac{1}{(z+t_h)^2} + \frac{1}{(z+t_h+2b)^2} - \frac{2}{(z+t_h+b)^2}\right) \tag{15}$$

instead of Eq. (13), where t_h is related to the thickness of the headgroup of the lipid bilayers (it should be less than twice the physical thickness of the headgroups of each layer, but larger than zero). In this case, the van der Walls attraction does not diverge, even for $z = 0$, and its magnitude in the vicinity of the surface is drastically affected by the value of t_h. Since the membranes are typically confined closer to the minimum of the potential, the decrease of the absolute value of this minimum allows the membrane to wander in more distant regions of the potential. Consequently, the decrease in the van der Waals interaction in the vicinity of the interface has a strong effect even at large separations, because of the thermal undulations, and is felt by the system as a strong long-range repulsion. This effect is depicted in Fig. 6, where the pressure is plotted as a function of distance for $K_C = 5.8 \times 10^{-20}$ J and various values for t_h.

M. Manciu, E. Ruckenstein / Journal of Colloid and Interface Science 309 (2007) 56–67 63

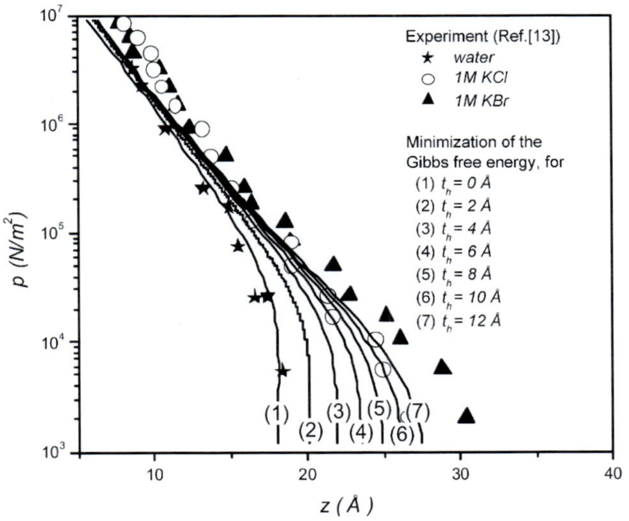

Fig. 6. Experimental values for the osmotic pressure as a function of separation distance from Ref. [13] (stars: water; circles: 1 M KCl; triangles: 1 M KBr) are compared with calculations for constant parameter values for hydration ($A_H = 1.6 \times 10^8$ N/m², $\lambda_H = 2.1$ Å) and $K_C = 5.8 \times 10^{-20}$ J, but various values for t_h for the van der Waals interactions (with $H = 9.2 \times 10^{-21}$ J and $b = 39$ Å).

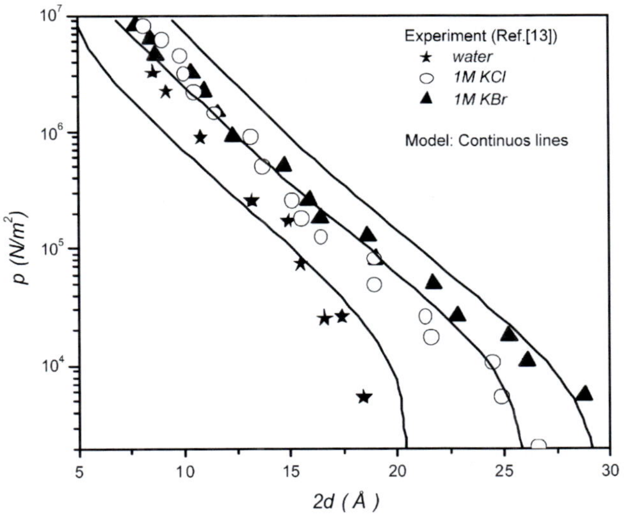

Fig. 7. Experimental values for the osmotic pressure as a function of separation distance from Ref. [13] (stars: water, circles: 1 M KCl; triangles: 1 M KBr) are compared with calculations that combine the polarization model for hydration forces with the statistical treatment of the undulation, for the following values of the parameters: $\frac{p_\perp}{\varepsilon'} = 9$ Debyes, $A = 60$ Å², $\Delta' = 1$ Å, $\sigma = 0$, $\lambda_m = 6$ Å, $d_{\text{cut-off}} = 2.5$ Å, $K_C = 4 \times 10^{-20}$ J, $t_h = 6$ Å, $H = 9.2 \times 10^{-21}$ J, $\Delta W_K = 0.5kT$ for $d < 4$ Å, $\Delta W_{Cl} = 0$ and $\Delta W_{Br} = 1kT$ for $d < 4$ Å, ΔW_K, ΔW_{Cl} and $\Delta W_{Br} = 0$ for $d > 4$ Å.

4. Discussion and conclusions

If one considers that the thermal undulations generate a constant, additive interaction, independent of salt concentration, the polarization model cannot describe the hydration interactions between neutral lipid bilayers, because for $\lambda_m = 14.9$ Å the interaction is too strong at large separations and large ionic strengths, and has a too large decay length at small separations and high ionic strengths. On the other hand, for smaller values of the dipole correlation lengths (e.g., $\lambda_m = 4$ Å) the polarization model predicts a very week hydration force between surfaces, much smaller than the van der Waals attraction.

However, it was noted in a previous section that the repulsion due to thermal undulations should not be treated as an additive, constant interaction. Whenever the interactions between membranes (double layer, hydration, van der Waals) change, so does the contribution due to thermal fluctuation, even if the bending rigidity K_C remains the same (which is not completely true). In addition to the traditional interactions (double layer, hydration and van der Waals) one has to account also for the Born-like repulsion between surface headgroups, as well as for the possible decrease of the van der Waals attraction between bilayers, due to the polar nature of their headgroups. While both of these effects affect the interaction potential between rigid surfaces only at very short separations (less than 10 Å), the undulations of the membranes propagate them at much larger separations. In the present treatment of the thermal undulation, it is possible to associate the swelling of the membrane with an increase in the hydration forces provided by the polarization model, even if one considers that the bending rigidity K_C and the Hamaker constant are not affected at all by the addition of salt. For example, in Fig. 7 we calculated the interactions as function of the distance, using for the interaction potentials the hydration force provided by the polarization model (Eq. (A.21)) and the van der Waals interaction (Eq. (15)) using the parameters $\frac{p_\perp}{\varepsilon'} = 9$ Debyes, $A = 60$ Å², $\Delta' = 1$ Å, $\sigma = 0$, $\lambda_m = 6$ Å, $d_{\text{cut-off}} = 2.5$ Å, $t_h = 6$ Å, $H = 9.2 \times 10^{-21}$ J, $\Delta W_K = 0.5kT$ for $d < 4$ Å, $\Delta W_{Cl} = 0$ and $\Delta W_{Br} = 1kT$ for $d < 4$ Å). The effect of the thermal undulations has been calculated as suggested above, by the minimization of the Gibbs free energy of the system, using $K_C = 4 \times 10^{-20}$ J.

While one can, in principle, explain the swelling of bilayers upon addition of salt even by assuming that the Hamaker constant does not depend on the electrolyte concentration, this was not the purpose of this paper. Note that the value $\frac{p_\perp}{\varepsilon'} = 9$ Debyes employed in the previous calculations is too large for surface dipoles. We fully agree with Petrache et al. that the Hamaker constant is decreased by the addition of salt, maybe not as much as 75% but nevertheless by a measurable quantity. What we tried to emphasize here is that there are so many unknown quantities in the interactions between lipid bilayers, that it is very difficult to obtain reliable information for the dependence of the Hamaker constant on electrolyte concentrations from this type of experiments.

Acknowledgments

We are indebted to Horia Petrache for valuable discussions and for providing us the data published in Refs. [13] and [14].

Appendix A. Review of the polarization model

The interactions of ions with all the charges contained between two planar surfaces, separated by a distance $2d$, can be described by a mean field potential $\psi(z)$ which obeys the Poisson equation:

$$\frac{\mathrm{d}}{\mathrm{d}z}\big(\varepsilon_0 E(z) + P(z)\big) = \rho_{\mathrm{el}}(z), \tag{A.1}$$

where ε_0 is the vacuum permittivity, $E \equiv -\frac{\mathrm{d}\psi}{\mathrm{d}z}$ is the electric field, P the polarization and ρ_{el} the charge density.

At microscopic scale, a water molecule is subjected to a "local" electric field, which is the sum between a "macroscopic field" generated by sources from outside the medium, and a field generated by the other water molecules surrounding it [26]. For a uniformly polarized medium, the latter corresponds to the field generated in a spherical cavity in a medium, due to the polarization of the whole medium (the Lorenz field $E_{\mathrm{Lorenz}} = \frac{P}{3\varepsilon_0}$ [26]). Therefore, the local field acting on a water molecule in an uniformly polarized medium is given by:

$$E_{\mathrm{local}} = E + \frac{P}{3\varepsilon_0} \tag{A.2}$$

and the average dipole moment of the water molecule, $m = Pv_0$, where v_0 is the volume occupied by a water molecule, by [18]:

$$\begin{aligned} m &= \gamma E_{\mathrm{local}} = \gamma\left(E + \frac{m}{3\varepsilon_0 v_0}\right) \\ &= \frac{\gamma}{1 - \frac{\gamma}{3\varepsilon_0 v_0}} E \equiv \varepsilon_0 v_0(\varepsilon - 1)E, \end{aligned} \tag{A.3}$$

where γ is the microscopic polarizability.

If the polarization is nonuniform, an additional field E_P is generated by the neighboring dipoles. A simple way to calculate this field [15] is to neglect the field generated by the remote dipoles, because it is screened by the intervening water molecules, and to consider only the field generated by the neighboring dipoles, screened by an effective dielectric constant ε'', much lower than the bulk value, $\varepsilon = 80$. In order to calculate the additional field E_P, the local structure of water must be known. A simple assumption is that the water is organized in ice-like layers of thickness Δ in the vicinity of the surface (see Fig. A.1). The water molecules near the surface dipoles (gray area in Fig. A.1) are assumed to be bound to the surface dipoles and the distance between the center of the first ice-like water layer and the center of the surface dipoles is denoted Δ'. The additional field in layer "i," $E_{P,i}$ is generated by the neighboring water molecules from the same layer "i," as well as by the neighboring water molecules from the adjacent layers, "$i-1$" and "$i+1$." The average dipole moment of a water molecule, m_i, is constant in each layer, because the layers are assumed parallel to the surface, and m is a function of the distance from the surface z alone. Since the field generated by the dipoles is proportional to their dipole moments, the additional field can be written as:

$$E_{P,i} = C_1 m_{i-1} + C_0 m_i + C_1 m_{i+1}, \tag{A.4}$$

where [15]:

$$C_0 = -\frac{3.766}{4\pi\varepsilon_0\varepsilon'' l^3}, \qquad C_1 = \frac{1.827}{4\pi\varepsilon_0\varepsilon'' l^3}, \tag{A.5}$$

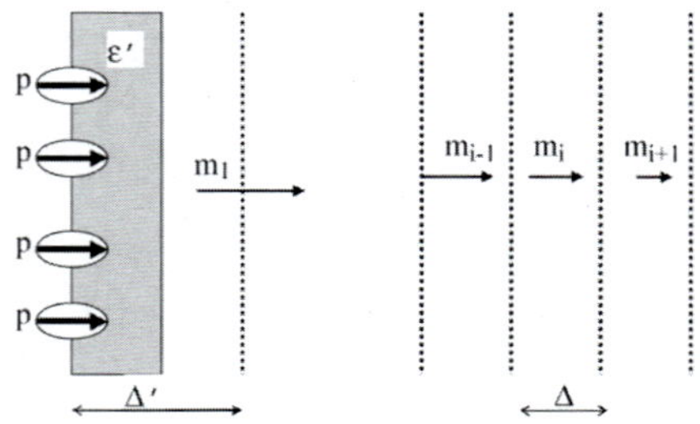

Fig. A.1. A sketch of water layering in the vicinity of a dipoles-bearing surface. In the vicinity of the surface dipoles p, the bounded water (gray) has the dielectric constant ε', the center of the first water layer is at the distance Δ' from the center of surface dipoles.

l is the distance between the centers of two adjacent water molecules, and ε'' the dielectric constant for the interaction between neighboring molecules, which is expected to be much lower than 80. For a nonuniform polarization, the local field is therefore given by:

$$E_{\mathrm{local},i} = E + \frac{m_i}{3\varepsilon_0 v_0} + C_0 m_i + C_1(m_{i-1} + m_{i+1}) \tag{A.6}$$

and the average dipole moment of a water molecule by:

$$\begin{aligned} m_i &= \gamma E_{\mathrm{local}} = \gamma\left(E + \frac{m}{3\varepsilon_0 v_0} + E_P\right) = \frac{\gamma}{1 - \frac{\gamma}{3\varepsilon_0 v_0}}(E + E_P) \\ &= \varepsilon_0 v_0(\varepsilon - 1)(E + E_P) \\ &= \varepsilon_0 v_0(\varepsilon - 1)\big(E + C_0 m_i + C_1(m_{i-1} + m_{i+1})\big). \end{aligned} \tag{A.7}$$

By expanding in series the average dipole moment, Eq. (A.6) becomes:

$$\begin{aligned} E_P(z) &= C_0 m(z) + C_1\left(2m(z) + \Delta^2 \frac{\mathrm{d}^2 m(z)}{\mathrm{d}z^2} + \cdots\right) \\ &\cong C_1 \Delta^2 \frac{\mathrm{d}^2 m(z)}{\mathrm{d}z^2}, \end{aligned} \tag{A.8}$$

where Δ is the distance between the centers of two adjacent water layers. Equations (A.7) and (A.8) lead to:

$$\lambda_m^2 \frac{\mathrm{d}^2 m(z)}{\mathrm{d}z^2} = m(z) + \varepsilon_0 v_0(\varepsilon - 1)\frac{\mathrm{d}\psi(z)}{\mathrm{d}z} \tag{A.9}$$

with

$$\lambda_m = \Delta\big(\varepsilon_0 v_0(\varepsilon - 1)C_1\big)^{1/2} \tag{A.10}$$

representing a "dipolar correlation length." For $\varepsilon = 80$, $\Delta = 3.68$ Å, $v_0 = 30$ Å^3, $l = 2.76$ Å, $\varepsilon'' = 1$, Eq. (10) provides $\lambda_m = 14.9$ Å.

The chemical potential of an ion of species "i" and of charge q_i in the liquid is given by:

$$\mu_i(z) = \mu_{0i} + kT\ln\big(\gamma_i c_i(z)\big) + q_i\psi(z) + W_i(z), \tag{A.11}$$

where k is the Boltzmann constant, T the absolute temperature, and μ_{0i} represents the standard chemical potential of the ions of species "i," of charge q_i and concentration $c_i(z)$. The activity coefficient γ_i of the ions of species "i" will be assumed unity. The interaction free energy $W_i(z)$ should include all the other interactions of the ion with the medium, not accounted for by the mean field electrical potential ψ, such as the image forces [27], excluded volume effects [28], ion-dispersion [29] or ion-hydration forces [19]. Assuming that the system is in contact with an infinite reservoir in which the ion concentrations are c_{i0}, the total charge density is provided by:

$$\rho_{\mathrm{el}}(z)=\sum_i q_i c_{0i}\exp\left(-\frac{q_i\psi(z)+\Delta W_i(z)}{kT}\right), \tag{A.12}$$

where $\Delta W_i(z)$ is calculated with respect to the bulk.

Equations (A.1) and (A.12) can be combined to yield:

$$\varepsilon_0\frac{\mathrm{d}^2\psi(z)}{\mathrm{d}z^2}=-\sum_i q_i c_{0i}\exp\left(-\frac{q_i\psi(z)+\Delta W_i(z)}{kT}\right)+\frac{\mathrm{d}m(z)}{v_0\,\mathrm{d}z}. \tag{A.13}$$

Equations (A.9) and (A.13) represent a system of coupled nonlinear equations, which can be solved (under appropriate boundary conditions) for $\psi(z)$ and $m(z)$.

In a homogeneous medium a surface dipole density does not generate an electric field above the surface, and consequently it is ignored by the traditional theory of the double layer, in which the double layer is generated by surface charges alone. In the present treatment, both the surface dipole density and the surface charge density generate the double layer. The first boundary condition at the surface is provided by integrating the Poisson equation (A.1) over the volume of a flat box, which includes the surface, with the large sides parallel to the surface and a vanishingly thin width. After using the Gauss theorem, one obtains:

$$\varepsilon_0\frac{\partial\psi(z)}{\partial z}\bigg|_{z=-d}-\frac{m(z)}{v_0}\bigg|_{z=-d}=-\sigma, \tag{A.14}$$

where $z=-d$ indicates one of the surfaces, the separation distance between the identical surfaces being $2d$. The second boundary conditions at the surface is provided by the polarization of the water molecules from the vicinity of the surface, which is caused by the macroscopic field $E=-\frac{\partial\psi(z)}{\partial z}\big|_{z=-d}$, the field E_S generated by the surface dipoles and the field E_{SP} due to the neighboring water dipoles [15]:

$$\begin{aligned}m(0)&=\gamma\left(E+\frac{m(0)}{3\varepsilon_0 v_0}+E_{\mathrm{SP}}+E_S\right)\\&=\varepsilon_0 v_0(\varepsilon-1)(E+E_{\mathrm{SP}}+E_S)\end{aligned} \tag{A.15}$$

with [15]:

$$E_{\mathrm{SP}}=C_0 m(0)+C_1 m(\Delta), \tag{A.16}$$

and [15]:

$$E_S=\left(\frac{p_\perp}{\varepsilon'}\right)\frac{1}{2\pi\varepsilon_0}\frac{1}{(\frac{A}{\pi}+\Delta'^2)^{3/2}}, \tag{A.17}$$

where A is the area occupied by each surface dipole. For two identical surfaces immersed in an electrolyte separated by a distance $2d$, two additional boundary conditions are provided by the symmetry with respect to the middle distance, $z=0$:

$$\frac{\mathrm{d}\psi(z)}{\mathrm{d}z}\bigg|_{z=0}=0, \tag{A.18}$$

$$m(z)\big|_{x=0}=0. \tag{A.19}$$

Once the system of Eqs. (A.9) and (A.13) is solved for the boundary conditions ((A.14), (A.15), (A.18) and (A.19)), the total free energy of the system can be calculated by adding various contributions, due to the electric field, entropy of ions, chemical energy and interactions between ions and surfaces.

The electrostatic free energy includes the usual macroscopic free energy density $\frac{1}{2}ED$ (with the displacement field $D=\varepsilon_0 E+P$) and also the interaction of the dipoles with the additional field E_P due to the nonuniform polarization [15]:

$$\begin{aligned}f_{\mathrm{el}}(2d)&=\int_{-d}^{d}\left[\frac{1}{2}E(\varepsilon_0 E+P)-\frac{1}{2}PE_P\right]\mathrm{d}z\\&=\frac{1}{2}\int_{-d}^{d}\left[\left(\varepsilon_0\frac{\mathrm{d}\psi(z)}{\mathrm{d}z}-\frac{m(z)}{v_0}\right)\frac{\mathrm{d}\psi}{\mathrm{d}z}\right.\\&\qquad\left.-C_1\Delta^2\frac{m(z)}{v_0}\frac{\mathrm{d}^2m(z)}{\mathrm{d}z^2}\right]\mathrm{d}z.\end{aligned} \tag{A.20a}$$

The entropic contribution of all the ion species present in the electrolyte, with respect to a homogeneous system of constant ion concentrations, is given by [30]:

$$\begin{aligned}f_{\mathrm{ent}}(2d)&=-T\Delta S\\&=kT\sum_i\int_{-d}^{d}\left(c_i\ln\left(\frac{c_i}{c_{i0}}\right)-c_i+c_{i0}\right)\mathrm{d}z,\end{aligned} \tag{A.20b}$$

where "i" runs over all the ions species, and the change in the chemical energy, with respect to infinite separation, due to the charging of the surface is provided by [31]:

$$\Delta f_{\mathrm{ch}}(2d)=-\int_{\sigma(\infty)}^{\sigma(l)}\psi_S(\sigma)\,\mathrm{d}\sigma, \tag{A.20c}$$

where σ is the surface charge density and ψ_S the surface potential. The free energy due to the interactions between the ions and the surface is given by [19]:

$$\begin{aligned}&f_{\text{ion-surf}}(2d)\\&=\int_{-d}^{d}\left(\Delta W_A(z)c_E\exp\left(\frac{q_A\psi-\Delta W_A}{kT}\right)\right.\\&\qquad\left.+\Delta W_C(z)c_E\exp\left(\frac{-q_C\psi-\Delta W_C}{kT}\right)\right)\mathrm{d}z.\end{aligned} \tag{A.20d}$$

The force per unit area is consequently obtained from the derivative:

$$p(2d)=-\frac{\mathrm{d}(f_{\mathrm{el}}(2d)+f_{\mathrm{ent}}(2d)+f_{\mathrm{ch}}(2d)+f_{\text{ion-surf}}(2d))}{\mathrm{d}(2d)}. \tag{A.21}$$

Appendix B. Thermal undulation of bilayers

There is only one case in which the thermal undulations of a membrane can be evaluated analytically, namely when one membrane undulates in a harmonic potential. Let us assume that the position, along the z direction, of each point of the membrane is described by the function $u(x,y)$, with $u(x,y) \equiv 0$ corresponding to a nonundulating membrane (see Fig. B.1).

Assuming that the total area of the membrane is constant, the Hamiltonian is given by [4,32]:

$$\mathrm{H} = \int \mathrm{d}x\,\mathrm{d}y \left(\frac{1}{2} K_C \left(\frac{\partial^2 u(x,y)}{\partial x^2} + \frac{\partial^2 u(x,y)}{\partial y^2} \right)^2 + \frac{1}{2} B u^2(x,y) \right), \tag{B.1}$$

where K_C is the bending modulus of the membrane and B is the compression modulus (the spring constant, per unit area of the bilayer of the harmonic potential of interaction). Denoting by $\tilde{u}(q_x, q_y)$ the Fourier transform of $u(x,y)$, the average energy of each mode for a membrane of unit area was obtained from the equipartition principle as [4,32]:

$$\langle \tilde{u}^2 \rangle \left(\frac{1}{2} K_C q^4 + \frac{1}{2} B \right) = \frac{1}{2} kT. \tag{B.2}$$

Hence the mean square fluctuation is given by:

$$\sigma_{\text{und}}^2 = \langle u^2 \rangle = \sum_{q_x, q_y} \langle \tilde{u}^2 \rangle = \frac{kT}{(2\pi)^2} \int_0^\infty \frac{2\pi q\,\mathrm{d}q}{K_c q^4 + B} = \frac{kT}{8\sqrt{K_C B}} \tag{B.3}$$

and the distributions of the positions of the membrane is Gaussian:

$$\rho_{\text{und}}(u) = \frac{1}{\sqrt{2\pi}\,\sigma_{\text{und}}} \exp\left(-\frac{u^2}{2\sigma_{\text{und}}^2} \right). \tag{B.4}$$

This distribution is formally identical with the normal (Boltzmann) distribution of small independent, planar membranes of area s in the potential $\frac{1}{2}Bu^2$ [24]:

$$\rho_{\text{und}}(u) = \sqrt{\frac{sB}{2\pi kT}} \exp\left(-s \frac{Bu^2}{2kT} \right), \tag{B.5}$$

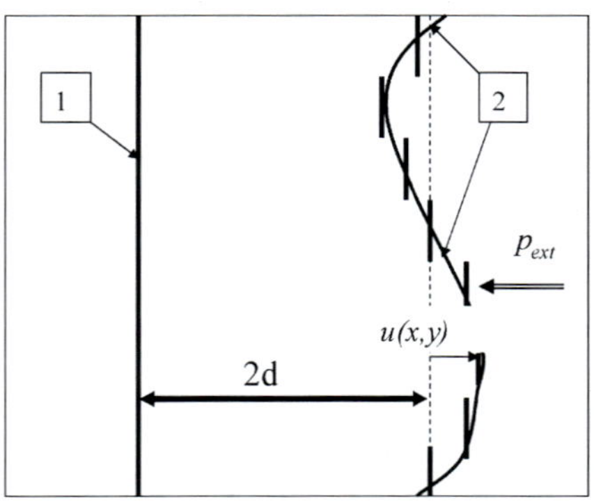

Fig. B.1. The distribution of distances between a thermal undulating membrane (2) in the harmonic potential generated by a rigid membrane (1), which is the same as that of small, independent, planar pieces of a suitable-chosen area S, Boltzmannian distributed in the same potential.

if the area s of the small planar pieces of the membranes, independently moving but nonundulating, is selected as:

$$s = \frac{kT}{B\sigma_{\text{und}}^2} = 8\sqrt{\frac{K_C}{B}}. \tag{B.6}$$

A useful extension of this method consists in assuming that the membrane acquires a Boltzmann-like distribution of type (B.5) even for an arbitrary, nonharmonic potential $f(u)$, which has a minimum at $u = 0$ [24,25]. When an external pressure p_{ext} is applied to the undulating membrane, the relevant statistical potential is the enthalpy $h(u) = f(u) + p_{\text{ext}}u$ and the membrane distribution is provided by:

$$\rho_{\text{und}}(u) = \frac{1}{N} \exp\left(-s \frac{(f(u) + p_{\text{ext}}u)}{kT} \right), \tag{B.7}$$

where the constant N can be obtained by normalization,

$$N = \int_u \rho_{\text{und}}(u)\,\mathrm{d}u. \tag{B.8}$$

The problem is to calculate the proper value for the area s, since Eq. (B.6) is not valid for arbitrary (inharmonic) potentials. Let us return to the elastic membrane in a harmonic potential. The partition function of the canonical ensemble in the Fourier space can be integrated [4]:

$$Z = \int \prod \mathrm{d}\tilde{u} \left(\exp\left(-\beta \sum_{q_x, q_y} \frac{1}{2} \tilde{u}^2 \left(K_C \vec{q}^4 + B \right) \right) \right) = \prod_{q_x, q_y} \sqrt{\frac{2\pi kT}{K_c q^4 + B}}. \tag{B.9}$$

The difference between the free energies of the free membrane (which formally corresponds to $B \to 0$) and that which interacts with the rigid wall via the harmonic potential $\frac{1}{2}Bu^2$ is given by [4]:

$$\Delta f = kT \frac{1}{(2\pi)^2} \int_0^\infty \mathrm{d}q\, 2\pi q \ln \sqrt{\frac{K_C q^4 + B}{K_C q^4}} = \frac{kT}{8} \sqrt{\frac{B}{K_C}}. \tag{B.10}$$

The entropic term of the free energy (per unit area) due to the confinement is obtained by subtracting the interaction energy per unit area from Eq. (B.10), a result which is essentially due to Helfrich [4]:

$$\Delta f_{\text{ent}} = \frac{kT}{8} \sqrt{\frac{B}{K_C}} - \frac{1}{2} B \langle u^2 \rangle_c = \frac{(kT)^2}{128 K_C \sigma_{\text{und}}^2}. \tag{B.11}$$

For the case when two undulating membrane interact with each other (instead of one undulating in the potential created by the other rigid membrane), the entropic contribution is provided by [16]:

$$\Delta f_{\text{ent}} = \frac{(kT)^2}{64 K_C \sigma_{\text{und}}^2}. \tag{B.12}$$

Let us now assume that the functional form for the entropic contribution, Eq. (B.12), holds even when the membrane is subjected to an arbitrary potential, $f(u)$. The total Gibbs free energy of the membrane at a constant external pressure p_{ext} is given by:

$$G(s) = \int_u \rho_{\text{und}}(u)\big(f(u) + p_{\text{ext}}u\big)\,du + \frac{(kT)^2}{64K_C\sigma_{\text{und}}^2}, \qquad \text{(B.13)}$$

where $\rho_{\text{und}}(u)$ is provided by Eqs. (B.7) and (B.8), while

$$\sigma_{\text{und}}^2 = \int_u \big(u - \langle u\rangle\big)^2 \rho_{\text{und}}(u)\,du. \qquad \text{(B.14)}$$

Minimization of the Gibbs free energy with respect to s provides the value s_0 as a function of the external pressure, which in turn provides the average distance between undulating membranes, for each value of the external pressure [25]:

$$\langle u\rangle = \frac{\int_u u \exp\big(-s_0 \frac{(f(u)+p_{\text{ext}}u)}{kT}\big)\,du}{\int_u \exp\big(-s_0 \frac{(f(u)+p_{\text{ext}}u)}{kT}\big)\,du}. \qquad \text{(B.15)}$$

References

[1] M. Bloom, E. Evans, O.G. Mouritsen, Q. Rev. Biophys. 24 (1991) 293.
[2] E.J. Verwey, J.Th.G. Overbeek, Theory of Stability of Lyophobic Colloids, Amsterdam, Elsevier, 1948.
[3] R. Rand, V.A. Parsegian, Biochim. Biophys. Acta 988 (1989) 351.
[4] W. Helfrich, Z. Naturforsch. 33a (1978) 305.
[5] M. Mutz, W. Helfrich, Phys. Rev. Lett. 62 (1989) 2881.
[6] T. Honger, K. Mortensen, J.H. Ipsen, J. Lemmich, R. Bauer, O.G. Mouritsen, Phys. Rev. Lett. 72 (1994) 3911.
[7] N. Chu, N. Kucerka, Y. Liu, S. Tristram-Nagle, J.F. Nagle, Phys. Rev. E 71 (2005) 041904.
[8] C. Pabst, H. Amenitsch, D.P. Kharakoz, P. Laggner, M. Rappolt, Phys. Rev. E 70 (2004) 021908.
[9] R. Lipowsky, S. Leibler, Phys. Rev. Lett. 56 (1986) 2541;
S. Leibler, R. Lipowsky, Phys. Rev. B 35 (1987) 7004.
[10] B. Deme, M. Dubois, T. Zemb, Biophys. J. 82 (2002) 215.
[11] S.S. Korreman, D. Posselt, Eur. Biophys. J. 30 (2001) 121.
[12] J. Mahanti, B. Ninham, Dispersion Forces, Academic Press, New York, 1976.
[13] H.I. Petrache, S. Tristram-Nagle, D. Harries, N. Kucerca, J.F. Nagle, V.A. Parsegian, J. Lipid Res. 47 (2006) 302.
[14] H.I. Petrache, T. Zemb, L. Belloni, V.A. Parsegian, Proc. Natl. Acad. Sci. 103 (2006) 7982.
[15] E. Ruckenstein, M. Manciu, Langmuir 18 (2002) 7584.
[16] M. Manciu, E. Ruckenstein, Langmuir 17 (2001) 2455.
[17] M.A. Bevan, D.C. Prieve, Langmuir 15 (1999) 7925.
[18] M. Manciu, E. Ruckenstein, Adv. Colloid Interface Sci. 112 (2004) 109.
[19] M. Manciu, E. Ruckenstein, Adv. Colloid Interface Sci. 105 (2003) 63.
[20] M. Manciu, O. Calvo, E. Ruckenstein, Adv. Colloid Interface Sci. 127 (2006) 29.
[21] J.J. Vale-Delgado, J.A. Molina-Bolivar, F. Galisteo-Gonzalez, M.J. Galvez-Ruiz, A. Feiler, M.W. Rutland, J. Chem. Phys. 123 (2005) 034708.
[22] E. Evans, V.A. Parsegian, Proc. Natl. Acad. Sci. 83 (1986) 7132.
[23] R. Podgornik, V.A. Parsegian, Langmuir 8 (1992) 557.
[24] M. Manciu, E. Ruckenstein, Langmuir 18 (2002) 4179.
[25] M. Manciu, E. Ruckenstein, Langmuir 20 (2004) 1775.
[26] J.D. Jackson, Classical Electrodynamics, Wiley, New York, 1975.
[27] B. Jonsson, H. Wennerstrom, J. Chem. Soc. Faraday Trans. II 79 (1983) 19.
[28] E. Ruckenstein, D. Schiby, Langmuir 1 (1985) 612.
[29] B.W. Ninham, V. Yaminsky, Langmuir 13 (1997) 2097.
[30] J.T.G. Overbeek, Colloids Surf. 51 (1990) 61.
[31] M. Manciu, E. Ruckenstein, Langmuir 19 (2003) 1114.
[32] D. Sornette, N. Ostrowsky, in: W.M. Gelbart, A. Ben-Shaul, D. Roux (Eds.), Micelles, Membranes, Microemulsions and Monolayers, Springer-Verlag, New York, 1994.

ACADEMIC PRESS

Available online at www.sciencedirect.com

SCIENCE DIRECT

Journal of Colloid and Interface Science 263 (2003) 156–161

Journal of Colloid and Interface Science

www.elsevier.com/locate/jcis

The effect of surface dipoles and of the field generated by a polarization gradient on the repulsive force

Haohao Huang, Marian Manciu, and Eli Ruckenstein *

Department of Chemical Engineering, State University of New York at Buffalo, Buffalo, NY 14260, USA

Received 9 September 2002; accepted 14 January 2003

Abstract

Double-layer and hydration interactions have been coupled into a single set of equations because both are dependent on the polarization of the water molecules. The coupled equations involve the electric fields generated by the surface charge and surface dipoles, as well as the field due to the neighboring dipoles in water. The dipoles on the surface are generated through the counterions' binding to sites of opposite charge. The equations obtained were employed to explain the restabilization observed experimentally at large ionic strengths for colloidal particles on which protein molecules were adsorbed. Polar molecules adsorbed on a charged surface of colloidal particle can generate a field either in the same direction as that generated by the charge or in the opposite direction. The effect of the sign of the dipole of the adsorbed polar molecules on the interaction between surfaces was also examined.

Keywords: Hydration force; Double-layer interaction; Surface dipoles; Restabilization

1. Introduction

The traditional double-layer theory combines the Poisson equation with the assumption that the polarization is proportional to the macroscopic electric field, and uses Boltzmann distributions for the concentrations of the ions. The potential of mean force, which should be used in the Boltzmann distribution, is approximated by the mean value of the electrical potential. The macroscopic field E and the polarization P are related via the Poisson equation

$$\varepsilon_0 \frac{dE}{dz} = \rho - \frac{dP}{dz}, \tag{1}$$

where ρ is the charge density due to the ions, ε_0 is the vacuum permittivity, and z is the distance from the middle between two surfaces, which are considered planar and parallel. In the traditional treatment, E and P are assumed to be proportional:

$$P = \chi E, \tag{2}$$

χ being the electrical susceptibility. Equation (2), however, can be used only when the electric field is sufficiently uniform, hence when the polarization gradient is sufficiently small. When this condition is not satisfied, the polarization gradient generates an additional electric field, because the fields acting on a water molecule due to the neighboring water dipoles do not compensate each other any longer. Only the fields induced by the neighboring dipoles are important because those caused by the more remote ones are screened by the intervening water molecules. Indeed, the dielectric constant is unity between two neighboring dipoles, but very large for the remote ones [1].

A large polarization gradient is generated in an aqueous electrolyte solution, either when the surface charge density is large or/and when there are dipoles on the surfaces. A general theory accounting for the above effects was developed in a previous article [2]. Two applications of this theory are examined in the present article. First we consider a colloidal dispersion or an emulsion stabilized by the adsorption of an ionic surfactant or a protein. In this case, the dissociation of the surfactant or protein molecules generates a charge that in the traditional theory is responsible for the stability of the system. The classic theory also predicts that, as a result of screening, the repulsive double-layer force decreases as the ionic strength increases. There are, however, experimental results [3] that show that, as the concentration of electrolyte increases, the stability ratio decreases and passes through a

* Corresponding author.
E-mail address: feaeliru@acsu.buffalo.edu (E. Ruckenstein).

0021-9797/03/$ – see front matter © 2003 Elsevier Science (USA). All rights reserved.
doi:10.1016/S0021-9797(03)00070-5

minimum, after which it increases. In other words, the colloidal system is restabilized at sufficiently high electrolyte concentrations. To explain the restabilization, we examine a colloidal system stabilized with adsorbed protein molecules in the presence of NaCl as electrolyte. The increase in electrolyte concentration displaces the dissociation equilibria to smaller degrees of dissociation. While the positive and negative surface charges are thus decreased, each of the neutralized charges is replaced by an ion pair (dipole). In addition, when the polarization gradient is sufficiently large, the neighboring dipoles in water generate a strong electric field. At high electrolyte concentrations the repulsion is dominated by the field generated by the surface dipoles, which propagates through water because of the local interactions between its molecular dipoles; this field is responsible for the restabilization.

Second we examine the case in which polar molecules present in water are adsorbed on the surface. They can polarize the water molecules nearby in the same direction or the direction opposite of that generated by the charge. This means that the adsorption of the polar molecules is expected to affect the repulsion.

The basic equations derived in previous papers [1,2] are first summarized; this is followed by examination of the above-described two cases. In contrast to the previous article in which the calculations were carried out in the linear approximation, nonlinear equations are employed in the present article.

2. Theoretical framework

Let us consider that near a surface the water is composed of a succession of ice-like layers, parallel to the surface, which are assumed to be planar. The electric field E_p that acts on a water molecule of layer i and is generated by the neighboring water molecules through their dipoles can be expressed as [2,4]

$$E_p(z_i) = C_1 m_{i-1} + C_0 m_i + C_1 m_{i+1}, \tag{3}$$

where m_i is the average dipole moment in layer i, C_0 and C_1 are interaction coefficients given by

$$C_0 = -\frac{3.766}{4\pi\varepsilon_0\varepsilon'' l^3}, \qquad C_1 = \frac{1.827}{4\pi\varepsilon_0\varepsilon'' l^3}, \tag{4}$$

l is the distance between the centers of two adjacent water molecules, and ε'' is the dielectric constant for the interaction between neighboring molecules, taken as unity in the present calculations. Equation (3) implies that only the neighboring molecules of layers $i \pm 1$ and layer i provide contributions. This assumption is reasonable because, as already mentioned, the more remote dipoles are screened by the intervening water molecules.

Expanding in series and neglecting higher-order terms, one obtains

$$E_p \approx (C_0 + 2C_1)m + C_1\Delta^2\frac{\partial^2 m}{\partial z^2} \approx C_1\Delta^2\frac{\partial^2 m}{\partial z^2}, \tag{5}$$

where Δ is the distance between the centers of two adjacent water layers, and $m(z)$ is the dipole moment as a function of z.

In traditional Lorentz–Debye theory, the local field that acts on a water dipole is given by $E + P/3\varepsilon_0$, where E is the macroscopic field and P is the polarization [5]. In the present case, the local field E_{local} should also include the field E_p generated by the neighboring dipoles of a water molecule.

Consequently,

$$\begin{aligned} E_{\text{local}} &= E + E_p + P/3\varepsilon_0 \\ &= E + C_1\Delta^2(\partial^2 m/\partial z^2) + P/3\varepsilon_0. \end{aligned} \tag{6}$$

The average dipole moment of a water molecule is given by $m = \gamma E_{\text{local}}$, where γ is the polarizability. Using for γ the Clausius–Mossotti equation,

$$\gamma = 3\varepsilon_0 v_0(\varepsilon - 1)/(\varepsilon + 2),$$

where v_0 is the volume of a water molecule and ε is the dielectric constant of water, and because $P = m/v_0$, one finally obtains

$$m(z) = \varepsilon_0 v_0(\varepsilon - 1)E(z) + \varepsilon_0 v_0(\varepsilon - 1)C_1\Delta^2\frac{\partial^2 m}{\partial z^2}. \tag{7}$$

Assuming Boltzmann distributions for the ions, one can write

$$\rho = -2c_E e \sinh(e\Psi/kT), \tag{8}$$

where c_E is the bulk electrolyte concentration. By combination of Eqs. (1) and (8), the Poisson equation becomes [2]

$$\frac{\partial^2\Psi(z)}{\partial z^2} = \frac{2ec_E}{\varepsilon_0}\sinh\left(\frac{e\Psi}{kT}\right) + \frac{1}{\varepsilon_0 v_0}\frac{\partial m(z)}{\partial z}. \tag{9}$$

Equations (7) and (9) constitute a system of equations for Ψ and m. To obtain particular solutions of Eqs. (7) and (9), boundary conditions must be provided.

When there are dipoles on the surface, the average field that they generate and that is acting on the first water layer is given by [4]

$$E_s = (p/\varepsilon')/\left[2\pi\varepsilon_0(A/\pi + \Delta'^2)^{3/2}\right], \tag{10}$$

where p is the normal component of the dipole moment of the surface dipoles, ε' is the local dielectric constant of the medium, A is the surface area per dipole of the surface ($1/A$ is the surface dipole density), and Δ' is the distance between a surface dipole and the center of the first layer of water molecules.

The electric field that is acting on the first water layer and is induced by the nearby water molecules can be obtained from Eq. (3) in which the term containing m_{i-1} is absent:

$$E_{p1} = C_0 m_1 + C_1 m_2. \tag{11}$$

Consequently, the dipole moment in the layer of water closest to the surface is given by

$$m|_{z=-d} = \gamma E_{\text{local}} = \gamma(E + E_s + E_{p1} + P/3\varepsilon_0) = \varepsilon_0 v_0(\varepsilon - 1)(E + E_s + E_p)|_{z=-d} = -m|_{z=d}, \tag{12}$$

where $2d$ is the distance between the two plates.

The electroneutrality condition provides a second boundary condition:

$$\sigma = -\frac{1}{2}\int_{-d}^{d} \rho\, dz = \frac{1}{2}\int_{-d}^{d}\left(\varepsilon_0 \frac{\partial^2 \Psi}{\partial z^2} - \frac{1}{v_0}\frac{\partial m}{\partial z}\right) dz = \frac{1}{2}\left[\varepsilon_0\left(\frac{\partial \Psi}{\partial z}\bigg|_{z=d} - \frac{\partial \Psi}{\partial z}\bigg|_{z=-d}\right) - \frac{1}{v_0}(m|_{z=d} - m|_{z=-d})\right] = \varepsilon_0 \frac{\partial \Psi}{\partial z}\bigg|_{z=d} - \frac{1}{v_0} m|_{z=d}. \tag{13}$$

Expressions for the surface charge σ and dipole densities are derived in the next section, where the dissociations of the surface groups are taken into account.

The free energy of the system per unit area contains a number of contributions,

$$F = F_{\text{chem}} + F_{\text{ele}} + F_{\text{entropy}}, \tag{14}$$

where F_{chem} is the chemical, F_{ele} the electrostatic, and F_{entropy} the entropic contributions. $F_{\text{chem}} = -2\sigma\Psi_s$ at constant surface potential Ψ_s and 0 at constant surface charge [6,7]. A general expression that is free of any constraint has the form [8]

$$F_{\text{chem}} = -2\int_{\sigma(\infty)}^{\sigma(2d)} \Psi_s\, d\sigma. \tag{15}$$

Equation (15) can be established by decomposing the pathway of Ψ_s versus σ in a succession of infinitesimal changes at constant charge followed by changes at constant Ψ_s.

The electrostatic free energy due to the electric field, which includes the interactions with the neighboring dipoles, is given by

$$F_{\text{ele}} = \frac{1}{2}\int_{-d}^{d}\left(\left(\varepsilon_0 E(z) + \frac{m(z)}{v_0}\right)E(z) - \frac{m(z)}{v_0}E_p(z)\right) dz. \tag{16}$$

The excess entropy contribution to the free energy is [7]

$$F_{\text{entropy}} = kT\sum_i \int_{-d}^{d}\left(c_i \ln\left(\frac{c_i}{c_{i,0}}\right) - c_i + c_{i,0}\right) dz = -c_E kT \sum_{i=1,2}\int_{-d}^{d}\left[\frac{(-1)^i e\Psi}{kT}\exp\left(\frac{(-1)^{i+1}e\Psi}{kT}\right) + \exp\left(\frac{(-1)^{i+1}e\Psi}{kT}\right) - 1\right] dz. \tag{17}$$

The free energy of the surface layer formed by the surface dipoles and the water molecules between them is assumed independent of $2d$.

3. Colloidal restabilization in protein–latex systems

Molina-Bolivar and Ortega-Vinuesa have studied the stability of polystyrene colloidal particles stabilized with an adsorbed layer of protein (IgG), and observed that the stability ratio decreased with increasing ionic strength of NaCl and passed through a minimum at sufficiently large electrolyte concentrations, after which the system was restabilized [3]. This anomalous restabilization of the colloidal system at high electrolyte concentrations cannot be explained in the framework of classic DLVO theory. As the concentration of electrolyte increases, DLVO theory predicts that the colloidal system becomes increasingly unstable. In what follows it is shown that the present theory, which takes into account the effect of the fields generated by the surface dipoles and the neighboring water molecules in the liquid, can explain the restabilization.

The surface charge is generated through the dissociation of the acidic and basic sites of the protein molecules. However, as the concentration of electrolyte increases, the dissociation equilibria are displaced in the direction of lower dissociations and the charges are replaced by ion pairs (dipoles).

Neglecting the adsorption of Cl^- anions onto the basic sites, the surface charge can be calculated by considering the equilibria

$$\mathrm{NaCl} \rightarrow \mathrm{Na}^+ + \mathrm{Cl}^-,$$

$$\mathrm{S}_A \leftrightarrow \mathrm{S}_A^- + \mathrm{H}^+,$$

$$\mathrm{S}_B \leftrightarrow \mathrm{S}_B^+ + \mathrm{OH}^-,$$

$$\mathrm{S}_A^- + \mathrm{Na}^+ \leftrightarrow \mathrm{S}_A\text{–Na},$$

where S_A is an acidic site, S_A^- a dissociated acidic site, S_B a basic site, S_B^+ a dissociated basic site, and S_A–Na an ion pair formed on an acidic site.

The following equilibrium equations can therefore be written:

$$K_{\mathrm{H}} = \frac{(1 - w - x)[\mathrm{H}^+]_S}{x}, \tag{18}$$

$$K_{\mathrm{OH}} = \frac{(1 - y)[\mathrm{OH}^-]_S}{y}, \tag{19}$$

$$K_{\mathrm{Na}} = \frac{(1 - w - x)[\mathrm{Na}^+]_S}{w}. \tag{20}$$

Here x is the fraction of acidic sites occupied by hydrogen ions, w is the fraction of acidic sites occupied by sodium ions, y is the fraction of basic sites occupied by OH^- ions, and $[H^+]_S$, $[Na^+]_S$, and $[OH^-]_S$ are the concentrations of the corresponding ions at the surface.

The ionic concentrations at the surface can be related to their bulk concentrations via the Boltzmann expressions

$$[H^+]_S = [H^+]\exp\left(\frac{-e\Psi_S}{kT}\right), \tag{21}$$

$$[Na^+]_S = [Na^+]\exp\left(\frac{-e\Psi_S}{kT}\right), \tag{22}$$

$$[OH^-]_S = [OH^-]\exp\left(\frac{e\Psi_S}{kT}\right), \tag{23}$$

where e is the protonic charge, and Ψ_S is the surface potential. Consequently, the surface charge density is given by

$$\sigma = e\left[N_B(1-y) - N_A(1-x-w)\right], \tag{24}$$

where N_B is the surface density of the basic sites and N_A is the surface density of the acidic sites. Because we assume that surface dipoles are formed through the adsorption of Na^+ only, the dipole density on the surface $(1/A)$ is given by wN_A.

The isoelectric point of IgG is between 6.0 and 8.0 [3]. By selecting its value as 7.0, the surface charge becomes zero at pH 7 and Eq. (24) leads to

$$N_B(1-y) = N_A(1-x-w). \tag{25}$$

Combining Eqs. (25), (18)–(20), one obtains

$$K_{OH} = \frac{10^{-7}}{\frac{10^{-7}N_B}{K_H N_A} - \frac{N_A - N_B}{N_A}}, \tag{26}$$

which allows calculation of K_{OH} in terms of K_H and the ratio N_B/N_A.

By use of the Derjaguin approximation, the repulsive interaction energy between two identical spherical particles of radius a is given by

$$V_R = \pi a \int_{H_0}^{\infty} (F - F_\infty)\, dH, \tag{27}$$

where H_0 is the minimum distance between the surfaces of the two spheres, and F_∞ is the repulsive free energy per unit area between two infinitely separated plates. The van der Waals attractive interaction between two particles has the form [10]

$$V_A = -\frac{A_H}{6}\left[\frac{2a^2}{H_0(4a+H_0)} + \frac{2a^2}{(2a+H_0)^2} + \ln\frac{H_0(4a+H_0)}{(2a+H_0)^2}\right], \tag{28}$$

A_H being the Hamaker constant. When $H_0 \ll a$, V_A can be approximated by $V_A = -A_H a/12H_0$. The total interacting free energy V_T between the two spheres is therefore given by

$$V_T = V_A + V_R. \tag{29}$$

The stability ratio W, which constitutes a measure of the stability of the colloidal system, is defined as

$$W = \frac{k_r}{k_s}, \tag{30}$$

k_r being the rate constant for rapid coagulation and k_s the rate constant for slow coagulation. W, which was determined experimentally via the low-angle scattering method [3,9], is related to the interaction potential via the expression

$$W = \frac{\int_0^\infty \frac{\beta(u)}{(u+2)^2}\exp\left(\frac{V_T}{kT}\right)du}{\int_0^\infty \frac{\beta(u)}{(u+2)^2}\exp\left(\frac{V_A}{kT}\right)du}, \tag{31}$$

where β is a hydrodynamic correction factor, $u = H_0/a$, V_T is the total interaction free energy between two colloidal particles, and V_A is the interaction free energy due to van der Waals attractive forces. The hydrodynamic correction factor β is taken as unity.

The calculations were carried out for the conditions under which the experiments were carried out: pH 8.0 and $D = 2a = 204$ nm. The values of the parameters are not known. Assuming for the distance between two acidic sites or two basic sites a value of 7 Å, one obtains $N_A = N_B = 2 \times 10^{18}$ m^{-2}. Typical values for the pK_H of amino acid residues of globular proteins are in the range 1.95–9.5 [11], and the value of 8×10^{-10} M was selected for K_H ($pK_H = 9.1$). For the K_{Na} of human erythrocytes, values of the order of 1 M or larger fitted the experimental data reasonably well [12]. The value of 1 M was therefore selected as the K_{Na} of IgG. The value $A_H = 3.2 \times 10^{-20}$ J, which has the right order of magnitude, was selected for the Hamaker constant.

By use of the above equations and values of the parameters, the nonlinear equations were solved numerically. The boundary value problem (Eqs. (7) and (9)), with the appropriate boundary conditions (Eqs. (12), (13) and the symmetry condition for Ψ) was solved using the standard shooting method and the bvp4c Matlab routine. Initial guesses were provided by the exact solutions in the linear approximation. The free energy F was obtained by adding the van der Waals interaction to the double-layer interaction calculated using Eqs. (14)–(17) and (27). The stability ratio was calculated by taking 300 Å as the largest value of H_0. The results are compared with the experimental data in Fig. 1 for three values of p/ε'. There are three regions in the figure: the DLVO stabilization region, the rapid aggregation region, and the restabilization region. DLVO theory can explain the behavior in the first two regions, in which the stability of a colloidal system decreases with increasing salt concentration until the concentration reaches the so-called critical coagulation concentration (ccc). Then the colloidal system aggregates rapidly, because the repulsive electric field is suppressed by the high salt concentration.

DLVO theory fails to explain the restabilization. However, by coupling the electric fields induced by the surface charge and surface dipoles and by the neighboring dipoles in water, one can obtain agreement with the experimental

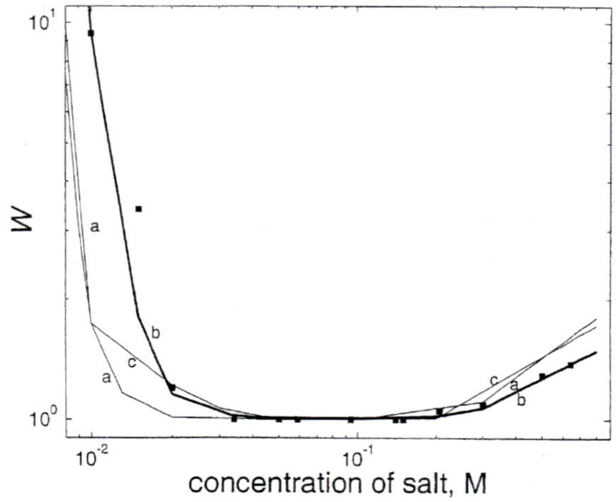

Fig. 1. Stability ratio W versus salt concentration. The experimental data (■) are from Fig. 6 of Ref. [3]. pH 8.0, $D = 2a = 204$ nm, $N_A = N_B = 2 \times 10^{18}$ m^{-2}, $K_H = 8 \times 10^{-10}$ M, $K_{Na} = 1$ M, $A_H = 3.2 \times 10^{-20}$ J. (a) $p/\varepsilon' = 1$ Debye; (b) $p/\varepsilon' = 1.8$ Debye; (c) $p/\varepsilon' = 3$ Debye.

results and provide a possible explanation for this phenomenon. The adsorption of Na^+ ion on the surface, which is stimulated by higher salt concentrations, increases the density of surface dipoles. The surface dipoles polarize the nearby water molecules, which in turn polarize the next layer and so on. The force generated by the overlap of the polarization layers of the two plates, often called hydration force, is responsible for the restabilization of the colloidal system as soon as the concentration of the electrolyte becomes higher than a critical stabilization concentration (csc).

4. Effect of the dipoles adsorbed on the surface

Some polar molecules present in water can be adsorbed onto the surface of colloidal particles. They can polarize nearby water molecules and this polarization propagates from layer to layer. It is assumed that the surface has a fixed charge and that the only surface dipoles are those of the adsorbed polar molecules. Figure 2 illustrates two different kinds of adsorption. In the first one (A), the adsorbed polar molecules expose to the liquid the same charge as that of the surface. The polarization generated by the dipoles of the adsorbed molecules is in this case, in the same direction as that generated by the charge. Consequently, the polarization induced in water is larger than that generated by the surface charge alone. As a result, the repulsion between the two plates is expected to be stronger than that generated by the surface charge alone.

In the second kind of adsorption (Fig. 2B), the dipole exposes to the solution a charge opposite of that of the surface. Correspondingly, the dipole generates a field opposite of that generated by the surface charge, and the adsorption is expected to decrease the repulsive force between the two plates. However, if the surface charge is small, then the dipoles of the adsorbed molecules alone generate repulsion.

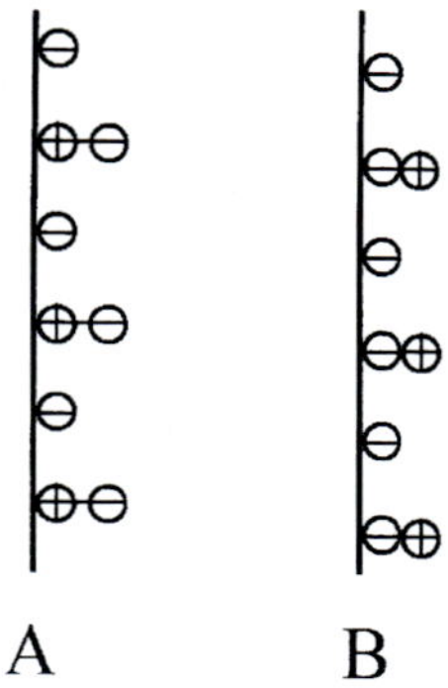

Fig. 2. Two different kinds of adsorption of dipoles onto a negatively charged surface.

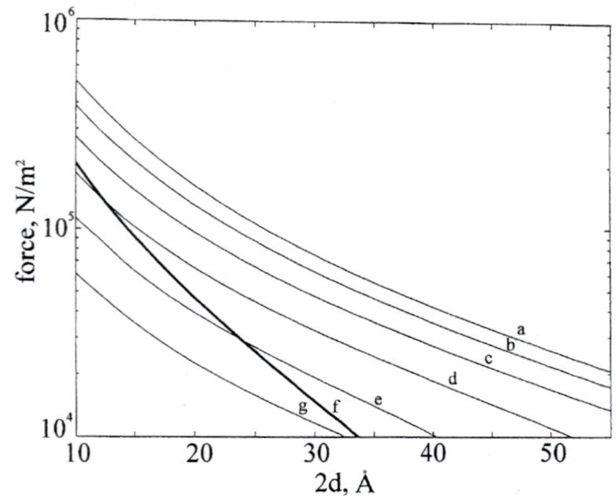

Fig. 3. Effect of the kind of adsorption on the force between two plates. $\sigma = 0.01$ C/m^2, Debye length $\lambda_D = 10$ Å, $A = 100$ Å^2. (a) $p/\varepsilon' = 3$ Debye; (b) $p/\varepsilon' = 2$ Debye; (c) $p/\varepsilon' = 1$ Debye; (d) $p/\varepsilon' = 0$; (e) $p/\varepsilon' = -1$ Debye; (f) DLVO theory; (g) $p/\varepsilon' = -2$ Debye.

Figure 3 presents the results of the calculations. They show that at large distances the repulsive force is always stronger than that predicted by DLVO theory, regardless the sign of the surface dipoles. This occurs because the gradient of the polarization generated by the charge induces a field in the same direction as that of the charge, and this makes the interactions to be of longer range than those predicted by the traditional Poisson–Boltzmann equation. At short distances which, however, should not be smaller than two layers of water molecules (~8 Å) for the calculation to be physically meaningful, the repulsion is stronger than that predicted by DLVO theory when the surface dipoles act in the same direction as the charge and weaker in the opposite case.

5. Conclusions

The kind of adsorption of polar molecules onto the surface of particles can affect interaction between particles. If the electric field generated by the adsorbed dipoles has the same direction as the electric field generated by the charge, the adsorbed dipoles increase the repulsion. If the electric field generated by those dipoles is in a direction opposite that generated by the charge, the adsorbed dipoles weaken the repulsion.

The theory could successfully explain the restabilization of protein-stabilized polymer latexes; the increase in surface dipole density with increasing electrolyte concentration, generated by the binding of the counterions to sites of opposite charge, is in this case responsible for this effect. This occurs because the repulsive force generated by the surface dipoles more than compensates for the decrease in repulsion caused by screening.

References

[1] D. Schiby, E. Ruckenstein, Chem. Phys. Lett. 95 (1983) 435.
[2] E. Ruckenstein, M. Manciu, Langmuir 18 (2002) 7584.
[3] J.A. Molina-Bolivar, J.L. Ortega-Vinuesa, Langmuir 15 (1999) 2644.
[4] M. Manciu, E. Ruckenstein, Langmuir 17 (2001) 7582.
[5] J.D. Kraus, Electromagnetics, 4th ed., McGraw–Hill, New York, 1992.
[6] E.J.W. Verwey, J.Th.G. Overbeek, Theory of the Stability of Lyophobic Colloids, Elsevier, Amsterdam, 1948.
[7] J.Th.G. Overbeek, Colloids Surf. 51 (1990) 61.
[8] M. Manciu, E. Ruckenstein, Langmuir 19 (2003) 1114.
[9] J.A. Molina-Bolivar, F. Galisteo-Gonzalez, R. Hidalgo-Alvarez, Phys. Rev. E 55 (1997) 4522.
[10] R.J. Hunter, Foundations of Colloid Science, Vol. 1, Oxford Science, Oxford, 1987.
[11] S.N. Timasheff, in: A. Veis (Ed.), Biological Polyelectrolyte, Dekker, New York, 1970.
[12] D.C. Prieve, E. Ruckenstein, J. Theor. Biol. 56 (1976) 205.

Langmuir **2003,** *19,* 3049–3055 3049

Colloid Restabilization at High Electrolyte Concentrations: Effect of Ion Valency

Eli Ruckenstein* and Haohao Huang

Department of Chemical Engineering, State University of New York at Buffalo, Buffalo, New York 14260

Received October 22, 2002. In Final Form: January 3, 2003

The scope of this paper is to explain the restabilization of colloidal dispersions at sufficiently high electrolyte concentrations. This restabilization could not be explained in the framework of the traditional double layer theory. An improved theory which accounts for the field generated in water by neighboring dipoles (the polarization gradient) as well as the field induced by surface dipoles could, however, explain the experimental results. The latter dipoles are generated by the adsorption of the ions of the electrolyte upon the surface charges with opposite sign formed through the dissociation of acidic and basic sites. This adsorption changes not only the surface charge but also generates ion pairs (dipoles), which induce a field in the neighboring water molecules that propagates further in the liquid. A higher valency has an important effect upon the restabilization because it makes the field induced by the neighboring water molecules stronger and the change of the surface charge with electrolyte concentration greater. Experimental results from literature regarding the stability of a dispersion of protein-covered latexes are explained on the basis of the new theory.

1. Introduction

Experiment revealed that, in most cases, the stability of a colloidal dispersion is decreased as the electrolyte concentration increases. The DLVO theory could explain, sometimes quantitatively, these experimental findings. However, many decades ago, Voet[1] prepared stable sols of metals (Pt, Pd) and salts (sulfides, halides) in highly concentrated solutions of H_2SO_4, H_3PO_4, and $CaCl_2$. Their dilution with water induced coagulation. Similar results have been reported more recently by Healy et al.,[2] who observed that amphoteric latex particles did not coagulate at high ionic strength of $LiNO_3$ (>1 M). Finally, Molina-Bolivar and Ortega-Vinuesa[3] have observed that the stability ratio of a dispersion of polystyrene particles stabilized with adsorbed protein molecules (Ig-G) decreased with increasing electrolyte concentration, passed through a minimum, after which it increased. Consequently, at sufficiently large electrolyte concentrations, the dispersion was restabilized.

The traditional theory could not explain the above observations. In recent papers concerned with monovalent electrolytes,[4,5] the restabilization was attributed to the following two effects: (i) the reassociation of charges on the surface and their replacement by ion pairs (surface dipoles), and (ii) the fields generated by the surface dipoles and in the bulk by neighboring dipoles. While the surface charge density formed through the dissociation of acidic and basic sites is decreased by the adsorption of counterions, the charges are replaced by ion pairs (dipoles) which polarize the water molecules nearby. This polarization propagates, and the overlap of the polarization layers generates a repulsive force (the hydration force) which becomes dominant at high electrolyte concentrations and is responsible for the restabilization. The previously developed theory[6] assumed additivity between the traditional double layer force and the hydration force generated by the surface dipoles. Because both forces depend on polarization, a new unitary theory was formulated in which they were coupled in a single repulsive force[4] in which the two effects were no longer additive. It is well-known that the valency of the counterion ν has a strong effect on colloidal stability,[7] since the critical coagulation concentration is proportional to ν^{-6}. The theoretical confirmation of this Schulze–Hardy rule was one of the major successes of the DLVO theory.[8] The scope of the present paper is to examine the effect of the valency of the electrolyte ions on colloidal restabilization. The valency is expected to be important for at least two reasons: (i) the strong effect which it has on the screening of the electric field; (ii) the adsorption of multivalent ions on surface charges of opposite sign which can easily cause a charge inversion on the surface.

The presentation is organized as follows: First, an expression for the surface charge density will be derived by assuming that the surface charge is generated by the dissociation of acidic and basic groups and the adsorption of cations and anions of the electrolyte upon the surface charges of opposite sign. Because intuition suggests that a charge reversal should occur with increasing electrolyte concentration and this reversal is compatible with a minimum in the stability ratio, the traditional double layer theory was first employed to verify if it can predict restabilization. The conclusion was that the traditional theory cannot predict restabilization when the electrolyte concentration becomes very large. Further, it will be shown that a unitary theory, which

* To whom correspondence should be addressed: Telephone: (716) 645-2911 ext. 2214. Fax: (716) 645-3822. E-mail: feaeliru@acsu.buffalo.edu

(1) Voet, A. Thesis, University of Amsterdam, 1935; see also: Kruyt, H. R. *Colloid Science*; Elsevier: New York, 1952.
(2) Healy, T. W.; Homola, A.; James, R. O.; Hunter, R. J. *Faraday Discuss. Chem. Soc.* **1978**, *65*, 156.
(3) Molina-Bolivar, J. A.; Ortega-Vinuesa, J. L. *Langmuir* **1999**, *15*, 2644.
(4) Ruckenstein, E.; Manciu, M. *Langmuir* **2002**, *18*, 7584.
(5) Huang, H.; Manciu, M.; Ruckenstein, E. *J. Colloid Interface Sci.*, in press.
(6) Manciu, M.; Ruckenstein, E. *Langmuir* **2001**, *17*, 7582.
(7) Overbeek, J. Th. G. In *Colloid Science*; Kruyt, H. R. Eds.; Elsevier: New York, 1952; Vol, 1.
(8) Verwey, E. J.; Overbeek, J. Th. G. *Theory of the Stability of Lyophobic Colloids*; Elsevier: Amsterdam, 1948.

10.1021/la026729y CCC: $25.00
Published on Web 02/27/2003

involves the charge of the surface, the surface dipole density, and the fields caused by the surface dipoles and by the neighboring water dipoles in the water layer, can provide reasonable results. A comparison with experiment will conclude the paper.

2. Dissociation and Adsorption of Cations and Anions on Protein Stabilized Latexes

Molina-Bolivar and Ortega-Vinuesa examined experimentally the stability of a dispersion of polystyrene latexes with chemisorbed protein (Ig-G) molecules upon their surface.[3] A latex suspension was introduced into a buffer containing protein molecules and the latter molecules reacted with the chloromethyl groups of the latex surface. After the protein binding, the stability ratio W, the ratio of the rate constants of rapid and slow coagulations, of the latexes in various electrolyte solutions was determined using the low-angle light-scattering technique.

Let us first derive an expression for the surface charge generated through the dissociation of the acidic and basic groups of the protein and adsorption of the ions of the electrolyte on the opposite charges of the surface.[9] The dissociations of the protein can be represented by the equilibria

$$S_A \leftrightarrow S_A^- + H^+ \tag{1a}$$

$$S_B \leftrightarrow S_B^+ + OH^- \tag{1b}$$

where S_A is an acidic site, S_A^- is a dissociated acidic site, S_B is a basic site, and S_B^+ is a dissociated basic site.

Since the cations and anions of the electrolyte can be adsorbed upon the dissociated acidic and, respectively, basic sites, the following adsorption equilibria should be included

$$S_A^- + M^{\nu+} \leftrightarrow (S_A{-}M^{\nu-1}) \tag{1c}$$

$$S_B^+ + N^- \leftrightarrow S_B{-}N \tag{1d}$$

where $M^{\nu+}$ is a cation of valency ν, $(S_A{-}M^{\nu-1})$ is an acidic site occupied by cation $M^{\nu+}$ and $S_B{-}N$ is a basic site occupied by anion N^-. Equilibrium 1c implies that a cation of valency ν is adsorbed on a single monovalent site.

Neglecting the interactions between the adsorbed species, one can write the following equilibria

$$K_H \equiv \frac{(1-\theta_M-\theta_H)[H^+]_S}{\theta_H} \tag{2a}$$

$$K_M \equiv \frac{(1-\theta_M-\theta_H)[M^{\nu+}]_S}{\theta_M} \tag{2b}$$

$$K_{OH} \equiv \frac{(1-\theta_{OH}-\theta_N)[OH^-]_S}{\theta_{OH}} \tag{2c}$$

$$K_N \equiv \frac{(1-\theta_{OH}-\theta_N)[N^-]_S}{\theta_N} \tag{2d}$$

where K_H, K_{OH}, K_M, and K_N are equilibrium constants for H, OH, and the cation and anion of the electrolyte, respectively, θ_H and θ_M denote the fractions of acidic sites occupied by hydrogen ions and cations $M^{\nu+}$, θ_{OH} and θ_N denote the fractions of basic sites occupied by hydroxide ions and anions N^-, and $[H^+]_S$, $[M^{\nu+}]_S$, $[OH^-]_S$, and $[N^-]_S$ are the concentrations of the corresponding ions near the surface.

The surface charge density is obtained by subtracting the negative charge density from the positive one

$$\sigma = e[N_B(1-\theta_{OH}-\theta_N) + (\nu-1)N_A\theta_M - N_A(1-\theta_H-\theta_M)] = e[N_B(1-\theta_{OH}-\theta_N) - N_A(1-\theta_H-\nu\theta_M)] \tag{3}$$

where N_B and N_A represent the numbers of basic and acidic sites per unit area, respectively, and e is the protonic charge.

Combined with eq 2a–d and Boltzmann distributions, eq 3 becomes

$$\sigma = e\left(\frac{N_B K_{OH} K_N}{K_{OH}K_N + [OH^-]\exp\left(\frac{e\Psi_s}{kT}\right)K_N + [N^-]\exp\left(\frac{e\Psi_s}{kT}\right)K_{OH}} + \frac{N_A(\nu-1)K_H[M^{\nu+}]\exp\left(-\frac{\nu e\Psi_s}{kT}\right) - N_A K_H K_M}{K_H K_M + K_M[H^+]\exp\left(-\frac{e\Psi_s}{kT}\right) + K_H[M^{\nu+}]\exp\left(-\frac{\nu e\Psi_s}{kT}\right)}\right) \tag{4}$$

where $[H^+]$, $[OH^-]$, $[M^{\nu+}]$, and $[N^-]$ are concentrations in the bulk, Ψ_s is the surface potential, k is the Boltzmann constant, and T is the temperature in K

If the adsorption of anions is neglected (such an approximation is reasonable for a negatively charged surface or a sufficiently large K_N), eq 4 reduces to

$$\sigma = e\left(\frac{N_B K_{OH}}{K_{OH} + [OH^-]\exp\left(\frac{e\Psi_s}{kT}\right)} - \frac{N_A(\nu-1)K_H[M^{\nu+}]\exp\left(-\frac{\nu e\Psi_s}{kT}\right) - N_A K_H K_M}{K_H K_M + K_M[H^+]\exp\left(-\frac{e\Psi_s}{kT}\right) + K_H[M^{\nu+}]\exp\left(-\frac{\nu e\Psi_s}{kT}\right)}\right) \tag{5}$$

3. Calculation of the Stability Ratio on the Basis of the Traditional Double Layer Theory

The free energy of the system due to repulsion was calculated using the Derjaguin approximation, hence by expressing the repulsive free energy between two spherical particles in terms of that between two parallel plates.

The Poisson–Boltzmann equation in one dimension has the form

$$\frac{d^2\Psi}{dz^2} = -\frac{1}{\epsilon\epsilon_0}\sum_i n_i^0 \nu_i e \exp(-\nu_i e\Psi/kT) \tag{6}$$

where Ψ is the electrical potential, z the distance from the middle distance between plates, n_i^0 the number of ions of species i per unit volume in the bulk solution, ϵ the dielectric constant of the medium, and ϵ_0 the vacuum permittivity.

(9) Prieve, D. C.; Ruckenstein, E. *J. Theor. Biol.* **1976**, *56*, 205.

The electroneutrality provides the boundary conditions

$$\left.\frac{d\Psi}{dz}\right|_{z=-d} = -\left.\frac{d\Psi}{dz}\right|_{z=d} = -\frac{\sigma}{\epsilon\epsilon_0} \tag{7}$$

where $2d$ is the distance between the two plates.

The repulsive free energy of the system is the sum of three contributions[10]

$$F = F_{ele} + F_{entropy} + F_{chem} \tag{8}$$

where F_{ele}, $F_{entropy}$, and F_{chem} are the electrostatic, entropic, and chemical contributions to the free energy, respectively.

The quantity needed in the calculation of the stability ratio is the repulsive free energy per unit area and hence the work per unit area needed to bring the two plates from infinity to the distance $2d$

$$\Delta F = F(2d) - F(2d{=}\infty) \tag{9}$$

The electrostatic and entropic contributions to the free energy can be calculated using the following expressions[10]

$$F_{ele} = \frac{1}{2}\epsilon\epsilon_0 \int_{-d}^{d} \left(\frac{d\Psi}{dz}\right)^2 dz \tag{10}$$

$$F_{entropy} = kT\sum_i \int_{-d}^{d} \left(n_i \ln\left(\frac{n_i}{n_i^0}\right) - n_i + n_i^0\right) dz \tag{11}$$

where n_i is the number ions of species i per unit volume.

The chemical contribution to the free energy can be calculated by decomposing the pathway from infinity to $2d$ into infinitesimal steps at constant charge, for which the chemical contributions are zero,[8] each followed by an infinitesimal step at constant potential, for which the chemical contributions are $-2\Psi_s\Delta\sigma$.[8] The summation over the entire path leads to[11]

$$\Delta F_{chem} = -2\int_{\sigma(\infty)}^{\sigma(2d)} \Psi_s \, d\sigma \tag{12}$$

The free energy of the surface layer formed of the surface dipoles and the water molecules between them is considered independent of the distance $2d$.

If the radius of the spherical colloidal particles a is much larger than the shortest distance between the surfaces of two particles, the repulsive free energy between two identical spherical particles at a distance H_0 of shortest approach is given in the Derjaguin approximation by[12]

$$V_R = \pi a \int_{H_0}^{\infty} \Delta F \, dH \tag{13}$$

where $H = 2d$.

The van der Waals attractive energy between two spheres of radius a has the form[8]

$$V_A = -\frac{A_H}{6}\left[\frac{2a^2}{H_0(4a + H_0)} + \frac{2a^2}{(2a + H_0)^2} + \ln\frac{H_0(4a + H_0)}{(2a + H_0)^2}\right] \tag{14}$$

where A_H is the Hamaker constant.

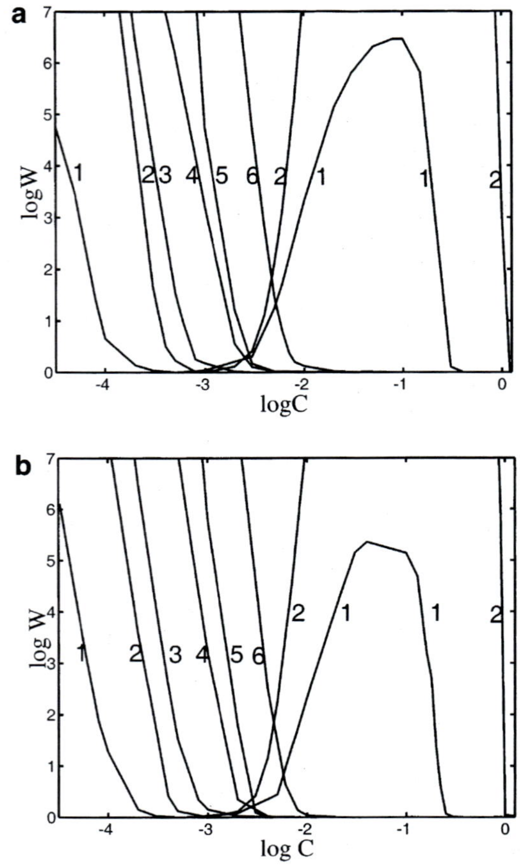

Figure 1. (a) Stability ratio W against electrolyte concentration C (in M). The calculations were carried out with the traditional theory. $N_A = N_B = 2 \times 10^{18}$ m^{-2}, $K_H = 6 \times 10^{-10}$ M, $A_H = 1 \times 10^{-20}$J, pH = 8.0, a = 102 nm, and $K_N = 6 \times 10^{-5}$ M. Key: (1) 3:1 electrolyte with K_M = 0.001 M; (2) 2:1 electrolyte with K_M = 0.001 M; (3) 3:1 electrolyte with K_M = 0.01 M; (4) 3:1 electrolyte with K_M = 0.1 M; (5) 2:1 electrolyte with K_M = 0.01 M; (6) 2:1 electrolyte with K_M = 0.1 M. (b) Stability ratio W against electrolyte concentration C (in M). The calculations were carried out with the traditional theory. The adsorption of anions is neglected. $N_A = N_B = 2 \times 10^{18}$ m^{-2}, $K_H = 6 \times 10^{-10}$ M, $A_H = 1 \times 10^{-20}$J, pH = 8.0, a = 102 nm, and $K_N = \infty$. Key: (1) 3:1 electrolyte with K_M = 0.001 M; (2) 2:1 electrolyte with K_M = 0.001 M; (3) 3:1 electrolyte with K_M = 0.01 M; (4) 3:1 electrolyte with K_M = 0.1 M; (5) 2:1 electrolyte with K_M = 0.01 M; (6) 2:1 electrolyte with K_M = 0.1 M.

The colloidal stability can be expressed through the stability ratio W, which is the ratio between the rate constants of rapid and slow coagulations[13,14]

$$W = \frac{\int_0^{\infty} \frac{\beta(u)}{(u+2)^2} \exp\left(\frac{V_T}{kT}\right) du}{\int_0^{\infty} \frac{\beta(u)}{(u+2)^2} \exp\left(\frac{V_A}{kT}\right) du} \tag{15}$$

where β is a hydrodynamic correction factor, $u = H_0/a$, V_T is the sum of the repulsive V_R and the van der Waals attractive V_A free energies between two spherical particles. The hydrodynamic correction factor β was taken as unity.

Parts a and b of Figure 1 examine the effect of the valency on the stability ratio in the framework of the

(10) Overbeek, J. Th. G. *Colloids Surf.* **1990**, *51*, 61.
(11) Manciu, M.; Ruckenstein, E, *Langmuir*, in press.
(12) Hunter, R. J. *Foundations of Colloid Science*;Clarendon Press: Oxford, England, 1986; Vol. 1.
(13) Honig, E. P.; Roebersen, G. J.; Wiersema, P. H. *J. Colloid Interface Sci.* **1971**, *36*, 97.
(14) Derjaguin, B. V.; Muller, V. M. *Dokl. Akad. Nauk SSSR (Engl. Transl.)* **1967**, *176*, 738.

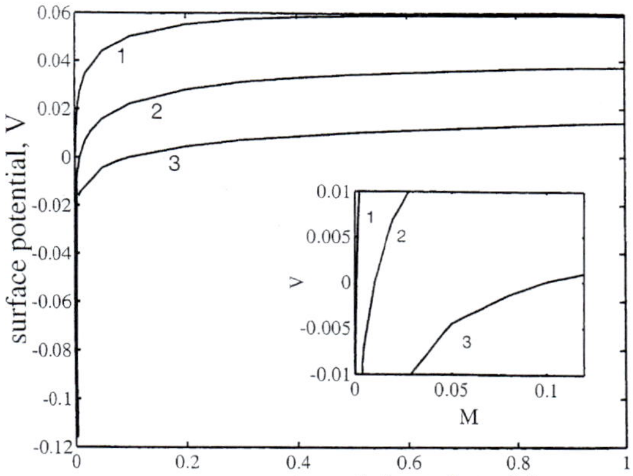

Figure 2. Reversal of the surface potential for a single plate. The calculations were carried out with the traditional theory. Key: (1) 2:1 electrolyte with $K_M = 0.001$ M; (2) 2:1 electrolyte with $K_M = 0.01$ M; (3) 2:1 electrolyte with $K_M = 0.1$ M. The other parameters are as in Figure 1a.

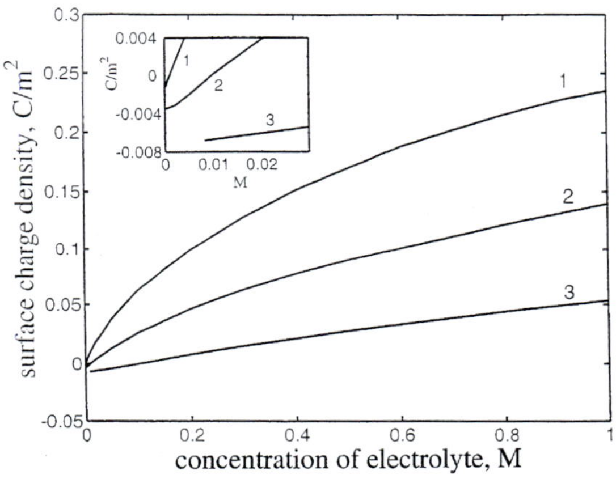

Figure 3. Reversal of the surface charge for a single plate. The calculations were carried out with the traditional theory. Key: (1) 2:1 electrolyte with $K_M = 0.001$ M; (2) 2:1 electrolyte with $K_M = 0.01$ M; (3) 2:1 electrolyte with $K_M = 0.1$ M. The other parameters are as in Figure 1a.

classical theory. In Figure 1a, the anion adsorption equilibrium constant $K_N = 6 \times 10^{-5}$ M, whereas in Figure 1b, $K_N = \infty$. There is a small difference between the two figures as a result of a low anion adsorption in Figure 1a and zero anion adsorption in Figure 1b. For strong cation adsorption ($K_M = 0.001$ M), the stability ratio passes through a minimum, followed by a maximum, after which it decreases with increasing electrolyte concentration. However, for low cation adsorptions ($K_M = 0.01$ or 0.1 M), the stability ratio decreases monotonically with increasing electrolyte concentration.

The above behavior is a result of two competing factors: one of them is the screening effect of the electrolyte ions, and the other is the change in the sign of the surface charge. At relatively low ionic strengths, the stability decreases due to ion screening. When the electrolyte concentration becomes sufficiently large, the multivalency and strong adsorption easily generate a change in the surface charge from negative to positive, and the system is restabilized. The stability decreases, however, at sufficiently large electrolyte concentrations because the screening becomes again dominant. It is important to emphasize that for monovalent ions W decays monotonically as the electrolyte concentration decreases. This occurs because the effect of charge inversion is much less important for monovalent than multivalent ions. Figures 2 and 3, in which the surface potential and the surface charge density are plotted against the electrolyte concentration, confirm the above interpretation. When the cation adsorption is weak, the screening effect of the electrolyte ions is dominant at all electrolyte concentrations. The above calculations indicate that at sufficiently high ionic strengths the traditional double layer theory predicts the screening of the repulsive force. In contrast, experiment appears to indicate that at high ionic strengths a restabilization should occur.

4. The Unitary Theory and Its Application to the Calculation of the Stability Ratio

It is well-known that the traditional double layer theory is valid in a limited range of concentrations for monovalent electrolytes, but is much less valid for higher valency electrolytes.[15] The traditional theory starts from the Poisson equation

$$\epsilon_0 \frac{dE}{dz} = \rho - \frac{dP}{dz} \tag{16}$$

where ρ is the charge density, E is the macroscopic field, and P is the polarization. The main assumption in that theory is that

$$P = \chi E \tag{17}$$

where χ is the electrical susceptibility, which is considered constant. Equation 17 is however valid only for a uniform field and hence at most when the gradient of the field is sufficiently small. Because, particularly for high valency electrolytes, the variation of the field with the distance is relatively steep, the above expression has to be modified by including the field E_p induced upon a molecule of water by the neighboring water dipoles. In the traditional Lorentz approach, on which the traditional double layer theory is based, the local field that acts upon a dipole is given by $E + (P/3\epsilon_0)$.[16] In the present case the local field has to contain the additional field E_p induced by the neighboring dipoles

$$E_{local} = E + (P/3\epsilon_0) + E_p \tag{18}$$

where

$$E_p(z_i) = C_1 m_{i-1} + C_0 m_i + C_1 m_{i+1} \tag{19}$$

In eq 19, the field is calculated by assuming that the water molecules are organized in icelike layers, m_i being the average dipole moment of water in layer i and C_0 and C_1 being interaction coefficients given by[6]

$$C_0 = -\frac{3.766}{4\pi\epsilon_0\epsilon'' l^3};\quad C_1 = \frac{1.827}{4\pi\epsilon_0\epsilon'' l^3} \tag{20}$$

In eq 20, l is the distance between the centers of two adjacent water molecules, and ϵ'' is the dielectric constant for the interaction between two neighboring molecules.

Only the neighboring water molecules are assumed to induce a field upon a water molecule of layer i, because the effect of the more distant molecules is screened by the intervening ones (the dielectric constant $\epsilon'' = 1$ for neighboring molecules and around 80 for the more distant ones, at room temperature).

(15) Levine, S.; Bell, G. M. *Discuss. Faraday Soc.* **1966**, *42*, 69.
(16) Frankl, D. R. *Electromagnetic Theory*; Prentice Hall: Englewood Cliffs, NJ, 1986.

Expanding eq 19 in series, one obtains

$$E_p \approx (C_0 + 2C_1)m + C_1\Delta^2 \frac{\partial^2 m}{\partial z^2} \approx C_1\Delta^2 \frac{\partial^2 m}{\partial z^2} \quad (21)$$

where Δ is the distance between the centers of two adjacent water layers.

The average dipole moment of a water molecule is given by

$$m = \gamma E_{local} \quad (22)$$

where γ is the polarizability for which, as in the traditional double layer theory, the Clausius−Mossotti equation

$$\gamma = 3\epsilon_0 v_0(\epsilon - 1)/(\epsilon + 2) \quad (23)$$

where v_0 is the volume of a water molecule, will be employed.

Combining eqs 18, 21, 22, and 23, yields

$$P(z) = \epsilon_0(\epsilon - 1)E(z) + \epsilon_0 v_0(\epsilon - 1)C_1\Delta^2 \frac{\partial^2 P}{\partial z^2} \quad (24)$$

Using Boltzmann expressions for the ion distributions, the Poisson−Boltzmann equation becomes

$$\frac{d^2\Psi}{dz^2} = -\frac{1}{\epsilon_0}\sum_i n_i^0 \nu_i e \exp(-\nu_i e\Psi/kT) + \frac{1}{\epsilon_0 v_0}\frac{dm(z)}{dz} \quad (25)$$

Let us now derive the boundary conditions.

The adsorption of the ions of the electrolyte changes the charge of the surface and generates simultaneously surface ion pairs (dipoles). Let us first assume that only the cation is adsorbed. The electric field induced by a surface dipole on the first layer of water molecules is given by[6]

$$E_s = (p/\epsilon')/2\pi\epsilon_0 (A/\pi + \Delta'^2)^{3/2} \quad (26)$$

where p is the normal component of the dipole moment of a surface dipole, ϵ' is the local dielectric constant near the surface, A is the surface area per surface dipole ($1/A$ is the surface dipole density), and Δ' is the distance between a surface dipole and the center of the first layer of water molecules.

The surface area per dipole, and hence the number of dipoles per unit area, is related to the surface potential and electrolyte concentration through the following equation

$$A = \frac{1}{N_A\theta_M} = \frac{K_M K_H + K_M[H^+]_s + K_H[M^{\nu+}]_s}{K_H N_A [M^{\nu+}]_s}$$
$$= \{K_M K_H + K_M[H^+]\exp(-e\Psi_s/kT) + K_H[M^{\nu+}] \times \exp(-\nu e\Psi_s/kT)\}/\{K_H N_A [M^{\nu+}]\exp(-\nu e\Psi_s/kT)\} \quad (27)$$

By inclusion of the field induced by the surface dipole into the local field that is acting upon the first layer of water molecules, the dipole moment of the first layer of water molecules near the surface is given by

$$m|_{z=\pm d} = \gamma E_{local} = \gamma(E + E_s + E_p + P/3\epsilon_0)|_{z=\pm d} \quad (28)$$

where

$$E_p|_{z=\pm d} = \mp(C_0 m_1 + C_1 m_2) \quad (19a)$$

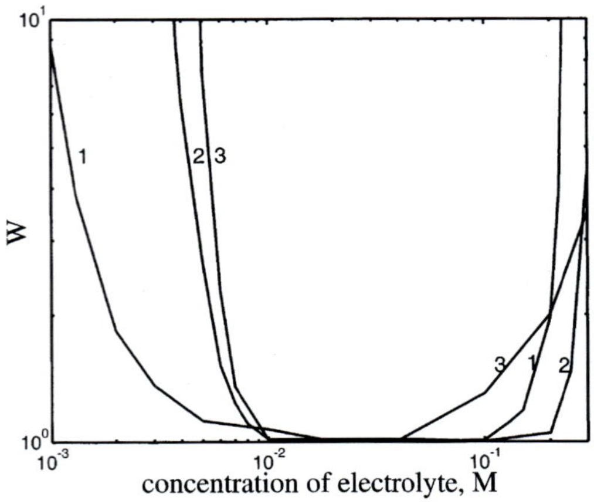

Figure 4. Stability ratio W against electrolyte concentration C (in M). The calculations were carried out with the new theory. The adsorption of anions is neglected. $N_A = N_B = 2 \times 10^{18}\ m^{-2}$, $K_H = 6 \times 10^{-10}$ M, pH = 8.0, $A_H = 1 \times 10^{-20}$J, and a = 102 nm. Key: (1) 3:1 electrolyte with K_M = 0.1 M; (2) 2:1 electrolyte with K_M = 0.1 M; (3) 1:1 electrolyte with K_M = 0.1 M.

Equation 19a is obtained from eq 19 by taking into account that for the first layer of water near the surface the layer $i - 1$ is missing.

Consequently, since $P = m/v_0$

$$m|_{z=\pm d} = \epsilon_0 v_0(\epsilon - 1)(E + E_s + E_p)|_{z=\pm d} \quad (29)$$

The electroneutrality condition provides a second boundary condition

$$\sigma = -\frac{1}{2}\int_{-d}^{d} \rho\, dz = \frac{1}{2}\int_{-d}^{d}\left(\epsilon_0\frac{\partial^2\Psi}{\partial z^2} - \frac{1}{v_0}\frac{\partial m}{\partial z}\right)dz$$
$$= \frac{1}{2}\left[\epsilon_0\left(\frac{\partial\Psi}{\partial z}\Big|_{z=d} - \frac{\partial\Psi}{\partial z}\Big|_{z=-d}\right) - \frac{1}{v_0}(m|_{z=d} - m|_{z=-d})\right]$$
$$= \epsilon_0\frac{\partial\Psi}{\partial z}\Big|_{z=d} - \frac{1}{v_0}m\Big|_{z=d}$$
$$= e\left(\frac{N_B K_{OH}}{K_{OH} + [OH^-]\exp\left(\frac{e\Psi_s}{kT}\right)} - \frac{N_A(\nu - 1)K_H[M^{\nu+}]\exp\left(-\frac{\nu e\Psi_s}{kT}\right) - N_A K_H K_M}{K_H K_M + K_M[H^+]\exp\left(-\frac{e\Psi_s}{kT}\right) + K_H[M^{\nu+}]\exp\left(-\frac{\nu e\Psi_s}{kT}\right)}\right) \quad (30)$$

In addition, due to the symmetry of the system

$$m = 0 \quad \text{and} \quad \frac{d\Psi}{dz} = 0 \quad \text{for } z = 0 \quad (31)$$

Using eqs 29−31 as boundary conditions, eqs 24 and 25 can be solved to provide the profiles of the electric potential Ψ and the dipole moment $m = Pv_0$.

Because of the interactions with neighboring dipoles, which generate the field E_p, the electrical contribution to the free energy contains an additional term when compared to the traditional theory

$$F_{ele} = \frac{1}{2}\int_{-d}^{d}\left(\left(\epsilon_0 E(z) + \frac{m(z)}{v_0}\right)E(z) - \frac{m(z)}{v_0}E_p(z)\right)dz \quad (32)$$

Figure 4, in which the stability ratio is plotted against the electrolyte concentration, contains three domains: the DLVO domain, the rapid aggregation domain, and the restabilization domain. The calculations revealed that the surface dipoles and the additional contribution to the local field generated by the neighboring dipoles in the water layers can account for the restabilization. With increasing electrolyte concentration, the density of surface dipoles increases and the colloidal dispersion becomes increasingly more stable for electrolyte concentrations higher than the critical stabilization concentration.

Because of the surface charge reversal as the electrolyte concentration increases, it is more realistic to also include the adsorption of anions among the adsorption equilibria, since they can have a relatively high concentration near a positively charged surface.

The adsorption of anions generates surface dipoles with a sign opposite to that generated by the adsorption of cations. The total electric field induced by the surface dipoles is obtained by adding the electric fields induced by the two kinds of ion pairs (surface dipoles)

$$E_s = (p_a/\epsilon')/2\pi\epsilon_0 (A_a/\pi + \Delta'^2)^{3/2} + (p_b/\epsilon')/2\pi\epsilon_0 (A_b/\pi + \Delta'^2)^{3/2} \quad (33)$$

where p_a, p_b are the normal components of the surface dipole moments generated by the adsorptions of anions and cations, respectively, and the surface area per dipole A_a and A_b can be calculated using the following expressions

$$A_a = \frac{1}{N_A\theta_M} = \frac{K_MK_H + K_M[H^+]_s + K_H[M^{\nu+}]_s}{N_AK_H[M^{\nu+}]_s}$$
$$= \{K_MK_H + K_M[H^+]\exp(-e\Psi_s/kT) + K_H[M^{\nu+}] \times \exp(-\nu e\Psi_s/kT)\}/\{N_AK_H[M^{\nu+}]\exp(-\nu e\Psi_s/kT)\} \quad (34)$$

$$A_b = \frac{1}{N_B\theta_N} = \frac{K_NK_{OH} + K_N[OH^-]_s + K_{OH}[N^-]_s}{N_BK_{OH}[N^-]_s}$$
$$= \frac{K_NK_{OH}\exp(-e\Psi_s/kT) + K_N[OH^-] + K_{OH}[N^-]}{N_BK_{OH}[N^-]} \quad (35)$$

where $[N^-]_s$ and $[N^-]$ are the concentrations of N^- near the surface and in the bulk.

In this case the boundary conditions acquire the forms

$$\sigma = -\frac{1}{2}\int_{-d}^{d}\rho dz = \frac{1}{2}\int_{-d}^{d}\left(\epsilon_0\frac{\partial^2\Psi}{\partial z^2} - \frac{1}{v_0}\frac{\partial m}{\partial z}\right)dz = \epsilon_0\frac{\partial\Psi}{\partial z}\bigg|_{z=d} - \frac{1}{v_0}m\bigg|_{z=d}$$

$$= e\left(\frac{N_BK_{OH}K_N}{K_{OH}K_N + [OH^-]\exp\left(\frac{e\Psi_s}{kT}\right)K_N + [N^-]\exp\left(\frac{e\Psi_s}{kT}\right)K_{OH}} + \frac{N_A(\nu-1)K_H[M^{\nu+}]\exp\left(-\frac{\nu e\Psi_s}{kT}\right) - N_AK_HK_M}{K_HK_M + K_M[H^+]\exp\left(-\frac{e\Psi_s}{kT}\right) + K_H[M^{\nu+}]\exp\left(-\frac{\nu e\Psi_s}{kT}\right)}\right) \quad (36)$$

and

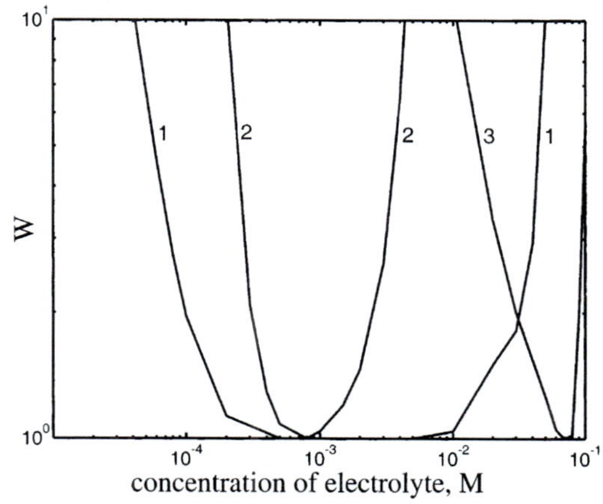

Figure 5. Stability ratio W against electrolyte concentration C(in M). The calculations were carried out with the new theory. The adsorption of anions is taken into account. $N_A = N_B = 2 \times 10^{18}$ m^{-2}, $K_H = 8 \times 10^{-10}$ M, pH = 8.0, $A_H = 1 \times 10^{-20}$J, a = 102 nm, and $K_N = 6 \times 10^{-5}$ M. Key: (1) 1:1 electrolyte with K_M = 0.001 M; (2) 2:1 electrolyte with K_M = 0.001 M; (3) 3:1 electrolyte with K_M = 0.001 M.

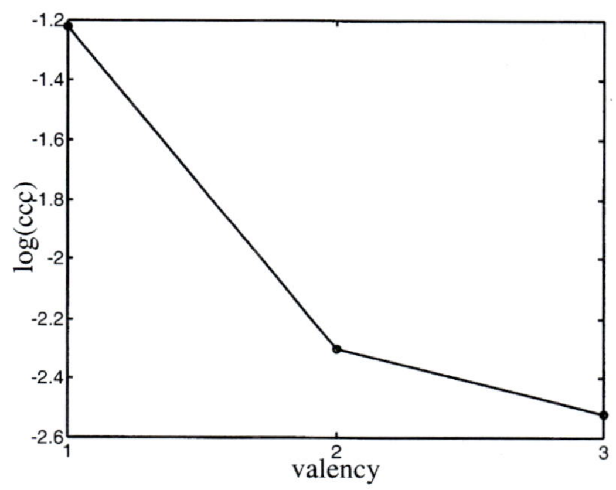

Figure 6. Relationship between the critical coagulation concentration and valency for the parameters of Figure 5.

$$m|_{z=-d} = \epsilon_0 v_0(\epsilon - 1)(E + E_p + (p_a/\epsilon')/2\pi\epsilon_0 (A_a/\pi + \Delta'^2)^{3/2} + (p_b/\epsilon')/2\pi\epsilon_0 (A_b/\pi + \Delta'^2)^{3/2})|_{z=-d} \quad (37)$$

Figure 5, in which the stability ratio is plotted against electrolyte concentration, for cations of valencies 1, 2, and 3 and monovalent anions, shows that the critical coagulation concentration increases with decreasing cation valency (Figure 6). This rule is not valid for the critical stabilization concentration.

5. Comparison with Experiment

The experiments carried out by Molina-Bolivar and Ortega-Venuesa with a monovalent cation NaCl and a divalent cation $CaCl_2$ electrolytes, described in the Introduction, are compared with calculated results in Figure 7.

The adsorption equilibrium constant K_{OH} was calculated in term of K_H using the expression

$$K_{OH} = \frac{10^{-7}}{\frac{10^{-7}N_B}{K_HN_A} - \frac{N_A - N_B}{N_A}} \quad (38)$$

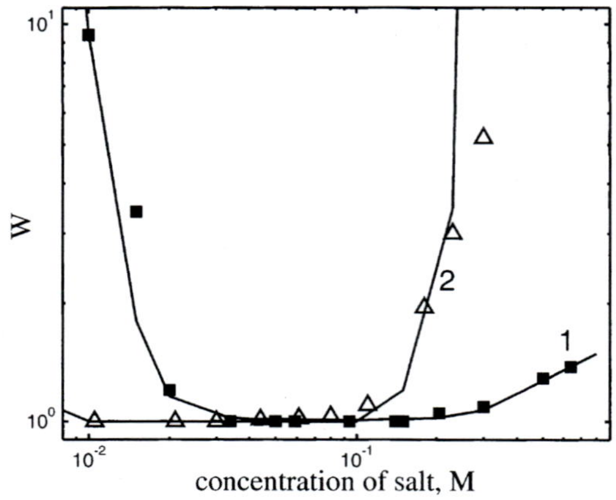

Figure 7. Comparison between experimental data and calculations for the dependence of the stability ratio W on the electrolyte concentration C (in M). The calculations were carried out with the new theory. $N_A = N_B = 2 \times 10^{18}$ m^{-2}, $K_H = 8 \times 10^{-10}$ M, pH = 8.0, $A_H = 1 \times 10^{-20}$J, $D = 2a = 204$ nm, and $K_{Cl} = 6 \times 10^{-5}$ M. Key: (1) NaCl with $K_{Na} = 1$ M; (2) $CaCl_2$ with $K_{Ca} = 0.02$ M. ■ and △ are experimental data for NaCl and $CaCl_2$ taken from Figure 6 of ref 3.

which was obtained from eq 4 at the isoelectric point and in the absence of electrolyte. The value of the pH at the isoelectric point was selected 7.0, on the basis of the experimental values, which are in the range 6.0−8.0.

Selecting a distance of 7 Å between two acidic or basic sites, one obtains $N_A = N_B = 2 \times 10^{18}$ m^{-2}. The pK_H of amino acid residues of globular proteins are in the range of 1.95−9.5,[17] and, for this reason, a value of 8×10^{-10} M was selected for K_H (pK_H = 9.1). The value of 0.02 M was selected for the K_{Ca} of Ig-G. The value $A_H = 1 \times 10^{-20}$J, which has the right order of magnitude, was selected for the Hamaker constant.[12] The values of the other parameters employed in the calculations were: $K_{Cl} = 6 \times 10^{-5}$ M, $p_a/\epsilon' = 3$ D for Ca^{2+}, $p_a/\epsilon' = 1.8$ D for Na^+, $p_b/\epsilon' = -1$ D. In these experiments, the diameter of the polystyrene particles was 204 nm, and the pH of the solution was 8.0.

The calculations demonstrate that the surface dipoles and the field caused by neighboring molecules in water play important roles in colloid stability and can explain the restabilization.

6. Conclusions

In the framework of the traditional double layer theory, when there is a strong adsorption of a multivalent cation of the electrolyte, the stability ratio of a colloidal dispersion first decreases, passes through a minimum, followed by a maximum, after which it decreases as the concentration of the (multivalent) electrolyte increases. When the adsorption is weak, the stability ratio decreases monotonically with increasing electrolyte concentration. For a monovalent electrolyte, the stability ratio calculated in the framework of the classical double layer theory decreases monotonically with increasing electrolyte concentration. Consequently, in the traditional theory, the ion screening becomes dominant at high ionic strength, whereas experiment appears to show that the colloidal dispersion is restabilized.

A new theory was proposed in which the adsorption of the ions of the electrolyte generates surface ion pairs (dipoles), which induce an electric field. In addition, the neighboring dipoles in water also generate a field in water. The restabilization of the colloidal dispersion at sufficiently high ionic strengths appears to be an effect of these fields. The charge inversion produced by the adsorption of the ions of the electrolyte plays also a role particularly for the high valency ions.

LA026729Y

(17) Timasheff, S. N. *Biological Polyelectrolyte*; Veis, A., Ed.; Marcel Dekker: New York, 1970.

94 *Langmuir* **2005,** *21,* 94–99

On the Restabilization of Protein-Covered Latex Colloids at High Ionic Strengths

Haohao Huang, Marian Manciu,[†] and Eli Ruckenstein*

Department of Chemical and Biological Engineering, State University of New York at Buffalo, Buffalo, New York 14260

Received July 20, 2004. In Final Form: October 6, 2004

Recent experiments on restabilization of protein-covered latex colloids at high ionic strengths reported by Lopez-Leon et al.[1] revealed strong specific anion effects. The same authors also emphasized that a recent polarization model, which involves both hydration and double layer forces, can account only for some of their experimental results but are in disagreement with other experimental results. The aim of the present paper is to show that most experimental results of ref 1 can be described, more than qualitatively, when the association equilibria for all the ions (with both the acidic and basic sites of the protein) are taken into account. As the traditional Poisson–Boltzmann approach, the polarization model neglects additional interactions between ions, and ions and surfaces, not included in the "mean field" electrical potential; therefore, a complete quantitative agreement should not be expected. While many of the discrepancies between calculations and experiment occur at low ionic strengths (10^{-4}–10^{-2} M), in the range of validity of the traditional DLVO theory, the latter can neither explain them. It is suggested that the structural modifications of the protein configuration induced by the electrolyte are responsible for some of the disagreements between experiment and calculations.

1. Introduction

The traditional Derjaguin–Landau–Verwey–Overbeek (DLVO) theory regards the stability of colloids as a balance between a repulsive double layer interaction due to the surface charges and an attractive van der Waals interaction.[2] At high electrolyte concentrations, the electrostatic interactions are screened and the van der Waals attraction leads to the coagulation of colloids. While describing qualitatively the behavior of most colloids, this theory is in disagreement with numerous experimental results. For example, it is known that the neutral lipid bilayers[3] and the strong hydrophilic colloids (e.g., silica)[4] are stable regardless of the ionic strength and an additional repulsion is required to explain those experiments. This repulsion was sometimes associated with the structuring of water in the vicinity of an interface (hence called "hydration force"),[4] but there is still no consensus regarding its microscopic origins.

One of the models for the hydration force, the polarization model,[5] assumes that the hydration force is generated by the local correlations between neighboring dipoles present on the surface and in water. The macroscopic continuum theory, in which water is assumed to be a homogeneous dielectric, predicts that there is no electric field above or below a neutral surface carrying a uniform dipolar density. However, at microscopic level the water is hardly homogeneous, and the electric interactions between neighboring molecules are much less screened than the interactions between remote molecules; consequently, a polarization is induced in water in the vicinity of dipolar surfaces.[5–7] It should be noted that Molecular Dynamics simulations reported also a polarization of water in the presence of surface dipoles,[8] which is not proportional to the electric field,[9] as assumed in the traditional macroscopic theory. Both results support the predictions provided by the polarization model regarding the hydration interactions.

The surface dipoles are sometimes generated by the association of the cations and anions with the negative and positive sites, respectively, of the surface. These associations decrease the surface charge but increase the surface dipole densities, and the polarization induced by them has been used[10,11] to explain the experimental data regarding the restabilization of protein-covered polystyrene colloids at high ionic strengths reported by Molina-Bolivar and Ortega-Vinuesa.[12]

The present article was stimulated by the recent experimental data on protein-covered latex colloidal systems immersed in various electrolyte solutions: NaCl, $NaNO_3$, NaSCN and $Ca(NO_3)_2$, which showed strong specific anionic effects on the restabilization curves.[1] In the opinion of López-León et al.,[1] the above polarization model for double layer/hydration forces could explain only some of their experiments, but not all of them. However, they assumed that at pH = 10 the adsorption of anions was negligible; hence "specific anion effects" could not be predicted by their association with the positive sites of the surface. Furthermore, at pH = 4 they assumed the

* Corresponding author: e-mail address, feaeliru@acsu.buffalo.edu; phone, (716) 645 2911/2214; fax, (716) 645 3822.

† Permanent address: Physics Department, University of Texas at El Paso.

(1) López-León, T.; Gea-Jódar, P. M.; Bastos-González, D.; Ortega-Vinuesa, J. L. *Langmuir* **2005**, *21*, 87–93.
(2) Deryaguin, B. V.; Landau, L. *Acta Physicochim. URSS* **1941**, *14*, 633. Verwey, E. J.; Overbeek, J. Th. G. *Theory of stability of lyophobic colloids*; Elsevier: Amsterdam, 1948.
(3) Rand, R. P.; Parsegian, V. A. *Biochim. Biophys. Acta* **1989**, *988*, 351.
(4) Israelachvili, J. *Intermolecular and Suface Forces*; Academic Press: New York, 1992.
(5) Schiby, D.; Ruckenstein, E. *Chem. Phys. Lett.* **1983**, *95*, 435.
(6) Manciu, M.; Ruckenstein, E. *Langmuir* **2001**, *17*, 7061.
(7) Ruckenstein, E.; Manciu, M. *Langmuir* **2002**, *18*, 7584.
(8) Perera, L.; Essmann, U.; Berkowitz, M. L. *Prog. Colloid Polym. Sci.* **1997**, *103*, 107.
(9) Faraudo, J.; Bresme, F. *Phys. Rev. Lett.* **2004**, *92*, 236102–1.
(10) Huang, H.; Manciu, M.; Ruckenstein, E. *J. Colloid Interface Sci.* **2003**, *263*, 156.
(11) Ruckenstein, E.; Huang, H. *Langmuir* **2003**, *19*, 3049.
(12) Molina-Bolivar, J. A.; Ortega-Vinuesa, J. L. *Langmuir* **1999**, *15*, 2644.

10.1021/la0481795 CCC: $30.25
Published on Web 12/08/2004

adsorption of cations to be negligible and the "specific cation effects", observed experimentally, could not be predicted.[1]

Strong specific anion effects were reported particularly at low electrolyte concentrations (10^{-4}–10^{-2} M),[1] a range in which the DLVO theory is considered accurate. However, as shown later, the present experimental data cannot be reproduced by the traditional theory in this range of electrolyte concentrations. In the past, no agreement could be obtained, on the basis of the traditional theory, because small changes in the values of the parameters, caused by the nonuniformity of the particles, affected strongly the stability ratio.[13] The polarization model provides similar results in the above range of electrolyte concentrations, when the dipole densities are sufficiently low and cannot explain the data.

It is clear that a perfect agreement with experiment cannot be provided by a theory which ignores the additional interactions between ions, and ions and surfaces, not included in the mean field potential (such as image forces,[14] excluded volume effects,[15] and ion-dispersion[16] or ion-hydration forces[17]). However, it will be shown that the experimental results reported by López-León et al.[1] can be more than qualitatively reproduced for uni-univalent electrolytes by the present polarization model for hydration/double layer forces, if one accounts for the association equilibria with the surface sites for all the ions present in the electrolyte (H^+, OH^-, anions, and cations).[11] Some additional reasons for the quantitative disagreements, involving the structural modifications of the adsorbed protein layer and the nonuniformity of the colloidal particles, will be also noted.

2. Review of the Theoretical Framework

If one assumes that all the interactions between ions are mediated by a "mean field" potential ψ and that the ions are Boltzmannian distributed, the charge density ρ is given by

$$\rho = -e(c_E + c_{OH})\exp\left(\frac{e\psi}{kT}\right) + e(c_E + c_H)\exp\left(-\frac{e\psi}{kT}\right) \quad (1)$$

where e is the protonic charge, c_E is the bulk electrolyte concentration (assumed for the sake of simplicity uni-univalent), c_{OH} and c_H are the concentrations of OH^- and H^+ in the bulk, respectively, k is the Boltzmann constant, and T is the absolute temperature. The macroscopic electric field $E \equiv -\nabla\psi$ obeys the Poisson equation

$$\nabla(\epsilon_0 E + P) = \rho \quad (2)$$

where ϵ_0 is the vacuum permittivity and $P = m/v_0$ is the polarization, with m being the average dipole of a water molecule of volume v_0. Equations 1 and 2 can be combined into

$$\epsilon_0 \frac{\partial^2\psi(z)}{\partial z^2} = e(c_E + c_{OH})\exp\left(\frac{e\psi(z)}{kT}\right) - e(c_E + c_H)\exp\left(-\frac{e\psi(z)}{kT}\right) + \frac{1}{v_0}\frac{\partial m(z)}{\partial z} \quad (3)$$

In a typical "macroscopic" assumption of proportionality between polarization and applied electric field, $P = \epsilon_0(\epsilon - 1)E$, where ϵ is the dielectric constant, and eq 3 reduces to the traditional Poisson–Boltzmann equation (the concentrations c_H and c_{OH} being in general much smaller than c_E). However, if the correlations between neighboring dipoles are taken into account, the following constitutive equation relating the polarization to the "macroscopic" electric field is obtained[7]

$$\lambda_m^2 \frac{\partial^2 m(z)}{\partial z^2} = m(z) + \epsilon_0 v_0(\epsilon - 1)\frac{\partial\psi(z)}{\partial z} \quad (4)$$

where z is the direction normal to the planar surface containing the dipoles. If the "dipole correlation length" $\lambda_m = 0$, the usual macroscopic constitutive relation $P = \epsilon_0(\epsilon - 1)E$ (and, consequently, the traditional Poisson–Boltzmann theory) is recovered.

The solutions for $\psi(z)$ and $m(z)$ of the system of eqs 3 and 4 depend not only on the surface charge density but also on the surface dipole densities. The boundary conditions are related to the surface charge and the surface dipoles generated by the association of the cations and H^+ with some of the N_A acidic surface sites per unit area of the surface and of the anions and OH^- with some of the N_B basic surface sites per unit area. The details are given elsewhere,[11] and the results will only be briefly reviewed here.

The surface charge density is given by

$$\sigma = \frac{eN_B}{1 + \left(\frac{c_{OH}}{K_{OH}} + \frac{c_E}{K_A}\right)\exp\left(\frac{e\psi_S}{kT}\right)} - \frac{eN_A}{1 + \left(\frac{c_H}{K_H} + \frac{c_E}{K_C}\right)\exp\left(-\frac{e\psi_S}{kT}\right)} \quad (5)$$

where K_A and K_C are the association-equilibria constants for anions and cations, respectively. The average areas occupied by anions, A_A, and cations, A_C, on the surface are provided by

$$A_A = \frac{1 + \frac{K_A}{K_{OH}}\frac{c_{OH}}{c_E} + \frac{K_A}{c_E}\exp\left(-\frac{e\psi_S}{kT}\right)}{N_B} \quad (6a)$$

$$A_C = \frac{1 + \frac{K_C}{K_H}\frac{c_H}{c_E} + \frac{K_C}{c_E}\exp\left(\frac{e\psi_S}{kT}\right)}{N_A} \quad (6b)$$

(13) Prieve, D. C.; Ruckenstein, E. *J. Colloid Interface Sci.* **1980**, *73*, 539.
(14) Jönnson, B.; Wennerström, H. *J. Chem. Soc., Faraday Trans.* **1983**, *79*, 19.
(15) Levine, S.; Bell, G. M. *Discuss. Faraday Soc.* **1966**, *42*, 69; Ruckenstein, E.; Schiby, D. *Langmuir* **1985**, *1*, 612;
(16) Boström, M.; Williams, D. M. R.; Ninham, B. W. *Phys. Rev. Lett.* **2001**, *87*, 168103.
(17) Manciu, M.; Ruckenstein, E. *Adv. Colloid Interface Sci.* **2003**, *105*, 63; Ruckenstein, E.; Manciu, M. *Adv. Colloid Interface Sci.* **2003**, *105*, 177.

The two boundary condition at the surface ($z = 0$) are given by

$$\epsilon_0 \frac{\partial\psi(z)}{\partial z}|_{z=0} - \frac{m(z)}{v_0}|_{z=0} = -\sigma \tag{7a}$$

and

$$(1 - \epsilon_0 v_0(\epsilon - 1)(C_0 + C_1))m(z)|_{z=0} - \epsilon_0 v_0(\epsilon - 1)\Delta C_1 \left.\frac{\partial m}{\partial z}\right|_{z=0} + \epsilon_0 v_0(\epsilon - 1) \left.\frac{\partial \psi}{\partial z}\right|_{z=0} = \epsilon_0 v_0(\epsilon - 1)\left[\left(\frac{p_A}{\epsilon'}\right)\frac{1}{2\pi\epsilon_0}\frac{1}{\left(\frac{A_A}{\pi} + \Delta'^2\right)^{3/2}} + \left(\frac{p_C}{\epsilon'}\right)\frac{1}{2\pi\epsilon_0}\frac{1}{\left(\frac{A_C}{\pi} + \Delta'^2\right)^{3/2}}\right] \tag{7b}$$

where p_A/ϵ' and p_C/ϵ' represent the effective dipoles generated by the association of the anions and cations with the positive and negative sites of the surface (note that the fields generated by the two types of dipoles are oriented in opposite directions), respectively, ϵ' is an effective dielectric constant of water in the vicinity of the surface, $\Delta = 3.68$ Å is the distance between two icelike water layers in the liquid, $\Delta' = 2$ Å is the distance between the surface dipoles and the first water layer, and the coupling parameters C_0 and C_1 are given by

$$C_0 = -\frac{3.766}{4\pi\epsilon_0\epsilon'' l^3} \tag{8a}$$

$$C_1 = \frac{1.827}{4\pi\epsilon_0\epsilon'' l^3} \tag{8b}$$

$l = 2.76$ Å being the shortest distance between two water molecules and $\epsilon'' = 1$ the effective dielectric constant for the electrostatic interactions between neighboring water molecules. The boundary condition (7b) takes explicitly into account the polarization generated by the surface dipoles, while eq 7a is a consequence of the Poisson equation (2).

For two identical surfaces immersed in an electrolyte separated by a distance $2d$, the symmetry provides the other two boundary conditions at the middle distance $z = d$

$$\left.\frac{\partial\psi(z)}{\partial z}\right|_{z=d} = 0 \tag{7c}$$

$$m(z)|_{z=d} = 0 \tag{7d}$$

Once the system of eqs 3 and 4 is solved under the boundary conditions (7a–d), the total free energy of the system can be calculated by adding the van der Waals interactions between the surfaces to the double layer free energy composed of electrostatic, entropic, and chemical contributions,[11] and the stability ratio can be calculated in the usual manner.[18]

(18) Hunter, R. J. *Foundations of Colloid Science*; Oxford Science Publications: New York, 1987.

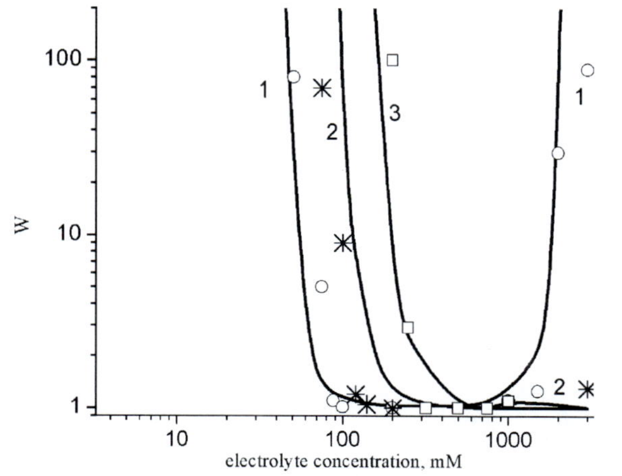

Figure 1. Experimental values of the stability ratio of protein-covered latex particles as a function of electrolyte concentration, at pH = 4.0, reported by Lopez-Leon et al.,[1] compared to those calculated from the polarization-based hydration model, for the following parameter values: $N_A = 1.2 \times 10^{18}$ sites/m², $N_B = 2.31 \times 10^{18}$ sites/m², $A_H = 0.5 \times 10^{-20}$ J, $K_H = 10^{-6}$ M, $K_{OH} = 8.95 \times 10^{-8}$M, $K_{Na} = 0.021$ M, $(p/\epsilon')_{Na} = 1.8$ D; (1) $K_{SCN} = 0.65$ M, $(p/\epsilon')_{SCN} = -1.65$ D; (2) $K_{NO_3} = 0.62$ M, $(p/\epsilon')_{NO_3} = -1.8$ D; (3) $K_{Cl} = 0.76$ M, $(p/\epsilon')_{Cl} = -2.3$ D; circles, NaSCN; squares, NaCl; stars, $NaNO_3$.

3. Coagulation of Protein-Covered Latex Particles

Let us first analyze the stability ratio for particles immersed in an electrolyte of pH = 4. The densities of acidic and basic sites, as well as their dissociation constants, are unknown. While one of these parameters could be determined as a function of the others from the isoelectric point (IP), even the latter is uncertain: $6.1 < \text{IP} < 8.5$.[19] The following values have been selected for the parameters: $N_A = 1.2 \times 10^{18}$ sites/m², $N_B = 2.31 \times 10^{18}$ sites/m², $K_H = 10^{-6}$ M, and $K_{OH} = 8.95 \times 10^{-8}$ M. The calculated stability ratio is plotted in Figure 1, using for the cation (Na^+) $K_{Na} = 0.021$ M, $(p/\epsilon')_{Na} = 1.8$ D, while for the various anions: $K_{Cl} = 0.76$ M, $(p/\epsilon')_{Cl} = -2.3$ D, $K_{NO_3} = 0.62$ M, $(p/\epsilon')_{NO_3} = -1.8$ D, and $K_{SCN} = 0.65$ M, $(p/\epsilon')_{SCN} = -1.65$ D. As shown in Figure 1, a reasonable agreement between experiment and calculations was obtained for uni-univalent electrolytes by selecting a rather low value for the Hamaker constant of the van der Waals interactions between colloidal particles, $A_H = 0.5 \times 10^{-20}$ J.

The calculations at pH = 10.0 are compared in Figure 2 with the experimental data provided by ref 1 for NaN0₃ and NaCl. The only modifications in the parameter values selected above are for the number of basic sites per unit area and for the Hamaker constant, which were selected in this case as $N_B = 1.62 \times 10^{18}$ sites/m² and $A_H = 0.9 \times 10^{-20}$ J. Again, qualitative agreement was obtained. There is no doubt that better agreement with experiment could have been obtained by other fittings of the values of the parameters to the experimental data. However, this is not very meaningful. The large number of parameters of unknown values involved in calculations raises doubt whether a good fit can be obtained only for the appropriate values of the parameters. As a matter of fact, because of the large number of parameters, a previous paper reported a good fit for the stability ratio of one of the above systems for different parameter values.[10]

In Figure 3, the experimentally determined stability ratio[1] is plotted for low concentrations of NaSCN and pH

(19) López-León, T.; Jódar-Reyes, A. B.; Ortega-Vinuesa, J. L.; Bastos-González, D. Submitted for publication in *J. Colloid Interface Sci.*

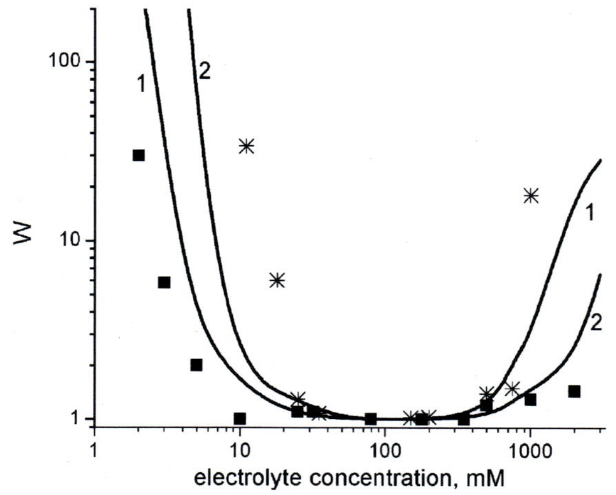

Figure 2. Experimental values of the stability ratio of protein-covered latex particle as a function of electrolyte concentration, at pH = 10.0, reported by Lopez-Leon et al.,[1] compared to those calculated from the polarization-based hydration model, for the following parameter values: $N_A = 1.2 \times 10^{18}$ sites/m^2, $N_B = 1.62 \times 10^{18}$ sites/m^2, $A_H = 0.9 \times 10^{-20}$ J, $K_H = 10^{-6}$ M, $K_{OH} = 8.95 \times 10^{-8}$ M, $K_{Na} = 0.021$ M, $(p/\epsilon')_{Na} = 1.8$ D; (1) $K_{Cl} = 0.76$ M, $(p/\epsilon')_{Cl} = -2.3$ D; (2) $K_{NO_3} = 0.62$ M, $(p/\epsilon')_{NO_3} = -1.8$ D; stars, $NaNO_3$; squares, NaCl.

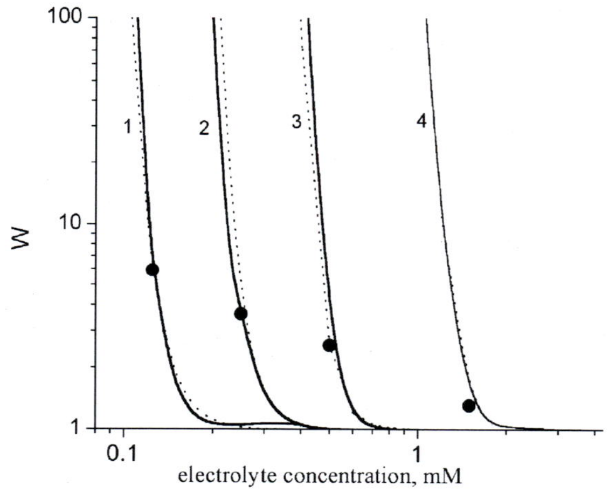

Figure 3. Stability ratios at low concentrations of NaSCN at pH = 10. The calculations have been carried out for $N_A = N_B = 0.5 \times 10^{18}$ sites/m^2 and various dissociation constants K_{Na} based on the present model (thick lines) and on the DLVO theory, with different values for K_{Na} (dotted lines), predict almost identical results but varies much more rapidly with electrolyte concentration than the experimental values reported in ref 1 (circles). $K_H = K_{OH} = 10^{-6}$ M, $(p/\epsilon')_{Na} = 1.8$ D, $K_{SCN} = 0.8$ M, $(p/\epsilon')_{SCN} = -0.8$ D, $A_H = 0.1 \times 10^{-20}$ J. (1) polarization model, $K_{Na} = 12.5 \times 10^{-6}$ M; DLVO, $K_{Na} = 2.5 \times 10^{-6}$ M. (2) polarization model, $K_{Na} = 23.1 \times 10^{-6}$ M; DLVO, $K_{Na} = 5.83 \times 10^{-6}$ M. (3) polarization model, $K_{Na} = 54.7 \times 10^{-6}$ M; DLVO, $K_{Na} = 12.6 \times 10^{-6}$ M. (4) polarization model, $K_{Na} = 164 \times 10^{-6}$ M; DLVO, $K_{Na} = 42 \times 10^{-6}$ M.

= 10. One can see that this ratio varies extremely slowly over a wide range of electrolyte concentrations. The low values of electrolyte concentrations, at which relatively rapid coagulation occurs (e.g., 1.25×10^{-4} M), suggest a low surface charge density. Therefore, in the calculations based on the polarization model (solid lines), the values $N_A = N_B = 0.5 \times 10^{18}$ sites/ m^2 were selected, and the value of the association constant for Na, K_{Na}, was varied to fit at least one experimental point, all the other parameter values remaining unchanged. All the calculations predict a stability ratio that varies much more rapidly with the ionic strength than the experimental one. It is therefore unlikely to find a set of parameters that can reproduce accurately this experimental behavior.

At low ionic strengths, the polarization model predicts results which are qualitatively similar to those obtained from the traditional DLVO theory. The main differences are due to the different boundary conditions, generated by the surface dipoles (ignored in the traditional theory). To obtain a quantitative agreement between the two theories, an additional surface charge should be employed in the DLVO calculations, to compensate for the field generated by the surface dipoles. The calculation of the stability ratio was repeated in the framework of the DLVO theory (dotted lines), with the K_{Na} selected to fit each of the experimental data points. As shown in Figure 3, there is good agreement (at low ionic strengths) between the results of the two treatments, and none of them can predict the slow variation of the stability ratio observed experimentally. It should be noted that at large ionic strengths the results of the two theoretical models are drastically different and that the traditional one cannot predict restabilization.

One might expect the additional interactions, not included in the electrical potential, between the large structure-breaking SCN^- and interface to be responsible for this effect. However, it was shown that these additional interactions (such as ion-dispersion[16] or ion-hydration[17] forces) are, in general, important only at sufficiently high electrolyte concentrations (above about 0.05 M) but negligible below about 0.01 M. A modality to include in the framework of the polarization model the additional interactions between ions and surfaces, not accounted for by the mean field potential ψ, was recently proposed.[20] The previous conclusion, that these additional interactions are negligible at low electrolyte concentration,[17] is supported by the new model.[20]

We are therefore inclined to believe that the non-DLVO behavior reported in the recent experiments at low ionic strengths[1] are mainly due to the nonuniformity of the colloidal particles and to the configurational changes of the adsorbed protein, triggered particularly by the added electrolyte. Let us first emphasize that the nonuniformity of the particles affects strongly the stability ratio.[13] The proteins were adsorbed on latex particles at pH = 6, then the pH of the system was changed to either pH 4 or 10 and stored up to 1 week before the light scattering experiments.[19] It was observed that the amount of protein adsorbed decreased with time,[19] and even if the latex particles were uniformly covered initially (which is unknown), one would expect some “voids” to be generated until the light scattering experiments. The stability ratio depends strongly on the coverage of the latex particles at the points of closest approach, and experiment provides only a statistical average over a polydisperse system. The statistical average might vary slowly with the electrolyte concentration, whereas the stability ratio calculated for identical particles (with constant numbers of acidic and basic sites), which are plotted in Figure 3, varies much more rapidly.

An important issue is whether the structure of the adsorbed protein is modified by the addition of electrolyte. The electrolyte is added only at the beginning of the light scattering experiment,[1] and one might expect that during the experiment (about 2 min), and at low electrolyte concentrations, not much modification in the structure of the adsorbed layer can occur. This expectation, however, is probably not accurate. The strong specific ionic effects

(20) Manciu, M.; Ruckenstein, E. *Adv. Colloid Interface Sci.*, **2004**, *112*, 109.

determined experimentally at low electrolyte concentrations, which cannot be explained by the existing theories, can be easily understood if one assumes that each type of electrolyte triggers different structural configurations of the adsorbed protein layer. Moreover, electrophoresis experiments, performed on the same system, revealed a rather low IP for the protein-covered latex particles immersed in low ionic strength electrolyte solutions ($c_E = 6 \times 10^{-3}$ M).[19] When a protein with an IP ~ 7 is adsorbed on a polystyrene particle with an IP ~ 10.5, one expects an intermediate IP to result. The experimental values obtained for the IP (between 5 and 6 for the electrolytes used in the experiment)[19] suggest either that many anions collapsed on the surface (perhaps on other places than on the empty basic sites) or that the structure of the protein was in such a manner modified that much more basic sites were exposed to water. A lowering by two units of the IP indicates a quite large change (by orders of magnitude) of the number of surface acidic and basic sites, and such large changes are possible only if structural modifications of the adsorbed layer occur.

It should be noted that the Hamaker constant of the van der Waals interaction between particles is strongly affected by the properties of the external layer of the particles. Because the colloidal particles interact through water, the incorporation of water molecules into the adsorbed protein layer decreases strongly the van der Waals interactions between particles. It was previously reported[21,22] that large decreases in the effective Hamaker constant (by orders of magnitude) could be achieved if water penetrates the external layer of the particles. Therefore, the relatively low value for the Hamaker constant in our calculations might suggest that the layer adsorbed on the latex particle contains a large amount of water.

If the structure of the adsorbed layer is modified during experiment, then a more accurate model is required to take into account the complex structure of the particles. Therefore, one should not expect a perfect match between calculations and experimental results.

An important issue that has to be emphasized is that the experimentally determined dependence of the stability ratio on electrolyte concentration, at low ionic strengths, exhibits (at least for NaSCN) a strongly non-DLVO behavior, in a range in which the DLVO theory is considered fairly accurate. Therefore, we are inclined to believe that the electrolyte (even at low ionic strength) induces indeed structural modifications of the adsorbed protein layer at least near the interface.

Another issue which has to be clarified is the criticism in ref 1 that the hydration/double layer model fails because it assumes an icelike layered structure of water in the vicinity of an interface, while the structure-breaking ions (such as SCN^-) probably destroy this layered structure.[1] First, the hydration/double layer model assumes only correlations between neighboring dipoles, not necessarily an icelike structure. The correlations are perhaps stronger when the water is organized in icelike layers (and might even lead to an oscillatory behavior of the polarization)[23] but do not vanish in less organized structures; they are only weaker. In fact, the value for the dipole correlation length λ_m was derived assuming that only a cluster of about 5 Å around the central water molecule preserves locally the structure of ice.[7] The influence of disorder on dipole correlation length was discussed elsewhere.[23]

The destruction of the icelike ordering by SCN^- might lower the hydration force, although such an effect was not reported for other systems (e.g., neutral lipid bilayers in water). However, the decrease in hydration due to the destruction of the icelike layers cannot explain the experiments of López-León et al.[1] The rapid coagulation at low electrolyte concentrations ($10^{-4}-10^{-3}$ M) is not due to the decrease in the hydration force: at these ionic strengths, even a small surface charge controls the coagulation. The decrease in the hydration force should affect coagulation at high ionic strength; however, at high ionic strength, the system was restabilized by adding NaSCN.

Furthermore, the restabilization achieved by the addition of NaSCN is even stronger than that obtained through the addition of NaCl. However, the opposite behavior should have been observed, because the stronger structure-breaking SCN^- is expected to destroy the icelike structure more than Cl^- and, consequently, would lower more the hydration repulsion. Of course, one might suspect that the strong restabilization by NaSCN might be due to an unusually large negative charge generated on the surface by the accumulation of SCN^-, but this hypothesis is invalidated by the electrophoretic experiments, which indicated an almost neutral surface at high ionic strengths.[19]

4. Conclusion

It was recently suggested[1] that some of the strong specific anionic effects observed regarding the coagulation of protein-covered latex particles cannot be predicted by a theory of hydration/double layer interaction,[5,7] which accounts for the correlation between neighboring dipoles and the formation of dipoles on the surface.[10,11]

Here it is shown that the above theory can in most cases predict at least qualitatively the experimental behavior. The unusual strong specific anion effects observed for SCN^- at low ionic strengths are in disagreement with both the present and the traditional DLVO theory. The present theory is an extension of the traditional theory and predicts qualitatively similar results for sufficiently charged surfaces at low ionic strengths. Notable differences occur between the two treatments either at low surface charges or at high ionic strengths, at which the double layer is screened.

The strong discrepancy between experiment and the traditional DLVO theory at low ionic strengths (where the latter theory is considered to be accurate) cannot be explained by additional interactions between ions and surfaces, because they are negligible below 0.01 M. Therefore, we are inclined to believe that the structural modification of the adsorbed protein by the addition of a structure breaking ion, such as SCN^- is mainly responsible for the quantitative disagreement between experiment and model calculations. The nonuniformity of the colloidal particles may be also responsible for the disagreement.

It is well-known that the structure-breaking anions generate strong specific ion effects for proteins (while the type of cations does not matter much).[24] In contrast, the structure-making cations (e.g., K^+, Li^+, Cs^+) have strong specific effects for the interactions between latex[25] or mica[26] surfaces. An explanation for this apparent paradox might be that the structure-breaking ions are more likely to be

(21) Johnson, R. E.; Morrison, W. H. *Adv. Ceram.* **1987**, *21*, 323.
(22) Ruckenstein, E. *Colloids Surf.* **1993**, *63*, 271.
(23) Manciu, M.; Ruckenstein, E. *Langmuir* **2001**, *17*, 7582.
(24) Cacace, M. G.; Landau, E. M.; Ramsden, J. J. *Q. Rev. Biophys.* **1997**, *30*, 241.
(25) Healy, T. W.; Homola, A.; James, R. O.; Hunter, R. J. *Faraday Discuss. Chem. Soc.* **1978**, *65*, 156.
(26) Pashley, R. M. *J. Colloid Interface Sci.* **1981**, *83*, 531.

expelled from the bulk water toward the surface, whereas the structure-making ions are in general depleted from the interface.[17] The unstable surfaces, such as those of the proteins, which can be easily modified, are particularly sensitive to the ions that can approach them closely, such as SCN^-, and modify the water properties near them. In contrast, the interactions between stable surfaces are particularly sensitive to the counterions depletion, because the screening of the double layer is much weaker in this case. Therefore, for negatively charged stable surfaces, one expects strong specific effects for the structure-making cations, such K^+, Li^+, or Cs^+. The strong specific effects obtained experimentally for structure-breaking ions,[1] such as SCN^-, suggest that the structure of the surface is modified by the ions; the structural modifications are also expected to occur for multivalence ions, such as Ca^{2+}. Therefore, we suggest that the departure of the experimental data reported in ref 1 from the predictions of DLVO theory (at low ionic strengths) and from the polarization model (at all ionic strengths) are mainly due to the structural modification of the adsorbed protein layer, induced by electrolyte. This conclusion is also supported by electrophoretic experiments performed on the same systems, which showed a drastic change in the isoelectric points of the particles, dependent on the nature of the added electrolyte.[19]

LA0481795

ELSEVIER

Advances in Colloid and Interface Science 112 (2004) 109–128

ADVANCES IN COLLOID AND INTERFACE SCIENCE

www.elsevier.com/locate/cis

The polarization model for hydration/double layer interactions: the role of the electrolyte ions

Marian Manciu[1], Eli Ruckenstein*

Department of Chemical and Biological Engineering, State University of New York at Buffalo, United States

Available online 10 November 2004

Abstract

The interactions between hydrophilic surfaces in water cannot be always explained on the basis of the traditional Derjaguin–Landau–Verwey–Overbeek (DLVO) theory, and an additional repulsion, the "hydration force" is required to accommodate the experimental data. While this force is in general associated with the organization of water in the vicinity of the surface, different models for the hydration were typically required to explain different experiments. In this article, it is shown that the polarization-model for the double layer/hydration proposed by the authors can explain both (i) the repulsion between neutral lipid bilayers, with a short decay length (~2 Å), which is almost independent of the electrolyte concentration, and, at the same time, (ii) the repulsion between weakly charged mica surfaces, with a longer decay length (~10 Å), exhibiting not only a dependence on the ionic strength, but also strong ion-specific effects. The model, which was previously employed to explain the restabilization of protein-covered latex particles at high ionic strengths and the existence of a long-range repulsion between the apoferritin molecules at moderate ionic strengths, is extended to account for the additional interactions between ions and surfaces, not included in the mean field electrical potential. The effect of the disorder in the water structure on the dipole correlation length is examined and the conditions under which the results of the polarization model are qualitatively similar to those obtained by the traditional theory via parameter fitting are emphasized. However, there are conditions under which the polarization model predicts results that cannot be recovered by the traditional theory via parameter fitting.

Keywords: Polarization model; Hydration forces; Double layer interaction; Ion-specific effects; Restabilization

Contents

* Corresponding author. Tel.: +1 716 645 2911/2214; fax: +1 716 645 3822.
E-mail address: feaeliru@acsu.buffalo.edu (E. Ruckenstein).
[1] Permanent address: Physics Department, University of Texas at El Paso.

0001-8686/$ - see front matter
doi:10.1016/j.cis.2004.09.001

1. Introduction

Long ago, Gouy [1] and Chapman [2] suggested that the distribution of ions in the vicinity of a charged surface immersed in an electrolyte can be obtained by assuming that the ions interact with all the other charges in the system via a "mean field" electrical potential ψ that obeys the Poisson equation. Assuming that the ions obey Boltzmann statistics, the "mean field" ψ and the ion distributions can be calculated for given boundary conditions (typically, relations for the charge or potential on the surfaces). This "Poisson–Boltzmann" formalism was employed later by Langmuir [3] to explain the force between two charged surfaces immersed in an electrolyte, and remained since then the basic equation for the behaviour of colloids.

It is perhaps curious that such a simple theory is still surviving the criticism that it received for almost a century. One reason is that some parameters involved (the surface charge and surface potential) are not accurately known and can be in most cases fitted to account for the experimental data. Sometimes, the results predicted are not achievable physically. A well-known example is the prediction by the traditional Poisson–Boltzmann formalism of a counter-ion density exceeding the available volume in the vicinity of a highly charged surface [4]. Stern corrected this paradox by considering a layer of immobile counter-ions in the vicinity of the interface [4]. Since then, many such corrections have been introduced in the traditional approach, to account for the changes in the dielectric constant of water due to the presence of ions [5], or to high electric field strengths [6], to account for the finite size of the ions [7], for image forces [8], for ion-dispersion [9] and ion-hydration [10] forces, and so on.

Another clear failure of the Poisson–Boltzmann approach was provided by the experiments regarding the force between neutral lipid bilayers [11]. The repulsion required to explain their stability was determined to have an almost exponential dependence, with a decay length of about 2–3 Å [11], and neither this decay length nor the magnitude of the interactions were dependent on the ionic strength. This interaction was initially attributed to the structuring of water near the surface (the "hydration" of the surfaces) and it is usually called "hydration force" [12]. The microscopic origins of this interaction are still under debate.

There are good reasons for these disagreements, as it will be noted in what follows. According to the traditional Derjaguin–Landau–Verwey–Overbeek (DLVO) theory [13], most colloids coagulate when the electrolyte concentration is sufficiently high to screen the repulsion generated by the surface charge; however, the hydration forces are strong enough to prevent the coagulation of silica particles regardless of the electrolyte concentration [12]. Therefore, one open question is why the hydration force is of paramount importance in some cases and negligible in others?

Amphoteric latex particles, at pH values larger than those corresponding to the isoelectric point, coagulated when the $CsNO_3$ concentration exceeded about 0.3 M, but remained stable up to concentrations exceeding 3 M of KNO_3 [14]. The additional repulsive interaction between surfaces, which could explain the stability, was initially related to the hydration of ions [14]: to achieve contact between surfaces, the ions must lose their hydrating water molecules, a process that is energetically unfavorable. A more involved statistical model of this interaction was suggested recently [15]. Since the energetic penalty is larger for K^+ than for Cs^+ ion, one could understand why the system is stable with KNO_3 but not with $CsNO_3$, and also why at low pH values the system is unstable at high ionic strengths. In the latter case, the anions are attracted near the positive surfaces. Because their hydration energies are lower than those of the cations, the repulsion generated by the dehydration of anions might not be sufficient to prevent coagulation. Also, because the cations are depleted near the positively charged interface, there were no specific cation effects observed in the experiments at pH values lower than the isoelectric point of the particles [14].

The above model for the "hydration force" assumes that the ions (particularly, the structure making cations) play a dominant role. This conclusion is supported by many experiments with charged surfaces, which indicate that the traditional "double layer" theory dominates the interactions at low and intermediate ionic strengths, while the "hydration" interactions occur only at large electrolyte concentrations, at which there are sufficient ions to be "dehydrated" and also to screen the double layer. The competition of ions for hydrating water molecules (an excluded-volume effect) was shown to generate an increase in repulsion, when compared to the traditional theory, and therefore to generate a "hydration force" at sufficiently large ionic strengths [7,16].

However, a different picture is provided by the experiments with neutral phospholipid bilayers in water, the stability of which (presumably due to the hydration force) is almost unaffected by the electrolyte concentration. This type of "hydration force" certainly cannot be due to "ion dehydration". Another open question is therefore whether or not the "hydration forces" are independent of the electrolyte concentration.

A number of additional questions are further raised by experiments regarding forces between mica surfaces [17]. At short distances between surfaces, the interactions are always dominated by the divergently large van der Waals attractions. However, the potential barrier seems to increase drastically after the electrolyte concentration exceeds a particular value, which indicates the emergence of a "hydration repulsion" at a critical electrolyte concentration. The minimum electrolyte concentration at which this "hydration force" occurs depends strongly on the nature of the electrolyte: 6×10^{-2} M for LiCl, 10^{-2} M for NaCl, 10^{-3} M for CsCl and 3×10^{-4} M for KCl [17]. The extremely low electrolyte concentrations required to prevent the jump of the surfaces in their primary minimum suggest that this hydration force is not necessarily related to the "dehydration" of cations: if a sufficiently large repulsion could be generated by 3×10^{-4} M KCl, then probably no colloidal system will coagulate at 1 M KCl, which is not the case. Furthermore, the difference of about 2 orders of magnitude between the minimum concentrations of LiCl and KCl required to generate the hydration interactions represents an enormous "specific ion effect", which is unlikely to be explained by the difference in the hydration energies between the Li^+ and K^+ cations. Even if such a strong effect could be explained on that basis, it would be difficult to explain why this strong effect does not occur in other colloidal systems.

Another interesting result of the experiments of Pashley [17] is that the decay length of the strong interaction measured at close separations is about 10 Å, in contrast to the value of about 2–3 Å reported for the interaction between lipid bilayers [11]. The range of hydration interactions is also questioned by other experiments regarding the restabilization of apoferritin molecules at large ionic strengths [18]. The large values of the second virial coefficient obtained at high electrolyte concentrations cannot be explained by the sum between the traditional double layer force and a short-range hydration force [19].

Some of the experiments cited above suggest that the ions do not affect the hydration force, whereas in other experiments the ionic strength is not only important, but strong cation-specific effects are present [14,17]. However, it is well known since Hofmeister [20] that the specificity of anions influences strongly the precipitation of proteins from electrolyte solutions, while the nature of cations plays in this case a less important role. While the list of contradictory experiments about the nature of the hydration force is much longer, we will only add here some recent results about the restabilization of protein-covered latex particles [21], which indicate that the restabilization at high ionic strength (presumably, due to the hydration forces) is strongly influenced by the nature of both the anions and cations involved. It should be, however, noted that the restabilization of electrosterically stabilized colloids at high ionic strength was not associated to the hydration forces, and other mechanisms were suggested instead [22].

In summary, the contradictory evidence brought by various experiments makes difficult to answer some fundamental questions about the hydration forces, such as:

(i) whether or not the hydration force depends on the electrolyte concentration?
(ii) if they do depend on the ionic strength, do specific ion effects occur for anions, cations, both or none of them?
(iii) what is the decay length of the hydration force and on which parameters does it depend (nature of the surface, nature of the electrolyte and electrolyte concentration)?
(iv) how do the surfaces affect the magnitude of the hydration forces?

It should be noted that many attempts have been made to explain the hydration force by microscopic models. For example, the disagreements between experiment and the traditional mean field theory were attributed to the ion–ion correlations (ignored by the mean field theory); one of the most successful theories is based on the anisotropic hypernetted chain method [23]. It was shown on the basis of that theory that an additional repulsion (similar to the hydration force) can indeed occur at small separations and large ionic strengths [24]. However, the results are so strongly dependent on the choice of the interaction potentials between the ion-pairs, that even a "double layer attraction" could be predicted [25]. In addition, the hydration force in the absence of an electrolyte, or at moderate ionic strength, could not be explained by this model.

Another model assumed the formation of a cage-like "network" of hydrogen bonds around a surface immersed in water. Because the bonds must be destroyed when two surfaces approach each other, a repulsion is generated. A model calculation was presented by Attard and Batchelor [26]. While intuitively appealing, this model cannot relate in a simple manner the hydration interactions to the nature of the surface or to the electrolyte concentration.

Another model [27] attempted to explain only the hydration interactions between neutral phospholipid bilayers, because those interactions appear to be quite different from the hydration forces determined in other systems: they have a much shorter decay length and are almost independent of electrolyte concentration [11]. The theory was based on a steric repulsion between lipid molecules, which protruding from one bilayer collide with the molecules of the opposite bilayer [27]. However, it was shown that the hydration interactions are not affected much by the polymerization of bilayers, which essentially exclude molecular protrusions [28].

The present article is structured as follows. In the first section, the physical basis of the polarization model for the hydration/double layer interactions proposed by the authors

[29,30] will be briefly reviewed, and the model will be extended to account for the additional interactions between ions, and ions and interfaces, not included in the mean field electrical potential. In the last part of the article, it will be shown that the polarization model for hydration/double layer interactions can provide in general a reasonable qualitative agreement with the experimental results listed above.

2. The polarization model for hydration/double layer interactions

2.1. The physical basis of the model

In some of the first models proposed to explain the microscopic origin of the hydration force, it was suggested that the hydration interactions are generated by the polarization of the water molecules located in the vicinity of a dipolar surface [29,31]. This hypothesis was supported by experiment, which revealed a direct correlation between the magnitude of the hydration force and the surface dipole density of phospholipid membranes [32], the latter being proportional to the experimentally accessible potential across the membrane (the so-called χ-potential). It is intuitively appealing to assume that high surface dipolar densities correspond to strongly hydrophilic surfaces and hence that between such surfaces large hydration forces are generated. However, before establishing a relation between the χ-potential and the polarization of water in vicinity of a surface, there are some difficulties that have to be addressed.

First, a continuous homogeneous dipolar density on a flat surface between two continuous media does not generate any electric field above the surface. The homogeneous dipolar surface might be thought as composed of two sheets of constant and equal surface densities of opposite charges; the electric field generated by such a charge distribution vanishes everywhere but between the sheets, hence it is not clear how a polarization can be generated in the vicinity of such an interface. A second difficulty was indicated by molecular dynamics simulations, which showed that the water near an interface is indeed polarized, but the average polarization oscillated with the distance from the interface [33]. The first polarization-based models [29,31] predicted, however, an exponential decay.

The first difficulty could be explained by observing that water is not homogeneous at a molecular scale [34]. The interactions with remote dipoles are screened by the intervening water molecules, because the effective dielectric constant of their interaction is comparable with the macroscopic dielectric constant (ε=80). In contrast, the interaction between neighboring dipoles is much less screened, because there is no intervening medium between them. While assuming a constant ε for all interactions would predict a vanishing electric field near the surface, the local value of an "effective" ε leads to a net electric field, which can polarize the water molecules above the surface.

The second difficulty can be removed if one assumes that in the vicinity of an interface the water is organized in ice-like layers. The electrical interactions between the water dipoles of successive layers lead indeed to an oscillatory behavior of the polarization [35]. If the surface is not perfectly flat, of if the water is not perfectly organized in water layers, the statistical average smooth out the polarization oscillations [35]. The latter results have been also supported by molecular dynamics simulation, in which the surface dipoles were allowed to move [36]. Let us now examine in detail how the correlation between neighboring dipoles occurs.

2.2. The correlation between neighboring dipoles

The electrical field always obeys the Poisson Equation [37]:

$$\nabla(\varepsilon_0 E + P) = \rho \tag{1a}$$

where ε_0 is the vacuum permittivity, $E \equiv -\nabla\psi$ is the electric field, ψ the electrical potential, P the polarization and ρ the charge density. In a typical "macroscopic" assumption of proportionality between polarization and applied electric field,

$$P = \varepsilon_0(\varepsilon - 1)E, \tag{2}$$

where ε is the dielectric constant, one obtains the well-known form of the Poisson equation

$$\Delta\psi = -\frac{\rho}{\varepsilon\varepsilon_0}, \tag{1b}$$

which, by assuming Boltzmannian distributions for the electrolyte ions, leads to the traditional Poisson–Boltzmann equation. However, at microscopic scales, the water is not homogeneous and a water molecule is subjected to a "local" electric field, which is the sum between a "macroscopic field" generated by the sources from outside the medium, and a field generated by the other water molecules surrounding it [37]. For a uniform polarized medium, the latter corresponds to the field generated in a spherical cavity in a medium, due to the polarization of the whole medium (the Lorenz field $E_{\mathrm{Lorenz}}=P/3\varepsilon_0$ [37]). Therefore, the local field acting on a water molecule is given by:

$$E_{\mathrm{local}} = E + \frac{P}{3\varepsilon_0} \tag{3}$$

and the average dipole moment of the water molecule, $m=P/v_0$, where v_0 is the volume occupied by a water molecule, by:

$$m = \gamma E_{\mathrm{local}} = \gamma\left(E + \frac{m}{3\varepsilon_0 v_0}\right) = \frac{\gamma}{1 - \frac{\gamma}{3\varepsilon_0 v_0}} E \equiv \varepsilon_0 v_0(\varepsilon - 1)E \tag{4}$$

where γ is the microscopic polarizability. The last equality was obtained using Eq. (2) because for an uniform polarized medium, the polarization is proportional to the macroscopic electric field, and corresponds to the Clausius–Mossotti relation between the microscopic molecular polarizability γ and the macroscopic dielectric constant ε.

If the polarization is non-uniform, an additional field E_P is generated by the neighboring dipoles. A simple way to calculate this field [34] is to neglect the field generated by the remote dipoles, because it is screened by the intervening water molecules, and to consider only the field generated by the neighboring dipoles, screened by an effective dielectric constant ε'', much lower than the bulk value, ε=80. In order to calculate the additional field E_P, the local structure of water must be known. A simple assumption is that the water is organized in ice-like layers in the vicinity of the surface, of thickness Δ (see Fig. 1). The water molecules near the surface dipoles (gray area in Fig. 1) are assumed to be bound by the dipoles and the distance between the center of the first ice-like water layer and the center of the surface dipoles is Δ'. The value of the additional field in layer i, $E_{P,i}$, is generated by the neighboring water molecules from the same layer i, as well as by the neighboring water molecules from the adjacent layers, $i-1$ and $i+1$. The average dipole moment of a water molecule, m_i, is constant in each layer, because the layers are assumed parallel to the surface, and m is a function of the distance from the surface x alone. Since the field generated by the dipoles is proportional to their dipole moments, the additional field becomes:

$$E_{P,i} = C_1 m_{i-1} + C_0 m_i + C_1 m_{i+1} \tag{5a}$$

The values of the coupling constants C_0 and C_1 were calculated by assuming that only the first 26 neighbors of a water molecule in layer i (12 from the same layer i and 7 from each neighboring layer, $i\pm1$), generated a sufficiently

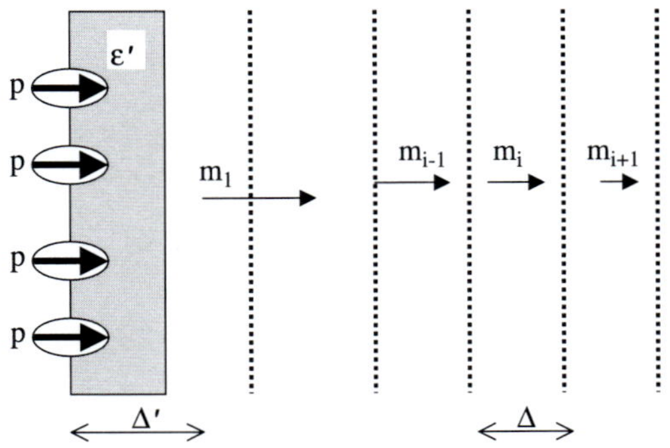

Fig. 1. A sketch of water layering in the vicinity of a dipoles-bearing surface. In the vicinity of the surface dipoles p, the bounded water (gray) has the dielectric constant ε', and the center of the first water layer is at the distance Δ' from the center of surface dipoles.

strong field, and the contributions of the more remote dipoles were neglected [34]. Under these circumstances,

$$C_0 = -\frac{3.766}{4\pi\varepsilon_0\varepsilon'' l^3} \qquad C_1 = \frac{1.827}{4\pi\varepsilon_0\varepsilon'' l^3} \tag{6}$$

with l the distance between the centers of two adjacent water molecules and ε'' the dielectric constant for the interaction between neighboring molecules, which is expected to be nearer to unity than to the dielectric constant of water, ε=80, because there are no intervening molecules to screen the interactions. For an uniform polarized medium, (m_i=m_{i-1}=m_{i+1}), the additional field E_P vanishes, because $C_0+2C_1 \cong 0$, and the traditional theory (Eq. (4)) is recovered. For a non-uniform polarization, the local field is given by

$$E_{\mathrm{local,i}} = E + \frac{m_i}{3\varepsilon_0 v_0} + C_0 m_i + C_1(m_{i-1} + m_{i+1}) \tag{7}$$

and the average dipole moment of a water molecule by:

$$\begin{aligned} m_i &= \gamma E_{local} = \gamma\left(E + \frac{m}{3\varepsilon_0 v_0} + E_P\right) \\ &= \frac{\gamma}{1 - \frac{\gamma}{3\varepsilon_0 v_0}}(E + E_P) = \varepsilon_0 v_0(\varepsilon - 1)(E + E_P) \\ &= \varepsilon_0 v_0(\varepsilon - 1)(E + C_0 m_i + C_1(m_{i-1} + m_{i+1})) \end{aligned} \tag{8}$$

where the relation between the molecular polarizability γ and macroscopic dielectric constant, Eq. (4), was employed. Eq. (8) relates the average dipole moment of the water molecules in one layer, m_i, to the average dipole moments of the molecules in adjacent layers, $m_{i\pm1}$. By expanding in series the average dipole moment, Eq. (5a) becomes:

$$\begin{aligned} E_P(x) &= C_0 m(x) + C_1\left(2m(x) + \Delta^2\frac{\partial^2 m(x)}{\partial x^2} + \ldots\right) \\ &\cong C_1\Delta^2\frac{\partial^2 m(x)}{\partial x^2} \end{aligned} \tag{5b}$$

where Δ is the distance between the centers of two adjacent water layers. Eqs. (8) and (5b) lead to:

$$\lambda_m^2\frac{\partial^2 m(x)}{\partial x^2} = m(x) + \varepsilon_0 v_0(\varepsilon - 1)\frac{\partial\psi(x)}{\partial x} \tag{9}$$

with

$$\lambda_m = \Delta(\varepsilon_0 v_0(\varepsilon - 1)C_1)^{\frac{1}{2}} \tag{10}$$

representing a "dipolar correlation length". When λ_m=0 or $\partial^2 m/\partial x^2$=0 the usual macroscopic constitutive relation $P=\varepsilon_0(\varepsilon-1)E$ is recovered. However, for ε=80, Δ=3.68 Å, v_0=30 Å^3, l=2.76 Å, ε''=1, Eq. (10) provides λ_m=14.9 Å, and, in general, the polarization and the electric field are not proportional to each other. It is of interest to note that such an "anomalous dielectric behaviour" of water in the vicinity of a surface was recently observed in molecular dynamics simulations of ionic Newton Black films [38].

The above expression of λ_m [30] is based on the assumptions that the local cluster surrounding a water molecule has an ice-like structure and is parallel to the surface. The first approximation is reasonable, because of the small size of the local clusters (with a radius twice the distance of closest approach between water molecules and containing only the first 26 neighbors). The second assumption is less accurate; however, a simple estimation showed that the "tilting" of the cluster with an angle up to 20° generated only a small decrease of the correlation length [35], which did not affect much the results. This insensitivity to the orientation of the cluster is due to the relatively high spherical symmetry of the first neighbors of a water molecule in an ice-like structure.

It should be emphasized that the assumption of an ice-like structure of water in the vicinity of the surface is only an approximation used to calculate the dipole correlation length λ_m (Eq. (10)). In fact, if the water would be organized in perfect ice-like layers parallel to the planar surface, the model would predict an oscillatory behaviour of the polarization in the vicinity of the surface [35].

2.3. The effect of disorder in the water structure on the correlation length λ_m

As will be shown later, the correlation length between dipoles is responsible for the hydration force, but also for the increase of the range of the interactions at large ionic strengths. The derivation of an expression for the correlation length assumed an ordered cluster around each water molecule. The local ordering is expected to be stronger in the vicinity of the surface and at high polarizations, and the correlation length to decrease with the distance to the surface and with decreasing fields. The electrolyte ions are also expected to affect the local ordering as well as the interactions between neighboring dipoles. While a microscopic theory to account for the local disorder in the structure of water is extremely difficult to develop, simple considerations can provide an estimate of the role of the local disorder on the double layer/hydration interactions predicted by the model.

The coupling between the dipoles of adjacent layers, described by the coupling constant C_1, is ultimately responsible for the propagation of the polarization through water. A simple manner to account for the decrease of the coupling with decreasing polarization is through an expression of the form:

$$C_1(x) = C_1\left(1 - \exp\left(-\frac{m(x)}{m_0}\right)\right) \tag{11a}$$

where m_0 is an empirical parameter, which controls the coupling. For large polarizations, $m \gg m_0$ and the coupling is strong: $C_1(x)=C_1$. For low polarizations, $m(x) \ll m_0$ and the coupling is vanishingly weak. Consequently, Eq. (8) becomes:

$$\begin{aligned} m(x) &= \varepsilon_0 v_0(\varepsilon-1)(E+E_P) \\ &= \varepsilon_0 v_0(\varepsilon-1)(E + C_0(x)m(x) + C_1(x)m(x-\Delta) \\ &\quad + C_1(x)m(x+\Delta)) \\ &= \varepsilon_0 v_0(\varepsilon-1)\Big(E + (C_0(x)+2C_1(x))m(x) \\ &\quad + C_1(x)\Delta^2\frac{\partial^2 m}{\partial x^2} + \dots\Big) \\ &\cong -\varepsilon_0 v_0(\varepsilon-1)\frac{\partial\psi}{\partial x} + \lambda_m^2(x)\frac{\partial^2 m}{\partial x^2} \end{aligned} \tag{11b}$$

with a dipole correlation length:

$$\begin{aligned} \lambda_m(x) &= \Delta(\varepsilon_0 v_0(\varepsilon-1)C_1(x))^{\frac{1}{2}} \\ &= \lambda_m\left(1-\exp\left(-\frac{m(x)}{m_0}\right)\right)^{\frac{1}{2}} \end{aligned} \tag{12}$$

which decreases with decreasing polarization.

2.4. The Boltzmann distribution of ions

The chemical potential of an ion of species i and of charge q_i in the liquid is given by [10]:

$$\mu_i(x) = \mu_{0i} + kT\ln(f_i c_i(x)) + q_i\psi(x) + W_i(x) \tag{13}$$

where k is the Boltzmann constant, T the absolute temperature and μ_{0i} represents the standard chemical potential of the ions of species i, of charge q_i and concentration $c_i(x)$. The activity coefficient f_i of the ions of species i will be assumed unity. The interaction free energy $W_i(x)$ should include all the other interactions of the ion with the medium (not accounted for by the mean field electrical potential ψ), such as the image forces [8], excluded volume effects [7], ion-dispersion [9] or ion-hydration forces [10]. Assuming that the system is in contact with an infinite reservoir in which the ion concentrations are c_{i0}, the total charge density is provided by:

$$\rho(x) = \sum_i q_i c_{0i}\exp\left(-\frac{q_i\psi(x)+\Delta W_i(x)}{kT}\right) \tag{14}$$

where $\Delta W_i(x)$ is calculated with respect to the bulk.

Eqs. (1a) and (14) can be combined to yield:

$$\varepsilon_0\frac{\partial^2\psi(x)}{\partial x^2} = -\sum_i q_i c_{0i}\exp\left(-\frac{q_i\psi(x)+\Delta W_i(x)}{kT}\right) + \frac{\partial m(x)}{v_0\,\partial x} \tag{15}$$

Eqs. (9) and (15) represent a system of coupled nonlinear equations, which can be solved (under appropriate boundary conditions) for $\psi(x)$ and $m(x)$.

There are various interactions between ions and surfaces. Here our focus will be on the interactions generated by the hydration of ions, which can be accounted for in a simple manner [10]. A structure-making ion is surrounded in bulk water by a sheath of hydrating water molecules. When approaching a surface at distances of the order of molecular size, the ion has to lose some of the hydrating molecules, hence the approach to the surface is unfavorable. The hydration energy of a structure-making ion is of the order of 10^2kT, and even a decrease of hydration by a few percent affects strongly the ion distribution in the vicinity of the surface. In fact, the increase of the surface tension of the water/air interface with increasing electrolyte concentration led Langmuir to suggest (based on the Gibbs adsorption equation) the existence of a layer of a few Ångstroms, completely depleted of ions [39]. Indeed, a large potential barrier with a width of a few Ångstroms for the structure-making ions can describe fairly well the behaviour of the air/water surface tension of the simple electrolytes [10]. The ions, however, are not completely depleted near a physical surface immersed in an electrolyte, because no recombination with the surface groups could take place in this case. Instead of an infinite potential barrier near the interface, it is reasonable to expect only a potential barrier with a height of the order of kT.

These repulsive ion-hydration forces acting on the structure-making ions decrease their concentration in the vicinity of the interface. In contrast, the structure-breaking ions have unfavorable interactions with the water molecules, because the bulk water molecules can better organize their hydrogen bonding network in their absence. Consequently, the structure breaking ions are expelled from the bulk toward the interface, where their concentration is increased [10].

The details of the interactions between ions and surfaces are not known, because they involve large-scale ab initio quantum mechanical calculations. A simple manner to account for the change in the free energy when the ions approach the interface is via a potential well (or a potential barrier) with a depth (height) $\Delta W(x)$ of the order of kT and with a width w of the order of a few Ångstroms [10]. The results are not qualitatively affected by the shape of the interaction potentials, a square or a triangular well providing similar results as a power-law interaction [10].

2.5. Boundary conditions for the polarization model

As already noted, in a homogeneous medium, a surface dipole density does not generate an electric field above the surface and, consequently, it is ignored by the traditional theory of the double layer. In the latter theory, the double layer is generated by surface charges alone. In the present treatment, both the surface dipole density and the surface charge density generate the double layer. Assuming that the surface charge is generated by the dissociation of the acidic and basic groups of the surface, while the surface dipole are generated by the association of the electrolyte ions with the corresponding sites, for a uni-univalent electrolyte the surface charge density is given by [40]:

$$\sigma = \frac{eN_B}{1+\left(\frac{c_{OH}}{K_{OH}}+\frac{c_E}{K_A}\right)\exp\left(\frac{e\psi_S}{kT}\right)} - \frac{eN_A}{1+\left(\frac{c_H}{K_H}+\frac{c_E}{K_C}\right)\exp\left(-\frac{e\psi_S}{kT}\right)} \tag{16}$$

where K_A and K_C are the association-equilibria constants for the anions and cations, respectively; N_A and N_B are the number of acidic and basic sites per unit area, respectively; c_H and c_{OH} are the bulk concentration of H^+ and OH^- ions, respectively; and c_E is the concentration of electrolyte in the bulk. The average areas, A_A, occupied by anions and A_C by cations on the surface are provided by:

$$A_A = \frac{1+\frac{K_A}{K_{OH}}\frac{c_{OH}}{c_E}+\frac{K_A}{c_E}\exp\left(-\frac{e\psi_S}{kT}\right)}{N_B} \tag{17a}$$

$$A_C = \frac{1+\frac{K_C}{K_H}\frac{c_H}{c_E}+\frac{K_C}{c_E}\exp\left(\frac{e\psi_S}{kT}\right)}{N_A} \tag{17b}$$

The first boundary condition at the surface is provided by integrating the Poisson Eqs. (1a) and (1b) over the volume of a flat box, which includes the surface, with the large sides parallel to the surface and a vanishingly thin width. After using the Gauss theorem, one obtains:

$$\varepsilon_0 \left.\frac{\partial\psi(x)}{\partial x}\right|_{x=0} - \left.\frac{m(x)}{v_0}\right|_{x=0} = -\sigma \tag{18a}$$

where x=0 indicates the surface. The second boundary conditions at the surface (Eq. (18b)) is provided by the polarization of the water molecules from the vicinity of the surface, which is caused by the macroscopic field $E=-\frac{\partial\psi(x)}{\partial x}|_{x=0}$, the field generated by the surface dipoles and the field E_{SP} due to the neighboring water dipoles:

$$E_{SP} = C_0 m(0) + C_1 m(\Delta), \tag{19}$$

which differs from Eqs. (5a) and (5b) because there are no water dipoles behind the surface ($m(-\Delta)$=0). The field generated by the surface dipoles is calculated by averaging the field generated by a dipole p in a medium of effective dielectric constant ε' at a distance Δ' above the center of the dipole over an area corresponding to the area occupied by a surface dipole (see Fig. 1). If the area occupied by the dipole on the surface is large, the field vanishes. Since the dipoles are formed through the association of the cations with the acidic surface groups, generating an effective dipolar moment p_C/ε', as well as by the association of anions with the basic surface groups, of an effective dipolar moment p_A/ε', the average field due to the dipoles is given by [40]:

$$E_S = \left[\left(\frac{p_A}{\varepsilon'}\right)\frac{1}{2\pi\varepsilon_0}\frac{1}{\left(\frac{A_A}{\pi}+\Delta'^2\right)^{\frac{3}{2}}}+\left(\frac{p_C}{\varepsilon'}\right)\frac{1}{2\pi\varepsilon_0}\frac{1}{\left(\frac{A_C}{\pi}+\Delta'^2\right)^{\frac{3}{2}}}\right] \tag{20}$$

Therefore, the second boundary condition at the surface is given by:

$$m(0)=\gamma\left(E+\frac{m(0)}{3\varepsilon_0 v_0}+E_{SP}+E_S\right)$$
$$=\varepsilon_0 v_0(\varepsilon-1)(E+E_{SP}+E_S) \quad (21)$$

which, by expanding $m(\Delta)=m(0)+\Delta\left.\frac{\partial m(x)}{\partial x}\right|_{x=0}$, becomes

$$(1-\varepsilon_0 v_0(\varepsilon-1)(C_0+C_1))m(x)|_{x=0} - \varepsilon_0 v_0(\varepsilon-1)\Delta C_1\left.\frac{\partial m}{\partial x}\right|_{x=0}+\varepsilon_0 v_0(\varepsilon-1)\left.\frac{\partial\psi}{\partial x}\right|_{x=0}$$
$$=\varepsilon_0 v_0(\varepsilon-1)\left[\left(\frac{p_A}{\varepsilon'}\right)\frac{1}{2\pi\varepsilon_0}\frac{1}{\left(\frac{A_A}{\pi}+\Delta'^2\right)^{\frac{3}{2}}}+\left(\frac{p_C}{\varepsilon'}\right)\frac{1}{2\pi\varepsilon_0}\frac{1}{\left(\frac{A_C}{\pi}+\Delta'^2\right)^{\frac{3}{2}}}\right] \quad (18b)$$

For a single surface immersed in an electrolyte, the other two boundary conditions necessary to solve the system of Eqs. (9) and (15) are provided by the vanishing of both the electric field and the polarization far away from the surface:

$$\left.\frac{\partial\psi(x)}{\partial x}\right|_{x=\infty}=0 \quad (22a)$$

$$m(\infty)=0 \quad (22b)$$

For two identical surfaces immersed in an electrolyte separated by a distance $2d$, these boundary conditions are provided by the symmetry with respect to the middle distance, $x=d$:

$$\left.\frac{\partial\psi(x)}{\partial x}\right|_{x=d}=0 \quad (23a)$$

$$m(x)|_{x=d}=0 \quad (23b)$$

2.6. The interaction between identical surfaces immersed in an electrolyte

Once the system of Eqs. (9) and (15) is solved for the boundary conditions (18a), (18b), (23a) and (23b), the total free energy of the system can be calculated by adding various contributions, due to the electric field, entropy of ions, chemical energy and interactions between ions and surfaces.

The electrostatic free energy includes the usual macroscopic free energy density (1/2)ED (with the displacement field $D=\varepsilon_0 E+P$ [37]) and also the interaction of the dipoles with the additional field E_P due to the non-uniform polarization [30]:

$$F_{el}(2d)=\int_0^{2d}\left[\frac{1}{2}E(\varepsilon_0 E+P)-\frac{1}{2}PE_P\right]dx$$
$$=\frac{1}{2}\int_0^{2d}\left[\left(\varepsilon_0\frac{\partial\psi(x)}{\partial x}-\frac{m(x)}{v_0}\right)\frac{\partial\psi}{\partial x}-C_1\Delta^2\frac{m(x)}{v_0}\frac{\partial^2 m(x)}{\partial x^2}\right]dx \quad (24a)$$

The entropic contribution of all the ion species present in the electrolyte, with respect to a homogeneous system of constant ion concentrations, is given by [41]:

$$F_{ent}(2d)=-T\Delta S$$
$$=kT\sum_i\int_0^{2d}\left(c_i\ln\left(\frac{c_i}{c_{i0}}\right)-c_i+c_{i0}\right)dx \quad (24b)$$

where i runs over all the ions species (including H^+ and OH^-), and the change in the chemical energy, with respect to infinite separation, due to the charging of the surface is provided by [42]:

$$\Delta F_{ch}(2d)=-\int_{\sigma(\infty)}^{\sigma(2d)}\psi_S(\sigma)d\sigma \quad (24c)$$

where σ is the surface charge density and ψ_S the surface potential. The free energy due to the interactions between the ions and the surface is given by [10]:

$$F_{ion-surf}(2d)=\int_0^{2d}\left(\Delta W_A(x)c_E\exp\left(\frac{e\psi-\Delta W_A}{kT}\right)+\Delta W_C c_E\exp\left(\frac{-e\psi-\Delta W_C}{kT}\right)\right)dx \quad (24d)$$

The force per unit area is consequently obtained from the derivative:

$$\Pi(2d)=-\frac{\partial(F_{el}(2d)+F_{ent}(2d)+F_{ch}(2d)+F_{ion-surf}(2d))}{\partial(2d)} \quad (25)$$

3. Consequences of the polarization model

3.1. Polarization skin effect

In the traditional double layer theory, the average dipole moment of a water molecule is proportional to the macroscopic field (the derivative of the electrical potential) because the local field is also proportional to the latter field. In the polarization model and for a uniformly polarized medium, the field generated by the neighboring dipoles also vanishes, $(C_0+2C_1)m\approx 0$, since the negative field generated by the dipoles from the same layer ($C_0<0$) is compensated by the positive fields generated by the dipoles from the adjacent layers ($C_1>0$), and the traditional theory is recovered. However, at the surface, one adjacent layer is missing, and there is no compensation anymore. The total field generated by the neighboring water molecules on the first layer is opposite in direction to the average dipole moment; hence, it induces a net depolarization, even in the absence of surface dipoles.

One of the differences between the polarization model and the traditional theory is that the local field at the surface, predicted only by the former model, opposes the polarization of the medium. The solutions of the system of Eqs. (9) and (15), for low electrolyte concentrations (c_E=0.001 M), with the boundary conditions (18a), (18b), (22a) and (22b), and no surface dipoles ($p_C/\varepsilon' = p_A/\varepsilon' \equiv 0$) are plotted in Fig. 2a. The depolarization field at the surface strongly affects the shape of the electrical potential and average dipole moment only within a few Ångstroms from the surface (a "skin" effect); however, a traditional Poisson–Boltzmann treatment for the same values of the parameters predicts different results at any distance from the surface. When a suitable charge is added on the surface to compensate for this field, the results of the polarization model and the traditional theory are almost identical far away from the surface (see Fig. 2a). This result is supporting the general agreement that the traditional theory is accurate at intermediate electrolyte concentrations [12], when the surface charge is considered a fitting parameter. However, at higher ionic strength (c_E=0.1 M, Fig. 2b), the polarization model and the traditional theory predict different behaviors for $\psi(x)$ and $m(x)$, which cannot be accounted for by changing the values of the surface charges.

Therefore, even in the absence of surface dipoles, the polarization model for the hydration/double layer predicts qualitatively different results from those of the traditional theory at moderate and high ionic strengths. At low electrolyte concentrations, the quantitative differences between the two models, far away from the surface, can be accounted for by suitable modifications of the surface charge. The shape of the electric field and polarization within a few Ångstroms from the surface, predicted by the two models, are however different at all electrolyte concentrations.

3.2. *Hydration forces between neutral lipid bilayers: independence of electrolyte concentration*

In the absence of an electrolyte, Eq. (15) becomes:

$$\varepsilon_0\frac{\partial^2\psi(x)}{\partial x^2}=\frac{1}{v_0}\frac{\partial m(x)}{\partial x} \qquad (26)$$

which can be integrated to lead to:

$$m(x)=\varepsilon_0 v_0\frac{\partial\psi(x)}{\partial x}+C \qquad (27)$$

The integration constant is selected in such a manner (C=0) for the polarization to vanish far away from the surface. From Eqs. (9) and (27), one obtains:

$$\frac{\lambda_m^2}{\varepsilon}\frac{\partial^2 m(x)}{\partial x^2}=m(x) \qquad (28)$$

The solution of Eq. (28) is provided by the sum of two terms, which are proportional to $\exp\left(-\frac{x}{\lambda_H}\right)$ and $\exp\left(\frac{x}{\lambda_H}\right)$, with the "hydration" decay length $\lambda_H=\frac{\lambda_m}{\sqrt{\varepsilon}}\approx 1.67\mathring{A}$. For a single surface, the boundary condition (22b) implies that the average dipole moment has the form:

$$m=m_1\exp\left(-\frac{x}{\lambda_H}\right). \qquad (29)$$

where m_1 is the value of the average dipole moment near the surface.

Since there are no surface charges, the magnitude of the polarization ($P=m/v_0$) depends only on the surface dipoles. The interaction between two planar surfaces is important only when the polarization regions overlap, and increases with the magnitude and density of the surface dipoles. This result is supported by the experiments with neutral phospholipid bilayers, for which the interactions are proportional with the potential across the membrane, the latter being proportional to the surface dipole density [32]. Despite the absence of a surface charge, there is a potential in the vicinity of the surface, generated by the gradient of polarization (see Fig. 3a). In the presence of an electrolyte, the range of the polarization is increased, indicating that the hydration force predicted by the polarization model depends on electrolyte concentration.

The force between neutral surfaces (with a surface dipole density) depends on the electrolyte concentrations, as shown in Fig. 3b, particularly at large separations. However, at small separations, the interaction appears to be well described by an exponential with a decay length λ_H. For neutral lipid bilayers, the equilibrium is reached at a distance of about 20 Å, at which the attractive van der Waals interaction balances the repulsive hydration and thermal undulation interactions [43]. The experiments regarding the forces between neutral lipid bilayers [11] sample the interactions at separations smaller than 20 Å, for which the dependence on ionic strength is much weaker. By adding to the total pressure a typical van der Waals disjoining pressure [12]:

$$\Pi_{vdW}(2d)=-\frac{A_H}{6\pi}\frac{1}{(2d)^3} \qquad (30)$$

where A_H is the Hamaker constant, one obtains the curves plotted in Fig. 3c. The value $A_H=0.2\times10^{-20}$ J was selected to provide the equilibrium distance at about 20 Å from the

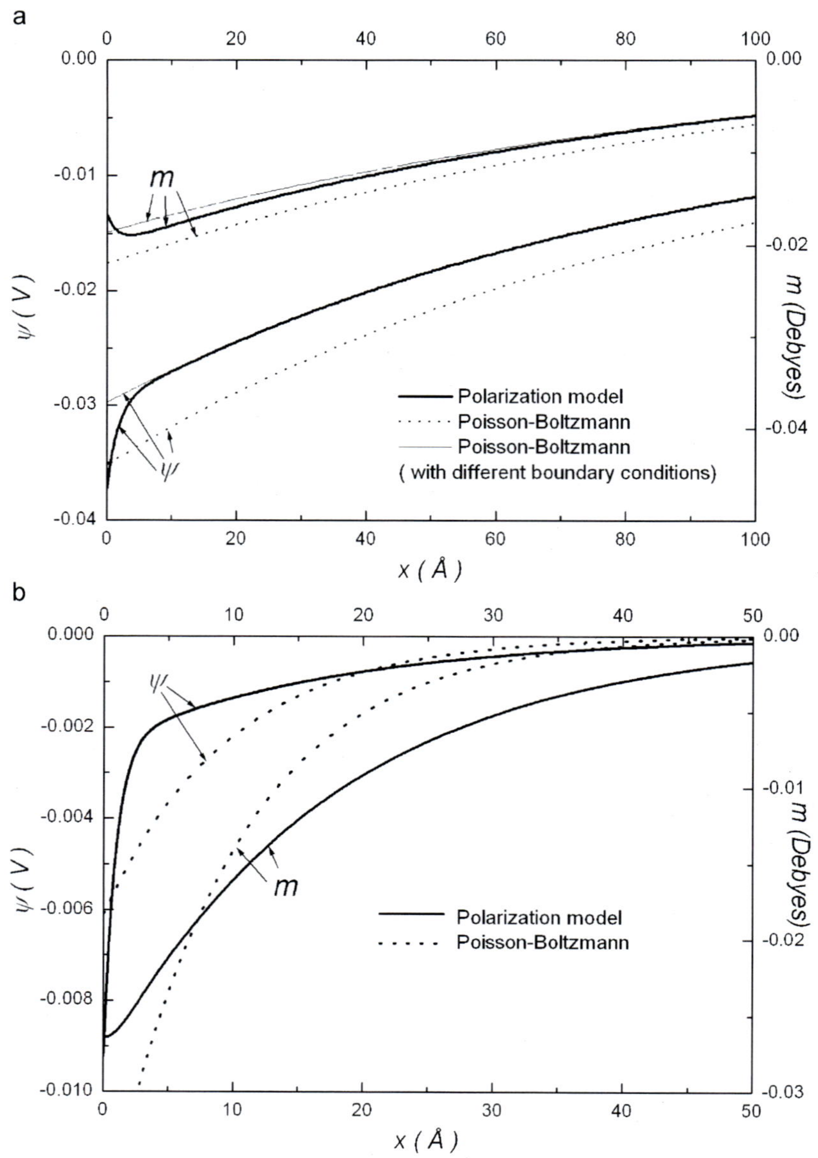

Fig. 2. (a) The potential and the average dipole moment of water molecules in the vicinity of a charged surface, calculated using the polarization model (thick lines) for the following values of the parameters: c_E=0.001 M, $N_A=1\times10^{17}$ sites/m^2, $N_B=1\times10^{17}$ sites/m^2, ΔW_C=0, ΔW_A=0, pH=8.0, $K_H=10^{-6}$ M, $K_{OH}=10^{-6}$ M, K_C=0.5 M, K_A=0.5 M, p_C/ε' =0, p_A/ε' =0, ε=80, λ_m=14.9 Å, m_0=0. While the calculation based on the traditional Poisson–Boltzmann theory (dotted line) provides quantitatively different results, a suitable slight adjustment of the value of the surface charge density in the Poisson–Boltzmann calculations (thin lines) led to almost identical results far away from the surfaces, for the two theories. Within a few Ångstroms from surface, the polarization model predicts qualitatively different results (a skin effect). (b) The potential and the average dipole moment of water molecules in the vicinity of a charged surface, calculated using the polarization model (thick lines) for the same values of the parameters as in (a), except now c_E=0. 1 M. The results predicted by the traditional theory (dotted line) are qualitatively different at any distance from the surface. For moderate and high electrolyte concentrations, the results of the two theories cannot be made to agree by a suitable adjustment of the surface charge density.

Fig. 3. (a) The potential and the average dipole moment of water molecules in the vicinity of a charged surface, calculated using the polarization model for a neutral surface (σ=0) with surface dipoles occupying an area A_C=50 Å^2, p_C/ε' =1 Debye, p_A/ε' =0, ε=80, λ_m=14.9 Å, m_0=0, ΔW_C=0, ΔW_A=0, for various electrolyte concentrations. (b) The force per unit area between two identical surfaces, calculated using the polarization model and the same values of the parameters as for Fig. 3a, is compared with an exponential with a decay length λ=2 Å. At large separations and high electrolyte concentrations, the results of the polarization model are not well described by an exponential decay with λ=2 Å. (c) The total force per unit area between two identical surfaces as in (b), calculated by adding to the results of the polarization model a typical van der Waals interaction between surfaces, with $A_H=0.2\times10^{-20}$ J. For most separations distances, smaller than the equilibrium separation (the only separations experimentally accessible), the repulsion is well approximated by the same exponential (with the decay length λ=2 Å), regardless of the electrolyte concentration. The polarization model for the hydration/double layer interactions predicts that the equilibrium separation between neutral lipid bilayers slightly increases with increasing electrolyte concentration.

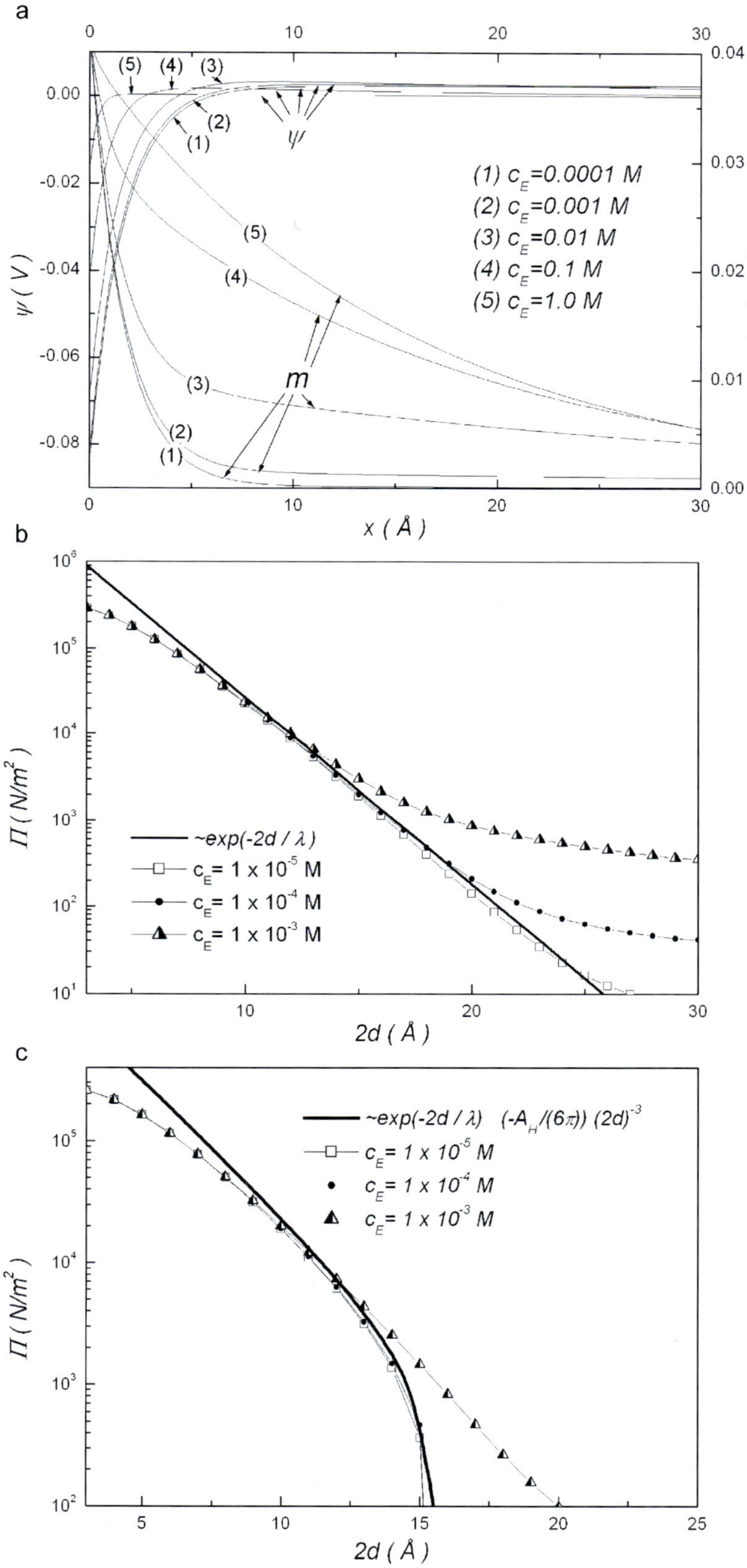
a
(1) c_E=0.0001 M
(2) c_E=0.001 M
(3) c_E=0.01 M
(4) c_E=0.1 M
(5) c_E=1.0 M
ψ (V)
m (Debyes)
x (Å)
ψ
m
b
~exp(-2d / λ)
c_E= 1 x 10^{-5} M
c_E= 1 x 10^{-4} M
c_E= 1 x 10^{-3} M
Π (N/m^2)
2d (Å)
c
~exp(-2d / λ) (-A_H/(6π)) (2d)$^{-3}$
c_E= 1 x 10^{-5} M
c_E= 1 x 10^{-4} M
c_E= 1 x 10^{-3} M
Π (N/m^2)
2d (Å)

Fig. 3.

surface. For most separation distances (smaller than the equilibrium separation), the pressure is apparently independent of the electrolyte concentration. However, the equilibrium separation distance predicted by the polarization model seems to increase by a few Angstroms with increasing electrolyte concentration. This result appears to be supported by recent preliminary experiments with neutral lipid bilayers, which indicate an increase of the hydration force with increasing electrolyte concentration [44].

3.3. Hydration forces between mica surfaces: ion-specific effects

In the previous section, it was shown that a hydration force with a short decay length and almost independent of the electrolyte concentration can be achieved with neutral lipid bilayers at short separation distances. At separations larger than about 20 Å, the attractive van der Waals interactions dominated the interactions.

A completely different situation occurs for mica surfaces, for which both charges and dipoles are present on the surface. For these surfaces, strong specific cation effects have been observed experimentally at relatively low electrolyte concentrations [17]. These specific ion effects cannot be explained by the additional interactions between ions (such as excluded volume effects) or ions and surfaces (such as ion-dispersion [9] or ion-hydration [10] forces), because these interactions are in general negligible for electrolyte concentrations smaller than about 0.05 M.

The experimental data reported by Pashley [17] raise two important issues. The first one is why the critical electrolyte concentration at which the hydration force emerges, depends so strongly on the type of electrolyte (6×10^{-2} for LiCl, 3×10^{-4} M for KCl). The second issue is why the decay length of the interactions at short separations is about 10 Å, about five times larger than that corresponding to the hydration force between lipid bilayers.

Let us first note that, when the surface charges and surface dipoles are generated via the dissociation of the moieties of the surface, and via the association of the electrolytes ions with the dissociated sites of the surfaces, respectively, the fields generated by the surface charges and surface dipoles, in the vicinity of the surface, are oriented in opposite directions. An interesting situation occurs when the field generated by surface charges and surface dipoles are comparable to each other; in this case, even small changes in the association constants or in the effective surface dipole p/ε' affect strongly the interactions. The mica surface is a complex one and many types of surface groups are present; it is therefore possible for the fields generated by the surface charges and dipoles to be oriented in the same direction and hence reinforce each other. The force between charged surfaces immersed in an electrolyte of concentration c_E=0.001 M is plotted in Fig. 4a, by assuming a constant surface charge density σ=−0.01 C/m^2, a constant area per dipole A=100 Å^2, and various values for the effective dipole moment, p/ε' of the surface dipoles. In all the cases investigated, the interactions at large separations are well approximated by an exponential decay, with the decay length roughly provided by the Debye–Hückel length [12]:

$$\lambda_{\mathrm{DH}} = \sqrt{\frac{\varepsilon\varepsilon_0 kT}{\sum_i q_i^2 c_{0i}}}, \tag{31}$$

while at short separations the decay length for the interaction is different. For the set of parameters employed, the later decay length is about 10 Å. When a suitable van der Waals disjoining pressure (Eq. (30), A_H=1×10^{-20} J) is added to the interactions, the total interactions for systems with p/ε' =0 or 3 Debyes appear to obey the traditional DLVO theory, with no additional "hydration" repulsion emerging at low separations. The interactions become attractive for separation smaller than about 20 Å and it appears that no hydration force exists (see Fig. 4b). However, when the field generated by the surface dipoles reinforces the field generated by the surface charge, with p/ε' =−5 or −7 Debyes, a "hydration force" occurs at low separations, with a decay length of the order of 10 Å.

At very low separations, the divergently large van der Waals attraction dominates; however, the maximum pressure that must be applied to fuse the mica surfaces is about two orders of magnitude larger in the latter case, than in the case when p/ε' =3 Debyes.

In summary, a "hydration force" is always predicted by the polarization model at low separations and its apparent decay length depends on the electrolyte concentration. However, if the magnitude of the repulsion is small, the total interaction (including the van der Waals attraction between surfaces) seems to be well described by the traditional DLVO theory [13]. In contrast, when the repulsion is large, a "hydration force" with a decay length of about 10 Å emerges in the vicinity of the interface. The transition between a "DLVO-like" result and a "DLVO plus a hydration force" result can be generated by relatively small modifications of the surface dipoles formed on the mica surface. Various cations have different association constants with the surface groups, hence can generate different surface charges and particularly different surface dipole densities. Because of the high sensitivity of the interactions on the latter density, strong ion-specific effect could be explained on this basis.

3.4. Long-ranged interactions at high ionic strengths: restabilization

When the surface charges and the surface dipoles are formed via the dissociation, and via the association with electrolyte ions of the moieties of the surface, respectively, the increase of ionic strength decreases the surface charge, but at the same time increases the surface dipole density. Since the electric fields induced in the vicinity of the surface by the surface charge and surface dipoles oppose each other,

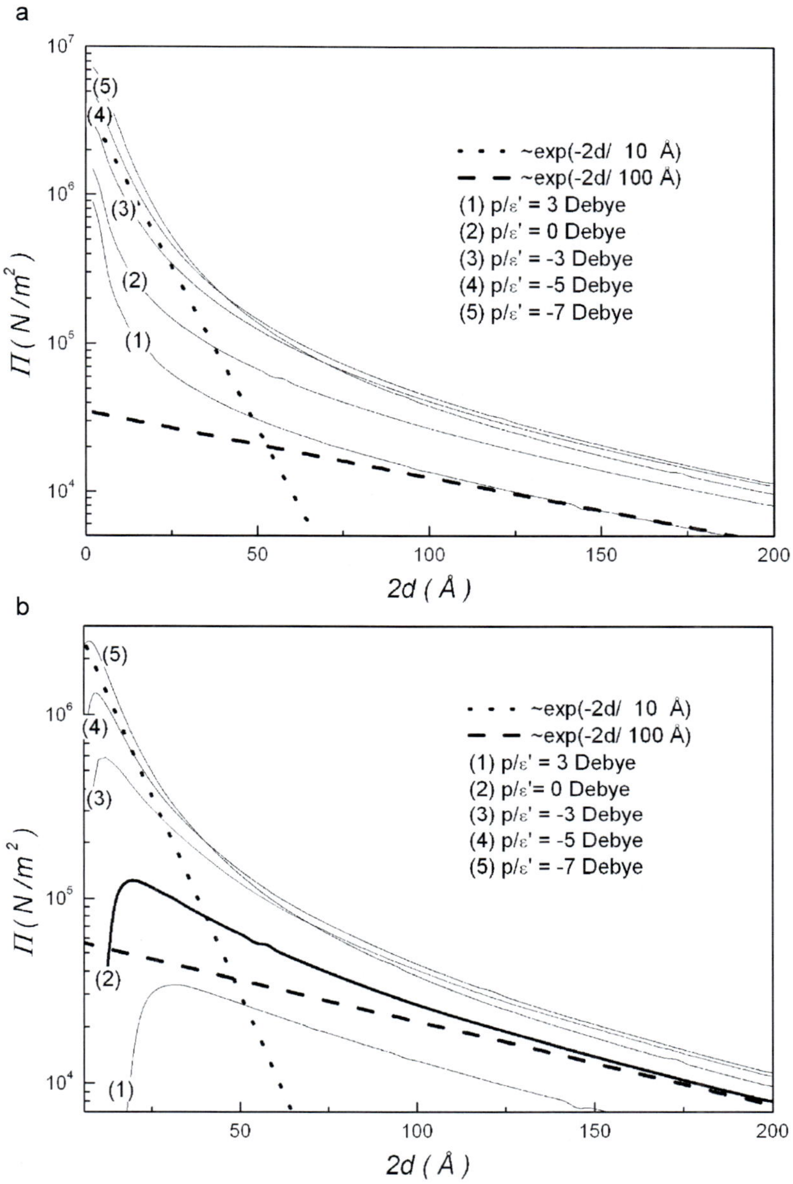

Fig. 4. (a) The repulsion force predicted by the polarization model between two identical surfaces, for c_E=0.001 M, σ=0.01 C/m^2, A_{dipole}=100 Å^2, ΔW_C=0, ΔW_A=0, ε=80, λ_m=14.9 Å, m_0=0) and various values for the effective dipole moment, p/ε' . In the vicinity of the surface, the force is well approximated by an exponential with a decay length λ~10 Å, while at large separations the force is well approximated by an exponential with the decay length λ_{DH} provided by the traditional theory. (b) The total interactions, obtained by adding to the repulsion force calculated in (a) a typical van der Waals attraction between surfaces, with a Hamaker constant $A_H=1\times10^{-20}$ J. For some values of the effective dipole moment (p/ε' =0 or 3 Debyes), the total interaction seems to obey the DLVO theory, while for the other values (p/ε' =−5 or −7 Debyes), a "hydration force" with a decay length λ~10 Å emerges at low separation distances.

the total interaction might run through a minimum at an electrolyte concentration for which these fields cancel each other. At low ionic strengths, the charges dominate the interactions, which decrease with increasing ionic strength, while at large electrolyte concentrations, the increase of ionic strength generates more dipoles on the surface, via the association of the ions with the dissociated surface groups, hence increasing the interactions. Therefore, at intermediate electrolyte concentrations, the repulsive double layer/hydration interactions predicted by the polarization model might be lower (in absolute value) than the attractive van der Waals interactions and the system will rapidly coagulate, while at both low and high ionic strengths the repulsive interactions might be sufficiently strong to prevent coagulation [34]. This mechanism was employed to explain the restabilization of protein-covered latex particles [40,45] observed experimentally [21].

The stability of some colloids at high ionic strengths can be explained by any theory that relates the hydration force to the surface dipole density. More difficult to explain are some experimental data regarding the second virial coefficient of the interactions between apoferritin molecules, which at low ionic strengths (0.01–0.075 M) decreases with increasing electrolyte concentrations, as expected from the traditional DLVO theory, but increases markedly with electrolyte concentration at large ionic strengths (~0.25 M) [18]. A short-range repulsion can prevent coagulation, if it is sufficiently strong; however, a short-range repulsion cannot lead to such a large increase in the second virial coefficient [19]. In fact, even an infinite hard-core repulsion for separations distances smaller than 50 Å between the surfaces of two apoferritin molecules is not sufficient to explain such large virial coefficients, while an exponentially decaying repulsion, with a decay length of about 2 Å, must have a huge magnitude to explain the experiment. This magnitude must be by about 10 orders of magnitude larger than that determined for neutral lipid bilayers and is clearly not realistic [19].

Therefore, the interaction between apoferritin molecules cannot be explained by a phenomenological model which simply assumes that the hydration force has a fixed decay length and a magnitude proportional to the surface dipole density.

A peculiar feature of the polarization model is the emergence of a long-ranged interaction at high ionic strengths. To understand qualitatively the behaviour of such an interaction, let us note than for small electrical potentials, $\frac{e\psi}{kT} \ll 1$, and the system of Eqs. (9) and (15) can be linearized and hence can be solved analytically. In the linear approximation, the system becomes [30]:

$$\frac{\partial^2\psi(x)}{\partial x^2} = \frac{\varepsilon}{\lambda_{\mathrm{DH}}^2}\psi + \frac{1}{\varepsilon_0 v_0}\frac{\partial m(x)}{\partial x}, \tag{32a}$$

$$\lambda_m^2\frac{\partial^2 m(x)}{\partial x^2} = m(x) + \varepsilon_0 v_0(\varepsilon-1)\frac{\partial\psi(x)}{\partial x} \tag{32b}$$

and its solutions are proportional to $\exp\left(\pm\frac{x}{\lambda_i}\right)$. The values of the characteristic decay lengths λ_i (i=1 or 2) are determined from the characteristic equation (obtained by introducing the solutions in the system):

$$\left(\frac{1}{\lambda^2}-\frac{\varepsilon}{\lambda_{\mathrm{DH}}^2}\right)\left(\frac{\lambda_m^2}{\lambda^2}-1\right)-\frac{\varepsilon-1}{\lambda^2}$$
$$=0 \Leftrightarrow \lambda^4 - (\lambda_{\mathrm{DH}}^2+\lambda_m^2)\lambda^2 + \frac{\lambda_{\mathrm{DH}}^2\lambda_m^2}{\varepsilon} = 0 \tag{33}$$

and are given by:

$$\lambda_{1,2} = \sqrt{\frac{(\lambda_{\mathrm{DH}}^2+\lambda_m^2) \pm \sqrt{\lambda_{\mathrm{DH}}^4+\lambda_m^4+2\lambda_{\mathrm{DH}}^2\lambda_m^2\left(1-\frac{2}{\varepsilon}\right)}}{2}} \tag{34}$$

Both solutions of Eq. (34) are real and positive, because the dielectric constant is in the range $1<\varepsilon<\infty$. The two characteristic lengths as a function of the Debye–Hückel length λ_{DH} are plotted in Fig. 4a. For $\lambda_{\mathrm{DH}} \gg \lambda_m (\lambda_{\mathrm{DH}} \rightarrow \infty)$, the characteristic lengths are given by:

$$\lambda_1 \cong \lambda_{\mathrm{DH}} \tag{35a}$$

$$\lambda_2 \cong \lambda_{\mathrm{H}} = \frac{\lambda_m}{\sqrt{\varepsilon}} \tag{35b}$$

which implies that two types of decay lengths coexist in the system: one roughly described by the traditional theory and the other corresponding to a short-range hydration force, independent of electrolyte concentration.

However, at high ionic strengths, $\lambda_{\mathrm{DH}} \ll \lambda_m (\lambda_{\mathrm{DH}} \rightarrow 0)$ the characteristic lengths are approximated by:

$$\lambda_1 \cong \lambda_m \tag{36a}$$

$$\lambda_2 \rightarrow 0 \tag{36b}$$

which implies than an intermediate decay length (since λ_m=14.9 Å), much larger than both λ_{DH} and λ_{H}, is present in the system at high electrolyte concentrations.

The contributions of each characteristic length to the solution is determined by the boundary conditions (18a), (18b), (22a) and (22b) for one surface or boundary conditions (18a), (18b), (23a) and (23b) for two identical surfaces. While the analytical solutions (and the characteristic decay lengths) are valid only for the linear approximation of the system, the behaviour of the non-linear system is qualitatively similar, at least for sufficiently small electrical potentials. Indeed, a calculation based on the polarization model, in which the interactions between spherical molecules were calculated using the Derjaguin approximation, showed that it is possible to obtain such long-ranged interactions between apoferritin molecules, which could explain the unusually strong increase of the second virial coefficient at large electrolyte concentrations [19]. The circles in Fig. 5b represent the experimental data reported in Ref. [18], while the continuos line are the results of the calculations for various effective surface dipolar moments, reported in Ref. [19]. The details of the calculations are provided in Ref. [19].

3.5. Role of ion-hydration forces in the polarization model

There are various manners in which ion-specific effects can be accounted for by the polarization model. The first one is related to the different association constants of the ions with the dissociated surface groups. The charging of the surface depends not only on the electrolyte concentration, but also on the type of electrolyte. These ion-specific effects are similar to those predicted by the traditional theory [46]. A typical consequence of such effects is illustrated by the dependence of the isoelectric point on the nature of electrolyte [21].

A second manner of predicting ion-specific effects by the polarization model is via the magnitude and density of the

M. Manciu, E. Ruckenstein / Advances in Colloid and Interface Science 112 (2004) 109–128 123

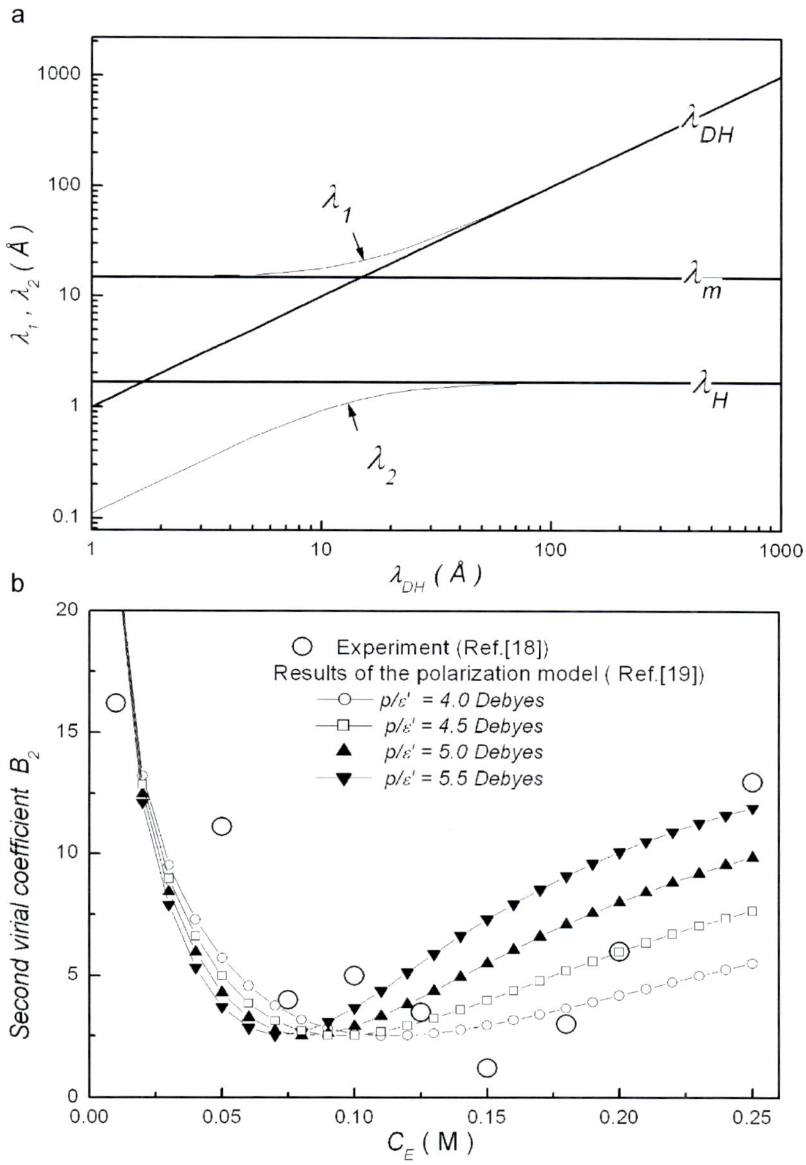

Fig. 5. (a) The dependence of the characteristic decay lengths λ_1, λ_2 of the system on the Debye–Hückel length λ_{DH} $\left(\lambda_m = 14.9\ \text{Å},\ \lambda_H = \frac{\lambda_m}{\sqrt{\varepsilon}} = 1.67\ \text{Å}\right)$. (b) The second virial coefficient for apoferritin molecules, obtained from experiment in Ref. [18] is compared with calculations based on the polarization model for hydration/double layer interactions for various values of the effective surface dipole moment. The details of the calculations are provided in Ref. [19].

dipoles formed on the surface. Even if different types of ions would have the same association constants, the magnitude of the dipoles formed on the surface should depend on the specificity of the chemical bonding between ions and surface groups, as well as on the hydration characteristics of the ions (the strongly hydrated, structure-making ions probably generating dipoles of larger magnitude, because they might retain some of the hydrating water molecules even after the association with surface groups). The surface dipoles generate a potential in the vicinity of the surface; therefore, the association equilibria of all ions are affected by the magnitude and density of the surface dipoles. The specific ion effects are particularly important when the electric fields generated by the surface charges and the surface dipoles are comparable in the vicinity of the surface. When they are oriented in opposite directions, small changes in the magnitude of the dipoles will affect strongly the total interaction at low separations, as shown in Fig. 4b, which can be well described either by the traditional DLVO interaction at low p/ε' values, or by the traditional DLVO interaction plus a short-range "hydration force", at large p/ε' values.

There is a third manner to introduce specific ion effects via the polarization model, which will be discussed in this section. While at low electrolyte concentrations the additional interactions between ions and interfaces can be in general neglected, at moderately high electrolyte concentrations, these additional interactions have to be taken into account.

A simple manner to account for the interactions between ions and surfaces, $\Delta W_i(x)$, is via a potential well (or a potential barrier) with a depth (height) of the order of kT and with a width w of the order of a few Ångstroms [10]. Such ion potentials have been selected because of the changes in the hydration energy of ions when they approach the interface: a structure-breaking ion disorganize the structure of bulk water, and it is expelled toward the interface, because water can better organize in its absence ($\Delta W(x)<0$). In contrast, a structure-making ion is strongly bound to their hydration water molecule. Since when approaching the interface, they have to lose some of these molecules, they prefer the bulk water ($\Delta W(x)>0$). While the details of these interactions are unknown, the results are not affected qualitatively by the shape selected for the potentials [10].

Most univalent anions (e.g. Br^-, NO_3^-, SCN^-) are structure breaking, while most of the simple univalent cations (e.g. K^+, Na^+, Li^+, Cs^+) are structure making. In the model calculations that follows, it will be assumed, for the sake of simplicity, that the anions are structure-breaking ($\Delta W_A(x)<0$), while the univalent cations are structure-making ($\Delta W_C(x)>0$). In Fig. 6a–d, the potential in the

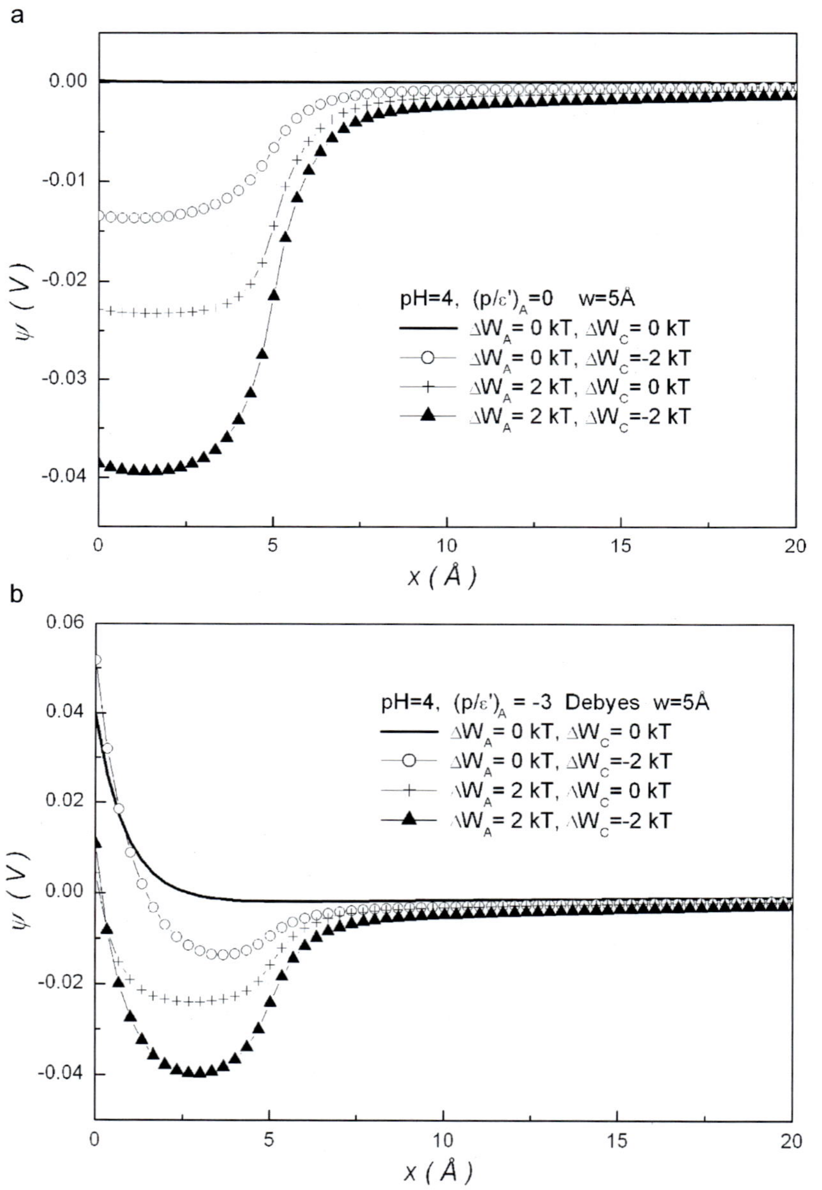

Fig. 6. The potential in the vicinity of a surface when ion-hydration interactions affect the anions (ΔW_A) and the cations (ΔW_C) The other parameters are c_E=0.1 M, $N_A=1\times10^{18}$ sites/m^2, $N_B=1\times10^{18}$ sites/m^2, $K_H=10^{-7}$ M, $K_{OH}=10^{-7}$ M, $K_C=10^{-4}$ M, $K_A=10^{-4}$ M, ε=80, λ_m=14.9 Å, m_0=0, w=5 Å, and (a) pH=4, p_A/ε' =0; (b) pH=4, p_A/ε' =−3 Debyes; (c) pH=10, p_C/ε' =0; (d) pH=10, p_C/ε' =3 Debyes. The values of p_C/ε' do not affect the results plotted in (a) and (b), while the values of p_A/ε' do not affect the results plotted in (c) and (d). (e) The force between two identical surfaces for the same parameter values as in (a–d). Only the magnitude of the force is strongly dependent on the value of the surface dipoles and ion-hydration forces, but not its functional behaviour.

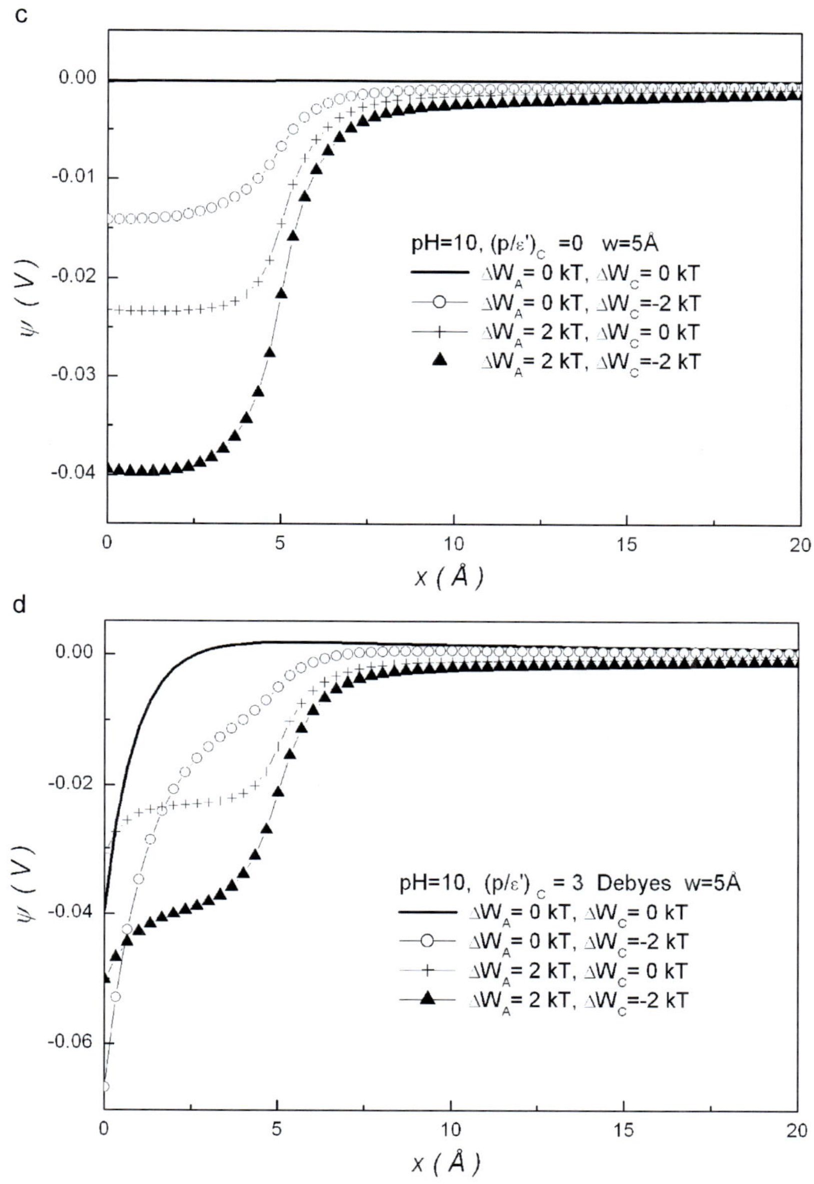

Fig. 6 (*continued*).

vicinity of the surface is calculated for various types of ion-hydration interactions. At small pH values (Fig. 6a,b), the association of anions is important, because most acidic sites are not dissociated. Consequently, changes in the effective dipole moment of the surface dipole formed by the association of cations do not affect much the potential, while changes in the effective surface dipole formed by the association of anions affect drastically the electric potential. However, the interactions of both cations and anions with the surface affect the potential.

In Fig. 6c,d, the calculations were repeated for large pH values, for which most of the basic sites of the surface are associated with OH^-, and consequently the association of anions is negligible. In this case, only specific cation effects can be predicted by changes in the effective dipole moment on the surface. As already noted, the interactions between ions and surface can account for both anion and cation specificity.

The interactions between two plates are plotted in Fig. 6e; the magnitude of the interaction is strongly affected by the surface charge and surface dipole densities. Therefore, specific ion effects can be accounted for through different association constants, different dipole moments of the ion pair formed on the surface, or different interactions between the ions and surfaces. However, the dependence of force on the distance appears to be much less affected by the surface conditions. In the next section, it will be shown that the forces at large separations are dominated by the largest characteristic decay length of the system. When the dipole correlation length, λ_m, is much smaller than the Debye–Hückel decay length, the latter approximates well the largest characteristic decay length of the system.

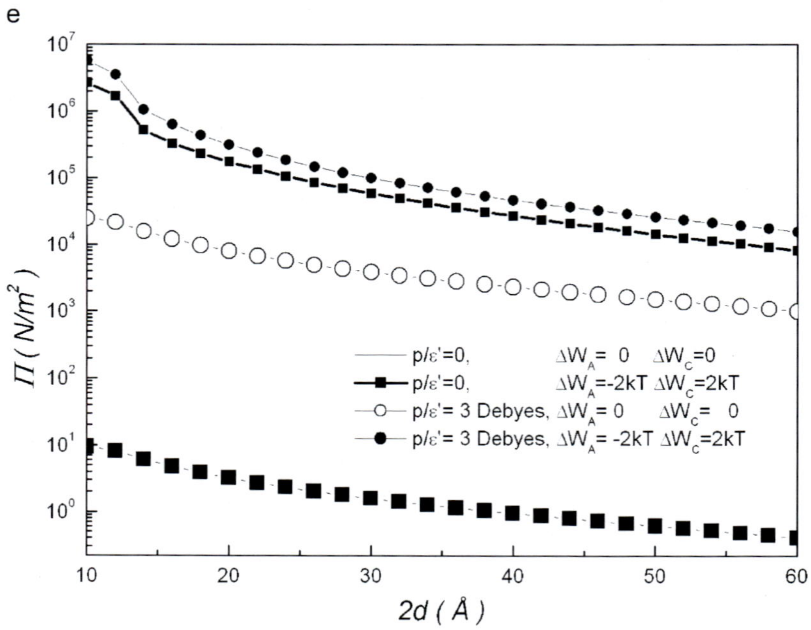

Fig. 6 (*continued*).

3.6. The effect of the correlation length λ_m on the interactions

In Section 3.3, it was shown that at low ionic strengths the interactions between surfaces, at large separations, are well described by an exponential with a decay length roughly provided by the traditional theory (c_E=0.001 M, λ_{DH}=100 Å). The magnitude of the interaction differs from that predicted by the DLVO theory; however, in the latter theory, the surface charge density is merely an unknown parameter, used to fit the experimental results. Therefore, the results of the polarization model at low ionic strength can in general be recovered (at large separations) by the DLVO theory when suitable values for the surface charge density are employed.

The reason for this behaviour is that, in the linear approximation, the potential and the polarization are a sum of two exponentially decaying terms, and at large separations the largest term dominates. At low ionic strengths, $\lambda_{DH} \gg \lambda_m \approx 15$ Å, and the largest characteristic length of the system is $\lambda_1 \approx \lambda_{DH}$. The situation is completely different when λ_{DH} is comparable to λ_m and λ_1 becomes considerable larger than λ_{DH}. In this case, the polarization model predicts, at large separations, an exponentially decaying interaction, but with a decay length completely different from λ_{DH}.

An important issue is whether or not this increase in the decay length occurs for any colloidal system. The derivation of the value for λ_m=14.9 Å assumed an ice-like order for the clusters around each water molecule [30]. When this order is lowered, the value of the dipole correlation length is expected to decrease. A phenomenological model for the decrease of λ_m with the decrease of the average dipolar moment of water molecules is proposed in Section 2.3. In this section, the consequences of the decrease of the dipole correlation length are estimated.

The potential in the vicinity of a charged surface, calculated for m_0=0.05 Debyes (in Eq. (12)) and for c_E=0.1 M, are plotted in Fig. 7a. For the sake of simplicity, the surface dipoles were neglected ($p_A/\varepsilon' = p_C/\varepsilon'$ =0). For large surface charge densities, a large polarization is generated in the vicinity of the surface; hence, the dipole correlation length is large and the results of the polarization model (open symbols) are very different form the results of the traditional theory (filled symbols). However, at large separations, the polarization decreases and the results of the two theories become comparable. For lower surface charge densities, the polarization in the vicinity of the interface is lower and the results of the two theories become comparable starting at smaller distances (see Fig. 7a).

The same behaviour is obeyed by the forces between surfaces, plotted in Fig. 7b. Note than the change in $C_1(x)$ (Eqs. (11a) and (11b)) affects not only the dipole correlation length λ_m, but also the boundary condition (18b) and the electrostatic free energy Eq. (24a). At lower surface charge densities, the force predicted by the polarization model (open symbols) roughly decays as those predicted by the traditional theory (filled symbols). However, for higher surface charge densities, the force predicted by the polarization model is much larger than the traditional one, with a decay length larger than that predicted by the Debye–Hückel theory.

In summary, when the Debye–Hückel λ_{DH} largely exceeds the dipole correlation length λ_m (either because of low ionic strength, or because of disorder), the polarization model predicts, at large separations, results similar to those obtained from the DLVO theory (with suitably adjusted

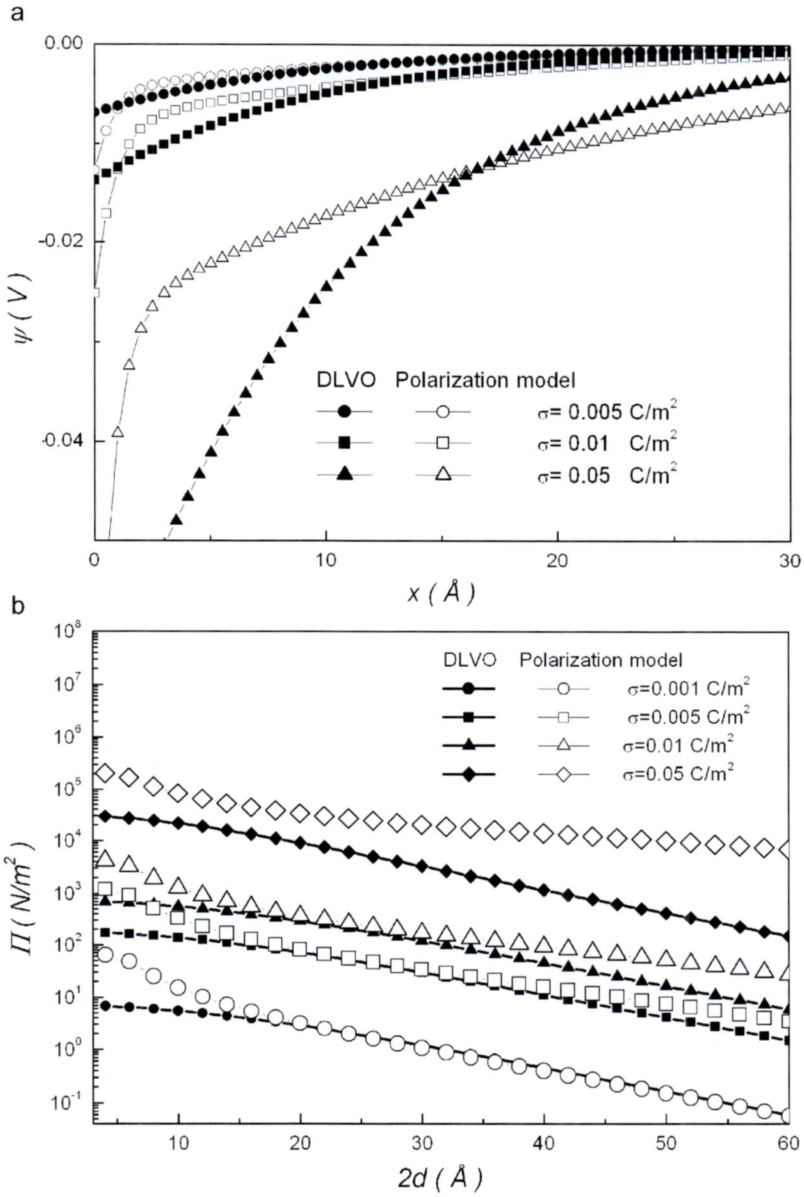

Fig. 7. (a) The potential in the vicinity of a surface for various surface charge densities (c_E=0.1 M, p_C/ε' =0, p_A/ε' =0) calculated using the polarization model (open symbols) and the traditional theory (filled symbols). The effect of disorder on the dipole correlation length is taken into account via an empirical parameter, m_0=0.05 Debyes. When the average dipole moment of the water molecule m is much larger than m_0 (at large surface charge densities and near the surface), the (disorder-free) polarization model is obtained. When $m \ll m_0$ (at low surface charge densities and far away from the surface), the DLVO theory is recovered. (b) The force between the identical surfaces of (a), immersed in an electrolyte of c_E=0.1 M, for various surface charge densities, calculated using the polarization model which includes disorder (m_0=0.05 Debyes) (open symbols) and the traditional theory (filled symbols). At low surface charge densities, the traditional theory is recovered, while at large densities the polarization model predicts an interaction completely different (in both magnitude and slope) from that obtained from the DLVO theory.

boundary conditions). However, when λ_{DH} is comparable with λ_m, qualitatively different results are predicted by the two theories.

4. Conclusion

The interactions between some colloidal surfaces immersed in an electrolyte can be well described by the traditional DLVO theory [13], at least in a certain range of electrolyte concentrations and by employing the surface properties (surface charge, potential or dissociation constants) as fitting parameters. However, the inter-particle interactions, as well as other aspects of colloidal behaviour of many other systems cannot be understood in terms of the traditional theory, and a supplementary force ("the hydration force") is sometimes employed to account for the experimental results.

While the hydration force was associated with the structuring of water in the vicinity of a lyophilic surface [12], there is no consensus about its microscopic origin. This incertitude is probably due to the apparent contradictory experimental results: for phospholipid bilayers, the hydration force is apparently independent of the electrolyte concentration and has a decay length of about 2 Å [11], while for mica surfaces the hydration is strongly dependent not only on the electrolyte concentration, but also on the nature of the cation (cation-specific effects) [17] and has a decay length of about 10 Å.

In this article, it is shown that the polarization model proposed earlier by the authors [30] can explain both these apparently contradictory experimental results [11,17]. The physical basis of the model [29,30] is reviewed and the conditions under which the polarization model predicts results qualitatively similar to those obtained from the traditional DLVO theory are discussed. The qualitative departures from the traditional theory are also examined, and it is noted that the polarization model can explain experimental results regarding the stability of some colloids at high ionic strengths [14], and the restabilization of protein-covered latex particles at large electrolyte concentrations [21,22], via the formation of surface dipoles due to the association of the electrolyte ions with surface groups. The polarization model can also predict the large virial coefficient of apoferritin molecules at high ionic strengths [18], because the correlation between neighboring dipoles increases the range of the interactions.

The polarization model is extended to account for the ion–ion and ion–surface interactions, not included in the "mean field" electrical potential. The role of the disorder on the dipole correlation length λ_m is modeled through an empirical relation, and it is shown that the polarization model reduces to the traditional Poisson–Boltzmann formalism (modified to account for additional interactions) when λ_m becomes sufficiently small.

References

[1] Gouy G. J Phys Radium 1910;9:457.
[2] Chapman DL. Phyl Mag 1913;25:475.
[3] Langmuir I. J Chem Phys 1938;6:893.
[4] Stern O. Z Elektrochem 1924;30:508.
[5] Spaarnay MJ. Recl Trav Chim Pays-Bas 1958;77:872.
[6] Gur Y, Ravina I, Babchin AJ. J Colloid Interface Sci 1978;64:333.
[7] Levine S, Bell GM. Discuss Faraday Soc 1966;42:69; Ruckenstein E, Schiby D. Langmuir 1985;1:612.
[8] Jonsson B, Wennerstrom H. J Chem Soc, Faraday Trans II 1983; 79:19.
[9] Boström M, Williams DMR, Ninham BW. Phys Rev Lett 2001;87: 168103.
[10] Manciu M, Ruckenstein E. Adv Colloid Interface Sci 2003;105:63; Ruckenstein E, Manciu M. Adv Colloid Interface Sci 2003;105:177.
[11] Rand RP, Parsegian VA. Biochim Biophys Acta 1989;988:351.
[12] Derjaguin BV, Churaev NV, Muller VM. Surface forces, 1987, Plenum.
[13] Deryagin BV, Landau L. Acta Physicochim URSS 1941;14:633; Verwey EJ, Overbeek JThG. Theory of Stability of Lyophobic colloids. Amsterdam: Elsevier; 1948.
[14] Healy TW, Homola A, James RO, Hunter RJ. Faraday Discuss Chem Soc 1978;65:156.
[15] Paunov VN, Kaler EW, Sandler SI, Petsev DN. J Colloid Interface Sci 2001;240:640.
[16] Manciu M, Ruckenstein E. Langmuir 2002;18:5178.
[17] Pashley RM. J Colloid Interface Sci 1981;83:531.
[18] Petsev DN, Vekilov PG. Phys Rev Lett 2000;84:1334.
[19] Manciu M, Ruckenstein E. Langmuir 2002;18:8910.
[20] Collinns KD, Washabaugh MW. Q Rev Biophys 1985;18:323.
[21] Molina-Bolivar JA, Galisteo-Gonzalez F, Hidalgo-Alvarez R. Phys Rev E 1997;55:4522; López-León T, Jódar-Reyes AB, Ortega-Vinuesa JL, Bastos-González D. J Colloid Interface Sci 2004 [submitted for publication].
[22] Stenkamp VS, McGuiggan PM, Berg JC. Langmuir 2001;17:637.
[23] Kjellander R, Marcelja S. Chem Scr 1985;25:73 Chem Phys Lett 1985;120:393.
[24] Marcelja S. Nature 1997;385:689.
[25] Otto F, Patey GN. J Chem Phys 2000;113:2851.
[26] Attard P, Batchelor MT. Chem Phys Lett 1988;149:206.
[27] Israelachvili JN, Wennerstrom H. Langmuir 1990;6:873.
[28] Binder H, Dietrich U, Schalke M, Pfeiffer H. Langmuir 1999;15:4857.
[29] Schiby D, Ruckenstein E. Chem Phys Lett 1983;95:435.
[30] Ruckenstein E, Manciu M. Langmuir 2002;18:7584.
[31] Gruen DWR, Marcelja S. Faraday Trans. II, (1983) 211 and 225.
[32] Simon SA, McIntosh TJ. Proc Natl Acad Sci U S A 1989;86:9263.
[33] Berkowitz ML, Raghavan K. Langmuir 1991;7:1042.
[34] Manciu M, Ruckenstein E. Langmuir 2001;17:7061.
[35] Manciu M, Ruckenstein E. Langmuir 2001;17:7582.
[36] Perera L, Essmann U, Berkowitz ML. Prog Colloid & Polym Sci 1997;103:107.
[37] Jackson JD. Classical Electrodynamics. New York: John Wiley & Sons; 1975.
[38] Faraudo J, Bresme F. Phys Rev Lett 2004;92:236102–1.
[39] Langmuir I. J Am Chem Soc 1917;39:1848.
[40] Ruckenstein E, Huang H. Langmuir 2003;19:3049.
[41] Overbeek JTG. Colloids Surf 1990;51:61.
[42] Manciu M, Ruckenstein E. Langmuir 2003;19:1114.
[43] Helfrich W. Z Naturforsch 1978;33a:305; Manciu M, Ruckenstein E. Langmuir 2002;18:4179.
[44] Petrache HI, Kimchi I, Harries D, Tristram-Nagle S, Podgornik R, Parsegian VA. Biophys J 2004;86:379A.
[45] Huang HH, Manciu M, Ruckenstein E. J Colloid Interface Sci 2003; 263:156.
[46] Prieve DC, Ruckenstein E. J Theor Biol 1976;56:205.

ELSEVIER

Advances in Colloid and Interface Science 127 (2006) 29–42

ADVANCES IN COLLOID AND INTERFACE SCIENCE

www.elsevier.com/locate/cis

Polarization model for poorly-organized interfacial water: Hydration forces between silica surfaces

Marian Manciu [a,*], Oscar Calvo [a], Eli Ruckenstein [b,1]

[a] *Department of Physics, University of Texas at El Paso, El Paso, TX, 79968, United States*
[b] *Department of Chemical and Biological Engineering, State University of New York at Buffalo, Buffalo, NY, 14260, United States*

Available online 3 October 2006

Abstract

The main goal of this paper is to review the theoretical models which can be used to describe the interactions between silica surfaces and to show that a model proposed earlier by the authors (the polarization model), which accounts concomitantly for double layer and hydration forces, can be adapted to explain recent experiments in this direction. When the water molecules near the interface were considered to have an ice-like structure, a strong coupling between the double layer and hydration forces (described by the correlation length between neighboring dipoles, λ_m) generates long range interactions, larger than the experimentally determined interactions between silica surfaces. Arguments are brought that a gel layer is likely to be formed on the surface of silica, which, by generating disorder in the interfacial water layers, can decrease strongly the value of λ_m. Since the prediction of λ_m involves a choice for the microscopic structure of water, which is often unknown, the polarization model is also presented here as a phenomenological theory, in which λ_m is used as a fitting parameter. Two extreme cases are considered. In one of them, the water molecules near the interface are considered to have an ice-like structure, whereas in the other they are considered randomly distributed. In the first case, the dipole correlation length $\lambda_m = 14.9$ Å. In the second limiting case, λ_m can be of the order of 1 Å. It is shown that, for $\lambda_m = 4$ Å, a more than qualitative agreement with the experiment could be obtained, for reasonable values of the parameters involved (e.g. surface dipole strength and density, dipole location, surface charge).

Keywords: Hydration force; Colloidal interactions; Silica

Contents

* Corresponding author. Tel.: +1 915 747 7531; fax: +1 915 747 5447.
E-mail addresses: mmanciu@utep.edu (M. Manciu), feaeliru@acsu.buffalo.edu (E. Ruckenstein).
[1] Tel.: +1 716 645 2911/2214; fax: +1 716 645 3822.

0001-8686/$ - see front matter © 2006 Elsevier B.V. All rights reserved.
doi:10.1016/j.cis.2006.08.001

1. Introduction

Most of the water-mediated interactions between surfaces are described in terms of the DLVO theory [1,2]. When a surface is immersed in water containing an electrolyte, a cloud of ions can be formed around it, and if two such surfaces approach each other, the overlap of the ionic clouds generates repulsive interactions. In the traditional Poisson–Boltzmann approach, the ions are assumed to obey Boltzmannian distributions in a "mean field" potential. In spite of these rather drastic approximations, the Poisson–Boltzmann theory of the double layer interaction, coupled with the van der Waals attractions (the DLVO theory), could explain in most cases, at least qualitatively, and often quantitatively, the colloidal interactions [1,2].

The limitations of this simplified model have been immediately recognized, and the first criticism [3] even preceded the full development of the DLVO theory. Since then many improvements of the theory have been proposed, to account for the finite size of the ions [4], image forces [5], dielectric corrections [6], ion correlations [7], ion-dispersion [8] and ion-hydration forces [9], to name only a few. Despite the many corrections brought to the traditional DLVO theory, there are some experiments, such as those regarding the stability of neutral lipid multilayers, which could still not be explained within this framework. It is therefore commonly accepted that an additional repulsion occurs when two surfaces approach each other at a distance shorter than a few nanometers. Because this force was initially related to the structuring of water near surfaces, it is commonly named "hydration force" [10].

1.1. Hydration force

At a simple phenomenological level, the "hydration force" can be described via an exponentially decreasing force, additive to (and independent of) the DLVO double layer and van der Waals forces [10]. An alternative phenomenological description is to consider the existence of an "order parameter" and a Landau-like expansion of the free energy in that parameter. When only some of the expansion terms are retained in the latter expansion, both phenomenological descriptions lead to similar behaviors for the hydration forces [11].

The debate started when various microscopic models for the origin of the "hydration force" have been proposed. One plausible possibility is that the "hydration force" might involve in fact a plethora of different phenomena, each having a different microscopic origin. For examples, the hydration interaction measured between mica surfaces [12] exhibits an oscillatory behaviour, in contrast with the hydration force measured between neutral lipid bilayers [10]. Even more puzzling, experiment shows that the latter forces seem to be almost independent of electrolyte concentration, while in other systems they depend not only on the ionic strength, but even on the kind of ions employed, exhibiting strong specific ion effects. For example, for the interactions between mica surfaces, deviations from the DLVO theory (hence a "hydration force") has been obtained only above some critical electrolyte concentrations, which were about $c_{E,\mathrm{critical}}=10^{-2}$ M for LiCl and NaCl, $c_{E,\mathrm{critical}}=10^{-3}$ M for CsCl, and $c_{E,\mathrm{critical}}=3\times10^{-4}$ M for KCl [13]. Also, amphoteric latex particles, at pH values above the isoelectric point, coagulated when the $CsNO_3$ exceeded about 0.3 M, but remained stable up to concentrations exceeding 3 M of KNO_3 [14], pointing out toward strong specific ionic effects associated with the hydration force.

The earlier attempts to explain hydration forces suggested merely corrections to the Poisson–Boltzmann approach, to account for the additional effects not included in the mean field [4–9]. The more recent models proposed that the free energy of the system contain not only the traditional DLVO free energy terms, but also an additional free energy term associated with the interface. Several possible microscopic origins of the additional term have been proposed, such as (i) the structuring of water near the interface, with the corresponding disruption of hydrogen bonding [15], (ii) the dipoles present in water and on surfaces [5,16,17], (iii) the repulsion generated by ion dehydration [14,18] and (iv) the steric effect [19,20].

The disruption of hydrogen bonding constitutes an appealing possible explanation of the hydration force. Attard and Batchelor presented a two-dimensional lattice model, and concluded that both the preexponential factor and the decay length of the hydration force are determined by only one unknown parameter, the concentration w of Bjerrum defects in the vicinity of the surface [15], which are proton disorders of the ice structure, consisting of either no H atom or two H atoms between two neighboring O atoms (L and D defects, respectively). It is clear that this model cannot predict the oscillatory behavior of hydration in some systems [12] or why the ions (or even the nature of the ions) seem to govern the coagulation of colloids in other systems [13], [14]. A more serious issue is that for neutral lipid bilayers, the preexponential factor of the hydration force depends on the nature of the lipid that forms the bilayers, while the decay length (2–3 Å) seems to be almost independent of it [10]. For this reason, only one parameter (w) is perhaps not sufficient to describe simultaneously both quantities.

An ion dehydration origin of hydration force has been initially proposed by Healy et al. [14] and a statistical model has been developed recently by Paunov et al. [18]. Healy et al. suggested that the approach at very short distances of two surfaces is possible only if the ions adsorbed on the surfaces have lost their hydration shell, therefore the ions with large hydration energies might generate hydration forces large enough to prevent coagulation. One problem with this model is that the range of the hydration interactions cannot exceed two diameters of the hydrated ion (about 14 Å) [18], whereas hydration forces have been currently identified at distances of up to 20–40 Å. Another difficulty is that the hydration force in this model is directly related to the ionic strength, whereas in many systems they are present regardless of the presence or absence of an electrolyte (e.g. for neutral lipid bilayers [10]).

The steric repulsion, caused e.g. by coating the colloidal particles with a polymer, is well known to generate a force large enough to prevent coagulation at any electrolyte concentration [19]. It was suggested by Israelachvili and Wennerström that the statistical overlap of lipids protruded from lipid bilayers might generate an exponentially decaying force, with a preexponential factor and decay length comparable to those experimentally measured for neutral lipid bilayers [20]. Clearly, this model cannot explain the hydration forces for other systems, such as those between mica or silica surfaces. Its validity was, however, questioned even for lipid bilayers, because the polymerization of the lipids within the bilayers, which was expected to decrease drastically the protrusion rate, did not affect the hydration force [21].

The initial microscopic models for the hydration force have been somehow related to the presence of permanent dipoles on the water molecules and on the immersed surfaces. This choice was natural, because the large dielectric constant of water (hence its solvation ability for polar molecules) is largely due to the permanent dipole moment of the water molecule. Jönsson and Wennerström suggested that the interactions between the surface dipoles and their image charges might be responsible for the hydration force [5]. However it turned out that in order to explain a typical experiment, the magnitude of the surface dipoles would have to be about two orders of magnitude larger than the normal molecular dipoles [22]. Gruen and Marcelja [17] suggested that the order parameter in the Marcelja–Radic treatment [11] might be the water polarization, and they conjectured that the Bjerrum defects might constitute the source of the polarization field, in analogy with the ions, which provide the source of the electric field [17]. The model considered the polarization and the electric fields as independent, and coupled equations for them have been obtained through a variational procedure. When a suitable value was selected for the concentration of Bjerrum defects, an exponential repulsion similar to the hydration force obtained experimentally was acquired. However, this theory did not explain how the concentration of Bjerrum defects in water is related to the physical properties of the immersed surface and it is not clear how the polarization of the surface induces the polarization of the water molecules (in a continuous model, a surface with a constant dipolar density does not generate any electric field in the fluid in which it is immersed). In addition, it was later suggested that the exponential decay of the hydration force predicted by the Gruen–Marcelja model was a result of an inconsistent choice for the non-local dielectric function [23]. The strongest critique of the model came, however, from Molecular Dynamics simulation, which showed that for some systems the water polarization might oscillate near an immersed surface [24,25], whereas the disruption of hydrogen bonding (the Bjerrum defects) varies monotonically [26]. Therefore, unlike the disruption of the hydrogen bonding, the polarization does not constitute a proper order parameter, in the sense of Landau's theory.

The various qualitative behaviors of the hydration force in different systems (either oscillatory [12] or monotonic [10] , with various decay lengths (2–3 Å [10] or about 10 Å [13]), either independent of electrolyte concentration [10] or exhibiting strong specific ion effects [14]) appear to point out toward the existence of a number of different microscopic origins for the short range repulsions between surfaces immersed in water, in excess to those accounted by the DLVO theory. On the other hand, there are some striking similarities between the hydration forces in different systems. For example, the Molecular Dynamics simulation showed an oscillatory behaviour of the polarization of water between two layers of sodium dodecylsulfate [25] and a corresponding oscillatory force [27]. The results are comparable with the earlier Molecular Dynamics simulations of Berkowitz et al., which indicated an oscillatory polarization in the vicinity of a neutral lipid bilayer [24]. However, when the headgroups of the bilayer surface have been allowed to move, the polarization oscillations have been smoothed away and therefore a monotonic behaviour of the hydration force has arisen [28]. Therefore, the oscillatory behaviour of the hydration force, for the flat mica surfaces, and the monotonic behaviour for fluid lipid bilayers might be explained by the same molecular mechanism, namely the polarization of the water molecules in the vicinity of the surfaces.

A model for this polarization was proposed by Schiby and Ruckenstein [16], who suggested that it might be a consequence of the discrete nature of the water molecules and surface dipoles [16]. A uniform, continuous surface dipolar density does not generate any electrical field in the adjacent continuous medium. In contrast, when discrete dipoles are placed on a surface lattice, an exponentially decaying field near the surface is generated [29]. Whereas this field (calculated for a continuous medium) is too weak to explain alone the hydration force, when the discrete nature of the water molecules is accounted in the dielectric response of water (which consequently becomes non-local), the correlation between the surface dipoles and the water dipoles becomes strong enough to generate a repulsion similar to the hydration force [30]. Moreover, when the water is organized in ice-like layers (presumably in the vicinity of a flat surface, like mica) the dipole correlation generates an oscillatory polarization in water, in the vicinity of the surface, and a corresponding oscillatory hydration force [31]. Within this "polarization model", the coupling between the "hydration", the "double layer" [32] and ion-hydration interactions [9] could explain why specific ion effects occur and why the decay length of the "hydration forces" might vary between 2 Å and about 10 Å [33].

1.2. Short range interactions between silica surfaces

The present paper has been stimulated by a recent article [22], which reported experiments on short range interactions between silica surfaces and explored possible explanations of the results by using various theoretical models for the hydration force. It is somewhat surprising that most theoretical models investigated provided a reasonable agreement with the experiment, while the best fit of the experimental data of the polarization model provided qualitatively different results (a short range attraction and a long range repulsion, instead of a short range repulsion and long range attraction). The surprise is due to the fact that the polarization model has been shown to describe the hydration forces for various systems [33] and also because it involves many unknown constants, which could be used as fitting parameters. Nevertheless, Valle-Delgado et al., concluded that the polarization model, with the values of the parameters employed for neutral lipid bilayers [32], cannot explain the hydration force measured for silica surfaces.

Two explanations for the discrepancy are possible. One of them is related directly and the other one indirectly to the peculiar structure of the silica surface, which increases its roughness upon immersion in water. A freshly broken quartz surface is composed of dangling silicon and oxygen bonds, which are quite unstable and can be hydroxylated within seconds if water is available, forming mostly silanol groups, which in turn interact with the neighboring water molecules, which are strongly organized by them [34]. Considering also the relatively large dissolution rate of silica surfaces [35] it is very unlikely that the surface of silica remains flat after its immersion in water. Even more drastically, it was suggested [36] that the hydroxilation of the siloxane groups (Si–O–Si) to silanol (Si–OH) groups might continue upon immersion in water, forming-$Si(OH)_2$–O–$Si(OH)_2$–OH polymeric chains, which in turn generate a silica gel with a thickness of about 10 Å that covers the silica surface [36]. Such a gel might generate a steric repulsion between two silica surfaces, and therefore might provide a microscopic explanation for the hydration forces measured in silica [36]. The latter suggestion was, however, challenged by Yoon and Vivek [37], who observed that the hydration force disappeared when 15% methanol was added to water; they argued that the adsorption of methanol on the silica surface cannot affect too much the roughness of the surface or the structure of the polymer-like silica hairs near the surface [37], but it is likely to disrupt the hydrogen bonding between the silica surface and the neighboring water molecules, hence the organization of water near the silica surface. Consequently Yoon and Vivek suggested that the organization of the interfacial water should be responsible for the hydration force between silica surfaces [37].

The second possible explanation, that relies indirectly on the roughness of the silica surface, is based on the interaction between silica and the interfacial water. It has been suggested that the water adsorbed on silica is strongly oriented by the surface and has characteristics (entropy, mobility, dielectric constant) drastically different from those of bulk water, for at least the first three water monolayers (about 10 Å) [38]. Because the interfacial water is more disorganized than the bulk water (or an ice-like layer of water), the correlation between neighboring dipoles, which is related to their relative position, is expected to be also weakened. The consequences of a weaker dipole correlation on the polarization model for hydration forces have been investigated briefly in a previous article [33]. In the present paper, a slightly different approach is employed. In Section 2, the basics of the polarization model as a phenomenological theory will be presented, which involves only a single parameter, the dipole correlation length λ_m. The reason for this approach is that the evaluation of λ_m requires knowledge of the local structure of water, which is often missing. By employing λ_m as an unknown (fitting) parameter, the problem is considerably simplified. In Section 3, a microscopic justification for the dipole correlation length is presented and the effect of the structural disorder of water on λ_m is examined. It will be shown that the disorder can generate values for λ_m much smaller than λ_m=14.9 Å, which was obtained for an ice-like organization of water in the vicinity of neutral lipid bilayers [32]. In Section 4, the results provided by the polarization model for the hydration forces are compared with the experimentally determined silica interactions [22]. Using the value λ_m=4 Å, it will be shown that a more than qualitative agreement with experiment can be obtained. This value is consistent with a random distribution of the nearest neighbors water molecules. The implication of the results obtained from the fit on the structure of silica surfaces will be discussed.

2. Polarization model for the hydration force: a phenomenological approach

2.1. Basic equations

In a continuous and homogeneous liquid, the polarization is related to the macroscopic electric field via a local dielectric constant $\varepsilon(r)$,

$$\vec{P}(r) = \varepsilon_0(\varepsilon(r)-1)\vec{E}(r), \tag{1}$$

where ε_0 is the vacuum permittivity, $\vec{P}(r)$ is the polarization, $\vec{E}(r)$ is the macroscopic electric field and $\varepsilon(r)$ is the dielectric constant at point r. In the polarization model, the average polarization of a water molecule depends not only on the macroscopic electric field applied, but also on the dipoles of the neighboring molecules. If the water is considered a continuous and homogeneous dielectric, with, $\varepsilon(r)\equiv\varepsilon$, then the field at the location of a water molecule, created by all the dipoles of all the other molecules, should vanish. However, if the interactions between neighboring dipoles are less screened than the interactions with the more remote dipoles, there will be a net remaining field at the location of the molecule, created by all the surrounding dipoles [30]. This field polarizes the molecule and will generate a correlation in the orientation of the neighboring dipoles. Assuming that the surface immersed in water is planar and that the coordinate z is normal to the surface, the polarization $m(r)=v_0P(r)$ of a water molecule of volume v_0 is provided by [32]:

$$m(z) = \varepsilon_0 v_0(\varepsilon-1)E(z) + \lambda_m^2\frac{d^2m(z)}{dz^2} \tag{2}$$

Eq. (2) accounts for the correlation of neighboring dipoles through only one parameter, the dipole correlation length λ_m. An examination of the microscopic origin of the parameter λ_m (and the justification of Eq. (2)) will be presented in Section 3; until then the dipolar coupling can be thought as a phenomenological fitting parameter of the polarization model. Clearly, when $\lambda_m\to 0$, Eq. (2) becomes the conventional relation between polarization and macroscopic electric field, Eq. (1).

The mean field potential $\psi(z)$, defined by $E(z) = -\frac{d\psi(z)}{dz}$, is related to the polarization and charge via the Poisson equation:

$$\frac{d^2\psi(z)}{dz^2} = -\frac{\rho(z)}{\varepsilon_0} + \frac{1}{\varepsilon_0 v_0}\frac{dm(z)}{dz} \tag{3}$$

Assuming that the charge distribution is due only to the ions of a dissociated electrolyte of concentration c_E, which obeys Boltzmannian distributions:

$$\rho(z) = c_E q_C \exp\left(-\frac{q_C\psi(z) + \Delta W_C(z)}{kT}\right) + c_E q_A \exp\left(-\frac{q_A\psi(z) + \Delta W_A(z)}{kT}\right) \tag{4}$$

where k is the Boltzmann constant, T is the absolute temperature, q_A and q_C are the anion and cation charges, and ΔW_A and ΔW_C are free energies of interactions with the interface, not included in the mean field, of the anions and cations, respectively. Some typical examples for the latter interactions are: the image forces [5], the ion-dispersion [8] or the ion-hydration forces [9].

Eqs. (1), (3) and (4) represent a Modified Poisson–Boltzmann approach, which, when the ions interacts only via the mean field potential $\psi(z)$ (e.g. $\Delta W_a = \Delta W_c \equiv 0$), reduces to the well-known Poisson–Boltzmann equation. The only change in the polarization model is the replacement of the constitutive equation Eq. (1) by Eq. (2), which accounts for the correlation in the orientation of neighboring dipoles. However, since $\psi(z)$ and $m(z)$ are now independent functions, four boundary conditions are needed to solve the system composed of Eqs. (2) (3) and (4). For only one surface immersed in an electrolyte, $\psi(z=\infty)=0$, $m(z=\infty)=0$ whereas for two identical surfaces, the symmetry of ψ and m requires that $\frac{d\psi(z=0)}{dz} = 0$ and $m(z=0)=0$ where $z=0$ represents the midpoint between surfaces. In a phenomenological approach the surface charge (or the surface potential) as well as the water polarization near the surface can be considered fitting parameters. In Section 3, in which a microscopic model for the dipole correlation length is presented, it will be shown that knowledge of the surface properties (e.g. surface charge density, surface dipolar density, location and magnitude of the surface dipoles) provides natural boundary conditions for the equations of the polarization model.

2.2. Free energy

In the Poisson–Boltzmann theory, the free energy per unit area of two identical planar surfaces placed at the distances $z=-d$ and $z=d$, respectively, is composed of an electrostatic energy [39],

$$U_{\mathrm{el}}(2d) = \frac{1}{2}\int_{-d}^{d} D(z)E(z)dz = \frac{1}{2}\int_{-d}^{d} \varepsilon_0\varepsilon E^2(z)dz \tag{5a}$$

(in which the displacement vector is considered proportional to the macroscopic electric field, $D=\varepsilon_0\varepsilon E$), an entropic term due to the ions [39],

$$U_{\mathrm{ent}}(2d) = -T\Delta S = kT\sum_i \int_{-d}^{d}\left(c_i \ln\left(\frac{c_i}{c_E}\right) - c_i + c_E\right)dz \tag{5b}$$

where the sum runs over the anions and cations present in the system, a "chemical" free energy [39,40]

$$U_{\mathrm{ch}}(2d) = -\int_{\sigma(\infty)}^{\sigma(d)} \psi_S(\sigma)d\sigma \tag{5c}$$

in which the integral is performed by considering that the surfaces approach from infinite distance to the final separation $2d$, and σ is the surface charge density at each separation, and a term due to the additional interactions between ions and interfaces, not included in the mean field [41]:

$$U_{\mathrm{int}}(2d) = \int_{z=-d}^{z=d} (c_A(z)\Delta W_A(z) + c_C(z)\Delta W_C(z))dz \tag{5d}$$

where c_A and c_C are the local concentrations of anions and cations, respectively, and ΔW_A and ΔW_C are defined after Eq. (4).

The electrostatic free energy (Eq. (5a)) accounts for the interaction of the charges and dipoles with the macroscopic field in the conventional theory. However, in the polarization model, there is an additional contribution to the energy, because the dipoles interact with the field that generates their correlation. As it will be shown in the next section, Eq. (2) can be rewritten as [32]:

$$m(z) = \gamma E_{\mathrm{total}} = \varepsilon_0 v_0(\varepsilon-1)(E(z) + E_m(z)) \tag{6a}$$

where γ is the molecular polarizability and E_m is the local field that aligns the neighboring dipoles, which is given by:

$$E_m = \frac{\lambda_m^2}{\varepsilon_0 v_0(\varepsilon-1)}\frac{d^2m(z)}{dz^2} \tag{6b}$$

Consequently, the electrostatic free energy in the polarization model becomes:

$$\begin{aligned} U_{\mathrm{el}}(2d) &= \int_{-d}^{d}\left[\frac{1}{2}E(\varepsilon_0 E + P) - \frac{1}{2}\frac{m(z)}{v_0}E_m\right]dz \\ &= \frac{1}{2}\int_{-d}^{d}\left[\left(\varepsilon_0 E(z) + \frac{m(z)}{v_0}\right)E(z) - \frac{1}{2}\frac{m(z)}{v_0} \times \frac{\lambda_m^2}{\varepsilon_0 v_0(\varepsilon-1)}\frac{d^2m(z)}{dz^2}\right]dz \\ &= \frac{1}{2}\int_{-d}^{d}\left[\left(\varepsilon_0 E(z) + \frac{2m(z)}{v_0}\right)E(z) - \frac{1}{2}\frac{m(z)^2}{\varepsilon_0 v_0^2(\varepsilon-1)}\right]dz \end{aligned} \tag{5e}$$

where Eq. (6a) was employed in the last substitution.

In summary, the polarization model represents an extended Poisson–Boltzmann approach, in which the "hydration" and the "double layer" are not independent interactions, but are intimately coupled to each other, via an electrostatic coupling between the fields $\psi(z)$ and $m(z)$. These fields can be calculated by solving Eqs. (2) (3) and (4), and the total free energy of the system can be obtained by summing up the terms provided by

Eqs. (5b) (5c) (5d) (5e). The constitutive equation of the polarization model contains only one unknown parameter, namely λ_m, which expresses the correlation between neighboring dipoles. However, in order to solve the system of equations, both the surface charge density σ (or the surface potential $\psi(z=-d)$) as well as the polarization of water near the surface have to be known. The latter can be related, in a microscopic model that will be examined in the next section, to the surface dipoles. In the limit $\lambda_m \to 0$, the polarization model reduces to the Poisson–Boltzmann approach, and the two boundary conditions become dependent on each other, because in this case

$$m(z)|_{z=-d} = -\varepsilon_0 v_0(\varepsilon-1)\frac{d\psi(z)}{dz} = v_0\frac{\varepsilon-1}{\varepsilon}\sigma \qquad (7)$$

The polarization model has been employed previously to show that, at low electrolyte concentrations, the total interaction can be indeed calculated as the sum between two apparent independent interactions, a traditional "double layer" and an exponentially decaying "hydration force" [32]. However, at high electrolyte concentrations the coupling between the "double layer" and "hydration" can generate an unusually long range repulsion, which is perhaps responsible for the restabilization of apoferritin molecules [42]. For various combinations of the surface charge and surface dipole density, interactions with decay lengths between 2 to about 10 Å could be predicted [33]. With the supplementary hypothesis that the water is structured as ice-like layers near the interface (note that in this case a discrete version of Eq. (2) had to be employed [31]), the polarization model could explain the oscillations of the polarization near the surfaces [31], [43], which were observed via Molecular Dynamics simulations [24,25] and the corresponding non-monotonic hydration force [27].

If no microscopic model is available to calculate the dipole correlation length, one might consider λ_m as a phenomenological parameter of the polarization model. However, before investigating the applicability of the polarization model to the interactions between silica surfaces, a more microscopic representation is used in the next section to obtain an expression for λ_m and to derive a relation between the surface dipoles and the polarization of water near the interface.

3. Microscopic origin of the polarization model

3.1. Dipole correlation length

Let us first review the basics of the Lorentz theory for polarization. If one assumes that a constant macroscopic field is applied to a homogeneous medium of dielectric constant ε, the polarization through the medium will be uniform. However, the polarization of a molecule is not proportional to the macroscopic electric field (created by sources external to the dielectric), but to the local electric field, which contains also the field generated by all the other molecules of the dielectric. To account for the latter, one can separate the medium in a spherical cavity (in which the central molecule and its molecular neighbors reside, see Fig. 1A) and the rest of the medium, which

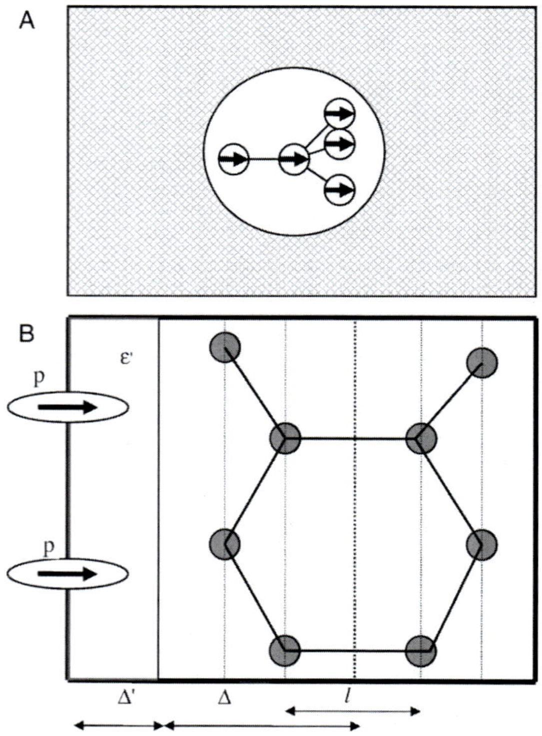

Fig. 1. A. A cavity in the dielectric, containing a central water molecule and its neighbors. The radius of the cavity is considered large enough, for the molecular details of the dielectric outside the cavity to be ignored. Consequently, the dielectric outside the cavity can be considered as a homogeneous and continuous medium. B. A schematic structure of the silica–water interface. Each surface dipole exerts an average field on the water molecules above him; the fields generated by the remote dipoles are neglected.

is far enough from the central molecule, to be able to ignore its molecular details. This means that the medium outside the cavity can be considered to behave as a homogeneous, continuous dielectric. The field acting on the central molecule is composed of three terms: the macroscopic field E, the Lorenz field $E_{\text{Lorenz}} = \frac{P}{3\varepsilon_0}$ created inside the spherical cavity embedded in a continuous dielectric with the uniform polarization P [44], and the field E_m created by the neighboring molecules:

$$E_{\text{local}} = E + \frac{P}{3\varepsilon_0} + E_m, \qquad (8)$$

It was shown by Lorentz [44] that, for a medium of constant polarization and with a sufficient symmetric distribution of the neighboring molecules around the central one, the field E_m vanishes. Then, the average dipole moment of the central molecule, $m(z)=v_0P(z)$, where v_0 is the volume occupied by a water molecule, is provided by [32]:

$$m = \gamma E_{\text{local}} = \gamma\left(E + \frac{m}{3\varepsilon_0 v_0}\right) = \frac{\gamma}{1-\frac{\gamma}{3\varepsilon_0 v_0}}E \equiv \varepsilon_0 v_0(\varepsilon-1)E \qquad (9)$$

where γ is the molecular polarizability. The last equality is equivalent to Eq. (1), because for a uniform polarized medium the polarization is proportional to the macroscopic electric field;

it provides the Clausius–Mossotti relation between the microscopic molecular polarizability γ and the macroscopic dielectric constant ε.

If the polarization is non-uniform, an additional field E_m is generated by the neighboring dipoles. A simple way to calculate this field [30] is to neglect the field generated by the remote dipoles, because it is screened by the intervening water molecules, and to consider only the field generated by the first "i" neighboring dipoles, screened by an effective dielectric constant ε'', much lower than the bulk value, $\varepsilon=80$.

Each neighboring molecule "i", with an average dipole moment m_i, located at the position r_i from the central molecule, will create at the location z_0 of the central molecule the field:

$$\vec{E}_{m,i}(z_0)=\frac{1}{4\pi\varepsilon_0\varepsilon''}\frac{3\vec{r_i}(\vec{m_i}\cdot\vec{r_i})-\vec{m_i}(\vec{r_i}\cdot\vec{ri})}{r_i^5} \tag{10a}$$

Since we are interested only in the dipole correlation along the z directions (it is assumed that $\psi(z)$ and $m(z)$ do not depend on the x and y coordinates in the plane of the surface; because of this planar symmetry the x and y components of $\vec{m}(z)$ vanish), the component of the field along the z direction has the form:

$$E_{m,i}(z_0)=\frac{m_i}{4\pi\varepsilon_0\varepsilon''}\frac{3\cos^2\theta_i-1}{r_i^3} \tag{10b}$$

where θ_i represents the angle between r_i and the axis z. Assuming that the water polarization is a continuous function of z, the average dipole moment can be expanded around $z=z_0$ to obtain:

$$m_i=m(z_0)+(z_i-z_0)\frac{dm(z_0)}{dz}+\frac{1}{2}(z_i-z_0)^2\frac{d^2m(z_0)}{dz^2}+\ldots. \tag{11}$$

Consequently the additional field is provided by:

$$E_m=\sum_i E_{m,i}=\frac{1}{4\pi\varepsilon_0\varepsilon''}\sum_i m_i\frac{3\cos^2\theta_i-1}{r_i^3} \tag{12}$$

The terms of higher order in Eq. (11) can be neglected. Indeed, the polarization of water near the surface is expected to decrease strongly and can be, in general, well approximated by an exponential, $m(z)=m_0\exp\left(-\frac{z}{\lambda}\right)$; consequently, the term of order n of the expansion is proportional to $\frac{1}{n!}\frac{(z_i-z_0)^n}{\lambda^n}$. Since the field generated by the more remote dipoles is negligible (because it is screened by the intervening water molecules), for the near neighbors $(z_i-z_0)<\lambda$ and the series is rapidly convergent.

In order to evaluate the additional field E_m, the local structure of water must be known. One simple assumption, which was made previously [30], was that the water is organized in ice-like layers in the vicinity of the surface, and that this structure is preserved at least for a distance $2l$ around a central molecule, where l is the distance of closest approach between two water molecules. Consequently, a cavity of radius $2l$ has been considered around the central water molecule, containing 27 water molecules. By considering in Eqs. (11) and (12) the positions of the first 26 neighbors of a water molecule as in an ice-like layer, one can write [32]:

$$\begin{aligned}E_m(z_0)&=\frac{m(z_0)}{4\pi\varepsilon_0\varepsilon'' l^3}(-3.77+2\cdot1.83)+\frac{1.8272\Delta^2}{4\pi\varepsilon''\varepsilon_0 l^3}\frac{d^2m(z_0)}{dz^2}\\&\approx\frac{1.83}{4\pi\varepsilon''\varepsilon_0 l}\frac{16}{9}\frac{d^2m(z_0)}{dz^2}\equiv\frac{\lambda_m^2}{\varepsilon_0 v_0(\varepsilon-1)}\frac{d^2m(z_0)}{dz^2}\end{aligned} \tag{13}$$

where $\Delta=(4/3)\,l$ is the distance between the centers of adjacent ice-like water layers. The last equality introduces a new parameter, the dipole correlation length.

$$\lambda_m=\frac{2}{3}\sqrt{\frac{1.83v_0(\varepsilon-1)}{\pi\varepsilon'' l}} \tag{14}$$

In Eq. (13), the term proportional to $m(z_0)$ is negligible (which shows that the field of the neighbors vanishes for a uniform polarization of the medium, as suggested long ago by Lorentz [44]), and the terms proportional to the derivatives of odd order vanish because of symmetry. Once the field E_m has been determined, Eq. (9) should be replaced by:

$$\begin{aligned}m&=\gamma E_{\text{local}}=\gamma\left(E+\frac{m}{3\varepsilon_0 v_0}+E_m\right)=\frac{\gamma}{1-\frac{\gamma}{3\varepsilon_0 v_0}}E\\&\equiv\varepsilon_0 v_0(\varepsilon-1)(E+E_m)=\varepsilon_0 v_0(\varepsilon-1)E+\lambda_m^2\frac{d^2m}{dz^2}\end{aligned} \tag{15}$$

The latter equation coincides with Eq.(2) of the previous section, employed by the polarization model. In the limit of low electrolyte concentrations $c_E\to0$, the integration of Eq. (2) (with $\rho(z)\equiv0$) leads to [32]:

$$m(z)=\varepsilon_0 v_0\frac{d\psi(z)}{dz}, \tag{16}$$

which was inserted in Eq. (3) provides:

$$m(z)=\frac{\lambda_m^2}{\varepsilon}\frac{d^2m(z)}{dz^2} \tag{17}$$

The solution of Eq. (17) is:

$$m(z)=m(0)\exp\left(-\frac{z}{\lambda_H}\right) \tag{18}$$

with

$$\lambda_H=\frac{\lambda_m}{\sqrt{\varepsilon}} \tag{19}$$

The largest possible value of λ_m in Eq. (14) is obtained when $\varepsilon''=1$ for all the dipoles within the cavity (no screening), which provides $\lambda_m=14.9$ Å [30], and Eq. (19) leads to $\lambda_H=1.67$ Å. This value is comparable with the decay length of the hydration force measured in neutral lipid bilayers [10].

However, there is no reason for the dipolar correlation length to acquire its maximum value in all the systems. We estimated that the tilting of the clusters with small angles (below 20°)

decreases λ_m only by a few percents [31]. The consequences of a weaker dipole correlation for the interactions between surfaces have been also investigated earlier [33]. As shown by Valle-Delgado et al. [22], the value $\lambda_m = 14.9$ Å is too large to explain the interactions between silica surfaces. Let us examine what happens in the opposite extreme case, when the water molecules near interface, instead of having the ordered structure of ice, have their neighboring water molecules at random orientations.

For a random orientation, a neighboring molecule at a distance r_i from the central one is situated with equal probability on a spherical shell of radius r_i. Therefore, the field generated by each neighbor located inside the dielectric cavity can be obtained by averaging the field given by Eq. (10b)

$$\begin{aligned}\bar{E}_{m,i} &= \frac{1}{4\pi r_i^2}\int_{\theta=0}^{\pi}\int_{\varphi=0}^{2\pi} E_{m,i} r_i^2 \sin\theta d\theta d\varphi \\ &= \frac{1}{4\pi\varepsilon_0}\frac{1}{4\pi}\int_{\theta=0}^{\pi}\int_{\varphi=0}^{2\pi}\frac{m_i}{\varepsilon''}\frac{3\cos^2\theta-1}{r_i^3}\sin\theta d\theta d\varphi \\ &= \frac{1}{8\pi\varepsilon_0\varepsilon'' r_i^3}\int_{\theta=0}^{\pi} m_i(3\cos^2\theta-1)\sin\theta d\theta \end{aligned} \tag{20}$$

Using Eq. (11), Eq. (20) becomes:

$$\begin{aligned}\bar{E}_{m,i} &= \frac{1}{8\pi\varepsilon_0\varepsilon'' r_i^3}\int_{\theta=0}^{\pi} m_i(3\cos^2\theta-1)\sin\theta d\theta \\ &= \frac{1}{8\pi\varepsilon_0\varepsilon'' r_i^3}\int_{\theta=0}^{\pi}\left(m(z_0) + r_i\cos\theta\frac{dm(z_0)}{dz} + \frac{1}{2}r_i^2\cos^2\theta\frac{d^2m(z_0)}{dz^2}\right)(3\cos^2\theta-1)\sin\theta d\theta \\ &= \frac{m(z_0)}{8\pi\varepsilon_0\varepsilon'' r_i^3}\int_{\theta=0}^{\pi}(3\cos^2\theta-1)\sin\theta d\theta + \frac{dm(z_0)}{dz}\frac{1}{8\pi\varepsilon_0\varepsilon'' r_i^2} \\ &\quad\times\int_{\theta=0}^{\pi}(3\cos^2\theta-1)\sin\theta\cos\theta d\theta + \frac{d^2m(z_0)}{dz^2}\frac{1}{16\pi\varepsilon_0\varepsilon'' r_i} \\ &\quad\times\int_{\theta=0}^{\pi}(3\cos^2\theta-1)\sin\theta\cos^2\theta d\theta = \frac{1}{30\pi\varepsilon_0\varepsilon'' r_i}\frac{d^2m(z_0)}{dz^2}\end{aligned} \tag{21}$$

In Eq. (21) the terms corresponding to $m(z_0)$ and its odd order derivatives with respect to z vanish (note that the field in Eq. (10b) is proportional to the Legendre polynomial of the first order, which explains the vanishings). Assuming that on each shell of radius r_j are N_j neighbors of order j, the total field is provided by:

$$\begin{aligned}E_m &= \sum_i \bar{E}_{m,i} = \sum_j N_j \bar{E}_{m,j} \\ &= \frac{1}{30\pi\varepsilon_0}\left(\sum_j \frac{N_j}{\varepsilon_j'' r_j}\right)\frac{d^2m(z_0)}{dz^2} \equiv \frac{\lambda_m^2}{\varepsilon_0 v_0(\varepsilon-1)}\frac{d^2m(z_0)}{dz^2}\end{aligned} \tag{22}$$

where ε_j'' depends on r_j (with large screenings for the remote dipoles and less screening for the neighboring dipoles). Eq. (22) defines λ_m:

$$\lambda_m = \sqrt{\frac{v_0(\varepsilon-1)}{30\pi}\left(\sum_j \frac{N_j}{\varepsilon_j'' r_j}\right)} \tag{23}$$

By assuming that the field of the first 4 neighbors of a water molecule is screened by a medium with ε'' close to unity, while the field of the other dipoles by a much larger dielectric constant ($\varepsilon_j'' \gg 1$), the dipole correlation length is provided by:

$$\lambda_m \cong \sqrt{\frac{4v_0(\varepsilon-1)}{30\pi l\varepsilon''}} \tag{24}$$

which, for typical values of the structural parameters of water ($v_0 = 32.4$ Å^3, $l = 2.76$ Å, $\varepsilon = 80$), becomes:

$$\lambda_m = \frac{6.27}{\sqrt{\varepsilon''}}\text{Å} \tag{25}$$

The last equation indicates that dipolar correlations length smaller than $\lambda_m = 15$ Å are expected to occur in disordered systems.

3.2. Boundary conditions

As already noted, in a homogeneous medium a surface dipole density does not generate an electric field above the surface, and consequently such a field is ignored in the traditional theory of the double layer, which takes into account only the surface charges. In the polarization model, both the surface charge and the surface dipole densities generate electrostatic interactions (commonly denoted as "double layer" and "hydration "forces").

The first boundary condition at the surface is provided by integrating the Poisson Eq. (3) over the volume of a cylindrical box, which includes the surface, with the flat sides parallel to the surface and a vanishingly thin height. After using the Gauss theorem, one obtains:

$$\varepsilon_0\left.\frac{d\psi(z)}{dz}\right|_{z=-d} - \left.\frac{m(z)}{v_0}\right|_{z=-d} = -\sigma \tag{26}$$

where $z = -d$ indicates the surface and σ is the surface charge density. The second boundary condition at the surface is provided by the polarization of the water molecules from the vicinity of the surface, which is caused by the macroscopic field $E = -\frac{d\psi(z)}{dz}|_{z=-d}$, the average field $\bar{E}_s$ generated by the surface dipoles and the field E_m due to the neighboring water dipoles.

The field generated by the surface dipoles is calculated by averaging the field generated by a dipole p in a medium of effective dielectric constant ε', at a distance Δ' above the center of the dipole over an area A corresponding to the area occupied by a surface dipole (see Fig. 1B). The fields generated by all the others (remote) surface dipoles are neglected, because they are screened by the intervening water molecules.

The electric field, normal to the surface, generated by a surface dipole, at a point whose position vector makes an angle θ with the normal, is given by:

$$E_s = \frac{p}{4\pi\varepsilon'\varepsilon_0 r^3}(3\cos^2(\theta)-1) \tag{27}$$

where $r = \frac{\Delta'}{\cos(\theta)}$. For the average field produced by a surface dipole one obtains [30]:

$$\begin{aligned}\bar{E}_S &= \frac{p}{\pi R_0^2}\int_0^{R_0}\frac{2\pi\rho d\rho}{4\pi\varepsilon'\varepsilon_0(\Delta^2+\rho^2)^{\frac{3}{2}}}\left(3\frac{\Delta^2}{\Delta^2+\rho^2}-1\right)\\ &= \frac{p}{2\pi\varepsilon_0\varepsilon'}\frac{1}{(R_0^2+\Delta^2)^{\frac{3}{2}}} = \frac{p}{2\pi\varepsilon_0\varepsilon'}\frac{1}{\left(\frac{A}{\pi}+\Delta^2\right)^{\frac{3}{2}}}\end{aligned} \tag{28}$$

where $A=\pi R_0^2$ is the average area occupied by a surface dipole. The field generated by the neighboring dipoles is provided by an expression similar to Eq. (12):

$$E_m = \sum_i E_{m,i} = \frac{1}{4\pi\varepsilon_0\varepsilon''}\sum_i m_i\frac{3\cos^2\theta_i-1}{r_i^3} \tag{29}$$

with the difference that in this case, the index "i" runs only over the neighbors which are above the interface (with $z>z_0$, z_0 being the position of the interfacial water molecule).

For water that is organized (at least locally) in ice-like layers in the vicinity of the surface, with the structure preserved for at least a distance $2l$ around a central interfacial molecule, the average field was evaluated previously as [30,33]:

$$\begin{aligned}E_m &= m(z_0)\left(-\frac{1.9}{4\pi\varepsilon_0\varepsilon'' l^3}\right)+\frac{dm(z_0)}{dz}\frac{1.83}{3\pi\varepsilon_0\varepsilon'' l^2}\\ &\quad+\frac{d^2m(z_0)}{dz^2}\frac{2\cdot 1.83}{9\pi\varepsilon_0\varepsilon'' l},\end{aligned} \tag{30}$$

which, after employing Eq. (3), becomes:

$$\begin{aligned}E_m &= m(z_0)\left(-\frac{1.9}{4\pi\varepsilon_0\varepsilon'' l^3}+\frac{3.66}{9\pi\varepsilon_0\varepsilon'' l\lambda_m^2}\right)\\ &\quad+\frac{dm(z_0)}{dz}\frac{1.83}{3\pi\varepsilon_0\varepsilon'' l^2}+\frac{d\psi(z_0)}{dz}\frac{3.66v_0(\varepsilon-1)}{9\pi\varepsilon'' l\lambda_m^2}\end{aligned} \tag{31}$$

In the other extreme case, in which the neighboring water molecules are randomly oriented, one can assume that the water molecules neighboring an interfacial water molecule are located on a hemisphere centered on that water molecule. As in Eq. (20) with the difference that θ runs now form 0 to $\pi/2$ instead of π, the average field generated by a neighbor "i" is provided by:

$$\begin{aligned}\bar{E}_{m,i} &= \frac{1}{2\pi r_i^2}\int_{\theta=0}^{\pi/2}\int_{\varphi=0}^{2\pi}E_{m,i}r_i^2\sin\theta d\theta d\varphi\\ &= \frac{1}{4\pi\varepsilon_0}\frac{1}{2\pi}\int_{\theta=0}^{\pi/2}\int_{\varphi=0}^{2\pi}\frac{m_i}{\varepsilon''}\frac{3\cos^2\theta-1}{r_i^3}\sin\theta d\theta d\varphi\\ &= \frac{1}{4\pi\varepsilon_0\varepsilon'' r_i^3}\int_{\theta=0}^{\pi/2}m_i(3\cos^2\theta-1)\sin\theta d\theta\end{aligned} \tag{32}$$

which becomes, after employing Eq. (11) with $z-z_0=r_i\cos\theta$:

$$\begin{aligned}\bar{E}_{m,i} &= \frac{1}{4\pi\varepsilon_0\varepsilon'' r_i^3}\int_{\theta=0}^{\pi/2}m_i(3\cos^2\theta-1)\sin\theta d\theta\\ &= \frac{1}{4\pi\varepsilon_0\varepsilon'' r_i^3}\int_{\theta=0}^{\pi/2}\left(m(z_0)+r_i\cos\theta\frac{dm(z_0)}{dz}\right.\\ &\quad\left.+\frac{1}{2}r_i^2\cos^2\theta\frac{d^2m(z_0)}{dz^2}\right)(3\cos^2\theta-1)\sin\theta d\theta\\ &= \frac{m(z_0)}{4\pi\varepsilon_0\varepsilon'' r_i^3}\int_{\theta=0}^{\pi/2}(3\cos^2\theta-1)\sin\theta d\theta+\frac{dm(z_0)}{dz}\frac{1}{4\pi\varepsilon_0\varepsilon'' r_i^2}\\ &\quad\times\int_{\theta=0}^{\pi/2}(3\cos^2\theta-1)\sin\theta\cos\theta d\theta+\frac{d^2m(z_0)}{dz^2}\frac{1}{8\pi\varepsilon_0\varepsilon'' r_i}\\ &\quad\times\int_{\theta=0}^{\pi/2}(3\cos^2\theta-1)\sin\theta\cos^2\theta d\theta=\frac{dm(z_0)}{dz}\frac{1}{16\pi\varepsilon_0\varepsilon'' r_i^2}\\ &\quad+\frac{d^2m(z_0)}{dz^2}\frac{1}{30\pi\varepsilon_0\varepsilon'' r_i}\end{aligned} \tag{33a}$$

which, after using Eq. (15), becomes

$$\begin{aligned}\bar{E}_{m,i} &= \frac{m(z_0)}{30\pi\varepsilon_0\varepsilon'' r_i\lambda_m^2}+\frac{dm(z_0)}{dz}\frac{1}{16\pi\varepsilon_0\varepsilon'' r_i^2}\\ &\quad+\frac{v_0(\varepsilon-1)}{30\pi\varepsilon'' r_i\lambda_m^2}\frac{d\psi(z_0)}{dz}\end{aligned} \tag{33b}$$

and the total field E_m at the location of the central interfacial molecule is given by:

$$E_m = \sum_i \bar{E}_{m,i} \tag{34}$$

where the index "i" runs only over the neighbors of the central molecule situated above the interface ($z>z_0$). Accounting only for the first neighbors, the field becomes

$$E_m = \frac{m(z_0)}{15\pi\varepsilon_0\varepsilon'' l\lambda_m^2}+\frac{dm(z_0)}{dz}\frac{1}{8\pi\varepsilon_0\varepsilon'' l^2}+\frac{v_0(\varepsilon-1)}{15\pi\varepsilon'' l\lambda_m^2}\frac{d\psi(z_0)}{dz}. \tag{35}$$

Finally, the second boundary condition at the surface is provided by:

$$\begin{aligned}m(z_0) &= \gamma\left(E(z_0)+\frac{m(z_0)}{3\varepsilon_0 v_0}+\bar{E}_S(z_0)+E_m(z_0)\right)\\ &= \varepsilon_0 v_0(\varepsilon-1)\left(-\frac{d\psi(z_0)}{dz}+\bar{E}_S(z_0)+E_m(z_0)\right)\end{aligned} \tag{36}$$

with $\bar{E}_s$ provided by Eq. (28) and E_m given either by Eq. (31) (for an ice-like structuring of water near the interface), or by Eq. (35) (for the random orientation case).

As noted in the previous section, for a single surface immersed in an electrolyte, the other two boundary conditions necessary to solve the system of Eqs. (2)–(4) are provided by the vanishing of both the electric field and the polarization far away from the surface:

$$\left.\frac{d\psi(z)}{dz}\right|_{z=\infty}=0 \tag{37}$$

$$m(z=\infty)=0 \tag{38}$$

whereas for two identical surfaces immersed in an electrolyte separated by a distance $2d$, the symmetry with respect to the middle distance, $z=0$, provides:

$$\left.\frac{d\psi(z)}{\partial z}\right|_{z=0}=0 \tag{39}$$

$$m(z)|_{z=0}=0 \tag{40}$$

Once the functions $\psi(z)$ and $m(z)$ are obtained, one can calculate the free energy of the system by adding the free energy terms provided by Eqs. (5b) (5c) (5d) (5e). A typical application is presented in the next section, in which we will examine whether or not the polarization model can be employed to represent the interactions between silica surfaces.

4. Interactions between silica surfaces

One of the reasons for the present investigation was that recent experiments on the interactions between silica surfaces seemed to be well described by most theoretical models [22], and, in contrast, the polarization model provided not only an unsuccessful fit, but also incoherent results, because it predicted a decrease of the surface dipole density with increasing electrolyte concentration [22]. We will try to explain why the polarization model provided such an unsuccessful fit.

Firstly, let us note that a good fitting of the experimental data does not imply necessarily a good theoretical model. For example, the reasonable agreement between experiment [22] and the Attard and Pattey model [45] (which is an extension of the Jönnson and Wennerström model [5]) has been obtained for surface dipoles between 150 and 800 Debyes [22] (note that the permanent dipole moment of the water molecule is 1.87 Debyes). While the Paunov et al. model [18] provided good agreement [22], it predicted a very short range for the hydration force (up to about 15 Å), whereas in most cases hydration forces have been identified for up to about 20–40 Å. In the Attard and Batchelor model [15], both the preexponential factor and the decay length of the hydration force depend only on one parameter (the concentration of Bjerrum defects w); therefore, in order to fit the data, a very steep hydration force was predicted, with a decay length about one order of magnitude smaller than the experimental one. The agreement of the Marcelja and Radic model [11] with the experiment is not surprising, because the model is very similar to adding an exponential repulsion (with a preexponential factor and decay length as fitting parameters) to the DLVO interactions. Such a procedure is almost always successful [29] (a notable exception being the oscillatory hydration forces measured between mica surfaces [12]).

Before discussing the polarization model, it is useful to present the experimental procedure employed in Ref. [22]. The force measurements have been performed using the colloid probe technique, a method pioneered by Ducker et al. [46]. A silica bead with a diameter of 5 μm was glued to a standard V-shaped tipless cantilever with a spring constant of 0.12 N/m, mounted on an Atomic Force Microscope (AFM) Nanoscope III. The cantilever (with the silica sphere attached) was pressed against a silica surface, with both the cantilever and the surface immersed in an electrolyte of controlled pH and ionic strength. The AFM recorded both the displacement of the cantilever base as well as the deflection of the cantilever tip (which is proportional to the force between the silica sphere and the flat silica surface), and thus the force between silica surfaces could be recorded as a function of the relative displacement between the surfaces. Using the Derjaguin approximation [29], the force between the spherical surface of radius R and the flat surface can be related to the free energy of interactions between two flat surfaces by:

$$\frac{F(2d)}{R}=2\pi(U(2d)-U(\infty)) \tag{41}$$

where $U(2d)$ represents the total free energy, per unit area, between two flat surfaces separated by a distance $2d$, which includes the attractive van der Waals interactions, as well as the repulsive double layer and "hydration" interactions. In what follows, for the repulsive interactions we will use the polarization model, in which the two repulsive interactions are coupled. Consequently,

$$U(2d)=U_{\mathrm{ent}}(2d)+U_{\mathrm{ch}}(2d)+U_{\mathrm{int}}(2d)+U_{\mathrm{el}}(2d)+U_{\mathrm{vdW}}(2d), \tag{42}$$

where the first four terms are provided by Eqs. (5b) (5c) (5d) (5e) respectively, and the last one represents the van der Waals free energy.

The inset of Fig. 2A shows that the experimental data from Ref. [22] (c_E=1 M, pH=9) for distances larger than 20 Å (region that provides information particularly on the van der Waals interactions) are comparable with the resolution of the measurements with the Atomic Force Microscope, which is (in this case) 10^{-2} mN/m. Therefore it is difficult to obtain accurate values for the Hamaker constant. Also, for large $\frac{F}{R}$ values (of the order of 1 mN/m), there is an uncertainty in the distance measurements of a few Å. Because it is unlikely that those uncertainties are due to the resolution of the Atomic Force Microscope (the typical vertical distance accuracy for the AFM being higher by at least one order of magnitude), one might be tempted to assign them to a higher roughness of the silica surfaces or perhaps even to the formation of a silica gel above the surface, as suggested by Vigil et al. [36]. The latter supposition is also supported by the presence of a strong and steep, hard-wall-like interaction, at short separations, followed by a slower, exponentially-like decreasing repulsion (with a decay length of about 10 Å). A hairy gel formed near the silica interface is very likely to behave in this manner.

A first step in the understanding of the hydration interaction would be a fit with an exponential repulsion plus a van der Waals attraction:

$$\frac{F(2d)}{R} = B_H \exp\left(-\frac{2d}{\lambda_H}\right) - \frac{A_H}{6(2d)^2} \tag{43a}$$

where A_H is the Hamaker constant for silica–silica interactions through water while B_H and λ_H are unknown fitting parameters. However, as the fit tentative shown in Fig. 2A, such a trial fails to provide a reasonable agreement with experimental data. The reason for the failure can be understood from Fig. 2B. As in the DLVO theory, the addition of an exponential repulsion to a power low attraction generates two minima separated by an intermediate maximum. When the cantilever of the Atomic Force Microscope exerts the maximum force possible between the silica surfaces (which is related to the force constant of the cantilever and its maximum recordable deflection), the separation between surfaces does not reach the primary minimum, but remains somewhere in between the maximum and the secondary minimum (see Fig. 2B). Therefore, the shortest separation that the microscope can reach does not provide the contact between the silica surfaces (zero separation). The separation between silica surfaces is unknown (up to an additive constant), and a more reasonable fitting function should be of the form

$$\frac{F(2d)}{R} = B_H \exp\left(-\frac{2d}{\lambda_H}\right) - \frac{A_H}{6(2d+2t)^2} \tag{43b}$$

where $2t$ represent the unknown distance between the point of contact of the silica surfaces and the minimum separation reached by AFM. A reasonable agreement with experiment, both at short separations (Fig. 2B, bottom inset) and near the secondary minimum (Fig. 2B, top inset) has been obtained by employing the values B_H=3.35 mN/m, λ_H=5.3 Å and $2t$=15 Å in Eq. (25) For the Hamaker constant, we used the value $A_H = 8.3\times10^{-21}$ J, proposed by Valle et al. [22]. The incertitude in the position of the point of contact explains why the predicted interactions for most theoretical models investigated by Valle-Delagdo et al. seem to be shifted from the experimental data by about 10 Å towards higher values for the separation distances (see Figs. 3, 5, 7 and 9 of Ref. [22]).

Let us now turn our attention to the polarization model. It was shown that, in the linear approximation, the repulsive interactions can be described by the sum of two exponentially decreasing terms. At low ionic strength, their decay lengths are

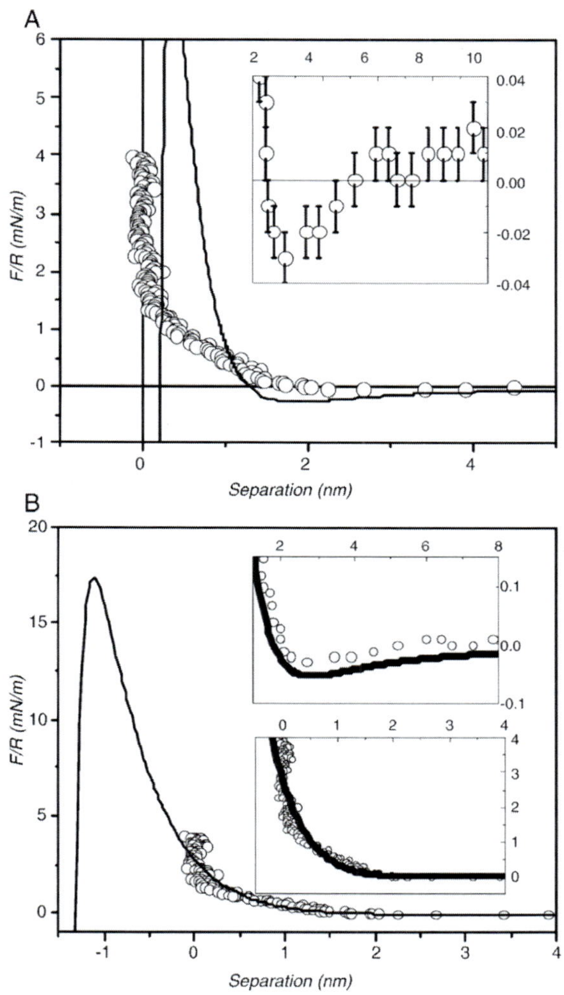

Fig. 2. A. The force between silica surfaces at c_E=1M and pH=9, determined experimentally in Ref. [22] (circles); the continuous line represents an unsuccessful fit with Eq. (43a). The origin of the x axis corresponds to the smallest separation distance between the two surfaces, attainable by AFM. The coordinate of the true point of contact between the surfaces cannot be obtained directly from experiment. The inset shows the region of the secondary minimum, where the magnitude of the interaction is comparable with instrumental resolution of 0.01 mN/m. B. The fit of the experimental data (c_E=1M, pH=9) with an exponential repulsion and a van der Waals attraction (Eq. (43b)). As in Fig. 2A, the origin of the x axis corresponds to the smallest separation attained in the experiment. The true point of contact between surfaces, obtained from fit, is located at a distance which is by $2t$=15 Å larger than the smallest separation recorded by AFM.

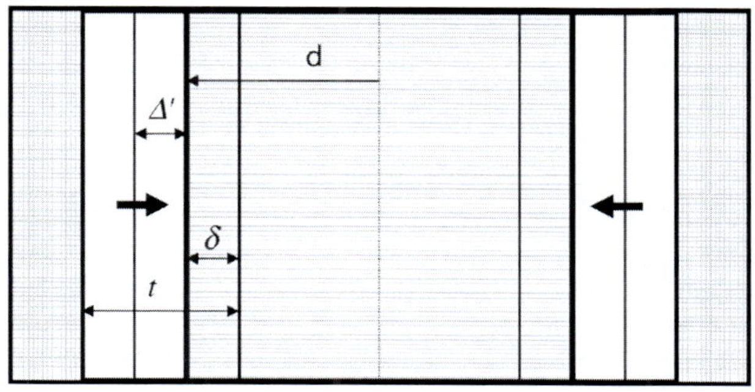

Fig. 3. The model employed for the calculations based on the polarization model in silica. The first water layer is at a distance d from the middle distance between silica surfaces. The surface dipoles of magnitude p=4 Debyes, each occupying an area A=50 Å^2, are located at a distance Δ'=1 Å below the first water monolayer (ε'=1). The shortest distance recorded experimentally corresponds to $d=\delta$. The surface of silica is at a distance $2t$=15 Å from the shortest experimental distance, as obtained from the best fit of the experiments with Eq. (43b), for A_H=8.3×10−^{21} J. If δ would be equal to t, the surface dipoles would be located on the silica surface. When $\delta<t$, the surface dipoles are located above the surface, at a distance $(t-\delta)$.

$\lambda_1 = \lambda_H = \frac{\lambda_m}{\sqrt{\varepsilon}}$ and $\lambda_2 = \lambda_{DH}$ (the Debye–Hückel length), but at high electrolyte concentrations, because of the coupling between "hydration" and "double layer", the largest decay length λ_2 becomes equal to λ_m [32]. By employing a value $\lambda_m = 14.9$ Å (corresponding to an ice-like layers organization of interfacial water), as performed in Ref. [22], it is very unlikely to obtain a good agreement of the polarization model and experimental results. However, the disorder induced in the interfacial water by the silica structure, as discussed in the previous sections, might reduce drastically the value of λ_m and a reasonable agreement might be obtained.

Since the structure of the silica interface is unknown, we will use the following simple model, sketched in Fig. 3. It will be assumed that the first water monolayer starts at $z=-d$. The surface dipoles of magnitude p, each occupying an area A on the surface, are located at a distance $\Delta'=1$ Å behind the first water monolayer ($z=-d+\Delta'$). Since the distance between the point of contact between silica surfaces and the smallest separation recorded by the AFM is unknown, it will be assumed that the shortest separation achieved experimentally corresponds to a distance through water of $2d=2\delta$.

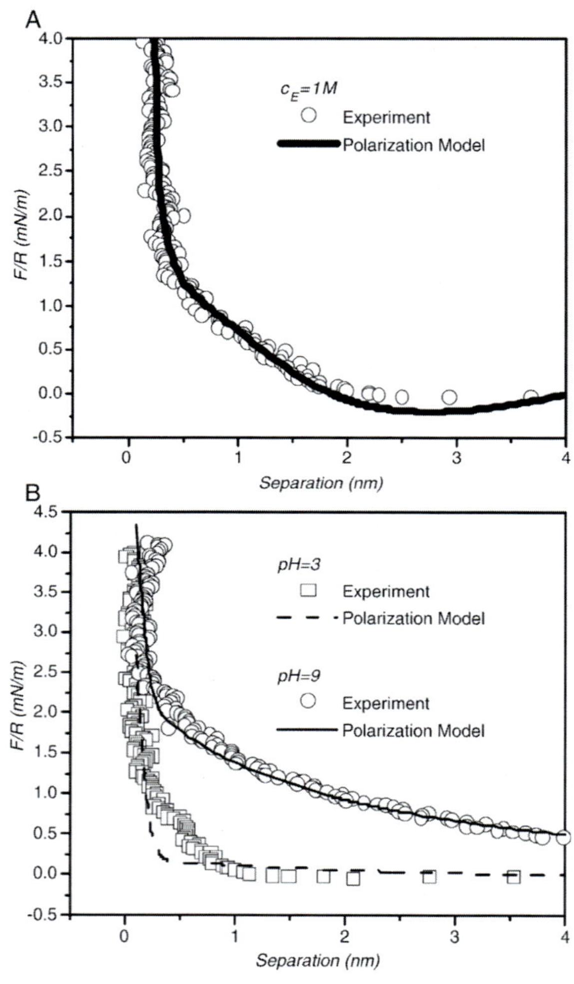

Fig. 4. A: Results of the polarization model (continuous line) are compared with the experimental data from Ref. [22] (circles), for c_E = 1 M and pH = 9. The values for the fitting parameters employed are: $\sigma=-0.1$ C/m^2, $2\delta=6$ Å. The values of all the other parameters are as before ($\lambda_m=4$ Å, $p=4$ Debyes, $\varepsilon'=1$, $\Delta'=1$ Å, $A=50$ Å^2, $2t=15$ Å and $A_H=8.3\ 10^{-21}$ J). B: Results of the polarization model are compared with the experimental data from Ref. [22], for c_E=0.01M, for pH = 3 (squares) and pH = 9 (circles). The values for the parameters derived from fitting are: $\sigma=-0.0027$ C/m^2, $\delta=8$ Å (pH = 3) and $\sigma=-0.0065$ C/m^2, $2\delta=6$ Å (pH = 9). The values of all the other parameters employed are as before ($\lambda_m=4$ Å, $p=4$ Debyes, $\varepsilon'=1$, $\Delta'=1$ Å, $A=50$ Å^2, $2t=15$ Å and $A_H=8.3\ 10^{-21}$ J).

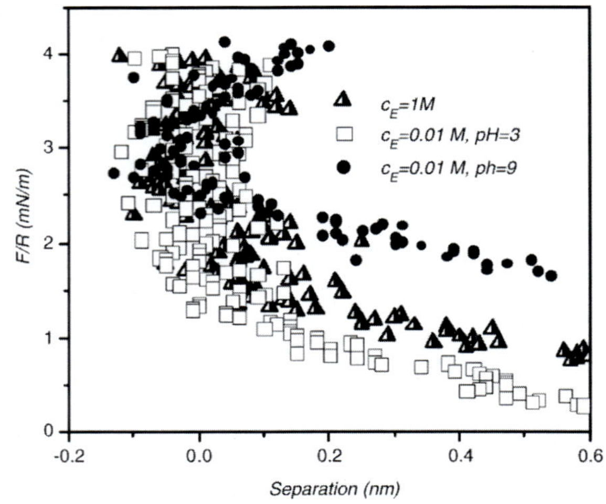

Fig. 5. All experimental data exhibit a hard-wall-like repulsion at very short separations, regardless of pH or ionic strength. Note that the hydration forces increases at higher ionic strength: c_E=1 M (triangles), compared with c_E=0.01 M, pH = 3 (squares). At pH = 9 (circles) and c_E=0.01 M, the surface charges create a long range double layer repulsion.

At this point, the polarization model contains a number of unknown parameters (e.g. λ_m, p, A, Δ', ε'', A_H, t, δ and σ). If the microscopic structure of the surface would be known, some of these parameters would be related to each other and to the experimental details (e.g. the surface charge density and the surface dipole densities would depend on the recombination equilibria of the surface groups with the electrolyte ions, which in turn depend on c_E, pH, and $\psi(-d)$). A fit with so many parameters would, however, not be very relevant, and in order to simplify the problem, we will employ some reasonable selected values for some of these parameters.

To be consistent with the experimental data for hydration forces measured between silica interfaces, λ_m has to be of the order of 4 Å, a value which is compatible with Eq. (24) for a random orientation of the neighboring water molecules. Therefore, in what follows, we will employ only the value λ_m=4 Å. For the van der Waals interactions, we will assume in all calculations $A_H=8.3\ 10^{-21}$ J and $2t=$ 15 Å, as obtained previously form the fit with Eq. (43b). The magnitudes of the surface dipoles will be considered 4 Debyes (about twice that of a water molecule) and it will be assumed that each dipole occupies on the surface an area of 50 Å^2 and that they are located at a distance $\Delta'=1$ Å below the first water monolayer.

Consequently, in what follows only two fitting parameters will be considered, the unknown surface charge densities σ and the unknown shortest separation obtained experimentally, 2δ. Figs. 4A and B shows a more than qualitative agreement with experimental data, obtained for the following fitting parameters: for the experiment at c_E=1 M and pH=9, the values σ=−0.1 C/m^2, 2δ=6 Å were employed (Fig. 4A); for the experiment at c_E=0.01 M and pH=3, the values σ=−0.0027 C/m^2 and 2δ=8 Å were employed; for the experiment at c_E=0.01 M and pH=9, the values σ=−0.0065 C/m^2 and 2δ=6 Å were employed. The values of the fitting surface charges indicate that the silica surface becomes more negatively charged at higher pH and higher ionic strengths, which is consistent with the expectation for recombination of the ionic moieties. At higher pH values, the surface becomes more negatively charged, because there are more OH^- available for recombination on the surface. At higher ionic strengths, the surface potential decreases drastically (because of the increase in the Debye–Hückel screening) and therefore the Boltzmann factor for the distribution of the OH^- anions near the surface is strongly increased.

The most important question is, however, whether or not the polarization of water is mainly responsible for the short range interactions between silica surfaces. The fittings shown in Fig. 4 indicate that the surface dipoles are situated at a distance 2δ of about 6 Å from the shortest separation obtained experimentally. In contrast, the fitting of the van der Waals interactions (Eq. (43b), Fig. 2B) implies that the flat silica surfaces are at a distance 2t of about 15 Å from the shortest separation obtained experimentally. If a value $t=\delta$ would have been obtained, it would indicate that the surface dipoles were actually situated on the flat silica surfaces. The results suggest instead that the surface dipoles, responsible for the polarization of the first water monolayer, are situated at a distance $t-\delta\approx 5$ Å above each of those surfaces. This conclusion would strongly support the formation of a silica gel, with a thickness of about 5 Å, on the silica surface. The possible formation of a gel is also supported by the experimental data obtained for various pHs and electrolyte concentrations, represented for convenience in Fig. 5, which exhibit an almost hard-wall-like repulsion at short separations.

On the other hand, in support of the hypothesis that the polarization of water might play an important role in the hydration force between silica surfaces, one should note that the polarization model predicts an increase in the hydration force at higher ionic strength [30], which can be indeed observed in Fig. 5, by comparing the experiments at c_E=0.01 M (pH=3) with those at c_E=1 M. While both the gel-induced steric repulsion and the polarization model are consistent with the present experimental data, a final decision about the microscopic origin of the hydration force in the case of silica should be postponed until more accurate data or additional information regarding the nature of the silica surface will become available.

5. Conclusions

This paper has been stimulated by recent experiments on the interactions between silica surfaces, which could not be represented by the polarization model proposed by the authors [32,33]. In that model, the water molecules near the surface were considered to be organized in ice-like layers and consequently a large dipole correlation length, which is a measure of the range of interaction between dipoles, λ_m=14.9 Å, was obtained. The main goal of this paper was to show how the dipole correlation length λ_m is affected by the disorder in the structure of the interfacial water, and values much smaller than 14.9 Å could be thus obtained. Since the evaluation of λ_m is based on a choice for the microscopic structure of water, which is often unknown, the polarization model was also presented as a phenomenological model, in which the dipole correlation length λ_m was used as a parameter. It was also shown that, for λ_m=4 Å, a more than qualitative agreement with the experiment can be obtained, for reasonable values of the parameters involved (e.g. dipole strength, dipole density, dipole locations, surface charges), which are consistent with the surface recombination models. The value λ_m=4 Å is consistent with the assumption of random orientations for the nearest neighbors of a water molecule.

The results suggest that a gel might form above the surface of silica, as proposed by Vigil et al. [36], which induces disorder in the interfacial water layers.

Acknowledgement

We are indebted to Valle-Delgado for kindly providing us the experimental data reported in Ref. [22] for the forces between silica surfaces.

References

[1] Deryagin BV, Landau L. Acta Physicochim. URSS 1941;14:633.
[2] Verwey EJ, Overbeek JThG. Theory of stability of lyophobic colloids. Amsterdam: Elsevier; 1948.
[3] Stern O. Z Eletrochem 1924;30:508.
[4] Ruckenstein E, Schiby D. Langmuir 1985;1:612.
[5] Jonsson B, Wennerstrom H. J Chem Soc, Faraday Trans II 1983;79:19.
[6] Spaarnay MJ. Recl Trav Chim Pays-Bas 1958;77:872.
[7] Kjellander R, Marcelja S. J Chem Phys 1985;82:2122.
[8] Ninham BW, Yaminsky V. Langmuir 1997;13:2097.
[9] Manciu M, Ruckenstein E. Adv Colloid Interface Sci 2003;105:63.
[10] Rand RP, Parsegian VA. Biochim Biophys Acta 1989;988:351.
[11] Marcelja S, Radic N. Chem Phys Lett 1976;42:129.
[12] Israelachvilli JN, Pashley RM. Nature 1983;306:249.
[13] Pashley RM. J Colloid Interface Sci 1981;83:531.
[14] Healy TW, Homola A, James RO, Hunter RJ. Faraday Discuss Chem Soc 1978;65:156.
[15] Attard P, Batchelor MT. Chem Phys Lett 1988;149:206.
[16] Schiby D, Ruckenstein E. Chem Phys Lett 1983;95:435.
[17] Gruen DWR, Marcelja S. Faraday Trans II 1983;79:211–25.
[18] Paunov VN, Kaler EW, Sandler SI, Petsev DN. J Colloid Interface Sci 2001;240:640.
[19] Napper DH. Steric stabilization of colloidal dispersions. London: Academic Press; 1983.
[20] Israelachvili JN, Wennerstrom H. Langmuir 1990;6:873.
[21] Binder H, Dietrich U, Schalke M, Pfeiffer H. Langmuir 1999;15:4857.
[22] Valle-Delgado JJ, Molina-Bolivar JA, Galisteo-Gonzalez F, Galvez-Ruiz MJ, Feiler A, Rutland MW. J Chem Phys 2005;123:034708.
[23] Attard P, Wei DQ, Patey GN. Chem Phys Lett 1990;172:69.
[24] Berkowitz ML, Raghavan K. Langmuir 1991;7:1042.
[25] Faraudo J, Bresme F. Phys Rev Lett 2004;92:236102.
[26] Kjellander R, Marcelja S. Chem Phys Lett 1985;120:393.
[27] Faraudo J, Bresme F. Phys Rev Lett 2005;94:077802.

[28] Perera L, Essmann U, Berkowitz ML. Prog Colloid & Polym Sci 1997;103:107.
[29] Israelachvili J. Intermolecular and surface forces. Academic Press; 1992.
[30] Manciu M, Ruckenstein E. Langmuir 2001;17:7061.
[31] Manciu M, Ruckenstein E. Langmuir 2001;17:7582.
[32] Ruckenstein E, Manciu M. Langmuir 2002;18:7584.
[33] Manciu M, Ruckenstein E. Adv Colloid Interface Sci 2004;112:109.
[34] Parks GA. J Geophys Res 1984;89:3997–4008.
[35] Rimstidt JD, Barnes HL. Geochim Cosmochim Acta 1980;44:1683.
[36] Vigil G, Xu Z, Steinberg S, Israelachvili J. J Colloid Interface Sci 1994;165:367.
[37] Yoon R-H, Vivek S. J Colloid Interface Sci 1998;204:179.
[38] Sermon PA. J Chem Soc, Faraday Trans I 1980;76:885.
[39] Overbeek JTG. Colloid Surf 1990;51:61.
[40] Manciu M, Ruckenstein E. Langmuir 2003;19:1114.
[41] Ruckenstein E, Manciu M. Adv Colloid Interface Sci 2003;105:177.
[42] Manciu M, Ruckenstein E. Langmuir 2002;18:8910.
[43] Manciu M, Ruckenstein E. Langmuir 2005;21:11749.
[44] Jackson JD. Classical electrodynamics. New York: John Wiley and Sons; 1975.
[45] Attard P, Patey GN. Phys Rev, A 1991;43:2953.
[46] Ducker WA, Senden TJ, Pashley RM. Langmuir 1992;8:1831.

Introduction to CHAPTER 8
Polymer brushes

8.1 B. Li, E. Ruckenstein: "*Statistical thermodynamics of end-attached chain monolayers,*" JOURNAL OF CHEMICAL PHYSICS 106:1 (1997) 280–288.

8.2 E. Ruckenstein, B. Li: "*Steric interactions between two grafted polymer brushes,*" JOURNAL OF CHEMICAL PHYSICS 107: 3 (1997) 932–942.

8.3 M. Manciu, E. Ruckenstein: "*Simple Model for Grafted Polymer Brushes,*" LANGMUIR 20 (2004) 6490–6500.

8.4 M. Manciu, E. Ruckenstein: "*On the Monomer Density of Grafted Polyelectrolyte Brushes and Their Interactions,*" LANGMUIR 20 (2004) 8155–8164.

8.5 H. Huang, E. Ruckenstein: "*Double-layer interaction between two plates with hairy surfaces*" JOURNAL OF COLLOID AND INTERFACE SCIENCE 273 (2004) 181–190.

8.6 H. Huang, E. Ruckenstein: "*Double layer interaction between two plates with polyelectrolyte brushes*" JOURNAL OF COLLOID AND INTERFACE SCIENCE 275 (2004) 548–554.

8.7 H. Huang, E. Ruckenstein: "*The bridging force between two plates by polyelectrolyte chains*" ADVANCES IN COLLOID AND INTERFACE SCIENCE 112 (2004) 37–47.

8.8 H. Huang, E. Ruckenstein: "*Steric and Bridging Interactions between Two Plates Induced by Grafted Polyelectrolytes,*" LANGMUIR 22 (2006) 3174–3179.

8.9 H. Huang, E. Ruckenstein: "*Effect of Steric, Double Layer, and Depletion Interactions on the Stability of Colloids in Systems Containing a Polymer and an Electrolyte,*" LANGMUIR 22 (2006) 4541–4546.

8.10 M. Manciu, E. Ruckenstein: "*Estimation of the available surface and the jamming coverage in the Random Sequential Adsorption of a binary mixture of disks,*" COLLOIDS AND SURFACES A-PHYSICOCHEMICAL AND ENGINEERING ASPECTS 232 (2004) 1–10.

A general theory, based on a simple cubic lattice model and matrix formalism is developed. The theory accounts for the chain stiffness, the nearest-neighbor bond correlations and the intermolecular interactions. The model allows one to calculate the density profiles of the lateral, forward and backward bonds as functions of the layer number and the orientational probability for a bond to be lateral, forward or backward as functions of the bond number counted from the attached end. It is shown that the effect of the bond correlations becomes less important with decreasing chain length [8.1]. The model is employed to calculate the steric interactions between two grafted polyelectrolyte brushes, by accounting for the bond correlations and the interdigitation between brushes, and it is shown that the results are in agreement with experiment. Comparison with other theoretical models (that neglect interdigitation) indicates that the interdigitation is not negligible and can increase the repulsive force by an order of magnitude [8.2].

A simple theoretical treatment of a polymer brush is proposed, which consists of constructing its partition function by considering all possible distributions of a random walk ending at every distance z from the surface. This distance is assumed to be the position of the last monomer of the chain. A free energy minimized with respect to the local volume fraction of the monomers, calculated using the Flory-Huggins expression, is associated to each of these distributions; it is also assumed that the number of distributions with energy close to the minimum is proportional to the total number of paths that end at the distance z. The model can be employed for both good and poor solvents, and is compatible with both a parabolic-like brush profile at moderate graft densities, and with an almost step-like density for highly stretched brushes. One concludes that in good and moderately poor solvents, the interactions between grafted polymer brushes are always repulsive, while in poor solvents the interactions are repulsive at small separations but attractive at intermediate separations, results in agreement with experiment [8.3]. The model is extended to charged polyelectrolyte brushes, and the dependence of the brush thickness on electrolyte concentration is obtained. One concludes that the trapping of a fraction of counterions in the brush strongly affects the thickness of the brush. Consequently, if two brushes approach each other more rapidly than the ions diffuse in the direction parallel to the surface, the trapping of the counterions between brushes can affect the interactions by orders of magnitude [8.4].

E. Ruckenstein, M. Manciu, *Nanodispersions*, DOI 10.1007/978-1-4419-1415-6_8, © Springer Science+Business Media, LLC 2010

When hairy chains are extended into the continuous phase, they orient the water molecules and therefore lower the local dielectric constant of water. The local change in the dielectric constant has strong effects on the ion distribution (via changes in the Born hydration energy and on the Poisson equation) and therefore on the repulsion between two plates [8.5]. A model for polyelectrolyte brushes is proposed, which assumes that the polyelectrolyte chain is a rigid cylinder, on whose surface charges are generated through the dissociation of ionizable sites and adsorption of the cations of the electrolyte. In the brush region, the electric potential is described by a two-dimensional Poisson-Boltzmann equation (in the plane of the surface), while in the region free of polyelectrolyte chains it is described by a one-dimensional Poisson-Boltzmann equation (in the direction normal to the surface). The effects of electrolyte concentration, *pH*, brush thickness and chain coverage density on the repulsion between plates are examined [8.6].

The chains of the grafted polyelectrolyte brushes can be adsorbed on the other plate, thus generating bridging (hence attraction) between plates. A self-consistent field approach is employed to investigate the effect of bridging, by taking into account the van der Waals interactions between the segments of the polyelectrolyte molecules and the plates, and the electrostatic and volume exclusion interactions [8.7]. An alternate approach to steric and bridging interactions also is proposed, which consists of calculating the free energy of the system by using a number of contributions (the Flory-Huggins mixing free energy for the grafted chains and liquid; the van der Waals interactions between segments and plates; the connectivity free energy of the segments; and the adsorption energy. For charged plates, the electrostatic free energy and the free energy of the electrolyte are also included). The free energy minimization with respect to segment concentration and electric potential provides expressions for them. Once these quantities are known, the interactions between plates with grafted polyelectrolyte molecules, possessing both steric and bridging forces, are calculated [8.8]. Experiments have shown that polystyrene latexes are restabilized at sufficiently high electrolyte concentrations in the presence of an amphiphilic block copolymer, but are again destabilized at even higher electrolyte concentrations. These phenomena occur because of steric repulsion at intermediate electrolyte concentrations and depletion attractive interactions at higher ionic strengths [8.9].

The area available to a disk (or to a sphere) to be adsorbed on a surface during a random sequential adsorption (RSA) of a binary mixture of disks (spheres) is in general estimated via Monte-Carlo simulations, which are very time consuming. A simple analytical expression is proposed, which provides good estimates of the jamming points for the RSA of a binary mixture of disks or spheres. The predictions of the simple formula have been compared with the Monte-Carlo simulations existing in the literature [8.10].

Statistical thermodynamics of end-attached chain monolayers

Buqiang Li[a)] and Eli Ruckenstein[b)]
Department of Chemical Engineering, State University of New York at Buffalo, Amherst, New York 14260

(Received 5 April 1996; accepted 25 September 1996)

A simple cubic lattice model and a matrix formalism are used to calculate the segment density distribution and structural properties of monolayers consisting of end-attached chains and solvent molecules. The chain stiffness, the nearest-neighbor bond correlations, and the intermolecular interactions are taken into account. The calculations are performed for chains of various lengths, from short to polymer brushes. The main difference between the present treatment and those already available consists in the more compact formulation of the basic equations. The incorporation of the nearest-neighbor bond correlations, or higher values of the interaction parameter results in a higher average segment density and a narrower distribution profile of the chain free ends. With decreasing chain length, the effect of the bond correlations becomes less important. The model allows us to calculate the density profiles of the lateral, forward, and backward bonds as a function of the layer number, and to calculate the orientational probability for a bond to be lateral, forward, or backward as a function of the bond number counted from the attached end. © *1997 American Institute of Physics.* [S0021-9606(97)50801-4]

I. INTRODUCTION

The statistics of chain conformations is of importance in the investigation of surfactant monolayers, bilayer membranes, micellar aggregates as well as polymer (adsorbed or grafted) monolayers, etc. Two cases are commonly encountered. In one of them, a compact hydrocarbon domain of uniform segment density is separated from water by polar headgroups. The hydrocarbon chains have a free end which is located in the hydrocarbon domain and another end located at the interface with water. The packing constraints result in the variation of the chain structural order within the hydrocarbon domain along a coordinate determined by the geometry of the system. In the other one, solvent molecules are present among the hydrocarbon chains and both the segment density and the chain conformation depend on the distance to the surface. Many theories[1–14] for the chain conformational statistics have been developed.

The early theories[1–8] involved a single chain mean-field approximation. Using the rotational isomeric state scheme, Gruen[1] calculated the probability distribution function of the chain conformation and the conformational free energy of the surfactant tails in an aggregate, assuming that the hydrocarbon tails form a liquidlike hydrocarbon core of uniform density. In a series of papers, Ben-Shaul *et al.*[2–4] improved the preceding approach and provided more detailed calculations. An extension[5] by the same authors considered a nonuniform segment density. Assuming a simple cubic lattice and using the generator-matrix formalism, Dill *et al.*[6–8] developed a chain statistics for the case free of bond backtracks. All the above approaches neglected the nearest-neighbor bond correlations.

Some recent treatments[9–14] have taken into account the effect of the nearest-neighbor bond correlations on the probability for a site to be occupied. The self-consistent field (SCF) theory of Scheutjens and Fleer[9] could generate, on the basis of a rotational isomeric state scheme, a full set of conformations, whose weights were provided by the Boltzmann statistics. Leermakers and Scheutjens[10] improved the preceding approach, generating the self-consistent anisotropic (SCAF) theory. In the SCAF theory, the probability for a site to be occupied by a segment depends not only on the overall segment concentration in the layer in which it will be located, but also on the density of bonds in the direction of the bond to which the considered segment is connected (nearest neighbor bond correlation). The presence of the bond correlations results in a significant modification of the chain conformational entropy. They have applied the theory to micelles[11] and to polymer brushes.[12]

In a paper regarding phase transitions in monolayers,[13] Cantor and McIlroy incorporated the bond correlations and intermolecular interactions using an approach similar to that of Ref. 10. Finally, Cantor has incorporated the bond correlations in the generator-matrix formalism to calculate the elastic properties[14] of films of athermal surfactant mixtures.

In this paper, a physical picture equivalent to those of Leermakers and Scheutjens[10,12] and Cantor[13] is employed. However, a more simple generator-matrix methodology is used to examine the effects of the intramolecular interactions, the nearest-neighbor bond correlations, and the intermolecular interactions on the structural properties of end-attached chain monolayers. While equivalent to that developed in Refs. 10 and 12, the method is more simple and compact, and hence more convenient from a computational point of view.

II. THEORY

A. Partition function

We consider a two-dimensional mixture which contains N end-attached chains and N_0 solvent molecules and employ the simple cubic lattice model for its representation. The lat-

a)Permanent address: Chemical Engineering Department, Beijing, University of Chemical Technology, Beijing 100029, People's Republic of China.
b)Author to whom correspondence should be addressed.

tice is divided into M layers parallel to the surface, numbered from the surface as $i=1,2,\ldots,M$, each lattice layer containing L sites. The chains consist of $\gamma+1$ segments which are connected through γ bonds and occupy $\gamma+1$ sites. The first segment of each chain is constrained to be located in the first lattice layer. Each solvent molecule is considered to occupy one site. The number of lattice layers, M, is taken equal to the number of segments of the chain; consequently, one can write the relation

$$N_0+N(\gamma+1)=L(\gamma+1). \tag{1}$$

It is also assumed that the bulk solution contains only the pure solvent molecules.

Denoting by N_c the number of chains in conformation c and by N_i^0 the number of solvent molecules in layer i, one can write that $N=\Sigma_c N_c$ and $N_0=\Sigma_i N_i^0$. For a specified set of conformations $\{N_c\}$, the number of chains in each conformation and the number of solvent molecules in each layer are given. When only the nearest neighbor interactions are taken into account and the segments and the solvent molecules are distributed at random in each of the layers, the system has the same energy for each of the possible arrangements. The canonical partition function $Q(T,L,M,\{N_c\})$ can be written as

$$Q(T,L,M,\{N_c\})=\frac{\Omega}{\Omega^0}\exp\left(-\frac{\Delta U}{kT}\right), \tag{2}$$

where Ω is the combinatorial factor for the specified conformational profile $\{N_c\}$, Ω^0 is the product of the combinatorial factors of the chains and solvent molecules in their reference states, and ΔU is the energy change of the chains and solvent molecules from their reference states to the specified conformational profile in the chain-solvent mixture.

The combinatoral factor, Ω depends on the number of ways (ω) in which one can place successively all the chains and then the solvent molecules on the lattice. ω can be factorized as follows:

$$\omega=\omega_c\omega_s\omega_b, \tag{3}$$

where ω_c is the number of ways in which one can place successively the chains and then the solvent molecules on the lattice, considering that the sites occupied by the previous segments are empty, hence, that they can be occupied again (no excluded volume interactions are, hence, involved); ω_s is the product of the vacancy probabilities for all the segments and solvent molecules to be placed one after another on the lattice; ω_b is the product of correction factors resulting from the effect of the nearest-neighbor bonds, already present in a given direction, on the vacancy probability for a considered segment to be a part of a bond in the same direction.

Denoting by g_c the number of arrangements that a chain in conformation c can have when its first segment is fixed, one can write

$$\omega_c=L^N\prod_c g_c^{N_c}\prod_{i=1}^{M} L^{N_i^0}. \tag{4}$$

In the preceding equation, g_c is given by the product, $g_c=\Pi_{s=1}^{\gamma} z_{s+1,s}^c$, where $z_{s+1,s}^c$ stands for the number of sites which segment $s+1$ of a chain in conformation c can occupy around segment s. Indeed, the second segment has $z_{2,1}^c$ possible locations when the first segment is fixed, the third segment has $z_{3,2}^c$ possible locations, and the last segment $\gamma+1$ has $z_{\gamma+1,\gamma}^c$ possible locations.

ω_s arises because a lattice layer is completely empty only before the first segment is placed in the layer. When a layer is only partly empty, a vacancy probability must be included for a segment to be placed in the layer. Denoting by v_i the number of segments already present in the layer, the vacancy probability for the subsequent segment to be placed in that layer is equal to $(L-v_i)/L$; consequently, for all the segments and all the solvent molecules to be placed on the lattice successively, one obtains

$$\omega_s=\prod_{i=1}^{M}\prod_{v_i=0}^{L-1}\left(\frac{L-v_i}{L}\right). \tag{5}$$

ω_b arises because the density of bonds already present in a given direction increases the vacancy probability for the end segment of a bond to be placed in the same direction. The number of bonds in direction ξ which have their starting segments located in layer i is denoted by $v_{\xi i}$. In the cubic lattice, ξ can be x or y for the lateral bonds, which are located in layer i in the direction x or y, and z for the vertical bonds between layers i and $i+1$. For a bond to be placed from layer i in the direction ξ, a correction factor, $L/(L-v_{\xi i})$ must be included in the vacancy probability because fewer sites than L are accessible. For all the chains to be placed successively on the lattice, one obtains that ω_b is given by

$$\omega_b=\prod_{i=1}^{M}\prod_{\xi}\prod_{v_{\xi i}=0}^{N_{\xi i}-1}\left(\frac{L}{L-v_{\xi i}}\right), \tag{6}$$

where $N_{\xi i}$ is the number of bonds in direction ξ with the starting segments in layer i.

Substituting Eqs. (4), (5), and (6) into Eq. (3) and accounting for the indistinguishability among the chains in the same conformation and among the solvent molecules in each layer, the combinatoral factor Ω is given by

$$\Omega=(L!)^M\left(\prod_c\frac{g_c^{N_c}}{N_c!}\right)\left(\prod_{i=1}^{M}\frac{1}{N_i^0!}\right)\left[\prod_{i=1}^{M}\prod_{\xi}\frac{(L-N_{\xi i})!}{L!}\right]. \tag{7}$$

In what follows, the ordered close-packed hydrocarbon monolayer, in which each chain has an all-*trans* conformation and is oriented normal to the surface, is taken as the reference state of the chains. In the reference state, both the chain conformational energy and entropy are taken equal to zero. The pure solvent, with $(L-N)$ solvent molecules in contact with the surface, is taken as the reference state of the solvent molecules. For the reference states, $\Omega^0=1$.

Using Stirling's approximation for the factorials in Eq. (7), the logarithm of Ω/Ω^0, corresponding to the entropy of the chain-solvent mixture, is given by the expression

$$\ln\left(\frac{\Omega}{\Omega^0}\right) = -\sum_c N_c \ln\left(\frac{N_c}{g_c L}\right) - \sum_{i=1}^{M} N_i^0 \ln(\Phi_i^0)$$
$$+\sum_{i=1}^{M} \sum_{\xi} (L - N_{\xi i}) \ln(1 - \Phi_{\xi i}), \qquad (8)$$

where $\Phi_i^0 = N_i^0 / L$ is the fraction of sites occupied by solvent molecules in layer i and $\Phi_{\xi i} = N_{\xi i}/L$ is the density of bonds in direction ξ with at least one segment in layer i.

For a given conformational profile $\{N_c\}$, the mixing energy between the chains and solvent molecules, ΔU, involves three contributions; (i) the change of the contact energies of the segments and solvent molecules with the surface (ΔU_s); (ii) the mixing energy (ΔU_{mix}) of the chains and solvent (intermolecular interactions); and (iii) the conformational energy change (ΔU_{con}) of the chains (intramolecular interactions). Consequently,

$$\Delta U = \Delta U_s + \Delta U_{\text{mix}} + \Delta U_{\text{con}}. \qquad (9)$$

The contact energy change of the segments and solvent molecules with the surface is simply given by

$$\Delta U_s = (N_1 - N) U_s + N_1^0 U_s^0 - (L - N) U_s^0, \qquad (10)$$

where N_1 and N_1^0 are the numbers of segments and solvent molecules in the first lattice layer, respectively, and U_s and U_s^0 are the contact energies of a segment and a solvent molecule with the surface, respectively. The first segment of a chain does not contribute to ΔU_s because it remains connected to the surface. The last term on the right-hand side of Eq. (10) represents the contact energy of $(L-N)$ solvent molecules with the surface in its reference state. Since $N_1 + N_1^0 = L$, Eq. (10) can be rewritten as

$$\frac{\Delta U_s}{kT} = -(N_1 - N)\chi_s, \qquad (11)$$

where $\chi_s = (U_s^0 - U_s)/kT$, and $U_s^0 - U_s$ represents the energy change of replacing a segment by a solvent molecule at the surface.

Considering only the nearest neighbor interactions for the mixing energy of the chain-solvent mixture, one obtains

$$\frac{\Delta U_{\text{mix}}}{kT} = \sum_{i=1}^{M} N_i \langle \Phi_i^0 \rangle \chi - N\lambda_1 \chi, \qquad (12)$$

where N_i stands for the number of segments in layer i, χ is the Flory–Huggins segment-solvent interaction parameter, and $\langle \Phi_i^0 \rangle$, the average fraction of sites occupied by the solvent molecules around a site located in layer i, is given by

$$\langle \Phi_i^0 \rangle = \lambda_1 \Phi_{i-1}^0 + \lambda_0 \Phi_i^0 + \lambda_1 \Phi_{i+1}^0. \qquad (13)$$

In Eqs. (12) and (13), λ_1 and λ_0 are the fractions of nearest-neighbor sites in an adjacent and the same layer, respectively. Since no solvent molecules are present in layers $i \leqslant 0$ and no segments are present in layers $i \geqslant \gamma+2$, $\Phi_0^0 = 0$, and $\Phi_{\gamma+2}^0 = 1$. In addition, $\langle \Phi_i^0 \rangle + \langle \Phi_i \rangle = 1 - \lambda_1$ for $i=1$ and $\langle \Phi_i^0 \rangle + \langle \Phi_i \rangle = 1$ for $i \geqslant 2$, where Φ_i is the fraction of sites occupied by segments in layer i.

The second term on the right-hand side of Eq. (12) represents the exchange energy between the chains in their reference state and the solvent molecules covering them. The summation in the first term already includes the term $N_{\gamma+1}\lambda_1\chi$, which accounts for the exchange energy between the last lattice layer $\gamma+1$ of the chain-solvent mixture and the bulk pure solvent.

The chain conformational energy change is given by the following expression:

$$\frac{\Delta U_{\text{con}}}{kT} = \sum_c \frac{N_c E_c}{kT}, \qquad (14)$$

where E_c stands for the conformational energy of a chain in conformation c.

Combining Eqs. (9), (11), (12), and (14), one obtains

$$\frac{\Delta U}{kT} = -(N_1 - N)\chi_s + \left(\sum_{i=1}^{M} N_i \langle \Phi_i^0 \rangle - N\lambda_1 \right) \chi$$
$$+\sum_c \frac{N_c E_c}{kT}. \qquad (15)$$

B. Equilibrium distribution

The mixing free energy of the chains and the solvent molecules for a given chain conformational profile $\{N_c\}$ is given by $F = -kT \ln Q$. The equilibrium conformational profile can be obtained by minimizing the free energy F subject to the following constraints:

$$\sum_c N_c \gamma_{ci} + N_i^0 - L = 0, \quad i = 1,2,\ldots,\gamma+1 \qquad (16)$$

and

$$N - \sum_c N_c = 0. \qquad (17)$$

In Eq. (16), γ_{ci} stands for the number of segments of a chain in conformation c located in layer i. The two equations express the obvious conditions that each lattice layer must be occupied and that the total number of chains is constant. The Lagrange multiplier method is used to calculate the minimum free energy subject to the above constraints. By introducing the multipliers α_i, for each of the constraints given by Eq. (16), and β for the constraint expressed by Eq. (17), one can write

$$f = \ln(Q) + \sum_{i=1}^{M} \alpha_i \left(\sum_c N_c \gamma_{ci} + N_i^0 - L \right)$$
$$+\beta \left(N - \sum_c N_c \right), \qquad (18)$$

where $\ln(Q)$ can be obtained from Eqs. (8) and (15). The equilibrium conformational profile of the chains and the equilibrium distribution of the solvent molecules can be calculated using the equations

$$\frac{\partial f}{\partial N_c}=\frac{\partial \ln(Q)}{\partial N_c}+\sum_{i=1}^{M}\alpha_i\gamma_{ci}-\beta=0 \tag{19}$$

and

$$\frac{\partial f}{\partial N_j^0}=\frac{\partial \ln(Q)}{\partial N_j^0}+\alpha_j=0,\quad j=1,2,\ldots,\gamma+1. \tag{20}$$

Introducing $\ln(Q)$ into the above two equations and observing that $\partial N_i/\partial N_c=\gamma_{ci}$ and $\partial N_{\xi i}/\partial N_c=\gamma_{c\xi i}$, one obtains

$$\ln\left(\frac{N_c}{Lg_c}\right)=-1-r-\beta-\frac{E_c}{kT}+\gamma_{c1}\chi_s+\sum_{i=1}^{M}\gamma_{ci}(\alpha_i-\langle\Phi_i^0\rangle\chi)-\sum_{i=1}^{M}\sum_{\xi}\gamma_{c\xi i}\ln(1-\Phi_{\xi i}) \tag{21}$$

and

$$\ln(\Phi_j^0)=-1+\alpha_j-\langle\Phi_j\rangle\chi,\quad j=1,2,\ldots,\gamma+1. \tag{22}$$

Eliminating α_j between the above two equations, yields

$$\frac{N_c}{L}=e^{-\beta}g_c\exp\left(-\frac{E_c}{kT}\right)\eta_s^{\gamma_{c1}}\left(\prod_{i=1}^{M}p_i^{\gamma_{ci}}\right)\times\left[\prod_{i=1}^{M}\prod_{\xi}(p_{\xi i})^{\gamma_{c\xi i}}\right], \tag{23}$$

where

$$\eta_s=\exp(\chi_s)=\exp\left(\frac{U_s^0-U_s}{kT}\right), \tag{23a}$$

$$p_i=\Phi_i^0\exp[(\langle\Phi_i\rangle-\langle\Phi_i^0\rangle)\chi], \tag{23b}$$

and

$$p_{\xi i}=\frac{1}{1-\Phi_{\xi i}}. \tag{23c}$$

In the preceding equations, η_s, p_i, and $p_{\xi i}$ are weight factors which account for the difference between the interactions of the segments and of the solvent molecules with the surface, for the intermolecular interactions between the segments and the solvent molecules and for the nearest-neighbor bond correlations, respectively.

Assuming that on average, half of the lateral bonds in layer i are in the x direction and the other half in the y direction, and denoting by Φ_{Li} the total lateral bond density in layer i, one can write

$$p_{xi}=p_{yi}=\frac{1}{1-\Phi_{Li}/2}\equiv p_{Li}. \tag{24}$$

According to Eq. (23), the equilibrium number of chains in conformation c depends on the number g_c of arrangements of a chain in that conformation, the conformational energy E_c, the surface parameter η_s, as well as the weight factors due to the chain-solvent interactions (p_i) and to the nearest-neighbor bond correlations ($p_{\xi i}$). In the above formulation of the problem, the intramolecular and intermolecular interactions as well as the bond correlations are taken into account.

$e^{-\beta}$ can be eliminated by combining Eqs. (17) and (23). One thus obtains the following equilibrium distribution:

$$\frac{N_c}{N}=\frac{W_c}{\Xi}, \tag{25}$$

where W_c represents the statistical weight of a chain in conformation c, given by

$$W_c=g_c\exp\left(-\frac{E_c}{kT}\right)\eta_s^{\gamma_{c1}}\left(\prod_{i=1}^{M}p_i^{\gamma_{ci}}\right)\left(\prod_{i=1}^{M}\prod_{\xi}p_{\xi i}^{\gamma_{c\xi i}}\right) \tag{26}$$

and Ξ is the chain conformation partition function defined as

$$\Xi=\sum_c W_c. \tag{27}$$

C. Chain conformational statistics

The chain width of a hydrocarbon molecule (4.6 Å) is taken as the linear dimension of a lattice site. Consequently, in the simple cubic lattice, a cube with a side equal to 4.6 Å represents the unit cell. Since the effective length per methylene group in an all-*trans* chain is 1.275 Å, 3.6 methylene groups are located in a cube and will be considered to represent a segment. In the lattice, a site has four nearest neighbors in the same layer and one in each of the adjacent layers. In addition, only two relative orientations of two consecutive bonds can occur, namely the collinear and the bent one at 90°. The bending energies of the collinear and bent bond pairs are taken zero and ϵ, respectively.

Denoting by $s\rightarrow s+1$ the bond which connects segment s to segment $s+1$, there are four conformations of the bond pair ($s-1\rightarrow s\rightarrow s+1$) (summarized in Table I). (1) The bond pair is bent, segment $s-1$ is located in a layer adjacent to segment s and segment $s+1$ has 4 possible locations ($z_{s+1,s}=4$); (2) the bond pair is bent, segments $s-1$, s, and $s+1$ are located in the same layer and segment $s+1$ has two possible locations ($z_{s+1,s}=2$); (3) the bond pair is bent, segments $s-1$ and s are located in the same layer and segment $s+1$ has only one possible location ($z_{s+1,s}=1$); (4) the bond pair is collinear and segment $s+1$ has only one possible location ($z_{s+1,s}=1$). Bond zero, which connects the first segment with the surface, is considered to be forward. When the first bond is lateral, $z_{2,1}=4$ and an energy ϵ_h which may differ from ϵ is involved. Denoting by n_{cb1},n_{cb2},n_{cb3} the numbers of bond pairs in conformation c corresponding to types 1, 2, and 3, respectively, and by n_{ch}(=0 or 1) the number of bent bond pairs ($0\rightarrow 1\rightarrow 2$) in conformation c, one can write

$$g_c=4^{n_{cb1}}2^{n_{cb2}}4^{n_{ch}},\quad E_c=(n_{cb1}+n_{cb2}+n_{cb3})\epsilon+n_{ch}\epsilon_h. \tag{28}$$

Substituting g_c and E_c into Eq. (26), the statistical weight of a chain in conformation c becomes

TABLE I. Types of bond pairs.

Type	Conformation	$z_{s+1,s}$	Bending energy
1	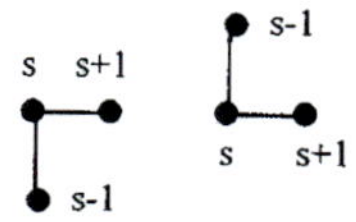	4	ε
2	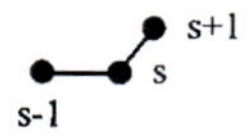	2	ε
3	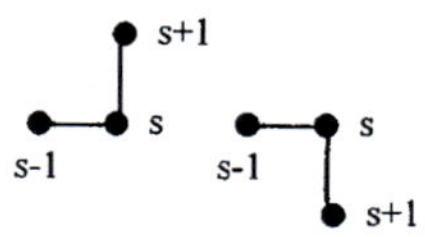	1	ε
4	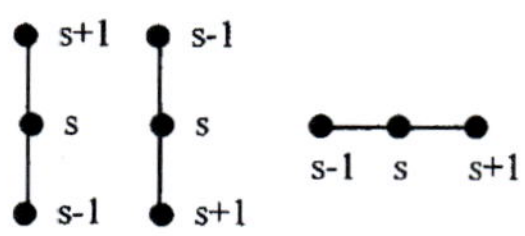	1	0

$$W_c=\alpha_h^{n_{ch}}\alpha^{n_{cb1}}\beta^{n_{cb2}}\omega^{n_{cb3}}\eta_s^{\gamma_{c1}}\left(\prod_{i=1}^{M} p_i^{\gamma_{ci}}\right)\times\left(\prod_{i=1}^{M}\prod_{\xi} p_{\xi i}^{\gamma_{c\xi i}}\right), \tag{29}$$

where $\alpha_h=4\omega_h$, $\alpha=4\omega$, $\beta=2\omega$. $\omega_h=\exp(-\epsilon_h/kT)$ and $\omega=\exp(-\epsilon/kT)$ are the weight factors of a pair of bent bonds relative to a pair of collinear bonds; ω characterizes the chain stiffness. A small value of ω (large bond bending energy ϵ) corresponds to a stiff chain.

Using the generator-matrix formalism of the chain conformational statistics, the chain conformation partition function Ξ can be expressed as a product of three matrices

$$\Xi=\sum_c W_c=\mathbf{A}_0\mathbf{G}^{\gamma}\mathbf{B}, \tag{30}$$

where $\mathbf{G}$ is the supermatrix

$$\mathbf{G}=\begin{pmatrix} \mathbf{q}_1 & \mathbf{p}_1 & \mathbf{0} & \cdots & \cdots & \cdots & \mathbf{0} \\ \mathbf{r}_2 & \mathbf{q}_2 & \mathbf{p}_2 & \cdots & \cdots & \cdots & \mathbf{0} \\ \cdots & \ddots & \ddots & \ddots & \cdots & \cdots & \cdots \\ \cdots & \cdots & \ddots & \ddots & \ddots & \cdots & \cdots \\ \mathbf{0} & \cdots & \cdots & \mathbf{r}_{\gamma-1} & \mathbf{q}_{\gamma-1} & \mathbf{p}_{\gamma-1} & \mathbf{0} \\ \mathbf{0} & \cdots & \cdots & \cdots & \mathbf{0} & \mathbf{q}_{\gamma} & \mathbf{p}_{\gamma} \\ \mathbf{0} & \cdots & \cdots & \cdots & \cdots & \cdots & \mathbf{0} \end{pmatrix}$$

whose elements are 2×2 submatrices given by

$$\mathbf{q}_1=\eta_s p_1 p_{L1}\begin{pmatrix} 0 & \alpha_h \\ 0 & \beta+1 \end{pmatrix};$$

$$\mathbf{q}_i=p_i p_{Li}\begin{pmatrix} 0 & \alpha \\ 0 & \beta+1 \end{pmatrix}, \quad i=2,3,\ldots,\gamma,$$

$$\mathbf{p}_i=p_i p_{zi} v_{pr}\begin{pmatrix} 1 & 0 \\ \omega & 0 \end{pmatrix}, \quad i=1,2,\ldots,\gamma,$$

$$\mathbf{r}_2=\eta_s p_2 p_{z1} v_{rp}\begin{pmatrix} 1 & 0 \\ \omega & 0 \end{pmatrix};$$

$$\mathbf{r}_i=p_i p_{z,i-1} v_{rp}\begin{pmatrix} 1 & 0 \\ \omega & 0 \end{pmatrix}, \quad i=2,3,\ldots,\gamma-1,$$

$$\mathbf{0}=\begin{pmatrix} 0 & 0 \\ 0 & 0 \end{pmatrix},$$

and $\mathbf{A}_0$ and $\mathbf{B}$ are the vectors

$$\mathbf{A}_0=(\mathbf{a}_1,\mathbf{0},\ldots,\mathbf{0});\ \mathbf{a}_1=(1,0),\ \mathbf{0}=(0,0),$$

$$\mathbf{B}=\mathrm{col}(\mathbf{c}_1,\mathbf{c}_2,\ldots,\mathbf{c}_{\gamma+1});\ \mathbf{c}_i=\mathrm{col}(p_i,p_i).$$

In the elements $\mathbf{p}$ and $\mathbf{r}$, the Heaviside functions v_{pr} and v_{rp} are included to avoid bond backfolding since a bond cannot be backward when the previous bond was forward and vice versa. In the supermatrix $\mathbf{G}$, $\mathbf{q}_i$ characterizes a lateral bond in layer i, $\mathbf{p}_i$ a forward bond starting from layer i, and $\mathbf{r}_i$ a backward bond starting from layer i. $\mathbf{q}_i$, $\mathbf{p}_i$, and $\mathbf{r}_i$ depend on three kinds of parameters. The parameters α_h, α, β, and ω arise from local chain stiffness and bond arrangements, p_i from the intermolecular interactions and p_{Li}, p_{zi} from the nearest-neighbor bond correlations.

For the chains whose last segment is located in layer i, the statistical weight, $W_{\gamma i}$, can be expressed as

$$W_{\gamma i}=\mathbf{A}_0\mathbf{G}^{\gamma}\mathbf{C}_i, \tag{31}$$

where

$$\mathbf{C}_i=\mathrm{col}(\mathbf{0},\ldots,\mathbf{0},\mathbf{c}_i,\mathbf{0},\ldots,\mathbf{0});\ \mathbf{c}_i=\mathrm{col}(p_i,p_i),\ \mathbf{0}=\mathrm{col}(0,0).$$

The probability $P_{\gamma i}$ for a chain to end in layer i is given by

$$P_{\gamma i}=\frac{N_{\gamma i}}{N}=\frac{W_{\gamma i}}{\Xi}=\frac{\mathbf{A}_0\mathbf{G}^{\gamma}\mathbf{C}_i}{\mathbf{A}_0\mathbf{G}^{\gamma}\mathbf{B}}, \tag{32}$$

where $N_{\gamma i}$ stands for the number of chains whose free ends are located in layer i.

For the chains with the kth bond lateral in layer i, the chains with the kth bond forward from layer i to $i+1$ and the chains with the kth bond backward from layer i to $i-1$, the probabilities, P_{kLi}, and P_{kFi}, P_{kBi} are given, respectively, by

$$P_{kLi}=\mathbf{A}_0\mathbf{G}^{k-1}\mathbf{Q}_i\mathbf{G}^{\gamma-k}\mathbf{B}/\Xi, \quad i=1,2,\ldots,\gamma, \tag{33}$$

$$P_{kFi}=\mathbf{A}_0\mathbf{G}^{k-1}\mathbf{P}_i\mathbf{G}^{\gamma-k}\mathbf{B}/\Xi, \quad i=1,2,\ldots,\gamma, \tag{34}$$

and

$$P_{kBi}=\mathbf{A}_0\mathbf{G}^{k-1}\mathbf{R}_i\mathbf{G}^{\gamma-k}\mathbf{B}/\Xi, \quad i=2,3,\ldots,\gamma-1, \tag{35}$$

where $\mathbf{Q}_i$, $\mathbf{P}_i$ or $\mathbf{R}_i$ is a supermatrix which has only the (i,i), $(i,i+1)$ or $(i,i-1)$ (i.e., $\mathbf{q}_i$, $\mathbf{p}_i$ or $\mathbf{r}_i$) as a nonzero element.

Since segment k is located in layer i if the bond k, forward, lateral or backward, starts from layer i, the probability for segment k to be in layer i is given by the sum of the probabilities for bond k to start from layer i,

$$P_{ki}=P_{kLi}+P_{kFi}+P_{kBi}$$
$$=\frac{1}{\Xi}\mathbf{A}_0\mathbf{G}^{k-1}(\mathbf{Q}_i+\mathbf{P}_i+\mathbf{R}_i)\mathbf{G}^{\gamma-k}\mathbf{B}. \quad (36)$$

The total number of lateral bonds in layer i (N_{Li}), forward bonds from layer i to $i+1$ (N_{Fi}) or backward bonds from layer i to $i-1$ (N_{Bi}) is given by the sum of the corresponding bonds over the bond number k,

$$N_{Li}=N\sum_{k=i}^{\gamma} P_{kLi}, \quad i=1,2,\ldots,\gamma, \quad (37)$$

$$N_{Fi}=N\sum_{k=i}^{\gamma} P_{kFi}, \quad i=1,2,\ldots,\gamma, \quad (38)$$

and

$$N_{Bi}=N\sum_{k=i+1}^{\gamma} P_{kBi}, \quad i=2,3,\ldots,\gamma-1. \quad (39)$$

Combining Eqs. (33) and (37), one obtains for the number of lateral bonds per site in layer i (lateral bond density, $N_{Li}/L=\Phi_{Li}$) the expression

$$\Phi_{Li}=\frac{\sigma}{\Xi}\mathbf{A}_0\sum_{k=i}^{\gamma}\mathbf{G}^{k-1}\mathbf{Q}_i\mathbf{G}^{\gamma-k}\mathbf{B}, \quad i=1,2,\ldots,\gamma, \quad (40)$$

where $\sigma=N/L$ represents the number of chains per site at the surface (surface chain density). The sum of N_{Fi} and N_{Bi+1}, obtained using Eqs. (34), (35), (38), and (39), provides the vertical bond density between layer i and $i+1$ $[(N_{Fi}+N_{B,i+1})/L=\Phi_{zi}]$,

$$\Phi_{zi}=\frac{\sigma}{\Xi}\mathbf{A}_0\left(\sum_{k=i}^{\gamma}\mathbf{G}^{k-1}\mathbf{P}_i\mathbf{G}^{\gamma-k}+\sum_{k=i+1}^{\gamma}\mathbf{G}^{k-1}\mathbf{R}_{i+1}\mathbf{G}^{\gamma-k}\right)\mathbf{B}, \quad i=1,2,\ldots,\gamma. \quad (41)$$

Equations (40) and (41) represent a system of 2γ equations which contain $3\gamma+1$ unknowns, namely Φ_{Li} and Φ_{zi} with $i=1,2,\ldots,\gamma$, and Φ_i with $i=1,2,\ldots,\gamma+1$. In order to obtain both the segment and bond density distributions, additional equations, which are derived below, are required.

For aliphatic chains, a methylene group ($-CH_2-$) is connected to two covalent bonds and a terminal group ($-CH_3$) to a single covalent bond. Hence, a methylene group has two internal connection points and a terminal group has only one. Correspondingly, a lateral bond in layer i provides two connection points in that layer, while a vertical bond, either between layer i and $i+1$ or between layer i and $i-1$, provides only one connection point in layer i. A mass balance of the internal connection points provided by the segments present in layer i leads to

$$N_{\gamma i}+2(N_i-N_{\gamma i})=N_{z,i-1}+N_{zi}+2N_{Li}, \quad i=1,2,\ldots,\gamma+1 \quad (42)$$

which can be rewritten as

$$2\Phi_i=P_{\gamma i}\sigma+\Phi_{z,i-1}+\Phi_{zi}+2\Phi_{Li}, \quad i=1,2,\ldots,\gamma+1. \quad (43)$$

Since the first segment is connected to the surface and no lateral and forward bonds are present in the last layer, $\Phi_{z0}=\sigma$ and $\Phi_{z,\gamma+1}=\Phi_{L,\gamma+1}=0$. Equations (40), (41), and (43) provide a complete system of equations for the $3\gamma+1$ unknown quantities. They can be numerically solved to obtain the distributions of segment and bond densities $\{\Phi_i\},\{\Phi_{Li}\},\{\Phi_{zi}\}$ as a function of layer number to the surface.

The distributions of both segment and bond densities allow us to calculate any structural property of the monolayer. The bond orientational probabilities (forward, lateral, and backward) as well as the bond order parameter of a chain as a function of bond number along the chain are given below.

The sum of P_{kFi}, the sum of P_{kLi} (both over all the layers from 1 to k) and the sum of P_{kBi} over all the layers from 2 to $k-1$ provide the probabilities for the kth bond to be forward (P_{kF}), lateral (P_{kL}) and backward (P_{kB}), respectively,

$$P_{kF}=\frac{1}{\Xi}\mathbf{A}_0\mathbf{G}^{k-1}\left(\sum_{i=1}^{k}\mathbf{P}_i\right)\mathbf{G}^{\gamma-k}\mathbf{B}, \quad k=1,2,\ldots,\gamma, \quad (44)$$

$$P_{kL}=\frac{1}{\Xi}\mathbf{A}_0\mathbf{G}^{k-1}\left(\sum_{i=1}^{k}\mathbf{Q}_i\right)\mathbf{G}^{\gamma-k}\mathbf{B}, \quad k=1,2,\ldots,\gamma, \quad (45)$$

and

$$P_{kB}=\frac{1}{\Xi}\mathbf{A}_0\mathbf{G}^{k-1}\left(\sum_{i=2}^{k-1}\mathbf{R}_i\right)\mathbf{G}^{\gamma-k}\mathbf{B}, \quad k=3,4,\ldots,\gamma. \quad (46)$$

The bond order parameter as a function of bond number to the attached end is given by the following expression:

$$S_k=\tfrac{3}{2}(1-P_{kL})-\tfrac{1}{2}. \quad (47)$$

III. RESULTS AND DISCUSSIONS

In this section, calculations are carried out to determine the segment and bond density profiles as a function of distance to the surface, as well as the bond orientational probability and order parameter as a function of bond number to the attached end. The effects of the chain lengths, the nearest-neighbor bond correlations, the intermolecular interactions and the bond backtracks are examined. In most of the calculations, a constant chain stiffness, $\omega=\omega_h=0.5$, corresponding to a bond bending energy of $kT\ln 2$, a constant chain surface density, $\sigma=0.1$, and a constant surface parameter, $\eta_s=1$, are used. The chain length is varied between 20 bonds (21 segments) and a polymer brush with 400 bonds (401 segments). Figure 1 provides a comparison between the present calculations and those based on the limited bond

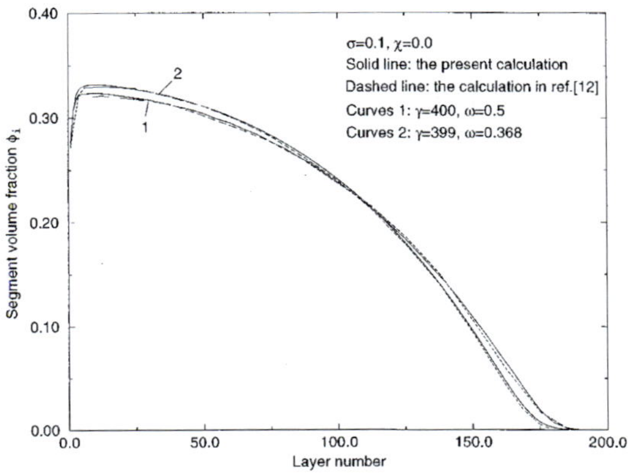

FIG. 1. A comparison between the present calculations and those of Wijmans *et al.* (Ref. 12) for two polymer brushes.

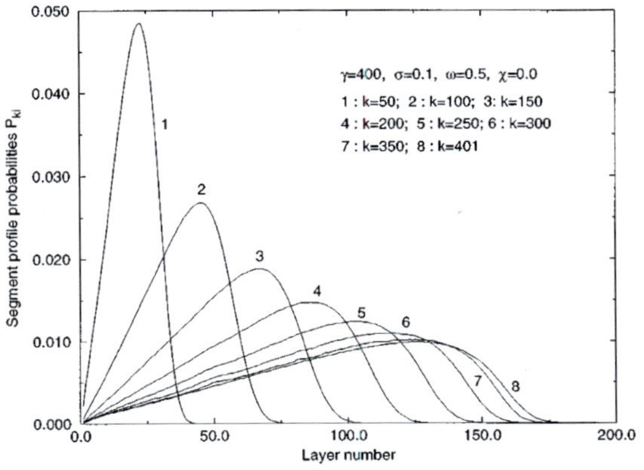

FIG. 3. The probability profile of a segment in the polymer brush with 400 bonds.

flexibility model of Wijmans *et al.*,[12] which is a cubic lattice version of the self-consistent anisotropic field (SCAF) theory of Leermakers and Scheutjens.[10] Two sets of curves are included. The first ignores the nearest-neighbor bond correlations and is for a polymer brush containing 401 segments with a chain stiffness $\omega=0.5$. The second one accounts for the bond correlations and is for a polymer brush containing 400 segments with a chain stiffness $\omega=0.368$ (hence a bond bending energy $\epsilon=kT$). The results of the present calculations and those of Wijmans *et al.*[12] almost coincide. In addition, the two sets of curves are near to one another, indicating that in those cases the increase in chain stiffness is compensated by the incorporation of the nearest-neighbor bond correlations.

IV. SEGMENT DENSITY PROFILE

The calculated segment volume fractions are presented in Fig. 2 as a function of layer number, for chains with 20 to 400 bonds. Two sets of theoretical curves are included. The solid curves account for the nearest-neighbor bond correlations, while the dashed ones do not involve them. One can see that the incorporation of the nearest-neighbor bond correlations results in a higher average segment density and consequently in a smaller brush height. The effect of the bond correlations becomes more significant as the chain length increases but is negligible for the short chain with 20 bonds. Regardless of whether the bond correlations are incorporated or not, the segment density first increases and subsequently decreases as a function of the distance to the surface.

Figure 3 presents the probability profiles for various segments in the polymer brush with $\gamma=400$ as a function of the layer number. The profile becomes wider with increasing segment number to the attached end, indicating a higher positional disorder for larger segment numbers. One can see that for segment numbers from $k=50$ to $k=300$, the profiles vary appreciably, while between $k=350$ and $k=401$ they are approximately the same.

V. BOND DENSITY PROFILE AND BOND ORIENTATION PROBABILITY

The density profiles of lateral, forward and backward bonds as a function of layer number to the surface are pre-

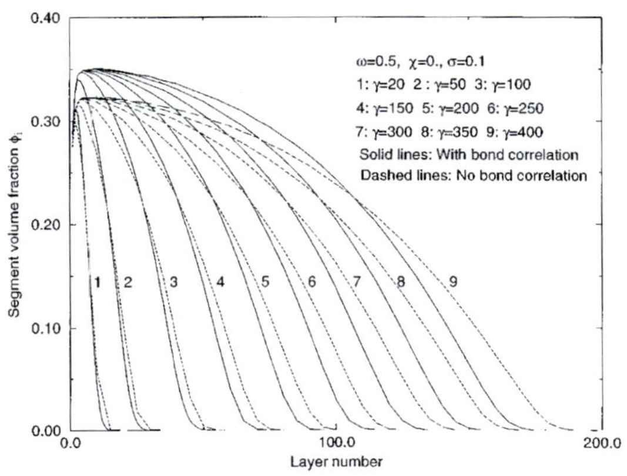

FIG. 2. The segment density profiles for chains of various length.

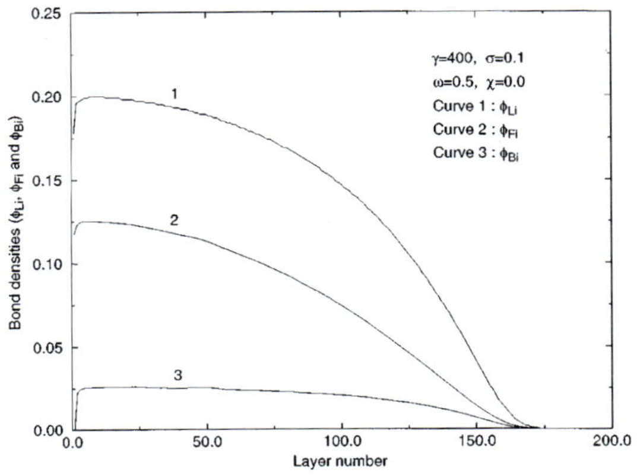

FIG. 4. The density profiles of forward, lateral, and backward bonds for the polymer brush with 400 bonds.

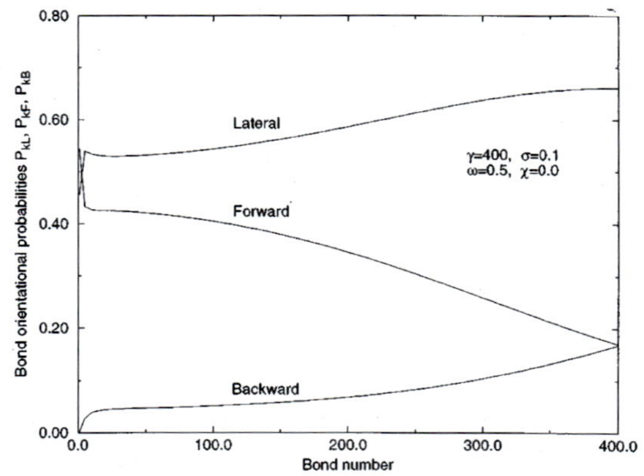

FIG. 5. Bond orientational probability as a function of bond number accounted from the attached end.

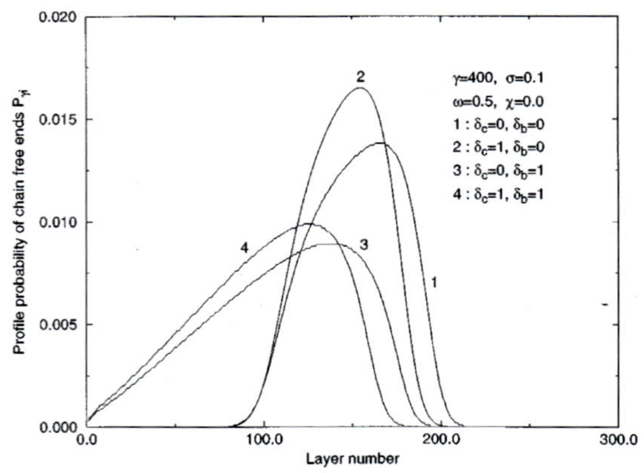

FIG. 7. The effect of bond correlations and bond backtracks on the profile probability of chain free ends for the polymer brush with 400 bonds.

sented in Fig. 4 for a polymer brush with 400 bonds. The sum of Φ_{Fi} and $\Phi_{B,i+1}$ provides the density of the vertical bonds between layers i and $i+1$. Because of the low segment density in the first layer (see Fig. 2), both the lateral bond density in the first layer and the forward bond density from layer 1 to 2 have low values. Φ_{B1} is equal to zero because no backward bond from the first layer is possible. The backward bond density has much lower values compared to those of lateral and forward bonds, indicating that most of the bonds are forward or lateral because of the constraint imposed by the surface. In contrast to the lateral or forward bond densities which decrease rapidly with increasing distance to the surface, the backward bond density remains approximately constant in a large range of layer numbers.

Bond orientational probabilities as a function of bond number to the attached end are plotted in Fig. 5 for a brush with 400 bonds. The first bond has a large forward probability. The orientational probability of a bond to be lateral increases rapidly while that of a bond to be forward decreases rapidly with increasing bond number. After a small critical layer number, the orientational probability for a bond to be lateral increases slowly while that to be forward decreases slowly. The orientational probability for a bond to be backward always increases. The last bond has the largest lateral probability and the same probabilities to be forward or backward.

VI. EFFECT OF BOND CORRELATIONS AND BOND BACKTRACKS

For the polymer brush with 400 bonds, the profiles of the segment density and of the chain free ends are plotted in Figs. 6 and 7 as a function of the distance to the surface to examine the effects of bond correlations and bond backtracks. Figure 8 presents the bond order parameter as a function of the bond number. Each of these figures contains four curves. $\delta_c=0$ in these curves indicates that no bond correlations are taken into account in the calculations and $\delta_c=1$ that they are taken into account. A similar convention, $\delta_b=0$ or 1, is used for the bond backtracks. The bond backtracks have a significant influence on the segment density, chain free end and order parameter profiles. In Figs. 6, 7, and 8, curves 1 and 2 show that when the bond backtracks are neglected, the segment density is uniform within a large distance to the surface (Fig. 6), the profile of the chain free ends is narrow

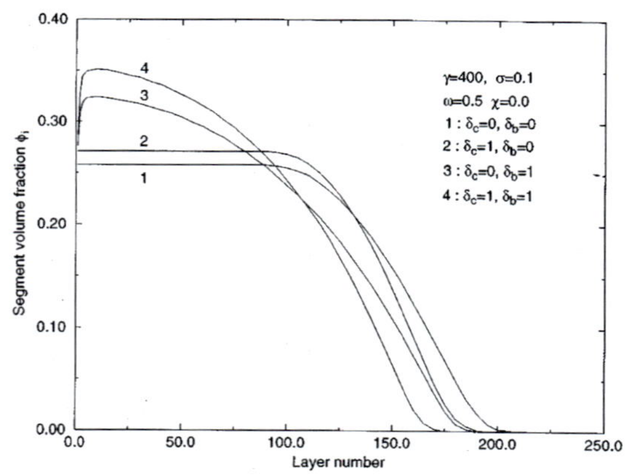

FIG. 6. The effect of bond correlations and bond backtracks on segment density profile for the polymer brush with 400 bonds.

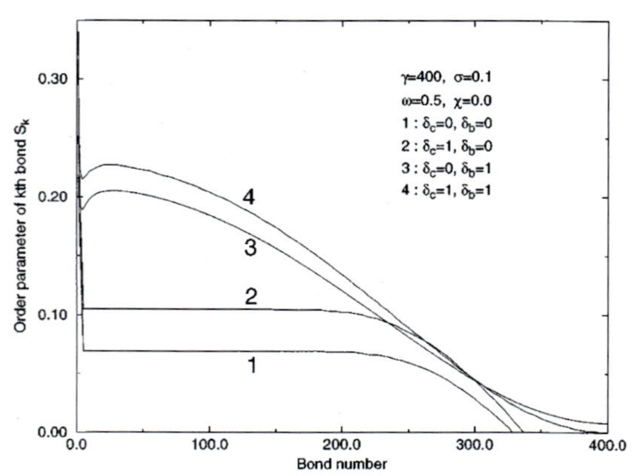

FIG. 8. The effect of bond correlations and bond backtracks on the bond order parameter for the polymer brush with 400 bonds.

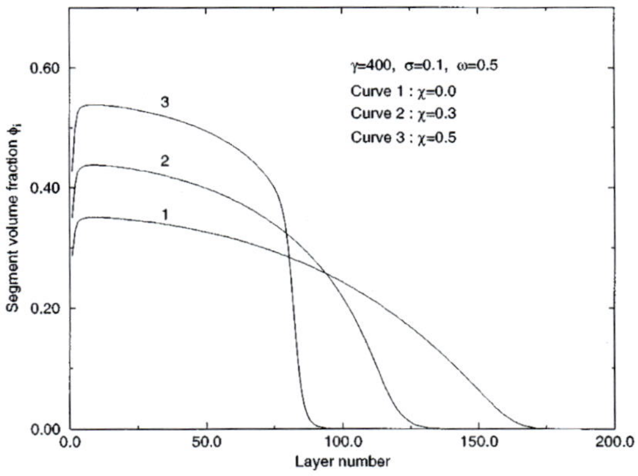

FIG. 9. The effect of intermolecular interaction parameter on the segment density profile for the polymer brush with 400 bonds.

(Fig. 7) and the bond order parameter exhibits a large horizontal transition region (Fig. 8). When the bond backtracks are taken into account (curves 3 and 4 in each of the Figs. 6, 7, and 8), the segment density (Fig. 6) and order parameter (Fig. 8) increase near the surface and then decrease, and the chain free ends have a wide profile. The effect of bond correlations is revealed by comparing curve 4 with 3 and curve 2 with 1. The incorporation of the bond correlations provides a higher average segment density and consequently, a narrower profile of chain free ends and higher order parameters. The order parameter decreases sharply from its maximum at the first bond to a much lower value at approximately the third bond.

VII. EFFECT OF INTERMOLECULAR INTERACTIONS

The effect of intermolecular interactions is characterized by the Flory–Huggins interaction parameter. For the polymer brush with 400 bonds, three values of the interaction parameter were selected in the calculations and the segment density profiles are plotted in Fig. 9. $\chi=0$ corresponds to a good solvent and $\chi=0.5$ to a poor one. With increasing value of χ, the height of the polymer brush decreases. For the poor solvent ($\chi=0.5$), the chain segments tend to be more uniformly distributed (curve 3).

[1](a) D. W. R. Gruen, Biochim. Biophys. Acta **595**, 161 (1980); (b) J. Colloid Interface Sci. **84**, 281 (1981); (c) J. Phys. Chem. **89**, 146 (1985); (d) **89**, 153 (1985).
[2](a) A. Ben-Shaul, I. Szleifer, and W. M. Gelbart, Proc. Natl. Acad. Sci. USA **81**, 4601 (1984); (b) J. Chem. Phys. **83**, 3597 (1985).
[3]I. Szleifer, A. Ben-Shaul, and W. M. Gelbart, J. Chem. Phys. **83**, 3612 (1985).
[4]I. Szleifer and A. Ben-Shaul, J. Chem. Phys. **85**, 5345 (1986).
[5]I. Szleifer, A. Ben-Shaul, and W. M. Gelbart, J. Phys. Chem. **94**, 5081 (1990).
[6](a) K. A. Dill, and P. J. Flory, Proc. Natl. Acad. Sci. USA **77**, 3115 (1980); (b) **78**, 676 (1980).
[7]K. A. Dill and R. S. Cantor, Macromolecules **17**, 380 (1984).
[8]R. S. Cantor and K. A. Dill, Macromolecules **17**, 384 (1984).
[9](a) J. M. H. M. Scheutjens and G. J. Fleer, J. Phys. Chem. **83**, 1619 (1979); (b) **84**, 178 (1980); (c) Macromolecules **18**, 1882 (1985).
[10]F. A. M. Leermakers and J. M. H. M. Scheutjens, J. Chem. Phys. **89**, 6912 (1988).
[11]F. A. M. Leermakers and J. Lyklema, Colloids and Surfaces **67**, 239 (1992).
[12]C. M. Wijmans, F. A. M. Leermakers, and G. J. Fleer, J. Chem. Phys. **101**, 8214 (1994).
[13]R. S. Cantor and P. M. McIlroy, J. Chem. Phys. **91**, 416 (1989).
[14](a) R. S. Cantor J. Chem. Phys. **99**, 7124 (1993); (b) **103**, 4765 (1995).

Steric interactions between two grafted polymer brushes

Eli Ruckenstein[a] and Buqiang Li[b]
Department of Chemical Engineering, State University of New York at Buffalo, Amherst, New York 14260

(Received 10 January 1997; accepted 8 April 1997)

A lattice model and a generator-matrix method are employed to calculate the interaction force profile between two grafted polymer brushes. The correlation between neighboring bonds and the interdigitation between the two brushes are taken into account. The calculations show that the effect of incorporating the bond correlations is equivalent to an increase in the value of the polymer–solvent interaction parameter when the bond correlations are ignored. The interdigitation between the two brushes decreases the free energy of the system and consequently results in a smaller steric repulsion. A complete interdigitation occurs at a separation close to half the separation between the two plates for which the interaction force is zero. The model is compared with the experimental interaction force profiles for ten systems which involve poly(2-vinylpyridine)-polyisoprene (PVP-PI), poly(2-vinylpyridine)-polystyrene (PVP-PS) block copolymers as well as end-functionalized polystyrenes (PS-X). For most of the systems, the theoretical predictions are in good agreement with experiment. In addition, the present results are compared with the equations proposed by de Gennes, based on the assumption of a step distribution function for the segment density, and by Milner *et al.*, based on the parabolic distribution of the segment density. Both equations neglected interdigitation. It is shown in this paper that the interdigitation is not negligible and that it can decrease by an order of magnitude the repulsive force. © *1997 American Institute of Physics.* [S0021-9606(97)50427-2]

I. INTRODUCTION

A very effective way of stabilizing colloids is to modify the surface by grafting a polymer.[1] In this manner, a high polymer density for which the chains can overlap to form a polymer brush can be obtained and this leads to a strong steric repulsion between the colloidal particles. In recent years, this problem has received attention both experimentally[2–10] and theoretically.[11–42] The theoretical investigations focused firstly on the density profile of a single brush on a flat or curved surface and secondly on the interactions between two brushes.

Most of the theories were based on the self-consistent mean field approximation.[12–32] The other ones include the scaling analysis,[33–37] molecular dynamics simulations[38,39] and Monte Carlo simulations.[40,41] The self-consistent mean field theories[12–32] were developed along the following three lines: (1) on the basis of a lattice model,[12–22] (2) on the basis of a diffusion type equation,[23–28] and (3) analytical approaches.[29–32]

The lattice-based self-consistent mean field theories[12–22] were developed mainly by the Wageningen group and can be traced back to the Scheutjens–Fleer polymer adsorption theory,[12,13] which is a lattice theory that extends the Flory–Huggins approach for homogeneous polymer solutions to inhomogeneous systems. In the lattice model, all possible conformations of the grafted (and also free) chains are generated on a lattice, each conformation being weighted by the appropriate Boltzmann factor. In the early treatments, the polymer molecules were described as freely jointed chains and the chain conformations treated as step-weighted random walks in a potential field. Some improvements were made by incorporating the chain stiffness[14] and the correlations between the nearest-neighboring parallel bonds.[15] Wijmans *et al.*,[17] using the rotational isomeric state scheme of Leermakers and Scheutjens[15] which involves the chain stiffness and the bond correlation, provided a detailed calculation for single brushes. Fleer and Scheutjens[19] extended their theory[12,13] for the adsorption of homopolymers to block copolymers. The block copolymer contains an adsorbing (or anchoring) block and a nonadsorbing block (or tail). The latter behaves like an end-grafted chain and at high densities, the end-grafted chains form a brush. The interaction free energy between two surfaces grafted with adsorbed di- and tri-block copolymers for completely flexible chains was also calculated. Recently, Wijmans, Leermakers, and Fleer,[21] developed a self-consistent model which enabled the calculation of the interaction potential between two flexible polymer-coated spherical particles.

The second kind of self-consistent theories[23–28] employed a formalism originally developed by Edwards,[23] in which the polymer was described as a freely-jointed chain consisting of segments of a given length and the conformation of a chain was treated as a random walk in the presence of a potential field. A fundamental quantity in these theories is the segment probability distribution function which is governed by a self-consistent diffusion equation. Whitmore and Noolandi[27] presented calculations regarding the adsorption on a single surface and on two parallel surfaces. The repulsive force acting between the two surfaces was calculated and the results were in good agreement with two sets of experiments. Recently, the diffusion self-consistent mean

[a]Author to whom the correspondence should be addressed.
[b]Permanent address: Beijing University of Chemical Technology, Beijing 100029, People's Republic of China.

 0021-9606/97/107(3)/932/11/$10.00

field theory was used by Lin and Gast[28] to calculate the steric repulsion between two spherical particles on which polymer brushes were grafted, using an approximation which improved the Derjaguin approximation, since the latter one can not be applied to small particles.

Analytical self-consistent mean field theories were developed independently by Zhulina *et al.*[29,30] and Milner *et al.*[31,32] They are based on the assumption that for large stretchings of the grafted chains with respect to their Gaussian dimension, one can approximate the set of conformations of a stretched grafted chain by a set of most likely trajectories, and predict for such cases a parabolic density profile. In the calculations of the interactions between the two brushes, the interdigitation between the chains was ignored.

The scaling analysis was suggested by Alexander[33] and de Gennes.[34,35] Patel and Tirrell[36] compared the scaling equations with their experimental force profiles for various di- and tri-block copolymers and Pincus[37] extended them to polyelectrolyte brushes. The scaling analysis assumed a step-function for the distribution of the polymer segment density. For a neutral brush, the interaction free energy involves two contributions; an osmotic term which depends on the segment density and an elastic one which arises because of chain stretching. For a polyelectrolyte brush, the Poisson–Boltzmann equation plays also a part. While there is good agreement between theory and experiment regarding the dependence of the steric repulsive force on distance, the equations contain two parameters which have to be provided by experiment.

In this paper, we calculate on the basis of a previously developed formalism for a brush[42] the interaction force profile between two identical polymer brushes grafted to two crossed cylindrical surfaces. As noted[42] previously, the method is equivalent but different formally from that developed in Refs. 15 and 17, being more simple and compact. A detailed comparison with recent experiments,[9,10] which involve ten systems, is carried out. The Derjaguin approximation is combined with the flat geometry results to calculate this force profile. Particular emphasis is on the effects of the interdigitation between the chains of the two brushes and of the nearest-neighboring bond correlation.

II. THEORY

A. Chain conformational statistics

We consider a system consisting of two identical polymer brushes (grafted to two parallel plates) and solvent molecules. The cubic lattice model is employed to represent the mixture of polymer chains and solvent (schematically shown in Fig. 1). Each polymer segment and each solvent molecule occupies one lattice site. The lattice is divided into $2m$ layers parallel to the plates, numbered from any of the two plates as $1, 2,\ldots, i,\ldots,m,\ m+1,\ldots,2m+1-i,\ldots,2m$. Each lattice layer contains L sites. Since the two brushes are considered identical, there exists a plane with respect to which the segment density and bond conformations are symmetrical. Thus, layers i and $2m+1-i$ with $i=1,2,\ldots,m$ contain the same numbers of segments, of solvent molecules and have the same

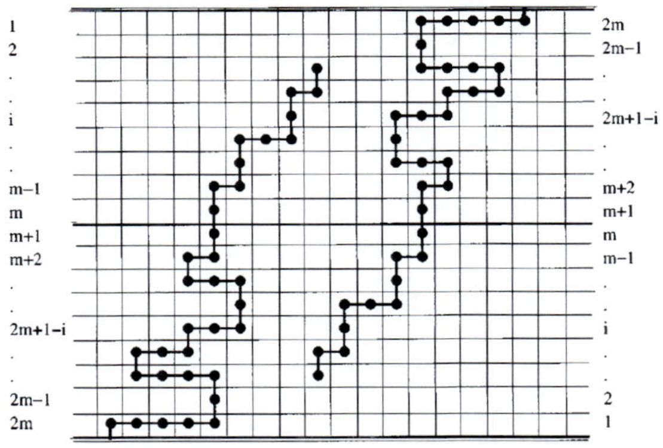

FIG. 1. Schematic representation of the lattice model.

bond conformations. We consider only monodisperse brushes. Each brush contains N chains and each chain γ segments which are connected through γ_b bonds and occupy γ sites ($\gamma=\gamma_b+1$). The first segment of each chain is constrained to be located in layer 1 or $2m$.

In the cubic lattice, only two relative orientations of two consecutive bonds can occur, namely the collinear and the bent one at 90°. The bending energies of the collinear and the bent bond pairs are taken to be zero and ϵ, respectively. The chain stiffness is characterized by the parameter $\omega=\exp(-\epsilon/kT)$. A small value of ω (large bond bending energy ϵ) corresponds to a stiff chain. Taking into account four types of bond pairs in the cubic lattice, depending on whether a bond pair is collinear or bent and on how many ways the bond pair can arrange on the lattice, the statistical weight of a chain in conformation c can be expressed[42] as

$$W_c=\alpha_h^{n_{ch}}\alpha^{n_{cb1}}\beta^{n_{cb2}}\omega^{n_{cb3}}\left(\prod_{i=1}^{2m} p_i^{\gamma_{ci}}\right)\left(\prod_{i=1}^{2m}\prod_{\xi} p_{\xi i}^{\gamma_{c\xi i}}\right) \qquad (1)$$

which differs from the corresponding equation for a single brush only because $2m$ replaces m. In Eq. (1), $n_{ch}=0$ or 1 when the first bond is forward or lateral in conformation c, n_{cb1}, n_{cb2}, and n_{cb3} denote the numbers of bond pairs in conformation c corresponding to three different bent bond pairs. The quantities α_h, α, and β are given by the expressions $\alpha_h=4\omega_h$, $\alpha=4\omega$, and $\beta=2\omega$ with $\omega_h=\exp(-\epsilon_h/kT)$ and $\omega=\exp(-\epsilon/kT)$, where ϵ_h stands for the bending energy of the first bond near the surface which may differ from ϵ. γ_{ci} stands for the number of segments in layer i provided by one chain in conformation c and $\gamma_{c\xi i}$ for the number of bonds in direction ξ with at least one segment in layer i provided by one chain in conformation c. In the cubic lattice, ξ can be x or y for the lateral bonds and z for the vertical one.

p_i is the weight factor due to the segment–solvent intermolecular interactions, given by

$$p_i=\Phi_i^0 \exp[(\langle\Phi_i\rangle-\langle\Phi_i^0\rangle)\chi+(\delta_{1i}+\delta_{2mi})\chi_s],$$

$$i=1,2,\ldots,2m, \qquad (2)$$

where Φ_i^0 and Φ_i are the fractions of sites occupied by the solvent molecules and segments, respectively, the brackets $\langle\,\rangle$ stand for the nearest neighbor average, χ is the polymer–solvent interaction parameter, $\chi_s=(U_s^0-U_s)/kT$ (with U_s^0 and U_s denoting the contact energies of the solvent molecules and segments with the surface, respectively), δ is the Delta function.

$p_{\xi i}$ is the weight factor due to the correlation between the nearest-neighboring parallel bonds, expressed as

$$p_{\xi i}=\frac{1}{1-\Phi_{\xi i}}, \quad i=1,2,\ldots,2m, \tag{3}$$

where $\Phi_{\xi i}$ is the number of bonds per lattice site (bond density) in direction ξ with at least one segment in layer i. Assuming that on average, half of the lateral bonds in layer i are in the x direction and the other half in the y direction, and denoting by Φ_{Li} the total lateral bond density in layer i, one can write for the lateral bonds,

$$p_{xi}=p_{yi}=\frac{1}{1-\Phi_{Li}/2}\equiv p_{Li}, \quad i=1,2,\ldots,2m. \tag{4}$$

It is worth noting that the surface parameter $\eta_s=\exp(\chi_s)$ which appears in the corresponding equation of the preceding paper[42] does not appear in Eq. (1) because in the present derivation, the effect of χ_s was included in the expression for the weight factor of a segment located in either layer 1 or $2m$.

Using the generator-matrix formalism of the chain conformational statistics, the chain conformation partition function, Ξ, can be expressed using a product of three matrices,[42,43,44]

$$\Xi=\sum_c W_c=\mathbf{A}\mathbf{G}^{\gamma_b}\mathbf{B}, \tag{5}$$

where $\mathbf{G}$ is a $\nu\times\nu$ supermatrix,

$$\mathbf{G}=\begin{bmatrix} \mathbf{q}_1 & \mathbf{p}_1 & \cdots & \cdots & \cdots & \mathbf{0} \\ \mathbf{r}_2 & \mathbf{q}_2 & \mathbf{p}_2 & \cdots & \cdots & \mathbf{0} \\ \cdots & \cdots & \cdots & \cdots & \cdots & \cdots \\ \mathbf{0} & \cdots & \cdots & \mathbf{r}_{\nu-1} & \mathbf{q}_{\nu-1} & \mathbf{p}_{\nu-1} \\ \mathbf{0} & \cdots & \cdots & \cdots & \mathbf{r}_\nu & \mathbf{q}_\nu \end{bmatrix}$$

whose elements are the 2×2 submatrices

$$\mathbf{r}_i=p_i p_{zi-1} \upsilon_{rp}\begin{bmatrix} 1 & 0 \\ \omega & 0 \end{bmatrix}, \quad i=2,3,\ldots,\nu,$$

$$\mathbf{q}_i=p_i p_{Li}\begin{bmatrix} 0 & \alpha \\ 0 & \beta+1 \end{bmatrix}, \quad i=1,2,\ldots,\nu,$$

$$\mathbf{p}_i=p_i p_{zi} \upsilon_{pr}\begin{bmatrix} 1 & 0 \\ \omega & 0 \end{bmatrix}, \quad i=1,2,\ldots,(\nu-1),$$

$$\mathbf{0}=\begin{bmatrix} 0 & 0 \\ 0 & 0 \end{bmatrix}.$$

$\mathbf{A}$ and $\mathbf{B}$ are vectors containing ν elements

$$\mathbf{A}=[\mathbf{a},\mathbf{0},\ldots,\mathbf{0}], \quad \mathbf{a}=[1,0], \quad \mathbf{0}=[0,0],$$

$$\mathbf{B}=\mathrm{col}[\mathbf{c}_1,\mathbf{c}_2,\ldots,\mathbf{c}_\nu], \quad \mathbf{c}_i=\mathrm{col}[p_i,p_i].$$

In the above equations, ν stands for the highest layer which can be reached by a chain. For the present case, $\nu=2m$. In the supermatrix $\mathbf{G}$, $\mathbf{r}_i$ represents a backward bond starting from layer i, $\mathbf{q}_i$ a lateral bond in layer i, and $\mathbf{p}_i$ a forward bond starting from layer i. In the elements $\mathbf{r}_i$ and $\mathbf{p}_i$, the Heaviside functions υ_{rp} and υ_{pr} are included to avoid bond-folding, since a bond can not be backward when the previous bond was forward and vice versa. The elements $\mathbf{r}_i,\mathbf{q}_i$, and $\mathbf{p}_i$ depend on three kinds of parameters. The first kind of parameters α, β, and ω arise from the local chain stiffness and bond arrangements, the second p_i from the segment–solvent interactions, and the third kind $p_{\xi i}$ from the correlations between nearest-neighboring parallel bonds.

The supermatrix $\mathbf{G}$ differs from that proposed by DiMarzio and Rubin,[43] and employed by Scheutjens and Fleer[12] and Cosgrove *et al.*[16] since it accounts for the bond correlation and chain stiffness and excludes via the Heaviside functions the bondfolding. Compared to the supermatrix of Dill,[44] it accounts, in addition, for the bond backtrack and bond correlation.

The statistical weight ($W_{\gamma i}$) and the probability ($P_{\gamma i}$) for a chain to end in layer i are given by

$$W_{\gamma i}=\mathbf{A}\mathbf{G}^{\gamma_b}\mathbf{C}_i \tag{6}$$

and

$$P_{\gamma i}=\frac{N_{\gamma i}}{N}=\frac{W_{\gamma i}}{\Xi}, \tag{7}$$

where

$$\mathbf{C}_i=\mathrm{col}[\mathbf{0},\ldots,\mathbf{c}_i,\ldots,\mathbf{0}], \quad \mathbf{c}_i=\mathrm{col}[p_i,p_i], \quad \mathbf{0}=\mathrm{col}[0,0],$$

and $N_{\gamma i}$ stands for the number of chains whose free ends are located in layer i.

For the chains with the kth bond lateral in layer i, the chains with the kth bond forward from layer i to $i+1$ and the chains with the kth bond backward from layer i to $i-1$, the probabilities P_{kLi}, P_{kFi}, and P_{kBi} are given, respectively, by

$$P_{kLi}=\frac{1}{\Xi}\mathbf{A}\mathbf{G}^{k-1}\mathbf{Q}_i\mathbf{G}^{\gamma_b-k}\mathbf{B}, \quad i=1,2,\ldots,\nu, \tag{8}$$

$$P_{kFi}=\frac{1}{\Xi}\mathbf{A}\mathbf{G}^{k-1}\mathbf{P}_i\mathbf{G}^{\gamma_b-k}\mathbf{B}, \quad i=1,2,\ldots,\nu-1, \tag{9}$$

and

$$P_{kBi}=\frac{1}{\Xi}\mathbf{A}\mathbf{G}^{k-1}\mathbf{R}_i\mathbf{G}^{\gamma_b-k}\mathbf{B}, \quad i=2,3\ldots,\nu, \tag{10}$$

where $\mathbf{Q}_i$, $\mathbf{P}_i$ or $\mathbf{R}_i$ is a supermatrix which has only the (i,i), $(i,i+1)$ or $(i,i-1)$ (i.e., $\mathbf{q}_i$, $\mathbf{p}_i$ or $\mathbf{r}_i$) as a nonzero element. Since no bond can be forward from layer ν and no bond can be backward from layer 1, $P_{kF\nu}$ and P_{kB1} are zero.

Because the kth segment of a chain is located in layer i if the kth bond of the chain is forward, lateral, or backward from layer i, the probability for the kth segment to be in layer i is equal to the sum of the probabilities for the kth bond to start from layer i,

$$P_{ki} \equiv \frac{N_{ki}}{N} = \frac{1}{\Xi} \mathbf{A}\mathbf{G}^{k-1}(\mathbf{Q}_i + \mathbf{P}_i + \mathbf{R}_i)\mathbf{G}^{\gamma_b - k}\mathbf{B}, \neg \quad i = 1,2,\ldots,\nu, \tag{11}$$

where N_{ki} represents the number of chains whose kth segment is located in layer i. Eq. (11) can be applied to the segments from 1 to γ_b. For the end segment (chain free end), Eq. (7) should be used.

According to Fig. 1, the number of segments provided by the lower brush in layer i can be obtained by summing N_{ki} from segment i to the end segment γ, while that provided by the upper brush is given by summing $N_{k2m+1-i}$ from segment $2m+1-i$ to the end segment γ. Consequently, the total number of segments in layer i, N_i, is given by

$$N_i = \sum_{k=i}^{\gamma} N_{ki} + \sum_{k=2m+1-i}^{\gamma} N_{k2m+1-i}. \tag{12}$$

Using Eqs. (7) and (11), Eq. (12) becomes

$$\Phi_i = \sigma(P_{\gamma i} + P_{\gamma 2m+1-i}) + \sigma\left(\sum_{k=i}^{\gamma_b} P_{ki} + \sum_{k=2m+1-i}^{\gamma_b} P_{k2m+1-i}\right). \tag{13}$$

Similarly, one can obtain, for the lateral bond density (Φ_{Li}) in layer i and for the vertical bond density between layers i and $i+1$ (Φ_{zi}), the following equations:

$$\Phi_{Li} = \sigma\left(\sum_{k=i}^{\gamma_b} P_{kLi} + \sum_{k=2m+1-i}^{\gamma_b} P_{kL2m+1-i}\right), \tag{14}$$

$$\Phi_{zi} = \sigma\left(\sum_{k=i}^{\gamma_b} P_{kFi} + \sum_{k=i+2}^{\gamma_b} P_{kBi+1} + \sum_{k=2m-i}^{\gamma_b} P_{kF2m-i} + \sum_{k=2m+2-i}^{\gamma_b} P_{kB2m+1-i}\right). \tag{15}$$

Since the two brushes are symmetrical, only the distributions of the segment density and bond densities along the distance from layer 1 to m, namely Φ_i, Φ_{Li}, and Φ_{zi} with $i = 1,2,\ldots,m$, are required. They can be obtained by numerically solving Eqs. (13), (14), and (15), each containing m equations. Instead of Eq. (13), a more simple set of equivalent equations is derived below. One can demonstrate that the new set of equations is a linear combination of Eqs. (13), (14), and (15).

In a linear chain, a middle segment is connected to two bonds and a terminal segment to a single bond; in other words a middle segment has two connection points and a terminal one only one. A lateral bond in layer i provides two connection points in that layer, while a vertical, either between layer i and $i+1$ or between layer i and $i-1$, provides only one connection point in layer i. In addition, a terminal segment located in layer i, whether connected to a lateral or to a vertical bond, provides only one connection point in that layer. Consequently, a mass balance of the connection points provided by the segments present in layer i provides the equation

$$2N_i - N_{\gamma i} - N_{\gamma 2m+1-i} = N_{zi-1} + N_{zi} + 2N_{Li}. \tag{16}$$

Dividing by L, the above equation becomes

$$2\Phi_i = \sigma(P_{\gamma i} + P_{\gamma 2m+1-i}) + \Phi_{zi-1} + \Phi_{zi} + 2\Phi_{Li}. \tag{17}$$

Equations (14), (15), and (17) will be used to calculate the quantities Φ_i, Φ_{Li}, and Φ_{zi}. The obtained distributions of the segment density and bond densities allow to calculate any structural property of the two interacting polymer brushes as well as the mixing free energy. The latter quantity is used to obtain the interaction force profile.

When the effect of bond correlations is neglected, only Eq. (13) is required to calculate the distribution of segment density from layer 1 to m; the $p_{\xi i}$'s, which are present in the elements $\mathbf{r}_i$, $\mathbf{q}_i$, and $\mathbf{p}_i$ to represent the weight factors due to the bond correlations, become in that case equal to unity.

B. Interaction free energy of two brushes

Denoting by N_c the number of chains in a brush in conformation c, one can write $N = \Sigma N_c$. Since the two brushes are symmetrical, the whole system has $2N_c$ chains in conformation c. For a specified set of conformations $\{N_c\}$ of the N chains of any brush, the canonical partition function $Q(T,L,2m,\{2N_c\})$ can be written as

$$Q(T,L,2m,\{2N_c\}) = \frac{\Omega}{\Omega_0} \exp\left(-\frac{\Delta U}{kT}\right), \tag{18}$$

where Ω is the combinatorial factor for the specified set of conformations $\{N_c\}$, Ω_0 is the product of the combinatorial factors of the chains and solvent molecules in their reference states, ΔU is the energy change of the chains and the solvent molecules from their reference states to the specified set of conformations. The ordered close-packed brush, in which each chain has an all-*trans* conformation (hence both the conformational energy and entropy are taken zero), is taken as the reference state of the chains, while the pure solvent as the reference state of the solvent molecules. For the reference states, $\Omega_0 = 1$.

Extending the corresponding equations of the preceding paper[42] for a single brush to the present case involving two interacting brushes, one can write

$$\Omega = (L!)^{2m}\left(\prod_c \frac{g_c^{N_c}}{N_c!}\right)^2 \left(\prod_{i=1}^{2m} \frac{1}{N_i^0!}\right) \times \left[\prod_{i=1}^{2m}\prod_{\xi} \frac{(L - N_{\xi i})!}{L!}\right], \tag{19}$$

$$\frac{\Delta U}{kT}=2\left[N_1^0\chi_s+\frac{(L-N)U_s}{kT}+\sum_{i=1}^{m}N_i\langle\Phi_i^0\rangle\chi+\sum_c\frac{N_cE_c}{kT}\right], \tag{20}$$

where g_c is the number of arrangements that a chain in conformation c can have when its first segment is fixed, N_i^0 is the number of solvent molecules in layer i, $N_{\xi i}$ is the number of bonds in direction ξ with at least one segment in layer i, $\langle\Phi_i^0\rangle$ is the average fraction of sites occupied by the solvent molecules around a site located in layer i and E_c is the conformational energy of a chain in conformation c.

At equilibrium, $N_c=NW_c/\Xi$. Introducing the later expression in Eqs. (19) and (20), one finally obtains for the segment-solvent mixing free energy $A=-kT\ln(Q)$, the expression,

$$\frac{A}{kT}=2N\ln\left(\frac{N}{L\Xi}\right)+L\sum_{i=1}^{2m}\ln(\Phi_i^0)+\sum_{i=1}^{2m}N_i\langle\Phi_i\rangle\chi-L\times\sum_{i=1}^{2m}\sum_{\xi}\ln(1-\Phi_{\xi i}). \tag{21}$$

Taking into account the symmetry of the two brushes, the mixing free energy per surface site ($f=A/2L$) becomes

$$\frac{f}{kT}=\sigma\ln\left(\frac{\sigma}{\Xi}\right)+\sum_{i=1}^{m}\ln(\Phi_i^0)+\sum_{i=1}^{m}\Phi_i\langle\Phi_i\rangle\chi-\sum_{i=1}^{m}\sum_{\xi}\ln(1-\Phi_{\xi i})+\frac{1}{2}\ln(1-\Phi_{zm}), \tag{22}$$

where $\sigma=N/L$ stands for the number of chains per surface site (surface density).

The mixing free energy f becomes independent of the separation D between the two surfaces for distances larger than $2L_0$ (twice the equilibrium thickness of an uncompressed brush L_0) for which the steric repulsion is zero. Denoting by f_0 the mixing free energy at $D\geqslant 2L_0$, the interaction free energy per unit area (E) between two grafted brushes is given by

$$E(D)=2(f-f_0)/l^2, \tag{23}$$

where l stands for the length of a site and l^2 is the area per site. The interaction free energy can be converted to the interaction force (F) between two crossed cylinders of radius R, which is measured experimentally, using the Derjaguin approximation,

$$F(D)/R=2\pi E(D). \tag{24}$$

Recently,[21,28] two kinds of self-consistent mean field models were developed to calculate the interaction potential between layers of flexible polymer chains grafted to spherical surfaces whose radii are comparable to the separation between the two surfaces. The calculations showed that for small particles the repulsion is less steep than that provided by the Derjaguin approximation, while with increasing radii, the interaction potential becomes close to that given by the Derjaguin approximation. For the present comparisons with experiment, the Derjaguin approximation is considered to be applicable, since the curvature radius of the surface (cylindrical surface) is much larger than the separation between the two surfaces (10^5 times).

III. RESULTS AND DISCUSSIONS

In this section, the theory is compared with the experimental data available regarding the interaction force profile between two crossed cylinders bearing two identical grafted polymer brushes. The results will be also compared with the equation proposed by de Gennes,[35] which involves a step function for the segment density distribution, and that proposed by Milner *et al.*[31] which involves a parabolic distribution. The interdigitation between the two brushes has been neglected in both preceding equations. In the present paper, the distributions of segment density, bond densities and chain free ends as well as the steric repulsion force are calculated taking into account the interdigitation between the two brushes and the correlations between the nearest-neighboring parallel bonds.

The comparison between theory and experiment is made for the following systems: end-functionalized polystyrenes [polystyrene-$(CH_2)_3N^+(CH_3)_2(CH_2)_3SO_3^-$ denoted as PS-X] of various molecular weights and block copolymers [poly(2-vinylpyridine)-polyisoprene denoted as PVP-PI and poly(2-vinylpyridine)–polystyrene denoted as PVP-PS] over a range of molecular weights of each of the blocks. The end group X or the PVP block serves as an ''anchor,'' which attaches the PS or PI chains to the surface forming grafted polymer brushes in a solvent (here toluene).

A. Chain characteristic properties and lattice parameters

The application of the lattice model to various systems requires a procedure to determine the lattice size and the chain bending energy, hence the number of segments and the chain stiffness in the lattice. Based on the equivalence between the contour length, volume, and gyration radius of a real polymer chain and that in the lattice model, the following equations were suggested:[22]

$$\gamma l=l_bM/M_s, \tag{25}$$

$$\gamma l^3=\frac{M}{\rho\bar{N}}, \tag{26}$$

and

$$pl^2\gamma/6=a^2M. \tag{27}$$

In the above equations, M is the molecular weight of the polymer, M_s is the molecular weight of the monomer, l is the lattice spacing, l_b the bond length between two monomers (estimated from the structure of the monomer), ρ the polymer density, $\bar{N}$ Avogadro's number, $a^2=R_g/M^{1/2}$ with R_g the gyration radius, and p is the ratio between the Kuhn length and the bond length l in the lattice model (lattice spacing). The ratio p is related to the bond bending energy ϵ through the following expression.[22]

TABLE I. Chain characteristics and interaction parameters.

Polymer	M_s (g/mol)	l_b (Å)	ρ^a (g/cm³)	ω	l (Å)	χ^a_{FH} (Φ=1.0–0.0)	χ'	χ''
Polystyrene¬	104¬	2.53¬	1.130¬	0.27	7.77¬	0.67–0.04¬	0.20	0.03
Polyisoprene¬	68¬	5.07¬	0.913¬	0.27	4.94¬	0.36–0.32¬	0.35	0.18

[a]Reference 45.

$$p=1+\frac{1}{2}\exp\left(\frac{\epsilon}{kT}\right)=1+\frac{1}{2\omega}, \quad (28)$$

where $\omega=\exp(-\epsilon/kT)$ is the chain stiffness parameter.

Using Eqs. (25)–(28), one obtains for the lattice spacing l, the chain stiffness parameter ω and the number of segments γ the expressions

$$l=\left(\frac{M_s}{\rho l_b \bar{N}}\right)^{1/2}, \quad (29)$$

$$\omega=\frac{1}{2}\left(\frac{6a^2M_s}{ll_b}-1\right)^{-1}, \quad (30)$$

and

$$\gamma=\frac{l_bM}{lM_s}. \quad (31)$$

Equations (29) and (30) show that l and ω are independent of the chain length; this is as expected. For the PS chains, $a=0.3$ Å,[22] $M_s=104$ g/mol, $l_b=2.53$ Å, and $\rho=1.13$ g/cm³;[45] hence l and ω are 7.77 Å and 0.27, respectively. For the PI chains, no value for a is available and R_g had to be used to calculate ω. The following expression for the R_g of a PI (also PS) chain is available,[10]

$$R_g=b(M/M_s)^{0.595} \quad (32)$$

with $b=1.79$ Å for the PI chain and 1.86 Å for the PS chain. Combined with Eq. (32), Eq. (30) yields

$$\omega=\frac{1}{2}\left[\frac{6b^2}{ll_b}\left(\frac{M}{M_s}\right)^{0.19}-1\right]^{-1}. \quad (33)$$

According to Eq. (33), the chain stiffness parameter ω decreases weakly with the chain length. Consequently, the polymer chain becomes slightly stiffer with increasing chain length, which is unreasonable. For this reason, an average of the values provided by Eq. (33) for the four PI chains employed was used. For the PI chains, $M_s=68$ g/mol, l_b = 5.07 Å, $\rho=0.913$ g/cm³;[45] consequently $l=4.94$ Å and $\omega=0.29$ (calculated as the arithmetic average). All the parameters for PS and PI chains are listed in Table I, which also includes three polymer–solvent interaction parameters for reasons explained below.

B. Interaction parameters and bond correlations

Two sets of theoretical curves are plotted in Fig. 2 to examine the effect of the interaction parameter on the theoretical interaction force profile between two identical PS-X (140 K) brushes. The first set was calculated by ignoring the

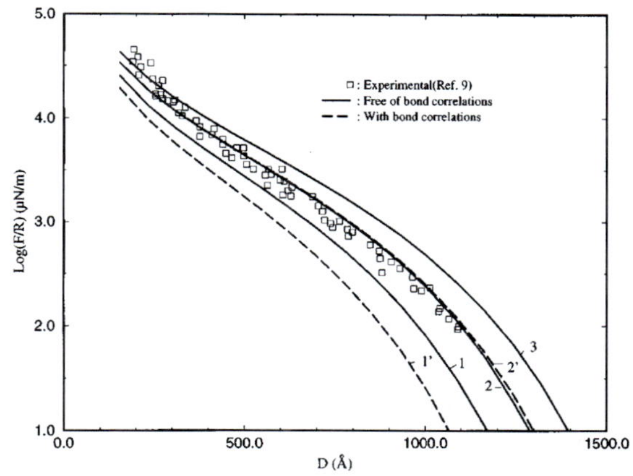

FIG. 2. The effect of bond correlations in the PS-X (140 K)-toluene system. Free of bond correlations (solid curves): 1, $\chi'=0.3$; 2, $\chi'=0.2$; 3, $\chi'=0.1$ and with the bond correlations (dashed curves): 1; $\chi''=0.2$; 2; $\chi''=0.03$.

bond correlations and the second by incorporating them. χ' and χ'' denote the interaction parameters in the two cases. In each of the two cases, the theoretical force profile curve is moved up with decreasing interaction parameter, indicating, as expected, that a good solvent results in a higher steric repulsion between the two brushes. By comparing the two sets of curves, one can note that for the same value of the interaction parameter, the incorporation of the bond correlations reduces the repulsion between the two brushes. This happens because the correlations between the nearest-neighboring parallel bonds lead to a more efficient packing of the segments, and as a result the segment density increases. Consequently, the bond correlations have the same effect as that caused by an increased value of the interaction parameter. Indeed, almost the same profiles of the interaction force, the segment density and the chain free ends were obtained by neglecting the bond correlations but by using a larger value for the interaction parameter($\chi''=\chi'+0.17$). Such a suggestion was made previously by Wijmans *et al.*,[17] who demonstrated that, for sufficiently low polymer concentrations, $\chi''=\chi'+0.17$. They applied this observation to single brushes.

For the PS-X (140 K) brushes and the two cases (without and with bond correlations), one can see from Fig. 2 that the theoretical force profile (solid curve 2) for the case free of bond correlations and $\chi'=0.2$. and that (dashed curve 2′) which incorporates the bond correlations and uses for χ'' the value 0.03 almost overlap. In addition, the theoretical curves 2 and 2′ fit well the experimental data of Taunton *et al.*[9]

In the present calculations, the interaction parameter (χ' or χ'') is the only undetermined parameter. For the PI chains–toluene interactions, χ' is selected 0.35, which is close to the experimental values of the Flory–Huggins interaction parameter[45] ($\chi_{FH}=0.32$–0.36 in the concentration range $\Phi=0$–1 and in the temperature range 25–55 °C). For the PS chains–toluene interactions, χ' is taken 0.2, which is in the lower range of the experimental Flory–Huggins interaction parameters[45] ($\chi_{FH}=0.04$–0.67 in the concentration

TABLE II. Molecular weights, number of segments, and surface densities.

Polymers	Molecular weight (M_{PI} or M_{PS}) ($\times 10^{-3}$ g/mol)	The number of segments (γ)	Surface density $\sigma \times 10^3$
PVP-PI 63-39¬	38.8¬	588¬	2.12
PVP-PI 26-50¬	50.4¬	754¬	3.05
PVP-PI 38-69¬	69.3¬	1041¬	2.44
PVP-PI 30-217¬	217.0¬	3274¬	0.83
PVP-PS 60-60¬	60.0¬	185¬	4.95
PVP-PS 60-90¬	95.0¬	292¬	5.31
PS-X (26 K)	26.3¬	81¬	24.43[a]
PS-X (58 K)	57.8¬	178¬	13.74[a]
PS-X (140 K)	139.8¬	430¬	7.79
PS-X (375 K)	374.8¬	1153¬	2.72[a]

[a]Value calculated as noted in the text.

range $\Phi=0-1$ and in the temperature range 162–229 °C). For either of the two kinds of chains (PI or PS), the difference $\chi'-\chi''=0.17$. The values of the interaction parameters are listed in Table I.

C. Comparison with experimental data

In this section, the calculated results are compared with experimental data for the PS-X polymers and for the PVP-PI and PVP-PS block copolymers. The two-dimensional surface density (number of chains per surface site) in the lattice model σ is proportional to the surface grafting density σ_e (number of chains per unit area), namely $\sigma=\sigma_e l^2$. For the systems involved in the calculations, the surface density σ and the number of segments γ are listed in Table II.

For the experiments carried out by Watanabe and Tirrel,[10] which involved the PVP-PI and PVP-PS block copolymers, we use the surface grafting densities of the PI and PS chains calculated by them. For the experiments performed by Taunton *et al.*[9] involving the PS-X polymers, only the surface grafting density of PS-X (140 K) (3 ± 0.5 mg/m^2) was given. The average value 3 mg/m^2 was therefore used in our calculations. For the other three PS-X polymers, the surface grafting densities were obtained by us on the basis of the theoretical analysis provided by the authors and we do not know how accurate the values are.

The comparison between theory and experiment is presented in Figs. 3–6. Since when the bond correlations are ignored and the interaction parameter χ' is used, or when the bond correlations are incorporated and the interaction parameter $\chi''=\chi'-0.17$ is employed, the theoretical curves almost overlap, we plot in the figures only the theoretical curves for the case with bond correlations. Since for the block copolymer PVP-PI 30-217, which has a long chain ($\gamma=3274$), the incorporation of bond correlations will require a time-consuming computation, only the calculations for the case free of bond correlations were performed, using, however, a larger interaction parameter.

For comparison, Figs. 3–6 include the theoretical curves based on de Gennes[35] and Milner *et al.*[31] equations. de Gennes equation is an extension to two interacting brushes of Alexander's ideas[33] regarding the single grafted brushes. In

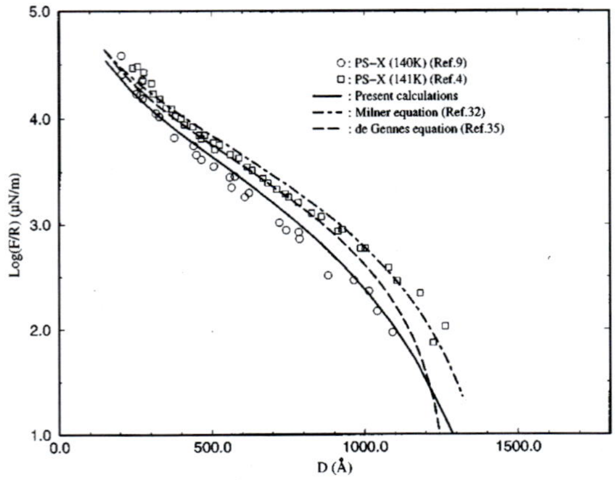

FIG. 3. A comparison among the present calculations, and the equations of de Gennes and Milner *et al.* The curve based on the de Gennes expression was obtained in Ref. 4 by fitting experimental results, while that based on the Milner *et al.* expression, which does not involve any fitting parameter, was provided in Ref. 32.

Alexander's treatment, the grafted brush is assumed to have a uniform segment density (step function distribution), and each chain to consist of connected semidilute "blobs." The osmotic repulsion between blobs tends to stretch the chains, while the elastic free energy of the chains has the opposite effect. For a single brush, the minimization of the overall free energy with respect to the brush thickness yields the equilibrium brush thickness L_0, given by

$$L_0=s(R_F/s)^{5/3}, \tag{34}$$

where s and R_F are the mean spacing between grafting points and the Flory radius, respectively. In the above equation, a numerical prefactor of order unity was ignored. The quantity s was calculated from the surface grafting density σ_e ($s=\sigma_e^{-1/2}$) and R_F ($R_F=6^{1/2}R_g$) from the radius of gyration R_g, given by Eq. (32).

Considering two contributions, a repulsive osmotic one due to the increase in the polymer concentration, and an

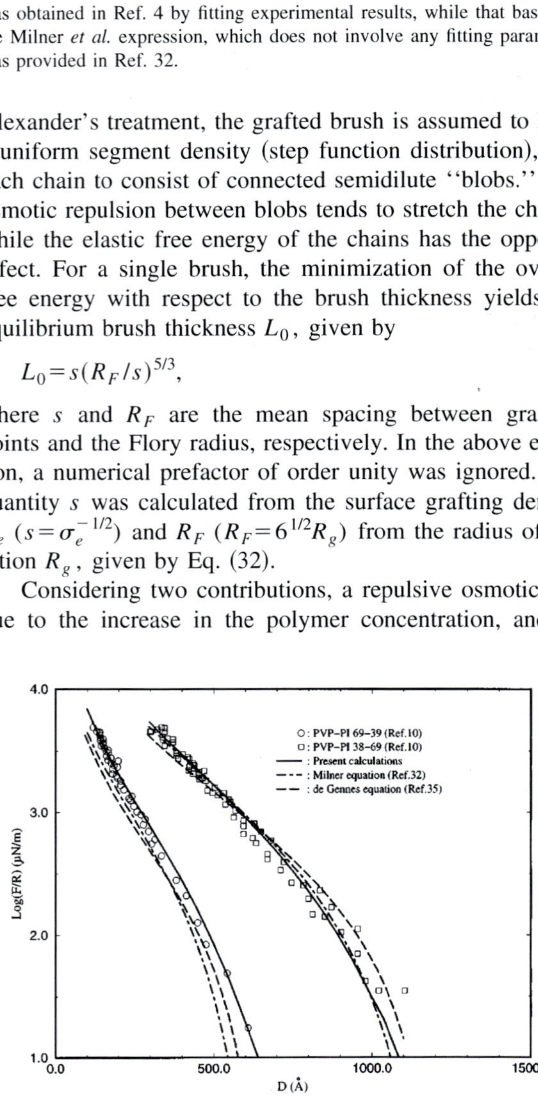

FIG. 4. Interaction force profiles for PVP-PI 69-39 and PVP-PI 38-69 in toluene. The parameters ϖ and v in the Milner *et al.* expression and the prefactors k_L and k_F in the de Gennes expression are obtained by fitting the experiments for PVP-PI 38-69.

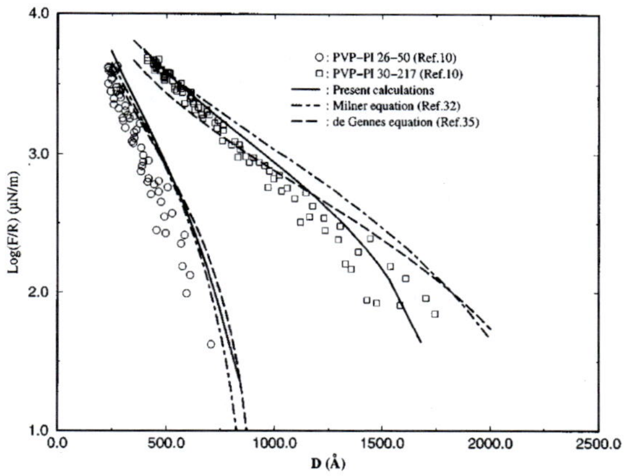

FIG. 5. Interaction force profiles for PVP-PI 26-50 and PVP-PI 30-217 in toluene. The parameters ϖ and υ in the Milner *et al.* expression and the prefactors k_L and k_F in the de Gennes expression are obtained by fitting the experiments for PVP-PI 38-69.

elastic one due to the decrease of the elastic free energy, de Gennes[35] suggested for the interaction force per unit area the following equation;

$$f_G(D)=\frac{kT}{s^3}\left[\left(\frac{2L_0}{D}\right)^{9/4}-\left(\frac{D}{2L_0}\right)^{3/4}\right]. \tag{35}$$

A prefactor in the above equation was ignored.

In order to test the validity of de Gennes′ scaling model, one can examine if the two prefactors in Eqs. (34) and (35) remain constant when the polymer system is changed. We introduce two prefactors k_L and k_F in Eqs. (34) and (35), respectively, and calculate the interaction free energy per unit area using the expression,

$$E(D)=\int_D^{2L_0} f_G(D')dD'. \tag{36}$$

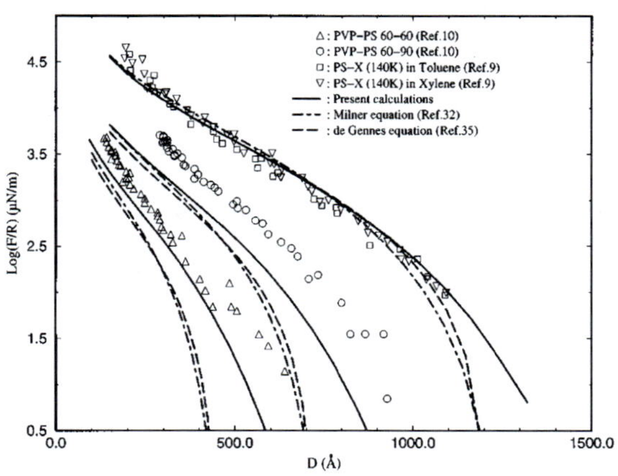

FIG. 6. Interaction force profiles for PVP-PS 60-60, PVP-PS 60-90, and PS-X (140 K) in toluene. The parameters ϖ and υ in the Milner *et al.* expression and the prefactors k_L and k_F in the de Gennes expression are obtained by fitting the experiments for PS-X(140 K).

The two prefactors, k_L and k_F, can be obtained by fitting the experimental force-distance data for a given system. The fitted results show that for the four PI chains considered in the calculations, k_L remains approximately constant (0.91–1.15), while for the six PS chains k_L has the values 0.74–0.81. However, k_F varies within a wider range. For the PI chains, k_F varies from 0.68 to 0.90 while for the PS chains, k_F varies between 0.98 and 2.82. These calculations show that k_L depends to some extent on the nature of the chain, and that k_F depends on both the nature of the chain and the chain length.

In contrast to de Gennes' model which involves a step function for the segment density distribution, the model of Milner *et al.*, based on a self-consistent mean field approach, predicts a parabolic distribution of the segment density distribution in either a free brush or in a brush interacting with another one. Neglecting the interdigitation, the free energy per unit area of a brush as a function of separation is given by

$$f_M(u)=\frac{5}{9}f_M^0\left(\frac{1}{u}+u^2-\frac{u^5}{5}\right), \tag{37}$$

where $u=D/2L_0$ and f_M^0 is the free energy per unit area of the free brush of thickness L_0 The quantities L_0 and f_M^0 can be calculated using the expressions

$$L_0=\left(\frac{12}{\pi^2}\right)^{1/3}\left(\frac{M}{M_s}\right)\left(\frac{\sigma_e\varpi}{\upsilon}\right)^{1/3}, \tag{38}$$

$$\frac{f_M^0}{kT}=\frac{9}{10}\left(\frac{\pi^2}{12}\right)\left(\frac{M}{M_s}\right)(\sigma^5\varpi^2\upsilon)^{1/3}, \tag{39}$$

where ϖ (with dimensions of length^3) is an excluded volume parameter and υ (with dimensions of length^{-2}) is a coefficient present in the chain elastic contribution. Milner[32] used the model to interpret the experimental force profile obtained by Taunton *et al.*[4] for the PS-X (141 K). ϖ and υ were calculated from the osmotic pressure $\pi(\Phi)$ and the gyration radius R_g using the expressions $\pi(\Phi)=\varpi\Phi^2/2$ and $2R_g^2=M/(M_s\upsilon)$. Using a surface grafting density of 3.5 mg/m^2 (the upper end of the experimental range 3 $\pm$0.5 mg/m^2), good agreement with experiment was obtained.

For the PS-X (140 K) chain, the present calculation and the equations of Milner *et al.* and de Gennes are plotted in Fig. 3. In the figure, two sets of experimental data obtained by the same authors[4,9] in two different papers are included. One can see that the model of Milner *et al.* and that of de Gennes fit well the early experiments, while the present calculation is in good agreement with the new ones. Regardless of the large difference between the two sets of experimental data, the models of de Gennes and Milner *et al.* fit well either the old or the new experiments by adjusting the parameters. Indeed, using $\varpi=45.2$ Å^3 and $\upsilon=0.069$ Å^{-2} instead of those calculated ($\varpi=33.7$ Å^3 and $\upsilon=0.052$ Å^{-2}), the model of Milner *et al.* fits well the new experiments. When the de Gennes model is fitted to the new experiments, a smaller prefactor k_F=0.8 replaces k_F=1.5.

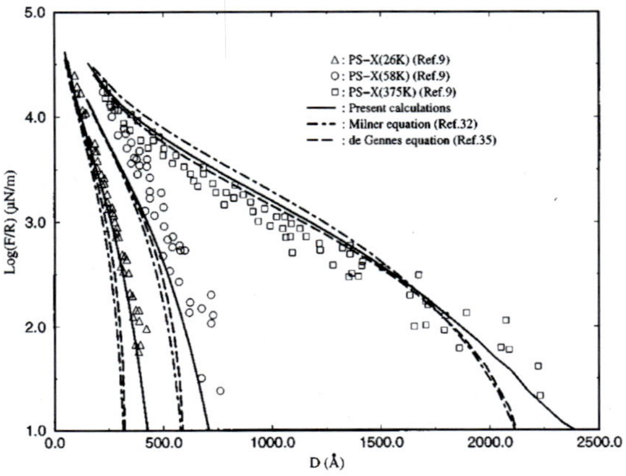

FIG. 7. Interaction force profiles for PS-X(26 K), PS-X(58 K), and PS-X(375 K) in toluene. The parameters ϖ and v in the Milner *et al.* expression and the prefactors k_L and k_F in the de Gennes expression are obtained by fitting the experiments for PS-X(140 K).

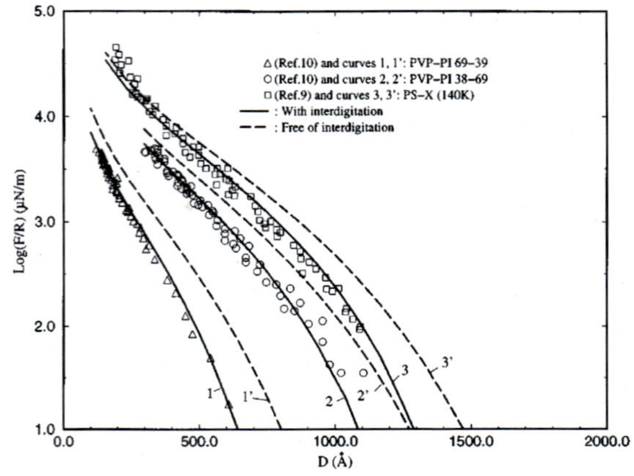

FIG. 8. The effect of the interdigitation between two brushes.

A comparison between the three models and experiments involving ten systems is made in Figs. 4–7. In the comparison, the parameters ϖ and v in the model of Milner *et al.* and the prefactors k_L and k_F in de Gennes' model are taken the same for the same kind of chains (PI or PS). For the PS chains, they are obtained by fitting the experiments for the PS-X (140 K), while for the PI chains, by fitting those for PVP-PI 38-69. The values of the parameters are listed in Table III.

Figures 4 and 5 present comparisons for four PI chains of various molecular weights (from 3.88×10^4 to 2.17×10^5). For the PVP-PI 38-69-toluene system which was selected to determine the optimum parameters in the Milner *et al.* and de Gennes' models, the present calculations, which do not involve any fitted parameters, are in very good agreement with experiment over the entire range of the force profile. For both the PVP-PI 69-39 with a short PI chain and the PVP-PI 30-217 with a long PI chain, the present calculations fit well the experimental force profiles, while the other two models provide lower values for the former system over the entire range of separation and higher values for the latter system in a wide separation range. For the PVP-PI 26-50, the three models provide higher values than experiment. For the PVP-PI systems, the models of Milner *et al.* and de Gennes provide similar results.

Figures 6 and 7 present a comparison for six PS chains of various molecular weights (from 2.63×10^4 to 3.75×10^5). For the PS-X (140 K)-toluene system which was selected to determine the optimum parameters in the models of Milner *et al.* and de Gennes, the present calculations and the other two models provide approximately the same results within a wide region, in good agreement with experiment; at larger separations, the other two models deviate from experiment. For the PVP-PS 60-60 and PVP-PS 60-90 block copolymers (see Fig. 6) and PS-X (58 K) (see Fig. 7), the theoretical force profiles provided by the present calculations are lower than those provided by experiment, but can be shifted along the (logarithmic) vertical axis to fit the experimental data. For the PS-X (26 K) containing 185 segments, the present calculations fit well the experimental results, while for the PS-X (375 K) containing 1153 segments, there is a slight deviation (Fig. 7). For all the PS chains, the equations of Milner *et al.* and de Gennes lead to larger deviations from the data than the present model. Particularly, at large separations, the interaction force given by the models of Milner *et al.* and de Gennes decrease rapidly with increasing separation D. As a result, the two models predict a smaller separation $2L_0$ at which the interaction force between the two brushes becomes zero than the experimental one. The comparisons plotted in Figs. 4–7 show that for most systems, the present model is in good agreement with experiment; for some systems, such as PVP-PI 26-50, PVP-PS 60-90, and PS-X (58 K), a good fitting with experiment can be obtained by parallel shifting up the theoretical force profile curves. This happens probably because of inaccurate data regarding the surface density.

TABLE III. Fitted parameters in the equations of Milner *et al.* and de Gennes.

Polymer	ϖ (Å)	v ($Å^{-2}$)	k_L	k_F
Polystyrene	45.2	0.069	1.15	0.42
Polyisoprene	48.3	0.029	0.76	0.80

D. The interdigitation between two brushes

The effect of the interdigitation between the polymer chains of two brushes on the interaction force profile is plotted in Fig. 8 for three polymers PVP-PI 69-39, PVP-PI 38-69, and PS-X (140 K). The solid curves present the results in which the interdigitation was taken into account, while the dashed ones those in which the interdigitation was neglected. For the former, the chain conformational partition function [Eq. (5)] involves a $2m\times 2m$ supermatrix **G** for which the chain free ends can be located in any lattice layer between the two surfaces. For the latter, the conformational partition function involves a $m\times m$ supermatrix **G** with chains con-

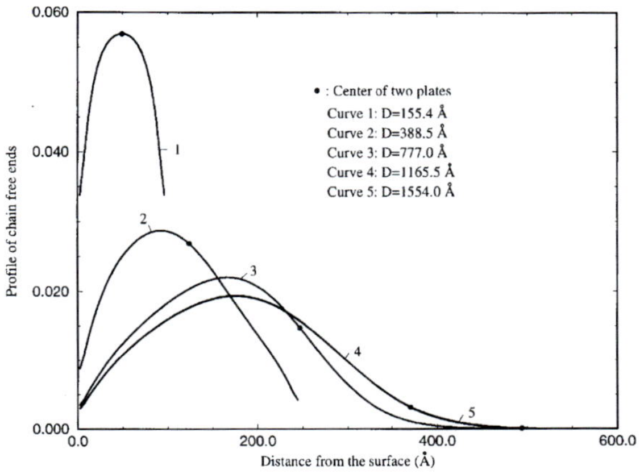

FIG. 9. The distribution of chain free ends in the PS-X(140 K)-toluene system.

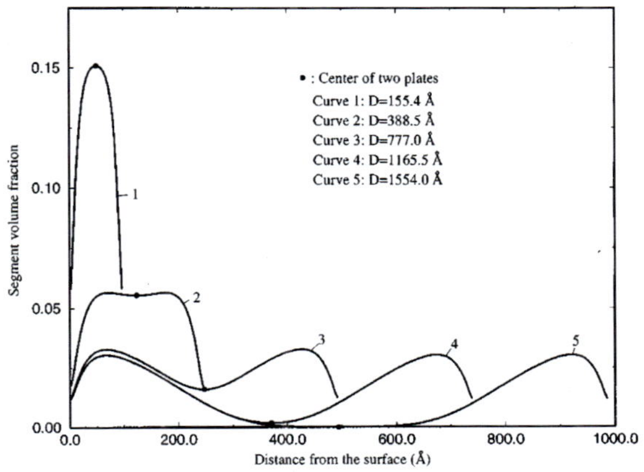

FIG. 10. The distribution of segment density in the PS-X(140 K)-toluene system.

fined between the middle plane and the surface on which the chains are grafted; the interaction energy between the two brushes is in this case the sum of the contributions of the two separate brushes.

One can see that the theoretical curves free of interdigitation are always located above those which account for the interdigitation, indicating that by neglecting interdigitation a much larger repulsion is obtained. The absolute deviation between the two theoretical results increases with decreasing separation. The interdigitation can not be neglected over the entire range of separations. It can decrease the repulsion by more than an order of magnitude. The equations of Milner *et al.* and de Gennes ignore the interdigitation, which is accounted only indirectly through the fitting parameters.

E. The distributions of chain free ends and segment density

For the PS-X (140 K) and five different separations, the distribution probabilities of the chain free ends are plotted in Fig. 9 as a function of the distance to the surface on which the chains are grafted. In the figure, the filled points denote the position of the middle plane. Therefore, the part of the theoretical curve at the right of the filled point provides the distribution of chain free ends of one brush in the other one. The thickness of the interdigitation region is equal to twice the distance from the middle plane to the point at which the distribution probability of chain free ends becomes zero. As expected, as the separation decreases from a large value (1554 Å) to a smaller one (155.4 Å), the filled point moves up and reaches a point located at the maximum of the distribution probability. There is a critical separation at which the chain free ends of one brush can reach the grafting surface of another brush and a full interdigitation occurs. For the PS-X (140 K) shown in Fig. 9, a full interdigitation occurs at separations smaller than 777 Å (curve 3), which is approximately half of the separation $2L_0$. With decreasing separation from 777 Å to 155.4 Å (see curves 3, 2, and 1), the distribution probability of the chain free ends located on the opposite surface increases starting from zero. For the small separation (155.4 Å), the theoretical distribution curve becomes symmetrical with respect to the middle plane.

For the same system and the same separation distances as in Fig. 9, the distributions of segment densities are given in Fig. 10. The filled point represents again the position of the middle plane. The segment density at the middle plane is close to zero at the large separation of 1554 Å (curve 5) and increases with decreasing separation. The separation at which the interactions start, which is twice the thickness of the single brush, is equal to 1554 Å. The above theoretical value is slightly larger than the experimental one 1300±100 Å. With decreasing separation, the segment density becomes more uniformly distributed. At a separation smaller than 388.5 Å (curve 2), one can achieve an almost uniform segment density distribution in the middle region. For all the separations, the segment density decreases close the surfaces, indicating a depletion of polymer chains at the surface.

IV. CONCLUSIONS

On the basis of a cubic lattice model and generator-matrix formalism, the interaction force profile between two identical polymer brushes grafted on two curved surfaces is calculated. This model allows to examine in detail the effect of the interdigitation between the two brushes. As expected, the presence of interdigitation results in a smaller interaction free energy and hence decreases the steric repulsion between the two brushes. The calculations show that complete interdigitation, at which the free end of a chain grafted on a surface reaches the opposite surface, is reached at a separation L_0, which represents the thickness of a single brush in thermodynamic equilibrium.

The effect of the correlations between the nearest neighboring parallel bonds is important. However, it is shown that the calculations can be simplified by ignoring these correlations and by replacing the interaction parameter employed when the correlations are included with a value larger by 0.17.

The calculations for four PI chains of various molecular weights, from 3.88×10^4 to 2.17×10^5, and six PS chains of various molecular weights, from 2.63×10^4 to 3.75×10^5, are in good agreement with experiment for most systems.

[1] D. Napper, *Polymeric Stabilization of Colloidal Dispersions* (Academic, London, 1983).
[2] A. Halperin, M. Tirrell, and T. P. Lodge, Adv. Polym. Sci. **100**, 31 (1992).
[3] G. Hadziioannou, S. Patel, S. Granick, and M. Tirrell, J. Am. Chem. Soc. **108**, 2869 (1986).
[4] H. J. Taunton, C. Toprakcioglu, L. J. Fellers, and J. Klein, Nature **332**, 712 (1988).
[5] H. J. Taunton, C. Toprakcioglu, L. J. Fetters, and J. Klein, Colloid Surf. **31**, 151 (1988).
[6] H. J. Taunton, C. Toprakcioglu, L. J. Fetters, and J. Klein, Polym. Prepr. **30**, 368 (1989).
[7] M. A. Ansarifar and P. F. Luckham, Polymer **29**, 328 (1988).
[8] J. Marra and M. L. Hair, Colloid Surf. **34**, 215 (1988).
[9] H. J. Taunton, C. Toprakcioglu, L. J. Fetters, and J. Klein, Macromolecules **23**, 571 (1990).
[10] H. Watanabe and M. Tirrell, Macromolecules **26**, 6455 (1993).
[11] S. T. Milner, Science **251**, 905 (1991).
[12] J. M. H. M. Scheutjens and G. J. Fleer, J. Phys. Chem. **83**, 1619 (1979).
[13] J. M. H. M. Scheutjens and G. J. Fleer, J. Phys. Chem. **84**, 178 (1980).
[14] F. A. M. Leermakers and J. M. H. M. Scheutjens, J. Chem. Phys. **89**, 3264 (1988).
[15] F. A. M. Leermakers and J. M. H. M. Scheutjens, J. Chem. Phys. **89**, 6912 (1988).
[16] T. Cosgrove, T. Heath, B. van Lent, F. A. M. Leermakers, and J. M. H. M. Scheujtens, Macromolecules **20**, 1692 (1987).
[17] C. M. Wijmans, F. A. M. Leermakers, and G. J. Fleer, J. Chem. Phys. **101**, 8214 (1994).
[18] C. M. Wijmans, J. M. H. M. Scheutjens, and E. B. Zhulina, Macromolecules **25**, 2657 (1992).
[19] G. J. Fleer and J. M. H. M. Scheutjens, Colloid Surf. **51**, 281 (1990).
[20] C. M. Wijmans and E. B. Zhulina, Macromolecules **26**, 7214 (1993).
[21] C. M. Wijmans, F. A. M. Leermakers, and G. J. Fleer, Langmuir **10**, 4514 (1994).
[22] C. M. Wijmans and B. J. Factor, Macromolecules **29**, 4406 (1996).
[23] S. F. Edwards, Proc. Phys. Soc. **85**, 613 (1965).
[24] A. K. Dolan and S. F. Edwards, Proc. R. Soc. London, Ser. A **337**, 509 (1974).
[25] A. K. Dolan and S. F. Edwards, Proc. R. Soc. London, Ser. A **343**, 427 (1975).
[26] M. Muthukumar, and J.-S. Ho, Macromolecules **22**, 965 (1989).
[27] M. D. Whitmore and J. Noolandi, Macromolecules **23**, 3321 (1990).
[28] E. K. Lin and A. P. Gast, Macromolecules **29**, 390 (1996).
[29] E. B. Zhulina, O. V. Borisov, and V. A. Priamitsyn, J. Colloid Interface Sci. **137**, 495 (1990).
[30] E. B. Zhulina, O. V. Borisov, and L. Brombacher, Macromolecules **24**, 4679 (1991).
[31] S. T. Milner, T. A. Witten, and M. E. Cates, Macromolecules **21**, 2610 (1988).
[32] S. T. Milner, Europhys. Lett. **7**, 695 (1988).
[33] S. Alexander, J. Phys. (Paris) **38**, 983 (1977).
[34] P.-G. de Gennes, Macromolecules **13**, 1069 (1980).
[35] P.-G. de Gennes, C. R. Acad. Sci. Paris **300**, 839 (1985).
[36] S. Patel, M. Tirrell, and G. Hadziioannou, Colloids Surf. **31**, 157 (1988).
[37] P. Pincus, Macromolecules **24**, 2912 (1991).
[38] M. Murat and G. S. Grest, Phys. Rev. Lett. **63**, 1074 (1989).
[39] M. Murat and G. S. Grest, Macromolecules **22**, 4054 (1989).
[40] A. Chakrabarti and R. Toral, Macromolecules **23**, 2016 (1990).
[41] P. Y. Lai and K. Binder, J. Chem. Phys. **95**, 9288 (1991).
[42] B. Li and E. Ruckenstein, J. Chem. Phys. **106**, 280 (1997).
[43] E. A. DiMarzio and R. J. Rubin, J. Chem. Phys. **55**, 4318 (1971).
[44] K. A. Dill and R. S. Cantor, Macromolecules **17**, 380 (1984).
[45] J. Bandrup and E. J. Immerght, *Polymer Handbook*, 3rd ed. (Wiley, New York, 1989).

6490 *Langmuir* **2004,** *20,* 6490–6500

Simple Model for Grafted Polymer Brushes

Marian Manciu and Eli Ruckenstein*

Department of Chemical Engineering, State University of New York at Buffalo, Buffalo, New York 14260

Received January 23, 2004. In Final Form: April 29, 2004

The first theories of grafted polymer brushes assumed a step profile for the monomer density. Later, the real density profile was obtained from Monte Carlo or molecular dynamics simulations and calculated numerically using a self-consistent field theory. The analytical approximations of the solutions of the self-consistent field equations provided a parabolic dependence of the self-consistent field, which in turn led to a parabolic distribution for the monomer density in neutral brushes. As shown by numerical simulations, this model is not accurate for dense polymer brushes, with highly stretched polymers. In addition, the scaling laws obtained from the analytical approximations of the self-consistent field theory are identical to those derived from the earlier step-profile-approximation and predict a vanishing thickness of the brush at low graft densities, and a thickness exceeding the length of the polymer chains at high graft densities. Here a simple model is suggested to calculate the monomer density and the interaction between surfaces with grafted polymer brushes, based on an approximate calculation of the partition function of the polymer chains. The present model can be employed for both good and poor solvents, is compatible with a parabolic-like profile at moderate graft densities, and leads to an almost steplike density for highly stretched brushes. While the thickness of the brush depends strongly on solvent quality, it is a continuous function in the vicinity of the Θ temperature. In good and moderately poor solvents, the interactions between surfaces with grafted polymer brushes are always repulsive, whereas in poor solvents the interactions are repulsive at small separations and become attractive at intermediate separation distances, in agreement with experiment. At large separations, a very weak repulsion is predicted.

1. Introduction

One of the first important applications of polymer brushes was to the stabilization of colloidal dispersions.[1] The brushes were generated through the grafting of polymers on the surface of colloidal particles. When two particles approach each other, the overlap of their polymer brushes generates a steric repulsion, which can prevent their coagulation. A simple theory for polymer brushes was proposed by Alexander,[2] who assumed that the grafted polymer is composed of a sequence of blobs. The sectional area of each blob was provided by the area occupied by each grafted polymer on the surface, and it was assumed that, inside the blobs, the relation between the chain length and the occupied volume is the same as for a free polymer in a solvent. The repulsion due to the overlap of two polymer brushes was subsequently evaluated by de Gennes[3] up to an arbitrary proportionality constant, from scaling considerations.

The simple Alexander–de Gennes theory, which assumed a steplike monomer density in the brush, captured the dependence of the interaction on the physical parameters (length of the polymer, density of grafting, quality of the solvent) and provided a satisfactory approximation for the calculation of the steric repulsion. However, new applications of grafted polymers on surfaces, such as the control of the catalytic selectivities of some chemical reactions by varying the thickness of a brush,[4] the prevention of the adsorption of proteins on surfaces (a condition required for biocompatibility),[5] or the control of a membrane selectivity by changing the pH of the solution[6] required a more accurate theory of polymer brushes.

The properties of grafted polymer brushes could be determined accurately via the traditional Monte Carlo[7] or molecular dynamics[8] simulations. Since one should account not only for the monomer–monomer interactions but also for the interactions with the solvent molecules, in general only calculations for systems of relatively small size could be performed. A modality to increase the calculation efficiency is to employ simplifying assumptions; one of the most successful was proposed by Dolan and Edwards,[9] who considered that the monomers of the grafted polymer are distributed according to a random walk in an external potential. This potential is generated by monomer–monomer and monomer–solvent interactions and depends on the local monomer concentration. Since the monomer concentration in the brush depends on the potential, which in turn depends on the monomer concentration, the potential can be determined only self-consistently via successive approximations. A convenient modality to implement self-consistent calculations[10] is based on a lattice model due to Scheutjens and Fleer.[11] The lattice-based self-consistent field theory was later improved to account for anisotropy, bond correlations, chain stiffness, and intermolecular interactions.[12–14]

* Corresponding author. E-mail: feaeliru@acsu.buffalo.edu. Telephone: (716) 645 2911/2214. Fax: (716) 645 3822

(1) Napper, D. H. *Polymeric stabilization of colloidal dispersion*; Academic Press: San Diego, CA, 1983.
(2) Alexander, S. *J. Phys. (Paris)* **1977**, *38*, 983.
(3) de Gennes, P. G. *Adv. Colloid Interface Sci.* **1987**, *27*, 189.
(4) Ruckenstein, E.; Hong, L. *J. Catal.* **1992**, *136*, 178.
(5) Carignano, M. A.; Szleifer, I. *Colloids Surf. B* **2000**, *18*, 169.
(6) Idol, W. K.; Anderson, J. L. *J. Membr. Sci.* **1986**, *28*, 269.
(7) Muthukumar, M.; Ho, J.-S. *Macromolecules* **1989**, *22*, 965.
(8) Seidel, C. *Macromolecules* **2003**, *36*, 2536.
(9) Dolan, A. K.; Edwards, F. R. S. *Proc. Royal Soc. London A* **1974**, *337*, 509.
(10) Hirz, S. Modeling of Interactions Between Adsorbed Block Copolymers. M.S. Thesis, University of Minnessota, Minneapolis, MN, 1988. Cosgrove, T.; Heath, T.; van Lent, B.; Leermakers, F.; Scheutjens, J. *Macromolecules* **1987**, *20*, 1692.
(11) Scheutjens, J. M. H. M.; Fleer, G. J. *J. Phys. Chem.* **1979**, *83*, 1619 and **1980**, *84*, 178.
(12) Leermakers, F. A. M.; Scheutjens, J. M. H. M. *J. Chem. Phys.* **1988**, *89*, 6912.
(13) Cantor, R. S.; McIlroy, P. M. *J. Chem. Phys.* **1989**, *91*, 416.

10.1021/la049781y CCC: $27.50
Published on Web 06/16/2004

Both the Monte Carlo and molecular dynamics simulations, as well as the numerical solutions of the self-consistent field (SCF) equations, are time-expensive and do not offer a clear insight regarding the relations between the parameters of the grafted polymers and the properties of the brushes. A major progress in the analytical approximation of the SCF solution was suggested by Milner et al. (the MWC theory),[15] based on the observation that, at high stretching, the partition function of the brush is dominated by the "classical path" (the most probable distribution). This approach is similar to the semiclassical approach of quantum mechanics, the Wentzel–Kramers–Brillouin (WKB) approximation. Under these assumptions, it turned out that, for equal-length polymer chains, the self-consistent field is parabolic, which leads to a parabolic distribution of the monomer density for neutral brushes. The simplicity of this result became extremely appealing, and most subsequent theories of the polyelectrolyte brushes (such as those of Miklavic and Marcelja,[16] Misra et al.,[17] and Zhulina and Borisov,[18] to name only a few of them) were based on the "parabolic field" approximation. A more intuitive model to describe the parabolic distribution of monomers in grafted polymer brushes was proposed by Pincus,[19] who considered that the profile of the brush is a result of the competition between an excluded-volume free energy and the decrease of the entropy of the polymer in the stretched configuration. Assuming that the excluded volume interactions are proportional to the square of the local monomer concentration and also a Gaussian elasticity of the polymer, this model led also to a parabolic distribution of the monomers in the brush.[19]

It should be noted that the "parabolic field" was derived by assuming a high stretching of the polymers; it provides, however, a better description of polymer brushes in the intermediate-stretching regime (good solvents and moderate graft densities). In contrast, the strongly stretched brushes, as obtained from molecular dynamics simulations,[8] seem to be better described by the Alexander step profile. As a matter of fact, the analytical approximations of the self-consistent field theory provided the same scaling laws for the brush thickness as those derived earlier by Alexander in the framework of the step-profile approximation.[19] These laws clearly fail at both very low graft densities (by predicting a vanishing small thickness of the brush) as well as at very high graft densities (by predicting a brush thickness larger than the length of the grafted polymer).[20] The limitation of the parabolic profile description only to brushes with moderate graft densities immersed in good solvents was also recognized by Shim and Cates, who proposed an ad hoc functional form for the free energy of stretching of the chain and consequently obtained a density profile much flatter than a parabola.[21]

Because the neutral polymer brushes are unlikely to be very strongly stretched, the MWC approach usually provides in this case an accurate approximation. The situation is however different for the highly charged grafted polyelectrolytes, which are almost completely stretched.[8,22] The purpose of this article is to present a simple method to calculate the thickness of the strongly stretched grafted polymer brushes and the interactions between surfaces covered by strongly stretched brushes. The calculations are based on an approximate Monte Carlo method, involving the assumptions that the configurations of minimum free energy are dominating the partition function and that the probability that the farthest monomer from the surface located at the distance z is approximated by the probability that the last monomer of the chain reaches the same distance. Under this hypothesis, in good solvents, the model predicts a constant density of monomers for each configuration that ends up at distance z, which is similar to the approximation successfully employed by Flory to explain the scaling laws of the free polymers in good solvents.[23]

2. The Profile of Neutral Polymer Brushes

The traditional Monte Carlo method implies the generation of all possible configurations of grafted polymers and the calculation of the following partition function for a polymer

$$Z = \sum_i \exp\left(-\frac{U_i}{kT}\right) \tag{1}$$

where k is the Boltzmann constant, T the absolute temperature, "i" runs over all possible configurations and U_i represents the total energy of a configuration, which includes the interactions between the monomers of the same and neighboring polymer chains and the interactions between the monomers and the grafting surface as well as their interactions with the solvent molecules. A usual modality to generate the configurations of a polymer composed of N monomers of length "a" is to assume that the polymer is composed of N_l independent pieces of persistence length "l" (where $lN_l \equiv aN$), that are connected but can assume any relative orientations. This model constitutes the statistical equivalent of a random walk. Since the length of the grafted polymers perpendicular to the surface is much larger than their size parallel to the surface, one can assume that the in-plane monomer concentration is constant. In the absence of any monomer interactions, the probability $\psi'(z)$ for the last monomer of the chain to reach the position z is well approximated (for large N) by the analytical solution of a noninteracting random walk with a perfect-reflecting wall at the origin (see Appendix)[24]

$$\psi'(z) \cong \sqrt{\frac{2}{\pi\sigma^2}} \exp\left(-\frac{z^2}{2\sigma^2}\right) \quad 0 < z < Na \tag{2a}$$

where $\sigma = l\sqrt{N_l}$ is the root-mean square distance of the random walk of N_l steps of length l. When additional interactions are present, the probability for a configuration to end at a distance z is multiplied by its corresponding Boltzmann factor

$$\psi(z) \propto \exp\left(-\frac{z^2}{2\sigma^2}\right) \sum_i \exp\left(-\frac{U_i(z)}{kT}\right) \tag{2b}$$

where the sum is performed over all possible configurations

(14) Li, B. Q.; Ruckenstein, E. *J. Chem. Phys.* **1997**, *106*, 280. Ruckenstein, E.; Li, B. Q. *J. Chem. Phys.* **1997**, *107*, 3.
(15) Milner, S. T.; Witten, T. A.; Cates, M. E. *Macromolecules*, **1988**, *21*, 2610.
(16) Miklavic, S. J.; Marcelja, S. *J. Phys. Chem.* **1988,** *92*, 6718.
(17) Misra, S.; Varanasi, S.; Varanasi, P. P. *Macromolecules* **1989**, *22*, 4173.
(18) Zhulina, E. B.; Borisov, O. V. *J. Chem. Phys.* **1997**, *107*, 5952.
(19) Pincus, P. *Macromolecules* **1991**, *24*, 2912.
(20) de Gennes, P. G. *Scaling Concepts in Polymer Physics*; Cornell University Press: Ithaca, NY, 1979.
(21) Shim, D. F. K.; Cates, M. E. *J. Phys. (Paris)* **1989**, *50*, 3535.
(22) Huang, H.; Ruckenstein, E. *J. Colloid Interface Sci.* **2004**, in press.
(23) Flory, P. *Principles of Polymer Chemistry*; Cornell University Press: Ithaca, NY, 1971.
(24) Chandrasekhar, S. *Rev. Mod. Phys.* **1943**, *15*, 3.

"i" whose last monomer of the chain is located at z and have the total free energies $U_i(z)$.

The random walk analogy was employed in the derivation of eq 2b only to provide the number of all possible configurations that end up at the distance z. The main difficulty is to calculate the free energies for each of the possible configurations. A first approximation, the mean field approach, consist in assuming that the total free energy density of a configuration (which includes all the interactions between monomers and solvent molecules as well as the chain entropy of mixing) can be described in term of the local monomer volume fraction ϕ by the Flory–Huggins mixing free energy density[20,23]

$$F_{\mathrm{FH}}(z) = \frac{kT}{a^3}\left(\frac{\phi}{N}\ln(\phi) + (1-\phi)\ln(1-\phi) + \chi\phi(1-\phi)\right) \cong \frac{kT}{a^3}\left((\chi-1)\phi + \left(\frac{1}{2}-\chi\right)\phi^2 + \frac{1}{6}\phi^3 + \frac{1}{12}\phi^4 ...\right) \quad (3a)$$

where a is the length of a monomer, χ is the Flory interaction parameter,[20,23] and the term proportional to $1/N$ was neglected because N is large. The term linear in ϕ, while nonnegligible for the mixing free energy density eq 3a, is not important for the total Flory–Huggins mixing free energy of the configuration, since its integral over the volume of the brush

$$s^2\int_0^{Na}\left(\frac{kT}{a^3}(\chi-1)\phi(\zeta)\right)\mathrm{d}\zeta = \frac{kTs^2}{a^3}(\chi-1)\int_0^{Na}\phi(\zeta)\,\mathrm{d}\zeta = \frac{kT}{a^3}(\chi-1)Nv = \text{const} \quad (3b)$$

provides (see eq 4b) the same constant for each possible configuration, and the same additive constant in the free energy of each configuration does not modify the result. For this reason, a convenient approximation of the Flory–Huggins free energy density is given by[20,25]

$$F_{\mathrm{FH}}(\phi) \cong \frac{kT}{a^3}\left(-\frac{1}{2}\tau\phi^2 + \frac{1}{6}w\phi^3\right) \quad (3c)$$

where τ is a dimensionless excluded volume parameter ($\tau = (2\chi - 1)$ with $\tau < 0$ for a good solvent and $\tau > 0$ for a poor solvent) and w is the third virial coefficient, which is positive and typically on the order of unity.[25] The latter coefficient should roughly account for the higher order terms in eq 3a (all of them being positive), as well as for the additional interactions between monomers, not included in the traditional Flory–Huggins mixing free energy density eq 3a.

The second approximation employed here is to consider that the configurations of the minimum energy provide most of the contributions to the average Boltzmann factor in eq 2b, the well-known saddle point approximation of statistical mechanics.

For a good solvent, the Flory–Huggins free energy density is minimized by the lowest possible volume fraction of monomers, which is in turn obtained for a completely stretched chain. In the stretched configuration that ends up at the distance $z < Na$ a random walker moves away from the surface until reaches a distance $z' > z$ and then returns to z (such as the length of the path is $z' + (z' - z) = Na$; hence, $z' = (Na + z)/2$. However, there is only one such configuration among the $2C_N^k = 2N!/k!(N-k)!$ configurations that ends up at $z = ka$, where the factor 2 multiplying the binomial coefficient accounts for the reflecting wall at the origin. While this particular configuration has the highest Boltzmann factor, its contribution to the sum in eq 2b is in most cases not significant, because of its low statistical weight. In what follows, a third approximation will be employed, namely that we will take into account only the configurations for which the last monomer of the chain is also the farthest from the surface. As will be shown below, under these assumptions, the minimum Flory–Huggins free energy in good solvents is provided by the configurations with a constant distribution of monomers. Such a distribution was successfully employed by Flory to derive a relation between the root-mean-square distance between the ends and the number of monomers of a polymer in good solvents,[23] and it provided the correct value for the scaling coefficient in one dimension and values within a percent of the most accurate numerical results in 2 and 3 dimensions.[20]

The Flory–Huggins free energy of a configuration for which the monomers are confined between 0 and z is given by

$$U_{\mathrm{FH}}(z) = s^2\int_0^z F_{\mathrm{FH}}(\zeta)\,\mathrm{d}\zeta = \frac{s^2kT}{a^3}\int_0^z\left(-\frac{\tau}{2}\phi^2(\zeta) + \frac{w}{6}\phi^3(\zeta)\right)\mathrm{d}\zeta \quad (4a)$$

where the monomer volume fraction ϕ obeys the normalization condition

$$\int_0^z \phi(\zeta)\,\mathrm{d}\zeta = \frac{Nv}{s^2} \quad (4b)$$

where N is the number of monomers of volume v in a chain, s^2 is the area corresponding to a grafted chain on the surface, and Na is the maximum length of a polymer.

The function $\phi(z)$ which minimizes the Flory–Huggins free energy (eq 4a) with the constraint (eq 4b) is provided by the extremum of the functional

$$\int_0^z \mathscr{F}(\phi)\,\mathrm{d}\zeta = \frac{s^2kT}{a^3}\int_0^z\left(-\frac{\tau}{2}\phi^2(\zeta) + \frac{w}{6}\phi^3(\zeta) + \lambda\phi(\zeta)\right)\mathrm{d}\zeta \quad (5a)$$

where λ is a Lagrange multiplier. The corresponding Euler–Lagrange equation

$$\frac{\partial}{\partial\zeta}\frac{\partial\mathscr{F}}{\partial\left(\frac{\partial\phi}{\partial\zeta}\right)} - \frac{\partial\mathscr{F}}{\partial\phi} = 0 \quad (5b)$$

leads to

$$-\tau\phi + \frac{w}{2}\phi^2 + \lambda = 0 \quad (5c)$$

The Euler–Lagrange equation (eq 5c) does not depend on ζ, and the continuous solution is a constant obtained from the constraint (4b):

$$\phi(\zeta,z) = \frac{Nv}{s^2 z} \quad (5d)$$

While (5d) is the only continuous solution that is an extremal of the functional (5a), a discontinuous solution, which corresponds to the separation into two homogeneous phases, might lead to a lower value of the free energy. Therefore, we will consider configurations with volume fraction ϕ_a on a domain (not necessarily connect) of length $z_a < z/2$ and with volume fraction ϕ_b on the remaining

(25) Ross, R. S.; Pincus, P. *Europhys. Lett.* **1992**, *19*, 79.

domain of length $z_b = z - z_a$. The restriction that $z_a < z/2$ does not reduce the generality, since the other case is obtained by interchanging the subscripts a and b. The Flory–Huggins free energy of such a configuration is given by

$$U(z,\phi_a,z_a) = \frac{s^2kT}{a^3}\left(z_a\left(-\frac{\tau}{2}\phi_a^{\,2} + \frac{w}{6}\phi_a^{\,3}\right) + (z - z_a) \times \left(-\frac{\tau}{2}\left(\frac{\frac{Nv}{s^2} - \phi_a z_a}{z - z_a}\right)^2 + \frac{w}{6}\left(\frac{\frac{Nv}{s^2} - \phi_a z_a}{z - z_a}\right)^3\right)\right) \quad (6)$$

where the normalization eq 4b ($z_a\phi_a + (z - z_a)\phi_b = Nv/s^2$) was taken into account. The extremal of $U(z,\phi_a, z_a)$ with respect to ϕ_a leads to two pair of solutions:

$$\text{(i)}\quad \phi_a = \frac{Nv}{s^2z},\quad z_a = z,\quad \text{which is equivalent to eq 5d} \quad (7a)$$

$$\text{(ii)}\quad \phi_a = \frac{\frac{2\tau}{w}z - \frac{2\tau}{w}z_a - \frac{Nv}{s^2}}{z - 2z_a},\quad \phi_b = \frac{\frac{Nv}{s^2} - \frac{2\tau}{w}z_a}{z - 2z_a} \quad (7b)$$

Introducing the pair of solutions (eq 7b) in eq 6, the free energy becomes a function of z_a, with its derivative with respect to z_a given by

$$\frac{\partial U(z,z_a)}{\partial z_a} = \frac{2}{3w^2}\left(\frac{\frac{Nv}{s^2}w - \tau z}{z - 2z_a}\right)^3 \quad (8)$$

For $\tau < 0$ (good solvent), the derivative is always positive; hence, the minimum corresponds to $z_a = 0$, which leads to $z_b = z$ and $\phi_b = Nv/s^2z$, which is the same as eq 5d.

For $\tau > 0$ (poor solvents), the derivative is positive for z smaller than

$$z_{\min} = \frac{w}{\tau}\frac{Nv}{s^2} \quad (9)$$

and negative for $z > z_{\min}$. In the first case ($z < z_{\min}$, poor solvent), the minimum free energy is obtained again for $z_a = 0$ (a constant distribution of monomers between 0 and z) while in the latter case ($z > z_{\min}$, poor solvent), the Flory–Huggins free energy is a monotonically decreasing function of z_a. However, the volume fraction ϕ_b in eq 7b, which also decreases with increasing z_a, cannot become smaller than that corresponding to a completely stretched chain:

$$\phi_{\min} = \frac{Nv}{s^2Na} = \frac{v}{s^2a}$$

The maximum value physically allowed for z_a, obtained from the condition $\phi_b = \phi_{\min}$,

$$\frac{\frac{Nv}{s^2} - \frac{2\tau}{w}z_a}{z - 2z_a} = \frac{v}{s^2a} \quad (10a)$$

is given by

$$z_{a,\max} = \frac{1}{2}\frac{Na - z}{\frac{\tau}{w}\frac{s^2a}{v} - 1} \quad (10b)$$

which employed in eq 7b leads to the volume fractions:

$$\phi_a = \phi_{\max} = \frac{2\tau}{w} - \frac{v}{s^2a} \quad \text{(on a domain of length } z_{a,\max}) \quad (10c)$$

$$\phi_b = \phi_{\min} = \frac{v}{s^2a} \quad \text{(on a domain of length } z - z_{a,\max}) \quad (10d)$$

Parts c and d of eq 10 simply state that the system separates into two phases, if the solvent is poor ($\tau > 0$) and the confinement is weak ($z > z_{\min} = (w/\tau)\, Nv/s^2$), but not when the confinement is strong ($z < z_{\min}$). While the diluted polymer in the bulk solvent undergoes a phase separation as soon as τ becomes positive (at the Θ temperature), only the configurations of the grafted polymer that are sufficiently long undergo a phase separation. This occurs because at high volume fractions, the excluded volume interactions (due to the positive third virial coefficient w) dominate the Flory–Huggins free energy and prevent the polymer collapse. If the grafting density is sufficiently high (small s) in eq 9, the minimum length $z_{\min}$ required for phase transition might exceed the maximum length of the polymer Na regardless of the value τ, therefore none of the physically allowed chain configurations undergoes phase separation.

In summary, the configurations of minimum Flory–Huggins free energy for monomers distributed between 0 and z correspond to the following.

(i) Good solvents ($\tau < 0$) and $0 < z < Na$:

$$\phi(\zeta,z) = \frac{Nv}{s^2z} \quad (5d)$$

(ii) Poor solvents ($\tau > 0$) and $0 < z < z_{\min} = (Nv/s^2)(w/\tau)$:

$$\phi(\zeta,z) = \frac{Nv}{s^2z} \quad (5d)$$

(iii) Poor solvents ($\tau > 0$) and $z_{\min} = (Nv/s^2)(w/\tau) < z < Na$:

$$\phi(\zeta,z) = \phi_{\max} = \frac{2\tau}{w} - \frac{v}{s^2a} \quad \text{on an interval of length}$$

$$z_{a,\max} = \frac{1}{2}\frac{Na - z}{\frac{\tau}{w}\frac{s^2a}{v} - 1} \quad (10c)$$

$$\phi(\zeta,z) = \phi_{\min} = \frac{v}{s^2a} \quad \text{on an interval of length } z\text{-}z_{a,\max} \quad (10d)$$

Therefore, the minimum total Flory–Huggins energy of a chain configuration confined between 0 and z is given by the following.

(i) Good solvent in the range $0 < z < Na$ or poor solvent in the range $0 < z < z_{\min}$:

$$U_{\mathrm{FH}}(z) = s^2\int_0^z F_{\mathrm{FH}}(\zeta)\, d\zeta = \frac{kT}{a^3}\left(-\frac{\tau N^2v^2}{2s^2z} + \frac{wN^3v^3}{6s^4z^2}\right) \quad (11a)$$

(ii) Poor solvent in the range $z_{\min} < z < Na$:

$$U_{FH}(z) = \frac{kTs^2}{a^3}\left(z_{a,\max}\left(-\frac{\tau}{2}\phi_{\max}{}^2 + \frac{w}{6}\phi_{\max}{}^3\right) + (z - z_{a\max})\left(-\frac{\tau}{2}\phi_{\min}{}^2 + \frac{w}{6}\phi_{\min}{}^3\right)\right) \quad (11b)$$

The probability for the last monomer of the chain to reach point z is consequently obtained from eq 2b:

$$\psi(z) = \frac{1}{C_1}\exp\left(-\frac{z^2}{2\sigma^2}\right)\exp\left(-\frac{U_{FH}(z)}{kT}\right), \quad 0 < z < Na \quad (12a)$$

where $\sigma = l\sqrt{N_l}$ is the root-mean-square length of the random walk of persistence length l and C_1 is a normalization constant which obeys

$$\int_0^{Na} \psi(z)\, dz = 1 \quad (12b)$$

The monomer volume fraction in the brush $\Phi(z)$ is the statistical average of the contributions of configurations ending up at different positions. In poor solvents, the configurations with $z > z_{\min}$ that minimize the Flory–Huggins energy do not have a constant monomer density. However, the monomers are equally like to accumulate at a volume fraction $\phi_{\max}$ in any region between 0 and z; therefore, an average over all the possible configurations (of the same minimum energy) would lead to a constant monomer density between 0 and z. For $z < z_{\min}$ (in poor solvents) or for good solvents, the configuration of minimum energy corresponds to a constant distribution of monomers between 0 and z, $\phi(\zeta,z) = Nv/s^2z$ for $0 < \zeta < z$ and $\phi(\zeta,z) = 0$for $\zeta > z$. Consequently, the chains that end up at distances smaller than z do not contribute to the volume fraction of the monomers at z, whereas the chains which end up at distances larger than z contribute $\phi(\zeta,z) = Nv/s^2z$ with the probability $\psi(z)$ to the total monomer fraction $\Phi(z)$. Therefore, the monomer volume fraction is given by

$$\Phi(z) = \frac{1}{C_1}\int_z^{Na} \psi(\xi)\phi(\zeta,\xi)\, d\xi = \frac{1}{C_1}\int_z^{Na} \frac{Nv}{s^2\xi}\exp\left(-\frac{\xi^2}{2\sigma^2}\right)\exp\left(-\frac{U_{FH}(\xi)}{kT}\right) d\xi \quad (13)$$

The results will be illustrated through this paper with model calculations for a grafted brush with the following values of the parameters: $s = 10$ Å, $a = 1$ Å, $N = 1000$, $l = 10$, $v = 1$ Å^3. In parts a and b of Figure 1, the probability for the last monomer of the chain to be at z (Figure 1a) and the local volume fraction of the monomers in the brush (Figure 1b) are plotted as a function of the distance from the surface for good solvents ($\tau < 0$). In the absence of interactions between monomers $\tau = w \equiv 0$, $\psi(z)$ is Gaussian and the profile of the brush can be roughly approximated by a parabola. The brush is swollen because of the repulsive Flory–Huggins interactions, which increase with increasing volume fraction ϕ. At low graft densities (large values of s, hence low values of ϕ) or for moderately good solvents (small absolute value of τ and w) the brush profile can be still approximated by a parabola. However, by increasing the graft density (small s) or solvent quality (large absolute value of τ) a strongly stretched brush is generated, with an almost constant density profile over most of the brush, but with an exponential-like tail, in agreement with the numerical simulations.[8] The most likely chain configuration in the brush is obtained from the condition of vanishing of the derivative of $\psi(z)$ with respect to z in eq 12a. Neglecting the second term in the Flory–Huggins interaction ($w = 0$) leads to

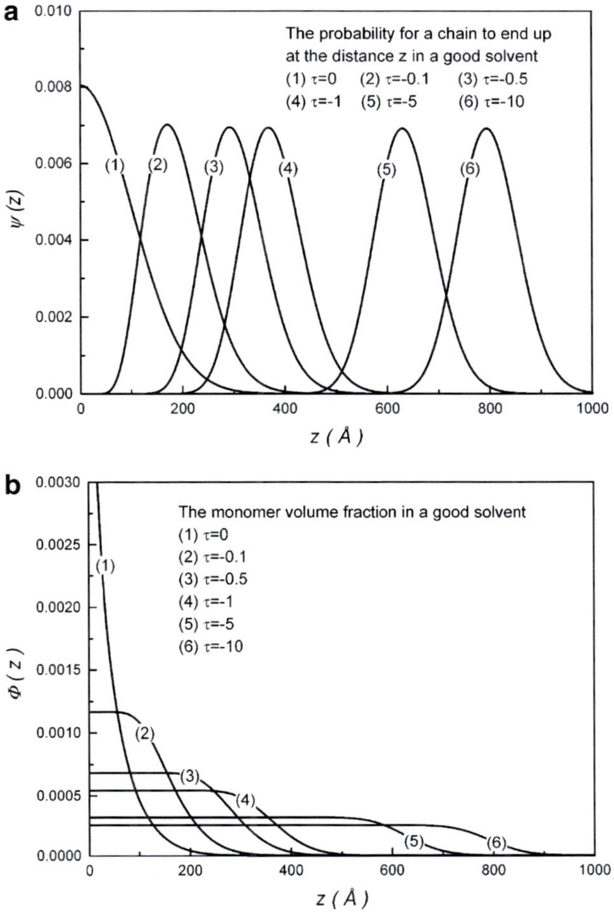

Figure 1. (a) Probability $\psi(z)$ for the last monomer of a grafted chain to be located at the distance z from the surface and (b) the local volume fraction of monomers in a good solvent, for $s = 10$ Å, $a = 1$ Å, $N = 1000$, $l = 10$ Å, $v = 1$ Å^3, and various solvent qualities.

$$\bar{z} = \left(-\frac{\tau l^2 N_l N^2 v^2}{2s^2a^3}\right)^{1/3} \quad (14)$$

If the length of the most probable chain configuration $\bar{z}$ is assumed to be equal to the thickness of the brush L, eq 14 provides the same dependence between the graft density and thickness as the Alexander theory. Whereas eq 14 accounts explicitly for the persistence length of a polymer (its degree of flexibility), with the simplified assumptions that $l \sim a$, $N_l \sim N$, $v \sim a^3$ the Alexander scaling law is completely recovered:

$$L \propto N\left(\frac{\tau v a^2}{2s^2}\right)^{1/3} \quad (15)$$

Here τv represents an "excluded volume" of the interaction.[19]

However, as shown in Figure 1a, the approximation of the thickness of the brush by the length of the most probable configuration is in many cases arbitrary, since configurations that end up at different distances can contribute notably to the brush thickness. In addition, the derivation of eq 14 implies that the random walk can be described by a Gaussian extended from $z = 0$ to $z = \infty$, while the random walk should not exceed the maximum chain length Na. Equations 12 and 13 do account for this inconsistency (by a cutoff in the integral limits) and do not predict a brush thickness that exceeds Na. Another feature of the present model is that whereas at vanishing Flory–Huggins interactions ($\tau = w = 0$) the most probable configuration ends up at $z = 0$, the brush does not collapse on the surface, as suggested by the scaling laws (eqs 14 or 15), because the configurations extended to $z > 0$ have nonvanishing probabilities.

The present model is based on several assumptions: (i) the possible configurations of the grafted chain are described by a random walk; (ii) their free energy densities are expressed as functions of the local monomer volume fraction alone; (iii) the configurations of minimum energy dominate the partition function of the system; (iv) only the configurations with monomers distributed between the surface and the position of the last monomer of the chain, assumed to be the farthest one, are taken into account. The latter assumption basically implies that the probability that the most distant monomer from the surface reaches the distance z is equal to the probability that the last monomer of the chain reaches this distance; this approximation clearly fails when z is in the vicinity of the surface. However, in swollen brushes the behavior of the monomers in the vicinity of the surface is less important than the behavior of the distant monomers, which are primarily responsible for the brush thickness and for the interactions between brushes.

As a matter of fact, in good solvents, the above assumptions predict that only configurations with constant monomer density between 0 and z contribute essentially to the partition function. This is equivalent to the Flory model for polymer swelling in a good solvent, which was remarkably successful, especially for one dimension.[20,23] It is therefore to be expected that the model is reasonably accurate in good solvents and fails in extremely poor solvents, because the configurations in the vicinity of the surface dominate the collapsed brush. It should be however noted that the sharp distinction between good and poor solvent at the Θ temperature occurs only at high dilutions, whereas for high volume fractions the Flory–Huggins free energy is dominated by the volume-exclusion interaction between monomers and increases with ϕ even for poor solvents. Therefore, at high grafting densities, the brush might be swollen even at temperatures below Θ; hence, the present model is applicable.

The probability $\psi(z)$ that the last monomer of the chain is located at the distance z, and the volume fraction $\Phi(z)$ of the monomers in the grafted brush immersed in a moderately poor solvent are plotted in Figure 2, parts a and b, using for the parameters the values selected above (s = 10 Å, a = 1 Å, N = 1000, l = 10, v = 1 Å^3), a fixed value for the third virial coefficient $w = 1$ and various values for the interacting parameter τ. It is of interest to note that the brush thickness is modified continuously when the quality of the solvent changes from good ($\tau < 0$) to poor ($\tau > 0$). This occurs because the Flory–Huggins free energy is minimized by a constant monomer distribution in both good solvents and poor solvent for configurations with $z < z_{\min} = Nv/s^2\, w/\tau$. For small values of τ, $z_{\min}$ (for which phase separation occurs) exceeds the maximum length of the chain Na. Therefore, in these cases, the polymer behaves as immersed in a good solvent, even below the Θ temperature. This uniform collapse of the brushes induced by changing the solvent quality is in agreement with the results reported by Ross and Pincus.[25]

The most probable configurations in Θ solvents ($\chi = 1/2$, $\tau = 0$), obtained from the extremum of eq 12, occurs at a distance given by

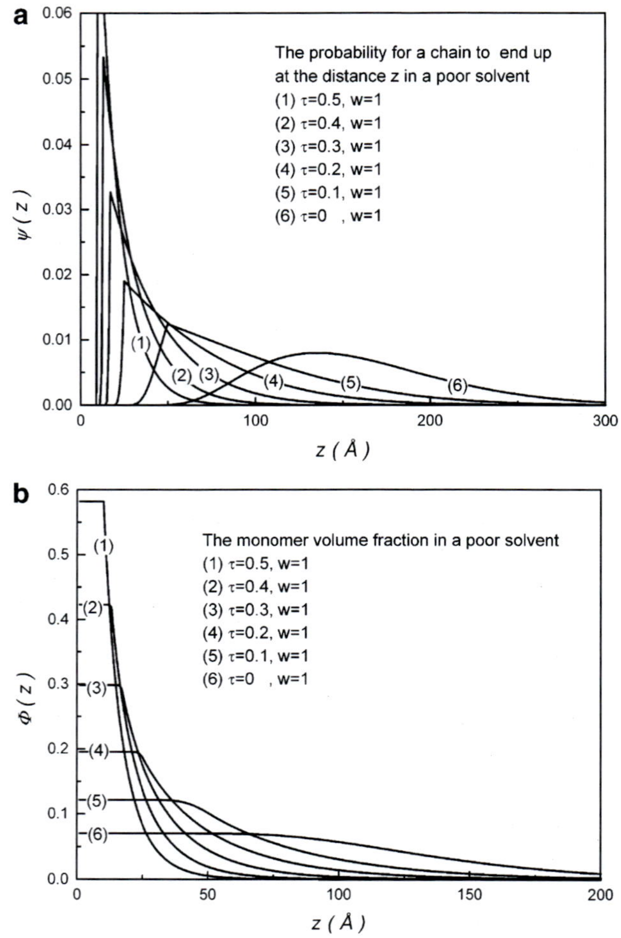

Figure 2. (a) Probability $\psi(z)$ for the last monomer of a grafted chain to be located at the distance z from the surface and (b) the local volume fraction of monomers in a poor solvent, for s = 10 Å, a = 1 Å, N = 1000, l = 10 Å, v = 1 Å^3, and various solvent qualities.

$$\bar{z} = \left(\frac{wN^3N_l v^3 l^2}{3a^3 s^4}\right)^{1/4} \tag{16}$$

which differs from the Alexander scaling law, eq 15.

In Figure 3, the length of the most probable chain configuration (which roughly accounts for the thickness of the brush), is plotted as a function of the quality of the solvent τ for various values of the third virial coefficient w. For small values of the third virial coefficient, there is a large variation in the thickness of the brush (albeit continuous) in the vicinity of the Θ temperature, as observed experimentally.[26]

3. The Interaction between Grafted Neutral Brushes

A. Good Solvents. When two surfaces with grafted polymer brushes approach each other, the overlap of the neutral brushes generates an interaction between surfaces. In good solvents, the Flory–Huggins mixing free energy density (eq 3c) increases with the monomer concentration; therefore, one expects that the overlap of the brushes would lead always to repulsion. In the present

(26) Kilbey, S. M.; Watanabe, H.; Tirrell, M. *Macromolecules* **2001**, *34*, 5249.

6496 *Langmuir, Vol. 20, No. 15, 2004* *Manciu and Ruckenstein*

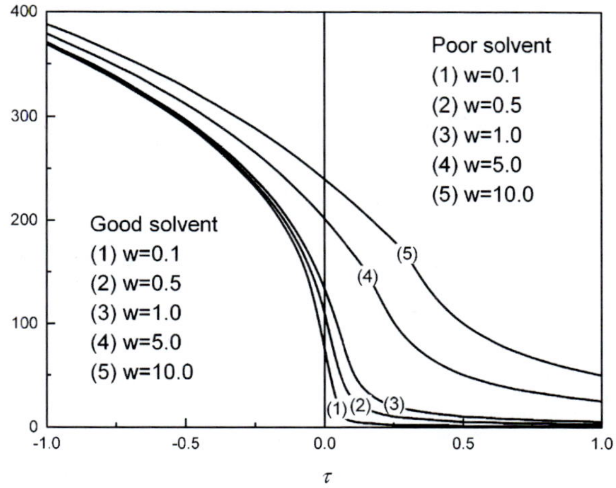

Figure 3. Length $L \equiv \bar{z}$ of the most probable chain configuration as a function of solvent quality for various values of the third virial coefficient w.

model, the profile of each brush depends not only on the physical properties of the individual brushes (number of monomers N, monomer length a, monomer volume v, persistence length l, excluded volume parameter τ, third virial coefficient w, grafting area s^2), but also on the distance $2d$ between surfaces. The probability for a chain grafted on surface "1" to end up at the distance z_1 from the first surface, whereas a chain which starts on surface "2" to end up at the position z_2 from the surface "2" is given by (see Appendix):

$$\psi(z_1,z_2) = \frac{1}{C_2}\psi_1'(z_1)\psi_2'(z_2)\sum_{i,j} \exp\left(-\frac{U_{FH}{}^{i,j}(z_1,z_2)}{kT}\right) \quad (17a)$$

where the sum is performed over all possible configurations which end up at z_1 and z_2, respectively, $\psi_k'(z_k)$ $(k = 1, 2)$ represents the probability that a random walk, constrained by two reflecting walls at 0 and $2d$, which starts on the surface "k" and ends up at the distance z_k from that surface (see Appendix), and the normalization constant C_2 is obtained by taking into account that both chains should end up somewhere between 0 and $2d$:

$$\int_0^{2d}\int_0^{2d} \psi(z_1,z_2)\, dz_1\, dz_2 = 1 \quad (17b)$$

Assuming, as previously, that the Boltzmann factor of a configuration is dominated by the configuration corresponding to a minimum energy $U_{FH}(z_1,z_2)$ with the constraints that all the monomers of the first chain are distributed between the first surface and z_1 while all the monomers of the second chain are distributed between the second surface and z_2, the total free energy of the overlapping brushes, per unit area, is the statistical average over all possible configurations (see Appendix)

$$F(2d) = \frac{1}{s^2}\int_0^{2d}\int_0^{2d} \psi(z_1,z_2)U_{FH}(z_1,z_2)\, dz_1\, dz_2 \quad (18)$$

and the force per unit area is provided by its derivative with respect to the distance $2d$.

The calculations are considerable simplified if only the symmetric configurations ($z_1 = z_2$) would be taken into account and the paths that reach z after reflections on the opposite walls would be ignored. In this case, eq 17a becomes

$$\psi(z) = \frac{1}{C_2}\exp\left(-\frac{z^2}{\sigma^2}\right)\exp\left(-\frac{U_{FH}(z)}{kT}\right) \quad (19a)$$

The Flory–Huggins free energy for the nonoverlapping configurations ($0 < z < d$) is given by

$$U_{FH}(z) = 2z\frac{kTs^2}{a^3}\left(-\frac{\tau}{2}\left(\frac{Nv}{s^2z}\right)^2 + \frac{w}{6}\left(\frac{Nv}{s^2z}\right)^3\right) \quad 0 < z < d \quad (19b)$$

which is the sum of the energies of the independent configurations that end up at $z_1 = z_2 = z$, while for $z > d$

$$U_{FH}(z) = 2d\frac{kTs^2}{a^3}\left(-\frac{\tau}{2}\left(\frac{Nv}{s^2d}\right)^2 + \frac{w}{6}\left(\frac{Nv}{s^2d}\right)^3\right) \quad d < z < 2d \quad (19c)$$

which implies that, for two overlapping configurations, the minimum Flory–Huggins free energy in good solvents is obtained for a constant distribution of monomers between 0 and $2d$. The free energy per unit area of the system is given by

$$F(2d) = \frac{1}{s^2}\int_0^{2d} \psi(z)U_{FH}(z)\, dz \quad (19d)$$

and the force per unit area

$$\Pi(2d) = -\frac{\partial F(2d)}{\partial(2d)} \quad (19e)$$

In Figure 4a, the force per unit area between two identical surfaces with grafted polymer brushes immersed in good solvents, with the parameters selected as above (s = 10 Å, a = 1 Å, N = 1000, l = 10 Å, v = 1 Å^3) is calculated for various solvent qualities using all possible configurations (continuous line) as well as the simplified version which accounts only for the symmetric configurations. The ratio between the forces calculated in the two different manners, represented in Figure 4b, shows that the simpler calculation is fairly accurate for high overlaps (strong interactions).

B. Poor Solvents. When the exponential tails of two brushes immersed in a poor solvent overlap, one might expect the local increase in the monomer volume fraction to lead to a more negative Flory–Huggins free energy and therefore to a lower total free energy of the system, which corresponds to an attraction between surfaces. However, as will be shown below, this is not always true.

Earlier experiments on polystyrene adsorbed on mica immersed in cyclohexane showed that below the Θ temperature the force was attractive for large separations between the mica plates, but became repulsive for small separations.[27] However, it was found, for the same system, that the attraction persisted even for temperatures higher than Θ. For this reason, its origin was attributed to the formation of polymer bridges between the surfaces.[28] Later, another experiment, in which the bridging was essentially eliminated, has shown that at large separations the force was attractive for $T < \Theta$ and repulsive for $T > \Theta$.[29]

First, let us note that the attractive interaction could not be predicted by theories which did not consider the interpenetration of the chains from opposing surfaces but

(27) Klein, J. *Nature* **1980**, *288*, 248.
(28) Israelachvili, J. N.; Tirell, M.; Klein, J.; Almog, Y. *Macromolecules* **1984**, *17*, 204.
(29) Hadziioannou, G.; Patel, S.; Granick, S.; Tirell, M. *J. Am. Chem. Soc.* **1986**, *108*, 2869.

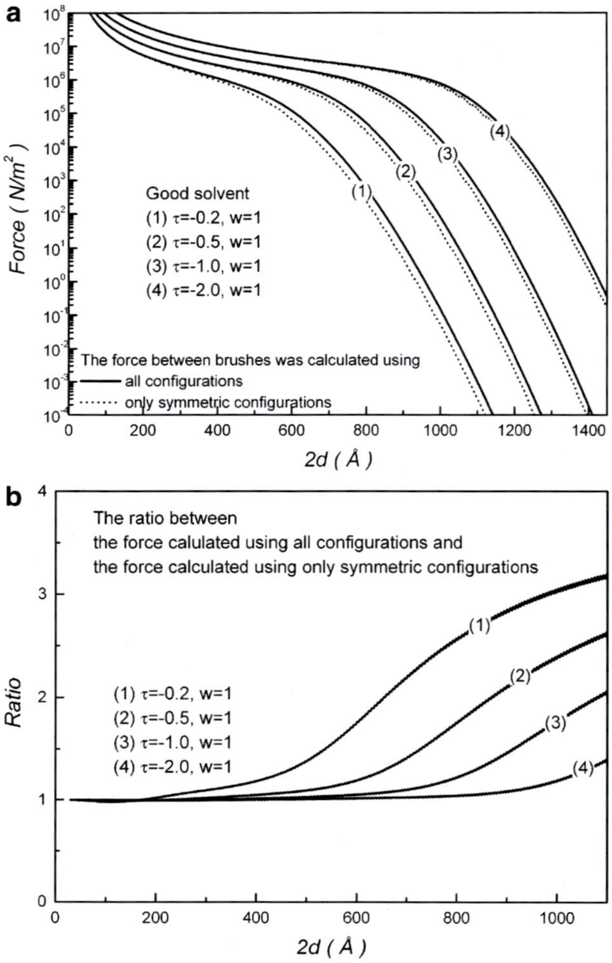

Figure 4. (a) Force per unit area between two identical surfaces with grafted polymers, separated by a distance $2d$ for $s = 10$ Å, $a = 1$ Å, $N = 1000$, $l = 10$ Å, $v = 1$ Å^3 in a good solvent, calculated using all possible chain configurations, or using only the symmetric configurations. (b) Ratio between the two forces, showing that the restriction to symmetric configurations is a good approximation, particularly at large interactions.

assumed instead that each brush was distributed between the surface and the middle distance.[3] Indeed, in the latter model, the total free energy of the compressed brush should be higher than that of the free brush; otherwise, the brush would have been adopted spontaneously a compressed configuration, even in the absence of the other surface. Therefore, the interactions between surfaces with non-interpenetrating brushes were always repulsive; in those cases, the attraction could be explained only if additional interactions (such as those due to polymer bridging) were taken into account.

Let us now analyze how the interactions between brushes occur in the present model. Our system is described by the statistical average of all possible configurations of chains that start from opposite surfaces and end up at the distances z_1 and z_2, respectively. For each possible path that is not reflected by the opposite wall, it is assumed that the monomers of the chains grafted on surfaces "1" and "2" are distributed between 0 and z_1 and between z_2 and $2d$, respectively. The paths that are reflected by the opposite walls have the monomers distributed everywhere between 0 and $2d$. Only the configurations that minimize the Flory–Huggins free energy are taken into account. Since a complete evaluation of all possible combinations is quite tedious, we will focus here only on the symmetric configurations, neglecting also the paths reflected on the opposite walls. As in the previous section, the limitation to symmetric configurations does not affect the results qualitatively.

For the highly confined configurations that end up at distances smaller than $z_{\min} = (w/\tau)(Nv/s^2)$ in a poor solvent, the minimum Flory–Huggins energy is obtained for an uniform (constant) monomer distribution. These configurations behave as immersed in a good solvent even below the Θ temperature (the monomer does not separate from solvent). For the confined configurations, with monomer volume fractions exceeding $\phi_{\max} = 2\tau/w$, corresponding to a separation $2d_{\min} = (w/\tau)(Nv/s^2)$, the increase of the local volume fraction due to the overlap of the brushes leads to an increase of the Flory–Huggins mixing free energy. These unfavorable interactions (dominated by the excluded-volume repulsion) are responsible for the strong, hard-wall-like repulsion, which occurs always at high overlaps ($2d < 2d_{\min}$), regardless of the quality of the solvent.

It is of interest to note that below (but in the vicinity of) the Θ temperature, the minimum confinement distance for which the overlap reduces the free energy might exceed the maximum length of the chains, $2d_{\min} > 2Na$. Because for $2d > 2Na$ there are no interactions between brushes, since the chains cannot overlap, the interactions occur in this case for $2d < 2Na < 2d_{\min}$ and hence are repulsive. Therefore, the brushes repel each other at all separations, as if they would have been immersed in a good solvent. This might explain the repulsion (and no attraction) between brushes observed recently in near-Θ solvents.[26]

When the separation between surfaces exceeds $2d_{\min}$, the Flory–Huggins mixing free energy decreases for all overlapping ($z > d$) configurations. Using the same procedure as in section 2, one can show that the Flory–Huggins free energy of a configuration whose chains end at z (with $z > d$) is minimized by a constant monomer distribution for $2d < 2z_{\min} = 2(w/\tau)(Nv/s^2)$ and for $2d > 2z_{\min}$ by two constant monomer distributions

$$\phi_a = \frac{\frac{2\tau}{w}2d - \frac{2\tau}{w}z_a - 2\frac{Nv}{s^2}}{z - 2z_a} \quad \text{(over a domain of length } z_a\text{)} \tag{20a}$$

$$\phi_b = \frac{2\frac{Nv}{s^2} - \frac{2\tau}{w}z_a}{z - 2z_a} \quad \text{(over a domain of length } z_b = z - z_a\text{)} \tag{20b}$$

and the condition $\phi_b = \phi_{\min} = v/s^2a$ leads to the solution

$$z_{2d} = \frac{1}{2}\frac{2Na - 2d}{\frac{\tau}{w}\frac{s^2a}{v} - 1} > 2z_{a,\max} = \frac{1}{2}\frac{2Na - 2z}{\frac{\tau}{w}\frac{s^2a}{v} - 1} \tag{20c}$$

The values of $\phi_{\max}$ and $\phi_{\min}$, obtained by introducing eq 20c in eq 20, parts a and b, are the same as those in section 2

$$\phi_a = \phi_{\max} = \frac{2\tau}{w} - \frac{v}{s^2a} \quad \text{(over a domain of length } z_{2d}\text{)} \tag{20d}$$

$$\phi_b = \phi_{\min} = \frac{v}{s^2a} \quad \text{(over a domain of length } z\text{-}z_{2d}\text{)} \tag{20e}$$

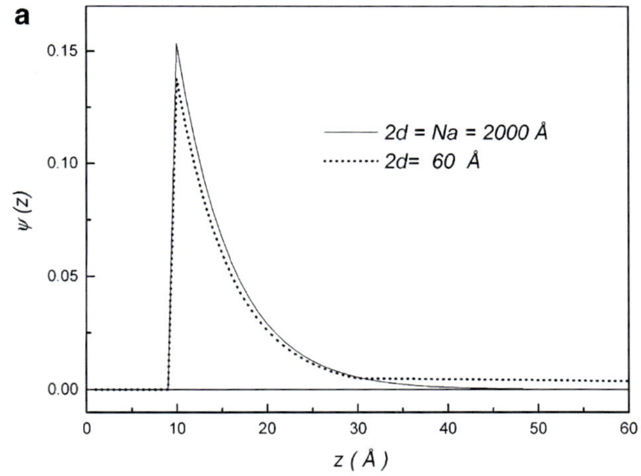

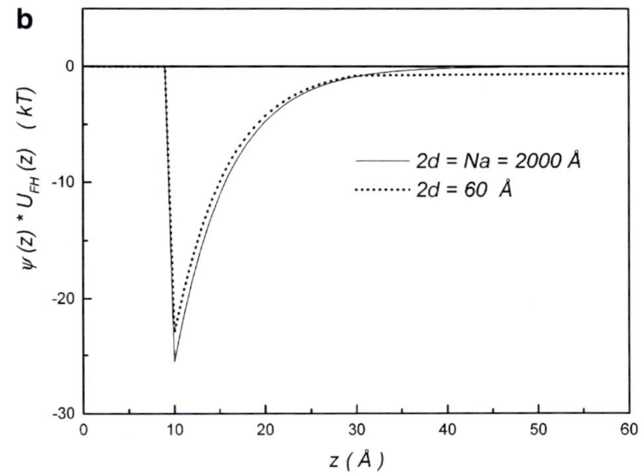

Figure 5. (a) Probability $\psi(z)$ for a chain to end up at distance z in a poor solvent differs in noninteracting brushes ($2d = 2Na = 2000$ Å) and overlapping brushes ($2d = 60$ Å). The swelling of brushes in poor solvents due to overlapping increases $\psi(z)$ at large z on the expense of decreasing $\psi(z)$ at moderate z. Because the former configurations have higher Flory–Huggins mixing free energies, they might increase the total free energy of the system. (b) Function $\psi(z)U_{FH}(z)$, the integral of which provides the total free energy. The balance between its increase (in absolute value) for large values of z and its decrease for small values of z provides the sign of the interactions between surfaces.

In this case, the energy of the overlapping configurations decreases because the length of the interval over which the monomers accumulate to the energetically favorable ϕ_{max} value increases. The separation of the monomers from the solvent is favored by the presence of the other brush.

Since the overlapping of the stretched configurations lowers their free energy, one expects the overlapping of the brushes to always decrease the total energy of the system at large separations. Surprisingly, this does not happen. In Figure 5a, the probability $\psi(z)$ for a configuration to end at the distance z is compared for large separation distances with the same probability at small separations, when the configurations overlap. The decrease of the energy of the configurations for large values of z leads to an increase in the probability for the chain to reach this distance, at the expense of decreasing $\psi(z)$ for small values of z. However, the "stretched" configurations have higher Flory–Huggins energies that the "collapsed" ones, and the increase of the number of stretched configurations might increase the total free energy of the system. The product $\psi(z)U_{FH}(z)$, whose

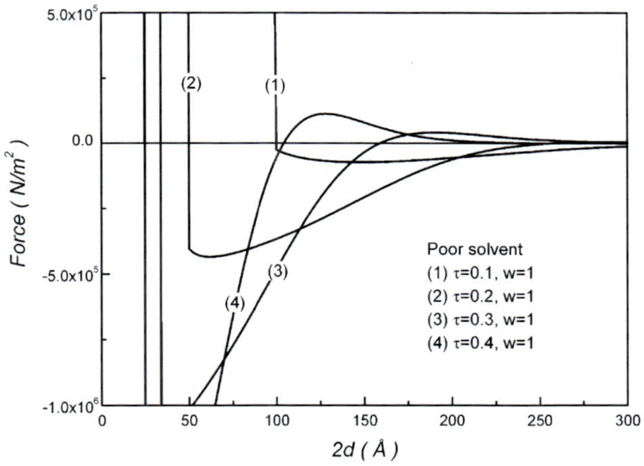

Figure 6. (a) Force per unit area between two identical surfaces with grafted polymers, separated by a distance $2d$ for $s = 10$ Å, $a = 1$ Å, $N = 1000$, $l = 10$ Å, and $v = 1$ Å^3 in a poor solvent, calculated using only the symmetric configurations. At small separations, there is a strong repulsion due to excluded-volume interactions (carried by the third virial coefficient). At intermediate separations there is an attractive region due to the decrease of the free energy of configurations due to the chain overlap. At large separations, the swelling of the brush due to chain overlap (the increase of $\psi(z)$ at large z) favors configurations with high Flory–Huggins mixing free energies, thus increasing the total energy of the system and therefore generating a repulsion. The absolute value of this repulsion is, however, smaller by orders of magnitude than the attraction that occurs at intermediate separations and smaller by many orders of magnitude than the strong repulsion that occurs at small separations.

integral with respect to z provides the total free energy,is plotted in Figure 5b.

In Figure 6, the force between brushes is calculated for various solvent qualities. At small overlaps, the decrease of $\psi(z)$ for small z dominates, and the interaction is repulsive. However, at intermediate separations an attraction between brushes occurs, followed by a much stronger repulsion at small separations, in agreement with experiment.[27] It should be noted that the very weak repulsion that occurs at large separations, which was not reported experimentally, is by a few orders of magnitude smaller than the attraction and by many orders of magnitude smaller than the strong repulsion at small separations. This weak repulsion is a consequence of the statistical average over possible configurations employed in this article.

4. Conclusions

The traditional analytical treatments of the distribution of monomers in a grafted neutral polymer brush and the interaction between two surfaces with grafted polymer brushes are based either on the assumption of a step profile brush or on a parabolic profile brush. The latter assumption is based on an approximate solution of the self-consistent field equation (MWC).[15] Both approaches predict the same scaling relations between the thickness of the polymer brush and the graft density, which fail at both very low and very high graft densities. In this article an alternate approach was suggested, based on an approximate Monte Carlo procedure. It is assumed that the most contributions to the Boltzmann factor of the configurations that end up at the distance z are provided by the configurations of minimum Flory–Huggins mixing free energy, with a monomer distribution restricted between the surface and the distance z. In a good solvent, these configurations correspond to a constant volume fraction of monomers, an approach that was successfully employed by Flory to predict the scaling laws of a polymer in a good solvent. The method is particularly valuable for strong-stretched polymers, for which the parabolic approximation is not accurate. Because at high volume fractions of monomers (typical for a high-density polymer graft), the Flory–Huggins free energy increases with the monomer volume fraction and the brushes are swollen even for poor solvent qualities, the method can be applied to high-density brushes immersed in both good and poor quality solvents. When the graft density is not very high, or when the interaction parameters (τ and w) are small, the profile of the brush can be well approximated by a parabola. At higher interactions (either higher graft densities or larger excluded-volumes), the monomer density is almost constant in the vicinity of the surface, but decays exponentially at large distances, in agreement with molecular dynamics simulations and the numerical solutions of the self-consistent field equations. By assuming that the thickness of the brush is equal to the length of the most probable chain configuration (a rather poor approximation), one can recover for good solvents the scaling laws derived by previous analytical treatments; however, the brush thickness obeys a different scaling law at the Θ temperature. The interactions between two surfaces with grafting polymer brushes immersed in good solvents are always repulsive, with a magnitude dependent on the quality of the solvent. For marginally poor solvents, the model predicts also repulsive interactions at any separation. However, for poor solvents, the present theory predicts a very weak repulsion at large separations, followed by a (much stronger) attractive minimum of the free energy at intermediate separation distances, and a strong repulsion at small separations.

Appendix

The probability for an one-dimensional, free and unconstrained random walker, starting from the origin, to reach the distance z after N_l steps of length l is well approximated (at large N_l) by

$$\psi^U(z) = \frac{1}{\sqrt{2\pi\sigma^2}}\exp\left(-\frac{z^2}{2\sigma^2}\right), \quad 0 < z < N_l l \qquad \text{(A1)}$$

where $\sigma = l\sqrt{N_l}$ is the root-mean square distance of the random walk.[24] The probability for a free random walker that starts at 0 to arrive at the distance z in the presence of a reflecting wall at the distance d_1 is equal to the sum of the probabilities for an unconstrained random walker to arrive at the point z and at its "image" point $2d_1 - z$, in the absence of the reflecting wall.[24] When the reflecting wall is located at the origin, the probability for the "reflected" walk to reach a distance $z > 0$ is simply twice the probability to reach the same distance in the absence of the reflecting wall (eq 2a).

The general solution of a random walk on a lattice between perfect reflecting walls was provided by Feller.[30] In what follows, a simpler approximate solution will be constructed using the method of images.[24] When the walk is confined between two reflecting walls, one at the origin and the other at the distance $2d$, the probability for the constrained walk to reach the distance z, $\psi'(z)$, can be written as a sum of probabilities ψ^U for the unconstrained walk to reach various "image" points, as follows.

$\psi^U(z)$: The possible paths either do not cross the points 0 and $2d$ or cross them an even number of times. The presence of reflecting walls at 0 or $2d$ does not modify the probability to reach z, because after two successive reflections on the same wall the unconstrained path is always recovered.

$\psi^U(-z)$: The possible paths cross an odd number of times the origin. Because of the reflecting wall at 0, the constrained walker arrives at z instead (some paths cross an even number of times $2d$, but inserting a reflecting wall there does not affect the probability, as noted above).

$\psi^U(4d - z)$: The possible paths cross the point $2d$ an odd number of times; because of reflections on that wall, the walker arrives at the point z instead.

$\psi^U(-4d + z)$: The possible paths cross the origin an odd number of times, followed by crossing the point at $-2d$ an odd number of times. Because of the reflection on the walls at origin and at $2d$, the constrained walker reaches the point z instead, and so on. Since the length of the walk cannot exceed $N_l l \equiv Na$, only the image points at shorter distances from origin than Na should be taken into account.

For the monomer distribution, a distinction has to be made between paths of type a, which are not reflected by opposite walls, and paths of type b, which return to z after reflecting at least once on the opposite wall. In the first case

$$\psi^a(z) = \psi^U(z) + \psi^U(-z) \qquad \text{(A2a)}$$

and in the calculation of the configuration of minimum energy it will be assumed that the monomers can be distributed between the starting point 0 and the end point z.

For the remaining configurations

$$\psi^b(z) = \psi^U(4d - z) + \psi^U(4d + z) + \psi^U(8d - z) + \psi^U(8d + z) + \ldots \qquad \text{(A2b)}$$

Since the chain reaches both walls at least once, it will be assumed that its monomers are distributed between the walls. The energy of the system of overlapping brushes, per unit area, becomes

$$\psi(z_1,z_2) = \frac{1}{C_2}\psi_1'(z_1)\psi_2'(z_2)\sum_{i,j}\exp\left(-\frac{U_{FH}{}^{i,j}(z_1,z_2)}{kT}\right) \qquad \text{(14a)}$$

$$\cong \frac{1}{C_2}\left[\psi^a(z_1)\psi^a(z_2)\exp\left(-\frac{U_{FH}{}^{aa}(z_1,z_2)}{kT}\right) + \psi^b(z_1)\psi^b(z_2)\exp\left(-\frac{U_{FH}{}^{bb}(z_1,z_2)}{kT}\right) + \psi^a(z_1)\psi^b(z_2) \times \exp\left(-\frac{U_{FH}{}^{ab}(z_1,z_2)}{kT}\right) + \psi^b(z_1)\psi^a(z_2)\exp\left(-\frac{U_{FH}{}^{ba}(z_1,z_2)}{kT}\right)\right] \qquad \text{(A3)}$$

where in eq 14a the sum is performed over all possible configurations i, j with chains that start on the surface "1" and "2" and end up at z_1 and z_2, respectively, while in eq A3 only the configurations of minimum Flory–Huggins free energy are taken into account. In the latter case, the minimum depends on whether the chain reaches the opposite wall (type b) or not (type a).

For the configurations which do not overlap, the minimum Flory–Huggins energy is obtained for constant monomer distributions between 0 and z_1, and between the location z_2 and $2d$, respectively. Figure 7 presents the

(30) Feller, W. *An Introduction to Probability Theory and its Application*; John Wiley & Sons: New York, 1968; Vol. 1, Chapter XVI.

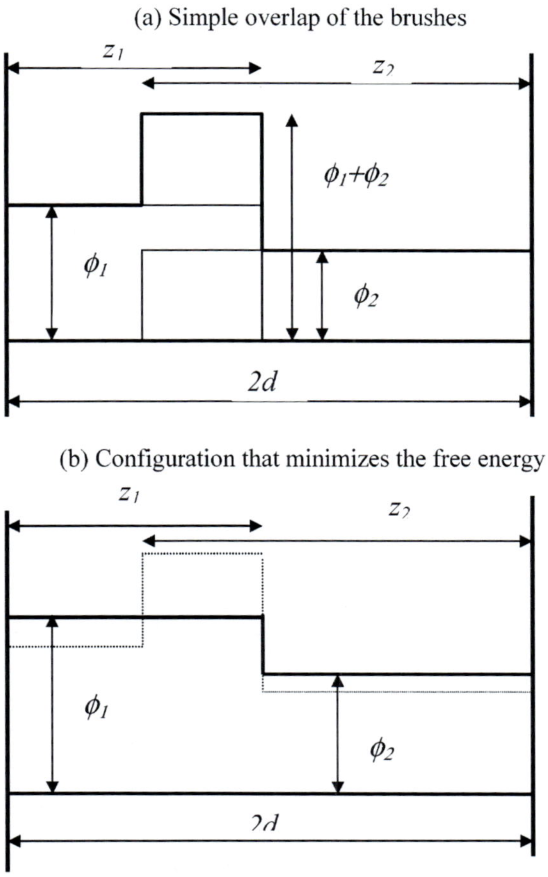

Figure 7. Local volume fractions of the monomers for two configurations of chains which end up at z_1 and z_2, respectively: (a) simple addition of the volume fractions of the brushes; (b) configuration that minimizes the Flory–Huggins mixing free energy for the conditions mentioned in the text.

monomer distribution obtained from the sum of the volume fractions of two overlapping configurations. This result clearly does not correspond to a minimum Flory–Huggins free energy, because of the high value of ϕ in the overlapping region. A smaller free energy is obtained for the configurations in which the excess volume fraction is redistributed to the nonoverlapping regions. In addition to the requirements that all the monomers of each brush should be located between 0 and z_1 or z_2 and $2d$, respectively, it will be assumed that in the overlapping region, the number of monomers belonging to each grafted brush is proportional to the volume fraction of those monomers in the nonoverlapping regions. The result is a constant volume fraction ϕ_1 between 0 and z_1 and a constant volume fraction ϕ_2 between z_1 and $2d$, where $\phi_1 > \phi_2$ (see Figure A.1b). The volume fractions in each region obey therefore the conservation equations:

$$(2d - z_2)\phi_1 + \alpha(z_1 + z_2 - 2d)\phi_1 = \frac{Nv}{s^2} \quad \text{(A4a)}$$

$$(2d - z_1)\phi_2 + \alpha(z_1 + z_2 - 2d)\phi_2 = \frac{Nv}{s^2} \quad \text{(A4b)}$$

$$\alpha\phi_1 + \alpha\phi_2 = \phi_1 \quad \text{(A4c)}$$

Equation A4, parts a–c, can be solved for α, ϕ_1, and ϕ_2.

For the case *bb*, when both chains undergo reflections on the opposite walls, the Flory–Huggins free energy is minimized by a constant distribution of monomers between walls. In the case *ab*, when the second chain is reflected by the opposite wall, but not the first chain, the proportional redistribution of the excess monomers leads to a configuration with volume density ϕ_1 between 0 and z_1 and ϕ_2 between z_1 and $2d$, which obey the equations

$$\alpha z_1\phi_1 = \frac{Nv}{s^2} \quad \text{(A5a)}$$

$$(2d - z_1)\phi_2 + \alpha z_1\phi_2 = \frac{Nv}{s^2} \quad \text{(A5b)}$$

$$\alpha\phi_1 + \alpha\phi_2 = \phi_1 \quad \text{(A5c)}$$

and are similar for the *ba* case. The average free energy density is consequently calculated from

$$F(2d) \cong \frac{1}{C_2}\int_0^{2d}\int_0^{2d} \sum_{i=a,b;j=a,b} U_{FH}^{i,j}(z_1,z_2)\psi^i(z_1)\psi^j(z_2) \exp\left(-\frac{U_{FH}^{i,j}(z_1,z_2)}{kT}\right) dz_1\,dz_2 \quad \text{(A.6)}$$

where the indices i and j run now over the two types of paths, a and b.

LA049781Y

Langmuir **2004,** *20,* 8155–8164 8155

On the Monomer Density of Grafted Polyelectrolyte Brushes and Their Interactions

Marian Manciu and Eli Ruckenstein*

Department of Chemical Engineering, State University of New York at Buffalo, Buffalo, New York 14260

Received March 31, 2004. In Final Form: June 30, 2004

Most of the modern theories of grafted polyelectrolyte brushes are valid only for moderate stretching of the polyelectrolyte. However, particularly at low ionic strength and high grafting densities, even a moderate charge of the polyelectrolyte can generate a strong stretching. A simple mean field model for strongly stretched grafted polyelectrolyte brushes is suggested, based on an approximate calculation of the partition function of a polyelectrolyte chain. It is shown that the average Boltzmann factor of a possible chain configuration can be approximated by the Boltzmann factor of a configuration with a constant monomer distribution, for which the free energy can be readily obtained. The monomer density in the brush and the interaction between two surfaces with grafted polyelectrolyte brushes could be calculated as a statistical average over all possible configurations. Some simple analytical results are derived, and their accuracy is examined. The dependence of the brush thickness on the electrolyte concentration is investigated, and it is shown that the trapping of a fraction of counterions in the brush influences strongly the thickness of the brush. When two surfaces with grafted polyelectrolyte brushes approach each other more rapidly than the ion diffusion parallel to the surface, the trapping of the counterions between the brushes can affect the interactions by orders of magnitude.

1. Introduction

The grafting of neutral polymers on colloidal surfaces is a well-known method to enhance the stability of colloids, since the steric repulsion between brushes is not highly sensitive to changes in ionic strength or pH, unlike the double layer repulsion induced by surface charges. The predictions of the monomer density distribution and of the thickness of neutral brushes have been challenging problems, the earlier scaling theories assuming a constant density of the monomers within the brush,[1] whereas the most recent ones, originating from a model due to Dolan and Edwards[2] in which the neutral polymer follows a random walk from the surface in a field generated by the polymer itself, suggested a parabolic density distribution of monomers.[3]

During the same period, the charged grafted polyelectrolyte chains received considerable less attention than the neutral ones. One of the reasons is that the sufficiently strong charged grafted polyelectrolyte chains are almost completely stretched because of the electrostatic repulsion between their charges, which implies that a step-profile density of the monomers is an accurate approximation.[4] However, the grafted polyelectrolytes received recently renewed attention because of a plethora of potential applications, such as the possibilities (i) to control the catalytic selectivity of some chemical reactions by varying the thickness of the brush,[5] (ii) to prevent the adsorption of proteins on surfaces,[6] and (iii) to design porous filters for pH-controlled gating.[7] The recent theories of grafted polyelectrolytes[8–10] are based on an extension to charged brushes of the approximate solution of the self-consistent field equations obtained for neutral brushes,[3] which predicted a parabolic shape for the self-consistent field. However, as pointed out by Shim and Cates,[11] the parabolic shape of the self-consistent field is accurate only for a moderate stretching but is not valid for very high and very small stretchings. It should be noted that the Milner–Witten–Cates theory[3] leads to the same scaling laws for the thickness of the brush as the step profile (the Alexander model)[12] and that these scaling laws predict vanishing thicknesses for small grafting densities and thicknesses that exceed the length of the grafted polymers at strong stretching.

However, the grafted polyelectrolytes are often strongly stretched, because such stretchings can be sometimes generated by only a moderate charging of the polyelectrolyte. This fact can be understood qualitatively using the following simple analysis. Let us assume that to each grafted chain corresponds an area s^2 on the surface and a parallelepipedic atmosphere with the same bottom area and a length equal to the thickness of the brush. The flux of the electric field through the sides of the parallelepiped is proportional to the inside charge ne. This charge is generated through the dissociation of the polyelectrolyte but also includes the fraction of the polyelectrolyte counterions trapped in the brush and the electrolyte ions.

* Corresponding author: e-mail address, feaeliru@acsu.buffalo.edu; phone, (716) 645-2911(extention 2214); fax, (716) 645-3822.

(1) Alexander, S. *J. Phys.* (*Paris*) **1977**, *38*, 983.
(2) Dolan, A. K.; Edwards, F. R. S. *Proc. R. Soc. London, Ser. A* **1974**, *337*, 509.
(3) Milner, S. T.; Witten, T. A.; Cates, M. E. *Macromolecules* **1988**, *21*, 2610.
(4) Huang, H.; Ruckenstein, E. *J. Colloid Interface Sci.* **2004**, *275*, 548.
(5) Ruckenstein E.; Hong, L. *J. Catal.* **1992**, *136*, 178.
(6) Carignano, M. A.; Szleifer, I. *Colloids Surf., B* **2000**, *18*, 169.
(7) Ito, Y.; Ochiai, Y.; Park, Y. S.; Imanishi, Y. *J. Am. Chem. Soc.* **1997**, *119*, 1619.
(8) Miklavic, S. J.; Marcelja, S. J. *Phys. Chem.* **1988**, *92*, 6718.
(9) Zhulina, E. B.; Borisov, O. V. *J. Chem. Phys.* **1997**, *107*, 5952.
(10) Misra, S.; Varanasi, S.; Varanasi, P. P. *Macromolecules* **1989**, *22*, 4173.
(11) Shim, D. F. K.; Cates, M. E. *J. Phys.* (*Paris*) **1989**, *50*, 3535.
(12) Pincus P. *Macromolecules* **1991**, *24*, 2912.

10.1021/la049168e CCC: $27.50
Published on Web 08/14/2004

The electric field vanishes on the lateral sides of the atmosphere because of the periodicity of the grafting and also on its bottom surface, because of the overall electroneutrality of the system. Therefore, a large electric field is generated at the top of the brush, of the order of $E = ne/\epsilon\epsilon_0 s^2$, where ϵ is the dielectric constant within the brush and ϵ_0 is the vacuum permittivity. The force required to stretch a polymer is of the order of kT/l, where l is the persistence length.[13] The force acting on an elementary charge e located at the end of a chain is given by $eE \cong ne^2/\epsilon\epsilon_0 s^2 \approx kT/l$, which for typical values of the parameters ($\epsilon = 80$, $s^2 = 1000$ Å^2, $l = 10$ Å) leads to a value of the order of unity for n. This result implies that a few charges within the atmosphere are in general sufficient to exert a sensible stretching of the grafted polyelectrolyte. It should be emphasized that the length of the parrallelepipedic atmosphere (the thickness of the brush) is in general much larger than its lateral length s, therefore the strong stretching is a result of the confinement of the flux of the electric field to a relatively small area. This strong stretching is in contrast to those of a free polyelectrolyte in a solvent or to a single grafted polyelectrolyte, cases in which much larger charges are required to stretch the polyelectrolyte.

In this paper, a simple modality to calculate the density distribution of the monomers in a grafted polyelectrolyte brush as well as the interaction between surfaces with grafted polyelectrolyte brushes is suggested. The model is an extension to charged polymer brushes of a treatment proposed recently for neutral polymer brushes.[14] It is based on a simplifying assumption applied to a standard Monte Carlo procedure, namely, that the average Boltzmann factor of a possible configuration is dominated by the configuration with the lowest free energy. For a neutral brush immersed in a good solvent, this configuration corresponds to a uniform density of monomers.[14] This constant monomer distribution approximation for each possible configuration was successfully employed by Flory to explain the scaling relation between the length and the gyration radius of a polymer in a good solvent.[15]

In polyelectrolytes, the configuration of minimum free energy would be provided by a constant monomer density only if the electrical potential is constant through the brush. Because, as it will be shown in what follows, this condition is almost satisfied, the approximation of a constant monomer distribution is accurate and simplifies considerably the calculations.

2. Monomer Density and Interaction between Polyelectrolyte Brushes

General Framework. The traditional approach to calculate the density distribution of a polyelectrolyte brush consists of generating all the possible configurations of the polyelectrolyte and calculating the total energy corresponding to each configuration. The probability that such a configuration occurs is proportional to its corresponding Boltzmann factor. A simple modality to generate the configurations of a polymer composed of N monomers of length a grafted on a surface is to assume that the polymer is composed of N_l independent pieces of persistence length l, which are connected but can assume any relative orientations. In the absence of any other interactions between monomers (apart from their mutual bonding), the probability that the last monomer of the chain reaches the distance z is provided by the solution of a random walk[16]

$$\Xi_0(z) = 0 \qquad z < 0 \tag{1a}$$

$$\Xi_0(z) = \left(\frac{2}{\pi\sigma^2}\right)^{1/2} \exp\left(-\frac{z^2}{2\sigma^2}\right) \qquad z > 0 \tag{1b}$$

where $\sigma = lN_l^{1/2}$ is the root-mean-square distance of a one-dimensional random walk of N_l steps of length l, if one assumes that the surface represents a perfect reflecting wall for the random walk.[16]

In the presence of additional interactions, the probability $\Xi_0(z)$ for a configuration to end up at a distance z must be multiplied by the corresponding Boltzmann factor

$$\Xi(z) = \frac{1}{C_1} \exp\left(-\frac{z^2}{2\sigma^2}\right)\left\langle \exp\left(-\frac{U(z)}{kT}\right)\right\rangle \tag{2a}$$

where $U(z)$ is the total free energy of a configuration and the average $\langle\ \rangle$ is performed over all possible chain configurations whose last monomer is located at z. The constant C_1 is provided by the normalization

$$\int_0^{Na} \Xi(z)\,dz = 1 \tag{2b}$$

where $Na = N_l l$ represents the maximum length of the polyelectrolyte.

The main difficulty in eqs 2 is to calculate the Boltzmann factor averaged over all possible configurations. As previously suggested,[14] it will be assumed that the configuration of lowest free energy dominates the partition function (the saddle-point approximation of statistical mechanics) and therefore the average Boltzmann factor of all the configurations which end up at a distance z can be approximated by a Boltzmann factor corresponding to the configuration of lowest free energy. A further approximation is to consider that the configuration of lowest free energy that ends up at the distance z corresponds to a constant density of monomers $\phi(\zeta,z)$ distributed between the surface and the distance z

$$\phi(\zeta,z) = \frac{N}{s^2 z} \qquad 0 < \zeta < z \tag{3a}$$

$$\phi(\zeta,z) = 0 \qquad \zeta > z \tag{3b}$$

where N is the number of monomers in a chain and s^2 is the area corresponding to a grafted polyelectrolyte on the surface. The above expressions provide a minimum free energy for neutral chains immersed in good solvents and for charged chains, if the electrical potential would be constant through the brush. Since the calculations show that the potential is almost constant through the brush (see eq 14a and Figure 1), the last condition is almost satisfied and eqs 3 are accurate approximations of the monomer distributions corresponding to the minimum of the free energy.

(13) Marko J. F.; Siggia E. D. *Macromolecules* **1995**, *28*, 8759.
(14) Manciu, M.; Ruckenstein, E. *Langmuir* **2004**, *20*, 6490.
(15) Flory, P. *Principles of Polymer Chemistry*; Cornell University Press: Ithaca, NY, 1971.
(16) Chandrasekhar, S. *Rev. Mod. Phys.* **1943**, *15*, 3.

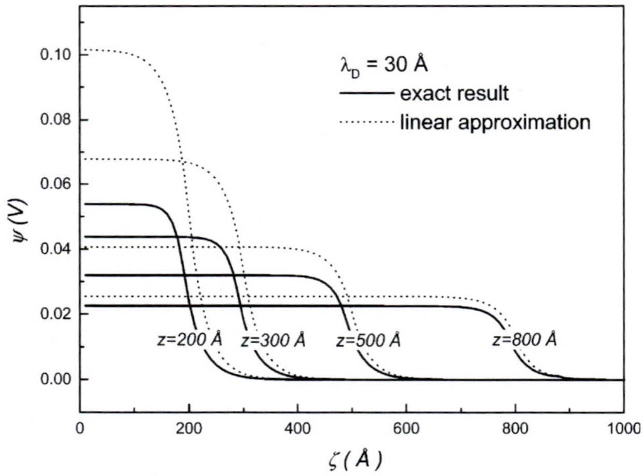

Figure 1. The electric potential as a function of the distance from the surface for various values of z, and $N = 1000$, $a = 1$ Å, $\eta = 0.005$, $c_E = 0.01$ M, $s^2 = 1000$ Å^2, $\epsilon = 80$: thick continuous line, exact result; dotted line, linear approximation.

Assuming that each monomer carries an average charge ηe, where η is the degree of dissociation and e is the elementary charge, and that the brush is immersed in an electrolyte solution in contact with an infinite reservoir of concentration c_E, the potential inside the brush obeys the Poisson equation

$$\frac{\partial^2 \psi}{\partial \zeta^2} = \frac{2ec_E}{\epsilon\epsilon_0} \sinh\left(\frac{e\psi}{kT}\right) - \frac{\eta e}{\epsilon\epsilon_0} \frac{N}{s^2 z} \qquad 0 < \zeta < z \tag{4a}$$

$$\frac{\partial^2 \psi}{\partial \zeta^2} = \frac{2ec_E}{\epsilon\epsilon_0} \sinh\left(\frac{e\psi}{kT}\right) \qquad \zeta > z \tag{4b}$$

where $\psi(\zeta)$ is the electrical potential, ζ the coordinate measured from the surface, ϵ the dielectric constant, ϵ_0 the vacuum permittivitty, k the Boltzmann constant, and T the absolute temperature and Boltzmannian distributions were assumed for the electrolyte ions. The solution of eqs 4 is obtained by assuming the continuity of the electric potential and its derivative at $\zeta = z$, and the boundary conditions

$$\left.\frac{\partial \psi}{\partial \zeta}\right|_{\zeta=0} = -\frac{\sigma_S}{\epsilon\epsilon_0} \tag{5a}$$

and

$$\psi|_{\zeta \to \infty} = 0 \tag{5b}$$

where σ_S represents the surface charge density. Once the potential has been determined, the total free energy corresponding to a polyelectrolyte configuration that occupies an area s^2 on the surface and whose last monomer is located at the distance z from the surface can be calculated as the sum between a Flory–Huggins free energy, due to the monomer–monomer and monomer–solvent interactions, which does not depend on the charging of the polyelectrolyte and a double layer free energy, due to the charging of the polyelectrolyte and of the grafted surface.

The latter one is composed of the contribution of the electric field[17]

$$F_{el}(z) = \frac{s^2}{2} \int_0^\infty \epsilon\epsilon_0 \left(\frac{\partial \psi}{\partial \zeta}\right)^2 d\zeta \tag{6a}$$

and the entropy of electrolyte ions of concentration $c_i(\zeta)$, F_{ent}[17]

$$F_{ent}(z) = kTs^2 \sum_i \int_0^\infty \left(c_i \ln\left(\frac{c_i}{c_E}\right) - c_i + c_E\right) d\zeta \tag{6b}$$

while the Flory–Huggins free energy is given by[12,18]

$$F_{FH}(z) = s^2 \frac{kT}{a^3} \int_0^z \left(-\frac{1}{2}\tau v^2 \phi^2 + \frac{1}{6} w v^3 \phi^3\right) d\zeta = \frac{kT}{a^3}\left(-\frac{\tau N^2 v^2}{2s^2 z} + \frac{wN^3 v^3}{6s^4 z^2}\right) \tag{6c}$$

where a is the length of a monomer, v its volume, τ is the dimensionless excluded volume parameter ($\tau < 0$ for a good solvent and $\tau > 0$ for a poor solvent), and w is the third virial coefficient (typically positive and of the order of unity). In this paper it will be assumed that the charges of the surface and polyelectrolyte do not depend on the separation between the two grafted polyelectrolyte brushes and, hence, that the chemical free energy of the double layer is a constant.[17]

The probability for a configuration to end up at the distance z is therefore given by

$$\Xi(z) = \frac{1}{C_1} \exp\left(-\frac{z^2}{2\sigma^2}\right) \exp\left(-\frac{F_{el} + F_{ent} + F_{FH}}{kT}\right) \tag{7}$$

The average monomer density $\Phi(z)$ of the brush is the statistical average of the contributions of all configurations that end up at all possible positions between z and Na. The contribution at the location ζ of a configuration which ends up at a distance z equals $\phi(\zeta,z) = N/s^2 z$ when $\zeta < z$ and $\phi(\zeta,z) = 0$ when $\zeta > z$; therefore the statistical average of configurations is given by[14]

$$\Phi(z) = \int_z^{Na} \Xi(\varsigma) \frac{N}{s^2 \varsigma} d\varsigma \tag{8}$$

When two surfaces with grafted polyelectrolyte approach each other, the probability for a chain grafted on surface "1" to end up at the distance z_1 from the first surface, whereas a chain which starts on surface "2" to end up at position z_2 from the surface "2" is given by

$$\Xi(z_1,z_2) = \frac{1}{C_2} \exp\left(-\frac{z_1^2}{2\sigma^2}\right) \exp\left(-\frac{z_2^2}{2\sigma^2}\right) \left\langle \exp\left(-\frac{U(z_1,z_2)}{kT}\right)\right\rangle_{z_1,z_2} \tag{9a}$$

where the average is performed over all possible configurations which end up at z_1 and z_2, respectively. The normalization constant C_2 is obtained by taking into account that neither chain can be longer than Na and

(17) Overbeek, J. T. G. *Colloids Surf.* **1990**, *51*, 61. Manciu, M.; Ruckenstein, E. *Langmuir* **2003**, *19*, 1114.

(18) de Gennes, P. G. *Scaling Concepts in Polymer Physics*; Cornell University Press: Ithaca, NY, 1979.

both of them should end up somewhere between 0 and $2d$

$$\int_0^{\xi}\int_0^{\xi} \Xi(z_1,z_2)\,\mathrm{d}z_1\,\mathrm{d}z_2 = 1 \qquad (9b)$$

where ξ is the shortest of the distances Na and $2d$.

Assuming, as previously, that the partition function is dominated by the configurations with constant monomer density between each of the surfaces and the position z_1 and z_2, respectively, eq 9a becomes

$$\Xi(z_1,z_2) = \frac{1}{C_2}\exp\left(-\frac{z_1^{\,2}}{2\sigma^2}\right)\exp\left(-\frac{z_2^{\,2}}{2\sigma^2}\right)\exp\left(-\frac{F_{\mathrm{el}}+F_{\mathrm{ent}}+F_{\mathrm{FH}}}{kT}\right) \qquad (10)$$

where the integrals of the electric (6a), entropic (6b), and excluded volume contributions (6c) are this time integrated between 0 and $2d$. The total free energy of the interacting brushes, per unit area, is the statistical average over all possible configurations, given by

$$F_{\mathrm{brushes}} = \frac{1}{s^2}\int_0^{\xi}\int_0^{\xi}(F_{\mathrm{el}}+F_{\mathrm{ent}}+F_{\mathrm{FH}})\Xi(z_1,z_2)\,\mathrm{d}z_1\,\mathrm{d}z_2 \qquad (11a)$$

The force per unit area between surfaces can be obtained from the derivative of the free energy per unit area, eq 11a, with respect to the separation distance $2d$

$$\Pi(2d) = -\frac{\partial F_{\mathrm{brushes}}}{\partial(2d)} \qquad (11b)$$

3. Monomer Distribution in a Grafted Polyelectrolyte Brush Immersed in an Infinite Reservoir

A. Linear Approximation. The linear approximation of the Poisson–Boltzmann equation ($\sinh(e\psi/kT) \cong e\psi/kT$) is in general accurate only for small potentials ($\psi < kT/e$). However, it offers insight into the qualitative behavior, even at larger potentials. In the linear approximation, the Poisson–Boltzmann equations (4) become

$$\frac{\partial^2\psi}{\partial\zeta^2} = \frac{\psi}{\lambda_{\mathrm{D}}^{\,2}} - \frac{\eta e}{\epsilon\epsilon_0}\frac{N}{s^2 z} \qquad (12a)$$

$$0 < \zeta < z$$

$$\frac{\partial^2\psi}{\partial\zeta^2} = \frac{\psi}{\lambda_{\mathrm{D}}^{\,2}} \qquad (12b)$$

$$\zeta > z$$

where $\lambda_{\mathrm{D}} = (\epsilon\epsilon_0 kT/2e^2c_{\mathrm{E}})^{1/2}$ is the Debye–Hückel length, with the solutions

$$\psi(\zeta) = \frac{\eta e}{\epsilon\epsilon_0}\frac{N}{s^2 z}\lambda_{\mathrm{D}}^{\,2} + A_1\exp\left(\frac{\zeta}{\lambda_{\mathrm{D}}}\right) + A_2\exp\left(-\frac{\zeta}{\lambda_{\mathrm{D}}}\right) \qquad (13a)$$

$$0 < \zeta < z$$

$$\psi(\zeta) = A_3\exp\left(\frac{\zeta}{\lambda_{\mathrm{D}}}\right) + A_4\exp\left(-\frac{\zeta}{\lambda_{\mathrm{D}}}\right) \qquad (13b)$$

$$\zeta > z$$

Assuming a neutral grafted surface ($\sigma_{\mathrm{S}} = 0$), with the boundary conditions eqs 5a,b, and using the continuity of the electric potential and its derivative at $\zeta = z$, one obtains

$$\psi(\zeta) = \frac{\eta e}{\epsilon\epsilon_0}\frac{N}{s^2 z}\lambda_{\mathrm{D}}^{\,2}\left[1 - \cosh\left(\frac{\zeta}{\lambda_{\mathrm{D}}}\right)\exp\left(-\frac{z}{\lambda_{\mathrm{D}}}\right)\right] \qquad (14a)$$

$$\zeta < z$$

$$\psi(\zeta) = \frac{\eta e}{\epsilon\epsilon_0}\frac{N}{s^2 z}\lambda_{\mathrm{D}}^{\,2}\sinh\left(\frac{z}{\lambda_{\mathrm{D}}}\right)\exp\left(-\frac{\zeta}{\lambda_{\mathrm{D}}}\right) \qquad (14b)$$

$$\zeta > z$$

The electrical potential through the brush (eq 14a) is almost constant, since for $(z - \zeta)/\lambda_D \gg 1$, $\cosh(\zeta/\lambda_{\mathrm{D}})\exp(-z/\lambda_{\mathrm{D}}) \approx (1/2)\exp(-(z-\zeta)/\lambda_{\mathrm{D}}) \ll 1$ and varies notably only when $(z - \zeta)/\lambda_{\mathrm{D}}$ is of the order of unity. A typical solution of eqs 14 is compared in Figure 1 to the results obtained from the numerical integration of the Poisson–Boltzmann equations (4a,b) for various values of z (the maximum length of a configuration) and the following values of the parameters: $N = 1000$, $a = 1$ Å, $\eta = 0.005$, $c_{\mathrm{E}} = 0.01$ M, $s^2 = 1000$ Å^2, $\epsilon = 80$. The linear approximation is in general not accurate for small values of z, which correspond to large potentials; however, these configurations have also large double layer energies and, therefore, small contributions to the final statistical average.

In the linear approximation, the double layer energy of a configuration of distance z, obtained from the integration of eqs 6a and 6b, is given by

$$F_{\mathrm{DL}}(z) = \int_0^z (F_{\mathrm{el}} + F_{\mathrm{ent}})\,\mathrm{d}\zeta \cong s^2\frac{\epsilon\epsilon_0}{2}\int_0^z\left(\left(\frac{\mathrm{d}\psi(\zeta)}{\mathrm{d}\zeta}\right)^2 + \frac{\psi^2(\zeta)}{\lambda_{\mathrm{D}}^{\,2}}\right)\mathrm{d}\zeta = \frac{\lambda_{\mathrm{D}}^{\,2}\eta^2e^2N^2}{4\epsilon\epsilon_0 s^2 z^2}\left(4z + 4\lambda_{\mathrm{D}}\exp\left(-\frac{2z}{\lambda_{\mathrm{D}}}\right)\cosh^2\left(\frac{z}{\lambda_{\mathrm{D}}}\right) - 3\lambda_{\mathrm{D}} - \lambda_{\mathrm{D}}\exp\left(-\frac{4z}{\lambda_{\mathrm{D}}}\right)\right) \qquad (15)$$

For a charged polyelectrolyte brush, the double layer energy, eq 15 is in general much larger than the excluded volume interaction, eq 6c. The probability for a brush to end up at a distance z (eq 7) is represented in Figure 2a, for the above values of the parameters and $l = 10$ Å, $\sigma = lN_l^{1/2} = 100$ Å, $\tau = -1.0$, $w = 0.0$, $v = 1.0$ Å^3. Since the selected area corresponding to a grafted chain on a surface is relatively small ($s^2 = 1000$ Å^2), low dissociation fractions (of the order of a few percent) are sufficient to generate large electrical potentials and consequently strong stretching of the brush. A qualitative analysis starting from eq 14a shows that the potential at the grafting surface, for a strong stretching ($z \cong Na$), is of the order of

$$\psi(0) \cong \frac{\eta e}{\epsilon\epsilon_0}\frac{N\lambda_{\mathrm{D}}^{\,2}}{s^2 z} \approx \frac{\eta e}{\epsilon\epsilon_0}\frac{\lambda_{\mathrm{D}}^{\,2}}{s^2 a} \qquad (16)$$

This implies that even at moderate electrolyte concentrations ($c_{\mathrm{E}} = 0.01$ M, $\lambda_{\mathrm{D}} = 30$ Å) and small dissociation fractions η, $\psi(0) > kT/e$ and the linear approximation of the Poisson–Boltzmann equations fails. An approximation, which is more accurate at large potentials, is suggested below.

B. Nonlinear Approximation. The numerical solution of the Poisson–Boltzmann system of eqs 4a and b, as well as the analytical solutions (eqs 14) of its linear approximation (eqs 12) indicate that the potential is almost constant through each configuration, until a distance of

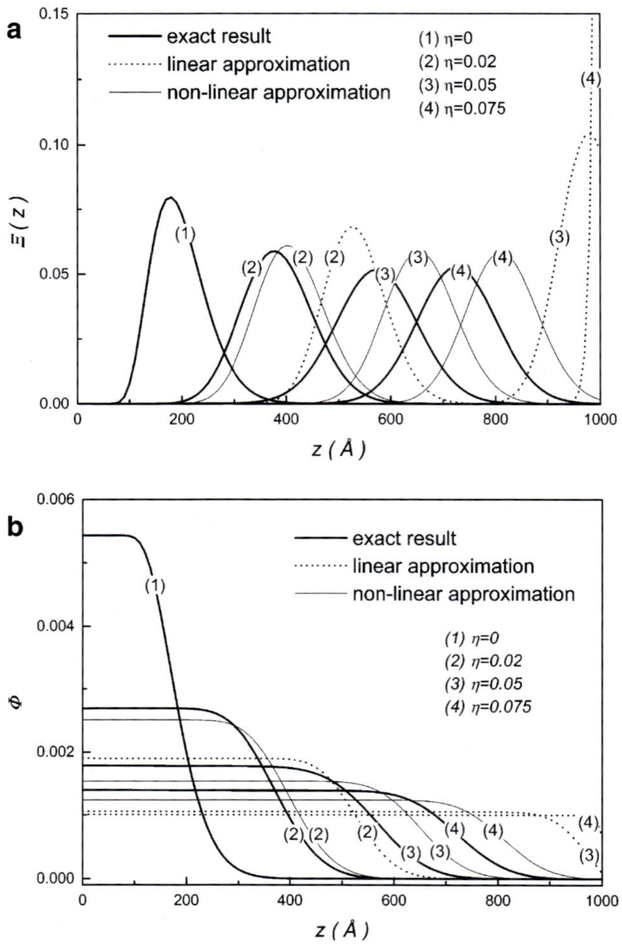

Figure 2. (a) The probability for a chain whose last monomer is located at a distance z for $N = 1000$, $a = 1$ Å, $c_E = 0.01$ M, $s^2 = 1000$ Å^2, $\epsilon = 80$, $\sigma = 100$ Å, $\tau = -1.0$, $w = 0.0$, $v = 1$ Å^3, and various degrees of dissociation η: thick continuous line, exact result; dotted line, linear approximation; continuous line, nonlinear approximation. (b) The monomer density in brushes with various degrees of dissociation for $N = 1000$, $a = 1$ Å, $c_E = 0.01$ M, $s^2 = 1000$ Å^2, $\epsilon = 80$, $\sigma = 100$ Å, $\tau = -1.0$, $w = 0.0$, and $v = 1$ Å^3: thick continuous line, exact result; dotted line, linear approximation; continuous line, nonlinear approximation.

the order of λ_D from the maximum length of the configuration (see Figure 1). A simple approximation consists of assuming a constant potential $\bar{\psi}$ between 0 and z. Outside the brush (for $z < \zeta < \infty$), the traditional Poisson–Boltzmann equation is valid and can be integrated to provide[19]

$$-\left.\frac{d\psi(\zeta)}{d\zeta}\right|_z = \frac{2kT}{\lambda_D e}\sinh\left(\frac{e\psi(z)}{2kT}\right) \tag{17a}$$

where the electric field at z is related to the charge distribution between 0 and z (due to the polyelectrolyte and electrolyte ions), which is given by

$$\int_0^z\left(\frac{\eta e N}{\epsilon\epsilon_0 s^2 z} - \frac{2ec_E}{\epsilon\epsilon_0}\sinh\left(\frac{e\bar{\psi}}{kT}\right)\right)d\zeta = \frac{\eta e N}{\epsilon\epsilon_0 s^2} - \frac{2ec_E z}{\epsilon\epsilon_0}\sinh\left(\frac{e\bar{\psi}}{kT}\right) \tag{17b}$$

By combining eqs 17a and b, one obtains

$$\sinh\left(\frac{e\bar{\psi}}{kT}\right) + \frac{2\lambda_D}{z}\sinh\left(\frac{e\bar{\psi}}{2kT}\right) = \frac{\eta N}{2c_E s^2 z} \tag{17c}$$

from which a numerical solution for $\bar{\psi}$ can be readily obtained. When the potential is sufficiently large, $\sinh(e\bar{\psi}/2kT) \ll \sinh(e\bar{\psi}/kT)$ (in many cases λ_D is also much smaller than z), and eq 17c reduces to

$$2c_E\sinh\left(\frac{e\bar{\psi}}{kT}\right) = \frac{\eta N}{s^2 z} \tag{17d}$$

which simply states that the constant potential $\bar{\psi}$ is such that the electrolyte ions compensate exactly the charge of the polyelectrolyte. The solution of eq 17d is provided by

$$\bar{\psi} \cong \frac{kT}{e}\ln\left(\left(\left(\frac{\eta N}{2c_E s^2 z}\right)^2 + 1\right)^{1/2} - \frac{\eta N}{2c_E s^2 z}\right) \tag{18}$$

As a matter of fact, the approximate surface potential predicted by the linear analysis can be derived using the same assumption of a vanishing charge distribution near the grafting surface. Indeed, $\eta N/s^2 z = 2c_E\sinh(e\psi(0)/kT) \cong 2ec_E\psi(0)/kT = \epsilon\epsilon_0\psi(0)/e\lambda_D^2$, which coincides with eq 16.

In the region of constant potential, the electrostatic energy vanishes (because $d\psi/d\zeta \equiv 0$) while the entropic contribution is given by

$$F_{DL}(z) = kTs^2\sum_i\int_0^z\left(c_i\ln\left(\frac{c_i}{c_E}\right) - c_i + c_E\right)d\zeta = c_E kTs^2 z\left(\frac{e\bar{\psi}}{kT}\exp\left(\frac{e\bar{\psi}}{kT}\right) - \frac{e\bar{\psi}}{kT}\exp\left(-\frac{e\bar{\psi}}{kT}\right) - \exp\left(\frac{e\bar{\psi}}{kT}\right) - \exp\left(-\frac{e\bar{\psi}}{kT}\right) + 2\right) \tag{19}$$

The contribution to the double layer free energy from the region between z and infinity can be calculated by solving the corresponding traditional Poisson–Boltzmann equation. However, the value of $\bar{\psi}$ given by eq 18 is an overestimate of the potential inside the brush (since it implies a potential large enough to cancel the charge density in the brush) and consequently eq 19 is also an overestimate of the free energy of the double layer between 0 and z. Therefore, eq 19 can be used as an approximation of the total double layer free energy for the configuration that ends up at the distance z.

In Figure 2a, the probability for a brush to end up at a distance z is calculated using the simple nonlinear approximation based on eqs 18 and 19; this approximation provides in general a much better agreement with the numerical calculations based on the nonlinear equations than the linear approximation at high potentials. The corresponding monomer density distributions, statistically averaged over all possible configurations, are plotted in Figure 2b.

The thickness of the grafted polyelectrolyte brush can be described roughly by the length of the most probable configuration, which is provided by the solution of

$$\frac{d}{dz}\left(\exp\left(-\frac{z^2}{2\sigma^2}\right)\exp\left(-\frac{F_{DL}(z) + F_{FH}(z)}{kT}\right)\right) = 0 \tag{20}$$

(19) Hunter, R. J. *Foundations of Colloid Science*; Clarendon Press: Oxford, 1987.

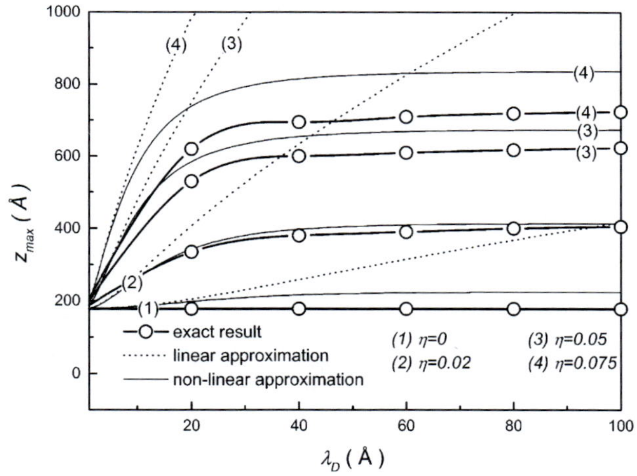

Figure 3. The thickness of the brush (the length of the most probable configuration) vs the Debye–Hückel length for various degrees of dissociation and $N = 1000$, $a = 1$ Å, $c_E = 0.01$ M, $s^2 = 1000$ Å^2, $\epsilon = 80$, $\sigma = 100$ Å, $\tau = -1.0$, $w = 0.0$, and $v = 1$Å^3.

The dependence of the length of the most probable configuration on the electrolyte concentration is plotted in Figure 3. The brush is strongly stretched for most ionic strengths. As already noted, this occurs because at lower electrolyte concentrations there are not many electrolyte counterions to screen the polyelectrolyte charges; therefore the potential in the brush is high, even for small degrees of dissociation.

4. Monomer Distribution in a Grafted Polyelectrolyte Brush Immersed in a Finite Reservoir

A. Linear Approximation. As noted in previous sections, even dissociation of the order of a few percent of the grafted polyelectrolyte brushes might be enough to stretch the chains, particularly at low electrolyte concentrations. In an infinite reservoir, the density of the counterions generated by the dissociation of the polyelectrolyte groups vanishes, and only the electrolyte ions are left to compensate the brush charge. An interesting situation occurs when the brush is immersed in a finite reservoir, because a fraction of the counterions of the polyelectrolyte remains trapped inside the brush, neutralizing a part of its charge. In this case, assuming a finite reservoir of length $L > Na$ (see Figure 4), the Poisson–Boltzmann equations become

$$\frac{\partial^2\psi}{\partial\zeta^2} = \frac{e\left(c_1\exp\left(\frac{e\psi}{kT}\right) - c_2\exp\left(-\frac{e\psi}{kT}\right)\right)}{\epsilon\epsilon_0} - \frac{\eta e}{\epsilon\epsilon_0}\frac{N}{s^2 z} + \frac{\alpha e N}{\epsilon\epsilon_0 s^2 L}\exp\left(\frac{e\psi}{kT}\right) \quad (21a)$$

$$0 < \zeta < z$$

$$\frac{\partial^2\psi}{\partial\zeta^2} = \frac{e\left(c_1\exp\left(\frac{e\psi}{kT}\right) - c_2\exp\left(-\frac{e\psi}{kT}\right)\right)}{\epsilon\epsilon_0} + \frac{\alpha e N}{\epsilon\epsilon_0 s^2 L}\exp\left(\frac{e\psi}{kT}\right) \quad (21b)$$

$$z < \zeta < L$$

where the constants c_1, c_2, and α are related to the concentrations of counterions of the polyelectrolyte and

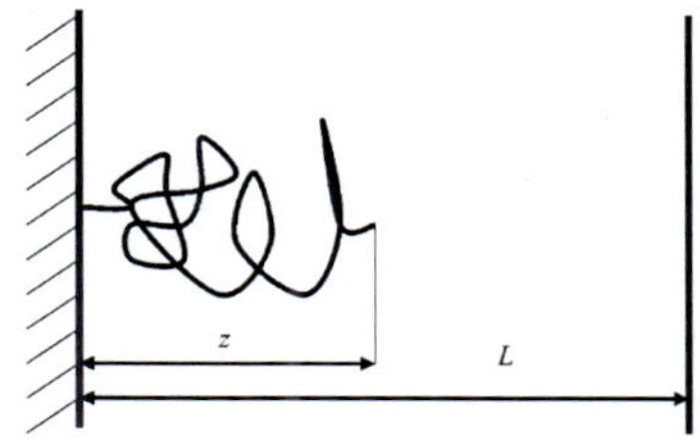

Figure 4. A configuration of a polyelectrolyte molecule, grafted on a surface, whose last monomer is located at the distance z, immersed in a finite reservoir of length L.

of the electrolyte ions through the expressions

$$\int_0^L c_1\exp\left(\frac{e\psi(\zeta)}{kT}\right)\mathrm{d}\zeta = c_E L \quad (22a)$$

$$\int_0^L c_2\exp\left(-\frac{e\psi(\zeta)}{kT}\right)\mathrm{d}\zeta = c_E L \quad (22b)$$

$$\int_0^L \alpha\exp\left(\frac{e\psi(\zeta)}{kT}\right)\mathrm{d}\zeta = \eta L \quad (22c)$$

Assuming as above that the grafting surface is not charged ($\sigma_S = 0$), the electroneutrality of the system leads to the boundary conditions

$$\left.\frac{\partial^2\psi}{\partial\zeta^2}\right|_{\zeta=0} = 0 \quad (23a)$$

$$\left.\frac{\partial^2\psi}{\partial\zeta^2}\right|_{\zeta=L} = 0 \quad (23b)$$

An analytical solution of the above system can be obtained in the linear approximation as follows. The Poisson–Boltzmann equations become

$$\frac{\partial^2\psi}{\partial\zeta^2} = a_1\psi + a_2 - a_3 \quad (24a)$$

$$0 < \zeta < z$$

$$\frac{\partial^2\psi}{\partial\zeta^2} = a_1\psi + a_2 \quad (24b)$$

$$z < \zeta < L$$

where

$$a_1 = \frac{e^2}{\epsilon\epsilon_0 kT}\left(c_1 + c_2 + \alpha\frac{N}{s^2 L}\right) \quad (25a)$$

$$a_2 = \frac{e}{\epsilon\epsilon_0}\left(c_1 - c_2 + \alpha\frac{N}{s^2 L}\right) \quad (25b)$$

$$a_3 = \eta\frac{eN}{\epsilon\epsilon_0 s^2 z} \quad (25c)$$

The solution of the linear system (24), which obeys the boundary conditions (23a,b) and has ψ and $\mathrm{d}\psi/\mathrm{d}\zeta$ continuous at z is determined up to an arbitrary constant potential, which is irrelevant for the problem. It should be, however, noted that eqs 22 imply that the ion concentrations reach the values c_1, c_2, and $\alpha N/s^2 L$ at the point where $\psi = 0$. Selecting this point to be at $\zeta = L$

(the far side of the reservoir), one obtains the solution

$$\psi(\zeta) = \frac{a_3}{a_1}\left(1 - \frac{\sinh(za_1^{1/2})}{\sinh(La_1^{1/2})} - \frac{\sinh((L-z)a_1^{1/2})}{\sinh(La_1^{1/2})}\cosh(\zeta a_1^{1/2})\right) \quad (26a)$$

$$0 < \zeta < z$$

$$\psi(\zeta) = \frac{a_3}{a_1}\frac{\sinh(za_1^{1/2})}{\sinh(La_1^{1/2})}(\cosh((L-\zeta)a_1^{1/2}) - 1) \quad (26b)$$

$$z < \zeta < L$$

Since in the linear approximation

$$\int_0^L \exp\left(\frac{e\psi(\zeta)}{kT}\right) d\zeta = L + \frac{e}{kT}\int_0^L \psi(\zeta)\, d\zeta = L + \frac{e}{kT}\frac{a_3}{a_1}\left(z - L\frac{\sinh(za_1^{1/2})}{\sinh(La_1^{1/2})}\right) \quad (27)$$

Equations 22 become

$$c_1 = \frac{c_E}{1 + \frac{e}{kT}\frac{a_3}{a_1}\left(\frac{z}{L} - \frac{\sinh(za_1^{1/2})}{\sinh(La_1^{1/2})}\right)} \quad (28a)$$

$$c_2 = \frac{c_E}{1 - \frac{e}{kT}\frac{a_3}{a_1}\left(\frac{z}{L} - \frac{\sinh(za_1^{1/2})}{\sinh(La_1^{1/2})}\right)} \quad (28b)$$

$$\alpha = \frac{\eta}{1 + \frac{e}{kT}\frac{a_3}{a_1}\left(\frac{z}{L} - \frac{\sinh(za_1^{1/2})}{\sinh(La_1^{1/2})}\right)} \quad (28c)$$

It should be again emphasized that the linear approximation is accurate only for $\psi \ll kT/e$. Whereas the solutions for the potential (eqs 26) are qualitatively (but not quantitatively!) valid even for $\psi > kT/e$, eqs 28 fail in this case and might even predict negative concentrations of co-ions.

Equations 28a–c can be solved for the constants c_1, c_2, and α as functions of η, c_E, z, and L. Then the double layer free energy of a configuration that ends up at the distance z in a reservoir of finite length L is given by

$$F_{DL}(z) = \frac{s^2\epsilon\epsilon_0}{2}\int_0^L \left(\frac{d\psi(\zeta)}{d\zeta}\right)^2 d\zeta + kTs^2\int_0^L \left[\frac{ec_1}{kT}\exp\left(\frac{e\psi}{kT}\right)\ln\left(\frac{c_1}{c_E}\right) - \frac{ec_2}{kT}\exp\left(-\frac{e\psi}{kT}\right)\ln\left(\frac{c_2}{c_E}\right) - c_1\exp\left(\frac{e\psi}{kT}\right) - c_2\exp\left(-\frac{e\psi}{kT}\right) + 2c_E + \frac{e}{kT}\frac{\alpha N}{s^2L}\exp\left(\frac{e\psi}{kT}\right)\ln\left(\frac{\alpha}{\eta}\right) - \frac{\alpha N}{s^2L}\exp\left(\frac{e\psi}{kT}\right) + \frac{\eta N}{s^2L}\right] \quad (29)$$

where it was taken into account that not only the electrolyte ions but also the counterions of the polyelectrolyte chains contribute to the entropy of the double layer, because, as a result of the double layer formation, their distribution differs from the average density $\eta N/s^2L$.

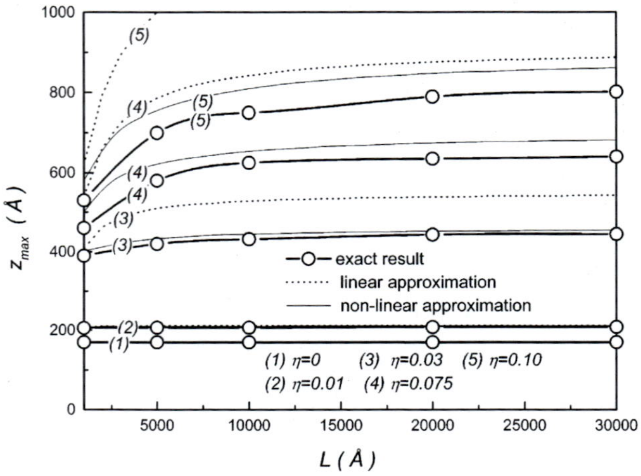

Figure 5. The dependence of the brush thickness (the length of the most probable configuration) on the finite reservoir size L for various degrees of dissociation and $N = 1000$, $a = 1$ Å, $c_E = 0.01$ M, $s^2 = 1000$ Å^2, $\epsilon = 80$, $\sigma = 100$ Å, $\tau = -1.0$, $w = 0.0$, and $v = 1.0$ Å^3.

B. Nonlinear Approximation. The linear approximation in general is not accurate for potentials larger than kT/e. A simple approximation, which is more accurate than the linear one at larger potentials, can be obtained by assuming that the electrical potential is constant along the configuration that ends at the distance z and vanishes in the rest of the reservoir. This average potential can be obtained, as in section 3B, from the condition of vanishing of the total charge density inside the brush

$$\eta\frac{N}{s^2z} + c_2\exp\left(-\frac{e\bar{\psi}}{kT}\right) - \left(c_1 + \alpha\frac{N}{s^2L}\right)\exp\left(\frac{e\bar{\psi}}{kT}\right) = 0 \quad (30a)$$

The conservation of the number of ions in the reservoir (eqs 22) leads to

$$c_1 = \frac{c_E L}{z\exp\left(\frac{e\bar{\psi}}{kT}\right) + L - z} \quad (30b)$$

$$c_2 = \frac{c_E L}{z\exp\left(-\frac{e\bar{\psi}}{kT}\right) + L - z} \quad (30c)$$

$$\alpha = \frac{\eta L}{z\exp\left(\frac{e\bar{\psi}}{kT}\right) + L - z} \quad (30d)$$

The solution of $\bar{\psi}$ from eqs 30a–d can be used to obtain the following approximate expression for the double layer free energy

$$F_{DL}(z) = kTs^2z\left[\frac{ec_1}{kT}\exp\left(\frac{e\bar{\psi}}{kT}\right)\ln\left(\frac{c_1}{c_E}\right) - \frac{ec_2}{kT}\exp\left(-\frac{e\bar{\psi}}{kT}\right)\ln\left(\frac{c_2}{c_E}\right) - c_1\exp\left(\frac{e\bar{\psi}}{kT}\right) - c_2\exp\left(-\frac{e\bar{\psi}}{kT}\right) + 2c_E + \frac{e}{kT}\frac{\alpha N}{s^2L}\exp\left(\frac{e\bar{\psi}}{kT}\right)\ln\left(\frac{\alpha}{\eta}\right) - \frac{\alpha N}{s^2L}\exp\left(\frac{e\bar{\psi}}{kT}\right) + \frac{\eta N}{s^2L}\right] \quad (31)$$

In Figure 5, the length of the most probable configuration is plotted as a function of the reservoir length L for

$N = 1000$, $a = 1$ Å, $l = 10$ Å, $\sigma = lN_l^{1/2} = 100$ Å, $\eta = 0.01$, $c_E = 0.01$ M, $s^2 = 1000$ Å^2, $\epsilon = 80$, $\tau = -1.0$, $w = 0.0$, and $v = 1.0$ Å^3. For small reservoirs, most counterions are not able to leave the brush; hence the overall charging is small. This situation is not unlike some molecular dynamics simulations,[20] where the counterions were explicitly taken into account, but the size of the system was limited (because of computational requirements). However, when the brush is embedded in an infinitely large reservoir and chemical equilibrium is reached, there are no counterions left in the brush, and a stronger stretching is achieved.

5. The Interactions between Surfaces with Grafted Polyelectrolyte Brushes: The Role of the Rate of Approach of the Surfaces

A. Thermodynamic Equilibrium (Slow Approach of the Surfaces). The general formalism for the interaction between two surfaces with grafted polyelectrolyte brushes, presented in this paper in section 2, implies that, for each separation distance $2d$, the Poisson–Boltzmann equations have to be solved for every possible configuration of the chains, ending up at z_1 and z_2, where $0 < z_1 < \xi$, $0 < z_2 < \xi$, ξ being the smallest between the maximum length of a chain, Na, and the distance $2d$ between surfaces. Since such calculations are tedious, here a much simpler analysis is used for the interaction between surfaces. It was shown previously[14] that for neutral brushes and good solvents, the symmetric configurations (for which $z_1 = z_2$) are the most important contributions to the free energy, and (at least at strong repulsion) the contributions of the other configurations can be neglected. Such an approximation is reasonable because for a single brush only the configurations around the most probable one have significant probabilities $\Xi(z)$ (see Figure 2a). For two brushes, $\Xi(z_1,z_2) \approx \Xi(z_1)\Xi(z_2)$ and for the asymmetric configurations ($z_1 \neq z_2$), at least one of the probabilities is small.

For the symmetric configurations with $z < d$, a constant monomer density between surface and z will be considered for each brush, while for $z > d$ (physical overlap of the brushes) it will be assumed that all the monomers are homogeneously distributed between the two surfaces.

In the absence of a surface charge density on the plates ($\sigma_S = 0$), the solution of the Poisson–Boltzmann equations (4a and b) are obtained for the boundary conditions

$$\left.\frac{\partial\psi}{\partial\zeta}\right|_{\zeta=0} = 0 \tag{32a}$$

$$\left.\frac{\partial\psi}{\partial\zeta}\right|_{\zeta=d} = 0 \tag{32b}$$

and the requirements of continuity of $\psi(\zeta)$ and $d\psi(\zeta)/d\zeta$ at $\zeta = z$. The free energy of each configuration that ends up at the distance z is obtained from eqs 6a–c, while the free energy of the system is obtained from eqs 9–11, by considering only the symmetric configurations

$$F_{\text{brushes}} = \int_0^{2d} (F_{\text{el}}(z) + F_{\text{ent}}(z) + F_{\text{FH}}(z))\Xi(z)\,dz \tag{33}$$

For small charges of the brush or large ionic strengths, one can employ the linear approximation of the Poisson–Boltzmann equations (eqs 12) with the boundary conditions (eqs 32). The solution that obeys the continuity conditions at $\zeta = z$ is given by

$$\psi(\zeta) = \frac{\eta e}{\epsilon\epsilon_0}\frac{N}{s^2 z}\lambda_D^2\left[1 - \cosh\left(\frac{\zeta}{\lambda_D}\right)\frac{\sinh\left(\frac{d-z}{\lambda_D}\right)}{\sinh\left(\frac{d}{\lambda_D}\right)}\right] \quad 0 < \zeta < z \tag{34a}$$

$$\psi(\zeta) = \frac{\eta e}{\epsilon\epsilon_0}\frac{N}{s^2 z}\lambda_D^2\cosh\left(\frac{d-\zeta}{\lambda_D}\right)\frac{\sinh\left(\frac{z}{\lambda_D}\right)}{\sinh\left(\frac{d}{\lambda_D}\right)} \quad z < \zeta < d \tag{34b}$$

Consequently the total double layer free energy is given by

$$F_{\text{DL}}(z) = \frac{\eta^2 e^2}{4\epsilon\epsilon_0}\frac{N^2\lambda_D^2}{s^2 z^2 \sinh^2\left(\frac{d}{\lambda_D}\right)}\left[(2z - \lambda_D)\sinh\left(\frac{2d}{\lambda_D}\right) + \lambda_D\left(\sinh\left(\frac{2d-2z}{\lambda_D}\right) + \sinh\left(\frac{2z}{\lambda_D}\right)\right) - 2z\right] \tag{35}$$

In Figure 6a, the force per unit area between surfaces with grafted polyelectrolyte brushes, plotted as a function of their separation distance $2d$, calculated in the linear approximation, is compared with the numerical solution of the nonlinear Poisson–Boltzmann equations, for a system with $N = 1000$, $a = 1$ Å, $c_E = 0.01$ M, $s^2 = 1000$ Å^2, $\epsilon = 80$, $l = 10$ Å, $\sigma = 100$ Å, $\tau = -1.0$, $w = 0.0$, $v = 1.0$ Å^3 and various degrees of dissociation. As expected, even moderate degrees of dissociation of the polyelectrolyte generate large potentials, for which the linear approximation fails.

B. Quasi-Thermodynamical Equilibrium (Rapid Approach of the Surfaces). The motion of the counterions along the distance perpendicular to the surfaces is in general sufficiently rapid to reach the thermal equilibrium along this direction during the interaction. If the approach of the surfaces is slow, the counterions of the polyelectrolyte chains can also reach thermal equilibrium with the infinite reservoir. However, if the surfaces are sufficiently large, and the approaching of the plates sufficiently rapid, the diffusion of the ions in the planes parallel to the surface toward the infinite reservoir might be much slower than the rate of approach of the surfaces. In this case, the ions reach thermal equilibrium along the direction perpendicular to the surfaces but can remain practically trapped between the surfaces. A simple modality to describe such a behavior is to consider a system with infinitely large surfaces, with all the ions (the counterions of the polyelectrolyte chains and the electrolyte ions) located between the surfaces. Each half of the system obeys the same equations as the brush embedded in a finite reservoir treated in the previous paragraph, if the reservoir thickness L is replaced by the (variable) half distance d between surfaces. In the linear approximation, the solution for the potential (up to an arbitrary constant) is provided by eqs 26a,b, while the corresponding double layer free energy is given by eq 29.

In Figure 6b, the force per unit area between surfaces with grafted polyelectrolyte brushes, plotted as a function of their separation distance $2d$, provided by the linear approximation is compared with the numerical result

(20) Seidel, C. *Macromolecules* **2003**, *36*, 2536.

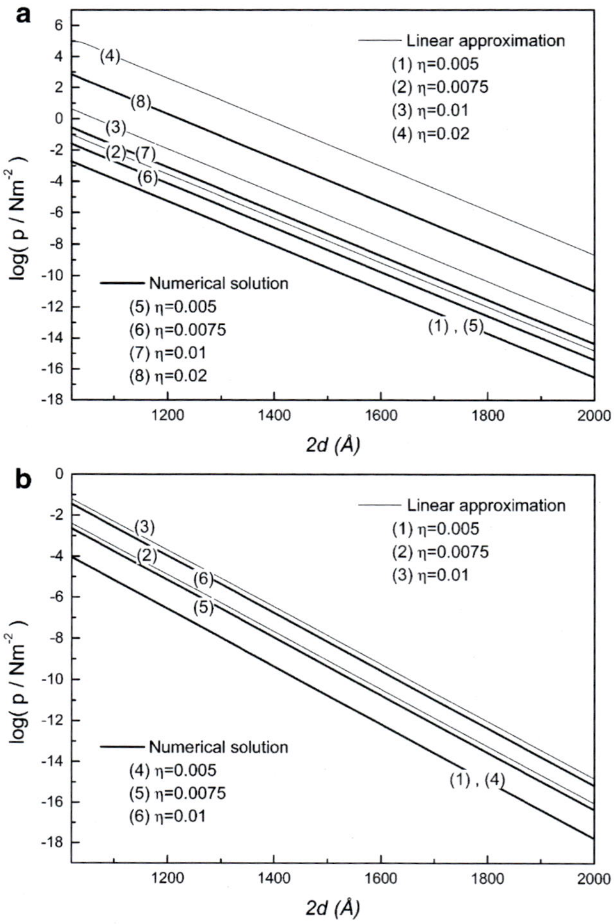

Figure 6. (a) The force per unit area between two identical plates with grafted polyelectrolyte brushes, as a function of the separation distance, for $N = 1000$, $a = 1$ Å, $c_E = 0.01$ M, $s^2 = 1000$ Å^2, $\epsilon = 80$, $\sigma = 100$ Å, $\tau = -1.0$, $w = 0.0$, and $v = 1.0$ Å^3 calculated in the linear approximation is compared to the numerical result obtained from the nonlinear equations, assuming that the counterions diffuse into an infinite reservoir (slow approach). (b) The force per unit area between two identical plates with grafted polyelectrolyte brushes, as a function of the separation distance, for $N = 1000$, $a = 1$ Å, $c_E = 0.01$ M, $s^2 = 1000$ Å^2, $\epsilon = 80$, $\sigma = 100$ Å, $\tau = -1.0$, $w = 0.0$, and $v = 1.0$ Å^3 calculated in the linear approximation is compared to the numerical result obtained from the nonlinear equations, assuming that the counterions remained trapped between the plates (rapid approach).

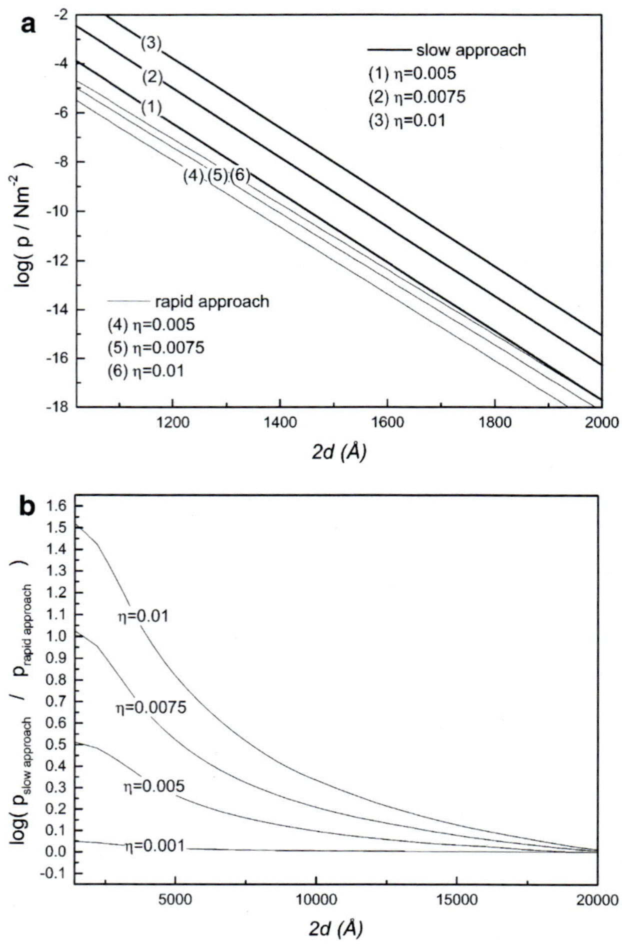

Figure 7. (a) The force per unit area between two identical plates with grafted polyelectrolyte brushes, as a function of the separation distance, for $N = 1000$, $a = 1$ Å, $c_E = 0.01$ M, $s^2 = 1000$ Å^2, $\epsilon = 80$, $\sigma = 100$ Å, $\tau = -1.0$, $w = 0.0$, $v = 1.0$ Å^3, and various degrees of dissociation, calculated by assuming that the counterions either diffuse into an infinite reservoir (slow approach) or remain trapped between the plates (rapid approach) (b) The ratio between the interactions calculated assuming that the ions reach thermal equilibrium with an infinite reservoir (slow approach of the surfaces) and those calculated assuming that all the ions remain trapped between the surfaces (rapid approach of large surfaces), for $N = 1000$, $a = 1$ Å, $c_E = 0.01$ M, $s^2 = 1000$ Å^2, $\epsilon = 80$, $\sigma = 100$ Å, $\tau = -1.0$, $w = 0.0$, $v = 1.0$ Å^3, and various degrees of dissociation. At small separation distances and even moderate charging, the interactions can differ by orders of magnitude. At large separations, the differences become negligible.

obtained from the integration of a nonlinear system for a system with the same parameter values as above ($N = 1000$, $a = 1$ Å, $c_E = 0.01$ M, $s^2 = 1000$ Å^2, $\epsilon = 80$, $l = 10$ Å, $\sigma = 100$ Å, $\tau = -1.0$, $w = 0.0$, $v = 1$ Å^3) and various degrees of dissociation. Because of the counterions trapping, the potentials are in general smaller than in the previous case; hence the linear approximation is a better one.

The most interesting results are the changes in the interactions between plates as a function of the rate of approach. The extreme cases, when all the counterions are trapped between plates (rapid approach of the surfaces) and when the counterions diffuse into the infinite reservoir (slow approach of the surfaces) are compared in Figure 7a for a system with the same parameter values as above ($N = 1000$, $a = 1$ Å, $c_E = 0.01$ M, $s^2 = 1000$ Å^2, $\epsilon = 80$, $l = 10$ Å, $\sigma = 100$ Å, $\tau = -1.0$, $w = 0.0$, $v = 1$ Å^3) and various degrees of dissociation, the force being calculated numerically in each case, using the nonlinear equations and only the symmetric configurations. Even for moderate degrees of dissociation, there is a difference of orders of magnitude between the two cases. The large difference is due mainly to the strong stretching of the brushes in the absence of the polyelectrolyte counterions. The double layer interaction roughly decays exponentially with a decay length $\lambda_D \cong 30$ Å (for $c_E = 0.01$ M), and variations in brush thicknesses of the order of 100 Å lead to changes of orders of magnitude in the interactions. In addition, when the counterions are trapped between plates and compensate a part of the polyelectrolyte charge, a smaller potential is generated. At very large separations between plates, as expected, the difference between the two cases vanishes, because the counterions have sufficient room to leave the brush (see Figure 7b).

6. Conclusion

A simple mean field theory for polyelectrolyte brushes is suggested, based on an approximate Monte Carlo-type calculation. The method can be applied at high polyelectrolyte stretchings, for which the traditional treatments based on the self-consistent-field fail. It is assumed that the Boltzmann factor for the configurations that end up at a distance z from the surface can be approximated by a Boltzmann factor of the configuration that ends at the same distance and has a constant density distribution of monomers. The free energy of the latter can be calculated by solving the appropriate Poisson–Boltzmann equations. Both the density distribution of the monomer in the brush, calculated as the statistical average of all possible configurations, as well as the interactions between surfaces can be provided by this method. Simple expressions for the free energy of a configuration have been obtained, both in the linear approximation (valid at small potentials) and in a nonliner approximation, which is more accurate at high potentials. The dependence of the brush thickness (defined as the length of the most probable configuration) on electrolyte concentration was investigated. A simple modality to calculate the interaction between surfaces with grafted polyelectrolyte brushes using only symmetric configurations has been suggested. The possibility for a part of the counterions to remain trapped in the brush was also examined, and it was shown that even at moderate degrees of dissociation of the polyelectrolyte chains, it might reduce the force between brushes by orders of magnitude at small separations. Since for plates approaching rapidly (or having large surfaces) the counterions might not have sufficient time to diffuse into the infinite reservoir, the trapping of counterions can play an important role in colloid coagulation.

LA049168E

Available online at www.sciencedirect.com
SCIENCE DIRECT
ELSEVIER
Journal of Colloid and Interface Science 273 (2004) 181–190
JOURNAL OF Colloid and Interface Science
www.elsevier.com/locate/jcis

Double-layer interaction between two plates with hairy surfaces

Haohao Huang and Eli Ruckenstein *

Department of Chemical Engineering, State University of New York at Buffalo, Buffalo, NY 14260, USA

Received 8 July 2003; accepted 29 August 2003

Abstract

In most theoretical treatments of colloidal particles with hairy surfaces, only the steric effect is taken into account. The steric force is a short-range interaction and acts only when the chains on different particles begin to interpenetrate each other. However, since the hairy chains are extended into the continuous phase, they constrain the orientation of the water molecules near the surface and, as a result, the dielectric constant in that region can become very different from that in the bulk. The low dielectric constant affects the distributions of ion concentrations and the gradient of the electric field. Therefore, the double-layer interactions between two plates with hairy surfaces cannot be calculated on the basis of the classical Gouy–Chapman theory, which involves a uniform dielectric constant in the Poisson–Boltzmann equation. A model which accounts for the difference in dielectric constants in the hairy region and outside that region is therefore proposed. The ion specificity is also taken into account by using Born's expression for the free energy of hydration of ions. The repulsive forces calculated via the Gouy–Chapman theory and via the new model are compared. The hairy region can have a long range effect on the repulsive double-layer interactions.

Keywords: Hairy surface; Double-layer interaction; Hydration energy; Specific ion effect

1. Introduction

Polymer molecules are often employed to stabilize colloids [1]. In most theoretical treatments of the effect of polymer adsorption [2–5], only the steric force is taken into account, and the steric force and the traditional double-layer force for particles devoid of "hairs" are assumed to be additive. The steric force is a short-range interaction which acts only when the chains on the surfaces of the two particles interpenetrate [6–8]. However, in addition to this short-range interaction, a hairy surface can also generate another effect, because it can change the dielectric constant in the vicinity of the surface.

Approaches have been developed which modify the traditional double-layer theory by assuming that the dielectric constant depends upon position. This dependence is considered to occur because the dielectric constant is affected by the electric field [9–12]. However, the dielectric constant can also be changed by the presence of hairs on the surface. Indeed, hairy chains on the surface constrain the orientation of the water molecules nearby, and, as a result, in the region near the surface the dielectric constant becomes lower than in the bulk. In fact, the water molecules are ordered very near the particle surface even in the absence of a hairy surface [13] and this ordered structure leads to a lower local dielectric constant near the surface than in the bulk. The thickness of the latter region is, however, very small. In what follows it will be shown that, by changing the dielectric constant near the surface, the hairy chains can have a long-range effect, much longer than the range of steric repulsion.

In this paper, a model is proposed that involves two different dielectric constants in the calculation of the electrical potential between two plates, one for the region near the surface and the other outside this region. The ion specificity will be also taken into account via Born's expression for the hydration free energy of ions. The repulsive force was calculated numerically at constant surface charge density at constant surface potential and for a surface charge generated through dissociation equilibrium. The effects of electrolyte concentration, of the dielectric constant in the hairy region, of the thickness of the hairy region, and of the nature of the electrolyte were investigated. The repulsive forces calculated via the Gouy–Chapman theory and those based on the new model are compared. An analytical expression for the repulsive force in the framework of the linear approximation is provided.

* Corresponding author.
E-mail address: feaeliru@acsu.buffalo.edu (E. Ruckenstein).

0021-9797/$ – see front matter
doi:10.1016/j.jcis.2003.08.060

2. Theoretical framework

The region near the surface (region I), of thickness δ, is considered to have a dielectric constant ε^{I} different from that in the remaining region between the plates (region II) $\varepsilon^{\mathrm{II}}$, due to the constraint on the water molecules induced by the hairy surface. For simplicity, the dielectric constants in both regions are assumed to be constant, with that in region II equal to the bulk dielectric constant.

The electrical potential distribution between the two plates is given by the Poisson equations

$$\frac{\partial^2 \Psi}{\partial x^2} = -\frac{\rho}{\varepsilon^{\mathrm{I}}\varepsilon_0} \quad \text{(for region I)} \tag{1}$$

and

$$\frac{\partial^2 \Psi}{\partial x^2} = -\frac{\rho}{\varepsilon^{\mathrm{II}}\varepsilon_0} \quad \text{(for region II)}, \tag{2}$$

where Ψ is the electrical potential, x is the distance from the middle between the two plates, ρ is the charge density, ε^{I} and $\varepsilon^{\mathrm{II}}$ are the dielectric constants in regions I and II, respectively, and ε_0 is the vacuum permittivity.

According to the Born theory, the hydration free energy of ion i can be expressed as

$$U_i^h = \frac{A_i}{\varepsilon}, \tag{3}$$

where A_i is a hydration constant, which is specific to each ion and could be evaluated from experimental hydration free energies [14,15], and ε is the dielectric constant of the medium in which ion i is located. The values of A_i for various ions used in the calculations are listed in Table 1.

The equalities between the chemical potentials in region I and the bulk and region II and the bulk lead to

$$kT \ln n_i^{\mathrm{I}} + ez_i\Psi + \frac{A_i}{\varepsilon^{\mathrm{I}}} = kT \ln n_i^0 + \frac{A_i}{\varepsilon^{\mathrm{II}}} \tag{4}$$

and

$$kT \ln n_i^{\mathrm{II}} + ez_i\Psi + \frac{A_i}{\varepsilon^{\mathrm{II}}} = kT \ln n_i^0 + \frac{A_i}{\varepsilon^{\mathrm{II}}}, \tag{5}$$

where n_i^0 is the bulk concentration of ion i, k is the Boltzmann constant, T is the absolute temperature, e is the protonic charge, z_i is the ion valency (positive for cations and negative for anions), and n_i^{I} and n_i^{II} are the concentrations of ion i in regions I and II, respectively. Consequently,

$$n_i^{\mathrm{I}} = n_i^0 \exp\left[-\frac{ez_i\Psi}{kT} - \frac{A_i}{kT}\left(\frac{1}{\varepsilon^{\mathrm{I}}} - \frac{1}{\varepsilon^{\mathrm{II}}}\right)\right] \quad \text{(for region I)} \tag{6}$$

and

$$n_i^{\mathrm{II}} = n_i^0 \exp\left(-\frac{ez_i\Psi}{kT}\right) \quad \text{(for region II)}. \tag{7}$$

Substituting Eqs. (6) and (7) into Eqs. (1) and (2), respectively, the Poisson–Boltzmann equations become

$$\frac{\partial^2 \Psi}{\partial x^2} = -\frac{e\sum_i n_i^0 z_i \exp\left[-\frac{ez_i\Psi}{kT} - \frac{A_i}{kT}\left(\frac{1}{\varepsilon^{\mathrm{I}}} - \frac{1}{\varepsilon^{\mathrm{II}}}\right)\right]}{\varepsilon^{\mathrm{I}}\varepsilon_0} \quad \text{(for region I)}, \tag{8}$$

$$\frac{\partial^2 \Psi}{\partial x^2} = -\frac{e\sum_i n_i^0 z_i \exp\left(-\frac{ez_i\Psi}{kT}\right)}{\varepsilon^{\mathrm{II}}\varepsilon_0} \quad \text{(for region II)}. \tag{9}$$

For surfaces that have a constant potential Ψ_0, the boundary condition at the surface is

$$\Psi|_{x=-d} = \Psi_0, \tag{10}$$

where $2d$ is the distance between the plates. For surfaces at constant charge density σ, the boundary condition at the surface becomes

$$\left.\frac{\partial \Psi}{\partial x}\right|_{x=-d} = -\frac{\sigma}{\varepsilon^{\mathrm{I}}\varepsilon_0}. \tag{11}$$

The symmetry condition provides an additional boundary condition,

$$\left.\frac{\partial \Psi}{\partial x}\right|_{x=0} = 0. \tag{12}$$

At the boundary between the two regions, the electric potentials should be the same,

$$\Psi|_{x=-d+\delta}^{\mathrm{I}} = \Psi|_{x=-d+\delta}^{\mathrm{II}}, \tag{13}$$

and if there is no charge accumulation, the Gauss law provides the equation

$$\varepsilon^{\mathrm{I}} \left.\frac{\partial \Psi}{\partial x}\right|_{x=-d+\delta}^{\mathrm{I}} = \varepsilon^{\mathrm{II}} \left.\frac{\partial \Psi}{\partial x}\right|_{x=-d+\delta}^{\mathrm{II}}, \tag{14}$$

where the superscripts I and II indicate regions I and II, respectively.

An equation of the form (8) was previously employed [11], in a different context.

Equations (8) and (9) and the boundary conditions (10) or (11) and (12)–(14) allow the profile of the electric potential to be calculated numerically.

The Langmuir equation cannot be used to calculate the repulsive force, because it implies ideal behavior of ions and the presence of the term $A_i(1/\varepsilon^{\mathrm{I}} - 1/\varepsilon^{\mathrm{II}})$ in Eq. (6) implies nonideality. Consequently, to obtain the repulsive force between two plates, it is necessary to calculate the free energy

Table 1
The value of A_i for different ions [11]

Ion	A_i (kcal/mol)
Li^+	127.1
Na^+	103.0
K^+	85.1
F^-	84.4
Cl^-	70.8
Br^-	63.8

H. Huang, E. Ruckenstein / Journal of Colloid and Interface Science 273 (2004) 181–190 183

of the system as a function of the distance $2d$ between them, its derivative with respect to $2d$ providing the force. The free energy of the system F is composed of four contributions,

$$F = F_{\text{ele}} + F_{\text{entropy}} + F_{\text{che}} + F_{\text{hyd}}, \tag{15}$$

where F_{ele} is the electrostatic, F_{entropy} the entropic, F_{chem} the chemical, and F_{hyd} the hydration contribution.

The electrostatic contribution comes from the electrical potential energy of the charge in the field [16]:

$$F_{\text{ele}} = \int_{-d}^{d} \frac{\varepsilon\varepsilon_0}{2}\left(\frac{\partial\Psi}{\partial x}\right)^2 dx$$

$$= 2\int_{-d}^{-d+\delta} \frac{\varepsilon^{\text{I}}\varepsilon_0}{2}\left(\frac{\partial\Psi}{\partial x}\right)^2 dx + \int_{-d+\delta}^{d-\delta} \frac{\varepsilon^{\text{II}}\varepsilon_0}{2}\left(\frac{\partial\Psi}{\partial x}\right)^2 dx. \tag{16}$$

The entropic contribution to the free energy is given by [16]

$$F_{\text{entropy}} = kT\sum_i \int_{-d}^{d} \left(n_i \ln\left(\frac{n_i}{n_i^0}\right) - n_i + n_i^0\right) dx. \tag{17}$$

The chemical contribution $F_{\text{chem}} = -2\Psi_s\Delta\sigma$ at constant surface potential Ψ_s and $= 0$ at constant surface charge. A universal expression free of any constraints has the form [17]

$$F_{\text{chem}} = -2\int_{\sigma(\infty)}^{\sigma(2d)} \Psi_s \, d\sigma. \tag{18}$$

Equation (18) was derived by decomposing the pathway of Ψ_s vs σ in a succession of infinitesimal changes at constant charge followed by changes at constant Ψ_s [17].

The hydration contribution is provided by the difference between the hydration free energies of ions in the region with low dielectric constant and in the bulk,

$$F_{\text{hyd}} = 2\sum_i \int_{-d}^{-d+\delta} n_i A_i\left(\frac{1}{\varepsilon^{\text{I}}} - \frac{1}{\varepsilon^{\text{II}}}\right) dx. \tag{19}$$

For $\varepsilon^{\text{I}} = \varepsilon^{\text{II}}$, the hydration contribution F_{hyd} becomes zero.

Since the dielectric constant and the ion distributions are provided by different expressions in the two regions, the electrostatic contribution and the entropic contribution should be calculated separately for the two regions, the total contributions being their sum.

The repulsive force f between the two parallel plates is given by the derivative of the free energy with respect to the distance $2d$ between them,

$$f = -\frac{\partial F}{\partial(2d)}. \tag{20}$$

In the numerical calculations, the force provided by the classical theory was obtained by assuming $\varepsilon^{\text{I}} = \varepsilon^{\text{II}} = 80$.

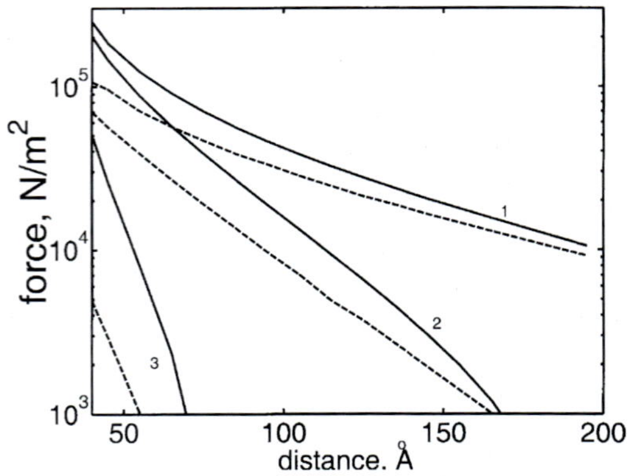

Fig. 1. The repulsive force between two plates at constant surface charge and various bulk NaCl concentrations. The following parameters were employed in the calculations: $\sigma = -0.01$ C/m^2, $\varepsilon^{\text{I}} = 10$, $\varepsilon^{\text{II}} = 80$, $\delta = 10$ Å. (1) 0.001; (2) 0.01; (3) 0.1 M NaCl. The solid lines are for the force predicted by the new model and the dashed lines for the force predicted by the Gouy–Chapman theory.

3. Results and discussion

3.1. Constant surface charge density

3.1.1. The effect of electrolyte concentration on double-layer interaction

The double-layer repulsive force calculated using Eqs. (7), (8), and (10)–(20) is plotted in Fig. 1 for various NaCl concentrations. In the calculations, the dielectric constant for region I was taken to be 10 and for region II 80; the thickness of region I was taken to be $\delta = 10$ Å and the surface charge density $\sigma = -0.01$ C/m^2. As expected, the force decreases with increasing electrolyte concentration. The difference between the forces calculated on the basis of the new model and predicted by the traditional theory increases with increasing electrolyte concentration. At concentration 0.1 M and a distance of 40 Å, the force is 10 times larger than predicted by the Gouy–Chapman theory. However, at 0.001 M and a distance of 40 Å, it is only about twice as large.

In Fig. 2, the ion distributions are plotted for various electrolyte concentrations. Compared with those predicted by the Gouy–Chapman theory, the ion concentrations in region I are extremely low. This occurs because the hydration free energies of the ions are inversely proportional to the dielectric constant (see Eq. (3)). As a result, the ions have higher affinity for region II than region I. Even though the surface is negatively charged, the concentration of cations is still lower near the surface than in the bulk. Fig. 2 also shows that at high salt concentrations the anion concentration is higher than that of the cation in a fraction of region I, even though the electrical potential is negative. This behavior is a result of the larger hydration constant of the cations than of the anions. For a dielectric constant of 10 in region I, the difference between the hydration free energies of Na^+ and Cl^- is $4.8kT$ (Table 1). Only for electrical potentials smaller than -0.062 V ($= (A_{\text{Cl}}/\varepsilon^{\text{I}} - A_{\text{Na}}/\varepsilon^{\text{I}})/2e$) can the difference between the electrostatic energies of the anions and cations compensate for the difference in hydration free energies, and the cation concentration becomes larger than that of the anion. However, for electrical potentials larger than −0.062 V, the total potential energy of the cation becomes larger than that of the anion. Consequently, in spite of the negative electrical potential, the cation concentration becomes lower than that of the anion for electrical potentials larger than −0.062 V. Since the electrical potential decreases with decreasing electrolyte concentration, the thickness of the fraction of region I with negative charge density decreases and becomes zero at a concentration of 0.001 M.

The electrical potential profile for various electrolyte concentrations is plotted in Fig. 3, which shows that its magnitude decreases with increasing electrolyte concentration. The surface potential is more negative than for a uniform dielectric constant between plates. The electrical field in region I is almost constant, and its magnitude is much larger than in region II, due to the lower charge density in the former region caused by the lower dielectric constant. The electric field in region I did not change as the electrolyte concentration increased because the surface charge density was considered constant and the electrical field depended little on position in region I.

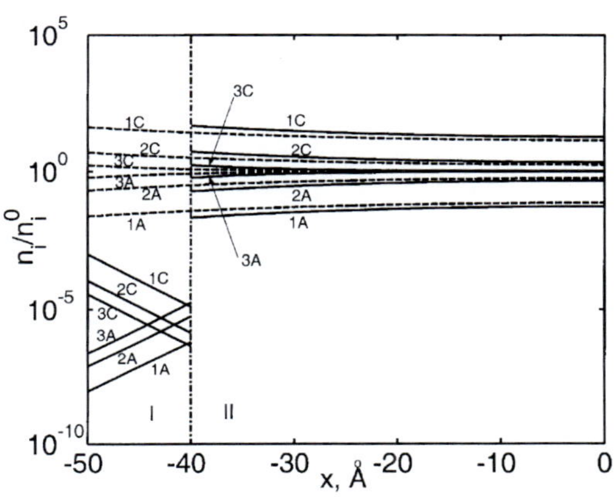

Fig. 2. The ion distributions between two plates for various bulk NaCl concentrations. The distance between the two plates is 100 Å. The other parameters employed are: $\sigma = -0.01$ C/m^2, $\varepsilon^{I} = 10$, $\varepsilon^{II} = 80$, $\delta = 10$ Å. (1) 0.001; (2) 0.01; (3) 0.1 M NaCl. In the figure, A and C indicate Cl^- and Na^+, respectively. The solid lines are for the ion distributions predicted by the new model and the dashed lines for those predicted by the Gouy–Chapman theory.

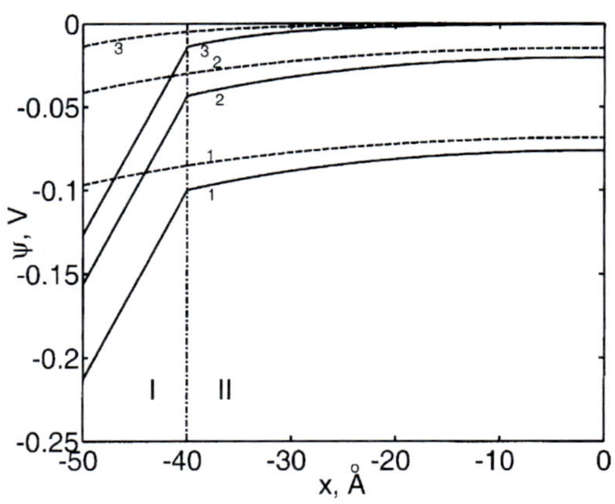

Fig. 3. The electrical potential distributions between two plates for various bulk NaCl concentrations. The distance between the two plates is 100 Å. The other parameters employed are $\sigma = -0.01$ C/m^2, $\varepsilon^{I} = 10$, $\varepsilon^{II} = 80$, $\delta = 10$ Å. (1) 0.001; (2) 0.01; (3) 0.1 M NaCl. The solid lines are for the electrical potentials predicted by the new model and the dashed lines for the electrical potentials predicted by the Gouy–Chapman theory.

3.1.2. *The effect of the dielectric constant on double-layer interaction*

Fig. 4 presents the effect of the dielectric constant of region I on the double-layer force. The force increases with decreasing dielectric constant. As the dielectric constant increases, the hydration free energies of the ions decrease, the ion concentrations in region I increase (Fig. 5), and the magnitudes of the surface potential and of the electric field in

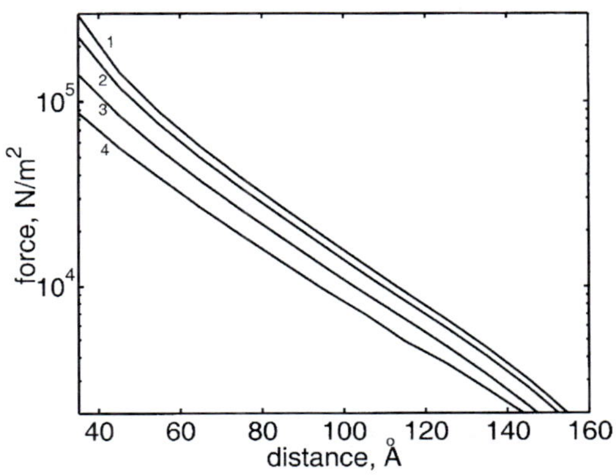

Fig. 4. The repulsive force between two plates for various values of the dielectric constant ε^{I}. The parameters employed in the calculations are: $\sigma = -0.01$ C/m^2, $\varepsilon^{II} = 80$, $\delta = 10$ Å, NaCl concentration = 0.01 M. (1) $\varepsilon^{I} = 10$; (2) $\varepsilon^{I} = 40$; (3) $\varepsilon^{I} = 60$; (4) $\varepsilon^{I} = 80$.

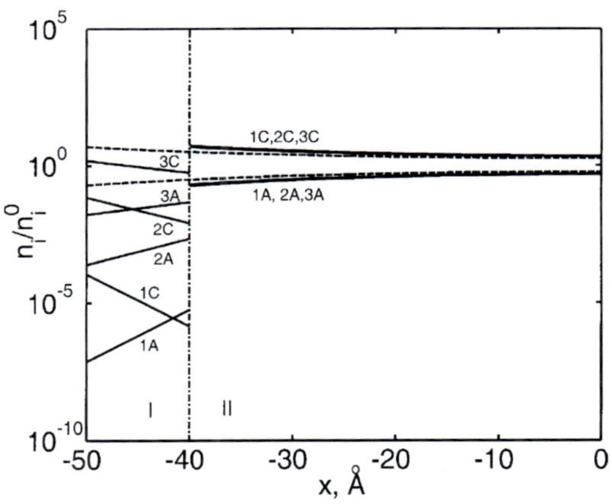

Fig. 5. The ion distributions between two plates for various values of the dielectric constant ε^{I}. The distance between the two plates is 100 Å. The parameters employed in the calculations are: $\sigma = -0.01$ C/m^2, $\varepsilon^{II} = 80$, $\delta = 10$ Å, NaCl concentration = 0.01 M. (1) $\varepsilon^{I} = 10$; (2) $\varepsilon^{I} = 20$; (3) $\varepsilon^{I} = 40$. In the figure, A and C indicate Cl^- and Na^+, respectively. The solid lines are for the ion distributions predicted by the new model and the dashed lines for those predicted by the Gouy–Chapman theory.

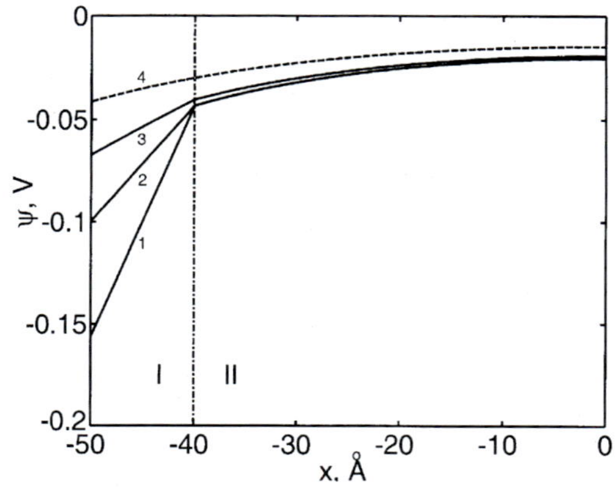

Fig. 6. The electrical potential distributions between two plates for various dielectric constants. The distance between the two plates is 100 Å. The other parameters employed are: $\sigma = -0.01\ C/m^2$, $\varepsilon^{II} = 80$, $\delta = 10$ Å, NaCl concentration = 0.01 M. (1) $\varepsilon^I = 10$; (2) $\varepsilon^I = 20$; (3) $\varepsilon^I = 40$; (4) $\varepsilon^I = 80$ (the Gouy–Chapman theory).

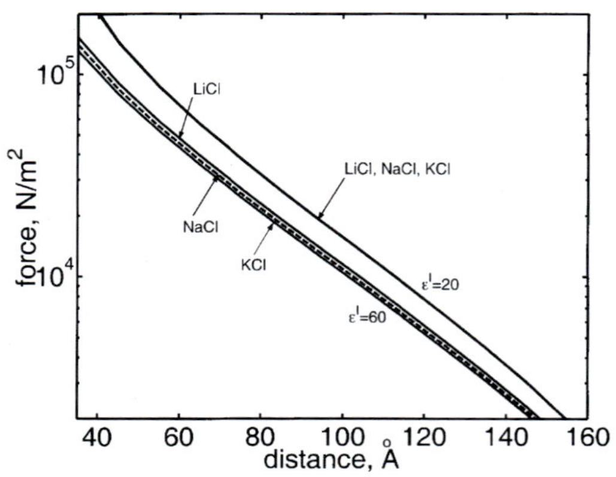

Fig. 7. The effect of the nature of cations on the force for low and high dielectric constants. The parameters employed in the calculations are: $\sigma = -0.01\ C/m^2$, $\varepsilon^{II} = 80$, $\delta = 10$ Å, electrolyte concentration = 0.01 M. For $\varepsilon^I = 20$, the three curves for LiCl, KCl, and NaCl are almost the same.

region I decrease (Fig. 6). The calculations have shown that the electrostatic and the entropic contributions to the free energy decrease with increasing dielectric constant. However, the hydration contribution first increases, because of the drastic increase of the number of ions in region I, and then decreases, because of the decrease of the hydration free energies with increasing dielectric constant. Because the magnitude of the surface potential decreases, the total free energy and the force decrease with increasing dielectric constant in region I.

3.1.3. The effect of the natures of ions on the force

Table 1 shows that different ions have different hydration free energies in media with the same dielectric constant.

Fig. 7 presents the dependence of the force on the distance between plates for various cations (LiCl, NaCl, and KCl) and for high and low dielectric constants in region I. When the dielectric constant is low (10 or 20), the differences between the forces for various cations are small, even

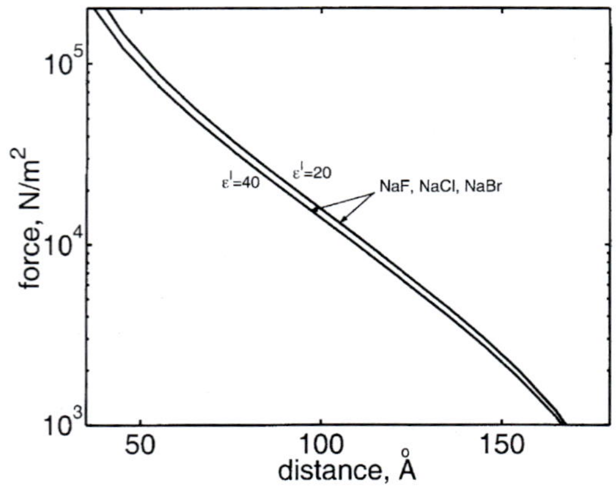

Fig. 8. The effect of the nature of the anion on the force for low and high dielectric constants. The parameters employed in the calculations are: $\sigma = -0.01\ C/m^2$, $\varepsilon^{II} = 80$, $\delta = 10$ Å, electrolyte concentration = 0.01 M. The curves for NaF, NaBr, and NaCl are almost the same for both $\varepsilon^I = 20$ and 40.

undetectable. However, for a dielectric constant of 60, the nature of the cation provides a specific effect on the force. Although the hydration free energy of each ion is larger at low dielectric constants, the total hydration free energy of the colloidal system is in such cases too small to affect the force between plates, because of the low ion concentrations in region I. This low total hydration free energy makes the system insensitive to specific ion effects. As the dielectric constant increases, the hydration free energy of each ion moderately decreases, whereas the ion concentrations in region I drastically increase. For this reason, the total hydration free energy of the system is not negligible and differences in the forces for various cations become noticeable, but small. Li^+ has a higher hydration free energy than Na^+ or K^+; furthermore, calculations have shown that the dielectric constant affected much more than the hydration constant the concentrations of ions in region I. For these reasons, the total hydration free energy is larger for Li^+ than for Na^+ or K^+ cations and the force for LiCl is larger than for NaCl or KCl.

Fig. 8 presents the effect of hydration free energy of the anion (NaF, NaCl, and NaBr) on the force between the plates. For both high and low dielectric constants, the differences between forces for various anions are negligible. This happens because the concentration of anions in region I is very small and the differences in the hydration constants of the anions are also small. Therefore, the free energy and the force are almost the same for these anions.

3.1.4. The effect of the thickness of region I on the force

Since region I has a different dielectric constant, its thickness will significantly affect the repulsive force between the two plates. Fig. 9 shows that the force increases as the thickness of region I increases. This occurs because the electrical potential becomes more negative with increasing δ and because of the larger contribution of region I to the total free energy.

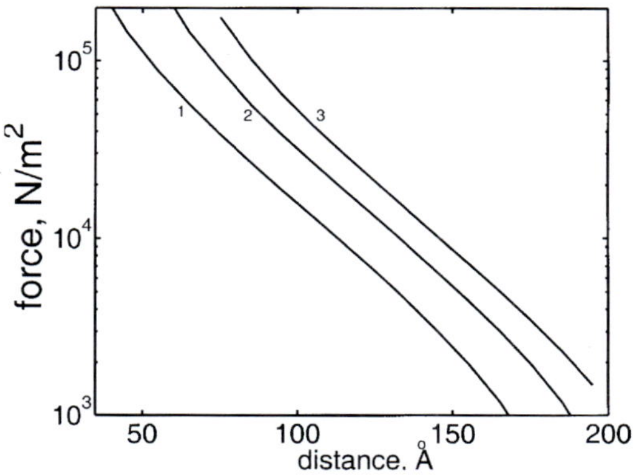

Fig. 9. The effect of the thickness of region I on the force. The parameters employed in the calculations are: $\sigma = -0.01$ C/m^2, $\varepsilon^{I} = 10$, $\varepsilon^{II} = 80$, NaCl concentration $= 0.01$ M. (1) $\delta = 10$ Å; (2) $\delta = 20$ Å; (3) $\delta = 30$ Å.

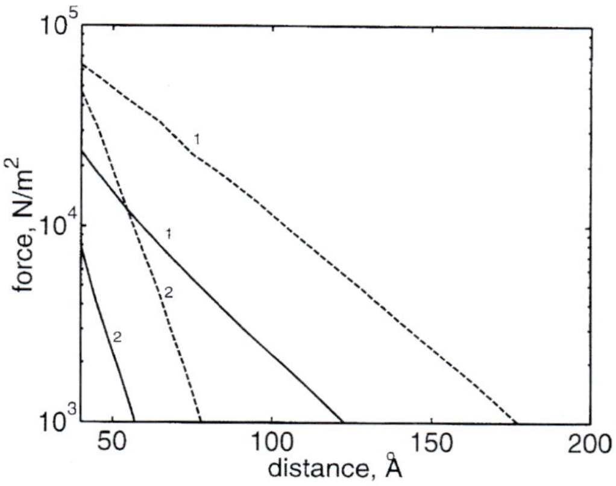

Fig. 10. The repulsive force between two plates at constant surface potential and various NaCl concentrations. The following parameters are employed in the calculations: $\Psi_0 = -0.05$ V, $\varepsilon^{I} = 10$, $\varepsilon^{II} = 80$, $\delta = 10$ Å. NaCl concentration: (1) 0.001 M; (2) 0.01 M. The solid lines are for the force predicted by the new model and the dashed lines are for the force predicted by the Gouy–Chapman theory.

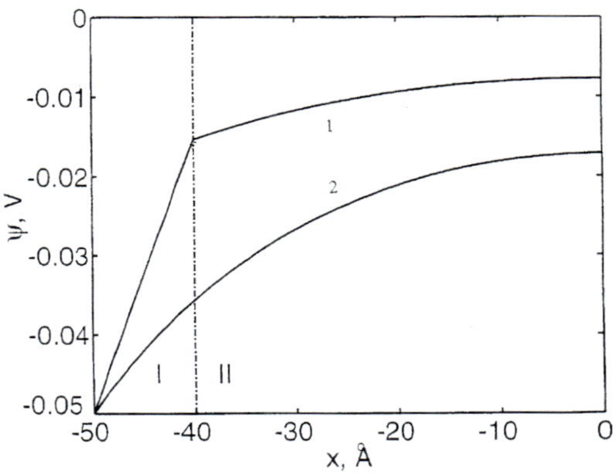

Fig. 11. The electrical potential distributions between two plates. The distance between the two plates is 100 Å. The other parameters employed are: $\Psi_0 = -0.05$ V, $\varepsilon^{I} = 10$, $\varepsilon^{II} = 80$, $\delta = 10$ Å, NaCl concentration $= 0.01$ M. (1) The new model; (2) the Gouy–Chapman theory.

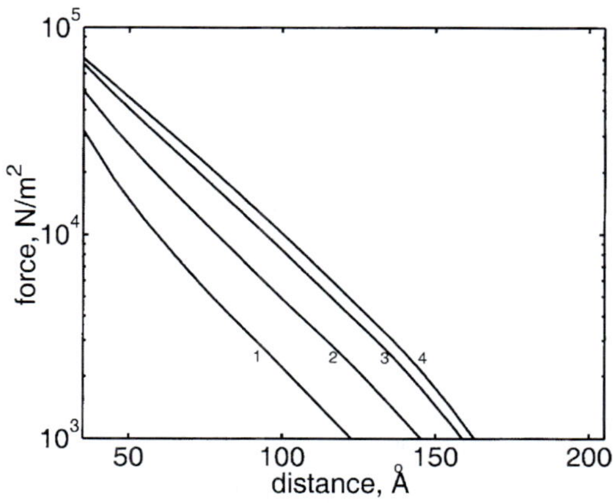

Fig. 12. The repulsive force between two plates for various dielectric constants ε^{I}. The parameters employed in the calculations are: $\Psi_0 = -0.05$ V, $\varepsilon^{II} = 80$, $\delta = 10$ Å, NaCl concentration $= 0.01$ M. (1) $\varepsilon^{I} = 10$; (2) $\varepsilon^{I} = 20$; (3) $\varepsilon^{I} = 40$; (4) $\varepsilon^{I} = 60$.

3.2. Constant surface potential

3.2.1. The effect of the dielectric constant on the force

In the case of constant surface charge density, the force was larger than predicted by the classical Gouy–Chapman theory. In contrast, for constant surface potential, as shown by Fig. 10, the force predicted by the new model is smaller than that provided by the classical theory. Because of the low dielectric constant in region I, the charge density near the surface is much smaller in the new model than in the classical one.[1] As a result, the electrical potential varies more rapidly with the distance to the surface (and its slope is larger near the surface) for the former model. However, because the ratio between the dielectric constant in region I and in the classical model is small, the surface charge is smaller in the new model, and the repulsion force lower. Indeed, Fig. 11 shows that the ratio of slopes in the new and classical models is 2.5, whereas the ratio of the dielectric constants is 1/8. Fig. 12 provides the dependence of the force on the dielectric constant in region I. In contrast to the case of constant surface charge density, at constant surface potential, the higher the dielectric constant of region I, the larger is the repulsion between plates.

3.2.2. The effect of ion nature on the force

At high or low dielectric constants, there are small differences among the forces for various cations or anions (Figs. 13 and 14). For a dielectric constant in region I of 40, the force between the plates at a distance of 45 Å is 4.83×10^4 for LiCl, 4.78×10^4 for NaCl, and 4.77×10^4 N/m^2 for NaF, hence almost the same.

3.2.3. The effect of region I thickness on the force

Fig. 15 shows that the force between the two plates decreases as the thickness of region I increases. This behavior occurs because the electrical potential between the plates,

[1] This can be demonstrated starting from the Poisson equation and taking into account that $\rho_{\text{new model}} \ll \rho_{\text{classical model}}$.

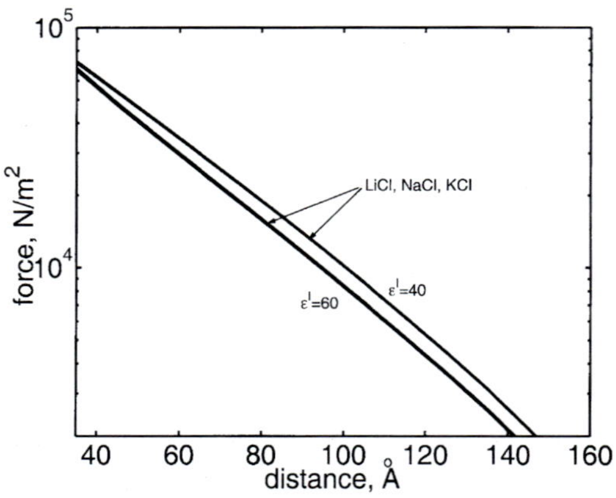

Fig. 13. The effect of the nature of cation on the force at low and high dielectric constants. The parameters employed in the calculations are: $\Psi_0 = -0.05$ V, $\varepsilon^{\mathrm{II}} = 80$, $\delta = 10$ Å, electrolyte concentration = 0.01 M. The curves for LiCl, KCl, and NaCl are almost the same.

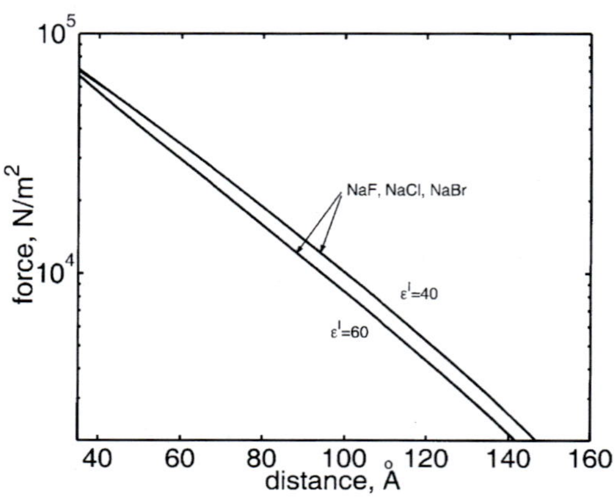

Fig. 14. The effect of the nature of the anion on the force at low and high dielectric constants. The parameters employed in the calculations are: $\Psi_0 = -0.05$ V, $\varepsilon^{\mathrm{II}} = 80$, $\delta = 10$ Å, electrolyte concentration = 0.01 M. The curves for NaF, NaBr, and NaCl are almost the same.

and hence the surface charge density, increases as δ increases. Therefore, the thickness of region I has a different effect on the force at constant surface charge density and constant surface potential.

3.3. The surface charge generated through the dissociation equilibrium

Let us assume that there is a dissociation equilibrium at the surface of the plate,

$$S^- + M^+ \leftrightarrow S\text{–}M, \tag{21}$$

where S^- is a negatively charged site on the surface, M^+ is a cation which can be adsorbed on a negatively charged site, and S–M is the surface site occupied by cation M^+.

At equilibrium,

$$K_{\mathrm{eq}} = \frac{1-\theta}{[M^+]_s\theta}, \tag{22}$$

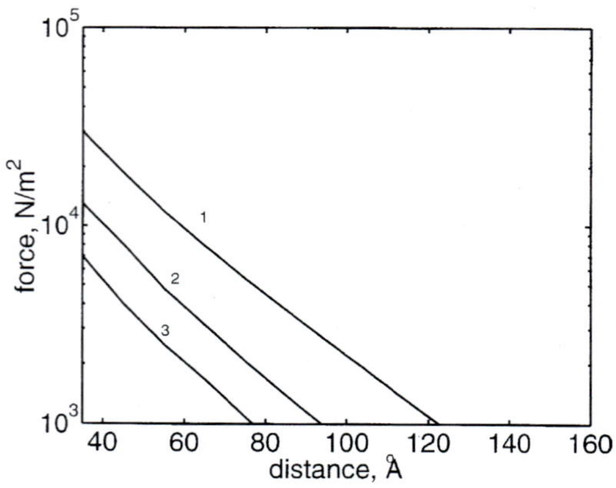

Fig. 15. The effect of the thickness of region I on the force. The parameters employed in the calculations are: $\Psi_0 = -0.05$ V, $\varepsilon^{\mathrm{I}} = 10$, $\varepsilon^{\mathrm{II}} = 80$, NaCl concentration = 0.01 M. (1) $\delta = 10$ Å; (2) $\delta = 20$ Å; (3) $\delta = 30$ Å.

where K_{eq} is the adsorption equilibrium constant, θ is the fraction of surface sites unoccupied by cation M^+, and $[M^+]_s$ is the concentration of M^+ near the surface, which can be related to the surface potential and the bulk concentration via the equality of the chemical potentials.

The surface charge density is given by

$$\sigma = \theta\sigma_0, \tag{23}$$

where σ_0 is the surface charge density when there is no adsorption of M^+ onto the (negative) surface sites.

Substituting Eq. (23) into Eq. (22), a boundary condition for the surface charge density generated through the dissociation equilibrium is obtained. Equations (8) and (9) can be solved numerically for the boundary conditions (11)–(14), (23), to calculate the force between plates.

3.3.1. The effect of electrolyte concentration on the force

As the electrolyte concentration increases, the magnitude of the surface charge density decreases, whereas the screening effect increases. However, because of the lower dielectric constant in region I, the magnitude of the surface charge density does not decrease as much as in the Gouy–Chapman theory and the screening effect is not so strong. Therefore, at high electrolyte concentrations, the force predicted by the new model is much larger than that provided by the Gouy–Chapman theory. In Fig. 16, which is for a concentration of 0.1 M NaCl, the force obtained is two orders of magnitude larger than that predicted by the Gouy–Chapman theory. It should be emphasized that the comparison was made for the same adsorption equilibrium constants, even though, in reality, they depend on the dielectric constant.

3.3.2. The effect of the dielectric constant on the force

As the dielectric constant of region I increases, the concentrations of ions near the surface increase, the magnitude of the surface charge density decreases, and the force between the two plates decreases (Fig. 17). Compared to the constant surface charge case, for which only the ion concentrations in the double layer increased with increasing dielec-

188 H. Huang, E. Ruckenstein / Journal of Colloid and Interface Science 273 (2004) 181–190

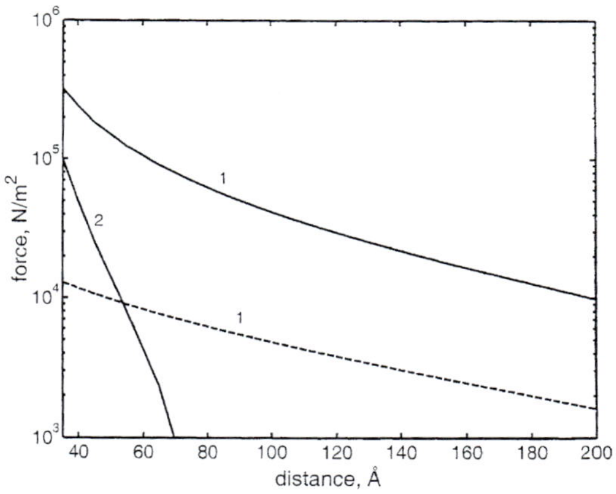

Fig. 16. The repulsive force between two plates with adsorption equilibrium at various NaCl concentrations. The following parameters are employed in the calculations: $\sigma_0 = -0.01\ \mathrm{C/m^2}$, $\varepsilon^{\mathrm{I}} = 10$, $\varepsilon^{\mathrm{II}} = 80$, $\delta = 10$ Å, $K_{\mathrm{eq}} = 1000$. NaCl concentration: (1) 0.001; (2) 0.01 M. The solid lines are for the force predicted by the new model and the dashed lines are for the force predicted by the Gouy–Chapman theory.

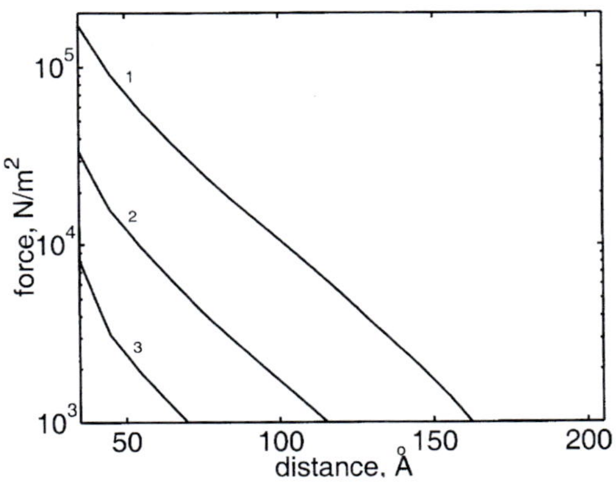

Fig. 17. The repulsive force between two plates for various dielectric constants ε^{I}. The parameters employed in the calculations are: $\sigma_0 = -0.01\ \mathrm{C/m^2}$, $\varepsilon^{\mathrm{II}} = 80$, $\delta = 10$ Å, NaCl concentration = 0.01 M. (1) $\varepsilon^{\mathrm{I}} = 20$; (2) $\varepsilon^{\mathrm{I}} = 40$; (3) $\varepsilon^{\mathrm{I}} = 60$.

tric constant, in the present case the magnitude of the surface charge density has also increased.

3.3.3. The effect of adsorption constant on the force

It is obvious that a larger adsorption constant will lead to a lower force. However, due to the low ion concentrations in region I, the adsorption constant has different effects on the equilibrium in the new model and in the Gouy–Chapman theory.

In the new model, the ion concentrations near the surface are very low. Therefore, only a large value of the adsorption constant can affect the surface charge density. For sufficiently small adsorption constants, the magnitude of the surface charge density acquires its maximum value σ_0. In the Gouy–Chapman theory, the cation concentration near the surface is comparatively large. If the adsorption constant is also large, the magnitude of the surface charge density will be very small (Fig. 18).

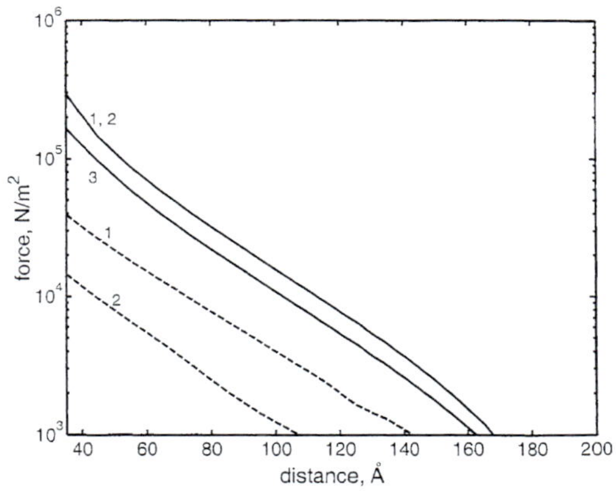

Fig. 18. The repulsive force between two plates for various equilibrium constants. The parameters employed in the calculations are: $\sigma_0 = -0.01\ \mathrm{C/m^2}$, $\varepsilon^{\mathrm{I}} = 10$, $\varepsilon^{\mathrm{II}} = 80$, $\delta = 10$ Å, NaCl concentration = 0.01 M. (1) $K_{\mathrm{eq}} = 20$; (2) $K_{\mathrm{eq}} = 100$; (3) $K_{\mathrm{eq}} = 1 \times 10^5$. The solid lines are for the forces predicted by the new model and the dashed ones for those predicted by the Gouy–Chapman theory.

4. Comparison with the Langmuir expression

As already noted, the Langmuir equation is not rigorously valid in the present case because of the nonideal term $(A_i/\varepsilon^{\mathrm{I}})$, which appears in the chemical potential. Nevertheless, we carried out calculations using that equation and obtained in all cases differences with respect to the new model of at most 5%. This can be explained by observing that the mentioned term appears only in region I and not in region II, where the ideal approximation is considered valid.

In what follows, a detailed comparison will be carried out in the Debye–Hückel approximation, for which simple analytical expressions can be derived.

Expanding the right-hand side terms of Eqs. (8) and (9) in series and neglecting the higher order terms, one obtains

$$\frac{\partial^2 y}{\partial x^2} = \frac{\kappa_{\mathrm{I}}^2}{2}\left(\left(\frac{1}{w_-} - \frac{1}{w_+}\right) + y\left(\frac{1}{w_-} + \frac{1}{w_+}\right)\right) \quad \text{(for region I)} \tag{24}$$

and

$$\frac{\partial^2 y}{\partial x^2} = \kappa_{\mathrm{II}}^2 y \quad \text{(for region II)}, \tag{25}$$

where $y = e\Psi/(kT)$, $1/\kappa_{\mathrm{I}}$ and $1/\kappa_{\mathrm{II}}$ are the Debye lengths in region I and II, respectively, and

$$w_- = \exp\left(\frac{A_-}{kT}\left(\frac{1}{\varepsilon^{\mathrm{I}}} - \frac{1}{\varepsilon^{\mathrm{II}}}\right)\right), \tag{26}$$

$$w_+ = \exp\left(\frac{A_+}{kT}\left(\frac{1}{\varepsilon^{\mathrm{I}}} - \frac{1}{\varepsilon^{\mathrm{II}}}\right)\right). \tag{27}$$

Equations (24) and (25) can be solved analytically for the boundary condition (10) or (11) and (12)–(14). The electrical potential distribution in region I is given by

$$y = \frac{w_- - w_+}{w_- + w_+} + C_1 \exp\left(\frac{\sqrt{w_- + w_+}}{\sqrt{2w_- w_+}} \kappa_I x\right) + C_2 \exp\left(-\frac{\sqrt{w_- + w_+}}{\sqrt{2w_- w_+}} \kappa_I x\right) \quad (28)$$

and in region II by

$$y = 2g \cosh(\kappa_{II} x), \quad (29)$$

where $2g$ is the electrical potential at the middle distance divided by kT/e. Expressions for the constants g, C_1, and C_2 are given in Appendix A.

An equation for the force is obtained by substituting Eqs. (A.1) or (A.4) into the Langmuir equation [18]:

$$f = 2n_0 kT\big(\cosh(2g) - 1\big). \quad (30)$$

For the case in which the surface charge is a result of dissociation equilibrium, one can use the same equation as for constant surface charge combined with the equilibrium equations (22) and (23) and the equality of the chemical potential at the surface and in bulk ($kT \ln[M^+]_s + ez_i\Psi_0 + A_i/\varepsilon^I = kT \ln[M^+] + A_i/\varepsilon^{II}$).

These calculations have shown that the differences between the exact numerical results and Eq. (30) are less than 5%.

5. Conclusions

In this paper, a model which takes into account the effect of a lower dielectric constant near a surface due to the presence of hairs on the surface is developed. At constant surface charge, the force between the two plates is larger than predicted by the Gouy–Chapman theory. As the dielectric constant increases, the force decreases. The nature of cations has a specific effect on the force between plates.

At constant surface potential, the force provided by the new model is smaller than predicted by the Gouy–Chapman theory. In contrast, at constant surface charge density, the dielectric constant and the thickness of region I have effects opposite to those at constant surface potential.

When the surface charge is generated through dissociation equilibrium, the difference between the forces calculated using the new model and the Gouy–Chapman theory is much larger than for constant charge density or constant surface potential.

Appendix A

For constant surface charge density

$$g = e^{\kappa_{II}(d-\delta)}\left[\lambda_2(e^{\delta\lambda_1} - 1)(w_+ - w_-) + 4\varepsilon_I(w_+ + w_-)\mu\sqrt{w_- w_+}e^{\delta\lambda_1/2}\right] \Big/ \Big\{(w_+ + w_-)\big[(\lambda_3 - \lambda_2)(e^{2\kappa_{II}(d-\delta)+\lambda_1\delta} - 1) + (\lambda_2 + \lambda_3)(e^{2\kappa_{II}(d-\delta)} - e^{\lambda_1\delta})\big]\Big\}, \quad (A.1)$$

$$C_1 = \Bigg\{\sqrt{2}\sqrt{w_- w_+}\exp\left(-\frac{\lambda_1(d-\delta)}{2}\right) \times \left[(\lambda_2 + \lambda_3)(w_- + w_+)\mu \exp\left(\frac{\lambda_1\delta}{2}\right) + (w_+ - w_-)\lambda_2\kappa_{II}\frac{\varepsilon_{II}}{\varepsilon_I}\big(\exp(2\kappa_{II}(d-\delta)) - 1\big)\right] + \sqrt{2}\sqrt{w_- w_+}\exp\left(-\frac{\lambda_1(d-\delta)}{2}\right)(\lambda_2 - \lambda_3) \times (w_- + w_+)\mu \exp\left(2\kappa_{II}(d-\delta) + \frac{\lambda_1\delta}{2}\right)\Bigg\} \Bigg/ \Bigg\{\kappa_I(w_- + w_+)^{3/2} \times \Big[(\lambda_2 - \lambda_3)\big(1 - \exp(2\kappa_{II}(d-\delta) + \lambda_1\delta)\big) + (\lambda_2 + \lambda_3)\big(\exp(2\kappa_{II}(d-\delta)) + \exp(\lambda_1\delta)\big)\Big]\Bigg\}, \quad (A.2)$$

$$C_2 = \Bigg\{\sqrt{2}\sqrt{w_- w_+}\exp\left(\frac{\lambda_1 d}{2}\right)(w_- - w_+)\frac{\lambda_2\kappa_{II}\varepsilon_{II}}{\varepsilon_I} \times \left[\exp\left(\frac{\lambda_1\delta}{2}\right) - \exp\left(2\kappa_{II}(d-\delta) + \frac{\lambda_1\delta}{2}\right)\right] + \sqrt{2}\sqrt{w_- w_+}\exp\left(\frac{\lambda_1 d}{2}\right)(w_- + w_+)\mu \times \Big[(\lambda_2 + \lambda_3)\exp\big(2\kappa_{II}(d-\delta)\big) + (\lambda_2 - \lambda_3)\Big]\Bigg\} \Bigg/ \Bigg\{\kappa_I(w_- + w_+)^{3/2} \times \Big[(\lambda_2 - \lambda_3)\big(1 - \exp(2\kappa_{II}(d-\delta) + \lambda_1\delta)\big) + (\lambda_2 + \lambda_3)\big(\exp(2\kappa_{II}(d-\delta)) + \exp(\lambda_1\delta)\big)\Big]\Bigg\}, \quad (A.3)$$

and for constant surface potential

$$g = \lambda_2\kappa_{II}(d-\delta)\Big[(e^{\delta\lambda_1} + 1)(w_- - w_+) + 2e^{\delta\lambda_1/2}\big((y_0 - 1)w_- + (y_0 + 1)w_+\big)\Big] \Big/ \Big\{(w_+ + w_-)\big[(\lambda_2 - \lambda_3)(e^{2\kappa_{II}(d-\delta)+\lambda_1\delta} + 1) + (\lambda_2 + \lambda_3)(e^{2\kappa_{II}(d-\delta)} + e^{\lambda_1\delta})\big]\Big\}, \quad (A.4)$$

$$C_1 = \Biggl\{ \exp\left(\frac{\lambda_1(-d+3\delta)}{2}\right)\bigl[(\lambda_2-\lambda_3)y_0(w_-+w_+) + (\lambda_3-\lambda_2)(w_--w_+)\bigr] + \exp\left[\frac{\lambda_1(-d+2\delta)}{2} - 2\kappa_{II}(d-\delta)\right] \times (\lambda_2+\lambda_3)\bigl[y_0(w_-+w_+) - w_- + w_+\bigr] + \Bigl\{ \exp\bigl[-2\kappa_{II}(d-\delta)\bigr]\lambda_3(w_--w_+) + \lambda_3(w_+-w_-)\Bigr\}\Biggr\} \Bigg/ \Biggl\{ (w_-+w_+) \times \Bigl\{(\lambda_2+\lambda_3)\bigl[\exp\bigl(-2\kappa_{II}(d-\delta)+\delta\lambda_1\bigr)+1\bigr] + (\lambda_2-\lambda_3)\bigl[\exp\bigl(2\kappa_{II}(-d+\delta)\bigr)+\exp(\lambda_1\delta)\bigr]\Bigr\}\Biggr\}, \tag{A.5}$$

$$C_2 = \Biggl[\Biggl\{ \exp\left[\frac{\lambda_1 d}{2} - 2\kappa_{II}(d-\delta)\right](\lambda_2-\lambda_3) + \exp\left(\frac{\lambda_1 d}{2}\right)(\lambda_2+\lambda_3)\Biggr\} \times \bigl[y_0(w_-+w_+) - w_- + w_+\bigr] + \exp\bigl[-2\kappa_{II}(d-\delta)+\lambda_1\delta\bigr]\lambda_3(w_+-w_-) + \exp(\lambda_1\delta)\lambda_3(w_--w_+)\Biggr] \Bigg/ \Biggl\{ (w_-+w_+) \times \Bigl\{(\lambda_2+\lambda_3)\bigl[\exp\bigl(-2\kappa_{II}(d-\delta)+\delta\lambda_1\bigr)+1\bigr] + (\lambda_2-\lambda_3)\bigl[\exp\bigl(2\kappa_{II}(-d+\delta)\bigr)+\exp(\lambda_1\delta)\bigr]\Bigr\}\Biggr\}, \tag{A.6}$$

where

$$\lambda_1 = -\sqrt{2(w_-+w_+)}\kappa_I/\sqrt{w_-w_+}, \tag{A.7}$$

$$\lambda_2 = \sqrt{2(w_-+w_+)}\kappa_I\varepsilon^{I}, \tag{A.8}$$

$$\lambda_3 = 2\kappa_{II}\varepsilon^{II}\sqrt{w_-w_+}, \tag{A.9}$$

$$\mu = \frac{e\sigma}{kT\varepsilon^{I}\varepsilon_0}, \tag{A.10}$$

$$y_0 = \frac{e\Psi_0}{kT}. \tag{A.11}$$

References

[1] D.H. Napper, Polymeric Stabilization of Colloidal Dispersions, Academic Press, New York, 1983.
[2] R.J. Hunter, Foundations of Colloid Science, Vol. 1, Oxford Science Publications, Oxford, 1987.
[3] H. Casanova, J. Chen, E. Dickinson, B.S. Murray, P.V. Nelson, M. Whittle, Phys. Chem. Chem. Phys. 2 (2000) 3861.
[4] N.S. Bell, J. Sindel, F. Aldinger, W.M. Sigmund, J. Colloid Interface Sci. 254 (2002) 296.
[5] Y.L. Ong, A. Razatos, G. Georgiou, M.M. Sharma, Langmuir 15 (1999) 2719.
[6] S. Alexander, J. Phys. 38 (1977) 983.
[7] P.G. de Gennes, Adv. Colloid Interface Sci. 27 (1987) 189.
[8] E. Ruckenstein, B. Li, J. Chem. Phys. 107 (1997) 932.
[9] B.E. Conway, J.O'M. Bockris, I.A. Ammar, Trans. Faraday Soc. 47 (1951) 756.
[10] M.J. Spaarnay, Rec. Trav. Chim. 77 (1958) 382.
[11] Y. Gur, I. Ravina, A.J. Babchin, J. Colloid Interface Sci. 64 (1978) 333.
[12] S. Basu, M. Sharma, J. Colloid Interface Sci. 165 (1994) 355.
[13] G. Nemethy, H.A. Scheraga, J. Chem. Phys. 36 (1962) 3382.
[14] R.M. Noyes, J. Am. Chem. Soc. 84 (1962) 513.
[15] R.H. Stokes, J. Am. Chem. Soc. 86 (1964) 979.
[16] J.Th.G. Overbeek, Colloids Surf. 51 (1990) 61.
[17] M. Manciu, E. Ruckenstein, Langmuir 19 (2003) 1114.
[18] I. Langmuir, J. Chem. Phys. 6 (1938) 893.

Available online at www.sciencedirect.com
SCIENCE DIRECT
Journal of Colloid and Interface Science 275 (2004) 548–554
JOURNAL OF Colloid and Interface Science
www.elsevier.com/locate/jcis

Double layer interaction between two plates with polyelectrolyte brushes

Haohao Huang and Eli Ruckenstein *

Department of Chemical and Biological Engineering, State University of New York at Buffalo, Buffalo, NY 14260, USA

Received 19 December 2003; accepted 24 February 2004
Available online 1 April 2004

Abstract

When polyelectrolyte chains are grafted to colloidal particles, the electric field between particles is affected by the charges of the chains. In some previous theoretical attempts, the charge density of the polyelectrolyte chains per unit length was considered constant, and its effect was accounted for by introducing an additional constant charge density into the unidimensional Poisson–Boltzmann equation, which was evaluated assuming that it is uniformly distributed in the polyelectrolyte volume of the brush. In this paper, a more detailed model is employed for the calculation of the electrical potential between two plates on which polyelectrolyte brushes are present. In this model, the polyelectrolyte chain is viewed as a rigid cylinder, on the surface of which charges are generated through the dissociation of ionizable sites and adsorption of the cations of the electrolyte. To each of the chains an atmosphere is attached which for simplicity is assumed cylindrical. In the brush region, the electrical potential is described by a two-dimensional Poisson–Boltzmann equation, while in the region free of polyelectrolyte chains by a unidimensional Poisson–Boltzmann equation. Such a model is physically suitable when the charges of the chains are sufficiently large for the repulsion they generate to ensure that the chains are fully extended. Such cases are quite frequent, because relatively low charges lead to an almost complete extension of the chains. In this paper, both the plate surface and the surface of the cylinders are considered charged. The effects of electrolyte concentration, pH, brush thickness and chain coverage density on the repulsion between plates are examined.

Keywords: Polyelectrolyte; Repulsion force; Cylindrical model

1. Introduction

Colloidal dispersions can be stabilized by attaching polymer chains to their surface [1–9]. When neutral polymer chains grafted on two parallel plates interpenetrate, a steric repulsion is generated. If the polymer chains grafted to the plates are charged, the double layer interaction between the two plates is also affected by the presence of the chains.

Using a self-consistent field theory, Misra et al. [10] examined the effect of the brush charge, electrolyte concentration, and surface charge density on the brush thickness. They extended the self-consistent field polymer brush theory suggested by Milner et al. [11] to the case of a polyelectrolyte brush. The theory involves a parabolic monomer concentration profile rather than the step-function suggested by Alexander [12] and de Gennes [13,14]. The repulsion force between two brushes in poor solvents was also calculated using a similar approach [15].

In some previous calculations [16–18] of the interactions involving polyelectrolyte chains grafted to two surfaces, the charge of the polyelectrolyte chains was assumed to be constant and this fixed charge density was introduced into a one-dimensional Poisson–Boltzmann equation to calculate the electrical potential profile. All the above treatments involved a unidimensional model.

In this paper, a two-dimensional model is proposed, in which the polyelectrolyte chains are treated as rigid cylinders normally grafted to each of the plates. On the surface of the cylinders, ionizable sites are uniformly distributed and the surface charge of the cylinder is generated through their dissociation. Using this two-dimensional model, the repulsive force between two plates is calculated as a function of electrolyte concentration, surface density of polyelectrolyte chains on the plate, thickness of the polyelectrolyte brushes, and bulk pH (denoted in what follows as pH_0).

* Corresponding author. Fax: +1-716-645-3822.
E-mail address: feaeliru@acsu.buffalo.edu (E. Ruckenstein).

0021-9797/$ – see front matter
doi:10.1016/j.jcis.2004.02.058

2. The model

Polyelectrolyte chains possessing two kinds of ionizable sites, such as the protein molecules, are considered to be grafted to a charged plate surface. The positive and negative charges on the chain surface arise from the dissociation of two kinds of ionizable sites. For sufficiently high charge densities, the polymer chains become fully extended and hence the chains can be viewed as cylinders normal to the plate surface. One end of the cylinder is located on the plate and its height is equal to the length of the polyelectrolyte chain. To each of the cylinders an atmosphere is associated by dividing the space between cylinders in equal parallelepipeds. To simplify the treatment, the external boundary of the atmosphere will be replaced by a cylinder of equal volume of radius R_2 (Fig. 1). The height h of this cylindrical atmosphere is equal to the completely extended length of the polyelectrolyte molecules. The space between the plates is divided into two regions. One of them, region I, contains the polyelectrolyte chains, whereas the other one, region II, is free of chains.

Because in region I, the boundary of the atmosphere around each polymer chain was approximated with a cylinder, the electrical potential is related to the electrolyte concentration via a Poisson–Boltzmann equation in cylindrical coordinates,

$$\frac{1}{r}\frac{\partial}{\partial r}\left(r\frac{\partial \Psi}{\partial r}\right)+\frac{\partial^2 \Psi}{\partial z^2}=\frac{2(c_0+c_h)eN_0\sinh(e\Psi/kT)}{\varepsilon\varepsilon_0}, \quad (1)$$

where Ψ is the electrical potential, r and z are the radial and normal coordinates, c_0 is the bulk electrolyte (NaCl) concentration, c_h is the bulk acid concentration which is employed to change the pH, N_0 is the Avogadro number, e is the protonic charge, k is the Boltzmann constant, T is the absolute temperature, ε is the dielectric constant of water, and ε_0 is the vacuum permittivity. Equation (1) will be solved for the cell between the two cylinders of radii R_1 and R_2 and height h using suitable boundary conditions which are provided below.

The boundary condition for the Poisson–Boltzmann equation on the chain surface is

$$\left.\frac{\partial \Psi}{\partial r}\right|_{r=R_1}=-\frac{\sigma(z)}{\varepsilon\varepsilon_0}, \quad (2)$$

where R_1 is the radius of the cylindrical chain, and σ is the surface charge density on the surface of the cylinder of radius R_1.

At the plate surface, the boundary condition is

$$\left.\frac{\partial \Psi}{\partial z}\right|_{z=0}=-\frac{\sigma_p}{\varepsilon\varepsilon_0}, \quad (3)$$

where σ_p is the free plate surface charge density which is assumed constant in the calculation.

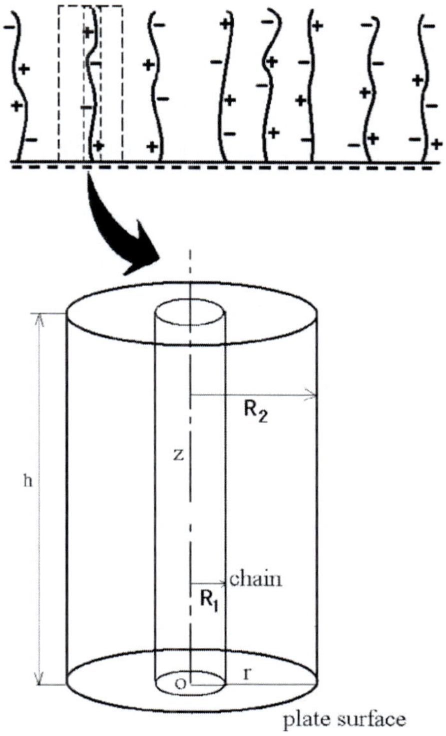

Fig. 1. Schematics of the stretched polyelectrolyte chains on the plate surface. The stretched polyelectrolyte chain is represented as a cylinder. The chains are assumed to have the same length.

For reason of symmetry, at the boundary of the cylindrical atmosphere,

$$\left.\frac{\partial \Psi}{\partial r}\right|_{r=R_2}=0. \quad (4)$$

The dissociation of the two kinds of ionizable sites provides the chain surface with positive and negative charges through the dissociation equilibria [19]

$$G_aH \leftrightarrow G_a^- + H^+, \quad (5)$$

$$G_bH^+ \leftrightarrow G_b + H^+, \quad (6)$$

$$G_aNa \leftrightarrow G_a^- + Na^+, \quad (7)$$

where G_aH and G_b are the ionizable sites which generate the negative G_a^- and positive G_bH^+ charges, respectively, and H^+ and Na^+ are the hydrogen and sodium ions, respectively.

By neglecting the interactions between species, one can write the equilibria

$$K_a \equiv \frac{(1-\theta_a-\theta_{Na})[H^+]_S}{\theta_a}, \quad (8)$$

$$K_b \equiv \frac{(1-\theta_b)[H^+]_S}{\theta_b}, \quad (9)$$

$$K_{Na} \equiv \frac{(1-\theta_a-\theta_{Na})[Na^+]_S}{\theta_{Na}}, \quad (10)$$

where K_a, K_b, and K_{Na} are equilibrium constants (the values $pK_a = 3.4$, $pK_b = 12$, and $K_{Na} = 1\times 10^8$ M, obtained in Ref. [19] by fitting the experimental data for proteins, are employed in the calculation), θ_a and θ_{Na} denote the fractions of sites G_a occupied by H^+ and Na^+, respectively, θ_b is the fraction of sites G_b occupied by H^+, and $[H^+]_S$ is the concentration of hydrogen ions near the surface of the chains.

The net charge density σ on the chain cylindrical surface is the difference between the positive and negative surface charge densities

$$\sigma = e\left[N_b\theta_b - N_a(1-\theta_a-\theta_{Na})\right], \tag{11}$$

where N_a and N_b are the numbers of ionizable sites G_aH and G_b per unit area, respectively.

Combined with Eqs. (8)–(10) and assuming Boltzmann distributions, Eq. (11) becomes

$$\frac{\sigma}{e} = \frac{N_b[H^+]_0\exp(-\frac{e\Psi_s}{kT})}{K_b + [H^+]_0\exp(-\frac{e\Psi_s}{kT})} - \frac{N_aK_aK_{Na}}{(K_{Na}[H^+]_0 + K_a[Na^+]_0)\exp(-\frac{e\Psi_s}{kT}) + K_aK_{Na}}, \tag{12}$$

where $[H^+]_0$ and $[Na^+]_0$ are the bulk concentrations of the corresponding ions and Ψ_s is the surface electrical potential.

In the second region, region II, which is outside the polyelectrolyte brush, the electrical potential will be assumed to be a one-dimensional function of z, independent of r. Two boundary conditions at the boundary between the two regions can be written. In one of them, the electrical potential at the boundary in the outside region is assumed to be equal to the average of the electrical potential at the boundary of the brush region. In the other one, the electric field at the boundary of the outside region is assumed to be equal to the average of the field at the boundary of the brush region,

$$\Psi\big|_{z=h}^{\mathrm{II}} = \frac{2}{R_2^2 - R_1^2}\int_{R_1}^{R_2} r\Psi\big|_{z=h}^{\mathrm{I}}\,dr, \tag{13}$$

$$\frac{d\Psi}{dz}\bigg|_{z=h}^{\mathrm{II}} = \frac{2}{R_2^2 - R_1^2}\int_{R_1}^{R_2} r\frac{\partial\Psi}{\partial z}\bigg|_{z=h}^{\mathrm{I}}\,dr, \tag{14}$$

where I and II represent the brush and the outside regions, respectively.

In reality, there is a transition region in which the potential in region II satisfies a two-dimensional Poisson–Boltzmann equation before becoming one-dimensional. However, for the sake of simplicity, the thickness of that region is considered small.

In the outside region, the Poisson–Boltzmann equation has the form

$$\frac{d^2\Psi}{dz^2} = \frac{2(c_0 + c_h)eN_0\sinh(e\Psi/kT)}{\varepsilon\varepsilon_0}. \tag{15}$$

Because of the symmetry of the system, there is a boundary condition in the middle between the two plates,

$$\frac{d\Psi}{dz}\bigg|_{z=d} = 0, \tag{16}$$

where $2d$ is the distance between the plates.

Using the boundary conditions (2)–(4), (13), (14), and (16), the Poisson–Boltzmann equations (1) and (15) can be solved numerically to obtain the electrical potential profile.

Using the electrical potential in the middle between the two plates, the force per unit area can be calculated using the Langmuir expression [20]

$$F = 2kT(c_0 + c_h)N_0\left[\cosh\left(\frac{e\Psi_0}{kT}\right) - 1\right], \tag{17}$$

where Ψ_0 is the electrical potential at the middle between the two plates.

3. Results and discussion

3.1. The effect of the electrolyte concentration on the force

The electrolyte concentration affects the repulsive force between the plates through the screening effect. As the electrolyte concentration increases, the magnitudes of the electrical potential in the brush and the outside regions decrease (Fig. 2). Therefore, as expected, the force decreases with increasing electrolyte concentration (Fig. 3). For comparison purposes, the figure also contains the force between two plates free of polyelectrolyte chains but with the same surface charge density σ_p (-0.01 C/m^2) on the plates as when the chains are present.

Figs. 4 and 5 present the chain surface electrical potentials Ψ_s and the chain surface charge density σ at various electrolyte concentrations, and $pH_0 = 7$, respectively. For the conditions employed, the charge of the polyelectrolyte

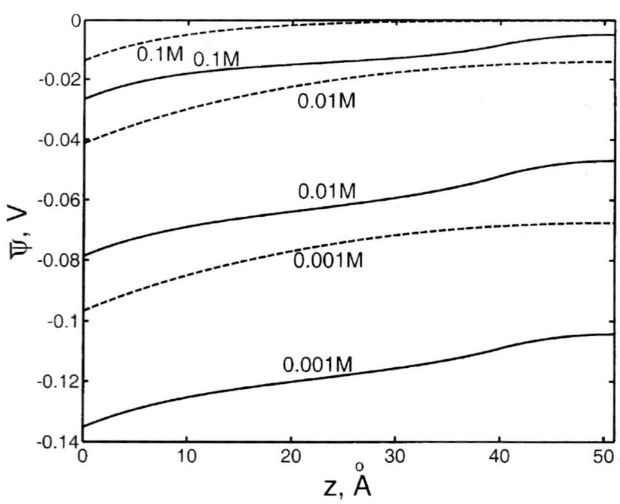

Fig. 2. The average electrical potential $\bar{\Psi} = \int_{R_1}^{R_2} 2\Psi\,dr/(R_2^2 - R_1^2)$ for various electrolyte concentrations c_0. The parameters used in the calculation are $pK_a = 3.4$, $pK_b = 12$, $N_a = 8\times10^{13}$ cm^{-2}, $N_b = 4\times10^{13}$ cm^{-2}, $pH_0 = 7$, $h = 40$ Å, $R_1 = 5$ Å, $R_2 = 25$ Å, $d = 51$ Å, $\sigma_p = -0.01$ C/m^2. The dashed lines are for two bare plates.

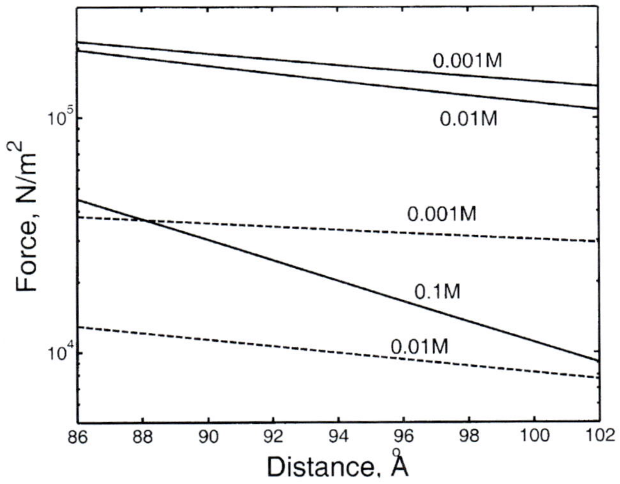

Fig. 3. The repulsive force between two plates at various electrolyte concentrations c_0. The parameters used in the calculation are $pK_a = 3.4$, $pK_b = 12$, $N_a = 8 \times 10^{13}\ \text{cm}^{-2}$, $N_b = 4 \times 10^{13}\ \text{cm}^{-2}$, $pH_0 = 7$, $h = 40$ Å, $R_1 = 5$ Å, $R_2 = 25$ Å, $\sigma_p = -0.01\ \text{C/m}^2$. The dashed lines are for two bare plates.

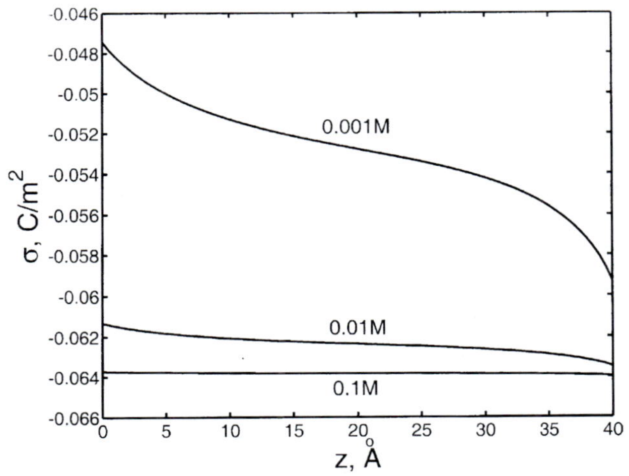

Fig. 5. The chain surface charge density for various electrolyte concentrations. The distance between the two plates is 100 Å. The other parameters are as in Fig. 2.

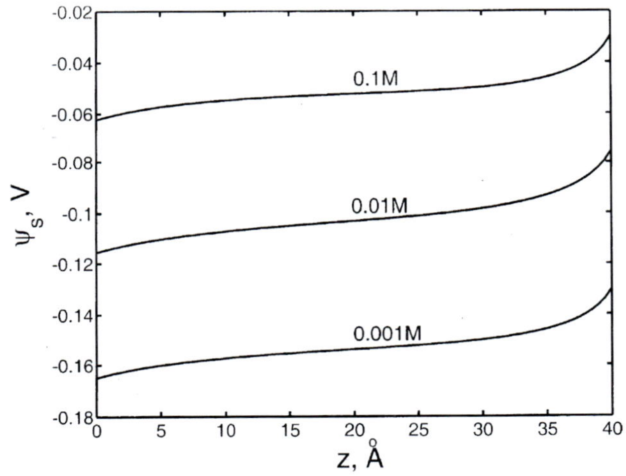

Fig. 4. The chain surface electrical potential at various electrolyte concentrations. The distance between the two plates is 100 Å. The other parameters are as in Fig. 2.

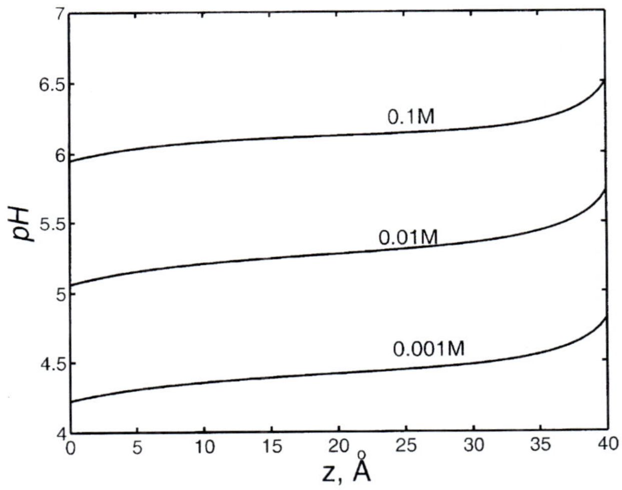

Fig. 6. The local pH near the chain surface. The distance between the two plates is 100 Å. The other parameters are as in Fig. 2.

chains and the sign of the surface potential of the chains are negative. As expected, the same sign of the chain charge as that of the plate increases the repulsive force between the two plates. For higher electrolyte concentrations, the chains have less negative surface electrical potentials. Because in our calculations the surface charge density of the plate σ_p is negative, the potential at the end of the chain near the plate is more negative than at the free end. The electrical potential being more negative near the plates, the concentration of H^+ ions near the chains in the vicinity of the plate are higher (i.e., lower local pH; see Fig. 6) than further from the plate. Through the dissociation equilibria, this provides lower dissociations of G_aH and G_bH^+ near the plates than near the free end. As a result, the chain surface charge density σ is not uniformly distributed, being more negative near the free end. This variation is stronger for low electrolyte concentrations.

3.2. *The effect of the bulk pH (pH_0)*

The pH_0 changes the surface charge density on the chain through the dissociation equilibria. Fig. 7 shows that when pH_0 increases from 2 to 7, the force between the two plates first decreases, passes through a minimum, and then increases. Figs. 8 and 9 allow to explain this behavior. Fig. 8 shows that the surface charge density σ on the chain is positive for $pH_0 = 2.0$ and $pH_0 = 3.0$, but becomes negative for $pH_0 = 5.0$ and 7.0. This occurs because the surface charge of the chain is generated through the dissociation of surface sites. As the pH_0 increases, the number of hydrogen ions dissociated from the sites increases. As a result, the surface charge becomes negative at high values of pH_0 and posi-

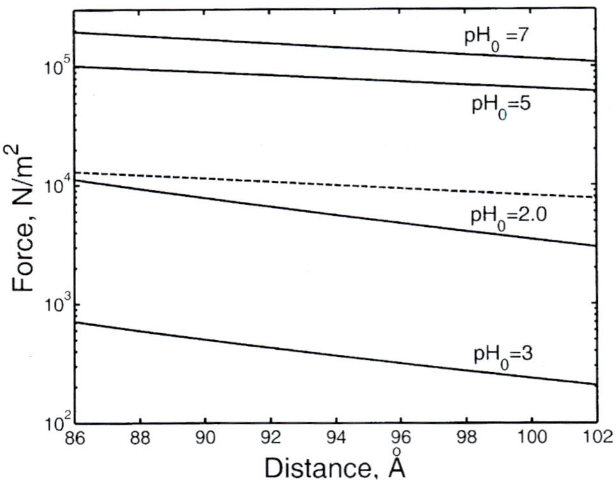

Fig. 7. The repulsive force between two plates for various values of pH_0. The parameters used in the calculations are $pK_a = 3.4$, $pK_b = 12$, $N_a = 8 \times 10^{13}\ cm^{-2}$, $N_b = 4 \times 10^{13}\ cm^{-2}$, $R_1 = 5$ Å, $R_2 = 25$ Å, $h = 40$ Å, $c_0 = 0.01$ M, $\sigma_p = -0.01\ C/m^2$. The dashed lines are for two bare plates.

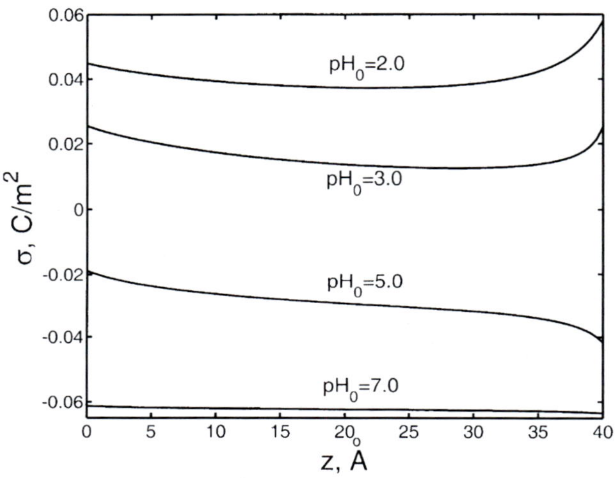

Fig. 8. The surface charge density along the chain for various values of pH_0. The distance between the two plates is 100 Å. The other parameters are as in Fig. 7.

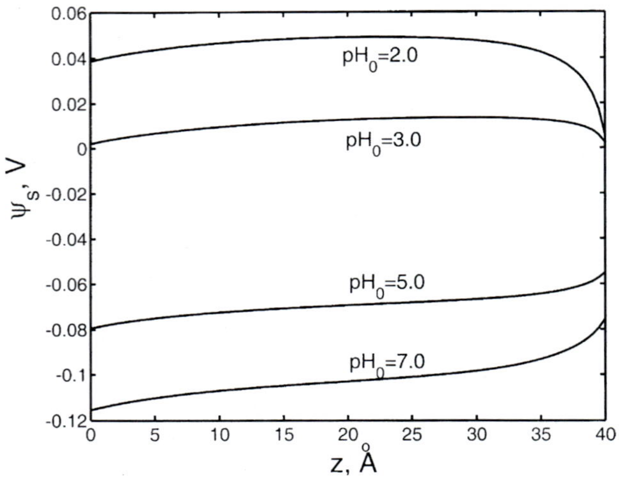

Fig. 9. The surface electrical potential along the chain for various values of pH_0. The distance between the two plates is 100 Å. The other parameters are as in Fig. 7.

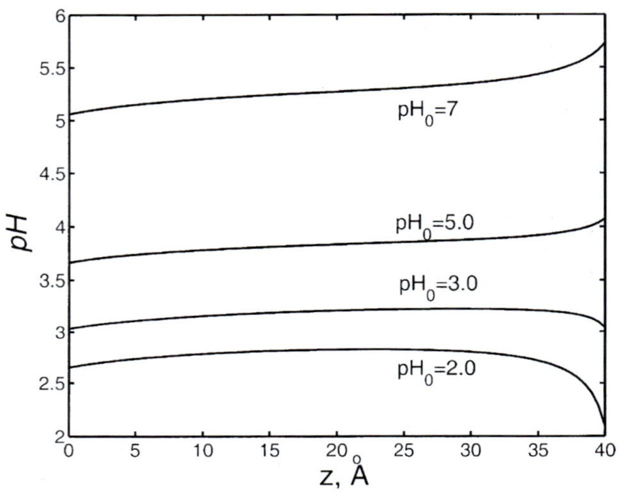

Fig. 10. The local pH near the chain surface. The distance between the two plates is 100 Å. The other parameters are as in Fig. 7.

tive at low values (see Eq. (12)). Consequently, the surface charge density σ of the chain is positive for both pH_0 2.0 and 3.0 (but larger for $pH_0 = 2.0$ than for $pH_0 = 3.0$) and becomes negative for $pH_0 = 5.0$ or 7.0. Because the charge of the plate is assumed negative, the force between the plates decreases at low pH_0 values and increases at larger values, as pH_0 increases. The repulsion force between two bare plates of charge density $-0.01\ C/m^2$ is located between those for $pH_0 = 2.0$ and $pH_0 = 5.0$ for plates covered with brushes.

The local pH (Fig. 10) along the chain near the plate is affected by the surface charge of the plates, which being negative attracts H^+ ions. The latter ions can associate to the chain which thus becomes less negatively charged near the plate when the chains are negatively charged (high pH_0), or more positively charged when the chains are positively charged (low pH_0). When the chains are negatively charged, the electrical potential at the free end of the chain is less negative than near the plate. This leads to a lower H^+ concentration and to a more negative surface charge density on the chains near the free end than near the plate.

It should be also noted that at high pH_0 the chain surface charge density decreases and the chain surface electrical potential increases along the chain monotonously. However, at low pH_0 values, the surface charge density on the chains first decreases and then increases, and the surface electrical potential first increases and then decreases. This occurs because the surface charge of the chains has the same sign as the plates at high pH_0 and the opposite sign at low values.

For low pH_0 values, the electrical potential of the chains is positive, and repulses the hydrogen ions. As a result, the local pH near the chain surface is larger than the bulk pH_0

H. Huang, E. Ruckenstein / Journal of Colloid and Interface Science 275 (2004) 548–554 553

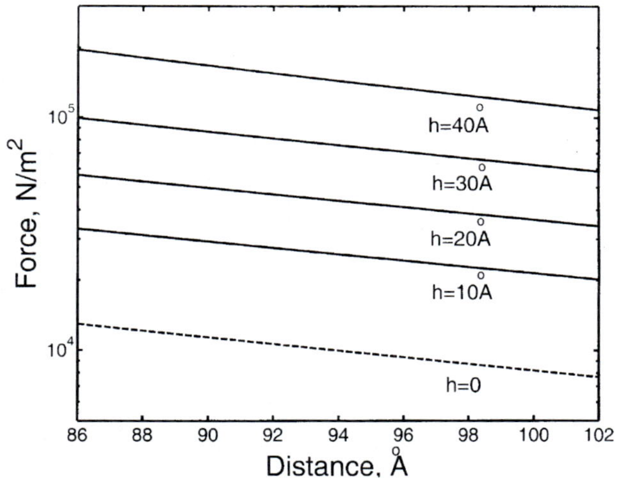

Fig. 11. The effect of the thickness of the polyelectrolyte brush h on the repulsive force. The parameters used in the calculations are $pK_a = 3.4$, $pK_b = 12$, $N_a = 8 \times 10^{13}\ \mathrm{cm}^{-2}$, $N_b = 4 \times 10^{13}\ \mathrm{cm}^{-2}$, $\mathrm{pH}_0 = 7$, $R_1 = 5$ Å, $R_2 = 25$ Å, $c_0 = 0.01$ M, $\sigma_p = -0.01\ \mathrm{C/m^2}$. The dashed line is for two bare plates.

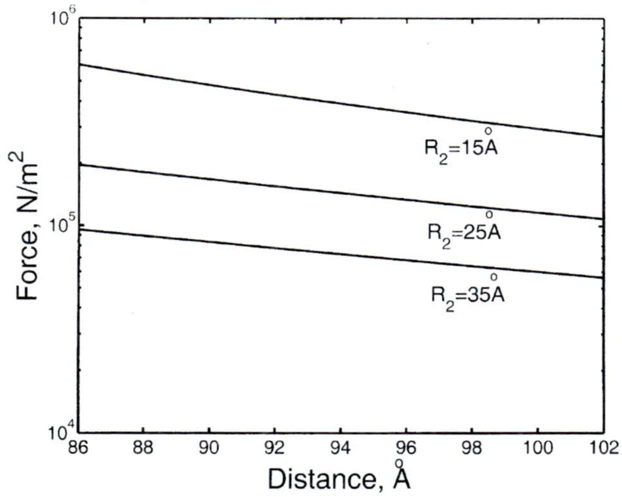

Fig. 12. The effect of the coverage density on the repulsive force. The parameters used in the calculations are $pK_a = 3.4$, $pK_b = 12$, $N_a = 8 \times 10^{13}\ \mathrm{cm}^{-2}$, $N_b = 4 \times 10^{13}\ \mathrm{cm}^{-2}$, $\mathrm{pH}_0 = 7$, $R_1 = 5$ Å, $h = 40$ Å, $c_0 = 0.01$ M, $\sigma_p = -0.01\ \mathrm{C/m^2}$.

for $\mathrm{pH}_0 = 2.0$ or 3.0. For $\mathrm{pH}_0 = 5.0$ or 7.0, the local pH near the chain surface is lower than the bulk pH_0, because the surface electrical potential is negative.

3.3. *The effect of the thickness of the polyelectrolyte brush*

Fig. 11 presents the effect of the thickness of the polyelectrolyte brush on the repulsive force for $\mathrm{pH}_0 = 7$ and $c_0 = 0.01$ M. For large thicknesses of the brush, the force is larger than for small thicknesses. The calculations also indicated that, for $h > 15$ Å and $\mathrm{pH}_0 = 2$ or 9, the force increases as the thickness of the brush increases.

3.4. *The effect of the chain density of the polyelectrolyte chains on the plates*

The effect of the density of the polyelectrolyte chains on the force is presented in Fig. 12 for $\mathrm{pH}_0 = 7$. As expected, the force between the two plates increases with increasing coverage density. This occurs because the decrease of R_2 increases the number of chains per unit area and the magnitude of the average electrical potential due to the interaction between the chains in the brush.

3.5. *Collapse of the polyelectrolyte brush*

Under certain conditions the polyelectrolyte brush has the tendency to collapse on the plate surface. In Figs. 8 and 9, one can see that the chain surface charge has a positive value near the plate for $\mathrm{pH}_0 = 2.0$ or 3.0; however, because the plate has a negative charge, the electric field in this region is negative. This means that the polyelectrolyte segments are attracted by the plates. If the repulsion between the monomers of the chain is weaker than the attractive force between the oppositely charged plate surface and monomers, the chains will collapse on the plate surface. The present model is no longer valid under such conditions.

Table 1
The chain surface charge density σ_a used in Eq. (19) to obtain the same force as the one calculated from the two-dimensional model

pH_0	$10^3\sigma_a$ ($\mathrm{C/m^2}$)
7.0	−69
5.0	−36
3.0	−7.5
2.0	0.98

3.6. *Comparison with the traditional treatment*

In the traditional treatment, the electrical potential is assumed unidimensional and a uniform charge density is introduced into the Poisson–Boltzmann equation to account for the presence of the charged chains. If the chain surface has an average charge density σ_a, the uniform charge density ρ_{un} can be evaluated as the total surface charge of a chain divided by the brush volume of the cylinder of radius R_2 and height h,

$$\rho_{un} = \frac{2\pi R_1 h \sigma_a}{\pi R_2^2 h} = \frac{2R_1\sigma_a}{R_2^2}. \tag{18}$$

Using this expression for the uniform charge density, the Poisson–Boltzmann equation in region I acquires the form

$$\frac{\partial^2 \Psi}{\partial z^2} = \frac{2(c_0 + c_h)eN_0 \sinh(e\Psi/kT) + 2R_1\sigma_a/R_2^2}{\varepsilon\varepsilon_0}. \tag{19}$$

By equating the force provided by the two-dimensional model to that obtained from Eq. (19), the values listed in Table 1 have been obtained for σ_a as a function of pH_0.

554 H. Huang, E. Ruckenstein / Journal of Colloid and Interface Science 275 (2004) 548–554

Table 2
The average chain surface charge density $\bar{\sigma}$ for various pH_0 values and various distances between the two plates[a]

	$10^3\bar{\sigma}$ (C/m^2)			
$2d$ (Å)	$pH_0 = 2.0$	$pH_0 = 3.0$	$pH_0 = 5.0$	$pH_0 = 7.0$
86	40.455	16.074	−28.176	−62.257
88	40.631	16.147	−28.394	−62.281
90	40.766	16.202	−28.585	−62.301
92	40.871	16.244	−28.755	−62.318
94	40.956	16.278	−28.908	−62.332
96	41.024	16.304	−29.047	−62.345
98	41.079	16.325	−29.174	−62.356
100	41.124	16.342	−29.290	−62.366
102	41.160	16.355	−29.397	−62.376

[a] The parameters used in the calculations are $c_0 = 0.01$ M, $pK_a = 3.4$, $pK_b = 12$, $N_a = 8 \times 10^{13}$ cm^{-2}, $N_b = 4 \times 10^{13}$ cm^{-2}, $h = 40$ Å, $R_1 = 5$ Å, $R_2 = 25$ Å, $\sigma_p = -0.01$ C/m^2.

Let us now compare the value of σ_a to the average values $\bar{\sigma}$ provided by the expression

$$\bar{\sigma} = \frac{1}{h}\int_0^h \sigma(z)\,dz, \tag{20}$$

where σ is the surface charge density on a chain calculated on the basis of the two-dimensional model. Table 2 lists the values obtained.

Comparing the two tables, one can see that σ_a is somewhat different from $\bar{\sigma}$. This indicates that the one-dimensional model approximation can be unsatisfactory. In the extreme case in which the sites are completely dissociated, one can easily evaluate σ_a, which becomes equal to $e(N_b - N_a)$.

4. Conclusions

In this paper, a two-dimensional model is employed to describe the repulsive interactions between two parallel plates to which polyelectrolyte chains are grafted. The chains are assumed to be completely extended. The charge is generated through the dissociation of the ionizable sites of the chains. The force between the two plates at distances larger than twice the thickness of the brush of chains was calculated. The calculation revealed that with increasing bulk pH_0, the force can pass through a minimum, and the larger the thickness of the polyelectrolyte brush and the coverage of the plate, the larger the repulsive force.

References

[1] D.H. Napper, Polymeric Stabilization of Colloidal Dispersions, Academic Press, New York, 1983.
[2] H. Casanova, J. Chen, E. Dickinson, B.S. Murray, P.V. Nelson, M. Whittle, Phys. Chem. Chem. Phys. 2 (2000) 3861.
[3] N.S. Bell, J. Sindel, F. Aldinger, W.M. Sigmund, J. Colloid Interface Sci. 254 (2002) 296.
[4] Y.L. Ong, A. Razatos, G. Georgiou, M.M. Sharma, Langmuir 15 (1999) 2719.
[5] S. Alexander, J. Phys. 38 (1977) 983.
[6] P.G. de Gennes, Adv. Colloid Interface Sci. 27 (1987) 189.
[7] E. Ruckenstein, B. Li, J. Chem. Phys. 107 (1997) 932.
[8] E. Zhulina, O. Borisov, V. Priamitsyn, J. Colloid Interface Sci. 137 (1990) 495.
[9] M. Muthukumar, J. Ho, Macromolecules 22 (1989) 965.
[10] S. Misra, S. Varanasi, P.P. Varanasi, Macromolecules 22 (1989) 4173.
[11] S.T. Milner, T.A. Witten, M.E. Cates, Macromolecules 21 (1988) 2610.
[12] S. Alexander, J. Phys. 38 (1977) 983.
[13] P.G. de Gennes, J. Phys. 37 (1976) 1443.
[14] P.G. de Gennes, Macromolecules 13 (1980) 1069.
[15] S. Misra, S. Varanasi, W. Mattice, Macromolecules 29 (1996) 6055.
[16] H. Ohshima, Colloid Polym. Sci. 277 (1999) 535.
[17] D.A. Hoagland, Macromolecules 23 (1990) 2781.
[18] D. Dean, J. Seog, C. Ortiz, A.J. Grodzinsky, Langmuir 19 (2003) 5526.
[19] D. Prieve, E. Ruckenstein, J. Theor. Biol. 56 (1976) 205.
[20] I. Langmuir, J. Chem. Phys. 6 (1938) 893.

ELSEVIER

Advances in Colloid and Interface Science 112 (2004) 37–47

ADVANCES IN COLLOID AND INTERFACE SCIENCE

www.elsevier.com/locate/cis

The bridging force between two plates by polyelectrolyte chains

Haohao Huang, Eli Ruckenstein*

Department of Chemical Engineering, State University of New York at Buffalo, 303 Furnas Hall, Buffalo NY 14260, USA

Available online 14 August 2004

Abstract

The interaction between particles in a colloidal system can be significantly affected by their bridging by polyelectrolyte chains. In this paper, the bridging is investigated by using a self-consistent field approach which takes into account the van der Waals interactions between the segments of the polyelectrolyte molecules and the plates, as well as the electrostatic and volume exclusion interactions. A positive contribution to the force between two plates is generated by the van der Waals interactions between the segments and the plates. This positive repulsive contribution plays an important role in the force when the distances between the plates are small. With increasing van der Waals interaction strength between segments and plates, the force between the plates becomes more repulsive at small distances and more attractive at large distances. When the surfaces of the plates have a constant surface electrical potential and a charge sign opposite to that of the polyelectrolyte chains, the force between the two plates becomes less attractive as the bulk polyelectrolyte concentration increases. This behavior is due to a higher bulk counterion concentration dissociated from the polyelectrolyte molecules. At short distances, the force between plates is more repulsive for stiffer chains. A comparison between theoretical and experimental results regarding the contraction of the interlayer separation between the platelets of vermiculite clays against the concentration of poly(vinyl methyl ether) was made.

Keywords: Bridging interaction; Polyelectrolyte; Self-consistent field theory; van der Waals interactions; Repulsive force

Contents

* Corresponding author. Tel.: +1 716 645 2911x2214; fax: +1 716 645 3822.
E-mail address: feaeliru@acsu.buffalo.edu (E. Ruckenstein).

0001-8686/$ - see front matter
doi:10.1016/j.cis.2004.06.003

1. Introduction

Polymers are often used to stabilize colloidal systems by grafting them on the particle surfaces to provide steric repulsion [1,2]. Polymers can also induce flocculation due to either "depletion" or "bridging" interactions [3].

The bridging interaction occurs because monomers of the same polymer chain are adsorbed on different particles and draw them together [4,5]. The bridging was extensively studied experimentally [6–8], theoretically [9–11] and by Monte Carlo simulations [12–16]. It was exploited practically for water purification by adding high molecular weight polyacrylamides to flocculate the small particles suspended in water [17]. However, the strength of the bridging could not be quantitatively expressed.

Recently, it was noted that the addition of poly(vinyl methyl ether) to a clay–salt–water system induced the contraction of the interlayer separation between the clay platelets [18]. The n-butylammonium vermiculite used in those experiments provided an ideal structure of regularly spaced and parallel charged colloidal platelets in a salt solution [19,20]. Swenson et al. [21–23] studied the effect of the addition of poly(ethylene oxide) (PEO) on the distance between the vermiculite layers and observed that, with increasing PEO concentration, the distance between the vermiculite layers decreased. They estimated that the bridging force per bridge was 1.4 pN.

Polyelectrolyte molecules affect in a more complicated manner than the neutral ones the bridging interactions between colloidal particles because they influence through their dissociation the electrical potential of the electrical double layer [24–27] which, in turn, affects the conformation of the chains and the overall electrostatic interaction.

A number of theoretical approaches which can describe the segment concentration distribution of polymers adsorbed on the surface of a plate or the surfaces of two interacting plates [28–44] were proposed. Edwards [28] employed a continuous self-consistent field theory to obtain the segment distribution in a solution of flexible neutral polymers adsorbed on a plate. Scheutjens and Fleer [32,33] developed a lattice model to examine the polymer adsorption on a plate. Skau et al. [42] compared the two approaches and obtained quantitative agreement for the segment density profile in a good solvent. Li and Ruckenstein [43] employed a lattice model to calculate the interaction between two grafted polymer brushes with the emphasis on the interdigitation between the chains of the two brushes [44]. Monte Carlo simulations were also carried out to investigate the effect of grafted polyelectrolyte on the interaction between two colloidal particles [45].

Using a self-consistent field theory, Varoqui et al. [46] calculated the segment density profile of polyelectrolyte molecules adsorbed on one plate and Podgornik [9] extended the procedure to the interaction between two plates.

Borukhov et al. [11] investigated the effect of polyelectrolyte adsorption on the intercolloidal forces. They assume that the adsorption of polyelectrolyte molecules on the surface of the plates was solely due to the electrostatic interactions between the negatively charged plates and the positively charged polyelectrolyte segments. However, there are additional interactions between the surfaces of the plates and the segments, such as the van der Waals interactions.

When the polyelectrolyte chains form bridges between two plates, the plates are pulled towards each other by the chains. If the van der Waals interactions between the segments of the chains and the surfaces are sufficiently strong, the increased polymer concentration in the gap between the plates can significantly affect the bridging force between them.

A self-consistent field theory was employed by us to calculate the bridging force between two plates immersed in a polyelectrolyte solution, following Borukhov et al.'s [11] approach. However, in the calculations, the van der Waals interactions between plates and segments were also taken into account. It was shown that the effect of the latter interactions is important.

2. Self-consistent field theory

Two infinite-size plates are immersed in a semidilute solution of polyelectrolyte in a good solvent which also contains the small ions of a salt. One of the plates is located at x=0 and the other one at x=D. The system is considered to be in contact with a large reservoir, which contains a polyelectrolyte/salt solution. In addition to the electrostatic interactions, the segments of the polymer have a van der Waals interaction $-UkT$ with the plates. In the mean field approach, the intra- and interchain interactions together with the electric field induced by the surface charges of the plates and polyelectrolyte molecules are expressed as an external potential. Within the mean-field approximation, the free energy of the system with respect to that in the reservoir can be expressed as the sum of three contributions, the polymer contribution F_{pol}, the salt ions contribution F_{ion} and the electrostatic field contribution F_{ele},

$$F=\int_0^D h\mathrm{d}x = F_{\text{pol}}+F_{\text{ion}}+F_{\text{ele}} \tag{1}$$

where h is the free energy density.

The polymer contribution F_{pol} includes the effect of the polymer chain connectivity and the short-range excluded volume interaction

$$\beta F_{\text{pol}}=\int_0^D\left[\frac{a^2}{6}\left(\frac{\mathrm{d}\varphi}{\mathrm{d}x}\right)^2+\frac{1}{2}v(\varphi^4-\varphi_b^4)-(\varphi^2-\varphi_b^2)\mu_p-U\varphi^2\right]\mathrm{d}x \tag{2}$$

H. Huang, E. Ruckenstein / Advances in Colloid and Interface Science 112 (2004) 37–47 39

where a is the Kuhn segment length of the chain, $\beta=1/kT$, k is the Boltzmann constant, T is the absolute temperature, v is the excluded volume parameter, φ is the "polymer order parameter," which is related to the local concentration $c_p(x)$ of the segments of the polymer chains through $c_p(x)=\varphi^2(x)$, $kT\mu_p$ is the chemical potential of the segments of the chains and the subscript b indicates the bulk. The first term in Eq. (2) is due to the connectivity of the segments of the polymer chain, the second term to the excluded volume interaction between segments, the third term to the concentration difference between a point in the gap and the bulk and the fourth term to the van der Waals interaction between plates and segments. A larger a means a stiffer chain. For a good solvent, v has a positive value, while for a poor solvent, v has a negative value, i.e., the interaction between segments becomes attractive. In the θ condition, v is zero. In the following calculations, v is assumed a positive constant, independent of the concentration of the salt in solution. The chemical potential $kT\mu_p$ can be obtained from the $kT\mu_p=vkT\varphi_b^2$.

The ionic contribution to the free energy of the system is due to the entropic increase of the small ions between the plates compared to that in the reservoir.

$$\beta F_{\text{ion}} = \int_0^D \left[\left(\sum_i c_i \ln\frac{x_i}{x_{i,b}}\right) + c_w \ln\frac{1-\sum_i x_i}{1-\sum_i x_{i,b}}\right]\mathrm{d}x$$
$$\approx \int_0^D \sum_i \left(c_i \ln\frac{c_i}{c_{i,b}} - c_i + c_{i,b}\right)\mathrm{d}x \quad (3)$$

where c_i is the concentration of ion i, c_w is the concentration of water and x_i is the mole fraction of ion i. In the following calculations, two kinds of small ions are present in solution: the ions of the salt and the counterions dissociated from the polyelectrolyte molecules which are assumed to be the same as one of the ions of the salt.

The electrostatic free energy contribution is the sum of the electrostatic energy of the solution and the chemical contribution of the surface, if the potential determining ions are in thermodynamic equilibrium with those in solution. At constant surface electrical potential Ψ_0, the chemical contribution is given by

$$F_{\text{chem}} = -2\sigma\psi_0 \quad (4)$$

where σ is the surface charge density.

Therefore, the electrostatic free energy contribution can be expressed as

$$F_{\text{ele}} = \frac{1}{2}\varepsilon\varepsilon_0 \int_0^D \left(\frac{\mathrm{d}\psi}{\mathrm{d}x}\right)^2 \mathrm{d}x - 2\sigma\psi_0 \quad (5)$$

where ε is the dielectric constant, ε_0 is the vacuum permittivity and ψ is the electrical potential.

Using the Gauss law and the Green theorem, the electrostatic contribution can be rewritten as

$$F_{\text{ele}} = \frac{1}{2}\varepsilon\varepsilon_0 \int_0^D \left(\frac{\mathrm{d}\psi}{\mathrm{d}x}\right)^2 \mathrm{d}x - \varepsilon\varepsilon_0 \int_0^D \left[\psi\frac{\mathrm{d}^2\psi}{\mathrm{d}x^2} + \left(\frac{\mathrm{d}\psi}{\mathrm{d}x}\right)^2\right]\mathrm{d}x$$
$$= -\varepsilon\varepsilon_0 \int_0^D \left[\psi\frac{\mathrm{d}^2\psi}{\mathrm{d}x^2} + \frac{1}{2}\left(\frac{\mathrm{d}\psi}{\mathrm{d}x}\right)^2\right]\mathrm{d}x \quad (6)$$

Because of the Poisson equation,

$$\frac{\mathrm{d}^2\psi}{\mathrm{d}x^2} = -\frac{\rho}{\varepsilon\varepsilon_0} \quad (7)$$

where ρ is the charge density in the solution, Eq. (6) becomes

$$F_{\text{ele}} = \int_0^D \left[e\psi\left(c^+ - c^- + \alpha\varphi^2\right) - \frac{1}{2}\left(\frac{\mathrm{d}\psi}{\mathrm{d}x}\right)^2\right]\mathrm{d}x \quad (8)$$

where c^+ and c^- are the concentrations of the cations and anions, respectively.

If the surface charge density changes with the distance between the plates through equilibrium, the chemical contribution to the free energy is given by [47]

$$F_{\text{chem}} = -2\sigma(D=\infty)\Psi(D=\infty) - 2\int_{\sigma(\infty)}^{\sigma(D)} \Psi_0 \mathrm{d}\sigma \quad (9)$$

The second term on the right-hand side of Eq. (9) represents the variation of the chemical contribution to the free energy when the distance between the plates is changed from infinity to D. It was obtained by decomposing the pathway from infinity to D into infinitesimal steps at constant charge, for which the chemical contributions to the free energy are zero, each followed by an infinitesimal step at constant potential with a change $\mathrm{d}\sigma$ of the surface charge density, providing each a contribution $-2\Psi_0\mathrm{d}\sigma$.

By minimizing the free energy with respect to ψ, and φ, one obtains

$$\frac{a^2}{6}\varphi'' = (\alpha\beta e\psi - U)\varphi + v\left(\varphi^3 - \varphi\varphi_b^2\right) \quad (10)$$

$$\frac{\mathrm{d}^2\psi}{\mathrm{d}x^2} = \frac{1}{\varepsilon\varepsilon_0}\left[2ce\sinh(\beta e\psi) - \alpha e\varphi^2 + \alpha e\varphi_b^2\exp(\beta e\psi)\right] \quad (11)$$

Eq. (10) represents the self-consistent field equation for the local segment density of the polymer chains subject to an external electrical potential ψ, a van der Waals interaction with the plates $-UkT$ and an excluded volume interaction. Eq. (11) is a modified Poisson–Boltzmann equation in which the first term accounts for the charges of the small ions of the salt, the second term for the charges of the polyelectrolyte chains and the third one for the charges of the ions dissociated from the polyelectrolyte molecules.

Combined with the appropriate boundary conditions, Eqs. (10) and (11) can be solved to obtain the distributions of the electrical potential and of the chain segment density. Both quantities are required to calculate the interaction force between plates.

If one assumes that no polymer segments are adsorbed onto the surfaces, the boundary condition commonly used [48,39,11,49] for the polymer order parameter at the surface is

$$\varphi(0) = 0 \tag{12}$$

This boundary condition implies that the chain segments are depleted in the region near the plates. This is an arguable boundary condition [50,51].

When the surfaces of the plates are assumed to be kept at constant surface potential ψ_0,

$$\psi(0) = \psi_0 \tag{13a}$$

If the surfaces of the plates are kept at constant charge density, the corresponding boundary condition becomes

$$\left.\frac{d\psi}{dx}\right|_{x=0} = -\frac{\sigma}{\varepsilon\varepsilon_0} \tag{13b}$$

At the midway between plates, the symmetry of the system provides two boundary conditions

$$\left.\frac{d\varphi}{dx}\right|_{x=D/2} = 0 \tag{14}$$

$$\left.\frac{d\psi}{dx}\right|_{x=D/2} = 0 \tag{15}$$

The force per unit area between the plates can be calculated as the derivative of the free energy with respect to the distance between plates

$$f = -\frac{d[F(D) - F(D=\infty)]}{dD} \tag{16}$$

Consequently, the force can be expressed as

$$\begin{aligned}\beta f = &-\alpha\beta\varphi_{D/2}^2 e\psi_{D/2} - \frac{v}{2}\left[\varphi_{D/2}^2 - \varphi_b^2\right]^2 + U_{D/2}\varphi_{D/2}^2 \\ &+ \alpha\varphi_b^2\left\{\exp\left[e\psi_{D/2}/(kT)\right] - 1\right\} \\ &+ 2c\left\{\cosh\left[e\psi_{D/2}/kT\right] - 1\right\}\end{aligned} \tag{17}$$

where $\Psi_{D/2}$ is the electrical potential at the middle between plates, $U_{D/2}$ is the van der Waals interaction between plates and segments at the middle and $\varphi_{D/2}^2$ is the segment concentration at the middle. When the electrical potential at the middle distance is positive, the first two terms of Eq. (17) are negative, and the last two terms are positive. It should be noted that, although the plates and the segments have an attractive interactions $-U$, the third term is positive because U is positive. At small distances, the third term has a significant positive contribution to f because $U_{D/2}$ is large at such distances. However, at large distances between the plates, the contribution of the third term is not important because $U_{D/2}$ drastically decreases with increasing distance between plates, and the first term is negative and dominant because the electrical potential at the middle is positive at large distances.

If only the small ions are present in the system, the above equation reduces to the Langmuir expression

$$f = 2ckT\{\cosh[e\psi(D/2)/kT] - 1\} \tag{18}$$

For the van der Waals interactions between segments and plates, one can write the expression [52]

$$U = \frac{A_H a'^3}{3kT}\left[\frac{1}{x^3} + \frac{1}{(D-x)^3}\right] \tag{19}$$

where A_H is the Hamaker constant and $\pi a'^3/6$ is the volume of a segment assumed spherical and of radius a'. Very close to the surface, there is a Born repulsion between plates and the polyelectrolyte chains, which will be included in the calculations by considering an excluded region of 3.5 Å near the plates.

3. Neutral polymer between two plates

If the plates and the chains are not charged, Eq. (10) becomes

$$\frac{a^2}{6}\varphi'' = -U\varphi + v(\varphi^3 - \varphi\varphi_b^2) \tag{20}$$

The second term on the right-hand side of Eq. (20) is negligible when the distance between plates is not too large. As an approximation, φ between two plates can be obtained from the superposition of the profiles of φ near the single plates. For a single plate, the equation $\frac{a^2}{6}\varphi'' = -\frac{A_H a'^3}{3kTx^3}\varphi$

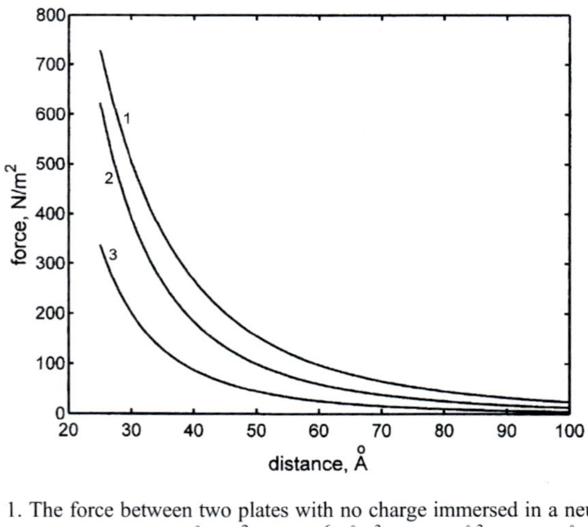

Fig. 1. The force between two plates with no charge immersed in a neutral polymer solution. a=5 Å, φ_b^2=1×10^{-6} Å^{-3}, v=50 Å^3, a'=5 Å. (1) A_H=0.5×10^{-20} J; (2) A_H=0.25×10^{-20} J; (3) A_H=0.1×10^{-20} J.

combined with the boundary conditions $\varphi(0)=0$ and $d\varphi/dx(\infty)=0$ has the solution

$$\varphi(x) = c_1 x^{1/2} J_1\left(2C^{1/2}x^{-1/2}\right) \tag{21}$$

where $C=2A_{\mathrm{H}}a'^{3}/(a^2kT)$, J_1 is the first-order Bessel function of the first kind and $c_1 \cong \varphi_b/\sqrt{C}$ is obtained from the segment concentration at infinite distance $\varphi(\infty)=\varphi_b$.

By assuming that the profiles for the two plates are additive, one obtains

$$\varphi(x) = c_1 x^{1/2} J_1\left(2C^{1/2}x^{-1/2}\right) + c_1(D-x)^{1/2} J_1\left(2C^{1/2}(D-x)^{-1/2}\right)$$

By expanding the Bessel function in series and neglecting higher order terms, φ at the middle between two plates can be expressed as

$$\varphi_{D/2} = 2c_1(D/2)^{1/2} J_0\left(2^{3/2}C^{1/2}D^{-1/2}\right) \cong 2\varphi_b\left[\frac{C}{D}\left(\frac{C}{3D}-1\right)+1\right] \tag{22}$$

Therefore, the force between two plates is given by

$$\beta f = -\frac{v}{2}\left[\varphi_{D/2}^2-\varphi_b^2\right]^2 + U_{D/2}\varphi_{D/2}^2 = -\frac{v\varphi_b^4}{2}\left\{4\left[\frac{C}{D}\left(\frac{C}{3D}-1\right)+1\right]^2-1\right\}^2 + \frac{64A_H a'^{3}\varphi_b^2}{3kTD^3}\left[\frac{C}{D}\left(\frac{C}{3D}-1\right)+1\right]^2 \tag{23}$$

The force is repulsive for neutral polymers (Fig. 1) because the second term in Eq. (23) is dominant. Because of the van der Waals interactions between the plates and the segments, the latter are attracted into the gap between the plates and the segment concentration $\varphi_{D/2}^2$ is larger than that in the bulk. However, the magnitude of the second term in Eq. (23), which provides a repulsive contribution to the force, is in most cases larger than the first term. Consequently, the force between the plates is repulsive. As the van der Waals interaction strength increases, the repulsive force increases. Numerical calculations based on Eq. (20) led to results which coincide with those obtained with Eq. (21).

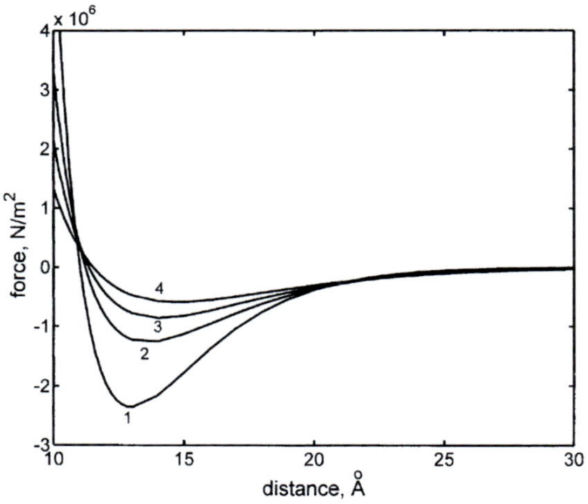

Fig. 2. The force between plates for various van der Waals interaction strengths. $\alpha=1$, $a=5$ Å, $\varphi_b^2=1\times10^{-6}$ Å^{-3}, $e\Psi_0/kT=-2$, $v=50$ Å^3, $c=0.001$ M, $a'=5$ Å. (1) $A_H=1\times10^{-20}$ J; (2) $A_H=0.5\times10^{-20}$ J; (3) $A_H=0.25\times10^{-20}$ J; (4) $A_H=0.05\times10^{-20}$ J.

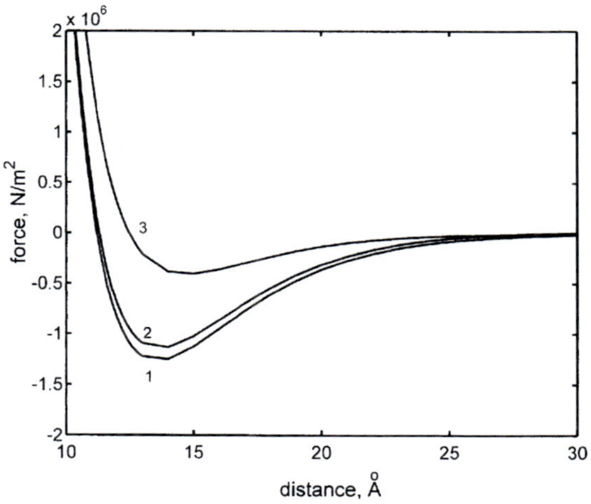

Fig. 3. The force between two plates at various salt concentrations. The values of the parameters used in the calculations are: $\alpha=1$, $a=5$ Å, $\varphi_b^2=1\times10^{-6}$ Å^{-3}, $e\Psi_0/kT=-2$, $v=50$ Å^3, $A_H=0.5\times10^{-20}$ J, $a'=5$ Å. (1) $c=0.001$ M; (2) $c=0.1$ M; (3) $c=1$ M.

4. Numerical results

If the polymers are charged, the segment concentration and the force between plates can be calculated numerically only.

4.1. The effect of the van der Waals interactions between plates and segments

Because of the van der Waals interactions, the plates attract a larger number of polymer segments into the gap between the plates than when there are only electrostatic interactions with the chain segments. As a result, the interaction between plates, with the latter at constant electrical potential, becomes more attractive at distances larger than about 12 Å than when the van der Waals interaction is neglected (Fig. 2). This behavior occurs because φ^2 increases and hence the number of bridges increases. The force between plates at distances smaller than 12 Å becomes, however, more repulsive when the van der Waals interaction is taken into account than when it is neglected. This increased repulsion can be attributed to the third term in Eq. (17), which provides a positive contribution to the force and becomes dominant at short distances.

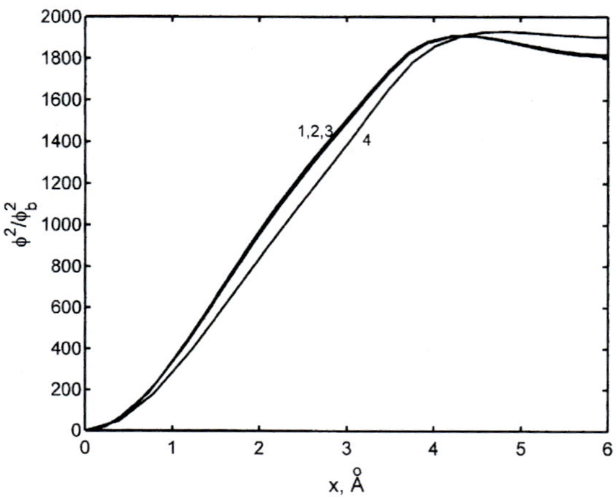

Fig. 4. The segment concentration distribution between plates for various salt concentrations. Only half space from one plate to the midway is plotted. D=12 Å, and the other parameter values used in calculations are as for Fig. 3. (1) c=0.001 M; (2) c=0.01 M; (3) c=0.1 M; (4) c=1 M.

As the distance between plates increases, the third term drastically decreases, and the first term, which provides a negative contribution when the electrical potential at the midway is positive, becomes dominant at larger distances. Hence, the van der Waals interaction between the plates and the segments of the chains plays an important role. As the van der Waals interaction between segments and plates becomes stronger, the force between plates becomes more attractive at moderate distances (D>about 12 Å) between them and more repulsive at D<12 Å.

4.2. The effect of salt concentration

The effect of salt concentration on the force is presented in Fig. 3 for a Hamaker constant of 0.5×10^{-20} J. As the salt concentration increases, the force between plates at constant surface potential becomes more attractive at large distances.

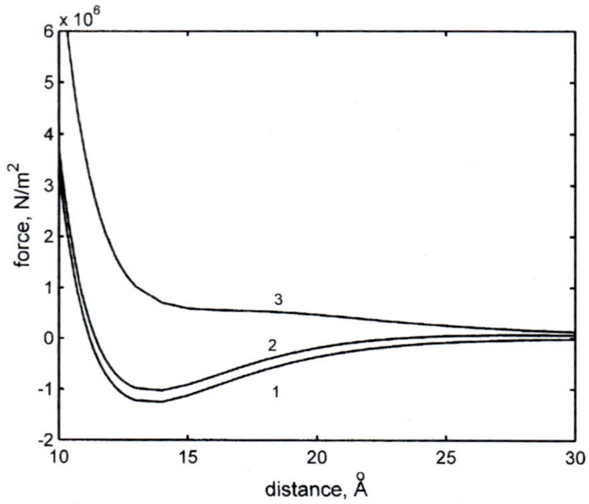

Fig. 6. The force between plates for various bulk polyelectrolyte concentrations. α=1, a=5 Å, c=0.001 M, $e\Psi_0/kT$=−2, v=50 Å^3, A_H=0.5×10^{-20} J, a' =5 Å. (1) φ_b^2=1×10^{-6} Å^{-3}; (2) φ_b^2=1×10^{-4} Å^{-3}; (3) φ_b^2=1×10^{-3} Å^{-3}.

The segment concentration profiles is presented in Fig. 4 for φ_b^2=1×10^{-6} Å^{-3}. In the gap between plates, there are two concentration peaks, each located near one of the surfaces. The concentrations of the monomer in the gap can be by 3 orders of magnitude larger than its bulk concentration. These segment concentrations are much larger than the bulk salt concentration. As a result, salt concentrations lower than 0.1 M have a small effect on the force. For φ_b^2=1×10^{-6} Å^{-3}, the electrical potential and the segment concentration at the midplane are almost independent of the salt concentration when the latter is lower than 0.1 M (Fig. 5). However, for salt concentrations higher than 0.1 M, the segment concentration at the midway increases and the electrical potential at the midway decreases with increasing salt concentration. As a result, the force becomes more

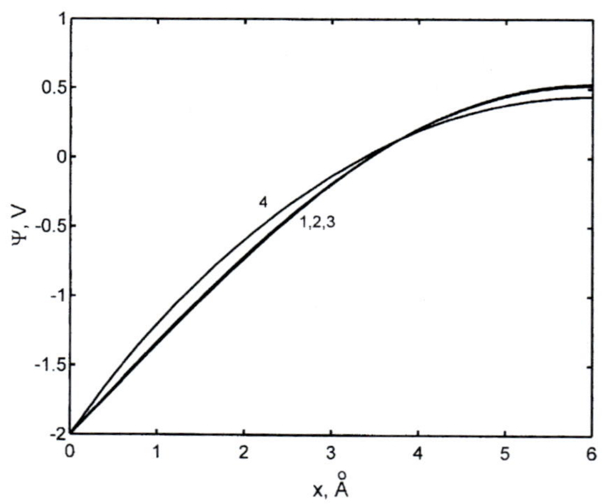

Fig. 5. The electrical potential distribution between plates for various salt concentrations. Only half space from one plate to the midway is plotted. D=12 Å, and the other parameter values used in calculations are as for Fig. 3. (1) c=0.001 M; (2) c=0.01 M; (3) c=0.1 M; (4) c=1 M.

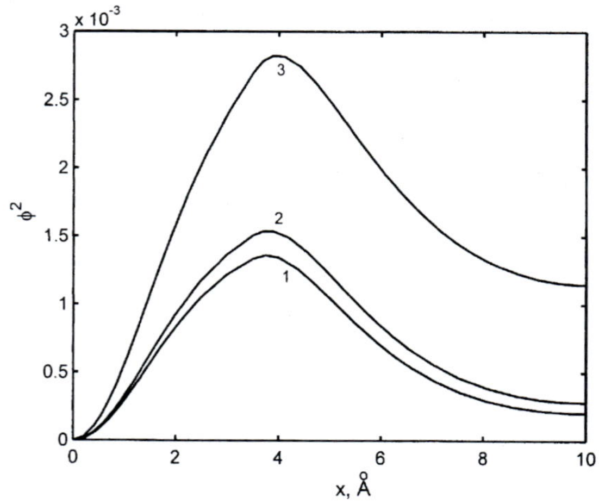

Fig. 7. The segment concentration distribution between the plates for various bulk polyelectrolyte concentrations. D=20 Å, and the other parameter values used in calculations are as for Fig. 6. (1) φ_b^2=1×10^{-6} Å^{-3}; (2) φ_b^2=1×10^{-4} Å^{-3}; (3) φ_b^2=1×10^{-3} Å^{-3}.

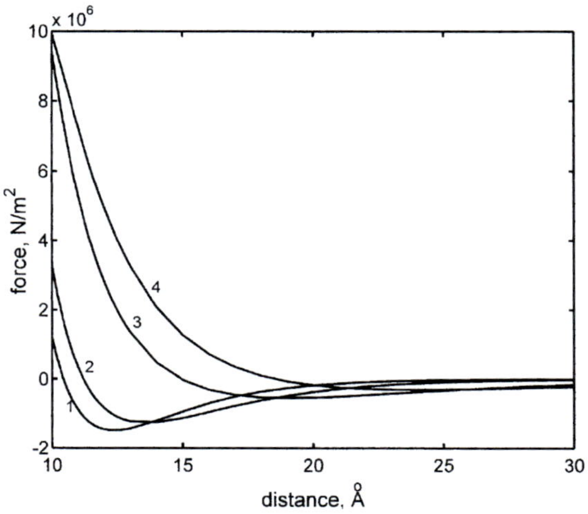

Fig. 8. The force between plates for various stiffnesses of polyelectrolyte chain. The parameter values used in calculations are: α=1, c=0.001 M, $\varphi_b^2=1\times10^{-6}$ Å^{-3}, $e\Psi_0/kT=-2$, v=50 Å^3, $A_H=0.5\times10^{-20}$ J, a' =5 Å. (1) a=4 Å; (2) a=5 Å; (3) a=8 Å; (4) a=10 Å.

repulsive at both short and large distances with increasing salt concentration.

4.3. The effect of the bulk polymer concentration

The bulk segment concentration affects the distributions of the electrostatic potential and osmotic pressure in solution. Fig. 6 shows that as the bulk segment concentration increases, the force becomes less attractive and at large distances and, for high bulk polymer concentrations, it becomes slightly repulsive.

The concentration of the segments is plotted in Fig. 7 for D=20 Å. As expected, the polymer segment concentration at the middle is larger for higher bulk concentrations. A higher segment concentration can lead to an increased attraction between plates because it increases the number of bridges.

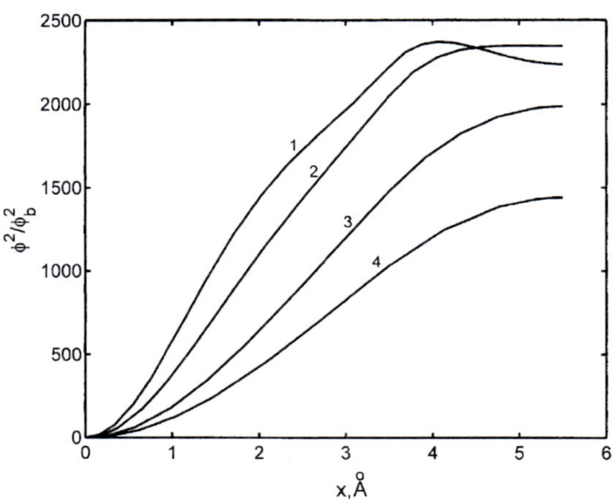

Fig. 9. The segment concentration distribution between plates at a small distance for various stiffnesses of the polyelectrolyte chains. D=11 Å, and the other parameter values used in the calculation are as for Fig. 8. (1) a=4 Å; (2) a=5 Å; (3) a=8 Å; (4) a=10 Å.

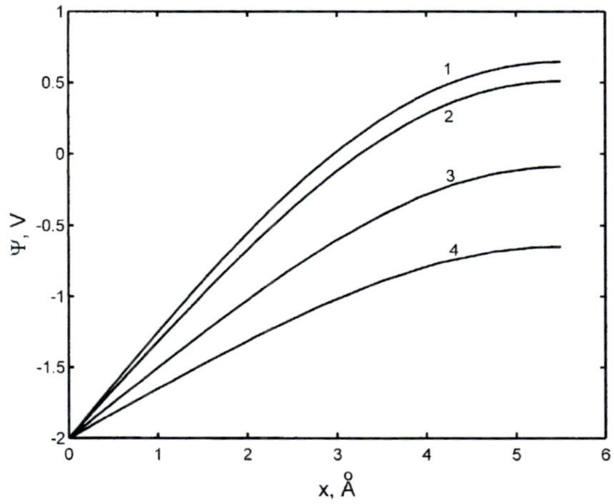

Fig. 10. The electrical potential distribution between plates at small distances for various stiffnesses of the polyelectrolyte chain. D=11 Å, and the other parameter values used in calculation are as for Fig. 8. (1) a=4 Å; (2) a=5 Å; (3) a=8 Å; (4) a=10 Å.

However, it also leads to a higher concentration of counterions dissociated from the polyelectrolyte that generate repulsion [the fourth term on the right-hand side of Eq. (17)]. This higher concentration can generate a higher repulsion between plates for the same electrical potential at the middle.

4.4. The effect of the stiffness of the chains

The stiffness of the chains, represented by the term $\frac{a^2}{6}\left(\frac{d\varphi}{dx}\right)^2$ in Eq. (2), together with the electrical and van der Waals interactions determine the conformation of the polyelectrolyte chains and hence their charge and segment concentration distributions in solution. The chains become more extended with increasing stiffness. As a result, at short distances, the force between plates is repulsive for

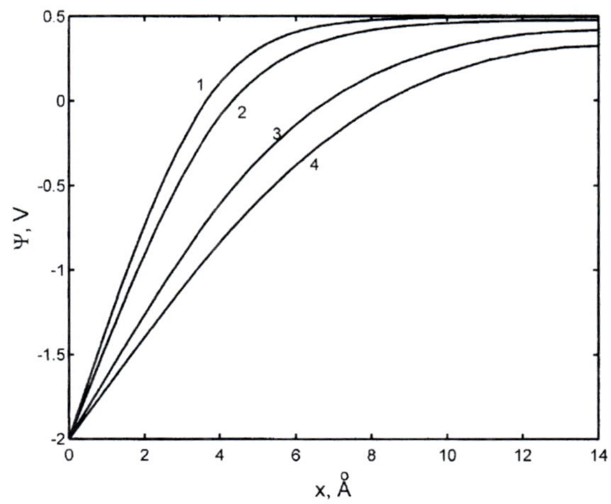

Fig. 11. The electrical potential distribution between plates at large distances for various stiffnesses of the polyelectrolyte chain. D=28 Å, and the other parameter values used in calculation are as for Fig. 8. (1) a=4 Å; (2) a=5 Å; (3) a=8 Å; (4) a=10 Å.

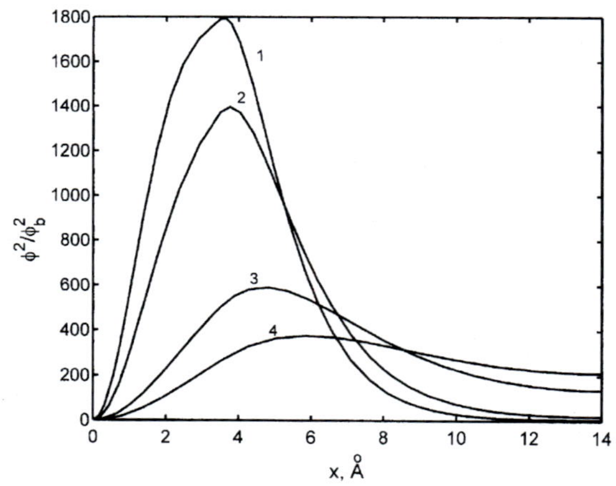

Fig. 12. The segment concentration distribution between plates at a large distance for various stiffnesses of the polyelectrolyte chains. D=28 Å, and the other parameter values used in the calculation are as for Fig. 8. (1) a=4 Å; (2) a=5 Å; (3) a=8 Å; (4) a=10 Å.

a=8 and 10 Å and decreases with decreasing stiffness (Fig. 8). For flexible chains, a larger number of chains are present between plates at small distances between them (Fig. 9) because they can be more easily located there than the stiffer chains. The increase of the electrical potential with the distance from the plates is larger for the flexible than stiffer chains (Fig. 10). For a=8 or 10 Å, the electrical potential at the middle is negative and leads to a positive contribution of the first term of Eq. (17) to the force. However, the positive electrical potential at the middle for flexible chains provides a negative contribution of the first term of Eq. (17) to the force. Consequently, the force is more repulsive for the stiffer than flexible chains at small distances between plates.

The electrical potential at the middle between plates increases with increasing distance between them. For stiffer chains, the electrical potential is negative at short distances, providing a repulsive contribution to the force, but becomes positive at large ones (Fig. 11), generating an attractive contribution. The flexible chains are more easily attracted by the plates and generate a concentration peak near each of the plates. The segment concentration at the middle between plates is low. However, at large distances between plates, the stiff chains provide a higher segment concentration at the middle than the flexible ones (Fig. 12). Consequently, the force becomes more attractive for the stiffer than flexible chains when the distances between plates become sufficiently large.

4.5. The effect of the charge density of the chains

Fig. 13 presents the effect of the charge density per segment, which is assumed uniform along the chains, at constant surface potential of the plates, on the force. The surface of the plates is considered negatively charged and the chains are considered to be positively charged. As the charge density of the chains increases, the location of the minimum of the attractive force is displaced to a smaller distance. At distances larger than 16 Å, the smaller the charge density of the segments α, the stronger the attraction between plates. As the charge density of the chains increases, the peaks of the segment concentrations are displaced towards the plate surface because of the stronger electrostatic attraction between segments and plates. At the middle between plates, the segment concentration is lower for higher charge densities of the segments because of the positive electrical potential. A lower segment concentration at the middle for high charge densities of the segments leads to a small negative contribution of the first term of Eq. (17), which is dominant at large distances. Therefore, the force at large distances is more attractive for lower charge densities of the segments.

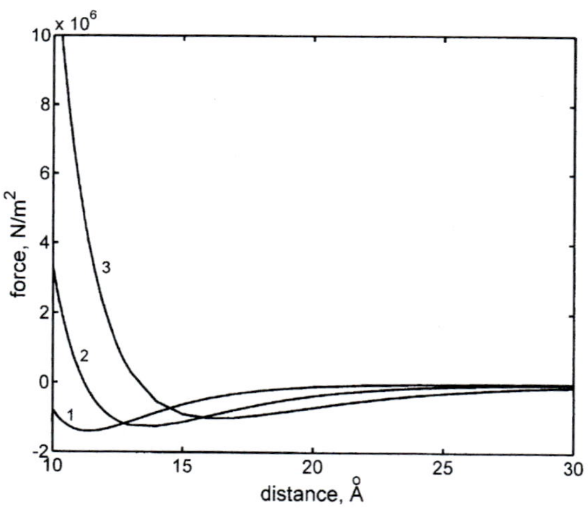

Fig. 13. The force between plates for various charge densities of the polyelectrolyte segments. The parameter values used in calculations are: a=5 Å, φ_b^2=1×10^{-6} Å^{-3}, $e\Psi_0/kT$=−2, v=50 Å^3, A_H=0.5×10^{-20} J, c=0.001 M, a' =5 Å. (1) α=2; (2) α=1; (3) α=0.5.

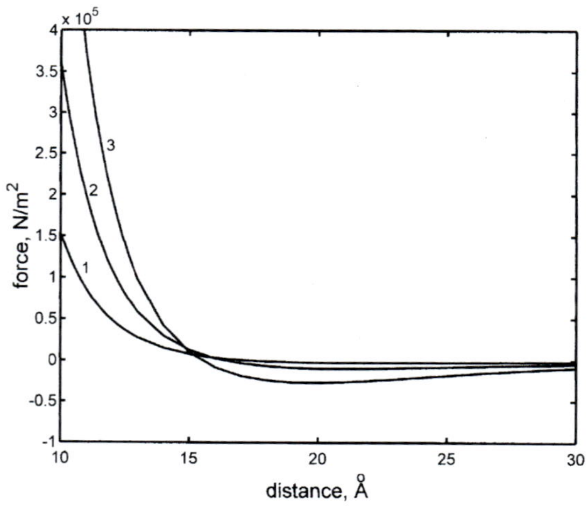

Fig. 14. The force between plates for various surface charge densities of the plates. The values of the parameters used in calculations are: α=1, a=5 Å, φ_b^2=1×10^{-6} Å^{-3}, $e\Psi_0/kT$=−2, v=50 Å^3, A_H=0.5×10^{-20} J, c=0.001 M, a' =5 Å. (1) σ=−0.002 C/m^2; (2) σ=−0.005 C/m^2; (3) σ=−0.01 C/m^2.

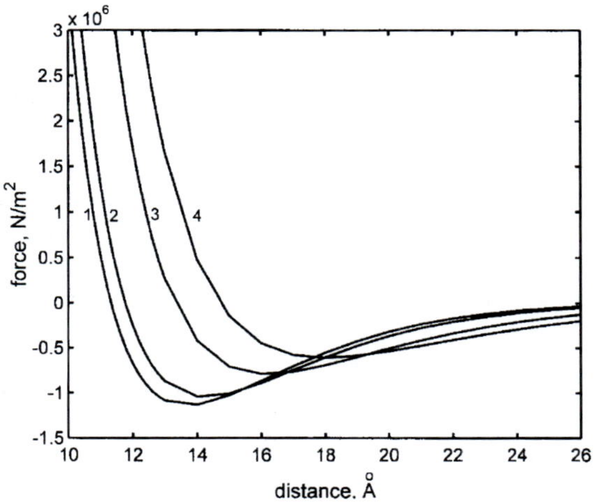

Fig. 15. The force between plates immersed in the solution of a polyelectrolyte which can generate charges through dissociation equilibrium. The values of the parameters used in calculations are: $\tau=1$, $a=5$ Å, $\varphi_b^2=1\times10^{-6}$ Å^{-3}, $e\Psi_0/kT=2$, $v=50$ Å^3, $A_H=0.5\times10^{-20}$ J, $c=0.1$ M, $a'=5$ Å. (1) $K=1\times10^{-2}$ Å^{-3}; (2) $K=5\times10^{-4}$ Å^{-3}; (3) $K=1\times10^{-4}$ Å^{-3}; (4) $K=5\times10^{-5}$ Å^{-3}.

4.6. The effect of the plate surface charge density

In Fig. 14, the force between plates is plotted against the distance D for various constant plate surface charge densities. The force is repulsive at short distances but becomes attractive at large ones. This behavior occurs because the first term on the right-hand side of Eq. (17), which is dominant, is positive at small distances because the electrical potential at the middle between plates is negative. At large distances, the electrical potential at the middle becomes positive, and the force becomes attractive because of the negative contribution of the first term on the right-hand side of Eq. (17), which is dominant.

4.7. Polyelectrolyte chains generate their charge through dissociation equilibrium

If the polyelectrolyte chains generate their charge through the dissociation equilibrium,

$$\mathrm{SNa} \leftrightarrow \mathrm{S}^- + \mathrm{Na}^+ \tag{24}$$

where SNa is an ionizable group of the polyelectrolyte chain and S^- is a dissociated group.

The fraction θ of the dissociated polyelectrolyte ionizable groups is given by

$$K \equiv \frac{\theta[\mathrm{Na}^+]}{1-\theta} \tag{25}$$

where K is the dissociation equilibrium constant and $[Na^+]$ is the sodium concentration near the surface of the chains.

Consequently,

$$\alpha = -\tau\theta = -\tau K/\left[K + c_b^+ \exp(-e\psi/kT)\right] \tag{26}$$

where τ is the number of ionizable groups of each segment and c_b^+ is the bulk concentration of Na^+ ions. The bulk concentration of sodium ions is provided by the sum of the electrolyte concentration in the bulk and of the counterion concentration of the polyelectrolyte in the bulk.

$$c_b^+ = c + \theta\tau\varphi_b^2$$

The force can be therefore expressed as

$$\begin{aligned}\beta f = {} & \beta\tau\varphi_{D/2}^2 e\psi_{D/2} K/\{K + c_b^+ \exp[-e\psi_{D/2}/(kT)]\} \\ & - v(\varphi_{D/2}^2 - \varphi_b^2)^2/2 + U_{D/2}\varphi_{D/2}^2 \\ & + \tau\varphi_b^2 K[\exp(-e\psi_{D/2}/kT) - 1]/(K + c_b^+) \\ & + 2c[\cosh(e\psi_{D/2}/kT) - 1]\end{aligned} \tag{27}$$

Fig. 15 presents the effect of the equilibrium constant K on the force. As K increases, the force between plates can change sign from repulsive to attractive. When the dissociation constant of the polyelectrolyte ionizable groups increases, the concentration of polyelectrolyte between the two plates decreases. Therefore, the dominant term for short distances, the third term of Eq. (17), becomes larger as the dissociation constant decreases, and the force less repulsive. At large distances, the first term of Eq. (17) is dominant and the higher the dissociation constant, the less attractive the force between plates.

5. Comparison with experiment

Jinnai et al. [18] measured by neutron-scattering the interlayer separation between the butylammonium vermiculite platelets as a function of the poly(vinyl methyl ether) (PVME) volume fraction in an aqueous butylammonium chloride solution. The n-butylammonium vermiculite swelled in the n-butylammonium chloride solution, and the interlayer separation of the platelets increased from 2 to 12 nm at $8<T<20$ °C. When the PVME was added to this

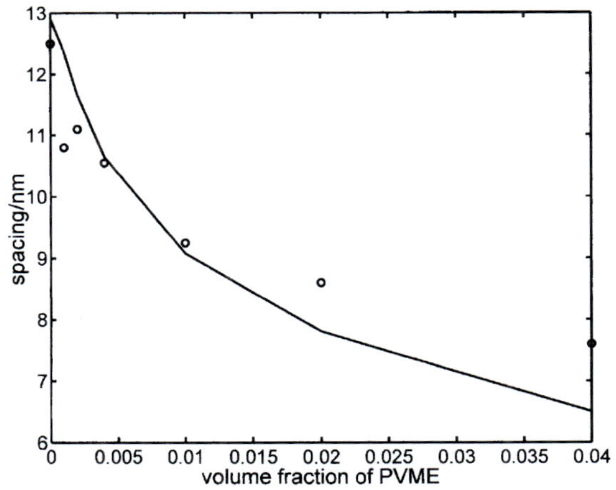

Fig. 16. The interlayer separation of n-butylammonium vermiculite against the volume fraction of PVME. The experimental data are from Fig. 4 of Ref. [18]. The values of the parameters employed in the calculation were: $T=8$ °C, the salt concentration is 0.1 M, $a=1.88$ nm, the Hamaker constant between the plates $A'_H=1\times10^{-19}$ J, the Hamaker constant between the segments and plates $A_H=1\times10^{-20}$ J, $\sigma=-0.035$ C/m^2, $v=300$ Å^3, $\varepsilon=83$.

clay–salt–water system, the interlayer separation decreased with increasing PVME volume fraction.

The latter behavior can be attributed to the bridging of the vermiculite layers by PVME. The experimental and theoretical results are compared in Fig. 16. The molecular weight of PVME is 18000, value which satisfies the condition for the chain to be sufficiently long imposed by the theory. In the calculations, the total free energy of the system is the sum of the free energy F and the van der Waals interaction energy F_{vdw} between plates, which is given by

$$F_{\mathrm{vdw}} = -\frac{A'_H}{48\pi}\left[\frac{1}{D^2} + \frac{1}{(D+\delta)^2} - \frac{2}{(D+\delta/2)^2}\right]$$

where A'_{H} is the Hamaker constant between the platelets and δ=1 nm is the thickness of the platelets.

The position of the secondary minimum of the total free energy plus the thickness of a platelet (1 nm) was considered as the equilibrium separation between layers. The surface charge density was assumed to be constant. The Kuhn segment length was evaluated using the expression

$$a = \langle R^2 \rangle / R_{\mathrm{exe}} \tag{28}$$

where $\langle R^2 \rangle$ is the mean square end-to-end distance and R_{exe} is the fully extended chain length, which is 77.5 nm for PVME with a molecular weight of 18000. For $\langle R^2 \rangle$ in solution, one can use the expression [53]

$$\langle R^2 \rangle^{1/2} = 900\mathrm{M}^{1/2} \times 10^{-4}\mathrm{nm} \tag{29}$$

which together with Eq. (28) leads to a=1.88 nm. The following values of the parameters were employed in the calculation: the Hamaker constant between platelets $A'_{\mathrm{H}} = 1\times10^{-19}$ J, the Hamaker constant between the segments and plates $A_{\mathrm{H}} = 1\times10^{-20}$ J, the charge density of the clay platelets σ=−0.035 C/m^2, the excluded volume v=300 Å^3, the temperature T=8 °C and the salt concentration c=0.1 M. For the dielectric constant, the value 83 was employed in the calculations for T=8 °C. The satisfactory agreement between experiment and calculations suggest that the contraction of the separation between the clay platelets can be the result of polymer bridging interactions.

6. Conclusions

The van der Waals interactions between the polyelectrolyte segments and plates are included into the mean-field approach together with the excluded volume and electrostatic interactions. A repulsive contribution of the force between plates is generated by the van der Waals interactions between the segments and the plates. With increasing van der Waals interactions between the segments and the plates, the force between the plates at constant surface potential becomes more repulsive at small distances between the plates and more attractive at large distances because of a large segment concentration in the gap. As the bulk concentration of the segment concentration increases, the force becomes less attractive because of the higher counterion concentration in the bulk. When the polyelectrolyte chains generate their charge through a dissociation equilibrium between their ionizable group and counterions, as the equilibrium constant increases, the concentration of the polyelectrolyte molecules between the two plates at constant potential increases and the force decreases. The experimental results of the contraction of the interlayer distance between layers of the n-butylammonium vermiculite induced by the poly(vinyl methyl ether) are in satisfactory agreement with the theoretical calculations.

References

[1] D.H. Napper, Polymeric Stabilization of Colloidal Dispersions, Academic Press, New York, 1983.
[2] P.G. de Gennes, Adv. Colloid Interface Sci. 27 (1987) 189.
[3] R.R. Netz, D. Andelman, Phys. Rep. 380 (2003) 1.
[4] R.K. Iler, J. Colloid Interface Sci. 37 (1971) 364.
[5] E. Dickinson, L. Eriksson, Adv. Colloid Interface Sci. 34 (1991) 1.
[6] P.F. Luckham, J. Klein, J. Chem. Soc., Faraday Trans. 80 (1984) 865.
[7] P.M. Claesson, B.W. Ninham, Langmuir 8 (1992) 1506.
[8] K. Lowack, C.A. Helm, Macromolecules 31 (1998) 823.
[9] R. Podgornik, J. Phys. Chem. 96 (1992) 884.
[10] R. Podgornik, J. Chem. Phys. 118 (2003) 11286.
[11] I. Borukhov, D. Andelman, H. Orland, J. Phys. Chem., B 103 (1999) 5042.
[12] T. Akesson, C. Woodward, B. Jonsson, J. Chem. Phys. 91 (1989) 2461.
[13] J. Jimenez, J. de Joannis, I. Bitsanis, R. Rajagopalan, Macromolecules 33 (2000) 7157.
[14] S. Misra, W.L. Mattice, Macromolecules 27 (1994) 2058.
[15] S. van Giessen, I. Szleifer, J. Chem. Phys. 102 (1995) 9069.
[16] T. Holzl, M. Wittkop, S. Kreitmeier, D. Goritz, J. Chem. Phys. 106 (1997) 7792.
[17] D.H. Everett, Basic Principles of Colloid Science, Royal Society of Chemistry, London, 1988.
[18] H. Jinnai, M.V. Smalley, T. Hashimoto, S. Koizum, Langmuir 12 (1996) 1199.
[19] M.V. Smalley, Langmuir 10 (1994) 2884.
[20] J. Swenson, M.V. Smalley, R.K. Thomas, R.J. Crawford, L.F. Braganza, Langmuir 13 (1997) 6654.
[21] M.V. Smalley, H.L.M. Hatharasinghe, I. Osborne, J. Swenson, S.M. King, Langmuir 17 (2001) 3800.
[22] J. Swenson, M.V. Smalley, H.L.M. Hatharasinghe, Phys. Rev. Lett. 81 (1998) 5840.
[23] J. Swenson, M.V. Smalley, H.L.M. Hatharasinghe, J. Chem. Phys. 110 (1999) 9750.
[24] H. Ohshima, Colloid Polym. Sci. 277 (1999) 535.
[25] D.A. Hoagland, Macromolecules 23 (1990) 2781.
[26] D. Dean, J. Seog, C. Ortiz, A.J. Grodzinsky, Langmuir 19 (2003) 5526.
[27] H.H. Huang, E. Ruckenstein, J. Colloid Interface Sci. 275 (2004) 548.
[28] S.F. Edwards, Proc. Phys. Soc. 85 (1965) 613.
[29] F.W. Wiegel, J. Phys., A 10 (1977) 297.
[30] P.G. de Gennes, Scaling Concepts in Polymer Physics, Cornell Univ. Press, Ithaca, 1979.
[31] P.G. de Gennes, Macromolecules 15 (1982) 492.
[32] J.M.H.M. Scheutjens, G.J. Fleer, J. Phys. Chem. 83 (1979) 1619.
[33] J.M.H.M. Scheutjens, G.J. Fleer, Macromolecules 18 (1985) 1882.

[34] M. Muthukumar, J. Chem. Phys. 86 (1987) 7230.
[35] X. Chatellier, J.-F Joanny, J. Phys. II Fr. 6 (1996) 1669.
[36] J.-F. Joanny, Eur. Phys. J., B 9 (1999) 117.
[37] R. Varoqui, J. Phys. II Fr. 3 (1993) 1097.
[38] E.M. Blokhuis, K.I. Skau, J.B. Avalos, J. Chem. Phys. 119 (2003) 3483.
[39] I. Borukhov, D. Andelman, H. Orland, Macromolecules 31 (1998) 1665.
[40] I. Borukhov, D. Andelman, H. Orland, Euro. Phys. J., B 5 (1998) 869.
[41] A. Shafir, D. Andelman, R.R. Netz, J. Chem. Phys. 119 (2003) 2355.
[42] K.I. Skau, E.M. Blokhuis, J. van Male, Macromolecules 37 (2004) 1969.
[43] B.Q. Li, E. Ruckenstein, J. Chem. Phys. 106 (1997) 280.
[44] E. Ruckenstein, B.Q. Li, J. Chem. Phys. 107 (1997) 932.
[45] M.K. Granfeldt, B. Jönsson, C.E. Woodward, J. Phys. Chem. 95 (1991) 4819.
[46] R. Varoqui, A. Johner, A. Elaissari, J. Chem. Phys. 94 (1991) 6873.
[47] M. Manciu, E. Ruckenstein, Langmuir 19 (2003) 1114.
[48] E.A. Di Marzio, J. Chem. Phys. 42 (1965) 2101.
[49] R.J. Allen, P.B. Warren, Langmuir 20 (2004) 1997.
[50] S.J. Miklavic, J. Phys. Chem. 97 (1993) 3924.
[51] R. Podgornik, J. Phys. Chem. 97 (1993) 3927.
[52] J.N. Israelachvili, Intermolecular and Surface Forces, Academic Press, New York, 1985.
[53] J. Brandrup, E.H. Immergut, Polymer Handbook, Wiley, New York, 1989.

3174 *Langmuir* **2006,** *22,* 3174−3179

Steric and Bridging Interactions between Two Plates Induced by Grafted Polyelectrolytes

Haohao Huang and Eli Ruckenstein*

Department of Chemical and Biological Engineering, State University of New York at Buffalo, Buffalo, New York 14260

Received October 17, 2005. In Final Form: January 3, 2006

If colloidal particles are grafted with a polymer, then the grafted chains can provide steric repulsion between them. If some of the grafted polymer chains are also adsorbed to a second particle, then a bridging force is generated as well. For uncharged plates and polymer, the following contributions to the free energy of the system have been taken into account in the calculation of the interaction force: (i) the Flory−Huggins expression for the mixing free energy of the grafted chains with the liquid; (ii) the entropy loss due to the connectivity of the polymeric segments; (iii) the van der Waals interactions between the segments and the plates; and (iv) the free energy of adsorption of the polymer segments of the grafted chains on the other plate. For charged plates, the electrostatic free energy as well as the free energy of the electrolyte are included in the total free energy of the system. By minimizing the free energy with respect to the segment concentration and, when it is the case, with respect to the electrical potential, equations for the segment number density distribution and for the electrical potential are obtained, on the basis of which the interactions between two plates grafted with polymer chains that can be also adsorbed on the other plate were calculated. The interaction thus obtained includes steric and bridging forces.

1. Introduction

When two plates with polymer chains grafted to their surface approach each other, a steric repulsion is generated between the two brushes as soon as the tips of the grafted chains begin to contact each other.[1,2] The repulsive steric interactions can increase the stability of a colloidal system. When the distance between the plates becomes less than twice the thickness of the polymer brush, the polymer chains on the two plates begin to interdigitate. The increase of the segment concentration in the interdigitation domain leads to repulsion. When opposite polymer brushes contact each other, the brushes become compressed in addition to their interdigitation. Therefore, the steric repulsion includes entropic and elastic contributions. However, if some of the grafted chains are also adsorbed on the second plate, then the polymers draw the plates toward each other, resulting in an attractive bridging force.[3,4] The bridging force may induce the coagulation of the colloidal particles. The bridging was extensively examined experimentally,[5−7] theoretically,[8−10] and by Monte Carlo simulations.[11−14] Of course, the steric and bridging forces are simultaneously present between the plates, and their effects are not expected to be additive.

Many theories were developed to describe the distribution of segments between two polymer brushes. On the basis of a lattice model, Scheutjens et al. developed a discrete self-consistent theory to describe the steric repulsion.[15−18] A similar model involving interdigitation was employed by Li and Ruckenstein.[19,20] The assumption of free jointed chains used by Scheutjens et al. was replaced by the latter authors with the assumption of a correlation between neighboring bonds.

Another kind of self-consistent theory used a continuum diffusion representation to describe the distribution of segments.[21−23] The segments were assumed to be subjected to an external potential of a mean field. An analytical approximation of the latter self-consistent theory was suggested by Milner et al. (the MWC model)[24] on the basis of the observation that at high stretching the partition function of the brush is dominated by the classical path as the most probable distribution. Under this assumption, it was found that the self-consistent field is parabolic and leads to a parabolic distribution of the monomer density. Similar theories for polyelectrolyte brushes[25−27] also adopted the parabolic distribution approximation.

The polyelectrolyte molecules affect in a more complex manner than the neutral ones the steric and bridging interactions between colloidal particles because they affect the electrical potential of the electrical double layer,[28,29] which in turn affects the conformation of the chains and the overall electrostatic interactions.

* To whom correspondence should be addressed. E-mail: feaeliru@acsu.buffalo.edu. Tel: (716) 645-2911 ext. 2214. Fax: (716) 645-3822.

(1) de Gennes, P. G. *Adv. Colloid Interface Sci.* **1987**, *27*, 189.
(2) Roan, J. R; Kawakatsu, T. *J. Chem. Phys.* **2002**, *116*, 7283.
(3) Iler, R. K. *J. Colloid Interface Sci.* **1971**, *37*, 364.
(4) Dickinson, E.; Eriksson, L. *Adv. Colloid Interface Sci.* **1991**, *34*, 1.
(5) Luckham, P. F.; Klein, J. *J. Chem. Soc., Faraday Trans.* **1984**, *80*, 865.
(6) Claesson, P. M.; Ninham, B. W. *Langmuir* **1992**, *8*, 1506.
(7) Lowack, K.; Helm, C. A. *Macromolecules* **1998**, *31*, 823.
(8) Varoqui, R. *J. Phys. II* **1993**, *3*, 1097.
(9) Podgornik, R. *J. Phys. Chem.* **1992**, *96*, 884.
(10) Borukhov, I.; Andelman, D.; Orland, H. *J. Phys. Chem. B* **1999**, *103*, 5042.
(11) Akesson, T.; Woodward, C.; Jonsson, B. *J. Chem. Phys.* **1989**, *91*, 2461.
(12) Jimenez, J.; de Joannis, J.; Bitsanis, I.; Rajagopalan, R. *Macromolecules* **2000**, *33*, 7157.
(13) Misra, S.; Mattice, W. L. *Macromolecules* **1994**, *27*, 2058.
(14) van Giessen, S.; Szleifer, I. *J. Chem. Phys.* **1995**, *102*, 9069.
(15) Scheutjens, J. M. H. M.; Fleer, G. J. *J. Phys. Chem.* **1979**, *83*, 1619.
(16) Leermakers, F. A. M.; Sheutjens, J. M. H. M. *J. Chem. Phys.* **1988**, *89*, 3264.
(17) Leermakers, F. A. M.; Sheutjens, J. M. H. M. *J. Chem. Phys.* **1988**, *89*, 6912.
(18) Wijmans, C. M.; Leermakers, F. A. M.; Fleer, G. J. *Langmuir* **1994**, *10*, 4514.
(19) Li, B. Q.; Ruckenstein, E. *J. Chem. Phys.* **1997**, *106*, 280.
(20) Ruckenstein, E.; Li, B. Q. *J. Chem. Phys.* **1997**, *107*, 932.
(21) Edwards, S. F. *Proc. Phys. Soc.* **1965**, *85*, 613.
(22) Borukhov, I.; Andelman, D.; Orland, H. *J. Phys. Chem. B* **1999**, *103*, 5042.
(23) Huang, H. H.; Ruckenstein, E. *Adv. Colloid Interface Sci.* **2004**, *112*, 37.
(24) Milner, S. T.; Witten, T. A.; Cates, M. E. *Macromolecules* **1988**, *21*, 2610.
(25) Miklavic, S. J.; Marcelja, S. *J. Phys. Chem.* **1988**, *92*, 6718.
(26) Misra, S.; Varanasi, S.; Varanasi, P. P. *Macromolecules* **1989**, *22*, 4173.
(27) Zhulina, E. B.; Borisov, O. V. *J. Chem. Phys.* **1997**, *107*, 5952.

10.1021/la0527947 CCC: $33.50
Published on Web 03/01/2006

A statistical model by Manciu and Ruckenstein,[30,31] which assumed that the chain distributions are generated by all possible random walks and employed an average free energy for all of the chain distributions that end at a certain distance x from the interface, showed that by accounting for the most likely chain distribution and discarding all other chain distributions one arrives at the same scaling law as does the MWC model. However, when all possible distributions (with their appropriate Boltzmann weights) have been taken into account, a departure from the MWC model results, most notably, at high grafting density, when the brush becomes more step-like than parabolic-like.[30]

With the assumption of a step function for the distribution of the segment density and using the ground-state dominance approximation, Alexander[32] and de Gennes[1] suggested a scaling analysis to calculate the interaction between two brushes.

An equation is derived in the present article for the free energy of the system by using various contributions to the free energy. For uncharged plates and polymer, the following contributions were considered: (i) the Flory–Huggins expression for the mixing free energy of the grafted chains with the liquid; (ii) the entropy loss due to the connectivity of the segments of the chains; (iii) the van der Waals interaction between segments and plates; and (iv) the free energy of adsorption of the segments of the grafted polymers on the other plate. For charged surfaces and polymers, the electrostatic free energy and the free energy of the electrolyte were also added to obtain the total free energy of the system. The minimum free energy with respect to the segment density and, when it was the case, with respect to the electrical potential provides two differential equations for the segment density and electrical potential. The minimum of the free energy with respect to the segment density of the polymer near the surface provided the concentration of the segments that are subject to adsorption.

Because both the steric and bridging interactions occur simultaneously, these two interactions are coupled and should not be calculated separately. Some polymer chains are grafted on only one plate and contribute to the steric repulsion; the chains that are also adsorbed on the interacting plate contribute to bridging.

The goal of the present article is to calculate the interaction between two plates grafted with polymer chains, the segments of which can be also adsorbed on the interacting second plate.

2. Theoretical Framework

Two plates grafted with polymer chains on their surface are immersed in an electrolyte solution. The two plates are considered to be located at $x = 0$ and D, respectively, and the system is considered to be in contact with a large reservoir containing an electrolyte solution free of polymer. Compared to the concentration of the ions of the electrolyte, the concentration of the counterions dissociated from the grafted polyelectrolyte is neglected.

The free energy F of the system contains a polymer (F_p), a double layer (F_d), and an adsorption (F_a) contribution.

$$F = F_p + F_d + F_a \tag{1}$$

The polymer contribution contains the Flory–Huggins mixing free energy of the segments with the solvent molecules, the van der Waals interaction $-UkT$ between segments and the two plates, and the entropy loss caused by the connectivity of the segments[8,23]

$$\beta F_p = \int_0^D \left[-\frac{1}{2l^3}\tau v^2 n^2 + \frac{1}{6l^3} w v^3 n^3 - Un + \frac{l^2}{6}\left(\frac{d\varphi}{dx}\right)^2 \right] dx \tag{2}$$

In eq 2, n and v are the local number density and the volume of a polymer segment, τ is the dimensionless excluded volume parameter, w is the second virial coefficient, $\varphi = \sqrt{n}$, l is the Kuhn length of a segment, $\beta = 1/kT$, k is the Boltzmann constant, and T is the absolute temperature.

For the van der Waals interactions between plates and segments in the solution, one can write the expression

$$U = \frac{A_H l'^3}{3kT}\left[\frac{1}{x^3} + \frac{1}{(D-x)^3}\right] \tag{3}$$

where A_H is the Hamaker constant and l' is the diameter of a segment that is assumed to be spherical. Close to the surface, there is a Born repulsion that we consider that generates an infinite repulsion over a distance of 3.5 Å.

The double-layer contribution F_d contains the entropic F_{ent} of the ions and the electrostatic contributions F_{ele}. The entropic contribution of the electrolyte to the free energy of the system is due to the change in the entropy of ions between the plates compared to that in the reservoir.

$$\beta F_{ent} = \int_0^D \left[\left(\sum_i c_i \ln \frac{x_i}{x_{i,b}} \right) + c_w \ln \frac{1 - \sum_i x_i}{1 - \sum_i x_{i,b}} \right] dx \approx \int_0^D \sum_i \left(c_i \ln \frac{c_i}{c_{i,b}} - c_i + c_{i,b} \right) dx \tag{4}$$

where c_i is the concentration of ion i, c_w is the concentration of water, x_i is the mole fraction of ion i, and the subscript b indicates the concentration in the bulk.

The electrostatic free-energy contribution is the sum of the electrostatic energy of the system and the chemical contribution of the surface. For constant surface potential ψ_0, the chemical contribution is given by

$$F_{chem} = -2\sigma\psi_0 \tag{5}$$

where σ is the surface charge density.

Therefore, the electrostatic free-energy contribution can be expressed as

$$F_{ele} = \frac{1}{2}\epsilon\epsilon_0 \int_0^D \left(\frac{d\psi}{dx}\right)^2 dx - 2\sigma\psi_0 = \frac{1}{2}\epsilon\epsilon_0 \int_0^D \left(\frac{d\psi}{dx}\right)^2 dx + 2\epsilon\epsilon_0\psi_0 \frac{d\psi}{dx}\Big|_{x=0} \tag{6a}$$

Using Green's theorem, eq 6a can be rewritten as

$$F_{ele} = \frac{1}{2}\epsilon\epsilon_0 \int_0^D \left(\frac{d\psi}{dx}\right)^2 dx - \epsilon\epsilon_0 \int_0^D \left[\psi \frac{d^2\psi}{dx^2} + \left(\frac{d\psi}{dx}\right)^2 \right] dx = -\epsilon\epsilon_0 \int_0^D \left[\psi \frac{d^2\psi}{dx^2} + \frac{1}{2}\left(\frac{d\psi}{dx}\right)^2 \right] dx \tag{6b}$$

where ϵ is the dielectric constant, ϵ_0 is the vacuum permittivity, and ψ is the electrical potential.

The optimization of the total free energy of the system with respect to the electrical potential provides the following relation

(28) Ohshima, H. *Colloid Polym. Sci.* **1999**, *277*, 535.
(29) Hoagland, D. A. *Macromolecules* **1990**, *23*, 2781.
(30) Manciu, M.; Ruckenstein, E. *Langmuir* **2004**, *20*, 8155.
(31) Manciu, M.; Ruckenstein, E. *Langmuir* **2004**, *20*, 6490.
(32) Alexander, S. *J. Phys.* **1977**, *38*, 983.

between the electrical potential and the charge density (Poisson equation)

$$\frac{d^2\psi}{dx^2} = -\frac{\rho}{\epsilon\epsilon_0} \tag{7}$$

where the charge density in the solution ρ includes the charges from the electrolyte ions and the polyelectrolyte

$$\rho = e(c^+ - c^- + \alpha\varphi^2) \tag{8}$$

c^+ and c^- are the concentrations of the cations and anions, respectively, and α is the charge of each segment.

Substituting eq 7 into eq 6b, one obtains

$$F_{ele} = \int_0^D \left[e\psi(c^+ - c^- + \alpha\varphi^2) - \frac{1}{2}\left(\frac{d\psi}{dx}\right)^2\right] dx \tag{9}$$

The adsorption contribution to the free energy is a result of the replacement of solvent molecules on the surface by adsorbed polymer segments. The adsorption contribution is given by

$$F_{ads} = \frac{2\Gamma\Delta G_{ads}}{N_0} \tag{10}$$

where Γ is the number of segments adsorbed per unit area, ΔG_{ads} is the molar adsorption free energy of the polymer, and N_0 is Avogadro's number.

Typical values for ΔG_{ads} are -13 kJ/mol for poly(acrylic acid) on alumina and -10 to -16 kJ/mol for polyacrylamide on talc.[33] Here the adsorption free energy of a polymer molecule is assumed to be the same as the adsorption free energy of a segment.

The number of segments adsorbed per unit area can be calculated using the Langmuir adsorption isotherm

$$\Gamma = \frac{(s_0 - s)K_{ads}n_s}{1 + K_{ads}n_s} \tag{11}$$

where s_0 is the total number of adsorption sites on the surface per unit area, s is the grafting density of the polymer chains. n_s is the segment concentration near the plates and K_{ads} is the adsorption constant ($RT \ln K_{ads} = -\Delta G_{ads}$, R being the gas constant and T the temperature in K taken 300 K).

All of the segments between the two plates belong to the grafted polymer chains, and the plates are in contact with an electrolyte solution free of polymer. Consequently, the total number of segments in the solution between the two plates is constant. If the grafting density of the polymer chains is denoted by s, then the total number of segments in the solution is $2sN$, where N is the number of segments per chain. This constraint must be taken into account in the minimization of the free energy with respect to the segment density. Because the number of segments in solution is much larger than the number of segments adsorbed, the number of the latter segments can be neglected. Hence, one can write

$$2sN \approx \int_0^D \varphi^2 \, dx \tag{12}$$

The minimization of the free energy with respect to φ under the constraint (eq 12) and considering that $v = l^3$ leads to

$$\frac{l^2}{6}\frac{d^2\varphi}{dx^2} = -\tau l^3\varphi^3 + \frac{1}{2}wl^6\varphi^5 - U\varphi + \alpha\beta e\psi\varphi + \lambda\varphi \tag{13}$$

where λ is a Lagrange multiplier, which is provided by eq 12.

The minimization of the free energy with respect to ψ leads to the equation

$$\frac{d^2\psi}{dx^2} = \frac{1}{\epsilon\epsilon_0}[2ce \sinh(\beta e\psi) - \alpha e\varphi^2] \tag{14}$$

Equation 13 provides the segment density of the polymer chains subjected to the external electrical potential, ψ, and the van der Waals interaction with the plates, $-UkT$. Equation 14 represents a modified Poisson–Boltzmann equation in which the first term on the right-hand side accounts for the charges of the small ions of the salt and the second term accounts for the charges of the polyelectrolyte chains. When the polymer is not charged, eqs 13 and 14 can be solved independently.

Because of symmetry, there are two boundary conditions at the middle distance between them

$$\frac{d\psi}{dx}(x = D/2) = 0 \tag{15}$$

$$\frac{d\varphi}{dx}(x = D/2) = 0 \tag{16}$$

At the surface of the charged plates, the boundary condition is

$$\psi = \psi_0 \tag{17}$$

for a constant potential of the plates.

The segment concentrations at $z = 0$ and D, which are the same, are selected by trial and error such as to minimize the free energy. The Lagrange multiplier is determined also by trial and error using eq 12.

The profiles of the segment concentration and the electrical potential for the minimum free energy allow one to calculate the interaction force f between the two plates, which is given by the expression

$$f = -\frac{\delta F}{\delta D} = kT\left(\frac{1}{2}l^3\tau\varphi_{D/2}^4 - \frac{1}{6}wl^6\varphi_{D/2}^6\right) + kTU_{D/2}\varphi_{D/2}^2 + 2c\left\{\cosh\left[\frac{e\psi_{D/2}}{kT}\right]\right\} + 2s_0K_{ads}RT \ln K_{ads}\frac{\partial(\varphi_D^2/(1 + K_{ads}\varphi_D^2))}{\partial D} \tag{18}$$

3. Calculation of the Steric Interaction

If the grafted polymer is not adsorbed on the second plate, then the grafted polymer provides only steric interaction between the plates. This repulsive steric interaction increases the stability of the colloidal system. To calculate the steric interaction between two plates, the adsorption constant K_{ads} in eq 11 should be taken to be zero in this case.

Figure 1 presents the steric interaction between two plates for various graft densities. As the grafting density increases, the segment density at the middle between plates increases. Therefore, the steric interaction between the two brushes increases.

As shown in Figure 2, the steric repulsion increases as the polymer chain becomes longer. As the length of the polymer

(33) Farrokhpay, S.; Morris, G. E.; Fornasiero, D.; Self, P. *J. Colloid Interface Sci.* **2004**, *274*, 33.

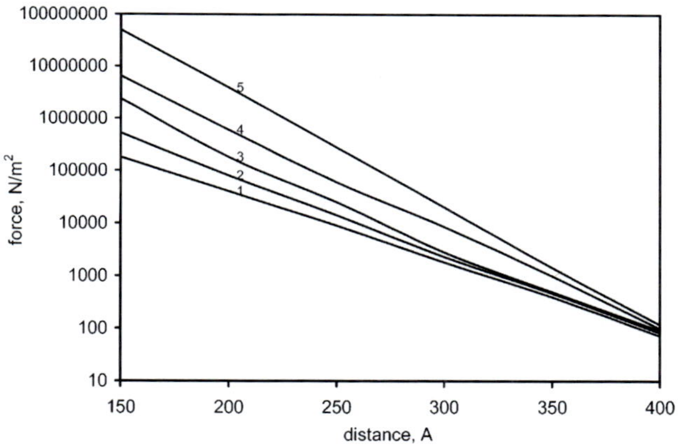

Figure 1. Steric interaction between two plates for various grafting densities. $\tau = 0.3$, $w = 1$, $N = 200$, $l = 5$ Å, $\psi_0 = 0.01$ V, $\alpha = 1e$, $c = 0.01$ M, and $s_0 = 0.8$ nm^{-2}. s/s_0 is equal to the following values: (1) 0.01; (2) 0.02; (3) 0.03; (4) 0.04; and (5) 0.05.

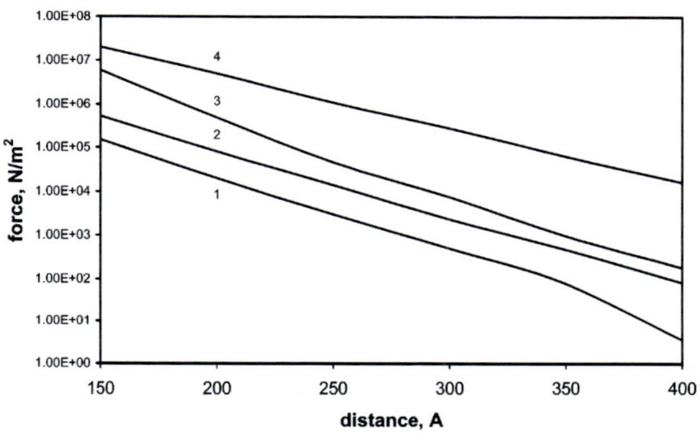

Figure 2. Steric interaction between two plates for various chain lengths. $\tau = 0.3$, $w = 1$, $l = 5$ Å, $\psi_0 = 0.01$ V, $\alpha = 1e$, $c = 0.01$ M, and $s_0 = 0.8$ nm^{-2}. The ratio of the number of grafting sites to the number of adsorption sites is 0.02. N is equal to the following values: (1) 150; (2) 200; (3) 300; and (4) 500.

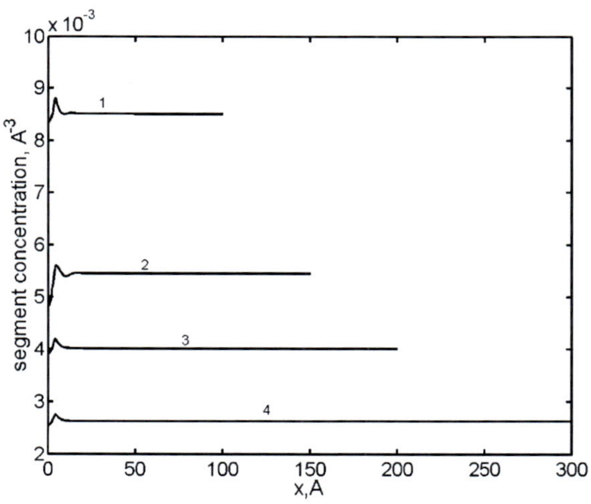

Figure 3. Distribution of polymer segments between two plates at various distances. $\tau = 0.2$, $w = 1$, $l = 5$ Å, $\psi_0 = 0.005$ V, $\alpha = 2e$, $N = 200$, $s_0 = 4$ nm^{-2}, and $c = 0.01$ M. The ratio of the number of grafting sites to the number of adsorption sites is 0.1. Only half the distance is plotted. (1) 200; (2) 300; (3) 400; and (4) 600 Å.

increases, the number of segments between the two plates increases, and this results in an increase in the free energy and in the repulsive force.

Figure 3 presents the distribution of polymer segments between the two plates for various distances. The distribution of the

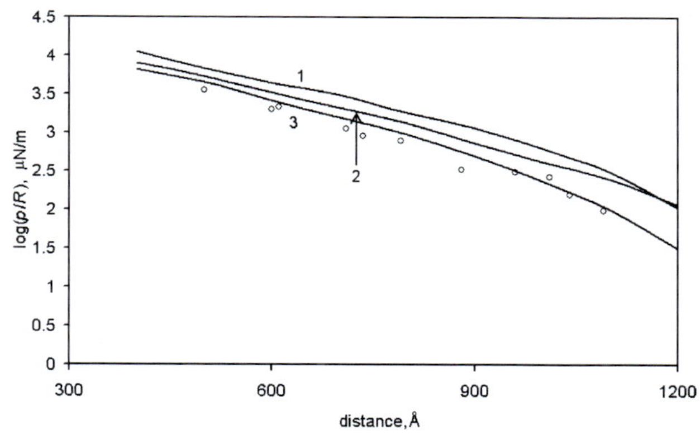

Figure 4. Steric interactions between two cross cylinders for $\tau = 0.3$, $w = 1$, $l = 7.77$ Å, $N = 430$, $s_0 = 1.66$ nm^{-2}, $\alpha = 0$, $\psi_0 = 0$, and $s/s_0 = 0.00779$. (1) de Gennes's model; (2) the present model; (3) Ruckenstein and Li's model. The experimental data (o) are from ref 34.

concentration of the segments against the distance to the surface of the plate can be well approximated by a step function. As the distance between the two plates increases, the concentration of the polymer segments decreases. Our model confirms one of the basic assumption of the Alexander–de Gennes theory, namely, that the segment distribution can be approximated by a step function.

For comparison purposes, the expressions for the interaction between two uncharged plates grafted with uncharged chains suggested by de Gennes[1] and Ruckenstein and Li[20] are given below and compared with results of the present calculations.

Two effects, the osmotic and the elastic contributions, were taken into account in the de Gennes analysis for the interaction between two parallel plates[1]

$$f \cong \frac{kT}{d^3}\left\{\left(\frac{2L_0}{D}\right)^{9/4} - \left(\frac{D}{2L_0}\right)^{3/4}\right\} \tag{19}$$

where d is the average separation between two graft sites and L_0 is the brush thickness.

Using a lattice model, Ruckenstein and Li derived a more rigorous expression for the steric interaction between two parallel plates.[20]

The interaction force p between two crossed cylinders of radius R, for which experimental data are available,[34] can be obtained from the interaction free energy F between two parallel plates using the Derjaguin approximation

$$\frac{p}{R} = 2\pi F \tag{20}$$

A comparison between the three expressions for the steric interaction p is made in Figure 4, which shows that for polystyrene the present model is in better agreement with the more rigorous approach of Ruckenstein and Li than the equation of de Gennes.

4. Steric Bridging Interactions Induced by a Polyelectrolyte

If the grafted polymer molecules are also adsorbed on the second plate, then the force between the two plates includes both bridging and steric interactions. In this case, the adsorption contributions to the total free energy must be included among the free-energy contributions. When the polymer is charged, eqs

(34) Taunton, H. J.; Toprakcioglu, C.; Fetters, L. J.; Klein, J. *Macromolecules* **1990**, *23*, 571.

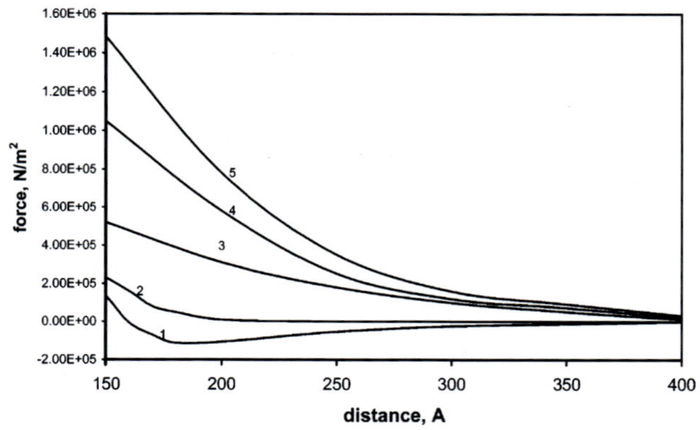

Figure 5. Interaction between two plates at various grafting densities. $\tau = 0.3$, $w = 1$, $N = 200$, $l = 5$ Å, $\psi_0 = 0.01$ V, $\alpha = 1e$, $c = 0.01$ M, $s_0 = 0.8$ nm^{-2}, and $\Delta G_{ads} = -13.25$ kJ/mol. s/s_0 is equal to the following values: (1) 0.01; (2) 0.02; (3) 0.03; (4) 0.04; and (5) 0.05.

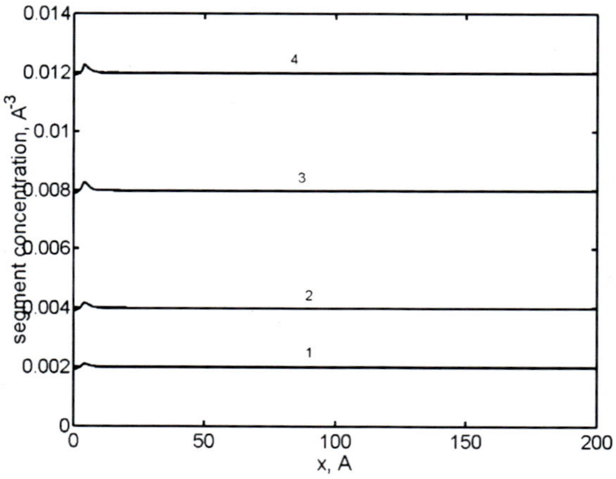

Figure 6. Distribution of polymer segments between two plates for various grafting densities. $\tau = 0.2$, $w = 1$, $l = 5$ Å, $\psi_0 = 0.005$ V, $\alpha = 2e$, $c = 0.01$ M, $N = 200$, and $\Delta G_{ads} = -13.25$ kJ/mol. s/s_0 is equal to the following values: (1) 0.05; (2) 0.1; (3) 0.2; and (4) 0.3 and $s_0 = 4$ nm^{-2}.

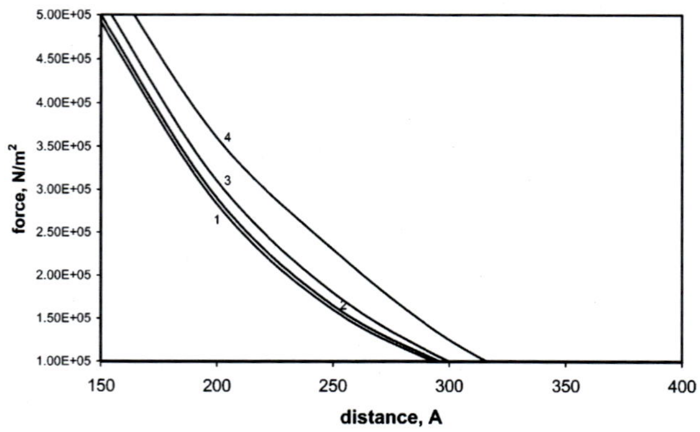

Figure 7. Interaction between two plates for various segment lengths l. $\tau = 0.3$, $w = 1$, $N = 200$, $\psi_0 = 0.01$ V, $\alpha = 1e$, $c = 0.01$ M, $s_0 = 0.8$ nm^{-2}, and $\Delta G_{ads} = -13.25$ kJ/mol. The ratio of the number of grafting sites to the number of adsorption sites is 0.02. l (nm) is equal to the following values: (1) 0.3; (2) 0.4; (3) 0.5; and (4) 0.8.

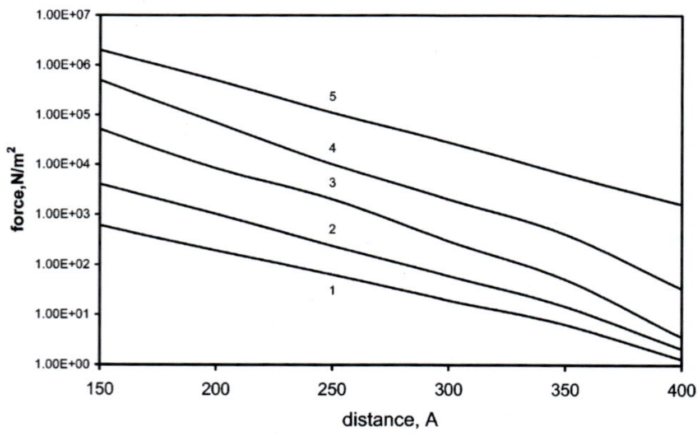

Figure 8. Interaction between two plates for various chain lengths. $\tau = 0.3$, $w = 1$, $l = 5$ Å, $\psi_0 = 0.01$ V, $\alpha = 1e$, $c = 0.01$ M, $s_0 = 0.8$ nm^{-2}, and $\Delta G_{ads} = -13.25$ kJ/mol. The ratio of the number of grafting sites to the number of adsorption sites is 0.02. N is equal to the following values: (1) 150; (2) 200; (3) 300; (4) 500; and (5) 800.

13 and 14 are not independent and have to be solved simultaneously. The segment concentration n_s at the surface was selected by minimizing the free energy of the system with respect to n_s.

4.1. Effect of the Grafting Density of the Polymer. The grafting density of the polymer affects the total number of segments and the adsorption of the segments on the plate surface. Figure 5 shows that as the grafting density of the segments increases the force becomes more repulsive. At relatively small values of the grafting density, the force decreases with increasing distance and passes through a minimum, after which it increases. Figure 5 shows that there are conditions under which the overall force becomes negative, and hence the bridging becomes dominant.

Figure 6 shows that the shape of the segment concentration distribution can be described by a step function and that the uniform concentration becomes larger for higher grafting densities. Consequently, higher bridging is generated, and the attractive bridging force increases. However, the higher segment concentration also increases the steric interaction. Because the latter dominates, the overall interaction that includes both the bridging and the steric forces becomes more repulsive with increasing grafting density, at least for the parameters employed in the present calculations.

4.2. Effect of the Stiffness of the Chains. The stiffness of the chains, represented by the term $l^2/6\,(d\varphi/dx)^2$ in eq 2, together with the electrical and van der Waals interactions determines the conformation of the polyelectrolyte chains and hence their charge and segment concentration distributions in solution. A larger value of l means a stiffer polymer chain. The chains become more extended with increasing chain stiffness, and the repulsive force between plates increases with increasing stiffness (Figure 7). For stiffer chains, the contribution of bridging is smaller than for flexible chains because the concentration of segments near the plate surface decreases with increasing stiffness.

4.3. Effect of the Length of the Polymer. As the length N of the chain increases, the repulsive force becomes stronger (Figure 8). Obviously, the steric repulsion is expected to increase with increasing N. The adsorption of the polymer segments is expected to increase as well. However, for the conditions employed, the steric repulsion increases more than the bridging attraction.

4.4. Segment Charge through Dissociation Equilibrium. If the polyelectrolyte chains generate their charge through dissociation equilibrium, then

$$SM \leftrightarrow M^- + S^+ \tag{21}$$

where SM is an ionizable group of the polyelectrolyte chain, M^- is the anion of the electrolyte and chain, and S^+ is a dissociated site.

The fraction θ of the dissociated ionizable groups is given by

$$K_d \equiv \frac{\theta[M^-]}{1-\theta} \tag{22}$$

where K_d is the dissociation equilibrium constant and $[M^-]$ is the anion concentration.

$$[M^-] = c\exp\left(\frac{e\psi}{kT}\right) \tag{23}$$

where c is the electrolyte concentration in the reservoir.

$$\alpha = \xi e\theta = \frac{\xi e K_d}{K_d + c\exp(e\psi/kT)} \tag{24}$$

where ξ is the number of ionizable groups per segment.

Figure 9 presents the effect of the equilibrium constant K_d on the force. As K_d increases, the force becomes more repulsive. This happens because as the dissociation constant of the polyelectrolyte ionizable groups increases the segments of the polyelectrolyte become more positively charged. Therefore, the segment adsorption on the surfaces that are also positively charged becomes smaller. As a result, the bridging contribution becomes smaller, and the interaction becomes more repulsive with increasing dissociation constant.

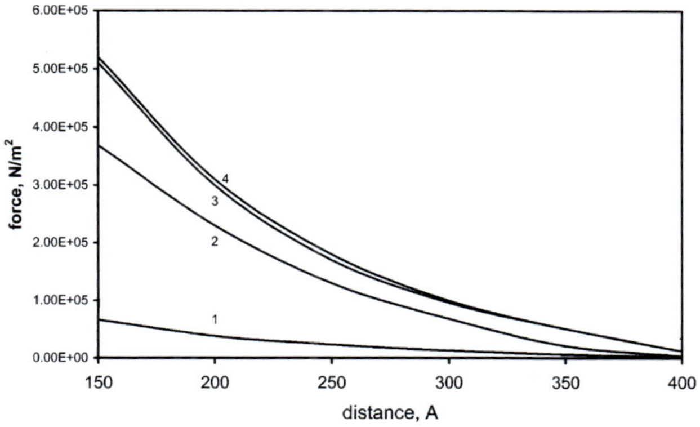

Figure 9. Interaction between two plates grafted with a polyelectrolyte for various dissociation equilibrium constants. $\tau = 0.3$, $w = 1$, $l = 5$ Å, $\psi_0 = 0.01$ V, $c = 0.01$ M, $s_0 = 0.8$ nm^{-2}, $\Delta G_{ads} = -13.25$ kJ/mol, and $\xi = 1$. The ratio of the number of grafting sites to the number of adsorption sites is 0.02. K_d is equal to the following values: (1) 1; (2) 10; (3) 100; and (4) 1000 mol/m^3.

5. Conclusions

The interaction between two plates grafted with charged polymer chains was calculated by taking into account both the bridging and steric interactions. The free energy of the system was calculated by accounting for a number of contributions: (i) the Flory–Huggins free energy for the grafted chains and liquid; (ii) the van der Waals interactions between the segments and plates; (iii) the connectivity free energy of the segments; and (iv) the adsorption free energy. For charged polymers and surfaces, an entropic contribution due to the electrolyte ions and an electrostatic contribution were included as well.

The overall interaction becomes more repulsive with increasing grafting density of the polymer on the surface and with increasing length of the polymer chains. When the plates and the segments have the same sign of charge, the interaction is more repulsive for higher charge densities of the segments because both the steric interaction and the double-layer force are stronger and the bridging weaker. The interaction between two plates is more repulsive for flexible charged chains than for stiff charged chains. There are conditions under which the bridging attraction dominates the steric repulsion. This occurs at sufficiently low grafting densities as well as sufficiently low segment charges (when the charges of the segments and the plate surfaces have the same sign).

LA0527947

Langmuir **2006,** *22,* 4541–4546 4541

Effect of Steric, Double-Layer, and Depletion Interactions on the Stability of Colloids in Systems Containing a Polymer and an Electrolyte

Haohao Huang and Eli Ruckenstein*

Department of Chemical and Biological Engineering, State University of New York at Buffalo, Buffalo, New York 14260

Received January 20, 2006. In Final Form: March 1, 2006

Experiments carried out by Stenkamp et al. [Stenkamp, V. S.; McGuiggan, P.; Berg, J. C. *Langmuir* **2001**, 17, 637.] have shown that polystyrene latexes can be restabilized at sufficiently high electrolyte concentrations in the presence of an amphiphilic block copolymer [poly(ethylene oxide)−poly(propylene oxide)−poly(ethylene oxide) (PEO−PPO−PEO)] At even higher electrolyte concentrations, the systems can again be destabilized. The present paper attempts to explain the restabilization through the dominance of steric interactions and the destabilization through the dominance of depletion interactions. Because of salting out, as the concentration of electrolyte increases, the polymer molecules are increasingly precipitated onto the surface of the latex particles and, at sufficiently high electrolyte concentrations, form, in addition, aggregates. The precipitation onto the latex particles generates steric repulsion, which is responsible for the restabilization, whereas the formation of aggregates generates depletion interactions, which are responsible for destabilization.

1. Introduction

The stability of a colloidal dispersion generally decreases as the electrolyte concentration increases.[2] At low electrolyte concentrations, the electrostatic repulsion is responsible for the stability of the colloidal system. However, at sufficiently high electrolyte concentrations, the thickness of the double layer is significantly decreased, and the electrostatic repulsion no longer contributes to the stability of the system.

The steric stabilization, which is imparted by polymer molecules grafted onto the colloidal particles, is extensively employed.[3] Amphiphilic block copolymers are widely used as steric stabilizers. The solvent-incompatible moieties of the block copolymer provide anchors for the polymer molecules that are adsorbed onto the surface of the colloidal particles, and the solvent-compatible (buoy) moieties extend into the solvent phase. When two particles with block copolymers on their surface approach each other, a steric repulsion is generated between the two particles as soon as the tips of the buoy moieties begin to contact, and this repulsion increases the stability of the colloidal system.[4−6] Polymers can also induce aggregation due to either "depletion"[7−11] or "bridging" interactions.[12−15]

Stenkamp et al.[1] experimentally determined the stability ratio of polystyrene latexes as a function of electrolyte concentration for various electrolyte solutions in the presence of the amphiphilic block copolymer poly(ethylene oxide)−poly(propylene oxide)−poly(ethylene oxide) (PEO−PPO−PEO). As the electrolyte concentration increases, the stability ratio of the colloidal system first decreases as a result of the decrease of the electrostatic stabilization. However, at sufficiently high electrolyte concentrations, the stability ratio increases with increasing electrolyte concentration. In some of the colloidal dispersions, Stenkamp et al. observed that the restabilization was followed by destabilization at higher electrolyte concentrations. Stenkamp et al. expressed the opinion that neither the additional adsorption of the block copolymer caused by decreased polymer solvency with increasing electrolyte concentration, nor ion binding, but rather the presence of micelles or aggregates is responsible for the restabilization. Another possible and more plausible explanation is that the steric interaction is responsible for the restabilization at sufficiently high electrolyte concentrations because of the salting out of the polymer onto the surface of the colloidal particles. The solubility of the polymer in water decreases as the electrolyte concentration increases. As a result, the hydrophobic PPO chains of the block copolymer are adsorbed upon the surface of the polystyrene particle, and the PEO chains stretch out, forming a brush that stabilizes the colloidal dispersion. Whereas, for not too large electrolyte concentrations, the polymer molecules precipitate upon the surface of the particles, at large concentrations, they generate, in addition, aggregates. The latter particles are responsible for the attractive depletion interactions and hence for the destabilization, which follows the maximum in the curve stability ratio against electrolyte concentration. The above explanation of the behavior of the complex dispersions investigated by Stenkamp et al. can be accurate only if, for not too low electrolyte concentrations, the concentration of the block copolymer is lower than the critical micelle concentration (CMC) and, for larger ones, greater than the CMC. The dependence of the CMC on the electrolyte concentration is not known for the present case. However, in pure water, at 25 °C, the CMC is very

* To whom correspondence should be addressed. Telephone: (716) 645-2911 ext. 2214. E-mail: feaeliru@acsu.buffalo.edu. Fax: (716) 645-3822.

(1) Stenkamp, V. S.; McGuiggan, P.; Berg, J. C. *Langmuir* **2001**, *17*, 637.
(2) Verwey, E. J.; Overbeek, J. T. G. *Theory of the Stability of Lyophobic Colloids*; Elsevier: Amsterdam, 1948.
(3) Napper D. H. *Polymeric Stabilization of Colloidal Dispersions*; Academic Press: New York, 1983.
(4) Alexander, S. *J. Phys.* **1977**, *38*, 983
(5) Roan, J. R.; Kawakatsu, T. *J. Chem. Phys.* **2002**, *116*, 7283.
(6) Bell, N. S.; Sindel, J.; Aldinger, F.; Sigmund, W. M., *J. Colloid Interface Sci.* **2002**, *254*, 296.
(7) Asakura, S.; Oosawa, F. *J. Chem. Phys.* **1954**, *22*, 1255.
(8) Asakura, S.; Oosawa, F. *J. Polym. Sci.* **1958**, *33*, 183.
(9) Vrji. A. *Pure Appl. Chem.* **1976**, *48*, 471.
(10) Joanny, J. F.; Leibler, L.; de Gennes, P. G. *J. Polym. Sci., Polym. Phys. E.* **1979**, *17*, 1073.
(11) Rao, I. V.; Ruckenstein, E. *J. Colloid Interface Sci.* **1985**, *108*, 389.
(12) Dickinson, E.; Eriksson, L. *Adv. Colloid Interface Sci.* **1991**, *34*, 1.
(13) Podgornik, R. *J. Phys. Chem.* **1992**, *96*, 884.
(14) Borukhov, I.; Andelman, D.; Orland, H. *J. Phys. Chem. B* **1999**, *103*, 5042.
(15) Huang, H.; Ruckenstein, E. *Langmuir* **2006**, *22*, 3174.

10.1021/la0602057 CCC: $33.50
Published on Web 04/11/2006

large (25.5 g/L for Pluronic L64 and 24.7 g/L for Pluronic L35 [16]), and at large electrolyte concentrations, intuition suggests that it is much smaller. Consequently, the proposed explanation might be accurate.

In the present paper, calculations are carried out to explain the restabilization followed sometimes by destabilization, as the electrolyte concentration increases, observed by Stenkamp et al. In these calculations, double-layer, steric, and depletion interactions are taken into account. The steric interaction at various electrolyte concentrations is calculated using the scaling theory. The surface density of the polymer chain was evaluated by using the Sechenov equation for the polymer solubility as a function of electrolyte concentration.

2. Theoretical Framework

The colloidal stability can be expressed as the stability ratio, which is given by the expression[17]

$$W = \int_2^\infty \frac{\exp(V_{\text{tot}}/kT)}{s^2}\mathrm{d}s \Big/ \int_2^\infty \frac{\exp(V_{\text{vdW}}/kT)}{s^2}\mathrm{d}s \tag{1}$$

where W is the stability ratio, V_{tot} is the total interaction potential between two spherical particles, k is the Boltzmann constant, T is the absolute temperature, s is the ratio between the center distance of two colloidal particles and the radius of the particles, and V_{vdW} is the van der Waals interaction potential.

Because the block copolymer is uncharged, the electrostatic repulsion, the steric interaction, and the depletion repulsion are independent quantities and are hence additive. The total interaction potential V_{tot} is therefore given by the sum

$$V_{\text{tot}} = V_{\text{d}} + V_{\text{steric}} + V_{\text{dep}} + V_{\text{vdW}} \tag{2}$$

where V_{d} is the double-layer potential between two spheres, V_{steric} is the steric potential, V_{dep} is the depletion potential, and V_{vdW} is the van der Waals interaction potential.

When the radius of the spherical colloidal particles a is much larger than the shortest distance between the surfaces of two colloidal particles, the free energy between two identical spherical particles, due to double-layer, steric, and depletion interactions, can be calculated using the Derjaguin approximation:[2]

$$V_i = \pi a \int_{H_0}^\infty [F_i(H) - F_i(\infty)]\mathrm{d}H \quad i = \text{d, steric, dep} \tag{3}$$

where $F_i(H)$ is the free energy of interaction between two parallel plates, H is the distance between plates, and H_0 is the shortest approach between the two spheres.

The solubility of a block copolymer in water decreases as the concentration of electrolyte increases. When the concentration of the polymer is larger than its solubility, the polymer molecules precipitate onto the surface or form aggregates that remain dispersed into the colloidal system or deposit on the wall of the vessel that contains the colloidal dispersion. Let us start from two parallel plates, and then calculate the stability ratio of the system for spheres using the Derjaguin approximation.

2.1. Double-Layer Interaction. When the copolymer PEO−PPO−PEO precipitates onto the surface of the particles, the PPO segments are attached to the surface, and the PEO segments extend into the solution. The dielectric constant near the surface of the plates is changed by the presence of the polymer chains because the latter have a lower dielectric constant. In addition, the polymer chains present on the surface constrain the orientation of the water molecules nearby. As a result, in the region near the surface, the dielectric constant is lower than in the bulk. The region near the surface (region I) of thickness δ is considered to have a dielectric constant ϵ^{I} different from that in the remaining region between plates (region II), ϵ^{II}, because of the lower dielectric constant of the polymer and the constraint induced by the chains on the water molecules. For simplicity, the dielectric constants in both regions are assumed to be constant, with that in region II being equal to the dielectric constant of water (80) and that in region I being equal to 60.

The electrical potential distribution between the two plates is given by the Poisson equations

$$\frac{\partial^2\Psi}{\partial x^2} = -\frac{\rho}{\epsilon^{\text{I}}\epsilon_0} \text{ (for region I)} \tag{4}$$

and

$$\frac{\partial^2\Psi}{\partial x^2} = -\frac{\rho}{\epsilon^{\text{II}}\epsilon_0} \text{ (for region II)} \tag{5}$$

where Ψ is the electrical potential, ρ is the local charge density, ϵ^{I} and ϵ^{II} are the dielectric constants in regions I and II, respectively, and ϵ_0 is the vacuum permittivity.

At the boundary between the two regions, the electric potentials should be the same:

$$\Psi|^{\text{I}}_{x=-d+\delta} = \Psi|^{\text{II}}_{x=-d+\delta} \tag{6}$$

where $2d$ is the distance between the two plates, and x is measured from the middle of the two plates.

If there is no charge accumulation, the Gauss law provides the equation

$$\epsilon^{\text{I}}\frac{\partial\Psi}{\partial x}\Big|^{\text{I}}_{x=-d+\delta} = \epsilon^{\text{II}}\frac{\partial\Psi}{\partial x}\Big|^{\text{II}}_{x=-d+\delta} \tag{7}$$

where the superscripts I and II refer to regions I and II, respectively.

The boundary condition at the surface has the form

$$\frac{\mathrm{d}\Psi}{\mathrm{d}x} = -\frac{\sigma}{\epsilon\epsilon_0} \tag{8}$$

where σ is the surface charge density.

The symmetry condition provides an additional boundary condition,

$$\frac{\partial\Psi}{\partial x}\Big|_{x=0} = 0 \tag{9}$$

Using boundary conditions 8 and 9, the Poisson equations 4 and 5 can be solved to calculate the distribution of the electrical potential for a given surface charge density.

For two plates, the double-layer free energy F_{d} of the interaction between two plates is the sum of three contributions:[18]

$$F_{\text{d}} = F_{\text{ele}} + F_{\text{entropy}} + F_{\text{chem}} \tag{10}$$

where F_{ele}, F_{entropy}, and F_{chem} are the electrostatic, entropic due to the small ions, and chemical contributions to the free energy, respectively.

(16) Lopes, J. R.; Loh, W. *Langmuir* **1998**, *14*, 750.
(17) Hunter, R. J. *Foundations of Colloid Science*; Clarendon Press: Oxford, England, 1986; Vol. 1.
(18) Overbeek, J. T. G. *Colloids Surf.* **1990**, *51*, 61.

The electrostatic contribution comes from the electrical potential energy of the charge in the field:

$$F_{\text{ele}} = \int_{-d}^{d} \frac{\epsilon\epsilon_0}{2}\left(\frac{\partial\Psi}{\partial x}\right)^2 \mathrm{d}x = 2\int_{-d}^{-d+\delta} \frac{\epsilon^{\mathrm{I}}\epsilon_0}{2}\left(\frac{\partial\Psi}{\partial x}\right)^2 \mathrm{d}x + \int_{-d+\delta}^{d-\delta} \frac{\epsilon^{\mathrm{II}}\epsilon_0}{2}\left(\frac{\partial\Psi}{\partial x}\right)^2 \mathrm{d}x \quad (11)$$

Because the two layers have different dielectric constants, the electrostatic contribution is the sum for those two layers.

The entropic contribution to the free energy due to the electrolyte ions is given by[18]

$$F_{\text{entropy}} = kT\sum_i \int_{-d}^{d} n_i \ln\left(\frac{n_i}{n_i^0}\right) - n_i + n_i^0)\mathrm{d}x \quad (12)$$

where n_i is the number of ions of species i per unit volume, and n_i^0 is its value in the bulk.

The chemical contribution ΔF_{chem} is 0 at constant surface charge density.

2.2. Steric Interaction. The steric interaction force f_{steric} between two plates containing attached polymer chains can be evaluated on the basis of the scaling theory[19]

$$f_{\text{steric}} \cong \frac{kT}{D^3}\left[\left(\frac{h_c}{2d}\right)^{9/4} - \left(\frac{2d}{h_c}\right)^{3/4}\right] \quad (13)$$

where D is the average distance between two attachment points, and $h_c/2$ is the length of the buoy chain in the solvent phase. (It is assumed to be equal to δ.) This equation is valid only for $2d < h_c$. For $2d > h_c$, the steric force is zero. Therefore, the steric interaction free energy between two plates is given by

$$F_{\text{steric}} \cong -\int_{h_c}^{2d} f_{\text{steric}} d(2d) = -\frac{4kTh_c}{D^3}\left[\frac{1}{5}\left(1 - \left(\frac{h_c}{2d}\right)^{5/4}\right) + \frac{1}{7}\left(1 - \left(\frac{2d}{h_c}\right)^{7/4}\right)\right] \quad (14)$$

The steric interaction energy increases as the average distance between the attachment points decreases. The electrolyte concentration affects the steric interaction through the distance D between the attachment points. To evaluate the distance between the attachment points at different electrolyte concentrations, we consider that the polymer begins to precipitate when the polymer concentration becomes larger than the solubility of the polymer in water. As the electrolyte concentration increases, the solubility of the polymer in the solution decreases. Therefore, the polymer chains dissolved in the solution are salted out onto the surface of the particles and/or form aggregates with increasing electrolyte concentration. As a result of the first effect, the steric interaction free energy increases. The second effect is responsible for depletion interactions.

2.3. Depletion Interaction. When the distance between two plates is smaller than the diameter l of the small particles (micelles), which, in the present case, are uncharged, the depletion force f_{dep} acting between two plates is given by[7,20,21]

$$f_{\text{dep}} = -n_m kT \quad (15)$$

(19) de Gennes P. G. *Adv. Colloid Interface Sci.* **1987**, *27*, 189.
(20) Walz, J. Y.; Sharma, A. *J. Colloid Interface Sci.* **1994**, *168*, 485.
(21) Huang, H.; Ruckenstein *Langmuir* **2004**, *20*, 5412

where n_m is the number of aggregates (micelles) per unit volume.

When the distance between the two plates is larger than the diameter of the small particles, the depletion force becomes zero because the small particles can be located in the gap.

Consequently, the depletion free energy between two plates is given by

$$F_{\text{dep}} = -\int_l^{2d} f_{\text{dep}} d(2d) = n_m kT(2d - l) \quad (16)$$

To calculate the surface density of the polymer on the particles and the number of micelles at various electrolyte concentrations, the solubility of the polymer in the electrolyte solution is required. The Sechenov equation can be employed to calculate the solubility of a polymer in an electrolyte solution. This equation was initially proposed to calculate the solubility of a gas,[22] but it is valid for a polymer solution as well.[23] It has the form

$$\log\frac{C}{c_w} = -k_c c_{\text{salt}} \quad (17)$$

where C is the solubility of the polymer in an electrolyte solution of concentration c_{salt}, c_w is the polymer solubility in pure water, and k_c is the Sechenov constant, which depends on both the electrolyte and solute. Tadros and Vincent[24] have observed that, by decreasing the polymer solvency, one can increase the adsorbed amount.

When the polymer concentration is larger than its solubility, the excess polymer is assumed to precipitate on the surface of the particles and/or to form micelles. This assumption allows us to write for the polymer surface density η the expression

$$\eta = \alpha\frac{C_0 - C}{4MN\pi a^2} + \eta_0 \quad (18)$$

where N represents the colloidal particle concentration, 2×10^{12} particles/L in the experiments considered;[1] C_0 (g/L) is the initial concentration of the block copolymer in solution; η_0 (mol/m^2) is the surface density of polymer before the addition of the electrolyte; α is the fraction of precipitated polymer adsorbed onto the surface; and M is the molecular weight of the polymer.

The average distance D between the attachment points decreases with increasing electrolyte concentration due to the precipitation caused by the salting out, and can be calculated from the surface density of the polymer:

$$D = \frac{1}{\sqrt{2\eta N_A}} \quad (19)$$

where the factor 2 accounts for the two PEO blocks of the copolymer chain, and N_A is Avogadro's number.

The ratio of the polymer deposited on the surface of the particles to that generating micelles was determined by fitting the experimental results. If the number of polymer chains in each micelle is m (which is taken as 50 in the present calculations[1]), the number density of the micelles is

$$n_m = (1 - \alpha)N_A\frac{C_0 - C}{mM} \quad (20)$$

The size l of an aggregate can be easily calculated. For $m = 50$, one obtains that $l \approx 100$ Å. The above equation will be used to calculate the depletion free energy between particles. In some

(22) Sechenov, M. *Z. Phys. Chem.* **1889**, *4*, 117.
(23) Borukhov, I.; Leibler, L. *Phys. Rev. E* **2000**, *62*, R41.
(24) Tadros, T. F.; Vincent, B. *J. Phys. Chem.* **1980**, *84*, 1575

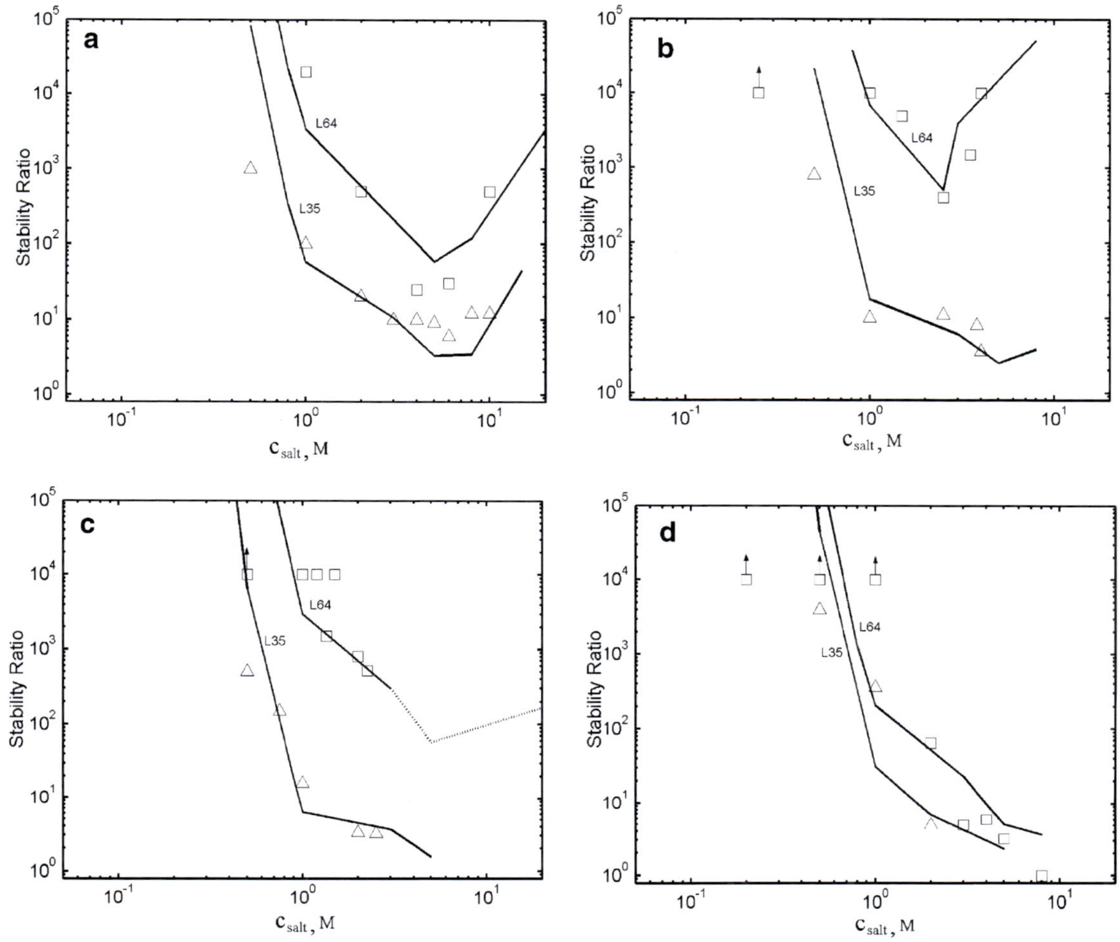

Figure 1. The stability ratio vs electrolyte concentration for various polymer lengths. The data are from ref 1. □: L64 ($M = 2900$); △: L35 ($M = 1900$). The initial polymer concentration is 60×10^{-3} g/L, $a = 110$ nm, $A_H = 1 \times 10^{-21}$ J, $h_c = 2$ nm, $\sigma = -0.028$ C/m^2. (a) The electrolyte is LiCl. Parameters used in the calculation: for Pluronic L35, $k_c = 5 \times 10^{-7}$ M^{-1} and $\eta_0 = 4.15 \times 10^{-8}$ mol/m^2; for Pluronic L64, $k_c = 6 \times 10^{-7}$ M^{-1} and $\eta_0 = 5.1 \times 10^{-8}$ mol/m^2. (b) The electrolyte is NaCl. Parameters used in the calculation: for Pluronic L35, $k_c = 1 \times 10^{-6}$ M^{-1} and $\eta_0 = 3.8 \times 10^{-8}$ mol/m^2; for Pluronic L64, $k_c = 3 \times 10^{-6}$ M^{-1} and $\eta_0 = 5.1 \times 10^{-8}$ mol/m^2. (c) The electrolyte is KNO_3. Parameters used in the calculation: for Pluronic L35, $k_c = 1 \times 10^{-6}$ M^{-1} and $\eta_0 = 3.5 \times 10^{-8}$ mol/m^2; for Pluronic L64, $k_c = 1 \times 10^{-7}$ M^{-1} and $\eta_0 = 5.1 \times 10^{-8}$ mol/m^2. The dotted lines represent the calculation results in a range for which no experimental data are available. (d) The electrolyte is KSCN. Parameters used in the calculation: for Pluronic L35, $k_c = 5 \times 10^{-7}$M^{-1} and $\eta_0 = 4 \times 10^{-8}$ mol/m^2; for Pluronic L64, $k_c = 1 \times 10^{-7}$ M^{-1} and $\eta_0 = 4.5 \times 10^{-8}$ mol/m^2.

cases, the stability ratio decreases with increasing electrolyte concentration after a minimum followed by a maximum. To explain this behavior, the depletion interaction was taken into account. In some cases, the stability ratio does not pass through a maximum after the minimum; the fraction of polymer that forms micelles is assumed in those cases to be zero.

2.4. van der Waals Interaction. The van der Waals interaction between two identical spherical particles of radius a has the form[16]

$$V_{\mathrm{vdW}} = -\frac{A_H}{6}\left[\frac{2a^2}{H_0(4a + H_0)} + \frac{2a^2}{(2a + H_0)^2} + \ln\frac{H_0(4a + H_0)}{(2a + H_0)^2}\right] \quad (21)$$

where A_H is the Hamaker constant, and H_0 is the closest distance between the surfaces of the two spheres.

Adding all interaction free energies together, the stability ratio can be calculated for various electrolyte concentrations.

3. Fitting the Experimental Results

In Stenkamp et al.'s experiments,[1] the particles consisted of polystyrene latex containing sulfate functional groups. In the calculations, we assumed a constant surface charge density. Complex dispersions were generated by the addition of electrolyte solutions to polystyrene latex dispersions containing the amphiphilic block copolymer PEO–PPO–PEO. The stability of such a dispersion was quantified by determining the stability ratio through photon correlation spectroscopy of the incipient aggregation. The stability ratio first decreases and then increases with increasing electrolyte concentration. However, in some cases, at sufficiently high electrolyte concentrations, the stability ratio decreased again after passing through a maximum with increasing electrolyte concentration. The restabilization and destabilization concentrations depend on the nature of the electrolyte. In the present paper, we attempt to explain the restabilization and destabilization using the above theoretical framework.

Our explanation implies that the excess polymer precipitates onto the surface of the particles when the polymer concentration becomes larger than its solubility, which decreases with increasing electrolyte concentration in the solution.

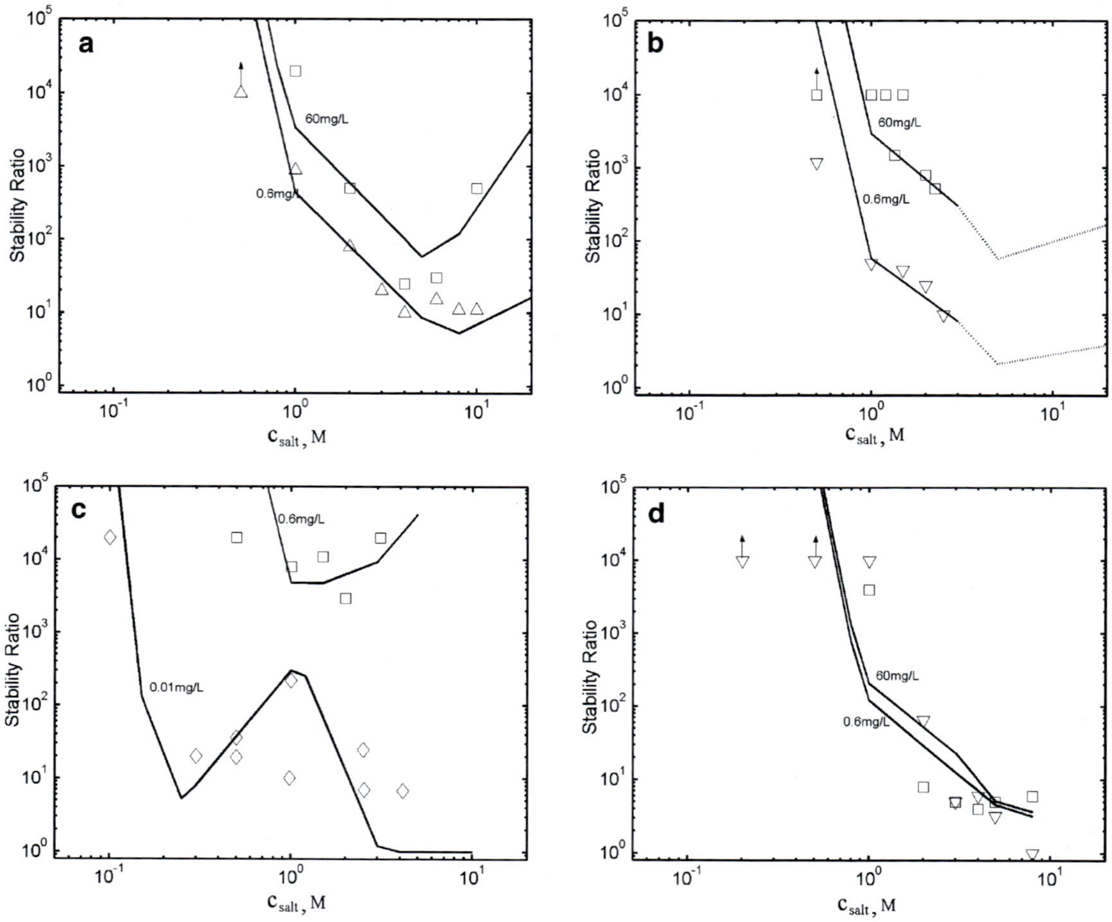

Figure 2. The stability ratio vs electrolyte concentration for various initial polymer concentrations. The data are from ref 1. Pluronic L64 is employed. $a = 110$ nm, $A_H = 1 \times 10^{-21}$ J, $h_c = 2$ nm, $\sigma = -0.028$ C/m^2, $l = 10$ nm. (a) The electrolyte is LiCl. Parameters used in the calculation: for $C_0 = 60 \times 10^{-3}$g/L, $k_c = 6 \times 10^{-7}$ M^{-1} and $\eta_0 = 5.1 \times 10^{-8}$ mol/m^2; for $C_0 = 0.6 \times 10^{-3}$ g/L, $k_c = 6 \times 10^{-7}$ M^{-1} and $\eta_0 = 4.8 \times 10^{-8}$ mol/m^2. (b) The electrolyte is KNO_3. Parameters used in the calculation: for $C_0 = 60 \times 10^{-3}$ g/L, $k_c = 1 \times 10^{-7}$ M^{-1} and $\eta_0 = 5.1 \times 10^{-8}$ mol/m^2; for $C_0 = 0.6 \times 10^{-3}$ g/L, $k_c = 1 \times 10^{-7}$ M^{-1} and $\eta_0 = 4.2 \times 10^{-8}$ mol/m^2. The dotted lines represent the calculation results in a range for which no experimental data are available. (c) The electrolyte is KBr. Parameters used in the calculation: for $C_0 = 0.01 \times 10^{-3}$ g/L, $k_c = 5 \times 10^{-6}$ M^{-1} and $\eta_0 = 1.2 \times 10^{-8}$ mol/m^2; for $C_0 = 0.6 \times 10^{-3}$g/L, $k_c = 5 \times 10^{-6}$ M^{-1} and $\eta_0 = 4.9 \times 10^{-8}$ mol/m^2. (d) The electrolyte is KSCN. Parameters used in the calculation: for $C_0 = 60 \times 10^{-3}$g/L, $k_c = 1 \times 10^{-7}$ M^{-1} and $\eta_0 = 4.5 \times 10^{-8}$ mol/m^2; for $C_0 = 0.6 \times 10^{-3}$g/L, $k_c = 1 \times 10^{-7}$ M^{-1} and $\eta_0 = 4.2 \times 10^{-8}$ mol/m^2.

The effect of the molecular weight of the polymer is presented in Figure 1a–d. The increase in the molecular weight of the polymer increases the stability ratio. In some cases, restabilization occurs at sufficiently high electrolyte concentrations. The lowest molecular weight Pluronic L35 does not induce restabilization, while Pluronic L64 induces restabilization for the electrolytes LiCl, NaCl, and KNO_3. For Pluronics L35 and L64 in KSCN, the stability ratio decreases to 1 at high electrolyte concentrations without passing through a minimum, or a minimum and a maximum.

Another experimental aspect investigated by Stenkamp et al. was the effect of the concentration of the polymer initially dissolved in solution (Figure 2). In general, by decreasing the initial concentration of the polymer, the probability for restabilization to occur decreases. For LiCl and KNO_3, a decrease from 60×10^{-3} to 0.6×10^{-3} g/L of the initial concentration of the polymer causes the loss of restabilization. KCl had a stronger propensity for restabilization and could restabilize the colloidal dispersion for a 0.6×10^{-3} g/L initial polymer concentration, but not for a 0.01×10^{-3} g/L concentration.

The hydrodynamic diameter of the latex particles was also determined by Stenkamp et al. The results revealed that the average diameter of the restabilized latexes was close to that of the polymer-coated latex particles in distilled water. However, this does not disprove that a significant additional adsorption is one of the possible causes of restabilization because the increase in the polymer coverage of the surface with increasing electrolyte concentration does not mean that the hydrodynamic diameter changed appreciably. Because the hydrophilic moiety of a polymer chain extends into the water phase, more adsorption of the hydrophobic moieties onto the particle surface due to the salting out will not lead to an appreciable increase in the thickness of the adsorption layer. On the contrary, the thickness of the adsorption layer has the tendency to decrease because of the poor solvency of the polymer at high electrolyte concentrations.

The more than qualitative agreement between the experimental data and the calculations shows that the steric interaction can indeed contribute to restabilization at high electrolyte concentrations below the CMC. Of course, if the salting-out effect of the electrolyte is not large enough to precipitate enough polymer on the surface (small k_c), the restabilization alone or restabilization followed by destabilization will not occur.

4. Conclusions

The stability of the polystyrene latex in electrolyte solutions containing an amphiphilic block polymer first decreases because of the decrease in the electrostatic repulsion. However, at higher electrolyte concentrations, the polystyrene latex could be restabilized in the presence of amphiphilic block copolymers (Pluronics), which, because of salting out, precipitate onto the particles. This occurs because, with increasing surface coverage of the polymer, the steric interaction between two latex particles increases. When, because of salting out, aggregates are also formed, an attractive depletion force is also generated, which decreases the stability ratio at very high electrolyte concentrations. The calculations suggest that the steric interaction and the depletion force can explain the restabilization followed sometimes by destabilization with increasing electrolyte concentration.

LA0602057

ELSEVIER

Colloids and Surfaces A: Physicochem. Eng. Aspects 232 (2004) 1–10

COLLOIDS AND SURFACES A

www.elsevier.com/locate/colsurfa

Estimation of the available surface and the jamming coverage in the Random Sequential Adsorption of a binary mixture of disks

Marian Manciu, Eli Ruckenstein*

Department of Chemical Engineering, State University of New York at Buffalo, Buffalo, NY 14260, USA

Received 11 August 2003; accepted 2 October 2003

Abstract

A simple analytical expression is proposed for the area available to a disk on a surface for a Random Sequential Adsorption (RSA) of binary mixtures of disks. The expression was obtained by combining the low-order terms of the density expansion of the available area with the asymptotic behaviour of the surface coverage near the jamming point. Comparison with Monte Carlo simulations shows that this approach provides a fair estimation of the jamming coverage for both kinds of disks.

Keywords: Random Sequential Adsorption; Jamming; Disk mixture

1. Introduction

Long ago, Langmuir suggested that the rate of deposition of particles on a surface is proportional to the density of particles in the vicinity of the surface and to the available area on the surface [1]. However, the calculation of the available area is still an open problem. In a first approximation, one can assume that the available area is the total area of the surface minus the area already occupied by the adsorbed particles [1]. A better approximation can be obtained if the adsorbed particles, assumed to have the shape of a disk, are in thermal equilibrium on the surface, either because of surface diffusion and/or of adsorption/desorption kinetics. In this case, one can use one of the empirical equations available for the compressibility of a 2D gas of hard disks, calculate the chemical potential in excess to that of an ideal gas [2] and then use the Widom relation between the area available to one particle and its excess chemical potential on the surface (the particle insertion method) [3]. The method is accurate at low densities of adsorbed particles, where the equations of state are accurate, but, in general, poor at high concentrations. The equations of state for hard disks are based on the virial expansion and only the first few coefficients of this low-density expansion are known [4]; they do not provide sufficient information for the high density behaviour.

A typical example is a fortunate continuation of the virial expansion of a hard sphere gas, the Carnahan–Starling equation of state [5], which predicts that the divergence of the compressibility occurs when the total volume of spheres is equal to the total volume available, while at close packing the spheres occupy only about 74% of the total volume. Since the density at which the compressibility diverges is known for monodisperse rigid spheres, it can be incorporated "a posteriori" in the equation of state [6], or one can test which one of the Padé analytic continuations of the virial series provides a better agreement [7]. However, there is no reliable procedure to predict "a priori" the compressibility divergence using only the first virial coefficients. As a consequence, there is no procedure which can predict even roughly the divergence (phase transition) for a binary mixture of hard spheres. While accurate empirical equations of state are known at low densities for binary mixtures [8], they predict divergence of the compressibility only when the total volume is occupied. The situation is similar for a 2D binary mixture of hard disks.

The non-equilibrium problem is even more complicated. The large particles can have surface binding energy much larger than kT and in this case they neither diffuse nor desorb from the surface. The Random Sequential Adsorption (RSA) model [9] assumes that a particle, which arrives at a random location on a surface, is adsorbed only if there is

* Corresponding author. Tel.: +1-716-645-2911/2214; fax: +1-716-645-3822.
E-mail address: feaeliru@acsu.buffalo.edu (E. Ruckenstein).

0927-7757/$ – see front matter © 2003 Elsevier B.V. All rights reserved.
doi:10.1016/j.colsurfa.2003.10.001

sufficient room for it around that particular location (that is, it does not overlap with any other adsorbed particles). Only one continuum RSA model, the 1D random parking of cars, received an analytical solution [10]. This problem implies that points are selected randomly along a line of length L. If there is enough space around the point to fit a "car" of length l, the car is parked; otherwise, another random point is selected, and so on. For a $L \to \infty$ line, when the density of cars is about $N_{\text{RSA}} l/L \cong 0.76$ there is no more room to place a car. This result can be contrasted to the "equilibrium" problem of parking cars, for which cars can be parked until close compaction $N_{\text{E}} l/L = 1$ (the 1D gas of hard rods does not have a phase transition [11]).

Widom [9] realized the importance of this problem for statistical mechanics and showed that the centers of the particles of a hard disk gas, in an equilibrium position, are not uniformly random distributed. The available area for a new particle $\Phi(\theta)$ can be written as a power series in particle density $\theta = N\pi r^2/A$, where N is the number of adsorbed particles, r their radius and A the total area of the surface. The coefficients of the series terms are identical up to the second power of θ for the equilibrium and the RSA models. The differences in the higher powers coefficients lead for RSA to jamming for $\theta_{\text{C}} = 0.76$, 0.547 and 0.38 for the 1D (segments on a line), 2D (disks on a surface) and 3D (spheres on a volume), respectively, while for the equilibrium configurations the close-packing occurs at $\theta = 1$, 0.91 and 0.74, respectively.

The RSA model received renewed attention after Feder [12] observed that the adsorption on the surface of apoferritin molecules (large iron-storage proteins with a diameter of about 10 nm), which adsorb irreversibly, reached saturation at a coverage $\theta_{\text{C}} = 0.518$. Monte Carlo simulations of Random Sequential Adsorption of disks on a surface last prohibitively long in the vicinity of the jamming point; however Feder [12] noted that in the vicinity of the jamming coverage, θ has a power-law dependence on time:

$$\theta_{\text{C}} - \theta \propto \frac{1}{\sqrt{t}}. \quad (1)$$

By extrapolating this scaling law, an accurate value for the jamming coverage, $\theta_{\text{C}} = 0.547$ was obtained. The power-law (1) was later demonstrated theoretically, and it was shown that it is accurate not only in the immediate vicinity of jamming, but in a broader θ range [13,14].

Schaaf and Talbot [15] continued Widom's analytical approach by calculating the available area for the RSA of disks and obtained the coefficient of θ^3, which is different in RSA and equilibrium models. The next coefficient was obtained independently by Dickman et al. [16] and Given [17]. The first five terms of the series are not, however, very useful for the calculation of the jamming point θ_{C}. Indeed, using the five known terms, there is no jamming, because $\Phi(\theta) = 0$ has no solution for $0 < \theta < 1$. Furthermore, almost all analytical continuations based on Padé approximants $P[i, j]$ provided the same conclusion, that no jamming occurs (except $P[1, 4]$, which predicted an unsatisfactory value $\theta_{\text{C}} \sim 0.4$). A modality to employ the scaling law (4) in the analytical continuation of the series was proposed by Dickman et al. [16]. They calculated the first terms of the time expansion of the coverage $\theta(t)$, transformed it to a new variable with an appropriate asymptotic behavior $y = 1 - (1/\sqrt{1 + bt})$ and found, when the Padé [3,2] approximant was employed, an excellent agreement for the jamming coverage (within about 0.15%) by selecting $b = 3$. The use of a different value for b or of another Padé approximant deteriorated, however, dramatically the agreement. Later, Wang [18] suggested to estimate b from the convergence of the jamming coverages predicted by various Padé approximants. While the method seems to work well for the RSA of oriented squares (for which the first nine terms of the low-density expansion are known), Wang concluded than there are not sufficient known terms in the RSA expansion for disks to reach an accurate prediction [18].

The estimation of the jamming coverage for the RSA of monodisperse disks is not an important issue, because its value is already accurately known from Monte Carlo simulations [12]. However, it is of interest to develop a procedure that can predict the available area and the jamming coverage for a mixture of disks, for which much less information is available. Even at equilibrium, for which reasonable accurate equations of state for binary mixtures of hard disks are known for low densities [19,20], the available area vanishes only for the unphysical total coverage $\theta = \theta_{\text{S}} + \theta_{\text{L}} = 1$ (where the subscripts S and L stand for small and large disk radii, respectively), hence there is no "jamming". Exact analytical expressions are known only for the first three virial coefficients of a binary mixture of disks [21]. The fourth and fifth coefficients were computed numerically for some diameter ratios and molar fractions for an equilibrium gas [22]. However, there are no such calculations for the RSA model.

Let us first briefly review the recent conclusions about the jamming in a RSA model of binary mixtures. For the 1D model, an exact solution for the random parking of "cars" of two different sizes predicted that the coverage is always larger for a binary mixture than for unisized cars [23]. However, another solution for the 1D model predicted a smaller coverage for binary than for unisized cars in a continuum model, but a larger coverage for binary in a lattice model [24]. For the 2D model, a Monte Carlo simulation showed that a binary mixture of disks always covered the surface better than monodisperse disks [25]. However, another Monte Carlo simulation of deposition of larger spheres on a surface randomly precovered with smaller ones indicated that the total coverage is always smaller for binary mixtures than for unisized disks [26].

These puzzling results can be understood qualitatively. In a mixture of large and small disks, if the large disks are adsorbed first, they can cover up to a fraction θ_{C} of the surface. The jamming then is reached for the large disks, but the small disks can still be adsorbed, and the binary mixture covers a larger fraction of the surface than the monodisperse disks. In the limit of very large and very small disks, the total coverage is clearly $\theta_C + (1-\theta_C)\theta_C \sim 0.795$. However, if some small disks are adsorbed first without reaching jamming, there is enough room for the small disks, but might not be for the larger ones adsorbed later. In this case, the coverage is lower for the binary mixture. Therefore, the available area and the jamming depend not only on the concentration of the particle already adsorbed on the surface, θ_S and θ_L, but also on the order of their adsorption.

The only approximate analytical solution for the RSA of a binary mixture of hard disks was proposed by Talbot and Schaaf [27]. Their theory is exact in the limit of vanishing small disks radius $r_S \to 0$, but fails when the ratio $\gamma = r_L/r_S$ of the two kinds of disk radii is less than 3.3; its accuracy for intermediate values is not known. Later, Talbot et al. [28] observed that an approximate expression for the available area derived from the equilibrium Scaled Particle Theory (SPT) [19] provided a reasonable approximation for the available area for a non-equilibrium RSA model, up to the vicinity of the jamming coverage. While this expression can be used to calculate accurately the initial kinetics of adsorption, it invariably predicts that the abundant particles will be adsorbed on the surface until $\theta = 1$, because the Scaled Particle Theory cannot predict jamming.

The evaluations of the jamming coverage and of the available area for a binary mixture is of interest in many practical applications, such as protein and cell separation (affinity chromatography), biocompatibility of biomaterials and separation of toner and ink particles [29]. A simple approach to estimate the available area to a disk and the jamming points for binary mixtures of disks will be proposed in what follows. The approach combines the power-law dependence of the available area in the vicinity of the jamming point with the known dependence at low coverage densities. For monodisperse disks, an excellent approximate expression can be constructed in this manner by employing information about the behaviour near the jamming point provided by Monte Carlo simulations. However, reasonable results can be obtained without using the latter information. To calculate the area available to a large disk on a surface for binary mixtures $\Phi_L(\theta_S, \theta_L)$, it will be assumed that the small disks are deposited first until θ_S is reached and then the large disks are adsorbed until jamming. To calculate the area available to small disks, "large" and "small" should be interchanged. The available area $\Phi_L(\theta_S, 0)$ can be calculated using one of the available approximations, that are accurate at low densities. The approximations for $\Phi_L(\theta_S, 0)$ are in most cases accurate, even when the same approximation for $\Phi_L(\theta_S, \theta_L)$ fails. The available area $\Phi_L(\theta_S, \theta)$ is a function of θ alone and can be continued analytically taking into account its asymptotic behavior, which is the same as that corresponding to monodisperse disks. The predictions of the jamming points, obtained from $\Phi_L(\theta_S, (\theta_L)(\theta_S))_C) \equiv 0$ are compared with the Monte Carlo simulations available in [25,26].

2. RSA of monodisperse disks

The rate of adsorption of monodisperse disks on a surface, $d\theta/dt$, is proportional to the total surface available to a disk, Φ:

$$\frac{d\theta}{dt} = \Phi(\theta) \tag{2}$$

where the surface coverage $\theta = N\pi r^2/A$, N being the number of adsorbed disks, r the radius of a disk, A the total area and t a dimensionless time. Monte Carlo simulations indicated that, in the vicinity of jamming,

$$\theta = \theta_C - \frac{K}{\sqrt{t}} \tag{3}$$

where $\theta_C = 0.547$ and $K = 0.236$ [12].

From Eq. (3) one obtains that

$$\frac{d\theta}{dt} = \frac{K}{2\sqrt{t^3}}$$

which combined with Eqs. (2) and (3), leads for the available area in the vicinity of the jamming point to the expression:

$$\Phi^F(\theta) = \frac{1}{2K^2}(\theta_C - \theta)^3 \tag{4}$$

The first terms of the series expansion of the available area for low surface coverages are also known [15–17]:

$$\Phi(\theta) = \sum_{i=0} c_i\theta^i = 1 - 4\theta + 3.307973\theta^2 + 1.406876\theta^3 + 0.7205656\theta^4 \tag{5}$$

An approximate expression, which can interpolate between Eqs. (4) and (5), has the form [15]:

$$\Phi^i(\theta) = (\theta_C - \theta)^3(a_1 + a_2\theta + a_3\theta^2 + a_4\theta^3 + \cdots) \tag{6}$$

where the coefficients a_i are determined from the conditions to obey the limit law (4) near θ_C:

$$a_1 + a_2\theta_C + a_3\theta_C^2 + a_4\theta_C^3 + \cdots = \frac{1}{2K^2} \tag{6a}$$

and to match the terms of the θ expansion (5):

$$\theta_C^3 a_1 = 1 \tag{6b}$$

$$\theta_C^3 a_2 - 3\theta_C^2 a_1 = -4 \tag{6c}$$

$$\theta_C^3 a_3 - 3\theta_C^2 a_2 + 3\theta_C a_1 = 3.308, \ldots \tag{6d}$$

The best agreement with the limiting laws Eqs. (4) and (5) is obtained using only four terms in Eq. (6). Then, Eq. (6a)–(6d) lead to

$$a_1 = \frac{1}{\theta_C^3} = 6.107 \tag{7a}$$

$$a_2 = \frac{3-4\theta_C}{\theta_C^4} = 9.059 \quad (7b)$$

$$a_3 = \frac{3.308\theta_C^2 - 12\theta_C + 6}{\theta_C^5} = 8.669 \quad (7c)$$

$$a_4 = \frac{(1/2K^2) - 3.308\theta_C^2 + 16\theta_C - 10}{\theta_C} = -28.56 \quad (7d)$$

Only the first three terms of expansion (5) were employed in Eq. (7a)–(7d). These terms coincide to those of the equilibrium expansion. The use of the next coefficients does not improve, however, the accuracy, a situation that is not uncommon in the analytical continuation of a virial expansion [4].

The interpolating approximation (6) is compared with the limiting laws, namely the series expansion Eq. (5) and Feder law Eq. (4) in Fig. 1a. The accuracy of the interpolation can be better seen in Fig. 1b, where the ratios between Eq. (6) and the limiting laws are plotted. The interpolating function (6) satisfies both limiting laws (in their ranges of validity) within a few percents, practically representing the Monte Carlo simulations within their numerical error.

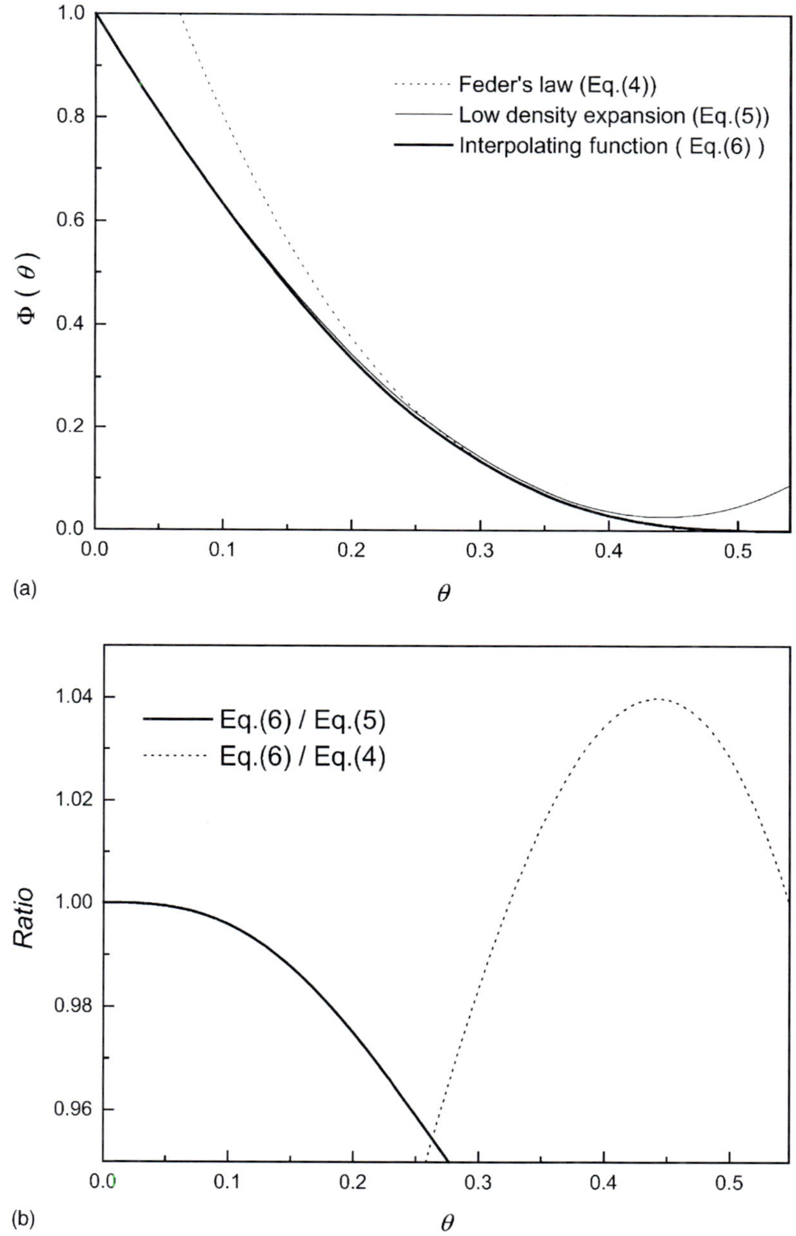

Fig. 1. (a) The area available to a new disk as a function of surface coverage. (b) The ratios of the interpolating function, Eq. (6) to the limiting laws (Eqs. (4) and (5)) as functions of surface coverage. The limiting law (4) is valid at high surface coverages (near jamming), while Eq. (5) is valid at low coverages.

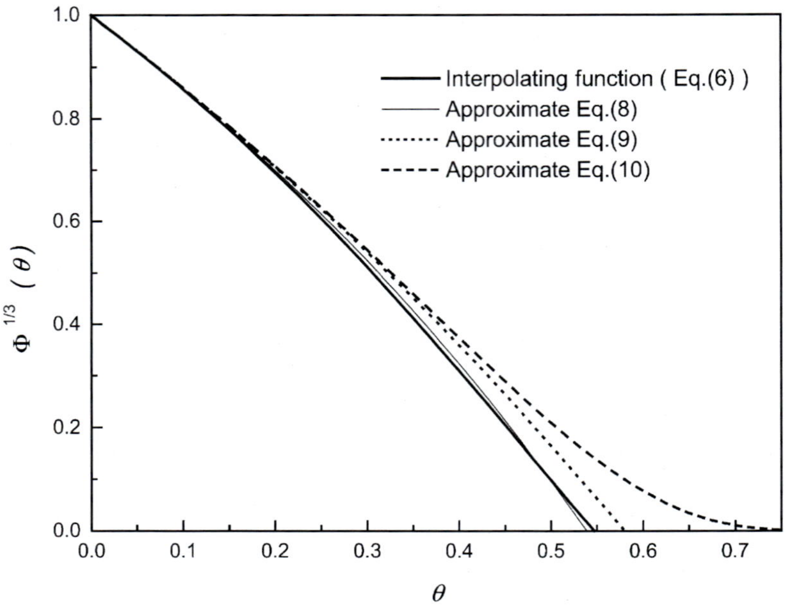

Fig. 2. Predictions for the available area $\Phi(\theta)$.

The jamming point can also be evaluated without using the information provided by the Monte Carlo simulations [30]. While this approach is not useful for monodisperse disks, for which a much better interpolating expression can be obtained using the accurate values for θ_C and K predicted by Monte Carlo simulations, it can be employed to estimate the available area and the jamming coverage for binary mixtures of disks, for which in general the values of θ_C and K are not known.

In Fig. 2, we represent the interpolating function in a natural representation, $(\Phi(\theta))^{1/3}$ versus θ, which becomes linear near the jamming point. The interpolating function can be approximated by an expression of the type

$$(\tilde{\Phi})(\theta))^{1/3} = (a_0 - \theta)\left(\sum_{i=0} a_{i+1}\theta^i\right) \tag{8}$$

with the coefficients provided by the matching of the first terms of the low-density expansion Eq. (5). The best approximate was obtained by using the first four coefficients, which are given by

$$a_0 a_1 = \sqrt[3]{c_0} \tag{8a}$$

$$a_2 a_0 - a_1 = \sqrt[3]{c_0}\frac{c_1}{3c_0} \tag{8b}$$

$$a_3 a_0 - a_2 = \sqrt[3]{c_0}\left(\frac{c_2}{3c_0} - \left(\frac{c_1}{3c_0}\right)^2\right) \tag{8c}$$

$$a_3 = \sqrt[3]{c_0}\left(\frac{2c_1c_2}{9c_0^2} - \frac{5c_1^3}{81c_0^3}\right) \tag{8d}$$

where c_i are the coefficients of the low-density expansion (Eq. (5)). The approximate expression predicts a jamming at $a_0 = 0.539$, which is not far from the value obtained from Monte Carlo simulations, $\theta_C = 0.547$.

An even simpler approximation can be constructed by matching only the first three terms of expansion (5), with those of expression:

$$(\tilde{\Phi}(\theta))^{1/3} = b_1(b_0 - \theta) + b_2(b_0 - \theta)^2 \tag{9}$$

whose coefficients are obtained from

$$b_0 b_1 + b_2 b_0^2 = \sqrt[3]{c_0} \tag{9a}$$

$$b_1 + 2b_2 b_0 = -\sqrt[3]{c_0}\frac{c_1}{3c_0} \tag{9b}$$

$$b_2 = \sqrt[3]{c_0}\left(\frac{c_2}{3c_0} - \left(\frac{c_1}{3c_0}\right)^2\right) \tag{9c}$$

providing jamming for $\theta_C = b_0 = 0.578$.

It was recently noted by Talbot et al. [28] that an approximate equation derived for thermal equilibrium of disks from the Scaled Particle Theory, namely [19]

$$\Phi^{\mathrm{SPT}}(\theta) = (1 - \theta)\exp\left(-\frac{3\theta}{1-\theta} - \frac{\theta^2}{(1-\theta)^2}\right) \tag{10}$$

provided good agreement for the available area in the non-equilibrium RSA model, up to the vicinity of the jamming coverage. In Fig. 2, the interpolating function (6), is compared with the approximations (8) and (9), which are more accurate than the SPT prediction (10), particularly at high θ values.

3. RSA of binary mixture of disks

When two kinds of disks of radii r_S and $r_L \equiv \gamma r_S$ (with $\gamma > 1$) are deposited randomly on a surface, the problem becomes much more involved. It was shown that the Feder law holds for monodisperse disks [12–14]. In contrast, for the RSA of binary mixtures, the jamming of the large particles is reached exponentially and then the density of small particles obeys Feder's power law (Eq. (4)) until their jamming [31]. However, the Monte Carlo simulations of an alternate deposition of particles, namely: first all the small ones and then the large ones, revealed that the approach to jamming by the large particles also obeys Feder's law [26]. The jamming coverage θ_C and the proportionality constant K depend on the ratio $\gamma = r_L/r_S$ and on the initial area occupied by the small particles (the precovered area), θ_S.

To seek a reasonable accurate analytical approximation for the available area, as a function of $\theta_S = N_S\pi r_S^2/A$ and $\theta_L = N_L\pi r_L^2/A$ one should have accurate values for a reasonable number of coefficients in the low-density expansion of the binary RSA model, which is not a trivial task. Even for binary mixtures of disks at equilibrium, a problem that received much more attention than RSA, analytical expressions are known only for the first three terms of the virial expansion [21]. The values of the fourth and fifth terms, obtained using laborious numerical calculations, were reported only for a few values of γ and molar fractions of the two types of disks [22]. In the non-equilibrium RSA of binary particles, one should take into account, when calculating the higher terms of the series, not only various γ and molar fractions, but also the order of deposition of particles. Furthermore, as already noted, it is not clear whether the involved calculations needed to obtain the next unknown terms of the low-density expansion would improve much the accuracy of estimating the jamming coverage.

An alternate possibility is to obtain approximations for the coefficients of the low-density expansion using analytical approximations of the available surface and to extend analytically these approximations. The expressions for available area, such us those derived from Scaled Particle Theory, $\Phi_i^{SPT}(\theta_S, \theta_L)$ [19]:

$$\Phi_S^{SPT}(\theta_S, \theta_L) = (1-\theta_S-\theta_L) \times \exp\left(-\frac{3\theta_S + (\theta_L/\gamma)(2+(1/\gamma))}{1-\theta_S-\theta_L} - \frac{(\theta_S+(\theta_L/\gamma))^2}{(1-\theta_S-\theta_L)^2}\right) \quad (11a)$$

$$\Phi_L^{SPT}(\theta_S, \theta_L) = (1-\theta_S-\theta_L) \times \exp\left(-\frac{3\theta_L + \gamma(2+\gamma)\theta_S}{1-\theta_S-\theta_L} - \frac{(\theta_S+\gamma\theta_L)^2}{(1-\theta_S-\theta_L)^2}\right) \quad (11b)$$

were already shown by Talbot et al. [28] to be excellent approximations at low densities. However, by predicting a total final coverage of $\theta_S + \theta_L = 1$, they fail in the vicinity of the jamming point. Therefore, they fail to provide both the final coverage and the final ratio of adsorbed species at jamming.

In the present approach, the area available for large disks, $\Phi_L(\theta_S, \theta_L)$, will be approximated as follows (interchange everywhere "S" and "L" for the area available to small disks). It will be assumed as an approximation that only the small disks are deposited until the surface coverage θ_S is reached, and then only the large particles are adsorbed until their jamming. Along the path $(\theta_S = 0, \theta_L = 0) \to (\theta_S, 0)$ it is assumed that the initial area available to the large disks, $\Phi_L(\theta_S, 0)$ can be approximated by $\Phi_L^{SPT}(\theta_S, 0)$, providing that this it is not in the vicinity of the jamming point (which implies that $\Phi_L^{SPT}(\theta_S, 0)$ is sufficiently large). Along the path $(\theta_S, 0) \to (\theta_S, \theta)$, the SPT approximation fails at large θ, in the vicinity of the large disks jamming point. However, $\Phi_L(\theta_S, \theta)$ is a function of θ alone, and obeys the asymptotic behaviour (4) (with the constants θ_C and K dependent on γ and θ_S). Therefore, an expression $\tilde{\Phi}_L(\theta_S, \theta)$ of the type (8) or type (9) can be constructed starting from the first coefficients of the low-density expansion:

$$\Phi_L(\theta_S, \theta) = \sum_{k=0} \frac{1}{k!} \left.\frac{\partial^k \Phi_L^{SPT}(\theta_S, \theta)}{\partial \theta^k}\right|_{\theta=0} \theta^k \quad (12)$$

The approximation $\tilde{\Phi}_L(\theta_S, \theta_L)$ will provide an estimate for the available area for large disks on a surface upon which the adsorbed small and large disks cover the fraction θ_S and θ_L of the surface, respectively. While no data are available in literature for $\Phi_L(\theta_S, \theta_L)$, the accuracy of the estimation $\tilde{\Phi}_L(\theta_S, \theta_L)$ can be verified by comparing the jamming of large disks obtained analytically from the conditions

$$\tilde{\Phi}_L(\theta_S, (\theta_L)_C) = 0; \qquad \tilde{\Phi}_S((\theta_S)_C, \theta_L) = 0 \quad (13)$$

with the jamming predicted by Monte Carlo simulations.

Let us first examine whether Eq. (8) or Eq. (9) provides better results, by calculating the total jamming coverage, $\theta_C \equiv \theta_S + (\theta_L)_C$ as a function of θ_S in the simple case $\gamma = 1$ (monodisperse disks), for which accurate results are known from Monte Carlo simulations ($\theta_C = 0.547$).

The low-θ_L expansion is given by

$$\Phi_L(\theta_S, \theta) = \sum_{k=0} \frac{1}{k!} \left.\frac{\partial^k \Phi(\theta)}{\partial \theta^k}\right|_{\theta=\theta_S} (\theta-\theta_S)^k \quad (14)$$

were we selected for $\Phi(\theta)$ either the interpolating function (6) or the SPT expression (10). The interpolating function (6) was preferred to expression (5) because the former is accurate over the entire θ range, while the latter only for low θ values.

The jamming of the "large particles" estimated in this manner for various initial coverages θ_S are compared in Fig. 3 with the exact result, $\theta_C = 0.547$. As expected, both approximations based on Eqs. (8) and (9) are better at large θ_S values, where Feder's law is obeyed, when the interpolating function is employed for the low-θ_L expansion. In

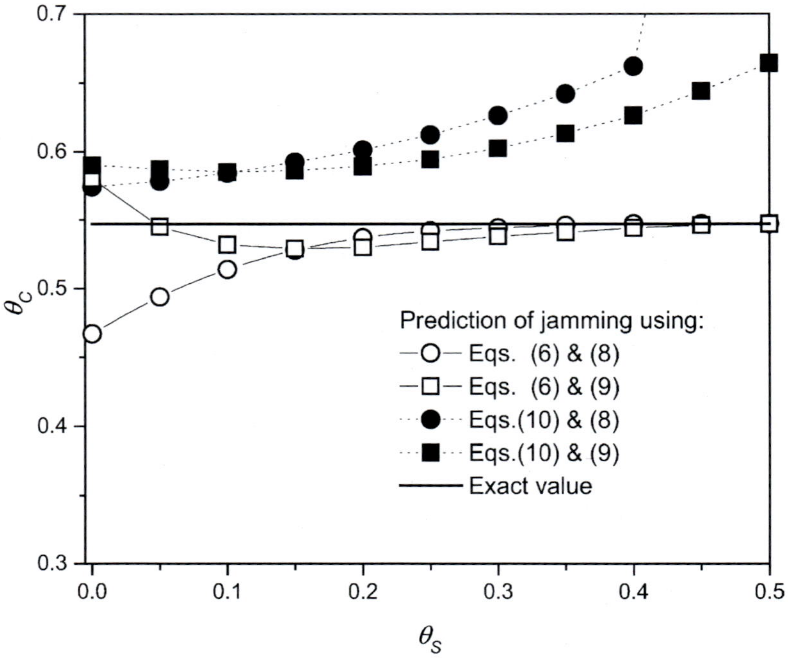

Fig. 3. The total jamming coverage for a "mixture" of monodisperse disks ($\gamma = 1$) predicted using either Eq. (8) or Eq. (9), combined with the low-θ_L expansion terms calculated from either Eq.(6) or Eq. (10), as a function of the "small" disks coverage, θ_S, are compared to the accurate jamming point, $\theta_C = 0.547$.

contrast, when the SPT expression (10) is used in Eq. (14), the jamming point estimation deteriorates at high θ_S values, mainly because Eq. (10) ceases to be a good approximation in that region. While both approximations (8) and (9) overpredict systematically the jamming, when the SPT expression is employed for the low-θ expansion (14), Eq. (9) appears to be a better one and will be used in what follows.

Let us compare the approach described above with the existing Monte Carlo simulations. While the values for the available area $\Phi_S(\theta_S, \theta_L)$ are not available in literature, we

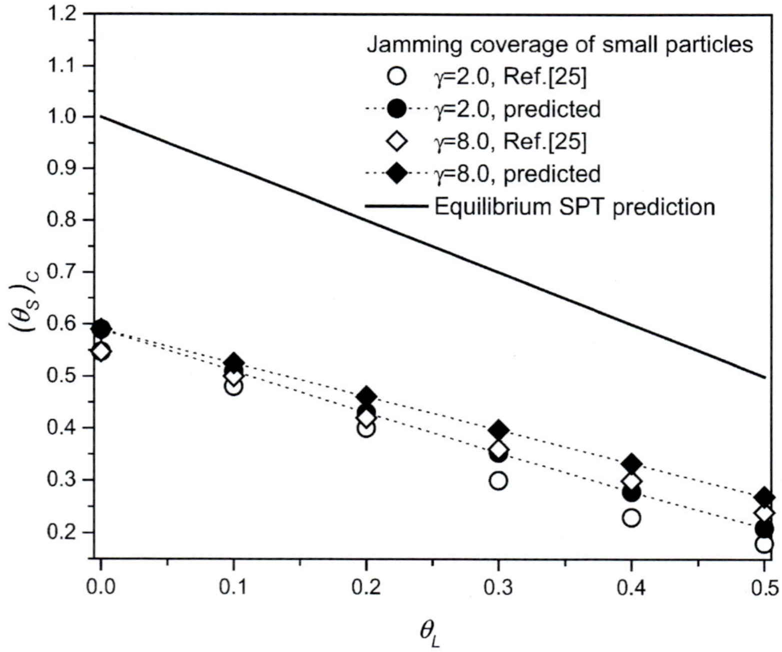

Fig. 4. Predicted jamming point for small disks, as a function of the surface coverage by large disks, θ_L, compared with the results of Monte Carlo simulations reported in [25], for $\gamma = 2.0$ and 8.0.

will use the results obtained by Meakin and Jullien [25] for the RSA of a binary mixture of hard disks of different radii and different concentrations. While both kinds of disks are present in the vicinity of the surface, the large ones reach exponentially the jamming and then practically only the small ones are adsorbed until they reach their jamming point. The values of θ_L and $(\theta_S)_C$, which correspond to the jamming of small particles, $\Phi_S(\theta_S, \theta_L) = 0$, derived from Monte Carlo simulations [25] are compared in Fig. 4 with the estimates based on the approach described above, which employs Eqs. (9), (11a), (12) and (13). Except for the systematic overprediction of jamming by about 0.05, due mainly to the use of the SPT approximation, the estimates are in a fair agreement with the simulations.

It is of interest to test the accuracy of the present approach by estimating the jamming of large disks in the presence of small ones. When both kinds of disks are present in the vicinity of the surface, the large ones reach rapidly their

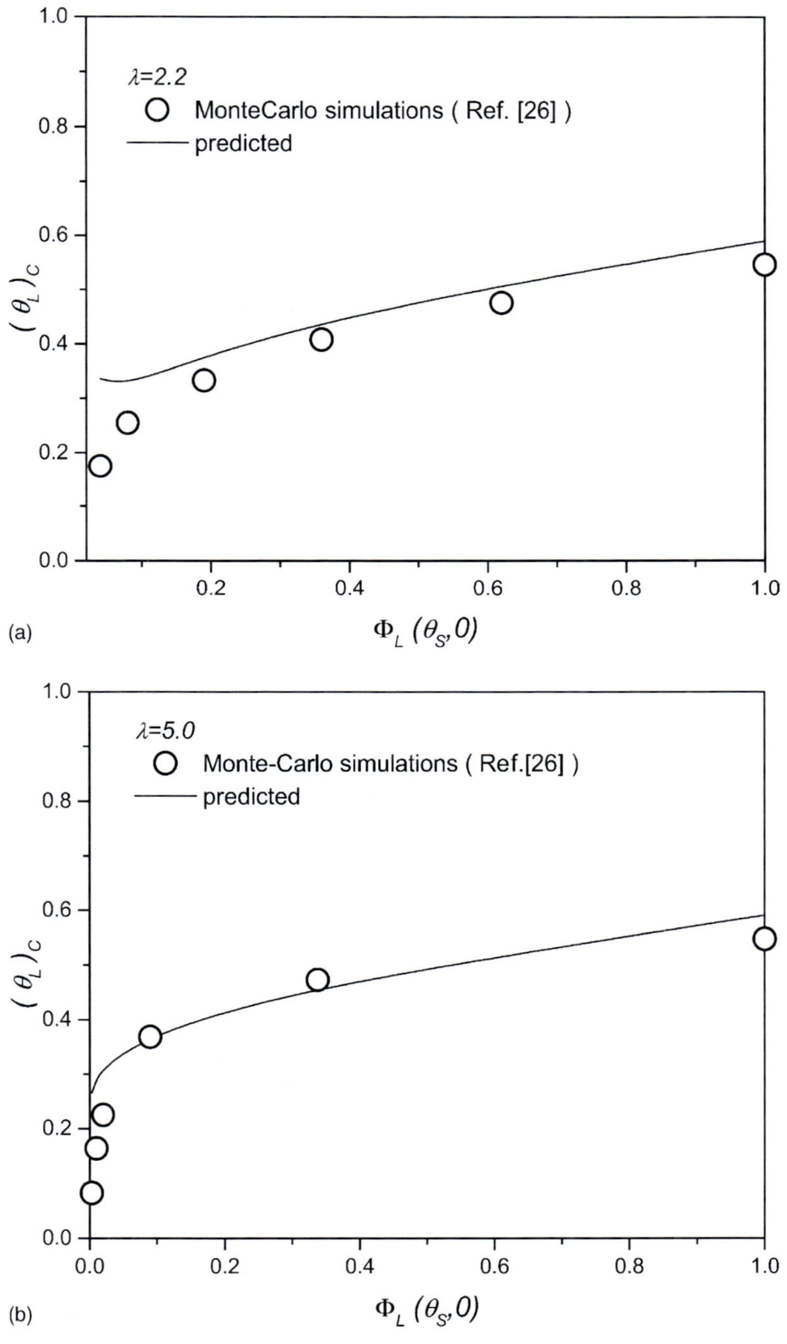

Fig. 5. Predicted jamming point for the large spheres as a function of the initial surface available to the large spheres, $\Phi_L(\theta_S,0)$, compared with the results of the Monte Carlo simulations reported in [26]: (a) $\lambda = 2.2$; (b) $\lambda = 5.0$; (c) $\lambda = 10.0$.

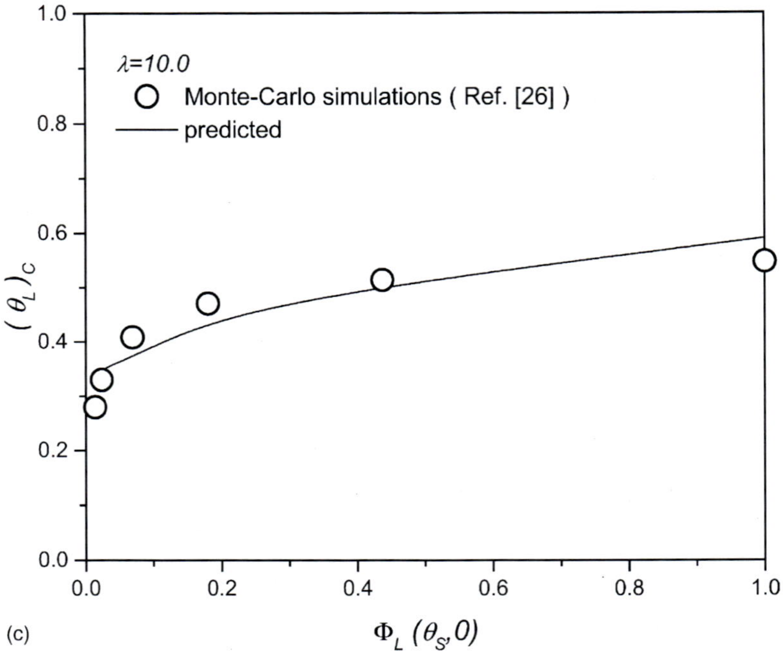

Fig. 5. (*Continued*).

jamming, whereas the small disks continue to be adsorbed on the surface for a much longer time. Because the small disks coverage changes very much in the region in which the large disks are almost at their jamming (see Fig. 2 of [25]), the value of θ_S, at which the jamming of large particles occurs, cannot be determined accurately (the precision being of the same order of magnitude as the estimation employed here). Accurate values could be obtained from Monte Carlo simulations of successive depositions of small and large disks. The only simulations reported in literature, however, refer to a related, but different problem, namely the RSA of large spheres adsorbed on a surface precovered with small spheres [26].

As noted by Talbot and Schaaf [27], the surface available for a large sphere of radius $a_L = \lambda a_S$ on a surface precovered with small spheres is the same, in the limit $\theta_L \to 0$, as the surface available to large disks of radius $r_L = \gamma r_S$ for

$$\gamma = 2\sqrt{\lambda} - 1. \qquad (15)$$

Unfortunately, since the mapping of spheres into disks fails for $\theta_L \neq 0$, the derivatives with respect to θ_L cannot be calculated accurately from Eqs. (11a) and (11b). Since these values are needed to approximate Φ, this inaccuracy generates additional errors.

An additional difficulty is that the SPT approximation is excellent for low surface coverages, (when $\Phi(\theta_S, \theta_L) \sim 1$), but fails at large surface coverages ($\Phi(\theta_S, \theta_L) \sim 0$). While the small particles might not occupy a large area on the surface, they can exclude most of the surface to the large ones. Since the SPT approximation fails when the available area is small, we plot in Fig. 5a–c the jamming coverage of large particles, $\Phi_L(\theta_S, (\theta_L)_C)$ as a function of the initial area available to them, $\Phi_L(\theta_S, 0)$, to better identify the range of validity of the approximation. There is a good agreement between the Monte Carlo simulations and the present approach for not too low values of $\Phi_L(\theta_S, 0)$.

4. Conclusions

Despite the attention received by the Random Sequential Adsorption model of particles, only the one-dimensional problem (Renyi's car parking problem) was solved analytically. Accurate results could be obtained from Monte Carlo simulations in higher dimensions. Simple and accurate expressions for the available area can be constructed from the low-density expansion, combined with the information provided by Monte Carlo simulations for the behaviour near the jamming point. However, reasonable estimates can be made without using the Monte Carlo results. In the present paper, a procedure was suggested to estimate the available area for a Random Sequential Adsorption of binary mixture of disks, for which the Monte Carlo calculations are prohibitively long. It was assumed that first all the small (large) disks were deposited until a value θ_S (θ_L) was reached, then the other type of disks were adsorbed until they reached their jamming. Then, the approximate Eqs. (11a) and (11b) were employed to calculate the low-density expansion coefficients of the available area as a function of $\theta_L(\theta_S)$ alone. Finally, an approximate expression (9) with appropriate asymptotic behaviour near the jamming point was constructed. Whereas this method provided only estimates of the jamming points, they represent a significant improvement over the predictions of the Scaled Particle Theory, Eqs. (11a) and (11b).

These estimates were compared to the Monte Carlo simulations available in literature for the Random Sequential Adsorption of a binary mixture of disks and spheres.

References

[1] I. Langmuir, J. Am. Chem. Soc. 40 (1918) 1361.
[2] M.K. Khoshkbarchi, J.H. Vera, Fluid Phase Equilib. 142 (1998) 131.
[3] B. Widom, J. Chem. Phys. 39 (1963) 2808.
[4] E.J.J. van Rensburg, J. Phys. A 26 (1993) 4805.
[5] N.F. Carnahan, K.E. Starling, J. Chem. Phys. 51 (1969) 635.
[6] K.R. Hall, J. Chem. Phys. 57 (1972) 2252.
[7] F.H. Ree, W.G. Hoover, J. Chem. Phys. 40 (1964) 939.
[8] G.A. Mansoori, N.F. Carnahan, K.E. Starling, T.W. Leland, J. Chem. Phys. 54 (1971) 1523.
[9] B. Widom, J. Chem. Phys. 44 (1966) 3888.
[10] A. Renyi, Publ. Math. Inst. Hung. Acad. Sci. 3 (1958) 109.
[11] L. Tonks, Phys. Rev. 50 (1936) 955.
[12] J. Feder, J. Theor. Biol. 87 (1980) 237.
[13] R.K. Swendson, Phys. Rev. A 24 (1981) 504.
[14] Y. Pomeau, J. Phys. A. 13 (1980) L193.
[15] P. Schaaf, J. Talbot, J. Chem. Phys. 91 (1989) 4401.
[16] R. Dickman, J.S. Wang, I. Jensen, J. Chem. Phys. 94 (1991) 8252.
[17] J.A. Given, Phys. Rev. A 45 (1992) 816.
[18] J.-S. Wang, Colloids Surf. A 165 (2000) 325.
[19] J.L. Lebowitz, E. Hefand, E. Praestgaard, J. Chem. Phys. 34 (1965) 1037.
[20] C. Bario, J.R. Solana, J. Chem. Phys. 115 (2001) 7123.
[21] J.L. Barrat, H. Xu, H.J.-P. Hansen, M. Baus, J. Phys. C: Solid State Phys. 21 (1988) 3165.
[22] E. Enciso, N.G. Almarza, M.A. Gonzalez, F.J. Bermejo, Phys. Rev. E 57 (1998) 4486.
[23] M.K. Hassan, J. Schmidt, B. Blasius, J. Kurths, Phys Rev. E 65 (2002) 045103R.
[24] B. Bonnier, Phys. Rev. E 64 (2001) 066111.
[25] P. Meakin, R. Jullien, Phys. Rev. A 46 (1992) 2029.
[26] Z. Adamczick, P. Weronski, J. Chem. Phys. 108 (1998) 9851.
[27] J. Talbot, P. Schaaf, Phys. Rev. A 40 (1989) 422.
[28] J. Talbot, X. Jin, N.-H. Wang, Langmuir 10 (1994) 1663.
[29] Z. Adamczyk, Adv. Colloids Interf. Sci. 100 (2003) 267.
[30] J.W. Evans, Phys. Rev. Lett. 62 (1989) 2641.
[31] M.C. Bartelt, V. Privman, Phys. Rev. A 44 (1991) R2227.

MIX
Papier aus verantwortungsvollen Quellen
Paper from responsible sources
FSC® C105338

Printed by Books on Demand, Germany